AN INTEGRATED APPROACH TO
ENVIRONMENTAL MANAGEMENT

AN INTEGRATED APPROACH TO ENVIRONMENTAL MANAGEMENT

Edited by

DIBYENDU SARKAR
Professor of Environmental Geochemistry
Founding Director of the PhD Program in Environmental Management
Montclair State University, New Jersey

RUPALI DATTA
Associate Professor of Environmental Remediation
Department of Biological Science
Michigan Technological University, Michigan

AVINANDAN MUKHERJEE
Dean, College of Business
Clayton State University
Georgia

ROBYN HANNIGAN
Founding Dean, School for the Environment
University of Massachusetts at Boston
Massachusetts

Copyright © 2016 by John Wiley & Sons, Inc. All rights reserved

Published by John Wiley & Sons, Inc., Hoboken, New Jersey
Published simultaneously in Canada

No part of this publication may be reproduced, stored in a retrieval system, or transmitted in any form or by any means, electronic, mechanical, photocopying, recording, scanning, or otherwise, except as permitted under Section 107 or 108 of the 1976 United States Copyright Act, without either the prior written permission of the Publisher, or authorization through payment of the appropriate per-copy fee to the Copyright Clearance Center, Inc., 222 Rosewood Drive, Danvers, MA 01923, (978) 750-8400, fax (978) 750-4470, or on the web at www.copyright.com. Requests to the Publisher for permission should be addressed to the Permissions Department, John Wiley & Sons, Inc., 111 River Street, Hoboken, NJ 07030, (201) 748-6011, fax (201) 748-6008, or online at http://www.wiley.com/go/permissions.

Limit of Liability/Disclaimer of Warranty: While the publisher and author have used their best efforts in preparing this book, they make no representations or warranties with respect to the accuracy or completeness of the contents of this book and specifically disclaim any implied warranties of merchantability or fitness for a particular purpose. No warranty may be created or extended by sales representatives or written sales materials. The advice and strategies contained herein may not be suitable for your situation. You should consult with a professional where appropriate. Neither the publisher nor author shall be liable for any loss of profit or any other commercial damages, including but not limited to special, incidental, consequential, or other damages.

For general information on our other products and services or for technical support, please contact our Customer Care Department within the United States at (800) 762-2974, outside the United States at (317) 572-3993 or fax (317) 572-4002.

Wiley also publishes its books in a variety of electronic formats. Some content that appears in print may not be available in electronic formats. For more information about Wiley products, visit our web site at www.wiley.com.

Library of Congress Cataloging-in-Publication Data:

An integrated approach to environmental management / edited by Dibyendu Sarkar, Rupali Datta, Avinandan Mukherjee, Robyn Hannigan.
  pages cm
 Includes bibliographical references and index.
 ISBN 978-1-118-74435-2 (cloth)
1. Environmental management. I. Sarkar, Dibyendu, editor. II. Datta, Rupali, editor. III. Mukherjee, Avinandan, editor. IV. Hannigan, Robyn, editor.
 GE300.I583 2015
 333.7–dc23
                                                                                                                                 2015006397

Set in 10/12pt Times by SPi Global, Pondicherry, India

Printed in the United States of America

10 9 8 7 6 5 4 3 2 1

1 2016

CONTENTS

About the Editors		ix
Contributors		xi
Preface		xiii
Endorsements		xv
SECTION I	**ENVIRONMENTAL MANAGEMENT: THE NATURAL SCIENCE AND ENGINEERING PERSPECTIVE**	1
1	**Geology in Environmental Management** *Michael A. Kruge*	3
2	**Biology in Environmental Management** *Audrey L. Mayer*	47
3	**Soil Science in Environmental Management** *Nadine Kabengi and Maria Chrysochoou*	75
4	**Green Chemistry and Ecological Engineering as A Framework for Sustainable Development** *Shyam R. Asolekar, R. Gopichandran, Anand M. Hiremath and Dinesh Kumar*	97
5	**Green Energy and Climate Change** *R. Gopichandran, Shyam R. Asolekar, Omkar Jani, Dinesh Kumar and Anand M. Hiremath*	127
6	**Engineering in Environmental Management** *Yang Deng*	151

| 7 | Green Architecture in Environmental Management | 173 |

Jason Kliwinski and Amy Ferdinand

SECTION II ENVIRONMENTAL MANAGEMENT: THE BUSINESS AND SOCIAL SCIENCE PERSPECTIVE — 193

| 8 | Business Strategies for Environmental Sustainability | 195 |

Avinandan Mukherjee, Naz Onel and Rosita Nuñez

| 9 | Green Marketing Strategies | 231 |

Mehdi Taghian, Michael Jay Polonsky and Clare D'Souza

| 10 | Role of Environmental, Social, and Governance (ESG) Factors in Financial Investments | 255 |

A. Seddik Meziani

| 11 | The Role of Public Relations and Organizational Communication in Environmental Management | 277 |

Ricard W. Jensen

| 12 | The Economics of Environmental Management | 289 |

David Timmons

| 13 | Law and Policy in Environmental Management | 305 |

Dianne Rahm

| 14 | Environmental Ethics | 337 |

Tatjana Višak

SECTION III ENVIRONMENTAL MANAGEMENT: THE METHODS AND TOOLS PERSPECTIVE — 363

| 15 | Participatory Approaches in Environmental Management | 365 |

Stentor B. Danielson

| 16 | Statistics in Environmental Management | 383 |

Jennifer A. Brown

| 17 | Remote Sensing in Environmental Management | 397 |

Mark J. Chopping

| 18 | Geographic Information Systems in Environmental Management | 423 |

Danlin L. Yu and Scott W. Buchanan

| 19 | Life Cycle Analysis as a Management Tool in Environmental Systems | 441 |

Dimitrios A. Georgakellos

| 20 | Environmental Audit in Environmental Management | 465 |

Ian T. Nicolson

| 21 | Risk Assessment as a Tool in Environmental Management | 503 |

Kofi Asante-Duah

Appendix A: Supplemental Readings   521

Appendix B: Model Syllabus   549

Appendix C: Model Environmental Management Curricula (BS, MS, PhD)   579

Index   587

ABOUT THE EDITORS

Dr. Dibyendu Sarkar is a professor of environmental geochemistry and the founding director of the environmental management PhD program (2009–2015) at Montclair State University, New Jersey. Prior to joining Montclair State, Dibs served as an assistant and associate professor and associate dean of Graduate Studies and Research at the University of Texas at San Antonio (2000–2008), after graduating with a PhD in geochemistry from the University of Tennessee (December 1997) and working as a postdoctoral researcher in Soil and Water Science at the University of Florida (1998–2000). Between 2000 and 2015, he advised 10 PhD students and 15 MS students and trained 14 postdoctoral research associates. Dibs has so far published over 300 journal articles, book chapters, conference proceedings, and technical abstracts. He has authored a research monograph, edited two books, and has generated more than $5 million in grant funding to support his research activities and those of his students/postdocs. His research has appeared in a wide range of environmental journals, such as Environmental Science and Technology, Journal of Hazardous Materials, Journal of Environmental Quality, Soil Science Society of America Journal, Chemosphere, Environmental Pollution, Journal of Colloid and Interface Science, etc. Dibs is a member of several scientific and professional organizations, including Geological Society of America, American Society of Agronomy, American Geophysical Union, and American Association of Petroleum Geologists, and has received many research and teaching awards from them. He has served on numerous committees and organized many symposia and theme sessions for these organizations. Dibs is a Fellow of the Geological Society of America and a principal of SIROM Scientific Solutions, LLC. He is the editor-in-chief of a Springer journal, Current Pollution Reports; the technical editor of another Springer journal, International Journal of Environmental Science and Technology; and an associate editor of Geosphere (online journal of the Geological Society of America), Environmental Geosciences (quarterly journal of the Division of Environmental Geosciences of the American Association of Petroleum Geologists), and Soil Science Society of America Journal. Dibs serves on the editorial board of the Environmental Pollution, an Elsevier journal, and as a reviewer for more than 60 journals and several grant funding agencies, including NSF and NIH.

Dr. Rupali Datta is an associate professor and the graduate program director of the Department of Biological Sciences at Michigan Technological University. Rupali's primary research interest lies in the application of plant biochemistry, genetics, molecular biology, and microbiology in solving environmental problems, using phytoremediation, plant–microbe interactions, and bioremediation. Her research involves the study of interactions between plant, soil, microbial, and water systems to understand the mechanisms of uptake and detoxification of specific environmental contaminants in biota from two broad angles—biochemistry and genetics. She has close to 100 research publications and more than 150 technical abstracts and conference proceedings and has generated more than two million dollars in research funding. Her research is strongly related to student experiential learning. So far, she has graduated four PhD students and eight MS students and supervised four postdoctoral fellows. Prior to joining Michigan Tech, Rupali was an assistant professor in the Department of Earth and Environmental Sciences at the University of Texas at San Antonio. Rupali is an associate editor of two Springer journals—namely, *Current Pollution Reports* and *International Journal of Environmental Science and Technology*—and serves as a reviewer for more than 50 journals and several grant funding agencies. She is also a principal of SIROM Scientific Solutions, LLC.

Dr. Avinandan Mukherjee is professor of marketing and international business and dean of the College of Business at Clayton State University, Metro Atlanta, Georgia. Avi is a doctoral faculty in the PhD program in environmental management at Montclair State University. He is also the editor-in-chief of the *International Journal of Pharmaceutical and Healthcare Marketing*. Prior to joining Clayton State, Avi was professor and department chair of marketing at the School of Business in Montclair State University. Other than Montclair State, Avi has taught at Penn State University, Rutgers University, and New Jersey Institute of Technology (United States), University of Bradford (United Kingdom), Nanyang Technological University (Singapore), and Indian Institute of Management Calcutta (India). He has authored more than 100 articles in refereed journals, conference proceedings, and edited books and has more than 1000 citations of his published work. His research has appeared in the *Journal of Retailing, Journal of Business Research, Service Industries Journal, Communications of the ACM, Journal of the Operational Research Society, Journal of Services Marketing, Journal of Marketing Management, International Journal of Advertising, International Journal of Bank Marketing*, etc. Avi has been guest editor for *European Journal of Marketing* and *Journal of Services Marketing*, editorial board member of *Hospital Topics* and *Asia-Pacific Journal of Marketing and Logistics*, and ad hoc reviewer for a variety of journals. Avi has so far advised 5 PhD students and a host of master's degree students and postdoctoral scholars over his academic career.

Dr. Robyn Hannigan is professor and founding dean of School for the Environment at the University of Massachusetts at Boston. A graduate of the University of Rochester, Robyn's research centers on coastal ecosystem and resource management and evaluation of past climate to inform adaptation in coastal systems. Specifically, her work focuses on the impact of ocean acidification on fish and shellfish and the reconstruction of ocean acidification events in Earth's deep past. Robyn has published over 100 peer-reviewed articles across fields of geochemistry and environmental science. She is a fellow of the American Association for the Advancement of Science and Geological Society of America and an Aldo Leopold Leadership fellow. She served as the chair of the Consortium of Universities for the Advancement of Hydrologic Sciences, Inc. In addition to Robyn's academic achievements, she and her students hold several patents in areas of sample introduction technologies for mass spectrometric identification of important metals in biological samples. Robyn started a company with former students, GeoMed Analytical, which uses geochemical methods to study human health and seafood resource issues such as food sourcing and metals in disease treatment and diagnosis.

The editors were assisted in the Social Science subsection by **Dr. Neeraj Vedwan**, associate professor of anthropology at Montclair State University.

CONTRIBUTORS

Asante-Duah, Kofi, Ph.D., Chief Science Advisor—Risk Assessment/Toxicology, Environmental Protection Administration, District Department of the Environment, Washington, DC, USA

Asolekar, Shyam R., Ph.D., Professor, Centre for Environmental Science and Engineering, Indian Institute of Technology Bombay, Powai, Mumbai, India

Brown, Jennifer A., Ph.D., School of Mathematics and Statistics, University of Canterbury, Christchurch, New Zealand

Buchanan, Scott W., M.S., Ph.D., Student, Department of Natural Resources Science, University of Rhode Island, Kingston, RI, USA

Chopping, Mark J., Ph.D., Professor, Department of Earth and Environmental Studies, Montclair State University, Montclair, NJ, USA

Chrysochoou, Maria, Ph.D., Associate Professor, Department of Civil and Environmental Engineering, University of Connecticut, Storrs, CT, USA

Danielson, Stentor B., Ph.D., Assistant Professor, Department of Geography, Geology and Environment, Slippery Rock University, Slippery Rock, PA, USA

Deng, Yang, Ph.D., Associate Professor, Department of Earth and Environmental Studies, Montclair State University, Montclair, NJ, USA

D'Souza, Clare, La Trobe Business School, Melbourne, VIC, Australia

Ferdinand, Amy, Ph.D., Director, Environmental Safety and Sustainability, Montclair State University, Montclair, NJ, USA

Georgakellos, Dimitrios A., Ph.D., Professor of Technology, Energy and Environmental Management, Department of Business Administration, School of Economic, Business and International Studies, University of Piraeus, Piraeus, Greece

Gopichandran, R., Ph.D., Director, Vigyan Prasar, Department of Science and Technology, Government of India, New Delhi, India

Hiremath, Anand M., M.Tech., Ph.D., Candidate, Centre for Environmental Science and Engineering, Indian Institute of Technology Bombay, Powai, Mumbai, India

Jani, Omkar, Ph.D., Principal Research Scientist, Solar Energy Research Wing, Gujarat Energy Research and Management Institute, Innovation and Incubation Centre (GERMI-RIIC), Gandhinagar, Gujarat, India

Jensen, Ricard W., Ph.D., Professor, Department of Marketing, Montclair State University, Montclair, NJ, USA

Kabengi, Nadine, Ph.D., Assistant Professor, Department of Geosciences and Department of Chemistry, Georgia State University, Atlanta, GA, USA

Kliwinski, Jason, B.Arch., Director of Sustainable Design at Parette Somjen Architects and Partner, Green Building Center, New York, NJ, USA

Kruge, Michael A., Ph.D., Professor, Earth & Environmental Studies Department, Montclair State University, Montclair, NJ, USA

Kumar, Dinesh, Ph.D., Candidate, Centre for Environmental Science and Engineering, Indian Institute of Technology Bombay, Powai, Mumbai, India

Mayer, Audrey L., Ph.D., Associate Professor of Ecology and Environmental Policy, School of Forest Resources and Environmental Science and Department of Social Sciences, Michigan Technological University, Houghton, MI, USA

Meziani, A. Seddik, Ph.D., Professor, Department of Economics & Finance, School of Business, Montclair State University, Montclair, NJ, USA

Mukherjee, Avinandan, Ph.D., Dean, College of Business, Clayton State University, Morrow, GA, USA and School of Business, Montclair State University, Montclair, NJ, USA

Nicolson, Ian T., Ph.D., Honorary Assistant Professor, The Kadoorie Institute, The University of Hong Kong, Pokfulam, Hong Kong

Nuñez, Rosita, Ph.D., Candidate, Global Director, Marketing, Lorza Group, Rochester, NY, USA

Onel, Naz, Ph.D., School of Business, Stockton University, Galloway, NJ, USA

Polonsky, Michael Jay, Ph.D., Professor, Department of Marketing, Faculty of Business and Law, Deakin University, Melbourne, VIC, Australia

Rahm, Dianne, Ph.D., Professor, Department of Political Science, Texas State University, San Marcos, TX, USA

Taghian, Mehdi, Ph.D., Senior Lecturer, Department of Marketing, Faculty of Business and Law, Deakin University, Melbourne, VIC, Australia

Timmons, David, Ph.D., Assistant Professor, Department of Economics, University of Massachusetts, Boston, MA, USA

Višak, Tatjana, Ph.D., Lecturer, Department of Philosophy and Business Ethics, Mannheim University, Mannheim, Germany

Yu, Danlin L., Ph.D., Associate Professor, Department of Earth and Environmental Studies, Montclair State University, Montclair, NJ, USA

PREFACE

The last few decades have seen a tremendous increase in the anthropogenic pressures on the environment. The result has been widespread environmental degradation, along with increasing negative social and economic impacts, especially on the vulnerable sections of the population, not only in the United States but all around the globe. Although environmental management concepts have been around since the 1970s, it is mostly in recent years that they have become ubiquitous. Thus far, the efforts to combat environmental problems have been predominantly piecemeal and top-down; consequently, both the reception of environmental policies and their effectiveness have been less than satisfactory in most cases. Today, increasing concerns about the environment are leading legislators, regulators, communities, corporations, and consumers to make new choices in terms of scientific research priorities, engineering investments, infrastructure and regulations, purchasing patterns, product usage, and disposal behavior, to name a few of many. The emerging field of environmental management takes a holistic view of environmental problems, recognizing the interconnectedness of social, economic, political, and environmental processes. The emphasis on primary benefits of environmental awareness and sensitivity to sustainability paradigms is critical to winning over the mainstream population in this global race to manage the most important issues where environment plays a critical role, such as climate change and human and ecological health. Accordingly, the field adopts an eclectic and integrative approach to environmental problem solving, combining theories, research methods, and analytical techniques drawn from a gamut of disciplines, as far apart as geochemistry and political science. The heterogeneity inherent in the subject of environmental management makes the field an exciting arena of discovery while also producing challenges in terms of coherence and consistency.

Our main goal in writing the book was to provide a single one-stop treatise on the rapidly evolving multidisciplinary subject of environmental management. The idea came to us back in fall 2008 when the senior editor (DS) joined Montclair State University to build its PhD program in environmental management and soon realized the challenges in putting together a comprehensive tertiary-level program in this highly interdisciplinary field. The definition of environmental management varied from person to person, depending on their individual areas of specialization. It was almost like John Godfrey Saxe's classic poem "The Blind Men and the Elephant." People's concept of environmental management depends upon which part of the elephant they are handling. To complicate things further, there wasn't any comprehensive textbook on environmental management that approaches the subject from various disciplinary viewpoints with an objective to bring them all together in the end. Thus, came the idea of this book and an interdisciplinary team of editors and a contributor from the broad areas of science, engineering, social science, and business, who approached highly qualified authors to discuss the roles of their individual disciplines in environmental management.

The book is intended to be the first and only scholarly book that will be positioned as a leading textbook, as well as a specialist reference resource of academic information and analysis on environmental management from multiple perspectives, highlighting cutting-edge research, new concepts and theories, and fresh practical ideas and initiatives that can be readily applied in societies and organizations. The book will mitigate a major deficiency in the field of environmental management, as currently envisioned: namely, it's fracturing along the disciplinary boundaries, with the result that presents a fragmented picture, with its different components seemingly at odds with each other. The book—by simultaneously adopting multiple perspectives and by striving to unify them, through a rigorous examination of the underlying interconnections—will provide a much-needed integrative thrust that will clarify and crystallize the subject of environmental management. This will be particularly useful for practitioners and students of policy who need to be keenly aware of and sensitive to the distinctive aspects of environmental management that can both be overlapping and even competing at times.

The 21 chapters in the book are divided into three perspective sections, namely, (I) Natural Science and Engineering, (II) Business and Social Science, and (III) Methods and Tools. In Section I, we discuss how the principles of geology, biology, soil science, chemistry, and engineering are applied in environmental management research and practices. We also deal with green energy, climate change, and green architecture. The first half of Section II is devoted to the elements of sustainable business, such as green marketing, corporate social responsibility, socially responsible investing, environmental economics, and the role of public relations and organization communications in environmental management. Social science aspects are discussed in the second half of Section II, specifically focusing on participatory approaches, environmental ethics, and environmental law and policy. Section III focuses on the common methods and tools that we use in environmental management research and practice, including statistics, geographic information systems, remote sensing, as well as life cycle analysis, environmental auditing, and environmental risk assessment. Each chapter is accompanied by a detailed list of supplementary readings to guide the interested reader to relevant literature on the topic (Appendix A). The chapters in the book are meant to be primers, which can be extended to develop individual courses. Appendix B contains model syllabi for the benefit of both instructors who may want to offer the courses and students who would get some idea on what a full course in those individual topics might look like. Finally, in Appendix C, we present three model curricula for environmental management degrees in all three levels, bachelors, masters, and doctoral.

We sincerely hope that the book justifies our aim to provide a convenient one-stop compendium consisting of theories, analytical techniques, and applications drawn from natural sciences and engineering, social sciences and policy, and business and economics, which are constitutive of the contemporary environmental management. We hope that it is deemed useful by both students and practitioners in the field, particularly the students finding the broad coverage of the field a useful starting point for pursuing their specific interests. The bird's-eye view afforded by the book hopefully enables students, especially at the postgraduate level (both master's and doctoral) to better locate their own interests and research in the evolving trajectory of the field. Hopefully, the practitioners and policy makers will also be able to appreciate the tremendous possibilities in the field of environmental management that are often obscured by the academic treatments focused on a "nuts-and-bolts" approach alone. Here, the multiple perspectives will be particularly useful in encouraging lateral thinking and in promoting cross-fertilization across disciplinary boundaries.

This book would not have been possible without the contributions of the many authors who worked diligently with us as we tried to assemble this complex volume of interrelated topics, which complement one another in the process of developing a holistic platform to encompass the intricacies of this highly interdisciplinary, continuously evolving field of study. Michael Leventhal of Wiley provided apt guidance on publication requirements and kept us on time. The family members of the four editors (DS, RD, AM, RH) made sacrifices to make the project successful.

January 2015

DIBYENDU SARKAR
RUPALI DATTA
AVINANDAN MUKHERJEE
ROBYN HANNIGAN
(with contributions from Neeraj Vedwan)

ENDORSEMENTS

An Integrated Approach to Environmental Management—edited by Drs. Dibyendu Sarkar, Rupali Datta, Avinandan Mukherjee, and Robyn Hannigan—makes a strong case for the necessity of and the complexity of environmental management. The book identifies the interplay of technical elements and business and societal elements. An important part of environmental management is defining the context of the problem being managed—geological, biological, chemical, and structural (soil). The book discusses how the technical context defines engineering parameters needed to attain sustainable solutions. The second main aspect of integrated management is defining and managing the societal context including economics, legal, ethics, and policy. The goal is to define participatory management approaches. Finally, the book provides a cogent overview of new methods and tools for integrated management. An Integrated Approach to Environmental Management provides the framework and details needed to fulfill its promise of an integrated approach to environmental management. It is a tremendous resource for environmental professionals.

DR. RICHARD A. BROWN
Technical Fellow
Environmental Resources Management, Inc.

This is a long-awaited textbook in environmental management. The authors masterfully combine the latest scientific discoveries, technological advances, and social and economic aspects to present an in-depth analysis and state-of-the-art methodologies for the implementation of innovative solutions and management practices toward a sustainable world. The sustainment of mankind and advancement of our civilization depend on our ability to properly utilize the planet's resources and promulgate policies and practices that address the social and economic impacts of anthropogenic activities. This book is written with particular focus on these pressing issues, and it discusses them with lucidity and erudition. It constitutes an invaluable reference for the student, teacher, and practitioner of environmental management and sustainability.

DR. CHRISTOS CHRISTODOULATOS
Professor and Vice Provost of Innovation and Entrepreneurship
Stevens Institute of Technology

Environmental management degrees are now being offered at various levels in universities and are increasingly in demand. However, most textbooks on this topic concentrate on the scientific and engineering aspects, ignoring the economic, business, and social aspects of environmental management. This book is a very laudable attempt on the part of the editors to provide an integrated perspective on environmental management, for the first time. All readers, whatever their field of expertise or intended expertise in environmental management, will find much to spark their interest in this book. Its breadth and scope, the variety of subject matter explored, and detailed nature of the chapters will provoke creative thoughts in both students and educators alike. The book is brilliantly conceived and organized; the chapters are well thought out and written by experts; the editors are all highly qualified and experienced and have all done a great job in identifying and assembling the perfect contents under the leadership of the senior editor, Dr. Sarkar, who directs one of the very few PhD programs in environmental management in the United States. As an educator, I particularly appreciate the teaching tools the book provides, in terms of model syllabi and further reading suggestions.

DR. NURDAN S DUZGOREN-AYDIN
Professor and Chairperson, Department of Geoscience and Geography
New Jersey City University

With *An Integrated Approach to Environmental Management*, Professor Sarkar and his coeditors have developed a timely and much needed treatment of the highly multidisciplinary and rapidly changing environmental management field. As an environmental engineer with more than 25 years of experience from the diverse perspectives of industry, consulting, and academia, I applaud the editor's efforts to bridge the disparate but interrelated aspects of environmental management from the scientific bases (e.g., geology, biology, and chemistry) to the business elements (e.g., sustainability, laws and policy, and public relations) and applications. This book provides an effective framework by which environmental professionals can rationalize and engage with complex anthropogenic environmental challenges in our increasingly interconnected global society. Moreover, it helps to reinforce the fact that solutions to such vexing environmental challenges as climate change, the proliferation of surface water and groundwater contamination in fast-growing industrial economies, and Brownfields redevelopment will invariably require multidisciplinary perspectives, experience, and paradigms. I believe that An Integrated Approach to Environmental Management will prove to be a "go-to" asset on every environmental professional's bookshelf.

Dr. Daniel W. Elliott
Senior Consultant and Environmental Engineer
Geosyntec, Inc.

This unique book provides readers with an approach to environmental management that synthesizes all the different areas under which current analyses and policies are conducted. Much of the literature that pertains to this subject is disjointed in its approach, largely ignoring the business aspects to environmental management. As Adam Smith observed, as long as the economic and financial incentives are not in line with the desired environmental objectives, no amount of regulation or exhortation can induce businesses as well as individuals to adopt the required goals.

This volume, however, remedies the challenge of disjointed analyses by presenting us with an amazing "one-stop shop." The finance/business/economics synthesis pertaining to environmental management presented in this book, coupled with its unique mix of theory and practice, makes it a veritable game changer in the field of environmental management.

Dr. Farrokh Langdana
Director, Executive MBA Program, and
Professor of Finance/Economics
Rutgers Business School

This book coauthored by faculty from physical sciences as well as business school brings about a unique integration of the many diverse topics that would normally be found in eight different books. The responsibility for developing a sustainable future lies on the shoulder of every citizen, professional, and most importantly everyone contributing to the cause. This unique book brings us together and provides the platform to work as a group.

Dr. Somenath Mitra
Executive Director, Otto York Center for
Environmental Engineering and Science
Distinguished Professor, Chemistry and
Environmental Science
New Jersey Institute of Technology

Environmental management is a rapidly developing field of interdisciplinary pedagogy. Many universities now offer bachelor's and master's degrees in environmental management, and a few offer doctoral degrees. However, there is a serious lack in availability of comprehensive textbooks in environmental management. The book edited by Sarkar, Datta, Mukherjee, and Hannigan fills that void. It provides a clear panoramic view of the field, with expert authors providing their disciplinary viewpoints and commentaries, which have been nicely organized in 3 major perspectives: natural science and engineering, business and social science, and methods and tools. The highly qualified editors have done an extremely good job in pulling all these various perspectives together in a single volume, thus making the book a true one-stop shop for students interested in pursuing a career in environmental management. Faculty members will appreciate the model syllabi accompanying each chapter, along with the comprehensive further readings list.

Dr. Max Seel
Provost and Vice President for Academic Affairs
Michigan Technological University

Many advanced academic text and reference books do a good job in providing a great deal of detailed information on the topic at hand, primarily drawing from the work in one discipline and perhaps a few closely related ones. The complexity of the topic in this book, environmental management, requires us to draw on many fields to understand the problems better and, more importantly, to formulate solutions to the multiple issues involved. I am particularly impressed with the philosophy of the book that breaks down our stereotypical academic silos and thus provides an enriched and enhanced perspective on this important topic. There is indeed a significant need for An Integrated Approach to Environment Management, which includes research in the natural sciences, engineering, the social sciences, and business and management, and I am glad to see that the authors have addressed this need in their book.

Dr. Beheruz N. Sethna
President Emeritus and Regents' Professor
of Business
The University of West Georgia

This book provides a refreshingly integrated reference to environmental management encompassing natural science, engineering, economics, statistics, remote sensing, and a broad array of multidisciplinary facets. Varieties of conceptual bases presented in this book offer an opportunity to the readers to derive cumulative wisdom by connecting the lessons learned from different chapters. There is a well-proportionate combination of theoretical and practical aspects of environmental management. The comprehensive treatise of both fundamental and applied components of environmental studies under one cover makes the book unique. Inclusion of local as well as global issues across wide geographical diversity furthers the universal appeal of this book.

DR. SHANKAR SHARMA
Senior Environmental Scientist of Energy and
Scientific Development Adviser
State of California

(Disclaimer: This is not an endorsement by the State of California and solely represents Dr. Sharma's academic observation.)

This book represents a fresh and balanced perspective on environmental management. There is growing urgency to understand and measure environmental outcomes at all levels of the global economy. The integration of a science and engineering with sustainability, policy, and business will provide students and practitioners with a comprehensive range of tools and ideas for advancing environmental practices. The model curricula will support programs developing the environmental leaders of tomorrow. As a scientist having supported environmental services in the private sector for nearly 30 years, I appreciate the importance of balancing good science with economic reality.

DR. TAMARA L. SORELL
Chief Scientist/National Risk Practice Lead
Brown and Caldwell

Finally, there is a comprehensive textbook in the interdisciplinary field of environmental management! I am very impressed by the breadth of the book (covering topics in natural science and engineering as well as business and social science), thanks to the diverse expertise of the editors (environmental geochemistry/biology/science and business/marketing). This is not only a very useful textbook for students but also a valuable resource book for practitioners in the environmental management field.

DR. PENGFEI ZHANG
Professor and Chair, Department of Earth and
Atmospheric Sciences
City College of New York, CUNY

SECTION I

ENVIRONMENTAL MANAGEMENT: THE NATURAL SCIENCE AND ENGINEERING PERSPECTIVE

1

GEOLOGY IN ENVIRONMENTAL MANAGEMENT

MICHAEL A. KRUGE

Earth & Environmental Studies Department, Montclair State University, Montclair, NJ, USA

Abstract: From the geological perspective, the two overriding environmental management concerns are the destructive impact of hazardous natural events on human health and property and the deleterious impact of human activity on the natural environment. The knowledge derived from the geological sciences serves as the basis for a more enlightened approach to the reduction of unnecessary risk involved in the siting and construction of buildings and transportation networks, as well as the extraction of natural resources and waste management. Armed with such knowledge along with political sensitivity, environmental managers will have opportunities for positive social impact in negotiating the challenges as they weigh costs, risks, and benefits. When considering natural resource and energy issues, environmental managers should foster science-based solutions to maximize resource utilization while minimizing harmful impacts, bearing in mind externalities and long-term consequences.

The chapter provides an overview of key geological aspects of environmental management, illustrating fundamental principles via representative examples. The main geological subjects addressed include volcanic eruptions, earthquakes, coastal processes, freshwater resources, waste management, and fossil fuel resources. They are discussed in tandem with their associated environmental problems and risks.

Keywords: volcanic hazards, lahar, earthquake hazards, tsunami, seismic safety, liquefaction, slope instability, coastal hazards, barrier island, flooding hazards, eutrophication, saline lake, groundwater overdraft, sinkhole, solid waste disposal, nuclear waste disposal, coal mining, acid mine drainage, petroleum system, Deepwater Horizon oil spill, hydraulic fracturing ("fracking").

1.1 Introduction	4	
1.2 Volcanic Hazards	4	
1.2.1 Mt. Vesuvius: Ancient Pompeii and Modern Naples	4	
1.2.2 Volcanic Eruptions and Aviation	6	
1.2.3 Lahar Hazards: Northwestern Washington State	7	
1.3 Earthquake-Related Hazards	7	
1.3.1 The Great East Japan Earthquake and Tsunami	7	
1.3.2 California Seismic Safety Standards and Preparedness	10	
1.3.3 Liquefaction Hazards in San Francisco	10	
1.3.4 Earthquake-Induced Slope Failures in the San Francisco Bay Area	10	
1.4 Coastal Processes and Environmental Management	13	
1.4.1 Coastal Sediment Transport Dynamics and Engineered Shore Structures	13	
1.4.2 Coastal Sediment Transport Dynamics in Response to Major Storms and Sea Level Rise	18	
1.5 Environmental Management of Rivers and Lakes	19	
1.5.1 Flooding Hazards	19	
1.5.2 Eutrophication: "Too Much of a Good Thing"	21	
1.5.3 Lakes in Arid Regions	23	
1.6 Groundwater Management and Karst Hazards	24	
1.6.1 Groundwater Overdraft: Impact on Agriculture and Municipal Water Supplies	24	
1.6.2 Karst Topography and Sinkhole Hazards	27	
1.7 Geological Factors Impacting Waste Management	29	
1.7.1 Municipal Solid Waste Disposal and Landfill Leachate	29	
1.7.2 Nuclear Waste Disposal	31	
1.8 Energy Resource Extraction and its Environmental Consequences	32	
1.8.1 Coal and Coal Mining	32	
1.8.2 The Petroleum System	36	
1.8.3 Petroleum and Natural Gas Extraction: Environmental Considerations	37	
1.8.4 Consideration of Externalities	40	
1.9 Concluding Remarks	40	
Appendix 1.A Geographic Coordinates of the Examples Presented.	42	
References	42	

An Integrated Approach to Environmental Management, First Edition. Edited by Dibyendu Sarkar, Rupali Datta, Avinandan Mukherjee, and Robyn Hannigan.
© 2016 John Wiley & Sons, Inc. Published 2016 by John Wiley & Sons, Inc.

1.1 INTRODUCTION

From the geological perspective, there are two overriding environmental management concerns: (i) the destructive impact of hazardous natural events on human health and property and (ii) the deleterious impact of human activity on the natural environment. This two-way flow of undesirable influence creates a complex web in which the ambitions of groups of people from differing economic and social circumstances come into conflict. The dynamism of Earth's near-surface processes has often caught people unaware, leading to sudden loss of life and livelihood. The knowledge derived from the geological sciences serves as the basis for a more enlightened approach to the reduction of unnecessary risk involved in the siting and construction of buildings and transportation networks, as well as the extraction of natural resources and waste management.

When facing the possibility of a natural disaster—volcanic eruption, earthquake, landslide, and flood—the avoidance of hazardous areas is perhaps the simplest and surest protective measure. However, in many cases, homes, infrastructure, and even entire cities are built in the line of fire. The aesthetic appeal of a seaside house, favorable farming conditions, and ingrained traditions all motivate residents to remain in potentially hazardous areas. Engineers are able to devise complex and effective defenses, but these are expensive. Since these measures may provide protection for disasters that have not occurred within historic memory, decision makers may be reluctant to mandate them. The tensions between the scientific/technical and the political sides of environmental issues play out in *cost–benefit* and, especially, *risk–benefit* analyses (Petroski, 2013), by which the possible rewards to be derived from an activity or policy are weighed against the expense or hazard entailed. One management question that must be considered is: Who pays for damages to private property caused by natural disasters?

This chapter provides an overview of the geological underpinnings of several key environmental concerns. It is neither comprehensive nor encyclopedic, but rather employs illustrative examples to highlight the major environmental management issues presented. It takes the point of view that environmental management should (i) protect populations from undue exposure to natural hazards and (ii) protect natural systems from undue *anthropogenic* pressure.

Several of the examples presented are illustrated with aerial or satellite imagery. If the reader is not acquainted with the benefits of and insights derived from the "bird's-eye view," observation of familiar places with web-based utilities such as Google Earth is recommended. As an invitation to further exploration, the appendix to this chapter provides the coordinates of all places discussed herein, so that they may be virtually visited at the reader's leisure. Recognizing that this book will be read by specialists from different disciplines, the geological, ecological, and engineering technical terms are flagged with italic type upon first use. The appended reference list to this chapter includes many publications that merit further reading, including several peer-reviewed general review articles along with contributions from journalists succinctly summarizing the political and social implications of the technical issues raised.

1.2 VOLCANIC HAZARDS

We will examine two of the principal types of volcanic hazards afflicting urbanized areas. One is the direct, often violent, eruption of *pyroclastic* materials. Unlike liquid lava, pyroclastic debris solidifies at the same time that the volcano ejects it, so that it may coat the ground as a blanket of fine volcanic ash or roar down the volcano's slope as a hot flow impossible to outrun. In an explosive eruption, ash is typically propelled high into the atmosphere and may be transported long distances by prevailing winds. The other hazard type is a *lahar*, in which erupted material such as ash mixes with water, usually during a volcanic disturbance, and destructively flows down river valleys like a fast-moving wall of wet concrete.

1.2.1 Mt. Vesuvius: Ancient Pompeii and Modern Naples

The disastrous eruption of Mt. Vesuvius in ancient Roman times (AD 79) began to capture public imagination during the eighteenth century, when serious excavations started there. The fascination continued in modern times (Fig. 1.1), and the ruined Roman cites of Pompeii and Herculaneum remain popular tourist destinations, having details remarkably preserved by the eruption's massive volcanic deposits. The environmental management aspect of the tragedy leads one to consider why people would build cities near a volcano and how they may best prepare themselves for a possible future eruption, in other words to contemplate a classic risk–benefit dilemma.

Central western Italy, a fertile region with a pleasant climate and good harbors, would certainly have been an attractive place to settle, then as now. Peoples of the ancient world did not have the benefits of our scientific knowledge base nor the instrumentation that we use to study active volcanoes. Nonetheless, contemporary historical accounts indicate that the AD 79 eruption was preceded by strong earthquakes, and it could be presumed that inhabitants in cities near Vesuvius did feel some concern for their safety from time to time. While archeologists estimate that the death toll within Pompeii proper may have been as high as 2000, it is more difficult to estimate how many survived (or at least managed to flee beyond the city limits), perhaps between 6 and 20 thousand. The eruption occurred in two phases. During the initial phase, the city was blanketed with ash fallout (*pumice lapilli*) propelled by a pressurized steam (*phreatic*) explosion. This was followed by several waves of flow deposits (*pyroclastic density currents*) possibly as hot as 400°C. Many of the victims during the initial phase were sheltering indoors and killed as roofs

1.2 VOLCANIC HAZARDS 5

FIGURE 1.1 Plaster casts of victims of the disastrous AD 79 eruption of Vesuvius. (Photo: Lancevortex; Wikipedia, 2014a).

collapsed under the weight of the accumulating ash, particularly flat or low-angle roofs. The hot pyroclastic currents of the second phase were more thoroughly destructive, killing those who had managed to hide safely during the ash fall. Walls facing the oncoming flow collapsed, while those standing parallel to the flow direction tended to endure. The force of the flow diminished as it progressively demolished wall after wall in its path. Since the "milder" initial phase of the eruption was nonetheless deadly, we may conclude that structures built near volcanoes prone to ash eruptions should, at a minimum, have steeply sloping roofs (Luongo et al., 2003a, b).

It is instructive to view Mt. Vesuvius and the surrounding urban areas in their spatial context. A map prepared from a satellite image (Fig. 1.2) shows that Pompeii is only about 8 km southeast of the volcanic vent. During the AD 79 eruption, prevailing winds blew the initial ash cloud toward the south, burying Pompeii and hindering evacuation prior to the arrival of the more destructive pyroclastic flows. Although Herculaneum is even closer to the summit, it lies to the west and was spared the brunt of the ash fall, giving most of its inhabitants time to flee in advance of the flows which ultimately engulfed both cities (Luongo et al., 2003a).

To acquire the image used in Fig. 1.2, the satellite's sensors recorded *visible light* as well as *infrared radiation*. The colors were added later during image processing and were chosen to

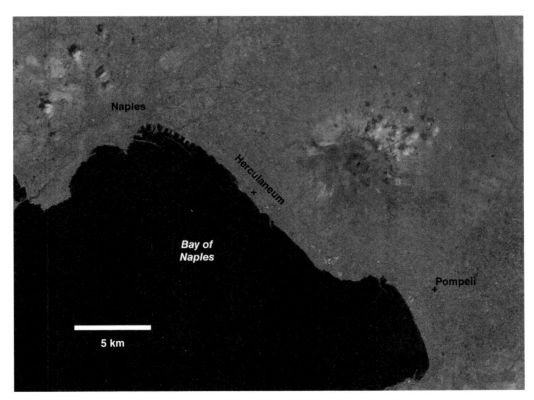

FIGURE 1.2 Map showing the proximity of the Mt. Vesuvius volcanic crater to the ruins of Pompeii and Herculaneum, as well as the present Naples metropolitan area. Base: natural color Landsat image from 1999, US Geological Survey. North is at the top of all maps in this chapter unless otherwise indicated.

give a "natural" look, with vegetation appearing green and urban areas tan to mauve. Water appears black and the few clouds present are white wisps. Close examination of Fig. 1.2 allows us to appreciate how near the present-day Naples metropolitan area lies to a still-active volcano and to imagine what might happen if Vesuvius erupts again, this time with a much larger population in the danger zone. While the volcano is intensively monitored for eruption precursors such as earthquakes and ground swelling, prompt and effective evacuation would obviously present a daunting logistical challenge. Mobilization due to a false alarm would lead to widespread consternation; thus, decision makers would be under intense pressure while attempting to make the correct call. An excess of caution may also have consequences, as an Italian court sentenced seven scientists to prison for manslaughter for failing to issue a safety warning in advance of the deadly 2009 earthquake in L'Aquila, Italy (Povoledo and Fountain, 2012).

1.2.2 Volcanic Eruptions and Aviation

While the authorities in Naples are prudent to prepare for a nearby eruption, those in tectonically quiet northern European capitals such as London, Paris, and Berlin do not normally worry about such matters. That changed in April 2010, during the eruption of Eyjafjallajökull in a remote part of Iceland more than 2000 km to the northwest (Fig. 1.3). While this eruption was considered to be of moderate intensity, prevailing air currents swept a large mass of fine pyroclastic ash over most of Europe, paralyzing commercial air traffic for several weeks. The concern was that the tiny ash particles (<63 µm in diameter) could cause jet engines to stall in flight, if they are present in sufficient concentration in the atmosphere (>2 mg m^{-3}). The 2010 eruption began with a phreatic phase, as melting glaciers on the summit of Eyjafjallajökull provided water for the pressurized steam. The resulting explosion propelled pyroclastic debris as high

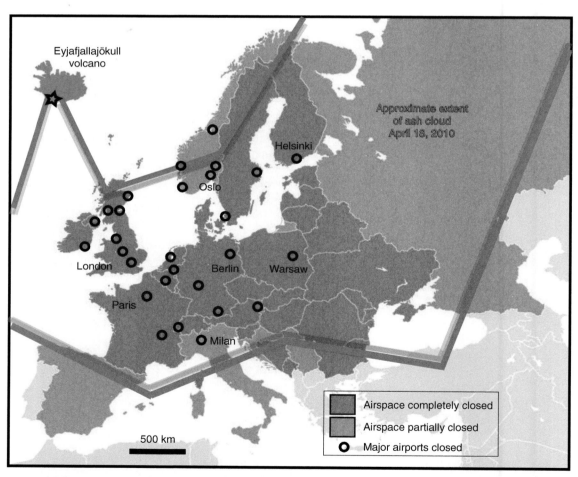

FIGURE 1.3 Map showing disruption to European air travel during the 2010 eruption of Eyjafjallajökull on Iceland. Green lines enclose area impacted by the ash cloud on April 18, 2010 (Met Office, 2014). On that day, airspace was completely closed in countries colored red and partially closed in those in orange (Wikipedia, 2014b). Black circles mark major airports closed on the previous day (New York Times, 2010).

as 11 km into the air, which is roughly the cruising altitude of commercial airliners. Eruptions continued for the next 40 days, with some pulses of ejecta reaching nearly that altitude again. Aside from the immediate concern for passenger and aircraft safety, the environmental management issues in this case are primarily financial, as this eruption caused thousands of flight cancelations and over a billion dollars in direct losses to airlines, with attendant costs rippling through the world economy. Health issues are another factor, as the finer ash particles provoke respiratory ailments at sufficiently high concentrations in air. Nine Volcanic Ash Advisory Centers (VAACs) have been established to provide timely alerts should an eruption anywhere in the world pose a hazard to aviation or human health, based on numerical modeling of eruptive and atmospheric conditions (Langmann et al., 2012).

1.2.3 Lahar Hazards: Northwestern Washington State

Major active *stratovolcanoes*, including Mt. Rainier near the Seattle–Tacoma (Washington) metropolitan area, are formed by alternating eruptions of liquid lava and solid pyroclastics such as ash. Mt. Rainier's last major eruption was about 1000 years ago, but *radiometric* dating methods exploiting the known *half-lives* of radioactive *isotopes* naturally present in its rocks have recognized numerous eruptive episodes that occurred at the present volcanic edifice over the past one million years. While a lava or pyroclastic eruption of Mt. Rainier would be sufficient to trigger governmental disaster response efforts by itself, the flanks of this volcano are also prone to hazardous mudflows (Fig. 1.4). These lahars form as hot erupting material mixes with snow and ice in the mountaintop glaciers, creating a dense, but fast-moving and deadly slurry. Some of the material on the peak of Mt. Rainier has been *hydrothermally* altered during past eruptions in the presence of hot, chemically reactive fluids. The alteration has weakened these rocks, making them prone to catastrophic collapse during an eruptive disturbance, thus forming another source of lahar material (John et al., 2008).

We can see in Fig. 1.4 that past lahars follow existing stream valleys as they descend the mountain, initially radially in all directions, but then turning westward and northwestward toward the coast, unfortunately the most densely populated part of the region. The map distinguishes between major lahar events with a 500–1000 years *recurrence* interval and flows that are smaller, but likely to occur more frequently (every 100–500 years). Both the magnitude and recurrence intervals are important in assessing risk factors when planning for lahars (or any other type of natural disaster). The destructive power inherent in lahars was evident on Mt. Ruapehu, New Zealand, in 2007, during an event that was triggered by a landslide of previously erupted pyroclastic material (Fig. 1.5).

Lahars on Mt. Rainier may be "hot," spawned by active eruptions, or "cold," rarer events, triggered by landslides. In either case, the downstream communities would be at risk, necessitating early warning systems and evacuation plans. Precursor earthquake activity would herald an eruption and thus also provide a timely warning of an impending lahar. On the other hand, sensors placed along the stream valleys would likely be the only signal of an approaching lahar resulting from a cold landslide, leaving little time for evacuation. In the Seattle–Tacoma region, educating the public about the dangers of lahars and effective emergency procedures forms an integral part of disaster preparedness plans. In addition to residents in their homes, planners must account for workers at their job sites, tourists, and patients in hospitals, as well as commuters and travelers in transit at the time of an event (Wood and Soulard, 2009). Such complexities pose a serious challenge to environmental managers and public safety officials.

1.3 EARTHQUAKE-RELATED HAZARDS

The *tectonic* forces jostling Earth's *lithospheric plates* are the drivers of explosive eruptions at volcanoes such as Vesuvius, Eyjafjallajökull, and Rainier. They are also responsible for earthquakes, which principally occur at the boundaries between tectonic plates. In some regions, plates collide, crumpling and twisting into nonvolcanic mountain ranges such as the Himalayas and the Alps. In other areas, the results of a collision will entail one plate *subducting*, diving down beneath its overriding opponent. In still other parts of the globe, plates are observed to pull away from each other, forming an ever-widening *rift* in the intervening gap. In a fourth tectonic style, opposing plates slip past one another in a *strike–slip* or lateral motion. We will review several representative examples of earthquake-related hazards provoked by subduction (Japan) and lateral motion (California), along with attendant environmental management considerations.

1.3.1 The Great East Japan Earthquake and Tsunami

Northern Japan sits perilously close to the plate boundary at which the Pacific plate is subducting beneath the far western extension of the North American plate at an average rate of about 9 cm year^{-1}. In 2011, the powerful Great East Japan earthquake registered a slip of more than 20 m at the plate boundary in a matter of seconds, at its *focus* under the Pacific Ocean about 130 km off the coast of the city of Sendai. Having an extraordinarily high *moment magnitude* (M_w) of 9.0, it was one of the strongest earthquakes in recorded history, killing about 17,500 people, as well as destroying coastal municipalities and infrastructure, notably the Fukushima nuclear power station (Lin et al., 2012; Wang et al., 2012). The statistics reflect the extreme misfortune. In addition to

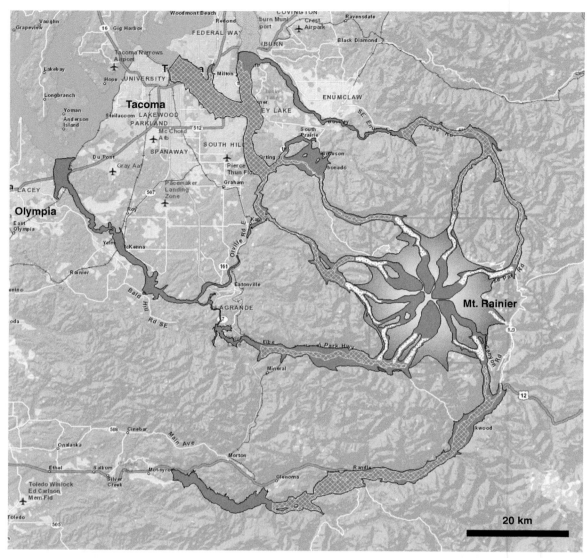

FIGURE 1.4 Map showing the lahar hazards in the region south of Seattle, Washington, in the event of a volcanic eruption at Mt. Rainier. Solid pink, large lahars with a 500–1000-year recurrence interval; cross-hatched pink, moderate lahars with a 100–500-year interval; speckled blue, old lahars buried by younger sediment; gray, lava or pyroclastic flows. *Source:* Washington State Geology and Earth Resources Division. (Washington State Department of Natural Resources, 2013.)

the deaths, about a half million residents were rendered homeless, millions lost electrical service, transportation systems were paralyzed, wide areas were contaminated by radioactive materials from the Fukushima plant, and damage costs were expected to exceed $200 billion (Davis et al., 2012).

The sudden, strong vertical motion on the seafloor triggered a powerful *tsunami* with *run-up* heights from 3 to 35 m, depending on the location along the Japanese coast. Roughly 92% of the fatalities due to this earthquake are directly attributable to the tsunami (Lin et al., 2012). Figure 1.6 dramatically illustrates the awesome destructive force involved, which overwhelmed the protective measures that had been in place. Regional planning accounted for strong earthquakes of only up to M_w 7.7. While it is not possible to forecast earthquakes days or hours in advance (as can be done with major storms), long-term predictions did foresee an event stronger than M_w 8.0, which perhaps should have motivated planners to further strengthen engineering standards and coastal defenses (Davis et al., 2012). (Note that the magnitude scale is *logarithmic*, such that an M_w 8.0 event releases nearly triple the energy of an M_w 7.7 earthquake and an M_w 9.0 is almost 90 times stronger.)

Disaster management, as a critical component of environmental management, requires long-term planning in

FIGURE 1.5 The 2007 Ruapehu lahar in New Zealand. For scale, note the green picnic table in the lower left corner. Photo courtesy of Geoff Mackley.

FIGURE 1.6 The powerful tsunami following the 2011 Tohoku earthquake stranded a ferry boat atop a building in Otsuchi, Japan. AP Photo/The Yomiuri Shimbun.

advance of a disaster, alertness prior to an event (to the extent that warnings are possible), emergency relief in the immediate aftermath, and continuing recovery efforts. The long-term processes before and after a disaster particularly benefit from environmental awareness, as they involve defensive engineering that should do more good than harm, land use planning to ideally keep the most vulnerable zones free of buildings and infrastructure, and the establishment of the safest possible evacuation routes (Nakanishi et al., 2013).

1.3.2 California Seismic Safety Standards and Preparedness

The Pacific plate slips northward relative to the North American plate along coastal California from the Mexican border to Mendocino County north of San Francisco. The *right-lateral* strike–slip San Andreas Fault delineates this unusual plate boundary and passes through or near to the state's two major urban clusters, each with millions of inhabitants. Major destructive earthquakes have occurred here in historic times, notably the M_w 7.8 San Francisco earthquake of 1906. Since the motion on a strike–slip fault is primarily horizontal rather than vertical, a major co-occurring tsunami is unlikely. Nonetheless, prudent residents remain alert to the possibility of a tsunami originating in other parts of the tectonically active Pacific Rim (Fig. 1.7).

The California Geological Survey (CGS) has been charged by the state legislature with the implementation of its Seismic Hazards Mapping Act. As part of the risk assessment, probabilistic projections constitute a major portion of this effort. Since the future cannot be precisely predicted, modeling the likelihood of a particular event becomes the sensible means for planning in the face of uncertainty. This is a general principle applicable to risk–benefit studies involving all types of natural threats. For example, seismic hazard maps should "show ground shaking levels which have a 10% probability of being exceeded in 50 years" (California Geological Survey, 2003). In addition to specifying probable occurrence, the regulations stipulate the consideration of seismic source type, earthquake frequency (based on the *paleoseismic* record or geological evidence of past earthquakes), range of earthquake magnitudes, seismic wave *attenuation* or loss of strength over distance at a site, and the extent to which a particular building site will amplify ground motion. All of this is done in the context of the Uniform Building Code and land use policy (California Geological Survey, 2004).

1.3.3 Liquefaction Hazards in San Francisco

An additional factor specified in the California seismic regulations (CGS, 2004) is *liquefaction* or the temporary loss of cohesion in water-saturated soils during an earthquake (Holzer et al., 2010). The potential for dangerous liquefaction is evaluated by geologic mapping, examination of groundwater data, drilling of geotechnical boreholes to sample subsurface soils and sediments, and seismic data collection (CGS, 2004). Areas of greatest concern tend to be low lying and underlain by recently deposited sediments and particularly by artificial fill often used in the past to convert wetlands into building sites. The results of such studies are summarized on maps, such as the one prepared for San Francisco (Fig. 1.8), on which the zones colored green are at risk for liquefaction. One of these vulnerable areas is the Marina District, where buildings did indeed collapse during the 1989 M_w 6.9 Loma Prieta earthquake (Fig. 1.9). Buildings of a particular size and construction were the most severely affected, apparently as they were by chance most closely attuned to the frequency of the seismic waves and because they were constructed before the advent of modern seismic standards (Sivathasan et al., 2000). Such a map is useful in determining which properties are candidates for seismic retrofitting, how stringent the engineering requirements should be for new structures, and which properties might be too risky to purchase.

1.3.4 Earthquake-Induced Slope Failures in the San Francisco Bay Area

The last criterion specified in the state regulations concerns landslides caused by earthquakes. The objective in this case is to consider slope stability and to compute the *factor of safety* for a given location based upon soil properties (*shear strength*), slope angle, and likely nature of seismically induced ground motion. Mapping of old landslides is essential, made more effective with the advent of *aerial photography* and other forms of *remote sensing* such as those involving radar. Recent landslides with historical records are the most useful for these purposes. The results of

FIGURE 1.7 A street sign near the Pacific Ocean beach in San Francisco, California, indicating the tsunami evacuation route heading inland toward higher ground. Photo: M.A. Kruge.

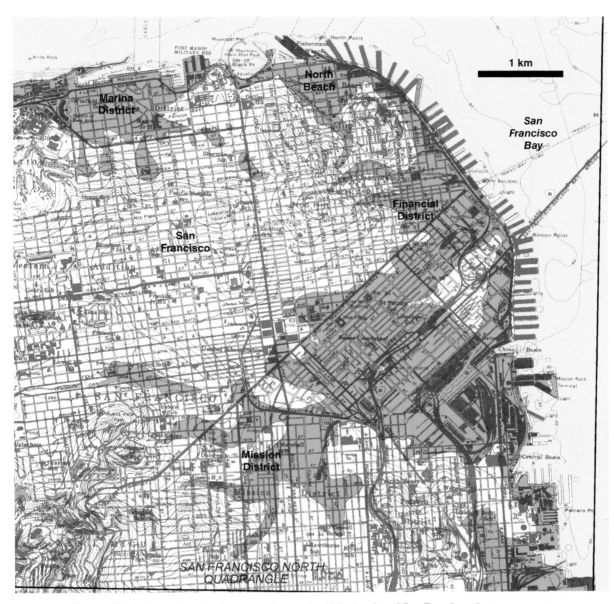

FIGURE 1.8 Map showing the zones (in green) within the city of San Francisco that are prone to liquefaction during an earthquake. Blue zones are prone to landslides. *Source:* Modified from California Division of Mines and Geology (2000).

the investigations can then be summarized in map form, such as the blue-shaded zones marking landslide-prone areas in San Francisco in Fig. 1.8. As might be expected, the likelihood of seismically induced landslides increases with increasing earthquake magnitude, proximity to *epicenter*, and local shaking intensity (as measured on the *Mercalli scale*). Rock fall and slides, as well as soil slides, slumps, and avalanches, are among the most common types of earthquake-related landslides (Keefer, 2002).

Housing developments in Daly City, California (just to the south of San Francisco on the Pacific coast), are vulnerable to earthquake-induced slope failures as they were built within or near the San Andreas Fault Zone (SAFZ). A sequence of three aerial images acquired years apart depict the hazards involved (Fig. 1.10). The first image, taken in 1956 prior to the real estate development, provides the context (Fig. 1.10a). The SAFZ trace, shown in orange, trends northwest–southeast across the image. Since this is a right-lateral strike–slip fault, the western block is moving northward relative to the eastern block, expressing a major tectonic feature—the boundary between the Pacific and North American plates at this location. The epicenter of the M_w 7.8

FIGURE 1.9 Aid workers carrying supplies past a collapsed building in San Francisco's Marina District in the aftermath of the 1989 Loma Prieta earthquake. Soils in the Marina District are prone to liquefaction (Fig. 1.8). Photo: California Conservation Corps (2013).

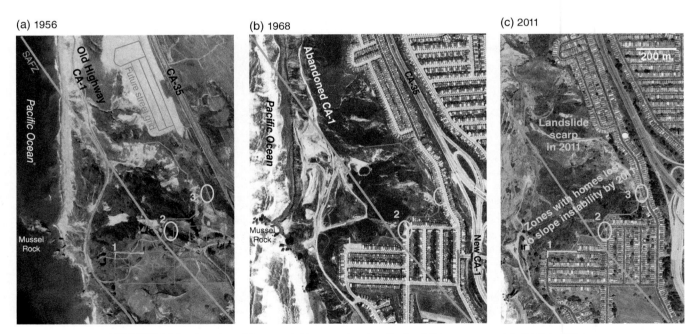

FIGURE 1.10 A series of three aerial images showing the landslide-prone coastal district of Daly City, California, where the San Andreas Fault Zone (SAFZ, traced approximately in orange) intersects the Pacific coast. (a) Taken in 1956, prior to the construction of the housing developments. The major landslide along the SAFZ southeast of the fault's intersection with the ocean appears dark gray. Zones 1–3 mark the sites of future problems seen in (c). (b) The view taken in 1968 shows the densely clustered single-family homes that were recently completed. (c) In the view from 2011, houses are missing in zones 1, 2, and 3 (along with a portion of the street in zone 1) as the landslide scarp (in blue) encroached upon the development. Base images: US Geological Survey. SAFZ location: Bonilla et al. (1998).

San Francisco quake in 1906 was in the SAFZ only about 5 km north of this location (Chourasia et al., 2008). In 1956, California Highway 1 still ran close to the sea, but was obviously at risk, as a large landslide zone is clearly visible around it. The blue line in Fig. 1.10a traces the position of the *scarp* at the head of the landslide as it appeared in 2011. The large light gray patch in the upper center of the image is land that was being prepared for housing construction, which was already underway to the immediate north. The red lines trace the position of the future street grid, perilously close to the scarp in the SAFZ at the bottom of the photo.

Housing construction continued into the 1960s, filling what had been open land with hundreds of closely spaced homes, evident in the 1968 image (Fig. 1.10b). By this time, the coastal stretch of Highway 1 had been long abandoned due to slope instability and rerouted to the east. In the view from 2011, the three zones marked in yellow indicate places where homes were lost due to slope instability (Fig. 1.10c). An entire half block at zone 1 has been abandoned and demolished, along with the street (traced in yellow), and gaps in the house sequences at zones 2 and 3 are apparent. Looking more closely at site 1 in 2011, the missing section of street (in yellow) and the eerie absence of houses in the fault zone are evident (Fig. 1.11a). A similarly close view of zone 3 shows the gap, wide enough for five houses, their lots consumed by the advancing scarp (Fig. 1.11b). The house marked with the red X at the edge of the gap (Fig. 1.11b) is also marked in Fig. 1.12, which shows the details on the ground at zone 3. Close examination of the aerial image from 1968 (Fig. 1.10b) reveals that eight homes planned for the northwestern corner of zone 1 were prudently never constructed, while 24 houses that were built there had to be demolished over the years.

Plagued by cracked walls, collapsed roadways, the potential for ruptured gas and water lines, and worse, the municipality was ultimately obliged to condemn properties at these sites. Conditions worsened after heavy winter rains (particularly intense during episodic *El Niño Southern Oscillation or ENSO* years) saturated the soil, reducing its load-bearing abilities. Stabilization measures would have proven to be too costly and ultimately likely ineffective. The environmental management process involved a search for funds to buy out long-time property owners as well as working with the residents on a one-to-one basis (Pence, 2000). It should have been clear in the 1950s just from the aerial photograph (Fig. 1.10a) that landslide-prone sites near a major active fault zone were unsuitable construction sites. This raises the environmental management concerns regarding land use, building codes, and financial responsibility for damages due to natural hazards.

In a relevant European case, owners of a Spanish hotel and adjacent homes destroyed in a 1994 landslide sued local and regional governments for damages. While the buildings had been constructed in a known slide-prone area, the structures had nonetheless been granted building permits that did not require prior geotechnical investigations. The suit against the governmental bodies might have been successful, except that the landslide was triggered by an intense, record-breaking rainstorm, which the court ruled to be *force majeure* (an "act of God"), thereby relieving the defendants of responsibility in this instance (Montoro Chiner, 1997).

1.4 COASTAL PROCESSES AND ENVIRONMENTAL MANAGEMENT

Coastal regions not threatened by major volcanic or seismic events may nonetheless be exposed to significant risks. Urbanization may encroach upon zones vulnerable to storm damage, exacerbated by the prospect of rising sea levels due to *global climate change* (Gornitz et al., 2002). Narrow highways on populated *barrier islands* and *spits* may prove to be inadequate evacuation routes in the face of an approaching major storm. The population of shore areas may swell with vacationers and temporary residents during warmer months, complicating emergency procedures should dangerous weather events require them. In spite of potential hazards, the beauty of coastal areas makes them attractive to visitors and property owners alike. Inevitably, complex environmental management issues arise, such as the conflict between the need for public open space and private property rights.

1.4.1 Coastal Sediment Transport Dynamics and Engineered Shore Structures

Frequent visitors to a coastal area may casually notice changes at a familiar beach, such as losses of sand after a winter storm. Residents of seaside vacation homes often demand governmental protection of their private property, by means of *beach nourishment* (sand replenishment), *dune* construction, or installation of durable *groins, jetties, sea walls, breakwaters,* and *revetments*.

The Atlantic shores of Long Island (New York) and New Jersey offer good examples of human interaction with dynamic coastal processes. For instance, in Atlantic Beach, New York, a series of groins was constructed along the beach to protect it from marine *erosion* (Fig. 1.13). Each groin was formed by piling large boulders in a line about 70 m long reaching out into the sea, spaced at intervals of about 200 m. At this location on the south shore of Long Island, the *longshore current* moves from east to west (right to left in the image), in turn transporting sediment (*longshore drift*) along the beach in the same direction. The groins, affixed perpendicularly to the shore, interfere with this process, and thus, sand accumulates on their *updrift* (east) sides, while sand is removed from the *downdrift* sides. If groins are correctly sized and spaced, they should have the intended overall effect of capturing the laterally migrating sand, protecting the beach as a whole. However, adjacent unprotected

FIGURE 1.11 Enlargements of the 2011 aerial image (Fig. 1.10c) showing details of the Daly City, California, landslide area in the San Andreas Fault Zone. Smaller houses are ca. 8×14 m. (a) Zone 1 showing missing houses and street (traced in yellow) at the head of the scarp. (b) Zone 3 showing a gap (yellow arrow) wide enough for five houses and the encroaching scarp. House X is also marked in Fig. 1.12. Base image: US Geological Survey.

FIGURE 1.12 Photographs (Daly City, California, 2013) showing the area of missing houses at zone 3 in Figs. 1.10c and 1.11b. For reference, the house marked with the red X is the same in all three photos. (a) View from the street looking southwestward toward the ocean showing an odd gap in the pattern of closely spaced homes. (b) View at the site itself showing the advancing head of the scarp. (c) View of the gap (marked in yellow) looking upward and eastward from the base of the landslide. Photos: M.A. Kruge.

beaches, particularly those downdrift of the groin field, may suffer increased erosion, a case of "robbing Peter to pay Paul" (Kana, 1995; Gornitz et al., 2002).

The *sediment budget* is of great concern to coastal planners, particularly along the eastern seaboard of the United States, who are in many instances compelled to repeatedly undertake expensive artificial beach nourishment operations when the natural sand supply fails to keep pace with erosion. Montauk Point, New York, forms the easternmost tip of Long Island's South Fork, jutting out like the prow of a ship into the

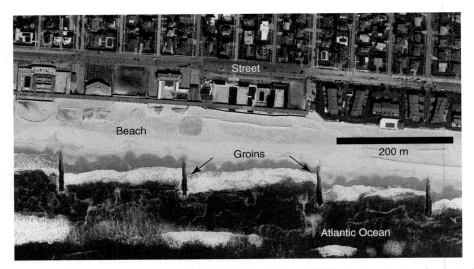

FIGURE 1.13 Aerial view of Atlantic Beach, New York, in 2010 showing how groins perturb patterns of sand deposition on the beach. The four groins in this photo are the dark, linear features perpendicular to the shoreline about 70 m long, constructed by piling large boulders. Sand is being transported westward by the longshore current (from right to left in the photo) and accumulates preferentially on the east sides of the groins, while the west sides are correspondingly starved of sand. Base image: US Geological Survey.

open Atlantic Ocean. This narrow peninsula is an eroding *headland* and is a principal natural source of sediment for the Long Island beaches to the west (McBride and Moslow, 1991). Like all of Long Island, Montauk is comprised of poorly consolidated *glacial till* and *outwash*, abandoned in place as the leading edge of the continental ice sheet began to melt back, some 20,000 years ago during the Pleistocene Epoch (Kana, 1995). The eighteenth-century lighthouse at Montauk Point, commissioned by President George Washington, is an important historical landmark and continues to provide navigational guidance. Its site is highly vulnerable to erosion by deep ocean waves approaching from the east and south, so environmental engineers undertook to preserve it by cloaking the soft glacial bluffs at its base with a massive, high revetment of large boulders (Fig. 1.14). Revetments are not a permanent solution in dynamic coastal settings, and they require periodic maintenance and replacement of displaced boulders (Yang et al., 2012).

A prominent feature of the Atlantic Coast from Long Island to Florida is the presence of numerous long, narrow barrier islands and spits oriented parallel to the mainland. Both are similar in their appearance, sediment dynamics, and environmental issues, except that spits are peninsulas attached at one end to the mainland. Atlantic Beach (Fig. 1.13) sits on Long Beach, one such barrier island. To the east lies Fire Island, measuring some 50 km from east to west, the longest barrier island in the New York/New Jersey region (Fig. 1.15). A barrier island separates the open ocean from a *lagoon* (here the Great South Bay), on the other side of which sits the mainland. As is typical of a barrier island or spit, Fire Island is very narrow: a half kilometer or less along much of its extent. The sea connects with the lagoon via inlets at both ends of the island, which are important conduits for tidal transport of sediments in both directions (Leatherman, 1985; Kana, 1995; Lentz et al., 2013). Inlets are ephemeral features and will tend to migrate laterally along the island as a function of sediment transport, in this instance the predominantly westward longshore drift. Historical records document that Fire Island Inlet migrated 8 km toward the west between 1825 and 1940, at which time engineers constructed a rock jetty at the western margin of the island to "lock" the channel in place for navigational convenience. As with other such artificial hardening measures, this produced unintended consequences and created the need for continual maintenance, in this case dredging the sand deposited by the westward drift (Leatherman, 1985). Sediments also migrate westward through the lagoon, entering via Moriches Inlet and exiting at Fire Island Inlet. There they accumulate as an *ebb tidal delta*, left behind by the falling tide just seaward of the inlet. In addition to the headlands to the east, Pleistocene glacial deposits a short distance offshore provide a natural source of sediment (Lentz et al., 2013).

Fire Island constitutes part of the Gateway National Seashore, which entails a complex framework of land management issues, as the island has wilderness areas; federal, state, and county parkland; and private communities. While new hardening measures are forbidden, beach replenishment and *beach scraping* (to obtain sand to enlarge protective dunes in residential zones) are ongoing. Dredging of inlets and offshore *borrow pits* provides the material for

1.4 COASTAL PROCESSES AND ENVIRONMENTAL MANAGEMENT   17

FIGURE 1.14 At Montauk Point, New York, the historical landmark lighthouse is protected from the Atlantic Ocean by a massive revetment of large boulders. (a) Aerial view from a 2010 high-resolution orthophoto (US Geological Survey). (b) View looking northward from the beach in 2011. Photo: M.A. Kruge.

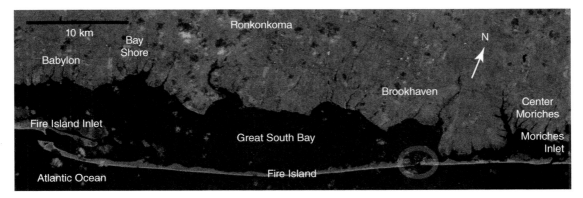

FIGURE 1.15 Satellite view of south-central Long Island, New York, in 2013 showing Fire Island. This barrier island was breached (where marked by yellow oval) in 2012 during Hurricane Sandy. Natural color Landsat image: US Geological Survey.

beach nourishment (Lentz et al., 2013). These artificial means are employed since the sand budget reflects a deficit, that is, the rate of erosion for the system as a whole exceeds the natural sediment supply (Kana, 1995). Sand suitable for replenishment is becoming increasingly difficult to acquire along the US Atlantic Coast, leading in some cases to conflict between neighboring communities. Officials in Florida have in fact proposed using crushed recycled glass as a sand substitute (Dean, 2012; Alvarez, 2013).

1.4.2 Coastal Sediment Transport Dynamics in Response to Major Storms and Sea Level Rise

Coastal barrier islands are impermanent features, subject to landward migration ("rollover") in response to sea level rise. In the absence of a human population, the ocean beach would continue to exist, progressively repositioned closer to the mainland, as the lagoon either shrank or *transgressed* in turn upon the shore behind it. Environmental management issues arise when buildings and infrastructure are placed on an ephemeral entity, particularly one that might appear relatively stable for years (Gornitz et al., 2002). Barrier island rollover is episodic, most dramatically in evidence after a major storm event, during which the narrow island may experience *overwash* in which powerful, storm-driven waves completely swamp a segment of the island. These waves erode the ocean beach and transport the sediment all the way across the island to the lagoon, in effect shifting the island's position bit by bit (Lentz et al., 2013).

Hurricane Sandy struck coastal New York and New Jersey in 2012 while it was in its *posttropical cyclone* phase, having migrated northward from its tropical zone of origin retaining tremendous strength. In part due to seabed morphology, the eastern portion is the most overwash-prone section of Fire Island (Leatherman, 1985; Lentz et al., 2013). Indeed, Hurricane Sandy opened a new inlet there some 260 m wide, at a locality unsurprisingly called Old Inlet (yellow oval, Fig. 1.15). It remained open in 2013, noticeably improving water quality in the lagoon due to increased water circulation (Foderaro, 2013). Engineers normally close new inlets quickly, since they counteract beach nourishment measures, transporting sand from the ocean beach to the lagoon to form *flood tide deltas* on the bay side of such inlets (Kana, 1995).

While there were no structures damaged during the formation of the new Fire Island inlet, that is unfortunately not always the case. During a 1992 *extratropical cyclone* (locally known as a *nor'easter*), overwash on the barrier island at Westhampton (just east of Fire Island) destroyed 60 homes. The resulting breach was soon filled and new homes were built on the very spot. That location lies immediately downdrift of a large groin field, which may increase its vulnerability (Gornitz et al., 2002). In Mantoloking, situated on a narrow spit along the New Jersey shore, Hurricane Sandy overrode beach dunes, heavily damaging

FIGURE 1.16 Oblique low-altitude airphoto looking westward at Mantoloking, New Jersey, in the aftermath of Hurricane Sandy, 2012. The houses in the foreground face the Atlantic Ocean. The streets between the rows of houses are flooded and littered with debris. The flooded houses in the background are on the lagoon (Barnegat Bay) side of the narrow peninsula, only about 200 m wide at this location. Photo: US Air Force.

beachfront properties and flooding homes on the lagoon side (Fig. 1.16). While the adjacent municipality of Bay Head was also flooded, the destructive force of the storm waves was moderated by a forgotten 130-year-old stone seawall hidden beneath the town's coastal dunes, illustrating the advantage of well-built hard coastal defenses at least for the properties directly protected by them (Irish et al., 2013).

Large-scale hard defenses, such as those in the Netherlands or the MOSE project in Venice, Italy, represent one extreme in coastal environmental management, with its advocates proposing a similar approach for the New York region (Ghezzo et al., 2010; Kleinfield, 2012). Property owners often clamor for smaller-scale hard structures (stone groins, seawalls, etc.) to protect their slice of the coast. However, armoring may have deleterious side effects, increasing the erosion rates on adjacent beaches as was discussed previously. Beach nourishment and dune enhancement are effective, softer measures, but are only temporary and must be repeated. Stricter building codes are obviously beneficial in stipulating the elevation of structures and mechanical systems for new construction as well as retrofits. Building owners in New York City who voluntarily exceeded existing code requirements fared better in Hurricane Sandy (Navarro, 2012). While some may consider New Jersey and Long Island beaches to be functionally "infrastructure" due to their importance to local economies (and thus worthy of costly maintenance), others advocate a *strategic retreat* from shore areas, cognizant of an expensive and ultimately futile battle in which nature has the advantage (Pilkey, 2012; Dean, 2012).

All of this is occurring in the context of ongoing sea level rise in the New York City region, which locally totals about 2.7 mm year^{-1} (half of which is due to global sea level rise, the remainder to local land *subsidence*). The rate is expected to increase throughout the twenty-first century according to global climate projections. Even if major storms do not increase in intensity or frequency, they will wreak greater destruction in low-lying areas, since a higher base sea level will increase the likelihood of flooding. Costs for coastal protective measures, be they hard or soft, will likely increase under this scenario (Gornitz et al., 2002). The legal challenges involved will likely be complex and divisive. An example is the proposal by Titus (1998) for *rolling easements* by which shore property owners would not be allowed to employ durable measures against sea level rise, thus preserving public access to the water as the private lots receded. In the United States, this would no doubt lead to a contest between those claiming the primacy of the constitutional proscription of "takings" and those favoring the common law tradition of open waterfronts. As in regions confronting volcanic and seismic hazards, effective environmental management of vulnerable coastal areas will demand broad-based knowledge, wisdom, and political skills.

1.5 ENVIRONMENTAL MANAGEMENT OF RIVERS AND LAKES

Rivers and lakes are economically important as sources of drinking and irrigation water, as navigation routes, as fisheries, and as tourist destinations. Their banks and shores host numerous human settlements, large and small. Rivers have commonly been a source of hydropower, from traditional water wheels to large, modern hydroelectric plants. Artificial canals and reservoirs may also be included in this category.

Episodic flooding by rivers has caused great loss of life and destruction of property. Droughts will reduce river flows and lake levels, impairing the activities enumerated earlier. All types of terrestrial surface water bodies may be afflicted by industrial and urban pollution. *Eutrophication* (excess of aquatic nutrients) of lakes, reservoirs, and lagoons can produce noxious algal blooms.

In the case of these important water resources, effective environmental management, as always, would both protect the environment from undue anthropogenic pressure and reduce environmental risks to inhabitants.

1.5.1 Flooding Hazards

In 1955, Hurricanes Connie and Diane struck the northeastern United States several days apart. The *runoff* from the storms surged down river systems in Connecticut. The heavily industrialized Naugatuck River Valley was among those strongly impacted. Factories including rubber, brass, and steel rolling mills dotted the river's *flood plain*, and many were quickly overwhelmed by the rising waters (Fig. 1.17). This event provides us with a compact case study of engineered responses to flooding hazards. As with the California landslide problem, it is instructive to view aerial images in sequence, in this case before and after the 1955 flood (Fig. 1.18). The views, taken in 1949 and 1972, show the same 1.5 km stretch of the Naugatuck River in the small industrial city of Ansonia, depicting roads, railways, factories, and housing. The flood plain ("u"; Fig. 1.18a) on the inside of the river's broad *meander* had been wisely left largely vacant prior to the flood, but other low-lying areas (e.g., "t" across the river) were densely developed with commercial and residential structures. These buildings were heavily damaged by the flood, so they were all condemned, demolished, and replaced by a shopping center, smaller commercial establishments, and a new street pattern ("t"; Fig. 1.18b). The roadway bridge ("v") at the bottom of the images was replaced by a more substantial one after the event. Most notably, an intensive engineering intervention drastically modified the river along this stretch.

FIGURE 1.17 Oblique low-altitude aerial view showing the flooded industrial zone along the Naugatuck River in Derby, Connecticut, in the aftermath of back-to-back Hurricanes Connie and Diane in August 1955. Photo courtesy of the Derby Public Library.

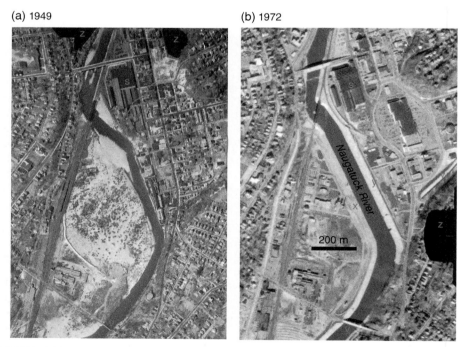

FIGURE 1.18 Aerial images of Ansonia, Connecticut, showing (a) the conditions in 1949, before the devastating 1955 Naugatuck River flood, and (b) the same area in 1972, after completion of the flood control projects undertaken in response to the 1955 flood. River flow is from north to south (top to bottom in these images). r, railroad; s, large factories; t, low-lying zone demolished and redeveloped as a shopping center and industrial park after the flood; u, flood plain largely vacant before 1955, but later the site of a sewage treatment plant; v, bridge over the Naugatuck River that was replaced after the flood; w, small tributary creek with a flood wall constructed after 1955; x, west bank of the river armored with a concrete and rock revetment after the flood; y, massive concrete flood wall on the river's east bank built after 1955; z, clips used during photographic processing. Base images: US Geological Survey.

FIGURE 1.19 View of the Naugatuck River in Ansonia, Connecticut, in 2001, looking upriver, just north of the area depicted in Fig. 1.18, showing the continuation of the rock and concrete revetment on the west bank and the high concrete flood wall on the east bank. Photo: M.A. Kruge.

A straightened and widened channel, a massive concrete and stone revetment on the west bank ("x"), and high concrete *flood wall* on the east bank ("y") are clearly evident in Fig. 1.18b. The small *tributary* stream on the east side ("w") was armored with its own flood wall. A modern sewage treatment plant was established in zone "u" in the flood plain, now protected by the revetment (Fig. 1.18b). More recently, the factories ("s") were demolished, and new retail, commercial, and light industrial facilities were constructed in zones "t" and "u," with the confidence provided by the hard flood control structures (Fig. 1.19).

Despite engineered flood control measures instituted on rivers across the United States, flooding continues to exact a toll in lost lives and property. Expert consensus indicates that one of the best remedies is to restrict construction on hazardous flood plains, yet such development persist (Dierauer et al., 2012). Large rivers (e.g., Mississippi and Rhine) are extensively lined with *dikes* and *levees*, hard raised structures designed to confine the flow to the channels during high water events. Channel volume is nonetheless finite, and *overtopping* may still occur during extreme cases. One solution is levee *setback*, by which the barriers are rebuilt farther back from the riverbed. This would increase the maximum channel volume (m^3) available during a high-flow event and consequently permit a higher flow rate ($m^3 s^{-1}$), allowing excess *discharge* to pass safely without flooding. However, structures currently protected could find themselves on the wrong side of a new relocated levee, meaning that *buyouts* would have to be considered in the cost–benefit analysis. Modeling that considers terrain characteristics, historic discharge rates, flood recurrence intervals, and land use would be effective as environmental managers weigh the expense of new infrastructure versus the cost of flood damage that would otherwise be incurred. Since many existing levees along the Mississippi River are aging and in need of repair, now is an opportune moment to consider the alternative of repositioning them (Dierauer et al., 2012).

1.5.2 Eutrophication: "Too Much of a Good Thing"

Nitrogen (N) and phosphorus (P) are nutrients essential for plant growth and are liberally applied to soils as fertilizers in the course of agricultural and horticultural activities. Excess nutrients not absorbed by the cultivated plants may escape via runoff and produce eutrophication in lakes within the *drainage basin*. This nourishment is essential for the growth not only of terrestrial plants, but for aquatic algae and vascular plants as well, masses of which may swell into noxious blooms in waters overloaded with nutrients. Rivers and marine systems can be similarly afflicted, particularly marine and *estuarine* lagoons behind barrier islands in urban areas (Kruge, 2013). In addition to applied fertilizers, the combustion of fossil fuels produces nitrogen emissions into the atmosphere, some of which returns to the surface. Phosphorus has been an important ingredient in household detergents, but many jurisdictions have now mandated reduction in its concentration or elimination from cleaning products. Municipal wastewater effluents contribute both N and P, as do animal manures accumulating in large feedlots. Eutrophic lakes have total N and P concentrations in excess of 650 and 30 mg m^{-3}, respectively, while for marine systems, the threshold is somewhat lower, and for the moving waters of a river, it is somewhat higher (Smith et al., 1999).

The sources of N and P (as well as other pollutants) to aquatic systems can be *point* or *nonpoint*. Point sources are single, discrete locations, such as wastewater treatment plants, waste disposal sites, animal feedlots, mines, and *combined sewer overflows* (CSOs). In environmental management terms, point sources are more amenable to engineered control than diffuse, widespread nonpoint sources, such as runoff from cultivated lands, pastures, suburban lawns, poorly functioning septic systems, and construction sites, as well as atmospheric deposition (Smith et al., 1999). CSOs are a particularly vexing problem in cities like New York, which have older systems in which storm sewers are interconnected with the sanitary sewers. During a heavy rainstorm, runoff will overwhelm the system, resulting in the discharge of raw sewage into local water bodies.

In the United States, eutrophication is the most common water quality issue and is not merely an aesthetic concern or a nuisance. Drinking water may develop taste and odor problems or even health risks, while water intake systems may clog with biomass. Fisheries and tourist-dependent enterprises may suffer losses. In the extreme, blooms may be due to highly toxic species of *cyanobacteria*. To prevent or reverse eutrophic conditions, reducing nutrient discharges at known major point sources (e.g., by upgrading wastewater treatment plants) is an important and obvious first measure. Since both N and P are limiting (essential) nutrients, it may be sufficient to reduce only one of them below the critical threshold (Smith et al., 1999). However, long-term success may be elusive. For example, Lake Erie (the Great Lake situated between the Canadian province of Ontario and the US state of Ohio) is once again plagued by massive summer algal blooms after decades of successful suppression efforts on

FIGURE 1.20 Satellite view of Mono Lake, California, and environs in July 1999. In the false-color scheme employed, water appears black, the bare desert surface appears tan shading to mauve in the dry outer margins of the lake bed, vegetation appears green, patches of snow lingering on the high Sierra Nevada mountain peaks along the west side are turquoise blue, and clouds appear white. The shrinkage of the lake due to declining water levels is most visible on its northeastern margin. Landsat image: US Geological Survey.

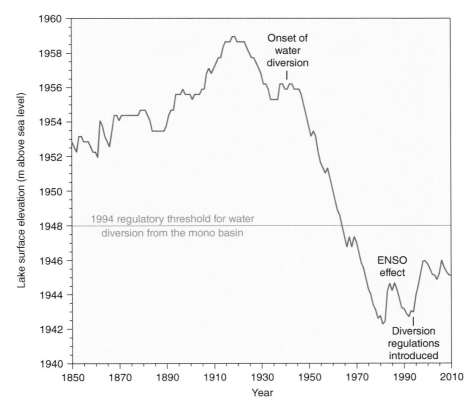

FIGURE 1.21 Fluctuations in Mono Lake water levels from 1850 to 2010. Data source: Mono Lake Committee (2013).

both sides of the international border (Wines, 2013a). *Beneficial use* may provide a way to profit from what otherwise is problematic. In one innovative example, algal biomass from blooms in Orbetello Lagoon (Italy) has been harvested as a feedstock for *biofuel* (Bastianoni et al., 2008).

1.5.3 Lakes in Arid Regions

A *desert* is a region in which the sparse rainfall (generally <250mm year^{-1}) is largely lost to evaporation, with little accumulating in streams, in lakes, and underground (Pipkin et al., 2008). Contemporary desert lakes in the western United States are often relicts of much larger freshwater bodies that formed during the cooler, wetter Pleistocene Epoch. As *desertification* proceeds, lake waters evaporate and precipitate mineral salts (*evaporites*), while the residual waters becomes increasingly saline or even *hypersaline*. An example is Mono Lake, a fairly large (160 km^2) hypersaline lake in the high desert of eastern California, the low point in an enclosed drainage basin with no outlet to the sea (Fig. 1.20). It has high concentrations of sodium, chloride, and carbonate *ions*, among others, and an alkaline *pH* of about 10 (Jellison and Melack, 1993). Close examination of Fig. 1.20 reveals concentric bands parallel to the present lake shore marking higher past strand lines, most visible extending for several kilometers to the northeast.

In 1941, the city of Los Angeles, some 400 km to the south, began to divert freshwater streams in the Mono Basin to enhance its municipal water supply (Jellison et al., 1996). Mono Lake water levels soon began a precipitous decline, dropping 14 m over the next 30 years (Fig. 1.21) and reducing lake surface area by about one-third. The ensuing popular outcry eventually had an effect in 1994, when the California State Water Resources Control Board instituted limits on water withdrawals applicable if lake levels fall below 1948 m (Jellison et al., 1998). Since that time, water levels have risen, but they remain below the regulatory threshold, which in turn is well below the historical lake level maximum recorded during the 1920s (Fig. 1.21).

The higher-than-normal precipitation of the 1982–1983 ENSO event freshened the uppermost layer of lake water (*epilimnion*), while the deeper waters (*hypolimnion*) remained saline and led to a temporary lake level rise of about 2 m (Fig. 1.21). This *meromictic* state, in which the less dense, fresher waters buoyantly remained "floating" above the denser, saltier water below, induced a *stratification* of the lake waters that persisted for several years thereafter.

During this period, the layers in the lake coexisted essentially unmixed and isolated from one other. Nutrients, particularly N, remained in the upper waters, prompting algal blooms (Jellison and Melack, 1993). Ironically, the subsequent rise in lake level due to the water diversion restrictions has triggered a recurrence of meromixis (Jellison et al., 1998). The decision that mandated the 1948 m minimum lake level was evidently the result of political compromise attempting to balance the water needs of Los Angeles with the ecological requirements of a healthy lake system. Although the Mono Lake area is sparsely populated with limited agricultural activity, human influences indirectly perturb the natural system in unexpected ways. It remains an interesting and important challenge for environmental managers.

1.6 GROUNDWATER MANAGEMENT AND KARST HAZARDS

Groundwater (subsurface water flowing slowly through *porous* and *permeable* soil, sediment, and rock) is one of the largest sources of freshwater available for use by humankind (Pipkin et al., 2008). Assuring adequate supplies of groundwater, discouraging its overuse, and safeguarding the resource from contamination are all important environmental management concerns. *Karst* topography develops in areas underlain by the more soluble types of *bedrock*, namely, *limestone* and evaporites. Groundwater can slowly dissolve such rocks to form caverns. A cavern may grow large enough to undermine the ground above, which can collapse without warning and form a *sinkhole* that can be large enough to swallow a building.

1.6.1 Groundwater Overdraft: Impact on Agriculture and Municipal Water Supplies

Some of the water falling as precipitation on the surface infiltrates into the ground and thereby *recharges* the groundwater resources below. Groundwater migrates in the subsurface, but slowly, on the order of centimeters or even millimeters per day, as it passes through tiny, convoluted *pore* space networks within soil, sediment, or suitable rock such as *sandstone*. The uppermost layer of material encountered generally has both air and water within its pore spaces and is termed the *vadose or unsaturated zone*. Below that, the available pore spaces are saturated with groundwater within the *saturated zone*. The *water table* is the boundary surface between the unsaturated zone above and the saturated below, moving up or down within the body of the porous, permeable *aquifer* over time as a function of the recharge rate. During periods of heavy rainfall, the water table (i.e., the top of the saturated zone) will rise and may even reach the surface. During *droughts*, the water table will fall and water wells

(a) 1991

(b) 2010

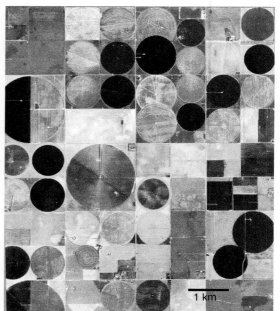

FIGURE 1.22 Two aerial images documenting the increased use of center pivot irrigation on southwestern Kansas farms (near Haskell, about 83 km southwest of Dodge City) with water withdrawn from the High Plains aquifer. For reference, the same three irrigated circles are marked in yellow on both images. (a) View in 1991. (b) View of same area in 2010. Base images: US Geological Survey.

may go dry if they are not deep enough. Ideally, recharge by precipitation or melting snow will keep pace with groundwater withdrawal from wells. Otherwise, conditions known as *overdraft*, *overpumpimg*, or *groundwater mining* may develop, leading to depletion of the resource and possible scarcity.

The productivity of the vast grain fields of the American Middle West has been increasingly sustained by irrigation, which taps into the High Plains aquifer system (also known as the Ogallala aquifer) stretching from Texas to South Dakota. *Central pivot irrigation* systems imprint the landscape in these states with large green circles of growing crops on the order of a kilometer in diameter, easily visible from high-flying aircraft. The increased use of these systems is readily apparent when comparing aerial images taken of the same part of western Kansas two decades apart (Fig. 1.22). In 1991, there were only three circles (highlighted in yellow on the black and white image) irrigated by central pivot mechanisms, whereas by 2010 there were 39 in the same area, some with double the diameter of the earlier fields. A central pivot system operates by withdrawing water from a well drilled in the center of the field. Water is delivered to the crops via perforated piping on wheels that slowly rotates around the well (Fig. 1.23a). The effect on the High Plains aquifer of water withdrawals in Kansas over a 66-year period is a steep and continuing drop in the water table of about 10 m at this location (Fig. 1.23b), indicating that groundwater consumption is outstripping the recharge capabilities of the system. The drier southern areas overlying the High Plains aquifer are being particularly impacted, forcing the reduction or even cessation of irrigation activities and switching from maize to less water intensive crops such as sorghum or to livestock (Wines, 2013b).

Central pivot irrigation is employed even in the arid southwestern United States (Fig. 1.24). This aerial image depicts several irrigated circles (about the same size as the ones in Kansas) under cultivation in the Nevada desert. They are sited in a river valley that remains dry for much of the year, but where the water table is evidently high enough to be reached by wells. Timely groundwater recharge in dry country is not to be taken for granted. Farmers there must be particularly careful to employ *sustainable* practices and avoid groundwater mining, as the water resources that they are exploiting were likely emplaced during the wetter climate of the Pleistocene ice age and thus will essentially be irreplaceable for the foreseeable future (Pipkin et al., 2008).

The US northeastern states normally have ample rainfall; nonetheless, if withdrawals exceed recharge, groundwater overdraft problems will occur there as well. On western Long Island, New York, early European settlers produced

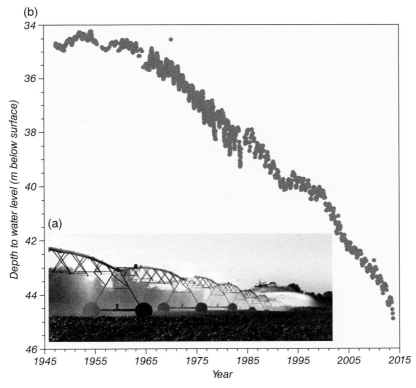

FIGURE 1.23 (a) Central pivot irrigation system in operation. Photo: US Dept. of Agriculture. (b) High Plains aquifer monitoring well data documenting a progressive 10 m drop in water level from 1947 to 2013, Colby, Kansas. Data source: US Geological Survey (2013).

groundwater from shallow wells in the Upper Glacial aquifer (Fig. 1.25). The aquifer was recharged by precipitation and infiltration from septic systems, but the latter also caused groundwater contamination, which became more severe as the population increased. To alleviate the pollution problem, municipal wells were drilled to tap the deeper Jameco and Magothy aquifers, shielded from surface contamination by the overlying impermeable Gardiners Clay *aquiclude*. The aquiclude also shields the deeper aquifers from ready groundwater recharge from above, resulting in a *hydrologic* imbalance in which deep well withdrawal rates exceeded recharge. The recharge deficit was exacerbated as individual septic systems were replaced by large municipal wastewater treatment plants (which release their effluent directly to the sea precluding recharge) and as suburban developments superseded farmlands, impeding water infiltration with impervious paved surfaces and buildings (Cohen et al., 1968).

As a further complication, freshwater aquifers in coastal regions are particularly vulnerable to saltwater intrusion (Oude Essink, 2001). A cross section running northward 14 km from Atlantic Beach on the Long Beach barrier island depicts southwestern Long Island's four main freshwater aquifers in light blue (Fig. 1.25). The drawing shows the status in the mid-twentieth century when overpumping was drawing saltwater into the freshwater aquifers from the south, intruding as three wedges shown in dark gray and advancing northward as rapidly as 100 m year^{-1} during the 1950s.

FIGURE 1.24 Oblique aerial photograph of an irrigated zone within a dry river valley in the Nevada desert, 38 km northwest of Tonopah, September 2013. The large center pivot field (marked by green arrow) is about 800 m in diameter. Note the large thermal solar electrical generating station (red arrow) under construction in the background. Photo: M.A. Kruge.

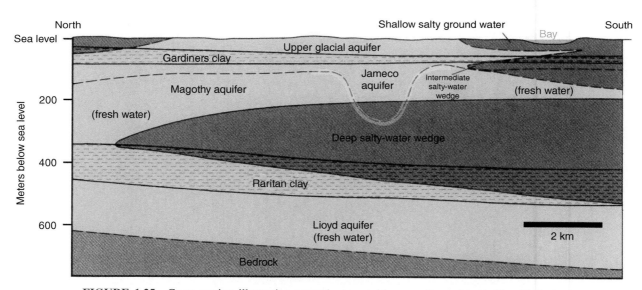

FIGURE 1.25 Cross section illustrating groundwater problems on Long Island, New York, in 1961. Three saltwater wedges (shallow, intermediate, and deep, shown in dark gray) invaded the Upper Glacial, Jameco, and Magothy freshwater aquifers (light blue) due to overpumping, which also resulted in a cone of depression (highlighted in yellow). The section goes from Atlantic Beach (Fig. 1.13) northward to Valley Stream. *Source:* Modified from Cohen et al. (1968). Courtesy of the New York State Department of Environmental Conservation.

Overpumping also created a *cone of depression* in the Jameco aquifer (highlighted in yellow) such that the wells at that location began to produce saline water, detected during water quality testing as increasing chloride ion contents (Cohen et al., 1968). Queens County was ultimately connected to the New York City surface water reservoir system, thus alleviating their groundwater problems, whereas Nassau County still relies on groundwater and must vigilantly work to avoid contamination and overpumping problems.

1.6.2 Karst Topography and Sinkhole Hazards

Much of the US state of Florida is underlain by limestone bedrock, which is predominately composed of *calcite*, a calcium carbonate mineral. Carbonate rocks are susceptible to slow dissolution by groundwater, the source of which is rainwater that is normally slightly acidic. Sinkholes produced by collapse of caverns in this karst terrain are particularly prevalent in central Florida. A satellite view of the city of Winter Haven is typical, showing dozens of large and small sinkholes (flooded and appearing deceptively as picturesque lakes) within the ≈40 km^2 area of the image (Fig. 1.26). Seen from the air, sinkhole lakes have characteristically rounded margins, collectively presenting a "Swiss cheese" appearance in map view. Sinkhole formation mechanisms include direct roof collapse, slippage of unconsolidated soil or sediment into a cavernous void below, and a passive sagging of the surface materials (Gutiérrez et al., 2008a).

The sudden ground failure forming a new sinkhole can be frightening, costly, and even fatal to residents. In 2013,

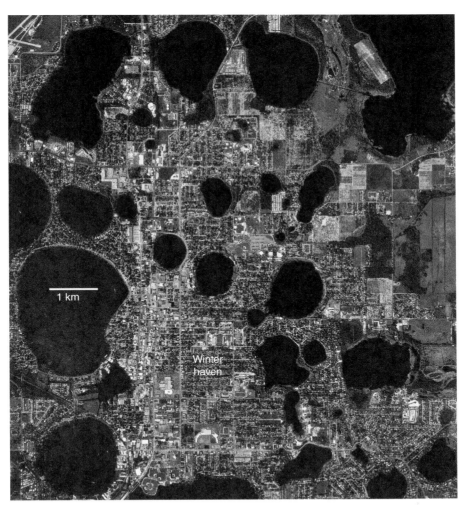

FIGURE 1.26 Satellite image of Winter Haven, Florida, taken in 1994. The "Swiss cheese" landscape is pockmarked with flooded sinkholes forming lakes of varying sizes. In this false-color image, water appears dark gray to black and vegetation appears red, while streets and larger buildings appear pale bluish gray. Base image: US Geological Survey.

FIGURE 1.27 This building at the Summer Bay Resort in Clermont, Florida, suddenly collapsed in 2013 when a sinkhole opened beneath it. Photo courtesy of the Orlando Sentinel.

a recently built multiunit structure at a Clermont, Florida, resort complex some 40 km north of Winter Haven collapsed without warning, fortunately with no loss of life (Liston, 2013). The ruined building (Fig. 1.27) was sited upon terrain that is clearly karstic with many sinkhole lakes in the vicinity (Fig. 1.28). Within the frame of the aerial view, roughly one square kilometer in area, numerous small sinkholes are clearly visible in the image from 1952, prior to any real estate development (Fig. 1.28a). The large lake in the center may have formed by the coalescence of several sinkholes, but further evidence would be needed for confirmation. Six decades later, the resort complex is occupying a swath of higher ground between lakes in the eastern portion of the image (Fig. 1.28b). Some of the sinkholes are less evident, partially obscured by vegetation or fill, but they are still discernable, especially when the two images are viewed side by side. The site of the collapsed structure (still intact in Fig. 1.28b) is marked by a red rectangle on both images. The environmental management questions become (i) to what extent should the developers have anticipated further sinkhole formation and (ii) to what extent the responsible government entities should have restricted construction based on the risk. This then leads to considerations of appropriate building codes, structural retrofitting, insurance, and liability in the event of damages (Zisman, 2013)—essentially the same concerns evoked by earthquakes, flooding, and the other risks presented earlier.

Evaporites are even more prone to sinkhole formation than carbonate rocks. Common evaporite minerals such as *gypsum* and particularly *halite* (calcium sulfate and sodium chloride, respectively) are considerably more water soluble than calcite with lower mechanical strength, producing sinkholes that are more active and diverse (Gutiérrez et al., 2008a). Evaporite bedrock underlies about 7% of Spain, where the Zaragoza metropolitan area is particularly vulnerable, having evaporites overlain by permeable *alluvial* aquifers (Gutiérrez et al., 2008b, c). In a recent instance, sinkhole-induced damage to a relatively new apartment building in the Zaragoza area led municipal authorities to condemn the structure and relocate the inhabitants, due to imminent risk to their safety (Portella, 2013). Ironically, the sinkhole in this zone was well studied, having created problems for decades leading to the demolition of a factory on a neighboring property. The successor building at that site is a department store, constructed on a foundation of deep pilings reaching to solid bedrock below and thus remains undamaged. The affected apartment building was constructed with only a concrete slab foundation (Gutiérrez et al., 2009).

All geotechnical means of sinkhole risk evaluation have their limitations, and thus, investigations are best done using a combination of methods, using essentially the same tools as are used in paleoseismic and landslide studies. Examination of a series of aerial photographs taken over a period of decades (e.g., Fig. 1.28) and of maps prepared over the years is a logical early step. *Boreholes* should be drilled and the vertical sequence of materials characterized and *logged*. *Geophysical* surveys are nonintrusive and relatively inexpensive. These include *ground-penetrating radar* and *electrical resistivity* measurements. These are particularly effective in conjunction with trenching and together permit 2-D (Fig. 1.29) and even 3-D visualization of the zone (Gutiérrez et al., 2009; Zisman et al., 2013). Another approach employs satellite-based *synthetic*

(a) 1952

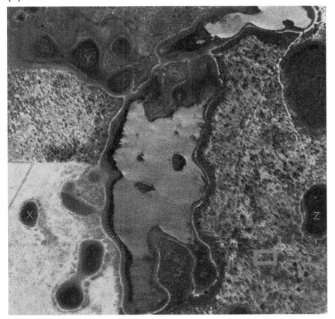

(b) 2010

FIGURE 1.28 Aerial images showing the distribution of sinkholes on the land surface in the vicinity of the Summer Bay Resort collapse site (red rectangle) in Clermont, Florida (Fig. 1.27). For reference, three smaller sinkholes (here, labeled X, Y, and Z) are marked on both images. (a) In 1952, the area was almost entirely undeveloped. A large lake (W) with two embayments on its southern end dominates the center of the image. (b) In 2010, while much of the land remains undeveloped, the Summer Bay Resort complex occupies the eastern third of the image, constructed upon what appeared to be a solid swath of land in 1952 (1.28a). The large lake (W) has shrunk and the sinkholes X, Y, and Z are less apparent in the later image. Base images: US Geological Survey.

aperture radar, with data from repeated passes of the satellite over the same area processed by differential interferometry to produce high-resolution imagery and quantitative measurements of subsidence (Tomás et al., 2014). If sinkhole damage to an existing building is suspected, a forensic evaluation of the structure is also warranted to confirm that the damage is not due to other unrelated factors, which could lead to unnecessary litigation and provide opportunities for unscrupulous building and repair contractors (Zisman, 2013; Zisman et al., 2013). As with all the hazards discussed previously, the safest and least expensive risk reduction strategy is to not build in such unstable areas in the first place (Gutiérrez et al., 2009).

1.7 GEOLOGICAL FACTORS IMPACTING WASTE MANAGEMENT

Human populations worldwide generate an ever-increasing amount of municipal solid waste, estimated at 1.7 billion tonnes per day in 2008, some 95% of which was being consigned to *landfills* (Foo and Hameed, 2009). High-level, long-lived nuclear wastes, although produced in much smaller quantities than municipal wastes, pose special and particularly difficult problems of their own. In both cases, potential groundwater and surface water contamination are of paramount concern.

1.7.1 Municipal Solid Waste Disposal and Landfill Leachate

Until recently, the uncontrolled disposal of municipal solid waste by filling low-lying areas or dumping on lands considered to be of little value was a common practice (Foo and Hameed, 2009). A case in point is the now-abandoned Malanka landfill in the Hackensack Meadowlands of New Jersey (Fig. 1.30). Precipitation can freely infiltrate and percolate through the buried waste, leaching contaminants into the groundwater system and surrounding estuary. To avoid the migration of the polluting *leachate* into the environment, it is important to design and construct a properly engineered landfill (Fig. 1.31). When initially siting a landfill, the first considerations include the nature of the host materials and the groundwater dynamics. The landfill should be placed within clay-rich sediments or soil, which functions in the groundwater system as low-permeability, passive barrier to the movement of leachate. However, it should be noted that even very thick, low-permeability, and competent clay units may have very small secondary pathways such as fractures or root burrows that act as significant conduits for contaminant flow (Pankow and Cherry, 1996). A porous and permeable sand layer would be a worse choice, although it could provide limited

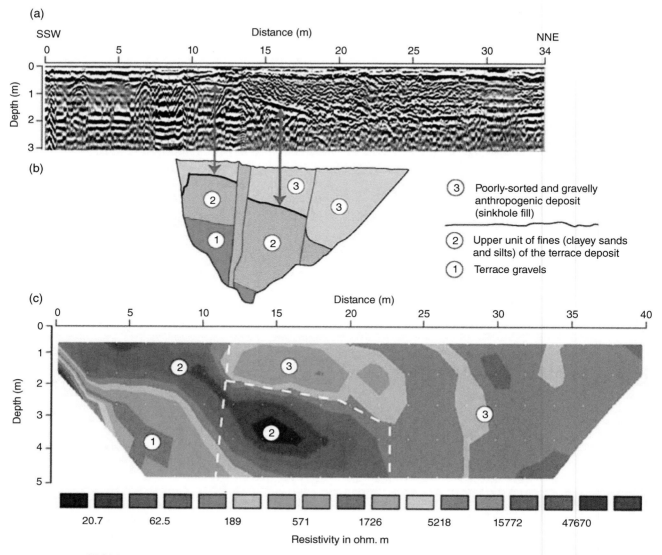

FIGURE 1.29 Geophysical methods used in the investigation of a large active sinkhole, Zaragoza, Spain. (a) Ground-penetrating radar profile. (b) Profile of excavated trench. (c) Electrical resistivity profile. *Source:* Gutiérrez et al. (2009); used with permission.

natural filtration of the leachate. Karst terrain would perform even more poorly, as contaminated groundwater would pass readily through its open subsurface conduit network with little filtration (Lindsey et al., 2010). Once chosen, the site should be excavated and lined with clay and synthetic barrier layers as additional insurance and then with a porous sand layer and piping to collect the leachate. After the landfill has been filled to capacity with waste, it should be sealed with an impermeable upper layer, covered with topsoil and revegetated. As leachate forms, it accumulates at the base of the landfill to be withdrawn via the pipe network for treatment. Wells around the perimeter monitor groundwater quality and for any fugitive leachate (Pipkin et al., 2008).

Aerobic degradation is the initial phase of waste decomposition in a landfill, producing volatile *fatty acids* that *biodegrade* relatively readily. Two of the parameters routinely measured during landfill monitoring (as well as more generally in other environmental investigations) are *biological oxygen demand (BOD)* and *chemical oxygen demand (COD)*. The aerobic phase is characterized by a relatively high BOD/COD ratio. Once the available oxygen has been consumed, the *anaerobic* degradation phase begins, favoring *methanogenic* bacteria. The BOD/COD ratio falls and high molecular weight *humic substances* become the dominant organic components (Kurniawan et al., 2006). Microbial methane accumulations may be hazardous, but the gas can be withdrawn via wells in the landfill and beneficially used as fuel.

FIGURE 1.30 Abandoned Malanka Landfill, Secaucus, New Jersey. Photo: M.A. Kruge.

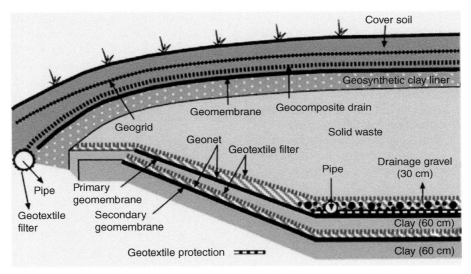

FIGURE 1.31 Cross section of a landfill lined with clay and synthetic materials (geogrid, geomembrane, geotextiles, etc.) to prevent groundwater contamination, with piping systems engineered to collect leachate for treatment. Ersoy et al. (2013); used with permission. In addition, the decomposing waste can be tapped to produce methane gas, which can be used beneficially as a fuel.

Associated pollutants in landfill leachate pose additional concerns. These include ammoniacal nitrogen, *heavy metals* and metalloids (such as As, Cd, Cr, Co, Cu, Hg, Ni, Pb, Zn), and *persistent organic compounds* (including aromatic hydrocarbons, phenols, halogenated compounds, pesticides). The collected leachate must be chemically and physically treated to stabilize these contaminants, employing methods including coagulation–flocculation; chemical precipitation; ammonium stripping; micro-, ultra-, and nanofiltration; activated carbon absorption; and ion exchange. The concentrated sludge produced during these steps must in turn be sequestered or destroyed safely (Kurniawan et al., 2006; Foo and Hameed, 2009). From an environmental management perspective, it would be sensible to minimize the need for complex, expensive engineered landfills through composting of organic wastes, recycling, and more sustainable production–consumption systems.

1.7.2 Nuclear Waste Disposal

Nuclear waste presents special disposal problems, as some of the constituent isotopes may have very long *half-lives* and thus remain dangerously radioactive for hundreds or even thousands of years. Radioactive wastes are generated by nuclear power stations (directly and by the associated fuel processing), in weapons production, at research facilities, in industrial processes, and by medical usage. Since there is a great diversity of nuclear waste types, a classification scheme is employed to properly guide the disposal process, based on the level of radioactivity (high, medium, or weak) and the half-life (long, short, or very short). Wastes may include relatively pure substances, contaminated components from decommissioned facilities, or protective clothing and tools that were in contact with nuclear materials (ANDRA, 2012).

The long-term fate of spent nuclear fuel and high-level radioactive waste in the United States remains unresolved, the subject of both scientific research and political controversy. Yucca Mountain, Nevada, was chosen by the US Congress in 1987 as the single repository in the country for all such waste, after a decade of site evaluation. It is a ridge of volcanic *tuff* (consolidated pyroclastic ash) situated in the sparsely populated Nevada desert roughly 120 km northwest of Las Vegas (Fig. 1.32a). The potential migration of *radionuclides* via dissolution in groundwater was a foremost concern during the evaluation. Having currently a semiarid climate, precipitation at the site is minimal and thus so is groundwater recharge. The thick vadose zone would isolate the subterranean repository high above the present water table, in galleries excavated deep into the mountain (Fig. 1.32b). In addition to these natural impediments to radionuclide migration, the waste would be packaged in robust, impermeable materials to create an engineered barrier to migration (Bodvarsson et al., 1999). Specific factors evaluated included the possibility of future *magmatic* activity at the site, earthquakes, climate change leading to more abundant rainfall, and human intrusion (Rechard et al., 2014a).

The initial standard for a disposal site was to limit radionuclide leakage to a 5 km radius over 10,000 years. Later, a more stringent 18 km over 1,000,000 year limit was set (Rechard et al., 2014b). Since 1983, $15 billion has been spent on the evaluation of the Yucca Mountain site, and although the scientific knowledge base was greatly enhanced, political opposition ultimately led to suspension of the project. Evaluation of a new site will not proceed unless it is first welcomed by its host community and sanctioned by local authorities (Rechard et al., 2014c).

1.8 ENERGY RESOURCE EXTRACTION AND ITS ENVIRONMENTAL CONSEQUENCES

Our civilization is strongly dependent on the combustion of vast quantities of *fossil fuels* (*coal, petroleum, natural gas*). One result of their utilization on such a scale is the continuing increase in atmospheric concentrations of the *greenhouse gas* carbon dioxide, in turn observed to provoke global climate change (IPCC, 2007). However, the emphasis in this section will be geological setting of the resources and the environmental consequences of their extraction.

1.8.1 Coal and Coal Mining

Coal *seams* are layered *sedimentary* rocks composed mostly of fossil organic matter, usually derived from ancient *terrestrial* swamp and bog plants. Microscopic examination of coal reveals fossil wood, charcoal, spores, pollen, roots, bark, tree resins, leaves, and algae, along with minor amounts of mineral matter (Killops and Killops, 2005). The ample *bituminous* coal resources in the eastern United States

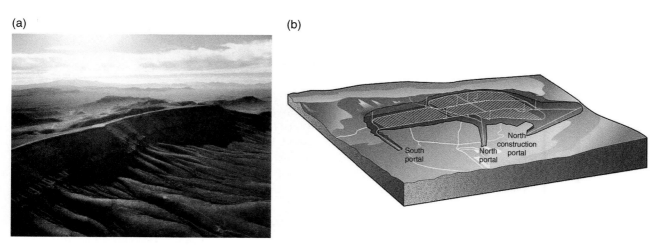

FIGURE 1.32 Proposed high-level nuclear waste storage site, Yucca Mountain, Nevada. (a) Oblique aerial photograph of the mountain. (b) Plan of the underground storage site. Images: Los Alamos National Laboratory.

are predominantly of *Upper Carboniferous* age (roughly 300 million years old) with large, economically significant deposits in the Appalachian and Illinois Basins. In addition to more traditional underground mining, coal is also extracted less expensively from large *open-pit* or *strip mines* on the surface in places where the seam is not too deeply buried. Enormous *draglines* remove the *overburden* to reveal the coal, which is then excavated and hauled out of the mine (Fig. 1.33a). Regulations stipulate that after the mine has been closed, the site should be *reclaimed*, with the ground surface restored to its original topographic contour and revegetated (Pipkin et al., 2008). Prior to the institution of these policies, mine operators simply abandoned worked-out strip mines, leaving barren piles of *tailings* and flooded pits (Fig. 1.33b).

Some coal seams may contain up to several percent sulfur. These high-sulfur coals were formed when their precursor coastal peat swamps were submerged during periods of sea level rise and the dissolved sulfate in the seawater subsequently reacted with the peat. The coal thus produced contains microscopic crystals of *pyrite* and other iron sulfide minerals, along with various organic sulfur forms (Gluskoter and Simon, 1968). The abandoned tailings of high-sulfur coal mines produce *acid mine drainage*, in which the sulfide

FIGURE 1.33 Open-pit coal mine, Wilmington, Illinois. (a) Active mining in 1938. (b) "Moonscape" of barren spoils piles and flooded pit after the mine was abandoned shortly thereafter. Photos courtesy of George Langford.

minerals are oxidized to sulfate in the presence of the oxygenated surface water:

$$FeS_2 + 7/2O_2 + H_2O \Rightarrow Fe^{2+} + 2SO_4^{2-} + 2H^+$$

Similar problems arise from the tailings of mines producing copper, lead, zinc, and other metals with sulfide ores (Johnson and Hallberg, 2005; Akcil and Koldas, 2006). The abandoned Will Scarlet open-pit coal mine in southern Illinois is a case in point, flooded with multihued acidic waters (Fig. 1.34). It has been deemed one of the worst cases of acid mine drainage in the United States, with streams nearby essentially devoid of aquatic life. The Land Reclamation Division of the Illinois Department of Natural Resources has been empowered to treat such abandoned properties, but funding is insufficient (Fitzgerald, 2012). Treatment strategies include physical, chemical, and biological methods. For older sites that were never reclaimed by their original operators, diverting streams or even groundwater flow may be helpful, although difficult. Chemical treatments seek to raise the pH and precipitate the iron compounds. Lining a streambed with limestone is a hybrid physical/chemical approach (Akcil and Koldas, 2006). Biological treatments include the creation of wetlands, permeable reactive barriers, and iron-oxidation bioreactors. In an instance of beneficial use, iron-rich sludge from an abandoned coal mine was used to produce paint pigment (Johnson and Hallberg, 2005).

Mountaintop removal–valley fill is a more recently developed surface coal mining method, on a dramatic scale comparable to or greater than open-pit mining, especially in West Virginia (Fig. 1.35). To accomplish this, the higher elevation overburden is removed with explosives to expose the coal and then dumped over the edge of the cut to fill adjacent valleys, obliterating or altering the stream drainage and groundwater infiltration systems therein (Griffith et al., 2012). This has accelerated rates of production of highly marketable coal from low-sulfur seams in West Virginia (Fedorko and Blake, 1998). (In addition to its potential to produce acid mine drainage, high-sulfur coal generates sulfate upon combustion in coal-fired power plants. *Acid rain* will result if the exhaust gases are not *scrubbed* to remove the sulfate. Low-sulfur coal reduces the need for scrubbers.) Arsenic, chromium, manganese, nickel, lead, and other trace elements are present in West Virginia coals in mean concentrations ranging from less than 1 to about $22\,mg\,kg^{-1}$ (Fig. 1.36). Mountaintop removal mining–valley fill operations may lead to increase *bioavailability* of some of these elements in downstream ecosystems and a possible increase in human birth defects (Pumure et al., 2010; Ahern et al., 2011).

The extent of old subsurface coal mines is often poorly known, a casualty of inaccurate or missing documentation, some mines having been abandoned generations ago. Surface subsidence above such old mines is a continuing environmental issue in coal regions, creating problems reminiscent of

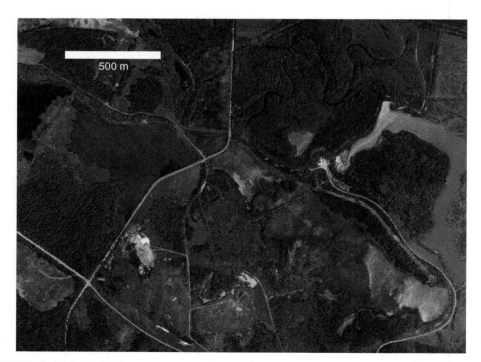

FIGURE 1.34 Aerial image of the abandoned Will Scarlet open-pit coal mine in Williamson County, Illinois. The unreclaimed pits have flooded with water, multicolored due to the high acidity. Forests and fields appear dark green in contrast. High-resolution orthophoto: US Geological Survey.

FIGURE 1.35 Oblique aerial photo of a "mountaintop removal" coal mining operation in January 2006, Kayford Mountain, West Virginia. Photo courtesy of Vivian Stockman.

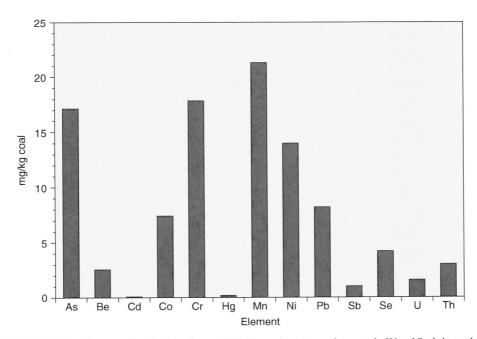

FIGURE 1.36 Average distribution of potentially hazardous trace elements in West Virginia coals. Data: West Virginia Geological and Economic Survey (2013).

FIGURE 1.37 Subsidence of an abandoned underground coal mine in an Indiana farm field. Photo courtesy of the Indiana Geological Survey (Harper, 2011).

those plaguing karst terrains (Fig. 1.37). The Indiana Geological Survey has spent decades creating maps to help combat the hazard, which has damaged homes, schools, and other structures (Blackford, 2012; Harper, 2011). Once a subsiding mine site has been recognized, it can be monitored for continuing movement by instrumentation (O'Connor and Murphy, 1997).

1.8.2 The Petroleum System

Commercial *petroleum* accumulations develop within large, layered *sedimentary basins*. The *petroleum system* is the conceptual framework used currently to understand the occurrence of conventional and nonconventional oil and gas resources in these basins (Hunt, 1996). It has five components or phases: aquatic biomass formation, biomass preservation in sediments, generation of petroleum, migration of petroleum, and trapping.

Whereas coal deposits derive most frequently from fossilized terrestrial plant remains, petroleum begins with microorganisms, especially algae, floating in the shallow *photic zone* in oceans and large lakes. If natural eutrophic conditions develop, algae will bloom in the surface waters, and it is their biomass that provides the essential raw material for petroleum and natural gas formation. If the blooms are sufficiently massive, the responding scavengers, particularly aerobic bacteria, may consume the available dissolved oxygen in the water faster than it can be replenished by marine currents, suppressing the scavenging processes. The biomass will then be able to settle to the bottom sediments before it can be fully recycled. There, it will be buried and preserved as sedimentation proceeds, completing the second phase of the process. As more sediment accumulates above it, the third phase begins. Progressively deeper burial leads to *geothermal* heating of the preserved *organic matter*, transforming the biomass into *kerogen* via the process of *diagenesis*. Then, with continued heating at yet higher temperatures (roughly 100–120°C, as are encountered typically at burial depths of several thousand meters in a sedimentary basin), the process known variously as *catagenesis*, *thermal maturation*, or *oil generation* cracks the kerogen into the constituent molecules of petroleum (mostly *hydrocarbons* containing 5–40 or more carbon atoms) or natural gas (predominantly methane with other small hydrocarbons). The clay-rich sediments that mixed with the biomass on the seafloor have by now hardened into *shale* and *mudstone* rich in organic matter and are called the *source rock* (Tissot and Welte, 1984; Hunt, 1996). Within the complex mixture of compounds present in petroleum, it is remarkable that molecules of obvious biological origin (such as *steroids*) are still preserved (Ourisson et al., 1984).

Once the oil and gas have been generated, the fourth phase begins, which is the *expulsion* out of the impermeable source rock into an adjacent porous, permeable rock such as a sandstone. There, it will interact with the water already present within the pore network and gradually rise, since gas and many types of petroleum are less dense than water. The hydrocarbons will continue to *migrate* upward through the permeable strata until their path is blocked by an overlying layer of impermeable rock, such as another shale. The oil and gas will then accumulate in the permeable *reservoir rock* beneath the *trap* in the fifth and final phase. To develop a conventional oil or gas deposit, all five phases of the petroleum system must be completed in sequence, a process that requires millions of years (Hunt, 1996).

1.8.3 Petroleum and Natural Gas Extraction: Environmental Considerations

A reservoir rock beneath a trap is the classic target sought for petroleum exploitation, whether the ancient sedimentary basin is presently onshore or offshore. Onshore exploration and production tends to be lower in cost, particularly if the location is not a remote one with a harsh climate. Offshore development of the resource began modestly, in shallow waters. Gradually, methods became more sophisticated and capable, with taller platforms that could stand in deeper waters. More recently, the development of floating platforms permits work in even deeper waters beyond the continental shelf, but at proportionately greater risk. In 2010, the *Deepwater Horizon* offshore oil platform caught fire in the Gulf of Mexico and sank with the loss of 11 crew members, as the well was being closed pending later production (Fig. 1.38a). The platform was operating in a water depth of roughly 1600 m with the target oil reservoir at about 4000 m below the sea bed; thus, the work was at such great depths that it was pushing the technological limits. Contributing factors to the disaster include a poor job installing and cementing the steel *casing* that lined the well bore and the failure of the *blowout preventer*, a device designed to quickly close the well in the event of an emergency. The fundamental problem leading to the fatal loss of control was the failure of oversight and supervision to keep pace with advances in a technology that had not yet been fully proven. While the tragedy aboard the platform would have been bad enough, there was more to come. The

FIGURE 1.38 (a) The Deepwater Horizon offshore oil platform caught fire and sank with the loss of 11 crew members in 2010, as the well was being closed pending later production. Photo: US Coast Guard. (b) Cumulative oil spill map for the Deepwater Horizon incident for which the darker gray colors indicate more days of sea surface oiling reported. *Source:* Modified from National Oceanic and Atmospheric Administration (2014).

well immediately began to leak oil into the Gulf, ultimately spilling an estimated 800,000 m^3 over a 5-month period before desperate attempts to staunch the flow finally succeeded. This caused grave disruption to coastal ecosystems and fisheries in Louisiana and neighboring states (Fig. 1.38b) (Graham et al., 2011).

A disaster of the magnitude of the *Deepwater Horizon* story is fortunately rare, but there are fundamental issues that a society dependent on petroleum for much of its energy needs must face. While petroleum continues to be generated in the subsurface, the process is a very slow one, such that the present rate of consumption far surpasses the natural rate of replenishment. Thus, petroleum is for all practical purposes a *nonrenewable resource*. The questions regarding how much extractable oil remains and how long it will last remain elusive, particularly as new technologies such as offshore drilling increase *proven reserves*. An early attempt (Hubbert, 1956) to predict the trajectory of future global output concluded that peak oil production would occur in about the year 2000 followed by decades of slow decline (Fig. 1.39). This approach was based on extrapolation of the behavior of individual oil wells and fields over their productive lifetimes, which showed initial increases up to a maximum and then a gradual decrease. Petroleum engineers developed ever more sophisticated methods for *secondary* and *tertiary recovery*, dramatically extending oil field lifespans. Nevertheless, production decline inevitably takes its toll. The Hubbert approach was recently revisited (Nashawi et al., 2010), with surprisingly similar results (Fig. 1.39). They predict global "peak oil" in about 2015, followed by a century of decline, although cumulative production in their scenario is several times that of the Hubbert original. If these models are reasonably correct and peak production is imminent, the search for alternatives to petroleum should be given greater impetus.

Nashawi and coworkers (2010) did not consider the implications of the latest technical developments in oil and gas extraction from the so-called tight formations (rocks of low permeability) such as the *Marcellus Shale* of the northeastern United States (Fig. 1.40a). Hydrocarbons can be produced from such rocks after artificially inducing fracturing by forcing large volumes of a mixture of water and other substances into the well bore under high pressure

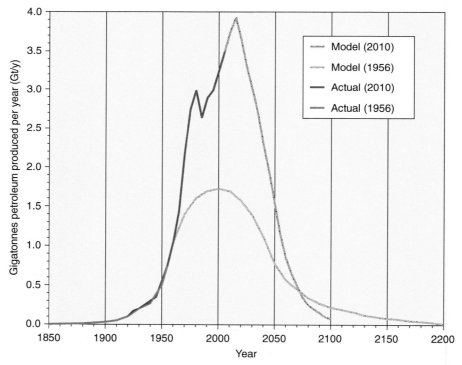

FIGURE 1.39 World petroleum production history and predictions showing peak oil production occurring in about the year 2000 (in red, as predicted by Hubbert, 1956) and in about 2015 according to more recent modeling (green, Nashawi et al., 2010). Dark colors indicate actual production data. Light colors indicate model predictions.

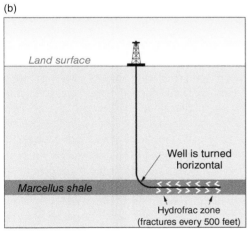

FIGURE 1.40 Marcellus Shale "tight gas" exploitation. (a) Map of the northeastern United States showing the extent of shales of Devonian age in the Appalachian Basin (outlined in green), of which the Marcellus Shale (in gray) constitutes a major part. (b) Schematic of the horizontal drilling and hydraulic fracturing ("fracking") techniques used to exploit natural gas deposits in the Marcellus Shale. *Source:* After Soeder and Kappel (2009).

("fracking"), thereby increasing the permeability. Applying a second innovation, *horizontal drilling*, in tandem with hydraulic fracturing further increases production from tight shales (Fig. 1.40b). This new technique permits the well operator to change the direction of the drillbit to remain within the target formation, following it laterally to expose more of the productive horizon to the well bore (Soeder and Kappel, 2009). If one considers this approach within the conceptual framework of the petroleum system, it is essentially an intervention in the system after the third phase, obviating the need for migration and trapping. In essence, the source rock also functions as the reservoir rock. There are many sedimentary basins in the world with mature source rock but without the geological components necessary for migration and trapping. Thus, there are now many more possibilities for hydrocarbon exploration and development.

High-volume hydraulic fracturing is not without its problems. The first is the large amount of water needed, particularly problematic in dry areas already straining to meet agricultural demand for water (Section 1.6.1), such as western Texas where the Eagle Ford Shale is a principal drilling objective (Barringer, 2013; Galbraith, 2013). More vexing is the question of water contamination (Urbina, 2011; Joyce, 2012), particularly from the *flowback* of the hydraulic fluids to the surface and the natural *produced waters* from deep underground that rise up the well with the hydrocarbons. Both classes of water are extremely saline, with up to $300\,g\,l^{-1}$ of total dissolved solids, creating concerns over their proper disposal (Rahm et al., 2013). The precise nature of the hydraulic fluids is guarded as a trade secret by operators and that in itself creates mistrust in the community. Chloride and bromide are among the major components present, for which Sun and others (2013) have recently proposed wastewater treatment methods. Cross-contamination of overlying freshwater aquifers is also a concern, whether by the brines or by methane (Vengosh et al., 2013). There have been reports of small earthquakes possibly induced by fracturing via reactivation of faults (Fountain, 2011; Rutqvist et al., 2013). In the United States, Texas and Pennsylvania have seen the most widespread use of high-volume hydraulic fracturing, while other states such as New York as well as other countries have been proceeding more cautiously as they develop their regulatory frameworks (Rahm, 2011; Eaton, 2013). Relatively, clean-burning natural gas, including that potentially producible in large volumes via hydraulic fracturing, may serve as an intermediate bridging step to a future state with a greater reliance on *renewable* energy sources (Eaton, 2013). The potential for deleterious effects of the fracturing process should encourage environmental managers to foster a science-based arrangement to maximize resource utilization while minimizing harmful consequences.

Another example of massive unconventional petroleum exploitation is the *oil sand* deposit in Alberta, Canada. Viewed within the conceptual framework of the petroleum system, this is a case for which the fifth phase, the trap, is missing. The permeable sandy reservoir is exposed at the

FIGURE 1.41 Oil sand refinery on the banks of the Athabasca River, Alberta, Canada. Dikes separate the river from the ponds of oily water used in the refining process. *Source:* Aerial photo courtesy of Peter Essick.

surface, where the oil has been converted into a viscous tar by microbial biodegradation. Hence, the oil sands must be strip mined like coal and then steam treated and refined to create a marketable liquid product (Fig. 1.41). The energy consumed in the mining and thermal treatment is costly, but oil prices are currently high enough to make this economically feasible. Of great concern is that when the carbon dioxide generated by mining and treatment is included, the net greenhouse gas emissions from the exploitation of this resource are nearly triple that of conventionally produced petroleum (Kunzig, 2009). Public objections arising from these worries have delayed the completion of a pipeline designed to bring the Alberta oil into the United States (Frosch, 2013).

1.8.4 Consideration of Externalities

Cost–benefit analyses performed by the energy industry regarding the utilization of nuclear power, coal, conventional petroleum and natural gas, hydrocarbons from hydraulic fracturing, and oil sands often ignore *externalities*. These are the damages to the environment and burdens placed upon society that are not reflected in the market price of the energy (Schleisner, 2000). Long-term planning decisions should be made with externalities in mind when comparing the economics of fossil energy systems with those of renewable energy sources, be they solar (Figs. 1.24 and 1.42), hydropower, wind, tidal, biomass, or waste to energy.

1.9 CONCLUDING REMARKS

In the course of this overview, the emphasis has been upon the geological basis of a number of major environmental management issues. Throughout, the perspective has been grounded in the belief that populations should be protected from undue exposure to natural hazards and that natural systems should likewise be shielded from undue anthropogenic pressure. Whether the threats were volcanic eruptions, earthquakes, floods, landslides, or subsidence, it has been emphasized that geological and environmental methodologies and knowledge can serve to provide a sensible means for risk reduction. Given that some human settlements will likely continue to be situated in hazardous areas, engineered approaches to protection of life and property will be beneficial if properly applied. Armed with both knowledge and political sensitivity, environmental managers will have opportunities for positive social impact in negotiating the challenges as they weigh costs, risks, and benefits. When considering natural resource and energy issues, environmental managers should foster science-based solutions to maximize resource utilization while minimizing harmful consequences, bearing in mind externalities and long-term impacts. As with the debate over strategies to cope with natural hazards, the course of future energy policy will be determined by the complex and contentious interplay of the scientific/technical, economic, and political drivers. At these critical junctures, it is evident that environmental managers have an important role to play.

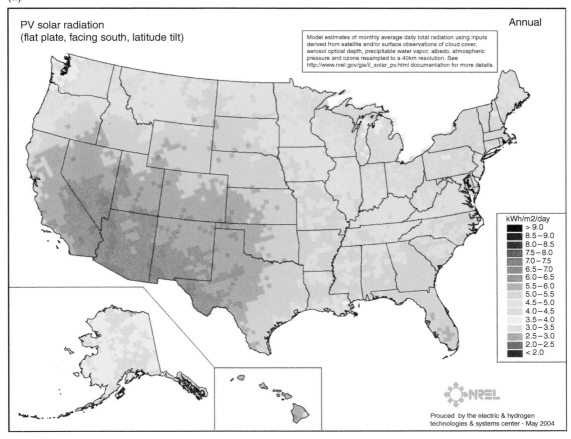

FIGURE 1.42 (a) Map of solar energy potential in the United States. The desert southwest has the greatest potential, as indicated by its red and orange colors on the map. (b) Tower and surrounding field of mirrors at the experimental thermal solar electrical generating plant "Solar Two" in the Mojave Desert, California. Images: National Renewable Energy Laboratory, US Dept. of Energy.

Appendix 1.A Geographic Coordinates of the Examples Presented

This table presents the geographic coordinates of the examples given in the text for further exploration by the reader, keyed to figure number. The digital degree format is used, such that south latitude and west longitude values are written as negative numbers. This is generally simpler to use than the degrees–minutes–seconds format and can be employed with Google Earth or other geographic information utility if the preferences are set to accept them.

Location	Figures	Latitude (°N)	Longitude (°E)
Mt. Vesuvius, Italy	2	40.8196	14.4281
Eyjafjallajökull Volcano, Iceland	3	60.6308	−19.6008
Mt. Rainier, Washington	4	46.8537	−121.7589
Mt. Ruapehu, New Zealand	5	−39.2836	175.5667
Otsuchi, Japan	6	39.3582	141.8994
San Francisco, California	7–9	37.7749	−122.4194
Daly City, California	10–12	37.6672	−122.4890
Atlantic Beach, New York	13, 25	40.5852	−73.7255
Montauk Lighthouse, New York	14	41.0708	−71.8570
Fire Island, New York	15	40.7231	−72.8907
Mantoloking, New Jersey	16	40.0438	−74.0487
Ansonia/Derby, Connecticut	17–19	41.3362	−73.0787
Mono Lake, California	20–21	38.0070	−119.0122
Haskell, Kansas	22	37.5262	−100.9067
Colby, Kansas	23	39.3914	−101.0672
NW of Tonopah, Nevada	24	38.3670	−117.4299
Winter Haven, Florida	26	28.0222	−81.7328
Clermont, Florida	27, 28	28.3494	−81.6584
Zaragoza, Spain	29	41.6427	−0.9356
Landfill, Secaucus, New Jersey	30	40.7534	−74.0855
Yucca Mountain, Nevada	32	36.8524	−116.4288
Mine, Wilmington, Illinois	33	41.2923	−88.1863
Will Scarlet Mine, Illinois	34	37.6525	−88.7232
Mine, Kayford Mountain, West Virginia	35	37.9698	−81.3818
Deepwater Horizon site, Gulf of Mexico	38	28.7381	−88.3659
Refinery, near Fort McMurray, Alberta	41	57.0050	−111.4816
Solar Two project, Barstow, California	42	34.8719	−116.8342

REFERENCES

Ahern, M.M., M. Hendryx, J. Conley, E. Fedorko, A. Ducatman, and K.J. Zullig. 2011. The association between mountaintop mining and birth defects among live births in central Appalachia, 1996–2003. *Environmental Research* 111: 838–846.

Akcil, A. and S. Koldas. 2006. Acid mine drainage (AMD): Causes, treatment and case studies. *Journal of Cleaner Production* 14:1139–1145.

Alvarez, L. 2013. Where sand is gold, the reserves are running dry. *The New York Times*. August 24, p. A14.

ANDRA. 2012. *Inventaire national des matières et déchets radioactifs—Rapport de synthèse. Agence nationale pour la gestion des déchets radioactif*. Châtenay-Malabry, France.

Barringer, F. 2013. Spread of hydrofracking could strain water resources in west, study finds. *The New York Times*. May 2, p. A12.

Bastianoni, S., F. Coppola, E. Tiezzi, A. Colacevich, F. Borghini, and S. Focardi. 2008. Biofuel potential production from the Orbetello lagoon macroalgae: A comparison with sunflower feedstock. *Biomass and Bioenergy* 32:619–628.

Blackford, N. 2012. A century later, abandoned coal mines pose serious risk to property. Evansville Courier and Press, June 2. Available at http://www.courierpress.com/news/local-news/54pt-hed4-15-of-story-x. Accessed October 15, 2013.

Bodvarsson, G.S., W. Boyle, R. Patterson, and D. Williams. 1999. Overview of scientific investigations at Yucca Mountain—The potential repository for high-level nuclear waste. *Journal of Contaminant Hydrology* 38:3–24.

Bonilla, M.G., C. Wentworth, M. Lucks, H. Schoonover, S. Graham, and T. May. 1998. Preliminary geologic map of the San Francisco south 7.5' quadrangle and part of the Hunters Point 7.5' quadrangle. U.S. Geological Survey Open-File Report 98–354. A Digital Database, San Francisco Bay Area, CA.

California Conservation Corps. 2013. Earthquakes. Available at http://www.ccc.ca.gov/emer/HistoricalResponse/Pages/earthquakes.aspx. Retrieved October 8, 2013.

California Division of Mines and Geology. 2000. State of California seismic hazard zones—City and County of San Francisco—Official map.

California Geological Survey. 2003. Seismic hazards zonation program. Article 10. Seismic hazards mapping. Available at http://www.conservation.ca.gov/cgs/shzp/Pages/article10.aspx. Retrieved October 6, 2013.

California Geological Survey. 2004. *Recommended Criteria for Delineating Seismic Hazard Zones in California*, Special Publication 118, revised. California Geological Survey, Sacramento, CA.

Chourasia, A., S. Cutchin, and B. Aagaard. 2008. Visualizing the ground motions of the 1906 San Francisco earthquake. *Computers & Geosciences* 34(12):1798–1805.

Cohen, P., O.L. Franke, and B.L. Foxworthy. 1968. An atlas of Long Island's water resources. *New York Water Resources Commission Bulletin* 62:117.

Davis C., V. Keilis-Borok, V. Kossobokov, and A. Soloviev. 2012. Advance prediction of the March 11, 2011 Great East Japan Earthquake: A missed opportunity for disaster preparedness. *International Journal of Disaster Risk Reduction* 1:17–32.

Dean, C. 2012. Costs of shoring up coastal communities. *The New York Times*. November 5, p. D1.

Dierauer J., N. Pinter, and J.W.F. Remo. 2012. Evaluation of levee setbacks for flood-loss reduction, Middle Mississippi River, USA. *Journal of Hydrology* 450–451:1–8.

Eaton, T.T. 2013. Science-based decision-making on complex issues: Marcellus shale gas hydrofracking and New York City water supply. *The Science of the Total Environment* 461–462: 158–169.

Ersoy, H., F. Bulut, and M. Berkün. 2013. Landfill site requirements on the rock environment: A case study. *Engineering Geology* 154:20–35.

Fedorko, N. and M. Blake 1998. A geologic overview of mountaintop removal mining in West Virginia. Executive Summary of a Report to the Committee on Post-Mining Land Use and Economic Aspects of Mountaintop Removal Mining. West Virginia Geological and Economic Survey, Morgantown, WV.

Fitzgerald, S. 2012. Former strip mine one of worst sites in Midwest. *The Southern Illinoisian*. July 15. Available at http://thesouthern.com/news/local/former-strip-mine-one-of-worst-sites-in-midwest/article_207b052c-ce1d-11e1-90cd-001a4bcf887a.html. Accessed October 15, 2013.

Foderaro, L.W. 2013. Breach through Fire Island also divides opinions. *The New York Times*. April 5, p. A13.

Foo, K.Y. and B.H. Hameed. 2009. An overview of landfill leachate treatment via activated carbon adsorption process. *Journal of Hazardous Materials* 171:54–60.

Fountain, H. 2011. Add quakes to rumblings over gas rush. *The New York Times*. December 12, p. D1.

Frosch, D. 2013. Amid pipeline debate, two costly cleanups forever change towns. *The New York Times*. August 11, p. A18.

Galbraith, K. 2013. As fracking increases, so do fears about water supply. *The New York Times* March 8, p. A21.

Ghezzo, M., S. Guerzoni, A. Cucco, and G. Umgiesser. 2010. Changes in Venice Lagoon dynamics due to construction of mobile barriers. *Coastal Engineering* 57:694–708.

Gluskoter, H.J. and J.A. Simon. 1968. *Sulfur in Illinois Coals*. Illinois State Geological Survey Circular 432. Illinois State Geological Survey, Urbana, IL.

Gornitz, V., S. Couch, and E.K. Hartig. 2002. Impacts of sea level rise in the New York City metropolitan area. *Global and Planetary Change* 32(1):61–88.

Graham, B., W.K. Reilly, F. Beinecke, D.F. Boesch, T.D. Garcia, C.A. Murray, and F. Ulmer. 2011. Deep water: The Gulf oil disaster and the future of offshore drilling. Report to the President. National Commission on the BP Deepwater Horizon Oil Spill and Offshore Drilling, Washington, DC.

Griffith, M.B., S.B. Norton, L.C. Alexander, A.I. Pollard, and S.D. LeDuc. 2012. The effects of mountaintop mines and valley fills on the physicochemical quality of stream ecosystems in the central Appalachians: A review. *Science of the Total Environment* 417–418:1–12.

Gutiérrez, F., A.H. Cooper, and K.S. Johnson. 2008a. Identification, prediction, and mitigation of sinkhole hazards in evaporite karst areas. *Environmental Geology* 53:1007–1022.

Gutiérrez, F., J.M. Calaforra, F. Cardona, F. Ortí, J.J. Durán, and P. Garay. 2008b. Geological and environmental implications of the evaporite karst in Spain. *Environmental Geology* 53:951–965.

Gutiérrez, F., J. Guerrero, and P. Lucha. 2008c. Quantitative sinkhole hazard assessment. A case study from the Ebro Valley evaporite alluvial karst (NE Spain). *Natural Hazards*. 45:211–233.

Gutiérrez, F., J.P. Galve, P. Lucha, J. Bonachea, L. Jordá, and R. Jordá. 2009. Investigation of a large collapse sinkhole affecting a multi-storey building by means of geophysics and the trenching technique (Zaragoza city, NE Spain). *Environmental Geology* 58:1107–1122.

Harper, D. 2011. Geologic hazards—Mine subsidence in Indiana. Indiana Geological Survey. Available at http://igs.indiana.edu/Hazards/Subsidence.cfm. Accessed November 7, 2013.

Holzer, T.L., A.S. Jayko, E. Hauksson, J.P.B. Fletcher, T.E. Noce, M.J. Bennett, C.M. Dietel, and K.W. Hudnut. 2010. Liquefaction caused by the 2009 Olancha, California (USA), M5.2 earthquake. *Engineering Geology* 116:184–188.

Hubbert M.K. 1956. *Nuclear Energy and the Fossil Fuels*. Shell Development Company, Exploration and Production Research Division, Publication No. 95, Shell Development Co., Exploration and Production Research Division, Houston, TX.

Hunt J.M. 1996. *Petroleum Geochemistry and Geology*, 2nd ed. Freeman, New York.

IPCC. 2007. Climate change 2007: The Physical science basis. In, S. Solomon, D. Qin, M. Manning, Z. Chen, M. Marquis, K.B. Averyt, M. Tignor, and H.L. Miller, eds., *Contribution of Working Group I to the Fourth Assessment Report of the Intergovernmental Panel on Climate Change*, Cambridge University Press, Cambridge/New York.

Irish, J.L., P.J. Lynett, R. Weiss, S.M. Smallegan, and W. Cheng. 2013. Buried relic seawall mitigates Hurricane Sandy's impacts. *Coastal Engineering* 80:79–82.

Jellison, R. and J.M. Melack. 1993. Algal photosynthetic activity and its response to meromixis in hypersaline Mono Lake, California. *Limnology and Oceanography* 38(4):818–837.

Jellison, R., R.F. Anderson, J.M. Melack, and D. Heil. 1996. Organic matter accumulation in sediments of hypersaline Mono Lake during a period of changing salinity. *Limnology and Oceanography* 41(7):1539–1544.

Jellison, R., J. Romero, and J.M. Melack. 1998. The onset of meromixis during restoration of Mono Lake, California: Unintended consequences of reducing water diversions. *Limnology and Oceanography* 43(4):706–711.

John, D.A., T.W. Sisson, G.N. Breit, R.O. Rye, and J.W. Vallance. 2008. Characteristics, extent and origin of hydrothermal alteration at Mount Rainier Volcano, Cascades Arc, USA: Implications for debris-flow hazards and mineral deposits. *Journal of Volcanology and Geothermal Research* 175:289–314.

Johnson D.B. and K.B. Hallberg. 2005. Acid mine drainage remediation options: A review. *Science of the Total Environment* 338:3–14.

Joyce, C. 2012. With gas boom, Pennsylvania fears new toxic legacy. National Public Radio. Available at http://www.npr.org/2012/05/14/149631363/when-fracking-comes-to-town-it-s-water-water-everywhere. Accessed May 14, 2012.

Kana, T.W. 1995. A mesoscale sediment budget for Long Island, New York. *Marine Geology* 126:87–110.

Keefer, D.K. 2002. Investigating landslides caused by earthquakes—A historical review. *Surveys in Geophysics* 23(6):473–510.

Killops, K. and V. Killops. 2005. *Introduction to Organic Geochemistry*, 2nd ed. Blackwell Publishing, Oxford.

Kleinfield, N.R. 2012. After getting back to normal, big job is facing new reality. *The New York Times*. November 4, p. A1.

Kruge, M.A. 2013. Oil pollution in water bodies of restricted circulation. In, M. Salgot, ed., *Stagnant Water Bodies Pollution*. Atelier, Barcelona, pp. 63–80.

Kunzig, R. 2009. The Canadian oil boom. *National Geographic Magazine*. March 2009. Available at http://ngm.nationalgeographic.com/2009/03/canadian-oil-sands/kunzig-text/1. Retrieved March 5, 2013.

Kurniawan, T.A., W-H. Lo, and G.Y.S. Chan. 2006. Physicochemical treatments for removal of recalcitrant contaminants from landfill leachate. *Journal of Hazardous Materials* B129:80–100.

Langmann, B., A. Folch, M. Hensch, and V. Matthias. 2012. Volcanic ash over Europe during the eruption of Eyjafjallajökull on Iceland, April–May 2010. *Atmospheric Environment* 48:1–8.

Leatherman, S.P. 1985. Geomorphic and stratigraphic analysis of Fire Island, New York. *Marine Geology* 63:173–195.

Lentz, E.E., C.J. Hapke, H.F. Stockdon, and R.E. Hehre. 2013. Improving understanding of near-term barrier island evolution through multi-decadal assessment of morphologic change. *Marine Geology* 337:125–139.

Lin, A., R. Ikuta, and G. Rao. 2012. Tsunami run-up associated with co-seismic thrust slip produced by the 2011 M_w 9.0 Off Pacific Coast of Tohoku earthquake, Japan. *Earth and Planetary Science Letters* 337–338:121–132.

Lindsey, B.D., B.G. Katz, M.P. Berndt, A.F. Ardis, and K.A. Skach. 2010. Relations between sinkhole density and anthropogenic contaminants in selected carbonate aquifers in the eastern United States. *Environmental Earth Science* 60:1073–1090.

Liston, B. 2013. Guests saved as Florida resort building falls into sinkhole. Available at http://www.reuters.com/article/2013/08/12/us-usa-florida-sinkhole-idUSBRE97B0D520130812. Retrieved September 24, 2013.

Luongo, G., A. Perrotta, and C. Scarpati. 2003a. Impact of the AD 79 explosive eruption on Pompeii, I. Relations amongst the depositional mechanisms of the pyroclastic products, the framework of the buildings and the associated destructive events. *Journal of Volcanology and Geothermal Research* 126:201–223.

Luongo, G., A. Perrotta, C. Scarpati, E. De Carolis, G. Patricelli, A. Ciarallo. 2003b. Impact of the AD 79 explosive eruption on Pompeii, II. Causes of death of the inhabitants inferred by stratigraphic analysis and areal distribution of the human casualties. *Journal of Volcanology and Geothermal Research* 126:169–200.

McBride, R.A. and T.F. Moslow. 1991. Origin, evolution, and distribution of shoreface sand ridges, Atlantic inner shelf, U.S.A.. *Marine Geology* 97:57–85.

Met Office. 2014. Volcanic Ash Advisory from London—Issued graphics, 20100418/1800Z. Available at http://www.metoffice.gov.uk/aviation/vaac/data/VAG_180542.png. Retrieved July 3, 2014.

Mono Lake Committee. 2013. Mono basin clearinghouse. Available at http://www.monobasinresearch.org/. Retrieved October 21, 2013.

Montoro Chiner, M.J.. 1997. Sobre la reclamació d'indemnització instada contra l'Administració de la Generalitat de Catalunya per diversos propietaris d'habitatges i edificacions afectades pels esllavissaments de terres soferts a la urbanització Cap de la Barra de l'Estartit. Dictamen 115/97. Generalitat de Catalunya, Comissió Jurídica Assessora, Memòria d'Activitats, Catalonia.

Nakanishi, H., K. Matsuo, and J. Black. 2013. Transportation planning methodologies for post-disaster recovery in regional communities: The East Japan Earthquake and tsunami 2011. *Journal of Transport Geography* 31:181–191.

Nashawi, I.S., A. Malallah, and M. Al-Bisharah. 2010. Forecasting world crude oil production using multicyclic Hubbert model. *Energy & Fuels* 24:1788–1800.

National Oceanic and Atmospheric Administration. 2014. ERMA deepwater Gulf response. Environmental Response Management Application. Available at http://gomex.erma.noaa.gov/erma.html. Retrieved July 3, 2014.

Navarro, M. 2012. After storm, dry floors prove value of exceeding city code. *The New York Times*. November 23, p. A15.

New York Times. 2010. Tracking airport status. Available at http://www.nytimes.com/interactive/2010/04/15/world/europe/airport-closings-graphic.html. Retrieved April 17, 2010.

O'Connor, K.M. and E.W. Murphy. 1997. TDR monitoring as a component of subsidence risk assessment over abandoned mines. *International Journal of Rock Mechanics and Mining Sciences* 34(3–4): 230.

Oude Essink, G.H.P. 2001. Improving fresh groundwater supply problems and solutions. *Ocean & Coastal Management* 44:429–449.

Ourisson, G., P. Albrecht, and M. Rohmer, 1984. The microbial origin of fossil fuels. *Scientific American* 251:44–51.

Pankow, J.F. and J.A. Cherry. 1996. *Dense Chlorinated Solvents and Other DNAPLs in Ground Water-History, Behavior, and Remediation*. Waterloo Press, Portland, OR, 522pp.

Pence, A. 2000. Cliff hangers: 17 houses on Daly City block sliding toward ocean plunge. *San Francisco Chronicle*, February 16. Available at http://www.sfgate.com/default/article/Cliff-Hangers-17-houses-on-Daly-City-block-3239855.php. Accessed October 15, 2013.

Petroski, H. 2013. The stormy politics of building. *The International New York Times*. October 22. Available at http://www.nytimes.

com/2013/10/23/opinion/international/the-stormy-politics-of-building.html. Accessed November 11, 2008.

Pilkey, O.H. 2012. We need to retreat from the beach. *The New York Times*. November 14, p. A35.

Pipkin, B.W., D.D. Trent, R. Hazlett, and P. Bierman. 2008. *Geology and the Environment*, 5th ed. Thompson Brooks/Cole. Belmont, CA.

Portella, G.P. 2013. El Ayuntamiento demolerá el edificio afectado por una dolina en Valdefierro y realojará a sus vecinos. Available at http://www.aragondigital.es/noticia.asp?notid=106279. Retrieved October 5, 2013.

Povoledo, E. and H. Fountain. 2012. Italy orders jail terms for 7 who didn't warn of deadly earthquake. *The New York Times*. October 23, p. A4.

Pumure, I., J.J. Renton, and R.B. Smart. 2010. Ultrasonic extraction of arsenic and selenium from rocks associated with mountaintop removal/valley fills coal mining: Estimation of bioaccessible concentrations. *Chemosphere* 78:1295–1300.

Rahm, D. 2011. Regulating hydraulic fracturing in shale gas plays: The case of Texas. *Energy Policy* 39:2974–2981.

Rahm, B.G., J.T. Bates, L.R. Bertoia, A.E. Galford, D.A. Yoxtheimer, and S.J. Riha. 2013. Wastewater management and Marcellus Shale gas development: Trends, drivers, and planning implications. *Journal of Environmental Management* 120: 105–113.

Rechard, R.P., G.A. Freeze, and F.V. Perry. 2014a. Hazards and scenarios examined for the Yucca Mountain disposal system for spent nuclear fuel and high-level radioactive waste. *Reliability Engineering and System Safety* 122:74–95.

Rechard, R.P., T.A. Cotton, and M.D. Voegele. 2014b. Site selection and regulatory basis for the Yucca Mountain disposal system for spent nuclear fuel and high-level radioactive waste. *Reliability Engineering and System Safety* 122:7–31.

Rechard, R.P., H.H. Liu, Y.W. Tsang, and S. Finsterle. 2014c. Site characterization of the Yucca Mountain disposal system for spent nuclear fuel and high-level radioactive waste. *Reliability Engineering and System Safety* 122:32–52.

Rutqvist J., A.P. Rinaldi, F. Cappa, and G.J. Moridis. 2013. Modeling of fault reactivation and induced seismicity during hydraulic fracturing of shale-gas reservoirs. *Journal of Petroleum Science and Engineering* 107:31–44.

Schleisner, L. 2000. Comparison of methodologies for externality assessment. *Energy Policy* 28(15):1127–1136.

Sivathasan, K., X.S. Li, K.K. Muraleetharan, C. Yogachandran, and K. Arulanandan. 2000. Application of three numerical procedures to evaluation of earthquake-induced damages. *Soil Dynamics and Earthquake Engineering* 20:325–339.

Smith, V.H., G.D. Tilman, and J.C. Nekola. 1999. Eutrophication: Impacts of excess nutrient inputs on freshwater, marine, and terrestrial ecosystems. *Environmental Pollution* 100:179–196.

Soeder, D.J. and W.M. Kappel. 2009. *Water Resources and Natural Gas Production from the Marcellus Shale*. U.S. Geological Survey Fact Sheet 2009–3032, U.S. Department of the Interior, U.S. Geological Survey, Reston, VA.

Sun, M., G.V. Lowry, and K.B. Gregory. 2013. Selective oxidation of bromide in wastewater brines from hydraulic fracturing. *Water Research* 47:3723–3731.

Tissot B.P. and D.H. Welte 1984. *Petroleum Formation and Occurrence*, 2nd ed. Springer-Verlag, Berlin.

Titus, J.C. 1998. Rising seas, coastal erosion, and the takings clause: How to save wetlands and beaches without hurting property owners. *Maryland Law Review* 57(4):1279–1399.

Tomás, R., R. Romero, J. Mulas, J.J. Marturià, J.J. Mallorquí, J.M. Lopez-Sanchez, G. Herrera, F. Gutiérrez, P.J. González, J. Fernández, S. Duque, A. Concha-Dimas, G. Cocksley, C. Castañeda, D. Carrasco, and P. Blanco. 2014. Radar interferometry techniques for the study of ground subsidence phenomena: A review of practical issues through cases in Spain. *Environmental Earth Sciences* 71(1):163–181.

U.S. Geological Survey. 2013. Groundwater levels for Kansas. Available at http://nwis.waterdata.usgs.gov/ks/nwis/gwlevels/. Retrieved October 7, 2013.

Urbina, I. 2011. Regulation lax as gas wells' tainted water hits rivers. *The New York Times*. February 27, p. A1.

Vengosh, A., N. Warner, R. Jackson, and T. Darrah. 2013. The effects of shale gas exploration and hydraulic fracturing on the quality of water resources in the United States. *Procedia Earth and Planetary Science* 7:863–866.

Wang, Z., W. Huang, D. Zhao, and S. Pei. 2012. Mapping the Tohoku forearc: Implications for the mechanism of the 2011 East Japan earthquake (Mw 9.0). *Tectonophysics* 524–525:147–154.

Washington State Department of Natural Resources. 2013. Natural hazards. Geology & Earth Resources Division. Available at https://fortress.wa.gov/dnr/geology/?Theme=natural_hazards. Retrieved October 6, 2013.

West Virginia Geological and Economic Survey. 2013. Trace elements in West Virginia coals. Available at http://www.wvgs.wvnet.edu/www/datastat/te/index.htm. Retrieved October 8, 2013.

Wikipedia. 2014a. Garden of the fugitives, Pompeii. Available at http://commons.wikimedia.org/w/index.php?title=File:Pompeii_Garden_of_the_Fugitives_02.jpg&oldid=121772188. Retrieved 23 December, 2014.

Wikipedia. 2014b. Air travel disruption after the 2010 Eyjafjallajökull eruption. Available at http://en.wikipedia.org/wiki/Air_travel_disruption_after_the_2010_Eyjafjallajökull_eruption. Retrieved July 3, 2014.

Wines, M. 2013a. Spring rain, then foul algae in ailing Lake Erie. *The New York Times*. March 15, p. A1.

Wines, M. 2013b. Wells dry, fertile plains turn to dust. *The New York Times*. May 20, p. A1.

Wood, N. and C. Soulard. 2009. Variations in population exposure and sensitivity to lahar hazards from Mount Rainier, Washington. *Journal of Volcanology and Geothermal Research* 188: 367–378.

Yang, B., C. Hwang, and H.K. Cordell. 2012. Use of LiDAR shoreline extraction for analyzing revetment rock beach protection: A case study of Jekyll Island State Park, USA. *Coastal Management* 69:1–15.

Zisman, E.D. 2013. The Florida sinkhole statute: Its evolution, impacts and needed improvements. *Carbonates and Evaporites* 28:95–102.

Zisman, E.D., M. Wightman, and J. Kestner. 2013. Sinkhole investigation methods: The next step after special publication no. 57. *Carbonates and Evaporites* 28:103–109.

2

BIOLOGY IN ENVIRONMENTAL MANAGEMENT

AUDREY L. MAYER
School of Forest Resources and Environmental Science and Department of Social Sciences, Michigan Technological University, Houghton, MI, USA

Abstract: Taking the perspective that it is the human activities that require management rather than the biological systems that we disturb, this chapter will describe an adaptive management approach to environmental management. Using technological advancements such as remote sensing and geographic information systems, ecologists can determine the most successful ways to harmonize human disturbances with natural ones and identify feasible biological targets for the system of interest. Understanding the scale of the system and its processes is critical, and using large-scale data sets along with remote sensing can help ecologists determine where the system is, where it needs to be, and whether preservation is the right management decision or if more active restoration or rehabilitation is necessary. Indicators (measurable variables) are used to monitor system movement toward targets and responses to management actions. The adaptive management process is iterative, so that new information on the response of the ecosystem to our management activities is used to improve the next round of decisions. The environmental management of the Florida Everglades is used throughout as an illustrative case study of how ecologists have used an adaptive management approach to restore and preserve an internationally famous ecosystem.

Keywords: adaptive management, conservation, ecology, Everglades, GIS, habitat, landscapes, monitoring, natural disturbances, pollution, reclamation, remote sensing, resilience, restoration, targets.

2.1	Introduction	47
2.2	Stage 1: Define the System	50
	2.2.1 Landscapes and Scale	50
	2.2.2 Abiotic Conditions	51
	2.2.3 Biodiversity and Ecosystem Processes	52
	2.2.4 Natural and Human-Induced Disturbances	53
2.3	Stage 2: Identify Target Conditions	53
	2.3.1 Setting Targets	54
2.4	Stage 3: Collect Data and Determine Distance to Target	55
	2.4.1 Land Classification Systems	56
	2.4.2 Continent-Scale Field Data Collection Networks	57
2.5	Stage 4: Preservation or Restoration?	59
	2.5.1 Preservation	59
	2.5.2 Restoration	63
	2.5.2.1 Reclamation	65
	2.5.2.2 Rehabilitation	65
	2.5.2.3 Restoration	66
	2.5.3 Restoring Natural Fire Disturbance Regimes	66
2.6	Stage 5: Monitoring and Adaptive Management	67
References		68

2.1 INTRODUCTION

Biology is "the study of life" and has grown from a relatively contained area of study in the seventeenth and eighteenth centuries into an immensely diverse collection of disciplines, ranging from genetics and microbiology (studies at the organismal level, such as bacteria and viruses) to biogeography (studying the evolution of species and communities at the global scale). Some of this diversification in biological studies has been driven by new technologies that allow us to observe mechanisms and phenomena that are beyond the scope of our own senses (e.g., electron microscopes, low-orbit satellites). And some of the diversification has to do with a paradigm shift

---

An Integrated Approach to Environmental Management, First Edition. Edited by Dibyendu Sarkar, Rupali Datta, Avinandan Mukherjee, and Robyn Hannigan.
© 2016 John Wiley & Sons, Inc. Published 2016 by John Wiley & Sons, Inc.

in how we, as humans, view our place in the world. While most of these disciplines in biology have applications to environmental management, this chapter will focus primarily on advances in the subdisciplines of ecology, which include conservation biology, urban ecology, landscape ecology, ecosystem ecology, biogeography (or macroecology), and restoration ecology.

Environmental management from a biological perspective includes the central tenet that often it is the human activities that place stress on the environment that require managing more so than the environment itself. Wali et al. (2010) offer six elements of environmental management that will be discussed throughout this chapter:

- *Sustainability of ecological processes that produce ecosystem goods and services*: Ecosystems provide products that human societies use, such as wood for fuel and construction from forests, and medicinal plants from wetlands and prairies. They also provide services that our societies depend upon, such as trees sequestering carbon in their tissues to regulate our climate. When human actions damage ecosystems, the capacity of these ecosystems to provide these goods and services are diminished (Millennium Ecosystem Assessment, 2005). Globally, these goods and services are worth many times that of the products and services traded in our markets (Costanza et al., 1997).
- *Systems thinking—interdependence, interconnectedness, and dynamism of ecological, social, and economic systems*: An emerging trend in many scientific disciplines is to move away from a reductionist methodology (examining the role and functioning of system parts) and toward holistic methodologies that examine how the entire system responds to disturbances or shocks (Bennett and McGinnis, 2008). Complexity theory gives scientists a framework to understand how feedbacks among systems and emergent behavior within them can create surprising behavior, such as collapses and catastrophes, even with very small changes (Scheffer, 2009; Mitchell, 2011). In complex systems, studying individual parts of the system does not allow one to predict how the system might respond to changes or disturbances (Meadows, 2008). For example, an unforeseen impact of the trophic cascade initiated by the reintroduction of wolves to Yellowstone National Park management was a dramatic improvement in the riparian zone of the Yellowstone River, as well as the geomorphology of the river itself (Beschta and Ripple, 2012; Ripple and Beschta, 2012). Decisions that rely on a reductionist approach to system knowledge therefore often fail to meet objectives, as the system does not respond as managers expect.
- *Hierarchy and heterogeneity of spatial and temporal scales*: Disciplines such as landscape ecology are primarily focused on how patterns and processes change with scale (Turner et al., 2010). Scale includes the attributes of *resolution*, the smallest level at which an observation is made (e.g., every minute, every square meter), and *extent*, the largest level at which a process or phenomena can be observed (e.g., over the past century, over the entire continent). Thus, scale is both spatial and temporal. Depending upon the scale at which you are working, an ecosystem will be affected by processes that operate at larger scales, which affect those processes that operate below it. For example, managing flooding within a watershed is subject to regional rainfall patterns, and management decisions will impact the heterogeneity (e.g., diversity) of in-stream habitat available to aquatic insects, some of which might prefer calm pools to fast rapids.
- *Ecologically compatible boundaries*: Too often in the past, environmental management has been confined within political boundaries; these boundaries are meaningless to ecosystems. One county may try to reduce flooding by planting trees along a river to widen the riparian zone, but if the county upstream from it in the watershed destroys its riparian buffer, the water in the river will still overflow its banks in the county attempting to manage the flooding. One of the most obstinate barriers to restoration of the Florida Everglades is the fact that the boundaries of the Everglades National Park and associated protected areas do not include the entire hydrologic system, which extends much further north to Lake Okeechobee and the Kissimmee River (Davis and Ogden, 1997; Stoneman, 1997; Richardson, 2008; Fig. 2.1). The importance of boundary choice and intentional engagement of all of the relevant players in environmental management and restoration cannot be overemphasized (Lackey, 1998).
- *Adaptive management to deal with ecosystem dynamics as well as uncertainty*: Adaptive management refers to a process of designing management interventions as if they were experiments (Allen and Stankey, 2008). A prairie manager might use prescribed fire in one area of a preserve to test how well fire might be used to control an invasive grass species. If the invasive shows no signs of recovery, that fire regime might then be used in other areas of the preserve. In this way, managers use scientific methods to understand how an ecosystem might respond to different management or restoration treatments (or no treatments) and measure the level of uncertainty in treatment outcomes. However, as sensible as this approach may sound, it has yet to be widely used in environmental management and restoration due to the difficulties involved in its practice and extrapolation of results (Lynch et al., 2008; Westgate et al., 2013). However, adaptive management offers a

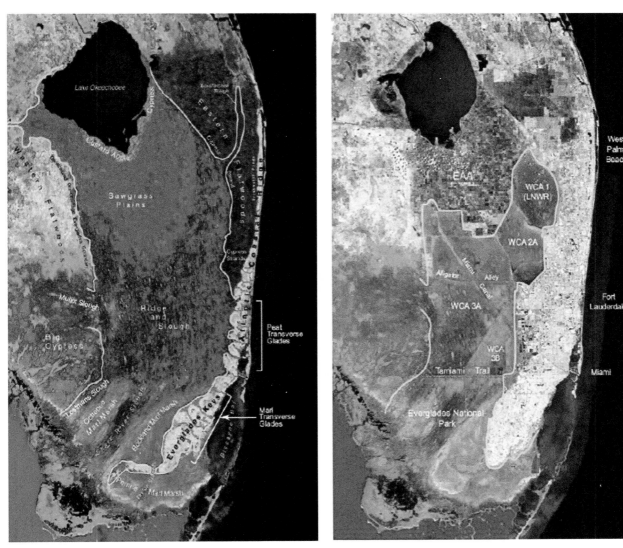

FIGURE 2.1 A historical and current view of the Florida Everglades ecosystem. Note the significant land use/land cover change from wetland habitats to agricultural (Everglades Agricultural Area (EAA)) and urban land uses and the drainage canals partitioning the Water Conservation Areas (WCA) into isolated fragments.

research management framework and is a practical way of studying the impacts of environmental management at realistic scales.

- *Collaborative and participative process of decision-making*: Decisions to manage or restore ecosystems, and the condition they should be managed or restored to, inherently involve social and economic decisions (Society for Ecological Restoration (SER), 2002; Sklar et al., 2005). For example, the decision to restore viable populations of wolves (*Canis lupus*) and protect these top predators in Yellowstone National Park was complex and controversial because of the numerous interested parties and stakeholders involved in the Greater Yellowstone Ecosystem (Bennett and McGinnis, 2008; Lynch et al., 2008). Wolves help control the number of grazers such as elk (*Cervus canadensis*), which were preventing forest regeneration through overbrowsing tree saplings, and are highly valued by recreationalists who travel there to observe them. However, wolves also present a danger to livestock in the area and hence exact an economic toll on ranchers (Licht et al., 2010; Lee et al., 2012; Eisenberg et al., 2013). Effective environmental management must include dialogue with communities that are likely to see both costs and benefits as a result of management decisions. Management must also be informed by adequate scientific information (data) and techniques (Lackey, 1998).

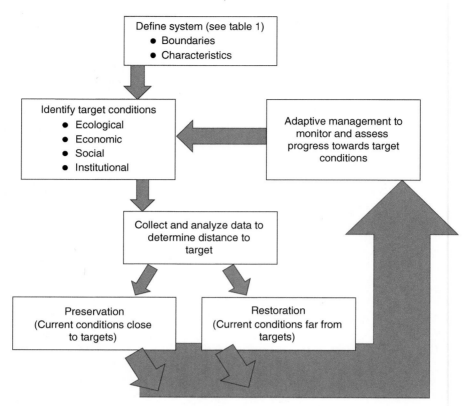

FIGURE 2.2 Stages of environmental management.

This chapter will present information on environmental management in the stages that are commonly used for its practice (Fig. 2.2). Management plans generally follow a five-stage cycle: first, the system to be managed is defined; next, the desired state of the system is identified; then data are collected and analyzed to determine the distance between the current system state and the target; next, best management practices (or best practices, described in Section 2.5) are utilized to either conserve or restore the system (depending upon what is needed to reach the target); and finally, the system is monitored to determine if and when additional actions are required.

2.2 STAGE 1: DEFINE THE SYSTEM

Although this may seem like an obvious first step, many of the failures in environmental management or restoration projects can be traced back to vague or ill-considered system boundaries and its characteristics or initiating management activities before the current conditions of the system (relative to the target conditions) are known. If management activities are initiated before the system is fully documented, any changes that are observed may not be attributable to the management activities. For example, if a dam is breached on a salmon (*Salmo* spp. and *Oncorhynchus* spp.) spawning river before data can be collected on the river's structure and hydrology, and before a baseline characterization of salmon population structure and demographics, then any increase in spawning salmon the next year cannot be attributed to the dam's removal with certainty. However, waiting for a system to be completely assessed and understood often risks ongoing damage to a system, further complicating management and restoration efforts. For many systems, "rapid assessment" approaches have been employed using specific methodologies that identify the more critical or pertinent information related to common management targets. Some of these targets and associated data might include high native biodiversity, low numbers of invasive species, well-functioning ecosystem processes like nutrient cycling, and relatively little anthropogenic degradation from transportation surfaces (O'Dea et al., 2004; Fennessy et al., 2007; Mayer et al., 2008; Sifneos et al., 2010).

2.2.1 Landscapes and Scale

While there is a great diversity in the size and scope of management and restoration projects, most projects are targeted to a specific landscape. Ecologists use the term "landscape" in a technical manner to identify the area over which there is significant heterogeneity in the pattern or process of interest (Turner et al., 2010; Jones et al., 2013). For example, a landscape in a tropical oak savanna might be considerably smaller than a landscape in a boreal forest,

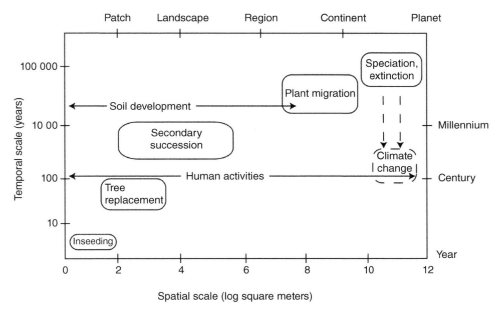

FIGURE 2.3 Scales of ecological processes. Note that human activities have increased the rate of climate change in the recent millennium. This is especially apparent with greenhouse gas emissions from the use of fossil fuels and land use/land cover change. Modified from Delcourt and Delcourt (1988).

given that boreal forest ecosystems have low species diversity and tend to occur in very large stands or patches. Furthermore, heterogeneity is not necessarily defined by a human perspective; a landscape for a field mouse (e.g., *Peromyscus* spp.) will naturally be much smaller than one for a human, since a mouse will perceive many more small details (e.g., fallen branches, a flowering plant) than a human will. Scale has both spatial and temporal dimensions, and processes such as disturbances, succession, and human influence operate at characteristic scales (Fig. 2.3). The resources that an individual or species needs also occur at different scales and in different patterns; for example, bald eagles (*Haliaeetus leucocephalus*) in eastern North America are very sensitive to the canopy structure of a riparian forest immediately around their nest, yet are also sensitive to human activity over a much longer distance (Thompson and McGarigal, 2002). The landscape boundaries and the scale at which data are collected are therefore highly dependent upon the targets or goals of the management or restoration project (Jones et al., 2013). Although the determination of these boundaries should theoretically be driven by knowledge of known ecological processes, most often, they are set in accordance to choices made for purely sociopolitical reasons and hence can hamper effective management (Blomquist and Schlager, 2005).

2.2.2 Abiotic Conditions

Characterizing the abiotic conditions of an area will help clarify the limits of what can be done to manage or restore it. Nutrient flows, parent rock material (which governs soil characteristics), climate, topography, and other features must be of sufficient extent or quality to support the target biological conditions to be attained. Some of these conditions (such as topography and geology) will be relatively unchanged by human activities (although humans are now altering climate and strip mining large areas in mountainous regions (Fig. 2.4)), but others such as nutrient cycles and soils are readily changed by many kinds of human interventions. Human land uses such as industry and agriculture have a large impact on carbon, nitrogen, and phosphorus cycles, among others (Wali et al., 2010). Understanding the abiotic (i.e., the nonliving components of an ecosystem) history of a site is critical to developing feasible management or restoration strategies. For example, the diversity and ecosystems of the Everglades evolved in an "oligotrophic" system, one with very low levels of nitrogen and phosphorus. As nutrient levels were increased through the use of agricultural fertilizers, plant species such as cattails (*Typha* spp.) that competed poorly with sawgrass (*Cladium jamaicense*) and muhly grass (*Muhlenbergia filipes*) in oligotrophic conditions were able to take over these higher-nutrient areas, changing habitat conditions for many native Everglades species (Richardson, 2008). Also, the southern half of Florida has little topographic relief, prompting water overflows from Lake Okeechobee during the rainy season to spread out in a wide, shallow river down through southern Florida. As transportation surfaces (roads) and water resources were rerouted and their flows shifted from their native drainage patterns, the overdry soil in drained areas eroded, creating large, low-lying areas where water pooled in the wet season while causing other areas to experience drought conditions. The altered

FIGURE 2.4 Images of strip mining (top) and mountaintop removal (bottom). Note how this remotely sensed data can detail the change in both land cover and topography. Photos courtesy of NASA and Jeffrey Gerard, respectively, both released into the public domain.

hydrology not only affected habitat distributions but also changed the fire disturbance regime over large areas of land (Richardson, 2008).

2.2.3 Biodiversity and Ecosystem Processes

The word "biodiversity" is a vague term that can refer to many different kinds of diversity at different organizational levels: genetic, population, species, etc. (Marcot, 2007). Functional diversity is also important and can be measured in terms of species membership in a feeding guild (e.g., herbivores, carnivores) or processes (e.g., carbon sequestration, soil stabilization) that contribute to the stability or functioning of the system (Petchey and Gaston, 2006; de Souza et al., 2013). For this reason, it is important to clarify at what level (e.g., population, species, community) one is using the term. For example, one of the most common ways to define and measure biodiversity is the Shannon diversity index, $H' = -\sum_{i=1}^{R} p_i \ln p_i$, where R is the total number of species in the community, and p_i is the proportion of individuals of ith unit, such as species (Magurran, 2003). When using the Shannon index, therefore, it is important to clarify which level of diversity i will represent. In this chapter, biodiversity will refer to diversity at the species level, unless otherwise noted.

Another important quality of an ecosystem is "endemism." Endemic species are those that are unique to a certain area; long-term observation at global scales tells us with some certainty that kangaroos and koala bears are only found in Australia. Ecosystems that support high biodiversity and high endemism are often prioritized for conservation or restoration. These ecosystems are sometimes referred to as "hot spots," which typically refer to areas at the global scale where conservation efforts have focused (Myers et al., 2000). The hot spot approach of identifying special areas of concern can be used at any scale to prioritize management and restoration activities.

Finally, the concept of "ecosystem services" (particularly in relation to their relationship to biodiversity) has become a central tenet in environmental management. These services are categorized into four main types: provisioning, regulating, supporting, and cultural (Millennium Ecosystem Assessment, 2005). If we consider deciduous forest ecosystems as an example, these forests provide wood for heat and construction, regulate the global climate through sequestering carbon in their biomass, support the nitrogen (N) cycle by using N to produce its biomass (and thereby preventing it from polluting nearby streams), and offer recreational activities that are culturally valued. These services are diminished as tree species are eliminated from ecosystems and as land use converts natural ecosystems to human-governed ones.

Ecosystem services are also tightly dependent upon ecosystem processes, such as the flow of energy from the sun through species in food webs (which allows for the buildup of biomass and enrichment of soil), and the migration of populations and species, which helps distribute nutrients over large areas. Some animal species use migration as a key strategy to maintain the availability of resources (especially due to seasonal changes in temperature and/or precipitation), and collectively, they can move a considerable amount of biomass and nutrients among regions (Reimchen and Fox, 2013; Bauer and Hoye, 2014). With the exception of deep-sea thermal ecosystems (which use chemosynthesis to convert chemicals into biomass), all ecosystems rely upon plants to convert sunlight (via photosynthesis) and water into biomass. This process is known as "primary productivity" and forms the base of most food webs (a large number of species connected through predator–prey relationships). The primary productivity of an area is generally a function of the available solar radiation and precipitation, modulated by nutrient availability. For this reason, deserts support less biomass than rainforests (due to low precipitation), and arctic grasslands grow much more slowly than tropical ones (due to low or highly seasonal sunlight). The predictability of biomes based on sunlight and precipitation is such that ecologists can determine which kinds of species and ecosystems are likely to flourish or flounder in any particular region. The definition of these biomes generally set practical limits on what can be managed for or restored.

2.2.4 Natural and Human-Induced Disturbances

Disturbances are a characteristic of all systems, and depending on their type and extent, can change the relationships within ecosystems, which then triggers a need for restoration or management. Disturbance can be roughly categorized into "pulse" and "press" types (although this is relative to the size of the system). Pulse disturbances are those that occur over a short period of time, such as the damaging winds from a tornado or a downpour that temporarily floods a stream. Press disturbances occur over a long period of time, long enough to change the functioning of the ecosystem or species composition of an area. Multiyear droughts would fall into this category, as they can substantially alter the vegetation and species diversity of an area, and this effect can last long after the drought subsides. The change that disturbances cause and the ecosystem's recovery time depend upon the frequency and intensity of the disturbance, along with its duration and extent. Each type of disturbance (both natural and anthropogenic) tends to operate at a characteristic scale (Fig. 2.2).

Although humans have been components of ecosystems since their evolution 100,000 years ago, human disturbances are often considered to be qualitatively different than other kinds of disturbances. Anthropogenic disturbances either deviate from the frequency or intensity of natural disturbances (e.g., increased flooding behind dams) or are qualitatively different from natural disturbances (e.g., persistent pollution from polychlorinated biphenyls (PCBs), dichlorodiphenyltrichloroethane (DDT), and other man-made chemicals). Indeed, the scope and intensity of humanity's alteration of the global environment have led many climatologists and geoscientists to deem our current epoch the "Anthropocene" (Jones, 2011).

2.3 STAGE 2: IDENTIFY TARGET CONDITIONS

As all other species, humans have altered their environments throughout their evolution, and these alterations have aided human population increase and expansion, although sometimes in the long term to our own detriment. Through the use of fire (for managing game animals or in shifting cultivation) and hunting and gathering activities, early humans changed the species composition (both plant and animal) around them and thereby habitats at a landscape scale (Willis et al., 2004; Willis and Birks, 2006; Pearce, 2013). As human societies advanced, so did the reach of human technologies; the area over which we altered our environments and the intensity of these alterations increased. For this reason, ecologists now refute the concept of "pristine" ecosystems, as evidence accrues that humans have introduced species, changed disturbance regimes, and impacted soils for millennia on most continents. Therefore, managers should acknowledge the difficulty of setting management targets for the conservation or restoration of ecosystems that preclude all human impact.

One of the most significant impacts on environmental systems comes from anthropogenic land cover change and land use (Foley et al., 2005; Table 2.1). Land cover refers to surface cover types such as grassland, forest, permafrost, or water; classification of cover types is context specific and can be very detailed (e.g., oak–maple deciduous forest (*Quercus* spp. and *Acer* spp.)) or very general (e.g., forest). Land use refers to the

TABLE 2.1 Land Cover Change by Biome[a]

Biome Type	Land Cover Change by 1950 (%)	Land Cover Change by 1990 (%)	Projected Land Cover Change by 2050 (%)
Mediterranean forests, woodlands, and scrub	−68	−69	−70
Temperate forest, steppe, and woodland	−67	−72	+5
Temperate broadleaf and mixed forests	−57	−57	−60
Tropical and subtropical dry broadleaf forests	−48	−56	−71
Flooded grasslands and savannas	−45	−53	−65
Tropical and subtropical grasslands, savannas, and shrublands	−42	−51	−70
Tropical and subtropical coniferous forests	−36	−40	−68
Deserts	−24	−28	−30
Montane grasslands and shrublands	−23	−25	−41
Tropical and subtropical moist broadleaf forests	−22	−25	−42
Temperate coniferous forests	−19	−20	−29
Boreal forests	−3	−4	−4
Tundra	<1	<1	−3

Positive numbers indicate a gain in land cover area; negative numbers indicate a loss in area. Much of the land cover types were converted to urban or agricultural land uses.

[a] From Millennium Ecosystem Assessment (2005).

dominant human activity in or use of that area, and can also be described in very general or very specific terms. Common land use categories are urban industrial, agriculture, recreation, or nature conservation. Trends in land cover and land use are closely related, as land cover usually dictates the optimal uses of an area, and land use has a large impact on the kinds of land cover found there. Increasingly, remote sensing data (that gathered from land-based or satellite sensors) are used to develop land use/land cover maps and analyses; however, historic land use/land cover information can also be gathered from spatially explicit data such as government land surveys in the United States (Manies and Mladenoff, 2000; Horning et al., 2010; Jones et al., 2013). Land suitability analysis (or land suitability evaluation) is undertaken to determine the most appropriate or productive use of an area, given climatic, topographic, and soil conditions, while giving consideration to multiple and possibly competing land use demands for the same area or region (Pereira and Duckstein, 1993; Randolph, 2003). These analyses therefore include economic, social, and environmental impact factors to determine the most optimal land use among various choices with respect to economic and noneconomic values (Munda et al., 1994; Guhathakurta, 2003).

Two new fields to emerge from land suitability analysis are "alternative futures scenarios" and "land degradation mapping." In the former, land use policies such as "greenbelts" (which limit urban sprawl and preserve agricultural areas near cities) are tested for their effectiveness for meeting land use/land cover and environmental quality targets, using computer models and spatially explicit data to project land use change into the future, often with geographic information systems (GIS) software (Millennium Ecosystem Assessment, 2003; Baker et al., 2004; Beardsley et al., 2009; Jones et al., 2013). Land degradation mapping also uses GIS to integrate different spatially explicit data sets and generate maps of land degradation risks, based on land use/land cover change, natural and human disturbances, and other pressures (Jabbar and Chen, 2006; Bojórquez-Tapia et al., 2013). As with land suitability analysis, land degradation is also highly context specific and depends upon who defines "degraded" land (Warren, 2002; Reed et al., 2013).

2.3.1 Setting Targets

In theory, the list of targets generated will be a measurable, representative set of characteristics of the reference ecosystem, and these targets will then guide the selection of indicators to determine when the managed or restored system has reached the targets. However, it is often difficult (if not impossible) to measure these attributes with the frequency and accuracy that are desirable for management decision making. Furthermore, the targets may give no indication of the successional pathway or processes necessary to maintain or restore a system to a state similar to that of the reference system (if one exists), and so additional information is needed on the dynamics of the system. For example, fire is often required to maintain a prairie (by preventing trees from establishing and shading out the grasses); however, these prairies require fires to occur within a range of specific frequency or intensities (Mayer and Henareh Khalyani, 2011). Therefore, the set of indicators used by managers is by necessity larger and more comprehensive than the actual set of management targets.

Indicators used for environmental management or restoration must meet several important criteria (Dale and Beyeler, 2001; Doren et al., 2009; Schultz et al., 2012). They must:

- Be relevant to the ecosystem
- Respond to system variability at an appropriate scale
- Be feasible to implement
- Be measurable
- Be sensitive to system drivers in a way that is predictable
- Be interpretable by multiple users
- Respond consistently with respect to desired targets
- Remain unaffected by irrelevant processes
- Be scientifically defensible

- Allow for assessment of distance to target
- Have a high degree of specificity (particularly related to management actions)
- Provide early warnings

Once a set of indicators meeting these criteria are developed, they will guide the data that are collected and the methodology used to collect them.

Another approach that can be taken at this stage is called "ecological risk assessment." Risk assessment is typically performed when there are known stressors, such as heavy metal pollutants, and when the probability that these stressors will impact specific species, ecological processes, or functions can be measured (U.S. Environmental Protection Agency, 1998). Characteristics of each type of stressor influence the level of toxicity, persistence, and impact that the stressor is likely to have in ecosystems. For example, pollutants (such as DDT) that are fat soluble tend to be stored in the fatty tissues of animals (as opposed to water-soluble pollutants that are routinely purged from the system). As larger organisms eat smaller ones, the pollutant can bioaccumulate through the food chain, so that top predators eventually receive lethal or damaging doses of a pollutant, even though each prey item contained only a small amount of the toxin. DDT exposure therefore poses a much greater risk to top predators such as bald eagles than to the fish species the eagle prey upon (Hinck et al., 2009). The assessment produces a level of expected or probable damage to ecological systems as well as an estimate of the uncertainty surrounding these predictions.

2.4 STAGE 3: COLLECT DATA AND DETERMINE DISTANCE TO TARGET

Once the management or restoration targets have been determined and a set of indicators has been developed to help managers guide the system toward those targets, information must be gathered to determine where the system is relative to those targets. Ideally, these data will be the same set used to monitor the system over the long term, as environmental data can be quite costly to obtain, especially over large areas and long periods of time. The types of data commonly used in environmental management include field observations collected on surveying missions, historical maps or field journals, automated monitoring stations, or sensor data from satellites and other low-orbit or atmospheric platforms (e.g., weather balloons). Environmental management is also increasingly using "citizen science" data, collected by large numbers of volunteers who make observations using a standardized protocol and report their observations through a webpage or mobile phone. The North American Breeding Bird Survey (managed by the US Geological Survey (USGS)) and the "Project FeederWatch" program at the Cornell Lab of Ornithology are two good examples of this type of monitoring data set (Bonney et al., 2009; Ziolkowski et al., 2010; Fig. 2.5).

The importance of GIS as a primary tool for environmental inventories, management, monitoring, and assessment activities cannot be overstated (Droj, 2012). GIS software allows users to combine many types of data as layers on a map, which can then be analyzed to identify areas at risk of degradation, probable water pollution sources, feasible restoration areas,

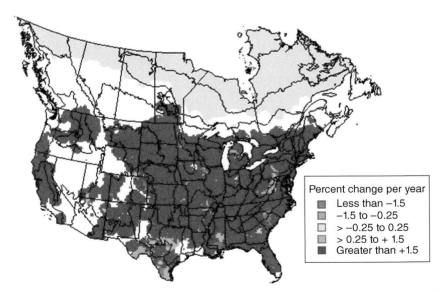

FIGURE 2.5 A trend map (years 1966–2011) was developed from volunteer observations for the North American Breeding Bird Survey (Sauer et al., 2012) for wild turkey (*Meleagris gallopavo*) populations around the United States and Canada. Across most of its range, the wild turkey has been quickly increasing in numbers.

and a variety of other common environmental management issues. GIS systems "improve our ability to:

1. analyze spatial relationships among human actions, their environmental impacts, and vulnerability of humans and ecosystems
2. assess the effectiveness of management strategies taken
3. monitor and detect environmental risks
4. classify ecosystems, and land uses/land covers; and
5. predict future implications of environmental changes on ecosystem productivity and health" (Wali et al., 2010).

Most GIS software tools also allow for the combination of different kinds of spatially explicit data: vector or raster. Vector data are points, lines, and polygons and can originate from sources such as hand-drawn maps, transportation infrastructure layers, river systems, private property boundaries, governmental jurisdiction boundaries, and citizen science programs, which prompt volunteers to enter observations at a single location (such as a rare plant sighting) into a website. Raster data are grids of attributes or data, such as satellite images and photographs. Often, these two kinds of data need to be combined to address management issues. For example, a project identifying recent forest harvests on property owned by nonindustrial private forest owners combined a raster-based satellite image (illustrating forest cover) with a vector-based property boundary file (indicating nonindustrial private property). The combination of the two layers yielded those properties where a forest harvest has been detected, indicated by the difference in image values for the pixels within a property boundary versus those around it, from 1 year to the next (Fig. 2.6).

Remote sensing provides a vast diversity of information on the environment due to the variety of sensors that have been mounted on satellites and other platforms. These data include passive data (e.g., solar energy reflected from the surface of the Earth as recorded in different wavelengths) to active data that uses radar or other tools to measure the path of a beam or signal after it returns back to the sensor (light detection and ranging (LiDAR) is one example). Remote sensing data have been used to measure and monitor ecosystem processes, vegetation structure and function, and wildlife habitat, in both terrestrial and marine environments (Franklin, 2010).

Remote sensing data can be used with minimal field data (although some ground-based data are necessary to validate the classification of the remote sensing images); however, the combination of remote sensing and field survey data is becoming standard practice (Franklin, 2010; Horning et al., 2010). This combination is made possible by the use of a Global Positioning System (GPS) tool that allows field data to be tagged with exact location information. Prior to GPS tools, field data would be mapped by hand and then digitized into a GIS system; this is still necessary when using old maps or other nonspatially explicit data.

2.4.1 Land Classification Systems

The base layer map for many environmental management projects is a classification of the land use/land cover of the area. This map can include the potential (natural) land cover based on biome, geology, or topography; the current land use; the desired land use; or any other combination of information on what has been present, what is present, what would have been present, and what can be present in the future. Although managers may come up with their own classification scheme to fit the project objectives more closely, there are many land classification systems that have been developed and used for a variety of projects (Loveland et al., 2000; Neumann et al., 2007). Many of these are free and available for public use and have incorporated remote sensing imagery so that they can be updated as new imagery becomes available (Horning et al., 2010):

- *National Land Cover Data system*: Originated by Anderson et al. (1976), this two-level hierarchical classification scheme categorizes land use/land cover into nine major classes, which are then split into more specific classes (Table 2.2); typically, the more broad Level I classes are used over very large areas, while Level II classes are used over smaller areas where more specificity is required. The classification system was used with Landsat data in 1992 and again in 2001 and 2006 by the Multi-Resolution Land Characteristics Consortium (MRLC), which has maintained an ongoing classification program for freely available land use/land cover data (Homer et al., 2007; Chander et al., 2009; Jin et al., 2013). The data have a 30 m resolution (as dictated by Landsat sensors) and are available for the entire United States.
- *Global Land Cover facility*: In addition to Landsat imagery, the GLC uses MODIS and AVHRR remote sensing data to aid users in identifying land use and land cover change (Horning et al., 2010). Like the NLCD, GLC data are free and available to the public and are routinely updated. The GLC data set is available in preclassified formats and allows users to develop their own classification schemes (e.g., Latifovic et al., 2004; Tateishi et al., 2011; Table 2.2). Due to the global scale of the database, much of the data are often at a coarser resolution than those available in the NLCD set, although the use of Landsat imagery can allow for more fine resolution analysis (Bartholome and Belward, 2005; Yu et al., 2013).
- *Land Cover Classification System*: Planned by the United Nations' Food and Agriculture Organization, the LCCS uses a three-level hierarchical system to classify areas first into "primarily vegetated" and "primarily nonvegetated" land cover types and allows for additional modular (i.e., nonhierarchical) classification within

FIGURE 2.6 Detection of harvests on nonindustrial private forests is illustrated with Landsat imagery, which is overlain with private property boundaries. Black pixels indicate a large negative change in biomass (decrease in the Normalized Difference Vegetation Index) from 1 year to the next, indicating intensive harvesting. Black lines are property boundaries. Top, harvests between years 2005 and 2006; bottom, harvests between years 2007 and 2008 (Tortini and Mayer, unpublished data).

Level III to produce very detailed land use/land cover maps (Di Gregorio and Jansen, 2000; Hüttich et al., 2011). The classification system used by the LCCS has been standardized in an International Standards Organization (ISO) format (ISO 19144-1:2009).

2.4.2 Continent-Scale Field Data Collection Networks

For most management and restoration projects, field survey and monitoring data protocols will be established for the site or area with variables and indicators specific to the management or restoration targets. However, it can be helpful to have field data from the surrounding area or region, and over a longer time period (especially historically), to understand how reasonable the targets are relative to what is and has been observed in the area. Data from nearby long-term monitoring stations can help determine whether targets have been set to conditions that are likely to be supported in the area. There are many such long-term monitoring systems around the world, including several well-supported ones in the United States:

- *National Oceanographic and Atmospheric Administration (NOAA) Weather Monitoring Stations*: The NOAA has supported a global network of local weather monitoring stations (both land and water

TABLE 2.2 Anderson et al.'s (1976) Land Use/Land Cover Classification Scheme for the United States as Compared to Tateishi et al.'s (2011) Global Land Cover Classes

Anderson Level I	Anderson Level II	Tateishi GLCNMO Class
1. Urban or built-up land	11. Residential 12. Commercial 13. Industrial 14. Transportation, communications, and utilities 15. Industrial and commercial complexes 16. Mixed urban or built-up land 17. Other urban or built-up land	18. Urban
2. Agricultural land	21. Cropland and pasture 22. Orchards, groves, vineyards, nurseries, and ornamental horticultural areas 23. Confined feeding operations 24. Other agricultural land	11. Cropland 12. Paddy field 13. Cropland/other vegetation mosaic
3. Rangeland	31. Herbaceous rangeland 32. Shrub and brush rangeland 33. Mixed rangeland	7. Shrub 8. Herbaceous 9. Herbaceous with sparse tree/shrub 10. Sparse vegetation
4. Forest land	41. Deciduous forest land 42. Evergreen forest land 43. Mixed forest land	1. Broadleaf evergreen forest 2. Broadleaf deciduous forest 5. Mixed forest 4. Needleleaf evergreen forest 3. Needleleaf deciduous forest 6. Tree open
5. Water	51. Streams and canals 52. Lakes 53. Reservoirs 54. Bays and estuaries	20. Water bodies
6. Wetland	61. Forested wetland 62. Nonforested wetland	14. Mangrove 15. Wetland
7. Barren land	71. Dry salt flats 72. Beaches 73. Sandy areas other than beaches 74. Bare exposed rock 75. Strip mines, quarries, and gravel pits 76. Transitional areas 77. Mixed barren land	17. Bare area, unconsolidated 16. Bare area, consolidated (gravel, rock)
8. Tundra	81. Shrub and brush tundra 82. Herbaceous tundra 83. Bare ground tundra 84. Wet tundra 85. Mixed tundra	
9. Perennial snow or ice	91. Perennial snowfields 92. Glaciers	19. Snow/ice

based) for decades. Data from weather stations and buoys are available from the NOAA's National Climatic Data Center, as are paleoclimate data from tree rings, ice cores, and lake and ocean sediment cores.

- *Long-Term Ecological Research (LTER)*: The LTER network of field laboratories was initiated by the National Science Foundation (NSF) in 1980 to encourage a deeper understanding of ecosystems and human–environment interactions. LTER sites are located across the United States (including Hawai'i and Alaska), and two have been established in Antarctica. The sites exemplify all of the major natural biomes in the United States, plus areas dominated by urban, suburban, and agricultural land use.

- *USGS and US Environmental Protection Agency (USEPA) Water Stations*: The USGS and the USEPA support two networks that collect data on water availability, quantity, and quality in water bodies throughout the United States. Originally available only through their individual websites, these data are now jointly available through the Water Quality Portal at http://www.waterqualitydata.us/.

- *National Ecological Observatory Network (NEON)*: The NEON is a new project, and like LTER, is supported by the NSF. The project aims to establish multiple stations that are strategically placed around the United States. However, whereas LTER sites may collect different kinds of data depending upon what the research activities at the site demand, NEON sites all collect the same monitoring data using the same methodology and equipment, so that the data can be pooled and used to detect patterns and trends at the continent scale.

2.5 STAGE 4: PRESERVATION OR RESTORATION?

Once the conditions of the system and their distance to management targets have been ascertained, managers must then determine whether preservation (keeping the system where it is) or restoration (returning the system to a previous state) is required. The term "preservation" can have very specific meanings in certain subfields, but here, it refers to the goal of preserving existing biodiversity, ecosystem functions, and processes because they are at or near to the targets set out in the management plan. As discussed in the following, restoration is also meant here as a very broad term that encompasses many approaches to return or drive a system toward a target state.

However, neither preservation nor restoration is a simple task, as straightforward linear relationships between disturbances or stressors and ecosystem responses are rare. One additional fire per decade may produce a nonlinear outcome, such as an order of magnitude reduction in invasive species abundances, for example, instead of simply halving them. Emerging research over the past several decades has revealed a startling ubiquity of alternative stable states in ecosystems, which complicate both management and restoration efforts (Beisner et al., 2003; Folke et al., 2004; Hobbs and Suding, 2007; Horan et al., 2011; Mayer and Henareh Khalyani, 2011; Barnosky et al., 2012). Alternative stable states (e.g., grassland vs. forest, also known as multiple stable states or regimes; Mayer and Rietkerk, 2004) are basins of attraction for a system, between which the system transitions nonlinearly or very suddenly. The structure of the feedbacks within the system drive these sudden transitions (or catastrophes) when the system crosses certain thresholds. These thresholds can be very difficult to detect before a system crosses them, although there are certain behaviors that can be detected in monitoring data that may indicate that the system is approaching a threshold (such as "critical slowing down," increased autocorrelation, or other statistical signals; Guttal and Jayaprakash, 2008; Biggs et al., 2009; Scheffer et al., 2009). These feedbacks often contribute to maintaining a system in a particular dynamic state, such that significant effort is required to push the system back into a preferred or restored state; this is called the "resilience" of the system (Gunderson, 2000). Once a system gains resilience in a state, if that state is undesirable to managers, it will require a great deal of effort to overcome this resilience and may require the restoration of conditions and processes far beyond where they were when the system was first degraded (i.e., a "hysteresis"; Beisner et al., 2003; Van Nes and Scheffer, 2005).

For example, the conditional boundaries between forests (dominated by trees), savannas (a mix of trees and grass), and grasslands (no trees) are not gradual; forests are a resilient state and do not slowly degrade into savannas (Sternberg, 2001; Mayer and Rietkerk, 2004; Hirota et al., 2011; Mayer and Henareh Khalyani, 2011; Staver et al., 2011). Depending upon precipitation and fire regimes, existing forest area, and the seasonality in precipitation of an area, either a forest, a savanna, or a grassland will persist unless one of those conditions are changed. For example, an increase in fire frequency will favor grass over trees and cause a forest to quickly transition to savanna or savanna to grassland (Fig. 2.7). When the location of these boundaries is known, particularly for fires and deforestation (both are anthropogenically generated disturbances in most of these systems), actions to maintain or restore these habitats are more likely to be successful.

In many environmental management fields (e.g., ecology, agriculture, forestry), managers use "best management practices" or "best practices" to achieve project targets or goals. Best practices refer to a collection of tools and methods that have been developed and tested over time in a variety of ecosystems and situations to achieve these targets successfully. Best practices are often used and revised in the process of "adaptive management," which is given further treatment later in the chapter.

2.5.1 Preservation

Until recently, the word "conservation" has usually referred to the management action of setting aside and protecting areas of high biodiversity, endemism, or uniqueness. However, more consideration is being paid to preserving ecosystem functions and services, even in areas that are not particularly diverse. For example, decades of discussion have taken place in the United States on the decision to open the Arctic National Wildlife Refuge in Alaska to fossil fuels exploration (Walker et al., 1987; Snyder, 2008; Sovacool, 2008). While this area does not support high biodiversity relative to other areas of the world, the area does provide key ecosystem goods and services (e.g., fisheries, caribou migration habitat, carbon storage) that would be damaged or destroyed by fossil energy extraction. In another example, the city of New York set aside a large area of the Adirondacks through a series of agreements with landowners to prevent significant land cover change (conversion to agriculture) and protect the ecosystem functions that are responsible for the quality and quantity of the city's water supply (Postel and Thompson, 2005). For the purposes of this chapter, "preservation" will encompass a variety of environmental

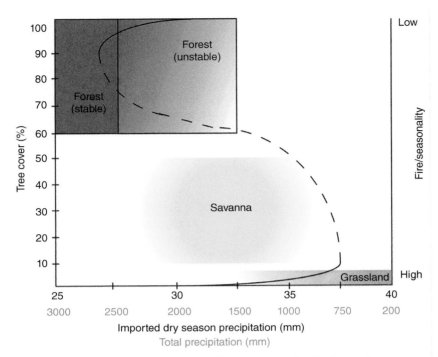

FIGURE 2.7 Hysteresis can occur between different biomes. In this figure, forest, savanna, and grassland states are dynamic due to the interplay between forest area, fire regimes, seasonality, and precipitation (overall and during the dry season). Figure adapted from Sternberg (2001) and Mayer and Henareh Khalyani (2011). Grasslands are not considered in Sternberg's (2001) original analysis, and there is likely a second hysteresis between savanna and grasslands to the right of his original hysteresis curve.

management actions that endeavor to maintain an ecosystem or area in a state that is similar to what can currently be found.

Once management targets have been identified and the state of the system is established, it is then necessary to determine which areas and how much area should be preserved and how this preservation should be maintained. One of the most commonly used best practices for this purpose is "gap analysis." Gap analysis is used in many disciplines to identify areas that require more attention or data support. For environmental management, it generally starts with GIS layers of areas that already receive some legal protection, plus layers that identify valuable ecological resources (e.g., distributions of rare species, old growth forest stands, etc.; Rodrigues et al., 2004; Dietz and Czech, 2005; Langhammer et al., 2007). When these layers are combined, areas of high value that are not currently protected are highlighted.

Preserving an area (land or water) for species, habitat, or ecosystem processes and functions requires an understanding of the types of human activities that are compatible with the ecological resources that are being preserved. While national parks and preserves are perhaps the most well known of the types of set-aside areas that exist, there is a suite of protection actions that can be taken that do not ban all human activity (beyond tourism) in an area, making preservation less contentious. For example, the IUCN has developed a range of protected area categories that take into account the level of anthropogenic disturbance to the system and the intensity of protection of "naturalness" desired (Table 2.3). These protected areas almost always have permanent, fixed boundaries; however, "floating reserves" (with boundaries that change over time) that account for disturbance and other dynamic processes are becoming a popular alternative (Cumming et al., 1996; Ramage et al., 2013). Given the impact that climate change is likely to have on species ranges (as species shift to find climatic envelopes that they are adapted to live in), the preserve areas set up today to protect rare and endangered species and habitats are not likely to perform this function in the long term.

Along with the level of protection, the size, shape, and configuration of a protected area or network of areas must be determined. Contributions from the field of biogeography can help managers determine these characteristics (Shafer, 1990; Lomolino et al., 2010). Biodiversity typically increases with area, so that a larger area will support more kinds of species than a smaller one—often, this relationship is predictive so that a "species-area curve" can be developed for biomes and habitats (Rosenzweig, 1995; Tjorve, 2010). As habitat is destroyed and fragmented into smaller patches, species will disappear from the area; generally, the rarer species disappear first, as they have fewer individuals and

TABLE 2.3 Protection Categories and Objectives for Preservation Areas Developed by the International Union for the Conservation of Nature[a]

IUCN Category	Definition	Management Objectives
1a: Strict nature reserve	"Strictly protected areas set aside to protect biodiversity and also possibly geological/geomorphological features, where human visitation, use and impacts are strictly controlled and limited to ensure protection of the conservation values. Such protected areas can serve as indispensable reference areas for scientific research and monitoring.	• To conserve regionally, nationally or globally outstanding ecosystems, species (occurrences or aggregations) and/or geodiversity features: these attributes will have been formed mostly or entirely by non-human forces and will be degraded or destroyed when subjected to all but very light human impact • To preserve ecosystems, species and geodiversity features in a state as undisturbed by recent human activity as possible • To secure examples of the natural environment for scientific studies, environmental monitoring and education, including baseline areas from which all avoidable access is excluded • To minimize disturbance through careful planning and implementation of research and other approved activities • To conserve cultural and spiritual values associated with nature
1b: Wilderness area	Protected areas are usually large unmodified or slightly modified areas, retaining their natural character and influence, without permanent or significant human habitation, which are protected and managed so as to preserve their natural condition.	• To protect the long-term ecological integrity of natural areas that are undisturbed by significant human activity, free of modern infrastructure and where natural forces and processes predominate, so that current and future generations have the opportunity to experience such areas • To provide for public access at levels and of a type which will maintain the wilderness qualities of the area for present and future generations • To enable indigenous communities to maintain their traditional wilderness-based lifestyle and customs, living at low density and using the available resources in ways compatible with the conservation objectives • To protect the relevant cultural and spiritual values and non-material benefits to indigenous or non-indigenous populations, such as solitude, respect for sacred sites, respect for ancestors, etc. • To allow for low-impact minimally invasive educational and scientific research activities, when such activities cannot be conducted outside the wilderness area
II: National park	Protected areas are large natural or near natural areas set aside to protect large-scale ecological processes, along with the complement of species and ecosystems characteristic of the area, which also provide a foundation for environmentally and culturally compatible spiritual, scientific, educational, recreational and visitor opportunities.	• To protect natural biodiversity along with its underlying ecological structure and supporting environmental processes, and to promote education and recreation • To manage the area in order to perpetuate, in as natural a state as possible, representative examples of physiographic regions, biotic communities, genetic resources and unimpaired natural processes • To maintain viable and ecologically functional populations and assemblages of native species at densities sufficient to conserve ecosystem integrity and resilience in the long term • To contribute in particular to conservation of wide-ranging species, regional ecological processes and migration routes • To manage visitor use for inspirational, educations, cultural and recreational purposes at a level which will not cause significant biological or ecological degradation to the natural resources • To take into account the needs of indigenous people and local communities, including subsistence resource use, in so far as these will not adversely affect the primary management objective • To contribute to local economies through tourism
III: Natural monument or feature	Protected areas are set aside to protect a specific natural monument, which can be a landform, sea mount, submarine cavern, geological feature such as a cave or even a living feature such as an ancient grove. They are generally quite small protected areas and often have high visitor value.	• To protect specific outstanding natural features and their associated biodiversity and habitats • To provide biodiversity protection in landscapes or seascapes that have otherwise undergone major changes • To protect specific natural sites with spiritual and/or cultural values where these also have biodiversity values • To conserve traditional spiritual and cultural values of the site

(Continued)

TABLE 2.3 (*Continued*)

IUCN Category	Definition	Management Objectives
IV: Habitat/ species management area	Protected areas aim to protect particular species or habitats and management reflects this priority. Many category IV protected areas will need regular, active interventions to address the requirements of particular species or to maintain habitats, but this is not a requirement of the category.	• To maintain, conserve and restore species and habitats • To protect vegetation patterns or other biological features through traditional management approaches • To protect fragments of habitats as components of landscape or seascape-scale conservation strategies • To develop public education and appreciation of the species and/or habitats concerned • To provide a means by which the urban residents may obtain regular contact with nature
V: Protected landscape/ seascape	A protected area where the interaction of people and nature over time has produced an area of distinct character with significant ecological, biological, cultural and scenic value; and where safeguarding the integrity of this interaction is vital to protecting and sustaining the area and its associated nature conservation and other values.	• To protect and sustain important landscapes/seascapes and the associated nature conservation and other values created by interactions with humans through traditional management practices • To maintain a balanced interaction of nature and culture through the protection of landscape and/or seascape and associated traditional management approaches, societies, cultures and spiritual values • To contribute to broad-scale conservation by maintaining species associated with cultural landscapes and/or by providing conservation opportunities in heavily used landscapes • To provide opportunities for enjoyment, well-being and socio-economic activity through recreation and tourism • To provide natural products and environmental services • To provide a framework to underpin active involvement by the community in the management of valued landscapes or seascapes and the natural and cultural heritage that they contain • To encourage the conservation of agrobiodiversity and aquatic biodiversity • To act as models of sustainability so that lessons can be learnt for wider application
VI: Protected areas with sustainable use of natural resources	Protected areas conserve ecosystems and habitats, together with associated cultural values and traditional natural resource management systems. They are generally large, with most of the area in a natural condition, where a proportion is under sustainable natural resource management and where low-level non-industrial use of natural resources compatible with nature conservation is seen as one of the main aims of the area.	• To protect natural ecosystems and use natural resources sustainably, when conservation and sustainable use can be mutually beneficial • To promote sustainable use of natural resources, considering ecological, economic and social dimensions • To promote social and economic benefits to local communities where relevant • To facilitate inter-generational security for local communities' livelihoods—therefore ensuring that such livelihoods are sustainable • To integrate other cultural approaches, belief systems and world-views within a range of social and economic approaches to nature conservation • To contribute to developing and/or maintaining a more balanced relationship between humans and the rest of nature • To contribute to sustainable development at national, regional and local level (in the last case mainly to local communities and/or indigenous peoples depending on the protected natural resources) • To facilitate scientific research and environmental monitoring, mainly related to the conservation and sustainable use of natural resources • To collaborate in the delivery of benefits to people, mostly local communities, living in or near to the designated protected area • To facilitate recreation and appropriate small-scale tourism"

[a]From IUCN; Dudley (2008).

therefore higher risk of extirpation. Therefore, managers often attempt to include as much area as possible when designating protected areas.

The shape of an area will impact the number and kinds of species that are likely to persist there (Shafer, 1990). Some species are quite general in their habitat requirements and will persist in "edge" habitats where two kinds of habitats meet, such as the shrubby boundary separating a grassland or abandoned pasture from a forest. However, other species require "interior" or "core" habitat and will avoid edge areas either because the microclimate conditions are too harsh (e.g., more windy, warmer) or because predators are more likely to be encountered there. These species will need large areas of habitat with minimal edge, which are more likely to be provided by round or square patches than by long, thin corridors. Corridors may be used by these species to travel from one large patch to another, so corridors are also protected along with large interior patches, even though they are not likely to provide breeding habitat. For example, the threatened northern spotted owl (*Strix occidentalis caurina*) requires large stands of old growth forest along the western coast of the United States, as their large territories and specific requirements for certain structural characteristics of forests are only supplied by core (or central) areas of these large stands (Marcot et al., 2013). However, young owls dispersing from their parents' territory may use short forested corridors to find other large stands of old growth forest to establish their own territory.

Finally, the configuration of protected areas and the habitats within them is also important (Shafer, 1990). Protecting entire ecosystems is usually not feasible (either due to cost or conflict with local communities), so networks of smaller protected areas may be established. These networks can be advantageous if they protect a diversity of geological, climate, and habitat types (Dudley, 2008). Protected area networks are not islands, but interact with the surrounding "matrix" habitats and land use at the landscape scale (Shafer, 1990). For this reason, managers are shifting their perspective to the landscape scale when setting up protected area networks, acknowledging that the ecosystems both inside and outside the networks are dynamic and connected (Vandermeer et al., 2010; Van Teeffelen et al., 2012; Willis et al., 2012; Kukkala and Moilanen, 2013). GIS and remote sensing data are very effective for evaluating landscape value for preservation (e.g., Willis et al., 2012).

Once the level of protection and optimal network design has been determined, then programs and policies must be used to preserve these areas. Governments can purchase or secure the areas as public space and place them in a protected status indefinitely; however, the appropriation of land (and marine) resources in this manner often creates a large amount of conflict with local communities who depend upon those resources (Naughton-Treves et al., 2005). One solution to mitigate this conflict is to compensate local communities for the lost resources (e.g., Pechacek et al., 2013), although the cost for compensation schemes in addition to the cost of preserving and managing the area can be prohibitive. Schemes that reduce only those activities that are ecologically damaging, and allow less-damaging activities that are critical to support local livelihoods, can be less costly and can minimize local resistance to protected areas (Naughton-Treves et al., 2005). Protected areas can even be welcomed by local communities if citizens are able to participate in their design or if a market-based system is set up to allow for those landowners who are willing to forego economic land uses for ecological conservation to be reasonably compensated by others who do not want to curb economic activity. Programs of this type in the United States include the Conservation Reserve Program for marginal agricultural land (which was originally intended to preserve soil fertility but has helped preserve critical breeding habitat for many grassland birds; Herkert, 2009) and wetland mitigation banking to halt the loss of wetland habitats (Burgin, 2010). Although the resulting reserve network may provide a less than optimal level of protection for biodiversity and ecosystem services, managers may face less resistance to the network and the land use restrictions it requires.

2.5.2 Restoration

Restoration is a broad term that encompasses many different kinds of activities or projects, some of which involve more severe or particular kinds of damaging activities. Ecological restoration is "an intentional activity that initiates or accelerates the recovery of an ecosystem with respect to its health, integrity and sustainability.... [it] is the process of assisting the recovery of an ecosystem that has been degraded, damaged, or destroyed" (SER, 2002). All restoration projects require indicators, targets, and a reference system. Indicators are sets of data that are collected and used to monitor key attributes of the system for improvement (and to adjust management plans when improvement does not occur). These indicators should represent as many of the system characteristics outlined in Table 2.4 as possible and reflect compositional, structural, and functional features that can be easily compared across systems or over time (Jaunatre et al., 2013). A reference system is a system that helps establish feasible targets for the project (SER, 2002); if no natural wetlands in the region support a particularly rare plant, for example, or if the damaged wetland itself has no history of supporting that species, then it is unlikely that it will do so in a restored state. A dynamic reference target (i.e., a changing system for which processes and succession trajectories are known) is highly preferable to a static target, given that restoration activities must often factor in or contend with ongoing external forces along with the system's own successional tendencies.

Restoration plans must include (SER, 2002):

1. A clear rationale as to why restoration is needed
2. An ecological description of the site designated for restoration
3. A statement of the goals and objectives of the restoration project
4. A designation and description of the reference system
5. An explanation of how the proposed restoration will integrate with the landscape and its flows of organisms and materials

TABLE 2.4 Boundaries and Attributes of System to Be Managed or Restored, Illustrated by the Florida Everglades Ecosystem[a]

Dimension	Characteristic	Examples
Boundaries	Spatial scale	46,000 km^2 watershed (from the Kissimmee River basin to Lake Okeechobee and to Gulf of Florida)
	Temporal scale	Processes: Annual hydrology patterns, seasonal disturbances (fires, hurricanes), decadal urbanization
Landscape	Major land use/land cover types	Urban (residential, commercial, industrial), agriculture (sugarcane, row crops), natural (see next row)
	Natural habitat types	Sawgrass slough, marl prairies, pine woodlands, cypress stands, mangrove forest
	Landscape composition	Current: Overabundance of cattail marsh, patches of invasive species; agricultural area in center and urban encroachment on boundaries
		Goal: Increased abundance and heterogeneity of natural habitat patches
	Landscape pattern and structure	Current: Disconnected hydrological flow by water management and transportation infrastructure
		Goal: Increased connectivity of hydrology and habitat
Biodiversity and ecosystem processes	Ecosystems and communities	Trophic structure: Condition of crocodiles and alligators (apex predators)
		Community dynamics: Highly dependent upon hydroperiod, animal migration, and breeding patterns
	Species and populations	Population dynamics: Increased breeding and abundance of American alligators (*Alligator mississippiensis*)
		Habitat suitability: Amount of muhly prairies for endangered Cape Sable seaside sparrow (*Ammodramus maritimus mirabilis*) breeding territories
	Ecosystem services	Carbon sequestration by pine forests, sloughs, and prairies
		Water storage by sawgrass sloughs
	Ecosystem processes	Energy flow
		Migration
Abiotic: Nutrients, soil, climate, hydrology, and geomorphology	Nutrient concentrations	Reduce nitrogen and phosphorus concentrations (e.g., [P] reduced to >10 µg/L)
	Surface and groundwater flows	Hydroperiod (number of days inundated, depth of inundation)
		Salinity (an indicator of saltwater intrusion into aquifers, occurs due to overpumping)
	Dynamic structural characteristics	Extent/distribution of connected floodplain
	Soils	Increased deposition/accretion of peat soils in sloughs and marl substrate in marl prairies
Natural disturbance regimes (e.g., hurricanes)	Frequency	Annual (mostly in June–November)
	Intensity	Highly variable (i.e., Saffir–Simpson scale, rainfall)
	Extent	Regional
	Duration	Days to weeks
Human-induced disturbance regimes (e.g., water pollution)	Frequency	Continuous
	Intensity	Seasonably variable (e.g., pulses after rainstorms wash fertilizer and pesticides into surface waters)
	Extent	Concentrated in Everglades agricultural area, water conservation areas, and eastern and western boundaries with urban areas
	Duration	Chronic

[a]This is not a complete list of characteristics, indicators, and goals.
Sources: Davis and Ogden (1997), Young and Sanzone (2002), Sklar et al. (2005), Gunderson and Light (2006), NRC (2007), Richardson (2008), Mazzotti et al. (2009), Lo Galbo et al. (2013), The Comprehensive Everglades Restoration Plan (www.evergladesplan.org).

6. Explicit plans, schedules, and budgets for site preparation, installation, and postinstallation activities, including a strategy for making prompt midcourse corrections
7. Well-developed and explicitly stated performance standards, with monitoring protocols by which the project can be evaluated
8. Strategies for long-term protection and maintenance of the restored ecosystem

If only minor ecological degradation needs to be repaired, restoration plans can be of short duration and specific targets can be identified. However, if the damage is extensive, decades to centuries may be needed to restore a system, and achieving narrow targets may not be possible given the dynamics of the system and the degree that conditions can change over that time (Clewell and Aronson, 2007).

Given the level and period of commitment required for the restoration of very degraded and/or large systems, the inclusion of the community and all stakeholders in the restoration planning and process is vital to the project's success (SER, 2002). Even after restoration, if the disturbances or forces that led to the system's degradation are still present, the system is not likely to maintain itself in the restored (or desired) state for long. Restoration targets often focus on the most noticeable differences in a system from its previous (desired) state or between the system and a reference site, including increased populations of specialized species and decreased populations of generalist species, a lower density and diversity of invasive species, enhanced functional and structural diversity, microclimate changes (including an increase in the number of kinds of microclimates available), soil improvements, retention of nutrients in the system, and restored disturbance regimes (Clewell and Aronson, 2007). Generally speaking, restoration targets can range from fairly superficial changes to an area to extensive manipulation of abiotic and biotic conditions.

2.5.2.1 Reclamation

Reclamation describes a type of restoration that is usually focused on sites after mining activities, which damage soils (through heavy metal pollution and salination), alter topography (especially mountaintop removal and open-pit mining methods), and acidify and pollute aquatic ecosystems with runoff from the mined area itself or the slurries and tailings left behind (SER, 2002; Wali et al., 2010). While one might argue that mining areas can be restored to a natural state if given enough time, generally, the amount of alteration to the abiotic and biotic conditions of a mining site is so extensive that only reclamation is possible. Mine reclamation usually involves restoring the area to a state that would allow for a limited number of productive uses, such as recreation.

In the past, when the majority of mining activity was conducted underground, the environmental impacts were relatively limited to polluted runoff from tailings areas (tailings are the waste materials that remain after the coal or ore has been removed) and groundwater pollution from liberation of heavy metals inside the abandoned mine shafts. Reclamation of tailings areas is usually accomplished using bioremediation and/or phytoremediation to break down toxic pollutants into less toxic compounds or to bind the toxins into biomass, which can then be harvested and disposed of property (Tordoff et al., 2000; Wali et al., 2010). Bioremediation uses organisms such as bacteria, fungi, or yeast, and phytoremediation uses different plant species; often particular plant species are chosen because they show a high affinity for organic pollutants or heavy metals (Atlas and Philp, 2005; Pilon-Smits, 2005). In some cases, plants can be used to remove valuable metals (such as gold, nickel, and aluminum) from the soil and concentrate them in plant tissues, which could then be harvested and recovered (Sheoran et al., 2009). However, and far more commonly, these plants are used to stabilize the polluted soil (and prevent heavy metals from becoming airborne or released into runoff and polluting larger areas), rather than to remove or break down the pollutants (Frérot et al., 2006).

However, given the labor and expense involved in underground mining methods, in the twentieth century, mining companies switched to surface mining methods, which include strip mining, open-pit mining, and mountaintop removal (Fig. 2.4). These methods remove all of the vegetation, soil, and rock above the seam of coal or mineral and use heavy machinery to excavate the seam. Soil and rock are called "fill" or "overburden" or "spoil" and are supposed to be stored in contained areas where they may be retrieved and used to reclaim the area once mining has ceased. However, these materials are more commonly disposed of in valleys where the fill causes still more ecological damage in the landscape (Tongway and Ludwig, 2011; Griffith et al., 2012). Particularly in sulfide metal mines, water runoff from the area can acidify receiving waters and introduce heavy metals and other pollutants that are difficult to mitigate (Hopkins et al., 2013); lime or other chemicals can be added to raise the pH in streams, but the treatment must continue until the mine runoff is contained (Potgieter-Vermaak et al., 2006; Sahoo et al., 2013). Aquatic organisms are highly sensitive to water pH, and acidic conditions cause a loss of biodiversity and ecosystem functioning that can be difficult to repair (Simon et al., 2009).

In the United States, once a seam of ore (coal or metal) is exhausted, the Surface Mining Control and Reclamation Act (SMCRA) of 1977 (plus recent amendments) requires that the area be rehabilitated to certain standards (U.S.C. §§ 1234–1328). Prior to mining, the SMCRA requires mine operators to first remove the topsoil by layers (A horizon and B horizon) and store and maintain them separately to preserve their microorganisms and fertility (often growing native vegetation on them). Spoil should also be sorted and maintained in piles that minimize erosion by wind or water, which can damage surrounding ecosystems; this spoil can be used to rebuild the topography after mining operations (Tongway and Ludwig, 2011). Once mining operations are completed and the topography rebuilt, the soil layers should be replaced (B first, then A) and native vegetation should be reseeded. Optimally, the vegetation should not require irrigation or fertilization after an initial establishment period (Tongway and Ludwig, 2011). The length of the monitoring period after restoration varies by region but usually from 5 to 10 years post-vegetation establishment.

2.5.2.2 Rehabilitation

Rehabilitation is more narrowly focused on restoring ecological processes and productivity, with less emphasis on biological diversity or community structure (SER, 2002). Rehabilitation is often undertaken with specific land use goals in mind, such as timber production or recreation (Wali et al., 2010). The rehabilitation of brownfields (former industrial sites in urban areas) to allow for their reuse commonly employs rehabilitation instead of a full restoration, since the sites are often intended to be reused as commercial or residential areas and not natural areas (Levine, 2002). With regards to goals and methods, the rehabilitation of brownfields is often more akin to reclamation than restoration.

2.5.2.3 Restoration

Restoration, as ecologists use the word, implies that a site will be returned to a highly resilient state similar or equivalent to a natural reference state, either one based on the history of the site or one based on a similar ecosystem somewhere else. The indicators and targets of the project often include high biodiversity, low or no invasive species, and a community structure that is representative of all of the ecological functions one would expect in a natural community (e.g., top predators, pollinators, etc.). Increasingly, restoration targets also include ecosystem services, although different end states can represent different trade-offs among services (such as carbon sequestration or native species biodiversity) that require discussions with stakeholders to prioritize (Bullock et al., 2011). The targets usually do not include land uses other than recreation or aesthetic values.

The Society for Ecological Restoration (SER, 2002) outlines nine characteristics of successfully restored (or "recovered") systems:

1. The restored ecosystem contains a characteristic assemblage of the species that occur in the reference ecosystem and that provide appropriate community structure.
2. The restored ecosystem consists of indigenous [native] species to the greatest practicable extent. In restored cultural ecosystems, allowances can be made for exotic [non-native] domesticated species and for non-invasive ruderal and segetal species that presumably co-evolved with them. Ruderals are plants that colonize disturbed sites, whereas segetals typically grow intermixed with crop species.
3. All functional groups necessary for the continued development and/or stability of the restored ecosystem are represented or, if they are not, the missing groups have the potential to colonize by natural means.
4. The physical environment of the restored ecosystem is capable of sustaining reproducing populations of the species necessary for its continued stability or development along the desired trajectory.
5. The restored ecosystem apparently functions normally for its ecological stage of development, and signs of dysfunction are absent.
6. The restored ecosystem is suitably integrated into a larger ecological matrix or landscape, with which it interacts through abiotic and biotic flows and exchanges.
7. Potential threats to the health and integrity of the restored ecosystem from the surrounding landscape have been eliminated or reduced as much as possible.
8. The restored ecosystem is sufficiently resilient to endure the normal periodic stress events in the local environment that serve to maintain the integrity of the ecosystem.
9. The restored ecosystem is self-sustaining to the same degree as its reference ecosystem, and has the potential to persist indefinitely under existing environmental conditions. Nevertheless, aspects of its biodiversity, structure and functioning may change as part of normal ecosystem development, and may fluctuate in response to normal periodic stress and occasional disturbance events of greater consequence. As in any intact ecosystem, the species composition and other attributes of a restored ecosystem may evolve as environmental conditions change.

These conditions represent the optimal targets, at least from the perspective of ecologists and environmental managers. Yet, depending upon the social and environmental context in which the restoration is taking place, they may not be feasible. For example, climate change may alter regional rainfall patterns such as some rare species that are "characteristic" of the community but are drought insensitive and may be lost from the regional ecosystem entirely. If managers are using the historical conditions at the site as a reference, it may not be obvious that this target condition (restoration of a characteristic assemblage of species) cannot be fully achieved. Managers may need to find concurrent reference systems in the region to determine whether target conditions cannot be met due to new constraints brought about by climate change.

2.5.3 Restoring Natural Fire Disturbance Regimes

Restoring characteristic species to an area may not always entail reseeding or transplanting them to an area. In some cases, a disturbance regime must be reestablished before any other restoration actions will be successful. Fire is an important natural disturbance in many ecosystems (McKenzie et al., 2011). Fire moves nutrients that were locked up in biomass (living and dead) back to the soil, creates areas for early successional species dependent upon sunlit space, and maintains landscape-scale heterogeneity. Some plant species (such as pines in the Southeastern United States and chaparral plants in California; Sugihara et al., 2006; Keeley et al., 2011; Latham, 2013) require fire to promote seed germination; the hot temperatures induce the tough seed coating to burst or burn away. Humans have altered fire regimes for thousands of years, using it to clear openings for game management and later suppressing it to prevent the loss of timber and property (Sugihara et al., 2006). As a result, many of these "pyrogenic" ecosystems (those that require fire disturbance) have very different habitat structure and landscape-scale patterns when fire is suppressed (Moritz et al., 2011; Swetnam et al., 2011). Invasive species can also interrupt or alter fire regimes and the patterns they create in the landscape (Duncan and Schmalzer, 2004; McKenzie et al., 2011). Managing or restoring these systems requires that fire regimes be restored to their natural characteristics, which include sufficient temperature or intensity, residence time (how long an area burns), patchiness of a burn, burning season, and germination promotion (Clewell and Aronson, 2007).

2.6 STAGE 5: MONITORING AND ADAPTIVE MANAGEMENT

Once management or restoration practices are underway, continuous monitoring will be necessary to determine whether the practices are having their intended effect. Monitoring will focus on indicators, which will help managers ascertain whether the system is improving or maintaining a status close to established targets. However, once this attainment stage is reached, it is not uncommon for additional indicators to be added to the monitoring regime, or for some indicators to be dropped, if new information is needed to monitor ecosystem responses that were not intended or expected when the initial set of indicators was developed. These modifications are particularly common when a system of adaptive management is used for preservation and restoration projects.

In the past, managers have used a trial-and-error (or worse, crisis and response; Gunderson and Light, 2006) approach to management and restoration. Sometimes enough data would be collected on these sites to conduct a meta-analysis to identify broad patterns in successes and failures, although generally managers would keep trying different approaches until one approach worked. However, this approach is quite costly (and wasteful) in terms of time and resources and rarely contributes to a broader understanding of how ecosystems work. Adaptive management has emerged over the past several decades to provide a more systematic approach that allows managers to learn from previous attempts and be more predictive for future ones (Gregory et al., 2006; Williams et al., 2007; McFadden et al., 2011; Zedler et al., 2012; Jones et al., 2013). This adaptive approach is particularly well suited when management issues are complex and entail considerable risk and uncertainty.

Allen and Gunderson (2011) define adaptive management as "a form of structured decision-making… [which] uses management actions as experiments to provide data supporting, or failing to support, competing hypotheses when there is uncertainty regarding the response of ecological systems to management activities…." Knowledge is gained from the process of using data to assess multiple hypotheses (posing testable questions regarding how the system might respond to certain actions) that are constructed collaboratively among the management team and stakeholders, as well as from clear evaluation of the experiments as to whether the hypotheses were rejected or not (Gunderson and Light, 2006; Zedler et al., 2012). However, adaptive management is not appropriate in all situations; for example, if uncertainty is high but the manager's ability to control portions of the system (such as test plots) is low, then adaptive management is not feasible and another management approach should be used instead (Gregory et al., 2006).

However, even when adaptive management is appropriate, Allen and Gunderson (2011) caution that there are nine obstacles that can hinder its successful application:

1. *Lack of stakeholder engagement*: … can lead to stakeholders rejecting results that vary from their expectations.
2. *Experiments are difficult*: [t]he reluctance to experiment can… be manifest as a need for control.
3. *Surprises are suppressed*: surprises… should be embraced as opportunities to learn rather than as externalities [to minimize].
4. *Prescriptions are followed*: Adaptive management processes that are too complex in their internal organization and too complex and fragile in their stakeholder network are apt to stick to the prescription no matter what.
5. *Action procrastination: learning and discussion remain the only ingredients*: Although no action is in fact a management action… situations where action is desired but prevented by obstructionists to the process… [are] often political in nature.
6. *Learning is not used to modify policy and management*: This pathology is similar to the one described above, but here what is learned is critical and important, but is shelved because the management actions identified and necessary are too politically, economically, or logistically difficult.
7. *Avoid hard truths: decision makers are risk averse*: To avoid hard truths small-scale management experiments are conducted, which may improve management and the state of the resource, around the edges. [see also Rosser, 2001]
8. *The process lacks leadership and direction*: [When a] very vocal or influential… stakeholder… hijacks the process… [it] can lead to a process to address a specific agenda other than learning how to best manage.
9. *Focus on planning, not action*: adaptive management programs can become stuck in a planning loop.

Therefore, although adaptive management may seem like a management technique that should always be used over trial and error as standard practice, applying it takes considerable more planning, effort, and resources and is not guaranteed to provide better results than other alternatives.

The decades of effort dedicated to managing and then restoring the Florida Everglades provides a cautionary tale of the difficulty that managers face, particularly for large, complex ecosystems with heavy demands on ecosystems services from surrounding local communities. Once regional managers conceded that there was a need to shift from trial-and-error management of the Everglades to an adaptive approach to its restoration (Walters et al., 1992; Gregory et al., 2006; Doren et al., 2009), numerous challenges began to emerge. Primary among these was the historical partition of the Everglades ecosystem into sections with boundaries that were dictated by political boundaries rather than ecological ones. With these political boundaries came a multitude of federal, state, tribal, and local agencies with jurisdictions and interests in how the Everglades would be managed and restored.

However, one of the greatest difficulties in the Everglades restoration project is mitigating the effects of the massive drainage infrastructure on the flows of water through different parts of the ecosystem, and especially seasonal timing (hydroperiod), which affected the habitats of the ecosystem (Light and Dineen, 1997; Lo Galbo et al., 2013; Fig. 2.1). The altered hydroperiod patterns across the system have not only changed the vegetation (and hence habitat) but have also changed disturbance regimes (e.g., more fires in drier areas, slower recovery from hurricanes in wetter areas) and allowed the establishment of invasive species (e.g., Australian *Melaleuca* trees in the eastern Everglades; DeAngelis, 1997; Mayer and Pimm, 1998; Mazzotti et al., 2009; Center et al., 2012; Lo Galbo et al., 2013; Comprehensive Everglades Restoration Plan; see www.evergladesplan.org). Indeed, of the five "critical components" of Everglades restoration, four address the hydrology and water supply: "water storage capacity, water delivery, seepage, water quality, and habitats" (NRC, 2007).

Ultimately, restoration cannot succeed without managing the human communities and land use in the system. Increasing urban populations continue to drive residential and commercial sprawl, converting natural habitats and agricultural land to urban land cover, resulting in more impervious surface area that impacts the Everglades' hydrology. Concurrently, these populations also place greater demands on ecosystem services such as freshwater provisions, flood protection, and recreational opportunities that further strain the Everglades system and make restoration more difficult. Ultimately, the extent to which the Everglades can be restored as an ecosystem has as much to do with the reformation of the politics and governance in the area as it does with returning the biological communities to a natural balance (Gunderson and Light, 2006). Human demands on the ecosystem will have to be brought in line and managed within ecological limits, simultaneously with the restoration of key parts of the ecosystem.

REFERENCES

Allen, C.R. and L.H. Gunderson. 2011. Pathology and failure in the design and implementation of adaptive management. *Journal of Environmental Management* 92:1379–1384.

Allen, C. and G.H. Stankey. 2008. *Adaptive Environmental Management: A Practitioner's Guide*. Springer, Dordrecht. 352pp.

Anderson, J.R., E.E. Hardy, J.T. Roach, and R.E. Witmer. 1976. A Land Use and Land Cover Classification System for Use with Remote Sensor Data. Geological Survey Professional Paper 964, U.S. Geological Survey. U.S. Government Printing Office, Washington, DC. Available at: http://landcover.usgs.gov/pdf/anderson.pdf (accessed March 5, 2015).

Atlas, R.M. and J.C. Philp. 2005. *Bioremediation: Applied Microbial Solutions for Real-World Environmental Cleanup*. ASM Press, Washington, DC.

Baker J.P., D.W. Hulse, S.V. Gregory, D. White, J. Van Sickle, P.A. Berger, D. Dole, and N.H. Schumaker. 2004. Alternative futures for the Willamette River Basin, Oregon. *Ecological Applications* 14:313–324.

Barnosky, A.D., E.A. Hadley, J. Bascompte, E.L. Berlow, J.H. Brown, M. Fortelius, W.M. Getz, J. Harte, A. Hastings, P.A. Marquet, N.D. Martinez, A. Mooers, P. Roopnarine, G. Vermeij, J.W. Williams, R. Gillespie, J. Kitzes, C. Marshall, N. Matzke, D.P. Mindell, E. Revilla, and A.B. Smith. 2012. Approaching a state shift in Earth's biosphere. *Nature* 486:52–58.

Bartholome, E. and A.S. Belward. 2005. GLC2000: A new approach to global land cover mapping from Earth observation data. *International Journal of Remote Sensing* 26:1959–1977.

Bauer, S. and B.J. Hoye. 2014. Migratory animals couple biodiversity and ecosystem functioning worldwide. *Science* 344: 54–62.

Beardsley, K., J.H. Thorne, N.E. Roth, S.Y. Gao, and M.C. McCoy. 2009. Assessing the influence of rapid urban growth and regional policies on biological resources. *Landscape and Urban Planning* 93:172–183.

Beisner, B.E., D.T. Haydon, and K. Cuddington. 2003. Alternative stable states in ecology. *Frontiers in Ecology and the Environment* 1:376–382.

Bennett, D. and D. McGinnis. 2008. Coupled and complex: Human–environment interaction in the Greater Yellowstone Ecosystem, USA. *Geoforum* 39:833–845.

Beschta, R.L. and W.J. Ripple. 2012. The role of large predators in maintaining riparian plant communities and river morphology. *Geomorphology* 157:88–98.

Biggs, R., S.R. Carpenter, and W.A. Brock. 2009. Turning back from the brink: Detecting an impending regime shift in time to avert it. *Proceedings of the National Academy of Sciences of the United States of America* 106:826–831.

Blomquist, W. and E. Schlager. 2005. Political pitfalls of Integrated Watershed Management. *Society and Natural Resources* 18:101–117.

Bojórquez-Tapia, L.A., B.M. Cruz-Bello, and L. Luna-González. 2013. Connotative land degradation mapping: A knowledge-based approach to land degradation assessment. *Environmental Modelling and Software* 40:51–64.

Bonney, R., C.B. Cooper, J. Dickinson, S. Kelling, T. Phillips, K.V. Rosenberg, and J. Shirk. 2009. Citizen science: A developing tool for expanding science knowledge and scientific literacy. *BioScience* 59:977–984.

Bullock, J.M., J. Aronson, A.C. Newton, R.F. Pywell, and J.M. Rey-Benayas. 2011. Restoration of ecosystem services and biodiversity: Conflicts and opportunities. *Trends in Ecology and Evolution* 26:541–549.

Burgin, S. 2010. 'Mitigation banks' for wetland conservation: A major success or an unmitigated disaster? *Wetlands Ecology and Management* 18:49–55.

Center, T.D., M.F. Purcell, P.D. Pratt, M.B. Rayamajhi, P.W. Tipping, S.A. Wright, and F.A. Dray. 2012. Biological control of *Melaleuca quinquenervia*: An Everglades invader. *Biocontrol* 57(2):151–165.

Chander, G., C. Huang, L.M. Yang, C. Homer, and C. Larson. 2009. Developing consistent Landsat data sets for large area applications: The MRLC 2001 Protocol. *IEEE Geoscience and Remote Sensing Letters* 6:777–781.

Clewell, A.F. and J. Aronson. 2007. *Ecological Restoration: Principles, Values, and Structure of an Emerging Profession*. Island Press, Washington, DC. 216pp.

Costanza, R., R. d'Arge, R. de Groot, S. Farber, M. Grasso, B. Hannon, K. Limburg, S. Naeem, R.V. O'Neill, J. Paruelo, R.G. Raskin, P. Sutton, and M. van den Belt. 1997. The value of the world's ecosystem services and natural capital. *Nature* 387:253–260.

Cumming, S.G., P.J. Burton, and B. Linkenberg. 1996. Boreal mixedwood forests may have no "representative" areas: Some implications for reserve design. *Ecography* 19:162–180.

Dale, V.H. and S.C. Beyeler. 2001. Challenges in the development and use of ecological indicators. *Ecological Indicators* 1:3–10.

Davis, S. and J.C. Ogden. 1997. *Everglades: The Ecosystem and Its Restoration*. CRC Press, Boca Raton, FL, 860pp.

DeAngelis, D. 1997. Synthesis: Spatial and temporal characteristics of the environment. pp. 307–322 in Davis, S.M. and J.C. Ogden (eds.), *Everglades: The Ecosystem and Its Restoration*. St. Lucie Press, Boca Raton, FL.

Delcourt, H.R. and P.A. Delcourt. 1988. Quaternary landscape ecology: Relevant scales in space and time. *Landscape Ecology* 2:23–44.

Di Gregorio, A. and L.J.M. Jansen. 2000. Land Cover Classification System (LCCS): Classification Concepts and User Manual. UNFAO, Rome. Available at: http://www.fao.org/docrep/003/x0596e/x0596e00.htm (accessed March 5, 2015).

Dietz, R.W. and B. Czech. 2005. Conservation deficits for the continental United States: an ecosystem gap analysis. *Conservation Biology* 19:1478–1487.

Doren, R.F., J.C. Trexler, A.D. Gottlieb, and M.C. Harwell. 2009. Ecological indicators for system-wide assessment of the Greater Everglades ecosystem restoration program. *Ecological Indicators* 9S:S2–S16.

Droj, G. 2012. GIS and remote sensing in environmental management. *Journal of Environmental Protection and Ecology* 13:361–367.

Dudley, N. (ed.). 2008. Guidelines for Applying Protected Area Management Categories. IUCN, Gland. X + 86pp. Available at: http://data.iucn.org/dbtw-wpd/edocs/paps-016.pdf (accessed March 5, 2015).

Duncan, B.W. and P.A. Schmalzer. 2004. Anthropogenic influences on potential fire spread in a pyrogenic ecosystem of Florida, USA. *Landscape Ecology* 19:153–165.

Eisenberg, C., S.T. Seager, and D.E. Hibbs. 2013. Wolf, elk, and aspen food web relationships: Context and complexity. *Forest Ecology and Management* 299:70–80.

Fennessy, M.S., A.D. Jacobs, and M.E. Kentula. 2007. An evaluation of rapid methods for assessing the ecological condition of wetlands. *Wetlands* 27:543–560.

Foley, A.J., R. DeFries, G.P. Asner, C. Barford, G. Bonan, S.R. Carpenter, F.S. Chapin, M.T. Coe, G.C. Daily, H.K. Givvs, J.H. Helkowski, T. Holloway, E.A. Howard, C.J. Kucharik, C. Monfreda, J.A. Patz, I.C. Prentice, N. Ramankutty, and P.K. Snyder. 2005. Global consequences of land use. *Science* 309:570–574.

Folke, C., S. Carpenter, B. Walker, M. Scheffer, T. Elmqvist, L. Gunderson, and C.S. Holling. 2004. Regime shifts, resilience, and biodiversity in ecosystem management. *Annual Review of Ecology, Evolution, and Systematics* 35:557–581.

Franklin, S.E. 2010. *Remote Sensing for Biodiversity and Wildlife Management: Synthesis and Applications*. The McGraw-Hill Companies, New York.

Frérot, H., C. Lefèbvre, W. Gruber, C. Collin, A. Dos Santos, and J. Escarré. 2006. Specific interactions between local metallicolous plants improve the phytostabilization of mine soils. *Plant and Soil* 282:53–65.

Gregory R., D. Ohlson, and J. Arvai. 2006. Deconstructing adaptive management: Criteria for applications to environmental management. *Ecological Applications* 16:2411–2425.

Griffith, M.B., S.B. Norton, L.C. Alexander, A.I. Pollard, and S.D. LeDuc. 2012. The effects of mountaintop mines and valley fills on the physicochemical quality of stream ecosystems in the central Appalachians: A review. *Science of the Total Environment* 417:1–12.

Guhathakurta, S. (ed.). 2003. *Integrated Land Use and Environmental Models: A Survey of Current Applications and Research*. Springer-Verlag, Berlin Heidelberg.

Gunderson, L.H. 2000. Ecological resilience—in theory and application. *Annual Review of Ecology and Systematics* 31:425–439.

Gunderson, L. and S.S. Light. 2006. Adaptive management and adaptive governance in the Everglades ecosystem. *Policy Sciences* 39:323–334.

Guttal, V. and C. Jayaprakash. 2008. Changing skewness: An early warning signal of regime shifts in ecosystems. *Ecology Letters* 11:450–460.

Herkert, J.R. 2009. Response of bird populations to farmland set-aside programs. *Conservation Biology* 23:1036–1040.

Hinck, J.E., C.J. Schmitt, K.A. Chojnacki, and D.E. Tillitt. 2009. Environmental contaminants in freshwater fish and their risk to piscivorous wildlife based on a national monitoring program. *Environmental Monitoring and Assessment* 152:469–494.

Hirota, M., M. Holmgren, E.H. Van Nes, and M. Scheffer. 2011. Global resilience of tropical forest and savanna to critical transitions. *Science* 334:232–235.

Hobbs, R.J. and K.N. Suding. 2007. *New Models for Ecosystem Dynamics and Restoration*. Island Press, Washington, DC.

Homer, C., J. Dewitz, J. Fry, M. Coan, N. Hossain, C. Larson, N. Herold, A. McKerrow, J.N. VanDriel, and J. Wickham. 2007. Completion of the 2001 National Land Cover Database for the conterminous United States. *Photogrammetric Engineering and Remote Sensing* 73:337–341.

Hopkins, R.L., B.M. Altier, D. Haselman, A.D. Merry, and J.J. White. 2013. Exploring the legacy effects of surface coal mining on stream chemistry. *Hydrobiologia* 713:87–95.

Horan, R.D., E.P. Fenichel, K.L.S. Drury, and D.M. Lodge. 2011. Managing ecological thresholds in coupled environmental–human systems. *Proceedings of the National Academy of Sciences of the United States of America* 108:7333–7338.

Horning, N., J.A. Robinson, E.J. Sterling, W. Turner, and S. Spector. 2010. *Remote Sensing for Ecology and Conservation*, Techniques in Ecology and Conservation Series. Oxford University Press, New York. 467pp.

Hüttich, C., M. Herold, B.J. Strohback, and S. Dech. 2011. Integrating in-situ, Landsat, and MODIS data for mapping in Southern African savannas: experiences of LCCS-based land-cover mapping in the Kalahari in Namibia. *Environmental Monitoring and Assessment* 176:531–547.

Jabbar, M.T. and X.L. Chen. 2006. Land degradation assessment with the aid of geo-information techniques. *Earth Surface Processes and Landforms* 31:777–784.

Jaunatre, R., E. Buisson, I. Muller, H. Morlon, F. Mesléard, and T. Dutoit. 2013. New synthetic indicators to assess community resilience and restoration success. *Ecological Indicators* 29:468–477.

Jin, S.M., L.M. Yang, P. Danielson, C. Homer, J. Fry, and G. Xian. 2013. A comprehensive change detection method for updating the National Land Cover Database to circa 2011. *Remote Sensing of Environment* 132:159–175.

Jones, N. 2011. Human influence comes of age. *Nature* 473:133.

Jones, K.B., G. Zurlini, F. Kienast, I. Petrosillo, T. Edwards, T.G. Wade, B.L. Li, and N. Zaccarelli. 2013. Informing landscape planning and design for sustaining ecosystem services from existing spatial patterns and knowledge. *Landscape Ecology* 28:1175–1192.

Keeley, J.E., J. Franklin, and C. D'Antonio. 2011. Fire and invasive plants on California landscapes. pp. 193–222 in McKenzie, D., C. Miller, and D.A. Falk (eds.), *The Landscape Ecology of Fires (Ecological Studies)*. Springer, Dordrecht, 350pp.

Kukkala, A.S. and A. Moilanen. 2013. Core concepts of spatial prioritisation in systematic conservation planning. *Biological Reviews* 88:443–464.

Lackey, R.T. 1998. Seven pillars of ecosystem management. *Landscape and Urban Planning* 40:21–30.

Langhammer, P.F., M.I. Bakarr, L. Bennun, T.M. Brooks, R.P. Clay, W. Darwall, N. De Silva, G.J. Edgar, G. Eken, L.D.C. Fishpool, G.A.B da Fonseca, M.N. Foster, D.H. Knox, P. Matiku, E.A. Radford, A.S.L Rodrigues, P. Salaman, W. Sechrest, and A.W. Tordoff. 2007. Identification and Gap Analysis of Key Biodiversity Areas: Targets for Comprehensive Protected Area Systems (Best Practice Protected Area Guidelines). World Conservation Union, Gland. Available at: http://data.iucn.org/dbtw-wpd/edocs/pag-015.pdf (accessed March 5, 2015).

Latham, D. 2013. *Painting the Landscape with Fire: Longleaf Pines and Fire Ecology*. University of South Carolina Press, Columbia. 224pp.

Latifovic, R., Z.-L. Zhu, J. Cihlar, C. Giri, and I. Olthof. 2004. Land cover mapping of North and Central America—Global Land Cover 2000. *Remote Sensing of Environment* 89:116–127.

Lee, Y., J.L. Harrison, C. Eisenberg, and B. Lee. 2012. Modeling biodiversity benefits and external costs from a keystone predator reintroduction policy. *Journal of Mountain Science* 9:385–394.

Levine, A.S. 2002. The Brownfields Revitalization and Environmental Restoration Act of 2001: The Benefits and the Limitations, 13 Vill. Envtl. L. J. 217. Available at: http://digitalcommons.law.villanova.edu/elj/vol13/iss2/1 (accessed April 30, 2015).

Licht, D.S., J.J. Millspaugh, K.E. Kunkel, C.O. Kochanny, and R.O. Peterson. 2010. Using small populations of wolves for ecosystem restoration and stewardship. *BioScience* 60:147–153.

Light, S.L. and J.W. Dineen. 1997. Water control in the Everglades: A historical perspective. pp. 47–84 in Davis, S.M. and J.C. Ogden (eds.), *Everglades: The Ecosystem and its Restoration*. St. Lucie Press, Boca Raton, FL.

Lo Galbo, A.M., M.S. Zimmerman, D. Hallac, G. Reynolds, J.H. Richards, and J.H. Lynch. 2013. Using hydrologic suitability for native Everglades slough vegetation to assess Everglades restoration scenarios. *Ecological Indicators* 24:294–304.

Lomolino, M.V., B.R. Riddle, R.J. Whittaker, and J.H. Brown. 2010. *Biogeography*, 4th Edition. Sinauer Associates, Inc., Sunderland, MA.

Loveland, T., B. Reed, J. Brown, D. Ohlen, Z. Zhu, L. Yang and J.W. Merchant. 2000. Development of a global land cover characteristics database and IGBP DISCover from 1 km AVHRR data. *International Journal of Remote Sensing* 21:1303–1330.

Lynch, H.J., S. Hodge, C. Albert, and M. Dunham. 2008. The greater yellowstone ecosystem: Challenges for regional ecosystem management. *Environmental Management* 41:820–833.

Magurran, A.E. 2003. *Measuring Biological Diversity*. Wiley-Blackwell, Malden, MA. 264pp.

Manies, K.L. and D.J. Mladenoff 2000. Testing methods to produce landscape-scale presettlement vegetation maps from the U.S. public land survey records. *Landscape Ecology* 15:741–754.

Marcot, B.G. 2007. Biodiversity and the lexicon zoo. *Forest Ecology and Management* 246:4–13.

Marcot, B.G., M.G. Raphael, N.H. Schumaker, and B. Galleher. 2013. How big and how close? Habitat patch size and spacing to conserve a threatened species. *Natural Resource Modeling* 26:194–214.

Mayer, A.L. and A. Henareh Khalyani. 2011. Grass trumps trees with fire. *Science* 334:188–189.

Mayer, A.L. and S.L. Pimm. 1998. Integrating endangered species protection and ecosystem management: The Cape Sable sparrow as a case study. pp. 53–68 in Mace, G.M., A. Balmford, and J.R. Ginsberg (eds.), *Conservation in a Changing World: Integrating Processes into Priorities for Action*. Cambridge University Press, New York.

Mayer, A.L. and M. Rietkerk. 2004. The dynamic regime concept for ecosystem management and restoration. *BioScience* 54:1013–1020.

Mayer, A.L., A.H. Roy, M. White, C.G. Maurice, and L. McKinney. 2008. Quick Assessment Protocols for measuring relative ecological significance of terrestrial ecosystems. US Environmental Protection Agency, Office of Research and Development, Cincinnati, OH, EPA/600/R-08/061, 37pp.

Mazzotti, F.J., G.R. Best, L.A. Brandt, M.S. Cherkiss, B.M. Jeffery, and K.G. Rice. 2009. Alligators and crocodiles as indicators for restoration of Everglades ecosystems. *Ecological Indicators* 9S:S137–S149.

McFadden, J.E., T.L. Hiller, and A.J. Tyre. 2011. Evaluating the efficacy of adaptive management approaches: Is there a formula for success? *Journal of Environmental Management* 92:1354–1359.

McKenzie, D., C. Miller, and D.A. Falk (eds.). 2011. *The Landscape Ecology of Fires (Ecological Studies)*. Springer, Dordrecht. 350pp.

Meadows, D.H. 2008. *Thinking in Systems – A primer*. Chelsea Green Publishing, White River Junction, VT.

Millennium Ecosystem Assessment. 2003. *Ecosystems and Human Well-Being: A Framework for Assessment*. Island Press, Washington, DC. 245pp.

Millennium Ecosystem Assessment. 2005. *Ecosystems and Human Well-Being: Synthesis*. Island Press, Washington, DC. 160pp.

Mitchell, M. 2011. *Complexity: A Guided Tour*. Oxford University Press, New York. 368pp.

Moritz, M.A., P.F. Hessburg, and N.A. Povak. 2011. Native fire regimes and landscape resilience. pp. 51–88 in McKenzie, D., C. Miller, D.A. Falk (eds.), *The Landscape Ecology of Fires (Ecological Studies)*. Springer, Dordrecht, 350pp.

Munda, G., P. Nijkamp, and P. Rietveld. 1994. Qualitative multicriteria evaluation for environmental management. *Ecological Economics* 10:97–112.

Myers, N., R.A. Mittermeier, C.G. Mittermeier, G.A.B da Fonseca, and J. Kent. 2000. Biodiversity hotspots for conservation priorities. *Nature* 403:854–858.

Naughton-Treves, L., M.B. Holland, and K. Brandon. 2005. The role of protected areas in conserving biodiversity and sustaining local livelihoods. *Annual Review of Environment and Resources* 30:219–252.

Neumann, K., M. Herold, A. Hartley, and C. Schmullius. 2007. Comparative assessment of CORINE2000 and GLC2000: Spatial analysis of land cover data for Europe. *International Journal of Applied Earth Observation and Geoinformation* 9:425–437.

NRC. 2007. *Progress Toward Restoring the Everglades: The First Biennial Review – 2006*. The National Academies Press, Washington, DC, 250pp. Available at: www.evergladesrestoration.gov (accessed April 30, 2015).

O'Dea, N., J.E.M. Watson, and R.J. Whittaker. 2004. Rapid assessment in conservation research: A critique of avifaunal assessment techniques illustrated by Ecuadorian and Madagascan case study data. *Diversity and Distributions* 10:55–63.

Pearce, F. 2013. True Nature: Revising Ideas on What is Pristine and Wild. Yale Environment 360 May 16, 2013. Available at: http://e360.yale.edu/feature/true_nature_revising_ideas_on_what_is_pristine_and_wild/2649/ (accessed March 5, 2015).

Pechacek, P., G. Li, J.S. Li, W. Wang, X.P. Wu, and J. Xu. 2013. Compensation payments for downsides generated by protected areas. *Ambio* 42:90–99.

Pereira, J.M.C. and L. Duckstein. 1993. A multiple criteria decision-making approach to GIS-based land suitability evaluation. *International Journal of Geographical Information Systems* 7:407–424.

Petchey, O.L. and K.J. Gaston. 2006. Functional diversity: Back to basics and looking forward. *Ecology Letters* 9:741–758.

Pilon-Smits, E. 2005. Phytoremediation. *Annual Review of Plant Biology* 56:15–39.

Postel, S.L. and B.H. Thompson. 2005. Watershed protection: Capturing the benefits of nature's water supply services. *Natural Resources Forum* 29:98–108.

Potgieter-Vermaak, S.S., J.H. Potgieter, P. Monama, and R. Van Grieken. 2006. Comparison of limestone, dolomite and fly ash as pre-treatment agents for acid mine drainage. *Minerals Engineering* 19:454–462.

Ramage, B.S., J. Kitzes, E.C. Marshalek, and M.D. Potts. 2013. Optimized Floating Refugia: A new strategy for species conservation in production forest landscapes. *Biodiversity and Conservation* 22:789–801.

Randolph, J. 2003. *Environmental Land Use Planning and Management*. Island Press, Washington, DC.

Reed, M.S., I. Fazey, L.C. Stringer, C.M. Raymond, M. Akhtar-Schuster, G. Begni, H. Bigas, S. Brehm, J. Briggs, R. Bryce, S. Buckmaster, R. Chanda, J. Davies, E. Diez, W. Essahli, A. Evely, N. Geeson, I. Hartmann, J. Holden, K. Hubacek, A.A.R. Ioris, B. Kruger, P. Laureano, J. Phllipson, C. Prell, C.H. Quinn, A.D. Reeves, M. Seely, R. Thomas, M.J. Van Der Werff Ten Bosch, P. Vergunst, and L. Wagner. 2013. Knowledge management for land degradation monitoring and assessment: An analysis of contemporary thinking. *Land Degradation and Development* 24:307–322.

Reimchen, T.E. and C.H. Fox. 2013. Fine-scale spatiotemporal influences of salmon on growth and nitrogen signatures of Sitka spruce tree rings. *BMC Ecology* 13:38.

Richardson, C. (ed.). 2008. *The Everglades Experiments: Lessons for Ecosystem Restoration*. Springer, New York. 702pp.

Ripple, W.J. and R.L. Beschta. 2012. Trophic cascades in Yellowstone: The first 15 years after wolf reintroduction. *Biological Conservation* 145:205–213.

Rodrigues, A.S.L., H.R. Akcakaya, S.J. Andelman, M.I. Bakarr, L. Boitani, T.M. Brooks, J.S. Chanson, L.D.C. Fishpool, G.A.B. Da Fonseca, K.J. Gaston, M. Hoffmann, P.A. Marquet, J.D.

Pilgrim, R.L. Pressey, J. Schipper, W. Sechrest, S.N. Stuart, L.G. Underhill, R.W. Waller, M.E.J. Watts, and X. Yan. 2004. Global gap analysis: Priority regions for expanding the global protected-area network. *BioScience* 54:1092–1100.

Rosenzweig, M.L. 1995. *Species Diversity in Space and Time*. Cambridge University Press, Cambridge/New York. 460pp.

Rosser, Jr. J.B. 2001. Complex ecologic-economic dynamics and environmental policy. *Ecological Economics* 37:23–37.

Sahoo, P.K., K. Kim, S.M. Equeenuddin, and M.A. Powell. 2013. Current approaches for mitigating acid mine drainage. *Reviews of Environmental Contamination and Toxicology* 226:1–32.

Sauer, J.R., J.E. Hines, J.E. Fallon, K.L. Pardieck, D.J. Ziolkowski Jr., and W.A. Link. 2012. *The North American Breeding Bird Survey, Results and Analysis 1966–2011. Version 07.03.2013*. USGS Patuxent Wildlife Research Center, Laurel, MD.

Scheffer, M. 2009. *Critical Transitions in Nature and Society*, Princeton Studies in Complexity. Princeton University Press, Princeton, NJ. 384pp.

Scheffer, M., J. Bascompte, W.A. Brock, V. Brovkin, S.R. Carpenter, V. Dakos, H. Held, E.H. van Nes, M. Rietkerk, and G. Sugihara. 2009. Early-warning signals for critical transitions. *Nature* 461:53–59.

Schultz, E.T., R.J. Johnston, K. Segerson, and E.Y. Besedin. 2012. Integrating ecology and economics for restoration: Using ecological indicators in valuation of ecosystem services. *Restoration Ecology* 20:304–310.

Shafer, C.L. 1990. *Nature Reserves: Island Theory and Conservation Practice*. Smithsonian Institution, Washington, DC. 189pp.

Sheoran, V., A.S. Sheoran, and P. Poonia. 2009. Phytomining: A review. *Minerals Engineering* 22:1007–1019.

Sifneos, J.C., A.T. Herlihy, A.D. Jacobs, and M.E. Kentula. 2010. Calibration of the Delaware Rapid Assessment Protocol to a comprehensive measure of wetland condition. *Wetlands* 30:1011–1022.

Simon, K.S., M.A. Simon, and E.F. Benfield. 2009. Variation in ecosystem function in Appalachian streams along an acidity gradient. *Ecological Applications* 19:1147–1160.

Sklar, F.H., M.J. Chimney, S. Newman, P. McCormick, D. Gawlik, S.L. Miao, C. McVoy, W. Said, J. Newman, C. Coronado, G. Crozier, M. Korvela, and K. Rutchey. 2005. The ecological-societal underpinnings of Everglades restoration. *Frontiers in Ecology and the Environment* 3:161–169.

Snyder, B. 2008. How to reach a compromise on drilling in AWNR. *Energy Policy* 36:937–939.

Society for Ecological Restoration (SER). 2002. The SER Primer on Ecological Restoration. Science and Policy Working Group, SER, Tucson, AZ. Available at: http://www.ser.org/docs/default-document-library/english.pdf (accessed March 5, 2015).

de Souza, D.M., D.F.B. Flynn, F. DeClerck, R.K. Rosenbaum, H. de Melo Lisboa, and T. Koellner. 2013. Land use impacts on biodiversity in LCA: Proposal of characterization factors based on functional diversity. *International Journal of Life Cycle Assessment* 18:1231–1242.

Sovacool, B.K. 2008. Spheres of argument concerning oil exploration in the Arctic National Wildlife Refuge: A crisis of environmental rhetoric? *Environmental Communication: A Journal of Nature and Culture* 2:340–361.

Staver, A.C., S. Archibald, and S.A. Levin. 2011. The global extent and determinants of savanna and forest as alternative biome states. *Science* 334:230–232.

Sternberg, L.D.L. 2001. Savanna-forest hysteresis in the tropics. *Global Ecology and Biogeography* 10:369–378.

Stoneman, D.M. 1997. *The Everglades: River of Grass*. Pineapple Press, Sarasota, FL, 480pp.

Sugihara, N.G., J. van Wagtendonk, K.E. Shaffer, J. Fites-Kaufman, and A. Thode. 2006. *Fire in California's Ecosystems*. University of California Press, Berkeley 612pp.

Swetnam, T., D.A. Falk, A.E. Hessl, and C. Farris. 2011. Reconstructing landscape pattern of historical fires and fire regimes. pp. 165–192 in McKenzie D., C. Miller, D.A. Falk (eds.). *The Landscape Ecology of Fires (Ecological Studies)*. Springer, Dordrecht, 350pp.

Tateishi, R., B. Uriyangqai, H. Al-Bilbisi, M.A. Ghar, J. Tsend-Ayush, T. Kobayashi, A. Kasimu, N.T. Hoan, A. Shalaby, B. Alsaaideh, T. Enkhzaya, H. Gegentana, and P. Sato. 2011. Production of global land cover data – GLCNMO. *International Journal of Digital Earth* 4:22–49.

Thompson, C.M. and K. McGarigal. 2002. The influence of research scale on bald eagle habitat selection along the lower Hudson River, New York (USA). *Landscape Ecology* 17:569–586.

Tjorve, E. 2010. How to resolve the SLOSS debate: Lessons from species-diversity models. *Journal of Theoretical Biology* 264:604–612.

Tongway, D.J. and J.A. Ludwig. 2011. *Restoring Disturbed Landscapes: Putting Principles into Practice*. Island Press, Washington, DC. 189pp.

Tordoff, G.M., A.J.M. Baker, and A.J. Willis. 2000. Current approaches to the revegetation and reclamation of metalliferous mine wastes. *Chemosphere* 41:219–228.

Turner, M.G., R.H. Gardner, and R.V. O'Neill. 2010. *Landscape Ecology in Theory and Practice: Pattern and Process*. Springer, New York. 406pp.

U.S. Environmental Protection Agency. 1998. Guidelines for Ecological Risk Assessment. Report No. EPA/630/R-95/002F, U.S. Environmental Protection Agency, Washington, DC. Available at: http://www.epa.gov/raf/publications/pdfs/ECOTXTBX.PDF (accessed March 5, 2015).

Van Nes, E.H. and M. Scheffer. 2005. Implications of spatial heterogeneity for catastrophic regime shifts in ecosystems. *Ecology* 86:1797–1807.

Van Teeffelen, A.J.A., C.C. Vos, and P. Opdam. 2012. Species in a dynamic world: Consequences of habitat network dynamics on conservation planning. *Biological Conservation* 153:239–253.

Vandermeer, J., I. Perfecto, and N. Schellhorn. 2010. Propagating sinks, ephemeral sources and percolating mosaics: Conservation in landscapes. *Landscape Ecology* 25:509–518.

Wali, M.K., F. Evrendilek, and M.S. Fennessy. 2010. *The Environment: Science, Issues, and Solutions*. CRC Press, Taylor & Francis Group, Boca Raton, FL.

Walker, D.A., P.J. Webber, E.F. Binnian, K.R. Everett, N.D. Lederer, E.A. Nordstrand, and M.D. Walker. 1987. Cumulative impacts of oil-fields on northern Alaskan landscapes. *Science* 238:757–761.

Walters, C., L. Gunderson, and C.S. Holling. 1992. Experimental policies for water management in the Everglades. *Ecological Applications* 2:189–202.

Warren, A. 2002. Land degradation is contextual. *Land Degradation and Development* 13:449–459.

Westgate, M.J., G.E. Likens, and D.B. Lindenmayer. 2013. Adaptive management of biological systems: A review. *Biological Conservation* 158:128–139.

Williams, B.K., R.C. Szaro, and C.D. Shapiro. 2007. Adaptive Management: The U.S. Department of the Interior Technical Guide. Adaptive Management Working Group, U.S. Department of the Interior, Washington, DC. Available at: http://www.doi.gov/initiatives/AdaptiveManagement/TechGuide.pdf (accessed March 5, 2015).

Willis, K.J. and H.J.B Birks. 2006. What is natural? The need for a long-term perspective in biodiversity conservation. *Science* 314(5803):1261–1265.

Willis, K.J., L. Gillson, and T.M. Brncic. 2004. How "virgin" is virgin rainforest? *Science* 304(5669):402–403.

Willis, K.J., E.S. Jeffers, C. Tovar, P.R. Long, N. Caithness, M.G.D. Smit, R. Hagemann, C. Collin-Hansen, and J. Weissenberger. 2012. Determining the ecological value of landscapes beyond protected areas. *Biological Conservation* 147:3–12.

Young, T.F. and S. Sanzone (eds.). 2002. A Framework for Assessing and Reporting on Ecological Condition. Report EPA-SAB-EPEC-02-009, United States Environmental Protection Agency, Washington, DC.

Yu, L., J. Wang, and P. Gong. 2013. Improving 30 m global land-cover map FROM-GLC with time series MODIS and auxiliary data sets: A segmentation-based approach. *International Journal of Remote Sensing* 34:5851–5867.

Zedler, J.B., J.M. Doherty, and N.A. Miller. 2012. Shifting restoration policy to address landscape change, novel ecosystems, and monitoring. *Ecology and Society* 17(4): 36.

Ziolkowski, Jr. D., K. Pardieck, and J.R. Sauer. 2010. On the road again for a bird survey that counts. *Birding* 42:32–40.

3

SOIL SCIENCE IN ENVIRONMENTAL MANAGEMENT

NADINE KABENGI[1] AND MARIA CHRYSOCHOOU[2]

[1] Department of Geosciences and Department of Chemistry, Georgia State University, Atlanta, GA, USA
[2] Department of Civil and Environmental Engineering, University of Connecticut, Storrs, CT, USA

Abstract: The thin crust of soil that covers most of the Earth's land surface is vital to life on Earth. Without it, there would be no food to eat, fiber for clothes, timber for building and fuel, and foundation for construction. Soil is where terrestrial life interfaces with water, geology, and the atmosphere and as such has numerous and diverse interactions with the biosphere, hydrosphere, and atmosphere. Soil science is the study of the physical, chemical, and biological processes that control soil functions and interactions with other environmental media. In a global perspective, human life depends on soil quality, and soil quality is greatly influenced by human life. Soils are an intricate part of the equation and a necessary component to the solution. This chapter will begin by providing a general perspective on soils and soil science. The second part presents an overview of the basic composition; fundamental physical, chemical, and biological properties; and key processes of soil formation, development, and behavior in the environment. The third part illustrates the relationship of soils with environmental management through four case studies selected to represent pressing environmental issues ranging from food production and waste management to sustaining clean resources and global climate change.

Keywords: soil science, soil functions, soil minerals, soil organic matter, soil physical properties, soil acidity and alkalinity, ion exchange, sorption, soil organisms soil formation, soil profiles, soil processes, chemical movement, soil fertility, integrated nutrient management, soil contamination, contaminant fate and transport, soil remediation, pesticide pollution, soil carbon cycle.

3.1	Introduction	75
3.2	Compositions, Properties, and Processes	76
	3.2.1 Compositions	76
	3.2.1.1 Mineral (Inorganic) Constituents	77
	3.2.1.2 Soil Organic Matter	77
	3.2.1.3 Soil Solution	78
	3.2.1.4 Soil Air	78
	3.2.2 Properties	78
	3.2.2.1 Physical Properties	78
	3.2.2.2 Chemical Properties	80
	3.2.2.3 Biological Properties	81
	3.2.3 Processes	81
	3.2.3.1 Soil Formation and Development	81
	3.2.3.2 Chemical Movement and Sorptive Processes	83
	3.2.3.3 Soil Transport Processes	85
3.3	Soils in Environmental Management	87
	3.3.1 Case Study 1: Soil Fertility Management for Sustainability	87
	3.3.2 Case Study 2: Contamination of Soil, Sediment, and Water	88
	3.3.3 Case Study 3: Pesticide Pollution	91
	3.3.4 Case Study 4: Global Warming and Soil Management	93
References		94

3.1 INTRODUCTION

In the metaphor that compares our planet to an orange, the crust—the outer rock layer of the Earth—is often times ascribed the role of the orange peel to highlight that its average thickness makes up less than 1% of the total earth radius. Soils, with an approximate worldwide average depth of 2m, make up less than 0.05% of the crust. Mathematically speaking, soils are barely significant digits; yet, in their absence, life as we know it may have been profoundly different, if not totally nonexistent. Without soils, there would be no food to eat, fiber for

---

An Integrated Approach to Environmental Management, First Edition. Edited by Dibyendu Sarkar, Rupali Datta, Avinandan Mukherjee, and Robyn Hannigan.
© 2016 John Wiley & Sons, Inc. Published 2016 by John Wiley & Sons, Inc.

clothes, timber for building and fuel, and foundation for construction: soils sustain life on Earth.

For at least six millennia, old civilizations busied themselves with productively tilling, irrigating, and draining their soils while accumulating, mostly by trial and error, a wealth of knowledge on successful management practices. In fact, some have argued that the rise and fall of empires has been influenced by the upsurge and decline of the overall quality of their soils. Yet despite the essential role soils played in the first known civilizations, it wasn't until much later (ca 1860s–1880s) that the idea of soil as a valuable natural body to be studied in connection with environmental conditions emerged from the work of V.V. Dokuchaev, generally considered the father of modern pedology. Along with many of his pupils, most notably N.M Sibirtsev and K.G. Klinka, Dokuchaev ushered soil science as a new branch of natural sciences. Around the same time, in the United States, Professor E.W. Hilgard began to develop basic concepts of soil formation and systematic methods of studying soils similar to those put forth in Russia. This early interest in the broad area of soil science, in both the United States and Russia, stemmed from the need to develop solutions to practical problem of agriculture. This early emphasis on soil science was the launching pad for the development of other soil divisions as distinguishable branches of sciences. Early on, these divisions ranged from ameliorative pedology, which studied theoretical ways to improve infertile soils to agronomical chemistry, the science of fertilizers. Nowadays, soil science consists of a large number of divisions and subdivisions, such as soil chemistry and physical chemistry, soil physics, soil fertility, soil microbiology, taxonomy, soil classification, geography, soil mineralogy, genesis of soils, etc. Over the years, a vividly dynamic interplay has been documented between the properties of soils developed naturally and in response to particular combinations of agricultural and environmental practices. Not only are soils complex, but they are consistently evolving and changing. A clear understanding of soils and how best to use and manage them requires an appreciation, consideration, and knowledge of meteorology, climatology, ecology, biology, hydrology geomorphology, geology, and other earth sciences.

Simply put, soil is where terrestrial life interfaces with water, geology, and the atmosphere, and soil science is the study of the physical, chemical, and biological processes that control soil functions and interactions with other environmental media.

One can broadly identify six general key roles: First, soils serve as media for plant growth by supporting root growth and supplying essential nutrients. Second, soils provide a habitat for a wide range of living organisms from large animals to tiny organisms. Third, soil properties control the quality and fate of water by affecting many hydrological processes such as filtration, contamination, water loss, and utilization. Fourth, soils act as recycling system, as dead bodies of plants, animals, and people are decomposed and assimilated into soils and their basic elements made available for reuse. Fifth, soils modify the atmosphere by absorbing and emitting gases such as carbon dioxide, methane, and water vapor. Finally, soils serve as an engineering medium for human-built systems. The diversity of these roles highlights the numerous interactions soils have with the biosphere, hydrosphere, and atmosphere and stipulates that a fundamental knowledge of soils is essential to a myriad of real-world problems related to environmental protection and sustainable management of natural resources. From a global perspective, human life depends on soil quality, and soil quality is greatly influenced by human life. From food production and waste management to sustaining clean resources and deciphering global climate systems, soils are an intricate part of the equation and a necessary component to the solution. This chapter will provide an overview of some fundamental concepts of soil science and will illustrate its relationship with environmental management through four case studies, which constitute selected pressing environmental issues.

3.2 COMPOSITIONS, PROPERTIES, AND PROCESSES

3.2.1 Compositions

Soils have various roles, ranging from being a medium to plant growth, a habitat for soil organisms, a system for recycling nutrients and wastes, a system for water supply and purification, and lastly a building material and foundation for engineered structures. As such, soil's definitions have been varied and focused on a specific function in a particular disciplinary context (soil science, geology, engineering, etc.).

In the context of environmental management and according to the US Department of Agriculture (USDA), soil is "a natural body comprised of solids, liquids, and gases that occur on the land surface, occupies space and is characterized by one or both of the following: horizons or layers that are distinguishable from the initial material as a result of additions, losses, transfers, and transformation of energy and matter, or the ability to support rooted plants in a natural environment." Based on this formal definition, not all of Earth's land is covered by soils, and land surfaces that do not have horizons and will not grow plants are classified as nonsoil.

Soils are composed of solids, liquid, and air, the relative proportions of which greatly influence soil behavior in the environment. The solids in soils can be either mineral or organic substances. With the exception of organic soils, most of soil's solids are minerals, that is, inorganic substances of defined chemical composition and crystal structure. Air and water usually occupy the total volume of pore spaces, which may represent up to 50% of a soil volume. Both soil moisture and aeration are essential to the growth of plants and organisms.

3.2.1.1 Mineral (Inorganic) Constituents

Mineral constituents of soils are derived from the rocks and minerals of the Earth's crust through weathering, which refers to their physical and/or chemical breakdown. They vary greatly in size, from fragments of rocks to gravel and fine sands, and in chemical composition. The mineralogical composition of soils is thus largely determined by the parent material and the degree of weathering. A mineral is defined as "a naturally occurring homogeneous solid, inorganically formed, with a definite chemical composition and an ordered atomic arrangement" (Mason and Berry, 1968). Thus, it is possible to have compounds with identical chemistry but different mineralogy, for example, quartz and cristobalite share the chemical formula SiO_2 but have a different crystal structure. The temperature and pressure of the formation environment dictate the stable mineral form. Minerals in soils are categorized into primary and secondary. Primary minerals are formed from molten lava with little composition change, for example, quartz, mica, and feldspars. Secondary minerals are formed by extensive breakdown and weathering of other soil minerals. Examples of secondary minerals include the wide range of silicate clays, iron oxides, carbonates, and gypsum. Silica (SiO_2) and alumina (Al_2O_3) groups are the primary building blocks to form the majority of the hundreds of possible soil minerals that may be present in soils. Minerals may be classified according to the type of structural arrangements of the silica and alumina groups, while the specific chemistry may be slightly different. For example, clay minerals are built based on alternating rows of sheets of silica tetrahedral and alumina octahedral; these sheets may have various degrees of substitution by other elements, such as Fe and Mg, and different ions in the interlayers between the sheets. Overall, eight elements (Si, Al, Fe, Ca, Na, K, Mg, and O) form more than 90% of the inorganic soil constituents. In addition to the aluminosilicates, iron oxyhydroxides and carbonates are important mineral phases in soils. The mineralogy of soils is dynamic, that is, it is subjected to constant weathering; however, most weathering processes occur over geological time frames, and soil mineralogy may be considered constant for most environmental management applications. Exceptions to this are carbonate minerals, which easily dissolve and precipitate if environmental parameters such as pH change; extreme conditions, such as acid mine drainage, can also induce relatively rapid mineralogical changes in soils.

Soil mineralogy is important in a variety of environmental management issues, including management of soil contamination, erosion control, eutrophication, and others, because it controls many of the physicochemical properties of the soil. For example, the sorption capacity for metals is related to the amount and type of clay minerals present in the fine fraction; the water holding capacity of a soil is also largely dependent on soil mineralogy, for example, a smectitic soil with 10% clay can absorb a lot more water compared to a kaolinitic soil. Thus, an understanding of soil mineralogical properties is of interest to a variety of environmental management issues.

3.2.1.2 Soil Organic Matter

Soil organic matter (SOM) encompasses both living organisms and nonliving organic components composed of decomposed remains of plants, animals, and microorganisms as well as other organic compounds produced by soil metabolic processes. SOM is highly carbonaceous material with more carbon stored in the world's soils than all of the world's vegetation and atmosphere combined. The SOM content of soils ranges from 0.5 to 5% in most mineral soils but can be up to 100% of organic soils (Sparks, 2003). Because SOM decomposes over time to carbon dioxide that returns to the atmosphere, it is important to replenish SOM content to maintain its beneficial impact; practices such as manuring aim to restore SOM in agricultural active soils.

SOM is structurally and functionally complex and varied. In addition to carbon, which comprises 50–60% of the SOM, O (35–40%), N (~4%), and H (3–5%) are the main constituents, along with lower amounts of P and S. The most chemically and physically active form is humus, a dark amorphous material that is relatively resistant to decay and hence tends to accumulate in soils. The chemical structure of humus has been difficult to define, yet it can be divided into various fractions based on their solubility and molecular weight (e.g., fulvic acid, humic acids, humin). Humic acids are high molecular weight (3,000–1,000,000 Da) compounds that are soluble in alkali (base), fulvic acids are soluble in both acid and base and have lower molecular weight (500–5,000 Da), and humin is insoluble in either. Humic substances do not have well-defined physical or chemical characteristics nor are they associated with particular classes of organic compounds. They still contain common functional groups such as phenolic, ketonic, and carboxylic groups, but their overall structure is not known. In addition to these, soils may contain nonhumic substances, such as lipids, carbohydrates, proteins, and amino acids. Typically, humic substances comprise 33–75% of the total SOM, whereby humic acid dominates in grassland soils and fulvic acid in forest soils (Sparks, 2003).

Although SOM represents a small fraction of the total soil mass, it greatly influences soil physical and chemical properties such as aggregation, water holding capacity, and soil fertility. Additionally, it provides an energy source for soil microorganisms and a matrix for sorption and degradation of organic contaminants. Thus, SOM plays an important role in many environmental management issues. However, there is still limited qualitative and quantitative understanding of its properties and role in many soil processes, and our knowledge is largely empirical.

3.2.1.3 Soil Solution

The liquid phase in soils is not pure water but contains significant quantities of solutes and dissolved gases and is hence best referred to as soil solution. The solutes in soil solution can be soluble inorganic compounds, such as nitrate (NO_3^-) and phosphate (PO_4^{3-}), which supply essential elements for plant growth. Typical major ions in soil solution are Na^+, K^+, Ca^{2+}, Mg^{2+}, Al^{3+}, HCO_3^- H_4SiO_4, Cl^-, NO_3^-, PO_4^{3-}, and SO_4^{2-}. Dissolved organic matter is also commonly present in soil solution, especially in the upper soil horizons. There also can be undesirable inorganic (e.g., heavy metals such as Pb) and/or organic contaminants (e.g., petroleum hydrocarbons) whose presence and movement need to be managed. The soil solution/solid interface plays a central role in key soil processes such as mineral dissolution, ion exchange, and sorption, which in turn control important soil processes such as nutrient availability and fate and transport of contaminants. A critical property of the soil solution is its pH as it governs many of the abovementioned processes. Thermodynamic descriptions of reactions in soil solution and solution/solid interactions have been developed and incorporated in software codes such as Visual MINTEQ (Gustafsson, 2010), PHREEQC (Appelo and Parkhurst, 1998), and others. These may be used to understand and/or predict changes in soil solution and solubility of compounds of interest, for example, under different scenarios of soil management practices.

3.2.1.4 Soil Air

Air and water share the pore space in soil. Hence, the volume of air in soil is inversely related to the soil water content and is affected by the same physical properties. The composition of soil air is different than that of the atmosphere: it contains a higher concentration of carbon dioxide and a lower concentration of oxygen. This is due to the metabolic activity or respiration of soil organisms and plant roots that consume oxygen. While the CO_2 concentration in the air is 0.035%, it typically ranges from 0.35% up to 10% in soil horizons; the deeper the soil horizon, the higher the CO_2 concentration because of slow O_2 diffusion. The rate of O_2 diffusion in the soil pores is strongly dependent on the water content, since diffusion in the gas phase is approximately 10,000 faster compared to diffusion in water. Thus, saturated soils typically have very low O_2 concentrations and can easily become anaerobic. The oxygen concentration in soil pores directly influences the redox potential, which is a controlling parameter for redox reactions. This has important implications for various processes including degradation of organic compounds and mobility of metals, including Fe and Mn.

Organic decomposition slows down and is incomplete in anaerobic soils, while metal mobility increases sharply, an undesirable effect. Improved soil aeration is thus an important consideration in soil management practices.

3.2.2 Properties

3.2.2.1 Physical Properties

Soil Color
One of the first noticeable soil physical properties is its color. While it doesn't affect behavior and use, soil color provides valuable clues for the composition, development, and environmental conditions. Soils display a wide variety of colors ranging from brown, red, yellow, and green to even black and white. Colors also differ in their intensity and brightness. A standard system for accurate color description has been developed, namely, the Munsell color charts, whereby a chunk of soil is taken and compared to standard color chips in the Munsell chart. Each color is described by three characteristics of color: the hue or dominant spectral color (usually redness or yellowness), the value or the degree of light/dark in a color in relation to neutral gray scale (a value of 0 refers to a back color), and the chroma or the strength of the hue, which usually describes the intensity or brightness.

Agents of Soil Colors—The most common agents of soil color are organic matter and iron oxides. Organic matter tends to darken the soil and oftentimes masks soil natural colors. Generally, iron oxides impart a reddish color in soils but may turn green/grayish under reduced conditions and are hence used as redoximorphic features or indicators that soils are or have been saturated and anoxic. Other less common color agents are manganese oxides (black), glauconite (green), and carbonates (whitish).

Soil Texture and Particle Size Distribution
Soil texture is a basic soil property that is critical for environmental management especially because it typically doesn't change much in most relevant applications. Soil texture refers to the relative amounts of three fractions: sand, silt, and clay. These fractions refer to well-established size ranges based on classification systems, of which the most common has been established by the USDA. The USDA system defines sand as particles with a diameter between 0.2 and 0.05 mm, silt as particles with a diameter between 0.05 and 0.002 mm, and clay as particles with a diameter less than 0.002 mm. Particle size distribution is related to several important physicochemical properties of soils, that is, it may be used as a proxy to evaluate some other important environmental and/or geotechnical properties. An important parameter is specific surface area, which increases as particle size decreases. Surface area constitutes the interface between a soil particle and its surrounding environment in relation to weathering, water retention, gas and dissolved chemical

interactions, as well as interactions with microorganisms. Hence, smaller particles such as clay tend to have a much bigger influence on soil properties.

Textural Classes—The knowledge of the sand, silt, and clay percentages allows soil scientists to assign to the soil a soil textural class, the name of which conveys a sense of the particle size distribution and general soil physical properties. This is done using the textural triangle, a triangular graph whose sides correspond to the percentages of each soil separates (Fig. 3.1).

Determination of Textural Classes—There are two methods for the determination of soil texture: texture-by-feel method and laboratory particle size analyses. Estimating soil texture by feel is mostly utilized in the field and involves manipulating moist soil samples through a series of squeezing and kneading the sample to estimate the proportions of sand, sit, and clay based on plasticity (clay), smoothness (silt), and grittiness (sand). The laboratory determination of soil texture and particle size distribution is based on sedimentation, that is, the rate at which particles settle out of aqueous solutions, which is directly proportional to their size. This relationship is referred to as Stokes' law. Briefly, the method involves dispersing a known amount of soil in an aqueous suspension and measuring the amount still in suspension after various settling times by measuring the density of the suspension using a hydrometer or a pipette.

Soil Structure

Soil structure refers to the arrangement of soil particles into aggregates or peds, which have a size and shape that are characteristics of the soil. The pattern of aggregates greatly influences movement of water, air, and heat. It also is an important factor in soil cultivation and root growth. The soil can range from being structureless, when soil particles are unattached to one another, to massive, when all mineral particles are packed tightly together. Soil structure is described in terms of shape (spheroidal, platy, prism-like, and block-like), size (fine, medium, coarse), and grade or the degree to which the structure is developed (strong, moderate, or weak). Soil structure related to the pore space is discussed in the next section.

Soil Density and Porosity

Soil bulk density and porosity are key properties for environmental and agricultural management of soils. These values are informative regarding the potential movement of water, air, compaction, and plant root growth.

Soil Bulk Density—Recall that soils contain solids and voids; then depending on the amounts of the latter, the mass of a particular volume of soil will vary. The mass of a unit volume of dry soil is defined as soil bulk density D_b. This measurement will include both solids and pores; thus, the greater the amount of pore spaces, the lower the weight and bulk density. Bulk densities range from $0.1 \, g/cm^3$ for histosols to $2.2 \, g/cm^3$ for compacted glacial tills but with more common values around $1.0–1.6 \, g/cm^3$. Bulk density is usually measured by driving a container of known volume into the soil to obtain an undisturbed soil volume, drying it to remove water, and weighing the dry mass.

Soil Particle Density—In contrast, soil particle density D_p is defined as the mass per unit volume of soil solids, that is, excluding the pores. Particle densities for most mineral soils narrowly vary between 2.60 and $2.75 \, g/cm^3$, with $2.65 \, g/cm^3$ the most commonly adopted value. It represents the density of quartz particles, which represent a good average for soil mineral particles, especially when most of the soil solid phase is made of quartz particles. Quartz (SiO_2) is the most common constituent of sandy soils and is a mineral, a concept that will be discussed in a subsequent section.

Porosity—Soil total porosity is defined as the ratio of the volume of pores to the total soil volume. It is related to the bulk and particle density through the following formula: $P = (1 - D_b/D_p) \times 100$. Equally important to the total porosity is the pore size distribution. Generally, pores are grouped by size into macropores (larger than about 0.08 mm), mesopores (between 0.03 and 0.08 mm), and micropores (between 0.005 and 0.03 mm). The size of the pore determines the function it will have in the soil in relation to the movement of air, water, plant roots, and microorganisms. For example, macropores in sandy soils will drain water faster. Overall, soils with clay as the dominant textural class will have smaller pores but greater overall porosity than soils with a sandy textural class.

Water Holding Capacity

The ability of soils to store and release water greatly influences all environmental systems and is fundamental to sound environmental management. In soils, water retention

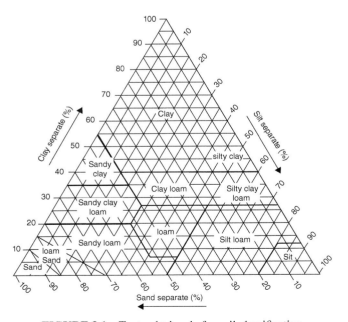

FIGURE 3.1 Textural triangle for soil classification.

and movement occur because of the combined effect of three forces: adhesion of water to soil solids, cohesion between water molecules, and gravity, which attracts water downward through the soil profile. Water movement in soils is intimately related to porosity and pore size distribution, as well as soil texture. Small pores will exert a greater tension or suction on soil water. For example, clayey soils will hold water to a greater degree and more tightly than sandy soils. Soil water content is measured by weighing a soil sample before and after water has been removed by drying. It may be expressed as percent by weight or by volume. Another way is to apply incrementally greater suction to the soil, gradually removing water from larger to smaller pores. Not all soil water is available to plants. The point at which plants cannot extract water from the soil is known as the wilting point. Two more important water benchmarks are saturation, that is, when all pores are filled with water, and field capacity, that is, when the soil has been drained by gravity and holds the remaining water through surface tension.

Engineering Properties

The most important soil properties that control the engineering behavior of soils are strength and compressibility, both of which are a function of fundamental properties such as grain size distribution and mineralogy, as well as *in situ* properties including void ratio and moisture content (Holtz et al., 2010).

Strength—Soil strength is typically defined as the maximum load that a soil can sustain without substantial deformation or failure. The mathematical description of soil strength relies on the concept of soil stress, which is defined as the load over the applied area. Soils sustain normal stresses that act perpendicular to the plane of interest (e.g., ground surface) and shear stresses that act parallel to the plane of interest. A critical combination of normal and shear stresses may cause soil failure, that is, large deformations. The purpose of various strength tests is to determine the magnitude of these critical loads and, if possible, the direction of failure. Triaxial strength tests on undisturbed soil samples are the most reliable tests developed for that purpose.

Compressibility—Consolidation is the process of slow water expulsion and vertical compression from saturated clay soils upon application of a normal load, for example, a building, or a reduction in effective stress due to lowering of the groundwater table, which in turn reduces the pore water pressure in the soil. Because of the very low hydraulic conductivity of clay soils, this process takes place over long periods of time, which may span decades or even hundreds or years. Soil compressibility is evaluated through consolidation tests, which simulate the progressive application of vertical loads on soils and measure the deformation over time.

3.2.2.2 *Chemical Properties*

Soil Acidity and Alkalinity

The soil pH is a master variable that controls many important agricultural and environmental aspects of soil science, as it relates to nutrient availability and contaminant fate and transport. Soil pH significantly affects the solubility of elements in soils and hence their potential availability as nutrients to plants and/or their movement. It is often one of the first chemical property to be determined whether in field or lab analyses. Soil pH ranges from about pH 4 (extremely acidic) to about pH 10 (very strongly basic or alkaline).

Acidity—The two main sources of soil acidity are the hydrogen (H^+) and aluminum (Al^{3+}) cation. Al^{3+} is considered acidic because it produces H^+ when it reacts with water. Generally, soil acidity is divided into three groups: active acidity, which is due to the H^+ and Al^{3+} ions in the soil solution; exchangeable acidity or salt-replaceable acidity, which corresponds to the H^+ and Al^{3+} that are released into the soil solution by other cations such as K^+; and residual acidity associated with the H^+ and Al^{3+} that are tightly bound to organic matter and silicates and are nonexchangeable. The main problem associated with extremely acidic soils is that aluminum and manganese can reach toxic levels and negatively affect crop growth and microbial activity. Trace metals such as Zn and Pb can also become increasingly mobile in acidic soils.

Alkalinity—The main source of alkalinity in soils are the base-forming cations, that is, Ca^{2+}, Mg^{2+}, K^+, and Na^+, which under arid and semiarid regime dominate the exchange processes and increase the OH^- concentration in soil solution. Other important alkalinity agents are carbonates (CO_3^{2-}) and bicarbonates (HCO_3^-) of several cations. The problems associated with strongly alkaline soils are low solubility of iron, zinc, copper, and manganese, which are essential micronutrients, and the resulting phosphorus deficiency, which tends to precipitate with calcium.

Buffering Capacity—An immediate implication of soil acidity and alkalinity on its management is the tendency of soils to resist changes in the pH of the soil solution. This is called soil buffering capacity. It ensures that soils are generally stable against wide pH fluctuations that might have negative consequences on plant and soil organisms. It also influences the amount and types of amendments required to have a desirable change in pH. Soil buffering capacity is a function of the type of acidity and alkalinity agents present but mostly depends on the cation exchange capacity (CEC) that will be described in the next section. Two examples of environmental issues related to acidification are the effect of acid rain and acid mine drainage on soil quality.

Ion Exchange Capacity

One of the most fascinating chemical properties of soils is the presence of an electrical charge on certain mineral and organic components. This electrical charge imparts soils the ability to attract and retain elements (e.g., Ca, Mg, K, Cu, Cl) and compounds (e.g., pesticides, hormones) with various degrees of selectivity and strength, allowing some to be released in the soil solution and hence made available to plants. This property is called soil exchange capacity and is applicable for both cations when the soil bears a net negative charge and anions when soils bears a net positive charge.

CEC—The main sources of negative charge in soils are clay minerals and organic matter (humus), as described in a subsequent section. The total negative charge is balanced by cations held at the surface through electrostatic forces. Other cations in solution can approach the cations held and the surface and replace them, in a process called cation exchange. The quantity of exchangeable cations per unit weight of dry soils is termed the CEC, expressed in centimoles of charge per kilogram of dry soils. The CEC is measured in the laboratory by displacing all the cations off the exchange sites by an "index" cation and counting those that were removed.

Anion Exchange Capacity—The basic mechanism of anion exchange is similar to that of cation exchange, and it occurs when a positive charge develops. The main sources of positive charge in soils are kaolinite, allophane, iron, and aluminum hydrous oxides that may develop a positive charge at low pH by adsorption of protons. Hence, the anion exchange capacity (AEC) of soils tends to decrease with increasing pH.

3.2.2.3 Biological Properties

Soil is a host to a large number and variety of organisms, both flora (plants) and fauna (animals).

These interact dynamically with one another, constituting a highly rich ecosystem that both affects and depends on soils. Soil is thus a biologically active medium.

Organism Diversity

A teaspoon of soil can contain millions of organisms representing a wide suite of phyla. A simple way to classify and organize these organisms is using their size into three categories: macro (larger than 2 mm), meso (medium size, between 2 and 0.2 mm), and micro (too small to be seen without a microscope, <0.2 mm). Selected examples of each are shown in Table 3.1.

Like in all ecosystems, maintaining diversity is essential to the health and proper functioning of the ecosystem, as each species carries a vital function. Organism diversity is regarded as an indicator of overall soil quality.

Organism Biomass

Soil biota depends on the soil environment, mineral, water, and air constituents for their energy and nutrient supply. Soil biomass is the total mass of the living fraction of the soil per unit volume. It is influenced by physical and chemical properties of the soil as well as competition and predation from other organisms. Biomass is related to SOM, the decomposed tissues of all soil biota.

Metabolic Activity

Soil organisms require carbon to build their cell constituents. Some organisms may obtain it by either breaking down or decomposing organic materials and are called heterotrophs. Others, referred to as autotrophs, obtain their necessary carbon from inorganic substances such as carbon dioxide (CO_2) or carbonate minerals using either light or chemical energy. Metabolic activity refers to the amount of carbon dioxide released from soil organisms' respiration and depends generally on the soil biomass. Because of their high numbers, microorganisms play a dominant role in soil metabolic activity.

3.2.3 Processes

3.2.3.1 Soil Formation and Development

Soil is formed over long periods of time (10,000–100,000 years) from the slow breakdown of rocks brought to Earth's outer surface by geological processes. Soil formation can be *in situ*, that is, formed in place of the original rocks, or *ex situ*, whereby fragmented rocks and minerals are transported from where the rocks originally occurred. This transport can be facilitated by gravity, water, ice, or wind. Soils formed from rocks deposited through these transport modes are referred to as colluvial, fluvial or alluvial, glacial, and eolian, respectively. Soils may also form from the accumulation of plant debris, as in the case for organic soils. In addition to the rock parent material, four other factors have been recognized to control and contribute to the soil formation process: climate, organisms (including human), physical relief or topography, and time. The combined effect of these five soil forming factors in a particular landscape results in different types of soils with

TABLE 3.1 Examples of Soil Organisms and Their Classification Using Size

Biota	Macro	Meso	Micro
Fauna (animals)	Moles, snakes, ants, snails, and earthworms	Mites, pot worms, and insect larvae	Nematodes
Flora (plants)	Mosses and feeder roots		Bacterial flora, fungi, and algae

distinctive physical, chemical, and biological properties. These characteristics, along with the presence and/or absence of many other diagnostic features, form the basis of soil taxonomy, an extensive but detailed soil classification system. The soil formation process continues through the subsequent transformations a soil continues to experience. These may include additions and losses of materials and chemical transformation such as acidification, oxidation, and reduction reactions. These changes impart further distinctive characteristics to older soils, separating them from younger, newly formed ones.

Weathering of Rocks and Minerals—Weathering is defined as the physical and chemical breakdown of rocks and minerals and includes destruction, modification, and synthesis of rocks and minerals. It is generally distinguished into physical and chemical weathering or decomposition.

Physical weathering breaks down rocks into smaller rocks without changing their chemical composition. This decrease in size is accompanied by an increase in surface area, exposing them to further physical and chemical weathering. Ultimately, physical disintegration results in the three constituents that form the inorganic portion of soils: sand, silt, and clay. Physical weathering is the result of many factors acting simultaneously. Temperature fluctuation throughout the day or the year causes the rock to expand and contract. These volume alterations weaken the rocks and break it apart. Wind, water, and ice are powerful abrasion agents that slowly wear down the rocks and cut them into even smaller pieces. Plant roots find their way into cracks, forcing them open and creating wider channels and more surface area. Physical weathering tends to dominate in dry, cool regions.

Through the action of water, oxygen, and acids, chemical weathering decomposes the rocks chemically, removing soluble materials and reprecipitating and synthesizing new (secondary) minerals with either slightly or completely different chemical composition than the primary minerals they originated from. Minerals present various degrees of resistance to chemical weathering, with end products such as oxides being the most resistant, as most other constituents in them have been removed and either washed away or reprecipitated into another mineral. Soil mineralogical composition is hence indicative of its formation and development stage. Chemical weathering dominates in hot and humid regions because of the ample water and higher temperatures that accelerate the rate of chemical reactions.

Soil Profile Formation—In addition to the combined influence of the five soil forming factors, each soil continues to experience additions, losses, transfers, and transformations of matter and energy. These formation and development processes create layers or horizons that are distinctively different from each other and from the original material. The vertical sequence of the horizons is specific to each soil and is termed as soil profile (Fig. 3.2). Generally, exposing a soil profile for a detailed examination is the first step for

FIGURE 3.2 Typical soil profile. *Source:* Courtesy of Dr. Brad Lee, Plant and Soil Sciences Department, University of Kentucky.

describing, classifying, and understanding a soil. Not all soil profiles contain the same horizons or layers, and not all layers and horizons have to be present in a specific soil profile. Soil scientists have delineated five master soil horizons (O, A, E, B, C). Sequentially from the top, these are:

O—humus or organic, mostly made of decomposing leaves

A—topsoil, mostly inorganic or minerals from parent material

E—eluviated or leached, layer mostly leached of clay, minerals, and organic matter

B—subsoil, rich in mineral and materials that have moved downward and accumulated

C—parent material, the deposit from which the soil developed

Most soils have three major horizons (A, B, C), and others may have an organic horizon or an E horizon. The latter is often found in older soils that had time to develop and experience significant leaching.

Other sublayers or subordinate horizons may occur and are further differentiated using lowercase letters, for example,

Bt or Oa whereby "t" refers to a layer of silicate accumulations and "a" to a sublayer of highly decomposed organic matter. There are around 26 possible designations of subdistinctions within master horizons.

Soil Development Processes—Many processes continuously influence soil formation and profile development and bring significant changes to soil physical and chemical properties. These changes occur on short but also longer timescales and are important in shaping the resulting soil profile. Some of the most common ones are briefly explained here. Leaching is the removal of material, usually soluble from the soil and soil solution, by percolating water that moves throughout the soil profile. This is especially important as rainwater is acidic even when it is unpolluted and may cause an overall acidification of the soil, upon which basic cations such as calcium, magnesium, sodium, etc. are washed away and lost. Physically, leaching may cause the downward translocation of the finer clay particles. In the long run, this results in zones of depletion or elluviation and zones of accumulation or illuviation in some horizons and throughout the soil profile. In the absence of enough water, soils experience an accumulation of cations such as sodium, calcium, and magnesium and salts such as sulfates and chloride. These processes, referred to as salinization and alkalization, also occur when salts are brought in with irrigation water, especially under arid and hot conditions. The pH of the soils may then increase to above 8, significantly affecting proper cultivation as the overall texture and fertility of soils deteriorate. Another shaping process for soil development is the physical removal of soil particles or erosion through the actions of wind and/or water flow. Soil erosion may become a serious environmental problem as not only soil fertility may be lost but also redeposition of particles may pollute waterways and rivers.

3.2.3.2 Chemical Movement and Sorptive Processes

Soils have the capacity to retain the essential nutrients and chemicals necessary for adequate plant growth. Hence, they can also bind chemicals that may be harmful to plants and pollute water environments. This property is mainly attributable to soil colloids, which are extremely small size particles, both inorganic and organic in nature, with a large surface area per unit mass. Both external and internal colloid surfaces bear a negative and/or positive charge, which attracts compounds of opposite charge. This retention can have various degrees of strength, with some chemicals available for further exchange and/or movement and other irreversibly bound to soil colloids and hence almost completely immobilized.

Soil Charge—Two types of charges are encountered on soil minerals and organic colloids: a permanent charge and a pH-dependent charge. The permanent charge, which is predominantly negative, occurs on clay minerals. These are designated as aluminosilicates, as their crystalline structure is built on a framework of aluminum and silicate sheets. The development charge occurs due to isomorphic substitution in the crystalline structure, that is, a replacement of a cation by another of the same size but of a different charge, for example, Al^{3+} replacing Si^{4+}. This substitution creates a charge imbalance. Like its name indicates, the pH-dependent charge is governed by the soil solution pH, and the same charged site can bear either a negative or positive charge. The pH-dependent charge arises at the surface and edges of clay minerals from unsatisfied charge in their structure. It also occurs on acidic functional groups (hydroxyl, carboxylic, sulfonic, etc.) in humus and hydrous metal oxides, which may either accept or release a proton (H^+) depending on the solution pH and their dissociation constants, thus creating either a positive or negative charge. The total charge created by these two mechanisms is generally satisfied by cations and anions present in the soil solution in the vicinity of the surface. The spatial distribution of these ions depends on the magnitude of the surface charge; on the chemical characteristics of the ions, that is, size, hydration sphere, and electronegativity; as well as on the concentrations of various ions in the soil solution. Counterions are concentrated close to the surface, with a gradient of decreased concentration away from the charge into the soil solution. This region is called the electrical double layer or diffuse double layer. It governs the overall charging behavior of colloids and influences almost all processes of relevance to chemical movement in soils. A great deal of effort has been invested by scientists to understand and characterize the diffuse layer structure and characteristics.

Ion Exchange—As discussed previously, the charge on clays and colloids warrants ions to counterbalance it. A small part of these ions enters the clay structure to satisfy the permanent negative charge inside the crystalline structure. These cations are held very tightly between the layers sheets of the aluminosilicate structures and become unavailable to the soil solution. These are termed nonexchangeable. Nevertheless, the majority of the counterions that balance the remaining of the soil charge on humus and colloid surfaces are not held very tightly to the surface, but only experience an electrostatic attraction and remain positioned outside the structure. These ions are coined exchangeable as they may be replaced by other ions in the soil solution and are hence available for plant uptake and movement with the soil solution. As defined in the section on chemical properties of soils, this process is called ion exchange, with CEC and AEC as the controlling chemical properties. Cation exchange is more prevalent than anion exchange, as clays and humus have overwhelmingly more negative charge. The ion exchange process has the following distinguishable characteristics: (i) it is rapid, (ii) it is reversible, and (iii) the exchange is chemically equivalent, that is, for the same number of total charge. It is important to note, however, that not all exchangeable ions are held with the same strength but that some cations are strongly preferred over others. Ion selectivity is an important concept in managing the types and amounts of ions desired on the surface of colloids and those present in the soil solutions for plant uptake and potential movement. The strong affinity for some cations does not originate from a different mechanistic bonding, but still remains an electrostatic attraction to the charged sites. In the event that the bonding process is not electrostatic, the process is no longer termed ion exchange but is referred to as chemisorption, adsorption, or even simply sorption.

Ion Sorption—Broadly defined, sorption is defined as any loss of chemical species from an aqueous solution phase to a contiguous solid phase. It thus encompasses three separate processes: absorption, adsorption or chemisorption, and surface precipitation. Absorption, which is sometimes also referred to as partitioning, is the process whereby the chemical species migrate or diffuse into the three-dimensional framework of the solid; it is thus not a strictly surface process. Adsorption or chemisorption is a strictly surface process, whereby selective bonding results in the net accumulation of a dissolved chemical species (the adsorbate) at the interface between the surface (the adsorbent) and the aqueous solution phase. Precipitation is defined as the growth of a new three-dimensional phase in solution or at the surface. Distinguishing between adsorption and sorption has been the topic of much research and debate. Yet, unless a new solid phase can be detected, the onset of precipitation and termination of adsorption is not easily pinpointed. Hence, the general term "sorption" is invoked where the specific retention mechanism cannot be identified.

The sorption process can be distinguished from ion exchange in many ways: (i) a high degree of selectivity toward particular cations and anions, (ii) a higher tendency for irreversibility or slower desorption, and (iii) a change in the surface charge of the colloid. Sorbed species are described in terms of their spatial position and proximity to the electrical diffuse layer and surface (outer sphere vs. inner sphere), the number of attachment sites unto the surface (one or monodentate, two or bidentate, and three or tridentate), and their strength and reversibility (reversible vs. irreversible). Ion sorption is the most important of the physicochemical processes responsible for the retention of inorganic and organic substances in the soil. This retention is extremely relevant to environmental management, as it often restricts the compound mobility and bioavailability, potentially creating soil and water pollution. Therefore, many efforts are exerted to quantify sorption of a wide range of essential and hazardous chemical compounds and determine their exact sorption mechanism to potentially predict the intensity, extent, and potential movement as a function of environmental conditions.

Analytically, ion sorption is measured by reacting a soil or the individual reactive components with a solution of known concentration of the chemical species for a fixed time and subsequently measured by the amount that has been removed from the solution. In tandem, the solid phase is further inspected and analyzed in an effort to determine the exact mechanism of complexation. Typically, such experiments are performed over several aqueous concentrations in order to produce a sorption isotherm (Fig. 3.3). Sorption isotherms can then be modeled using equations such as

$$\text{Freundlich}\left[C_{\text{soil}} = K_d \times C_{\text{aq}}^n \right] \text{ or}$$

$$\text{Langmuir}\left[C_{\text{soil}} = \frac{K \times C_{\text{aq}}}{1 + K \times C_{\text{aq}}} \right]$$

where:
C_{soil} is the sorbed amount in mg/kg.
C_{aq} is the dissolved concentration at equilibrium in mg/l.
n, K, and K_d are coefficients obtained through fitting the experimental data.

K_d is called the distribution coefficient and is the parameter most frequently used to model sorption reactions in transport modeling. The limitation of this approach is that K_d is an empirical parameter determined for a particular soil and compound under controlled conditions of pH and ionic strength; thus, any change in the geochemical environment requires determination of an appropriate K_d. In recent years, efforts have focused to develop thermodynamic descriptions of sorption reactions, described as surface complexation models (Goldberg et al., 2007); such models have the ability to describe sorption in a variety of environmental conditions.

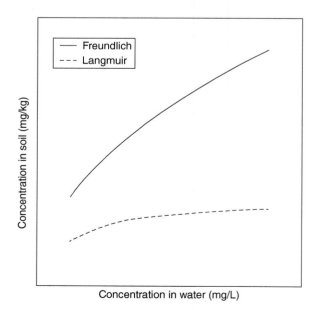

FIGURE 3.3 Example of Freundlich and Langmuir isotherms.

3.2.3.3 Soil Transport Processes

Soil is a three-phase system, comprising solid, water, and air, and interacts with other environmental media, that is, the atmosphere and water bodies. Given the heterogeneity of soil and the variable properties of these media, there is constant transport of mass and energy from and to the soil. Figure 3.4 illustrates the various transport processes that take place within a typical soil profile. Specifically, there are four types of transport: water, gas, solutes, and heat.

Water is exchanged between the unsaturated zone and the atmosphere through the processes of infiltration and evapotranspiration. Infiltration can be defined as the entry and vertical transport of water into the unsaturated soil profile. The maximum rate at which a soil can absorb water is called infiltration capacity, and it depends on several factors, including the intensity of precipitation, soil type and structure, soil moisture conditions, temperature, land use, and surface properties. In general, a soil with high content in fines and clay particles, low porosity, and high vegetative cover will have a much lower infiltration capacity compared to a sandy soil with high porosity and thin vegetation. Additionally, the infiltration rate decreases exponentially with increasing moisture content, reaching a constant value as the moisture content approaches saturation. When the infiltration rate is exceeded during a precipitation event, surface runoff is generated. Infiltration rates may be determined experimentally through infiltrometer tests or hydrograph analysis methods. Quantitative models have been developed in order to describe the influence of various factors on the generation of runoff, the most popular of which is the Soil Erosion Service Runoff Curve Number method developed by the USDA (USDA Soil Conservation Service, 1972). Williams et al. (2012) present a comprehensive overview of the method and its evolution and application examples. Detailed treatment of infiltration, evapotranspiration, and runoff may be found in hydrology textbooks, such as Chow and Mays (2013) and Viessman and Lewis (2002). The driving force for water movement in the subsurface has two components: changes in elevation (gravity) and changes in pressure. The Bernoulli equation for fluid flow, assuming an incompressible fluid and negligible viscous forces, states that for any point applies

$$\frac{v^2}{2} + gz + \frac{p}{\rho} = \text{constant} \quad (3.1)$$

where v is the fluid velocity, z the elevation of the point above a reference plane, g the acceleration of gravity, p the pressure, and ρ the fluid density. In porous media, the fluid velocity is very low and can be neglected. Normalizing the constant quantity in Equation 3.1 by dividing with g, we obtain the hydraulic head h. The hydraulic head is a measure of the fluid energy at any point in the subsurface and includes two components, the elevation head h_z and the pressure head h_p (Eq. 3.2), as illustrated in Fig. 3.5:

$$h = h_z + h_p = z + \frac{p}{\rho g} \quad (3.2)$$

The elevation head is measured above an arbitrary reference plane, while the pressure head corresponds to the height of the free water surface that will rise in a piezometer installed at the point of reference.

Changes in the hydraulic head induce fluid flow, which may be described by Darcy's law in one dimension and under a set of simplified conditions:

$$v = ki = k\frac{\Delta h}{\Delta l} \quad (3.3)$$

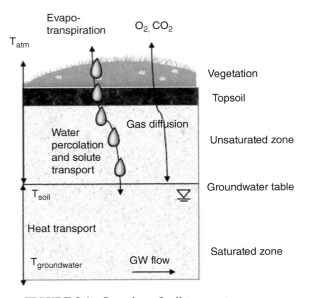

FIGURE 3.4 Overview of soil transport processes.

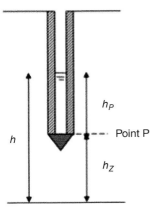

FIGURE 3.5 Total, pressure, and elevation head at any point in the subsurface.

where v is the fluid velocity, k the hydraulic conductivity of the porous medium, Δh the change in hydraulic head between two points, and Δl the distance between them. Darcy's law is valid when the flow is saturated, laminar (low flow velocity without turbulence), and constant (no temporal variations, constant hydraulic conductivity). Darcy's law may be extended to describe flow in the unsaturated zone as well, where the hydraulic conductivity is no longer constant but a function of the soil moisture content. Additionally, in the unsaturated zone, the pore water pressure is lower than the atmospheric pressure when the air forms a continuous phase, so that the total pressure head is negative and described as suction. The relationship between the pressure head, moisture content, and hydraulic conductivity in unsaturated soil is typically nonlinear and will dictate the specific approach to solve Darcy's equation, which turns into the Darcy–Buckingham equation (Raats and van Genuchten, 2006)

$$q = k(h)\frac{\partial h}{\partial z} \qquad (3.4)$$

where q is the specific flux and $k(h)$ the unsaturated hydraulic conductivity. Various models have been developed for the relationship between k and moisture content. Rumynin (2011) provides a basic overview of unsaturated flow transport, with more complete descriptions available by Stephens (1995) and Selker et al. (1999).

Groundwater flow in three dimensions is described by the Laplace equation, which cannot be solved analytically:

$$k_x \frac{\partial^2 h}{\partial x^2} + k_y \frac{\partial^2 h}{\partial y^2} + k_z \frac{\partial^2 h}{\partial z^2} = 0 \qquad (3.5)$$

However, it constitutes the basis for groundwater flow models, which solve the equation numerically and utilize field observations to calibrate flow directions and velocities. There is a wide variety of both commercial and freely available software codes, a comprehensive list of which may be found on the website of the US Geological Survey (USGS) (URL: http://water.usgs.gov/software/lists/groundwater/).

The movement of water in the subsurface results in the transport of dissolved chemical compounds, both vertically in the unsaturated zone and horizontally in the aquifer. Fluid flow or convection is the most important mechanism for solute transport in the subsurface; however, diffusion may also become important when the flow velocities are extremely low, for example, within clay layers. The basis for the mathematical description of solute transport in porous media is the advection–dispersion equation

$$\frac{\partial c}{\partial t} = \nabla(D\nabla c) - \nabla(\vec{v}c) + R \qquad (3.6)$$

where c is the concentration of the solute, D the dispersion coefficient, v the flow velocity, and R a term that describes the reactions of the solute along the flow path. The dispersion coefficient D describes two phenomena in the porous medium: molecular diffusion and dispersion due to flow velocity variations at the pore scale. Typically, the dispersion effect is orders of magnitude higher compared to molecular diffusion. The effect of dispersion is the spread of a solute both parallel and perpendicular to the groundwater flow direction assuming a point source, as shown in Fig. 3.6.

The advection–dispersion equation may be solved analytically for a small subset of problems. Walton (2008) provides solutions and examples for some typical transport problems in porous media, while Clark (2009) offers a more rigorous mathematical presentation of transport modeling in environmental systems. Similar to groundwater modeling, there are several software codes available to conduct fate and transport modeling of solutes in the subsurface, which range from simple one-dimensional models with simple reaction terms to complex codes that include biological reactions, mineral dissolution/precipitation, and unsaturated flow. In addition to the numerous commercial software packages available, there are free codes available through the USDA, US Department of Energy, USGS, and US Environmental Protection Agency (USEPA). Goldberg et al. (2007) provide a comprehensive overview of computer codes used by various federal agencies to this end.

Transport of gases in soil is also important for environmental management, given that they affect water quality and also that gases are implicated in important environmental issues, such as eutrophication and global warming. Gas transport occurs through advection and diffusion in the vapor phase, given that diffusion in liquid is four orders of magnitude slower (Scanlon et al., 2001). Advection is caused by pressure differences, and studies have shown that even very small pressure gradients can cause advective flux that much greater than the diffusive flux (Massmann and Farrier, 1992). Darcy's law may be used to describe advective gas flux using an appropriate expression for gas conductivity, and there are several different models (Fick's law, dusty gas model) to describe gas diffusion. Scanlon et al. (2001) provide a comprehensive overview of the mathematics and models used to describe gas flow in the subsurface. Special applications of

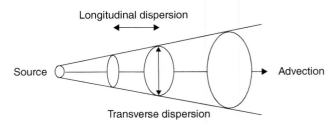

FIGURE 3.6 Illustration of spatial distribution of a contaminant subject to advection and dispersion.

environmental management such as soil vapor extraction and air sparging for remediation require more rigorous models to predict multicomponent gas flow in the unsaturated zone.

3.3 SOILS IN ENVIRONMENTAL MANAGEMENT

3.3.1 Case Study 1: Soil Fertility Management for Sustainability

The United Nations projections for the world population estimate an additional three billion people by year 2050. The Food and Agriculture Organization (FAO) projects that the world will have to produce 70% more food to feed itself. This is a daunting task even without the added challenges of safeguarding our environment and sustaining our natural resources and genetic diversity. More recently, at the Rio+20 summit in 2012, two important goals for the world community were launched—achieving Zero Hunger Challenge and Zero Net Land Degradation by 2030. It was an effort to combine hunger reduction with sustainable development efforts but also as reaffirmation of the importance of both to global issues such as food security, poverty reduction, and sustainable resources and environment management. These are all pressing global concerns that require significant and serious changes in agricultural practices, technological infrastructures, economic and institutional policies, as well as social reforms. In the agriculture sector, strides have been made in a number of areas: agricultural research continues to lead to improved technologies and practices, and investments in infrastructures and human capital significantly ameliorated the overall conditions for agriculture in many developing nations. The end goal remains to increase the global agriculture productivity, which is directly related to soil science. The way we manage and use our soils will be a crucial piece in solving these grand challenges, namely, maximizing soil productivity.

Soil fertility is defined as the capacity of soils to sustain crops, that is, provide the necessary nutritional requirement to plants. For adequate growth, plants require between 21 and 24 elements, separated into primary and secondary nutrients. The primary nutrients are nitrogen, phosphorus, and potassium and are required in significant amounts; otherwise, plant growth is severely limited. Less intensively needed are the secondary nutrients (e.g., calcium, sulfur) and micronutrients (chlorine, iron, molybdenum) whose main function is to ensure proper plant metabolic functions. Not only do plants require these nutrients in the proper amounts and correct proportions but also in a usable form and at the right time. The capacity of a soil to meet plant demand is a function of all the physical, chemical, and biological properties discussed in the previous section. For example, an acidic soil pH (<4.5) may hinder the bioavailability of some elements such as nitrogen and calcium while causing others like iron and manganese to be available in undesirably high concentrations.

A soil with a compact structure, such as a clayey textural class, will not allow the development of an extensive root system, which may limit yield if such a system is necessary. Thus, while some soils are inherently fertile, others may never be; maintaining a soil's fertility depends on the way soils are cultivated and managed.

A decline in soil fertility occurs in a variety of ways: depletion of nutrients from intense agriculture without proper replenishment and soil erosion via wind and water but also because of added tilling that leaves the topsoil bear and susceptible and other ill practices that lead to the physical and chemical deterioration of soils, for example, compaction, overirrigation, and accumulation of harmful substances.

Nutrient Resources—The pool of soil nutrients changes over time in what is known as the nutrient cycle (National Research Council, 1993), a complex soil–plant–atmosphere interactive system based on physical, chemical, and biological interactions that keep the inputs and outputs to the system in flux. Nutrient inputs to the soils can be from internal and external sources. Internal sources are those that supply nutrients from within the ecosystems and include weathering of mineral and rocks in a soil profile to release nutrients, atmospheric deposition and fixation, as well as recycling of plant residues (composting, cover crops, crop rotations) and animal manure. The main external sources of nutrients are fertilizers both in inorganic and organic forms. Nutrients are depleted from the soil through harvested plants, leaching, erosion, and atmospheric volatilization. The nutrient balance is the overall difference between all inputs and outputs to the system (Gruhn et al., 2000). A positive nutrient balance when inputs exceed output is usually an indication that the farming system is not efficient and more yields can be harvested. The demand for higher agricultural outputs intensified farming systems, with outputs now far exceeding the inputs. This results in a negative nutrient balance, where nutrients are mined and the overall fertility of soils is in decline. In that case, nutrients have to be replenished, and inorganic synthetic fertilizers are oftentimes the inexpensive source to use. A heavy use of fertilization may eventually lead to environmental pollution as large amounts of those nutrients find their way into water bodies causing *eutrophication*—an excess accumulation of nutrients (mostly nitrogen and phosphorus) that stimulates excessive plant growth such as algae. The excessive algae growth interferes with aquatic ecosystems and food chains as it blocks the sunlight from underwater. The bacterial decomposition of dead algae monopolizes the oxygen supply to the fish. The atmospheric volatilization of some of the excess nitrogen leads to the formation of nitrous oxides and eventually acid rain. Eutrophication is not solely the result of excessive or inefficient fertilizer use. Intensive industries and animal production along with other human activities such as detergent use often exacerbate the problem.

Integrated Nutrient Management—Scientists, researchers, and farmers commonly agree that the overall strategy to soil fertility or nutrient management should be a holistic approach that integrates organic and inorganic sources of fertilization with physical and biological measures of soil and water conservation. The goal of nutrient management is then to optimize the nutrient balance to have nutrient input matching as closely as possible all expected outputs. This not only avoids environmental contamination but also results in a cost-effective productivity, where nutrient resources are judiciously used without endangering the soil productivity of future generations. These principles comprise the integrated nutrient management (INM) or integrated plant nutrition systems (IPNS) approach. It is a flexible way that relies on a combination of best methods, principles, and practices referred to as best management practices (BMPs) to maximize soil fertility while taking into consideration local soil characteristics and climatic conditions, as well as respecting socioeconomic traditions. It differs in the rigidity of its stance from both low-input and high-input approaches that strongly favor one particular use of fertilizers.

Fertilizer Sources—INM promotes a judicious combination of inorganic fertilizers with organic sources in recognition of three important principles. First, while the release of nutrients from inorganic fertilizers is fast and hence can meet plant needs quickly, relying solely on inorganic fertilizers to maintain soil fertility is not sustainable. Inorganic fertilizers can have an overall acidifying effect on the soils. They are subject to global and local market fluctuations regarding prices and availability. Also, their cost efficiency in drylands is contingent on rainfall, which is often a risky proposition in arid climates. Second, organic sources alone may not suffice to meet the requirements of even a moderate yield and the increased demand for food. Also, organic fertilizers provide limited amounts of phosphorus and varying quantities of nitrogen as their quality in that regard differs greatly and their use should be adjusted accordingly. Third, mixed applications of inorganic and organic fertilizers are not only complementary but also highly synergistic. Organic matter increases water holding capacity and CEC and acts as a buffer against changes in soil pH. On the other hand, inorganic fertilizers enhance total biomass production in both root and shoot systems, in turn increasing the SOM. These organic sources can be locally produced on the farm from composting crop residues and animal manures and turning existing crop cover. They can also be from nonagricultural sources such as town reuse and industrial wastes. They may be cheaper and help mitigate the total cost of production.

Nutrient Conservation—Another critical component of INM is nutrient conservation, which can be achieved in two ways: first, making nutrient uptake more efficient and, second, developing soil conservation practices that prevent the physical loss of soil and nutrients. Ways to increase the efficiency of nutrient uptake include synchronizing the application to when nutrients are needed the most by the crops, placing fertilizers deeper in the soil to minimize loss of nitrogen to atmospheric volatilization, or using coated fertilizers to slow the release of nutrients susceptible to leaching, runoff, or volatilization. Furthermore, it is recommended to periodically test the soil to assess its fertility status and base nutrient applications on the crop needs accordingly. These and other similar practices are expected to improve nutrient uptake efficiency by as much as 20% in developing countries and 30% in the developed world (Bumb and Baanante, 1996).

Soil conservation techniques fall in two broad categories. The first general category of approaches aims to change the local physical properties to prevent soil and nutrients from being carried away by wind and water erosion. These practices include building physical structures such as terraces, which reduce soil erosion by changing the field physical characteristics (steepness and length of slopes). A more recent emphasis has been placed on practicing low-till or no-till farming, which seeks to maintain soil aggregation, surface characteristics, and water holding capacity, parameters that enable soils to better tolerate erosion. This is especially important since wind and water selectively carry finer soil particles that tend to be the richer in plant nutrients. The second category of soil conservation techniques involves developing physical barriers against erosion; mulch application, cover crops, and vegetation strips aim to diminish the direct impact of wind and rainfall intensity. A key step in soil conservation is to reduce or completely eliminate the time the land is under fallow, that is, left uncultivated for natural plants to replenish its fertility.

The promotion and adoption of INM in different parts of the world, and particularly in rural areas of developing countries, constitute a key step in soil fertility management. The consequences of soil fertility loss will reverberate across economic, social, and political sectors as higher food prices, increased rural poverty and famine relief, as well as threats of political unrests loom. Other necessary components toward increased global food productivity and food security include major institutional changes and effective extension and participation of all stakeholders. This constitutes a prime example of the interdisciplinary effort required to address a serious global issue and the central role soil science plays in it.

3.3.2 Case Study 2: Contamination of Soil, Sediment, and Water

Soil contamination is defined by the U.S. Environmental Protection Agency as the presence of hazardous substances in the naturally occurring soil. In Europe, the European

Commission has proposed the following definition of "contaminated site": a site where there is a confirmed presence, caused by human activities, of hazardous substances to such a degree that they pose a significant risk to human health or the environment, taking into account land use (Commission Proposal COM (2006) 232). Soil contamination typically occurs when there is improper use of chemical and/or waste handling, storage, and disposal. These substances cause soil contamination directly or indirectly by leaching in water that percolates through the soil and transports the contaminants both vertically and horizontally. It is also possible that atmospheric pollutants are deposited in the soil, causing an increase in their concentration; this is also described as "diffuse source" of soil contamination. Typically, point sources of soil contamination are more important and the focus of environmental management policies and actions; pesticide pollution is an important exception, addressed in a separate chapter.

The most important soil pollutants in Europe by frequency of occurrence according to the European Environment Agency are:

- Heavy metals (including As, Pb, Cd, Cr, Cu, Hg, Ni, and Zn)
- Mineral oil
- Polycyclic aromatic hydrocarbons (PAH)
- Benzene, toluene, ethylbenzene, and xylene (BTEX)
- Chlorinated hydrocarbons

In the United States, the Substance Priority List (SPL) contains the most frequent chemicals identified in the National Priorities List, which in turn contains the sites that enter the Superfund program.

This list is compiled by the USEPA and the Agency for Toxic Substances and Disease Registry (ATSDR) and updated yearly. Comparing the SPL with the five groups presented by the European Union (EU), it is apparent that the most frequent contaminants are identical in the industrialized world, given that they resulted from similar activities and practices.

Soil contamination results in the degradation of soil qualities and poses a risk to human health and the environment; these vary depending on the type of contamination present as well as soil properties. Potential effects include soil acidification, infertility, stunted plant growth, reduced crop yield, reduced microbial diversity, and others. Animals and humans may be affected through ingestion or dermal contact with the soil or with water contaminated by the soil pollutants. Regulation of contaminant levels in soil is typically based on the potential toxicity evaluated through toxicological studies. In the United States, the limits for soil pollutants are typically established as remediation standards by state environmental agencies and are specific to the end use of the site. For example, Connecticut regulations distinguish between Direct Exposure Criteria to protect human health when there is the possibility of direct contact (i.e., surface soil) and Pollutant Mobility Criteria for soil that has the potential to contaminated groundwater. These criteria vary depending on whether the site has residential or commercial/industrial use. In Europe, there are currently no EU-wide established criteria for soil contamination; however, various countries have established relevant thresholds.

Management of soil contamination is typically done on a site-specific basis, excluding pesticides and agricultural sources of contamination. However, there are also some cases where management is required at the regional scale for the so-called megasites; examples include the Hanford Site in Washington state and the Bitterfeld region in Germany. There are regulations that govern the investigation and remediation of contaminated sites, depending on the jurisdiction; for example, Superfund sites are regulated by the USEPA, while most other sites are regulated by state environmental protection agencies, both in the United States and in Europe. However, the development of site investigation and remedial action plans, as well as their evaluation in the broader socioeconomic context, is a complex process that requires in-depth knowledge of various disciplines, of which soil science is perhaps the most prominent.

Soil pollutants can interact with the soil in various ways, which may mitigate their impacts on the ecosystem and which directly influence the remedial strategies selected. Figure 3.7 provides an overview of soil–contaminant interactions and related examples. Soil can provide reactive compounds for the formation of insoluble precipitates. For example, soil minerals can exchange OH^- with the soil solution to form metal hydroxides when the pH is higher than approximately 5; phosphate present in the soil as apatite ($Ca_5(PO_4)_3(F,Cl,OH)$) can sequester metals either as insoluble phosphates such as pyromorphite ($Pb_5(PO_4)_3(F,Cl,OH)$) or by providing sorption sites. Sorption can retain cationic metals (Pb^{2+}, Zn^{2+}) in clay minerals for a wide pH range, while iron oxides typically complex cations at alkaline pH and oxyanions (CrO_4^{2-}, SO_4^{2-}, AsO_4^{-3}) at acidic and near-neutral pH. Sorption of organic contaminants depends on their hydrophobicity; typically, natural organic matter is a more efficient sorbent for common contaminants compared to soil minerals.

Soil may also contain reducing compounds, such as sulfide (S^{2-}) and ferrous iron (Fe^{2+}), if the redox environment allows for their stability. Reducing environments in soils may be encountered at contaminated sites with organic contamination, given that the biological degradation of the organics cause quick consumption of oxygen, which cannot always be replenished in the deeper soil horizons. In such cases, the presence of sulfides may cause sequestering of heavy metals as insoluble sulfides (e.g., CdS, PbS) or change

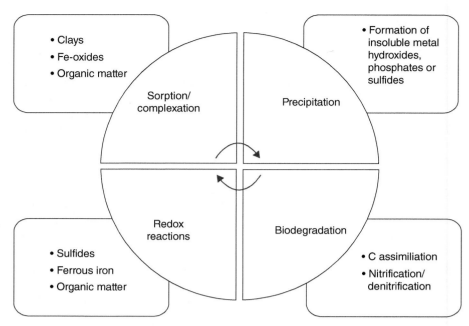

FIGURE 3.7 Overview of soil–contaminant interaction processes.

in the redox state of redox-active elements such as Cr and As. When hexavalent chromium is present as contaminant, the presence of naturally occurring organic matter, sulfide, and ferrous iron causes its transformation to nontoxic trivalent chromium, effectively constituting a natural attenuation process (Kozuh et al., 2000). Conversely, the reduction of As(V) to As(III) is an unwanted reaction, since As(III) is the more toxic and mobile form of arsenic in the environment (Brannon and Patrick, 1987).

Finally, soil is an important bacterial host and facilitates the transformation of organic contaminants to innocuous forms, notably carbon dioxide and water. However, in certain cases, biodegradation may increase the toxicity of contamination, for example, the reductive dechlorination of tetrachloroethene (PCE) and trichloroethene (TCE) may result in the persistence of vinyl chloride, a highly toxic compound (Siegrist et al., 2011). The rate and extent of biodegradation processes depend on a variety of factors: soil pH, redox potential, moisture content, availability of nutrients for bacterial growth, and the composition of the microbial community.

Collectively, these processes influence the fate and transport of contaminants in groundwater and also the remediation strategy chosen for a particular site. This is a very important aspect of the management of contaminated sites, which is oftentimes overlooked; in other words, the choice of remedial action does not depend only on the properties of the contaminant but also on the properties of the soil. Soil remediation strategies include chemical, physical, and biological processes, as shown in Fig. 3.8. The separation between physical and chemical treatments is artificial, given that the physical processes have a pronounced effect on soil

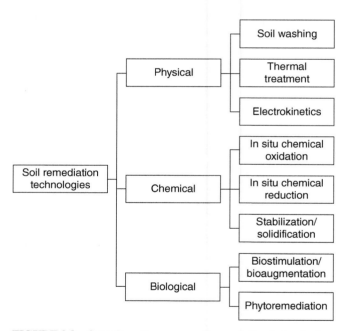

FIGURE 3.8 Overview of common soil remediation technologies.

chemistry; the distinction was based primarily on agent used for treatment, which does not directly involve chemicals for the physical processes. It is beyond the scope of this overview to provide details on each remediation technology; Sharma and Reddy (2004) and Lehr (2004) provide comprehensive descriptions of soil and water remediation technologies; the USEPA and the European Environment Agency websites are also important resources for related information.

Here, we will emphasize the fact that the chemistry of the native soil is influenced by the choice of the remedial action and that treatment success is highly dependent on soil physicochemical properties. As example for the former, *in situ* chemical reduction and the creation of an anoxic environment may lead to the mobilization of indigenous iron, which is then transported in the form of colloids and can mobilized other metals adsorbed on the colloidal fraction (Gschwend and Reynolds, 1987). Soil acidification and mobilization of secondary contaminants from the use of chemical treatment agents are also a common concern. Conversely, soil pH may prohibit the use of certain agents, for example, alkaline soils derived from limestone or dolomitic deposits are poor candidates for the use of iron-based chemicals such as zerovalent iron because of the tendency of iron to precipitate in alkaline environments.

Overall, understanding of soil processes is a key component in developing efficient and effective management strategies for contaminated soils. Not only is soil chemistry a fundamental aspect of designing remedial options but also a key parameter to take into account when considering the impacts of soil contamination and of the management practices themselves.

3.3.3 Case Study 3: Pesticide Pollution

Pollution by the widespread application of pesticides has been one of the major environmental quality issues since the 1960s, when the publication of *Silent Spring* by Rachel Carson raised awareness about the harmful biological effects of DDT and other pesticides. Although pesticide pollution concerns a variety of environmental media and is not strictly related to soil quality, soil plays a fundamental role in the fate and transport of pesticide compounds in the environment, and its properties should be taken into account when considering the negative impacts of pest control.

Pesticides are substances used to kill or control insects, weeds, fungi, rodents, bacteria, or other unwanted organisms (United States Geological Survey (USGS), 2007). Depending on the target organism, pesticides may be broadly classified into herbicides, insecticides, and fungicides, along with nematicides, molluscicides, fumigates, and a few other minor groups. Herbicides are by far the most widely applied group in developed countries, and their application exploded between 1965 and 1985 in the United States and other developed countries (USGS, 2007). At the same time, the use of insecticides dropped dramatically, both because several compounds proved to be acutely toxic and due to the development of genetic resistance by various organisms (Rathore and Nollet, 2012). However, insecticide use is still widespread in several developing countries, in order to combat diseases such as malaria. Overall, North America and Europe accounted for approximately half of all pesticide sales in 2008, while Latin America and Asia increased their market share from 11 and 16% in 1994 to 21 and 23% in 2008, respectively (Rathore and Nollet, 2012).

In addition to the differences in quantity, there are vast differences in the quality of the substances used in various countries. In the developed world, pesticides are heavily regulated due to the acute and chronic toxicity exhibited by various classes of compounds. The EU recently updated the Directive 98/8/EC on placing biocidal products on the market, and the current list of approved active substances contains only 42 compounds. In the United States, the Pesticide National Synthesis Project of the USGS listed 464 substances that were used as pesticides nationwide. The most widely used active substances in the United States are the herbicide glyphosate (*N*-(phosphonomethyl)glycine) that accounted for 32% of all pesticides used in agriculture in 2007 according to EPA estimates and the herbicide atrazine ($C_8H_{14}ClN_5$) with 13.5%; interestingly, atrazine is banned in the EU.

The USEPA also maintains a list of banned or severely restricted pesticides that currently includes 63 substances. The Stockholm Convention that entered into force in 2004 and is ratified by 160 governments has banned the use of 21 persistent organic pollutants (POPs), the majority of which are pesticides such as DDT, aldrin, chlordane, dieldrin, heptachlor, mirex, toxaphene, and lindane (as of 2009). Despite the bans, many of these chemicals are still used in African and Asian countries that have no or poorly enforced regulations and acute problems with transmittable diseases; some argue that the use of DDT to combat malaria is warranted if the relative impact is taken into account. Consequently, when Jaga and Dharmani (2003) compared DDT levels in human tissue globally, they observed higher levels in Africa and Asia compared to Western countries.

The health and ecological impacts of pesticides are varied. As mentioned previously, insecticides are typically more often acutely toxic compared to herbicides. There are four major classes of insecticides that vary in human toxicity and persistence in the environment: organochlorine and carbamate compounds are highly toxic and persistent, organophosphates are highly toxic but degradable, while pyrethroids have low mammalian toxicity (Rathore and Nollet, 2012). The toxic effects on humans range from simple skin irritations to lung and liver damage and damage to the nervous, reproductive, or immune system, depending on the particular compound. The US ATSDR contains detailed information about the toxicological profile of several pesticide substances, while the World Health Organization published a proposed classification of pesticides by hazard in 2010 (WHO, 2010). Richter (2002) reported an estimated 220,000 annual deaths and 26,000,000 cases of acute poisoning worldwide. In developed countries, exposure is mostly focused on the agricultural workforce, while in developing countries attempted suicides account for 2/3 of all acute pesticide poisonings (Richter, 2002).

In addition to human toxicity, pesticides have ecological impacts due to their toxicity to variety of organisms, from microbes all the way up the food chain to mammals. Given their ability to bioaccumulate, most pesticides persist in living organisms, disrupting ecosystem balances and threatening biodiversity. Pesticides may be inhaled, absorbed through the skin of humans or animals, persist as residue on plants and crops, or dissolve in water; thus, several exposure routes are viable for humans and other organisms.

Pesticide transport in the environmental occurs both through the atmosphere and the waterways, and certain compounds have the ability to travel large distances, finding their way even to the Arctic (Zhong et al., 2012). Bailey et al. (2000) showed that organochlorine pesticides (DDT, chlordane) could be transported through atmospheric circulation over the Pacific Ocean; wet deposition and bioaccumulation of such compounds can endanger the water quality and ecosystem health even in remote areas, where pesticides have never been used.

The transport behavior of pesticides in soils and waters is governed mainly by three processes: dissolution (solubility), sorption on soil and sediment particles, and (abiotic and biotic) degradation. The aqueous solubility of pesticides depends on the chemical structure, and values ranging from 10^{-6} to 10^6 mg/l have been reported in the literature (Arias-Estevez et al., 2008). Solubility is, in general, inversely proportional to sorption capacity, especially for neutral compounds that interact primarily with SOM. Sorption capacity is quantified through the soil–water partition coefficient K_{oc}, which is the distribution coefficient normalized over the organic carbon content of the soil $K_{oc} = K_d / TOC$. Arias-Estevez et al. (2008) showed that a near-perfect linear regression was observed between K_{oc} and aqueous solubility for nonionic pesticides, while compounds with basic or acidic functional groups (e.g., phenols, organic acids) have sorption behavior that is also dependent on pH and ionic strength. Such compounds present a much narrower range of solubility, with values between 10 and 1000 mg/l, that is, they are highly soluble in water.

The soil–water partition coefficient varies for a given compound and different soils; even though it is independent of the organic matter content, which is the controlling factor for sorption, there are other soil properties that influence its value. Kookana et al. (2008) reported atrazine K_{oc} values between 30 and 680 l/kg for 31 Australian soils. This has been attributed to various factors, including variation in the chemical properties of the SOM, the contribution of soil minerals to sorption, and organomineral interactions (Kookana et al., 2008). The analyses of Kookana et al. (2008) were performed on soils of similar mineralogy, highlighting the profound influence of the varying nature of SOM on sorption properties.

Degradation of pesticides can occur via hydrolysis, photolysis, or biodegradation (Rathore and Nollet, 2012). Hydrolysis is the reaction with water molecules, while photolysis is the reaction with UV radiation, both of which lead to breakdown of the original structure. Photolysis is only important in the soil surface, reaching only approximately 0.2–0.7 mm depth (Hebert and Miller, 1990). Biodegradation occurs through enzymatic breakdown of the compound by microbes that utilize it as carbon source. Biodegradation is largely dependent on sorption, since sorption has been observed to decelerate or inhibit microbial degradation for several compounds (Arias-Estevez et al., 2008). Additionally, this process has been observed to be time dependent, with pesticides becoming more resistant to degradation with time; this has been attributed to several aging mechanisms, including slow diffusion into soil pores and nanopores and irreversible sorption (Arias-Estevez et al., 2008). Biodegradation is generally described as an exponential function of time $C = C_0 \exp(-kt)$ from which the half-life of the compound may be expressed as $T_{1/2} = 0.693 / k$. The parameter k is empirical and is not a constant for a given compound, but it depends on the soil, temperature, water content, and other parameters. The half-life corresponds to the necessary time for 50% of the original concentration to be degraded; even though it is not a constant, it provides an order-of-magnitude estimation of the persistence of a particular pesticide in the environment. The half-life of some of the more persistent compounds may extend to years, for example, DDT has an average half-life of 2000 days (Hornsby et al., 1995). In comparison, atrazine has an average half-life of 10 days. As a result, legacy contamination of pesticides that have been banned for many years may still be present in waters and soils.

The USGS (2007) provides data on the presence of pesticides in US streams and groundwater in the period 1992–2001. 97% of water samples obtained over time in 186 streams had at least one pesticide compound present for agricultural and urban areas. Groundwater was observed to be less impacted, with approximately 60% detection in the same areas. In the same study, banned organochlorine compounds were detected with greater than 90% frequency in fish tissue and 60–80% detection in bed sediments. Most (>90%) of the detections in water involved concentrations lower than human health benchmarks. Agricultural streams located in the Corn Belt and Mississippi River Valley accounted for most concentrations that exceeded benchmarks, all involving atrazine (5 sites), cyanazine (4 sites), or dieldrin (2 sites), or 9 sites out of 83 streams sampled. Dieldrin was banned years before the sampling period, indicating the persistence of this organochlorine compound in natural waters. Overall, the likelihood of human exposure to significant concentrations of pesticides was found to be low. However, benchmarks of aquatic life and wildlife were violated in a very high number of cases, especially in agricultural areas; specifically, 57% of streams violated aquatic life benchmarks, and 87% of fish tissue violated wildlife benchmarks. Chlorpyrifos, azinphos-methyl, atrazine, and alachlor were the most frequent substances in those cases.

Konstantinou et al. (2006) summarized data from pesticide detection in various rivers across the EU, and herbicides such as atrazine were also the most frequently detected compounds, while lindane was the most frequent insecticide. Reported concentrations were generally in 100–1000 ng/l range in most cases for the reported compounds; as a measure of comparison, the drinking water limit in the United States is 3000 ng/l for atrazine, 2000 ng/l for alachlor, and 200 ng/l for lindane.

There are several BMPs in order to reduce environmental pollution from pesticide application. In the United States, state agencies typically publish fact sheets with suggested BMPs for farmers. Examples of suggested BMPs are:

- Application of lowest labeled pesticide rate.
- Rotation between pesticides within the same chemical family to avoid pest resistance.
- Avoid application of volatile chemicals at high temperatures, and in general, avoid windy and extreme weather conditions.
- Establish buffer zones in the vicinity of surface water and wells.
- Choose pesticide to be compatible with local soil conditions, topography, and climate.

3.3.4 Case Study 4: Global Warming and Soil Management

Soil is an important component in the global carbon cycle. Soil contains approximately 1500 Gt of organic carbon, which is double the amount of carbon in the atmosphere (750 Gt) and triple the amount of biotic C (560 Gt). Soil organic C undergoes various transformation processes, as illustrated in Fig. 3.9, most of which interact with atmospheric CO_2. The most important of these processes is soil respiration, which results in between 68 and 100 Gt of annual CO_2 emissions (Rustad et al., 2000). Soil respiration is counterbalanced by photosynthesis and carbon deposition as SOM due to litterfall, plant detritus, and plant and root exudates (Smith et al., 2008). Thus, changes in the carbon balance between the soil and the atmosphere can have substantial effect on the CO_2 concentration in the atmosphere (Kirschbaum, 2000; Lal, 2004; Smith et al., 2008).

Soil respiration is influenced by global warming due to a variety of factors, the most important of which are the effect of temperature and the effect of elevated CO_2 on plant growth and decomposition (Schlesinger and Andrews, 2000). These results in opposing trends; the increase in temperature results in an increase in the carbon decomposition rate based on kinetic theory, thus causing a net carbon loss. Conversely, plant grown in an elevated CO_2 warmer environment results in increase biomass production, and this increases return of C debris to the soil (Schlesinger and Andrews, 2000). Several climate change models have been developed to capture the entire carbon cycle (called C4 models); however, the uncertainty in contributions from soil and plant carbon remains quite high (Smith et al., 2008), resulting in differences in CO_2 concentration as large as 250 ppm (Friedlingstein et al., 2006). Changes in land use are also crucial in influencing the carbon balance, given that both the amount and the stability of the carbon input depend on the type of land use (Smith et al., 2008). The net effect of temperature change on soil respiration and whether soil is a net source or sink of CO_2 due to climate change are the subjects of heated debate, with dozens of studies devoted to them and a review paper published almost every year (e.g., Davidson and Janssens, 2006; Smith et al., 2008; Conant et al., 2011).

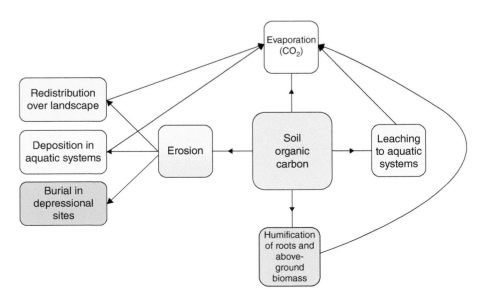

FIGURE 3.9 Soil organic carbon transformations. After Lal (2004).

Organic carbon from plant residues decomposes at different rates, depending on several factors, including plant residue composition, moisture, temperature, and soil biota (Smith et al., 2008). Forested areas tend to have large pools of organic C that is resistant to degradation, with grasslands having intermediate resistance and croplands minimum resistance to organic C decomposition (Guo and Gifford, 2002). Thus, different carbon pools have different residence times and variable response to changes in environmental factors. While an increase in temperature generally increases the rate of decomposition for all carbon pools, it is unclear to what extent these changes are significant in terms of the amounts released over space and time (Conant et al., 2011). In addition, changes in temperature and SOM properties cause changes in plant species and soil biota (Blankinship et al., 2011), whose feedback effects are currently poorly understood. Land management practices can also impact decomposition rates; for example, tillage increases carbon degradation rates due to breaking of macroaggregates that are subsequently exposed to increased weathering and microbial growth (Dawson and Smith, 2007). Intense agricultural activity results in significant carbon losses over time, the cumulative amount of which is estimated at 55–78 Gt C globally (Lal, 2004).

The spatial distribution of the impacts of global warming on soil C is also highly variable. The largest soil C pools (~50% of the total) are located in the Arctic, and 90% of it is buried in permafrost deposits (Tarnocai et al., 2009). Central America and certain areas in South America and Southeast Asia also contain large carbon deposits (FAO/IIASA/ISRIC/ISSCAS/JRC, 2012). Thus, the majority of soil carbon is located in areas with minimal direct anthropogenic influence. However, there are still several soil management strategies that could globally result in C sequestration.

Lal (2004) proposed that carbon sequestration in soils could result in C storage of 1 Gt/year globally to reach approximately 50–66% of the cumulative total C loss to date. Lal (2004) suggests that this is a necessary strategy irrespective of climate change, imposed by the large C losses that have caused degradation of soil quality in regions such as Central and South Asia including China, the Caribbean, Africa, and South America. Degradation in soil quality results in poor agricultural yield, hunger, and impoverishment. Carbon depletion in soil can be mitigated through the application of several management practices, including but not limited to:

- Reduced tillage of cropland soils
- Crop diversification and replacement by grassland and forest
- Manuring
- Grazing, fire, and nutrient management of range and grasslands
- Erosion control
- Rational water management and conservation

The overall purpose of these strategies is to increase carbon storage and soil quality by adding high amounts of biomass, minimizing disturbance of soil aggregates, enhancing diversity of fauna and flora, and strengthening elemental cycling mechanisms (Lal, 2004). Particular attention has been paid to the nitrogen cycle and its impact on the carbon cycle, and it has been suggested that nitrogen fertilization of forests may be an effective management strategy to stimulate carbon deposition (Holland et al., 1997). However, there is no evidence that this strategy may effective on a large scale, especially because the feedback on soil respiration is not well understood (Schlesinger and Andrews, 2000).

Overall, there is considerable uncertainty with respect to the interaction of soil organic carbon and climate change due to the complexity of soil physicochemical and biological processes and the large variability in soil properties across the world. While carbon sequestration in areas with intense agricultural activity appears to be an effective and feasible management strategy to combat the effects of both climate change and degradation of soil quality, the role of the large carbon deposits in northern peatlands and permafrost remains the subject of controversy and a field with great need for future research (Treat and Frolking, 2013).

REFERENCES

Appelo, C.A.J. and D.L. Parkhurst. 1998. Enhancements to the geochemical model PHREEQC—1D transport and reaction kinetics. Arehart, G.B. and J.R. Hulston. (eds). *Proceedings of the 9th international symposium on water-rock interaction, Taupo, New Zealand, 30 March–April 3, 1998*. Balkema, Rotterdam, p. 873–876.

Arias-Estevez, M., E. Lopez-Periago, E. Martinez-Carballo, J. Simal-Gandara, J.-C. Mejuto, and L. Garcia-Rio. 2008. The mobility and degradation of pesticides in soils and the pollution of groundwater resources. *Agriculture Ecosystems and Environment* 123:247–260.

Bailey, R., L.A. Barrie, C.J. Halsall, P. Fellin, and D.C.G. Muir. 2000. Atmospheric organochlorine pesticides in the western Canadian Arctic: evidence of transpacific transport. *Journal of Geophysical Research, [Atmospheres]* 105(D9):11805–11811.

Blankinship, J.C., P.A. Niklaus, and B.A. Hungate. 2011. A meta-analysis of responses of soil biota to global change. *Oecologia* 165:553–565.

Brannon, J. and W.H. Patrick. 1987. Fixation, transformation and mobilization of Arsenic in sediments. *Environmental Science and Technology* 21:450–459.

Bumb, B. and C. Baanante. 1996. *The role of fertilizer in sustaining food security and protecting the environment to 2020*, 2020 Vision Discussion Paper 17. IFPRI, Washington, DC.

Chow, V.T. and L. Mays. 2013. *Applied hydrology*, 2nd Edition. McGraw Hill, New York.

Clark, M.M. 2009. *Transport modeling for environmental engineers and scientists*, 2nd Edition. John Wiley & Sons, Inc., Hoboken, NJ.

Conant, R.T., M.G. Ryan, G.I. Agren, H.E. Birge, E.A. Davidson, P.E. Eliasson, S.E. Evans, S.D. Frey, C.P. Giardina, F.M. Hopkins, R. Hyvönen, M.U.F. Kirschbaum, J.M. Lavallee, J. Leifeld, W.J. Parton, J.M. Steinweg, M.D. Wallenstein, J.Å.M. Wetterstedt, and M.A. Bradford. 2011. Temperature and soil organic matter decomposition rates – synthesis of current knowledge and a way forward. *Global Change Biology* 17:3392–3404.

Davidson, E.A. and I.A. Janssens. 2006. Temperature sensitivity of soil carbon decomposition and feedbacks to climate change. *Nature* 440:165–173.

Dawson, J.J.C. and P. Smith. 2007. Carbon losses from soil and its consequences for land-use management. *Science of the Total Environment* 382(2–3): 165–190.

FAO/IIASA/ISRIC/ISSCAS/JRC. 2012. *Harmonized World Soil Database (version 1.2)*. FAO/IIASA, Rome/Laxenburg.

Friedlingstein, P., P. Cox, R. Betts, L. Bopp, W. von Bloh, V. Brovkin, P. Cadule, S. Doney, M. Eby, I. Fung, G. Bala, J. John, C. Jones, F. Joos, T. Kato, M. Kawamiya, W. Knorr, K. Lindsay, H. D. Matthews, T. Raddatz, P. Rayner, C. Reick, E. Roeckner, K.-G. Schnitzler, R. Schnur, K. Strassmann, A.J. Weaver, C. Yoshikawa, and N. Zeng, 2006. Climate–carbon cycle feedback analysis: results from the C4MIP model intercomparison. *Journal of Climate* 19:3337–3353.

Goldberg, S., L.J. Criscenti, D.R. Turner, J.A. Davis, and K.J. Cantrell. 2007. Adsorption-desorption processes in subsurface reactive transport modeling. *Vadose Zone Journal* 6:407–435.

Gruhn, P., F. Goletti, and M. Yudelman. 2000. *Integrated nutrient management, soil fertility and sustainable agriculture: current issues and future challenges*, Food, Agriculture and the Environment Discussion Paper 32. IFPRI, Washington, DC.

Gschwend, P.M. and M.D. Reynolds. 1987. Monodisperse ferrous phosphate colloids in an anoxic groundwater plume. *Journal of Contaminant Hydrology* 1:309–327.

Guo, L.B. and R.M. Gifford. 2002. Soil carbon stocks and land use change: a meta-analysis. *Global Change Biology* 8:345–360.

Gustafsson, J.P. 2010. Visual MINTEQ ver. 3.0. Available at http://www2.lwr.kth.se/English/OurSoftware/vminteq/index.htm. Verified April 16, 2015.

Hebert, V. and G. Miller. 1990. Depth dependence of direct and indirect photolysis on soil surfaces. *Journal of Agricultural and Food Chemistry* 38(3):913–918.

Holland, E.A., B.H. Braswell, J.-F. Lamarque, A. Townsend, J. Sulzman, J.-F. Muller, F. Dentener, G. Brasseur, H. Levy, J.E. Penner, and G.-J. Roelofs. 1997. Variations in the predicted spatial distribution of atmospheric nitrogen deposition and their impact on carbon uptake by terrestrial ecosystems. *Journal of Geophysical Research* 102:15849–15866.

Holtz, R.D., W.D. Covacs, and T. Sheehan. 2010. *Introduction to geotechnical engineering*, 2nd Edition. Prentice Hall, Upper Saddle River, NJ.

Hornsby, A.G., R.D. Wauchope, and A.E. Herner. 1995. *Pesticide properties in the environment*, Springer-Verlag, New York.

Jaga, K. and C. Dharmani. 2003. Sources of exposure to and public health implications of organophosphate pesticides. *Revista Panamericana de Salud Pública* 14(3):171–185.

Kirschbaum, M. 2000. Will changes in soil organic carbon act as a positive or negative feedback on global warming? *Biogeochemistry* 48:21–51.

Konstantinou, I., D. Hela, and I. Albanis. 2006. The status of pesticide pollution in surface waters (rivers and lakes) of Greece. Part I. Review on occurrence ad levels. *Environmental Pollution* 141:555–570.

Kookana, R., L. Janik, M. Forouzangohar, and S. Forrester. 2008. Prediction of Atrazine sorption coefficients in soils using mid-infrared spectroscopy and partial least squares analysis. *Journal of Agricultural and Food Chemistry* 2008(56):3208–3213.

Kozuh N., J. Stupar, and B. Gorenc. 2000. Reduction and oxidation processes in soils. *Environmental Science and Technology* 34:112–119.

Lal, R. 2004. Soil carbon sequestration impacts on climate change and food security. *Science* 304:1623–1627.

Lehr, J.H. 2004. *Wiley's remediation technologies handbook*. John Wiley & Sons, Inc., Hoboken, NJ.

Mason, N. and L.G. Berry. 1968. *Elements of mineralogy*. W.H. Freeman & Co Ltd, New York.

Massmann, J. and D.F. Farrier. 1992. Effects of atmospheric pressures on gas transport in the vadose zone. *Water Resources Research* 28:777–791.

National Research Council. 1993. *Soil and water quality: an agenda for agriculture*. National Academy Press, Washington, DC.

Raats, P.A.C, M.Th. van Genuchten. 2006. Milestone in soil physics. *Soil Science* 171:S21–S28.

Rathore, H. and L. Nollet. 2012. *Pesticides: evaluation of environmental pollution*, CRC Press, Taylor & Francis Group, Boca Raton, FL.

Richter, E.D. 2002. Acute human pesticide poisonings. In: *Encyclopedia of pest management*. Edited by D. Pimentel CRC Press, Boca Raton, FL.

Rumynin, V.G. 2011. *Subsurface solute transport models and case histories, theory and applications of transport in porous media 25*. Springer Science+Business Media B.V., Dordrecht.

Rustad, L.E., T.S. Huntington, and R. Boone. 2000. Controls on soil respiration: impacts for climate change. *Biogeochemistry* 48:1–6.

Scanlon, B.R., J.P. Nicot, and J.M. Massmann. 2001. Soil gas movement in unsaturated systems. In: *Soil physics companion*. Edited by A.W. Warrick. CRC Press, Boca Raton, FL.

Schlesinger, W.H. and J.A. Andrews. 2000. Soil respiration and the global carbon cycle. *Biogeochemistry* 48:7–20.

Selker J.S., J.T. McCord, and C.K. Keller. 1999. *Vadose zone processes*. Lewis Publishers, CRC Press, Boca Raton, FL.

Sharma, H.D. and K. Reddy. 2004. *Geoenvironmental engineering: site remediation, waste containment and emerging waste management technologies*, John Wiley & Sons, Inc., Hoboken, NJ.

Siegrist, E.L., M. Crimi, and T.J. Simpkin. (eds). 2011. *In situ chemical oxidation for groundwater remediation.* Springer, New York.

Smith, P., C. Fang, J. Dawson, and J.B. Moncrieff. 2008. Impact of global warming on soil organic carbon. *Advances in Agronomy* 97:1–43.

Sparks, D.L. 2003. *Environmental soil chemistry,* 2nd Edition. Academic Press, San Diego, CA.

Stephens, D.B. 1995. *Vadose zone hydrology.* Lewis Publishers, CRC Press, Boca Raton, FL.

Tarnocai, C., J.G. Canadell, E.A.G. Schuur, P. Kuhry, G. Mazhitova, and S. Zimov. 2009. Soil organic carbon pools in the northern circumpolar permafrost region. *Global Biogeochemical Cycles* 23:GB2023.

Treat, C.C. and S. Frolking. 2013. A permafrost carbon bomb? *Nature Climate Change* 3:865–866.

USDA Soil Conservation Service. 1972. *National engineering handbook, Section 4: Hydrology,* Chapters 4–10. Soil Conservation Service, U.S. Department of Agriculture. Washington, DC, 15-7–15-11.

USGS (United States Geological Survey). 2007. *Pesticides in the Nation's streams and ground water, 1992–2001,* Circular 1291. U.S. Geological Survey, Reston, VA.

Viessman Jr, W. and G.L. Lewis. 2002. *Introduction to hydrology,* 5th Edition. Prentice Hall, Upper Saddle River.

Walton, J.C. 2008. *Fate and transport of contaminants in the environment,* 1st Edition. College Publishing, Glen Allen, VA.

World Health Organization (WHO). 2010. *The WHO classification of pesticides by hazards, and guidelines to classification 2009.* Wissenschaftliche Verlagsgesellschaft mbH, Stuttgart.

Williams, J.R., N. Kannan, X. Wang, C. Santhi, and J.G. Arnold. 2012. Evolution of the SCS runoff curve number method and its application to continuous runoff simulation. *Journal of Hydrologic Engineering* 17(11):1221–1229.

Zhong, G., Z. Xie, M. Cai, A. Moller, R. Sturm, J. Tang, G. Zhang, J. He, and R. Ebinghaus. 2012. Distribution and air–sea exchange of current-use pesticides (CUPs) from East Asia to the high Arctic Ocean. *Environmental Science and Technology* 46(1):259–267.

4

GREEN CHEMISTRY AND ECOLOGICAL ENGINEERING AS A FRAMEWORK FOR SUSTAINABLE DEVELOPMENT

SHYAM R. ASOLEKAR[1], R. GOPICHANDRAN[2], ANAND M. HIREMATH[1] AND DINESH KUMAR[1]

[1] *Centre for Environmental Science and Engineering, Indian Institute of Technology Bombay, Powai, Mumbai, India*
[2] *Vigyan Prasar, Department of Science and Technology, Government of India, New Delhi, India*

Abstract: This chapter addresses the need for green chemistry, green engineering and ecological engineering to enable sustainable development. The highpoint of this chapter is the real-life case studies it provides to substantiate the above These deal with material efficiency and use of condemned resources to minimize consumption of non-renewable resources as raw materials; essential to sustain development. The nexus between trade and environment in the context of sustainable development and finally some strategies for transitioning to a sustainable future are also articulated.

Keywords: green chemistry, green engineering, ecological engineering, material efficiency, waste to energy, ship recycling, natural treatment systems, rejuvenation of lakes and rivers, trade and environment, social engineering, sustainable development, landfill mining, wastewater treatment and reuse, risks and hazards, principles of green chemistry, application of green engineering, organic food, constructed wetland, duckweed ponds, ship breaking.

4.1	Green Chemistry	97		
	4.1.1 Principles of Green Chemistry	98		
	4.1.2 Engineering Applications of Concepts from Green Chemistry	99		
	4.1.3 Reduction of Risk and Hazard	99		
	4.1.4 Green Chemistry and Sustainability Parameters	99		
4.2	Green Engineering	102		
	4.2.1 Principles of Green Engineering	102		
	4.2.2 Significance of Wastewater Treatment and Reuse	103		
	4.2.3 Need for Application of Green Engineering in Agriculture and Food Processing	103		
	4.2.4 Case Examples Illustrating Significance of GC&E	106		
4.3	Material Efficiency	106		
4.4	Unlocking Condemned Resources	110		
	4.4.1 Landfill Mining	110		
	4.4.2 Recycle of Low Chemical Potential Substances	111		
	4.4.3 Eco-Friendly Energy Generation from Biomass and Green Chemistry	113		
4.5	Ecological Engineering and Social Sustainability	113		
4.6	Case Study 1: Wastewater Treatment Systems as Microcosms of Eco-System	114		
4.7	Case Study 2: Integrating Natural Treatment Systems for Upgradation of Aquatic Systems	117		
4.8	Case Study 3: Sustainable Recycling of Obsolete Ships Through Ecological Engineering	118		
4.9	Nexus Between Trade and Environment	122		
4.10	Closure: Strategies for Transitioning to Sustainable Futures	122		
	Acknowledgments	123		
	Excercise	123		
	References	123		

4.1 GREEN CHEMISTRY

The concept of "sustainable development" was first put forth by the *Bruntland Commission* in 1987 (UN Documents, 1987). It was articulated and disseminated through the deliverables produced during the first *Earth Summit* titled: World Summit for Sustainable Development (WSSD). Then Summit took place at Rio-de Jenerio, Brazil in June, 1992 under the auspices of United Nations. Concurrently, the chemical sector grew leaps and bounds all over the world—especially in India and China. Steep competition to sell products at the cheapest price forced the chemical sector to develop newer chemical synthesis routes, optimize production processes and manufacture large quantities of cheaper

---

An Integrated Approach to Environmental Management, First Edition. Edited by Dibyendu Sarkar, Rupali Datta, Avinandan Mukherjee, and Robyn Hannigan.
© 2016 John Wiley & Sons, Inc. Published 2016 by John Wiley & Sons, Inc.

products. The chemical sector also came under increased scrutiny of environmental regulators and increasing criticism by community and environmental activists. Clearly, the times have changed since an era where production trumped everything! When the cost of pollution outweighs the benefits of economic growth brought about by the chemical sector, it is no longer ignored. Growth and development at the cost of environment and increasing ecological degradation is not appreciated anymore (Arceivala and Asolekar, 2006, 2012; Asolekar et al., 2014).

There are two basic reasons behind this enlightened approach to development. First, owing to the seminal contribution of system science of earth and ecology and in the light of several fundamental insights into the "working of nature", ecology and bio-diversity are being valued far more than any short term profits. In fact, in the *Supreme Court of India* as well as in the *International Court* at The Hague, Netherlands, the Courts have consistently upheld the "precautionary principle" in environmental matters and reinforced the superiority and priority of protection of ecology and bio-diversity without paying heed to the economics of pollution control. Second, it is now clear that subjecting the earth system and associated eco-systems to exploitative consumption of resources as well as indiscriminate chronic and episodic disposal of wastes creates irreversible damage to the system as a whole. (Asolekar and Gopichandran, 2005).

Clearly, a sustainable solution is one that minimizes consumption of resources without compromising growth. This challenge of slowing resource depletion and pollution of environment can be addressed using the approach of green chemistry and ecological engineering.

UNEP (2004a & b) defined green chemistry as "the development of greener technology to convert new and renewable resources into valuable products in a sustainable manner". Some of the related management implications are reflected in the Strategic Approach to International Chemicals Management (SAICM) endorsed by the World Summit on Sustainable Development in 2002. Rainer (2004) defined the role of producers, distributors, end users, government and other stakeholders for successful implementation of SAICM with a special emphasis on chemical safety. The REACH (Registration Evaluation & Authorization of Chemicals) concept was a logical complement (Hansson and Rudén, 2004) followed by the Globally Harmonized System (GHS) (Lotter, 2004) and the Design for Environment (DfE) approach (Ferat, 2004).

Green chemistry is the key to sustainable development as it will lead to new solutions to existing problems. It is the way to design chemicals with lesser environmental hazards and production process more sustainable. Green chemistry is applied from the very first stage of selection of raw materials throughout the product life cycle and includes design, manufacture, use, and ultimate disposal. Moreover, it presents opportunities for new processes and products with scientific and technological innovation.

4.1.1 Principles of Green Chemistry

Anastas and Warner in their book *Green Chemistry: Theory and Practice (1998)* defined green chemistry as the utilization of a set of principles that reduces or eliminates the use or generation of hazardous substances in the design, manufacture and application of chemical products.

Over the last few decades, environmental regulations as well as actions of environmental lobbyists forced industries to look at "green chemistry," "Clean Technology" and "Ecological Engineering" to minimize industrial pollution and ensure a sustainable future. Conventional end-of-pipe solutions reduce environmental impact of industrial activity without addressing the underlying causes. In comparison, "clean technology" solutions emphasize economically advantageous optimization of material and energy use and avoid wastes and emissions, to reduce environmental impact (Lancaster, 2002).

Ecological engineering is equally applicable in ecological as well as constructed simulation of natural system to achieve the engineering goals (Teal as cited by Kangas, 2003). Ecological engineering ecosystems are designed, constructed and operated to solve environmental problems instead of using conventional technologies. The goal of environmental engineering is to combine ecology with technologies for betterment of nature and human welfare. It is apparent that green chemistry, clean technology and ecological engineering are quite interconnected with each other and represent a comprehensive approach for sustainability.

During the 1990s, the US Environmental Protection Agency (EPA) strongly advocated use of green chemistry for pollution prevention; especially to industry and academia. In association with American Chemical Society the EPA developed 12 Guiding principles:

Principle 1. Prevention: It is better to prevent waste than to treat or clean up waste after it has been created.

Principle 2. Atom Economy: Synthetic methods should be designed to maximize incorporation of all materials used in the process into the final product.

Principle 3. Less Hazardous Chemical Syntheses: Wherever practicable, synthetic methods should be designed to use and generate substances that possess little or no toxicity to human health and the environment.

Principle 4. Designing Safer Chemicals: Chemical products should be designed to affect their desired function while minimizing their toxicity.

Principle 5. Safer Solvents and Auxiliaries: The use of auxiliary substances (e.g., solvents, separation agents, etc.) should be made unnecessary wherever possible and innocuous when used.

Principle 6. Design for Energy Efficiency: Energy requirements of chemical processes should be recognized for their environmental and economic impacts and should be minimized. If possible, synthetic methods should be conducted at ambient temperature and pressure.

Principle 7. Use of Renewable Feedstocks: A raw material or feedstock should be renewable rather than depleting whenever technically and economically practicable.

Principle 8. Reduce Derivatives: Unnecessary derivatization (use of blocking groups, protection/deprotection, temporary modification of physical/chemical processes) should be minimized or avoided if possible, because such steps require additional reagents and can generate waste.

Principle 9. Catalysis: Catalytic reagents (as selective as possible) are superior to stoichiometric reagents.

Principle 10. Design for Degradation: Chemical products should be designed so that they break down into innocuous degradation products at the end of their function and do not persist in the environment.

Principle 11. Real-time Analysis for Pollution Prevention: Analytical methodologies need to be further developed to allow real-time, in-process monitoring and control prior to the formation of hazardous substances.

Principle 12. Inherently Safer Chemistry for Accident Prevention: Substances and the form of a substance used in a chemical process should be chosen to minimize the potential for chemical accidents, including releases, explosions, and fires.

Warner et al. (2004) presented a comprehensive review of the origin of green chemistry and the applications of the 12 principles stated above.

4.1.2 Engineering Applications of Concepts from Green Chemistry

It is understood at the outset that describing theory of green chemistry alone may not be sufficient to illustrate the significance and range of potential engineering applications. Thus, in order to visualize engineering applications of some selected green chemistry concepts and to showcase their implications in day-to-day practices the following concepts along with their working principles and the corresponding potential applications are listed in Table 4.1.

The above table identifies possible applications of a few select technologies in conjunction with the principles of green chemistry. These help maximize resource conservation, improve health and safety attributes for workplace, increase material efficiency, and encourage use of by-products.

4.1.3 Reduction of Risk and Hazard

"Risk" may be defined as the probability of occurrence of an accident and its consequence. Conventionally, it is expressed as the product of the probability of occurrence (p) and its consequence (c). Further, hazard is any situation or condition, which has the potential to cause harm. The chemical and process industry sectors produce several essential as well as lifestyle chemicals, by-products and intermediates that are further used in synthesizing chemicals and products to make our day-to-day life interesting. It is to be noted that many of these chemicals and their raw materials, by-products and intermediates are hazardous in nature. It would not be pragmatic to take a stand that these chemicals should be banned from manufacturing. The only alternative in such situations is to manufacture these hazardous chemicals in a safe manner with minimum risk to work place and surrounding environment.

You may recall that one such tragedy that took place in a pesticide manufacturing petro-chemical plant named Union Carbide India Limited situated in City of Bhopal in central India in the State of Madhya Pradesh in the early hours of December 3, 1984. Thousands of people succumbed to a lethal gas after the accident in a vast area adjoining the premises of the manufacturing plant and many thousands were affected for years to come. Detailed accounts of the events leading to the accidents have been reported and discussed widely (Eckerman, 2001, 2005, 2013 as cited by URL 06, 2014; Lancaster, 2002; Asolekar and Gopichandran, 2005). This infamous accident has been the most expensive lesson learnt by the chemical process industry in a world that produces chemical products with the help of many hazardous chemicals and raw materials.

On the basis of a seminal investigation conducted by Eckerman (2001, 2005, 2013 as cited by URL 06, 2014) the cause of accident was identified as the release of approximately 42 tons of methyl iso-cynate (MIC) gas from the storage tank. Poor maintenance of the tank resulted in creation of an explosive mixture and the resulting cloud of toxic gas travelled in the downwind direction and dispersed. The possible composition of the gas cloud could have been phosgene, hydrogen cyanide, carbon monoxide, hydrogen chloride, oxides of nitrogen, mono-methyl amine (MMA) and carbon dioxide, either produced in the storage tank or in the atmosphere.

Clearly, the risk to the environment and human life is related to hazard and exposure to hazard. If the producer knows the hazard beforehand (e.g. toxicity of MIC), then reducing the exposure through protective equipment, safety devices and other methods of controlling the risk could have been possible. The concept of inherently safer design dictates that instead of controlling the exposure of the harmful chemical; the focus should be on removing the hazard from the process itself. In the case of the Bhopal tragedy, such a design involving large storages of toxic gases and thus the accident could have been avoided. Phosgene could have been made on site (using the "Just in Time" concept) and the use of MIC could have been avoided by initially reacting naphthol with phosgene and subsequently reacting the so-formed chloroformate with methylamine (Lancaster, 2002).

4.1.4 Green Chemistry and Sustainability Parameters

It is well known that chemistry deals with the science of matter and its transformation. It is the heart of most processes as well as the bridge between physics, material sciences and life sciences (Scholz-Boettcher et al., 1992). Raw materials and resources are used to create products of value to society.

TABLE 4.1 Different Concepts Utilized in Green Chemistry and Their Respective Principles and Potential Applications

Sr. No.	Concept	Principles	Potential Applications
1	Continuous flow stirred tank reactor	A continuous flow reactors fall into two categories, In the continuous flow stirred tank reactor (CFSTR); contents are stirred uniformly, and the concentration of various components remains same and decreases with time.	This technology dominates the field of wastewater treatment; in petrochemical industry and for drug discovery (Watts and Haswell, 2003), p-Iodinations in hydrocarbon media with the application of continuous flow reactor, (Slocuma et al., 2011), Production of amines (pharmaceuticals) from alcohols in a continuous flow fixed bed catalytic reactor (Lamb et al., 2010), Synthesis of heterocyclic compounds using continuous flow reactors (Watts and Wiles, 2013).
2	Micro-channel reactor	Number of small, parallel channels in μm range, short distance to wall, high surface/volume ratio, enhanced heat and mass transfer properties, especially suited to economical production on a small scale (URL 05, 2014).	Micro-channel reactors whereby reaction volumes are kept small and scale is highly flexible thus reducing hazards and risk. Examples of its utilization are: methanol synthesis, simulation of exhaust gas, reforming of propane in a heat exchange, integrated micro-channel reactor, preparation of mono-disperse biodegradable magnetic microspheres using a T-shaped micro-channel reactor (Liua et al., 2014).
3	Intensive processing systems		Intensive processing systems such as spinning disc reactors, which combine the benefits of low reaction volumes with excellent heat transfer and mixing characteristics. Process intensification has emerged as a promising field, which can effectively tackle the challenges of significant process enhancement, whilst also offering the potential to diminish the environmental impact presented by the chemical industry (Boodhoo and Harvey, 2013).
4	Electro-coagulation		EC systems provide environment-friendly and cost-effective results for sustainable water reuse, it is disconcerting to realize that many places in the world continue to use freshwater for cleaning purposes. Using dimensionally stable anode (Ti/Ru0.3Ti0.7O_2)—contrary to EC which uses sacrificial anodes—and iron cathode, proved that a simple electrochemical method may be effective for turbidity removal prior to reverse osmosis for seawater desalination (Rodriguez et al., 2007).
5	Advanced oxidation processes	Complete/partial degradation of recalcitrant organic compounds by hydroxy (HO.) or peroxy radical (HO_2). It does not generate secondary waste as in phase transfer processes (Andreozzi et al., 1999).	It has been used in oxidant chemicals for a range of water treatment purposes. To replace or reduce chlorine because of the concern that leads to formation of halogenated by-product compounds. The predominant oxidant applied so far has been ozone. There have been successful applications of chlorine dioxide and potassium permanganate in the treatment of surface waters. Ozone has been of particular interest because of its ability to degrade pesticide compounds and other organic micro-pollutants. However, typical water treatment conditions limit the effectiveness of ozone treatment by minimizing the generation of highly reactive radical species. from ozone. Research interest is currently focused on methods of enhancing radical formation, including combinations of ozone with either hydrogen peroxide, UV-irradiation, metal catalysts or activated carbon. Other treatment chemicals that combine both oxidation and coagulation/precipitation capabilities are also under active study at present.

TABLE 4.1 (Continued)

Sr. No.	Concept	Principles	Potential Applications
6	Sono-chemistry	Sono-chemistry is the use of ultrasonic non-hazardous acoustic radiation on chemical reactions.	Sonication enables the rapid dispersion of solids, decomposition of organics including biological components, as well as the formation of porous materials and nano-structures; It is applied in rate and yield improvements of diverse type of reactions, easy generation of reactive species, safer reaction conditions, preparation of micro materials and nanostructures (Cintas and Luche, 1999).
7	Microwave	Microwave heating is based on non-contacted irradiation of electromagnetic waves in a frequency range from 300 MHz to 300 GHz which activate dipole molecules depending upon their dielectric constants and microwave sensitizers which induce electron.	Solvent free methods like thermolysis of starch, amidation of carboxylic acids, isomerisation of eugenol etc. (Li et al., 2013). Pressurized microwave systems like etherification. High temperature reactions like hydrolysis of cellulose. Reaction with high temperature water like addition of water to olefins etc. Metal-catalyzed process. Application of microwave technology for utilization of recalcitrant biomass, Solvent-free microwave-extraction techniques of bioactive compounds from natural products, Solvent-free microwave organic synthesis.
8	Green catalyst	Green catalysis is a fundamental part of green chemistry, the design of chemical products and processes that diminish or eradicate the use and generation of hazardous substances and provide environmental protection and economic benefit (Anastas et al., 2001).	Alumina-sulfuric acid catalyzed eco-friendly synthesis of xanthenediones, Reduction with cadmium and diazo-coupling reaction for nitrate determination in waters, supercritical carbon dioxide to decaffeinate coffee.
9	Bio-catalysts	Biocatalysts are enzymes or whole cells which used for biocatalysis in industrial synthetic chemistry.	Biocatalysts are used to know about reaction kinetics, substrate specificity and operational stability, probes immobilization, production of acetic acid from ethanol with an immobilized Acetobacter strain, synthesis of 3,4-dihydroxylphenyl alanine (DOPA) with the catalysis of an oxidoreductase in pharmaceutical industry (DOPA is used in Parkinson's disease), synthesis of oligosaccharides with the help of glycosyltransferases bio-catalyst, use of hydrolases bio-catalyst in the organic synthesis like pesticide synthesis.
10	Phase transfer catalysts	Phase transfer catalysis (PTC) uses catalytic amounts of phase transfer agents which facilitate interphase transfer of species, making reactions between reagents in two immiscible phases possible.	PTC is used widely in the synthesis of various organic chemicals in both liquid-liquid and solid-liquid systems. The use of PTC combined with other rate enhancement techniques like sonochemistry, microwaves, electroorganic synthesis, and photochemistry, is being increasingly explored.
11	Membrane filtration	Membrane separation is practiced on feed streams ranging from gases to colloids. Microfiltration (MF) membranes are used to retain colloidal particles as large as several micrometers. For specific applications even ultrafilteration (UF) and nanofilteration (NF) membranes are used.	Membrane reactors that can maintain separation of aqueous and non-aqueous phases, hence simplifying the normally waste intensive separation stages of a process, separation of CO_2/CH_4 from gas stream by the use membrane separation technology, Removal of vapors from gas/vapor mixtures, separation of hydrocarbons (benzene)/metals for selective recovery of valuable metals etc.
12	Reverse osmosis membrane	Reverse osmosis.	Primarily used for separation of dissolved solids and molecules from solutions (Garud et al., 2011; Lee et al., 2011).

One of the sustainable parameters in green chemistry is careful optimization of resources to generate maximum efficiency. Application of this concept will lead to continued development of sustainable products.

Solvent substitution is an important green chemistry application. Some of the best sources of information on solvent substitution include Cue and Zhang (2009), Peric et al. (2012) on ecotoxicity and biodegradability of ionic liquids, Talaviya and Majmudar (2012), Macmillan et al. (2013), initiatives by Merck (2012) in pharmaceuticals, the textile dyeing operations (Gandhi et al., 2013) the Nike Materials Sustainability Index through its Sustainable Apparel Coalition (NIKE, Inc., 2012), use of enzyme in industrial production systems (Gaber, 2012; Jegannathan and Nielsen, 2012). The work at the Finnish Centre of Excellence in White Biotechnology on biopolymers and related efficient production of materials and chemicals from renewable natural resources is yet another example of green chemistry in action. In this context "White"/industrial biotechnology is aligned with the goals and approaches of green chemistry to develop sustainable production processes that twin environmental and energy conservation approaches.

Research and innovation are central to developing processes that fulfill the requirements of environmental and economic sustainability. Sustainability-related sensitivity and responsible thinking needs to be integrated into the problem-solving framework from the very beginning so that innovative technological solutions for the desired product or service can be developed.

4.2 GREEN ENGINEERING

In the recent years, people have started focusing on minimizing the negative impacts of anthropogenic activities on planet earth to complement its ability to sustain life. One of the aims of environmental science could be to incorporate green chemistry and create newer processes and products with the help of newer designs, materials and alternative energy sources for sustainable development. Such an endeavor can certainly be included under the umbrella of "Green Engineering". In addition, several approaches have been adopted including waste minimization, recycling of intermediates, by-products, solvents, water, metals and materials as well as gravitating to sustainable consumption. These approaches are also lately referred to as green engineering. The real challenge is the identification of scientific, social, economic, ecological and environmental variables that impact the system. Therefore, some principles need to be articulated and followed in a systematic manner to produce desired end results.

4.2.1 Principles of Green Engineering

As stated earlier, the application of the green chemistry principles would indeed help in achieving the objectives set forth by green engineering. What are the other tenets of green engineering? The study commissioned by the US EPA has made noteworthy contributions in articulation of the principles (URL 07) and subsequently by Anastas and Warner (1998) and Anastas and Zimmerman (2003). Those 12 principles of green engineering are reproduced verbatim below:

Principle 1. Designers need to strive to ensure that all materials and energy inputs and outputs are as inherently nonhazardous as possible.

Principle 2. It is better to prevent waste than to treat or clean up waste after it is formed.

Principle 3. Separation and purification operations should be designed to minimize energy consumption and materials use.

Principle 4. Products, processes, and systems should be designed to maximize mass, energy, space, and time efficiency.

Principle 5. Products, processes, and systems should be "output pulled" rather than "input pushed" through the use of energy and materials.

Principle 6. Embedded entropy and complexity must be viewed as an investment when making design choices on recycle, reuse, or beneficial disposition.

Principle 7. Targeted durability, not immortality, should be a design goal for products. After useful use of a product to disintegrate under natural conditions.

Principle 8. Design for unnecessary capacity or capability (e.g., "one size fits all") solutions should be considered a design flaw.

Principle 9. Material diversity in multi component products should be minimized to promote disassembly and value retention.

Principle 10. Design of products, processes, and systems must include integration and interconnectivity with available energy and materials flows.

Principle 11. Products, processes, and systems should be designed for performance in a commercial "afterlife."

Principle 12. Material and energy inputs should be renewable rather than depleting.

Anastas and Warner (1998) articulated these principles and methodologies to accomplish the goals of environmentally friendly and cost-effective design. It is well known that the 12 principles listed above are the most powerful enablers on our journey to sustainable futures because on one-hand they address optimization at molecular level and on other-hand the principles address system level reengineering. In other words, the key to achieving sustainability lies in the design of molecules, systems, processes and products through logical integration of the 12 green engineering principles. According to Allen and Shonnard (2001), this logical integration of green engineering principles will eventually bring benefit to the surrounding environment, economy and society.

It cannot be overemphasized that green chemistry and other complementary approaches are likely to generate change in all walks of life, particularly with respect to consumption of resources. Among all resources, water is unique because it is an essential driver for industrialization as well as agriculture. Water availability and access have been the principal bottlenecks in implementing projects of significance in free India and countries with comparable circumstances of growth—especially over the past two decades. Globalization and free trade practices have significantly increased water demand. It has now become clear that there is no alternative to treating municipal and industrial wastewaters to a high degree and recycling and reusing the treated effluents. The following section, therefore, has been included to highlight the significance of treatment technologies so that reuse of wastewater can be achieved in a sustainable manner. What better than green chemistry and ecological engineering to help achieve this goal?

4.2.2 Significance of Wastewater Treatment and Reuse

Nearly half the world's population does not have access to clean water. Also, water scarcity is a problem faced by communities worldwide today. It is popularly quoted that if the third world-war breaks out in the near future, it will be probably overwater! Water or "Blue Gold" is probably the only commodity in the world with a universally high value in all communities.

Green engineering plays a major role in treatment and recycling of wastewater generated by households, industries and institutions. There are many technologies available in the market, capable of treating industrial and municipal wastewaters to the highest possible degree. However, owing to the lack of cost-effective and environmental friendly technologies, the goal of recycling and reusing of treated wastewaters still remains a dream to be fulfilled. It is recognized that significant additional effort is required in connecting the buyers of recycled waters with the treatment facilities—so that the revenue earned through the sales of high quality treated wastewaters can subsidize the cost of primary and secondary treatment of the respective wastewaters. Some of the available wastewater treatment technologies capable of producing high quality treated effluents are listed in Table 4.2 below.

4.2.3 Need for Application of Green Engineering in Agriculture and Food Processing

One of the difficult problems associated with pollution of water resources such as lakes, rivers, ground water, reservoirs and man-made ponds is the so-called "non-point source pollution." For example, the surface runoff emerging from agricultural fields and residential areas (rural as well as urban) is known to leach and transport fertilizers, pesticides and herbicides into water bodies. Thus, dozens of extremely toxic farm chemicals are getting mobilized in the natural aquatic environment each year. These pollutants individually and collectively damage the eco-system and also impact public health through food chain. The fertilizers thus mobilized eventually cause eutrophication in surface waters and enhance the primary productivity in water body. Such deterioration of water resources and the impacts on corresponding aquatic life lead to destruction of flora and fauna. Similarly, pesticides and herbicides enter the food chain through aquatic processes and produce toxic effects in aquatic and human life. In response to these challenges, a new movement in agriculture is gaining stronger roots in the recent years, which is popularly called as "organic farming". If all agriculture in the world goes organic then it would be a different matter. However, in the interim period while much of the farm produce is based on the use of farm chemicals, there is a co-existence of organic farm produce with foods laced with toxic chemicals. To be able to solve this gigantic problem of replacing the contaminated foods of over six billion people in the world with cleaner organically grown foods is yet to be addressed in the twenty-first century. It can be argued that a little relief can possibly gained through application of appropriate downstream processing of farm produce and making deliberate efforts of infusing green engineering in this sector. Examples of such efforts include simplification of supply chain by consolidating food production and processing as well as connecting farmers directly to consumers/exporters.

The presence of many pesticides and toxic materials in routinely consumed foods has made food processing a concern worldwide as it directly affects the human health. Kaushik et al. (2009) argued that even though the organic farming is being promoted to tackle the problem, the process to promote it and its acceptance has been very slow. Ecobichon (2001) states that non-patented, more toxic, environmentally persistent and inexpensive chemicals are used extensively in developing nations, creating serious acute health problems and local and global environmental impacts. Winteringham (1971) noted that after all the remarkable progress in pesticide production, only a small fraction of used pesticides break down and a very large fraction of applied pesticides persist in residual form. The residual pesticides in the food chain further cause contamination and even get magnified through the food chain in trophic-levels. Food processing treatments including washing, peeling, canning or cooking may lead to a significant reduction of pesticide residues and therefore should be viewed as one of the short term solutions to impact on residues on humans.

Baking is the technique to prepare bread, pastries, pies and quiches *etc.* by dry heat in an oven for long time. A study conducted by Habiba and co-workers (1992) to find out the effect of baking on pesticide residues in potatoes; the team applied profenofos to potatoes 1 month before to its harvesting in Egypt. The study found that the level of pesticides reduces from 11.48 ppm in fresh potatoes to 0.22 and 0.19 ppm in microwave-baked and oven-baked potatoes, respectively. Sharma et al. (2005) claimed that the reduction in the pesticides concentration in baking process is because of evaporation, co-distillation and thermal degradation which take place due to heating in baking. The reduction depends on the thermal stability and chemical nature of pesticides.

The above discussion emphasizes the fact that it is important to address the concern of food safety through suitable processing techniques and appropriate storage period that

TABLE 4.2　Most Advance Wastewater Treatment Technologies

Sr. No	Technology	Principle	Typical Treatment Train and Potential Application
1	Activated carbon and sorption based technologies	Activated carbon works by the process of adsorption. Adsorption is when one material adheres to the surface of another material by means of physical and/or chemical attraction between the materials (Dabrowski, 2001; URL 01, 2014).	Among various water purification and recycling technologies, adsorption is a fast, inexpensive and universal method. Adsorption process may be classified as: cyclic batch, continuous counter-current, and chromatographic. **Bulk separation processes**: Normal paraffins, isoparaffins, aromatics, N_2/O_2, CO, CH_4, CO_2, N_2, acetone/vent streams, ethylene/vent streams, water/ethanol, p-xylene, o-xylene, m-xylene, p diethylbenzene/isomers mixture, fructose, glucose, detergent range olefins/paraffins. **Purification processes:** Organics/vent streams, water/natural gas, air, syngas, sulfur compounds/natural gas, hydrogen, LPG, solvents/air, odors/air, Nox/N_2, SO_2/vent streams, water/organics, oxygenated organics, chlorinated organic, odor, taste bodies/drinking water, sulfur compounds/organics, decolorizing petroleum fractions, sugar syrups, vegetable oils, etc. **Potential application:** For heat storage of solar energy (Cavalcante, 2000). For solar cooling and heat pumping. Solid-gas sorption refrigeration.
2	Ion-exchange	The ion exchange process comprises of the interchange of ions between a solution and an insoluble solid, *that is*, polymeric or mineralic ion exchangers such as ion exchange resins (functionalized porous or gel polymer), natural or synthetic zeolites, montmorillonite, clay, etc.	Ion exchange has numerous applications for industry as well as laboratory research. It is used in production of various acids, bases, salts. Ion exchange processes are commonly used in treatment of drinking and wastewater in commercial and industrial applications such as water softening, demineralization and decontamination. Phosphate, nitrate, ammonia, which appear in various types of agricultural, domestic and industrial wastewaters or heavy metals discharged in effluent from electroplating plants, metal finishing operations, as well as a number of mining and electronics industries can be removed by the process. Ion exchange is considered attractive because of the relative simplicity of application and in many cases is proven to be economic and effective technique to remove ions from wastewaters, particularly from diluted solutions.
3	Molecular sieves	Molecular sieves operate on the size exclusion principle. Smaller molecules that fit into the pores are adsorbed while larger molecules pass through. Molecular sieves have regular pore openings of 3, 4, 5, or 10Å. Polarity of the molecules matter because the highly polarized molecules are adsorbed more readily into the pores than non-polar molecules. Molecules are selectively removed in this manner. When a gas or liquid is passed through the mole sieve, smaller pieces are adsorbed while larger molecules pass through (URL 02, 2014).	Drying of air, Natural gas, alkane and refrigerant. Moisture removal in Polyurethane (PU) or Paint, Natural Gas Drying. Drying Cracked GasStatic drying of insulating glass units, whether air filled or gas-filled. Dehydration of highly polar compounds, such as methanol and ethanol. Dehydration of unsaturated hydrocarbons (e.g. ethylene, propylene, butadiene). Generation and purification of argon. Type 4A molecular sieve is typically used in regenerable drying systems to remove water vapor or contaminants which have a smaller critical diameter than 4Å.

TABLE 4.2 (*Continued*)

Sr. No	Technology	Principle	Typical Treatment Train and Potential Application
4	Solvent extraction		Solvent extraction for phosphorus. Recovery, liquid-liquid extraction for biotechnology-extraction of valuable products from fermentation broth, removal of high boiling organics from wastewater-such as phenol, aniline and nitrated aromatics, urification of heat sensitive materials, Recovery of products from reactions *etc*.
5	Photochemical and other advanced oxidation	Solvent extraction method of separation requires that the constituents have different relative solubilities in two immiscible, or only partially miscible liquid solvent.	Separation for close boiling liquids. Separation of liquids of poor relative volatility. As a substitute for vacuum distillation. As a substitute for evaporation. As a substitute for evaporation. As a substitute for fractional crystallization. Separation of heat sensitive materials. Separation of mixtures that forms azeotropes. As a substitute for more expensive chemical methods.
6	Ozonation	OZONE is a very powerful oxidizing agent, only next to OH radicals. How effective ozone is, will depend entirely on the nature of the contaminant and is directly dependant in the chemistry involved in the process. Many other oxidation agents are often used in combination with ozone to provide increased efficacy. Agents such as peroxides, UV, and conditions of high pH assist ozone in the oxidation process.	Ozone during preliminary stage is used for detoxification. Ozone at secondary stage is used for sludge reduction, and ozone during the tertiary stage is more common and used for disinfection, micro pollutant removal, COD reduction and decoloration. The location of ozone is dependent on the goal of ozonation. Decolorization and deodorization. De toxification. Disinfection. COD/BOD reduction. Sludge reduction.
7	Wet-air oxidation	WAO involves the liquid phase oxidation of organics or oxidizable inorganic components at elevated temperatures (125–320°C) and pressures (0.5–20 MPa) using a gaseous source of oxygen (usually air). Enhanced solubility of oxygen in aqueous solutions at elevated temperature and pressure provides a strong driving force for oxidation.	Wet air oxidation (WAO) is a well-established technique for wastewater treatment particularly toxic and high concentration organic wastewater. The efficiency of WAO can be improved by various means, such as adding a catalyst or using a stronger oxidant. Removal of chemical oxygen demand (COD), total organic carbon (TOC), theoretical oxygen demand (ThOD), biochemical oxygen demand (BOD).
8	Nano-technology	Nanotechnology, the engineering and art of manipulating matter at the nanoscale (1–100 nm), offers the potential of novel nanomaterials for the treatment of surface water, groundwater and wastewater (URL 03, 2014) contaminated by toxic metal ions, organic and inorganic solutes and microorganisms. Due to their unique activity toward recalcitrant contaminants and application flexibility, many nanomaterials are under active research and development.	Nanobiotechnology for the detection of microbial pathogens Nanofibers and nanobiocides in water purification Nanozymes for biofilm removal Nanofiltration for water and wastewater treatment Electrospinning nanofibers for water treatment Potential risks of using nanotechnology in water treatment are on human health.

(*Continued*)

TABLE 4.2 (Continued)

Sr. No	Technology	Principle	Typical Treatment Train and Potential Application
9	Reverse osmotic Membranes	The fundamental principle of reverse osmosis membrane technology is the use of pressure to separate soluble ions from water through a semi-permeable material.	Reverse osmosis has traditionally been utilized for producing low TDS water from sea water or brackish water sources.
		The membrane is usually a thin film composite material and is manufactured in a spiral configuration as opposed to a flat sheet or tube geometry. The predominant model used today for industrial applications is the spiral configuration (URL 04, 2014).	Heavy industry and power generating sectors now utilize reverse osmosis as a standard technology to produce high quality water. Membrane developments have lead to higher removal of soluble ions coupled with a reduction in operating pressures which has resulted in reverse osmosis becoming an economical technology for soluble ion removal.
10	Membrane filtration	Although there are a number of different methods of filtration that incorporate membrane technology, the most mature is pressure driven membrane filtration. This relies on a liquid being forced through a filter membrane with a high surface area. There are four basic pressure driven membrane filtration processes for liquid separations. These are, in ascending order of size of particle that can be separated: reverse osmosis, nanofiltration, ultrafiltration and microfiltration.	The use of reverse osmosis is well-established for desalination of drinking water and the production of deionized water for process use. Membrane filters are used to remove microbiological contaminants. Membrane filters are used to remove both dissolved and particulate inorganic substances. Membrane filters are used to remove organic compounds.

enhances food safety. Common and simple processing techniques acquire significance for reducing the harmful pesticide residues in food. Freezing as well as juicing and peeling are necessary to remove the pesticide residues in the skins. Cooking of food products helps to eliminate most of the pesticide residues. Removal of residues in food by processing is affected by type of food, insecticide type and nature and severity of processing procedure used. Hence a combination of processing techniques would suitably address the current situation in food safety.

4.2.4 Case Examples Illustrating Significance of GC&E

From apedagogic point of view, efforts have been made to include case studies and case examples in this Chapter that would illustrate the practical applications envisaged or already implemented in the field. At the end of this Chapter, three case studies have been presented with the intention of providing more details and insights into the specific interventions. However, such a detailed presentation could not be included on dozens of locations due to limitations of space and therefore, case examples also have been listed. Table 4.3 as presented below includes case examples, their highlights and significance of application of green chemistry, green engineering as well as GC&E practiced worldwide. On several occasions it was felt that it was neither possible to distinguish between application of chemistry, engineering or both nor was it easy to distinguish between the expression and communication of a mere possibility versus evidence of proven successful experiment. Thus, for the sake of completeness, case examples as well as the relevant literatures on application has been included in Table 4.3.

The above Table highlights the relevance of green chemistry and engineering principles in various industrial operations around the world. A closer inspection of the above Table reveals that application of green chemistry and engineering essentially enhances the quality of products in terms of cost-effectiveness, minimum waste generation, control of pollution etc. Additionally, it also enhances work place safety and sustainability of the production process through minimization of carbon footprint and impact on immediate surroundings including eco-systems.

4.3 MATERIAL EFFICIENCY

A snapshot of developments in the field of green chemistry presented below reveals an integrated perspective that has the twin benefits of optimal transformations of material and energy. Unit processes are targeted predominantly with an

TABLE 4.3 Case Examples and Literature on Green Chemistry, Green Engineering as well as GC&E Applications

Sr. No.	Description of the Case Example	Highlights	Principle	Reference
1	Cleaner production and eco-efficiency initiatives from the year 1996 to 2004 in Western Australia	The developments in the promotion and implementation of cleaner production (CP) and eco-efficiency (EE) in Western Australia in four stages: groundwork (1996–1999), experimentation (1999–2002), roll out (2002–2004) and re-orientation (2004 onward).	Green engineering	van Berkel (2007)
2	Application of environmental management tools for improving cleaner production in China	Factors such as difficulty in mainstreaming cleaner production in industries, constraints in financial and technical resources of small and medium-sized enterprises have slowed down the extensive adoption of cleaner production in China.	Green engineering	Hicks and Dietmar (2007)
3	Impact of cleaner production on health hazards in the workplace	Cost effectiveness of cleaner production and the role played by cleaner production in minimizing the health hazards of pollutants in the workplace.	Green chemistry and engineering	Unnikrishnan and Hegde (2006)
4	SekaBalikesir pulp and paper mill, Turkey: cleaner production opportunities	The benefits of the identified waste reduction options were analyzed for increasing production efficiency and achieving target raw effluent pollution loads from the pulp and paper mill in Turkey.	Green chemistry and engineering	Avsar and Demirer (2008)
5	Cleaner production and toxics use reduction in Massachusetts for preventing workers health and safety	Effect of cleaner production-pollution prevention in the form of toxics use reduction (TUR) on worker health and safety at three printed wire board facilities covered under the Massachusetts Toxics Use Reduction Act. Explored the relationship between worker health and safety and environmental protection within the corporate structure; Identified the factors driving companies to reduce toxics both inside and outside of their plants.	Green chemistry and engineering	Armenti et al. (2011)
6	A cleaner production of denim garment using amylase/cellulase/laccase	Using enzymes such as amylase, cellulase, laccase and their combinations applied on denim garment to conduct one step bio-desizing and bio-washing producing old-look appearance garment is proposed and analyzed.	Green chemistry and engineering	Ali and Gupta (2007)
7	Production of hydrogen from renewable and sustainable sources: A review	The industrial and emerging hydrogen production technologies including steam methane reformation, partial oxidation, auto thermal reforming, steam iron, plasma reforming, thermo-chemical water splitting and biological processes.	Green chemistry	Chaubey et al. (2013)

(*Continued*)

TABLE 4.3 (*Continued*)

Sr. No.	Description of the Case Example	Highlights	Principle	Reference
8	Cleaner production in the mining and minerals industry: An Industrial ecology framework	Industrial ecology and cleaner production as environmental management practices in the mining and minerals industry.	Green engineering	Basu and van Zyl (2006)
9	Eco-efficiency guiding micro-level actions: Ten basic steps towards sustainability	Compatibility between technological improvements at the micro-level and sustainability at the macro-level.	Green chemistry	Huppes and Ishikawa (2009)
10	Bio-ethanol fuel production and utilization trends	Production of bio-ethanol from biomass for reducing both consumption of crude oil and environmental pollution. Using bio-ethanol blended gasoline fuel for automobiles for significant reduction in petroleum use and exhaust greenhouse gas emission.	Green chemistry	Balat et al. (2009)
11	Use of bio-fuels for running the gas turbine: A review	The use of bio-fuels to run gas turbine reduces use of fossil-fuels and environmental concerns. The use of liquid and gaseous fuels from biomass will help to fulfill the Kyoto targets concerning global warming emissions.	Green chemistry and engineering	Gupta et al. (2010)
12	Eco-friendly/biodegradable lubricants: An overview	Highlight recent developments in biodegradable synthetic ester base stocks for formulation of new generation lubricants. The developed products can be used as automotive transmission fluids, metal working fluids, cold rolling oils, fire resistant hydraulic fluids, industrial gear oils, neat cutting oils and automotive gear lubricants either alone or in combination.	Green chemistry	Nagendramma and Kaul (2012)
13	Critical aspects of bio-based materials—reviewing methodologies and deriving recommendations	The treatment of biogenic carbon storage for quantifying the greenhouse gas emissions of bio-based materials in comparison with petrochemical materials.	Green chemistry and engineering	Pawelzik et al. (2013)
14	Low cost, eco-friendly layered $Li_{1.2}(Mn_{0.32}Ni_{0.32}Fe_{0.16})O_2$ nano particles for hybrid super capacitor applications	$Li_{1.2}(Mn_{0.32}Ni_{0.32}Fe_{0.16})O_2$ (LMNFO) nano particles with and without a chelating agent (adipic acid) were synthesized by sol–gel method.	Green chemistry and engineering	Karthikeyan et al. (2013)
15	Natural product based agents derived from industrial plants in textile applications: A review	Textile applications of environmental friendly plant-based products such as fibers, polysaccharides, dyes and pigments, poly-phenols, oils and other biologically active compounds. Also focus on plant derived bioactive agents with antimicrobial properties and application of these agents to the textiles.	Green chemistry and engineering	Shahid-ul-Islam et al. (2013)

increasing tendency to extract untransformed or partially transformed substrates from waste streams. Value addition to waste is yet another manifestation as part of eco–industrial development approaches across the world. These are essentially based on the principles of resource optimization seen in nature and inspire such transitions in artificially mediated processes.

The concept of "material efficiency" has been recently put forth by Professor J. M. Allwood and his co-workers in their recent White Paper published from the Cambridge University, Cambridge, UK (UK, Allwood et al., 2011). This concept essentially addresses engineering materials used in manufacturing one unit product (or one unit of economic output) in the context of manufacturing goods, consumer products or consumable durables. Nearly all engineering materials will be lost by human civilization through the process of locking them away in municipal and toxic hazards landfills over the coming decades. As a result, there will likely be a scarcity of precious metals in particular and engineering construction materials in general. This new form of externality will emerge in spite of taking full responsibility for the presently best practice of land-filling industrial and municipal solid wastes in a conscientious manner. Therefore, the time has come when the regulators and governing institutions will have to formulate resource consumption policies for responsible consumption, minimal use of materials during manufacturing and remanufacturing so as to achieve the so-called "material efficiency."

There are two schools of thoughts, however. The Allwood School advocates the use of "winning technologies" to improve material efficiency. On the other hand, the Soderholm and Tiltonb (2012) hypothesis argues the significance of implementing "winning policies" to improve material efficiency.

The Allwood School of thought emphasises on reducing use of virgin materials so that these materials are available for use by subsequent generations. The author further argued that global stockpiles of such materials as aluminium and steel, used in building, infrastructure and equipment, are enough to meet growing demands but environmental impact of production and processing new materials is of critical concern. Thus, it is not energy efficiency alone but the material efficiency that should be our main target in order to achieve sustainability. It is to be noted that material efficiency pertains to conquering the same level and types of services with less primary production. The ideas suggested by the Allwood et al. (2011) to improve material efficiency include extension of the lives of products, use of less metal by design, reduction in yield losses, divert manufacturing scrap, re-use old components before recycling and reduction of demand.

The demand for materials in a developed economy is not because of shortage of materials. It is because of low recycle and reuse of materials and the high repair cost of used materials compared to the cost of purchasing new ones. For example, In U.K., over the years 2000–2005, the number of garments purchased per consumer increased 33% and it was cheaper to buy a new pair of trousers than to repair a hole in their pockets (Allwood et al., 2006). Researchers recommend modification and reuse of products instead of replacing them altogether, and urge the development of adaptable designs that would help this process.

Allwood et al. (2006) further argued that big savings can be secured by designing more suitable and efficient products because industries presently use products with more material than is required. For example, by optimising beam designs to suit their use, a 30% material weight can be saved, which is further accompanied with a similar reduction in pollution.

Although, life extension is good example of extending material efficiency within the product in which they were originally used, another approach is to separate used products into their components, only to use them to manufacture new products. Such case studies have been reported by Asolekar (2006) regarding re-rolling of steel plates generated from ship-breaking activity in India. Scrap metals are usually sent for melting, which is an energy intensive process, whereas instead of melting they can be used directly in other activities.

Allwood et al. (2011) advocate re-use rather than recycling because in case of re-use, all the energy involved in melting, casting and re-rolling old steel can be avoided. However, old components of any product are recycled instead of reuse. This is true of beams from dismantled buildings. Figure 4.1 depicts the role of users, producers, designers as well as regulations in making any product efficient.

In industrialised nations, material efficiency strategies have not received much attention, mainly because of economic, regulatory and social barriers. However, evidence from waste management and the pursuit of energy efficiency suggests that these barriers might be overcome. It must be recognized that a single strategy is not equally effective for different materials.

Solderholm and Tiltonb (2012) address two types of material scarcity. One is short-term that persists for months or years but not longer than a decade. It was argued that unexpected sudden causes like war, export restrictions, strikes or business accidents might cause short term scarcity. The second type of material scarcity is caused by materials depletion. The availability of daily consumable materials, produced from renewable and non-renewable resources, is governed in turn by the availability of their resources from which they are extracted.

The presence of these types of behavioural failures will also motivate the implementation of information based policy measures. A potentially effective policy measure could be the use of performance standards that are common across products, that is, whether manufactured from primary or secondary materials. Such standards can prove to be beneficial to individuals and corporations in equal measure,

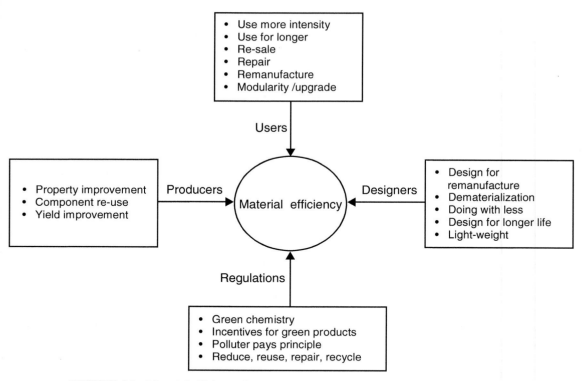

FIGURE 4.1 Material efficiency flow chart. *Source:* Adapted from Allwood et al. (2011).

providing a basis to compare off-setting capital and material operating costs.

The above illustrates that economically efficient use of materials may require public policy measures to address relevant market (and behavioural) failures. It should be clear from this that the actual amount of materials produced and processed will be endogenously determined, and efficient levels of production and use should also change as, for instance, relative prices, preferences, and incomes. Thus, according to economic theory, increased material efficiency does not represent a desirable end in itself, and policy measures that explicitly promote material efficiency will not necessarily (perhaps most often not) be economically efficient. Nevertheless, policy instruments that target environmental and information externalities may induce additional private measures to improve material efficiency, but it is very difficult *ex ante* to assert the quantum and the means.

4.4 UNLOCKING CONDEMNED RESOURCES

The generation of piles of solid wastes resulting from used and discarded consumer goods has been the flipside of automation and mass production—not to mention the huge quantities of air pollution and wastewater generated during manufacturing as well as consuming disposable products and lifestyle chemicals. No wonder, the growth rates of our stock piles of municipal and industrial solid wastes are outpacing capacities of municipal and industrial solid waste landfills. How can such a lifestyle be sustainable?

The need of the hour therefore is to reclaim and recycle useful metals and materials and recycle them into production systems as "recycled resources" which were otherwise deemed as "wastes" and condemned to landfills. It is a fact that millions of tons of partially depleted and reclaimable waste piles possessing several useful metals and materials are sequestered in landfills in almost every city and industrial cluster world over. This section highlights the significance of opportunities to reuse and recycle in the context of landfills as receptors of municipal and industrial wastes.

4.4.1 Landfill Mining

While improving the material efficiency might appear as the panacea for countering unsustainable use of engineering materials; there could be a variety of reasons impeding its implementation. Therefore, several researchers are exploring various aspects of "landfill mining."

Land-filling for the disposal of wastes has been used for a very long time. It is often the most preferred method due to low maintenance cost and availability of land at a reasonable price. The ever-increasing waste quantity and its complex composition have seriously impacted humans and environment. Many recent research papers have pointed out the implications of landfilling and nowadays, it is not being considered a sustainable way of disposing wastes. Increased land cost, unavailability of land, global warming gas emissions from landfill have compelled scientists and engineers to consider other scientific and engineered options of wastes disposal instead of land filling. Since, the landfilling has been practiced for many years and being used in many parts of world, numerous valuable materials have been locked inside landfill. Krook et al. (2012) estimates that the amount of copper deposited in landfills worldwide is comparable with available stock at present and an analysis shows that the other materials (other than metals) deposited in landfills in Sweden are enough to provide electricity to a district for a decade.

Several definitions of landfill mining have been put forth by different researches. One most widely referred to is given by Cossu et al. (1996). It states verbatim that *"the excavation and treatment of waste from an active or inactive landfill for one or more of the following purposes: conservation of landfill space, reduction in landfill area, elimination of a potentially contamination source, mitigation of an existing contamination source, energy recovery from excavated waste, reuse of recovered materials, reduction in waste management system costs and site re-development"*. The above definition emphasised more on holistic waste management rather than only recovery of materials. Savage et al. (1993) argue landfill mining as excavation, processing, treatment and recovery of materials. According to Krook et al. (2012) the three major objectives of landfill mining are extension of landfill lifetime, consolidation of landfill area facilitating final closure and remediation.

In their critical review on "landfill mining," Krook et al. (2012) describe many aspects of landfill mining. The authors have reviewed 39 research papers covering a decade (1988–2008) and argue that the field of landfill mining has not seen adequate research. The authors found only 12 out of 35 articles published in scientific journals in 20 years and these papers did not have a clear frame work on landfill mining. They accordingly emphasised such such research challenges as technology innovation, addressing the underlying conditions for realization and developing standardized framework for evaluating performance.

Bosmans et al. (2013) have given a comparative study of many thermal wastes to energy valorization technologies with their respective potentials. In figure, a flow sheet is given to describe many thermal WtE technologies. The technologies for valorization include:

1. Incineration: Full oxidative combustion
2. Gasification: Partial oxidation
3. Pyrolysis: Thermal oxidation of organic materials in the absence of organic materials
4. Combination Process: and
5. Plasma based technologies: (i) Plasma Pyrolysis and, (ii) Plasma Gasification and Vitrification

The potential of landfill mining depends on many factors that vary across sites. It depends upon the age, degree of mixing, type of landfill, location and initial composition of waste etc. Valorisation of landfill waste will depend on its quantity and quality. Bosmans et al. (2013) cites the feasibility of converting excavated MSW conversion into RDF in Belgium if the composition is similar to that of fresh MSW. The author argues that some WtE technologies accept raw MSW as input while others accept processed MSW because they are quite sensitive to the chemistry and fluid dynamics of waste. Figure 4.2 depicts the Framework for Integrated Waste to Energy.

4.4.2 Recycle of Low Chemical Potential Substances

Feng (2004) argued that a circular economy offers the best strategy to utilize resources, that are being rapidly depleted with development and modernization. The author further emphasised "reuse, recycle and reduce" as the principles which sustain a circular economy. Wen et al. (2007), state that unidirectional use of low potential chemicals produced from process industry is causing resource depletion, health problem and environment pollution. The use of low potential chemicals like $CaSO_4$, CO_2, H_2O, $NaCl$, $CaCO_3$ etc. will accelerate the circular economy and release the pressure from fresh resource consumption.

Low chemical potential substances are generated during the production of potential chemicals in phase transformation. These include production of phosphorus gypsum ($CaSO_4$) during the production of phosphoric acid and CO_2 generation from industrial processing including refining and the cement industry. Figure 4.3 depicts the ecology integration technology of $CaSO_4$ decomposition.

Phosphorus gypsum ($CaSO_4$) is an example of a substance with low chemical potential, produced during the extraction of phosphorus ore with sulphuric acid (Wen et al., 2007). Using traditional technology transformation of phosphorus gypsum produced during the process into a valuable product is not possible and much material is wasted. In addition, the land area requirement is very large and a fresh supply of sulphuric acid is required consistently.

The authors also described a technology developed in China, ecology integration technology, which has decomposition capacity of phosphorus gypsum and the technology has solved the gypsum pollution problem and enhanced the resource utilization potential.

Reaction	Heating Value (kJ/mol)
$2C + O_2 \leftrightarrow 2CO$	+246.4
$C + O_2 \leftrightarrow CO_2$	+408.8
$CH_4 + H_2O \leftrightarrow CO + 3H_2$	−206
$CH_4 + 2H_2O \leftrightarrow CO_2 + 4H_2$	−165
$C + CO_2 \leftrightarrow 2CO$	−172
$C + H_2O \leftrightarrow CO + H_2$	−131

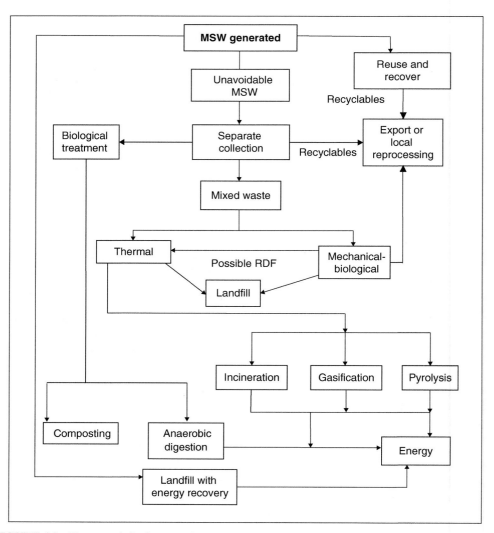

FIGURE 4.2 Framework for integrated waste to energy. *Source:* Adapted from Bosmans et al. (2013).

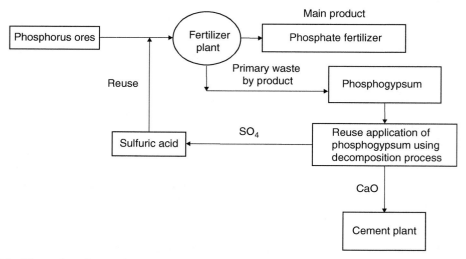

FIGURE 4.3 The ecology integration technology of $CaSO_4$ decomposition. *Source:* Adapted from Wen et al. (2007).

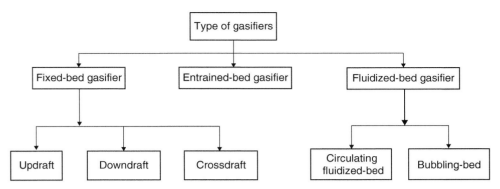

FIGURE 4.4 Types of gasifiers. *Source:* Adapted from Panwar et al. (2012).

4.4.3 Eco-Friendly Energy Generation from Biomass and Green Chemistry

Plants and algae generate biomass predominantly through photosynthesis. Biomass is an eco-friendly material, which can be efficiently utilized to enhance resource sustainability. Nurgul et al. (2001) argued that thermochemical processes are the most widely used methods for biomass conversion into high value fuel. There are many other processes, such as gasification, pyrolysis and carbonization, for biomass conversion into high-energy fuel. The following section highlights the significance of gasification process.

Gasification: According to Hanne et al. (2011) and McKendry (2002) gasification is an eco-friendly way of converting biomass into gaseous fuel at an elevated temperature of 800–1300°C. Gasification is done in the partial presence of oxygen. Partial combustion of solid fuel and producer gases generated during gasification contain high-energy components such as H_2, CH_4 along with impurities as CO, CO_2, nitrogen, sulfur, alkali compounds and tars (Balat et al., 2009 and Damartzis and Zabaniotou, 2011). Figure 4.4 depicts the type of gasifiers. The basic gasification reactions given below (source: Panwar et al., 2012).

4.5 ECOLOGICAL ENGINEERING AND SOCIAL SUSTAINABILITY

Ecological engineering was first defined by Odum (1963) and subsequently other alternative definitions have been given by other ecologists and engineers all over the world. Ecological engineering is a bridge between society, economics and ecology. Mitsch (1998) argued that ecological engineering is quite different from other disciplines of engineering and is based on the two pillars of ecology and engineering. Ecological engineering is a self-designed system modeled on the behavior of biological systems; and its ultimate goal is to create sustainable ecosystems. Xu and Li (2012) claimed that for the sake of sustainable development ecological engineering should be the basis for engineering management. The authors further stated that there is a lack of systematic concept of *ecological engineering based engineering management*. Therefore, there is a need to integrate ecology with engineering to achieve desired sustainable solution. With the goals of ecosystem restoration and sustainable ecosystem development, social sustainability lies at the heart of ecological engineering.

Social sustainability has been widely accepted as one of the three pillars of sustainable development along with economic and environmental aspects. Social sustainability focuses on the association of individuals, communities and societies, and societal provisions and expectations for (i) individual autonomy and realization of personal potential, (ii) participation in governance and rule making, (iii) citizenship and service to others, (iv) justice, (v) the propagation of knowledge, and (vi) resource distributions that affect the ability of that society to flourish over time. Vallance et al. (2011) argued categorization of social sustainability into three categories: *"development sustainability"*—which addresses poverty and equity, *"bridge sustainability"* with its concerns about changes in behavior so as to achieve biophysical environmental goals and *"maintenance sustainability"*— which refers to the preservation of socio-cultural patterns and practices in the context of social and economic change.

Various approaches of achieving social sustainability using ecological engineering have been applied world over. Three such approaches have been discussed in the following sections:

Case Study 1: Wastewater Treatment Systems as Microcosms of Eco-system,

Case Study 2: Integrating Natural Treatment Systems for Upgradation of Aquatic Systems and

Case Study 3: Sustainable Recycling of materials from Obsolete Ships through Ecological Engineering.

They illustrate a variety of roles played by application of integrated approaches (ecology with engineering) to solve the problems of great significance for India and countries with comparable challenges. It is noteworthy that there are two lessons to be learnt from these case studies. First, ecology-based engineering approach leads to the sustainable solution. Second, emphasis on ecological and environmental rejuvenation, conservation and up-gradation enhances long-term profitability and therefore willingness of the community to own and adapt it to their socio-cultural aspirations.

4.6 CASE STUDY 1: WASTEWATER TREATMENT SYSTEMS AS MICROCOSMS OF ECO-SYSTEM

The Water Act was promulgated in the year 1974 in India to address wastewater treatment and disposal related concerns. One of the challenges in those days was associated with regulating disposal of sewages and industrial trade effluents in the receiving bodies including lakes, rivers, estuaries and creaks as well as in the ocean. Several landlocked areas did not have the opportunity to dispose their treated effluents into surface waters and they demanded land application of this treated effluents either in agriculture or in commercial agro-forestry. Two considerations were critical in this context. First, the regulatory agencies including the Central Pollution Control Board (CPCB) governed by the Ministry of Environment Forests (MoEF), Government of India (GOI) or by the State Pollution Control Boards (SPCBs) in respective states in the union of India were expected to decide the constitution of "acceptable receiving body of water" suitable for disposal of treated sewages and effluents. Second, it was equally important to articulate and legally notify the minimum quality standards for treated sewages and industrial effluents that shall be deemed fit for disposal in the legal receiving bodies. Tables have now turned, in the today's context (after nearly 40 years), and the number of challenges associated with disposal of treated sewages and effluents has grown multi-fold. Today, a third important consideration has entered into the picture. Nearly every community is thirsty for water in India. The quality and quantity of water available for drinking, sanitation, industrial applications, commercial application including construction and irrigation are in short supply. One of the strategies being advocated by municipalities and communities is the reclamation and reuse of treated sewages. This strategy is currently reshaping the face of sewage treatment technologies and management in India.

Initially, the aim of a typical wastewater treatment plant was to improve the overall quality of the receiving water body into which treated or partially treated wastewaters were to be discharged. But, increasing demand for water by various sectors has mandated the reuse of treated effluents. Therefore, the Government of India has prescribed different kinds of sector-based regulations for wastewater treatment and disposal into water bodies with the hope that industries and communities will attempt to reclaim wastewaters and recycle them after appropriate treatment. Apart from the conventional mechanized wastewater treatment plants, natural treatment systems (NTSs) have also been implemented in India to cater to this growing need.

In the light of shortages of water in several parts of the world and especially in India; communities are looking for alternatives that are less power intensive and less expensive in providing some kind of primary and secondary treatment (Asolekar and Hiremath, 2013). The NTSs typically fill the gap in the sense that they need relatively low operation and maintenance (O&M) costs and far low power to run them when compared with conventional primary, secondary treatment alternatives—especially such as activated sludge process, trickling filter or extended aeration system.

A class of sewage treatment technologies that mimic natural processes such as interaction of soil-micro-organisms with pollutants as well as interaction of plants and other life in natural settings with pollutants in wastewaters are called as Natural Treatment Systems (NTSs). It is well known that the engineered natural wastewater treatment systems including river banks, wet-zones and their modified versions such as Constructed Wetlands (CWs), Waste Stabilization Ponds (WSPs), Sewage Fed Aquaculture ponds (SFA), Duckweed Ponds (DPs) or algal bacterial systems render effective treatment of biodegradable carbonaceous matter and successfully separate suspended particulates. However, it should be noted that among these wastewater treatment systems, not all are particularly effective in removing nitrogen and phosphorus.

In spite of their limitations, NTSs have attracted attention of environmental engineers and scientists by the virtue their abilities of treating sewages and wastewaters at phenomenally low O&M costs. They have been favourably looked upon in developing countries, especially because of their low power requirement. Conventional mechanized wastewater treatment systems are relatively expensive in terms of installation, operation and maintenance costs. Also, the conventional wastewater treatment systems, are efficient in removing COD but unable to reduce the nutrient pollution effectively—particularly nitrogen and phosphorus (Wu et al., 2011).

The cost of treating wastewater to conform with the high microbiological standards is often unaffordable to many developing countries, while the use of secondary treated wastewater is common. Newer solutions should be such that the peri-urban and small communities should be able to own and operate their wastewater treatment systems (Asolekar and Hiremath, 2013). Interestingly, in the recent past, communities seem to accept NTSs capable of providing adequate treatment to wastewaters and simultaneously supply fish and other sources of nutrition to fishing communities that manage the systems and generate adequate water for irrigation of farms and agro-forests. Moreover, they demand minimum energy and maintenance and provide a higher degree of treatment compared to conventional mechanised treatment systems of small communities in the last few years

TABLE 4.4 Types of NTSs Practiced for Wastewater Treatment and Reuse in India: Case Studies from India

S. No.	Technology	Location	Capacity	Performance	Designated use of Treated Wastewater
1	CW Ropar	Ropar, Punjab	0.5 MLD	Satisfactory	Pisciculture
2	WSP Mathura	Mathura, Utter Pradesh	20 MLD	Satisfactory	Irrigation
3	SFA Karnal	Karnal, Haryana	8 MLD	Satisfactory	Irrigation
4	DP Ludhiana	Ludhiana, Punjab	0.5 MLD	Satisfactory	Pisciculture
5	PP Karnal	Karnal, Haryana	40 MLD	Satisfactory	Pisciculture
6	KT Ujjain				

PLATE 4.1 (1) 0.5 MLD CW, Ropar, India, (2) 14.5 MLD WSP, Mathura, India, (3) 8 MLD SFA, Karnal, India (4) 0.5 MLD DP, Ludhiana, Punjab, India, (5) 14 MLD PP, Agra, India, and (6) 1.79 MLD KT Ujjain, India.

(Mara et al., 1992; Brix, 1994; Vymazal, 2002; Puigagut et al., 2007; Zimmels et al., 2008; Rana et al., 2011).

The types of NTSs mostly practiced for wastewater treatment and reuse in India include, constructed wetlands, waste stabilization ponds, sewage fed aquaculture, duckweed ponds and Karnal Technology for on-land disposal of wastewater. The summary and images of different case studies of NTSs from India are presented in Table 4.4 and Plate 4.1, respectively.

Constructed Wetland, Ropar, India: A decentralised CW-based wastewater treatment system of capacity 0.5 MLD was constructed in Ropar, Punjab, India in year 2006 for treating the domestic wastewater of village community. This system was established by the local village community with financial assistance from the Punjab State Government. The system was established to (i) improve the sanitation settings in the village community, and (ii) get a continuous supply of water for rejuvenating the adjacent pond used traditionally for fishing. In the past few years, the system has performed satisfactorily and created employment through pisci-culture in the rejuvenated pond.

The annual income that was recorded in the past few years happened to be about USD4500 to USD5000. The local community reported that, before establishment of CW, there were many small ditches in which wastewater would stagnate and led to mosquito breeding and foul odors in the community. The system has helped noticeably improve sanitation in the community. The treatment system is socially sustainable because it created job opportunities for the local people with appropriate treatment of wastewater. The system does not require any mechanized or delicate instrumentation. Since highly skilled manpower is not required for operating the system, the local population is able to manage the system effectively. The excess treated water from the fishpond is utilised to irrigate the nearby field crop.

Waste Stabilization Pond, Mathura, India: The Yamuna Action Plan (YAP) was put into motion by the Government of India to address the problem of severe pollution in the Yamuna River, one of the major rivers in India, under the umbrella of a larger National River Conservation Plan. In the year 2000, a 14.5 MLD sewage treatment plant (STP) consisting of a waste stabilization pond (WSP) system was constructed to treat domestic wastewater before discharge into the Yamuna River under Phase I of the YAP (YAP-I) in the city of Mathura, Uttar Pradesh, India. Shortly after, the STP was handed over by the state government of Uttar Pradesh to the Mathura Municipality for operation and maintenance. The primary purposes for establishment of this treatment plant were to limit excessive waste load on the Yamuna River by treating the domestic wastewater that was previously discharged directly to the river without treatment and to provide the treated wastewater to the peri-urban community for irrigation. A part of the treated wastewater is currently being reused for irrigation of agriculture fields adjoining the STP and the remainder is being discharged into the Yamuna River. In wastewater-irrigated fields, different types of crops are being produced. The most cultivated crops include Brinjal (eggplant), Colocasia, Cucumber, Rice, Wheat, Maize, Millet, Barley, Jute, Cotton, Sugarcane and Oil Seeds etc. The treated wastewater is well received by the farmers because of two major advantages compared to using bore-well water. First, the treated wastewater is available to farmers at a lower price compared to bore-well water, and second, it has reduced the fertilizer demand of the irrigated fields substantially without any drop in agricultural yields. Farmers therefore prefer using treated wastewater rather bore-well water. In future, the STP is expected to create employment opportunities for the nearby community with increasing agriculture activities if treated wastewater is available for irrigation. It was also observed that the communities involved in agricultural activities are poor and not able to afford the bore-well water for irrigating their fields. Therefore, the treatment plant has rapidly become the backbone for economic development of the poor community involved in agriculture activities in that area.

Sewage Fed Aquaculture, India: A waste stabilization pond (WSP) of 8 MLD was constructed in 1999 under the Yamuna Action Plan to treat domestic wastewater from the city of Karnal in the state of Haryana in northern India. The WSP system is being admirably managed by the Karnal Municipal Corporation in association with the local communities for the treatment of wastewaters. Generation of revenue has been achieved through practices of simultaneous wastewater treatment, pisciculture and irrigation. Fish breeding is practiced in the facultative and maturation ponds of the treatment system, where suitably treated wastewater provides a rich source of nutrition to the fishes and gets further purified in the process. The nutrient rich treated wastewater, is sold to farmers to support irrigation, serving the production of a wide variety of crops including rice, wheat, maize, millet, barley, jute, cotton, oilseeds and sugarcane. The fish bred in the treatment units are sold in the local market, generating reported revenue of around USD10000–15000 per year. The revenue generated from pisciculture and supply of treated wastewater to irrigation is utilized to collect wastewater and operation and maintenance of the treatment system. Needless to mention, employment opportunities that have been created by the initiative and has improved standards of living among the local communities.

Duckweed Pond, Ludhiana, India: The Duckweed Pond of 0.5 MLD was established in the year 2004 in the village Saidpur, Ludhiana, Punjab to address the problem of wastewater generated from the village community. Prior to establishment of treatment plant, the wastewater from the village community was disposed around the village in near low-land areas and created many problems like wastewater logging, odour problems, mosquitoes and many other unaesthetic issues. The wastewater treatment system generates revenue, which is utilized by the Village Council for sewage collection, operation and maintenance of treatment system. The wastewater is treated sequentially in a duckweed pond followed by a fish pond. Duckweed is harvested from duckweed pond and transferred to the fish pond as food stock. The excess treated water from fish pond is used to irrigate adjoining fields. The treatment plant generates three benefits for the operators, (i) revenue from pisiculture, (ii) revenue from orange trees planted on the edge and (iii) treated wastewater to the agriculture fields. It was observed that system generated about INR 50,000–70,000 per year, which is adequate for operation and maintenance for the system.

Polishing Pond, Agra, India: AUASB process with polishing ponds was constructed in Agra a small town in north India along the Cis-Yamuna area under YAP-I to treat 14 MLD wastewater in the year 2009. The overall performance of sewage treatment plant was found to be satisfactory and was able to meet the wastewater standards for disposal into the Yamuna River. The good quality of treated effluent made it attractive for reuse in irrigation. Plant operators reported that farmers adjoining the treatment plant have asked for the treated wastewater to irrigate their field but because of lack of a wastewater distribution system, the treated wastewater is not being currently reused. Currently, chlorine is used to disinfect the effluent before it is discharged into the Yamuna River. While chlorine may be toxic to river ecosystem, it is acceptable for application on agriculture land because the risk of exposure of farmers to pathogens is eliminated. The treatment plant and associated agriculture fields are close to the city of Agra, therefore the practice of irrigating vegetables that are not eaten in raw form may be allowed after secondary treatment followed by chlorine disinfection.

Karnal Technology, Ujjain, India: A 1.79 MLD Sewage Treatment Plant, Barogarh, Ujjain, Madhya Pradesh, central India situated on the bank of Shipra River is based on Karnal Technology (KT). The treatment plant was established in year 2002 through funding from the National River Conservation Directorate (NRCD) to minimize direct wastewater discharge into the Shipra River. The total daily sewage treatment (consumption) capacity is around 1.79 MLD. The treatment plant receives only domestic wastewater and the applied wastewater is completely absorbed by the soil-plant-bed. The applied wastewater does not appear to affect the growth and health of the planted trees since there is no indication of their mortality. The trees appear to be mature and regulatory body is planning to harvest the tree crop. The nearby community has warmly accepted the treatment plant as it gives a very pleasant green look to the area.

Opportunities and Constraints of Natural Treatment Systems: Eco-technologies are still in their infancy, to date and, thousands of eco-technology projects are proceeding throughout the world because scientists are continuously looking for new applications for eco-technology. Utilizing nature's potential to remediate contaminated lands, ground water and surface water, and to treat wastewater may work out to be inexpensive compared to conventional options.

Installation costs of most eco-technologies are typically lower because these installations use standard or slightly modified earthen structures and practices. Because the primary energy input is solar, operating costs are also low. Although, the low cost is attractive, eco- technology has only just begun to gain some commercial acceptance in India in the past few years based on encouraging findings from field-scale research. The rate and extent of remediation of many contaminants in different matrixes in the Indian context, however, is yet to be determined.

There is a widespread intuitive agreement among the today's scientific community that a site covered in vegetation seems less hazardous than an abandoned lot. Public acceptance of an eco-technology project using subsurface flow constructed wetland can be very high, because it looks like a garden (commonly called wastewater garden) park, providing shade, dust control, and serving as a habitat for birds and wildlife. The wastewater garden can achieve the same result as a septic system, but is far more attractive from an aesthetic standpoint. It is a zero discharge system for the complete utilization of wastewater, resulting in the end products of plant biomass, evaporated water, carbon dioxide, and heat. The plants, when harvested, may be composted (the natural method of recycling), used, or even sold for profit.

Eco-technology based treatment schemes, however, may not achieve regulatory treatment goals, since many biological treatments have endpoints above analytical detection limits. As a consequence, widespread use of eco-technologies is not found worldwide. The clean-up time can be longer with natural processes than with some physical or chemical processes—usually of the order of several growing seasons.

Eco-technologies seem to be affordable mostly when contaminants are within the reach of biological component of the system. For example, in the case of a phytoremediation project, the technology is only viable when contaminants are in the root zone of the plants (top 3–6 ft). For sites with contamination spread over a wide area, eco-technology may be the only economically feasible technology.

4.7 CASE STUDY 2: INTEGRATING NATURAL TREATMENT SYSTEMS FOR UPGRADATION OF AQUATIC SYSTEMS

The JalMahal Tourism Project, Jaipur envisages restoration of JalMahal monument in the Man Sagar Lake together with ecological restoration of 130 hectares Man Sagar Lake water along with its precincts and the development of adjoining 100 acres of prime lake front land to develop Tourism and Entertainment Projects. The site is located in the world renowned Jaipur-Amer Tourist Corridor which attracts approximately 650,000 national and 175,000 international tourists every year.

Prior to the rejuvenation effort, Mansagar Lake suffered from problems of siltation and settled deposits, contamination from inflow of wastewaters, decrease in water surface area due to artificial land formation, decline in spread area due to the outflow of waters for downstream irrigation, and other factors. All these problems have been overcome in time through a carefully planned eco-restoration program. In order to ensure that the Mansagar Lake was stocked with enough water, Jaipur City wastewater was treated using activated sludge process followed by tertiary treatment using a constructed wetland system and then discharged into the lake. After the restoration of Mansagar Lake in a step-wise manner, more than 20 species of birds are now nesting on the artificial islands created from dredging material from the lake. The Mansagar Lake happens to be a perfect example of restoration through integrating natural treatment systems for upgradation of aquatic systems in India (pictures are shown in Plate 4.2).

Since time immemorial, mankind has continued to exploit the nature and natural resources for their basic needs and prosperity. The overexploitation of natural resources, however, has introduced lethal contaminants into our environment and the matter has become an agenda for discussion among professionals from all arena including academics, politics, and researchers. Experts on environment, ecology, and development are now sending alarming signals to current and future generation to take appropriate action. Efforts are underway to develop technologies to avoid the further damage and correct our past mistakes. Indeed these efforts have been able to control the rate of further damage but the path to remedy of past activities is still not clear.

PLATE 4.2 Upgradation of aquatic systems throuh integrating natural treatment systems at Jal Mahal monument, Jaipur, India. (1) Jal Mahal monument before restoration, (2) Jal Mahal monument after restoration, (3) Natural Treatment system installed for fulfillment of daily Lake water requirements, and (4) Some of identified bird species in Lake habitat.

Interestingly, scientists and technocrats have now started postulating about appropriate and safe solutions to these pollution problems with substantial help from nature and natural resources. These could be simpler and less expensive—attractive solution for present and emerging environmental management problems. In sum, the above approaches (as described in two case studies) can be broadly labeled as the "applications of ecological engineering". These case studies also illustrate how ecosystems can be used in a sustainable manner for the benefit of communities. Moreover, recently a India-wide study completed on natural treatment system summarizes that the Indian Government has implemented more than one hundred facilities aimed at treatment and reuse of domestic wastewater as well as rejuvenation of lakes and rivers.

4.8 CASE STUDY 3: SUSTAINABLE RECYCLING OF OBSOLETE SHIPS THROUGH ECOLOGICAL ENGINEERING

"Ship recycling" is the ensemble of activities which involves recycling/reuse of ferrous/non-ferrous metals and materials derived from the process of scrapping (or recycling or dismantling) of obsolete ships. Ship recycling allows materials from ships, especially steel to be given a new life because more than 90% of the material recovered from ship recycling is steel and the remaining 10% are primarily reusable materials. Thus, ship recycling reduces the load on mining activities for the demand of steel. Research carried out by Asolekar (2006) shows that carbon foot print of steel produced by recycling obsolete vessels is 2.78 times less compared to the steel produced through mining activities.

Recycling of obsolete vessels is being practiced for more than four decades in India along the coast of Alang in the state of Gujarat (Plate 4.3). India is the leading ship recycling country in the world, dismantling nearly half of the end of life ships every year. The 12 km stretch of Alang beach has a very high tidal range (average 13 m) and gentle slope of 10° as well as consolidated sandy bottom—which makes the beach at Alang an ideal location for ship recycling. The beaching method of ship recycling in Alang helps a ship to use its own power to beach in the yard during high tide without using any external source of energy for pulling the ship to water-beach interface due to Alang's unique natural setting. Thus, the very first and vital step of beaching ships for recycling in Alang yards starts with the application of ecological engineering.

In other words, natural tidal forces and water stage is used to achieve the engineering goal of beaching. The competing technology practiced elsewhere in India and by other ship dismantling countries requires dry docks. This distinction of dismantling yards in Alang has made them competitive as well as environment friendly through application of

PLATE 4.3 Alang ship recycling yard.

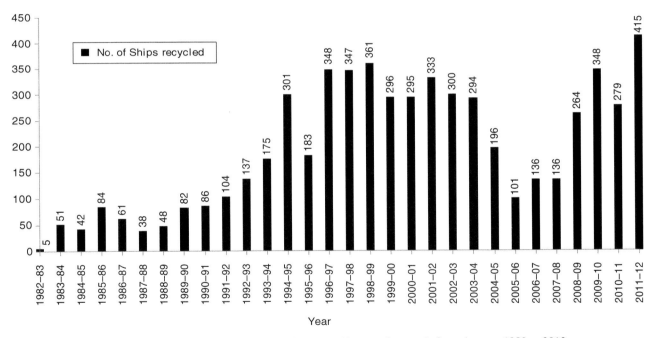

FIGURE 4.5 Number of ships recycled in Alang ship recycling yards from the year 1982 to 2012.

principles of ecological engineering. All other methods have proven to be expensive in terms of energy consumption, construction and maintenance of dry docks, slip ways, floating platforms and service tugs as well as from the point of view of the corresponding damage to ecosystem and environment.

Alang Port has more than 160 active yards engaged in ship recycling along its 12 km stretch beach. Figures 4.5 and 4.6 show the number of ships dismantled and the quantity of steel recycled in Alang Port over 30 years (1982–2012). This sector provides direct employment to around 60,000 people in the Alang ship recycling yards and more than 100,000 jobs in service industry and ancillaries dependent on recycling yards. In addition, nearly 400 small and medium scale steel and ferrous scrap metal re-rolling mills and electric arc furnances at the out-skirts of Alang Port reprocess the recycled steel produced by the yards in Alang and supply the market with value-added steel construction materials and plates. Also, more than 800

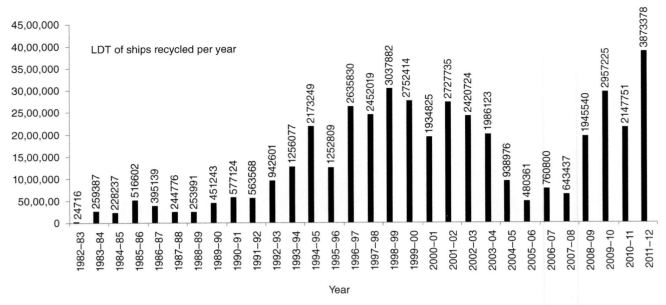

FIGURE 4.6 Quantity of steel recycled from obsolete ship recycling in Alang ship recycling yards from the year 1983 to 2012. 1 Light Displacement Tonnage (LDT) = 1000 Metric Tons.

PLATE 4.4 Plate cutting activity in Alang ship recycling yards using hand-held Oxygen-LPG torch.

shops selling refurbished second-hand materials are located in the vicinity of the ship recycling yards—which sell reuasable and recyclable objectsrecovered that are carefully separated during the ship dismantling process.

The ship recycling operation at Alang combines traditional ship recycling practices to save resources and energy with the application of modern engineering and management knowledge to organize the dismantling processes in a sustainable manner. For example, manual cutting of steel plates using hand-held Oxy-LPG torch is the most energy efficient, cost-effective and environmental friendly way of dismantling ships.

Plate 4.4 depicts the plate cutting activity in Alang ship recycling yards. A closer inspection of Plate 4.4 reveals that the entire ship breaking activity is carried out in open environment, thus minimizing workers' exposure to hazardous situations in enclosed spaces. Also, the refrigerants such as chloflorocarbons and other ozone layer depleting gases recovered from the obsolete ships during breaking are handed over to the authorized service agencies for reuse (if the chemical can be legally reused) or thermal destruction.

The existing ship recycling industry in Alang is based on a know-how gradually developed over the past 40 years by adopting social engineering practices. It uses the natural energy of tidalwaves combined with advantage from some unique geographical and coastal conditions existing in Alang Port. By putting together the obvious advantage of the existence of the unique natural settings with the presence of a well-nurtured market that recognizes the commercial value of refurbished and used second-hand materials, gadgets and components; the ship recycling industry in Alang has proven to be by far the most environment friendly way of ship recycling in the World. As reported by Deshpande et al. (2012) 1–5% of the entire steel demand in India is satisfied by the Alang ship recycling industry alone. Plate 4.5 shows the reusable materials recovered from obsolete ships before breaking and stored in the yard. Thus, India is the world's leading ship recycling country which works on the principle of waste to wealth.

Ships that need to be dismantled today were built 25 or more years ago when toxic components were not banned and could be used without any legal barriers during building and repairing of the ships. When refitting and repair becomes uneconomical, the ship is sent for recycling or for scrap. Ship owners usually sell their ships through brokers operating from different parts of the world such as London, Dubai, Singapore and Hamburg for recycling. Auction of "end-of-life" ship is mainly based on the amount of steel in the ship and usually measured in terms of Light Displacement Tonnage (LDT) (Deshpande et al., 2012, 2013; Asolekar, 2012; Asolekar and Hiremath, 2013).

The process of ship recycling is complicated due to the structural complexity of ships and many environmental, safety and health issues related to dismantling. The hazardous wastes output from ship recycling varies from less than 0.5 to 10% of the ship's total weight but this seemingly small percentage generates enormous quantities of hazardous and nonhazardous wastes because of the huge weight of a typical ship. These hazardous and nonhazardous wastes can potentially create pollution of the surrounding sea, land and air due to inappropriate dismantling practices. These wastes include hazardous materials such as asbestos, thermocol, glass wool, oily sludges, oily rags, bilge water, garbage as well as non-hazardous solid wastes.

Several interventions have been suggested by Asolekar and co-workers to handle these wastes generated from obsolete ship breaking in a scientific way. Some of them are listed below:

1. Scraping of paint on the surface of ship or steel plate before cutting to minimize exposure of workers to toxic heavy metals present in the coated paint.
2. Application of solidification and stabilization principle to handle paint chips generated during plate cutting operation.
3. Appropriate techniques for safe and environment friendly removal and disposal of asbestos and glass-wool are being experimented at laboratory scale.

In sum, ship recycling by beaching method, as carried out in the Indian subcontinent is an environmentally and economically sound practice and safe for workers. The industry is labour and capital intensive. Therefore, ship recycling solutions that are economically viable for all stakeholders and that are fair and equitable for worker's considering the socio-economic situation in the region is the key for long term success.

There is no option as attractive as "complete recycling" because any alteration leading to a lesser extent of recycling and reuse will certainly generate a larger environmental

PLATE 4.5 Reusable materials recovered from obsolete vessels before breaking and stored in the yard.

footprint and associated adverse health impact. Complete recycling of materials, objects and metals from the ship at its end-of-life is the most important feature of ship recycling in Alang, India.

4.9 NEXUS BETWEEN TRADE AND ENVIRONMENT

There is a complex relationship between trade and environment. Growing international trade has been recognized as one of the major contributors to environmental damage in the past few decades. The optimum and best utilization of resources can only be possible by practicing principles of sustainable development. At the most basic level, environment is the endless reserve of not only the basic inputs for trade (including metals, minerals, soil, forests and fisheries) but also the energy required to process these raw materials. Environment has also been treated as a dump yard for the huge volume of wastes generated as a result of numerous economic activities.

With an increasing number of countries joining the World Trade Organization (WTO), liberalization of trade and investment has become a global trend owing to the benefits of Foreign Direct Investment (FDI) inflows, which boost economic performance and growth of the host country. Conversely, this liberalization can also lead to an increase in pollution, especially for developing countries, which might have less stringent environmental regulations.

As a result of recognition of the environmental concerns associated with trade, global markets have undergone a fundamental shift from consumerism to "green consumerism", leading to a substantial increase in market demands for greener goods and services. In this view, technology transfer has assumed a key role to play in making a move from environmentally intensive production to cleaner production. Environmental regulations are also one of the major factors affecting trade patterns. Environmental assessments of agreements have been witnessed to hold an increased importance in mitigation of environmental impacts related to trade agreements. Trade policies and market based approaches have been identified as significant regulators of environmental impacts of trade, with technology transfer having a vital role to play (Asolekar and Gopichandran, 2005).

4.10 CLOSURE: STRATEGIES FOR TRANSITIONING TO SUSTAINABLE FUTURES

Anthropogenic activities are already becoming drivers for irreversible and immediate changes in earths system which intern impact on human wellbeing (Crutzen, 2002; Steffen, 2007). It is estimated that there will be an additional two billion people to the seven billion already existing on the Earth by 2050. It is to be noted that the regional problems will eventually cast their shadows on the world community. It is now proved by scientists all over the world that by providing sufficient evidence communities and nations can be urged to rapidly transition to global sustainability. Global sustainability will soon become prerequisite for human wellbeing in coming decades.

Tim Kasten (2012) defined the Global Chemicals Outlook in the context of changing chemical production over an eight year horizon till 2020 with significant increases in developing countries in particular. This also calls for focused action to tackle environmental effects including health effects on humans and other bio resources etc. UNEP lays special emphasis on its links with the WHO in particular through its UNEP WHO Health & Environment Strategic Alliance, the UNDP UNEP Partnership Initiative for integration of Sound Management of Chemicals (SMC) into development planning processes and the UNEP Guidance on the Development of Legal and Institutional Infrastructures for the Sound Management of Chemicals and Recovery of Costs of National Administration (LIRA). While the overall chemical output is valued at more than four trillion USD as of 2010, the costs of inaction on human health and environment also appear to be significant. The estimated value of green chemistry market by 2020 is about USD 100 billion. These aspects are further highlighted in the UNEP Global Chemicals Output (2013) and the growing impetus of green chemistry in several sectors. This approach is manifest in best practices, technologies and products only to displace hazardous materials and methods. A recent assessment by the Department of Innovation, Industry, Science and Research, Australia (2012) highlighted the scope for multi-pronged approaches involving green chemistry, bio—mimicry, green nanotechnology, whole system design, and sustainable energy interventions to ensure greater environmental sustainability. These are in response to a felt need to transition to low carbon economy pathways of development. The US President Green Chemistry Awards (Centi and Perathoner, 2009) are conferred on such aspects as greener synthetic pathways, designing greener chemicals and establishment of greener reaction conditions to acknowledge and further stimulate transitions.

There is a need to develop acknowledge based framework to address or to respond to the risks and opportunities of global environmental change. This knowledge can be developed using a new kind of research by integrating social and natural sciences—which will address the key human development objectives such as shelter, clothes, food, water, air as well as risk reduction at work places, energy equity and security and most importantly health of human beings and surrounding environment. Table 4.5 shows the potential strategies for transitioning to sustainable futures. It would be pointless to overemphasize the significant role green chemistry and ecological engineering will play in our endeavour of building sustainable world of tomorrow.

TABLE 4.5 Strategies for Transitioning to Sustainable Futures

Sr. No.	Strategy	Interventions
1	Innovation in science & technology	IGC & E, waste to energy, green reagents, alternate feedstock, safer design, nano-technology, bio-based chemicals
2	Preventive environmental management	Toxicity reduction, solvent recovery, reduced hazards, sustainable recycle and reuse, cleaner production, eco-industrial development, zero liquid discharge
3	Frugal technology	Ecoscale, bio-energy, bio catalyst, green processing, pinch technique
4	Incremental development	Holistic design, material efficiency, energy efficiency, sustainable consumption
5	Incorporate society/community aspirations	Use of renewable, sustainable disposal of waste
6	Ecological engineering	Bio-energy, biotechnology, bio-fuel, eco-friendly solvents

This chapter places special emphasis on an inclusive technological framework twinning natural conversions with artificially mediated processes. This is especially significant for value-added treatment and safer disposal. This emerging perspective is critical for developing countries that are yet to mainstream smart production. Large quantities of wastes generated lend themselves to these integrated treatment interventions with implications for reducing externalities significantly. This also creates the context for preventive practices through empirical evidence on the qualitative and quantitative profiles of waste streams. Waste mining could follow waste minimization and reinforces the systems approach in applied environmental management. Bilateral and multilateral institutions help establish an enabling environment to up-scale such interventions through locally adapted practices. This interface has significant public policy implication for sustainable management of resources and externalities, central to sustainable development.

ACKNOWLEDGEMENTS

The authors acknowledge the partial financial support provided by European Commission funded projects entitled: "SaphPani" (for our research in natural treatment systems), "DIVEST" (for our research in ship recycling), Gujarat Maritime Board funded project entitled: "Green Alang Initiative" (for our research in ship recycling) and Indian Institute of Technology Bombay (for our research in natural treatment systems and ship recycling). The authors are indebted to Sachin Kumar Pandey and Parmatma Prasad for their contribution in the field research and timely help.

The following list of exercises covers some of the important areas of green chemistry and related applications. These could be used in formal and informal learning settings.

EXERCISE

4.1 "Green chemistry principles can be used as a preventive environmental management tool" justify the statement. Explain the criteria which are responsible for sustainable solution.

4.2 Using the green chemistry principles develop a framework for sustainable ecological system. Articulate the relationship between green chemistry and green engineering.

4.3 Develop a concept map on "green engineering design." Write this term at the top center of your large sheet of paper, then brainstorm related ideas on sticky notes, arrange the notes so that related terms are closer to each other and draw links between terms that you think are related.

4.4 Generate a material efficiency diagram for the ship recycling sector and connect it with ecological engineering and social sustainability.

4.5 Investigate nutrient enrichment and eutrophication in natural waters. Articulate the effects of high inputs of nutrients from anthropogenic sources; explain what these sources may be and the effects on both the water quality and the aquatic biota. Is there a way to reverse this process? How can we prevent this occurring? What are the implications for inputs of nutrients to the sea?

4.6 What factors lead consumers to buy organic food? How much more expensive is organic food? What arguments are people using for and against buying organic foods?

4.7 Briefly explain the social sustainability, green engineering, ecological engineering and material efficiency aspects of the case studies discussed in this chapter. Also, conduct a brainstorming session on solving a real life problem using the approaches suggested in this chapter.

REFERENCES

Ali, I. and V.K. Gupta 2007. Advances in water treatment by adsorption technology. *Nature Protocols.* 1:2661–2667.

Allen, D.T. and D.R. Shonnard 2001. *Green engineering: Environmentally conscious design of chemical processes.* Prentice Hall, New York.

Allwood, J.M., S.E. Laursende, C.M. Rodriguez, and N.M.P. Bocke 2006. *Well dressed? The present and future sustainability of clothing and textiles in the United Kingdom.* University of Cambridge, Cambridge.

Allwood, J.M., M.F. Ashby, G. Timothy, G. Gutowaski, and E. Worrel 2011. Material efficiency: A white paper. *Resource, Conservation and Recycling.* 55:362–381.

Anastas, P.T. and J.C. Warner 1998. *Green chemistry: Theory and practice.* Oxford University Press, Oxford.

Anastas, P.T. and J.B. Zimmerman 2003. Design through the 12 principles of green engineering. *Environmental Science and Technology.* 37(5):95–101.

Anastas, P.T., M.M. Kirchhoffand, and T.C. Williamson 2001. Catalysis as a foundational pillar of green chemistry. *Applied Catalysis A: General.* 221:3–13.

Andreozzi, R., V. Caprio, A. Insola, and R. Marotta 1999. Advanced oxidation processes (AOP) for water purification and recovery. *Catalysis Today.* 53:51–59.

Arceivala, S.J. and S.R. Asolekar 2006. *Wastewater treatment for pollution control.* Tata McGraw Hill Education (India) Pvt. Ltd., New Delhi.

Arceivala, S.J. and S.R. Asolekar 2012. *Environmental studies: A practitioner's approach.* Tata McGraw Hill Education (India) Pvt. Ltd., New Delhi.

Armenti, K.R., R. Moure-Eraso, C. Slatin, and K. Geiser 2011. Primary prevention for worker health and safety: Cleaner production and toxics use reduction in Massachusetts. *Journal of Cleaner Production.* 19:488–497.

Asolekar, S.R. 2006. Status of management of solid hazardous wastes generated during dismantling of obsolete ships in India. In Proceedings of the first international conference on dismantling of obsolete vessels, Glasgow, UK.

Asolekar, S.R. 2012. Greening of ship recycling in India: Upgrading facilities in Alang. In Proceedings of 7th annual ship recycling conference, London, UK.

Asolekar, S.R. and R. Gopichandran 2005. *Preventive environmental management—An Indian perspective.* Foundation Books Pvt. Ltd., New Delhi.

Asolekar, S.R. and A.M. Hiremath 2013. India's contribution in preventive environmental management of obsolete vessels. *Indian Ports and Infrastructure Magazine.* 2012:5–7.

Asolekar S.R., P.P. Kalbar, M.K.M. Chaturvedi, and K.Y. Maillacheruvu 2014. Rejuvenation of Rivers and Lakes in India: Balancing societal priorities with technological possibilities. p. 181–229. In Ahuja S. (ed.) *Comprehensive water quality and purification.* Elsevier, Boston, MA.

Avsar, E. and G.N. Demirer 2008. Cleaner production opportunity assessment study in SEKA Balikesir pulp and paper mill. *Journal of Cleaner Production.* 16:422–431.

Balat, M., M. Balat, E. Kırtay, and H. Balat 2009. Main routes for the thermo-conversion of biomass into fuels and chemicals: Gasification systems. *Energy Conversion and Management.* 50:3158–3168.

Basu, A.J. and D.J.A. van Zyl 2006. Industrial ecology framework for achieving cleaner production in the mining and minerals industry. *Journal of Cleaner Production.* 14:299–304.

van Berkel, R. 2007. Cleaner production and eco-efficiency initiatives in Western Australia 1996–2004. *Journal of Cleaner Production.* 15:741–755.

Boodhoo, K.A. and A.P. Harvey 2013. *Process intensification technologies for green chemistry: Engineering solutions for sustainable chemical processing.* 1st ed. John Wiley & Sons, Ltd, Chichester.

Bosmans, A., I. Vanderreydt, D. Geysen, and L. Helsen. 2013. The crucial role of waste-to-energy technologies in enhanced landfill mining: A technology review. *Journal of Cleaner Production.* 55:10–23.

Brix, H. 1994. Constructed wetlands for municipal wastewater treatment in Europe. p. 325–334. In Mitsch, W.J. (ed.) *Global wetlands: Old world and new.* Elsevier, Amsterdam.

Cavalcante Jr., C.L. 2000. Industrial adsorption separation processes: Fundamentals, modeling and applications. *Latin American Applied Research.* 30:357–364.

Centi, G. and S. Perathoner 2009. From green to sustainable chemistry. p. 1–72. In Cavani, F., G. Centi, S. Perathoner, and F. Trifiro (eds.) *Sustainable industrial processes.* Wiley-VCH, Weinheim.

Chaubey, R., S. Sahu, O.O. James, and S. Maity 2013. A review on development of industrial processes and emerging techniques for production of hydrogen from renewable and sustainable sources. *Renewable and Sustainable Energy Reviews.* 23:443–462.

Cintas, P. and J.L. Luche 1999. Green chemistry: The sonochemical approach. *Green Chemistry.* 1:115–125.

Cossu, R., W. Hogland, and E. Salerni 1996. Landfill mining in Europe and the USA. ISWA Year Book. 107–114.

Crutzen, P.J. 2002. Geology of mankind: The Anthropocene. *Nature.* 415:23.

Cue, W. and J.B. Zhang 2009. Green process chemistry in the pharmaceutical industry. *Green Chemistry Letters and Reviews.* 2(4):193–211.

Dabrowski, A. 2001. Adsorption-from theory to practice. *Advances in Colloid and Interface Science.* 93:135–224.

Damartzis, T. and A. Zabaniotou 2011. Thermochemical conversion of biomass to second generation biofuels through integrated process design—A review. *Renewable and Sustainable Energy Reviews.* 15:366–378.

Department of Innovation, Industry, Science and Research, Australia. 2012. Trends in manufacturing to 2020: A foresighting discussion paper [online]. Available at http://www.innovation.gov.au/industry/futuremanufacturing/FMIIC/Documents/TrendsinManufacturingto2020.pdf (accessed on November 25, 2013 and verified on November 27, 2013) Future manufacturing. Industry Innovation Council, Australia.

Deshpande, P.C., A. Tilwankar, and S.R. Asolekar 2012. A novel approach to estimating potential maximum heavy metal exposure to ship recycling yard workers in Alang, India. *Science of the Total Environment.* 438:304–311.

Deshpande, P.C., P.P. Kalbar, A.K. Tilwankar, and S.R. Asolekar 2013. A novel approach to estimating resource consumption rates and emission factors for ship recycling yards in Alang, India. *Journal of Cleaner Production.* 59, 15:251–259.

Eckerman, I. 2001. Chemical Industry and Public Health—Bhopal as an example. Essay for MPH, A short overview, 57.

Eckerman, I. 2005. *The Bhopal Saga—Causes and Consequences of the World's Largest Industrial Disaster*. Universities Press, New Delhi.

Eckerman, I. 2013. Bhopal gas Catastrophe 1984: Causes and consequences. p. 302–316. In *Encyclopedia of environmental health*. Elsevier, Burlington.

Ecobichon, D.J. 2001. Pesticide use in developing countries. *Toxicology*. 160:27–33.

Feng, Z.J. 2004. *Introductory remarks of circular economy*. Beijing People's Publishing Agency, Beijing.

Ferat, M. 2004. Implementation of design for the environment (DFE) in a Mexican chemical group. *UNEP Industry & Environment*. 27(3):47–51.

Gaber, Y. 2012. Hydrolases as Catalysts for Green Chemistry and Industrial Application, Esterase, Lipase and Phytase. PhD Diss. Lund University, Lund.

Gandhi, R.R., J. Suresh, S. Gowri, S. Selvam, and M. Sundraraja 2013. Ultrasonic dyeing of enzyme treated organic cotton using *Nyctanthes Arbor-Tristis*. *Chemical Science Transactions*. 2(2):642–664.

Garud, R.M., S.V. Kore, V.S. Kore, and G.S. Kulkarni 2011. A short review on process and applications of reverse osmosis. *Universal Journal of Environmental Research and Technology*. 1:233–238.

Gupta, K.K., A. Rehman, and R.M. Sarviya 2010. Bio-fuels for the gas turbine: A review. *Renewable and Sustainable Energy Reviews*. 14:2946–2955.

Habiba, R.A., H.M. Ali, and S.M. Ismail 1992. Biochemical effects of Profenofos Residues in potatoes. *Journal of Agricultural and Food Chemistry*. 40(10):1852–1855.

Hanne, R., K. Kristina, K. Alexander, B. Arunas, S. Pekka, R. Matti, K. Outi, and N. Marita 2011. Thermal plasma-sprayed nickel catalysts in the clean-up of biomass gasification gas. *Fuel*. 90:1076–1089.

Hansson, S.O. and C. Rudén 2004. A science-based strategy for chemicals control. *UNEP Industry & Environment*. 27(3):12–15.

Hicks, C. and R. Dietmar 2007. Improving cleaner production through the application of environmental management tools in China. *Journal of Cleaner Production*. 15:395–408.

Huppes, G. and M. Ishikawa 2009. Eco-efficiency guiding micro-level actions towards sustainability: Ten basic steps for analysis. *Ecological Economics*. 68:1687–1700.

Jegannathan, K.R. and P.H. Nielsen 2012. Environmental assessment of enzyme use in industrial production: A literature review. *Journal of Cleaner Production*. 42:228–240.

Kangas, P.C. 2003. *Ecological engineering: Principles and practice*. Lewis Publishers, Boca Raton, FL.

Karthikeyan, K., S.H. Kim, K.J. Kim, and S.N. Lee 2013. Low cost, eco-friendly layered Li1.2 (MnO.32NiO.32FeO.16) O_2 nanoparticles for hybrid supercapacitor applications. *Electrochimica Acta*. 109:595–601.

Kasten, T. 2012. Towards a non-toxic environment: A UNEP perspective [online]. Available at http://www.kemi.se/Documents/Giftfri%20milj%C3%B6/Forum/2012/Presentationer/TimKasten.pdf (accessed on January 04, 2014 verified on January 08, 2014) UNEP, Nairobi, Kenya.

Kaushik, G., S. Satya, and S.N. Naik 2009. Food processing a tool to pesticide residue dissipation—A review. *Food Research International*. 42:26–40.

Krook, J., N. Svensson, and M. Eklund 2012. Landfill mining: A critical review of two decades of research. *Waste Management*. 32:513–520.

Lamb, G.W., F.A.A. Badran, J.M.J. Williams, and S.T. Kolaczkowski 2010. Production of pharmaceuticals: Amines from alcohols in acontinuous flow fixed bed catalytic reactor. *Chemical Engineering Research and Design*. 88:1533–1540.

Lancaster, M. 2002. *Green chemistry: An introductory text*. Royal Society of Chemistry, London.

Lee, K.P., T.C. Arnot, and D. Mattia 2011. A review of reverse osmosis membrane materials for desalination—Development to date and future potential. *Journal of Membrane Science*. 370:1–22.

Li, Y., A.S. Fabiano-Tixier, M.A. Vian, and F. Chemat 2013. Solvent-free microwave extraction of bioactive compounds provides a tool for green analytical chemistry. *Trends in Analytical Chemistry*. 47:1–11.

Liua, P., Y. Zhong, and Y. Luo 2014. Preparation of monodisperse biodegradable magnetic microspheres using a T-shaped microchannel reactor. *Materials Letters*. 117:37–40.

Lotter, L.H. 2004. Integrated chemical management: Dream or reality in the developing world? *UNEP Industry & Environment*. 27(3):19–22.

MacMillan, D.S., J. Murray, H.F. Sneddon, C. Jamiesona, and A.J.B. Watson 2013. Evaluation of alternative solvents in common amide coupling reactions: Replacement of dichloromethane and *N,N*-dimethylformamide. *Green Chemistry*. 15:596–600.

Mara D.D., S.W. Mills, H.W. Pearson, and G.P. Alabaster 1992. Waste stabilization ponds: A viable alternative for small community treatment systems. *Water and Environment Journal*. 6:72–78.

McKendry, P. 2002. Energy production from biomass (part 1): Overview of biomass. *Bioresource Technology*. 83:37–46.

Merck. 2012. Case study: A Journey to revolutionize drug manufacturing in an industry charged with helping the world's population be well: Green Chemistry is a critical tool [online]. Available at http://www.merckresponsibility.com/wp-content/uploads/2013/07/green_chemistry.pdf (accessed on December 15, 2013 verified on December 17, 2013) Merck, Whitehouse Station, New Jersey.

Mitsch, W.J. 1998. Ecological engineering—The 7-year itch. *Ecological Engineering*. 10(2):119–130.

Nagendramma, P. and S. Kaul 2012. Development of ecofriendly/biodegradable lubricants: An overview. *Renewable and Sustainable Energy Reviews*. 16:764–774.

NIKE, Inc. 2012. NIKE materials sustainability index [online]. Available at http://www.apparelcoalition.org/storage/Nike_MSI_2012_0724b.pdf (accessed on 11 December, 2013 and verified on 12 December, 2013) NIKE, Beaverton, Oregon.

Nurgul, O., E.P. Ayse, and P. Ersan 2001. Structural analysis of bio-oils from pyrolysis and steam pyrolysis of cottonseed cake. *Journal of Analytical and Applied Pyrolysis*. 60:89–101.

Odum H.T. 1963. *Experiments with engineering of marine ecosystems*. Publication of the Institute of Marine Science of the University of Texas, Port Aransas, TX. 374–403.

Panwar, N.L., R. Kothari, and V.V. Tyagi 2012. Thermo chemical conversion of biomass—Eco friendly energy routes. *Renewable and Sustainable Energy Reviews.* 16:1801–1816.

Pawelzik, P., M. Carus, J. Hotchkiss, and R. Narayan 2013. Critical aspects in the life cycle assessment (LCA) of bio-based materials-Reviewing methodologies and deriving recommendations. *Resources, Conservation and Recycling.* 73:211–228.

Peric, B., M. Esther, S. Jordi, C. Robert, and M.G. Antonia 2012. Green chemistry: Ecotoxicity and biodegradability of ionic liquids. p. 89–113. In *Recent Advances in Pharmaceutical Sciences II*. Transworld Research Network, Kerala.

Puigagut, J., J. Villaseñor, J.J. Salas, E. Bécares, and J. García 2007. Subsurface-flow constructed wetlands in Spain for the sanitation of small communities: A comparative study. *Ecological Engineering.* 30:312–319.

Rainer, K. 2004. Global strategy on chemicals management: Opportunities and risks. *UNEP Industry & Environment.* 27(3):7–8.

Rana, S., J. Jana, S.K. Bag, S. Mukherjee, J.K. Biswas, S. Ganguly, D. Sarkar, and B.B. Jana 2011. Performance of constructed wetlands in the reduction of cadmium in a sewage treatment cum fish farm at Kalyani, West Bengal, India. *Ecological Engineering.* 37:2096–2100.

Rodriguez, J., S. Stopic, G. Krause, and B. Friedrich 2007. Feasibility assessment of electrocoagulation towards a new sustainable wastewater treatment. *Environmental Science and Pollution Research.* 14(7):477–482.

Savage, G.M., C.G. Golueke, and E.L. von Stein 1993. Landfill mining: Past and present. *Biocycle.* 34:58–61.

Scholz-Boettcher, B.M., M. Bahadir, and H. Hopf 1992. *Angewandte Chemie.* 104:477–479; Angewandte Chemie International Edition England, 31: 443–444.

Shahid-ul-Islam T.W., M. Shahid and F. Mohammad 2013. Perspectives for natural product based agents derived from industrial plants in textile applications: A review. *Journal of Cleaner Production.* 57:2–18.

Sharma, J., S. Satya, V. Kumar, and D.K. Tewary 2005. Dissipation of pesticides during bread making. *Journal of Chemical Health and Food Safety.* 17–22.

Slocuma, D.W., K.C. Tekin, Q. Nguyen, P.E. Whitley, T.K. Reinschelda, and B. Fouzia 2011. p-Iodinations in hydrocarbon media: Continuous flow reactor application. *Tetrahedron Letters.* 52:7141–7145.

Soderholm, P. and J.E. Tiltonb 2012. Material efficiency: An economic perspective. *Resources, Conservation and Recycling.* 61:75–82.

Steffen, W.E. 2007. The Anthropocene: Are humans now overwhelming the great forces of Nature? *Ambio.* 36:614.

Talaviya, S. and F. Majmudar 2012. Green chemistry: A tool in Pharmaceutical Chemistry. *NHL Journal of Medical Sciences.* 1(1):7–13.

UN Documents. 1987. Our Common Future, Report of the World Commission on Environment and Development, World Commission on Environment and Development. Published as Annex to General Assembly document A/42/427, Development and International Co-operation: Environment August 2, 1987.

UNEP. 2004a. The chemical industry and international cooperation to manage chemical risks; facts and figures. *UNEP Industry and Environment.* 27(3):4–6.

UNEP. 2004b. Balancing the benefits of chemicals with their health and environmental risks. *UNEP Industry and Environment.* 27(2–3):4–6.

UNEP Global Chemicals Output. 2013. Towards sound management of chemicals [online]. Available at http://www.unep.org/hazardoussubstances/Portals/9/Mainstreaming/GCO/The%20Global%20Chemical%20Outlook_Full%20report_15Feb2013.pdf (accessed on December 05, 2013 and verified on December 12, 2013) UNEP, Nairobi, Kenya.

Unnikrishnan, S. and D.S. Hegde 2006. An analysis of cleaner production and its impact on health hazards in the workplace. *Environment International.* 32:87–94.

URL 01; Principle of operation – simple solutions http://industrialodorcontrol.com/mm5/pop_ups/Pop_up_33.pdf (accessed on January 08, 2014).

URL 02; Molecular sieve desiccants – nutec overseas fze http://www.nutecoverseas.com/molecular-sleeve.html (accessed on January 08, 2014).

URL 03; Nanotechnology in water treatment – highveld.com http://www.highveld.com/molecular-biology/nanotechnology.html (accessed on January 08, 2014).

URL 04; Reverse osmosis skids – degremont technologies http://www.degremont-technologies.com/dgtech.php?article458 (accessed on January 08, 2014).

URL 05; Microchannel reactor http://www.sintef.no/project/Trondheim_GTS/Presentasjoner/Scale-up%20of%20Microchannel%20Reactors%20for%20small%20scale%20GTL%20Proceses.pdf (accessed on January 08, 2014).

URL 06; Bhopal disaster http://en.wikipedia.org/wiki/Bhopal_disaster (accessed on January 09, 2014).

URL 07; USEPA 2000. Terminology Reference System (TRS 2.0), September 11, 2000, http://oaspub.epa.gov/trs/prc_qry.keywordhtm.

Vallance, S., H.C. Perkins, and J.E. Dixon 2011. What is social sustainability? A clarification of concepts. *Geoforum.* 42: 342–348.

Vymazal, J. 2002. Plants used in constructed wetlands with horizontal subsurface flow: A review. *Hydrobiologia.* 674:133–156.

Warner, J.C., A.S. Cannon, and K.M. Dye 2004. Green chemistry. *Environmental Impact Assessment Review.* 24:775–779.

Watts, P. and S.J. Haswell 2003. Continuous flow reactors for drug discovery. *Research Focus.* 8:586–593.

Watts, P. and C. Wiles 2013. Synthesis of heterocyclic compounds using continuous flow reactors. p. 531–548. In Scriven, E.F.V. *Pyridines: From lab to production.* Academic Press, Riverport Lane, MO.

Wen, C.F., Y.L. Zhaoa, and R.Z. Liang, 2007. Recycle of low chemical potential substance. *Resources, Conservation and Recycling.* 51:475–486.

Winteringham, F.P.W. 1971. Some global aspects of pesticide residue problems. *Israel Journal of Entomology.* 6:171–181.

Wu, H., J.Z.P. Li, J.Z.H. Xie, and B. Zhang, 2011. Nutrient removal in constructed microcosm wetlands for treating polluted river water in northern China. *Ecological Engineering.* 37:560–568.

Xu, J. and Z. Li, 2012. A review on Ecological Engineering based Engineering Management. *Omega.* 40(3):368–378.

Zimmels, Y., F. Kirzhner, and A. Malkovskaja 2008. Application and features of cascade aquatic plants system for sewage treatment. *Ecological Engineering.* 34:147–161.

5

GREEN ENERGY AND CLIMATE CHANGE

R. Gopichandran[1], Shyam R. Asolekar[2], Omkar Jani[3], Dinesh Kumar[2] and Anand M. Hiremath[2]

[1] *Vigyan Prasar, Department of Science and Technology, Government of India, New Delhi, India*
[2] *Centre for Environmental Science and Engineering, Indian Institute of Technology Bombay, Powai, Mumbai, India*
[3] *Solar Energy Research Wing, Gujarat Energy Research and Management Institute, Innovation and Incubation Centre (GERMI-RIIC), Gandhinagar, Gujarat, India*

Abstract: The present chapter defines the unique characteristics of the renewable energy sources and the climate benefits secured by their use. These are aligned with the commitments made by many countries to use these alternatives and help to avoid carbon dioxide and equivalent emissions to tackle challenges posed by climate change. Several multilateral institutions and national governments have joined the fray to help fulfill the goals of climate-friendly energy transitions. The chapter is intended to help the reader with a snapshot of all the aspects indicated previously. The chapter presents information regarding the emission reduction goals envisaged through the use of some predominantly proposed renewable energy/green energy alternatives. The status of commitments, results achieved, and emerging deliberations within the framework of the climate change convention is also presented. Based on the literature, the role played by multilateral institutions and financial institutions in understanding the green energy systems and its implications is highlighted. Reference is also made to other information sources concerning initiatives taken by countries around the world with respect to related enabling circumstances including regulations, fiscal and nonfiscal regulations, and pilot projects to demonstrate the feasibility of such interventions.

Some typical cases in point are the CAFÉ/GHG reporting program of the United States, the ecodesign directive of the EU, and the Top Runner Program of Japan. A special emphasis is placed on India's related initiatives with the hope that countries with a similar development status will be able to draw useful lessons from the information resources cited. A brief account of some predominant barriers is stated toward the end. The mechanics of their coming together are briefly covered with examples of their interventions. Important sources of information and a curriculum that will enable essential learnings in this area of study are also presented.

Keywords: green energy, wind, solar, geothermal, biomass, nuclear, smart grid, renewable sources, climate change, law and policy, regulations, conventions, UNEP, UNFCCCDM, solar panels, energy reports, global warming, CO_2, fuel cell, emission reduction.

5.1	Introduction to Important Attributes and Types of Green Energy	128	5.7 Some Important Precautions	146
5.2	Green Energy and Public Policy	128	5.8 The Way Forward	147
5.3	Emission Reduction Imperatives	129	References	147
5.4	Structural and Functional Dynamics of Renewable Energy Systems with Respect to Climate Change	130	Recommended Curriculum: Green Energy: Systems and Climate Efficiency	149
5.5	Industry Forums Active on Renewable Energy	141	Further Reading	150
5.6	Commitments and Collective Action to Mainstream Renewables	143		

An Integrated Approach to Environmental Management, First Edition. Edited by Dibyendu Sarkar, Rupali Datta, Avinandan Mukherjee, and Robyn Hannigan.
© 2016 John Wiley & Sons, Inc. Published 2016 by John Wiley & Sons, Inc.

5.1 INTRODUCTION TO IMPORTANT ATTRIBUTES AND TYPES OF GREEN ENERGY

The expression green energy is used interchangeably with renewable energy, and it rightly imposes an environment-friendly image of energy systems. Primary energy carriers are substances that have not undergone any technical conversion. These include coal, crude oil, natural gas, biomass, etc. Primary energy refers to energy content of the primary energy carriers and related energy flows. Secondary energy and secondary energy carriers are derived from the primary through conversions. Typical of these are petrol, electrical energy per se, etc. The energy form and application finally consumed by the end user represent the final energy carrier and the energy stream. Useful energy is the actual energy available that meets requirements and is a function of the losses incurred in these stages of conversions. Renewable energy represents primary energy that could be regarded as inexhaustible. Renewable energy sources, coupled with storage capacity, are consistently available via such natural phenomena as solar, wind, geothermal, and tidal systems. Accordingly, a significant variety of renewable energy forms and sources exists and must be derived depending on the end use. A wide range of technical processes are currently available, and likely numerous others will be identified in the future to allow these sources to be used appropriately.

Some of the important features of this green attribute are (i) sustainable sources, (ii) easy access to large quantities of energy, and (iii) the process of converting primary manifestations to end use. Importantly, when these systems are tapped and put to optimal use, they help to establish alternative livelihoods under many challenging socioeconomic circumstances with significantly lesser externalities. The main systems considered green are solar, biomass, and wind, while hydro too is to a certain extent. Geothermal and such emerging systems as heat pumps, solar hydrogen, and fuel cells add to the diversity of green energy options.

In case of solar energy systems, such aspects as spectral range, direct and diffuse impingement, and angle of receivers, measurements, absorption, temperature, emission, and transmission, become important. Some related aspects in the case of thermal solar systems include the efficiency of systems and solar fraction savings, types of collectors, heat stores, sensors, transfer medium, pumps, etc. Several system design concepts including those without circulation and varied circulation patterns and applications determine economic and environmental gains. For example, solar tower power stations, parabolic trough power plants, dish/stirling systems, updraft tower power plants, and solar pond power plants represent the large gamut of applications. These are in addition to photovoltaic (PV) systems and the implication of grid-connected or grid-independent functions.

These energy alternatives are distinctly different from fossil carbon energy systems for which the massive release of greenhouse gases (often referred in terms of carbon dioxide equivalents) can be attributed. Initiatives to mainstream the green energy alternatives are gaining momentum the world over. This heightened initiative is in response to the growing realization that irreversible damages to the planet may be resulting from the flooding of the atmosphere with carbon dioxide and other greenhouse gases. A synthesis of appropriate technologies, regulations, and market instruments enables the transition from fossil fuels to renewable energy sources. Importantly, the United Nations' "Framework Convention on Climate Change" (UNFCCC) embeds protocols that confer carbon credits through its clean development mechanism portfolio in particular and stimulates such transitions. Industry too has exhibited significant leadership and strengthened the case for larger-scale change over.

5.2 GREEN ENERGY AND PUBLIC POLICY

Three important facets of green energy dominate public law and policy across countries. The first and the foremost need is to mainstream alternative energy production and consumption systems in place of carbon-based systems to reduce the generation of greenhouse gases and hence tackle externalities generated in this process. A significant body of information and knowledge is available on the materials of construction, related infrastructure, unit processes and operations, related production, distribution, and efficient end use. Figure (on investments, UNEP 2013) presents a snapshot of increasing proportions of renewable in energy mix complemented with the interplay of policy considerations (Table 5.1). The need to quantify emission reduction benefits and therefore the opportunity to mainstream them emphatically have been substantiated by approved methodologies relevant to the renewable energy sector. (CDM methodology booklet, Nov 2012 Consolidated baseline methodology for grid-connected electricity generation from renewable sources, UNFCCC CDM Executive Board.)

The second facet concerns benefits for communities and economies including the development of alternative and mutually reinforcing livelihood opportunities with significant implications for local-level socioeconomic development. Quantifications of these benefits have also been periodically placed in the public domain by the initiatives of several bilateral and multilateral institutions.

The third facet involves coupling energy efficiency enhancement and emission reduction gains within the power generation and end-use sectors and the benefits for communities and countries as a whole through national agendas. Most importantly, this third aspect is manifested in commitments made by countries to sustain transitions to

TABLE 5.1 Policy Considerations[a]

FIT Design Issue	Investor Security	Energy Access	Grid Stability	Policy Costs	Price Stabilization	Electricity Portfolio Diversity	Administrative Complexity	Economic Development
Integration with policy targets	✓						✓	
Eligibility		✓	✓	✓		✓		✓
Tariff differentiation		✓		✓		✓	✓	✓
Payment based on	✓			✓	✓	✓	✓	
Payment duration	✓			✓	✓			
Payment structure	✓			✓	✓			
Inflation	✓				✓			
Cost recovery	✓			✓				
Interconnection guarantee	✓		✓					
Interconnection costs	✓		✓	✓				
Purchase and dispatch requirements	✓			✓				
Amount purchased	✓						✓	
Purchasing entity	✓						✓	
Commodities purchased	✓			✓			✓	
Triggers and adjustment	✓		✓	✓	✓		✓	
Contract issues	✓							
Payment currency	✓				✓			
Interaction with other incentives	✓		✓					

[a] From UNEP (2013).

climate-efficient and climate-resilient development pathways wherein economically and technologically robust management of renewable energy mixes within the larger energy portfolio is achieved. The major driver for green development is to work toward energy security and self-reliance as a hedge against future risks of fossil fuel supply. These three facets highlight the importance of initiatives that could be verified and projected for cumulative benefit across local, national, and global levels.

Some of the initiatives taken so far are as follows:

TERI, May 2010; Roadmap for upscaling and mainstreaming renewables, Ministry of New and Renewable Energy (MNRE), Dec 2010; Strategic plan for new and renewable energy sector for the period 2011–2017, MNRE, Feb 2011; Feed-in tariffs (FiT) as a policy instrument for promoting renewable energies and green economies in developing countries, UNEP; A citizen's guide to energy subsidies in India TERI and IISD, March 2012; Pradeep S. Mehta, 2012; S.P. Raghuvanshi, A.K. Raghav, and A. Chandra, July 2007; Emerging opportunities and challenges, India Energy Congress, 2012; Renewable energies—Investing against climate change: Renewable energies offer investors broad diversification opportunities with manageable risks, Allianz Global Investors, Nov 2012; Volker Krey and Leon Clarke, 2011; Energy and climate change British, Friends of the Earth.

5.3 EMISSION REDUCTION IMPERATIVES

Emission reduction is interconnected and has to be guided by well-adapted institutional, fiscal, and nonfiscal, regulatory, and related capacity building functions of institutions so that mutually reinforcing mechanisms and outputs will only sustain transitions in a positive and well-coordinated manner. Interestingly, many countries around the world have established locally adapted systems as stated earlier with varying degrees of the extent and depth of applications, aligned with their respective developmental aspirations. This conclusion is evident from some of the most recent publications (energy reports) from agencies worldwide including those of the International Energy Agency (2013); the Department of Industry, Innovation, Climate Change, Science, Research, and Tertiary Education, Australia (2013); the EC (2013); Frankfurt School–UNEP; the European Investment Bank; Yale Project on Climate Change Communication 2013; the Department of Energy and Climate Change, United Kingdom, 2013; the G8 countries; and the European Renewable Energy Council. These are further substantiated by recent sectoral investigation reports by Ernst and Young (2013); the Congressional Research Service Publication by Lattanzio (2013); Richard K. Lattanzio, FY2010-FY2014, Deloitte (Allen, 2013; Atul Kumar, 2013; Working with India to Tackle Climate Change, 2012; Energy-Efficient Cookers for Kitchens in India, 2013),

Bioenergy Times (Apr 2013), BioEnergy Council of India, the International Renewable Energy Agency (IRENA) *Handbook on Nationally Appropriate Mitigation Actions*, 2012 (IRENA Handbook on Renewable Energy, 2012), the World Resources Institute and WWF 2013 (World Resources Institute, 2013), and the World Energy Council of India 2012 (Emerging Opportunities and Challenges, India Energy Congress—2012; Tester (2005); Leiserowitz (2013); Milne and Field (2012); India Energy Congress (2013)).

Some of the crosscutting aspects of sustainability indicated in all the above-cited reports of the agencies worldwide include the cost of technologies, including the materials of construction, standards and best practices to sustain performance levels, quality and quantitative profiles of the proposed renewable energy resources, sustained access to these resources, and capability of entities that generate renewable energy out of such energy substrates to efficiently produce and distribute energy. On the other hand, the preparedness of markets and related institutions to secure the carbon-friendly services produced becomes equally important. At a much higher level of convergence and aggregation, it is essential to ask if countries have aligned their resources and development plans to mainstream alternatives in a systematic manner. These aspects of management of renewable energy resources are of paramount importance because they are embedded within the core framework of emission reduction goals.

It is needless to emphasize that oil reserves are reported to last just another four decades, and coal may last a few decades more than oil reserves. The oil reserves in many countries are already exploited beyond their optimal levels, and for these countries, the alternatives are to import increasingly costly (and diminishing) fossil fuels or to turn to an increased percentage of green energy production and use. Policies that promote renewables are the only option to sustain access and use to meet increasing demands for growth. On the other hand, a huge segment of the world's population does not have access to power grids. It is essential to establish a distributed electricity supply system to meet their needs while improving the quality of life. The essential needs include lighting for basic life and community functions and communication systems, especially in areas well off the grid.

The combustion of coal, oil, and natural gas releases CO_2, a predominant input causing global warming. Production of electricity entails release of more than a third of all carbon emissions globally. An important complementing reason is the low efficiency of power plants in converting fuels into power. It is important to enhance investments considerably to ensure sustainability of energy production, access, and efficient use. This will ultimately bring more renewable sources of energy into picture and thus will help in enhancing the proportion of renewables. Cogeneration is an interesting option wherein waste heat from power generation is used. Significantly, high efficiencies are achieved in these processes twinning economic and environmental gains. This is especially from conventional steam turbines and calls for a close proximity of end user systems including conurbations, community services such as hospitals.

Several cost–benefit models (UNEP, 2012) tend to reinforce larger economic benefits in the longer run in addition to lesser dependence on fossil fuels. Importantly, the latter will only become increasingly costlier. Some of the instruments proposed include FiT that modulate prices for consumers. These models could also phase out less efficient conventional systems as renewable systems are expected to gain ground.

One of the most interesting options as a long-term energy source is the sun, the main source of energy for our planet. The sun provides several hundredfold greater energy resources compared to our current global energy consumption. A significant portion of this quantum is received by the oceans and results in evaporation of billions of cubic meters of water every hour from the oceans. Energy from the sun can be used to generate heat and electricity. These are termed solar thermal and solar power, respectively. It is well known that temperature differential results in fluctuation of atmospheric pressure generate wind and influence weather change. These in turn influence the occurrence and distribution of vegetation and hence access to biomass. Hydro systems are obviously related to rain and hence the intricate link between biomass, wind power, and hydropower as manifestations of solar energy. Importantly, these systems are inexhaustible with reference to our scales of consumption and are deemed renewable forms of energy.

5.4 STRUCTURAL AND FUNCTIONAL DYNAMICS OF RENEWABLE ENERGY SYSTEMS WITH RESPECT TO CLIMATE CHANGE

Seifried and Witzel (2010) presented some interesting figures in this context. These pertain to the extensive land area needed to host solar energy systems and the benefits that can accrue to meet energy needs at the global scale. Some solar energy systems need direct sunlight, but most can utilize both direct and diffuse sunlight.

Solar Energy Systems: Among the renewable energy technologies available today, solar energy holds a special status primarily due to its large scale of potential, as well as the justification, that all most all energy, renewable, or nonrenewable, are a consequence of solar energy stored in our geological and atmospheric system. Figure 5.1 shows potential use of solar energy, and the volume of each box in the figure is proportional to the amount of energy available or consumed.

Solar energy utilization can be philosophically classified into two broad groups: passive and active.

A "passive solar energy system," often interchangeable with "passive solar design," does not utilize any active

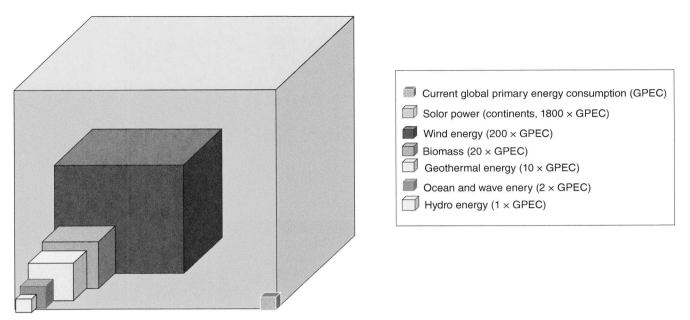

FIGURE 5.1 A schematic symbolizing the potential of solar energy. *Source:* Dr.-Ing. Joachim Nitsch, "Technologische und energiewirtschaftliche Perspektiven erneuerbarer Energien," 2007. http://www.dlr.de/Portaldata/41/Resources/dokumente/institut/system/publications/Leitstudie_2007_Toblach-18-10-07.pdf

mechanical systems to convert solar energy into a usable form and often may serve multiple purposes. By definition, passive solar systems are used to "collect, store and distribute energy by natural radiation, conduction and convection." A layman's example of utilizing passive solar energy is drying of clothes in the sun, which also has a tremendous economic value. This economic value is realized by the argument that if there was no such solar energy available to dry clothes, then there would be tremendous energy consumption using clothes dryers around the world! A more sophisticated application of passive solar energy is through designs of buildings by architects, where the designs are optimized based on the location and climate in order to use natural light and control indoor climate through reflection of radiation and routing of the available wind resource. In fact, evidence of passive solar designs dates back a couple of thousand years in the Native American as well as the Greek civilizations. Today, the maturity of solar passive designs is evident through several global and national-level building codes and standards, as well as certifications such as the Leadership in Energy and Environment Design (LEED) by the US Green Building Council and the Green Buildings Rating System India (GRIHA) by TERI, India.

Active solar energy systems are again broadly classified into two technologies based on the method of utilization of the available solar energy: solar thermal and solar PV. Solar thermal systems utilize the "heat" aspect of solar radiation to obtain the desired output. Simple solar thermal applications range from solar cooking to solar water heaters, where solar energy is collected and trapped either by flat plate collectors or evacuated tubes carrying water or other working fluids. While temperatures up to 150°C can be attained through simple absorption, higher temperatures are achieved by concentrating light mainly with the help of mirrors. The solar thermal energy-generating systems, also known as concentrated solar thermal (CST) systems, are classified based on the mechanism of concentration of solar radiation on the receiver and the subsequent temperatures achieved. Common solar thermal energy technologies that involve parabolic troughs, parabolic dishes, linear Fresnel reflectors, or central receiver towers are shown in Figs. 5.2, 5.3, 5.4, and 5.5, respectively.

The working fluids can attain temperatures in the range of 400°C through parabolic troughs and linear Fresnel concentrators; synthetic oils are often used as working fluids at these temperatures. Parabolic dishes and tower technologies achieve higher temperatures in excess of 550°C, where molten salts are used as the working fluids to transfer heat. The working fluid can either be stored for a few hours or immediately transported to heat exchangers, where the heat is used to transform water to steam to run steam turbines and generate electricity.

Solar PV technologies utilize the "light" aspect of solar radiation and directly convert the radiation energy into electricity through a device called a solar cell. PV is a direct and instantaneous method of electricity generation through the transfer of energy of a photon, that is, light to an electron; solar thermal technologies are an indirect method because

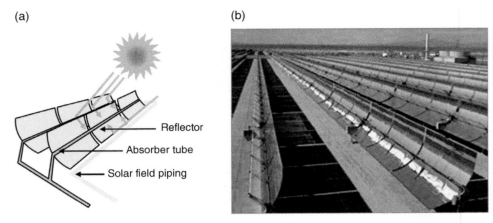

FIGURE 5.2 Solar parabolic trough: (a) concept and (b) photograph of an actual plant. *Source:* (a) International Energy Agency, "Technology Roadmap: Concentrating Solar Power," 2010. http://www.iea.org/publications/freepublications/publication/csp_roadmap.pdf. (b) http://commons.wikimedia.org/wiki/File:Solar_Array.jpg

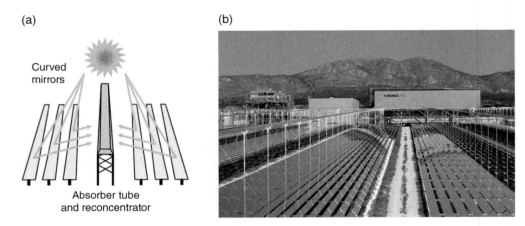

FIGURE 5.3 Linear Fresnel lens: (a) concept and (b) Puerto Errado 2 (PE2) by Novatec Solar, Spain. *Source:* (a) International Energy Agency, "Technology Roadmap: Concentrating Solar Power," 2010. http://www.iea.org/publications/freepublications/publication/csp_roadmap.pdf. (b) Author Novatec Solar. http://commons.wikimedia.org/wiki/File:Novatec_Solar_Puerto_Errado_2_BoP_PI.jpg

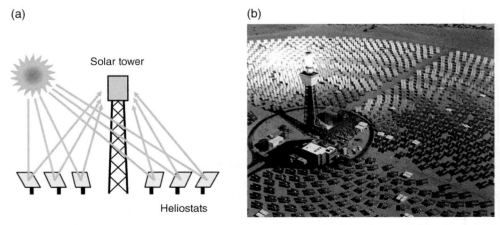

FIGURE 5.4 Central receiver technology: (a) concept and (b) Solar One at California, United States. *Source:* (a) International Energy Agency, "Technology Roadmap: Concentrating Solar Power," 2010. http://www.iea.org/publications/freepublications/publication/csp_roadmap.pdf. (b) http://commons.wikimedia.org/wiki/File:Solaronepowerplant.jpg

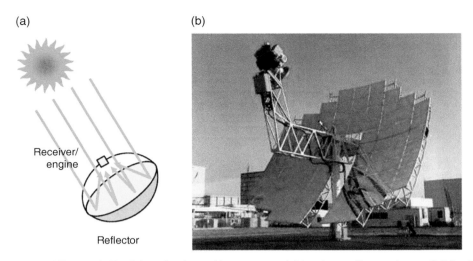

FIGURE 5.5 Parabolic dish technology: (a) concept and (b) solar sterling engine at California, United States. *Source:* (a) International Energy Agency, "Technology Roadmap: Concentrating Solar Power," 2010. http://www.iea.org/publications/freepublications/publication/csp_roadmap.pdf. (b) http://commons.wikimedia.org/wiki/File:SolarStirlingEngine.jpg

the radiation energy is first converted into mechanical energy of a turbine, which is then converted into electrical energy.

One of the major advantages of PV technologies is its versatility of scale, wherein solar cells of a milliwatts of capacity can be used to power small portable electronics such as a watch or a calculator; a few watts can be used to power lanterns, a few tens of watts to power streetlights, a few kilowatts to power homes, and megawatts to generate electricity at a power plant. Figure 5.6 depicts the application of solar energy in day-to-day appliances.

Solar technologies offer several advantages. First of all, the primary fuel, sunlight, is free and practically infinitely available. PV systems are simple and can be designed without any moving parts. As a result, the operation and maintenance costs of such plants are extremely low, less than even 1% of the capital cost. Such plants do not emit any harmful radiation, any emission or combustion, any high temperature or pressure components, or any waste disposal issues. Such systems are modular, where the capacity can be increased by integrating more PV modules, structures, and electronics. Also, solar energy systems generate energy during the daytime, when the demand for energy is also high. And finally, solar energy systems have an excellent safety record and a high public acceptance as a noble technology.

In spite of the advantages of solar technologies, the main challenge faced by the solar industry is the capital cost of investment. The high capital cost translates into a high tariff, which was typically two to three times higher than conventional electricity cost until recently. However, the cost of solar energy, particularly through PV, has been progressively decreasing due to multitude of factors such as fundamental understanding and research breakthroughs, learning through optimizing manufacturing processes, and sheer economies of scale. Figure 5.7 shows the cost trends in PV technology.

The phenomenon where the cost of solar electricity becomes equivalent to that of conventional electricity is known as "grid parity." Grid parity is a relative term and depends on many factors, both intrinsic and extrinsic to the solar technology. An example of intrinsic factors is the levelized cost of solar electricity from a plant installed in India would be less than a similar solar plant in Germany due to lower cost of labor and higher sunshine availability in India. An example of extrinsic factors is the cost of conventional electricity in island states and countries (e.g., Hawaii and Japan, respectively), is higher, and hence, they would reach grid parity faster than countries with fossil fuel reserves with lower electricity costs.

Even though the cost of solar energy is higher than conventional energy costs, the annual PV installations over the last 10 years have been increasing steadily at an average rate of 59%/year. Policy drivers have been the most effective tools for global solar market growth. Germany, the largest solar market today, utilizes the FiT model, where the utilities pay a higher solar tariff for the electricity generated by a solar developer and fed into the grid. Several countries have followed this model with minor variations. A second model is the net-metering model, initially launched in countries like the United States and Japan, where the energy meter of a consumer accounts for the energy consumed from the grid as well as the solar energy generated and fed into the grid; at the end of the billing cycle, the consumer is billed by the utility only on the "net" energy consumption. However, in order to make net metering viable, such schemes have to be supported with capital subsidies and other tax incentives. One of the major drivers at various state and national levels is the Renewable Purchase Obligation (RPO). Such obligations are typically fixed by the relevant electricity regulator, where the utilities and large consumers are mandated to purchase a

FIGURE 5.6 Possible scales of photovoltaic applications: (a) a calculator using milliwatt-scale solar cells, (b) a barn using a solar lantern at Kharaghoda, Gujarat, using 10 kW photovoltaic module, (c) a solar streetlight using two photovoltaic modules, (d) a rooftop solar photovoltaic installation on an office building in Gandhinagar, Gujarat, (e) a kilowatt-scale grid-connected ground-mounted 1 MW solar photovoltaic power plant at Gandhinagar, Gujarat, (f) a view of the multideveloper solar park at Charanka, Gujarat, which hosts more than 300 MW. *Source:* (a) http://pixabay.com/en/calculator-radhakrishnan-363215/. (c) http://commons.wikimedia.org/wiki/File:Solar_Street_Light.png

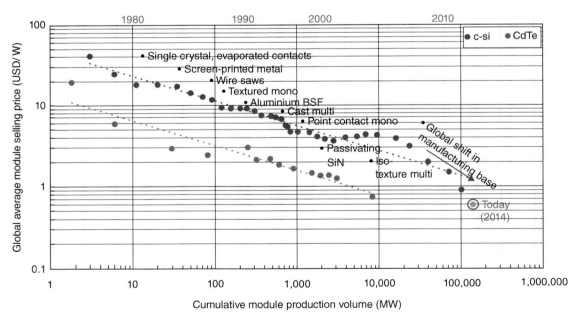

FIGURE 5.7 Cost trends in photovoltaic technology as a function of production capacity. Sources: Prof. Emanuel Sachs, Massachusetts Institute of Technology/1366 Technologies, Inc.; David Feldman, NREL data sources; for 1980–1984, "Large Quantity Buyers," Navigant Consulting (2006), Photovoltaic Manufacturer Shipments 2005/2006, Report NPS—Supply I (August 2006); for 1985–2011, "Large Quantity Buyers" Navigant Consulting (2011) Photovoltaic Manufacturer Shipments 2010/2011, Report NPS—Supply VI (April 2011); for inflation, Implicit Price Deflators for Gross Domestic Product, Bureau of Economic Analysis (September 29, 2011); for UBS Module ASP '11: UBS Global Solar Industry Update 2011 Volume 11 (June 2011); for Sep. '11, Chinese c–Si Spot Price: UBS Global Solar Industry Update 2011 Volume 13 (September 2011); Nemet, G.F. (2006), "Beyond the learning curve: factors influencing cost reductions in photovoltaics." Energy Policy 34(17): 3218–3232.

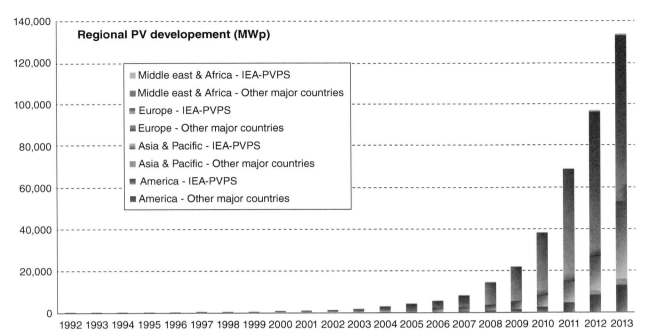

FIGURE 5.8 Cumulative photovoltaic market growth, 1992–2013. *Source:* International Energy Agency–Photovoltaic Power Systems Programme, "A Snapshot of Global PV 1992–2013," 2014. http://www.iea-pvps.org/fileadmin/dam/public/report/statistics/PVPS_report_-_A_Snapshot_of_Global_PV_-_1992-2013_-_final_3.pdf

certain fraction of renewable energy in their energy mix. RPOs can be set up as a general clean energy obligation irrespective of the source of fuel or a very specific solar energy purchase obligation. As a result, such utilities either develop their own solar or renewable energy plants or buy solar-generated electricity from independent power producers (IPP) through power purchase agreements (PPA). Figure 5.8 depicts the cumulative growth in PV market from the year 1992 to 2012.

Several countries have developed their indigenous renewable energy certificate (REC) mechanisms, where RPOs can be met by purchasing RECs through a common exchange or similar platform. RECs typically represent the "green" attribute of renewable energy, which are issued to a renewable energy generator when it is selling the "physical" attribute, that is, electricity, at nonpreferential tariffs. RECs would usually be issued by any central agency in charge of accounting power in the grid such as a state, regional, or national load dispatch center. It is also observed that in addition to RPOs, RECs are also driven through voluntary markets motivated with the purpose of going green.

Another major driver, especially in developing countries, is basic electrification. With the cost of solar energy declining, it is often observed by utilities and governments that it is more economic to invest in microgrids with distributed and renewable energy generation sources rather than to extend the high-voltage transmission infrastructure to a remote location. Business models targeting abatement in diesel generators such as in remote telecommunication towers are now turning out to be cost-effective due to the rising fuel costs and reducing solar costs. Figure 5.9 shows the solar energy technology marketing trends over the years.

It is evident that currently, we are in the early adoption stages of solar energy technologies but soon approaching grid parity. Until grid parity, such a technology would be heavily dependent on government policies and incentives. However, similar to the maturity trends of any other technology, once the parity is reached, solar technologies are expected to become self-driven and self-sustainable and enter the ramp-up stage of deployment. The ramp-up stage would witness a market growth in an order of magnitude, resulting into a more stable and mature sector. Figure 5.10 highlights the future trend of PV electricity generation trends (OECD/IEA, 2010).

There have been several studies undertaken to estimate the share of solar energy in the long term [M. M. Hand et al., "Renewable Electricity Futures Studies," National Renewable Energy Laboratory (NREL/TP-6A20-52409) 2012; CCSP, 2007: Scenarios of Greenhouse Gas Emissions and Atmospheric Concentrations (Part A) and Review of Integrated Scenario Development and Application (Part B). A Report by the US Climate Change Science Program and the Subcommittee on Global Change Research; International Energy Agency's "Technology Roadmap: Solar Photovoltaic Technology"; "Roadmap 2050: A Practical Guide to a Prosperous, Low-Carbon Europe," philanthropic European

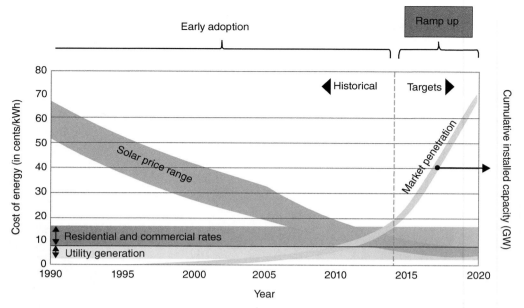

FIGURE 5.9 Market penetration trend of solar energy technologies. *Source:* US Department of Energy, "Multi-Year Program Plan: 2008–2012," 2008. https://www1.eere.energy.gov/solar/pdfs/solar_program_mypp_2008-2012.pdf

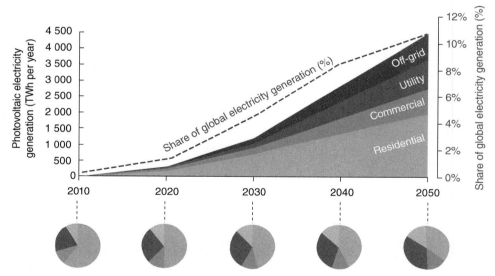

FIGURE 5.10 Evolution of photovoltaic electricity generation by end-use sector, 2010–2050. *Source:* International Energy Agency, "Technology Roadmap: Solar Photovoltaic Energy," 2010. http://www.iea.org/publications/freepublications/publication/pv_roadmap.pdf

climate foundation (ECF); Lynn Orr, Changing the World's Energy Systems, Stanford University Global Climate and Energy Project (after John Edwards, American Association of Petroleum Geologists); and SRI Consulting]. These studies take multiple factors into account including deployment of renewable energy systems, grid operation and stability, expansions in the transmission system, costs trends of various fuels, technological developments, environmental costs and implications, demand and supply gaps, and energy markets, as well as various policy scenarios.

All such studies, based on various assumptions and scenarios, suggest solar energy penetration anywhere between 10 and 30%. Further, solar energy deployment would also be distributed among residential, commercial, utility-scale, and off-grid sectors. Wind availability and velocity are central to the success of wind turbines. Also, it is essential to overcome the turbine's frictional resistance to increase efficiency. At optimal speeds, the turbine turns to generate electricity. The turbine's output is a function of the velocity, increasing power output by a power of three under optimal conditions. Turbines are switched off during storms. Worldwide, the installed wind capacity has reportedly grown approximately seven- to eightfold through the past decade with the main markets in Europe, the United States, and China. India too is catching up in this sector. Offshore turbines deliver higher yields due to higher wind velocities. Such systems are however subject to significant infrastructure hassles. Figure 5.11 depicts the global capacity to generate wind power.

Biomass: Plants absorb sunlight and through the process of photosynthesis, generate biomass. Biodiesel is an important spin-off from plant substrates and generates much lesser sulfur dioxide on combustion. This is often an ester derivative and can be used in diesel engines. In some countries, biodiesel is derived from rapeseed, and the rest of the plant parts serve as animal feed or manure for crop systems. Canola and sugarcane are also used to derive biodiesel. Ethanol derived from sugar beet could be blended and used as a fuel. Flex-fuel vehicles (FSV) can run mixtures of gasoline and ethanol. Synthetic fuels have emerged recently represented by biomass to liquid (BTL). Biomass is converted to a synthetic gas leading to an alcohol substrate.

Herbivores transform these energy and material substrates for their own physiological benefits and in the process generate waste. This waste too is an important biomass energy substrate lending itself to further biochemical decomposition that releases energy products. In this context, biomass can be divided into three categories in terms of energy use.

Freshly cut plants and manure that constitute wet biomass in particular help generate biogas on fermentation in an anaerobic environment. This biogas in turn is used to generate electricity. In many cases, direct heat applications are also evident.

Biogas plants can also be run as cogeneration units. Waste heat from the generator can be used again in the fermentation system if needed, and the additional power generated can be sent to the grid if connected suitably. The residual organic waste can be further fermented. Interestingly, dry biomass is represented by wood and straw predominantly used in the tropics as fuel. Energy plantations represent the third dimension of biomass for energy including electricity and heat. These energy substrates are homogenized to start with and transferred into a fermentation tank. A consortium of bacteria or specialized cultures is added to transform these substrates through extra cellular biochemical conversion pathways. It is well known that methane is the predominant component of biogas followed by carbon dioxide. The calorific value of the gas is determined by the composition of the substrate and the duration of the fermentation process. Fermentation in biogas plants improves the value of the manure. Interestingly, organic waste from kitchens including wastewater carries significant volumes of energy substrates and can augment other biomass.

Geothermal systems are based on the fact that heat flows consistently to the crust of the Earth from the core. This geothermal energy can be an important part of our renewable energy systems for sustainable energy supply. Importantly, this is emerging in a large number of countries around the world especially in regions with active volcanoes. Some of the leaders are the Philippines, Indonesia, Japan, Iceland, Italy, and California in the United States. Buildings are heated

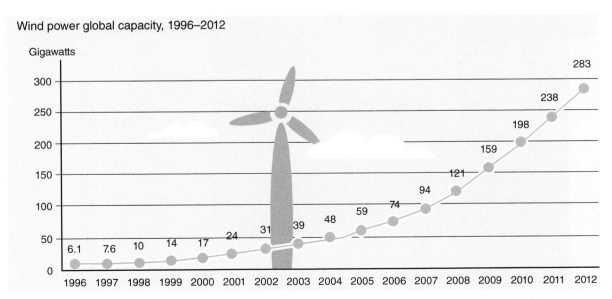

FIGURE 5.11 Wind power global capacity, 1996–2012. *Source:* REN21 Renewables 2013 Global Status Report. http://www.ren21.net/portals/0/documents/resources/gsr/2013/gsr2013_lowres.pdf

using down-hole heat exchangers. Water is injected to absorb underground heat and returns warmer to the surface. In hydrothermal systems that relate to significant depths under the earth, high temperatures are encountered. Water in the aquifer is pumped to the surface to interface with a heat exchanger.

Nuclear energy is used to produce electricity. Heat generated from the splitting of uranium atoms in a process known as fission is used to produce steam. This steam in turn powers turbines, which are used to produce the electricity that supplies the surrounding community. Nuclear power stations are set up in a multiple-step process that has been designed to help contain the energy and many of its negative by products. This process alone is the base of several advantages and disadvantages for this energy source.

Nuclear energy has numerous advantages over conventional as well as nonconventional energy sources. These include lower greenhouse gas emissions with lowest environmental impact of all energy sources; higher efficiency and reliability as compared to other alternate energy sources; reduced costs owing to long plant lives as well as price volatility owing to higher reliability, independence from natural forces; huge available reserves, which are expected to last another 100 years; and requirement of small quantity of fuel leading to low fuel cost and ease of transportation. Despite the numerous advantages, disadvantages of nuclear energy are inevitable and even disastrous. Most prominent drawback of nuclear energy is the safety concerns involved. From the initial stage of the process of energy generation, that is, mining transportation and storing of uranium, significant safety measures are required since uranium is a naturally unstable element and releases harmful radiations.

Further, both the nuclear waste and retired nuclear plants are a life-threatening legacy for hundreds of future generations. Leaks are also a major concern at nuclear power plants. Nuclear reactors are built with several safety systems designed to contain the radiation given off in the fission process. When the safety systems are not maintained, have structural flaws, or were improperly installed, a nuclear reactor could release harmful amounts of radiation into the environment during the process of regular use. During the operation of nuclear power plants, radioactive waste is produced, which in turn can be used for the production of nuclear weapons. In addition, the same know-how used to design nuclear power plants can to a certain extent be used to build nuclear weapons (nuclear proliferation). The time frame needed for formalities, planning, and building of a new nuclear power generation plant is in the range of 20–30 years in the western democracies. Even the cooling water used at nuclear power plants may be a source of heavy metals and other pollutants into the environment. Figure 5.12

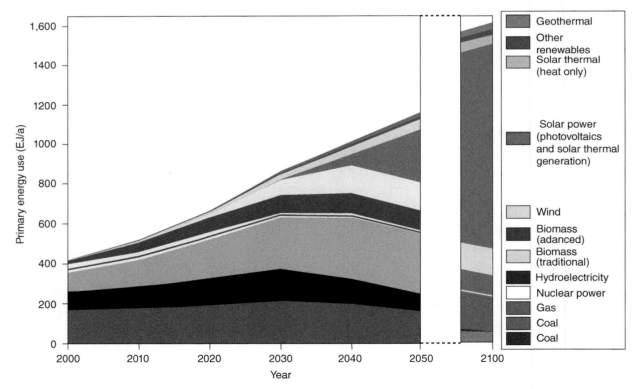

FIGURE 5.12 Project share by source of annual global energy production. *Source:* German Advisory Council on Global Change (WBGU), "World in Transition: Towards Sustainable Energy Systems," 2004. http://www.wbgu.de/fileadmin/templates/dateien/veroeffentlichungen/hauptgutachten/jg2003/wbgu_jg2003_engl.pdf

depicts project share by source of annual global energy production

The specific challenges and unique features of existing renewable energy sources are summarized in Table 5.2.

Several benefits are associated with the use of renewable energy technologies, including very low or no greenhouse gas emissions, making them a key component in any climate change mitigation strategy (IPCC, 2011). In the New Policies Scenario, total CO_2 savings across all sectors from renewables is expected to be about 4.1 gigatonnes (Gt) by 2035. (US Department of Energy National Renewable Energy Laboratory (NREL). "PV FAQs." December 2004. http://www.nrel.gov/docs/fy05osti/37322.pdf). Biofuels reduce emissions from oil in the transport sector by an estimated 0.4 Gt in 2035 but only if their production does not result in increases in emissions from direct or indirect land use changes. Bioenergy electricity could bring 1.3 Gt CO_2 equivalent (CO_2 eq.) emission savings per year in 2050, in addition to 0.7 Gt/year from biomass heat in industry and buildings, if the feedstock can be produced sustainably and used efficiently, with very low life cycle GHG emissions. Impacts of geothermal developments on natural geothermal features such as geysers, hot springs, and fumaroles are documented. The average rate of emissions for a coal-fired power plant is approximately 12 times greater than that of a geothermal power plant and about 6 times greater than a geothermal power plant for a natural gas-fired power plant.

Ocean energy does not directly emit CO_2 during operation; however, GHG emissions may arise from different aspects of the life cycle of ocean energy systems, including raw material extraction, component manufacturing, construction, maintenance, and decommissioning.

Smart grid primarily implies a modernized electricity grid system that integrates the power infrastructure with information and communication technology to make it more reliable, flexible, efficient, clean, safe, and customer friendly. The power infrastructure today is based on the same philosophy as when it was established more than a century ago, consisting of power generation, transmission, distribution, and retail sectors.

There are several drivers for smart grid development, which can be broadly classified based on the region implementing it. Most of the North American and West European countries are driven by grid efficiency and a desire to automate the grid and develop more competitive markets. These countries are also aiming to reduce peak demands, successfully integrate nondispatchable renewable energy systems, and ultimately strive toward energy independence. On the other hand, most of the Asian and Latin American countries are targeting the reliability of the grid, enhancement of the growth capacity of existing grids, reduction of electricity thefts, and improved access to its consumers.

It is through this layer of communication that intelligence or "smartness" can be added into the existing electricity grid.

TABLE 5.2 Opportunities and Challenges in Using Resources of Energy

Type	Opportunities and Challenges
Bioenergy	Biofuels have also been criticized for competing with food supply and contributing to deforestation. Not all bioenergy forms have equal direct greenhouse gas offset effects. Generally, grain-based ethanol provides the least offsets followed by cellulosic substrates, biodiesel, and electricity
Solar	Although the production of solar panels requires some inputs of raw materials and energy, solar power's environmental impact is minimal. The technology does not produce carbon, methane, or particulate emissions that fossil fuels emit, and it doesn't demand large-scale mining or drilling operations
Geothermal	Geothermal developments have minor environmental impacts. The disposal of wastewater containing small quantities of chemicals (boron and arsenic) and gases (H_2S and CO_2) is an important issue, however. Various methods are used to deal with this challenge. These include total reinjection of separated water, condensate and gases; chemical treatment; and mineral extraction. CO_2 emissions from low-temperature resources are negligible (0–1 g/kWh). Geothermal energy also takes up very little surface land—it has among the smallest footprint per kilowatt (kW) of any power generation technology, including coal, nuclear, and other renewables
Ocean energy	Ocean energy does not directly emit CO_2 during operation; however, GHG emissions may arise from different aspects of the life cycle of ocean energy systems, including raw material extraction, component manufacturing, construction, maintenance, and decommissioning
Wind	Careful location selection is needed to avoid conflicts on habitat preferences of fauna, and they do not need fossil fuel. They generate little environmental pollution during their manufacture, operation, and decommissioning
Nuclear energy	Nuclear energy has numerous advantages over conventional as well as nonconventional energy sources. These include lower greenhouse gas emissions with lowest environmental impact of all energy sources, higher efficiency, and reliability as compared to other alternate energy sources. Despite the numerous advantages, disadvantages of nuclear energy are inevitable and even disastrous. The most prominent drawback of nuclear energy is the safety concerns involved. From the initial stage of the process of energy generation, that is, mining transportation and storing of uranium, significant safety measures are required since uranium is a naturally unstable element and releases harmful radiations

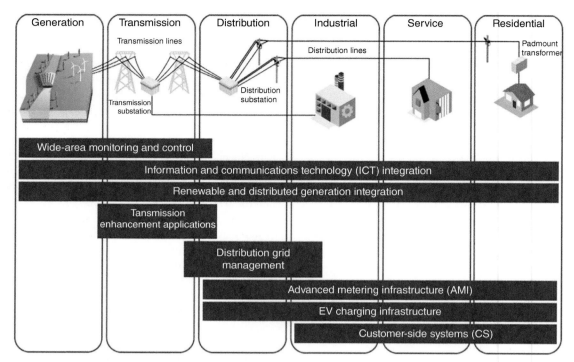

FIGURE 5.13 Smart grid technology areas. *Source:* International Energy Agency, "Technology Roadmap: Smart Grids," 2011. http://www.iea.org/publications/freepublications/publication/smart grids_roadmap.pdf

Smart grids enable increased demand response and energy storage technologies to stabilize the electricity system. The "smartening" of grids is already happening; it is not a one-time event. The first initiatives in the lines of the smart grid technology was through smart metering, wherein the utilities targeted remote meter reading to reduce meter reading costs. The next initiatives included two-way communication with meters, where remote meters could be remotely disconnected with the intention to reduce losses and thefts. The subsequent focus was on peak shifting, which was driven through dynamic tariffs; these initiatives also witnessed advanced functionalities such as remote firmware upgrades of meters. The latest trends are to standardize communication protocols of various devices as the volumes of deployment have been rapidly growing.

The physical and institutional complexity of electricity systems make it unlikely that the market alone will implement smart grids on the scale that is needed. Governments, the private sector, and consumer and environmental advocacy groups must work together to define electricity system needs and determine smart grid solutions. Figure 5.13 shows the areas where smart grid technology can be used. Information and communication technology is the backbone for a smart electricity infrastructure.

Various communication technologies including radio frequency (RF) wireless, general packet radio service (GPRS), and broadband and narrowband power line communication (PLC) as well as communication protocols are under standardization and optimization depending on the nature of information transferred and required reliability as well as the cost. These technologies are graded based on their reliability, interoperability, scalability, and security metrics.

One of the primary goals of a smart grid is to establish an advance metering infrastructure (AMI), which enables a two-way information flow providing customers and utilities with data on electricity price and consumption, including the time and amount of electricity consumed, and much more associated functionalities. Another major implication of smart grids will be in the success of integrating electric mobility with the utility grid, where the vehicles can also act as distributed energy storage facilities. Various demand response pilots are already ongoing around the globe, where participating consumers are incentivized to reduce their demand when the availability of power is low.

Such programs would be critical when large amounts of wind and solar energy systems, the generation of which cannot be scheduled, will be integrated into the grid. Above all, various pricing and economic models are expected to develop around such an advanced infrastructure.

Fuel cells are themselves not a source of renewable energy. They are clean since emissions are avoided and serve as efficient means of generating electricity and heat with significant implications for solar energy applications. They generate electricity from hydrogen and oxygen with high electric efficiency. Fuel cells allow a wide variety of fuels including natural gas, methanol biogas, and hydrogen. They could be used in cogeneration systems, the transport sector, etc.

5.5 INDUSTRY FORUMS ACTIVE ON RENEWABLE ENERGY

Some of the important industry forums active in this interface include the following:

a. The REN Alliance: The Programme for 50% Renewable Energy by 2035 cites the IPCC's 5th Assessment Report, "Climate Change 2013: The Physical Science Basis," released in September 2013. With increased use of renewable energy for electricity, heating, cooling, and transport fuels, well over 50% of the energy supply can be provided by renewable energy by 2035. REN Alliance members support authentic studies (e.g., REN21's Renewables Global Futures Report (2013)). It argues that a 50% or more renewable energy supply by 2035 is clearly achievable based on current and projected rates of technology deployments resulting from innovative policies. Growth in decentralized energy supplies, strengthened transmission networks, and increased uptake of small-scale renewable in rural areas disconnected from grids could reinforce this transition. Large-scale deployment of mass energy storage capacity can compensate for the variability of wind and solar and ensure reliability of supply.

b. The International Hydropower Association (IHA) is a nonprofit organization, with the mission to build and share knowledge on the role of hydropower in renewable energy system, responsible freshwater management, and climate change solutions. IHA champions continuous improvement in the hydropower sector through dialogue with all stakeholders. http://www.hydropower.org

c. The International Solar Energy Society (ISES) aims to achieve 100% renewable energy for all including efficiency and wise use. http://www.ises.org

d. WWEA is an international nonprofit association focusing on the wind sector worldwide. It also provides a platform to communicate with stakeholder groups interested in wind energy. http://www.wwindea.org.

e. The IGA is a nonpolitical, nonprofit, nongovernmental organization that encourages research and development and utilization of geothermal resources worldwide. This is through the publication of scientific and technical information among the geothermal specialists, the business community, governmental representatives, UN organizations, civil society and the general public. http://www.geothermal-energy.org

f. The World Bioenergy Association (WBA) supports and represents the wide range of actors in the bioenergy sector to promote the increasing utilization of bioenergy globally in an efficient, sustainable, economic, and environmentally friendly ways. http://www.worldbioenergy.org

Some of the predominant forms and functions of renewable in this perspective include provision of solar heat through passive systems including architectural measures to use solar energy, solar thermal heat provision through such active systems as solar thermal collectors, solar thermal electrical power systems including solar towers/farms and dish/chimney systems, PV systems that convert solar radiation into electrical energy, wind turbines, hydropower plants, use of ambient air and shallow geothermal energy to provide heat by using heat pumps, use of deep geothermal energy to provide heat or power, and use of biomass. In this context, efficiency is defined as the ratio of useful electricity or heat derived to the power input, say, solar radiation, geothermal energy, etc. Utilization ratio is the ratio of the total output of useful energy to total energy input and with reference to specific timescales.

Wind vanes are used to define direction, and plate/cup/impeller anemometers and thermal anemometers are also important components. The different types of geothermal heat reservoirs include shallow layers with about 20°C, mainly influenced by solar radiation up to a depth of around 10–20 m. This is also due to heat conductivity of the soil and by the circulating groundwater heated by solar energy (Table 5.3).

Warm and hot water hydrothermal low-pressure reservoirs; wet steam and hot or dry steam reservoirs are characterized by temperatures. These are also related to heat in water or steam-bearing rocks. 150–250°C occurs at depths where magma rocks are uplifted from the deep underground. High-pressure hydrothermal reservoirs with methane mix, hot dry rocks, and magma deposits are the other varieties. Geothermal reservoirs that help access more than 130°C can be used to generate electricity. Other related parameters in the case of passive systems include transmission coefficient, secondary heat flow, energy transmittance factor, thermal transmittance coefficient, and equivalent thermal transmittance coefficient along with transmission losses. Solar ponds are power plants. They utilize the impact of water stratification, typically filled with brine that functions as collector and heat storage.

TABLE 5.3 Presents UNEP Global Trends in Energy Investments on Renewable Energy Technologies

	Category	Year Unit	2004 $bn	2005 $bn	2006 $bn	2007 $bn	2008 $bn	2009 $bn	2010 $bn	2011 $bn	2012 $bn	2011–2012 Growth %	2004–2012 CAGR %
1	**Total investment**												
1.1	New investment		39.6	64.7	100.0	146.2	171.7	168.2	227.2	279.0	244.4	−12	26
1.2	Total transactions		48.4	90.7	135.6	204.7	231.0	232.5	285.8	352.5	296.7	−16	25
2	**New investment by value chain**												
2.1	Technology development												
2.1.1	Venture capital		0.4	0.6	1.2	2.2	3.2	1.6	2.5	2.6	2.3	−15	25
2.1.2	Government R&D		2.0	2.1	2.3	2.7	2.8	5.2	4.7	4.7	4.8	3	12
2.1.3	Corporate RD&D		3.0	2.9	3.3	3.6	4.0	4.0	4.6	4.8	4.8	−1	6
2.2	Equipment Manufacturing												
2.2.1	Private equity expansion capital		0.3	1.0	3.0	3.7	6.8	2.9	3.1	2.6	1.4	−46	20
2.2.2	Public markets		0.3	3.8	9.1	22.2	11.6	12.5	11.8	10.6	4.1	−61	41
2.3	Projects												
2.3.1	Asset finance		24.8	44.0	72.1	100.6	124.2	110.3	143.7	180.1	148.5	−18%	25
2.3.2	Of which reinvested equity		0.0	0.1	0.7	3.1	3.4	1.8	5.5	3.7	1.5	−60	—
2.3.3	Small distributed capacity		8.9	10.5	9.8	14.3	22.5	33.5	62.4	77.4	80.0	3%	32
	Total financial investment		25.8	49.3	84.7	125.6	142.4	125.5	155.6	192.2	154.8	−19	25
	Gov't R&D, corporate RD&D, small projects		13.8	15.4	15.3	20.6	29.3	42.7	71.7	86.8	89.6	3%	26
	Total new investment		39.6	64.7	100.0	146.2	171.7	168.2	227.2	279.0	244.4	−12	26
3	**M&A transactions**												
3.1	Private equity buyouts		0.8	3.8	1.8	3.6	5.5	2.5	1.9	3.0	2.4	−19	14
3.2	Public markets investor exits		0.0	1.4	2.7	4.2	1.0	2.6	4.7	0.1	0.4	200	41
3.3	Corporate M&A		2.4	7.9	12.7	20.4	18.0	21.5	18.0	29.5	7.1	−76	14
3.4	Project acquisition and refinancing		5.4	12.8	18.4	30.4	34.9	37.7	33.9	40.9	42.3	4	29
4	**New investment by sector**												
4.1	Wind		14.4	25.5	32.4	57.4	69.9	73.7	96.2	89.3	80.3	−10	24
4.2	Solar		12.3	16.4	22.1	39.1	59.3	62.3	99.9	158.1	140.4	−11	36
4.3	Biofuels		3.7	8.9	26.1	28.2	19.3	10.6	9.2	8.3	5.0	−40	4
4.4	Biomass and WtE		6.3	8.3	11.8	13.1	14.1	13.2	13.7	12.9	8.6	−34	4
4.5	Small hydro		1.5	4.6	5.4	5.9	7.1	5.3	4.5	6.5	7.8	20	22
4.6	Geothermal		1.4	0.9	1.4	1.8	1.8	2.7	3.5	3.7	2.1	−44	5
4.7	Marine		0.0	0.1	0.9	0.7	0.2	0.3	0.2	0.3	0.3	13	30
	Total		39.6	64.7	100.0	146.2	171.7	168.2	227.2	279.0	244.4	−12	26
5	**New investment by geography**												
5.1	United States		5.7	11.9	28.2	34.5	36.2	23.3	34.6	54.8	36.0	−34	26
5.2	Brazil		0.5	2.2	4.2	10.3	12.5	7.9	7.9	8.6	5.4	−37	34
5.3	AMER (excl. United States and Brazil)		1.4	3.4	3.4	5.0	5.6	5.9	11.5	8.3	9.5	14	27
5.4	Europe		19.6	29.4	38.4	61.7	72.9	74.7	101.3	112.3	79.9	−29	19
5.5	Middle East and Africa		0.6	0.6	1.2	1.7	2.7	1.7	5.0	3.5	11.5	228	46
5.6	China		2.6	5.8	10.2	15.8	25.0	37.2	40.0	54.7	66.6	22	50
5.7	India		2.4	3.2	5.5	6.3	5.2	4.4	8.7	13.0	6.5	−50	13
5.8	ASOC (excl. China and India)		6.7	8.3	8.9	11.0	11.5	13.2	18.1	23.8	29.0	22	20
	Total		39.6	64.7	100.0	146.2	171.7	168.2	227.2	279.0	244.4	−12	26

New investment volume adjusts for reinvested equity. Total values include estimates for undisclosed deals.
Source: UNEP, Bloomberg New Energy Finance.

5.6 COMMITMENTS AND COLLECTIVE ACTION TO MAINSTREAM RENEWABLES

The UNEP cautioned last month that the global community should not any further delay concerted action if the least cost path to keeping global temperature rise below 2° has to be exploited. This is increasingly evident in the growing gap in emission reduction with respect to set targets. This call for action was just before the CoP 19 at Warsaw.

The Emissions Gap Report 2013 of the UNEP involved 44 scientific groups in 17 countries. It emphasizes the need to avoid additional challenges after 2020 that will emerge if the gap is not narrowed immediately on a priority basis. These could be in the form of locked-in carbon-intensive infrastructure and dependence on unproven technologies in the medium to long term. This translates into the reality that 8–12 Gt of carbon dioxide equivalents would remain and therefore the need to choose alternative energy pathways in addition to capture and storage.

UNEP argues for a maximum of 44 Gt CO_2 eq by 2020, further cuts to 40 Gt by 2025, 35 Gt by 2030, and 22 Gt by 2050. This calls for renewed pledges and stronger commitment for action. Some of the important initiatives include a focus on energy efficiency enhancement that involves 50 countries in the case of the Global Efficient Lighting Partnership Programme, for instance. Renewables are expected to avoid 1–3 Gt of CO_2 eq by 2020. About USD 244 billion was invested in renewables in 2012, and 115 GW of new systems were installed worldwide. About 140 countries have set clean energy targets.

Mitigation benefits of actions quantify reduction in GHG emissions. These also include potential GHG emission reductions; expected spin-offs from emerging technology developments; and mitigative capacity encompassing social, political, institutional, and economic structures and conditions. Other benefits include indirectly induced related to sustainable development goals, including poverty eradication, reduction of local air pollution, or increased energy security and the avoided adaptation costs. Various types of government policies, the declining cost of many RE technologies, changes in the prices of fossil fuels, an increase of energy demand, and other factors have encouraged the continuing increase in the use of RE.

The UNEP (2013) has clearly articulated that countries strive to implement the Copenhagen Accord of the UNFCCC and recognize the possibility that GHG emissions will peak within the next decade. This calls for concerted efforts to reduce emissions substantially below the year 2000 levels by 2050. This translates into three major interrelated considerations. Countries need to meet the soaring global energy demand with primary energy in the order of 16 Gt of oil equivalent (Gtoe) in 2035 and around 21 Gtoe in 2050. This implies an increase in energy-related CO_2 emissions of 40% in 2030 and of 100% in 2050 relative to 2007 if adequate emission reduction policies are not implemented. Nuclear power is considered an option with the largest potential (1.88 Gt CO_2 equivalent (CO_2 eq.)) to mitigate GHG emissions at the lowest cost. This takes into account the externalities avoided through air pollution otherwise caused.

UNEP 2013 further reports that:

- At least 30 countries around the world already have shares of renewable energy above 20% and some as high as 50%.
- Already, grids can cope with 25–30% of intermittent RE in the mix.
- In many places:
  - Unsubsidized wind power is already competitive with conventional energy.
  - The cost of solar PV has reduced significantly recently and is set to reduce further through technology advances and grid parity before and by 2020.
  - 2012 saw the most dramatic shift yet in the balance of investment activity between developed and developing economies.
  - Outlays in developing countries reached US$ 112 billion representing 46% of the world total. In China, for example, wind power generation increased more than generation from coal. Wind generated 100.4 billion kWh in 2012, up 37% over 2011 and exceeding nuclear generation for the first time. In addition, more and more developing countries became technology developers and exporters.
  - Many regions, cities, towns, and communities are poised to achieve 100% renewable energy futures. These include Gussing (Austria), Dardesheim (Germany), Kuzumaki (Japan), and many medium-size cities as Malmo and San Francisco. Larger cities working that aim in time frames ranging from 2025 to 2050 include Copenhagen, Munich, and Sydney.

The IPCC 2011 in its Summary for Policymakers on Renewable Energy Sources and Climate Change Mitigation indicated that GHG emissions from consumption of fossil fuels and resultant CO_2 concentrations had increased to over 390 ppm by the end of the year 2010; an equivalent of 39% above preindustrial levels (Ottmar Edenhofer, Ramón Pichs-Madruga, and Youba Sokona, 2011).

A special emphasis is therefore on six renewable energy sources, namely, bioenergy, direct solar, geothermal, hydropower, ocean energy, and wind, that could play significant roles in present and future energy systems. Energy conservation and efficiency, fossil fuel switching, and nuclear and carbon capture and storage (CCS) are also integral to these transitions. Life Cycle Assessments (LCA) for electricity generation indicate that GHG emissions from RE technologies are nearly tenfold lower than those associated with fossil fuel options.

The report further indicates that despite global financial challenges, RE capacity continued to grow rapidly in 2009 (4–53%). This included wind, hydro, geothermal power, grid-connected PV, and solar hot water/heating. Advanced biofuels could provide higher GHG mitigation. Combining biomass conversion with CCS enhances GHG removal from the atmosphere in the long term. Life cycle assessments through majority of estimates for PV modules indicate a CO_2eq/kWh range of about 75% (30–80 g). This compares with levels for CSP-generated electricity at about 50% variation (14–32 g of CO_2eq/kWh). These emission levels are however nearly an order of magnitude lesser than those of natural gas-fired power systems. The emphasis is presently on solar thermal applications including active and passive and process heat for industry: PV systems, concentrating solar power, and solar energy to produce useful fuels. The main GHG emission from geothermal operations is CO_2, not created through combustion but emitted from naturally occurring sources.

The Ad Hoc Working Group on the Durban Platform for Enhanced Action (ADP) under workstream 2 invited Parties to submit information on actions and options to enhance mitigation goals, with a particular focus for the year 2013 (UNFCCC, May 2013). This was aligned with the emerging goal posts for 2015 on political coherence for mitigation, adaptation, and support for implementation. UNDP (2012a) pegged the gap at 8–13 Gt CO_2 eq in 2020; assuming that Parties implement their emission reduction pledges under the Convention and its Protocol. The technical potential for reducing emissions by 2020 is estimated to be about 17 ± 3 Gt CO_2 eq. The thematic areas and their mitigation potentials by 2020 include energy efficiency; renewable energy; fossil fuel subsidy reform; reduction of emissions from fluorinated greenhouse gases; reducing short-lived climate pollutants; all forms of transport, land use, including forestry and agriculture; and waste as substrates that can be transformed for energy.

Many Parties to the Convention made conditional and unconditional emission reduction pledges scheduled for 2020 under the Cancun Agreements. These pledges encompass quantified economy-wide emission reduction targets for developed countries under the Convention. The portfolios of nationally appropriated mitigation actions reflect the approaches of developing countries. Many Parties to the Convention cited the IEA World Energy Outlook 2012 that current policies on renewable energy can be enhanced to deliver emission reductions of around 1 Gt CO_2 eq by 2020 and 3 Gt CO_2 eq by 2030. The Emissions Gap Report suggests a potential of 1.5–2.5 Gt CO_2 eq from renewables considering their role in the production of about 4000 TWh of electricity. In the longer term, by 2050, the IPCC 2011 Special Report on Renewable Energy Sources and Climate Change Mitigation cited the above estimates that renewable energy could help to reduce by around one-third (220–560 Gt CO_2 eq) the projected cumulated fossil fuel CO_2 emissions (1530 Gt CO_2).

Almost all major economies have set themselves renewable energy targets. Several national emissions trading systems, offset mechanisms, and carbon taxes have provided further incentives to promote renewable energy. Some of the best-cited examples are those of Germany and Sao Paulo, Brazil. In the former, purchase and price of electricity generated by hydropower, wind energy, solar energy, landfill gas, sewage gas, and biomass are regulated and have achieved a fivefold increase in contribution over two decades. The latter has integrated its initiatives with municipal building codes to install solar water heating systems covering at least 40% of the energy used for hot water. By 2015, it is expected to be on target to allow several thousands of t CO_2 eq from the city's residential sector.

Denmark and Spain also have robust mechanisms and the IRENA supports countries in their transition to sustainable energy. Several of these policies foster best practices, set performance standards, eco-labeling, and other market-based mechanisms. Some of the other frequently mentioned initiatives include the Low Emission Development Strategies (LEDS) Global Partnership, the United Nations Collaborative Programme on Reducing Emissions from Deforestation and Forest Degradation in Developing Countries (UN-REDD Programme), the REDD-plus Partnership, the Secretary-General's Sustainable Energy for All, and the Climate and Clean Air Coalition to Reduce Short-Lived Climate Pollutants (CCAC). Many of these involve collaborating countries and stakeholders for optimal impacts.

The World Energy Outlook Special Report (2013) (International Energy Agency, June 2013) indicates that the energy-related CO_2 emissions reached a historic high last year. The non-OECD countries are responsible for a significant portion, China having used renewables to a significant extent, in addition to improvements and its energy and intensity. Several countries have also well-defined targets to mainstream renewables to tackle climate change impacts. This includes China, India, Brazil, the EU, etc. This is expected to enhance contribution of renewables to power generations by 7% in the next 7 years. The proportion varies across tens of scenarios that prompt assessments.

The 2030 framework for climate and energy policies stated by the EC (2013) (Brussels, March 2013) is a typical case in point. It is aligned with its low-carbon economy road map for 2050 with well-defined targets for GHG emission reduction. This includes a share of about 30% for renewables by the next 17 years and is guided by a 2020 target of 20% renewable energy portfolio with respect to final energy consumption. The recent report on global trends in renewable investments (McCrone, 2013) further substantiates in improving preparedness to enhance the role of renewable energy systems, notwithstanding the fact that investments have comparatively reduced in the present period. Importantly, however, investment patterns reportedly are increasing with developing economies quite significantly. The twinning of the use of renewable energy options and avoidance of GHG-intensive conventional sources have also helped to avoid several hundred megatons of CO_2. The report also presents interesting correlates of slowdown in economy and differential impact across the EU, the United States, and Australia, for instance, and the role played by increasing energy efficiency interventions and the use of renewables.

A recent review by Ernst and Young (2013) provides valuable insights into the dynamics of market-based mechanisms on the portfolio of renewable energy projects through the framework of country attractiveness indices. This is with special reference to the challenge imposed by shale gas, and the preponderance of biomass, geothermal energy, and other options. The milieu prevalent in China, Germany, India, France, the United Kingdom, Canada, South Africa, and Morocco is discussed and establishes the heterogeneous nature of determinants and success achieved. These dynamics are further substantiated by Deloitte (2013) (Allen, 2013; Trottier Energy Futures Project, 2013), providing hope for a larger role for renewables in the future.

Germany has increased the share of renewable electricity sources from 3.1% in 1990 to more than 25% today. With strong public and political support frameworks in place over the past decade, several towns and regions have committed to or already achieved—even surpassed—a 100% renewable electricity target. The country as a whole reached its 12.5% power target by 2010—3 years ahead of schedule—and 20% in 2011 and is on track to reach its renewable power goals of 35% by 2020 and 80% by 2050. Denmark mandates 100% RE especially in the heating and transportation applications by 2050. By 2011, wind, biogas, CHP waste, and a very small amount of PV solar contributed to the mix. In Austria, renewable resources represent almost 31% of energy consumption, and Upper Austria is set to commit 100% RE in the power and heat sectors by 2030.

Many commercial airlines use advanced biofuel blends on commercial flights. These include Air France–KLM that uses biofuel derived from waste cooking oil. British Airways plans to incorporate this renewable fuel in jet operations.

UNEP (2012) reported on the potential for renewable energy with reference to 15 countries and their emission reduction potential. The countries investigated are Angola, Belize, Burkina Faso, D.R. Congo, Fiji, Ghana, Haiti, Lesotho, Malawi, Mozambique, Myanmar, Rwanda, Sao Tomé and Principe, Senegal, and Trinidad & Tobago.

The report highlights the fact that there is very limited research on wind power in Africa. Some estimates on onshore wind resource indicate approximately 1750 GW potential. Wind resource varies significantly across the continent, much as some regions in particular enjoy relatively high-quality resources. These include Angola, Ghana, and Senegal. Angola targets 5,000 MW wind energy capacity by 2016, corresponding to 768,000 CERs/year. Ghana reports a potential of 5,600 MW. Myanmar could establish emission savings of up to 25 million tons of CO_2 based on 2500 full load hours.

Initiatives in all these countries could help eliminate 12 million tCO_2e at a conservative estimate. The African continent is reportedly endowed with significant hydropower potential, enough to meet all its energy needs. The continent could also produce 42 billion megawatt hours through solar systems including small- and large-scale technologies. The government of Lesotho has implemented codes of practice for some solar PV and solar heater installations. Belize has a high potential for residential solar water heating and large-scale solar PV. A nationwide solar water heater initiative is underway including the possibilities for developing a 50 MW solar PV plant. Fiji too is reportedly endowed with high potential for solar lighting of streetlights, residential solar water heating, and microscale solar PV. The country has already implemented different small-scale projects, such as solar streetlighting, and solar water pumps. This could benefit hundreds of households and generate around 10,000 CERs/year. Myanmar is equally endowed with significant solar resources that could benefit the large rural population.

Transitions in India: India's response to the call for collective action on mitigation and adaptation is reflected in her integrated approach driven predominantly by a proactive recognition of the cause impacts. India's energy plan on emission reduction portfolios reflects a highly adapted system of target settings and performance especially when she strives to meet the developmental aspirations of her citizens. While several of the developmental goals are well prioritized, she is versatile to the extent of initiating mutually reinforcing institutional, financial, and technological mechanisms to demonstrate the feasibility of mainstreaming several renewable energy systems. India's NATCOM I & II (Ministry of Environment & Forests, 2004, 2012) define her commitment to reduce the buildup of greenhouse gases with the special emphasis on energy and related systems.

The latter is especially aligned with her 12th 5-year plan reflecting the farsightedness of a resilient economy. Energy efficiency enhancement for significant cobenefits is embedded within her initiatives to capture opportunities even in the energy-intensive industries and the electricity sector, accounting for a very high proportion of emissions. The Perform, Achieve, and Trade scheme will enable enhanced performance only to help reduce a substantial quantity of CO_2 equivalent emissions within the next 2–3 years. This will contribute in a significant manner to the carbon-intensity reduction goals set for 2020 (Neelam Singh, 2013). The specific context of renewables has been recently elaborated by Deloitte (2013) with specific qualitative and quantitative assessments.

The links between energy security and sustainable extractions and use of energy from renewable sources is also substantiated duly recognizing the linkages across the missions of the National Action Plan on Climate Change. The report cites the Jawaharlal Nehru National Solar Mission and its targets over the next decade, the proportion of biomass, wind energy, and small hydropower in the energy mix. Yet another recent assessment has been reported by the Germanwatch and the Climate Action Network Europe (Jan Burck, Lukas Hermwille, and Laura Krings, Nov 2013). Importantly, the report presents developments across the world with reference to the Climate Change Performance Index citing the cases of Brazil, Denmark, Sweden, Portugal, Italy, Germany, the United States, China, India, Netherlands, Norway, Australia, Canada, Saudi Arabia, Qatar, and the EU as a whole. This index becomes important in the present context of understanding the role of renewables and that energy efficiency enhancement levels are not adequate enough to sustain transitions through renewable energy systems alone. A comprehensive review by Gupta and Anand (2013) argues that solar electricity supply systems have grown rapidly in India. A detailed statewise distribution has been presented with respect to the policy measures implemented at the local level. The options to enhance output especially through a National Solar Mission have also been presented.

The Global Wind Energy Council, the World Institute of Sustainable Energy, and the Indian Wind Turbine Manufacturers Association present a detailed analysis of the niche occupied by wind energy in India (2012) (Global Wind Energy Council, Nov 2012). They referred to the National Electricity Plan 2012 and the significant rise in new wind energy installations, recently in India and with reference to more than 12% shares of renewable energy of the total installed capacity in India as of 2012. Interestingly, wind power accounts for nearly 70% of this capacity and poised to enhance significantly in the couple of decades. The emission reduction correlates under various scenarios are also discussed at great length. The EU–India dialogue on energy, clean development, and climate change signifies synergies on several fronts, embodied in a joint action plan (Working with India to Tackle Climate Change, 2012). This includes several local-level initiatives, including the demonstration of clean coal technologies and carbon capture and storage interventions. A waste energy project focuses on anaerobic digestion for cofermentation leading to greenhouse capture and generation of bioenergy in addition to avoidance of greenhouse gas emission and recovery of nutrients.

Several other initiatives center on the management of municipal solid waste and evaluation of feedstock suitability for biogas production. This is in addition to establishment of solar PV plant built to scale and aimed at saving tens of thousands of tons of CO_2. Some of the applications include cooking using solar systems especially in poor households and integration of energy efficiency buildings with implications for sustainable livelihoods and offsetting several tons of biomass. Emission avoidance and reduction goals are also achieved in this process. The International Emissions Trading Association and the Environmental Defense Fund (Environmental Defense Fund, Sep 2013) refer to India's commitment to the Copenhagen Accord and her initiatives on installed PV and solar thermal capacity in addition to enhancement of energy efficiency leading to substantial CO_2 equivalent emission reduction.

A recent compilation by the MNRE and the UNDP (2012b) and of the MNRE 2010 (Ministry of New and Renewable Energy, 2010) presents an excellent overview of new and renewable energy initiatives in India covering bioenergy, microhydro, and solar energy applications. They have demonstrated multiple cobenefits for communities and the environment including universal access to energy services, improved energy efficiency, and a judicious mix of energy sources. These include energy plantations, biomass-based gasifier plants, biogas plants, etc. Related energy plantations have helped sequester thousands of tons of CO_2 in addition to improving related infrastructure services.

Some of the important spin-offs included pine-based and husk-based gasifiers in hilly regions, emission offsets, sustainable use of resources available in the vicinity, and improvement of energy efficiency using alternative biomass energy resources in industrial processes. Generation of biogas from wastes and lighting and cooking operations were also established. A large number of microhydro power plants to assist rural economies have also been established for successful electrification and other value-added applications in communities. Successful generation and use of solar power for multiple benefits for community have also been demonstrated. In addition to a micropower grid project and improvement of access to water through other system-related applications. These initiatives represent successful local adaptation through direct relevance for communities by enhancing the quality of life these connect that is central to sustainable management of alternative energy systems twinning the goals of mitigation and adaptation through alternative resources (V.K. Jain and S.N. Srinivas, 2012).

It is important to visualize these developments within the perspective of challenges faced by the energy sector in India. The International Energy Agency (Sun-Joo Ahn and Dagmar Graczyk, 2012) core capacities of entities engaged in the energy sector have to be significantly enhanced to help them tap emerging energy technologies. Pricing mechanisms have to harmonize commercial viability for the supplier and the end user, which is otherwise often subjected to the inclement market conditions and fiscal parameters. The scale of investments has to also be enhanced significantly, enabled by appropriate administrative processes.

5.7 SOME IMPORTANT PRECAUTIONS

While it is interesting to visualize the positive spin-offs, it is equally important to understand the major roadblocks in transitioning to an ideal renewable energy mix portfolio. Rasmussen (2010) deliberated on the environmental operational and economic attributes of wind and solar energy and highlighted the fact that the intermittent occurrence of wind makes it particularly difficult for the sector to harness efficiently. This in turn disrupts production and supply schedules and is therefore less reliable and stable. Critically, these parameters are determined by climatic forces including daily and seasonally temperature schedules. Unexpected bursts of wind further disrupt systems as a function of the volatility of the system. This has implications for poor performance and carbon efficiency of related energy systems based on the aforementioned; it is difficult to establish the dependency and consistency of wind-based systems. Solar systems also called for supplementary storage facilities to enhance reliability for supplying the grid. Operations and maintenance of solar panels continue to be significant problems in addition to the cost of infrastructure. It is equally important to synchronize system peak load periods with peak solar output, while this continuous to be a major challenge.

The Climate Parliament in India took stock of the wind energy scenario recently (2013) (Climate Parliament, July 2013) and asked for the immediate implementation of the generation-based incentive for wind energy. Related issues of withdrawal of accelerated depreciation benefit have reduced safeguards on long-term interests on this sector. Harmonized and validated data on wind potential are not available. Inadequate planning of transmission infrastructure further compounds related challenges. These are in addition to such aspects as incompatible turbine capacities and such system-related aspects as renewable energy procurement obligation not being implemented uniformly.

5.8 THE WAY FORWARD

It will be obvious from the preceding text that initiatives around the world are evolving continually, and it is essential to sustain the momentum of such transitions without any backsliding. While some of the barriers stated earlier appear to modulate the pace of large-scale integration, it is equally important to establish additional reinforcements.

A typical learning can be gathered from the successful implementation from the Montreal Protocol on Substances that deplete the ozone layer. This is with special reference to Compliance Assistance Programme of the Montreal Protocol that handholds countries and related institutions to understand alternatives and create the appropriate policy and plan framework to enable easy and targeted implementation programs. In this particular context of renewable energy context, it will be useful to design a technical assistance program that periodically helps counter recent and emerging tools and techniques that could be used for successful design and implementation This could pertain to science and technology regulations, institution mechanisms, fiscal and non-fiscal measures, and well-articulated and well-documented empirical evidences. As part of holistic management information support systems, this is especially in the case of rapidly developing economies to minimize the time taken in establishing such systems. Yet another important factor for the success of the Montreal Protocol was its ability to sustain investigations and successfully introduced alternatives. It is therefore essential to create and implement such enabling mechanisms including forecasts and technology leap from pathways. The opportunity to enhance performance across all the systems appears to be quite large.

REFERENCES

European Commission, 2013. A 2030 framework for climate and energy policies, Green Paper [Online]. Available at http://ec.europa.eu/energy/consultations/doc/com_2013_0169_green_paper_2030_en.pdf

Ahn, S. and D. Graczyk, 2012. Understanding energy challenges in India: Policies, players and issues [Online]. Available at https://www.iea.org/publications/freepublications/publication/India_study_FINAL_WEB.pdf (accessed February 27, 2015).

Allen, J., 2013. Alternative thinking 2013—Renewable energy under the microscope [Online]. Available at https://www2.deloitte.com/content/dam/Deloitte/global/Documents/dttl-er-AltThinking2013-07082013.pdf

Allianz Global Investors, (2012). Renewable energies – Investing against climate change: Renewable energies offer investors broad diversification opportunities with manageable risks. [Online]. Available at Market-Insights-Erneuerbare-Energien-Investieren-gegen-den-Klimawandel-EN.pdf

Australian National Greenhouse Accounts, National Greenhouse Accounts Factors, Department of Industry, Innovation, Climate Change, Science, Research, Tertiary Education, Australian Government [Online]. Available at http://www.climatechange.gov.au/sites/climatechange/files/documents/07_2013/national-greenhouse-accounts-factors-july-2013.pdf (accessed February 27, 2015).

Bioenergy Times, 2013. Policy initiatives [Online]. Available at http://www.abelloncleanenergy.com/Newsletters/Bio-Energy-Times/Pdf/April-2013.pdf (accessed February 27, 2015).

Burck, J., L. Hermwille, and L. Krings, 2013. The climate change performance index results 2013 [Online]. Available at http://germanwatch.org/en/ccpi_bame

CDM methodology booklet, United Nations Framework on climate change [Online]. Available at http://cdm.unfccc.int/methodologies/documentation/meth_booklet.pdf (accessed February 27, 2015).

Climate change and the path toward sustainable energy sources [Online]. Available at http://www.whitehouse.gov/sites/default/

files/docs/erp2013/ERP2013_Chapter_6.pdf (accessed February 27, 2015).

Climate Parliament, 2013. MPs key recommendations on wind energy [Online]. Available at http://www.climateparl.net/cpcontent/publications/MPs%20key%20Recommendations%20on%20wind%20Energy.pdf

Edenhofer, O., R. Pichs-Madruga, and Y. Sokona, 2011. Renewable energy resources and climate change mitigation special report of the Intergovernmental Panel on Climate Change for policymakers and technical summary [Online]. Available at http://www.ipcc.ch/pdf/special-reports/srren/SRREN_FD_SPM_final.pdf (accessed March 19, 2015).

Emerging opportunities and challenges, India Energy Congress—2012 [Online]. Available at http://www.pwc.in/assets/pdfs/publications-2011/wec-pwc-report.pdf (accessed February 27, 2015).

Energy and climate change, Friends of the Earth [Online]. Available at http://www.foe.co.uk/resource/factsheets/energy_climate_change.pdf (accessed February 27, 2015).

Energy-Efficient Cookers for Kitchens in India, Beyond Carbon [Online]. Available at https://www.myclimate.org/fileadmin/myc/klimaschutzprojekte/indien-7160/klimaschutzprojekt-indien-7160-project-story.pdf

Ernst & Young, 2013. Renewable energy country attractiveness indices. Issue 36. [Online]. Available at http://www.zonnekrachtcentrales.nl/assets/files/files/20121101%20CAI_issue-35_Nov-2012_DE0372.pdf

Estimated impacts of energy and climate change policies on energy prices and bills, Department of Energy and Climate Change, UK [Online]. Available at https://www.gov.uk/government/uploads/system/uploads/attachment_data/file/172923/130326_-_Price_and_Bill_Impacts_Report_Final.pdf (accessed February 27, 2015).

EUR 150 million loan to mitigate climate change in India, European Investment Bank [Online]. Available at http://www.eib.org/projects/press/2013/2013-028-eur-150-million-loan-to-mitigate-climate-change-in-india.htm (accessed March 26, 2015).

Feed-in tariffs as a policy instrument for promoting renewable energies and green economies in developing countries, United Nations Environmental Program [Online]. Available at http://www.unep.org/pdf/UNEP_FIT_Report_2012F.pdf (accessed February 27, 2015).

Gupta, S. K. and R. S. Anand 2013. Development of Solar Electricity Supply System in India: An Overview. *Journal of Solar Energy* 2013:1–10.

Inter-governmental Panel for Climate Change (IPCC) 2011. Special report of the Intergovernmental Panel on Climate Change renewable energy sources and climate change mitigation; summary for policymakers and technical summary. Available at https://www.ipcc.ch/pdf/special-reports/srren/SRREN_FD_SPM_final.pdf

India Second National Communication to the United Nations Framework Convention on Climate Change, Ministry of Environment & Forests [Online]. Available at http://moef.nic.in/downloads/public-information/India%20Second%20National%20Communication%20to%20UNFCCC.pdf. (accessed February 27, 2015).

India Wind Energy Outlook 2012, Global Wind Energy Council [Online]. Available at http://www.gwec.net/wp-content/uploads/2012/11/India-Wind-Energy-Outlook-2012.pdf (accessed February 27, 2015).

India: The World's carbon markets: A case study guide to emissions trading, Environmental Defense Fund [Online]. Available at http://www.ieta.org/assets/Reports/EmissionsTradingAroundTheWorld/edf_ieta_india_case_study_september_2013.pdf. (accessed February 27, 2015).

India's Initial National Communication to the United Nations Framework Convention on Climate Change, Ministry of Environment & Forests [Online]. Available at http://unfccc.int/resource/docs/natc/indnc1.pdf (accessed February 27, 2015).

India Energy Congress 2013. Securing tomorrow's energy today - policy and regulation [Online]. Available at http://mnre.gov.in/file-manager/akshay-urja/january-february-2013/EN/45.pdf

India Energy Congress 2012. Emerging opportunities and challenges [Online]. Available at http://www.pwc.in/assets/pdfs/publications-2011/wec-pwc-report.pdf

IRENA handbook on renewable energy Nationally Appropriate Mitigation Actions (NAMAs) for policy makers and project developers [Online]. Available at http://www.irena.org/DocumentDownloads/Publications/Handbook_RE_NAMAs.pdf (accessed February 27, 2015).

Jain, V. K. and S. N. Srinivas 2012. Empowering rural India the RE way—Inspiring success stories, Ministry of New and Renewable Energy [Online]. Available at http://s3idf.org/assets/2013/06/PICO_Biomass_MNRE_1_pdf (accessed February 27, 2015).

Krey, V. and L. Clarke 2011. Role of renewable energy in climate mitigation: A synthesis of recent scenarios. *Climate Policy* 11:1131–1158 [Online]. Available at http://www.climateaudit.info/pdf/ipcc/wg3/Krey%20and%20clarke%202011.pdf. (accessed February 27, 2015).

Kumar, A. 2013. India's Low Carbon Development Pathways to 2020, TERI [Online]. Available at https://workspace.imperial.ac.uk/climatechange/Public/Events/10.06.13%20India's%20emissions%20pathway%20to%202050/India%E2%80%99s%20Low%20Carbon%20Development%20Pathways%20to%202020%20-%20Atul%20Kumar.pdf

Lattanzio, R. K. 2013. The Global Climate Change Initiative (GCCI): Budget Authority and Request, FY2010-FY2014, Congress Research Service [Online]. Available at http://www.fas.org/sgp/crs/misc/R41845.pdf (accessed February 27, 2015).

Leiserowitz, A. 2013. Public Support for Climate and Energy Policies in April 2013, Centre for Climate Change Community [Online]. Available at http://environment.yale.edu/climate-communication/files/Climate-Policy-Report-April-2013-Revised.pdf (accessed February 27, 2015).

McCrone, A. 2013. Global trends in renewable energy investment 2013, Bloomberg New Energy Finance, Frankfurt School of Finance and Management [Online]. Available at http://www.unep.org/pdf/GTR-UNEP-FS-BNEF2.pdf (accessed February 27, 2015).

Meeting Renewable Energy Targets: Global lessons from the road to implementation, World Resources Institute [Online]. Available at http://awsassets.panda.org/downloads/meeting_renewable_energy_targets__low_res_.pdf (accessed February 27, 2015).

Mehta, P. S. 2012. *Mainstreaming Public Private Partnerships in India*, CUTS Institute for Regulation and Competition, New Delhi.

Milne, J. L. and C. B. Field. Assessment Report from the GCEP workshop on energy supply with negative carbon emissions, Global Climate and Energy Project, Stanford University [Online]. Available at http://gcep.stanford.edu/pdfs/rfpp/Report%20from%20GCEP%20Workshop%20on%20Energy%20Supply%20with%20Negative%20Emissions.pdf (accessed February 27, 2015).

Ministry of New and Renewable Energy (MNRE) 2010. Strategic plan for new and renewable energy sector for the period 2011-17, Ministry of New and Renewable Energy [Online]. Available at http://mnre.gov.in/file-manager/UserFiles/strategic_plan_mnre_2011_17.pdf

OECD/IEA, 2010. Energy Technology Roadmaps—Charting a low-carbon energy revolution. Available at http://www.iea.org/publications/freepublications/publication/pv_roadmap.pdf (accessed February 27, 2015).

Raghuvanshi, S. P., A. K. Raghav, and A. Chandra 2007. Renewable energy resources for climate change mitigation [Online]. Available at http://www.ecology.kee.hu/pdf/0604_015027.pdf (accessed February 27, 2015).

Rasmussen, K. 2010. A Rational Look at Renewable Energy and the implications of intermittent power [Online]. Available at http://carbon-sense.com/wp-content/uploads/2013/09/rational-look-at-renewable-energy.pdf

Redrawing the Energy-Climate Map, World Energy Outlook Special Report, International Energy Agency [Online]. Available at http://www.iea.org/publications/freepublications/publication/WEO_RedrawingEnergyClimateMap.pdf (accessed February 27, 2015).

REN21, 2013. Renewables 2013 Global Status Report (Paris: REN21 Secretariat). Available at: http://www.ren21.net/portals/0/documents/resources/gsr/2013/gsr2013_lowres.pdf (accessed February 27, 2015).

Road map for up scaling and mainstreaming renewables, Ministry of New and Renewable Energy [Online]. Available at http://www.direc2010.gov.in/pdf/DIREC-2010-Report.pdf (accessed February 27, 2015).

Seifried, D. and W. Witzel 2010. *Renewable energy – the facts*, Earthscan, London.

Singh, N. 2013. Creating market support for energy efficiency: India's Perform, Achieve and Trade Scheme [Online]. Available at http://cdkn.org/wp-content/uploads/2013/01/India-PAT_InsideStory.pdf (accessed February 27, 2015).

Tester, J. 2005. *Sustainable Energy: Choosing Among Options*, MIT Press, Cambridge, MA. p. 846.

Trottier Energy Futures Project, 2013. Low-carbon Energy Futures: A Review of National Scenarios [Online]. Available at http://www.davidsuzuki.org/publications/downloads/low-carbon%20Energy%20Futures.pdf (accessed March 19, 2015).

United Nations Environment Programme (UNEP 2013), Programme Performance Report 2012–2013, Available at: http://www.unep.org/annualreport/2013/docs/ppr.pdf#page=47

UNEP, 2012. A citizen's guide to energy subsidies in India, TERI and IISD [Online]. Available at http://www.iisd.org/gsi/sites/default/files/ffs_india_czguide.pdf (accessed February 27, 2015).

United Nations Development Programme (UNDP) 2012a. United Nations Environmental Protection, the emissions gap report 2012: a UNEP synthesis report [Online]. Available at http://www.unep.org/pdf/2012gapreport.pdf

United Nations Development Programme (UNDP) 2012b. United Nations Development Programme, Annual Report 2011/2012: The Sustainable Future We Want [Online]. Available at http://www.undp.org/content/dam/undp/library/corporate/UNDP-in-action/2012/English/UNDP-AnnualReport_ENGLISH.pdf

United Nations Development Programme (UNDP) 2013. United Nations Environmental Protection, the emissions gap report 2013: a UNEP synthesis report [Online]. Available at http://www.unep.org/pdf/UNEPEmissionsGapReport2013.pdf

Working with India to Tackle Climate Change, EU action against climate change in Europe and India [Online]. Available at http://eeas.europa.eu/delegations/india/documents/publications/working_with_india_to_tackle_climate_change_2012.pdf (accessed February 27, 2015).

RECOMMENDED CURRICULUM: GREEN ENERGY: SYSTEMS AND CLIMATE EFFICIENCY

1. Introduction and Structure
1.1 Forms and Functions of Energy
1.2 Energy Flows and Balance
1.2.1 Solar Radiation
1.2.2 Wind Flows
1.2.3 Ocean Energy
1.2.4 Rivers and Reservoirs
1.2.5 Geothermal
1.3 Green Energy Types and Characteristics
1.3.1 Solar
1.3.2 Wind
1.3.3 Tidal
1.3.4 Hydro-electric
1.3.5 Cogeneration
1.3.6 Biomass
1.3.7 Geothermal
1.3.8 Isolated and Grid Operations
1.3.9 Conversions, Losses and Coefficients
2. Solar Energy Principles and Systems
2.1 Architecture
2.2 Solar Thermal: Collectors, Heating Networks, Power Plants
2.3 Passive Solar Energy
2.4 Photovoltaic Power Generation: Arrays
2.5 Economic and Environmental Analysis
3. Biomass
3.1 Photosynthesis
3.2 Biogas and Cogeneration
3.3 Wood and other Substrates and Energy Plantations
3.4 Ethanol
3.5 Synthetic Fuels
3.6 Economic and Environmental Analysis

4. Hydro Electric Systems
4.1 Hydro Power Plants and Varieties
4.2 Setup, Intake, Turbines
4.3 Economic and Environmental Analysis
5. Geothermal Energy
5.1 Shallow Geothermal Sources and Utilization
5.2 Heat Supply, Hydrogeothermal and Deep Wells
5.3 Well Drilling
5.4 Heat Transfer
5.5 Leakage Monitoring
5.6 Sub-surface and Above Ground Systems
5.7 Economic and Environmental Analysis
6. Emerging Technologies
6.1 Fuel Cells
6.2 Solar hydrogen
7. Emission Offsets and Protocols
7.1 CDM Methodologies
7.2 Funding Parameters and Quantification of Paybacks
7.3 Carbon Markets
8. Policies
8.1 National and Local Imperatives and Agreements
8.2 Regional and Global Mechanisms
9. Green Energy Initiatives
9.1 Bilateral and Multilateral Processes
9.2 Successes and Challenges
9.3 Information Resources Including Databases
9.4 Industry Initiatives and Associations

FURTHER READING

Comprehensive renewable energy, vol 1: photovoltaic solar energy 2012. Elsevier, Amsterdam.

Comprehensive renewable energy, vol 2: wind energy 2012. Elsevier, Amsterdam.

Comprehensive renewable energy, vol 3: solar thermal systems: components and applications. 2012. Elsevier, Amsterdam.

Comprehensive renewable energy, vol 4: fuel cells and hydrogen technology 2012. Elsevier, Amsterdam.

Comprehensive renewable energy, vol 5: Biomass and biofuel production 2012. Elsevier, Amsterdam.

Comprehensive renewable energy, vol 6: hydro power 2012. Elsevier, Amsterdam.

Comprehensive renewable energy, vol 7: geothermal energy 2012. Elsevier, Amsterdam.

Comprehensive renewable energy, vol 8: ocean energy 2012. Elsevier, Amsterdam.

Cool energy: the renewable solution to global warming. 1990. Union of concerned scientists, Cambridge, MA.

Dieter, S. and Walter, W. 2010. *Renewable Energy, the facts*. Earthsean, London, p. 251.

Kicking the global warming and the case for renewable and nuclear energy 2006. Columbia University Press, New York.

Martin, K., Wolfgang, S. and Andreas, W. (Eds) 2007. Renewable energy, technology, economics and environment. Springer - Verlag, Berlin, p. 564.

Multi-criteria analysis in the renewable energy industry 2012. Springer-Verlag, London.

Policies and measures to reduce CO_2 emissions by efficiency and renewable 1996. WWF - Netherlands, Zeist, Zeist, The Netherlands.

Policy on Renewable Energy Development and Energy Conservation: Green Energy Imitative 2004. Ministry of Energy and Mineral Resource, Jakarta.

Proceedings of the IPCC scoping meeting on renewable energy sources 2008. IPCC, Germany.

Renewable energy and climate change 2010. John Wiley and Sons Ltd. West Sussex.

Renewable energy sources and climate change mitigation: special report of IPCC. 2012. Cambridge University Press, Cambridge.

Renewable energy sources and climate change mitigation: summary for policymakers and technical summary. 2011. Intergovernmental Panel on Climate Change, Geneva.

Renewable energy sources: a chance to combat climate change. 2009. Kluwer Law International, The Netherlands.

Renewables 2013. Global status report. http://www.ren21.net/Portals/0/documents/Resources/GSR/2013/GSR2013_lowres.pdf

The world wind energy association 2012. Annual report. http://www.wwindea.org/webimages/WorldWindEnergyReport2012_final.pdf

World Energy Outlook 2012. http://www.worldenergyoutlook.org/media/weowebsite/2012/WEO2012_Renewables.pdf

6

ENGINEERING IN ENVIRONMENTAL MANAGEMENT

YANG DENG

Department of Earth and Environmental Studies, Montclair State University, Montclair, NJ, USA

Abstract: To gain an appreciation for the basic engineering principles is critical to environmental management professionals due to the breadth and complexity of environmental problems. This chapter provides an introduction to engineering in environmental management, with an emphasis on environmental engineering. Following a brief discussion on engineering and environmental management, the concepts of materials and energy balances, which are the basic tools for understanding environmental processes and solving environmental problems, are first introduced. Thereafter, water resources, municipal solid waste management, and air resources in environmental engineering are discussed in details. Theory and design issues are integrated in the topics on physical, chemical, and biochemical operations and processes.

Keywords: engineering, engineers, environmental engineering, environmental management, materials balance, Reactors, energy balances, Pollutants, physical treatment, chemical treatment, biological treatment, water supply, drinking water, Wastewater, municipal solid waste, Landfills, Incineration, Recycling, air pollution.

6.1 Engineering and Environmental Management	151	
6.2 Materials and Energy Balances	152	
6.2.1 Materials Balance	152	
6.2.1.1 Reactors	154	
6.2.2 Energy Balance	154	
6.3 Water Resource Management System	155	
6.3.1 Water Supply Subsystem	155	
6.3.1.1 Water Pollutants	155	
6.3.1.2 Water Sources	155	
6.3.1.3 Water Treatment Systems	156	
6.3.1.4 Water Distribution System	159	
6.3.1.5 Drinking Water Standards	159	
6.3.2 Wastewater Disposal Subsystem	159	
6.3.2.1 Wastewater Pollutants	160	
6.3.2.2 Sanitary Sewers	160	
6.3.2.3 Wastewater Treatment Systems	161	
6.3.2.4 Wastewater Treatment Standards	163	
6.4 Municipal Solid Wastes	163	
6.4.1 Generation and Composition of MSW in the United States	163	
6.4.2 Integrated Solid Waste Management	164	
6.4.3 Landfilling	166	
6.5 Air Resource System	168	
6.5.1 Air Pollution Management	168	
6.5.2 Air Pollutants	168	
6.5.3 Stationary Source Control	169	
6.5.4 Mobile Source Control	170	
References	171	

6.1 ENGINEERING AND ENVIRONMENTAL MANAGEMENT

Engineering is a profession that applies mathematics and science to utilize the properties of matter and energy to create useful structures, machines, products, systems, and processes (Davis and Cornwell, 2008). The people who practice engineering are called *engineers*, and the licensed engineers in the United States are called *professorial engineers (PE)*. To become licensed, engineers must complete a 4-year college degree, work under a PE for at least 4 years, pass two intensive competency exams, and earn a license from their state's licensure boards. Then, PEs must continually maintain and improve their skills throughout their

---

An Integrated Approach to Environmental Management, First Edition. Edited by Dibyendu Sarkar, Rupali Datta, Avinandan Mukherjee, and Robyn Hannigan.
© 2016 John Wiley & Sons, Inc. Published 2016 by John Wiley & Sons, Inc.

TABLE 6.1 Fundamental Principles and Canons of Engineering

Fundamental principles	Engineers uphold and advance the integrity, honor, and dignity of the engineering profession by: 1. Using their knowledge and skill for the enhancement of human welfare and the environment 2. Being honest and impartial and serving with fidelity the public, their employers, and clients 3. Striving to increase the competence and prestige of the engineering profession 4. Supporting the professional and technical societies of their disciplines
Fundamental canons	1. Engineers shall hold paramount the safety, health, and welfare of the public and shall strive to comply with the principles of sustainable development in the performance of their professional duties 2. Engineers shall perform services only in areas of their competence 3. Engineers shall issue public statements only in an objective and truthful manner 4. Engineers shall act in professional matters for each employer or client as faithful agents or trustees and shall avoid conflicts of interest 5. Engineers shall build their professional reputation on the merit of their services and shall not compete unfairly with others 6. Engineers shall act in such a manner as to uphold and enhance the honor, integrity, and dignity of the engineering profession and shall act with zero tolerance for bribery, fraud, and corruption 7. Engineers shall continue their professional development throughout their careers and shall provide opportunities for the professional development of those engineers under their supervision

Source: Used with permission, American Society of Civil Engineers.

careers to retain their licenses (NSPE, 2013). The major engineering branches include chemical engineering, civil engineering, electrical engineering, and mechanical engineering, under which there exist numerous subdisciplinary and interdisciplinary categories. In fact, all facets of engineering, ranging from the materials processes implicitly designed into our creations to the residuals left behind in their manufacture and the resultant products of their use, can affect the environment (Lindeburg, 2003). The American Society of Civil Engineering (ASCE) outlines the fundamental principles and canons of engineering into its Code of Ethics, as shown in Table 6.1 (ASCE, 2013). The engineering type most associated with environmental management may be *environmental engineering*, a major subdiscipline of civil engineering. Environmental engineers use the principles of engineering, soil science, biology, and chemistry to develop solutions to environmental problems. They are involved in efforts to improve recycling, waste disposal, public health, and control of water and air pollution (Bureau-of-labor-statistics, 2008).

Engineering knowledge and practice provide a critically important tool to solve various problems in multiple environmental management settings. This allows us to recognize, select, and apply proper technological solutions that can prevent, treat, monitor, and remediate air, water, and soil pollution. Therefore, engineering plays an indispensable role in environmental management.

6.2 MATERIALS AND ENERGY BALANCES

Engineers utilize two key natural resources, materials and energy. Materials are useful due to their properties, which include strength, durability, ability to insulate or conduct, and unique chemical, electrical, or acoustical characteristics. Furthermore, energy can be transformed among different forms. The energy used in engineering includes fossil fuels (coal, petroleum, and natural gas) and alternative energies, such as wind, sunlight, hydropower, geothermal energy, tides, and nuclear fission. Material and energy balances are two vital tools that engineers use to quantitatively understand behaviors of our environmental systems. They serve as a method of accounting for the flow of materials and energy into and out of these systems (Davis and Cornwell, 2008).

6.2.1 Materials Balance

Materials balance is also called mass balance. The law of conservation of matter states that matter is neither created nor destroyed without nuclear reactions. However, materials balance does not mean that a specific substance existing in environmental systems cannot vary among different forms. For example, after organic wastes enter a landfill, its carbon may be microbiologically transformed into carbon dioxide (CO_2) that is released into air or dissolved into landfill leachate (a wastewater produced from landfills), exist as incompletely decomposed organic wastes, or remain in the residual organic wastes that are not subject to chemical or biochemical reactions. Although the carbons are in different forms, the amount originally existing in the incoming organic waste is equal to the sum of carbons in CO_2 and organic wastes following transformation.

To conduct an analysis of materials balance, we should first define a particular region of interest within a specific boundary. This region is called *control volume*. For example, when a researcher conducts laboratory-scale batch tests in a beaker to remove pesticides in water, the control volume is the beaker. When a scientist studies the fate of fine particulate matter (PM) (PM 2.5) in air within the City of New York, the city is the control volume.

Any substance entering a control volume may have four fates: (i) it leaves the system without change; (ii) it is reduced due to transformation to other substances; (iii) more substances are produced; and (iv) it is accumulated within the control volume. The general equation of materials balance is shown below:

$$\text{Accumulation rate} = \text{input rate} - \text{output rate} + \text{reaction rate} \quad (6.1)$$

The calculus-based equation can be expressed as follows:

$$\frac{dM}{dt} = \frac{d(M_{in})}{dt} - \frac{d(M_{out})}{dt} + rV \quad (6.2)$$

where:
M, the mass accumulated within the boundary
M_{in} and M_{out}, the mass flowing in and out, respectively
V, the system volume
r, reaction rate, which is "+" when the substance is generated and "−" when the substance decays

To correctly analyze materials balance, the following two issues need to be paid attention to:

1. *Conservative* substance versus *nonconservative* substance. If the substance of interest is not subject to any reaction (no production or decay), it is a conservative substance, and r in Equation 6.2 = 0. In contrast, if the substance is generated or decays in the system, it is a nonconservative substance. When the reaction follows a first-order pattern, $r = kC_0 e^{-kt}$ (generation) or $r = -kC_0 e^{-kt}$ (decay). It should be noted that the substance may be conservative under certain conditions, but nonconservative in others, depending on the environment. For example, chloride (Cl⁻) is usually inert (conservative) in water; however, the concentration of Cl⁻ in water dramatically drops in the presence of silver ion (Ag⁺) due to formation of insoluble silver chloride (AgCl, a white precipitate). Therefore, Cl⁻ is nonconservative in an environment where silver is present.

2. *Steady state* versus *unsteady state*. Steady state means that the concentration of a substance within the boundary does not change with time. In a steady state, the accumulation rate $(dM/dt) = 0$. In contrast, in an unsteady state, the concentration increases or decreases over time, and thus $dM/dt \neq 0$.

For example, a factory discharges its wastewater into a nearby stream, as shown in Fig. 6.1, and the flow rate of the stream is four times as that of the wastewater. The BOD_5 values of the stream and wastewater are 5 and 105 mg/l, respectively (BOD_5 represents the level of biodegradable organic wastes and will be discussed in detail in Section 6.3.2.1). It is assumed that the wastewater is rapidly and uniformly mixed with the stream. What is the BOD_5 after the mixing is completed?

Because wastewater is rapidly and completely mixed with the stream, where the wastewater merges into the stream is regarded as the control volume, and BOD_5 is conservative in this problem. Therefore, the materials balance equation is simplified as follows:

$$0 = \frac{d(M_{in})}{dt} - \frac{d(M_{out})}{dt} + 0 \quad (6.3)$$

When the substance of interest is water, Equation 6.3 can be rewritten as below:

$$0 = \frac{d(\rho V_{in})}{dt} - \frac{d(\rho V_{out})}{dt} \quad (6.4)$$

$$0 = \frac{dV_{in}}{dt} - \frac{dV_{in}}{dt} \quad (6.5)$$

$$0 = Q_{in} - Q_{out} \quad (6.6)$$

For the control volume, we have two inputs from the upstream and wastewater discharge and one output (the downstream):

$$0 = Q_{in} - Q_{out} \quad (6.7)$$

$$0 = (4Q + Q) - Q_m \quad (6.8)$$

$$Q_m = 5Q \quad (6.9)$$

When the substance of concern becomes BOD_5, the materials balance equation is below:

$$0 = \frac{d(M_{in})}{dt} - \frac{d(M_{out})}{dt} + 0 \quad (6.10)$$

$$0 = \frac{d(C_{in} V_{in})}{dt} - \frac{d(C_{out} V_{out})}{dt} \quad (6.11)$$

$$0 = \frac{C_{in} dV_{in}}{dt} - \frac{C_{out} dV_{out}}{dt} \quad (6.12)$$

$$0 = C_{in} Q_{in} - C_{out} Q_{out} \quad (6.13)$$

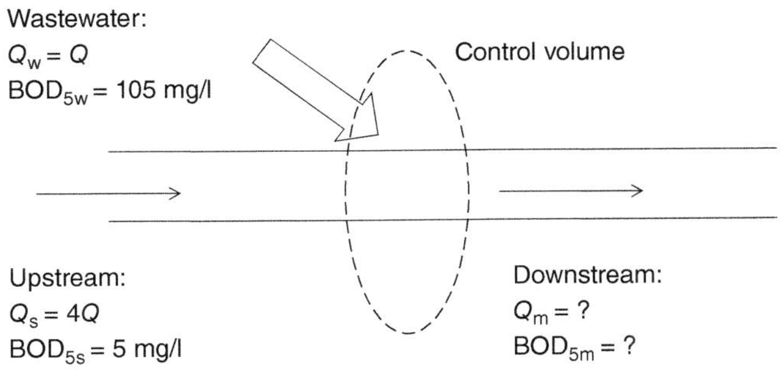

Wastewater:
$Q_w = Q$
$BOD_{5w} = 105$ mg/l

Upstream:
$Q_s = 4Q$
$BOD_{5s} = 5$ mg/l

Downstream:
$Q_m = ?$
$BOD_{5m} = ?$

FIGURE 6.1 Mass balance during dilution of wastewater in a stream.

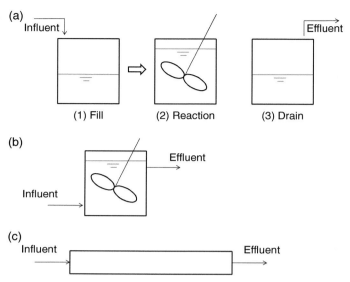

FIGURE 6.2 Types of basic reactors. (a) Batch reactor; (b) completely mixed flow reactor; and (c) plug flow reactor.

After the data given are inputted, the materials balance equation is

$$0 = (105 \times Q + 5 \times 4Q) - (BOD_{5m} \times 5Q)$$
$$BOD_{5m} = 25 \text{ mg/l}$$

6.2.1.1 Reactors

The concept of reactors is commonly used in engineering design. Reactors are the tanks in which certain physical, chemical, and biochemical reactions occur. They may be large square or rectangular basins, cylindrical tanks, pipes, long channels, columns, and towers (Crittenden et al., 2011). Selection and design of reactors are based on key stoichiometric and kinetic information of these reactions, in addition to knowledge in practical flow patterns. The concept of reactor models may be used to model natural systems as long as control volumes are appropriately selected.

There are three basic reactor types: (i) batch reactor (BR), (ii) completely mixed flow reactor (CMFR), and (iii) plug flow reactors (PFRs). BRs are operated in a fill-and-draw-type process as follows: First, materials are added to the reactor. Second, materials are completely mixed for sufficient time to allow the reaction to complete. Third, materials are drained after reaction. During the reaction within a BR, no flow enters or leaves, and the reaction rate proceeds at the identical rate everywhere within the reactor as a result of complex mixing. CMFR, also called continuous flow stirred tank reactors (CSTRs), are ideal when reactants and products continuously flow into and out of the reactors, and the reactions occur at a complete mixing state within the reactor. Within a CMFR, the reaction rate also proceeds at the identical rate everywhere. It should be noted that the concentration of any chemical in the effluent is the same as its concentration within the reactor. PFRs are ideal reactors in which flow continuously passes through in sequence. In an ideal state, complete mixing occurs only in the vertical direction of the flow, but no mixing occurs in the lateral direction (the fluid does not mix with the fluid elements in front of or behind it). The basic reactors types are shown in Fig. 6.2.

6.2.2 Energy Balance

Energy is defined as the capacity to do useful work. The unit of energy is joule (J). The first law of thermodynamics simply states that energy can neither be created nor destroyed without any nuclear reaction; however, it does not mean that the form of energy in a certain system cannot change. For example, the potential energy of water in a dam can be converted to mechanical energy by spinning turbines to generate electricity in a hydroelectric power plant.

Similar to materials balance, the system of study, like the control volume, is the first item to define. A system where both material and energy can flow through is called *an open system*, whereas the one in which only energy can flow across its boundary, but material cannot, is called *a closed system*. The first law of thermodynamics can be generally expressed by the following equations.

For an open system,

Net change of energy = total energy of mass entering system − total energy of mass leaving system + total energy crossing the system boundary   (6.14)

For a closed system,

$$\text{Net change of energy} = \text{total energy crossing the system boundary} \quad (6.15)$$

Here, the total energy (E) of a substance is the sum of its internal energy, kinetic energy, and potential energy. Note that the "total energy crossing the system boundary" may be heat or work. *Work* is done by a force, acting on a body, which results in a movement. It should be noted that energy has many forms that may be thermal, mechanical, kinetic, potential, electrical, or chemical.

6.3 WATER RESOURCE MANAGEMENT SYSTEM

The basic components in the water resource management system are presented in Fig. 6.3. We continuously withdraw water from the nature (mostly from fresh surface and groundwater). After appropriately treated, the water quality is improved to meet different intended purposes. Once the water is used, the water quality may degrade to different degrees. The wastewater should be properly treated prior to discharge; otherwise, pollution may occur. In arid areas (e.g., California), the treated wastewater may be collected, treated, and reused. Pathways 1–2, 3–4, and 5 represent a water supply subsystem, wastewater disposal subsystem, and water reuse subsystem, respectively. Water supply and wastewater disposal subsystems are discussed below.

6.3.1 Water Supply Subsystem

A flowchart of a water supply subsystem is shown in Fig. 6.4. The major functions of a water supply subsystem include collection, purification, transmission, and distribution of water. After raw water (the natural water that has not been purified and treated) is withdrawn from surface water (e.g., streams, lakes, and reservoirs) or pumped from groundwater through wells, it is treated in water treatment plants, where pollutants are significantly reduced. The finished water (the water that has been treated and ready for delivery) is then pumped into the pipe network in towns and cities. Excess finished water may be temporally stored in elevated water towers or ground-level reservoirs during low-demand periods (e.g., at night) and will reenter the distribution system as an additional supply source during peak water use times.

6.3.1.1 Water Pollutants

Undesirable substances exist in water sources naturally or due to pollution. Water pollutants of great concerns are discussed below.

Particles. Particles are finely divided solids that are larger than molecules, but are generally not distinguishable by the naked eye (Crittenden et al., 2011). Their sizes vary between 0.001 and 100 micrometers (μm). Settleable particles can be spontaneously removed through sedimentation; however, suspended particles cause cloudiness and cannot be removed

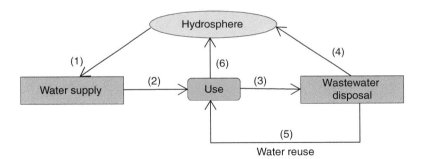

FIGURE 6.3 Water resource system.

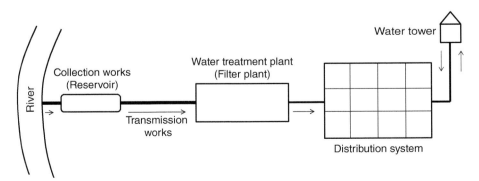

FIGURE 6.4 An example of water supply subsystem.

without treatment. These insoluble particulates impede the passage of light through water by scattering and absorbing light, thus contributing to the water's cloudiness, or *turbidity* (unit: nephelometric turbidity unit or NTU). Turbidity, primarily caused by colloids in water, is a major pollutant that must be lowered within surface water treatment plants. High turbidity can typically be found in raw water from rivers.

Color. Color in water results primarily from the presence of natural organic matter (NOM), particularly humic matter, causing a yellow-brown color (Andrew, 2005). Inorganic chemicals, like iron (Fe) for example, may contribute to color, causing groundwater to become brownish red.

Hardness. Total hardness is defined as the sum of calcium (Ca) and magnesium (Mg) concentrations (Andrew, 2005). Traditionally, it is expressed as mg/l $CaCO_3$. These cations (positively charged ions) are easily precipitated to form scale in pipes, reacting with soap to form difficult-to-remove scum. The hardness of water is typically classified as follows: 0 to <50 mg/l $CaCO_3$, soft; 50 to <100 mg/l $CaCO_3$, moderately hard; 100 to <150 mg/l $CaCO_3$, hard; and >150 mg/l $CaCO_3$, very hard (Crittenden et al., 2011). Groundwater frequently has a higher level of hardness than surface water.

Pathogens. Pathogenic microorganisms include bacteria, viruses, protozoa, and other organisms. They are excreted by people and animals with disease. Sufficiently high concentrations of pathogens make water unsafe for drinking, recreation, or fishing. Bacterial indictors (e.g., total coliform, *Escherichia coli,* and fecal coliform), which are bacteria that do not pose a health threat, can often indicate whether potentially harmful bacteria may be present in water. Viruses, which are excreted in feces of infected individuals, affect the human gastrointestinal tract. The US Environmental Protection Agency (USEPA) regulates *enteroviruses*, including polioviruses, coxsackieviruses, echoviruses, and other *enteroviruses.* Protozoa are microscopic, single-celled organisms, although they are relatively large compared to other microbes. Two types of protozoa, including *Cryptosporidium* and *Giardia lamblia*, are regulated in drinking water.

Salts. Salts and organic substances that do not evaporate contribute to total dissolved solids (TDS; units mg/l). Increased TDS in water sources may be caused by saltwater intrusion, the use of salt for snowmelt, or high salinity industrial discharge. The USEPA requires that the maximum contaminant level (MCL) of TDS in drinking water is 500 mg/l.

Emerging Contaminants. Emerging contaminants can be broadly defined as any synthetic or naturally occurring chemical or microorganism that is not commonly monitored in the environment; however, it has the potential to enter the environment and cause suspected adverse ecological and/or human health effects (USGS, 2013). These new contaminants have not previously detected at levels significantly different than expected. Examples of emerging contaminants include perfluorinated compounds, pharmaceuticals and personal care products (PPCPs), hormones, algal toxins, brominated flame retardants, and new disinfection by-products (DBPs) (e.g., NDMA) (Richardson and Ternes, 2011). During 1999 and 2000, the US Geological Survey (USGS) measured concentrations of 95 of pharmaceuticals, hormones, and other emerging organic wastewater contaminants (OWCs) in water samples from a network of 139 streams across 30 states of the United States. One or more OWCs were found in 80% of the streams. The most frequently detected compounds were coprostanol (fecal steroid), cholesterol (plant and animal steroid), *N,N*-diethyltoluamide (insect repellent), caffeine (stimulant), triclosan (antimicrobial disinfectant), tri(2-chloroethyl) phosphate (fire retardant), and 4-nonylphenol (nonionic detergent metabolite) (Kolpin et al., 2002). Of note, existing conventional water treatment processes are not specially designed or equipped with sufficient infrastructures to address these contaminants of emerging concern. As a consequence, numerous emerging contaminants are poorly removed in traditional water treatment facilities (Stackelberg et al., 2007; Vieno et al., 2007).

6.3.1.2 Water Sources

Sources of drinking water are groundwater or surface water. Generally speaking, groundwater is less contaminated (lower turbidity, lower or no color, and bacteriologically safe) with relatively constant composition; however, it often has high mineralization and hardness and sometimes contains high concentrations of iron (Fe) and manganese (Mn), as well as undesirable hydrogen sulfide (H_2S). Deep or shallow wells pump groundwater from aquifers to water treatment plants. Currently, more than 15.9 million water wells serve the United States, and approximately 500,000 new residential wells are constructed annually (NGWA, 2010).

Surface water (i.e., rivers, lakes, and reservoirs) is more readily contaminated and has higher turbidity and color than groundwater. Certain microorganisms are present with low mineralization and hardness, and tastes and odors are usually noticeable, especially during algal blooming. Among the different surface water types, raw water from reservoirs has the lowest water quality, followed by raw water from lakes and then rivers (Davis and Cornwell, 2008).

6.3.1.3 Water Treatment Systems

The location of a water treatment plant is influenced by three factors: (i) available resources, such as water source, power, and sewerage services; (ii) economic consideration, such as land cost, operating and maintenance (O&M) cost, and annual tax; and (iii) environmental factors, such as traffic (Lindeburg, 2003). A water treatment plant must be located above the flood plain with a 4–6 m elevation difference, which allows water to

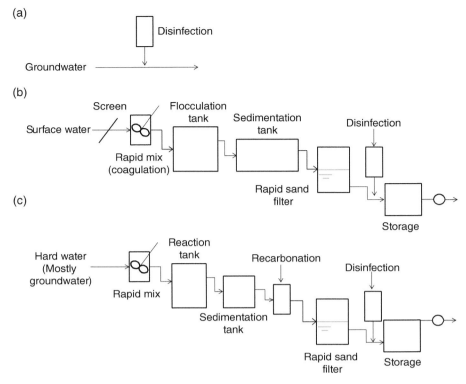

FIGURE 6.5 Flowcharts of three types of water treatment plants. (a) Simple disinfection; (b) filter plant; and (c) softening plant.

gravitationally flow between reactors. Lifetime of water treatment plant equipment typically ranges from 25 to 30 years.

Conventional water treatment systems are classified as simple disinfection, filter plants, and softening plants, as shown in Fig. 6.5. *Simple disinfection* (Fig. 6.5a) is usually used for high-quality groundwater. Addition of certain disinfectant guarantees that drinking water is microbiologically safe. Filter plants and softening plants are discussed below in detail.

Filter Plants (Fig. 6.5b). Filter plants are primarily used to treat surface water. A typical treatment train includes screens, rapid mixing to achieve coagulation, flocculation, sedimentation, filtration, and disinfection. This series of treatment can effectively remove turbidity and pathogens, color, NOM, and undesirable taste and odor.

Screen. Screening typically takes places prior to pumping. Coarse screens of vertical steel bars with 2–7 cm openings are used to protect pumps and mixing equipment from large objects. Following it, a finer screen (e.g., a traveling screen) can be installed to remove leaves, twigs, fish, etc. In addition, the construction of microstrainers can work to remove algae with 50–95% efficiency.

Coagulation. Coagulation is a process that destabilizes stable colloidal particles in raw water by the addition of certain chemicals, causing these particles to agglomerate. The added chemical agent is called coagulant, which can achieve destabilization through neutralization of colloidal surface charge (charge neutralization). This process is achieved through the following mechanisms: compression of colloidal electrical double layers, adsorption bridging, and netting sweeping (Edzwald, 2011).

Coagulants are typically classified into (i) hydrolyzing metal ions (e.g., aluminum and iron) and (ii) polymers. The hydrolyzing metal ion-based coagulants include aluminum sulfate ($Al_2(SO_4)_3 \cdot (14-16)H_2O$, commonly referred to as "alum"), ferric sulfate ($Fe_2(SO_4)_3 \cdot xH_2O$), and ferric chloride ($Fe_2Cl_3 \cdot xH_2O$). Since hydrolyzation of these coagulants can produce acids that reduce pH, raw water needs sufficient alkalinity to maintain an appropriate solution pH. Otherwise, lime (CaO) and soda ash (Na_2CO_3) need to be externally introduced. The addition of these trivalent cations significantly neutralizes negative charges on colloidal surfaces and compresses their unique dual electrical layers, thereby enhancing colloid aggregation. Polymers are long-chain organic compounds with numerous active adsorption sites. These active sites can adhere to particles, join them, and produce larger, tougher flocs (larger particles) with better settling capacity (adsorption bridging). Their molecular weights range from a few hundred to several million Daltons.

Polymers may be negative (anionic), positive (cationic), negative/positive (polyampholyte), or neutral (nonionic).

Since a coagulation process is completed within 0.1 s, the coagulation must be achieved by rapid and instantaneous mixing. Rapid mixing is accomplished within a tank using a vertical shaft mixer or a pipe with an in-line blender or a static mixer. Performance of rapid mixing is controlled by two operating factors, G (velocity gradient) and $G\theta$ (θ is the rapid mixing time). The levels of G and θ are fairly different during different mechanism-induced coagulations, for example, $G=3000-5000\,\text{s}^{-1}$ and $\theta=0.5\,\text{s}$ for adsorption–desorption coagulation, and $G=600-1000\,\text{s}^{-1}$ and $\theta=1-10\,\text{s}$ for netting sweep coagulation. The optimal pH (typically, pH 6–7 for alum, and pH 4–9 for ferric coagulant) and coagulant dose can be determined by classical jar tests.

Flocculation. Flocculation is a process that slowly brings the destabilized colloids into contact, resulting in larger and denser flocs that are readily settable. Flocculation is achieved with a paddle flocculator or a baffled chamber, and the most important operating factors controlling the process are G and $G\theta$. Because the formed flocs can be broken under a violent agitation, flocculation G is very low ($G=20-300\,\text{s}^{-1}$) relative to coagulation G. Moreover, the growth of floc during flocculation is a relatively slow process. Therefore, θ is high ($G\theta=36,000-400,000$).

Sedimentation. Sedimentation is the removal of solid particles (flocs) by gravity. Water typically stays in a sedimentation tank for a few hours during which flocs gradually settle down to the sludge zone. The removal efficiency of particles depend upon their settling speeds, particle distribution in terms of these settling speeds, flow rate, and surface area of the sedimentation tank. The produced sludge, called water treatment residuals (WTR), is an undesirable by-product and requires appropriate treatment and disposal. The effluent turbidity generally ranges between 1 and 10 NTU with a typical value of 2 NTU (Davis and Cornwell, 2008).

Filtration. Filtration is a process that removes fine particles that does not settle after water passes through a filter bed containing media. The most frequently used media is sand, and occasionally, dual (sand and anthracite) and triple (sand, anthracite, and garnet) media are used to enhance the treatment efficiency. In most cases, water flows down through media into filter by gravity. In modern water treatment plants, filters have a high loading rate of $1.4-6.8\,\text{l/s}\,\text{m}^2$. Filters are mostly square with a hydraulic head of 0.3–2.4 m. The filter heights are on the order of 3 m.

Filtration rate is significantly reduced, and/or the hydraulic head above filter media is dramatically increased when particles accumulate within the filter; therefore, filters need to be periodically cleaned. Cleaning, sometimes called *backwashing*, is completed in place by forcing clean water and fine bubbles backward through sands. The backwash expands the sand bed and removes fine particles from sands, after which the cleaned sands settle back into place. Generally, a treatment plant has at least three filters. Two can be operated to produce clean water when one is under backwash. Turbidity of filter effluent, which is discharged into a clear well storage reservoir, is dramatically reduced to less than 0.3 NTU (Davis and Cornwell, 2008).

Disinfection. Disinfection is a physical or chemical process that reduces pathogens (disease-producing microorganisms) to an acceptable level. Common disinfecting chemicals include chlorine, ozone, chlorine dioxide, and chloramine. Recently, ultraviolet (UV) irradiation has been also used for disinfection.

Chlorine is the most popular disinfectant in the United States due to its effectiveness, lasting disinfection capacity, and low cost. Chlorine can be added as a gas or as a liquid, at which point the Cl_2 hydrolyzes rapidly to form hypochlorous acid (HOCl):

$$Cl_2 + H_2O \leftrightarrow HOCl + HCl \qquad (6.16)$$

Chlorine in the form of HOCl and/or hypochlorite (OCl^-) is called *free available chlorine*. In the presence of ammonia, chlorine can react to form monochloramine, dichloramine, and trichloramine. In addition, chlorine reacts with organic nitrogenous compounds to form organic chloramine complexes. The chlorine existing in chemical combination with ammonia or organic nitrogen materials is defined as *combined available chlorine*. Because combined available chlorine is much less active than free available chlorine, breakpoint chlorination is used in practice during which chlorine dose is over a threshold level (breakpoint). This means that the residual chlorine is the more active form of free available chlorine, rather than combined chlorine.

The product of disinfection concentration (C) and time (T) (i.e., CT) is broadly used as criterion for microorganism inactivation. Generally, a high CT achieves a high pathogen removal. CT can be empirically expressed as follows:

$$CT = 0.9847 C^{0.1758} \text{pH}^{2.7519} \text{Temp}^{-0.1467} \qquad (6.17)$$

where $\text{pH} = -\log[H^+]$ and Temp = temperature in °C. To prevent regrowth of pathogenic microorganisms in a water distribution system, chlorine residual should be maintained until drinking water reaches users.

During disinfection, the added disinfectant may transform certain chemicals in water to numerous undesirable by-products. The so-called DBPs can be a great concern in public health. One well-known example is the combination of chlorine and DBP precursors, or NOMs, which react to form cancerogenic trihalomethanes (e.g., chloroform) and haloacetic acids. In the primary drinking water regulations, MCLs of total trihalomethanes (TTHMs) and five haloacetic acids (HAA5) are 0.080 and 0.060 mg/l, respectively. To control the DBP formation, the following strategies may be used: (i) change the application location of chlorine; (ii) effectively remove NOM prior to chlorination; (iii) reduce free chlorine residual; and (iv) apply alternative disinfectants. In fact, many alternatives for chlorine can be used, including ClO_2, O_2, and UV. However, these disinfectants are less effective for shorter time periods, so there is a risk of regrowth of microorganisms when water is present in water distribution systems. Moreover, usage of these alternative disinfectants may lead to formation of new DBPs. For example, cancerogenic bromate (BrO_3^-) can be produced when ozone is used to disinfect water containing bromide.

Softening Plants (Fig. 6.5c). The major function of softening plants is to remove hardness from raw water. Given that a high hardness level is usually found in groundwater, soften plants are frequently applied for groundwater treatment. In a softening plant, lime (CaO) and soda ash (Na_2CO_3) are added in a rapid mixer to react with Ca^{2+} and Mg^{2+} to form insoluble precipitates ($CaCO_3$ and $Mg(OH)_2$). Total hardness is the sum of carbonate hardness (i.e., the amount of hardness equal to the total hardness or the total alkalinity, whichever is less) and noncarbonate hardness (i.e., the total hardness in excess of alkalinity). These precipitates can be readily removed in subsequent sedimentation tanks. The added lime, also known as quick lime, forms hydrated lime ($Ca(OH)_2$) in water, which first removes dissolved CO_2 gas from the water, and then precipitates carbonate hardness as shown in the following equations:

$$Ca(HCO_3)_2 + Ca(OH)_2 \rightarrow 2CaCO_3 \downarrow + 2H_2O \quad (6.18)$$

$$Mg(HCO_3)_2 + 2Ca(OH)_2 \rightarrow 2CaCO_3 \downarrow + Mg(OH)_2 \downarrow + 2H_2O \quad (6.19)$$

To remove noncarbonate hardness due to Mg^{2+}, lime needs to be added:

$$Mg^{2+} + Ca(OH)_2 \rightarrow Mg(OH)_2 \downarrow + Ca^{2+} \quad (6.20)$$

To remove noncarbonate hardness due to Ca^{2+}, soda ash needs to be added:

$$Ca^{2+} + Na_2CO_3 \rightarrow CaCO_3 \downarrow + 2Na^+ \quad (6.21)$$

Of note, lime–soda ash softening cannot completely remove hardness from water in isolation. The minimum achieved Ca and Mg hardness are approximately 30 and 10 mg/l $CaCO_3$, respectively. To achieve a full removal of hardness, ion exchange may be used; however, it is not typically applied in a large-scale treatment plant.

6.3.1.4 Water Distribution System

Water distribution systems are designed to satisfy water quantity and pressure requirements for domestic, commercial, industrial, and firefighting purposes (Viessman et al., 2009).

The main elements in a distribution system include pipe systems, pumps, storage facilities, valves, and others (e.g., manholes, fire hydrants, and meters).

Pipes convey water. Pipelines may be grid, branching, or a combination of the two and are built of concrete, steel, cast iron, or plastic. Different *valves* in pipelines regulate water flow or pressure. For example, gate valves completely open or close pipes, check valves prevent backflow, pressure-reducing valves limit water pressure at specific points, and air release valves release trapped gas at high points of pipelines to prevent formation of a vacuum. *Distribution reservoirs* provide service storage to meet fluctuating water demands, to accommodate firefighting and emergency requirements, and to equalize operating pressures (Viessman et al., 2009). Distribution reservoirs include surface reservoirs, standpipes, and elevated tanks. The presence of these storage facilities increases reliability of the water supply and reduces the sizes of these pipes between them and water treatment plants, thereby lowering capital and operational costs. When water flows through pipes, valves, and junctions, the mechanical energy of water is reduced due to major and/or minor hydraulic losses. If needed, pumps are used to increase the hydraulic head of water to ensure an adequate water pressure in pipes. The most frequently used pumps are centrifugal and displacement. The former utilizes a rotating impeller to impart energy to water, whereas the latter uses a piston to withdraw water into a closed chamber and then expel water under pressure (Viessman et al., 2009). When more than one pump is used, they can be operated either in parallel or in series.

6.3.1.5 Drinking Water Standards

The USEPA established the National Primary Drinking Water Regulations (NPDWRs or primary standards) and National Secondary Drinking Water Regulations (NSDWRs or secondary standards). NPDWRs are legally enforceable standards that apply to public water systems and protect public health by limiting the levels of contaminants in drinking water. The regulated contaminants in NPDWRs are classified into the following categories: microorganisms (e.g., *Cryptosporidium*), disinfectants (e.g., chlorine), DBPs (e.g., TTHMs), inorganic chemicals (e.g., arsenic), organic chemicals (e.g., benzene), and radionuclides (e.g., uranium). In contrast, NSDWRs are nonenforceable guidelines regulating contaminants that may cause cosmetic effects (such as skin or tooth discoloration) or aesthetic effects (such as taste, odor, or color) in drinking water. The EPA recommends that secondary standards be applied to water systems, but does not require systems to comply.

6.3.2 Wastewater Disposal Subsystem

Wastewater, which may be industrial wastewater, domestic sewage, or both, should be safely disposed of to protect public health and prevent any occurrence of nuisances. A

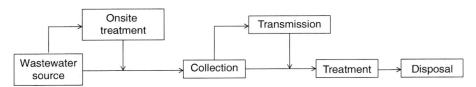

FIGURE 6.6 An example of wastewater disposal subsystem.

flowchart of wastewater disposal system is shown in Fig. 6.6. Certain pretreatment of industrial wastewater may be needed if some wastewater pollutants disturb the performance of municipal wastewater treatment plants (WWTP), also referred to as publicly owned treatment works (POTWs). When public or community sewer systems are not available, an on-site disposal system (e.g., septic tanks) needs to be considered. Characteristics and pollutant concentrations of industrial wastewater, and their treatment processes, are highly variable and depend heavily upon the properties of primary contaminants in wastewater and discharge requirements. Domestic sewer systems and treatment of municipal wastewater are discussed in detail in the following section.

6.3.2.1 Wastewater Pollutants

The major pollutants in municipal wastewater are discussed below.

Biochemical Oxygen Demand and Chemical Oxygen Demand. Biochemical oxygen demand (BOD) measures the amount of oxygen consumed by microorganisms in decomposing substances of wastewater and primarily indicates the level of aggregate biodegradable organic matter constituents. The most commonly used BOD indicator is a 5-day BOD (BOD_5) that measures oxygen consumption over a 5-day period. Chemical oxygen demand (COD) of wastewater is a measure of a maximum substance that can be oxidized and is an indicator of total organic matter. The ratio of BOD_5 to COD (BOD_5/COD) reflects the fraction of biodegradable organic matters in total organic waste.

Total Suspended Solids. Total suspended solids (TSS) are fine solid particles, organic or inorganic, which remain in suspension in wastewater. To measure TSS, a well-mixed wastewater sample is filtered through a preweighed standard glass fiber filter, and the residues retained on the filter are dried to a constant weight at 103–105°C. The subsequent increase in the filter weight represents TSS (Andrew, 2005). Because the nominal pore sizes of the filters employed in this technique can vary broadly from 0.45 to 2.0 μm, the TSS tests are arbitrary; however, its measurement provides an important effluent standard that portrays the performance of WWTP.

Nitrogen (N) and Phosphorus (P). Nitrogen (N) and phosphorus (P) are nutrients that can be present in wastewater. The most common N forms in wastewater include ammonia (NH_3), ammonium (NH_4^+), nitrite (NO_2^-), nitrate (NO_3^-), and organic nitrogen. The oxidation state of N in organic matter is −3.

Total nitrogen (TN) is comprised of organic nitrogen, ammonia nitrogen, nitrite nitrogen, and nitrate nitrogen. *Total Kjeldahl nitrogen* (TKN) is only the sum of organic and ammonia nitrogen. P is the other major nutrient in wastewater. It may exist in the forms of orthophosphate (PO_4^{3-}), polyphosphate, and organic phosphate. Municipal wastewater may contain 4–16 mg/l of phosphorus as P.

Color. Color roughly reflects the age of municipal wastewater. Fresh wastewater is light brown to gray. As the travel time of wastewater in the collection system increases and anaerobic condition is further enhanced, the color develops to dark gray and ultimately black (Tchobanoglous and Burton, 2003).

Pathogens. Pathogenic microorganisms in wastewater may be excreted by human beings and animals infected with disease or by those who are carriers of a particular infectious disease (Tchobanoglous and Burton, 2003). Pathogens can be classified into bacteria, viruses, protozoa, and helminths and may lead to different diseases such as gastroenteritis, leptospirosis, salmonellosis, typhoid fever, cholera, giardiasis, taeniasis, hepatitis, and so on. Therefore, it is crucial that disinfection be applied prior to discharge of treated wastewater.

6.3.2.2 Sanitary Sewers

A typical sanitary sewer capacity life varies between 25 and 50 years. This system is designed to handle peak and minimum wastewater flow without suspended solid sedimentation. Classification of sewers lines is also shown in Fig. 6.7. *Collectors* are comprised of collecting sewers and trunks (main sewers). Collectors transport wastewater from individual buildings to interceptors, carrying the hourly peak flow, including the wastewater carried and the flow infiltrated into the sewers. *Interceptors* are major sewers that carry wastewater from collectors to either the point of treatment (i.e., a treatment plant) or another interceptor.

Sewer pipes are made of concrete for large diameter sewer lines and of plastic (e.g., PVC) for small size lines. Vitrified clay, asbestos cement, steel, and cast iron were to be commonly used as sewer materials. The most common pipe is circular. Typically, airspace is needed for ventilation and to suppress production of hydrogen sulfide (H_2S). The wastewater flows in partially full sewer pipes, as the designed wastewater velocity should not be too high or too low. The minimum velocity (~0.6 m/s), also known as self-cleaning velocity, should be maintained to prevent the sedimentation of PM

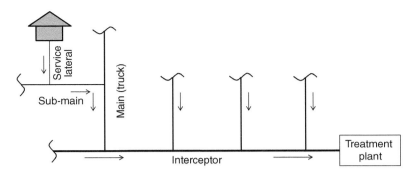

FIGURE 6.7 Sewer lines.

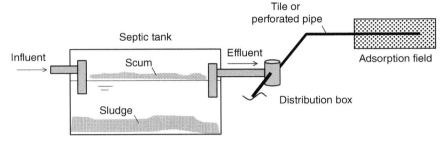

FIGURE 6.8 A conventional septic system.

within sewers. At a too high velocity (>3–4.5 m/s), pipes and manholes face erosion and displacement by shock hydraulic loadings (Lindeburg, 2003). The Manning equation is commonly used to size sewer pipes, considering flow rate, flow depth in the pipe, pipe grade, and pipe material. Usually, the designed pipe size should adequately transport peak flow at a depth that is 70% of the pipe diameter. *Manholes* are designed where flow directions, grades, and pipe sizes are changed or at intersection of sewer pipes. Manholes also allow for pipe cleaning, if needed. The maximum spacing between manholes is 120 m for less than 460 mm diameter pipes, 150 m for 460–1220 mm pipes, and 180–210 m for larger pipes, respectively (Lindeburg, 2003).

6.3.2.3 Wastewater Treatment Systems

Wastewater treatment aims to prevent pollution of the receiving water bodies or to reclaim wastewater for reuse. On-site treatment systems (e.g., septic tanks) are commonly applied in less densely populated areas because the treated wastewater typically percolates into surrounding soil. In contrast, WWTP are widely used in populated towns, cities, and metropolitan areas. Following treatment, the effluent is usually disposed of by dilution in rivers, lakes, groundwater, and the ocean.

On-Site (Decentralized) Disposal Systems. *Cesspools* are covered pits to which wastewater is discharged; however, leaching does occur in cesspools, allowing the waste liquid to seep into nearby soil and groundwater. Therefore, the cesspool on-site option is rarely used now.

Today, approximately 85–90% of on-site wastewater disposal systems are septic tanks, which consist of a vessel to store domestic sewage while sedimentation and anaerobic digestion occurs. A conventional septic system is composed of the septic tank, a distribution box, and an adsorption field, as shown in Fig. 6.8. In the septic tank, heavy particles settle by gravity and undergo anaerobic biological degradation, while grease floats on the surface and is then trapped. The typical detention time of wastewater ranges between 8 and 24 h, while the distribution box distributes the treated effluent into the subsequent adsorption field that includes a series of trenches with percolated pipes. A slime bactericidal layer, also known as clogging mat, can be formed at the bottom of trenches. When the effluent flows through the clogging mat and enters soil, pollutants can be further removed through various biochemical and physicochemical mechanisms. Only 30–50% of suspended solids are digested in septic tanks. The remaining solids must be periodically removed; otherwise, they will occupy the tank space and clog the tank. Approximately 50% (13.1 million) of total housing units with septic systems in the United States are in rural areas, while 47% (12.3 million) are in suburbs (USEPA, 2008). An estimated 10–20% of the US septic systems malfunction each year, however, causing local environmental pollution and creating a threat to public health.

Municipal (Centralized) Wastewater Treatment Systems. The flowchart of the treatment units in a traditional WWTP is shown in Fig. 6.9. Alternatives for the treatment fall within three classifications, including (i) primary treatment, (ii) secondary treatment, and (iii) tertiary (advanced) treatment. Each of the treatment degrees includes all the reactors in the previous treatment. For example, secondary treatment is comprised of a biological treatment

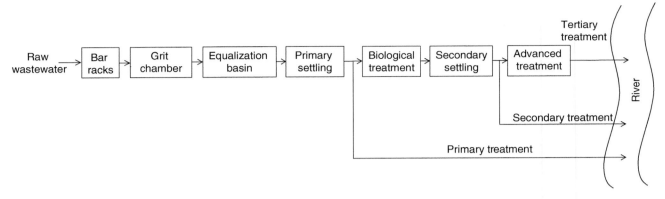

FIGURE 6.9 Degrees of different wastewater treatment.

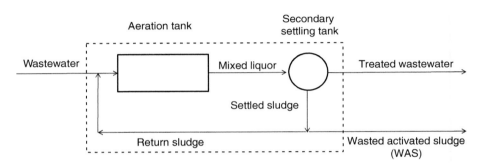

FIGURE 6.10 Schematic of an activated sludge wastewater treatment process.

unit and secondary settling unit, in addition to all of the units present in primary treatment.

Primary Treatment. The major purpose of primary treatment is to remove these wastewater pollutants as they either settle or float. The treatment is actually a mechanical process. As a result, approximately 50–60% of suspended solids are removed from wastewater. Meanwhile, 25–35% BOD_5 is reduced, though BOD removal is not the goal of the primary treatment. It would be noted that soluble BOD cannot be removed in this treatment step. Primary treatment alone was widely used in the United States; however, its isolated use is not acceptable now. Instead, primary treatment is commonly used prior to secondary treatment. Typical units in primary treatment include a bar rack, grit chamber, equalization basin, and a primary settling tank. Bar rack removes large objects and protects the subsequent pumps and other mechanical equipment. Grit chambers can slow wastewater, thereby allowing abrasive inert dense materials, such as sand, glass, silt, and pebbles, to settle out. Equalization basin is not a treatment, but can buffer inflow and load. Wastewater cannot enter a WWTP at a constant flow rate, and many WWTP treatment units must be sized with the maximum flow. Therefore, use of a flow equalization basin can dampen the variation of flow to ensure that wastewater will be treated at a nearly constant rate, thus significantly increasing the efficiency of the following treatment units. In a subsequent primary settling basin, some of suspended solids are removed from wastewater by gravity. The settled solid, called *raw sludge*, is then removed from the primary settling tank by mechanical scrapers and pumps. Light materials in the wastewater, such as grease and oil, float in the basin and can be collected and removed by a surface skimming system.

Secondary Treatment. Secondary treatment is mandatory in the United States. It utilizes biological processes to further treat the effluent from primary treatment for degradation of biodegradable organic matters. As a result, more than 85% of BOD_5 and suspended solids, in addition to 50% volatile solids, 25% of TN, and 20% of phosphorus, are removed. Accompanied with the BOD decomposition, the microbes gradually grow, subsequently removing biomass following biodegradation in a secondary settling tank. These biological processes may be operated under aerobic or anaerobic conditions, although aerobic biological treatment is more common than anaerobic decomposition for treatment of municipal wastewater. Various aerobic biological treatment processes have been developed, including a suspended growth type (e.g., activated sludge process) and an attached growth type (e.g., trickling filters and rotating biological contactors). Activated sludge process is discussed here in detail.

The *activated sludge process* is a secondary biological process in which a mixture of wastewater and sludge is agitated and aerated (Fig. 6.10). This sludge is an active mass of aerobic microorganisms (biological floc) under a starvation state—this is why it is called activated sludge. In practice, wastewater continuously flows into an aeration tank, into which air is injected to supply oxygen required for microorganisms to degrade organic waste. Aeration also ensures an adequate mixing of wastewater and activated sludge. Typically, the treatment process needs eight volumes of air for each volume of wastewater and lasts 6–8 h. Accompanied with microbial degradation of organic waste, the biomass progressively grows into a mixture of activated sludge and wastewater, called *mixed liquor*. It flows through the aeration tank and then enters the secondary sedimentation tank. Most of the settled sludge (return sludge), which makes up approximately 20–30% of wastewater, is returned to the aeration tank to maintain a high population of microorganisms in the aeration tank for rapid degradation of organic. The remaining sludge, or the *wasted activated sludge* (WAS), is then diverted for further treatment and disposal.

Tertiary (Advanced) Treatment. Tertiary treatment is targeted at specific wastewater pollutants that secondary treatment cannot effectively remove. Suspended solids, for instance, can be removed by granular filtration or microfiltration (MF) (a pressure-driven membrane process). Refractory organic matters or trace organic substances that resist biological decomposition can also be reduced through activated carbon (AC) adsorption or chemical oxidation (e.g., ozonation). Phosphorus can be removed by chemical precipitation in which aluminum, iron-based coagulants, or lime $(Ca(OH)_2)$ are added to transform dissolved P into an insoluble form. Ammonia can be removed by air stripping, breakpoint chlorination, or ion exchange. Soluble nutrient nitrogen can be transformed to nonnutrient nitrogen gas through a so-called nitrification/denitrification biological process. Lastly, undesirable inorganic ions can be removed by electrodialysis, ion exchange, or reverse osmosis (a high pressure-driven membrane process).

6.3.2.4 Wastewater Treatment Standards

WWTP must meet with all applicable wastewater quality standards for surface waters, drinking waters, and effluents of various types (Lindeburg, 2003). The Congress requires that municipalities and industries provide at least secondary treatment prior to discharging wastewater to natural water bodies. The USEPA evaluated performance data for POTWs practicing secondary treatment and established performance standards based on its evaluation. The secondary treatment standards are as shown in Table 6.2.

6.4 MUNICIPAL SOLID WASTES

The USEPA defines solid waste as any garbage, refuse, or sludge from WWTP, water supply treatment plants, or air pollution control facilities. It also refers to other discarded materials, including solid, liquid, semisolid, or contained gaseous material resulting from industrial, commercial, mining, agricultural operations, and community activities. The Resource Conservation and Recovery Act (RCRA) is a major federal statute governing solid waste. It delineates two types of wastes: *hazardous waste* and *nonhazardous waste*. The former is defined and regulated on RCRA Subtitle C, while the latter is addressed in RCRA Subtitle D.

Approximately 12 billion tons of nonhazardous waste, primarily composed of *industrial waste* and *municipal solid waste* (MSW), is generated every year within the United States. Industrial waste, defined by the EPA as nonhazardous materials that result from the production of goods and products, contributes to 2% of the nonhazardous waste (Masters and Ela, 2008). Most industrial waste is produced by via raw material extraction, material processing, and product manufacturing, such as construction and demolition debris, mining wastes, crude oil and natural gas waste, and medical waste.

MSW, more commonly known as trash or garbage, consists of everyday items that we use and then throw away. Not included are materials that also may be disposed of in landfills, but are not generally considered MSW, such as construction and demolition materials, municipal wastewater treatment sludge, and nonhazardous industrial wastes (USEPA, 2013). This section will focus on MSW.

6.4.1 Generation and Composition of MSW in the United States

Total MSW generation and per capita MSW generation in the United States are shown in Fig. 6.11. From 1960 until 2011, the total MSW generation has dramatically increased from 79.9 to 227.2 million tons, and the per capita MSU generation rate has almost doubled from 1.22 to 2.00 kg/person·day. Fractions of different types of MSU generated in 2011 are shown in Fig. 6.12. Paper and paperboard are the largest component of MSW (28.0% by weight), followed by food waste (14.5% by weight), yard trimmings (13.5% by weight), and plastics (12.7% by weight). Metals, rubber/leather/textile, wood, glass, and other MSW contribute to 8.8, 8.2, 6.4, 4.6, and 3.3%, respectively.

TABLE 6.2 Secondary Treatment Standards Established by the USEPA

Parameters	30-Day Average	7-Day Average
BOD_5	30 mg/l	45 mg/l
TSS	30 mg/l	45 mg/l
pH	6.0–9.0	—
Removal	85% BOD_5 and TSS	—

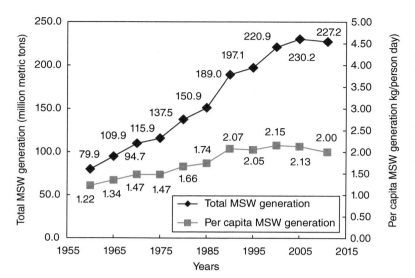

FIGURE 6.11 MSW generation rates in the United States during 1960–2011. *Source:* based on data from the USEPA (2013).

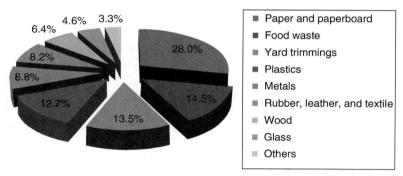

FIGURE 6.12 Composition of MSU generated within the United States in 2011 (by weight). *Source:* based on data from the USEPA (2013).

6.4.2 Integrated Solid Waste Management

Integrated solid waste management (ISWM) is a comprehensive waste prevention, recycling, composting, and disposal program. An effective ISWM system would prevent, recycle, and manage solid waste in the most effective manner, protecting both human health and the environment. The USEPA proposes a hierarchy of actions to implement ISWM: source reduction > recycling and composting > disposal.

Source Reduction. Source reduction, or waste prevention, designs products to reduce the amount of MSW generation and reduce waste toxicity levels. Typical source reduction strategies include selection of appropriate materials to be used, extension of material and/or product lifetime, reduction of amount and/or toxicity of materials, efficient management of manufacturing processes, and optimization of methods to package and transport products.

For example, chromated copper arsenate (CCA)-treated wood was the most widely used treated wood in the United States and represented about 80% of the wood preservation market through 2002. However, the toxic chromium (Cr), copper (Cu), and arsenic (As) impregnated into the treated wood have a potential to gradually leach out into soil and groundwater. Among the three toxic chemicals, As is of particular concern due to its extremely high toxicity. In 2000, approximately 4600 tons of As was estimated to leach out from in-service CCA-treated wood products (e.g., playground) in the state of Florida (Khan et al., 2006b). On March 17, 2003, the EPA signed an order in response to a voluntary request by wood preservative pesticide producers for cancellation of registration and termination of uses of certain CCA-treated wood products. This agreement required discontinued use of CCA-treated wood for most identified residential applications by December 31, 2003. Thereafter, new, low toxic, and alternative treated woods have become widely used (Khan et al., 2006a, b; Townsend et al., 2003).

Recycling and Composting. Recycling is a process to collect and process materials that would otherwise be thrown away as trash and instead turn them into new products. MSW recycling

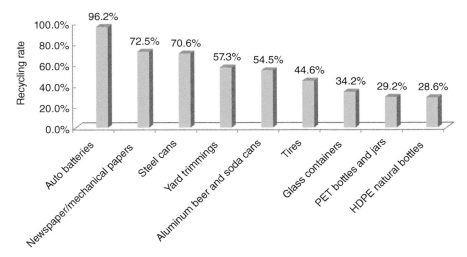

FIGURE 6.13 Recycling rates of different MSW products. *Source:* based on data from the USEPA (2013).

rates were below 10% prior to 1980 but dramatically increased from 10.1% in 1980 to 34.7% in 2011. Figure 6.13 shows the recycling rates of several commonly recycled MSW products in the United States during 2011. Auto batteries ranked number 1 (96.2%) in terms of the recycling rate, followed by newspaper/mechanical papers (72.5%) and steel cans (70.6%). Of note, 6.8 million tons of metals, including aluminum, steel, and mixed metals, were recycled from the US MSW pool in 2011, thereby reducing greenhouse gas emission equal to 20 million tons of carbon dioxide equivalent (CO_{2eq}) (USEPA, 2013).

In 2011, more than 9800 curbside recycling programs existed nationwide (USEPA, 2013). These recycling programs bring environmental and economic benefits, including, but not limited to, (i) reducing the amount of waste disposed of by landfills and incinerators; (ii) conserving natural resources such as timber, water, and minerals; (iii) reducing pollution caused by new raw materials; (iv) saving energy; (v) reducing greenhouse gas emissions contributing to global climate change; (vi) helping to sustain the environment for future generations; and (vii) helping to create new well-paying jobs in the recycling and manufacturing industries in the United States.

Composting is a process that produces humus-like material (compost) from aerobic biological stabilization of organic materials in solid waste. MSW that can be composted include yard trimmings, food wastes, manures, and wood chips. A full processing cycle of composting is 20–25 days, within which active degradation occurs over a 10–15 day period. The resulting compost is then used as a soil amendment or as a medium to grow plants. Humus, the major component in compost, is dark brown or black humus with a soil-like, earthy smell. After applied to soil, compost slowly decays, providing plants, animals, and microorganisms with minerals and nutrients. In 2011, about 3090 community composting programs were documented, a decrease from 3227 in 2002. Regardless, composting recovered over 18 million tons of MSW (USEPA, 2013) in total.

To successfully compost, five conditions should be considered:

1. *Feedstock and nutrient balance*. Different organic wastes should be appropriately mixed to obtain optimal balances in carbon and nutrients. The mass ratio of carbon to nitrogen (C:N) is typically between 20:1 and 25:1.
2. *Particle size*. Grinding, chipping, and shredding increase the MSW surface area where the microorganisms can feed. Moreover, smaller and uniform particles improve pile insulation to maintain an optimum reaction temperature. Typically, particle sizes are below 5 cm; however, if the particles are too small, little air can flow inside piles so that the compositing is inhibited.
3. *Oxygen*. Aerobic biodegradation is the principal mechanism of composting. It is vital to provide and maintain an adequate oxygen flow in piles. An aerobic condition can be maintained by turning the pile, placing the pile on a series of pipes, or including bulking agents (e.g., wood chips and shredded newspaper).
4. *Moisture content*. Moisture is another key element because microorganisms in compost piles survive only with adequate moisture. Additionally, water helps the transport of substances within the compost piles and makes the nutrients in organic wastes accessible to the microbes. Moisture may enter compost piles via composted wastes containing moisture, rainfall, or intended watering. Typically, moisture content is controlled between 50 and 60%.
5. *Temperature*. Biodegradation is an exothermic process. Well-operating compost requires a certain temperature range to optimize the microbial activity, and favorable temperatures can also promote rapid composting as well as destroy pathogens and weed seeds. The most frequently used temperature falls within 55–60°C.

The major benefits of composting include (i) producing useful compost, thereby reducing the need for water, fertilizers, and pesticides, and promoting higher yields of agricultural crops; (ii) removing solids, oil, grease, and heavy metals from stormwater runoff; (iii) capturing and destroying most of the industrial volatile organic chemicals (VOCs) in contaminated air; and (iv) serving as a marketable commodity and a low-cost alternative to standard landfill cover and artificial soil amendments. In contrast, a major drawback of composting is odor, which can be minimized by maintaining aerobic conditions and a proper cure time (Davis and Cornwell, 2008).

Disposal (Landfilling and Combustion). MSW that cannot be prevented or recycled should be properly disposed of by landfilling or incineration. Landfills in particular will be discussed in Section 6.4.3. Incineration (also called combustion) is a controlled burning of MSW waste, resulting in a volume reduction of 90% and a mass reduction of 75%. The final MSW incineration products include gases (e.g., carbon dioxide and water vapor) and remaining ash. The remaining residuals (ash) should be disposed of within landfills (Tchobanoglous and Kreith, 2002).

Incineration is an extreme chemical reaction where organic wastes are oxidized. The oxidizable elements in MSW include carbon (C) and hydrogen (H), in addition to, but at a much lesser extent, sulfur (S) and nitrogen (N). Following complete oxidation, C and H are oxidized to CO_2 and H_2O, respectively. Moreover, S is oxidized to sulfur dioxide (SO_2), and N may be oxidized to nitrous oxides (NO_x). To accomplish a successful incineration, the system should have excess oxygen with enough reaction time, the minimum temperature must be exceeded to initiate the combustion, and the turbulence in the combustion chamber must be ensured (Davis and Cornwell, 2008).

In a conventional incineration, MSW is simply combusted. In contrast, incineration at large installations is usually accompanied with energy recovery (Lindeburg, 2003). That is, incinerators convert water into steam to fuel heating systems or generate electricity. Facilities with boilers and electrical generators are referred to as *waste-to-steam plants* and *waste-to-energy (WTE) facilities*, respectively. *Mass burning* is the incineration of unprocessed MSW to generate steam. Following incineration, the ash accounts for approximately 27% of the MSW, including the final MSW oxidation products and noncombustible materials (e.g., glass and stone). The typical treatment capacity of a mass burning incinerator is 910–1800 Mg/day.

Over one-fifth of the US MSW incinerators use *refuse derived fuel* (RDF). RDF is the combustible portion of MSW, such as plastics and biodegradable waste, which is separated from noncombustible materials through shredding, screening, and air classifying. Modern RDF plants can retain more than 95% of the original combustible MSW and reduce the *mass yield* (mass yield = RDF mass: MSW mass) to less than 85%. The generated RDF has a heating value of 12–14 MJ/kg with approximately 24% moisture content and 12% ash content.

6.4.3 Landfilling

Sanitary landfills refer to land disposal sites that employ an engineered method of MSW disposal on land that minimizes public health and environmental impacts (Davis and Cornwell, 2008; Tchobanoglous and Kreith, 2002). During landfilling, MSW is spread to the smallest volume and covered by cover materials every day. All MSW landfills must comply with the federal regulations in 40 CFR Part 258 (Subtitle D of RCRA) or equivalent state regulations.

Site Selection, Operation, and Structure of Landfills. Site selection is the most difficult phase during development of an MSW landfill. Many factors must be carefully considered, including, but not limited to, cost and availability of the land, public opposition, direction and speed of wind, weather and climate, hydrology, geological conditions, and future growth in solid waste generation. Landfills are usually built in suitable geological areas away from faults, wetlands, flood plains, or other restricted areas. MSW fills landfills with two operation manners. The first is the *area method,* in which MSW is spread, compacted, and covered on flat ground at the end of each working day. This method is rarely limited by topography. The second option is the *trench method,* in which MSW is deposited in a trench and covered by the excavated trench soil. This method is suitable for a gently sloping land where the water table is low. Either of the operation methods can be used independently, or they can both be used in sequence or simultaneously.

Different components in a landfill are shown in Fig. 6.14. MSW is dumped in layers (typically 0.6–0.9 m thick) and then compacted by a crawler tractor or other equipment. Before the end of each working day, cover materials are placed on the compacted MSW with a density of 470–890 kg/m^3. The MSW and daily cover (the depth >0.15 m) form a *cell*. The cell has a typical height of 2.4 m and a slope ranging within 20–40°. The *daily cover* material, whose purpose is to prevent fire, odor, and scavenging, may be native soil or other approved materials (e.g., scrap tire chips and yard waste). The *lift* is the height of the covered layer. When the overall height of a landfill exceeds 15–20 m, *benches* are used to stabilize the landfill slope and provide a site for installation of landfill gas (LFG) collection pipes. After operation is completed, *final cover (cap)* with a minimum depth of 0.60 m is applied to reduce moisture entering the landfill. In addition, the final cover channels LFG and provides a place for vegetation to grow.

Environmental Consideration. When an MSW landfill is well operated, vectors, odor, and fire should not present complications. The two major environmental considerations of landfills are the (i) production of LFG and (ii) generation of landfill leachate.

LFG. The principal LFG produced as a result of microbial degradation of organic waste within landfills are carbon dioxide (CO_2) and methane (CH_4). Both gases account for approximately 50% of the total landfill in terms of volume, respectively. Other LFG are traceable, including oxygen gas

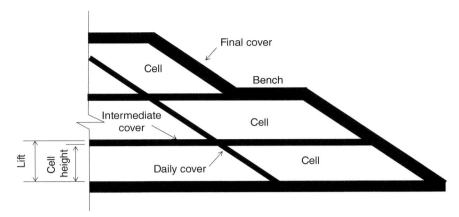

FIGURE 6.14 Cross section of a landfill.

FIGURE 6.15 Landfill leachate collected from a Caribbean MSW landfill.

(O_2), nitrogen gas (N_2), ammonia (NH_3), hydrogen gas (H_2), and CO (CO). Their generation rates depend heavily upon landfilling time. The gas generation occurs roughly in five sequential phases, including the initial adjustment phase, transition phase, acid phase, methane fermentation phase, and maturation phase (Tchobanoglous and Kreith, 2002). As the landfilling proceeds, the concentrations of oxygen (O_2) and nitrogen (N_2) gases trapped in the landfills gradually decrease, while CO_2 and CH_4 are slowly released as final products of anaerobic degradation of organic waste, as shown below:

$$C_aH_bO_cN_d + H_2O \xrightarrow{\text{anaerobic microbes}} CH_4 + CO_2 + NH_3 \quad (6.22)$$

Where $C_aH_bO_cD_d$ represents the general formula of MSW to be decomposed. LFG is problematic because both CO_2 and CH_4 are greenhouse gases, contributing to global warming. Particularly harmful is CH_4, which is more than 20 times stronger than CO_2 due to short atmospheric lifetime in terms of the greenhouse effect. In the United States, MSW landfills are the third largest source of human-related CH_4 emissions, accounting for approximately 17.5% of these emissions in 2011. The generated CH_4=(the effective ingredient of natural gas), however, may be utilized as a reliable, local, and renewable energy. It can be extracted and purified from landfills using a series of wells and a blower/flare (or vacuum) system. As a new energy source, LFG can generate electricity or replace traditional fuels in industries. Meanwhile, the captured LFG reduces odors accompanied with LFG emission and prevents CH_4 from contributing to smog and global climate change (USEPA, 2010).

Landfill Leachate. Landfill leachate is the liquid generated from landfill due to moisture oversaturation (Fig. 6.15).

This moisture originates from rainfall, surface drainage, groundwater, or the water present in the deposited MSW. The leachate volume in a landfill relies on the availability of water, landfill surface conditions, solid waste conditions, and underlying soil conditions (Lu et al., 1985). Chemical characteristics of landfill leachate are extremely variable due to variability in its generation rates over a wide time frame, nonuniformity of waste decomposition with respect to time and space, and influences related to environmental conditions and anthropogenic factors (Vagliasindi, 1995). Furthermore, its chemical composition is affected by landfill age, waste nature, moisture availability, temperature, pH, depth fills, compaction, and other factors (Viraraghavan and Singh, 1997).

Landfill leachate is highly contaminated and exhibits acute and chronic toxicity. As a complicated matrix, it contains high concentrations of organic and inorganic species. Of principal concern within leachates are dissolved organic matter and ammonia (Kjeldsen et al., 2002). Their concentrations may be tens to thousands times higher than those of municipal wastewater. Other undesirable leachate contaminants include toxic heavy metals and traceable persistent organic pollutants (e.g., pesticides). In addition, leachate typically has a yellow-brown color and strong odor. The uncontrolled leachate discharged directly into the aquatic environment has both acute and chronic impacts on the environment, which can severely diminish biodiversity and greatly reduce populations of sensitive species. Furthermore, public health may be seriously threatened by landfills that are near water sources. Old generation landfills were not designed with bottom liners, so certain landfills may experience greater impact than others. For example, New Jersey has the most Superfund sites (the most toxic sites) in the United States, and approximately 25% of Superfund sites in the state are contaminated by landfill leachate (Ezyske and Deng, 2012). The so-called *natural attenuation* (NA) landfills now apply native soil to reduce the concentrations of various leachate pollutants, and modern *containment landfills* possess a leachate collection system comprised of landfill liner and a leachate collection system. The liners may be synthetic membranes (e.g., HDPE, PVC, and CPE flexible membrane liners) or clay liners that can provide adequate protection with a 1.2–1.5 m thickness. The leachate collection system, consisting of a series of sloped terraces and percolated pipes at the bottom of landfills, does not allow leachate to pond in landfills and creates a significant hydraulic head on landfill liners. The collected leachate is thus removed from landfills and transported to holding facilities. Different leachate management strategies are available, including (i) spray irrigation on adjacent grassland, (ii) recirculation of leachate through the landfill, (iii) leachate evaporation using landfill-generated methane as fuel, (iv) cotreatment of sewage and leachate in POTWs, (v) biological or physical/chemical treatment, and (vi) wetland treatment. New attempts are being made to utilize landfill leachate for generation of electricity (Greenman et al., 2009).

6.5 AIR RESOURCE SYSTEM

6.5.1 Air Pollution Management

Air pollution may occur on a micro-, meso-, or macroscale. The first federal act associated with air pollution was the Air Pollution Control Act of 1955. This act established a program to federally sponsor air pollution research. The Clean Air Act was enacted in 1963, providing funds for state and local air pollution control agencies and allowing federal intervention to reduce interstate air pollution. Major amendments to the Clean Air Act were passed in 1970, 1977, and 1990. More specifically, the Amendments of 1970 required the EPA to set the National Ambient Air Quality Standards (NAAQS) and to issue the New Source Performance Standards (NSPS). The Amendments of 1977 relaxed the previous auto emission requirements and defined *prevention of significant deterioration* (PSD) areas. PSD applies to new major sources or major modifications at existing sources for pollutants where the area the source is located is in attainment or unclassifiable with the NAAQS. The Amendments of 1990 identified 189 hazardous air pollutants to be regulated, established SO_2 allowance for acid rain control, and imposed new requirements for auto emissions. Actions to implement the Clean Air Act have significantly reduced air pollution in the United States. For example, from 1970 to 2011, aggregate national emissions of six common pollutants (i.e., particles, ozone, lead, CO, nitrogen dioxide and SO_2) alone in the United States decreased by 68%, though the gross domestic product (GDP) grew by 212%.

6.5.2 Air Pollutants

The major air pollution source in the United States is fossil fuel combustion from both stationary and mobile sources. Air pollutants, which can be classified as primary and secondary, may be derived from wind erosion, industrial manufacture, and so on. *Primary air pollutants* are those that do not change their form in air after emission, while *secondary air pollutants* are formed in air from other emitted chemicals (Lindeburg, 2003). These air pollutants can cause negative health and/or environmental impacts. Several air pollutants of great concern are discussed below.

PM. PM is a complex mixture of extremely small particles and liquid droplets with a size of 0.05–100 μm, including acids (e.g., nitrates and sulfates), organic chemicals, metals, and soil or dust particles. PM is generated from industrial, agricultural, and construction-based activities, incomplete combustion from transportation, forest fires, and natural wind erosion. The sizes of these particles are directly associated with their potential health effect. Of particular concern are these particles with aerodynamic diameter less than 10 μm (PM 10) and less than 2.5 μm (PM 2.5). Exposure to such particles can affect both the lungs and heart.

Asbestos. Asbestos is an inert, strong, and incombustible fibrous silicate material. It was widely used in building materials, paper products, asbestos cement products, friction products, textiles, packing and gaskets, and asbestos-reinforced plastics. When asbestos is released into air from aged or damaged asbestos-containing materials, the long-term exposure via inhalation can cause lung cancer and mesothelioma (a consequence that has been well documented).

Lead (Pb). In US history, airborne lead (Pb) was largely released from fuels from on-road motor vehicles (such as cars and trucks) and industrial sources. After the USEPA removed Pb from vehicle gasoline, Pb emissions from the transportation sector declined dramatically by 99% from 1980 to 1999. Today, airborne Pb is mostly from ore and metal processing and piston-engine aircraft operating on leaded aviation gasoline. Once inhaled, Pb can distribute throughout the body via the blood and then accumulate in the bones. It can adversely affect the nervous system, kidney function, immune system, reproductive and developmental systems, and cardiovascular system.

Carbon Monoxide (CO). Carbon monoxide (CO) is a colorless, odorless gas emitted from various sources, such as motor vehicles, fossil fuel burning, industrial activities, and solid waste disposal. Motor vehicles account for more than 60% of the emission (Davis and Cornwell, 2008). If inhaled, CO can threaten one's health by reducing oxygen delivery to the organs and tissues. An extremely high level of CO can result in death.

6.5.3 Stationary Source Control

Cyclones. *Cyclone separators* are used to remove PM with aerodynamic diameters greater than 10–15 μm. The equipment is an inertial cone-shape collector in which particulate-laden gas is accelerated by a spiral motion. The particles are forced by the centrifugal force to impinge on the cyclone wall and slough off into a collection hopper. The cleaned air then moves out from the top of the cyclone.

Electrostatic Precipitators. *Electrostatic precipitators (ESP)* remove particular matter from gas. ESP are composed of alternating plates or wires that are under large direct current potential (30–75 V). When the gas passes through the charging fields, positively and negatively charged particulates are electrostatically attracted to negative and positive collection plates or wires, respectively. ESP is operated at a reasonable cost and has a high treatment efficiency (~99%); however, it is not applied for moist gas flow and requires heat during start-up and shutdown in case of acid gas condensation-induced corrosion.

Absorption. Absorption aims to remove gaseous pollutants. It is a mass transfer process in which gas is dissolved in liquid. Typically, absorption is achieved in a *spray chamber* or a *packed tower* through three sequential steps: (i) the gas-

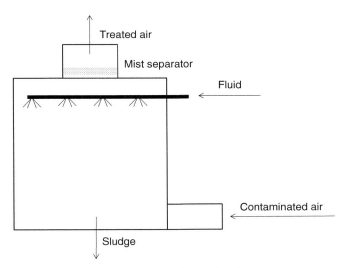

FIGURE 6.16 Spray chamber.

eous pollutants are diffused to the surface of the liquid; (ii) dissolution occurs (i.e., the gas crosses the interface of the gas and liquid); and (iii) the dissolved gas enters into the liquid. In a spray chamber, liquid droplets are formed to absorb the gases, as shown in Fig. 6.16. In a packed tower, the clean liquid flows down through the packing media, and the contaminated gas flows from bottom to top or from side to side. The plastic or ceramic packing media are designed to maximize the contact area of the gas and liquid surface and to thus increase the absorption efficiency. Absorption can remove SO_2 (SO_2), chlorine (Cl_2), ammonia (NH_3), hydrogen sulfide (H_2S), nitrous oxides (NO_x), and certain HCs.

Adsorption. Adsorption is a mass transfer process in which the pollutants (adsorbates) are physically and/or chemically bonded to a solid (adsorbent). The most frequently used adsorbent is AC, either in powdered or granular form. AC is characterized by a high specific surface area (up to 1000 m^2/g), which provides abundant active sites for adsorption of target contaminants. Adsorption is effective for removal of aromatic and aliphatic organic molecules (e.g., dioxins), heavy metals (e.g., Pb), and certain gases (e.g., SO_2, H_2S, and NO_2).

Combustion. Combustion is employed to remove combustible air pollutants such as CO and HCs. In contrast to solid waste incineration, combustion for air pollution control does not require following a strict emission control process. Generally, combustion is applied in three types. First, *direct flame incineration* is typically used for the air pollutants that have a net heating value (NHV) greater than 3.7 MJ/m^3, which adequately maintains combustion without any supplemental fuel source. Direct flame incineration does not produce any toxic combustion by-products. Second, *thermal incineration* is applied to these pollutants that require supplemental fuel (NHV = 40–750 KJ/m^3) to sustain combustion. Third, *catalytic incineration* requires a catalyst to produce combustion. This option is used for air pollutants similar to those suitably treated by thermal incineration; however, the catalytic incineration process significantly reduces the residence time and temperature.

6.5.4 Mobile Source Control

Vehicle-induced air pollution was noticed early in California, where traffic was responsible for smoggy skies over Los Angeles. In 1959, California adopted mobile source emission regulations to control HCs and CO. Standards have been widely expanded to include more air pollutants since then. The sources of mobile air pollution include cars, buses, trucks, trains, planes, and so on. The transportation sector in the United States contributes to significant fractions of all the criteria pollutants except for SO_2.

Institutional Issues and Regulation Strategies. The major solution to mobile source air pollution is the reduction of mobile miles driven by increasing mass transport, pedestrian access, bicycle use, as well as encouraging carpools. Internalization of the cost associated with mobile use also mitigates pollution, including toll roads, high parking fees, reserved lanes for mass transit and carpoolers, restricted access periods, and increased fuel taxes.

Regulations at the manufacture point and the use point can also effectively reduce the number of mobile sources. Under a preconsumer certification and enforcement program in the United States, automobiles can be sold only when they demonstrate compliance with the certification criteria. The manufactures are also required to provide a warranty on their emission control equipment. Some states (e.g., New Jersey) require automobile owners to periodically have their vehicles inspected for compliance with emissions restrictions and to make necessary repairs when noncompliance occurs.

The *Corporate Average Fuel Economy* (CAFE), enacted in 1975 in the United States, requires vehicle manufacturers to comply with the gas mileage, or fuel economy, standards set by the Department of Transportation (DOT). The CAFE values used to be computed using the city and highway fuel economy test results and the weighted average of vehicle sales (55% city and 45% highway). Tests were conducted in a laboratory by operating vehicles on a dynamometer. For example, if a car gets 18 mile/gallon (mpg) in the city and 26 mpg on the highway, its CAFE fuel efficiency could be obtained below:

$$MPG = \frac{1 \text{ mile}}{\frac{0.55 \text{ miles}}{18 \text{ miles/gallon}} + \frac{0.45 \text{ miles}}{26 \text{ miles/gallon}}} = 21 \text{ mpg}$$

In 2012, the CAFE standard has been expressed as a mathematical inverse linear formula with cutoff values. The CAFE values depend on vehicle footprint, which is defined as a measure of vehicle size determined by multiplying the vehicle's wheelbase by its average track width.

Emission Control. Combustion and evaporation are the major causes for emissions from gasoline engines. HCs can be volatilized from the carburetor, throttle body, and a vented gasoline tank. Noncombusted or incompletely oxidized HCs can be emitted from the crankcase. To control emissions, the vents from the air cleaner/gasoline tank and the crankcase are designed to capture volatilized HCs and noncombustible/partially oxidized HCs, respectively. The captured HCs can be adsorbed by AC contained in a canister in the engine. Criteria pollutants can also exist in the exhaust. An early approach to the exhaust control is to inject air, which enhances the oxidation of HCs and CO. It also recirculates a part of exhausted gas back to the incoming air/fuel mixture for control of nitrous oxides (NO_x). However, the currently preferred option is the three-way catalytic converter in which HC and CO are oxidized to CO_2, and NO_x is reduced to N_2 in the same catalyst bed.

Different emission controls are used for diesel engines. These engines are operated with oxygen-rich diesels and at a high temperature, thereby generating more NO_x and less HC and CO than gasoline engines. The removal efficiency of NO_x in the catalytic converter is relatively low due to the

---

MODEL SYLLABUS

Course Description

Introduction to the role of engineering in environmental management; materials balance and reactor analysis; energy balance; overview of water resources system, waste management, and air resources system; water pollution; water and wastewater treatment; solid waste engineering; air pollutants; and air pollution control technologies

Course Objective

By the end of this course, students will be able to:
- Understand the role of engineering in environmental management
- Define and identify the environmental issues and legal aspects
- Establish the concepts of water resource system, waste management, and air resources system
- Apply basic materials and energy balance tools to analyze environmental systems
- Understand sources and characteristics of traditional and emerging pollutants in water and wastewater, as well as water and wastewater treatment technologies
- Understand the strategies and methods of solid waste management
- Understand the sources and characteristics of air pollutants, as well as air pollution control technologies

presence of the rich oxygen mixture. A promising approach to diesel engine design is the homogeneous compression–ignition (HCCI) engine. An HCCI can achieve more complete combustion at a low temperature, thus significantly reducing the NO_x formation.

Alternative Fuels and New-Generation Automobiles. Numerous alternatives for gasoline are being investigated and attempted, including ethanol, methanol, biodiesel, propane, hydrogen, and electricity. Obviously, new energy sources will greatly reduce the automobile-induced air pollution, as well as our dependence upon petroleum, which is a limited traditional energy. Electric drive uses electric motors to augment, or even replace, internal combustion engines. *Hybrid electric vehicles* (HEVs) combine a conventional internal combustion engine and an onboard electric power system, including a generator, electric motor, and battery storage. The system can be designed in either a serial manner or a more popular parallel pattern. *Plug-in electric vehicles* (PEVs) are motor vehicles that can be recharged from any external source of electricity. PEVs allow the user to charge their batteries, mostly at night, by plugging the car into the utility grid. The power of these batteries adequately supports PEVs in their first few tens of miles running. After then, PEVs are operated as HEVs.

REFERENCES

Andrew, D. 2005. *Standard Methods for the Examination of Water and Wastewater*. American Public Health Association, Washington, DC.

ASCE. 2013. *Code of Ethics*. American Society of Civil Engineers (ASCE), Reston.

Bureau-of-labor-statistics. 2008. *Occupational Outlook Handbook*, Spring. US Department of Labor, Washington, DC.

Crittenden, J., R. Trussell, D. Hand, K. Howe, and G. Tchobanoglous. 2011. *Water Treatment: Principles and Design*. John Wiley & Sons, Inc., Hoboken, NJ.

Davis, M.L. and D.A. Cornwell. 2008. *Introduction to Environmental Engineering*. 4th ed. McGraw Hill, New York.

Edzwald, J.K. 2011. *Water Quality & Treatment: A Handbook on Drinking Water*. McGraw-Hill, New York.

Ezyske, C. and Y. Deng. 2012. *Landfill Management and Remediation Practices in New Jersey, United States*. Chapter 9, InTech.

Greenman, J., A. Gálvez, L. Giusti, and I. Ieropoulos. 2009. Electricity from landfill leachate using microbial fuel cells: comparison with a biological aerated filter. *Enzyme and Microbial Technology* 44:112–119.

Khan, B.I., J. Jambeck, H.M. Solo-Gabriele, T.G. Townsend, and Y. Cai. 2006a. Release of arsenic to the environment from CCA-treated wood. 2. Leaching and speciation during disposal. *Environmental Science & Technology* 40:994–999.

Khan, B.I., H.M. Solo-Gabriele, T.G. Townsend, and Y. Cai. 2006b. Release of arsenic to the environment from CCA-treated wood. 1. Leaching and speciation during service. *Environmental Science & Technology* 40:988–993.

Kjeldsen, P., M.A. Barlaz, A.P. Rooker, A. Baun, A. Ledin, and T.H. Christensen. 2002. Present and long-term composition of MSW landfill leachate: a review. *Critical Reviews in Environmental Science and Technology* 32:297–336.

Kolpin, D.W., E.T. Furlong, M.T. Meyer, E.M. Thurman, S.D. Zaugg, L.B. Barber, and H.T. Buxton. 2002. Pharmaceuticals, hormones, and other organic wastewater contaminants in US streams, 1999–2000: A national reconnaissance. *Environmental Science & Technology* 36:1202–1211.

Lindeburg, M.R. 2003. *Environmental Engineering Reference Manual for the PE Exam*. Professional Publications, Belmont, CA.

Lu, J., B. Eichenberger, and R.J. Stearns. 1985. *Leachate from Municipal Landfills; Production and Management*. Noyes Publications, Park Ridge, NJ.

Masters, G.M. and W. Ela. 2008. *Introduction to Environmental Engineering and Science*. Prentice Hall, Englewood Cliffs, NJ.

National Ground Water Association (NGWA). 2010. Groundwater facts. http://www.ngwa.org/fundamentals/use/documents/gwfactsheet.pdf. Accessed April 30, 2015.

National Society of Professional Engineers (NSPE). 2013. What is PE? http://www.nspe.org/resources/licensure/what-pe. Accessed April 30, 2015.

Richardson, S.D. and T.A. Ternes. 2011. Water analysis: emerging contaminants and current issues. *Analytical Chemistry* 83: 4614–4648.

Stackelberg, P.E., J. Gibs, E.T. Furlong, M.T. Meyer, S.D. Zaugg, and R.L. Lippincott. 2007. Efficiency of conventional drinking-water-treatment processes in removal of pharmaceuticals and other organic compounds. *Science of the Total Environment* 377:255–272.

Tchobanoglous, G. and F.L. Burton. 2003. Wastewater engineering. Management 7:1–4.

Tchobanoglous, G. and F. Kreith. 2002. *Handbook of Solid Waste Management*. McGraw-Hill, New York.

Townsend, T., H. Solo-Gabriele, T. Tolaymat, K. Stook, and N. Hosein. 2003. Chromium, copper, and arsenic concentrations in soil underneath CCA-treated wood structures. *Soil and Sediment Contamination* 12: 779–798

USEPA. 2008. Septic Systems Fact Sheet (EPA# 832-F-08-057). Office of Wastewater Management Decentralized Wastewater Program. http://water.epa.gov/aboutow/owm/upload/2009_06_22_septics_septic_systems_factsheet.pdf. Accessed April 30, 2015.

USEPA. 2010. Landfill Methane Outreach Program and Landfill Gas Energy: The Power of Partnership. http://www.epa.gov/lmop/documents/pdfs/lmopbro.pdf. Accessed April 30, 2015.

USEPA. 2013. *Municipal Solid Waste (MSW) in the United States: 2011 Facts and Figures*.USEPA, Washington, DC.

USGS. 2013. Emerging Contaminants in the Environment. http://toxics.usgs.gov/regional/emc/. Accessed April 30, 2015.

Vagliasindi, F.G.A. 1995. Landfill Leachate: Quantification, Characterization, and Treatment Options. International

Symposium on Groundwater Management – Proceedings, ASCE, New York, NY.

Vieno, N.M., H. Härkki, T. Tuhkanen, and L. Kronberg. 2007. Occurrence of pharmaceuticals in river water and their elimination in a pilot-scale drinking water treatment plant. *Environmental Science & Technology* 41:5077–5084.

Viessman, W., M.J. Hammer, and E.M. Perez. 2009. *Water Supply and Pollution Control*. Pearson Prentice Hall, Upper Saddle River, NJ.

Viraraghavan, T. and K.S. Singh. 1997. Anaerobic biotechnology for leachate treatment: A review. Proceedings of the Air & Waste Management Association's Annual Meeting & Exhibition, Toronto, Canada.

7

GREEN ARCHITECTURE IN ENVIRONMENTAL MANAGEMENT

Jason Kliwinski[1] and Amy Ferdinand[2]

[1] Green Building Center, New York, NY, USA
[2] Environmental Safety and Sustainability, Montclair State University, Montclair, NJ, USA

Abstract: The built environment (i.e., buildings and their occupants) consumes most of the world's energy and much of the earth's natural resources. The built environment also contributes to CO_2 and other greenhouse gas emissions that affect climate change. Climate change, resiliency and adaptation, sustainable design, and a host of other terms have emerged in the twentieth century as hot buttons in planning, codes, and critical thinking in every sector of the market place. Over 97% of world scientists agree that climate change is real, happening, and is accelerated by man's activities on this planet. Therefore, there should be no further discussion about whether or not we have a serious problem but, rather, what solutions to climate change environmental managers should explore. With ever-increasing disasters, whether natural or man-made, being the "new normal," there is a definite need among environmental managers, business leaders, and other stakeholders to become better informed on the radical changes the built environment must undergo to meet climate change targets. This chapter discusses green architecture as an environmental management tool to address the built environment's contribution to climate change.

Keywords: building performance, built environment, environmental assessment, environmental management, green architecture, green building, leadership in energy and environmental design, life cycle assessments, sustainable construction sustainable design, sustainable development.

7.1 Introduction	174	
7.1.1 The Impact of the Built Environment	174	
7.2 Integrated Design: The Key to Green Building Success	176	
7.2.1 Assemble the Right Team	178	
7.2.2 Set Your Sustainability Goals Early	178	
7.2.3 The Importance of Setting Goals Early	178	
7.2.4 Establish Open Communication and Conduct Charrettes	179	
7.2.5 Identify Synergies	179	
7.2.6 Maintain the Commitment to Sustainability	179	
7.3 The Team	179	
7.3.1 Do You Need a Green Champion?	179	
7.3.2 The Architect	180	
7.3.3 Site Analysis: The Civil Engineer and Landscape Architect	180	
7.3.4 The Structural Engineer	180	
7.3.5 The Mechanical, Electrical, and Plumbing (MEP) Engineering	181	
7.3.6 The Mechanical Engineer	181	
7.3.7 The Electrical Engineer	181	
7.3.8 The Plumbing Engineer	182	
7.3.9 The Commissioning Agent (CxA)	182	
7.3.10 Other Key Team Members	183	
7.4 Resource Use and Green Buildings	183	
7.4.1 Energy	183	
7.4.1.1 Reduce Demand	183	
7.4.1.2 Increase Efficiency of Systems	186	
7.4.1.3 Offset the Rest	186	
7.4.2 Water	188	
7.4.3 Materials	189	
7.4.4 Indoor Environmental Quality	189	
7.5 Conclusion	191	
References	191	

An Integrated Approach to Environmental Management, First Edition. Edited by Dibyendu Sarkar, Rupali Datta, Avinandan Mukherjee, and Robyn Hannigan.
© 2016 John Wiley & Sons, Inc. Published 2016 by John Wiley & Sons, Inc.

7.1 INTRODUCTION

7.1.1 The Impact of the Built Environment

Climate change, resiliency and adaptation, sustainable design and construction, life cycle costs, life cycle assessments, environmental assessments, and a host of other terms have emerged in the twentieth century as hot buttons in planning, codes, and critical thinking in every sector of the market place. Over 97% of world scientists agree that climate change is real, happening, and is accelerated by man's activities on this planet (Gore, 2006). Therefore, there should be no further discussion about whether or not we have a serious problem but, rather, what solutions to climate change environmental managers should explore.

The consequences of climate change have already affected the shores of the United States more in the last 10 years than at any time in our recorded history. "Superstorms" Katrina, in 2005, and Sandy, in 2012, have caused both billions of dollars in damage to property and substantial loss of life. The compounding effects of climate change on changing weather patterns and other climatic conditions are perhaps the most troubling. Warmer air holds more moisture and lowers water supply, creating droughts and unleashing violent, unexpected storms. We are now at 400 ppb of carbon in our atmosphere and rising, which is 100 ppb more than ever recorded in the last 800,000 years (Gore, 2006). With ever-increasing disasters—whether natural or man-made—being the "new normal," there is a definite need among environmental managers, business leaders, and other stakeholders to become better informed on the radical changes the built environment must undergo to meet climate change targets.

The built environment (i.e., buildings and their occupants) consumes most of the world's energy and much of the earth's natural resources. Buildings play a significant role in contributing to the current problem of climate change and are widely accepted as being responsible for nearly 40% of all carbon emissions in the United States alone, with transportation and industry making up the remaining 60% of emissions (USGBC, 2011). However, they also pose an unprecedented opportunity to help correct and even reverse their negative effects. In looking at our environmental portfolio, we must come to understand that its resources and assets are fixed. We have not found another planet like ours to date and are not optimistic that we will do so in our lifetime. Therefore, how we balance growth, current needs, and the limited resources and capacity of this planet are crucial to our very survival and way of life.

The concept of sustainable development (Green Building), introduced in 1987, at the World Commission on Environment and Development, is the path of progress, which meets the needs and aspirations of the present generation, without compromising the ability of future generations to meet their own needs (Brundtland, 1987). We now have the technology and solutions available to reverse our carbon emissions, curb global warming, and ultimately improve quality of life. Green buildings are a key element in the recipe for Carbon Neutrality and restorative action.

Many other definitions of green buildings and sustainable development exist, but they all include similar traits that address land, water, energy, and material consumption, as well as the creation of healthy indoor environments for the people who occupy them. For the purposes of this discussion, we define green buildings as those that have the following characteristics:

1. Produces at least as much energy as it uses from renewable sources, if not more (Fig. 7.1)

FIGURE 7.1 Solar canopy at TD bank.

FIGURE 7.2 Rainwater harvesting at the Holmes–Rulli residence.

FIGURE 7.3 Outdoor courtyard at Holmes–Rulli residence.

2. Collects and uses water on-site for all of its needs and thoroughly treats water leaving the site (Fig. 7.2)
3. Restores land and uses it thoughtfully to preserve views, provide access to daylight and natural breezes, and encourage its health and beauty (Fig. 7.3)
4. Conserves natural resources by utilizing materials that are local, sustainably harvested and produced, and recycled (Fig. 7.4)
5. Protects human health and well-being by providing nontoxic finishes, adequate ventilation, daylight, and a comfortable setting to perform the functions intended (Fig. 7.5)

Imagine a tree (Fig. 7.6), which lives off the amount of energy and water available to it, sequesters carbon while emitting oxygen, treats water through its root system before discharging it, prevents soil erosion, provides habitat and food for other species, adapts to the seasons and place to optimize its use of resources, and at the end of its useful life returns 100% of its resources to the environment benignly as nutrients. If our buildings could operate like a tree, then we would have a truly sustainable built environment. Sound impossible? It is not.

To achieve harmony with our environment, we must first determine on-site resource availability, understand the potential for each building to share resources regionally, and concurrently take great care not to exceed the building's local environmental capital. We often overlook this idea of integrating buildings with the environment. Nature does not operate in a vacuum. A tree would not be able to survive if its flowers were not pollinated by bees, its fruit not eaten by friendly residents, its leaves not composted for nutrients, or its access to sun and water restricted. For the last several decades, we have designed our buildings in a manner that ignores the environment and their neighbors and ensures that they function no matter where they are placed in relation to their surroundings. This practice assumes that all resources are unlimited. What happens when the grid fails, gas mains are emptied, and oil runs out? Will the building you are sitting in now still function and provide heating, cooling, ventilation, and light? The answer to this question will begin to tell you what is right or wrong with our buildings and might indicate what changes will lead toward a sustainable future.

FIGURE 7.4 Salvaged wood floors and mantel at the Morgan residence.

7.2 INTEGRATED DESIGN: THE KEY TO GREEN BUILDING SUCCESS

Nature and our example of the tree, specifically, accomplish very little in isolation. Rather, focusing on integration, synergy, and balance as the key traits of nature allows for a holistic system approach. The project team in today's construction industry is composed of highly specialized individuals and skill sets; however, each specialty, whether architectural, structural, mechanical, or others, must function as part of a cohesive team in order to develop truly cost-effective, sustainable projects. Many of the technologies and strategies employed by these teams of specialists on "green" building projects are readily available to almost any building type at no or little additional upfront costs, assuming a proper initial budget. Appropriate integration of technologies and development of synergy between building systems can lead to significant returns on investment socially, economically, and environmentally. The main reason green projects can be cost-effective, as opposed to ones that are not green or are less sustainable, is the availability of third-party rating systems such as LEED, which have become increasingly popular decision-support tool in recent years. The integrated design process (sustainable design and construction), which involves all the stakeholders from early in the planning and design process through completion of construction, is crucial to developing the most cost-effective, holistic green building projects. The team of highly specialized individuals must work collectively to identify synergies, opportunities, and cost/benefit scenarios while meeting the owner's programmatic and functional requirements. While sustainable/green design may or may not be a requirement or the primary driver of a project, meeting a

FIGURE 7.5 VOC free paint, daylighting, and natural ventilation at the Holmes–Rulli residence.

FIGURE 7.6 Tree and we.

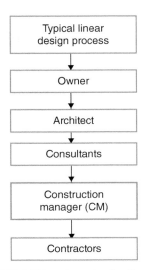

FIGURE 7.7 Linear Design Process.

building owner's program requirements, construction budget, project schedule, and particular aesthetic desires can all still be accomplished while cost-effectively integrating green design principles and strategies.

The integrated design process is not entirely new, but it is often less practiced given the nature of fast-paced deadlines and entrenched design processes. As Fig. 7.7 illustrates, the typical design process is often linear with little interaction among the team members. This process has historically proceeded as such: An owner develops a building program and gives it to his architect to prepare schematic plans and elevations, which determine the size, adjacencies, and aesthetics of spaces. The architect then gives this approved set of drawings to his team of engineers to figure out how to heat, cool, light, and power the occupied spaces. These efforts are coordinated to whatever extent possible, and then the final product is given to a construction manager and/or contractors to prepare final pricing to construct the building. There is very little two-way communication among team members, preventing the identification of system synergies and optimization of the project's energy, water, and resource efficiency. Energy modeling checks (if done at all) are completed after design to confirm building code compliance, rather than acting as a tool beforehand to inform decisions on orientation, window selection, building envelope design, and ventilation strategies. This is, in part, due to the rigorous design and construction schedule many projects are under, as well as the result of a high degree of specialization within the design/build industry. This high degree of specialization is both a benefit and detriment. While it ensures the individual aspects of a project are designed to the highest degree possible, it also fractures and compartmentalizes the design process.

Therefore, it is imperative that this highly specialized team meet early and often in the design process in order to understand the vision of the owner, establish environmental goals, identify synergies that achieve both the vision and the goals, and determine a plan that maintains a commitment to achieving those goals throughout the process. A project team typically includes the following: the owner, architect, engineers (which include structural, mechanical, electrical, plumbing, and civil at minimum on most projects), end users, contractors, and often a construction manager. In more recent projects, the commissioning authority (CxA) and energy modeler have also become an integral part of the sustainable design and construction team. Depending on the complexity and specialty of a building or development, additional specialists such as acoustical engineers, traffic consultants, environmental engineers, lighting designers, interior designers, etc. may be called upon by the architect to lend expert advice or design. No matter the size and complexity of the team, clear lines of communication and a process for decision-making are essential. Figure 7.8 illustrates the kind of circular nature of open communication among the team needed for a truly successful, cost-effective green project delivery. Instead of a linear process, in which the project is passed from one team member to another, the integrated process requires that the team come together to communicate their concerns, needs, and opportunities for collaboration among the specialties. This allows for swift identification and resolution of conflicts, communication, and reinforcement of goals, and maximization of opportunities for cost efficient, green building through collaborative processes.

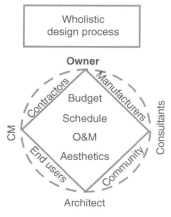

FIGURE 7.8 Wholistic/Holistic Design Process.

The design team must look at a building and its systems holistically to minimize waste and maximize efficiency. The strategies employed can vary greatly depending on the type of project, project team, owner goals, budget, and schedule. The integrated design process has several key steps, of which the project team must always address regardless of their sustainability goals:

- Assemble the right team for the project
- Set your sustainability goals early
- Establish open, consistent communication
- Conduct intense planning meetings (e.g., design charrettes) early in the design process (NREL, 2003)
- Identify synergies between building systems
- Track and maintain the commitment to green goals

7.2.1 Assemble the Right Team

It is imperative that the owners and project developers select qualified consulting, design, and/or building firms. This is largely determined by the "request for proposal" and the request for qualification' process. Embarking on a project with sustainability goals, even with an inexperienced project team or key team members, can have one of two likely results: (i) either the project will cost more to design and build than it needs to or (ii) it will not perform as intended to meet the sustainability goals. Identifying the key requirements of team members early in the proposal process, including (i) past sustainable project experience, (ii) particular expertise in areas of sustainable design, (iii) successful track record of on-time and on-budget project delivery, and (iv) experience in working with others in a team environment, is crucial. As the project teams gain experience in designing and delivering sustainable projects, the time required for design decreases, while the results and building performance likely improve. When this happens, instead of acting as an added service, sustainability becomes "just the way things are done" for both the design team and the owner.

7.2.2 Set Your Sustainability Goals Early

There are many third-party environmental assessment methods to identify and set a project's sustainability goals. As a nationally accepted standard of care in the industry, the Leadership in Energy and Environmental Design (LEED™) Green Building Rating System, as development by the US Green Building Council (USGBC, 2011), provides an excellent framework and checklist for establishing performance goals and exploring all aspects of a building to optimize its design. There are also similar systems, such as Living Building Challenge and Green Globes, which originated in Canada. Whatever system or third-party verification process is chosen, it is important to set goals early in the design process and to verify that these goals were effectively achieved.

Certification systems help ensure a holistic approach in all of the components of a building project, with categories including sustainable site design, water conservation, materials and resource conservation, energy and atmosphere conservation, and indoor environmental quality. Tools such as the LEED checklist are highly effective when used early on during the programming and goal-setting stage and throughout all the design and construction phases. Even if a project is not seeking certification through the US Green Building Council or others, active use of such guidelines during design can result in a significantly more sustainable project by encouraging the right conversations early in the process, requiring little to no additional upfront cost.

7.2.3 The Importance of Setting Goals Early

Usually, the questions of first cost and payback must be answered early in the process as scope and budgets are being finalized. They are early drivers on major decisions, affecting the building aesthetics, construction budget, technologies, and operational/maintenance costs. The idea of cost is both a short-term and long-term question, however. First (i.e., initial) cost and maintenance costs must be balanced; yet, in the typical 40-year life cycle cost of building ownership, operations/maintenance costs account for approximately 50%, while actual construction accounts for only 11%. When you understand this, it becomes apparent that decisions regarding construction, which has the smallest overall financial impact, can affect operation, the area with the largest financial impact. These decisions are made up front in design. Design (as an early phase of design and construction) accounts for an even smaller part of the total cost of ownership of a building, approximately 6–8% of the 11% initial construction cost. In short, all of the decisions that affect the entire life cycle cost of a building are made up front, when cost percentage is at its lowest on the building's timeline. This is why it is so critical to plan for high-performance design up front and to set your sustainability goals early in the process.

7.2.4 Establish Open Communication and Conduct Charrettes

While the use of tools, such as LEED, are effective in promoting integrated thinking, successful sustainable projects also require an open dialog between team members throughout the design process. Modern technology has enabled rapid, constant communication among team members through the use of e-mail, ftp websites, teleconferencing, texting, and online project management software. While these technologies are all wonderful tools in maintaining communication, it is imperative that all parties are consistently working toward the same goals. An excellent forum for establishing those goals is the design charrette (NREL, 2003).

The purpose of a design charrette planning meeting is to provide a forum that allows all of the stakeholders to communicate their concerns, visions, and expertise toward the design process. This intensive workshop, in which various stakeholders and experts from multiple disciplines are brought together, supports communication among the project team members, building users, and project management staff.

Most importantly, charrettes allow all participants to understand various goals for the facility, such as functionality, size, location, cost, aesthetics, health, safety, and future flexibility, as well as the integration of sustainable principals. The three most important results of a charrette should be as follows (NREL, 2003):

- An overall set of shared performance goals
- A project team with a shared vision
- An action plan for achieving a functional, aesthetically pleasing, cost-effective, and sustainable design

The format, content, length, and frequency of charrettes may vary. A good reference for planning charrettes may be found published by the National Renewable Energy Laboratory (2003).

7.2.5 Identify Synergies

The way to cost-effectively create green architecture is to optimize the synergies between building components. Building orientation, insulation, natural ventilation, and daylighting design can all have significant effects on the energy performance and indoor environmental quality of a project. Proper building orientation, for example, can reduce heat gain and energy costs by as much as 25% without adding any additional operational costs.

Additionally, designing for proper daylighting can reduce the need for artificial lighting from the hours of 10 A.M. to 3 P.M. in most locations. Lastly, including appropriate building insulation in walls and roofs, combined with appropriate glazing specifications and building orientation, can dramatically reduce heating and/or cooling loads, thus saving 30–60% in energy use.

While increased insulation may carry a nominal premium, the immediate first cost reduction in the size of heating/ventilation/air-conditioning (HVAC) equipment, including piping and fan sizes, as well as the quantifiable operational energy savings, makes the initial investment financially wise. As many energy models have shown, the general rule states that HVAC accounts for up to 75% of energy use in a typical building; therefore, concentrating a team's efforts on finding the best synergies between building orientation, daylighting, and envelope performance will have a significant return on investment. Moreover, it will likely result in a cost neutral move, as money is shifted from HVAC budgets to insulation or site development costs.

7.2.6 Maintain the Commitment to Sustainability

Sustainability goals can easily be lost in the design documentation process and even more so during construction. While green design and construction is not vastly different from conventional projects, it does often involve products, systems, or processes that are unfamiliar to many contractors and/or construction managers. Taking the time to educate contractors and construction managers in processes involved with sustainability is well worth the investment. For both the design team and owner, it is useful to continue using the same methodology throughout planning, design, and construction. This is where a performance goal setter and metric, such as the LEED rating system or others, comes into play once again. With an online interface that allows the design and construction team to track the project's sustainability goals during construction, rating systems like LEED are excellent tools for establishing and maintaining communication and expectations among team members (USGBC, 2013). It is imperative that sustainability goals are reviewed during the bidding process with prospective contractors as well as at each construction meeting. In the end, the best-laid plans are meaningless if the results do not measure up.

7.3 THE TEAM

As discussed thus far, the integrated design process is essential for attaining the most cost-effective, holistic solutions. As part of this process, highly specialized individuals are brought together. The sum of their collective expertise, within the confines of budget, schedule, maintenance, and aesthetics, is what dictates the project's level of sustainability. So, who are these team members and what are their roles?

7.3.1 Do You Need a Green Champion?

In order for any project, let alone a green project, to be successful, there needs to be a common vision that inspires and steers the decisions made by the team. This set of guiding principles is often best held when there is a champion or visionary that focuses project development through the lens of that sustainable vision. This role may be played by any number of team members or can also be an independent member not directly associated with actual design; however, a representative of the owner, architect, construction manager, or mechanical, electrical, and plumbing (MEP) engineer may play this role more effectively. In particular, the architect may be best suited to guide the sustainable vision of a project, as they already coordinate the project budget, schedule, aesthetics, and operational cost impact. Using the LEED rating system and checklist, the charrette, and modern communication practices, the Green Champion is responsible for guiding the team throughout each phase of the project and tracking the successful implementation of all sustainability goals through completion of construction. If a project seeks certification through a green building rating system, it may fall to this Champion to compile and submit the necessary documentation as part of their services.

7.3.2 The Architect

The architect has traditionally played many roles during the design process. Acting as an interior designer, team communication coordinator, meeting facilitator, technical building expert, code consultant, artist, programmer, cost estimator, and author of the construction documents, architects must now also be able to articulate the technical issues surrounding green design. These issues span everything from building orientation and envelope design to sustainable material selection and integration of performance mechanical, lighting, and renewable energy systems. More clients are demanding green buildings, and the only way to achieve this is if the architect through vision, coordination, and expertise successfully implements a holistic building design.

In defining a building program, schematic floor plan(s), preliminary budget, and orientation, the architect is responsible for establishing many of the fundamental synergistic opportunities, or lack, thereof, for the incorporation of cost-effective green design features very early in the process. In addition to being responsible for the early planning and proper sighting of buildings, architects also directly and dramatically affect the resource demands during both construction and operations. Perhaps the three most important factors the architect has prevue over, besides its aesthetics and functionality, are building orientation, building envelope design, and system integration. Additionally, architects also greatly affect indoor environmental quality in specifying interior finishes, furniture, and products such as paint, adhesives, flooring, seating, lighting, window treatments, wall covering, and daylighting controls. Using a rating system such as LEED, the architect is directly responsible for over 30 out of the minimum of 40 points needed for certification. Clearly, the architect's role is one of the most critical when it comes to ensuring the sustainability of a project.

7.3.3 Site Analysis: The Civil Engineer and Landscape Architect

Before considering the in-depth building design, it is critical to understand the project site, climate, and environment in which the building must function harmoniously with its environment. This affects the choices concerning orientation, insulation, glazing, and even HVAC system. Proper building orientation takes advantage of the resources readily available on-site, such as solar energy, wind, views, shading, water, and vegetation. Appropriate choices can be a low expense effort, yet still significantly create a holistic, green project. Specifically, vegetation selection can significantly impact water use for irrigation, energy use by the building through proper shading and protection, maintenance costs of the plants, ecological health of the site, and overall aesthetics. As we increase the use of ecologically engineered systems like living machines, constructed wetlands, and rainwater gardens, the roles of the civil engineer and landscape architect are integrated.

The Civil Engineer is generally responsible for site utilities design 5 ft beyond the building, along with connected services such as sewer, water, gas, and electric, all of which feed the site. In addition to the design and connection of site utilities, the Civil Engineer is responsible for mapping site topography of the site and designing pedestrian and vehicular access in and around the site. The Civil Engineer may also design specialty systems such as rainwater gardens, permeable pavement, or detention basins.

7.3.4 The Structural Engineer

The primary cost in materials and labor in many projects is the superstructure.

Designing the most efficient structure to perform the task will not only reduce first costs but also reduce the use of raw materials and, therefore, energy needed to process, transport, and install those materials (also known as embodied energy). Some of the most energy-intensive materials in a building project are in the manufacturing, transportation, and installation of items such as structural steel, concrete, and asphalt. When seeking to increase the sustainability of projects, the structural engineer must look for ways to minimize the number and size of structural members and substitute recycled and alternative products for energy-intensive raw materials. For example, substituting fly ash or slag, which is a by-product from coal-fired power plants, Portland cement in concrete not only reduces the greenhouse gas footprint of concrete, but it also increases the final strength of the concrete.

Concrete is also typically produced locally, as trucking concrete a long distance is not practical. Likewise, most steel is already made with 60–80% recycled content today. The structural engineer relies on materials that already have inherently sustainable features, such as local manufacturing and high recycled content. Ultimately, the structural engineer is responsible for calculations that ensure the building loads are safely transferred to and supported by the earth. No matter how sustainable a material may be, if it does not meet or ensure the structural requirements of a project, it will not be used by the structural engineer.

7.3.5 The Mechanical, Electrical, and Plumbing (MEP) Engineering

Buildings in the United States use approximately 65% of all of electricity produced. Within this, lighting and HVAC systems account for as much as 75% of energy used in buildings. In many green building rating systems today, such as LEED, high-efficiency HVAC and lighting are responsible for at least half of the LEED rating points required to obtain a certification as a sustainable building. Because MEP engineers are not typically involved early in the programming and schematic design process, they often inherit a building with its plans already confirmed by the owner. This oversight may handicap a MEP engineer as to how efficient they can get with off-the-shelf technologies.

7.3.6 The Mechanical Engineer

The primary responsibilities of the mechanical engineer include determining the ventilation rates required in spaces, determining heating and cooling loads based on building envelope insulation values provided by the architect, and sizing the equipment to meet the ventilation and heating/cooling requirements of the building. One other major responsibility of the mechanical engineer is to determine the distribution of these systems. In working with mechanical engineers, it is advisable to know what questions to ask and ideas to discuss, which should start with what technologies use the least energy to heat and cool air. Low energy strategies to consider include heat recovery, direct/indirect evaporative cooling, demand controlled ventilation, displacement ventilation, and geothermal energy. While not every solution is applicable to every project, various combinations of the aforementioned strategies, coupled with proper building orientation and insulation, can result in a cost-effective solution to reduce energy consumption of HVAC systems by 20–40% or more below the building code requirement. To determine the most cost-effective and efficient approach for a given project, the mechanical engineer should conduct a whole building energy model simulation. While some engineers are capable of providing this service, not all are. This model is a key tool in determining the most effective combination of technologies during the early design stages of a project. At a minimum, a whole building energy model should show the following:

- Energy use of all systems
- First cost of installation
- Operational costs
- Maintenance/operation costs
- Simple payback

Modeling allows the design team and the owner to weigh many factors simultaneously in the early stages of design. As an example, it is the general practice of most engineers to design an air conditioning system at a ratio of one ton of cooling to $250\,ft^2$ of floor area. This has been the conventional design for the last 30 years. In a high-performance project, this ratio should be closer to one ton of air conditioning per $500\,ft^2$. It is quite possible to go as high as $1000\,ft^2$, depending on the combination of technologies and load defining building parameters. Accurate energy modeling will help the design team refine the systems of the building and improve cost and efficiency through the simultaneous modeling of building envelope design, HVAC systems, lighting, and controls, as a whole.

Many HVAC systems, with combinations of technologies and delivery systems noted previously, also require more sophisticated automated controls than older systems. Most of these controls are known as direct digital controls or DDC for short. Coupling high-performance HVAC systems with precise controls that also link directly to occupant use through motion and CO_2 sensors will provide the most efficient results. The most complicated and difficult portion of high-performance HVAC systems is proper setup, testing, and burn-in of these controls and sensors. This stage is where commissioning of building HVAC systems and controls becomes critical to ensure proper installation and functionality. We will discuss the role of the commissioning agent later.

7.3.7 The Electrical Engineer

The electrical engineer, like the plumbing engineer, is often one of the last team members to begin work on projects. They must wait for the mechanical engineer to select equipment and the architect to determine the placement of key features, such as computers, lighting, and furniture. Regardless, the team must allocate proper time into the design schedule for completion of electrical work once the earlier steps are finished. The electrical engineer will then determine the electrical load of the building (i.e., how much power the facility will require to operate), the size of service and distribution systems, and distribution of life safety devices such as fire alarm and detection systems. The electrical load of the building is largely determined by the size of HVAC equipment, lighting levels, electrical equipment (i.e., computers), and hours of operation these systems are required to run.

The electrical engineer is responsible for the design and integration of technologies such as solar, wind, fuel cells, and combined heat and power systems. These systems, which often carry a cost premium but offer a relatively short return on investment, will also improve the results of the energy model as they reduce the electrical load of a building by providing power on-site through renewable sources. In order to leverage such technologies, building orientation and proper sighting play an important role.

As mentioned, lighting and HVAC constitute two of the major energy users in buildings. As the electrical engineer looks at the functions of spaces to determine appropriate lighting levels, it is important that they also understand the effect of daylighting on their approach. Their design strategies should incorporate daylighting controls that effectively respond to the available natural light. In addition to daylighting controls, further integration of artificial lighting with occupancy sensors will ensure that artificial lighting is not needlessly left on when spaces are unoccupied or adequate natural light levels exist. Effective use of occupancy sensors and integration with daylighting can reduce artificial lighting energy by 50% or more.

7.3.8 The Plumbing Engineer

The primary scope of work of the plumbing engineer includes the selection of plumbing fixtures, the sizing of supply piping and waste systems, and the design of life safety systems, such as building sprinklers. Within the context of sustainability, buildings account for 12% of all potable water use. With only about 2% of all the water on the planet being potable, the question is, "why do we flush drinking water?" The only plausible answer is simply because it has always been done this way, similar to why air-conditioning systems have traditionally had such poor tonnage to floor area ratio.

Using off-the-shelf lower flow fixtures for toilets, sinks, urinals, and showers, combined with motion sensors, a perceptive plumbing engineer can reduce the potable water use by as much as 30% without affecting the comfort of the building's occupants. In looking to improve this efficiency further, the plumbing engineer may recommend other technologies, such as rainwater catchment or waterless urinals. It has been found that in many buildings, toilet-flushing accounts for 60–80% of all potable water use. Using rainwater to flush our toilets would seem to be a logical, natural, and remarkably simple sustainable solution. In the event a rainwater system was to be used for this purpose, the engineer would need to know the annual rainfall of the area, building hours of operation, number of occupants, and number/type of fixture water use. With this information, the plumbing engineer can size the collection tank, pumps, filters, and distribution piping of a rainwater system and account for a percentage of storage in the event of drought. Within the LEED rating system, a single water-efficiency technology, such as a rainwater catchment system, can account for 5–10 points toward certification.

7.3.9 The Commissioning Agent (CxA)

Commissioning is still a relatively new term and process in the building industry. The commissioning process is intended to ensure that buildings operate as originally planned by design. For those of you in the design and construction community, you know that even the best-laid plans on paper can be lost in translation to an actual building project for a myriad of reasons. It is the job of this highly specialized building system integrator to test and confirm that the individual components of a building operate together as a holistic functioning system. Owners often question the need for commissioning, considering the cost unnecessary or the process one that can be performed using the existing design team. I like to make the analogy that the same owner who doubts the necessity for commissioning a building would never purchase a car if it had not passed a rigorous set of functional tests. Nevertheless, our buildings today are substantially more complex and expensive than a car, affecting the lives and well-being of far more than an individual driver. So shouldn't commissioning be a critical requirement? The second item to note is that our building code requires buildings over 50,000 square feet to have the HVAC systems and controls commissioned regardless, so in many cases, it is not optional.

The key function of the commissioning authority (CxA) is system integration testing. The CxA is a uniquely qualified individual because they usually need to have significant expertise in both the engineering and construction sides of the business. This cross-disciplinary experience is leveraged during both the design and construction phases to advise the design team of possible constructability and performance issues and to identify errors or flaws during construction, ultimately affecting the system's overall performance. ASHRAE Guideline 1-1996, the Building Commissioning Association (BCXA, 2013), and the National Environmental Balancing Bureau (NEBB, 2013) provide criteria for the commissioning process and the requirements for becoming a commissioning authority.

Green building rating systems, like LEED, take the requirements of ASHRAE to commission HVAC systems in buildings 50,000 square feet or larger a step further and require that all LEED projects are commissioned. Additionally, the building controls, lighting, and any specialty systems like a green roof, solar panels, etc. must also be commissioned. An excellent outline of the commissioning process can be found on the New Jersey Clean Energy Program website (NJCEP, 2013).

The commissioning process should ideally begin as early as possible in the design phase. By the middle of design development, as systems are beginning to be finalized and preliminary design completed, a commissioning review for functionality, efficiency, constructability, and maintenance can be a tremendous help to the design team and owner. Although involving the CxA during design is preferable, it is not required. At the least, commissioning requirements and a constructability review should be included in project specifications at the time of bid. Lastly, the scope of commissioning during construction may also include assembly of the project equipment and operations manual, coordination of owner operational training, compiling of system warranties, and a 10-month post construction inspection.

7.3.10 Other Key Team Members

The involvement of facility management staff, end users, the owner, and the contractor rounds out the integrated team. Each brings a unique perspective, to the table, which are important in formulating sustainable solutions. Facility managers are often not consulted during the design process; however, they have the most hands-on, day-to-day experience in dealing with the issues and challenges initial decisions may present over the life of the building. It is highly encouraged, therefore, that the facility team be involved during review and selection of systems for operational and maintenance concerns. Input from the end users of the building is equally important. After all, they are why the building is being built and for whom it must function best. Ensuring adequate size, adjacency, and functionality of space, while incorporating sustainable design goals, is critical to the long-term success of the building. Designing for future flexibility and easily adaptable space will also reduce renovation costs and allow the users to grow or change over time with the building. The needs of function can sometimes conflict with the goals of sustainability. For this reason, the input from the end user on functionality, buy-in, and sustainability goals is important.

The contractor, unfortunately, is often not brought into this equation until long after many or all of the design system decisions have been made. Typically, they are handed a set of finished construction documents for the first time and asked to put a guaranteed price on building what is proposed. The contractor's valuable input on constructability issues, cost, materials, and methods has been lost and ends up coming through at this stage as "value engineering" suggestions. This retroactive input can often radically alter the design, performance, and intent of a project unexpectedly for the owner, users, and design team. Therefore, it is far more cost-effective and timely include the contractor's input upfront during the design process as systems are being investigated and decisions are being made, rather than having to go back after the fact and to make major or changes to an entire set of construction documents. As you can see in Fig. 7.9, the opportunity to make changes early in the design phase is far less costly than if the changes are made later during construction.

FIGURE 7.9 Opportunity vs Cost: changes made in the design phase are less costly than during construction.

In publicly bid projects, receiving the contractor's input upfront may actually be deemed illegal. Recent trends in public bid processes use the design/build model because of the recognition of the added value, reduced costs, and expedient schedule that result from having the contractor on board as an active participant in the team during design.

7.4 RESOURCE USE AND GREEN BUILDINGS

7.4.1 Energy

We know buildings consume approximately 65% of all electric in the United States (USGBC, 2011) and that for every kWh of power used, it equates to approximately 1/2 pound of carbon emissions. We know that we are at the highest levels of carbon dioxide (CO_2) in our atmosphere ever recorded, and the consequences of just a few degrees of global warming equate to a meter more of sea level rise in high-density population zones. We have already witnessed billions of dollars of property damage, and a dangerous domino effect on food production, air quality, and regional climates, as a result of changing climate. Energy, therefore, is a major focus and impact of green buildings from an economic, environmental, and social standpoint, making it a key goal to reduce energy consumption as much as possible. Can we design net zero energy buildings today? Absolutely!

There are examples of this type of building all over the world, because the technology for a sustainable future already exists. There is a very simple formula for achieving net zero energy buildings (Fig. 7.10).

7.4.1.1 Reduce Demand

Breaking this formula down, reducing demand is where we need to start. We know that HVAC and lighting account for

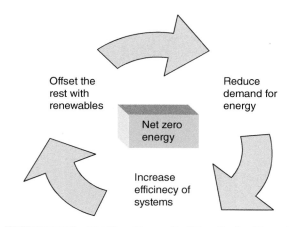

FIGURE 7.10 Net Zero Energy Building Design Formula.

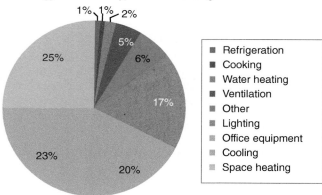

FIGURE 7.11 Energy end uses in typical office building (*Source:* USEPA green building working group).

FIGURE 7.12 Ideal building orientation (North eastern United States).

nearly 75% of all energy use in our buildings, but what most people do not consider is building orientation, building envelope efficiency, vegetation, shading, natural ventilation, daylight harvesting, and behavioral/operational management can affect building energy use.

We have the capability to reduce HVAC and lighting demand in our buildings by 50% or more by understanding these synergies. There is no one set combination of approaches, as different climates and conditions warrant different solutions and approaches. Understanding resources and site-specific conditions in order to develop solutions that optimize first costs and return on investment while achieving sustainability goals. In the Northeast United States, the ideal orientation by most standards is to have the long axis of the building running east/west (Figs. 7.11 and 7.12).

The reason for this is that it is much easier to control heat gain and glare from the south than from the east or west light and because northern light is the most constant. Northern sides of buildings tend to be the coldest in this climate, so care should be given to insulation values of walls and windows. Properly orienting the building will allow for abundant harvesting and control of daylighting, which can eliminate energy demand for artificial lighting during regular sunlight hours (Figs. 7.13 and 7.14).

In addition, looking at the building envelope, defining the ideal shape, and ensuring proper insulation levels and air sealing will reduce heating and cooling loads substantially. Related to daylighting and natural ventilation, trying to maintain a building depth of no more than 40 ft will allow for cross ventilation and adequate daylighting. When this is not possible, introduction of atriums, skylights, and/or solar tubes to provide lighting are all viable options. Insulation values are important in controlling loads. Generally, R30 walls, R50 roofs, and R5 windows provide a superior envelope and dramatically reduce the load as compared to code requirements. This affects the size of the mechanical system equally dramatically. The specific type of glass and the shading controls for glass will vary depending on the direction, elevation, and specific climate and site conditions. If you understand the inherent need to design to local conditions and specific orientation opportunities, then the obvious conclusion is that our buildings should not look the same on all four sides, which has been explored in many modern designs. In order to interact properly with its environment, a building's facades should all be designed to the specific conditions of each orientation. For example, clearer glass with higher insulation values and no sun control devices is appropriate on the north, while a darker glass and/or one with a very good solar heat gain coefficient and an exterior horizontal light shelf is appropriate on the southern facade. In turn, vegetation such as deciduous trees can provide shade in summer and drop their leaves in the winter, allowing for shade and solar gain on the southern and northern facades, respectively. This can significantly reduce loads on the building, as well. If we do not take advantage of capturing daylight and optimizing breezes onsite, then we immediately increase the requirement for energy consumption by an artificial mechanical system. The fact is, if properly designed and integrated, we can typically provide adequate daylight our buildings about 70% of the time and heat, cool, and ventilate our buildings 30–40% of the time without mechanical energy. The last component of demand reduction is the human factor. We can design and build these measures, but if the user chooses the use artificial light or mechanical ventilation, then we have essentially ignored the intentions of the sustainable design. It is, therefore, imperative that the occupants and facility managers

7.4 RESOURCE USE AND GREEN BUILDINGS   185

FIGURE 7.13 Light shelfs at School of the Future, Philadelphia, PA (LEED™ Gold Certified).

FIGURE 7.14 Light shelf at School of the Future, Philadelphia, PA.

of a building are part of the integrated project team from the beginning, understanding the goals and decisions. They should also be provided with the proper training in how to manage and operate the building as a holistic system. Taken in combination, these measures can reduce demand for energy by 25–50% from the start (Fig. 7.15).

7.4.1.2 Increase Efficiency of Systems

Once the team has taken measures to reduce demand as much as possible, it is time to start looking at the specific building systems such as heating, cooling, ventilation, lighting, and controls. There are literally dozens of options when it comes to these systems, and the choices can often be confusing. New technology is constantly developed, and it is important to consider and incorporate cutting edge system appropriately, as they are available. Using tools like whole building energy simulation, as defined by ASHRAE 90.1, can provide owners with information on first costs, operating costs, and return on investment, allowing them to make informed decisions on the right combination of these systems. Energy modeling is the key to understanding how buildings will perform with various system options and compared to one another. An important synergy to remember is that once you have optimized the orientation and building envelope and reduced the loads as much as possible, you can then reduce the size of equipment. This act will increase its efficiency without impacting your budget and, in some cases, actually reduce your budget. Some examples of high-performance HVAC mechanical systems and components include geothermal, variable refrigerant flow, demand controlled ventilation, evaporative cooling, heat recovery, displaced ventilation, entropy cooling, chilled beams, and solar thermal.

Lighting technology has progressed from the incandescent bulb to the compact fluorescent and to the LED technology of today. Fiber optics can be useful in distributing either natural or artificial light in a building, but this is less common today due to its cost and complexity. These technologies will undoubtedly change, improve, and evolve over time. Using energy modeling as the mechanism to evaluate these various strategies and combinations will allow design professionals and owners to make informed decisions and remain on the cutting edge of technological innovation; however, none of these systems works particularly well without proper controls and operation. As mechanical systems get more complicated, so too can their controls. The choices of building systems are also often affected by the ability of the owner's staff to understand and maintain a system. Controllability and integration with natural systems are key components of efficiently using the installed equipment. If you design a building for abundant daylighting, but do not install photo sensors to dim or shut the lights off, then you have lost the value of naturally lighting your building. The same holds true with mechanical systems. A high-performance system is one that does not know that the windows are open and, therefore, continues to run and waste energy. If we properly integrate controls, teach our occupants and facility managers how to use them, and rightsize the systems based on the dramatically reduced loads from the prior discussion, we can greatly increase the efficiency of the equipment itself while not increasing the construction budget. For example, if I have to buy a 100 ton air-conditioning system for a commercial office building instead of a 50-ton system because I did not properly size it based on reduced loads, then the system costs twice as much as it should, and I will likely have to buy a less efficient piece of equipment to afford it. Conversely, actually sizing the system to 50 tons based on the reduced loads of the building will allow me to greatly increase the efficiency of the equipment within the same cost parameters as the 100-ton unit.

This is part of the synergy and cost benefit of green architecture, at its best (Fig. 7.16).

7.4.1.3 Offset the Rest

At the end of the day, all buildings will continue to use energy of some kind to allow us to perform the tasks desired. No matter how much you reduce demand and increase the efficiency of equipment, some amount of power usage will

FIGURE 7.15 Thomas Edison State College, Trenton, NJ, daylit interior "great hall" with entropy cooling.

FIGURE 7.16 Morris County School of Technology, Denville, NJ, day-lit gymnasium with 100% outside air displacement ventilation system, FSC wood floors, and zero VOC paint.

FIGURE 7.17 Holmes–Rulli residence net zero electric solar.

remain. The question is how much, and can we reduce it enough to be able to cost-effectively offset the rest from renewable sources on-site? There are many forms of renewable energy. These typically include wind, solar, geothermal, some forms of hydro, and biofuels. Nuclear and gas are not renewable energy sources (Figs. 7.17 and 7.18).

The integration of these technologies in our buildings as part of the fabric of the building itself makes them more cost-effective than the typical approach of adding them on in addition to other building materials. For example, solar panels can be used as a waterproof roofing system, and in doing so, you can eliminate the need for other materials such as a roof deck and roof membrane. Depending on location, building orientation, shape, size, and local codes, different combinations of systems may be more appropriate than others. Each individual renewable energy technology has its own set of design criteria, benefits, costs, and downsides to consider, which we cannot discuss in detail in this chapter.

FIGURE 7.18 Microsoft School of the Future, Philadelphia PA, Solar electric glass (tinted areas) in the cafeteria.

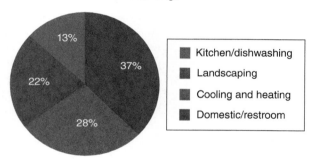

FIGURE 7.19 End uses of water in office buildings. *Source:* EPA WaterSense.

Suffice to say, the integration of these systems, after reducing demand and increasing efficiencies as much as possible, can readily and cost-effectively result in creating net zero energy projects (Fig. 7.19).

7.4.2 Water

As mentioned earlier, only about 2% of all water on our planet, which is composed of roughly 97% water, is actually potable. Of this, about 11% is in the form of surface water (reservoirs, lakes, rivers), with the remainder underground. Water efficiency has lagged behind energy largely because the cost of water is cheap by comparison. Similar to the approach discussed previously, reducing demand, increasing efficiency of equipment, and offsetting remaining use from

FIGURE 7.20 Microsoft School of the Future Rainwater Catchment for flushing 100% of all toilets ~ 56% of all water usage and over 1,000,000 gallons of water annually conserved.

renewable sources are the formula for achieving net zero water results. In the Northeast, for example, we get enough rainfall on average to flush our toilets nearly 3× over, yet we still typically pipe and flush drinking water in toilets. Toilets generally account for nearly 40% of all water usage in an average commercial building. Other types will have different water profiles, and this percentage can go as high as 60–80% in buildings such as schools (Fig. 7.20).

Following the concept of reducing demand first, the most practical approach is to select native plants around the building, therefore, requiring that no potable water be used for irrigation. You can also select toilet fixtures that eliminate or drastically reduce the need for potable water to flush, such as waterless urinals or composting/low flow toilets. Using the most efficient off-the-shelf technology and utilizing low flow toilets, faucets, showers, and appliances can radically reduce the amount of potable water required by a building by 30–50%. At minimum, fixtures bearing the EPA WaterSense label, which ensures a 20% reduction below code requirements, should be incorporated into any project (USEPA, 2013a).

Lastly, utilizing renewable resources to meet further needs would primarily entail rainwater harvesting. This technology can be used to flush toilets, irrigate lawns, and provide decorative water features for projects.

7.4.3 Materials

What qualifies a product as green? There are literally hundreds of different third-party certification systems surrounding the green product market, which is, at once, encouraging and utterly confusing. Green materials for the purposes of this discussion are generally those that reduce or eliminate their environmental impact during harvesting, manufacturing, installation, and at the end of its life. A true cradle-to-cradle approach where there is no such concept of waste, as embodied in the book "Cradle to Cradle" by William McDonough and Michael Bromgaurt, is the ideal situation. In this concept, all materials are either technological or organic food and are continuously cycled, back through the process without ever ending at a landfill. This requires a radical shift in our manufacturing process.

In short, green materials are often those that are local, contain recycled content, are sustainability harvested, and/or are rapidly renewable. There are third-party certification systems for each of these characteristics. LEED defines local as being harvested and manufactured within 500 miles of a project site. This greatly reduces the embodied energy and carbon footprint of materials arriving on a job site. Recycled content of products includes pre- and postconsumer. The Scientific Certification System (SCS) is one of the largest third-party certifiers of recycled content. Using recycled content products reduces the demand for raw materials and, simultaneously, the quantity of material finding its way to landfills. With over 40% of all raw materials globally consumed by buildings, this is an important characteristic to pay attention to, if we are to conserve our resources (Figs. 7.20 and 7.21).

Products like fiber cement siding, carpet, structural steel, gypsum wallboard, aluminum windows, porcelain tile, concrete, and ceiling tile can all contain high levels of recycled content.

Sustainably harvested materials are those that respect and even enhance the environment from which they are removed. The Forest Stewardship Council's certification program for

FIGURE 7.21 Morgan residence addition salvaged wood floors, fireplace mantel, and daylighting.

wood is a prime example of this (FSC, 2013). Their certification is a chain of custody certification that follows the wood from forest to mill to job site, ensuring that the ecosystem of the forest is being protected and enhanced as part of the process.

Rapidly renewable materials are those defined as having a 10-year or less growth cycle by LEED. Typically, these are plant-based products, such as bamboo-flooring and bamboo-wall paneling, cotton fiber insulation, wheat-based particle board, whey-based wood finish, cork-flooring or cork-wall covering, switch grass such as sisal wall covering, and carpeting that encompass species of quick growing wood, like Ash. Use of these products reduces the strain on long growth or high embodied energy materials. The raw materials are often self-regenerating or can be replanted and grown in short periods. Cork, for example, which comes from the cork tree in Portugal, is harvested from the bark every 8–10 years, which regrows on the tree without killing it. This can be done for generations without harming the tree (Figs. 7.22 and 7.23).

Many of these products are readily available at similar or same cost. They can be used in lieu of less environmentally friendly options without compromising aesthetics, cost, or durability. They just have to be specified as part of the project (Fig. 7.23).

7.4.4 Indoor Environmental Quality

Many do not understand the relation of buildings to human health. This is a growing concern when selecting what we put in our buildings. People spend 90% of their time in buildings, according to the (USEPA, 2013b). Surrounding ourselves with products that contain formaldehydes, volatile organic compounds, and chlorines that off-gas at room temperature, is not smart for the environment or for

FIGURE 7.22 Morgan residence addition salvaged wood floors, fireplace mantel, and daylighting.

FIGURE 7.23 Holmes–Rulli recycled glass backsplash by Oceanside, recycled glass and concrete counters by icestone, cork flooring, sustainably harvested wood cabinets, daylighting, and energy/water efficient appliances.

our health. These chemicals can cause a myriad of health issues, ranging from simple rashes and headaches to various forms of cancer, asthma, nervous system damage, and birth defects. The same products that we are looking at for environmentally friendly characteristics described previously need to be screened for chemical content as well. The Living Building Challenge (ILBI, 2013), a third-party certification system similar to LEED, has a category for materials that contains a chemical "red list" (ILBI, 2013). The building may not contain any of the chemicals in their materials or else it will not qualify for certification. Paints, carpets, adhesives and sealants, wood finish, furniture, fabrics, tile, and ceiling/wall materials can all often contain these harmful chemicals. Some places like the Green Living and Building Center have already screened their products for both environmentally friendly materials and indoor air quality issues. With a large number of different third-party certifications, misinformation, or lack of any information altogether from manufacturers, and disinterest from contractors, it can be challenging to source and properly install healthy materials in buildings. However, the human health impact on property value, risk management, and operational costs cannot be overlooked and remains to be an integral part of any sustainable environmental management program.

7.5 CONCLUSION

The old cliché, "There is no I in Team," rings true in sustainable design. It is truly a team effort requiring clear communication, coordination, and integration among a highly specialized multidisciplinary team. Setting goals early, establishing modes of communication, finding synergies between systems, and maintaining the commitment to these goals are the key elements for creating a cost-effective, environmentally responsible, healthily built environment.

Through the use of the integrated design process and a consistent metric of sustainability, such as LEED, a project can readily achieve substantial energy, water, and resource conservation while improving the indoor environment for occupants and protecting the natural environment for future generations. The current carbon footprint of our built environment is unsustainable and is already posing great risks realized by more frequent and devastating events across the globe. In looking at environmental management strategies, we cannot ignore the built environment and its impact on land, water, air, and resources. If everyone on the globe lived like individuals in the United States, we would need three to five planets to support life—and that is just simply not an option.

REFERENCES

BCXA. (2013). Building Commissioning Association. Retrieved October 23, 2013, from www.bcxa.org.

FSC. (2013). The Forest Stewardship Council. Retrieved October 23, 2013, from https://us.fsc.org/.

Gore, A. (2006). *An Inconvenient Truth: The Planetary Emergency of Global Warming and What We Can Do About It*. Emmaus, PA: Rodale Books.

ILBI. (2013). The Living Building Challenge. Retrieved October 23, 2013, from https://ilbi.org/lbc.

NEBB. (2013). National Environmental Balancing Bureau. Retrieved October 23, 2013, from http://www.nebb.org/.

NJCEP. (2013). New Jersey Clean Energy Program. Retrieved October 28, 2013, from http://www.njcleanenergy.com/commercial-industrial/programs/large-energy-users-program

NREL. (2003). *A Handbook for Planning and Conducting Charrettes for High-Performance Projects*. Golden, CO: National Renewable Energy Laboratory.

USEPA. (2013a). USEPA WaterSense. Retrieved October 28, 2013, from http://www.epa.gov/watersense/.

USEPA. (2013b). What are the Trends in Indoor Air Quality and Their Effects on Human Health?. Retrieved October 23, 2013b, from http://cfpub.epa.gov/eroe/index.cfm?fuseaction=list.listBySubTopic&ch=46&s=343.

USGBC. (2011). Public Policies Adopting or Referencing LEED. Accessed March 19, 2014, from http://www.usgbc.org/.

USGBC. (2013). LEED Online. Accessed March 19, 2014, from https://leedonline.usgbc.org.

SECTION II

ENVIRONMENTAL MANAGEMENT: THE BUSINESS AND SOCIAL SCIENCE PERSPECTIVE

8

BUSINESS STRATEGIES FOR ENVIRONMENTAL SUSTAINABILITY

AVINANDAN MUKHERJEE[1], NAZ ONEL[2] AND ROSITA NUÑEZ[3]

[1] College of Business, Clayton State University, Morrow, GA, USA
[2] School of Business, Stockton University, Galloway, NJ, USA
[3] Lorza Group, Rochester, NY, USA

Abstract: Environmental degradation caused by business activities is widely acknowledged all around the world. With the rising concerns over environmental destruction, individuals and businesses are becoming increasingly aware of the dangers that will occur if they do not carefully consider the needs of natural environment in every step they take. Thus, companies practice green initiatives at the firm or product level with the aim of introducing their efforts to help ease environmental impacts of their functions, such as reducing or eliminating ecologically harmful impacts of suppliers, productions, products, or end users. Developing successful business strategies in order to help sustain environmental well-being is an important way of adopting these attempts to satisfy needs of the stakeholders of business entities that can play a vital "bridging role" toward sustainability and today's necessary green lifestyles. This chapter elaborates on the growing importance of environmental sustainability for businesses and identifies different ways to adopt this necessary action for the survival of companies in today's evolving business era. Furthermore, the chapter explores the corporate social responsibility, which is also an important strategy for sustainable business success. In addition, the chapter looks at general business strategies and elaborates on the positioning businesses with sustainable practices. The chapter also includes a number of relevant important industry examples.

Keywords: environmental business sustainability, corporate social responsibility, ethics, sustainable supply chain management, green business, sustainable business, sustainable business strategy, greenwashing, green value chain, corporate sustainability reporting, Global Reporting Initiative (GRI), green consumption, green manufacturing, green technology, green finance, corporate social and financial performance.

8.1 Overview	196	
8.2 Introduction	196	
8.3 Michael Porter's Generic Business Strategies	197	
8.3.1 Cost Leadership	197	
8.3.2 Differentiation	198	
8.3.3 Focus or Niche Strategy	198	
8.3.4 Stuck in the Middle	198	
8.4 The Growing Importance of New Approach to Sustainability for Businesses	198	
8.4.1 Sustainability in the New Business Era	198	
8.4.2 Motivators for Companies to Become Sustainable	199	
8.5 Theories of Sustainable Business Strategies	200	
8.5.1 Corporate Social Responsibility (CSR)	200	
8.5.2 Shareholder Theory versus Stakeholder Theory Approach to Corporate Responsibility: Which One Is a Better Fit?	201	
8.5.3 Triple Bottom Line	203	
8.6 Sustainable Business Models and Implementation	203	
8.6.1 Ways to Green the Businesses	204	
8.6.2 Corporate Behavior on Sustainability	204	
8.6.3 Stakeholder Engagement in Sustainable Corporations	205	
8.6.4 Focus of Sustainable Corporations	205	
8.6.5 The Performance of Sustainable Corporations	205	
8.6.6 Relationship Between Corporate Social Performance and Corporate Financial Performance	205	

An Integrated Approach to Environmental Management, First Edition. Edited by Dibyendu Sarkar, Rupali Datta, Avinandan Mukherjee, and Robyn Hannigan.
© 2016 John Wiley & Sons, Inc. Published 2016 by John Wiley & Sons, Inc.

		8.6.7	Environmental Business Sustainability Development	207
8.7	Value Chain			208
	8.7.1	Michael Porter's Value Chain		208
	8.7.2	Greening the Value Chain		208
		8.7.2.1	Green Finance	208
		8.7.2.2	Green Technology	209
		8.7.2.3	Sustainable (Green) Supply Chain Management	210
		8.7.2.4	Green Manufacturing	211
		8.7.2.5	Green Marketing and Sales	211
		8.7.2.6	Green Consumption	212
8.8	Different Perspectives on Green Markets and Businesses			212
	8.8.1	Consumer Perspectives on Green Markets		212
	8.8.2	Company and Executive Perspectives on Green Markets		213
8.9	Corporate Sustainability Reporting			213
	8.9.1	Global Reporting Initiative (GRI)		214
		8.9.1.1	Introduction to GRI	214
		8.9.1.2	Industry Cases	214
8.10	Being Insincere: Greenwashing			220
8.11	Case Studies			222
	8.11.1	International Business Machines		222
		8.11.1.1	Conservation	222
		8.11.1.2	Climate Protection	222
		8.11.1.3	Product Stewardship	222
		8.11.1.4	Pollution Prevention	222
	8.11.2	Kraft Foods, Inc.		222
	8.11.3	Campbell Soup Company		223
		8.11.3.1	Campbell's Environmental Sustainability Policy	223
	8.11.4	Marks and Spencer		224
		8.11.4.1	Best Practices from Cases	224
8.12	The Future of Business Sustainability			226
8.13	Review Questions			226
References				226

8.1 OVERVIEW

The Earth does not have unlimited resources. Thus, it cannot provide for people's unlimited wants. With the rising concerns over environmental degradation, individuals are becoming increasingly aware of the dangers that will occur if they do not carefully consider the needs of mother Earth in every step they take. At the same time, companies that practice green initiatives—firm or product level—aim at introducing their efforts to help ease environmental impacts of their functions. These initiatives could be reducing or eliminating ecologically harmful impacts of suppliers, productions, products, and/or end users. Developing successful business strategies in order to help sustain environmental well-being is an important way of adopting these attempts to satisfy needs of the stakeholders of business entities (e.g., consumers, suppliers, stockholders, etc.). It plays a vital "bridging role" toward sustainability and today's necessary green lifestyles.

In this chapter, first, we will elaborate on the growing importance of environmental sustainability for businesses, and then, we will explore an important strategy for sustainable business success: corporate social responsibility. This chapter will also look at general business strategies as well as positioning businesses with sustainable practices. The chapter will also cover a number of relevant important industry examples.

8.2 INTRODUCTION

The best way to get people to take sustainability seriously is to frame it as it really is: not only a challenge that will affect every aspect of management but, for first movers, a source of enormous competitive advantage

Richard Locke
Director of the Thomas J. Watson Institute
for International Studies and
Professor of Political Science
Brown University
(Locke, 2009)

Environmental issues have slowly but surely moved up to the forefront of our social, political, and business agenda over the past few decades. Recent devastating disasters helped influence the environmental activists to revitalize the environmental movement with a faster pace. Hurricane Katrina, Superstorm Sandy, constant flooding in different regions of the world, water contamination, land degradation, and similar highly human-impacted environmental problems, in some cases less severe or more dreadful, have increasingly become a widespread social and fiscal subject matter.

The great challenge faced by nations today is to integrate economic growth with environmental sustainability and social welfare. Rapid growth of the middle class is causing fast-paced increase in consumption around the world. According to the World Economic Forum (2012), "each year until 2030, at least 150 million people will be entering the middle class. This will bring almost 60% of the world's population into a middle income bracket. Over the same period energy demand is projected to increase by 40%, and water demand is expected to outstrip supply by 40%." This tells us future human actions will be even more significant than today's in terms of impacting the planet. Recently, in response to alarming statistics, rapidly increasing number of business entities as well as consumers are showing a rising interest in sustainable initiatives all around the world to help ease the environmental impacts of elevated business practices and consumption habits.

The business aspect of environmentalism (corporate environmentalism) has constantly refined and matured from regulatory compliance to more sophisticated management and strategic concepts such as "pollution prevention," "industrial ecology," "life cycle analysis," "environmental justice," "environmental strategy," "environmental management," and, most recently, "sustainable development" (SD) (Bansal and Hoffman, 2012). Sustainability may be conceptualized as a social phenomenon that contributes to achieving development without deteriorating environmental conditions. It is a matter of a balanced approach. There is no doubt that development is necessary for the human beings to have better living conditions; however, this notion should not lead to a worsening of ecological conditions. If ecological conditions worsen or are unstable, then in actuality we would not be achieving development at all since living conditions would eventually become worse. For example, if we are achieving economic development and increasing the comfort in our lives by destroying the natural habitat of a certain species today, we would be adversely affected by this impact as a result of a chain effect that reaches into the future and worsens the living conditions of future generations. Subsequent generations would have to pay for we are doing today. In other words, if we are considerate about future generations in our pursuit of development, then we can say that we are being sustainable in our approach.

Various studies propose terms for sustainability and SD and provide insight into how they have become commonly used terms in today's industrialized world. According to Ebner and Baumgartner (2006, 2010), the origin of the term "SD" comes from the eighteenth century forestry. In that era, only a certain number of trees could be cut in order to guarantee a long-lasting tree population. This method ensured a continuous supply of wood without depleting resources for future generations.

The first scientific work on sustainability came from the not-for-profit organization, Club of Rome. After a meeting at MIT, a project was started to utilize computer-based modeling to describe various development paths. In 1972, the project published a report, "Limits to Growth," by Donnella Meadows, Dennis Meadow, and Jorgen Rander (Meadows et al., 1972). Eventually, the Club of Rome initiated an international discussion based on this report. In the course of discussion, an ecodevelopment approach was outlined that considered the protection of resources and the environment. This led to the SD mission statement that we have today (Baumgartner and Ebner, 2010).

In 1987, the United Nation (UN) World Commission on Environment and Development further defined SD as an ethical concept, and this definition has been widely adopted as follows: "Sustainable Development is a development that meets the needs of the present without compromising the ability of future generations to meet their own needs. It contains two key concepts: the concept of 'needs,' in particular the essential needs of the world's poor, to which overriding priority should be given; and the idea of limitations imposed by the state of technology and social organization of the environments ability to meet present and future needs" (p. 43)—cited in "Our Common Future" (World Commission on Environment and Development, 1987).

Both of these definitions have served as a foundation of the conceptualization of sustainability where SD balances economic growth with social equity and environmental protection. Thus, sustainability may be conceptualized as a social phenomenon that it contributes to achieving development without deteriorating environmental conditions. It is a matter of a balanced approach, in other words, "Enough, for all, forever." There is no doubt that development is necessary for the human beings to have better living conditions; however, this notion should not lead to a worsening of ecological conditions.

8.3 MICHAEL PORTER'S GENERIC BUSINESS STRATEGIES

Michael Porter's Generic (Competitive) Strategies
Michael Porter (1980) suggests that businesses can achieve a sustainable competitive advantage by adopting one of the three generic strategies: cost leadership, differentiation, and focus (or niche) strategy. There is also a "middle of the road" strategy, which has been considered to be the worst strategy for the long-term success of a company. In general, a company can adopt a strategy that is offensive or defensive with respect to competitive forces. A defensive strategy considers the structure of the industry as the way it is and positions the company accordingly by taking its strengths and weaknesses into account. On the other hand, an offensive strategy tries to alter the competitive environment by changing the underlying causes of each of the competitive forces. Although there are many specific strategies of each type (offensive/defensive), identifying which one would be the best for a specific company mostly depends on the circumstances. Porter suggests aforementioned three generic strategies for creating a defendable position for organizations to outperform competitors and become successful in the long run. Each of the three strategies is discussed below.

8.3.1 Cost Leadership

This strategy involves having the lowest per unit cost in the industry—that is, lowest cost relative to the industry rivals. For all production elements, such as materials and labor costs, the organization would aim to drive the costs as low as possible. Mostly, returns or profits would be low for these companies but nonetheless higher than competitors if it is a highly competitive industry. On the other hand, if there are only a few rivals, each firm can enjoy pricing power and high profits. Achieving cost leadership usually involves large-scale productions so that the companies can use the advantage of "economies of scale." A cost leadership business can create a competitive advantage and can be considered as a defendable strategy because it gives possibility to the company to defend itself against powerful buyers and suppliers. A good example to low-cost producer can be Walmart with its long-time cost leadership.

Achieving a low-cost position requires some important resources and skills for the companies, such as large up-front capital investment in new technologies, continued investment to maintain cost advantage through economies of scale and market share, innovations as well as development of cheaper and more effective ways to produce existing products, and monitoring the overheads and labor functions (incentive-based pay structure) very carefully.

8.3.2 Differentiation

Organizations can gain an edge over competitors by differentiating their offerings so that they appeal to customer segment or need more that other available products. According to Porter (1980), a differentiation strategy can focus on making the products stand out based on functionality, customer support, or quality. Differentiation can allow a firm to create or expand its market through offering a unique design, image, feature, or technology.

An organization that is engaged in a differentiation strategy can earn about average returns than its peers since differentiation helps to develop loyalty in customers and reduce the price sensitivity for products. Differentiation also can insulate the supply chain since the organization may be positioned to develop exclusive arrangements with suppliers and distributors.

Achieving a successful strategy of differentiation usually requires companies to have exclusivity, successful and strong marketing skills, product innovation as opposed to process innovation, applied R&D, high-level customer support, and less emphasis on incentive-based pay structure.

8.3.3 Focus or Niche Strategy

Organizations engaged in a focus strategy will direct all their efforts and resources toward one market segment, product segment, or geographical market and strive to become the leader in that category. The offering is often tailored to meet the needs specific to that market segment, to the extent where other firms may not be able to provide the same product as well. Examples can include Rolls-Royce and Bentley, which are both catering for the needs of people over the age of 50.

8.3.4 Stuck in the Middle

In the cases when businesses attempt to adopt all three strategies, cost leadership, differentiation, and niche (focus), those companies are called as "stuck in the middle." These kinds of companies usually do not have a clear business strategy and would want to cover everything to everyone. Unfortunately, this type of attempt usually increases running costs and cause confusion (it is rarely possible to please all sectors of a certain market). It has been widely suggested that these types of businesses that are stuck in the middle usually do the worst in their industry because of missing their critical business strengths.

In conclusion, to create a competitive advantage, a company should review its strengths very carefully and choose the most appropriate strategy for its success. This strategy could be cost leadership, differentiation, or focus. Whatever strategy a business decides to adopt, it needs to make sure that it is not stuck in the middle of the road. The main reason for this is that "one business cannot do everything well." Sustainability can contribute to competitive advantage by providing cost leadership, differentiation, or niche strategies for the companies.

8.4 THE GROWING IMPORTANCE OF NEW APPROACH TO SUSTAINABILITY FOR BUSINESSES

8.4.1 Sustainability in the New Business Era

As stated by Adam Werbach (2013) in his Harvard Business Press publication, the word "sustainability" was used by business leaders "to connote a company that had steady growth in its earnings" (p. 8) before it was cited in "Our Common Future" of UN World Commission on Environment and Development. But according to him, the meaning of the term has been diluted over time because of changing perceptions of the communities as well as business leaders. For instance, in some cases, the term "sustainability" is being used to describe organizations' philanthropic actions to save the environment. According to Werbach (2013), sustainable business can be described as "thriving in perpetuity." He also states that true sustainability should have four components with equal importance:

- Social: All members of the society should be considered carefully in every action. Society matters and true sustainability should take this into account, for example, considering public health, education, labor, and human rights.
- Economic: Actions should support operating in a profitable way. Economic needs of the people and businesses should be met. True sustainability helps with financial gain while protecting the other aspects of sustainability.
- Environmental: Actions and conditions that are related to ecological well-being should be considered. Protecting and restoring the ecosystem is as important as social, economic, and cultural aspects of sustainability.

- Cultural: Cultural diversity is an important merit of different societies. Protecting and valuing cultural diversity is another crucial component of true sustainability. Communities should be able to manifest their identities from generation to generation and carry on their traditions without losing them.

Sustainable strategies of the businesses should fulfill each one of these aspects, which are all integral to their long-term success. So, integrating environmental, social, ethical, and economic issues into all decisions of an organization can help achieve a balance of protecting the environment, creating economic development, increasing stakeholder value, and providing meaningful jobs, as part of creating a safe, just, and equitable society. These can be accepted as the core principles of today's business sustainability. Modern world demands for companies to carefully consider the finite resources on Earth, interdependence for the survival, need for minimizing inequalities in quality of life as well as social and economic well-being, necessity of respect for different cultures, and need for taking responsibility for leaving a legacy for future generations.

In order to be both sustainable and profitable, business strategy should be long term, recognize limits in resources, protect nature, transform current business practices, practice fairness, and enhance creativity. Sustainability can be incorporated into all aspects of business, including manufacturing, supply chain management, labor practices, transportation, facilities management, and more.

8.4.2 Motivators for Companies to Become Sustainable

There are several reasons why companies adopt sustainable business strategies. Many firms in different industries aim at gaining the early-mover advantage as they have to ultimately shift toward becoming green (Shafaat and Sultan, 2012). Some of the key benefits of sustainable business practices are listed below:

- *Reduced costs:* By reducing costs, it saves money for the companies (e.g., reducing wastes, eliminating unnecessary processes, or altering supply chain) in the long run. By using less natural resources, altering to a more efficient supply chain, using less chemical components, or reducing harmful wastes, companies can achieve substantial cost savings. Better utilizing company resources can put more money to company's bottom line that can be utilized in other functions of the company.
- *Market opportunity:* Greening of companies and products is a new market opportunity with fast-paced growth. It opens doors for new markets. It helps in accessing these new markets and getting all the benefits of competitive advantage. For example, in 2007, more than 36 million Americans spent over a quarter trillion dollars just on green products and services. This growing opportunity can be an advantage in terms of market share for businesses that can communicate their green initiatives successfully. Companies can grasp the opportunities involved in a "green" market estimated at US$209 billion (Shafaat and Sultan, 2012) by adopting virtuous green marketing strategies.
- *Source of differentiation:* The marketplace is increasingly competitive. Sustainability initiatives can provide a source of differentiation for companies, which they can slip off from their rivals and show how they are different. Companies can effectively differentiate themselves from the competitors in the marketplace by adopting various sustainability strategies.
- *Waste reduction:* Waste management can be problematic and cost sensitive for many companies. Adopting sustainable business actions can also help with eliminating the concept of waste, which can provide large amount of savings to companies.
- *Innovation and long-term growth:* It supports research, new ideas, and inventions—reinvention of business processes and the concept of products can be necessary to move toward green. These actions also help companies to stay tuned to new developments in the marketplace. It also ensures sustained long-term growth to the companies along with profitability.
- *Employee motivation:* It helps the company employees to feel connected, proud, and contented to be a part of an environmentally responsible and responsive company. It helps with engaging the team in a corporation and leads to a more motivated and productive employees. In general, employees are more willing to work for those companies that care for the environment.
- *Competitive advantage:* It leads to outperforming the industry average by gaining competitive advantage. This way, companies can outperform industry average. For example, recent research (Kearney, 2009) on 100 companies around the world showed that 16 of the 18 industries with sustainable business practices outperform their competitors by 15%. A company can do a lot with an extra 15% in its bottom line.
- *Environmental protection:* Survival of every company depends on healthy natural environment. Sustainable business models helps with preserving the ecological balance. Natural resource protection could be possible with widespread sustainable business activities, which can lead to better utilization of natural resources.
- *Increased corporate social responsibility (CSR):* Social responsibility performance of companies can increase with the help of green strategies. Many companies around the world are starting to realize the importance of achieving both environmental- and profit-related objectives. Green ideology adoption by a company can demonstrate company's willingness to engage in

socially responsible practices. It also shows that the company is meeting accountability demands of shareholders, the public, and governmental bodies. In some cases, companies move a step forward by undertaking more than what is expected by stakeholders, which helps generate consumer trust.

- *Increased consumer awareness:* Sustainable business actions can lead to increasing public awareness toward the planet and its conditions, as well as providing information on what each individual can do about it. This can lead to responsible consumption habits.
- *Responsiveness:* It demonstrates companies' willingness to respond consumer demands. It is crucial to be able to respond consumer demand promptly in the marketplace for companies to capture the market share and be successful. Environmental sustainability help companies to meet today's consumer demand in the context of green products and activities.
- *Corporate reputation and credibility:* Good intention would always bring good reputation for the companies. Sustainability actions can be a crucial step to improve reputation and gain customer trust. One of the most important aspects of sustainable business strategy adoption is making communication of genuine green actions and approaches by corporations visible to the public. This kind of direct communication methods would increase the companies' reputation observed by stakeholders, stockholders (investors prefer reputable companies), and government agencies. Elevated corporate reputation can change overall corporate pathway toward a more successful journey with its competitive advantage and increased profits.
- *Added value:* Companies can also add value by improving employee retention or motivation through sustainability activities or by raising prices or achieving higher market share with new or existing sustainable products. US-based Whole Foods Market, for instance, raised its sales by 13% a year from 2005 to 2009, in an economy experiencing single-digit growth.
- *Meeting regulations:* In recent years, increasing range of rules and regulations related to environmental and human health protection and/or fortification framed by the government to protect environment, consumers, and the society at large are putting into place. Many governmental bodies all around the world are working diligently to reduce or eliminate the production of harmful goods and by-products, which would lead to reduced manufacturing and consumption of destructive goods. Companies can meet these laws and regulations by adopting a sustainable pathway.
- *Building consumer trust and loyalty:* Similar to relationship marketing, green initiatives help with creating (i) trust, (ii) value, and (iii) confidence for customers. It demonstrates the company's level of involvement with society and moral obligation to be socially responsible, which are important aspects of gaining trust and loyalty.

8.5 THEORIES OF SUSTAINABLE BUSINESS STRATEGIES

8.5.1 Corporate Social Responsibility (CSR)

The concept of corporate social responsibility (CSR) has evolved since the 1970s. The changes occurred in our understanding of responsibility and also in the implementation by management (Moir, 2001). Frederick (1986, 1994) describes the evolution as starting with an assessment of firms' obligations to care for society's improvement—at this point, the question is whether or not the corporation had a responsibility to society. He calls this phase of CSR development CSR1. He notes that during the 1970s, corporations were moved to respond to social pressures, in a phase that he referred to as CSR2. The question became will the corporation respond and in what manner. Frederick expanded his description of CSR development to the current phase, which includes an ethical perspective in managerial decision making, which he refers to as CSR3. Frederick noted that the interaction between society and business needed to be viewed with an ethical focus to allow a systematic review of the impact of business on society.

Cannon (1992) examined the development of CSR in light of the legacy of the interaction between business, society, and government that became prevalent in the late twentieth century. The relationship between business and society was changed by new social norms and needs that required business to become involved with issues such as equal employment opportunities, pollution abatement, poverty alleviation, and others that had been in the domain of government in prior years. Cannon put forth that social value responsibility could be categorized as legal, moral, ethical, or philanthropic.

Harvard Business Review (2003) characterizes the current perspectives on CSR as "corporate philanthropy." Many large companies have jointly developed strategies to gain a competitive edge via their philanthropic activities. These strategies also help the corporations to gain brand recognition, boost productivity, reduce costs, and be viewed favorably by regulators.

Companies approach CSR differently to meet their varied perspectives on their responsibility to society. However, they are all using CSR as a means to concurrently manage business while having a positive impact on society (Samy et al., 2010).

There are different arguments on understanding of social responsibility by business executives. The most known one comes from Milton Friedman. Friedman supports the idea that the social responsibility of business only aims at increasing profits, and all the functions in terms of reflecting social responsibility of the company mainly try to achieve this aim. Friedman in his article "The Social Responsibility of Business is to Increase Its Profits" (Friedman, 1970) defines social responsibility as the spending of "corporate resources for socially beneficial purposes regardless of whether those undertaken expenditures are designed to help achieve the financial ends" (p. 298) of the firm. In this article published in the *New York Times Magazine*, he mainly argues that (i) executives of the corporations are only the employees of the business owners, the stockholders; (ii) these executives' job is mainly maximizing the profits of the company; (iii) stockholders authorize these executives only to do their job, which is stated "maximizing the profits"; (iv) these profits that they try to maximize belong to the owners (stockholders); (v) in general terms, "social responsibility" and "responsibility to increase profits" have different meanings; and (vi) if the executives spend company funds on socially responsible acts, this means that these executives are spending either owners', employees', or customers' money without their permission. According to Friedman, this kind of a spending is wrong and unacceptable. Consequently, these reasons lead him to advocate that those who support the idea of "social responsibility" of business executives in addition to maximizing the profits of the company are incorrect. In conclusion, he highlights the shareholder theory approach to be true by expressing that corporations do not have the obligation to exercise social responsibility. Friedman's approach typically undertakes the value maximization of the shareholders as the ultimate reference for corporate decision making.

To explain Milton Friedman's argument in more detail, we can look at the article from Brian P. Schaefer (2008), which presents a critique and broad analysis of Friedman's argument on understanding of social responsibility by business executives.

Schaefer in his article "Shareholders and Social Responsibility" states that Friedman constructs an incorrect argument as a consequence of focusing on the role of the managers and executives of the corporations, while he should be more concerned with the duties of shareholders. The author reveals Friedman's implied position on the duties of shareholders, which is considering shareholders to have "no duty" to guide the managers toward adopting social responsibility. Schaefer argues that this approach is irreconcilable with the foundations of some of the most widely respected, accepted, and influential moral theories in the West, namely, utilitarianism, or virtue ethics. He defines this as "convergence strategy, whereby multiple influential moral approaches are shown to align themselves against Friedman" (p. 297). According to Schaefer, this strategy illustrates that the position of Friedman lies on the outer fringes of mainstream Western moral beliefs "and that at least some of Friedman's professed adherents appear to offer incoherent moral views" (Schaefer, p. 297). Thus, when we approach the subject matter within the context of moral terms, Friedman does not appear to support a logical argument.

Although it was first stated over 30 years ago, Friedman's argument remains one of the most influential, and cited piece of work on shareholder theory, which is explained more in detail in the following section. And it seems that the debate will remain the discussion focus of many more scholars and researchers in the future.

8.5.2 Shareholder Theory versus Stakeholder Theory Approach to Corporate Responsibility: Which One Is a Better Fit?

There are two dominant business theories that often support conflicting views on corporate responsibility. According to the shareholder theory, the purpose of the corporation is to maximize the corporation's profits by realizing the specified ends of shareholders, which are identified as nondeceptive and legal (Schaefer, 2008). On the other hand, the stakeholder theory holds that the corporations should consider all the stakeholders' benefits despite whether doing so affects the profits positively (it may not maximize the profits). Corporations' stakeholders can be owners, management, employees, customers, and suppliers all of whom are crucial to the corporations' survival and success in the marketplace. The underlying concept of these two definitions is considered to be the moral issue (Schaefer, 2008), in which was not taken into account carefully by Friedman.

As mentioned earlier, Friedman expresses the shareholder theory by suggesting that corporations do not have the obligation or duty to exercise social responsibility. According to Hasnas (1998), social responsibility in this context is spending on socially beneficial purposes from corporate resources, whether it helps to achieve the maximization of profits or not. Then, it is easy to infer that Friedman's view does not support the stakeholder theory approach to corporate responsibility.

In the article "A Stakeholder Theory of the Modern Corporation," R. Edward Freeman discusses stakeholder theory and how it is put into practice in the modern world of business. In his argument, Freeman (2004) suggests that today's managers have more constraints placed on them in the way that they may pursue stockholders' interests at the expense of stakeholders' interests. Freeman's argument from a legal standpoint is that this is a result of pressures and changes in regulations, such as Clean Air Act and National Labor Relations Act. From an economic point of view, when management is cognizant of some stakeholder interests at the expense of other stakeholder's interests, it is possible for them not to be able to maximize the profits of the company to the greatest extent (e.g., tragedy of the commons, monopolies). Furthermore, he believes that the dominance of the stockholders is inaccurate, which is also the combined result of the first two arguments. He also suggests that all of the six stakeholders—stockholders, managers, suppliers, employees, customers, and the local communities—have the right not to be treated in a manner that would put them at a lower level than the others. This means that they have similar rights when it comes to consideration by management. Another important point he makes is that all the aforementioned stakeholder groups should take part in determining the direction of the company's future in which they have a stake. If different groups of stakeholders have conflicting stakes, it should be possible to resolve the conflict by different methods such as appealing to the Doctrine of Fair Contracts (i.e., normative core).

Hence, in the case of considering business functioning as a more balanced practice in terms of benefits gained, we could say that the stakeholder theory provides an opportunity for each of the stakeholders, such as stockholders, suppliers, employees, customers, and the local communities, to have their interests addressed. For instance, Freeman (2004) states that companies should be managed with the aim of balancing benefits of different stakeholders if they want to be successful. According to Freeman, balancing stakeholders' claims is one of management's top priorities and should constantly be evaluated over the course of the firms' existence (according to the Doctrine of Fair Contracts). This idea of a balanced approach echoes the stakeholder theory.

Under the discussion of the difference between the shareholder theory and stakeholder theory approach to corporate responsibility, Boatright (1994) follows a different approach. In his article, he talks about the logical gap between shareholders' property rights and the management's fiduciary duties toward the shareholders. After examining both sides of the argument, Boatright states that stakeholders are vulnerable because while stockholders can get rid of the stocks they own at any time, other stakeholders cannot end their relationship with the firm as easily and therefore are more likely to suffer due to bad circumstances. Furthermore, as determined by the corporate governance laws, shareholders' shares related to the business are offered on the basis of a "take it or leave it" rule, which gives them the option to move on if they are not happy with their shares. This gives them a privileged position compared to other stakeholders, who might not have the "luxury" of deciding how to proceed in the future. For instance, a local community that is being harmed by a firm's polluting activities might not have the option to relocate from where the firm operates.

Considering the aforementioned discussions on both sides, we suggest that the stakeholder theory approach is a better model of CSR compared to the shareholder theory. The shortcoming of the shareholder theory comes from considering stockholders as the only stakeholders who matter. That is the reason managers who adopt the shareholder view of the company cannot have a fair judgment when it comes to achieving corporate social responsibility. They would choose to increase the profits of the company no matter what happens to the rest of the stakeholders. Thus, we can say that the shareholder theory cannot function in parallel with social responsibility of corporations.

Shaefer's discussion of the stakeholder theory leads to the conclusion that is preferable to the shareholder model because of the general expectation of corporations' duty to exhibit social responsibility. We interpret this as doing what is morally right and is a valid reason to prefer the stakeholder theory to the shareholder theory, and so firms that operate on a shareholder model have good moral reasons to contemplate replacing it with a stakeholder model. As mentioned, briefly, the shareholder view holds that the purpose of the corporation is to realize the specified ends of shareholders, with the caution that those ends are legal and mainly nondeceptive. In practice, those ends are almost always to maximize the corporation's profits. On the other hand, the stakeholder theory holds that the firm should be managed for the benefit of all stakeholders, regardless of achieved profits. If a company wants to survive for the long term, it should not neglect its responsibilities toward all of its stakeholders.

Interestingly, Shaefer makes a suggestion regarding the shareholder theory that might be difficult to observe very often in real-life situations. He suggests that we must allow for the possibility that social responsibility can and may be sufficiently fulfilled in the shareholder model as well. Hence, he thinks that it may be better to leave open the possibility that some kind of shareholder theory will correspond to a morally acceptable framework for some corporations. This tells us that the decision on which theory would be a more appropriate (stakeholder or shareholder model) fit for the corporate responsibility will depend on the specific characteristics of the model that is adopted and may change on a case-by-case basis (e.g., moderate, nuanced versions of the stakeholder theory may be a better fit than a radical one).

One of the challenges that corporations face today is in maintaining a good image and reputation. Globalization and the widespread availability of technology have made it more difficult for corporations to manage information that is communicated externally (Wheeler et al., 2003). Potentially damaging information can be disseminated within minutes. In this case, the stockholders are affected adversely from the damaged reputation and its ensuing diminishing profits. Positive reputation and good image have emerged as valuable assets of a company (Roberts, 2003) that would not only produce a higher profit, but also help maintain the direction of corporate evolution. Therefore, today, an increasing number of companies are trying to incorporate more socially responsible models into their business strategies. In fact, according to Biloslavo and Trnavcevic (2009), increased company image and reputation can stimulate and change consumer's future purchase, loyalty, and demand behaviors.

Whether the stakeholder or the shareholder theory is dominant in a corporate strategy, adopting socially responsible models can contribute to improved reputation, goodwill, and added value for all groups of stakeholders in the long run.

8.5.3 Triple Bottom Line

It is increasingly apparent, following the aforementioned discussion, that we need to think in terms of building a new approach to CSR. In our argument, we suggest that none of the approaches (i.e., shareholder theory, stakeholder theory) answer the underlying need for a new approach to CSR. Therefore, there is a need to advance the "sustainability theory of business" as a response to this emerging need.

The sustainability theory of business encompasses "a balance between ecological, social and economic indicators that ensures the maintenance of the environment and society over time" (Boxer, 2007, p. 87). When doing this, financial gain is achieved as well. So, it meets the needs of stockholders, which is the main focus and aim of the shareholder theory. In this sustainability model, production and consumption of resources also need to be planned and managed in a way that produces not only an economic gain but also contribute to social and environmental well-being of the communities. This ideology of sustainability has a parallel notion with stakeholder theory. The approach we propose could be perceived as a similar notion to Elkington's (1998). His suggestion, named triple-bottom-line (TBL) (also known as the three pillars: people, profit, planet) approach, calls for three elements, people, profit, and the planet, to be considered in business activities to measure success.

Several corporations have taken up the call for incorporating the social equity factor into their sustainability and SD plans. Operating under the guise of the TBL approach, corporations have devised community-oriented strategies to participate in creating a more just and equitable society. According to Laff (2009), corporate TBL approaches include the design of new models that go beyond mere donations or one-time community service projects to broader community activating strategies that include educational grants for local schools, to responsibly sourcing, pricing, recycling, and reusing materials obtained from foreign suppliers. Corporations are placing an increased emphasis on the image they portray to the greater community and how they are viewed in the marketplace in general as opposed to strictly focusing on how they are financially performing within the marketplace (meaning that not supporting the shareholder theory). By extension, as mentioned earlier, corporations are recognizing that to improve their image, they must integrate social equity and justice factors into daily operations within the corporate structure. This necessitates an emphasis on employee satisfaction, involvement, and participation, which ensures that TBL and social equity factors specifically are ingrained in as opposed to merely appended to strategic plans after economic and environmental considerations have been taken into account. Therefore, a greater emphasis is placed on positively impacting the lives of employees and society at large (i.e., all the stakeholders) in ways that benefit not only the corporation, but local and global communities. This approach, thus, covers the part of the stockholder theory, as well as the stakeholder theory, which gives us a better model to execute.

There is no doubt that TBL as an approach can refine management insight on sustainability and, as a result, help corporations to develop business strategies to effectively address corporate social responsibility. Such an approach would strongly resonate with the suggestion by Garriga and Mele (2004), who stated that "business should be neither harmful to nor a parasite on society, but purely a positive contributor to the wellbeing of the society" (p. 62). This kind of a common good approach could be an effective source for long lasting achievement for companies.

8.6 SUSTAINABLE BUSINESS MODELS AND IMPLEMENTATION

Sustainability is now decisively on business management agendas worldwide, with corporate web pages featuring press releases and reports on sustainability performance. Some companies provide sustainability reports as separate documents to stakeholders, while some are integrating key sustainability performance metrics into their annual reports. Opponents to corporate engagement in sustainability suggest that it can be perceived as misappropriating and misallocating corporate resources from activities that increase shareholder value (Margolis and Walsh, 2003). Friedman (1970) criticized CSR as a form of theft and political subversion, with executives diverting resources away from shareholders, employees, and customers to accomplish social goals that should not be in the domain of privately owned enterprises, but should be addressed by government.

Despite these views, sustainable business models are positioned at the intersection of business that operates in a sustainable manner toward the environment and society and can also generate enough profit to remain viable and provide employment and goods and services. The primary concept behind this approach to business is that it is possible to do good while being successful.

Sustainable business models aim to reduce the negative social and environmental impacts of industry and enhance the positive. The natural capital model, proposed by Lovins et al. (1999), aims to move sustainable business management past efficiency improvements into areas that focus on increasing productivity of natural resources, increasing yields of biological production models, changing business models, and reinvesting in natural capital. However, in order for this approach to succeed, the established incentive systems that corporations use to measure performance, award employees, and set goals would need to be revised to incorporate sustainability accounting and reporting metrics that may focus more on goals such as natural resource conservation and improving work conditions.

Hart and Milstein (1999) also support the proposal by Lovins et al. They put forth the view that SD can be a driver for renewal and progress and that different types of economies require different business and sustainability strategies that match the production and consumption pattern of the marketplace.

For emerging and survival economies, social and technological leapfrog innovations are necessary to avoid the weaknesses in traditional business strategies and to develop sustainable business practices (Hart, 1997). For consumer economies in highly industrialized nations, Hart (1997) recommended that sustainable business models focus on minimizing social and ecological impacts and reducing environmental footprints.

These new sustainable business models can minimize negative social and environmental impacts while achieving SD. By incorporating sustainable practices into business models, industry can innovate and create value for the long term while caring for the natural world.

8.6.1 Ways to Green the Businesses

Firms can "green" themselves in three different ways: value-addition processes, management systems, and products:

1. *Value-addition processes*: Greening the value-addition processes could entail redesigning them, eliminating some of them, modifying technology, and/or inducting new technology—all with the objective of reducing the environmental impact (e.g., steel firm may install a new technology, thereby using less energy to produce steel).
2. *Management systems*: Adopting management systems that create conditions for reducing the environmental impact of value-addition processes (e.g., responsible care program of the chemical industry).
3. *Products*: The product level "greening" of firms could take place in the following ways:

    (i) Repair—Extend the life of a product by repairing its parts.
    (ii) Recondition—Extend the life of a product by significantly overhauling it.
    (iii) Remanufacture—The new product is based on old ones.
    (iv) Reuse—Design a product so that it can be used multiple times.
    (v) Recycle—Products can be reprocessed and converted into raw material to be used in another or the same product.
    (vi) Reduce—Even though the product uses less raw material or generates less disposable waste, it delivers benefits comparable to its former version or to competing products (Prakash, 2002).

8.6.2 Corporate Behavior on Sustainability

Eccles, Ioannou, and Serafeim, in their 2012 *Harvard Business Review* article "The Impact of a Corporate Culture of Sustainability on Corporate Behavior and Performance," classify firms into high sustainability and low sustainability types (Eccles et al., 2012).

High sustainability corporations are those that voluntarily adopted environmental and social policies many years ago. Ninety companies were identified in this research as high sustainability companies with a substantial number of environmental and social policies that have been adopted for a significant number of years since the early to mid-1990s (Eccles and Krzus, 2010)

Low sustainability companies exhibit fundamentally different characteristics from a matched sample of firms that adopted almost none of these policies.

The study shows that *high sustainability* firms perform better when we consider accounting rates of return, such as return on equity (ROE) and return on assets (ROA). Furthermore, according to the study, this outperformance is more pronounced for firms that sell products to individuals, compete on the basis of brand and reputation, and make substantial use of natural resources.

The emergence of a corporate culture of sustainability raises a number of fundamental questions for scholars of organizations:

- Does the governance of sustainable firms differ from traditional firms, and if yes, in what ways?
- Do sustainable firms have better stakeholder engagement and longer time horizons?
- How do their information collection and dissemination systems differ?

High sustainability firms use advanced environmental scanning and respond better to regulatory changes compared to low sustainability firms. Companies that actively invest in technologies to reduce their greenhouse gas (GHG) emissions make a bet on regulators imposing a tax on GHG emissions. Similarly, firms that invest in technologies that will allow them to develop solutions to reduce water consumption make a bet on water receiving a fair market price instead of being underpriced.

Another example comes from Intel Corporation. The company has invested more than $1 billion in the last decade to improve education globally. In 2010, in conjunction with US President Obama's "Educate to Innovate" campaign, Intel announced a $200 million commitment to advance math and science education in the United States.

8.6.3 Stakeholder Engagement in Sustainable Corporations

High sustainability firms will be more actively engaged and more likely to be held accountable for reviewing the environmental and social performance of the organization. Furthermore, *high sustainability* group are more likely to use monetary incentives to help executives focus on non-financial aspects of corporate performance that are important to the firm. For instance, Intel has linked executive compensation to environmental metrics since the mid-1990s, and since 2008, Intel links all employees' bonuses to environmental metrics.

Sustainable growth encompasses a business model that creates value consistently with the long-term preservation and enhancement of financial environmental and social capital. Also, studies show that adoption and implementation of sustainability policies reinforce a distinct type of corporate culture over the years. Plus, the firms that have instituted a culture of long-term relationships with key stakeholders can be better positioned to pursue the goals more efficiently. Additionally, high sustainability firms are more likely to provide feedback from their stakeholders directly to the board or other key departments. Therefore, firms with a culture of sustainability appear to be more proactive, more transparent, and more accountable in the way they engaged with their stakeholders.

8.6.4 Focus of Sustainable Corporations

High sustainability firms are focused more on a longer-term horizon in their communications with analyst and investors. A company communicates its norms and values both internally and externally. Therefore, high sustainability firms are significantly more likely to attract dedicated rather than transient investors.

Compared to those companies lack sustainability focus, sustainable firms are significantly more likely to measure execution of skill mapping and development strategy (54.1% vs. 16.2%), the number of fatalities in company (77.4% vs. 26.3%), and the number of "near missing" on serious accidents in company facilities (64.5% vs. 26.3%). Also, *high sustainability* firms focus on customers more than *low sustainability* firms. *High sustainability* firms have positive and significant relation between their employees and suppliers. Furthermore, the study from Harvard Business School (HBS) shows that 41.1% of the *high sustainability* firms have a global sustainability report compared to only 8.31% of the *low sustainability* firms. It is possible that *low sustainability* companies are not willing to share their actions on sustainability with a worldwide report.

Finally, the results of HBS's study showed *high sustainability* companies provide more social and environmental information to their customers and stakeholders. According to the results, 25.7% of the *high sustainability* firms integrate social information and 32.4% integrate environmental information. In contrast, only 5.4% of the *low sustainability* firms integrate social information and 10.8% integrate environmental information.

8.6.5 The Performance of Sustainable Corporations

High sustainability firms outperform *low sustainability* firms on the stock market significantly. For instance, investing $1 in the beginning of 1993 in a value-weighted portfolio of sustainable firms would have grown to $22.6 for *high sustainability firms* as opposed to $14.3 for *low sustainability firms* by the end of 2010, based on market price. In contrast, investing $1 in the beginning of 1993 in a value-weighted portfolio of traditional firms would have only grown to $15.4 for *high sustainability firms* as opposed to $11.7 for *low sustainability firms*. So, considering stock market, *high sustainability* firms were able to outperform the *low sustainability* group by 4.8% (Eccles et al., 2012).

Furthermore, it was found that sustainable firms outperform traditional ones based on accounting rates of return (ROE). Overall, *high sustainability* firms are able to significantly outperform compare to *low sustainability* firms.

The only condition assumed that *high sustainability* firms might underperform because they experience high labor costs by providing excessive benefits to their employees. For instance, companies with a culture of sustainability face tighter constraints in how they can behave. Since firms are trying to maximize profits subject to capacity constraints, tightening those constraints can lead to lower profits.

Still, the results by Eccles et al. (2012) showed that *high sustainability* companies exhibit lower volatility and obtain better ROA and ROE than the low sustainability companies.

8.6.6 Relationship Between Corporate Social Performance and Corporate Financial Performance

There are numerous studies in the literature that investigate the relationship between corporate social performance (CSP) and corporate financial performance (CFP). This relationship is important to businesses and policymakers to demonstrate that profitability and social responsibility do not have to be viewed as competing objectives, but can be achieved symbiotically.

Results of these studies have been mostly supportive of a positive CSP–CFP relationship; however, there have been inconsistencies. Some of the ambiguity has been attributed to the use of unreliable CSP measures, missing control variables, or assumed linearity without valid testing.

One of the most comprehensive studies of the CSP–CFP relationship, "Corporate Social and Financial Performance: A Meta-analysis" (Orlitzky et al., 2003), examines the issue with an integrative, quantitative study. The authors aimed to answer the question raised by Business Week in 1999, "Can business meet new social, environmental, and financial expectations and still win?" The meta-analysis included 52 studies, yielding a sample size of 33,878 observations. The specific objectives of the study were to (i) provide a statistical incorporation of the research that accumulated over the years on the relationship between CSP and CFP, (ii) assess the instrumental stakeholder theory in the context of CSP–CFP relationship, and (iii) evaluate various moderators.

The results of this meta-analysis showed that in the literature, across studies, CFP is positively correlated with CSP and that this relationship is not one way, but tends to be bidirectional and simultaneous. Furthermore, reputation is an important mediator of the relationship. In terms of cross-study variations, the authors found that stakeholder mismatching, sampling error, and measurement error could explain between 15 and 100% of the discrepancies in various subsets of CSP–CFP correlations.

Based on this meta-analysis integrating 30 years of research, the authors found that the answer to the question posed by Business Week is affirmative. The results of this meta-analysis show that there is a positive association between CSP and CFP across industries and across study contexts. In conclusion, the meta-analytic findings of the study suggested that corporate virtue in the form of social responsibility and, to a lesser extent, environmental responsibility is likely to pay off, although the operationalization of CSP and CFP also moderated a positive association. This meta-analysis, with a large-scale approach, established a greater degree of certainty with respect to the CSP–CFP relationship.

In one of the few longitudinal studies of the CSP–CFP relationship, Brammer and Millington (2008) examined the relationship within the context of corporate charitable giving, which is a particular component of CSP over different time horizons. The authors used a distinctive empirical approach where they chose this specific corporate philanthropic activity component, charitable giving, because it provides highly transparent insight into corporate strategy in the context of social responsiveness. It is thought that because philanthropic activities in most companies are subject to the control of the main board of directors who have a high degree of external visibility, this component plays a key role in shaping the perceptions of the company in the eyes of external stakeholders. Furthermore, these activities are neither intimately linked with operational functions of the company nor subject to legal compliance issues; therefore, they can be considered as a strong measurement of CSP. The empirical analysis covered a longitudinal data sample extracted from the annual reports of 537 UK companies quoted on the London Stock Exchange in 1999.

This study's contribution to the literature comes from utilizing four principal innovations in examining the link between CSP and CFP. First, the authors applied a two-stage empirical approach that builds on a method used widely in social science research. This method draws a distinction between the observed level of a given phenomenon and a level that might be expected on the basis of a range of predictors. This technique let the authors evaluate the degree to which a firm's charitable giving differs from that predicted on the basis of its characteristics, such as its size, industry, R&D, and advertising intensity. The authors then used these deviations from "normal" levels of charitable giving to construct a typology of firms. As a result, they were able to identify three groups of companies in terms of social responsiveness strategy: those firms with unusually high CSP, those firms with unusually low CSP, and those firms with normal CSP.

In the second stage of the analysis, the authors examined the financial performance characteristics of firms grouped according to their social responsiveness strategy. This two-stage methodology allowed them to examine the link between corporate social responsiveness strategy and financial performance while providing an efficient means of discriminating between the predictions of alternative conceptual models of the CSP–CFP relationship.

The basic model of the influences on corporate charitable giving developed in the study hypothesized that company contributions are a function of industry, firm size, labor intensity, and the availability of financial resources. For measuring financial performance, the authors used market-based performance measures. Their key findings concerned the significance of deviation from "expected" or "optimal" rates of social performance for a firm's financial performance and the longitudinal variability in the link between CSP and CFP. The results of the study showed that firms with "unusually high" and "unusually low" social performance have much higher financial performance than other firms. Furthermore, the results showed that firms with "unusually poor" social performance are doing best in the short run, and those with "unusually good" social performance are doing best over longer time periods.

The positive relationship between CSP and CFP is further supported by a study conducted by Callan and Thomas (2009). Though a majority of studies indicated that CSP is a determinant of CFP, some other aspects, such as how this affect occurs and which variables are the determinants, have been inconsistent.

The main finding of the study was a positive relationship between CSP and CFP. This study differed from earlier ones by specifying proper controls, such as capital investment, and measures of size, industry classification, research and development (R&D), and advertising. An important finding in this was that not all industries had consistent results in terms of financial returns to social responsibility activities. The authors found high variability across industries.

Soana (2011) further supports the suggestion that industry classification could influence the CSP–CFP relationship. An empirical analysis of 21 international and 16 Italian banks rated in 2005 by Ethibel (Ecolabel Index) and AXIA (an ethical rating agency operating in Italy), respectively, was conducted to examine the relationship. The global ethical rating helped the author to construct a relationship between the firms' financial performance and their global CSP. In contrast, analytical ethical ratings helped to examine each single component of ethics and the CFP relationship. For CFP, accounting and market measures were taken into account. The accounting measures were obtained from BankScope, which contains comprehensive information on banks throughout the world. From this database, the author gathered each bank's cost-to-income ratio, return on average equity (ROAE), and return on average assets (ROAA). While the cost-to-income ratio showed the efficiency of the company, the ROAE and ROAA measures provided information on the profitability. By taking the means of three market measures given as the market-to-book value, the price-to-book value, and the price–earnings ratio (gathered from Datastream), the author was able to come up with a combined market performance value.

The results of the study showed no statistically significant relationship between accounting ratios, market ratios, and global ethical ratings. Furthermore, the results did not support a statistically significant relationship between the ethical parameters included in the study (internal social policy, external social policy, economic policy, environmental policy) and accounting and market ratios for any of the banks. There were no negative correlations for these relationships. The author suggested that it may be possible that this specific sector (banking) resulted in these particular findings regarding the relationship between CSP and CFP, making a case for sector-specific evaluations of the CSP–CFP relationship.

This is supported by other researchers. For example, Beurden and Gossling's (2008) literature study on the possible correlation between corporate social and financial performance identifies industry as an influencing factor. Their literature review also determines several other factors, such as firm size, risk, and R&D to be important factors that influence the relationship between CSP and CFP. Therefore, it is possible to see varying results if a certain study focuses exclusively on small companies or high-risk companies, similar to focusing on only one industry (e.g., banks).

A question that can be raised is why is it that in some industries, CSP and CFP are not correlated. This may be due to the fact that dependence on stakeholders differs across industries and this causes different results. In fact, Mitchell et al. (1997) identified that each individual stakeholder group's power, urgency, and legitimacy influenced policies and practices of CSR differently for diverse industries. Thus, different industrial characteristics would shape CSR via variations in stakeholder group forces, and we can expect different results for various industries. This could explain why study of the banking industry did not show a significant correlation between CSP and CFP, as they likely had a different stakeholder group influence shaping banking CSR.

This brief review of the literature highlights the gap in terms of examining the CSP–CFP relationship from a sector-specific perspective, since industry membership can be a moderating effect. Another important issue is the need to conduct more meta-analyses. This approach would make it possible to develop a universally accepted theory, which at the moment seems a difficult task considering the varying outcomes by industry and various effects (e.g., size, risk, or measurement discrepancies). Therefore, meta-analyses should also focus on these elements in order to accurately explain the phenomenon. A unified theory might be difficult to achieve, but it is necessary, as recognized by several researchers (Simpson and Kohers, 2002; Marom, 2006; Soana, 2011). Although Marom (2006) tried to identify a unified theory, because of difficulties in measuring and mapping stakeholder utilities that depend on many variables, such as characteristics of a particular industry stakeholder group, he recommended first focusing on these areas to clarify the path toward a unified theory. Thus, we believe a meta-analysis of the relationship between CSP and CSF with a specific focus on different industries would contribute significantly to the literature.

8.6.7 Environmental Business Sustainability Development

Approaches to sustainable business development require a firm to identify where it is making a social and environmental impact and then evaluate its alternatives for reducing the impact while growing business and meeting customers' needs. As an example of this approach, Nestle launched a plan in 2009 that coordinates activities to promote sustainable cocoa: producing 12 million stronger and more productive plants over the next 10 years, teaching local farmers efficient and sustainable methods, purchasing beans from farms that use sustainable practices, and working with organizations to help tackle issues like child labor and poor access to health care and education.

Impact assessment plays a significant role in quantifying the impact, scope, and frequency of a firm's actions on the natural environment. Many corporations embark on sustainable business development by following a standardized framework, for example, the widely adopted ISO 14001, which provides guides to policy and procedures and standards and systems (Croner Publications Ltd, 1997).

In the early stages of sustainable business development, many corporations directed their efforts toward the everyday activities that could reduce environmental impact, such as recycling paper, improving production efficiencies, using light sensors, and reducing business travel. These were mainly inward-looking activities, and the benefit of these efforts could easily be seen in the bottom line.

As corporations and consumers became more aware and concerned about the state of the environment and the impact of industry on ecological systems, many companies were pressured both internally and externally to be more environmentally sensitive in their operations. This required looking beyond the direct internal efforts and at the sphere of influence of the organization and its products. At this point, firms had to identify their key stakeholders and assess how each group could contribute to reducing the environmental impact at each stage of the value chain. External stakeholders, such as suppliers, were required to provide sustainable alternatives, while customers were engaged in initiatives to recycle and reuse products. Internally, firms started to overhaul product design and optimize logistics in manufacturing and distribution to use energy and materials more efficiently. Life cycle assessment and analysis were performed to determine the impact of a product beyond the time horizon of a single owner or its useful life.

At this point, the use of standardized benchmarks, licensing, and certifications became necessary to allow companies to compare and differentiate suppliers and materials. Accountability to external stakeholders such as nongovernmental organizations (NGOs) and customers is addressed through voluntary reporting on sustainability performance. Standards on reporting have been developed to encourage transparency in sustainability assessments. Companies use auditing to ensure that its suppliers are meeting the sustainability targets and that the metrics being reported on are sound.

Environmental management systems in use today include an impact assessment to identify areas of vulnerability from environmental degradation as well as the use of standards and auditing to ensure that impacts are being monitored and environmental goals are being achieved. Companies can also benefit from audits since they can identify opportunities for efficiency improvements, cost recovery through recycling, and opportunities for sale of by-products and reduce the risk of being accused of negligence and liability (Barrow, 2006).

Sustainable business development is now focused on identifying opportunities for businesses to provide products that minimize carbon footprint while meeting consumers' needs, providing employment, and addressing concerns about the state of the environment and optimizing resource use.

The entire business value chain is impacted by sustainable environmental strategies. In the sections below, approaches to managing each step of the value chain from an environmental perspective will be discussed.

8.7 VALUE CHAIN

8.7.1 Michael Porter's Value Chain

Companies need to find ways to sustain their positions in the marketplace to survive. The crucial aspect to do this is to achieve a competitive advantage. According to Michael Porter (1980), this is only possible with superior performance, which can be achieved by having sustainably higher prices, lower costs, or both. This could mainly be the key to outperform the competitors.

In order to achieve competitive advantage by using aforementioned strategy of Porter, companies should manage their activities in every step of business actions very carefully. Collection of these essential business activities is also called "value chain." Magretta (2012) defines Porter's value chain framework as "the sequence of activities your company performs to design, produce, sell, deliver, and support its products" (p. 12). Figure 8.1 illustrates each sequential value chain steps.

8.7.2 Greening the Value Chain

Similar approach of Porter's value chain can be adapted to green business activities. Each step in the value chain can turn into green-focused activities that can help in achieving competitive advantage for companies. Value chain impacted by sustainable environmental strategies is shown in Fig. 8.2. Because of reasonable sense and resulting success of leading businesses, today, more and more sophisticated corporations are increasingly making green business decisions considering their environmental implications using similar type of green value chain. In the next sections, we will explore value-creating activities at each step in the green value chain introduced here.

8.7.2.1 Green Finance

As investors have become more aware and interested in opportunities to put their money to work to benefit the environment, banks have responded by moving beyond offering environmentally sensitive investments in equities

FIGURE 8.1 The value chain to create customer value. Source: Adapted from Magretta (2012).

FIGURE 8.2 Business actions to create customer value: greening the value chain.

to offering bonds that provide funds to support green initiatives. Some large banks, such as BNP Paribas, have created departments and leadership positions that are focused on sustainable capital markets (Gilbert, 2014). Banks are issuing bonds that provide funds to support green projects that have environmental improvement as their primary objective. Some corporations are also setting aside funds, separate from their capital budgets, that are earmarked for environmental projects within the organization.

Determining if a project meets the standards to be eligible for green financing can be ambiguous. In order to address this, several of the world's largest banks collectively drafted a guidance document, the Green Bond Principles, that includes guidelines and the criteria for issuing green bonds. One of the primary goals of the Green Bond Principles is to encourage transparency for projects that are funded by the green bond issues. These bonds support projects targeting environmentally sound infrastructure, renewable energy, development of technology to improve fuel efficiency and lower emissions, and factory renovations to reduce waste and water use.

The availability of green financing allows companies to obtain funding for environmental projects without placing a burden on internal funding, reducing competition within the firm, and encouraging development of products and technologies that benefit the environment. Investments and projects can be evaluated with the core principles of sustainability in mind, addressing concerns about our finite resources, interdependence of nations, inequalities, differences in cultures, and the responsibility of leaving a legacy for future generations.

Within the organization, management should be assessing green projects based on how well they address environmental issues such as climate change, energy use, water availability, risk of pollution and disease, and investor activism. Managers may have the opportunity to link sustainability efforts with benefits from energy reduction and efficiency improvement (will also reduce GHG emissions), the use of sustainable building design and land use, risk management in insurance practices, and efficient transportation options. Organizations also can improve their corporate image and brand profile and generate goodwill with customers and the community when they are able to demonstrate a commitment to the environment through green financing.

Green financing gives an organization the opportunity to evaluate projects with more than the established financial metrics such as payback and return on investment and to include nonfinancial criteria, for example, contribution to achieving environmental goals for reducing waste or resource consumption. Once a source of financing has been identified for a project, managers can move to the next phase of creating value for the corporation, by incorporating green technology into their product designs.

8.7.2.2 Green Technology

Green Innovations with Sustainable Technology
The search for sustainable technology has led to a number of green innovations that can support corporations in their efforts to reduce the environmental impact of business operations. GHG mitigation has been a leading focus of green technology development, since emissions have been a hot button topic related to global warming and its effects on climate change.

Technologies that utilize fuel more efficiently or use renewable energy have been at the forefront of developments to support environmental improvements and reduce GHG emissions. There is an expectation that there will be more regulation to drive industry toward a low-carbon operation, through policies such as cap and trade on emissions. Even in the absence of regulation, corporations are setting goals to achieve lower fossil fuel consumption. These goals not only reduce the negative environmental impacts but also contribute to savings on fuel and other energy use. Longer-life batteries, processes to recapture thermal energy, and carbon capture and storage are all technologies that have gained recent attention as solutions toward reducing resource consumption and emissions.

Another critical area that corporations have focused on to reduce their environmental impact is their resource use. Dire predictions have been made about future resource scarcity, for water, land, food, and nonrenewable materials. Some view resource scarcity as a driver for the third Industrial Revolution, as firms compete to find alternatives to fossil fuels and nonrenewable materials (Heck et al., 2014). Environmental managers within the corporate framework have strived to improve efficiencies so that less materials and water is being used during manufacturing as well as during product use. The use of recyclable or renewable materials, from packaging to feedstock and other raw materials, have been strongly encouraged and touted as an example of a firm's commitment to environmental improvement.

Environmental efforts to improve land use and reduce the impact on landfills have led corporations to seek out more biodegradable materials for use in packaging and products. Many companies have set sustainability goals that require new products to incorporate a greater percentage of biodegradable materials and have used these goals in their advertising campaigns to communicate their accomplishments to consumers. There has also been a trend by corporations toward reduced waste generation to limit the load on landfills. Companies have used technological processes to transform waste streams into refurbished or repurposed goods.

The use of less materials in packaging and manufacturing of goods also contribute to a better environmental profile from the logistics perspective, where there is less energy being used to transport goods to customers, resulting in lower emissions being released into the atmosphere.

Companies have also benefited from advances in materials science in their goal to reduce their impact on the natural environment. New materials with desirable conductance have been incorporated into electronics design to reduce the need to use harmful metals or toxic materials. The environmental impact from production, transport, and disposal of toxic materials is reduced or eliminated.

Technology is being viewed as a critical component toward achieving sustainability in business. Companies such as Tesla and Zipcar have provided solutions for concerns about fossil fuel use, emissions, and resource consumption by innovating and using advances in battery technology and information management. There are opportunities for firms to use new technologies to respond to consumer needs and reduce their environmental impact.

Developing Green Products and Services
One path to becoming a high sustainability organization is to develop green products and services while continuing to meet customers' needs. A firm may approach this by introducing new products that are environmentally sensitive or by redesigning its existing products. This approach is seen in many consumer durable companies, for example, in the automobile industry, where sector stalwarts are launching hybrid electric cars, while continuing to offer cars with combustion engines with improved efficiencies.

New product design can be made more environmentally friendly in a number of ways—using a lower level of nonrenewable materials, using less material, improving efficiencies, extending use life, eliminating toxic substances in products, or creating channels for end-of-life reuse. There are examples of each of these from mainstream consumer products. The consumer packaged goods industry have adopted each of these in one form or another, for example, by formulating more concentrated detergent products, resulting in lower chemical use, smaller amounts of packaging to be recycled, and fuel savings during transport of goods. By examining all aspects of delivering goods to consumers, from components to packaging and logistics, and extending the consideration to postconsumption fate, firms can identify opportunities to reduce the environmental impact of goods.

One challenge in developing green products and services is gauging the market's reception of these offerings. Foster and Green (2000) reported that customers have a limited understanding of the environmental impacts of some products and prefer green products only when there is no effect on cost or other performance measure. The challenge for R&D is to maintain product performance to meet customers' expectations while developing products that are better for the environment.

Perhaps, one of the most significant ways in which companies can reduce the environmental impact of their goods is by reducing the amount of material that terminates in landfills, where it can remain for years, sometimes leaching toxic substances into soil and groundwater. Programs that encourage consumer reuse or products that are designed to biodegrade quickly, without releasing toxic materials, are keys to improving the environmental profile of goods. Some companies allow consumers to return their used products for ecologically sensitive recycling or material recovery, sometimes even offering a discount toward future purchases.

There are many organizations that provide standards and certifications to guide companies in developing green products. Some of these are sponsored by governments and include incentives for consumers to purchase green goods. One example is the Environmental Protection Agency's (EPA) ENERGY STAR program that identifies home appliances that meet certain energy efficiency standards that allow the purchaser to receive a tax credit. NGOs also provide certifications that manufacturers can include on the labeling of their products.

Service companies can also green their processes by interacting with their customers in a manner that reduces their carbon footprint. The use of streaming media, along with other technologies, have been successfully used by cable television providers to deliver service and maintenance to customers, reducing in-person repair visits and eliminating trucks and their associated emissions.

Developing green products and services can be beneficial for the environment and for a firm's bottom line. Porter and Kramer (2006) argued that if a firm combined CSR activity, including environmental considerations, with core business strategy, then CSR can be a source of opportunity, innovation, and competitive advantage for the firm. The aim of developing green products and services can serve as a driver for R&D and marketing on the firm level and for an industry on the market level.

Green products can be launched as a new business initiative of an existing enterprise, allowing the firm to increase its green image while maintaining its traditional business. GE launched its Ecomagination initiative with the broad objectives of meeting environmental goals such as clean water, renewable energy, and reduced emissions, as well as increasing its investment in sustainable technologies and its revenues from sustainable products.

8.7.2.3 Sustainable (Green) Supply Chain Management

Sustainable supply chain management (SSCM) is essential to a firm's success in achieving its environmental management goals. The supply chain determines how the company acquires its raw materials that will be transformed into finished goods through its value-added activities. In the early stages of corporate environmental management, SSCM was largely a reaction to external stakeholder pressure from NGOs and customers. However, recent work has suggested that SSCM can also contribute to profitability, as it eliminates waste from all stages of a firm's operations (Kumar et al., 2012).

The main tenets of a SSCM program are to source materials so as to minimize environmental impact and to reduce waste. Reducing waste throughout the supply chain will allow the firm to be more profitable by lowering material costs and resource consumption. Low-impact sourcing forces the firm to identify where in its supply chain alternatives are required and to actively engage in finding alternatives. Walmart expects to generate $12 billion in global supply chain savings by 2013 through a packaging "scorecard" that could reduce packaging across the company's global supply chain by 5% from 2006 levels.

Firms can find help in greening the supply chain from nongovernmental and governmental certification bodies. Environmental standards and guidelines have been developed to identify lower hazard products, recycled-content products, sustainably grown and harvested products, water- and energy-efficient products, as well as alternatives for ozone-depleting chemicals. Suppliers in every sector have recognized the increased requirement for environmentally friendly substitutes and include them in their product offerings.

SSCM can also contribute to performance by improving an organization's reputation as a good citizen and raising its profile with target customers. Sustainable sourcing of ingredients, whether it's fair trade or organically harvested, have been a point of differentiation for many companies in the food sector.

SSCM also has the important role of managing products beyond the point of purchase and consumption. Reverse logistics is growing in importance, with firms developing policies for collection, refurbishing, recycling, and supply replenishment in target markets. Some companies provide attractive financial incentives to encourage customer to return items and increase the flow of remanufactured goods. The firm improves profitability by purchasing fewer new components, raises its sustainability profile by recycling, and develops a new market for its products.

Hunt and Davis (2008) suggested that supply chain management can contribute to an organization's competitive advantage in the marketplace through superior resource acquisition and identification of strategic supply partners. Purchasing strategy during times of resource scarcity can allow a firm to grow its operations when its competitors are unable to fill their supply pipelines. Proactive SSCM prior to resource shortages may require a firm to identify risks such as disruptions in supply and make accommodations or identify alternatives.

8.7.2.4 Green Manufacturing

Manufacturing is the system that companies use to transform their raw materials into finished goods. Green manufacturing incorporates practices into production that reduce the environmental impact of processes. Sustainability can be incorporated into the manufacturing phase of the value chain in three key areas—resource use, waste management, and logistics and design.

Firms can design their manufacturing systems to use less energy, use renewable energy where possible, or use energy more efficiently. The use of technology that improves efficiency can become part of the manufacturing system design. Techniques such as thermal energy recapture where heat from one process can be used in another step have been adopted in many chemical plants. Water is another critical resource that can be used more efficiently in manufacturing, and recycling water for use in other steps of a process can reduce consumption and waste water.

Green manufacturing is also focused on limiting waste, which lowers the environmental burden on landfills as well as emissions from transporting waste. Recycling manufacturing waste by identifying new markets for off specification products, and reworking goods or donating can reduce the waste from production. Some companies have developed innovative ways to use the waste generated at factories, and many are including zero landfill production as a sustainability goal. US-based global company SC Johnson has been working toward zero landfill plants for many years and now has seven sites that have successfully eliminated all materials going to landfill through reuse and recycling (www.scjohnson.com).

Green considerations can also be incorporated into the logistics and design of manufacturing systems. By producing close to key distribution points, companies can reduce the level of emissions and the energy used to transport goods to market. Factories can be laid out to transport raw materials and work in process to work stations in an efficient manner that can reduce energy use. Green manufacturing can also benefit from using updated materials management systems that accurately predict the amount of inputs that are required for a desired level of output. This would reduce waste generated, as well as the energy used to transport materials around a factory. The firm also benefits by reducing the amount of inventory, obsolete goods, and disposal costs associated with its operations.

Many of the steps that firms take to implement green manufacturing procedures also contribute to savings through improvements in efficiency and resource consumption. Green manufacturing can also improve the reputation of a company and its perception in the community, by reducing waste and emissions from production.

8.7.2.5 Green Marketing and Sales

Green marketing has also been called as "environmental marketing," "ecological marketing," "sustainable marketing," or "sustainability marketing." Prakash (2002) provides a definition that covers more recent approach of green marketing, which is "strategies to promote products by employing environmental claims either about their attributes or about the systems, policies and processes of the firms that manufacture or sell them."

Green marketing has become an important aspect of marketing management in the context of alleviating the human impacts on environment by adopting green consumption. Green marketing practices have evolved as an essential component for an organization's viability in the marketplace. Today, increasing concerns about the environment is causing consumers to change their shopping criteria, purchasing patterns, product usage, and disposal behavior greatly. As environmental concerns become more important to consumers, more and more companies are using "green" marketing as a way to improve their image with consumers and ultimately increase sales. The emphasis on primary benefits of green products by marketing managers is critical to winning over the mainstream consumers.

Sixty-one percent of American consumers believe that green goods perform less than conventional ones. Consumers also doubt the very greenness of green products. In this area, consumers trust scientists and environmental groups, not the government, the media, or businesses.

Educating consumers is an important aspect of green marketing because sometimes consumers are largely unaware of green products, and a business that sells them must see itself as an educator, not just a sales machine.

For example, Procter & Gamble's (P&G) Future Friendly campaign provided consumers with specific tips on how they can have a positive impact on the environment by making their homes more efficient, using less water, and reducing household waste.

8.7.2.6 Green Consumption

Emerging Green Consumers and Green Consumption
Today, there is an emergence and growth of markets for goods that are produced ethically, harmless to the environment, and traded fairly (Chan and Kotchen, 2014). These goods are called "green products," which are products "made with reduced amount of material, highly recyclable material, non-toxic material, do not involve animal testing, do not adversely affect protected species, require less energy during production or use, or have minimal or no packaging" (Wang, 2012, p. 165). Green products help reduce individuals' carbon footprints by impacting their aggregate consumption, such as hybrid and hydrogen-powered vehicles and recycled material products (Schoemaker and Day, 2011).

It is also important to note the interrelationship between green products and the lifestyle that promotes and encourages green consumption. Wang (2012) states that green consumption "involves replacing traditional products and services with products that are more effective and produce less pollution … [and it] based on the idea that the environmental damage cause by consumption can be reduced by adoption less damaging consumption style" (p. 165). In fact, there is no doubt that greening of the consumption is necessary to avoid jeopardizing future generations and to sustain basic needs and pursue a better quality of life.

Today, consumers in every parts of the world are demonstrating a growing interest in green and sustainable brands (Mukherjee and Onel, 2013). Buyers are steadily accepting the idea of green consumption, and even corporations are getting used to it (Esty and Winston, 2006; Bonini and Oppenheim, 2008a; Makower, 2009). As a sign of corporation adaptations to this new movement, green product and service development as well as green marketing by companies are becoming a visible trend in many industries, including food, auto, clothing, housing, and furniture (Mukherjee and Onel, 2013).

8.8 DIFFERENT PERSPECTIVES ON GREEN MARKETS AND BUSINESSES

In this section, we will discuss different points of view on green markets and businesses using examples from various industries. First, we will look at consumer perspectives, and then we will focus on company and executive perspectives in the context of sustainable businesses and green market actions.

8.8.1 Consumer Perspectives on Green Markets

Businesses seeking to succeed in the market for green goods need to manage their brands and images to gain trust with their target customer. Bloemers et al. (2001) reported on the market for green energy in McKinsey Quarterly. Results from focus groups indicate that 37% of consumers in Germany, and 46% of consumers in the United Kingdom, claimed that they would be willing to pay a 10% premium for green energy.

Companies around the world are also exploring the market for green energy by tapping into customers' interest in the environment. For example, in the United States, Greenmountain.com built a 100,000-customer base by selling "cleaner" energy through the Internet. The company buys energy from conventional producers but guarantees that its purchase mix is skewed toward environmentally friendly sources such as hydro and wind. In Germany, Greenpeace has tried selling green power to its 550,000 members through a combination of direct mail and conventional advertising. In the Netherlands, distribution companies such as Nuon and Essent sell green electricity with the verification and support of the Dutch World Wide Fund for Nature, which lends its name and logo to green energy suppliers.

These results may also apply to other business sectors. Focus group results suggest that organizations such as Friends of the Earth and Greenpeace, and even retail brands such as The Body Shop, IKEA, J. Sainsbury, and Virgin are perceived as credible stewards of the environment by customers. Companies may be better positioned to develop a market for green alternatives when there is an association with an environmentally friendly organization or label.

Bonini and Oppenheim (2008b) reports that consumers are concerned about climate change and may examine their consumption behavior and its impact on the environment. However, concern for the environment does not always carry through to purchase behavior. Data from a survey of 7751 consumers from Brazil, Canada, China, France,

Germany, India, the United Kingdom, and the United States indicate that 87% of consumers were worried about the environmental and the social impact of the products that they purchased (Bouton et al., 2010). However, less than 33% of those surveyed responded that they were ready to purchase or had purchased green products.

A similar outcome was seen in a 2007 Chain Store Age survey of 882 US consumers. Only 25% of respondents reported purchasing green products other than organic foods or energy-efficient lighting. Another survey, conducted in 2006, found that green laundry detergents and household cleaners accounted for less than 2% of US sales in their categories. In 2007, J.D. Power and Associates reported that hybrid cars made up little more than 2% of the US market.

An important challenge to business is to gauge consumers' needs for green products and to develop messaging that can successfully align environmental concerns with purchasing behavior.

8.8.2 Company and Executive Perspectives on Green Markets

Engaging in activities that address sustainability concerns can be beneficial to a company by improving its reputation with customers, employees, and investors; accommodating a focus on process improvements; and creating long-term shareholder value.

McKinsey (2010, 2011) has conducted surveys that gauge executives' views on the significance of addressing sustainability issues. In a 2010 survey, they found that 86% of executives agreed that engaging in sustainability contributed positively to shareholder value in the long term, versus 50% of respondents who felt that it led to short-term value creation. They also reported that companies that managed sustainability proactively were much likelier to seek and find value creation opportunities. Eighty-eight percent of the respondents to a McKinsey survey in 2011 at leading companies strongly agreed that their companies were actively seeking opportunities to invest in sustainability, versus 23% of all other respondents. The highest share of respondents (23%) affirmed that business portfolios that were being managed to address the trend in sustainability were adding significant value to companies in their industries.

According to McKinsey's 2010 survey, 55% of executives defined sustainability as the management of issues related to the environment (e.g., GHG emissions, energy efficiency, waste management, green product development, and water conservation); 48% of respondents agreed that sustainability included the management of governance issues, while 41% agreed that it also included the management of social issues, such as working conditions and labor standards. Just over a third of the respondents indicated that their companies had quantified the potential impact of environmental and social regulations on their businesses, with only 40% responding that they were prepared to deal with regulation in the next 3–5 years and confident about handling climate change issues.

A survey that was fielded online in 2011 by McKinsey to executives from multiple regions, industries, tenures, and company sizes received 3203 responses. Thirty-three percent more respondents indicated efficiency improvement and lowered operating costs as the top reasons for addressing sustainability at these companies, a 14% increase in the number of respondents choosing these reasons in 2010.

The survey also showed an increasing trend toward integrating sustainability across many processes, with 57% of respondents affirming that sustainability has been integrated into strategic planning at their companies and 29% indicating that it had been integrated into business practices. Sustainability objectives have been primarily integrated into the mission and values, followed by external communications by companies participating in the survey. Respondents indicated that there was still work to be done on addressing sustainability in the areas of supply chain management and budgeting.

Fifty-three percent of survey participants also responded that company performance on sustainability is at least somewhat important to attracting and retaining employees and those companies that take action on sustainability issues are more likely to gain an advantage in employee retention. McKinsey's 2010 survey of executives found that sustainability was considered as "extremely important" for managing company and brand reputation by 72% of the respondents, with 55% of respondents indicating that investments with a sustainability focus helped to build company reputation. More than a third of the respondents felt that building reputation was a key reason for addressing sustainability issues.

A survey of executives in the energy sector shed some light on sustainability in an environmentally sensitive industry. Ten percent of executives in this industry who responded to the survey indicated that addressing sustainability was the top priority on their CEOs' agendas, and 31% reported that it was a top three priority. Seventy-four percent of energy executives reported that sustainability was being incorporated into their companies' regulatory strategies, compared with 53% of respondents overall.

Research by McKinsey (2010, 2011) supports the view that companies are embracing sustainability with the intent of improving processes, pursuing growth, and adding value to their companies. Results from the energy sector indicate that an important step toward gaining recognition and improving the impact of sustainability activities could be to communicate better with investors and other stakeholders.

8.9 CORPORATE SUSTAINABILITY REPORTING

Besides engaging in responsible behavior, corporations must also consider if they will disclose on their social responsibility initiatives and to what extent. McIntosh et al. (1998) have made the argument that corporations have an ethical duty to disclose on the impact of their actions on society. Corporate governance, the systems of laws, rules, and factors that dictates how a company operates, allows management to pursue the objectives of the corporation and identify the metrics that will be used to determine success of those objectives (Samy et al., 2010).

8.9.1 Global Reporting Initiative (GRI)

8.9.1.1 Introduction to GRI

The Global Reporting Initiative (GRI) is a global nonprofit organization that provides guidelines on reporting on corporate social responsibility, providing information on economic, environmental, social, and governance performance. GRI has reworked some of its original key performance indicators (KPIs) since its inception in 1997 and is currently using the fourth generation, termed G4. The GRI has been noted as the standard for sustainability reporting for all organizations (O'Rourke, 2004).

The GRI website sets forth the vision for the 2007 set of guidelines as follows:

> …is that reporting on economic, environmental, and social performance by all organizations becomes as routine and comparable as financial reporting. GRI accomplishes this vision by developing, continually improving, and building capacity around the use of its sustainability reporting framework. (Global Reporting Initiative, 2007)

The GRI has been used as one of the variables when examining the relationship between CSR and financial performance. Organizations are required to provide disclosure on their activities in the following areas: economic, environmental, social, human rights, society, and product responsibility (Samy et al., 2010).

Companies using the GRI reporting framework are asked to declare the level to which they have applied the guidelines. GRI allows an organization to disclose the level of reporting it has chosen, which provides flexibility to companies to gradually adopt the GRI reporting guidelines. There are three levels of reporting, titled C, B, and A, with the reporting criteria in each level reflecting an increasing application of the framework. The report can be self-declared against a checklist, verified by an external third party, or checked by the GRI itself. If a report is externally verified, a (+) is added to the level of reporting designation, for example, A+, B+, and C+.

Section 8.9.1.2 provides cases of firms' reporting behaviors, taken from the GRI database for 2013 and companies' websites and reports. Three different GRI reporting frameworks were being used at this time—the third-generation framework, G3, G3.1, and, to a lesser degree, G4. G3.1 is an extension of the G3 guidelines, with a focus on community impacts, human rights, and gender equality. Reporting against the G3.1 guidelines was optional for companies—they could continue to report against G3 guidelines until the G4 guidelines were released in May 2013. These cases show that firms from all sectors can contribute to sustainability efforts.

8.9.1.2 Industry Cases

Agriculture
New Britain Palm Oil Limited New Britain Palm Oil Limited (**NBPOL**) is a large company in the agriculture sector, based in Papua New Guinea. NBPOL is the world's leading producer of sustainable palm oil certified in accordance with the Roundtable on Sustainable Palm Oil (RSPO) principles and criteria. The company is vertically integrated and active in all aspects of palm oil production, from planting to refining.

In 2013, NBPOL reported according to the GRI G3.1 guidelines, at a level of B. The report is self-declared. NBPOL has demonstrated a strong commitment to sustainability reporting by adopting the 3.1 guidelines ahead of schedule and disclosing its performance with respect to community impacts, human rights, and gender equality.

NBPOL's website has a link for sustainability that allows access to the companies' policies and positions on CSR issues, management structure, downloadable reports, information on stakeholder consultations, certifications, and community involvement. Besides furnishing the sustainability report, NBPOL has also provided a carbon footprint report, starting in 2011. NBPOL has been a leader in the oil palm sector with respect to sustainability reporting, releasing its first report in 2008. The company's website provides access to earlier reports, allowing stakeholders to assess progress on CSR performance.

NBPOL is an important player in an environmentally sensitive sector, where there is growing concern about land use, deforestation, use of pesticides, and labor violations. The company has addressed these issues by being transparent, proactive, and collaborative with stakeholders, leading change in the marketplace. These actions would lead to NBPOL being classified as a prospecting company, according to the Miles and Snow typology for management strategy.

Sourced and adapted from www.nbpol.com.pg

Automotive
Tata Motors India-based **Tata Motors** has been included in the list of companies reporting by GRI since 2008. The company's website provides communication on progress toward the UN Global Compact for each year since 2004, GRI reports since 2006, and social responsibility reports since 2006. There is a business responsibility report for the business year 2012–2013. Tata has shown a long-term and consistent approach to sustainability reporting.

Tata Motors' sustainability program has two key pillars—reduction and control of environmental pollution and restoration of ecological balance in the natural environment. The company has implemented programs in its manufacturing operations to conserve soil and water and has committed to efficient design to reduce emissions. Initiatives to restore ecological balance include design focused on efficient energy use, reduced emissions, and recycling materials. Tata Motors has also utilized renewable energy to reduce resource use and avoid carbon dioxide emissions from its plants. Water recycling programs during manufacturing has reduced water use and lowered the total water withdrawn for operations.

One important environmental issue that Tata Motors is addressing in India is the lack of an infrastructure for scrap car collection, treatment, dismantling, and recovery. Tata Motors is working with the Indian government to contribute to such a framework, borrowing from the end-of-life vehicle programs that have been implemented in the EU.

Tata Motors is committed to the communities where it operates, participating in projects to bring drinking water to villages, provide education, and reduce workplace injuries. The company has also provided safe driving education classes to its employees. In a sector that has the potential to determine the level of emissions and resource use, Tata Motors has demonstrated a strong commitment to conservation and sustainability and has received many corporate citizenship awards and recognitions for its efforts.

Sourced and adapted from www.tatamotors.com

Aviation

Qantas The aviation industry has the potential to be environmentally damaging due to emissions, hazardous materials, noise impacts, and energy consumption. Companies in this industry have been early adopters of sustainability initiatives and have seen the added benefit of reduced fuel costs and improved efficiencies.

Qantas, the Australian airline company, reviews its sustainability outcomes in its annual report. For 2013, the company reported a decrease in fuel consumption and carbon dioxide emissions for its operations. This was accomplished through improved fuel optimization and fleet renewal programs. Qantas also increased its use of biofuel and participates in a carbon offset purchase program. Fleet renewal also allowed Qantas to reduce its noise impact by using airplanes with newer and quieter engines.

Qantas also reported lower consumption of electricity and water and lower amounts of waste disposed to landfills. The company also has a sustainable sourcing program for the procurement of environmentally sensitive supplies and equipment.

Qantas did not use a GRI framework for its report; however, the sustainability statistics provided in its annual report were verified by an external auditor.

Sourced and adapted from www.qantas.com.au

Chemicals

Clariant The chemical industry has been the target of environmentalists for several decades and, as such, has been an early adopter of sustainability reporting. **Clariant**, a specialty chemical company based in Switzerland, has been included in the GRI reporting database since 2008. For 2013, Clariant reported using the GRI G3 framework at an A+ application level. The report was verified by GRI.

Clariant's 2012 sustainability report included its goals for reducing energy and water use, emissions, and waste by 2020. The company's sustainability strategy is clearly stated, including the voluntary commitments as part of the Global Responsible Care Charter and the Global Product Strategy. Clariant has achieved several certifications for environment, health, and safety for its worldwide sites.

Clariant has recognized its employees as key to achieving its sustainability goals, linking incentives such as bonuses to specific objectives in the area of sustainability. While recognizing that chemicals synthesis and transport can be potentially harmful to the environment, Clariant has taken steps to manage the risks. The company has implemented employee training programs to maintain safe practices and continuously monitors its processes and procedures. Clariant has also implemented rapid response teams for accidents and spills as part of its risk identification and risk management program. There are programs at Clariant's sites to address accidents, events, and complaints from the neighboring community. The company is also environmentally sensitive in its procurement programs, incorporating its sustainability standards into its purchasing strategies.

Clariant's CSR activities, reporting, and goals all demonstrate that the company is committed to operating in a sensitive manner, with the environment, safety, and health of its employees and communities as a priority.

Sourced and adapted from www.clariant.com

Commercial Services

Ernst & Young USA The commercial service sector is not generally considered as an industry with a significant environmental impact; however, firms in this sector are also actively engaged in sustainability efforts. **Ernst & Young USA**, a subsidiary of a global professional service firm, reported on sustainability in 2013 according to the GRI G3.1 framework. The report was verified by GRI.

The environmental stewardship program at Ernst & Young USA is focused on reducing the company's carbon footprint and waste stream. Direct emissions from electricity generation as well as indirect emissions from office energy consumption and business travel are measured and reported. Renewable energy is purchased for use in some of the company's data centers and office spaces. Many offices are LEED (i.e., Leadership in Energy & Environmental Design) and ENERGY STAR certified.

Ernst & Young USA is actively working on making large meetings more environmentally sensitive, from the sourcing of food to transportation of delegates and office supplies. There is also an initiative to increase the environmental awareness of employees, with the aim of encouraging more environmentally sensitive behaviors. This is accomplished through activities such as promoting the use of reusable beverage containers, supporting Earth Day activities, and providing sustainable choices in cafeterias. Ernst & Young USA is an example of a firm that has developed an environmental strategy even though it is not in an environmentally sensitive industry, demonstrating that everyone can help to improve the environment.

Sourced and adapted from www.ey.com

Computers

Lenovo The sustainability issues that have plagued the computer sector are labor practices, resource use, and end-of-life disposal. China-based **Lenovo** has made strides in addressing these issues and has pledged to improve its sustainability standards each year. Lenovo has been included in the GRI reporting list since 2009. In 2013, the company reported according to the GRI G3.1 framework, ahead of schedule, with an application level of A. The report was self-declared by the company with no external verification.

Lenovo has many programs in place to address the environmental challenges in its sector. The company has focused on minimizing the carbon impact of its operations and reducing the environmental impact of its supply chain. Life cycle management has been a target of Lenovo's environmental policy, with initiatives to support the use of environmentally sensitive materials and reduced packaging materials.

Recovery and recycling of materials has also been an area that Lenovo has made sustainability strides. During 2012, the company financed or managed the processing of more than 13,100 million pounds of consumer-returned and Lenovo-owned computer equipment. Through various initiatives such as waste-to-energy generation and parts recycling, only 2.7% of this material terminated in a landfill. Lenovo is involved in programs to continuously limit the amount of material, including postpurchase materials that go to landfills. Recycling and reusing are primary means to accomplish this, as well as designs to include more recycled plastic content in new products. In 2012, the company reports using more than 23 million pounds of recycled plastic, with almost half of that being net postconsumer and postindustrial plastics.

Sourced and adapted from www.lenovo.com

Construction Materials

Kawasaki Heavy Industries Japan-based **Kawasaki Heavy Industries** provides materials for the construction of transportation systems, energy systems, and industrial equipment. Kawasaki was included in the GRI database of reporting companies for the first time in 2013. The company did not use a GRI framework for its report, but the framework was referenced in Kawasaki's report.

Kawasaki's annual report for the year ending March 31, 2013, describes its commitment to the environment as "manufacturing that makes the Earth smile." The company's goals are to realize a low-carbon society and encourage recycling practices that can enhance society's ability to coexist with nature. Programs have been implemented to incorporate low environmental impact technologies into Kawasaki's operations. Kawasaki has also formulated green procurement guidelines that require its suppliers to implement environmental management systems and identify and monitor potential environmental risks from parts and materials.

Company report included details on the environmental management organization at Kawasaki, along with measurements on its progress toward reducing environmental impact of business activities. Some of the programs that were implemented during 2013 included those focused on reducing raw material, energy, and water use; minimizing harmful emissions; and recycling materials. The company also had initiatives to promote environmentally conscious logistics practices, to save energy and curb emissions during transportation of its goods and materials.

Sourced and adapted from www.khi.co.jp

Consumer Durables

LG Electronics The consumer durables sector can have a potentially significant environmental impact, since these goods often end their lives in landfills, use energy, and are used by almost everyone on a daily basis.

LG Electronics, based in the Republic of Korea, has pledged to contribute to a healthier and cleaner environment. This company produces televisions, mobile phones, computers, home appliances, and industrial equipment. The company is accomplishing its environmental goals by designing its products to be more energy efficient.

LG Electronics is also incorporating sustainability at the design phase, where it strives to limit the use of conflict minerals and include more recoverable material. Responsible sourcing and supply chain management also allows the company to meet its sustainability targets. Product recycling is encouraged to reduce the amount of material that goes to landfills. Design is being revamped to eliminate the use of hazardous materials so that when the products are disposed, there is reduced risk of soil and water contamination.

Internally, LG Electronics promotes sustainability initiatives through programs to increase employees' awareness of environmental and social issues. The company also supports efforts in the community to clean up rivers and plant trees.

LG Electronics' sustainability report for 2012–2013 was completed according to the GRI G3.1 guidelines. The report is available on the company's website. LG Electronics is also a member of the Dow Jones Sustainability Index and

RobescoSAM (i.e., investment company with sustainability focus).

Sourced and adapted from www.lg.com

Energy
Hess The energy sector has been one of the leaders in sustainability reporting, since it has been scrutinized for many decades for its environmental impact. US-based **Hess Corporation** reported on sustainability according to the GRI G3.1 framework at an application level of A+. The report was verified by a third party. Hess also utilized the GRI oil and gas sector supplement. Hess has been reporting according to the GRI framework since 2007. The company's website provides sustainability reports since 2002.

Some of the activities that Hess has engaged in to achieve its environmental goals include programs to reduce energy and water use, improve efficiencies to limit the level of emissions of GHG and other pollutants, prevent spills, and improve remediation and investment in green buildings.

Hess has also committed to improving the environment by conducting biodiversity screenings during site evaluation and selection for its operations. Environmental and social impact assessments are routinely used to mitigate environmental and biodiversity risks. Hess is working with industry committees to apply new technologies in drilling that can address environmental impacts.

Besides reporting according to the GRI framework, Hess also follows the guidance of the 2010 International Petroleum Industry Environmental Conservation Association, American Petroleum Industry, and the Oil and Gas Industry Guidance on Voluntary Sustainability Reporting. Hess has been consistent and dedicated to reporting on its environmental performance in compliance with legislative and industry initiatives.

Sourced and adapted from www.hess.com

Financial Services
Bank of America Bank of America Corporation is a US-based financial service company and one of the world's largest banks. The bank has grown over the past decades through mergers and acquisitions. The business can trace its roots to the Massachusetts Bank, which opened for business on July 5, 1784.

Bank of America reported according to the GRI G3.1 guidelines in 2013. The application level was B+, indicating that the company had its report verified by a third party. Bank of America did not use the sector-specific supplement for financial services. The company's website provides access to its CSR report, as well as an environmental sustainability report, both of which can be downloaded.

The bank has been at the forefront of environmental disclosure for the financial service sector, issuing its first goals for GHG reduction in 2004. The company has continued to implement programs at its facilities to conserve and reduce demand for natural resources. Bank of America's first report using the GRI framework was in 2006. In 2013, Bank of America issued the first "green bond" to provide financing for renewable energy and energy efficiency projects for residential, commercial, and public properties.

This organization has demonstrated leadership in the area of CSR and disclosure, by providing separate environmental reports, using the GRI G3.1 framework ahead of its implementation, having the report verified by a third party, financing environmental projects, and supporting the local communities.

Sourced and adapted from www.bankofamerica.com

Food and Beverage Products
Carlsberg The companies in the food and beverage products sector is perhaps the businesses that are most dependent on a healthy environment for their success. These companies must be good environmental stewards to ensure their present as well as future survival.

The Carlsberg Group, a Danish brewery, is a prime example of a company that is engaged in sustainable activities in this sector. This company has reporting according to the GRI G3 framework at a C application level in 2013. It was first included on the GRI reporting list in 2010.

The environmental focus areas of Carlsberg's sustainability program are energy and emissions, water conservation, and packaging reduction. The energy and emissions programs are implemented in the production and logistics of delivery. Production efficiency improvements to reduce energy use include heat capture and optimized equipment design. Carbon release is also reduced by efficiency improvements. Carlsberg is also reducing energy use and emissions through its logistics operations and alternative energy sources for transporting during distribution and sales.

Carlsberg is continuously working to conserve water and protect the quality, which is important to brewing. The amount of packaging is also being reduced, and reuse and recycling programs are being implemented to extend the life of packaging materials. The development of sustainable packaging is being assessed as another way of reducing materials use.

Sourced and adapted from http://www.carlsberggroup.com/csr/Pages/Default.aspx

Household and Personal Products
Kimberly–Clark Companies in the household and personal products sector have the opportunity to engage all consumers on sustainability issues by the advertising, packaging, and product design. **Kimberly–Clark Corporation**, based in the United States, has been reporting according to the GRI framework since 2008. Sustainability reports are available on the company's website from 2004. The 2012 report is available in several languages, including Spanish, Portuguese, Korean, and Chinese, reflecting the company's global reach.

Kimberly–Clark's environmental goals for its operations include reduction in water use, GHG, and waste to landfill. The company also has goals to improve the quality of water that is discharged, limiting negative impact on water quality in its communities. The supply chain is the focus of some of the sustainability programs, with goals to obtain material from certified suppliers.

Two areas where Kimberly–Clark has significant environmental impact are in the consumption and disposal of wood fiber for its disposable wipes and paper businesses and the use of corrugated cardboard and polyethylene for its packaging. Innovative approaches to packaging and product design has allowed Kimberly–Clark to reduce consumption of corrugate and polyfilm. The company has also monitored virgin wood fiber use by certifying suppliers and using new sources such as bamboo or wheat fiber.

Kimberly–Clark has also worked with hospitals and users of its professional products to reduce postconsumption waste by implementing programs to collect and compost disposable diapers and incontinence products in some locations.

Kimberly–Clark has received many recognitions for its environmental and sustainability efforts with regard to its operations, its workforce, and its community involvement.

Sourced and adapted from www.kimberly-clark.com

Logistics

Panalpina   The logistics sector is an industry with a potential to make a significant environmental impact with regard to its energy and materials use and GHG emissions. Simultaneously, logistics is an integral part of the global economy, with all businesses relying on companies in this sector to transport materials.

Switzerland based **Panalpina** is a large company in this sector that reported according to the G3 guidelines, at an application level of C since 2008. Panalpina did not provide a separate CSR report, but has integrated its sustainability reporting into its annual report. The environmental performance section of the annual report identifies focus areas that include electricity, fuel and water consumption, spillages, travel, paper, and toner cartridge use.

Additionally, the company's website provides a report on its integrated management system for quality, health and safety, and the environment. The report includes information on Panalpina's global environmental program, PanGreen. Panalpina also provides access to an innovative tool that customers can use to generate information on sources of carbon emission for their shipments, allowing them to assess transport efficiency and opportunities for reducing environmental impact.

Panalpina is an innovator in sustainability reporting. Integrating the CSR performance into the annual report demonstrates that sustainability is being addressed with the same importance as financial and governance performance. Panalpina is going further and providing its customers with tools to guide them in making more environmentally sensitive decisions in their businesses.

Sourced and adapted from www.panalpina.com

Mining

Aquarius Platinum Limited   Mining is generally viewed as an environmentally damaging industry, with a long history of chemical use, deforestation, and other practices that lead to erosion and pollution of soil and water. The sector is also associated with unsafe work conditions, with workers being required to spend many hours in confined areas, and potential exposure to blasting, heavy metals, and poor air quality.

Aquarius Platinum Limited is a Bermuda-based company that operates in South Africa and Zimbabwe, where it mines for platinum group metals. Aquarius reported according to the GRI G3 framework at an application level of C. The sector-specific supplement for the mining sector was not used for the report. Aquarius has provided a SD report annually since 2011 and has also reported according to the GRI framework since that year.

Aquarius's sustainability report highlights its performance on two critical issues—energy and water use and carbon emissions. Aquarius reported beyond the regulatory requirements for the sites that it operates, by submitting performance information to the Carbon Disclosure Project and the Water Disclosure Project. Ongoing monitoring and reporting on carbon emissions and water and energy usage is a priority for Aquarius's management. The company achieved its goal of zero major reportable environmental incidents for the year.

In recognition of the safety risks associated with mining, Aquarius has employed a highly mechanized approach to exploration and extraction in order to reduce risks and improve safety performance. The company has a target of zero harmful incidents for its operations, which it achieved for 2013.

Aquarius has also invested in the communities where it operates, by contributing to hospitals and schools. While operating in an industry that has a potential for a significant negative impact on the natural environment, Aquarius is taking proactive steps to manage responsibly and mitigate harm.

Sourced and adapted from www.aquariusplatinum.com

Railroad

Bombardier Transportation   With increasing concern about land use for highways, habitat fragmentation, and vehicular emissions, railroads have been viewed as a sustainable solution for mass transports and movement of goods. But this sector is not without sustainability issues, with noise pollution, energy use, and end-of-life impact being the dominant environmental concerns.

Bombardier Transportation is a German company and the only one in the world that manufactures both planes and trains, as well as provides transportation solutions across the globe. In 2013, Bombardier's sustainability report was completed in accordance with the GRI guidelines. The company's website provides news on its CSR activities as well as reports.

Bombardier's environmental strategy is focused on the design of its products. The company uses Design for the Environment and Design for Safety methodologies to optimize on environmental and safety considerations throughout the life cycle. This approach results in Bombardier designing energy-efficient trains with low noise and emissions, thus reducing the impacts on communities where they operate. By examining environmental impacts at production, use, maintenance, and disposal stages, Bombardier has been able to eliminate hazardous materials and increase the use of recycled material content. In 2013, Bombardier was well on its way to meet its goal of 100% use of renewable recycled materials for its products, with new rail vehicles with more than 95% of recoverable materials.

Bombardier has been releasing information on its products' environmental footprint since 1999, according to ISO standards, in order to allow its customers to benchmark its products against those from other companies.

Bombardier is also committed to reducing its environmental impact in its production operations, with goals to reduce resource use, increase the use of renewable energy, eliminate harmful substances, and reduce waste.

Sourced and adapted from www.bombardier.com

Retail
Hennes & Mauritz Hennes & Mauritz (H&M) is a global fashion retailer that is based in Sweden. CSR is featured broadly on the company website, with tabs for corporate governance and sustainability. In 2013, H&M reported against the G3.1 guidelines, at an application level of B. The aspects of sustainability that have challenged the retail sector are mostly with regard to resource use and labor practices.

H&M provides a separate sustainability report each year, which is available on the company's website. The first report was issued in 2002—all the reports are available for download, allowing stakeholders to measure progress against sustainability goals. The 2013 report provides information on the steps that H&M is taking to examine its supply chain to identify impacts, opportunities, and challenges to sustainability. H&M has also implemented programs to address postpurchase impacts by using fabrics that can be cleaned in an environmentally sensitive manner and allowing customers to return clothing to stores for recycling postuse, reducing the materials that go to landfills.

The H&M website describes the company as a responsible fashion company. H&M's reporting activities and business practices indicate that this company is strongly committed to sustainability in all aspects of its business. H&M proactively adopted the G3.1 guidelines ahead of the required implementation of May 2013.

Sourced and adapted from www.hm.com

Textiles and Apparel
Viyellatex Group Viyellatex Group, based in Bangladesh, is a vertically integrated garment manufacturer that was founded in 1996, which counts some of the fashion industry's biggest brands among its clients. In 2013, Viyellatex provided a sustainability report that applied the GRI G3.1 framework at an application level of B. The report was verified by GRI. Viyellatex has been reporting according to the GRI framework since 2011.

The company states on its website that it is committed to minimizing its carbon footprint and reducing energy and water use. It is achieving these goals by employing efficient technology to optimize material use and manufacturing output. Viyellatex became a signatory of the UN Global Compact and has provided annual reports on its progress with regard to human rights, labor standards, environment, and anticorruption. The company has made a commitment to the UN Private Sector Forum to be 25% more energy efficient by 2015 and to become carbon neutral by 2016.

Some activities that Viyellatex has engaged in to accomplish its environmental goals include reusing treated and utility water, harvesting rain water, recovering heat from manufacturing processes, and ensuring that its suppliers are following sound environmental policies.

Viyellatex Group's progress on CSR goals has earned the company numerous awards, recognitions, and certifications for sustainability, environmental compliance, and labor practices. The company is committed to improving the quality of life of its employees and communities in a country with challenging economic hurdles.

Sourced and adapted from www.viyellatexgroup.com

Tourism/Leisure
Royal Caribbean Cruises Limited The tourism and leisure industry has the potential to be a significant consumer of resources such as water and land and can have considerable impact on the natural environment. The companies in this sector that operate cruise ships are also likely to be major energy consumers and be scrutinized with regard to their waste management practices.

US-based **Royal Caribbean Cruises Limited** reported according to the GRI framework for the first time in 2013. Royal Caribbean reported using the GRI G3.1 framework at an application level of B. The company's website provides information on its environmental strategy and also access to its stewardship reports since 2008.

Royal Caribbean's environmental stewardship program is focused on cleaner seas and purer air. The company has initiatives to achieve these through environmentally friendly ship design, waste stream operational controls, and wastewater purification systems. Newer ships are being designed with solar technology, energy-efficient hull design, environmentally safe hull coatings, UV-filtering glass, and energy-efficient lighting that go beyond what is required by legislation. Royal Caribbean also works on reducing consumption by recycling, reusing, or donating a variety of materials including glass, paper, aluminum and steel cans, pallets, batteries, electronics, and kitchen grease.

Royal Caribbean has been recognized by wildlife and conservation organizations for its initiatives to protect the seas and the environment. In an industry where consumption and environmental impact is potentially high, Royal Caribbean has been proactive in adopting sustainable practices.

Sourced and adapted from www.royalcaribbean.com

Waste Management
Waste Management US-based **Waste Management** (WM), one of the world's largest companies in this sector, provided a sustainability report update in 2013, but did not use the GRI framework. The report included WM's sustainability goals for 2020 and KPIs for its progress since 2009. While WM did not apply the GRI framework, the calculations for its sustainability performance were completed using established protocols, such as the Solid Waste Industry for Climate Solutions protocol for determination of carbon storage and sequestration in landfills.

WM has made sustainability a priority, with a tab on its website that provides access to goals and reports since 2008. The focus of this company's sustainability program is the environment, where operations can have the largest impact. WM addresses working with businesses, communities, and residential customers to improve the environment and optimize sustainability.

Besides focusing on waste reduction, recycling, and safe disposal, WM is also developing sources of renewable energy, by recovering the naturally occurring gas inside landfills to generate electricity. In 2012, WM generated more than twice the amount of renewable electricity than the entire US solar industry.

WM is an example of a company that is tackling sustainability from different angles and arriving at solutions that benefit the community and the corporation. Even though the GRI framework is not being applied, the company has demonstrated that sustainability is a priority and is achieving its environmental goals.

Sourced and adapted from www.wm.com

8.10 BEING INSINCERE: GREENWASHING

There are numerous companies wanting to promote themselves as sustainable because green claims can be an essential promotional instrument to generate impact on consumers and, consequently, increase sales. Because business world is highly competitive, some corporations see an opportunity to capture the market share by representing themselves as environmentally concerned. Today, it is well known that consumer demand for greener products is changing the world (Terrachoice, 2010), and companies compete for the attention of these "green" consumers and eventually for the profits it would bring. In fact, for example, the "sins of greenwashing" research results show us that from 2009 to 2010, the number of "greener" products offered has increased by more than 70%, and from these increased numbers, more than 95% of the consumer products claiming to be green has been found committing at least one of the "sins of greenwashing"—that is, hidden trade-off, no proof, vagueness, worshipping false labels, irrelevance, lesser of two evils, and sin of fibbing (Terrachoice, 2010).

"Greenwash" is defined in the 10th edition of the Concise Oxford English Dictionary as the "Disinformation disseminated by an organization so as to present an environmentally responsible public image" (Oxford English Dictionary, 2014). According to Mohr (2005), greenwash is one of the main threats to the advancement of corporate sustainability, as well as one of the most taboos. He argues that businesses practice "greenwashing" for a variety of reasons, such as selling products with differentiation and enhancing reputation with environmental attitudes. Furthermore, many argue it is inevitable for the businesses to eventually practice greenwashing to enhance their position in the market. Examples in the marketplace illustrate "greenwashing" is concurrently an immense threat to the advancement of corporate sustainability.

Some argues the significance of marketing communications as a main contributor to greenwashing of the companies (Jadhi and Acikdilli, 2009). According to Jadhi and Acikdilli (2009), "most definitions of marketing invariably refer to customer need and want satisfaction," and to satisfy these needs and wants, marketing communications use all kinds of mediums in many different ways, such as "greenwashing." In her paper, Engel (2006) introduces a model to explain why some companies voluntarily engage in environmental issues in spite of higher costs and some others, on the other hand, practice greenwashing. She argues that in most cases quality of the green products—she refers to the environmental, social, health, or safety impacts of a product—cannot be observed by consumers after purchase and tries to understand this phenomenon with developing a model. Her results indicate when the quality of the product associated with the voluntary activity is large, monitoring intensity is high, and related cost increase is low, and additionally, when consumers believe the firm's honesty is high; the company is accepted to be a candid type that always prefers to produce high quality with genuine attempts. Typically, when these conditions are not met, she concludes, "we can expect greenwashing" (Engel, 2006).

According to Dahl (2010), the "greenwashing" has escalated because of constantly increasing competition between companies as well as declining US economy. The companies across different sectors are seeing the benefit of promoting themselves as "green," even if it is not a genuine attribute. However, confusion caused by greenwashing by various companies can be extremely harmful for consumers. According to Dahl, if this harm occurs, it can be a threat to the environment and public health. If the fact that consumers would not believe and become completely skeptical on all of the green claims, there can be a danger of losing a powerful tool for environmental improvements. Interestingly, in his article "Stemming the Tide of Greenwash," Gillespie (2008) brings a similar perspective to the issue with saying, "in the long term this may have a negative impact on public engagement with wider environmental issues." He stresses that intensified misleading environmental claims can cause consumers to undermine the environmental issues in the future. Therefore, the authors proclaim the necessity of a unified approach is to overcome these challenges.

Another reason asserted to be for "greenwashing" is to exhibit an image of compliance with demands of public, NGOs, and governmental agencies. In his paper, Alves (2009) states: "Green marketing and advertising are ever popular strategies to reconcile business interests with ecological interests, and more precisely, with the increased concern for sustainability issues." But obviously, as we mentioned earlier, not all companies are using green marketing legitimately. Some companies do green marketing to lessen the pressures for intensified standards. Especially recently, with the increasing regulations and rules by Federal Trade Commission (FTC), the companies are feeling more pressures to present themselves as more eco-friendly than they possibly are (Dahl, 2010). Alves (2009) suggests that as long as companies are not punished severely for contemptible practices, consequential progress toward sustainability will not be attained. This idea advocates the importance of private as well as publicly and independently monitored regulatory systems. Similarly, Gibson (2009) discusses the importance of FTC and EPA on setting the regulations and rules to limit the illegitimate activities of the corporations. However, he also adds that in spite of the great significance of the mentioned authorities' rules and limitations (named as Green Guides), these rules do not effectively control the increasing problem of "greenwashing" by themselves. He supports the idea of collaborative work to successfully deal with the problem, such as bringing the FTC and the EPA together. He concludes with stating: "With the environmental expertise of the EPA and the consumer knowledge of the FTC, the two agencies could ably protect consumers from deception in advertising and marketing, provide businesses with workable guidelines within which they must work, and ultimately, protect the environment by enabling consumers to choose truly green products and services" (Gibson, 2009). This perspective brings a unique approach to solve the problem of greenwashing.

Greenwashing can also be a way of increasing company reputation—unless it gets caught. Mohr (2005) argues that the company's reputation can be up to 60% of its total worth, which can easily cause corporations to try to find shortcuts. When this is the case, ultimately, greenwashing can be a way of augmenting one's reputation, especially in the short term. Selling products, in today's highly competitive markets, is extremely difficult without differentiation. This can be a hard mission for companies because of highly integrated markets and products. Therefore, differentiation on the basis of environmental attributes can be a way for businesses. If this approach is supported by the real attributes, then this can result with companies' financial gains, even though in some cases, disingenuous activities also can lead to similar results. Some scholars, such as Alves (2009) and Gibson (2009), argue the importance of regulations and rules to limit this kind of activities of the firms; but we believe these limitations, in a way, can increase the companies' efforts to look more environmentally friendly than they really are. This kind of culture also can lead to legal approval being the only point at which the issue of environmental claims is considered. Thus, we can say that there is a need for more voluntarily actions of companies supported by various authorities. Also, it is important to add that the collaborative actions of those authorities are imperative for reduction or, preferably, elimination of misleading claims.

In many cases, greenwashing may occur due to ignorance of the law and of societal expectations. In these kinds of instances, as Boiral (2009) argues, business culture and, more specifically, adopted behavior by company individuals play an important role. The managers, company teams, suppliers, consumers, and basically all involved stakeholders would like to hear the good news (Mohr, 2005); however, in the case of lacking transparency and honesty, it might be possible to perceive more illicit activities. Therefore, the importance of the corporate culture and behavior should not be disregarded when we investigate the reasons and possible solutions of this major problem. If there is a lacking trust and integrity in the corporation's body, indisputably, it would be difficult to implement any kind of regulations.

Although it seems difficult, as Aras and Crowther (2009) state: "sustainability is not only achievable but sustainable development is also a realistic possibility" for companies, especially, without "greenwashing." At large, corporations should balance the economic growth with social equity and environmental protection, which has been typically identified as business sustainability development. As Lovelock (1979) points out in Gaia hypothesis, viewing an organization as a part of a wider social and economic system is important. Hence, we believe any harmful act by them must be taken into account seriously for the future generations, since the negative costs can be increasingly evident as depleted natural resources, destroyed habitats and living conditions, and social conflicts.

8.11 CASE STUDIES

8.11.1 International Business Machines

Founded in 1911, the International Business Machines (IBM) Corporation is an American multinational technology and consulting corporation headquartered in New York, United States. The company revenue from continuing worldwide operations was $99.7 billion with a net income of $16.4 billion in 2013. IBM's worldwide number of employees was reported as 431,212 in 2013 (http://www.ibm.com/ibm/us/en/?lnk=fai-maib-usen).

In all of its business functions, from operations to design of its products and use of technology, IBM is committed itself to environmental leadership in the industry it serves worldwide. This is the reason the company was issued a corporate policy on environmental affairs in the very early years of environmental movement in 1971. The policy is supported by the company's global environmental management system, which is the key element of the company's efforts to achieve results consistent with environmental leadership. It also ensures the company to be vigilant in protecting the environment across all of its operations. The company's annual corporate environmental reporting first began in 1990 and has continued each year since.

In their latest corporate environmental affairs report in 2013, IBM listed some voluntary environmental performance goals and reported their results. According to this report, the company has four different types of environment protection goals as follows.

8.11.1.1 Conservation

- Energy conservation goal: Energy conservation savings equal to 3.5% of the company's total energy use annually. The company's energy saving was 6.5% of its total energy use in 2012.
- Water conservation goal: Water conservation savings equal to 2% of the company's total water use annually (at microelectronics manufacturing operations). The company achieved average annual water savings of 2.2% by the end of 2012.

8.11.1.2 Climate Protection

- CO_2 emissions reduction goal: IBM was able to reduce its CO_2 emissions by 40% between 1990 and 2005 based on its global energy conservation program. The company set an aggressive "second-generation" goal for itself as to reduce the emissions associated with its energy use 12% more between 2005 and 2012 via energy conservation and renewable energy procurement. By the end of 2012, their energy conservation and procurement of renewable energy resulted in a 15.7% reduction in CO_2 emissions from the 2005 base year goal. Through its annual energy conservation actions, the company was also able to save 6.1 billion kWh of electricity consumption (as well as $477 million) between 1990 and 2012.

8.11.1.3 Product Stewardship

- Recycled plastics goal: The goal is to ensure at least 5% of the total plastics procured annually by IBM and its suppliers to be from recycled plastics. In 2012, 12.6% of IBM's total weight of plastic purchases was recycled plastic, surpassing the corporate goal by 7.6%.
- Product recovery and recycling goal: The amount of product waste sent to landfills or incineration by the company for treatment should be less than 3% of the total amount processed. For this, reusing or recycling end-of-life products becomes crucial. In fact, in 2012, IBM's product end-of-life management operations processed end-of-life products for reuse or recycling worldwide and minimized its combined product landfill use and incineration rate by sending only 0.3% of the total to landfills or to incineration facilities for treatment.
- Product energy efficiency goals: This goal is to continually improving the computing power delivered for each kilowatt-hour (kWh) of electricity used with each product (new generation or model). The company was able to release new server products/models with 10–93% more computing power for each kWh of electricity used than the previous models/products in 2012.

8.11.1.4 Pollution Prevention

- Hazardous waste reduction goal: Reducing hazardous waste generation from the company's manufacturing processes. The 2012 reports showed that the company's hazardous waste generation indexed to output increased 2.9% (68 metric tons). Thus, this goal was not achieved.
- Nonhazardous waste recycling goal: The goal is to send an average of 75% of the generated nonhazardous waste for recycling. The company was successful in reaching this goal in 2012 by sending 87% of its nonhazardous waste to be recycled (sources: http://www.ibm.com/ibm/environment/; http://www.ibm.com/ibm/environment/annual/ibm_envkpi_2012.pdf).

8.11.2 Kraft Foods, Inc.

Kraft Foods Group is an American multinational confectionery, food, and beverage conglomerate. The US headquarters of the company is located in Northfield, in which they operate. They also encourage the suppliers to surpass these baseline requirements where possible so that these suppliers would have less impact on the environment.

Campbell gives responsibility to its employees to observe and advance its environmental sustainability policy listed above. Furthermore, company's environmental performance is watched by its Sustainability Leadership Team. This team along with the public affairs and corporate responsibility vice president reports progress regularly and takes comments and directions from the CEO and the board of directors of the company. The company promotes transparency and reports its sustainability performance regularly (http://www.campbellsoupcompany.com/).

8.11.4 Marks and Spencer

Marks and Spencer (also known as M&S, colloquially known as Marks and Sparks) is a major British retailer headquartered in Westminster, London, with 703 stores in the United Kingdom and 361 stores in more than 40 countries. Founded in 1884, M&S specializes in clothing and luxury food products (https://marksintime.marksandspencer.com/home).

The company launched Plan A in January 2007 with the aim of combating climate change, reducing waste, using sustainable raw materials, trading ethically, and helping its customers to lead healthier lifestyles. They set out 100 commitments that they aim to achieve in 5 years. Later, they increased the number of commitments to 180 to achieve by 2015. Their ultimate goal with these commitments and initiatives is to become the most sustainable major retailer in the world.

To meet the commitments stated in Plan A, the company is working with its customers and suppliers closely. For example, its customers give their clothes to the company and M&S helps those people in need. M&S also works with Oxfam to resell, reuse, or recycle unwanted clothes and helps support people living in poverty.

With the ads "Give Clothes a Future, Shwop: Old clothes shouldn't just be thrown out, they should have a future," the company gets attention of many concerned consumers all around the world. This attention helps with gathering clothes to Oxfam to resell online, in its stores or in international markets where there is demand (e.g., bras in Africa, warm clothing in Eastern Europe). If the items are not sold, then they could be recycled. As a result of this initiative, the company was able to collect 11 million donated items worth £8 million to the charity.

By shwopping, the company hopes to achieve a reduction in the number of clothes sent to landfills. They mainly want to collect as many clothes as they sell, which is another inspiring commitment of the company (http://corporate.marksandspencer.com/documents/policy-documents/plan-a-report-2013.pdf).

8.11.4.1 Best Practices from Cases

Company/Industry	Areas of Focus	Best Practices
New Britain Palm Oil Limited/agriculture	Reducing and managing its carbon footprint; addressing concerns about community impacts, human rights, and gender equality	Vertical integration of sustainable activities into all aspects of production, from planting to refining
Tata Motors/automotive	Working with government to develop a framework for scrap car collection, treatment, dismantling, and recovery	Addressing ecological impacts by implementing water recycling programs during manufacturing to reduce water use
Qantas/aviation	Improving fuel optimization and reducing noise impact through fleet renewal	Use of biofuel and participation in carbon offset purchase program; lower consumption of electricity and water and lower amounts of waste disposed to landfills
Clariant/chemicals	Responsible stewardship, programs to reduce incidents of spills and accidents, community involvement	Linking incentives such as employee bonuses to specific objectives in the area of sustainability
Ernst & Young USA/commercial services	Reducing carbon footprint and waste stream for office sites	Using renewable energy for data centers, LEED- and ENERGY STAR-certified offices
Lenovo/computers	Life cycle management, using more environmentally sensitive materials in products	Programs to promote product and parts recycling, reducing waste to landfills, using recycled postconsumer plastic in new designs
Kawasaki Heavy Industries/construction materials	Reducing resource consumption, minimizing harmful emissions, and recycling materials; initiatives to promote environmentally conscious logistics practices	Green procurement guidelines, use of low environmental impact technologies into operations

(Continued)

(*Continued*)

Company/Industry	Areas of Focus	Best Practices
LG Electronics/consumer durables	Designing energy-efficient products that limit the use of conflict minerals and hazardous components and include more recoverable materials	Internal programs to increase employees' awareness of environmental and social issues, product recycling programs
Hess Corporation/energy	Reducing water and energy consumption, limiting GHG emissions and the use of pollutants, implementing spill prevention initiatives	Conducting biodiversity screenings during site evaluation for operations, use of new drilling techniques to minimize environmental impacts
Bank of America Corporation/financial services	Reduction of GHG emissions at facilities, conservation of natural resources	Issuance of green bond to provide financing for renewable energy and energy efficiency projects
The Carlsberg Group/food and beverage products	Packaging reduction, water conservation, efficiency improvements; optimizing logistics; increasing use of renewable energy; reducing emissions	Programs to conserve water and protect water quality, increased use of recyclable packaging, development of sustainable packaging
Kimberly–Clark Corporation/household and personal products	Reduction in water consumption, GHG emission, and waste to landfill; supply chain management	Use of alternative sources of fiber for wipes business, reduction in postconsumption waste through collaboration with hospitals
Panalpina/logistics	Reducing electricity, fuel, and water consumption; preventative management for fuel spills	Providing tools to customers that allow them to assess transport efficiency and opportunities for reducing environmental impact
Aquarius Platinum Limited/mining	Minimizing energy and water use and limiting carbon emissions. Achieving zero major reportable environmental incidents and accidents	Providing reports to the Carbon Disclosure Project and Water Disclosure Project, using technology to limit safety risks
Bombardier Transportation/railroad	Designs that are energy efficient, with low levels of noise and emission; reducing use of hazardous material and nonrecyclable content	Applying design for the environment and design for safety methodologies to new products
H&M/retail	Reducing consumption use, incorporating textiles that can be cleaned in an environmentally sensitive manner	Managing its supply chain to identify social and environmental impacts, implementing programs to encourage recycling
Viyellatex Group/textiles and apparel	Minimizing its carbon footprint and reducing energy and water use	Reusing treated and utility water, harvesting rain water, recovering heat from manufacturing processes, ensuring that its suppliers are following sound environmental policies
Royal Caribbean Cruises Limited/tourism and leisure	Reducing ocean and air pollution via environmentally friendly ship design, waste water treatment systems, and reducing waste streams	Use of newer ships with solar technology and energy-efficient designs
Waste Management/waste management	Waste reduction, recycling, and safe disposal to minimize environmental impacts	Recovering naturally occurring gas inside landfills to generate electricity
International Business Machines (IBM)/technology services	Reducing emissions, recycling and product recovery, water and energy conservation, product efficiency improvements, pollution prevention	Releasing new server products/models with 93% more computing power for each kWh of electricity used than the previous models/products in 2012
Kraft Foods, Inc./food and beverage products	Reducing energy and water use; reducing CO_2 emissions, packaging, and waste; reducing mileage from its network; sourcing all coffee brands sustainably	Sustainable sourcing of agricultural commodities. Reaching new agricultural benchmarks by using third-party certifiers, such as Fairtrade, Rainforest Alliance, and 4C Association
Campbell Soup Company/consumer products and food manufacturing	Reducing energy and water use and waste generation, increasing recycling, sustainable packaging, sustainable agriculture, and supply chain optimization	Gives responsibility to its employees to observe and advance its environmental sustainability policy. Environmental performance is watched by its Sustainability Leadership Team
Marks & Spencer/retail and department stores	Waste reduction and recycling to minimize environmental impacts	Company tries to achieve a reduction in the number of clothes sent to landfills. They want to collect as many clothes with shwopping

8.12 THE FUTURE OF BUSINESS SUSTAINABILITY

The significance of business strategies to support environmental sustainability, or green strategies, is increasingly acknowledged by countless business executives, leaders, researchers, NGOs, consumers, and, most importantly, governments all around the world. In his landmark 2007 article in *The New York Times Magazine* titled "The Power of Green," noted economist Thomas L. Friedman provides us with an illuminating and thought-provoking treatise of the emerging green ideology, which "has the power to mobilise liberals and conservatives, evangelicals and atheists, big business and environmentalists around an agenda that can both pull us together and propel us forward" (Friedman, 2007). Green ideology, as Friedman states, has geostrategic, geoeconomic, capitalistic, and patriotic power that can reconnect and reunite America at home and restore its place as "hope and inspiration" around the world. A country that adopts living, operating, designing, manufacturing, and projecting in a new "green" way can have a unifying movement that can bring sustained success for the twenty-first century. This is a crucial opportunity that should not be neglected. In fact, the marketplace is already starting to realize the indisputable power of green. It is giving a new direction that is helping many companies, the environment, and their customers and will definitely help our grandkids, eventually.

Although slow moving and subdued, debate over climate–energy, environment–consumption, habitat–waste, and similar ones are out there and are leading to transformations widely, at small—and occasionally large—scales. Nevertheless, fast pace on moving toward green ideology can help achieve competitive global advantage for corporations. Green has to become part of a company's DNA, it has to hit Main Street, and, most importantly, at the individual level, it should mean more than a hobby and should outline the new way of survival for the company. This new way of life can bring a series of great opportunities toward success as well as solutions to countless environmental problems.

Integrating environmental, social, ethical, and economic issues into all decisions of an organization can help to achieve a balance of protecting the environment, creating economic development, increasing stakeholder value, and providing meaningful jobs, as part of creating a safe, just, and equitable society. These can be accepted as the core principles of today's business sustainability practices. As underlined in this chapter, the modern world requires companies to carefully consider the finite resources on Earth, interdependence for survival, need for minimizing inequalities in quality of life as well as social and economic well-being, necessity of respect for different cultures, and need for taking responsibility for leaving a legacy for future generations.

Sustainability practices can be incorporated into all phases of the value chain, from financing to postconsumption activities. The case studies included in this chapter illustrate the various ways that successful businesses in all industries are answering the need to build responsibility into their business strategies and practices. These companies have demonstrated that sustainability can be good for business, with results that support benefits in financial and nonfinancial aspects of performance. While there remains a lot of work to be done in improving our environment and righting social inequities, companies that address CSR issues in their strategy and business models are proving that it can be done.

Sustainability is not about cutting back or staying at a halt. It is creating a new way of lifestyle and inventing new industries that encourage innovative abundance for the next generations. Sustainability is about attaining dreams without destroying the Earth and becoming powerful by doing good.

8.13 REVIEW QUESTIONS

1. Is there a relationship between companies' socially responsible actions and their financial success?
2. What are the reasons that companies focus on environmentally responsible strategies? In other words, what are the motivations to act environmentally responsible for companies?
3. How can a company position itself by adopting a sustainable business strategy?
4. How can a company differentiate itself by adopting a sustainable business strategy?
5. List the value chain impacted by sustainable environmental strategies.
6. When compared, which approach to corporate responsibility is a better fit: shareholder theory or stakeholder theory?
7. What is Global Reporting Initiative (GRI)? How do the companies use it?
8. What is greenwashing? Why do some companies in the marketplace practice greenwashing?

REFERENCES

Alves, I.M. 2009. Green spin everywhere: how greenwashing reveals the limits of the CSR paradigm. *Journal of Global Change and Governance*. 2(1):1–26.

Aras, G., and D. Crowther. 2009. Corporate sustainability reporting: a study in disingenuity? *Journal of Business Ethics*. 87(1):279–288.

Bansal, P., and A.J. Hoffman. 2012. *The Oxford Handbook of Business and the Natural Environment*. Oxford University Press, Oxford.

Barrow, C.J. 2006. *Environmental Management for Sustainable Development*, 2nd edition. Routledge, London/New York.

Baumgartner, R.J., and D. Ebner. 2010. Corporate sustainability strategies: sustainability profiles and maturity levels. *Sustainable Development*. 18(2):76–89.

Beurden, P.V., and T. Gossling. 2008. The worth of values—a literature review on the relation between corporate social and financial performance. *Journal of Business Ethics*. 82:407–424.

Biloslavo, R., and A. Trnavcevic. 2009. Web sites as tools of communication of a "green" company. *Management Decision*. 47(7):1158–1173.

Bloemers, R., F. Magnani, and M. Peters. 2001. Paying a green premium. *The McKinsey Quarterly*. 3:15–17.

Boatright, J.R. 1994. What's so special about shareholders? *Business Ethics Quarterly*. 4(4):393–408.

Boiral, O. 2009. Greening the corporation through organizational citizenship behaviors. *Journal of Business Ethics*. 87:221–236.

Bonini, S., and J. Oppenheim. 2008a. Cultivating the green consumer. *Stanford Social Innovation Review*. 6:56–61.

Bonini, S., and J. Oppenheim. 2008b. Helping 'green' products grow. *The McKinsey Quarterly*. 3(2):1–8.

Bouton, S., J. Creyts, T. Kiely, J. Livingston, and T. Nauclér. 2010. *Energy Efficiency: A Compelling Global Resource*. McKinsey & Company, New York.

Boxer, L. 2007. Sustainability perspectives. *Philosophy of Management*. 6(2):86–98.

Brammer, S., and M. Millington. 2008. Does it pay to be different? An analysis of the relationship between corporate social and financial performance. *Strategic Management Journal*. 29:1325–1343.

Callan, S.J., and J.M. Thomas. 2009. Corporate financial performance and corporate social performance: an update and reinvestigation. *Corporate Social Responsibility and Environmental Management*. 16:61–78.

Cannon, T. 1992. *Corporate Responsibility*, 1st edition. Pitman Publishing, London.

Chan, N.W., and M.J. Kotchen. 2014. A generalized impure public good and linear characteristics model of green consumption. *Resource and Energy Economics*. 37:1–16.

Croner Publications Ltd. 1997. *Croner's Environmental Policy and Procedures*. Croner Publications, Kingston-upon-Thames, UK.

Dahl, R. 2010. Greenwashing: do you know what you're buying? *Environmental Health Perspectives*. 118(6):A246–A252.

Ebner, D., and R.J. Baumgartner. 2006. The relationship between sustainable development and corporate social responsibility. Corporate Responsibility Research Conference (CRRC), Dublin, Ireland. September, pp. 4–5.

Eccles, R.G., and M. Krzus. 2010. *One Report: Integrated Reporting for a Sustainable Strategy*. John Wiley & Sons, Inc., New York.

Eccles, R.G., I. Ioannou, and G. Serafeim. 2012. The impact of corporate sustainability on organizational processes and performance. National Bureau of Economic Research working paper. 17950:1–56.

Elkington, J. 1998. *Cannibals with Forks*. New Society Publishers, Gabriola Island, Canada.

Engel, S. 2006. Overcompliance, labeling, and lobbying: the case of credence goods. *Environmental Modeling and Assessment*. 11:115–130.

Esty, D.C., and A.S. Winston. 2006. *Green to Gold*. Yale University Press, New Haven.

Foster, C., and K. Green. 2000. Greening the innovation process. *Business Strategy and the Environment*. 9:287–303.

Frederick, W.C. 1986. Toward CSR3; why ethical analysis is indispensable and unavoidable in corporate affairs? *California Management Review*. 28:126–41.

Frederick, W.C. 1994. From CSR1 to CSR2. *Business and Society*. 33:150–166.

Freeman, R.E. 2004. A stakeholder theory of the modern corporation. p. 55–74. In T.L. Beauchamp and N.E. Bowie (eds.) *Ethical Theory and Business*. Pearson Prentice Hall, Upper Saddle River, NJ. Available at: http://academic.udayton.edu/lawrenceulrich/Stakeholder%20Theory.pdf

Friedman, M. 1970. The social responsibility of business is to increase its profits. *The New York Times Magazine*. September 13, 32–33, 122, 124, 126.

Friedman, T. 2007. The power of green. *The New York Times Magazine*, April 15, pg. 19. Accessible at: http://www.nytimes.com/2007/04/15/magazine/15green.t.html?pagewanted=all&_r=0 (accessed May 6, 2015).

Garriga, E., and D. Mele. 2004. Corporate social responsibility theories: mapping the territory. *Journal of Business Ethics*. 53:51–71.

Gibson, D. 2009. Awash in green: a critical perspective on environmental advertising. *Tulane Environmental Law Journal*. 22:423–440.

Gilbert, K. 2014. Investment firms and corporates embrace sustainable bonds [Online]. Available from http://www.institutionalinvestor.com/article/3328573/asset-management-fixed-income/investment-firms-and-corporates-embrace-sustainable-bonds.html#.VYpGwflViko (accessed on April 5, 2015). Institutional Investor April 8.

Gillespie, E. 2008. Stemming the tide of 'greenwash' [online]. Available from http://www.greenwashreport.org/downloads/stemming_the_tide_08.pdf. *Consumer Policy Review*. 18(3):79.

Global Reporting Initiative. 2007. Making the connection: the GRI guidelines and the global compact communication on progress [online]. Available from https://www.globalreporting.org/resourcelibrary/GRI-UNGC-Making-The-Connection.pdf (accessed on April 5, 2015). Global Reporting Initiative.

Hart, S.L. 1997. Beyond greening strategies for a sustainable world. *Harvard Business Review*. 75:66–76.

Hart, S.L., and M.B. Milstein. 1999. Global sustainability and the creative destruction of industries. *Sloan Management Review*. 41:23–33.

Harvard Business Review. 2003. *Business Review on Corporate Social Responsibility*. Harvard Business School, McGraw-Hill, London.

Hasnas, J. 1998. The normative theories of business ethics: a guide for the perplexed. *Business Ethics Quarterly*. 8:19–42.

Heck, S., M. Rogers, and P. Carroll. 2014. *Resource Revolution: How to Capture the Biggest Business Opportunity in a Century*. New Harvest, New York, NY.

Hunt, S.D., and D.F. Davis. 2008. Grounding supply chain management in resource-advantage theory. *Journal of Supply Chain Management.* 44(1):10–21.

Jadhi, K.S., and G. Acikdilli. 2009. Marketing communications and corporate social responsibility (CSR): marriage of convenience or shotgun wedding? *Journal of Business Ethics.* 88:103–113.

Kearney, A.T. 2009. Green winners: the performance of sustainability-focused companies during the financial crisis [online]. Available from http://www.atkearney.com/documents/10192/6972076a-9cdc-4b20-bc3a-d2a4c43c9c21 (accessed on April 5, 2015).

Kumar, S., S. Teichman, and T. Timpernagel. 2012. A green supply chain is a requirement for profitability. *International Journal of Production Research.* 50(5):1278–1296.

Laff, M. 2009. Managing organizational knowledge. "Triple bottom line: creating corporate social responsibility that makes sense" [online]. Available from http://www.astd.org/TD/Archives/2009/Feb/Free/0902_Triple_Bottom_Line.htm. Retrieved on January, 24, 2012.

Locke, R. 2009. Sustainability as fabric—and why smart managers will capitalize first. An interview with Richard M. Locke. MIT Sloan Management Review. January [online]. Available from http://sloanreview.mit.edu/article/sustainability-as-fabric-and-why-smart-managers-will-capitalize-first/ (accessed on February 28, 2015).

Lovelock, J.E. 1979. Gaia as seen through the atmosphere. *Atmospheric Environment.* 6(8):579–580.

Lovins, A.B., H.L. Lovins, and P. Hawken. 1999. A road map for natural capitalism. *Harvard Business Review.* May–June:145–158.

Magretta, J. 2012. *Understanding Michael Porter: The Essential Guide to Competition and Strategy.* Harvard Business Press, Watertown, MA.

Makower, J. 2009. *Strategies for the Green Economy.* McGraw Hill, New York.

Margolis, J.D., and J.P. Walsh. 2003. Misery loves companies: rethinking social initiatives by business. *Administrative Science Quarterly.* 48:268–305.

Marom, I.Y. 2006. Toward a unified theory of the CSP–CFP link. *Journal of Business Ethics.* 67:191–200.

McIntosh, M., D. Lepziger, K. Jones, and G. Coleman. 1998. *Corporate Citizenship: Successful Strategies for Responsible Companies.* Financial Times Management, Pitman, London.

McKinsey. 2010. How Companies Manage Sustainability. By Bonini, S., Gorner, S., and Jones, A. McKinsey Global Results. Retrieved from http://www.mckinsey.com/insights/sustainability/how_companies_manage_sustainability_mckinsey_global_survey_results (accessed on May 6, 2015).

McKinsey. 2011. The Business of Sustainability: Putting it Into Practice. By Bonini, S., and Gorner, S. McKinsey Global Survey Results. Retrieved from http://www.mckinsey.com/insights/energy_resources_materials/the_business_of_sustainability_mckinsey_global_survey_results (accessed on May 6, 2015).

Meadows, D.H., E.I. Goldsmith, and P. Meadow. 1972. *The Limits to Growth.* Volume 381. Earth Island Limited, London.

Mitchell, R.K., B.R. Agle, and D.J. Wood. 1997. Toward a theory of stakeholder identification and salience: defining the principle of who and what really counts. *Academy of Management Review.* 22(4):853–886.

Mohr, T. 2005. Reputation or reality? A discussion paper on greenwash & corporate sustainability [online]. Available from http://docs.google.com/viewer?a=v&q=cache:GzG4LAPTaHYJ:www.tec.org.au/greencapital/component/docman/doc_download/12-greenwashdiscussionpaperreputationorreality2005.html+argument+on+corporate+sustainability+greenwashing&hl=en&gl=us&pid=bl&srcid=ADGEEShUysKkMqLHAt2Ket GSGDgQThx2eYn3b3dr-kQkAl5aSJ6GzwBZgTd_H3LbO 10gVCYFSOihVKkrGMfDlojli7bGifSTVSMcV7TGJLdT3H PemsPt-F1YtM0ysWiqc-IJvfyYqLzP&sig=AHIEtbR1rGq7NF petYiRIUnj9ufdFPcRIw. Retrieved on November 21, 2010. Total Environment Centre. Green Capital Program. Sydney, Australia.

Moir, L. 2001. What do we mean by corporate social responsibility? *Corporate Governance.* 1(2):16–22.

Mukherjee, A., and N. Onel. 2013. Building green brands with social media: best practices from case studies. *Journal of Digital & Social Media Marketing.* 1(3):292–311.

O'Rourke, D. 2004. *Opportunities and Obstacles in CSR Reporting in Developing Countries.* University of California, Berkeley, CA.

Orlitzky, M., F.L. Schmidt, and S.L. Rynes. 2003. Corporate social and financial performance: a meta-analysis. *Organization Studies.* 24(3):403–441.

Oxford English Dictionary. 2014. Greenwash. Available from http://www.oxforddictionaries.com/us/definition/american_english/greenwash (accessed on February 28, 2015).

Porter, M.E. 1980. *Competitive Strategies.* Wiley, New York.

Porter, M.E., and M.R. Kramer. 2006. Strategy and society: the link between competitive advantage and corporate social responsibility. *Harvard Business Review.* 84:78–92.

Prakash, A. 2002. Green marketing, public policy and managerial strategies. *Business Strategy and the Environment.* 11(5):285–297.

Roberts, S. 2003. Supply chain specific? Understanding the patchy success of ethical sourcing initiatives. *Journal of Business Ethics.* 44(2/3):159–170.

Samy, M., G. Odemilin, and R. Bampton. 2010. Corporate social responsibility: a strategy for sustainable business success. An analysis of 20 selected British companies. *Corporate Governance.* 10(2):203–217.

Schaefer, B.P. 2008. Shareholders and social responsibility. *Journal of Business Ethics.* 81:297–312.

Schoemaker, P.J.H., and G.S. Day. 2011. Innovating in uncertain markets: 10 lessons for green technologies. *MIT Sloan Management Review.* 52(4):37–45. Available from http://search.proquest.com/docview/875531952?accountid=12536 (accessed on February 28, 2015).

Shafaat, F., and A. Sultan. 2012. Green marketing. *EXCEL International Journal of Multidisciplinary Management Studies.* 2(5):184–195.

Simpson, G.W., and T. Kohers. 2002. The link between corporate social and financial performance: evidence from the banking industry. *Journal of Business Ethics.* 35:97–109.

Soana, M. 2011. The relationship between corporate social performance and corporate financial performance in the banking sector. *Journal of Business Ethics.* 104:133–148.

Terrachoice 2010. Greenwashing Report 2010. The sins of greenwashing—home and family edition [online]. Available from http://sinsofgreenwashing.org/findings/greenwashing-report-2010. Retrieved on November 21, 2010 (accessed on February 28, 2015).

Wang, W. 2012. A study on consumer behavior for green products from a lifestyle perspective. *Journal of American Academy of Business*. 18(1):164–170.

Werbach, A. 2013. *Strategy for Sustainability: A Business Manifesto*. Harvard Business Press, Watertown, MA.

Wheeler, D., B. Colbert, and R.E. Freeman. 2003. Focusing on value: reconciling corporate social responsibility, sustainability and a stakeholder approach in a network world. *Journal of General Management*. 28(3):1–28.

World Commission on Environment and Development. 1987. *Our Common Future*. Volume 383. Oxford University Press, Oxford.

World Economic Forum. 2012. More with less: scaling sustainable consumption and resource efficiency. Available from https://www.cdp.net/en-US/News/Documents/more-with-less.pdf (accessed on April 5, 2015).

9

GREEN MARKETING STRATEGIES

MEHDI TAGHIAN[1], MICHAEL JAY POLONSKY[1] AND CLARE D'SOUZA[2]

[1] Department of Marketing, Faculty of Business and Law, Deakin University, Melbourne, VIC, Australia
[2] La Trobe Business School, Melbourne, VIC, Australia

Abstract: This chapter presents the fundamentals of "green" marketing by drawing on traditional marketing theory as well as research focused on green marketing context. It discusses five critical areas in green marketing. The first critical area stems from green marketing theory and practice that examines the logic for reducing the environmental impact of value creation and exchange. The second critical area highlights green marketing strategy that focuses on achieving organizational goals in ways that can reduce or eliminate negative impacts on the natural environment. The third critical area examines the green marketing mix that accounts for green products, green distribution, green pricing, and green promotion. By using traditional marketing concepts, the chapter identifies how the entire marketing mix elements should consistently provide a complete green product offering. Green products and processes need to be researched, designed, and manufactured to include environmentally safe ingredients and components. Products need to be strategically priced to reflect their green values, distributed in the green chain channels and displayed effectively to highlight their status, and accurately communicated to consumers and stakeholders. The fourth critical area illustrates governance and control. It shows how the holistic transformation toward greening the organization requires organizational culture change to gain support within and outside the firm to ensure environmental issues are appropriately considered. These can be assessed by using existing management mechanisms, such as environmental management systems and/or triple bottom line management, which ensure best practice and continuous improvements to occur. Lastly, the chapter discusses the future of green marketing and the direction that businesses need to take if they seek to be sustainable.

Keywords: green marketing theory, green positioning, green demand, green purchase decisions, stakeholders green sentiments, green products, green manufacturing processes, green prices, green logistics, green distribution, green distribution chain, green promotion, environmental logos, third-party certification, environmental management system, triple bottom line accounting, green R&D.

9.1 Overview	232	
9.2 Introduction	232	
9.3 Green Marketing Theory and Practice	233	
9.4 Green Marketing Strategy	234	
9.4.1 Green Market Positioning	235	
9.4.2 Green Consumers and Purchase Decisions	235	
9.5 Greening The Marketing Mix	236	
9.5.1 Green Products	237	
9.5.1.1 Developing New Greener Products	237	
9.5.2 Green Prices	238	
9.5.3 Green Logistics and Distribution	239	
9.5.3.1 Supply Issues	239	
9.5.3.2 Suppliers in the Green Chain	240	
9.5.3.3 Physical Distribution and Logistics	241	
9.6 Green Promotion	241	
9.6.1 Advertising	242	
9.6.2 Environmental Logos/Third-Party Endorsements	242	
9.6.3 Sales Promotion (Cause-Related Marketing)	243	

An Integrated Approach to Environmental Management, First Edition. Edited by Dibyendu Sarkar, Rupali Datta, Avinandan Mukherjee, and Robyn Hannigan.
© 2016 John Wiley & Sons, Inc. Published 2016 by John Wiley & Sons, Inc.

9.6.4	Personal Selling	243		9.7.2.1 Continuous Improvement	245
9.6.5	Publicity/Public Relations	244	9.7.3	Advantages of Using ISO Series	246
9.6.6	Social Media	244	9.7.4	Other Management Systems	246
9.7 Governance, Evaluation, and Control		244	9.7.5	The TBL	247
9.7.1	Implementation	244	9.8 The Future of Green Marketing		248
9.7.2	Governance	245	9.9 Review Questions		249
			References		249

9.1 OVERVIEW

Marketing practitioners and academics confront a changing, evolving, and uncertain environment. This means that they must adopt a different perspective to accommodate changes taking place in the natural environment. These changes include a dramatic and nonreversible increase in global warming and the depletion of nonrenewable natural resources (Albino, Balice, and Dangelico, 2009). The increasing global temperature is partly being blamed on increased human consumption and associated production (Ring, Lindner, Cross, and Schlesinger, 2012).

The opinions on the causes of global warming vary depending on who is delivering the message. For example, it is suggested that at times, misinformation has been distributed by both environmental scientists and nonscientists alike (Newport, 2010). While there are those who believe in the accuracy of the scientific reports about the human contribution to global warming, there are others who refute this argument. These "naysayers" suggest that the human contribution to the global warming issue is overstated and that global warming is largely independent from human activities. The debate and alternative perspectives on global warming creates substantial confusion in the minds of many within the community, which is sometimes further blamed on sensationalistic journalism (Adams, 2010).

The reality is that environmental change is a scientific truth, whatever its cause(s) (Ring et al., 2012). At issue is how the public and business should respond. Some people see business as being responsible for failing to exercise a duty of care for their customers' welfare and consumers' concern for the environment. Business is expected to be more proactive and careful about their initiatives creating efficiencies in production processes that reduce emissions and global warming (Porter and Kramer, 2006). Businesses need to actively design and introduce environmentally safe products, use more eco-friendly ingredients in their products, and adapt corporate processes and products to be less environmentally harmful. If human consumption is contributing to the degradation of the environment, then consumer choices need to be provided in a way that allows them to minimize their negative impact on the environment. This puts a substantial degree of responsibility on the shoulders of business to respond to consumers' expectations.

It also assumes businesses have a leadership role in promoting more sustainable consumption, by providing appropriate choices to consumers. To achieve this, businesses need to design and implement proactive organization-wide environmental strategies (Chen, 2011).

The literature documents many cases where the adoption of environmental strategies leads to improved business outcomes. Benefits to business include improved organizational processes and reduction in nonrenewable resource use (Albino et al., 2009) with overall cost reductions and improved profitability. Greening of activities results in improved corporate reputation (Griskevicius and Tybur, 2010). Greener corporate behavior enables consumers who are environmentally oriented to change their purchase behavior, resulting in increased market share for green organizations (Cronin, Smith, Gleim, Ramirez, and Martinez, 2010). As such, increased environmental sentiment has influenced businesses adoption of proactive environmentally focused activities (Dangelico and Pujari, 2010).

Given the overall business benefits mentioned above, business needs to take a leadership role in responding to consumers' environmental concerns. This is important because while there is scientific uncertainty about the environmental changes taking place, consumers are increasingly concerned about these changes. In this situation, business may need to apply the precautionary principle (Weitzman, 2009), where caution is practiced in all business decision making given the uncertainty about environmental change (Di Salvo and Raymond, 2010).

This chapter presents the fundamentals of "green" marketing within this uncertain context. In doing so, it draws on traditional marketing theory and research as well as theory and research focused on green marketing context.

9.2 INTRODUCTION

"Marketing is the activity, set of institutions, and processes for creating, communicating, delivering, and exchanging offerings that have value for customers, clients, partners, and society at large" (American Marketing Association, 2012). It is a discipline that is driven by the market's needs and sentiments, whether these are real, perceived or anticipated. Consumption in the quest to satisfy wants is being blamed for accelerating environmental degradation, which will be exasperated as the global population and the level of consumption rise. It is unproductive to assume that consumers will spontaneously realize that they are contributing to the world's environmental problems and thus automatically reduce their level of consumption (Sheth, Sethia, and Srinivas, 2011). Given all the alternative consumption information available, it is also unreasonable to expect that consumers would be fully informed of the environmental impact of all alternative choices. However, it may be that consumers will change their choices if information about the alternative "green" products were available (Young, Hwang, McDonald, and Oates, 2009.

Businesses have an ethical, economic, and social responsibility for providing environmentally safe "green" product alternatives (Thøgersen, Jørgensen, and Sandagerand, 2012). Green marketing is an extension of this responsibility. A range of alternative terms have been used in the literature to describe firms' behavior in this regard, including green marketing, ecological marketing, environmental marketing, sustainable marketing, responsible marketing, and pro-environmental marketing, all of which are a subset of the broader concept of prosocial marketing. We will use the term *green marketing* to include all these alternative terms, which is defined as being "the holistic management process responsible for identifying, anticipating and satisfying the needs of consumers and society, in a profitable and environmentally sustainable way" (Peattie, 2001, p. 18). This definition places a substantial level of responsibility on marketing to be the arbiter between environmental conditions and developing "green"-oriented corporate marketing strategies. Green marketing is not something that is nice or benevolent to do, but something that business is expected by their internal and external stakeholders to do, requiring a change in corporate culture and behavior. The concept needs to indicate long-term corporate commitment, overcoming any conflicting issues in planning, implementation, and evaluation of organizational strategy.

Technological developments, especially in process technologies and materials, have enabled superior responses to environmental risks, leading to reduced waste through improved work processes, which reduce costs and therefore improve profits. There is growing anecdotal evidence that green strategy pays off financially (Klassen and McLaughlin, 1996; Miles, Munila, and Russell, 1997; Soyez, 2012). Such positive monetary results encourage businesses to embrace green initiatives rather than treat them as a burden or to consider them a philanthropic "feel-good" exercise (Cronin et al., 2010). A green initiative can be viewed as a strategic approach in reforming the organization, aligning itself with doing the "right thing" as well as being responsive to the firm's environmental and social obligations (Babiak and Trendafilova, 2011). This philosophy goes far beyond a short-term profit orientation. Thus, while at the present greening may be optional, in the future, it is likely that greening will be mandatory (Cronin et al., 2010).

Businesses that fail to adapt to consumer and society's green expectations stand to lose their competitive edge (Zhang, Shen, and Wu, 2011). Moreover, it is important to recognize the marketing opportunities and challenges that are inherent in this environmental transformation that requires a redefinition of the working relationships between business and society.

What follows is a review of some of the key issues of interest in green marketing. These are classified, structured, and discussed as follows:

- Green marketing theory and practice
- Green marketing strategy
- Green marketing mix
- Governance and control
- Future of green marketing

9.3 GREEN MARKETING THEORY AND PRACTICE

The theory of green marketing examines the logic for reducing the environmental impact of value creation and exchange. Green marketing is one indication that a business is being market oriented and acting in a responsible manner. This is based on the assumption that market dynamics require businesses to have a positive response to consumers' concerns regarding the link between environmental change and human consumption (Sheth et al., 2011). Consumers may not be willing to consider reducing consumption, but rather are more selective in their purchases, giving preference to products with positive environmental attributes. Extant research suggests that some segments of consumers appear to be more influenced by environmental issues and product characteristics than others (D'Souza, Taghian, and Khosla, 2007).

Green marketing is practiced by businesses in a variety of ways, resulting in consumers being presented a variety of "green" and conventional products, simultaneously. "Green" products claim to have a range of benefits, including being produced using more "renewable" or less toxic ingredients, having less or biodegradable packaging or having a lower carbon footprint, etc. Such points of difference may lead to consumer perceptions of inferior quality or inflated prices for green products compared with their conventional counterparts (Rex and Baumann, 2007).

Several studies have shown that "green"-oriented consumers are willing to change their consumption behavior and actively participate in efforts to preserve the environment through their product selection (Tobler, Visschers, and Siegrist, 2011). Business customers are also integrating environmental performance into their procurement evaluation criteria (Thompson, Anderson, Hansen, and Kahle, 2010).

While consumers expect manufacturers to be more responsible, there is also substantial confusion about how to best meet the needs of these consumers (Albino et al., 2009). Informing and educating consumers is important although knowledge alone is not necessarily effective in driving green purchases (Polonsky, Vocino, Grau, Garma, and Ferdous, 2012). To assist in communicating the environmental impact of goods, a number of policy initiatives have emerged. For example, the "energy rating" label assists consumers to compare the energy efficiency of domestic appliances and provides incentives for manufacturers to improve the energy performance of appliances (Department of Resources, 2013). There are also fuel consumption guides of different vehicles (Department of Infrastructure and Transport, 2013). Nongovernment groups sometimes also provide information on environmental performance of firms and their green activities (Kollmuss, Zink, and Polycarp, 2008).

9.4 GREEN MARKETING STRATEGY

"Green" marketing strategy refers to adopting business "practices that contribute to environmental protection, ecological responsiveness and social responsibility" (Zhang et al., 2011, p. 158). It incorporates marketing practices, policies, and procedures that align with and respond to consumer and society concerns about the natural environment while pursuing the objective of creating wealth for the owners (Leonidou, Katsikeas, and Morgan, 2012). Therefore, "green" marketing strategy focuses on achieving organizational goals in ways that reduce or eliminate negative impacts on natural environment. To implement a green strategy usually forces the redesigning of the entire marketing process, because greening requires a total systems approach and organizational commitment (Zhang et al., 2011). A green strategy is centered on environmentally oriented activities throughout the value chain. It requires long-term organizational commitment and therefore represents a substantial shift in cultural, decision making, market evaluation, and changes in the planning and implementation of the marketing programs (Cronin et al., 2010). Greening strategy would lead to changes in material sourcing, design, production, logistics, and distribution as well as targeted marketing communication. Therefore, green marketing embraces the systemic review of the activities in the entire value chain and the restructuring of links that do not fit in with the environmentally oriented aspects of the green strategy. Change needs to be deep-rooted in all activities of the firm and therefore should be a top-down strategic flow, ensuring effective allocation of resources.

Research suggests that there are a growing number of firms adopting green marketing strategies (Dangelico and Pujari, 2010) and that "greening" can result in positive impacts on corporate financial performance (Klassen and McLaughlin, 1996; Molina-Azorín, Claver-Cortés, López-Gamero, and Tarí, 2009). Changes within organizational boardrooms (Taylor and Kay, 2011) and new generation of socially responsible managers are additional evidences that the greening of marketing strategy is occurring (Dief and Font, 2010). Being a "green" company can provide a competitive advantage (Miles and Covin, 2000).

While many firms view greening as a positive move, some firms engage in selective green practices by greening some of their activities, for example, by producing conventional products alongside "green" products (Darnall, Henriques, and Sadorsky, 2010). If the "green" strategy is product category driven rather than being organizational driven, the duality of the approach may create market confusion, as well as confusing internally strategy development (i.e., how green are we really?), both of which may distort the corporate image, fragment the market, and confuse internal and external stakeholders (Albino et al., 2009). To be consistent, "green" strategy should represent a complete integrated strategy that sets the corporate direction and the framework for "green" strategic orientation, which is implemented in all corporate activities. It is not possible to have a sustainable "green" marketing orientation in a firm that is not organizationally committed (Albino et al., 2009).

The role of marketing function in implementing the "green" strategy is key. Marketing ensures that the preferences of the consumers are reflected in corporate actions and can assist in balancing financial performance objectives and environmental product quality. The wider use of the triple bottom line (TBL) concept (discussed in Section 7.5) reflects this more balanced organizational responsibility imperative (De Giovanni, 2011).

A company's first task is to decide whether to adopt a green orientation and, if so, how it will implement the "green" strategy—that is, to what extent does the firm want to position itself as being "green" and what marketing activities need to be undertaken to support this strategy. The second task is to then evaluate the market dynamics and to identify the challenges and opportunities, which will direct the strategies to be undertaken (Zhang et al., 2011).

The decision on how green to be cannot be taken lightly. Whatever strategy is selected, it will require substantial investments to deliver the outcomes, and effective "green" organizational strategy is a long-term prospect (Babiak and Trendafilova, 2011).

Secondly, the implementation of a green strategy requires changes in a range of other actions and activities, such as finding new suppliers, sourcing new raw materials or ingredients, altering quality control procedures to include environmental issues, and c distribution system, including the selection of new channel members (Mollenkopf, Stolze, Tate, and Ueltschy, 2010). Depending on the level of greening, there may also need to be a change in organizational culture and values, which is a much harder task to achieve. "Green" marketing strategy is a holistic organizational shift, not simply using superficial environmental claims in communication activities.

9.4.1 Green Market Positioning

"Green" product/brand positioning, like any other positioning, is concerned with creating a distinctive image in the minds of key stakeholders. Positioning functions as a distinctive element of the "green" marketing strategy, because it sets up a broad set of expectations that must then be delivered on (Hunt, 2011). Positioning has been defined as the part of the brand identity and value proposition that is to be actively communicated to the target audience (Aaker and Joachimsthaler, 2000). It should incorporate the various marketing components and is established by using a deliberate strategy over time to establish the status of the desirable image for the product/brand.

"Green" positioning should be developed using perceptual mapping to identify the firm's market position relative to its competitors. Ideally, firms will pursue a unique market position, that is, one that has not yet been claimed by a competitor (Sharma, Lyer, Mehrotra, and Krishnan, 2010).

A firm's "green" positioning is the market status claimed by the firm, which is invested in and supported over the long term. "Green" market positioning needs to be constructed to create corporate identification, remove ambiguity in the mind of consumers, and strengthen the product's competitiveness. A key decision involves identifying the aspects of the brand that are more directly relevant to the consumers' concerns (Dangelico and Pontrandolfo, 2010). Some firms may choose to make "green" focal to their brand and activities, whereas others may use this as a secondary focus. No matter what position is selected, it should address the key dimensions of green brand equity, green brand image, green satisfaction, and the need for green trust (Chen, 2010) such that it has a differential advantage with targeted market segments.

The fundamental issue in "green" positioning is to establish credibility for the positioning. There is research that suggests that gaps may exist between what the business claims as "green aspects" and what they actually do (Hartmann, Ibanez, and Sainz, 2005). This gap between corporate intentions and actions creates consumers' mistrust and cynicism. As such, firms need to ensure that they can deliver on implicit and explicit green promises when selecting corporate positioning.

While consumer awareness of environmental issues is expected to increase in the future, at present, the demand for green goods is relatively small, with few consumers making their product selection primarily based on the environmental aspects of a product (Tucker, Rifon, Lee, and Reece, 2012). As will be discussed in the next section, this is a critical issue when identifying segments of green consumers to be targeted. There is also a gap between consumers' concern for the environment and their purchase behavior, referred to as the attitude–behavior gap (Boulstridge and Carrigan, 2000). If this attitude–behavior gap persists over time, the growth in the "green" market segment will be slow in the future. Therefore, marketers need to also adopt a strategy for converting "conventional" consumers to green consumers as well in order to increase the size of the green market segment.

9.4.2 Green Consumers and Purchase Decisions

One early rationale for firms to go green was that they initially wanted to target consumer segments that valued green attributes (Diamantopoulos, Schlegelmilch, Sinkovics, and Bohlen, 2003) and want green goods. Initially, green consumers were seen as only a small niche market. However, recent global studies suggest that large numbers of consumers purport to be "green." As can be seen in Table 9.1, according to the 2012 Greendex survey commissioned by the National Geographic (2012), the proportion of green consumers ranged from 73% in Mexico to 34% in South Korea. When the same consumers were asked what percentage of "other consumers" were green in their countries, the numbers dropped, varying from 40% in India to 25% in Russia. However, there is a moderate

TABLE 9.1 Greendex Results 2012—Percentage of Green Consumers and Green "Others"

Country	% Respondents Who Thought They Were Green	% Respondents Who Thought Others Were Green
Mexico	73	32
Spain	66	34
Brazil	66	33
Canada	65	38
India	63	40
Sweden	63	37
France	63	31
China	62	39
Argentina	62	30
Germany	58	33
Australia	53	36
America	52	35
Britain	52	33
Hungary	52	28
Japan	39	36
Russia	34	25
South Korea	32	30

correlation (0.429) between consumers' views of the two issues; thus, the greener the consumers, the more green they believe others are in their country.

Defining a "green consumer" is unfortunately difficult, since a diverse range of terms have been used to describe these consumers, including proenvironmental consumers, ecological consumers, and green consumers. Some researchers suggest that environmental issues fall within a wider sphere of responsible consumers or ethical consumers, that is, environmental issues are only part of the issues that consumers consider (Ballet, Bhukuth, and Carimentrand, 2014). In the absence of any commonly accepted definition, researchers face problems measuring the size of the green market.

Exacerbating the measurement problem is the observation that consumers frequently behave inconsistently (Thøgersen, 2004). That is, there is a discrepancy between green attitudes and actual purchase behavior. Some consumers may place a high value on selected environmental issues and consumption, but they often place a lower value on other environmental issues, and these are not reflected in their consumption. For example, people who purchase organic food for environmental reasons may not prefer to use public transportation. Thus, green consumers do not purchase green goods exclusively. For this reason, there is an inconsistency in the research around whether alternative marketing actions will trigger green behavior (Boulstridge and Carrigan, 2000).

One key factor driving green consumption behavior, as with all consumption, is that consumers must value environmental attributes sufficiently to warrant their inclusion in the decision criteria for purchases. This is especially important *if* green goods are sold at a premium price. Unfortunately, research on whether consumers are willing to pay more for environmental goods has been equivocal (Bloemers, Magnani, and Peters, 2001), with research suggesting that if there is willingness to pay a premium, they will only accept a small premium (D'Souza et al., 2007). Of course, like all issues, this too varies based on the good being purchased as in an increasing number of cases additional acquisition expenditures may be offset over time. For example, hybrid automobiles or long-life light bulbs may cost more to purchase but use less fuel and energy over time, and thus, it is possible to calculate a payback period for the additional expenditure (Camilleri and Larrick, 2014). Calculation of payback periods is unfortunately harder in other areas, such as where consumers spend more to purchase highly durable goods, that is, purchase fewer of this product over the consumer's life.

Research on green consumers traditionally focused on understanding how demographic factors impact behavior. It has been increasingly realized that demographics fails to capture the underlying complexity of proenvironmental behavior (Diamantopoulos et al., 2003). For example, factors such as environmental orientation (Cleveland, Kalamas, and Laroche, 2012), consumers' underlying values (Soyez, 2012), or even whether consumers take a long-term perspective (Milfont, Wilson, and Diniz, 2012) all have been found to impact on behavior.

Traditional approaches to segmentation for green consumers are more difficult. As with all segmentation activities, targeted segments must (a) be able to be measured in regard to size and growth potentials; (b) be profitable; (c) be accessible and marketers must be able to communicate with the segment, (d) have characteristics that can be measured; (e) respond to alternative marketing mixes differently from other segments; and (f) be relatively stable and thus enduring (Straughan and Roberts, 1999). Some studies have identified that psychographic criteria are more useful basis for "green" consumers' segmentation than demographic criteria (Akehurst, Afonso, and Gonçalves, 2012) and that consumers' perception of effectiveness and their altruistic attitudes provide better insight into "green" consumer behavior (Straughan and Roberts, 1999). Ginsberg and Bloom (2004) proposed that marketers can use the size of green segments and how differentiable the firms' green goods are from the competitors' goods to assist in designing an appropriate marketing strategy. As such, segmentation is a critical early step in determining the firm's environmental marketing strategy.

Of course, segmentation and strategy development becomes even more complicated when one considers the global segments of consumers and the dynamics of international business environments, as both vary simultaneously. Thus, marketers may have more difficulties identifying green segments globally, requiring adaptive green strategies within each country they operate.

9.5 GREENING THE MARKETING MIX

Given that green marketing requires an integrated set of activities, it is critical that firms know what they want to do and why they do it. All corporate actions need to support this consistent approach across the marketing mix (product, price, place, and promotion) since it is the mix components that demonstrates the firms' value proposition to the market. Consumers can easily identify when actions are inconsistent, which results in negative consumer views and potentially also negative publicity.

When firms have an environmental orientation as their key strategy focus, it will provide a blueprint for the other marketing decisions that will facilitate for them the implementation of the "green" activities across the entire organization. Every element of the marketing mix would be reviewed and restructured to comply with the green marketing approach. This will lead to the development of a consistent, verifiable, and meaningful "green" marketing mix elements.

9.5.1 Green Products

Developing "green" products for the market is a key decision that facilitates an environmental or green orientation. It is a means by which the company sets and delivers their environmental positioning along with the coordinated and integrated distribution, pricing, and promotion. It will establish their competitive characteristics in comparison to other key players (Dangelico and Pontrandolfo, 2010).

One critical aspect of green product development is that goods have environmental attributes, whether this is how they are made, used, or disposed. While there may still be some negative environmental impact for goods, ideally, this will be lower than competitors. For example, as reported by Datamonitor, it was estimated that 1570 green goods were introduced in 2009 (Greenbiz.com in Dangelico and Pujari, 2010).

There is debate about what constitutes a green product and what characteristics would qualify a product to be called "green." A definition for a green product suggested by Peattie (1995) indicates that a product may be called "green" when the key environmental aspects of its production and consumption are substantially improved in comparison to conventional products. Peattie (1995) also suggests that all products have some "negative" environmental impact, even green products, and thus, being "green" is a relative term. The issue then is how to reduce the negative environmental impact progressively, by improving the environmental performance both in manufacturing, use, and disposal. Obviously, any move away from negative environmental impact is beneficial. But at what level does a change in environmental impact qualify a product as being "green"?

Dangelico and Pontrandolfo (2010) provide a review of the various definitions of "green" products. The key features and characteristics that contribute to a "green" product relate to a long list of attributes that can improve the products' environmental performance. These attributes can be summarized as those that (i) cause no substantial environmental damage in the use, (ii) no substantial environmental damage in the production process, and (iii) substantial reduction in material usage and waste (Elkington and Hailes, 1988).

The key elements of being "green" relate to minimizing the material used in a product, creating efficiency in the manufacture of the product, and reducing the size and weight of the product components and choice of material used for packaging and the packaging format (Dangelico and Pontrandolfo, 2010). Green products are expected to provide good functional value; however, they may also have lower prices associated with more production efficiency and reduction of waste. Moreover, the extended use of renewable resources will also benefit the environment and society.

When discussing green goods, it is important to recognize that the term applies to the whole life cycle of the product and, therefore, embraces wider issues than only the physical product. When considering greening goods, changes can be made to any of three constituents, core components (product platform, product design, and functional feature), actual components (quality, styling, packaging, etc.), and support services components (installation, instructions, spare parts, etc.). The intention is to improve the product in order to reduce waste and material usage, enhancing recyclability and reusability (Albino et al., 2009). There may of course also be a new product that addresses the particular consumer need. When making this substantial change, firms potentially face more uncertain markets as they move through the product life cycle. Modifying existing goods or developing radical new goods is costly, while technological developments can potentially make production cheaper with new green innovations having high development costs (Lin and Chang, 2012).

Ideally, each approach to new product development will reduce the use of nonrenewable resources, avoid toxic material use, and increase the use of renewable materials (Straughan and Roberts, 1999). The attempt to use more of renewable resources and less of nonrenewables may be more sector focused. For example, in the energy and utilities industries, green product development is more directed toward developing more sophisticated technologies for energy efficiency and use of renewable resources, while in the industrial and consumer sectors, the emphasis is more in the use of recyclable materials and green packaging (Albino et al., 2009).

As long as green product attributes, benefits, and usage are consistent with conventional products, these are easy for consumers to interpret. It is only when new green attributes are sufficiently valued by consumers that their inclusions (or lack of inclusion) will shift consumption. Thus, at present, consumers seem to expect that green products have the same functional characteristics and similar prices.

9.5.1.1 Developing New Greener Products

There are various options for creating new greener products. Firstly, an existing conventional product may be modified (continuous innovation) to incorporate green features and attributes, for example, by changing some of the ingredients used in manufacturing the product—such as removing CFCs (Mishra and Sharma, 2010) or making changes in the packaging material used (moving to recycled packaging).

Secondly, it may be necessary to innovate and create a new product concept (discontinuous innovation), that is, a new way to solve an existing problem. New advanced technology is increasingly creating opportunities for new processes and products. For example, software packages that allow consumers to calculate their carbon footprint can be considered a new product (Polonsky and Taghian, 2010). Other product changes may influence the way consumers use a product requiring some adjustments to their behavior. For example, car sharing schemes (Meijkamp, 1998) such as

"Getgo Car Share" or "Flexicar" required consumers to plan their travel arrangements better than if they owned their own car (Luchs et al., 2011).

When designing a new "green" product, the structure of the product, the resources used to manufacture it, and how it is used need to be reviewed. These changes will consequently transform the product into a "green" good, sometimes without the end user being made aware of the changes. For example, the introduction of power painting reduced the waste of paint and avoided the use of harmful solvents associated with spray painting, without consumers being made aware of these changes (Polonsky and Taghian, 2010). Considerations for removing negative environmental impact drive the process of change, for example, Coke's new biodegradable plastic bottles were only achievable as technology advanced (Mülhaupt, 2013).

The product redesign may involve changing some aspect of the product that would have multiple benefits for the environment. For example, when soft drink manufacturers reduced the weight and thickness of the aluminum cans, it not only resulted in lower material requirements, but it also contributed to a lightweight finished product lighter leading to lower transportation costs. These packaging changes did not affect the brands or product, but provided substantial environmental benefits and did not in any way alter how consumers evaluated or used the product (Mülhaupt, 2013).

If changes in a product somehow influence the way consumers use the product or alter consumers' perceptions about product quality, value, and positioning of the product for the consumer, the changes should be carefully implemented.

9.5.2 Green Prices

Acquisition pricing for a "green" product is similar to pricing for conventional products in that prices need to be set with regard to specific marketing objectives and with consideration of key internal and external factors such as consumer expectations. The price set must be equal to or greater than the value consumers believe they acquire. Factors relevant to pricing depend on the cost of goods, industry, intensity of industry competition, technology, etc. For conventional products, the objectives of a price may range from optimizing profit to achieving specific return on investment and to creating and maintaining a certain image of the product supporting its market positioning.

Investments in green product innovation, as with all investments, may result in additional costs that need to be recouped over the product's life. A clear understanding of the costs components related to the new and cleaner production processes provides the basis for a green pricing strategy. If the product is "new," the initial costs may be relatively higher and its profitability may be lower; however, as sales increase over time, economies of scale will reduce costs and thus profitability may increase. In marketing, this is referred to as products moving through the product life cycle that has four stages:

1. Introduction—low sales, high costs, and low profitability.
2. Growth—sales increase, costs decline, and profitability increases.
3. Maturity—sales peak and begin to decline, costs continue to decrease, but profitability drops.
4. Decline—sales decline, costs may increase, and profits reduce further.

At all stages of the product life cycle, a comprehensive review of the costs components through the entire product's value chain and the logistic processes may identify and provide opportunities for cost saving and for increasing efficiency. For example, reformulation of the product and packaging modifications may identify opportunities for cost reductions.

An increasingly important issue relates to the cost of externalities, such as carbon produced. The issue is that the costs of pollution (i.e., externalities) are being subsidized by the general public. Market-based mechanisms (i.e., carbon taxes) are being imposed globally, which increase the costs of goods that are attributed to emitters and users of the products producing the pollution. The idea is that these market-based pricing mechanisms will stimulate new technological developments that reduce pollution and the associated costs; however, in the short term, the costs of these innovations still need to be born somewhere in the pricing of goods. Any consumer price increases associated with environmentally harmful goods should, according to the theory of supply and demand, reduce consumer demand for these goods.

Communicating the price–value relationship for some green products is more complex than for many conventional goods. Technological developments, along with new and improved product performance, create more durable or longer-life products. Thus, fewer goods are required to achieve the same long-term consumer benefit. For example, longer-life razor blades reduce the number of blades needed per consumer; thus, while consumers pay more for the initial blade, there is a lower cost per shave to the consumer (Chervev, 2012). This highlights that the green component of goods may in fact not be the core benefit, even though there is a significant environmental benefit. To attract consumers and ensure that they do not make unfair price comparisons, marketers need to encourage consumers to amortize the cost over the useful life of the product. For marketers, communicating these price–value relationships can be a challenge.

A pricing strategy that is adopted for some products may offer indirect environmental benefits through "voluntary carbon offsets." The scheme is designed to motivate consumers' contribution to lowering the carbon emissions through their purchase of a product. The carbon offset standards use the clean development mechanism (CDM) as a benchmark (Polonsky, Carlson, Prothero, and Kapelianis, 2002; Kollmuss et al., 2008). Some companies may offer opportunity to consumers to contribute to carbon emission reductions through charging additional costs (voluntarily paid) for their product and channeling this money to carbon reduction projects, such as the case with many airlines globally (MacKerron, Egerton, Gaskell, Parpia, and Mourato, 2009).

In pricing for a new green product, the traditional strategies of skimming (charging a premium for new goods, where there are high profits per item and low sales) or penetration pricing (charging a low price with low profits per item but high volumes) may also be considered, depending on the environmental benefits the product offers and the uniqueness of the technology that is used to produce the product. If the product offers substantial environmental features that the competition cannot replicate easily or quickly and if the demand for the product is strong, the opportunity for premium pricing and gaining higher profits using skimming price strategy exists. Therefore, green innovation may be cost-effective and a rewarding investment.

Adaptation of a conventional product to integrate environmental attributes may also occur. Firms' ability to charge for the added features will be dependent on whether consumers value these attributes. However, there is evidence that consumers' price sensitivity for new environmentally safe products' range is relatively small as compared to conventional products. Thus, the expectation that consumers will automatically choose a green product alternative because of its environmental benefits at premium prices may be optimistic (D'Souza et al., 2007). Many consumers continue to believe that green goods offer lower quality and relatively higher prices in comparison to conventional products; therefore, they offer a lower value for money (Lin and Chang, 2012). The challenge, therefore, may be to price competitively and to communicate the environmental benefits of the product to the consumers, particularly to the nonenvironmentally committed consumers.

To quickly achieve economies of scale firms may undertake a penetration strategy, which allows larger sales volumes. This low price strategy may also motivate more intense competition and efforts toward technological innovation. The competitive green market environment also partly dictates the firms' pricing strategy. A price can be (i) set at parity level with competition to maintain the status quo, (ii) set at lower than the market level to gain market share, or (iii) set at higher than the competition prices, projecting higher quality and higher relative value for the consumer. It does need to be remembered that in most cases, an increase in one firm's sales results in a decline in others; thus, competitors are likely to respond differently to the introduction of new green goods, no matter the pricing strategy used.

Another pricing consideration relates to whether firms produce green and conventional goods simultaneously. The pricing of a conventional brand and a new green alternative by the same firm may confuse consumers, especially if these two goods are positioned very differently. If both types of products are on offer providing consumer choice and targeting different segments (Rex and Baumann, 2007), these brands effectively compete with each other and cannibalize demand from each other with the potential to erode the growth of premium priced alternatives and potential to reduce long-term profitability.

9.5.3 Green Logistics and Distribution

It is critical that green goods are not only made and priced appropriately, but these need to be accessible to consumers, moving effectively through the distribution system. Thus, the environmental impact of how goods move from the supplier to the end consumer is an issue that also needs to be considered (Dyllick and Hockerts, 2002). Thus, many firms require managers to address environmental issues along the goods (green and nongreen) supply chain. Supply chain and logistics are deeply entrenched in creating value, and the movement of goods can have substantial environmental costs (Beamon, 1999). The challenge for distribution is not only about being efficient, timely, and cost-effective but also increasingly about reducing the environmental impact.

This section explores the supply issues and the physical distribution of goods with regards to their environmental impact. It draws on policy options to further recommend how firms can best achieve sustainable distribution.

9.5.3.1 Supply Issues

Greening the supply chain is a process that firms use to move their products from the supplier to the consumer in the least environmentally harmful way. Sustainable techniques and approaches are used by firms to efficiently incorporate their suppliers, manufacturers, agents, wholesalers, and retailers to distribute their goods and services to the end user. Similarly, logistics refers to the physical distribution of goods that include warehousing and transportation, which can greatly contribute to environmental degradation and pollution. Srivastava (2007, p. 54) defines green supply chain management as "integrating environmental thinking into supply-chain management, including product design, material sourcing and selection, manufacturing processes, delivery of the final product to the consumers as well as end-of-life management of the product after its useful life."

| Information availability | Performance measurement systems | Collaborative decision making | Process integration | Cost Benefit sharing |

Standardization of supply and distribution processes, planning approaches, and packaging

Consolidation of transports, reduction of empty runs through collaborative return freight, and collaborative use of logistics infrastructure

Collaborative optimization of volume and weight–based load efficiency

FIGURE 9.1 Success factors of cooperation for ecological sustainability in logistics. *Source:* TU Berlin and International Transfer Centre for Logistics.

In a supply chain, businesses go through the following steps:

Suppliers → Manufacturers → Wholesalers → Retailers → Consumers

At the firm level, the manager who procures materials and resources from a supplier is faced with the complex problems of identifying the environmental impacts of their supply chain members. This means that the manufacturer is faced with the intricate task of reviewing their supply chain from the supplier's end to the consumer's end, seeking to ensure that all firms in the supply chain minimize their environmental impacts. This may also mean encouraging suppliers to adopt different green strategies for their internal and external operations (Vachon and Klassen, 2006). One way to do this is to select those suppliers who are International Organization for Standardization (ISO) 14001 certified (Miles et al., 1997). Figure 9.1 shows a holistic view of ecological supply chain practices. According to Ing, Straube, and Doch (2010), a holistic view of the supply chain includes the areas of cooperation for ecological sustainability in logistics such as the standardization of supply and distribution, consolidation of transportation and logistics, and collaborative optimization of volume and weight, which are managerial processes. This involves internal and external partners and their relationships between emission/environmental impact levels, costs, and service quality. Ing et al. (2010) indicate that besides the optimization of the operational process of any supply chain, more emphasis should be placed on planning to ensure that systems operate effectively together. The areas of cooperation for environmental improvements include standardization of supply and distribution processes, consolidation of transportation modes, and collaborative optimization of the volume of goods. They suggest that high-quality planning will lead to greater sustainable logistics.

Thus, planning is a crucial stage for greening logistics and needs to be driven from top management, which requires that they (i.e., management) recognize the importance of a green approach. As such, firms need to set out clear purchasing criteria and processes to ensure that suppliers are assessed on their environmental performance (Drumwright, 1994).

9.5.3.2 Suppliers in the Green Chain

In general, suppliers are organizations who deliver goods and services to consumers or businesses. They may supply raw materials, components, services, or finished products. Green purchasing, sustainable procurement, or environmentally conscious purchasing from suppliers often known as green purchasing or environmentally responsible manufacturing (Handfield and Melnyk, 1996) is the process of incorporating environmental issues in purchasing activities (Handfield, Walton, Sroufe, and Melnyk, 2002). A green supplier incorporates policies aimed at cleaner processes that reduce or eliminate waste, minimize energy consumption, and improve resource efficiency and operational safety (Weissman and Sekutowski, 1991). In terms of the economic impact, green suppliers include environmental costs and balance these costs against alternatives. Similarly with social impacts, suppliers consider labor rights, health and safety issues, and community health and impact. Clearly, suppliers that are not socially responsible cannot be considered to be "green" suppliers (Miles and Munila, 2004). Lastly, when considering environmental impacts, suppliers look at renewable resources, environmental waste management, negative externalities, and other pollution precautionary measures.

There are several benefits in becoming more environmental, as green purchasing can significantly affect corporate performance (Handfield et al., 2002), in addition to providing better working conditions (i.e., employee benefits), improved brand reputation, and improved product quality.

9.5.3.3 Physical Distribution and Logistics

Physical distribution and logistics begin with the organization and its intermediaries. It involves integrated logistics management that involves transportation, materials handling, order processing, inventory control, warehousing, and packaging. Green logistics involves the process of planning and controlling the use of integrated logistics and the logistic systems' environmental impact. Environmental issues are evaluated within process inventory, finished goods, and transportation. In logistics, the economic impacts to be considered are activities such as traffic congestion and resource waste. Ecological impact considers greenhouse gas emission through transportation and waste product impacts (e.g., maintenance activities), and social impacts relate to the negative effects of pollution on public health (Beamon, 1999). Since logistics is the fundamental driver of economic growth, however, money invested in inventory is locked money (i.e., not generating productive capacity). Greening inventory involves keeping resources to a minimum and manages cycle stock to reduce costs through such things as lower lead times. There are several technologies that can assist in sustainable logistics. For example, these are inventory management software, radio-frequency identification (RFID) deployment, and Global Positioning System (GPS), which allow for transportation networks to be redesigned.

While one may improve the efficiency of transportation by having a low-carbon policy, transportation such as rail, sea, road, or air would depend on the infrastructure available. Intermodal transportation (using multiple modes) has become practical with containerization, and this has helped to reduce unloading time at ports and enhance the cost efficiency of international trade. Generally, intermodal transport allows products to stay in one container through the entire journey, reducing the carbon footprint of each mode of transportation (Sawadogo and Anciaux, 2011). Minimizing food miles are another example of where transportation can be addressed, by reducing the distance between where the food is manufactured and how far it travels to reach consumer's table. The greater the distance traveled, the more energy is used and more emissions are generated (Coley, Howard, and Winter, 2009). Thus, buying local produce is encouraged.

An increasingly important aspect of logistics is the issue of reverse logistics, which is the movement of waste—packaging, unwanted and postconsumption goods—from consumers back into the production system (Pokharel and Mutha, 2009). One way that this has been operationalized in firms is with "extended producer responsibility," which is a policy that promotes total life cycle environmental improvements of product systems by "extending the responsibilities of the manufacturer of the product to various parts of the entire life cycle of the product, and especially to the take-back, recycling and final disposal of the product" (Lindhqvist, 2000, p. 5).

It is increasingly recognized that in many cases "waste" has economic value, which if extracted will generate revenue as well as reduce the resource burden of extracting virgin raw materials. This integrated system is being used in a range of traditional "waste" type activities, such as reprocessing aluminum, glass, or paper. However, it allows materials to be used in one area such as clothing, to be reused by other consumers. In the case of clothing, the "reuse" has eliminated the need to produce tons of new clothing as well as reduced clothing going to landfills (Venkatesh, 2010). In other areas such as electronics, e-waste is a huge issue for material disposal but also extracting the value from materials that have finished their productive life (Widmer, Oswald-Krapf, Sinha-Khetriwal, Schnellmann, and Böni, 2005). Firms in a range of product systems are building in reverse logistics to minimize their environmental impact and allow firms to remanufacture some product component more cost-effectively than manufacturing from new components (Kerr and Ryan, 2001). Thus, reverse logistics can be a true environmental and financial, win-win situation.

In summary, green distribution therefore involves suppliers and the physical distribution of goods from the supplier's end to the consumer's end. In the area of suppliers, there is a need to manage the environmental impact from a supplier's perspective as firms focus on environmental procurement. The physical distribution examined how environmental issues can be incorporated by implementing green processes.

9.6 GREEN PROMOTION

When discussing green marketing, what comes to mind for most managers, regulators, and consumers is promotion and advertising. In reality, green promotion is much more. It is integrated communication whereby organizations communicate meaningful information to targeted audiences using a range of alternative communication tools (advertising, sales promotion, social media, personal selling, or public relations). We will discuss each of these tools briefly; however, before doing so, we want to highlight that the critical aspect of green promotion is that there needs to be something environmentally meaningful to say about the goods or service.

A number of countries around the world have adopted guidelines for the regulation of communication of environmental information (Kangun and Polonsky, 1995). The ISO has devised a set requirements in regard to "environmental labels and declarations" (ISO, 2012). While there is some variation in the wording used across regulation schemes, the requirements are generally similar in that claims should be "a) accurate and not misleading, b) substantiated and verifiable; c) unlikely to be misinterpreted" (ISO, 2012, p. 11). In 2012, the US Federal Trade Commission updated its green marketing guidelines (Federal Trade Commission, 2012), which, in addition to providing an overview to problems

TABLE 9.2 Part 260—Guides for the Use of Environmental Marketing Claims

Section	Title
260.1	Purpose, Scope, and Structure of the Guides
260.2	Interpretation and Substantiation of Environmental Marketing Claims
260.3	General Principles
260.4	General Environmental Benefit Claims
260.5	Carbon Offsets
260.6	Certifications and Seals of Approval
260.7	Compostable Claims
260.8	Degradable Claims
260.9	Free-Of Claims
260.10	Nontoxic Claims
260.11	Ozone-Safe and Ozone-Friendly Claims
260.12	Recyclable Claims
260.13	Recycled Content Claims
260.14	Refillable Claims
260.15	Renewable Energy Claims
260.16	Renewable Materials Claims
260.17	Source Reduction Claims

associated with green marketing, provides fourteen detailed descriptions of alternative types of claims, and these are explained in regard to how each might be misleading to consumers, as well as provided a discussion of how the claims could be effectively and accurately used (see, e.g., Sections 260.4–260.17) (Table 9.2).

Much of the concern related to green promotion arose early on its practice, where firms undertook what has come to be known as *greenwash* (Gillespie, 2008). That is making statements or claims that were exaggerated or meaningless, for example, "we care about the environment," "natural ingredients," or "XYZ-free" (when the product never contained XYZ). These "fuzzy" claims can easily lead to consumer confusion and skepticism as they add limited environmental information or consumer understanding in regard to how producers address environmental issues.

While there are a number of components of promotion, it is critical that all these are integrated (i.e., supporting a common message and image), referred to as integrated marketing communication. As such, many researchers talk about integrated marketing communication, rather promotion. For green or environmental issues, this means that environmental values are integrated through the firms' philosophy and strategy, as well as embedded in all marketing activities, not just those used in promotion (Holm, 2006).

9.6.1 Advertising

The American Marketing Association (2012) defines advertising as "the placement of announcements and persuasive messages in time [*i.e. on television or radio*] or space purchased," related to messages communicated by firms in print, radio, television, and online.

Early research into green advertising found widespread support that green "information" or claims were being misused. For example, the early work of Kangun, Carlson, and Grove (1991) found that out of 100 advertisements with environmental claims, 48% were viewed by "experts" to contain misleading claims and 68% were viewed by "average consumers" to contain misleading claims. Green claims were classified as being vague and ambiguous (51% experts; 70% average consumers), omissions (48% experts; 18% average consumers), and false/outright lies (1% experts; 12% average consumers). Later, international work by Polonsky, Carlson, Grove, and Kangun (1997) found misleading green claims appeared in advertising in the United States, the United Kingdom, Canada, and Australia; thus, "greenwash" appeared to be a global problem. Unfortunately, this trend is one that does not seem to have disappeared, and there are still concerns raised about the veracity of environmental claims even though international regulators have implemented rules relating to green marketing's use (Gillespie, 2008).

One of the difficulties with environmental claims is that there is no universal agreement as to the way these claims are communicated and multiple bodies "accredit" similar types of claims. For example, according to a World Wildlife Foundation study in 2008, there were at least eight alternative global organizations that certified firms' carbon offset schemes (Kollmuss et al., 2008), where each scheme has a different calculation of carbon footprints, as well as alternative criteria for including a project as an appropriate offset. This ambiguity makes it hard for consumers to assess alternative programs or to make effective decisions on which assessment method is meaningful (Polonsky, Garu, and Garma, 2010). Of course, consumers can interpret a range of information provided as firms being environmentally oriented, even if this is not the intent of the firm. For example, Polonsky et al. (2002) found that consumers were interpreting information such as "safe on your hands" as being environmentally focused. The consumers' rationale was that "if it does not hurt your hands, how can it hurt the environment." Whether this assessment is accurate or not is irrelevant, as it highlights that consumers interpret expressed and implicit claims potentially differently than firms intended.

9.6.2 Environmental Logos/Third-Party Endorsements

An overarching issue across promotional tools is that of environmental logos and third-party endorsements. Environmental logos are endorsements by an organization about the environmental attributes or qualities of a firm or its products. At one level, this should ideally make information more objective and thus more reliable, as it is "externally validated" (Mendleson and Polonsky, 1995). In practice, however, there are in fact multiple alternative bodies providing "certification" of environmental claims/information, which can include governmental bodies, commercial organizations, environmental nonprofits, industry bodies, international organizations, and even the firms. Research has found that the organization "endorsing" the claim may in fact influence consumers' assessment of the claim and endorsement (Dean and Biswas, 2001). For example, Ozanne and Vlosky (1997) found that in the area of environmental wood product certification schemes, nongovernmental environmental organizations and third-party organizations were more trusted than governmental and forestry industry schemes. One might question the objectivity logos and schemes if these are not rigorously and objectively evaluated. To partly address this, the ISO has developed a standard related to third-party certification schemes of environmental claims (ISO 14025) that builds on its generalized principles for all environmental claims, that is, there must be meaningful environmental impact that is objectively tested and where the criteria are readily available (ISO, 2012).

Working with third-party endorsement schemes should provide a systemic assessment of product's environmental attributes, thus enabling the firm to have meaningful information to promote. In some instances, third-party endorsements may create opportunities for additional input into the product development process (Mendleson and Polonsky, 1995), giving firms' added expertise on which to draw.

9.6.3 Sales Promotion (Cause-Related Marketing)

Promotions are incentives that are designed to encourage consumers to trial or purchase a good, for example, price specials, competitions, or added value offers. One of the most popular tools within promotion in regard to green marketing is the use of cause-related marketing (CRM) (not to be confused with customer relationship management also called CRM). Varadarajan and Menon (1988, p. 60) defined "cause related marketing as the process of formulating and implementing marketing activities that are characterized by an offer from the firm to contribute a specified amount to a designated charity when customers engage in profit-providing exchanges that satisfy organizational and individual objectives." Thus, firms give money to a charity or cause based on consumers' purchase (or other activity). For example, when you purchase product X, we will contribute 10% of the sales to charity Y.

CRM is a form of brand alliance (Lafferty, Goldsmith, and Hult, 2004) and seeks to leverage the value of the nonprofit brand with the for-profit organization. However, research has suggested that there may need to be some rational link between the two parties. For example, in the early 1990s, there was a jeans company that promoted that it would plant a tree for each pair of jeans sold, where there is no clear connection between the two (Mendleson and Polonsky, 1995). Consumers can interpret the links in CRM in many ways, for example, assuming that environmental nonprofits endorse the firm's actions; otherwise, they would not have CRM relationships. However, there is another research that suggests consumers interpret these CRM relationships differently if they are too closely linked and it may be seen as a firm's cynical attempt to create positive "spin" of their activities (Menon and Kahn, 2003). This is especially true when CRM is used in ways that is seen to counter a firm's negative performance, such as fuel companies promoting their environmental responsibility (Polonsky and Wood, 2001). Other criticisms have been leveled against the use of CRM, as more money is sometimes spent promoting the CRM program than is actually given to the cause.

CRM connections are designed to create augmented value to the purchase, which often is unrelated to the core value of the good. For example, giving 10% of the sale value to a charity does not increase the performance of the product or change the product in any way, other than how it is positioned in consumers' minds. The social value consumers get from purchasing a good with CRM is of course important and may serve as a significant differentiation among two otherwise "equal" products (Barone, Norman, and Miyazaki, 2007). Of course, CRM also is important in allowing firms to outwardly demonstrate their image and thus should align with the overall corporate values and image.

9.6.4 Personal Selling

Personal selling is a one-on-one interaction between the firm or its representative and the customer. It is a highly adaptive form of promotion as it can target the information to the specific customer. It is also generally more useful in regard to communicating complex information in which extensive explanation is required or the salesperson needs to assist the consumer in installing or using the good. This includes salespeople explaining the features of products to consumers and potential customers (in person, on phone, or synchronous online). Research in the area of personal selling and environmental goods is limited. There are, however, numerous alternative examples of where personal selling is critical in regard to green products:

- Consumers being contacted by electric utilities, promoting "green energy"
- Local municipalities contacting homeowners in regard to energy-efficient light bulbs, energy-saving devices, or water-saving devices
- For-profit organizations promoting the purchase of a range of devices (insulation, water tanks, solar panels, etc.)
- Service providers (such as architects) discussing how consumers (or businesses) provide design advice to

improve environmental efficiency or meet local or external standards
- Business markets where firms sell other organizations new technologies and processes to be integrated into existing activities (e.g., new water treatment activities or recycling systems)

In many of these programs, the salesperson needs to communicate the benefits (environmental and nonenvironmental) to consumers and customers. In cases where there are more substantive costs (such as new production processes or purchasing a solar energy system), it may be a matter of discussing the payback period of the product, that is, how long it takes for the acquisition costs to be offset by financial savings (Papadopoulos, Theodosiou, and Karatzas, 2002). Financial assessments of green product "investments" are potentially even more important in the business context, especially in the current cost-cutting economic environment (Kim, 2013).

Salespeople therefore need additional information and tools (such as software undertaking the relevant calculations) that allow them to adapt information for individual consumer and organizational circumstances. Salespeople also need to be able to communicate information in a way that leverages the relevant consumer or businesses interests and needs (Menegaki, 2012). As such, personal salespeople need to be integrated into the overall green promotional mix.

9.6.5 Publicity/Public Relations

Publicity/public relations are unpaid activities that generate media coverage about the product and/or organization. As it is not paid for, it cannot be controlled in terms of the amount of exposure or whether this is positive or negative. The firm does need to invest in activities that become newsworthy, as well as create "media" content (whether this is a press release or product launch).

Within green marketing, many opportunities exist to leverage positive initiatives. The development of an alliance between a firm and an environmental nonprofit may be something that can be promoted and could attract media attention. A new good with significantly improved environmental attributes may also attract media attention. Unfortunately, there are also unintended events as well, such as industrial accidents or product failures that also result in significant media attention. For example, a production disaster or accident can attract negative media attention, as might problems with the products or reports that marketing claims are excessively exaggerated (Roper, 2012).

9.6.6 Social Media

Social media is increasingly becoming important for marketing and green marketing in particular making the promotional mix more complicated, as this might include aspects of advertising, public relations, etc. (Mangold and Faulds, 2009). It is defined broadly as "internet applications … that allows for creation and exchange of user-generated content" (Kaplan and Haenlein, 2010, p. 61). These allow for greater degrees of engagement between the firm and its consumers, as well as between consumers. There are many social media tools available (Facebook, Twitter, blogs, etc.), which can either be driven or managed by the firm (Lodhia, 2012). However, there are also an increasing number of external social media vehicles that are driven by consumers or nongovernmental agencies where these individuals and groups can provide positive or negative feedback about a firm's green activities (Nwagbara, 2013). Social media is somewhat uncontrollable, especially when consumers can develop content or manage social media sources. For example, there are many "anti" corporate websites that discuss corporate environmental issues (real or perceived) in a negative light. While firms may try and spread positive messages with consumers through social media, this cannot be guaranteed, with consumers able to equally spread negative views about firms' activities. This is especially important in regard to environmental or quality "problems" that may arise, as social media networks spread information much more quickly than traditional media (Obar, Zube, and Lampe, 2012). Of course, like all promotional tools, firms need to spend resources effectively designing and managing social media, which may be more problematic given its 24/7 nature.

9.7 GOVERNANCE, EVALUATION, AND CONTROL

The adoption and implementation of green marketing strategy requires an organizational cultural transformation. The corporate management and environmental policies should detail intentions and objectives and state requirements for and implications on corporate resources to be used. Moreover, the expected performance levels need to be communicated to all key internal and external stakeholders, alongside periodic monitoring to ensure proper application of the strategy (Tang, Lai, and Cheng, 2012).

9.7.1 Implementation

The implementation of the "green" strategy begins with proper planning where green marketing strategy is at its core. Setting achievable, reasonable, and realistic objectives with clear understanding of the relevant market dynamics is a useful starting point. The adoption of the green marketing strategy introduces a substantial change in organizational attitude and culture. Like any other change, it may be resisted by some elements within the organization (Peattie and Crane, 2005). The internal environment and the possibility of resistance need to be identified and addressed through training and facilitating procedures. A review of the organizational structure, the skills, and the backgrounds of the key staff members responsible for implementing the new green strategy can suggest changes and adjustments in areas that could lubricate the process of successful implementation. Moreover, streamlining the activities of various sections and departments of the firm can motivate cooperation and collaboration among the key areas of the organization. Other issues may relate to the level of "green" strategy being incorporated into the firm's product portfolio. There may be a degree of confusion within the organization as well as in the marketplace if the company is characterized as following a dual strategy of providing both conventional and green alternatives in the same product category. This could also make implementation somewhat complicated.

The organization needs to undertake some objective assessment of whether it has successfully integrated environmental marketing into its activities. That is, identifying whether the changes in organizational behavior achieved the preestablished targets. First and foremost, firms need to assess changes in their environmental performance that will enable them to ensure that their green marketing activities/initiatives have brought about real improvements in the business and effectiveness regarding their environmental objectives, worthy of being communicated to stakeholders. A range of measures can be used to assist with this assessment such as evaluating changes in the company's carbon footprints or overall environmental impact assessments. Secondly, firms also need to evaluate how stakeholders evaluate their performance, as it is these stakeholders who will respond to green initiatives. The environmental and green marketing policy should be communicated to and be understood by their key internal and external stakeholders, including the employees, the public local residents, the community at large, the customers, the suppliers and regulators, the media, and the relevant unions.

9.7.2 Governance

Another mechanism available to assist in managing green marketing is accreditations or internationally recognized objective third-party certification. For example, the ISO has a range of environmental management and environmental marketing standards that can be used. Earlier in this chapter, we mentioned some of the voluntary standards in regard to environmental claims. There are also a number of certifiable standards related to environmental management, which would also include green marketing-related activities.

Environmental policy and green marketing can be maintained through using environmental management systems (EMS), which is the management of an organization's environmental programs in a comprehensive, systematic, planned, and documented manner. EMS includes the appropriate organizational structure to be used, planning of the environmental activities, detailing the resources needed for executing, and maintaining the policy for environmental protection (Gao, 2011). An EMS can also refer to a system, which monitors, tracks, and reports emissions information. In Australia and elsewhere, EMS are becoming web based (Boiral, 2007; COMM Engineering, 2013).

An EMS:

- Serves as a tool to improve environmental performance
- Provides a systematic way of managing an organization's environmental affairs
- Is the aspect of the organization's overall management structure that addresses immediate and long-term impacts of its goods, services, and processes on the environment
- Gives order and consistency for organizations to address environmental concerns through the allocation of resources, assignment of responsibility, and ongoing evaluation of practices, procedures, and processes
- Focuses on continual improvement of the system

ISO 14000 refers to a family of accredited standards and guidance to help organizations address environmental issues. Included in this series are standards for EMS, environmental and EMS auditing, environmental labeling, environmental performance evaluation, and life cycle assessment. ISO 14001 helps organizations to:

- Minimize how their operations (processes, etc.) negatively affect the environment (i.e., cause adverse changes to air, water, or land)
- Comply with applicable laws, regulations, and other environmentally oriented requirements
- Continually improve in the above

9.7.2.1 Continuous Improvement

Green marketing and environmental issues more generally require continuous improvement, which includes the cycle of planning for, implementing the environmental policy, reviewing the results achieved, and revising and/or preparing corrective actions. The EMS model includes a progressive review and revision of the environmental policy effectiveness.

An EMS follows the "Deming cycle" (Zwetsloot, 2003), plan–do–check–act (PDCA) cycle, as shown in Figure 9.2. The fundamental principle and overall goal of the ISO 14001 standard is the concept of continual improvement (Federal Facilities Council Report, 1999). ISO 14001 based on the PDCA methodology includes seventeen elements, grouped into five phases that relate to PDCA: environmental policy,

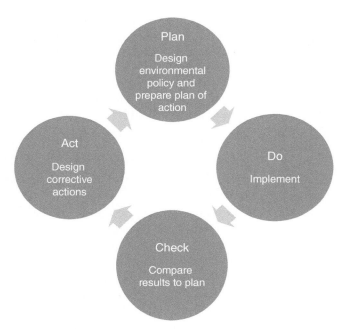

FIGURE 9.2 ISO 14001 plan–do–check–act model. *Source:* http://www.alexfert.com/new/content_ones/view/1/menuid:224

planning, implementation and operation, checking and corrective action, and lastly management review (Bennet and Martin, 1998).

The core requirement of a continual improvement process in ISO 14001 has three dimensions (Gastl, 2009):

- Expansion: more and more business areas get covered by the implemented EMS.
- Enrichment: more and more activities, products, processes, emissions, resources, etc. get managed by the implemented EMS.
- Upgrading: an improvement of the structural and organizational framework of the EMS, as well as an accumulation of know-how in dealing with business-related environmental issues.

Overall, the continual improvement process concept expects the organization to gradually move away from merely operational environmental measures toward a strategic approach on how to deal with environmental challenges.

9.7.3 Advantages of Using ISO Series

ISO 14001 was developed primarily to assist companies in reducing their environmental impact, but organizations can benefit from using it for a number of other economic benefits including higher conformance with legislative and regulatory requirements (Sheldon, 1997). These benefits may include (i) minimizing the risk of regulatory and environmental liability fines and improving the organization's efficiency (Delmas, 2001), leading to a reduction in waste and use of resources, which can lower the costs (ISO14001.com.au, 2010); (2) an internationally recognized standard, businesses operating in multiple locations across the globe can register as ISO 14001 compliant, eliminating the need for multiple registrations or certifications (Hutchens, 2010); (iii) a push by consumers, for companies to adopt stricter environmental regulations, making the incorporation of ISO 14001 a greater necessity for the long-term viability of businesses (Delmas and Montiel, 2009) and providing them with a competitive advantage against companies that do not adopt the standard. This, in turn, can have a positive impact on a company's asset value (Van der Veldt, 1997; Miles and Munila, 2004)) and can lead to improved public perceptions of the business, placing them in a better position to operate in the international marketplace (Sheldon, 1997). (iv) Finally, it can serve to reduce trade barriers between registered businesses (Miles et al., 1997; Van der Veldt, 1997), although it has also been argued that it may be a barrier to trade if it means that non-ISO-accredited firms, especially in developing countries, are cut out of the supply chain (Clapp, 1998; Miles and Covin, 2000).

9.7.4 Other Management Systems

While ISO is highly developed, a number of other approaches might be used to manage and implement green marketing-related activities (Gray, Owen, and Maunders, 1987; Gao, 2011), including:

- Natural resource accounting (a report that quantifies the costs and benefits of a company's operations in relation to the environment)
- Environmental audit (the practice of assessing, checking, testing, and verifying an aspect of environmental management)
- Corporate social responsibility (CSR) reports (a report that goes beyond economic performance and includes the company's responsibilities to other internal and external stakeholders)
- Social accounting (the process of communicating the social and environmental effects of organizational actions to particular interest groups)

Social accounting is increasingly being used and emphasizes the notion of corporate accountability for its societal impacts. Crowther (2000, p. 20) defines social accounting in this sense as "an approach to reporting a firm's activities which stresses the need for the identification of socially relevant behaviour, the determination of those to whom the company is accountable for its social performance and the development of appropriate measures and reporting techniques."

Social and environmental accounting, also sometimes referred to as corporate social reporting, CSR reporting, and nonfinancial reporting, is a report prepared by the directors of a company that details the costs and benefits of the organization's activities in relation to the society and to the natural environment. It presents these activities in monetary or financial terms. This report is used to communicate the social and environmental effects of economic actions to particular interest groups (Gray et al., 1987). It is typically used in the context of CSR. Social accounting is a response to the notion that financial accounting does not fully represent an organization's activities and outcomes. A company is concerned with more than only economic activities, and it is accountable to more stakeholder groups than only the owners. It has a wider purpose, responsibility, and accountability beyond what can be presented in a financial report. Companies influence both social and natural environments through their operations. Social accounting can be used for management control of the organization's activities and is therefore a tool that assists the management to keep track of their operation's social and environmental impacts. It facilitates the possibility of making informed choices and setting more appropriate objectives and allows for quantitative assessments of usually subjective elements of organizational impacts.

The benefits of social accounting in practice are many. It tends to (i) balance corporate economic activities with the company's wider responsibilities, (ii) allow for stakeholders' right of information, (iii) increase transparency of corporate activities, (iv) identify and assess the social and environmental impacts of the company's economic achievements, (v) establish corporate legitimacy, (vi) access more comprehensive information for decision making, and (vii) establish a socially responsible corporate image using a socially and environmentally targeted public relations activities.

The key criticism of social accounting is that in most cases it is a self-reported tool that is not independently audited for its accuracy and relevancy. However, the Social Accountability International system, SA8000, which is a series of behavioral standards for CSR, can provide a third-party "independent" certification (Miles and Munila, 2004). In most countries, there is no regulation accounting for socially relevant corporate activities (Gray, 2001). To overcome this problem, external social audits are sometimes considered, which are prepared by people independent of the organization and are conducted usually without the management's consent (Gray, 2001) with the intention to hold organizations accountable without their approval. There is also no standard format of social accounting reporting. However, the understanding is that this report should cover the company's activities that impact the environment, the employees, the ethical issues related to the consumers, and the community in general.

Environmental accounting is part of social accounting, which focuses on the environmental performance of a company. Environmental accounting reports the quantitative environmental data within the nonfinancial section of the annual report and details the damages made to the environment in the form of contamination of land and water or concerns about excessive pollution, etc. While environmental accounting is also a self-report, there are legislations for compulsory reporting in some form in some countries including Australia, Denmark, the Netherlands, and Korea.

9.7.5 The TBL

The TBL sometimes referred to as 3BL focuses on people, planet, and profit (or the three pillars). It captures an expanded spectrum of values and criteria for measuring organizational (and societal) contributions and/or performance (economic, ecological, and social) (Slaper and Hall, 2011).

In the private sector, a commitment to CSR implies a commitment to some form of TBL reporting. This is distinct from the more limited changes required to deal only with ecological issues (Bader, 2008). In practical terms, TBL accounting means expanding the traditional reporting framework to take into account ecological and social performance in addition to financial performance (Brown, Dillard, and Marshall, 2005).

The concept of TBL demands that a company's responsibility lies with stakeholders rather than shareholders. In this case, "stakeholders" refer to anyone who is influenced, either directly or indirectly, by the actions of the firm. According to the stakeholder theory, the business entity should be used as a vehicle for coordinating stakeholder interests, instead of maximizing shareholder (owner) profit.

The TBL is made up of "social, economic, and environmental" elements. The people, planet, and profit succinctly describe the TBL and the goal of sustainability.

The key arguments supporting TBL include (i) reaching untapped market potentials developing goods and services for underserved population, which are also profitable, and (ii) adapting to new business sectors and benefiting from business opportunities emerging in the area of social entrepreneurialism. For example, Fair Trade companies require ethical and sustainable practices from all of their suppliers and service providers (Sridhar, 2012).

While there is widespread support for the importance of ethical social conditions and preservation of the environment, some people argue about the appropriateness of the TBL for measuring business performance (Slaper and Hall, 2011; Sridhar, 2012; Sridhar and Jones, 2013). Criticism of the TBL mainly focuses on the notion that:

1. It diverts business activities away from the company's core competencies into areas that they have no expertise.
2. Effectiveness, as when a business attends to its focal business, it most effectively contributes to the improvements of all areas of society, social and environmental as well as economic.

3. The difficulty of its application in a monetary-based economic system, as there is no single way to measure the benefits of company activities to the society and environment as there is with profit; therefore, it would be misleading.
4. It may ultimately lead to firms avoiding legislation and mandatory measures and taxation, leading to a fictitious appearing people-friendly and eco-friendly image for some companies.

The ISO certification of "green" work procedures and third-party endorsements of "green" product characteristics provide objective verifications of the "green" nature of the business. However, the other control mechanisms available to the management through the various internally prepared reports collectively characterize an organization's planned actions and review of the status achieved. These reports, social accounting, environmental accounting, TBL, etc., can motivate periodic review and assessment of the actions being taken and provide a comparison between the intended and actual environmental performance levels, identifying and indicating areas that need changes and/or improvements.

9.8 THE FUTURE OF GREEN MARKETING

The importance of green marketing is most certainly likely to increase in the future. As the Earth's population continues to grow, it is clear that the environment cannot support the production for all 7+ billion consumers, consuming the same as Western societies consume today, especially as consumers in developing countries disposable income rises. As such, marketers, policymakers, and consumers need to redefine how we have our needs met (Polonsky, 2011). Public policy has already started to seek to address environmental issues in a range of countries. For example, Australia has banned incandescent light bulbs replacing them with long-life bulbs, although there are some who suggest these are not necessarily any less environmentally harmful (Elijošiutė, Balciukevičiūtė, and Denafas, 2012). As such, policy has been used to constrain consumer choice. In another broader example, governments around the world are putting in place housing standards that will reduce the energy requirements of new housing, thus changing the environmental impact of future homes (Kordjamshidi and King, 2009) and requiring people to build more environmentally friendly homes.

In other instances, market mechanisms have been used to encourage changes in behavior. For example, the carbon taxes imposed in different countries seek to allocate the costs of pollution to those creating it or using products arising from the production of pollution. The idea is that as prices rise traditional supply and demand factors will come into play and motivate consumers to change their behavior (Brännlund and Nordström, 2004). The supply and demand forces also recreate opportunities for green innovation, which takes advantage of increased consumer demand but also creates competitive advantage for more responsible goods, as well as the rising costs of resources (Porter and Van der Linde, 1995). For example, greater fuel efficiency of automobiles continues to reduce the harm associated with operating them, although this may unintentionally also stimulate more use (Small and Van Dender, 2007). There are other completely new product sectors arising, such as home solar systems to allow consumers to reduce the operating costs of their homes, but these frequently require significant initial incitements that may be difficult for some consumers to make (Zarnikau, 2003).

There does however also need to be a shift in consumer thinking and behavior to bring about environmental change. Mankind is part of nature and not separate from it. However, the dominant social paradigm takes a highly technological consumption focus (Kilbourne et al., 2009). As such, the question is how marketers and organizations more widely assist consumers in reducing their environmental impact. As has been described previously, there may be triggers imposed by governments, such as regulation of behavior or taxes that increase costs (and thus decrease demand). However, voluntary action to reduce consumer's environmental impact is also possible in a range of ways. One pressing issue is to develop (or examine) metrics that assess consumers' overall environmental impact, rather than focusing on individual behaviors. The limit with individual behaviors is that consumers are often inconsistent in activities (Thøgersen, 2004), that is, going green in one area and not another. Thus, a holistic assessment of consumer's environmental impact is something that needs to be considered, such as one's overall carbon footprint. A second issue for bringing about change is to get consumers to begin thinking more long term considering the full life cycle of costs (environmental and financial) of their behavior (Kaenzig and Wüstenhagen, 2010). The third challenge is to get consumers to change their views in regard to how to satisfy wants. This is not suggesting people go without or voluntarily simplify their lifestyles (Oates et al., 2008), although this is an option. Rather, the question is whether consumers can shift their emphasis to the use of want-satisfying attributes rather than acquiring the good that provides these attributes (Polonsky, 2011). A number of sharing schemes have arisen to allow consumers to do this (Meijkamp, 1998); however, "owing stuff" is still seen as a preferable option, suggesting consumers are more materially oriented.

There are a number of future challenges that transcend marketing and require system-wide changes, whether this be across an organization, economic system, or consumer

Illinois. Its mission is to be North America's best food and beverage company. The company has more than 20 brands, including Capri Sun, Jell-O, Planters, Miracle Whip, Oscar Mayer, Kraft, Maxwell House, Kool-Aid, Philadelphia, and Velveeta, with annual sales value of $19 billion. The company's five different main product groups are US Beverages, US Convenient Meals, Canada & NA Foodservice, US Cheese and Dairy, and US Grocery (http://www.kraftfoods group.com/about/index.aspx).

Kraft Foods' actions toward environmental sustainability are important because of its immense market penetration and share all over the world. This packaged food and snack giant introduced its first set of sustainability achievements for the years 2005–2010 as:

- %16 reduction in energy use
- %18 reduction in CO_2 emissions
- %30 reduction in incoming water
- %42 reduction in net waste
- 100,000 metric tons (200 million pound) reduction in packaging
- 16 million km (10 million road miles) elimination from its network

In addition to their prior sustainability focus areas of energy, carbon dioxide, water, waste, and packaging reduction, in 2011, the company added transportation and agricultural commodities to measurable fields. According to the company's sustainability executive Steve Yucknut, these new goals will help them do more. For instance, he states that their increased focus on sustainable agriculture will further boost their "scale to help accelerate long-range development in more communities and for more commodities than ever before."

The restructured sustainability goals, namely, Kraft Foods Big Sustainability Goals for 2015 timeframe, include:

- Sourcing all of the European coffee brands sustainably
- %25 increase in *sustainable sourcing of agricultural commodities*
- %15 reduction in energy use in manufacturing plants
- %15 reduction in energy-related CO_2 emissions in manufacturing plants
- %15 reduction in water consumption in manufacturing plants
- %15 reduction in waste at manufacturing plants
- Elimination of 50,000 metric tons (100 million pound) of packaging material
- Reduction of 80 million km (50 million miles) from transportation network (http://www.bloomberg.com/apps/news?pid=newsarchive&sid=aXVg.nq46bRI)

For Kraft Foods, the world's biggest purchaser of cocoa, coffee, and cashews, reaching these new agricultural benchmarks is possible by using third-party certifiers, such as Fairtrade, Rainforest Alliance, and 4C Association. Currently, the company is using these third-party certifiers extensively in their business activities. However, as criticized by a few (see Marc Gunther's discussion at http://www.marcgunther.com/how-kraft-sells-sustainability/), the company should also consider its total impact on Earth, not just as gains measured against its total production.

8.11.3 Campbell Soup Company

Campbell Soup Company, also called Campbell's, is an American canned soup and similar products producer based in New Jersey, United States. The company's products are sold worldwide in more than 120 countries. The company has been in business for more than 140 years and covers three divisions: the simple meals division, baked snacks division, and health beverage division.

8.11.3.1 Campbell's Environmental Sustainability Policy

The company is strongly committed to protecting its employees' health and safety and to conducting its operations in an environmentally responsible manner. This statement under the corporate governance policy is applicable to their businesses worldwide:

- In order to conserve natural resources, the company uses eco-efficient management strategies, performance metrics, and continuous improvement focused in five key areas that are important to Campbell's business success in the long term: (i) energy and water use, (ii) waste generation and recycling, (iii) sustainable packaging, (iv) sustainable agriculture, and (v) supply chain optimization.
- The company takes actions to reduce its GHG emissions continuously from its operations because it recognizes that climate change is a reality that demands business communities' attention and action.
- The company conducts its operations in accordance with currently available laws and regulations. It also engages in development of responsible standards with industry and public stakeholders and acts in voluntary initiatives.
- Campbell pays attention to providing safe workplace to its employees through a strong health and safety program. For this purpose, the company also trains its employees.
- The company wants its suppliers to meet the applicable environmental laws of the places (country, region, etc.)

community. While great advances have been made, companies need to confront both the challenges and opportunities of a changing environment, if the economic system as we know it is to continue to operate.

9.9 REVIEW QUESTIONS

1. Why may it be difficult to determine the facts from fiction regarding the causes of global warming?
2. Global warming is being caused mainly by consumption. In order to manage for environmental sustainability, businesses need to motivate consumers to consume less. Discuss.
3. All manufactured products need to be "green." They are expected to be environmentally and socially safe and provided at no extra cost to the consumer. Business needs to take a leading role in environmental sustainability through innovating in cleaner production. Discuss.
4. The big emerging economies are the main environmental polluters. They need to be technologically assisted by the industrialized economies to reduce their emissions. Discuss.
5. Is premium pricing for "green" innovations justifiable?
6. Give an example of how a marketer of "green" product can convincingly communicate the environmentally safe features of a product to consumers who are skeptics about environmental warming as a reality.
7. Environmental "green" marketing is a CSR, and therefore, it is at the discretion of managers to act on it. Comment.
8. Explain the differences between green product adaptation and green product innovation. Use examples to explain the differences.
9. Green product pricing strategies are fundamentally the same as those for pricing the conventional products. Discuss.
10. Businesses are responsible for "green" market development and for educating the consumers. Discuss.
11. Green marketing strategy focuses on achieving organizational goals in ways that reduce or eliminate negative impacts on natural environment. Discuss using examples from personal observations.
12. What is "green" market positioning?
13. How can "green" positioning credibility and claims be established?
14. The "green" business strategy leads to substantial reduction in waste and increases efficiency. Discuss using examples.
15. Explain the PDCA cycle of EMS.
16. Briefly outline the provisions of ISO 14001.

REFERENCES

Aaker, D.A. and E. Joachimsthaler. 2000. The brand relationship spectrum: The key to the brand architecture challenge. *California Management Review*. 42(4): 8–22.

Adams, B. 2010. Global warming: Sunny side up. The first-year papers (2010). Trinity College Digital Repository. Hartford, CT. http://digitalrepository.trincoll.edu/fypapers/2. Accessed March 19, 2015.

Akehurst, G., C. Afonso, and H.M. Gonçalves. 2012. Re-examining green purchase behaviour and the green consumer profile: new evidences. *Management Decision*. 50(5): 972–988.

Albino, V., A. Balice, and R.M. Dangelico. 2009. Environmental strategies and green product development: an overview on sustainability-driven companies. *Business Strategy and the Environment*. 18(1): 83–96.

American Marketing Association. 2012. Dictionary. http://www.marketingpower.com/_layouts/dictionary.aspx. Accessed March 19, 2015.

Babiak, K. and S. Trendafilova. 2011. CSR and environmental responsibility: motives and procedures to adopt green management practices. *Corporate Social Responsibility and Environmental Management*. 18(1): 11–24.

Bader, P. 2008. Sustainability, from principle to practice. Goethe-Institut. http://www.goethe.de/ges/umw/dos/nac/den/en3106180.htm. Accessed December 2014.

Ballet, J., A. Bhukuth, and A. Carimentrand. 2014. Child labor and responsible consumers: from boycotts to social labels, illustrated by the Indian hand-knotted carpet industry. *Business & Society*. 53(1): 71–104.

Barone, M.J., A.T. Norman, and A.D. Miyazaki. 2007. Consumer response to retailer use of cause-related marketing: is more fit better? *Journal of Retailing*. 83(4): 437–445.

Beamon, B.M. 1999. Designing the green supply chain. *Logistics Information Management*. 12(4): 332–342.

Bennet, J. and P. Martin. 1998. ISO 14031 and the future of environmental performance Evaluation. *Greener Management International*. 21: 71–86.

Bloemers, R., F. Magnani, and M. Peters. 2001. Paying a green premium. *McKinsey Quarterly*. 3: 15–17.

Boiral, O. 2007. Corporate greening through ISO 14001: a rational myth? *Organisation Science*. 18(1): 127–146.

Boulstridge, E. and M. Carrigan. 2000. Do consumers really care about corporate responsibility? Highlighting the attitude—behaviour gap. *Journal of Communication Management*. 4(4): 355–368.

Brännlund, R. and J. Nordström. 2004. Carbon tax simulations using a household demand model. *European Economic Review*. 48(1): 211–233.

Brown, D., J. Dillard, and S. Marshall. 2005. Strategically informed, environmentally conscious information requirements for accounting information systems. *Journal of Information Systems*. 19(2) 79–103.

Camilleri, A.R. and R.P. Larrick. 2014. Metric and scale design as choice architecture tools. *Journal of Public Policy & Marketing*. 33(1): 108–125.

Chen, Y. 2010. The drivers of green brand equity: green brand image, green satisfaction, and green trust. *Journal of Business Ethics*. 93(2): 307–319.

Chen, R.J.C. 2011. A review of 'Green to gold': how smart companies use environmental strategy to innovate, create value, and build competitive advantage. *Journal of Sustainable Tourism*. 19(6): 789–792.

Chervev, A. 2012. Rethinking Gillette's pricing with dollar shave's disruptive innovation. Bloomberg Business Week [Media and Marketing]. http://www.businessweek.com/articles/2012-04-10/rethinking-gillette-s-pricing-the-disruptive-innovation-of-a-dollar-shave. Accessed September 2013.

Clapp, J. 1998. The privatization of global environmental governance: ISO 14000 and the developing world. *Global Governance*. 4(3): 295–316.

Cleveland, M., M. Kalamas, and M. Laroche. 2012. It's not easy being green: exploring green creeds, green deeds, and internal environmental locus of control. *Psychology & Marketing*. 29(5): 293–305.

Coley, D., M. Howard, and M. Winter. 2009. Local food, food miles and carbon emissions: a comparison of farm shop and mass distribution approaches. *Food Policy*. 34(2): 150–155.

COMM Engineering. 2013. COMMtracker EMS, greenhouse gas reporting. http://www.commengineering.com/commtracker-ems.html. Accessed March 19, 2015.

Cronin, J.J., J.S. Smith, M.R. Gleim, E. Ramirez, and J.D. Martinez. 2010. Green marketing strategies: an examination of stakeholders and the opportunities they present. *Journal of the Academy of Marketing Science*. 39(1): 158–174.

Crowther, D. 2000. *Social and Environmental Accounting*. Financial Times Prentice Hall, London.

Dangelico, R.M. and M. Pontrandolfo. 2010. From green product definitions and classifications to the Green Option Matrix. *Journal of Cleaner Production*. 18(16–17): 1608–1628.

Dangelico, R.M. and D. Pujari. 2010. Mainstreaming green product innovation: why and how companies integrate environmental sustainability. *Journal of Business Ethics*. 95(1): 471–486.

Darnall, N., I. Henriques, and P. Sadorsky. 2010. Adopting proactive environmental strategy: the influence of stakeholders and firm size. *Journal of Management Studies*. 47(6): 1072–1094.

Dean, D. H. and A. Biswas. 2001. Third-party organization endorsement of products: an advertising cue affecting consumer prepurchase evaluation of goods and services. *Journal of Advertising*. 30(4): 41–57.

De Giovanni, P. 2011. Do internal and external environmental management contribute to the triple bottom line? *International Journal of Operations & Production Management*. 32(3): 265–290.

Delmas, P.D. 2001. Bone marker nomenclature. *Bone*. 28(6): 575–576.

Delmas, M. and I. Montiel. 2009. Greening the supply chain: when is customer pressure effective? *Journal of Economics & Management Strategy*. 18(1): 171–201.

Department of Infrastructure and Transport. 2013. Section 2: Outcomes and planned performance. http://www.infrastructure.gov.au/department/statements/2013_2014/budget/dit2a.aspx. Accessed March 19, 2015.

Department of Resources. 2013. The more stars the more savings! http://www.energyrating.gov.au. Accessed March 19, 2015.

Diamantopoulos, A., B.B. Schlegelmilch, R.R. Sinkovics, and G.M. Bohlen. 2003. Can socio-demographics still play a role in profiling green consumers? A review of the evidence and an empirical investigation. *Journal of Business Research*. 56(6): 465–480.

Dief, M. and X. Font. 2010. The determinants of hotels' marketing managers' green marketing behaviour. *Journal of Sustainable Tourism*. 18(2): 157–174.

Di Salvo, C.J.P. and L. Raymond. 2010. Defining the precautionary principle: an empirical analysis of elite discourse. *Environmental Politics*. 19(1): 86–106.

Drumwright, M.E. 1994. Socially responsible organizational buying: environmental concern as a noneconomic buying criterion. *The Journal of Marketing*. 58(3): 1–19.

D'Souza, C., M. Taghian, and R. Khosla. 2007. Examination of environmental beliefs and its impact on the influence of price, quality and demographic characteristics with respect to green purchase intention. *Journal of Targeting, Measurement and Analysis for Marketing*. 15(2): 69–78.

Dyllick, T. and K. Hockerts. 2002. Beyond the business case for corporate sustainability. *Business Strategy and the Environment*. 11(2): 130–141.

Elijošiutė, E., J. Balciukevičiūtė, and G. Denafas. 2012. Life cycle assessment of compact fluorescent and incandescent lamps: comparative analysis. *Environmental Research, Engineering and Management*. 61(3): 65–72.

Elkington, J. and J. Hailes. 1988. *The Green Consumer Guide*. Gollancz: London.

Federal Facilities Council Report. 1999. Environmental management systems and ISO 14001. http://www.nap.edu/openbook.php?record_id=6481. Accessed March 19, 2015.

Federal Trade Commission. 2012. Environmental Claims: Summary of the Green Guides. https://www.ftc.gov/tips-advice/business-center/guidance/environmental-claims-summary-green-guides. Accessed March 31, 2015.

Gao, Y. 2011. CSR in an emerging country: a content analysis of CSR reports of listed companies. *Baltic Journal of Management*. 6(2): 263–291.

Gastl, R. 2009. *CIP in Environmental Management*, 2nd Edition, 2009, VDF: Zurich. In Mann, T. 2012. An evaluation perspective on environmental management information systems software acquisition using a principle agency framework. Duke Environmental Leadership Masters of environmental Management, Nicholas School of the Environment.

Gillespie, E. 2008. Stemming the tide of greenwash. *Consumer Policy Review*. 18(3): 79–83.

Ginsberg, J.M. and P.N. Bloom. 2004. Choosing the right green marketing strategy. *MIT Sloan Management Review*. 46(1): 79–84.

Gray, R.H. 2001. Thirty years of social accounting, reporting and auditing: what (if anything) have we learnt. *Business Ethics: A European View*. 10(1): 9–15.

Gray, R.H., D.L. Owen, and K.T. Maunders. 1987. *Corporate Social Reporting: Accounting and accountability*. Prentice Hall: Hemel Hempstead, Hertfordshire.

Griskevicius, V. and J.M. Tybur. 2010. Going green to be seen: status, reputation, and conspicuous conservation. *Journal of Personality and Social Psychology*. 98(3): 392–404.

Handfield, R.B. and S.A. Melnyk. 1996. GreenSpeak. Purchasing Today July: 32–36.

Handfield, R.B., S.V. Walton, R. Sroufe, and S.A. Melnyk. 2002. Applying environmental criteria to supplier assessment: a study in the application of the Analytical Hierarchy Process. *European Journal of Operational Research*. 141(1): 70—87.

Hartmann, P., V.A. Ibanez, and F.J.F. Sainz. 2005. Green branding effects on attitude: functional versus emotional positioning strategies. *Marketing Intelligence & Planning*. 23(1): 9–29.

Holm, O. 2006. Integrated marketing communication: from tactics to strategy. *Corporate Communications: An International Journal*. 11(1): 23–33.

Hunt, S.D. 2011. Sustainable marketing, equity, and economic growth: a resource advantage, economic freedom approach. *Journal of the Academy of Marketing Science*. 39(1): 7–20.

Hutchens, S. 2010. Using ISO 9001 or ISO 14001 a competitive advantage. Intertek white paper. http://www.intertek.com. Accessed March 19, 2015.

Ing, D., F. Straube, and S. Doch. 2010. *How sustainability will foster a more collaborative logistics business. Delivering tomorrow: Towards Sustainable Logistics – How business innovation and green demand drive a carbon efficient industry*. Deutsche Post AG: Bonn.

International Organisation for Standardization (ISO). 2012. Event sustainability management systems – requirements with guidance touse.http://www.iso.org/iso/catalogue_detail?csnumber=54552. Accessed July 2013.

Kaenzig, J. and R. Wüstenhagen. 2010. The effect of life cycle cost information on consumer investment decisions regarding eco-innovation. *Journal of Industrial Ecology*. 14(1): 121–136.

Kangun N. and M.J. Polonsky. 1995. Regulation of environmental marketing claims: a comparative perspective. *International Journal of Advertising*. 11(1): 1–24.

Kangun, N., L. Carlson, and S.J. Grove. 1991. Environmental advertising claims: a preliminary investigation. *Journal of Public Policy & Marketing*. 10(2): 47–58.

Kaplan, A.M. and M. Haenlein. 2010. Users of the world, unite! The challenges and opportunities of Social Media. *Business Horizons*. 53(1): 59–68.

Kerr, W. and C. Ryan. 2001. Eco-efficiency gains from remanufacturing: a case study of photocopier remanufacturing at Fuji Xerox Australia. *Journal of Cleaner Production*. 9(1): 75–81.

Kilbourne, W.E., M.J. Dorsch, P. McDonagh, B. Urien, A. Prothero, M. Grunhagen, M.J. Polonsky, D. Marshell, J. Foley, and A. Bradshaw. 2009. The institutional foundations of materialism in Western societies: a conceptualization and empirical test. *Journal of Macromarketing*. 28(4): 259–278.

Kim, Y. 2013. Environmental, sustainable behaviors and innovation of firms during the financial crisis. *Business Strategy and the Environment*. 24(1): 58–72.

Klassen, R.D. and C.P. McLaughlin. 1996. The impact of environmental management on firm performance. *Management Science*. 42(8): 1199–1214.

Kollmuss, A., H. Zink, and C. Polycarp. 2008. *Making Sense of the Voluntary Carbon Market: A Comparison of Carbon Offset Standards*. WWF: Germany.

Kordjamshidi, M. and S. King. 2009. Overcoming problems in house energy ratings in temperate climates: a proposed new rating framework. *Energy and Buildings*. 41(1): 125–132.

Lafferty, B.A., R.E. Goldsmith, and G.T.M. Hult. 2004. The impact of the alliance on the partners: a look at cause–brand alliances. *Psychology & Marketing*. 21(7): 509–531.

Leonidou, C.N., C.S. Katsikeas, and N.A. Morgan. 2012. Greening the marketing mix: do firms do it and does it pay off? *Journal of the Academy of Marketing Science*. 41(1): 151–170.

Lin, Y. and C.A. Chang. 2012. Double standard: the role of environmental consciousness in green product usage. *Journal of Marketing*. 76(3): 125–134.

Lindhqvist, T. 2000. Extended producer responsibility in cleaner production: policy principle to promote environ-mental improvements of product systems, PhD Thesis, Lund University, Lund.

Lodhia, S. 2012. Web based social and environmental communication in the Australian minerals industry: an application of media richness framework. *Journal of Cleaner Production*. 25(1): 73–85.

Luchs, M., R. Walker, R.L. Rose, J.R. Catli, R. Gau, S. Kapitan, J. Mish, L. Ozanne, M. Phipps, B. Simpson, S. Subrahmanyan, and T. Weaver. 2011. Toward a sustainable marketplace: expanding options and benefits for consumers. *Journal of Research for Consumers*. 1(19): 1–12.

MacKerron, G.J., C. Egerton, C. Gaskell, A. Parpia, and S. Mourato. 2009. Willingness to pay for carbon offset certification and co-benefits among high-flying young adults in the UK. *Energy Policy*. 37(4): 1372–1381.

Mangold, W.G. and D.J Faulds. 2009. Social media: the new hybrid element of the promotion mix. *Business Horizons*. 52(4): 357–365.

Meijkamp, R. 1998. Changing consumer behaviour through eco-efficient services: an empirical study of car sharing in the Netherlands. *Business Strategy and the Environment*. 7(4): 234–244.

Mendleson, N. and M.J. Polonsky. 1995. Using strategic alliances to develop credible green marketing. *Journal of Consumer Marketing*. 12(2): 4–18.

Menegaki, A.N. 2012. A social marketing mix for renewable energy in Europe based on consumer stated preference surveys. *Renewable Energy*. 39(1): 30–39.

Menon, S. and B.E. Kahn. 2003. Corporate sponsorships of philanthropic activities: when do they impact perception of sponsor brand? *Journal of Consumer Psychology*. 13(3): 316–327.

Miles, M.P. and J.G. Covin. 2000. Environmental Marketing: a source of reputational, competitive and financial advantage. *Journal of Business Ethics*. 23(3): 299–311.

Miles, M.P. and L.S. Munila. 2004. The potential impact of social accountability certification on marketing: a short note. *Journal of Business Ethics*. 50(1): 1–11.

Miles, M.P., L.S. Munila, and G.R. Russell. 1997. Marketing and environmental registration: what industrial marketers should understand about ISO 14000. *Industrial Marketing Management*. 26(4): 363–370.

Milfont, T.L., J. Wilson, and P. Diniz. 2012. Temporal perspective and environmental engagement: a meta-analysis. *International Journal of Psychology*. 47(4): 325–334.

Mishra, P. and P. Sharma. 2010. Green marketing in India: emerging opportunities and challenges. *Journal of Engineering, Science and Management Education*. 3(1): 9–14.

Molina-Azorín, J., E. Claver-Cortés, M.D. López-Gamero, and J.J. Tarí. 2009. Green management and financial performance: a literature review. *Management Decision*. 47(7): 1080–1100.

Mollenkopf, D., H. Stolze, W. Tate, and M. Ueltschy. 2010. Green, lean, and global supply chains. *International Journal of Physical Distribution & Logistics Management*. 40(1/2): 14–41.

Mülhaupt, R. 2013. Green polymer chemistry and bio-based plastics: dreams and reality. *Macromolecular Chemistry and Physics*. 214(2): 159–174.

National Geographic. 2012. GreenDex 2012. National geographic. http://environment.nationalgeographic.com.au/environment/greendex. Accessed September 2013.

Newport, F. 2010. American's global warming concerns continue to drop. http://www.gallup.com/poll/126560/Americans-Global-Warming-Concerns-Continue-Drop.aspx?version=print. Accessed September 2013.

Nwagbara, U. 2013. The effects of social media on environmental sustainability activities of oil and gas multinationals in Nigeria. *Thunderbird International Business Review*. 55(6): 689–697.

Oates, C., S. McDonald, P. Alevizou, K. Hwang, W. Young, and L.A. McMorland. 2008. Marketing sustainability: use of information sources and degrees of voluntary simplicity. *Journal of Marketing Communications*. 14(5): 351–365.

Obar, J.A., P. Zube, and C. Lampe. 2012. Advocacy 2.0: an analysis of how advocacy groups in the United States perceive and use social media as tools for facilitating civic engagement and collective action. *Journal of Information Policy*. 2(1): 1–25.

Ozanne, L.K. and R.P. Vlosky. 1997. Willingness to pay for environmentally certified wood products: a consumer perspective. *Forest Products Journal*. 47(6): 39–48.

Papadopoulos, A.M., T.G. Theodosiou, and K.D. Karatzas. 2002. Feasibility of energy saving renovation measures in urban buildings: the impact of energy prices and the acceptable payback time criterion. *Energy and Buildings*. 34(5): 455–466.

Peattie, K. 1995. *Environmental Marketing Management: Meeting the Green Challenge*. Pitman: London.

Peattie, K. 2001. Towards sustainability: the third age of green marketing. *The Marketing Review*. 2(2): 129–146.

Peattie, K. and A. Crane. 2005. Green marketing: legend, myth, farce or prophesy? *Qualitative Research: An International Journal*. 8(4): 357–370.

Pokharel, S. and A. Mutha. 2009. Perspectives in reverse logistics: a review. *Resources, Conservation and Recycling*. 53(4): 175–182.

Polonsky, M.J. 2011. Transformative green marketing: impediments and opportunities. *Journal of Business Research*. 64(1): 1311–1319.

Polonsky, M.J. and M. Taghian. 2010. Green marketing, Chapter 6, *Australian Master Environment Guide*. CCH Australia Limited: Sydney.

Polonsky, M.J. and G. Wood. 2001. Can the over-commercialization of cause-related marketing harm society? *Journal of Macromarketing*. 21(1): 8–22.

Polonsky, M.J., L. Carlson, S. Grove, and N. Kangun. (1997. International environmental marketing claims: real changes or simple posturing? *International Marketing Review*. 14(4): 218–232.

Polonsky, M.J., L. Carlson, A. Prothero, and D. Kapelianis. 2002. A cross-cultural examination of the environmental information on packaging: implications for advertisers. *Advances in International Marketing*. 12(1): 153–174.

Polonsky, M.J., S.L. Garu, and R. Garma. 2010. The New Greenwash? Potential marketing problems with carbon offsets. *International Journal of Business Studies*. 18(1): 49–54.

Polonsky, M.J., A. Vocino, S.L. Grau, R. Garma, and A.S. Ferdous. 2012. The impact of general and carbon-related environment knowledge on attitudes and behavior of US consumers. *Journal of Marketing Management*. 28(3–4): 238–263.

Porter, M.E. and M.R. Kramer. 2006. Strategy and society. *Harvard Business Review*. 84(12): 78–92.

Porter, M.E. and C. Van der Linde. 1995. Green and competitive: ending the stalemate. *Harvard Business Review*. 73(5): 120–134.

Rex, E. and H. Baumann. 2007. Beyond ecolabels: what green marketing can learn from conventional marketing. *Journal of Cleaner Production*. 15(6): 567–576.

Ring, M.J., D. Lindner, E.F. Cross, and M.E. Schlesinger. 2012. Causes of the global warming observed since the 19th century. *Atmospheric and Climate Sciences*. 2(1): 401–415.

Roper, J. 2012. Environmental risk, sustainability discourses, and public relations. *Public Relations Inquiry*. 1(1): 69–87.

Sawadogo, M. and D. Anciaux. 2011. Intermodal transportation within the green supply chain: an approach based on ELECTRE method. *International Journal of Business Performance and Supply Chain Modelling*. 3(1): 43–65.

Sharma, A., G.R. Lyer, A. Mehrotra, and R. Krishnan. 2010. Sustainability and business-to-business marketing: a framework and implications. *Industrial Marketing Management*. 39(2): 330–341.

Sheldon, C. 1997. *ISO 14001 and Beyond, Environmental Management Systems in the Real World, Beyond ISO 14001: An Introduction to the ISO 14000 Series*. Greenleaf Publishing: Sheffield.

Sheth, J.N., N.K. Sethia, and S. Srinivas. 2011. Mindful consumption: a customer-centric approach to sustainability. *Journal of the Academy of Marketing Science*. 39(1): 21–39.

Slaper, T.F. and T. Hall. 2011. The triple bottom line: what is it and how does it work? *Indiana Business Review*. 86(1): 4–8.

Small, K.A. and K. Van Dender. 2007. Fuel efficiency and motor vehicle travel: the declining rebound effect. *The Energy Journal*. 28(1): 25–51.

Soyez, K. 2012. How national cultural values affect pro-environmental consumer behaviour. *International Marketing Review*. 29(6): 623–646.

Sridhar, K. 2012. The relationship between the adoption of Triple Bottom Line and enhanced corporate reputation and legitimacy. *Corporate Reputation Review*. 15(2): 69–87.

Sridhar, K and G. Jones. 2013. The three fundamental criticism of the Triple Bottom Line approach: an empirical study to link sustainability report in companies based in the Asia-Pacific region and TBL shortcomings. *Asian Journal of Business Ethics*. 2(1): 91–111.

Srivastava, S. 2007. Green supply-chain management: a state-of-the-art literature review. *International Journal of Management Reviews*. 9(1): 53–80.

Straughan, R.D. and J.A. Roberts. 1999. Environmental segmentation alternatives: a look at green consumer behavior in the new millennium. *Journal of Consumer Marketing*. 16(6): 558–575.

Tang, A.K.Y., K.H. Lai, and T.C.E. Cheng. 2012. Environmental governance of enterprises and their economic upshot through corporate reputation and consumer satisfaction. *Business Strategy and the Environment*. 21(6): 401–411.

Taylor III, P.L. and H.L. Kay. 2011. Green board as a climate-change imperative: appointing a climate-change expert to the audit committee. *University of Baltimore Journal of Environmental Law* 18: 215–261.

Thøgersen, J. 2004. A cognitive dissonance interpretation of consistencies and inconsistencies in environmentally-responsible behaviour. *Journal of Environmental Psychology*. 24(1): 93–103.

Thøgersen, J., A. Jørgensen, and S. Sandagerand. 2012. Consumer decision making regarding a "Green" everyday product. *Psychology and Marketing*. 29(4): 187–197.

Thompson, D.W., R.C. Anderson, E.N. Hansen, and L.R. Kahle. 2010. Green segmentation and environmental certification: insights from forest products. *Business Strategy and the Environment*. 19(5): 319–334.

Tobler, C., V.H.M. Visschers, and M. Siegrist. 2011. Eating green, consumers' willingness to adapt ecological food consumption behaviour. *Appetite*. 57(3): 674–682.

Tucker, E.M., N.J. Rifon, E.M. Lee, and B.B. Reece. 2012. Consumer receptivity to green ads. *Journal of Advertising*. 41(4): 9–23.

Vachon, S. and R. Klassen. 2006. Extending green practices across the supply chain: the impact of upstream and downstream integration. *International Journal of Operations & Production Management*. 26(7): 795–821.

Van der Veldt, D. 1997. Case studies of ISO 14001: a new business guide for global environmental protection. *Environmental Quality Management*. 7(1): 1–9.

Varadarajan, P.R. and A. Menon. 1988. Cause-related marketing: a coalignment of marketing strategy and corporate philanthropy. *Journal of Marketing*. 52(3): 58–74.

Venkatesh, V.G. 2010. Reverse logistics: an imperative area of research for fashion supply chain. *The IUP Journal of Supply Chain Management*. 7(1): 77–89.

Weissman, W.A and J.C. Sekutowski. 1991. Environmentally conscious manufacturing. *AT & T Technical Journal*. 70(6): 23–30.

Weitzman, M. 2009. On modelling and interpreting the economics of catastrophic climate change. *Review of Economics and Statistics*. 91(1): 1–19.

Widmer, R., H. Oswald-Krapf, D. Sinha-Khetriwal, M. Schnellmann, and H. Böni. 2005. Global perspectives on e-waste. *Environmental Impact Assessment Review*. 25(5): 436–458.

Young, W., K. Hwang, S. McDonald, and C.J. Oates. 2009. Sustainable consumption: green consumer behaviour when purchasing products. *Sustainable Development*. 18(1): 20–31.

Zarnikau, J. 2003. Consumer demand for 'green power' and energy efficiency. *Energy Policy*. 31(15): 1661–1672.

Zhang, X., L. Shen, and Y. Wu. 2011. Green strategy for gaining competitive advantage in housing development: a China study. *Journal of Cleaner Production*. 19(2–3): 157–167.

Zwetsloot, G. 2003. From management systems to corporate social responsibility. *Journal of Business Ethics*. 44(2): 201–208.

10

ROLE OF ENVIRONMENTAL, SOCIAL, AND GOVERNANCE (ESG) FACTORS IN FINANCIAL INVESTMENTS

A. Seddik Meziani
Department of Economics & Finance, School of Business, Montclair State University, Montclair, NJ, USA

Abstract: This chapter measures whether investing with Environmental, Social and Governance (ESG) issues in mind has turned into a compelling investment premise for fund managers. For this purpose, a series of market metrics were applied to all of the current ESG-based funds to measure whether they offer potential to satisfy a classical risk/return assessment of their performance. Contrary to studies that paint an utterly favorable picture of ESG investing, the results portrayed in this chapter are mixed. Although their yearly growth and their risk-adjusted returns in relation to the market benchmark used are notable, the same cannot be said of their performance in terms of the risk taken to achieve these returns and with regard to the important systematic risk they contribute.

Keywords: Environmental, social and governance (ESG) investments, responsible investing, sustainable investing, corporate governance, corporate ethics, green investing, Sarbanes-Oxley Act, ERISA plans, fiduciaries, faith-based funds, pension funds, mutual funds, Social Investment Forum Foundation (SIF), exchanged-traded funds, investment portfolio, portfolio-diversification, risk/return framework, Sharpe ratio, Beta, alpha, coefficient of determination, coefficient of variation, arbitrage, systematic risk, risk performance, return performance, abnormal returns, cumulative return, expense ratio.

10.1 Introduction	255	
10.2 History of ESG Investing: From Fringe to Mainstream	256	
10.3 Review of ESG Literature	257	
10.4 Review and Analysis of the ESG Market	257	
10.4.1 A Bird's-Eye View of the Overall ESG Market: Getting a Good Lay of the Land	258	
10.4.2 Case 1: The Faith-Based ESG Market	259	
10.4.2.1 Taking Stock of the Market	259	
10.4.2.2 Return Performance Statistics	261	
10.4.2.3 Risk Performance Statistics	261	
10.4.3 Case 2: ETFs and the ESG Market	261	
10.4.3.1 ESG ETFs: Return Performance Statistics	263	
10.4.3.2 ESG ETFs: Risk Performance Statistics	266	
10.5 Concluding Remarks	268	
Appendix: Exhibits	270	
References	276	

10.1 INTRODUCTION

In recent years, interest has increased in the so-called environmental, social, and governance (ESG) investments. Although ESG means different things to different people, it is above all about investing in companies that care about the environment and corporate responsibility, the latter referring to business processes that produce an overall positive impact on society.

Investors' increased concern for ESG issues has transformed such investment activity from a marginal type of investing to one of the fastest-growing areas on the investment spectrum. Its central tenet is that investors should be able to buy stocks and bonds in companies that pass ESG screens and avoid those that do not.

With people increasingly aligning their investment portfolios with their beliefs, ESG has also caught the attention of exchange-traded funds (ETFs) providers. Although socially

---

An Integrated Approach to Environmental Management, First Edition. Edited by Dibyendu Sarkar, Rupali Datta, Avinandan Mukherjee, and Robyn Hannigan.
© 2016 John Wiley & Sons, Inc. Published 2016 by John Wiley & Sons, Inc.

responsible ETFs still represent only a tiny portion of the $3 trillion ETF market, their share is expected to grow as interest in ESG continues to gain momentum.

The introduction of ESG factors into the ETF market is important for this type of investing for two main reasons. It represents a significant seal of approval by funds that have become hugely popular since their humble beginning in the early 1990s. ETFs' continuing success with the investing public has the potential to rub off on ESG investing by introducing the latter to a much larger group of investors. ETFs' much-hailed transparency could also reveal some interesting insights. It might show that these ESG ETFs do not consist only of companies at the vanguard of the social movement but also those that are all about profit while also taking into account sound ESG practices. They might be deemed inadequate by pure ESG investors, but that's not necessarily bad for the long-run success of this type of investing.

This chapter will outline the factors fueling the drive to ESG, and the ESG literature will be reviewed in some detail. Next, the incorporation of this type of investing into the ETF market will be charted and the impact it might have on ESG duly analyzed. In order to fully understand this impact, an explanation of the mechanics underlying ETFs also deserves some attention. Current concerns surrounding the liquidity of these funds and investors' apprehensions regarding their market performance could be more comprehensible following this explanation. Last but not least, this chapter will attempt to explain how ESG ETFs could be incorporated in an investment portfolio without compromising its overall risk-adjusted performance. After all, it's the only way to bring them out of the fringes of investing, where they have largely been languishing, and into a bigger investment arena.

10.2 HISTORY OF ESG INVESTING: FROM FRINGE TO MAINSTREAM

Responsible investing may date back as far as the eighteenth century, when the Religious Society of Friends or Quakers prohibited its members from investing in slavery and war (Schueth, 2003). That directive followed the opinion of John Wesley, the founder of Methodism, who emphasized to his followers that how money is used is the second-most-important subject of the New Testament teachings. The notion of responsible investing later evolved as religious groups added to their list of objections investments in businesses associated with the production of liquor, tobacco, and guns.

In the 1960s, responsible investing spread to more secular issues such as women's equality, civil rights, protests against the Vietnam War, and the subsequent concerns over the United States' increased reliance on nuclear power to satisfy its energy needs (Entine, 2003). In the 1970s, academic institutions and pension funds joined with faith-based groups in threatening to divest from large companies operating in South Africa under apartheid. The result was a negative flow of investment significant enough to ultimately force the country to end apartheid. More recently, international pressure was exercised to end genocide in the Darfur region of the Sudan. That pressure culminated with the US Congress passing the Sudan Accountability and Divestment Act of 2007,[1] which prohibited the federal government, along with state and local governments, from conducting business with companies doing business in the Sudan.

In the meanwhile, ESG has broadened its scope to include sustainable investing, which recognizes and rewards companies that incorporate environmental, social, ethical, and corporate governance guidelines in their investment approach. An unambiguous endorsement of this broadened version of ESG came from the Netherlands' two largest pension funds, Stichting Pensioenfonds ABP[2] and Stichting Pensioenfonds Zorg en Welzijn (known as PFZW)[3] as early as 1999, when both funds began applying environmental and other social screens to a portion of their assets. It was probably at the point that this form of investing became accurately referred to as ESG. In 2002, the California Public Employees' Retirement System (CalPERS), generally considered the largest public pension fund in the United States, followed suit by applying labor-standard screens to countries included in its emerging markets portfolio.

But ESG's biggest endorsement yet came from Norway, with the creation of the Petroleum Fund's Advisory Council on Ethics on November 19, 2004. The council's directive was to oversee the huge[4] Norway Government Pension Fund Global under a new mandate to incorporate environmental, social, and corporate governance factors into its investment strategies.[5]

The particular importance of corporate ethics in ESG investing was strengthened by the breakdown of Enron (in 2001) and WorldCom (2002) amid multibillion-dollar accounting scandals and large numbers of investors losing their life savings. Concerned that the general public might perceive those two significant corporate failings as a powerful indictment of lax government oversight, lawmakers reacted strongly in 2002 by passing the Sarbanes–Oxley Act in an effort to shore up confidence in Wall Street among individual investors.[6] However, Sarbanes–Oxley could not prevent the global financial meltdown of 2008–2009, initiated by the implosion of the

---

[1] http://www.gpo.gov/fdsys/pkg/BILLS-110s2271enr/pdf/BILLS-110s 2271enr.pdf

[2] Stichting Pensioenfonds ABP is the pension fund for employees in the government, public, and education sectors with invested capital of €292 billion as of March 31, 2013.

[3] Formerly known as PGGM, PFZW is the second-largest pension fund in The Netherlands, behind Stichting Pensioenfonds ABP with invested capital of €109 billion.

[4] A $783.3 billion (as of September 30, 2013) sovereign wealth fund established in 1990.

[5] See "Management Mandate for the Government Pension Fund Norway" at http://folketrygdfondet.no/images/Marketing/Rammeverk/Management%20 mandate%20for%20the%20Government%20Pension%20Fund%20 Norway%202011.pdf

[6] The Public Company Accounting Oversight Board was created by the Sarbanes–Oxley Act to enforce tougher auditing standards (http://pcaobus.org/Pages/default.aspx).

real estate market, which further eroded both consumers' and investors' confidence in corporate ethical behavior.

In spite of increasing public awareness of the issues that ESG has endeavored to promote with some level of success, it still faces key major challenges, one of which is the Employee Retirement Income Security Act of 1974 (ERISA).[7] ERISA's overriding goal of protecting employees' pensions continues to represent a major hurdle in realizing the goals promoted by ESG investing. Under Section 404(a) of ERISA, retirement plans' fiduciaries are required to act solely in the interest of plans participants and their beneficiaries.[8] As such, the fiduciaries must view the goals promoted by ESG as a secondary consideration, at best.

For example, fiduciaries who consider investing in the so-called green companies that promise to limit damage to the environment can do so only if they also consider nongreen companies. In other words, green investments will be incorporated into a plan only after its fiduciaries determine that such investments would provide a risk-adjusted return comparable to that of nongreen investments. But thankfully for ESG, defined benefit pension plans (ERISA plans) are steadily giving way to more flexible defined contribution plans, in which individuals can freely choose their investments. That should gradually lessen the adverse impact of fiduciary law on ESG investments.

10.3 REVIEW OF ESG LITERATURE

Relying heavily on the classical risk/return framework used extensively in portfolio management, academic studies have, in general, found little economic gain from ESG (Hickman et al., 1999).

As early as 1970, Milton Friedman stressed that the corporate responsibility is to increase profits. He termed ESG as no more than a "collectivist doctrine." Three decades later, Friedman's blunt assessment of ESG was echoed by the equally critical George Soros (2000), who suggested that "social justice is outside the competence of market economy." In 2001, David Henderson strongly criticized the proponents of ESG, warning that it was a new variation of capitalism whose "potentially damaging effects ... extend to economic systems as a whole, as well as to individual enterprises."

But some research has concluded that ESG can bring benefits to investors. For instance, Russo and Fouts (1997) analyzed the economic performance of a sample of 243 firms generated through the use of environmental screens over a period of 2 years. They concluded that a strong environmental stance correlates to better financial performance.

Van Dyck and Paulker (2001) observed that a strong social and environmental record of a corporation can provide valuable insights into the quality of a company's top management. Looking specifically at market returns from ESG, Bauer et al. (2005) and Galema et al. (2008) didn't find them significantly different from those of conventional investments. Although some consider the findings a victory for ESG, others consider them quite ambiguous, considering that the risk/return structure of classical analysis advises against making investment decisions based solely on returns without proper consideration of the risks involved.

In other research results favorable to ESG, Margolis et al. (2007) were able to identify a positive association between the use of social and environmental screens in investment analysis and market performance. Renneboog et al. (2008a, b) observed that institutional investors have subdivided newer ESG funds along environmental, social, corporate governance or ethical issues to accommodate individual investors' interests in specific issues. They noted that these subdivisions have helped bring new players into the market, ranging from institutional investors such as pension funds, insurance companies, and managed funds to organizational investors such as labor unions, international organizations, and academic institutions. A complete review of the literature dealing with the impact of using ESG screens on companies' financial performance can be found in Hoepner and McMillan (2009) and more recently in Capelle-Blancard and Monjon (2012).

There are also studies, such as Scholtens and Sievanen (2013), that showed how the ESG landscape has significantly changed for decades, increasingly attracting more interest from institutional investors, labor unions, international associations, and academic and governmental institutions. The studies note a key transition from a commitment by a cluster of engaged individuals in the 1960s and 1970s to a broader engagement of shareholders requiring more accountability from the firms they invested in.

Overall, the ESG literature indicates growing evidence that both individual and institutional investors are gradually willing to incorporate ethical, environmental, and social criteria in their investment decisions. It also indicates that with the increasing involvement of institutional investors, ESG is finally moving from fringe investing, where it has long languished, into the more mainstream investment environment.

The same literature also points out that the increasing use of investment criteria outside of the classical risk/return framework is also propelling ESG into a much bigger role in the investment world. The next section of this chapter will review the lay of the land for this type of investing and explore whether it's truly coming out of the cold and gradually becoming a noticeable player in the financial markets, as suggested by this literature.

10.4 REVIEW AND ANALYSIS OF THE ESG MARKET

Despite the growing importance of ESG funds in the investment field, skepticism still abounds. There is a common belief that ESG's market performance is impeded by the strict

---
[7] http://www.dol.gov/compliance/laws/comp-erisa.htm
[8] While some plan sponsors (employers) prefer to handle all the investment decisions related to their retirement plans, most prefer to outsource the management of the plan assets to a fiduciary. For a commission, the latter exercises effective control over the management or disposition of a plan's assets.

TABLE 10.1 ESG Funds in the United States (1995–2012)

Years	Total US-Domiciled Funds		Total US-Domiciled ESG Funds		
	Assets Under Management (in $Trillions)[a]	Biennial Growth (in %)	Assets Under Management (in $Billions)[b]	Share of Total US-Domiciled Assets (in %)	Biennial Growth (in %)
1995	7.00		12.00	0.17	
1997	13.70	95.71	96.00	0.70	700.00
1999	16.30	18.98	154.00	0.94	60.42
2001	19.90	22.09	136.00	0.68	−11.69
2003	19.20	−3.52	151.00	0.79	11.03
2005	24.40	27.08	179.00	0.73	18.54
2007	25.10	2.87	202.00	0.80	12.85
2010	25.20	0.40	569.00	2.26	181.68
2012	33.30	32.14	1013.00	3.04	78.03

[a] Nelson's Directory *of* Money Managers.
[b] The Forum for Sustainable and Responsible Investment. ESG funds include mutual funds, closed end funds, exchange-traded funds, investment fund, and other pooled products. Separate accounts *are* excluded.

social, ethical, and environmental screens typically used during the investment process. Doubters believe that these funds' ability to compete has effectively been limited by the screens' exclusion of too many high-performing companies.

10.4.1 A Bird's-Eye View of the Overall ESG Market: Getting a Good Lay of the Land

In response to a growing emphasis on responsible investing, a few organizations started reporting on ESG by analyzing its growth and the factors fueling it. The Social Investment Forum Foundation (SIF) is one of them. It began thoroughly informing the market on this type of investing via biennial surveys as early as 1995.[9]

Table 10.1 offers an overall summary of the results of SIF's surveys between 1995 and 2012. These data are subdivided into two distinct categories: total US-domiciled funds and total US-domiciled ESG funds. The first category refers to all US-based investments regardless of whether any criteria are applied to them, whereas the second one specifically measures US-domiciled assets believed to be based on responsible investment strategies.

A quick look at the data indicates that, in spite of a rapid growth, ESG funds remain a small portion of the total assets under management. In 2012, the year of SIF's latest survey release at the time of this chapter's writing, ESG funds totaled just over $1 trillion, out of a total $33 trillion for all US-domiciled funds. While some critics hastily conclude that ESG funds continue to be inadequate, a more detailed assessment of such investments could lead to a more balanced conclusion. That requires a closer look at Table 10.1.

Whereas it's true that ESG funds comprised only 3% of the total assets in 2012, it's important to take stock of their path since 1995 before drawing any meaningful conclusion.

That year, they stood at a meager 0.17% of the total. Their share has steadily increased, with institutional investors more widely embracing their key principles, even during the great recession of 2008–2009. Indeed, not only did these funds not suffer from the impact of this exceptional recession from 2007 to 2010, but on the contrary, they literally throve under unparalleled market stress, remarkably increasing 182%, from $202 to $569 billion. During the same period, the entire universe of US-domiciled funds increased by a stingy 0.40%, from $25.10 to $25.20 trillion. During the 2001–2003 period, also characterized by a recession, although a far-less remarkable one, ESG funds also exhibited a healthy growth of a little over 11%, from $136 to $151 billion, compared to a contraction of close to 4%, from $19.90 to $19.20 trillion, for the entire US population of investment funds.

Although more than 18 years of data and two recessions, albeit that last one was quite extreme, are needed to draw any meaningful conclusion, it's tempting nonetheless to consider that ESG funds may have proven themselves quite recession proof compared to the rest of the investment landscape. To be certain, their return performance, like that of other investments, was held down by the weight of the two recessions. But their supporters continued to pour in capital into them.

As to the factors fueling this growth, SIF noted in its various reports that the social screens used to identify ESG funds have not maintained the same importance from one survey period to the next. SIF observed in its earlier 1990s surveys that ESG investment strategies weighted tobacco, firearms, and human rights far more highly than environmental concerns. But in its 1999 survey,[10] SIF reported that the environment filter climbed sharply to first place with 79% of screened assets, up from the modest 38% reported in 1995.[11] SIF suggested that the environmental screen's impor-

---

[9] http://www.ussif.org/resources/pubs/
[10] http://www.ussif.org/files/Publications/99_Trends_Report.PDF
[11] http://www.ussif.org/files/Publications/95_trends_Report.pdf

tance in identifying ESG was boosted by alarming news on global warming and ozone depletion, bringing back to the public's mind the tragic accidents that happened in Bhopal and Chernobyl as well as the Exxon Valdez oil spill in Prince William Sound, Alaska.

As environmental pollution and the search for alternative energies capable of slowing climate change continued to fuel the news, socially responsible funds went on granting increased importance to the environment in their investment strategies throughout SIF's latest 2012 survey.[12] As a result, and according to SIF, ESG has now taken on the broader meaning of sustainable and responsible investing, where "sustainable" refers to investments seeking to deliver a positive impact on the environment.

10.4.2 Case 1: The Faith-Based ESG Market

As mentioned earlier, ESG finds its roots in the faith-based sector, which has long been committed to active engagement in screening, shareholder advocacy, and community investing. Table 10.2 highlights six major groups of faith-based funds where, practically speaking, some investors have found meaning by putting their money where their faith is. Stocks of companies that violate the creeds of an investor's religion or religious denomination are literally screened out of these funds. Exhibits 10.1 through 10.4 at the end of this chapter describe the 76 funds comprising these six main groups in more detail.

10.4.2.1 Taking Stock of the Market

Under the supervision of the Annuity Board of the Southern Baptist Church and founded for the purpose of providing investment management services for the evangelical marketplace, the GuideStone Funds is by far the largest faith-based group of funds, with over $22 billion in assets under management, as indicated in Table 10.2. They consist of 40 Christian-based mutual funds, the first of which, the GuideStone Equity Index Fund (GEQZX), was launched on August 27, 2001. These funds are socially screened according to the doctrine and practice of the Southern Baptist Church. Negative screens are used to exclude companies that violate the church's beliefs, such as those involved in embryonic stem cell research, military weapons, or adult entertainment.[13] Because the companies involved in activities deemed inappropriate by the fund manager are excluded, these funds only approximate the specific indices they were designed to track.

Of the two groups of funds that incorporate catholic values into their investment process, the Luther King Capital Management (LKCM) funds are the more prominent, twice the size of the Ave Maria funds (for the former, $3,136 million in assets under management spread over nine funds versus $1,365 million in six funds for the latter, as shown in Table 10.2). The LKCM Aquinas funds, founded on January 3, 1994, with the launch of LKCM Aquinas Value (AQEIX), LKCM Aquinas Small Cap (AQBLX), and LKCM Aquinas Growth (AQEGX), the oldest of the catholic funds, invest according to the US Conference of Catholic Bishops' Socially Responsible Investment Guidelines. These guidelines also incorporate environmental responsibility, affordable housing, and fair employment practices in their list of negative screens.[14]

The Ave Maria funds are relatively new compared to the LKCM funds. Their oldest mutual fund, Ave Maria Catholic Values (AVEMX), was launched on May 1, 2001, about 7 years after the first LKCM funds. The palm for the largest fund also goes to LKCM. The LKCM Small Cap Equity Advisor (LKSAX) holds $1.1 billion in assets under management, towering high over the $633 million in assets managed by the Ave Maria Rising Dividend (AVEDX), the largest of the Ave Maria funds.

Among other groups of faith-based funds, the Amana funds are the largest with about $7.4 billion in managed assets, followed by the Praxis funds with $1.4 billion and the Azzad funds, the smallest of the six groups with $96 million. The Amana funds were launched in 1984 by the North American Islamic Trust (NAIT). The Amana Income Investor (AMANX) fund began operations on June 23, 1986, and the Amana Growth Investor (AMAGX) fund, the largest of all faith-based funds with over $2 billion in gathered assets, started on February 3, 1994.

Note that many of the screens used to make investment decisions in accordance with Islamic principles are similar to those used by other faith-based funds, such as shunning companies involved with alcohol, tobacco, gambling, pornography, and abortion. Additionally, some of the Islamic funds may avoid excessive stock trading, considered a form of gambling. Since Islamic law explicitly forbids Muslims from practicing usury, Islamic funds may also screen out companies that charge interest rates on loans or carry too much debt on their balance sheets. Moreover, their screens eliminate companies that derive revenues from the sale of pork, consumption of which is forbidden by Islam.

Praxis' 13 funds managed under the stewardship of the US Mennonite Church are more modest in scope than Amana's funds. The largest of them, the Praxis Intermediate Income A (MIIAX), which also happens to have been the first fund in the group to be launched, holds about $345 million in managed assets as of November 15, 2013. As to Azzad's two funds, although the first, Azzad Ethical (ADJEX), was launched 13 years ago, Azzad, as noted earlier, remains far behind the other groups of faith-based funds with only $96 million in assets under management.

---

[12]http://www.ussif.org/files/Publications/12_trends_Report.pdf
[13]http://www.guidestonecapital.org/ProductsandServices/GuideStoneFunds
[14]http://www.usccb.org

TABLE 10.2 Faith-Based Funds/Summary Statistics

Fund Name	Orientation	Category	Benchmark	Fund Inception Date	Net Assets (in $Million)	Number of Funds	Expense Ratio (in %)	YTD Return (in %)	Five-Year Average Return (in %)	Sharpe Ratio	Beta
GuideStone	Southern Baptist				22,151	40					
First fund: GuideStone Equity Index Fund (GEQZX)		Large blend	S&P 500 index	Aug 27, 2001	332	1	0.54	13.07	11.60	1.04	1.07
Largest fund: GuideStone Funds International Eq GS2 (GIEYX)		Foreign large blend	Actively managed	Aug 27, 2001	1,660	1	0.38	24.89	15.00	1.31	0.99
Amana	Muslim				7,356	6				0.41	1.03
First fund: Amana Income Investor (AMANX)		Large blend	Actively managed	Jun 23, 1986	1,570	1	1.27	15.31	13.58	0.77	0.75
Largest fund: Amana Growth Investor (AMAGX)		Large growth	Actively managed	Feb 3, 1994	2,080	1	1.18	25.33	13.82	1.25	0.85
Ave Maria	Catholic				1,365	6	1.11	16.21	13.34	0.90	0.88
First fund: Ave Maria Catholic Values (AVEMX)		Mid-cap blend	Actively managed	May 1, 2001	238	1	1.24	19.30	14.12	1.10	0.84
Largest fund: Ave Maria Rising Dividend (AVEDX)		Large blend	Actively managed	May 2, 2005	633	1	1.48	19.85	16.35	0.95	1.11
Praxis	Mennonite				1,730	13	0.99	26.76	16.23	1.42	0.9
First Fund: Praxis Intermediate Income A (MIIAX)		Intermediate-term bond	Actively managed	May 12, 1999	345	1	0.92	16.26	13.43	1.16	0.96
Largest Fund: Praxis Intermediate Income A (MIIAX)		Intermediate-term bond	Actively managed	May 12, 1999	345	1	0.96	-0.92	6.80	1.10	0.84
LKCM	Catholic				3,136	9	Same	Same	Same	Same	Same
First fund: LKCM Aquinas Value (AQEIX)		Large blend	Actively managed	Jan 3, 1994	56	1	1.10	21.37	14.80	1.11	1.05
Largest fund: LKCM Small Cap Equity Advisor (LKSAX)		Small growth	Actively managed	Jun 5, 2003	1,060	1	1.50	25.95	15.22	1.06	1.12
Azzad	Muslim				96	2	0.94	26.93	18.27	1.08	1.18
First fund: Azzad Ethical (ADJEX)		Mid-cap growth	Actively managed	Dec 22, 2000	41	1	1.24	14.77	13.03	1.00	0.72
Largest fund: Azzad Wise Capital (WISEX)		Short-term bond	Actively managed	Apr 5, 2010	55	1	0.99	27.57	19.54	0.98	1.11
Vanguard 500 Index Inv (VFINX)		Large blend	S&P 500 index	Aug 31, 1976	155,290	1	1.49	1.96	N/A	1.01	0.33
							0.17	26.51	17.47	1.36	1.00

Source: Yahoo Finance as of November 15, 2013.

10.4.2.2 Return Performance Statistics

From a market performance perspective, Table 10.2 indicates that both the average year-to-date return and average 5-year returns of the faith-based funds are below those of the Vanguard 500 Index (VFINX), a mutual fund launched by Vanguard on August 31, 1976, for the specific purpose of tracking the returns of the S&P500 Index, widely considered by the market as a gauge for US equities. To be fair, however, their lagging performance was in many cases severely impacted by the less-than-stellar performance of their assets apportioned to international destinations, real estate, and bonds. Emerging markets were particularly volatile in 2013, pushing many concerned investors to withdraw their money, thereby impacting market returns. While returns on real estate assets have been at best lukewarm, given rising interest rates[15] and economic uncertainty, returns on fixed income funds noticeably lagged those of equities in 2013.

The comparison of faith-based funds to the VFINX will be far more equitable if we compare the performance of the latter to that of the former's US-domiciled equity funds. Many of these domestic funds show returns close to that of Vanguard 500. For instance, Table 10.2 indicates that the year-to-date and 5-year average returns of the GEQZX (24.89 and 15%) and AVEDX (26.76 and 16.23%) are in line with those of the Vanguard 500 (24.89 and 15%, respectively).[16]

Better yet, since the relationship between risk and return is a crucial concept in finance, investors comparing the performance of two funds must do it in terms of their risk-adjusted returns rather than on a pure return basis, in order to ensure that they are also adequately compensated for the risk they are assuming. The Sharpe ratio, which incorporates both components, is one of the most-used measures of risk-adjusted return. The higher the value of the Sharpe ratio, the more desirable the investment is, since investors can expect excess return for the extra risk they are exposed to. As a general rule, a Sharpe ratio of 1 is deemed good, 2 is great, and 3 is quite exceptional.

For the sake of a fair comparison and in light of the aforementioned issues experienced over the past few years by the real estate, emerging and bond markets, it's again preferable to compare the performance results of the VFINX with that of Table 10.2's domestic equity funds. Many of the individual faith-based funds show Sharpe ratios ranging between 1 and 1.5, indicating that, like those who invested in the Vanguard 500, their investors were similarly compensated for the additional risks they took.

10.4.2.3 Risk Performance Statistics

Beta is an important volatility statistic in the sense that it measures the systematic risk of a fund or the degree to which it responds to the volatility of the overall market. For example, a fund with a beta of two indicates that it's theoretically twice as volatile as the market, whereas a beta of 0.5 denotes that the fund is 50% less volatile. If the fund's beta is equal to 1, it means that it theoretically moves with the market. For example, the fact that the VFINX has a beta of 1 and provided investors with a return of 26.51% as of November 15, 2013, as shown in Table 10.2, means that the S&P500 Index, the index mirrored by the VFINX and used to represent the overall market, most likely generated about the same return over the same period.[17]

Hence, investors' use of beta as a measure of volatility helps identify the funds that meet their criteria for risk. Investors with a low tolerance for risk usually select funds with betas lower than 1, whereas those who seek risk may select those with higher betas, hoping to harvest higher returns from their investments. Many of the faith-based equity funds exhibit a beta higher than 1,[18] which should not come as a surprise considering that the use of screens to filter out investments that violate the dogmas of a specific religion requires active management, a process that can often lead to a higher beta. In comparison, the VFINX, being passively managed since its most important goal is to mimic its benchmark, has a beta of 1.

10.4.3 Case 2: ETFs and the ESG Market

ETFs have been in existence since the early 1990s, and the market demand for them has been rapidly increasing. The first ETF was the SPDR S&P500 ETF (SPY), launched on January 22, 1993, with the goal of tracking the performance of the S&P500 Index. Twenty years after its issuance, SPY remains the largest ETF, with nearly $164 billion in managed assets,[19] and a key indicative fund for the ETF market. As to the entire population of ETFs, ETF Global LLC estimates the assets held by all 1528 ETFs to be $1.7 trillion.[20]

In spite of substantial year-to-year asset growth for ETFs, the industry remains highly concentrated in the largest ETFs. Several products also appear to duplicate each other in both design and coverage, prompting some to suspect that more than a few ETFs were brought to market for the unique purpose of gaining visibility and market share for their issuers.

---

[15]The Fed already signaled its intention to begin pulling back on its monetary stimulus to the economy and letting rates rise in 2014. At its meeting on Wednesday, December 18, 2013, it decided to cut its monthly bond purchases from $85 to $75 billion beginning in January 2014 and is expected to continue paring back in regular increments at each meeting until the bond purchase program ends.

[16]More detailed data showing the performance statistics of the domestic faith-based funds under each of Table 10.2's six categories indicate that several of their domestic funds' returns compare favorably with those of the VFINX. These returns can be observed in Exhibits 10.1 through 10.4 at the end of the chapter.

[17]Moreover, VFINX's coefficient of variation, or R^2, is also equal to 100, indicating that all of its variations are explained by those of the S&P500 Index, the benchmark it tracks.

[18]Refer to Exhibits 10.1 through 10.4 at the end of the chapter.

[19]https://www.spdrs.com/index.seam

[20]http://www.etfg.com/

TABLE 10.3 A Quick View of the ESG ETF Sector

Ticker	Product Name	Issuer	Asset Class	Category	Focus	AUM of ETP (in $)
GIVE	AdvisorShares Global Echo ETF	AdvisorShares	Multiasset	Asset allocation	Target risk	6,072,000
DSI	iShares KLD 400 Social Index Fund ETF	BlackRock	Equity	Size and style	Large cap	272,945,000
ICLN	iShares S&P Global Clean Energy Index Fund ETF	BlackRock	Equity	Strategy	Theme	46,980,000
KLD	iShares KLD Select Social Index Fund ETF	BlackRock	Equity	Size and style	Large cap	241,693,200
FAN	First Trust ISE Global Wind Energy ETF	First Trust	Equity	Sector	Energy	80,017,523
FIW	First Trust ISE Water Index Fund ETF	First Trust	Equity	Sector	Natural resource	192,888,068
GRID	First Trust NASDAQ Clean Edge Smart Grid Infrastructure Index Fund ETF	First Trust	Equity	Strategy	Theme	12,622,122
QCLN	First Trust NASDAQ Clean Edge US Liquid Series Index Fund ETF	First Trust	Equity	Sector	Energy	94,080,036
CGW	Guggenheim S&P Global Water Index ETF	Guggenheim	Equity	Sector	Natural resource	310,786,000
TAN	Guggenheim Solar ETF	Guggenheim	Equity	Strategy	Theme	357,710,240
HECO	Huntington EcoLogical Strategy ETF	Huntington Strategy Shares	Equity	Broad equity	Broad equity	17,317,406
PBD	PowerShares Global Clean Energy Portfolio ETF	Invesco PowerShares	Equity	Strategy	Theme	86,602,250
PBW	PowerShares WilderHill Clean Energy Portfolio ETF	Invesco PowerShares	Equity	Strategy	Theme	221,094,000
PHO	PowerShares Water Resource Port ETF	Invesco PowerShares	Equity	Sector	Natural resource	1,012,092,000
PIO	PowerShares Global Water Portfolio ETF	Invesco PowerShares	Equity	Sector	Natural resource	248,325,000
PUW	PowerShares WilderHill Progressive Energy Portfolio ETF	Invesco PowerShares	Equity	Strategy	Theme	41,743,000
PZD	PowerShares Cleantech Portfolio ETF	Invesco PowerShares	Equity	Strategy	Theme	84,190,500
EAPS	ESG Shares Pax MSCI EAFE ESG Index ETF	Pax World	Equity	Broad equity	Broad equity	55,333,500
EVX	Market Vectors Environment Index ETF Fund ETF	Van Eck	Equity	Strategy	Theme	19,629,000
GEX	Market Vectors Global Alternative Energy ETF	Van Eck	Equity	Strategy	Theme	96,995,356
KWT	Market Vectors Solar Energy ETF	Van Eck	Equity	Strategy	Theme	29,120,000
Median	ex SPY					86,602,250
Average	ex SPY					168,011,248
High	ex SPY					1,012,092,000
Low	ex SPY					6,072,000
Total	AUM					3,528,236,201.00

Source: ETF Global LLC (as of December 31, 2013).

Most ETFs are considered wrappers priced via a real-time calculation, as their components are transparent. Large blocks of ETF shares can be created and redeemed through transactions between fund providers and large financial institutions known as authorized participants, or APs. This intricate process has contributed to keeping the price of a share more or less in line with the price of its underlying holdings through an intricate mechanism known as arbitrage: when an ETF becomes more (less) expensive than the sum of its underlying holdings, APs can sell (buy) its shares on the market, an arbitrage process that contributes to bringing its price back in line with the value of its underlying securities.

Since both the ETF and its underlying holdings are freely traded, experts agree that the process of creation and redemption has given ETFs' holdings an additional layer of liquidity. Hence, the issuance of an ETF to reflect a particular segment of a market potentially holds the prospect of added liquidity for the securities traded in that specific market sector. By extension, for the ESG market, its gradual inclusion into the ETF industry may well mean that (i) the market may be starting to look at it as a viable investment option, if not an intriguing one, and (ii) as importantly, its liquidity stands to be enhanced by the unique creation and redemption process underlying ETFs. Both considerations are very important to risk-wary but deep-pocketed market participants such as institutional investors, capable of finalizing ESG's transfer from the fringes of the market to its mainstream and positively impacting its liquidity in the process.

Table 10.3 shows a list of ESG ETFs generated through a query of the ETF Global LLC database.[21] It shows a total of 21 products representing $3.53 billion in managed assets. The table also indicates that this budding ETF sector is so far dominated in terms of assets under management by three major players: Invesco PowerShares with six products and a total of $1.69 billion in assets, BlackRock with three products and $561.62 million in assets, and First Trust with four products and $305.33 million in assets. With its PowerShares Water Resource Port ETF (PHO), Invesco also holds the largest of these products with $1.01 billion in managed assets. Alone, PHO hold a dominant 29% of ESG's total assets ($1.01B/$3.53B), demonstrating how concentrated this submarket is.

On the other side of the spectrum, AdvisorShares' only ETF (GIVE) is the smallest with $6.07 million in assets, representing a tiny 0.17% ($0.0061B/$3.53B) of the total. Of the three dominant sector players, BlackRock has the category's oldest ETF with its iShares KLD Select Social Index Fund, issued on January 28, 2005. Taken together, these numbers indicate that not only does the ESG ETF sector remain a very small portion of the overall ETF industry in spite of rising interest on the part of institutional investors, but it also remains highly concentrated.

Graph 1 shows the historical growth of ESG ETFs. Although their current number remains insignificant compared to the total number of ETFs available to US investors, as seen in Table 10.3, their growth is nonetheless observable, rising from 3 in 2005 to 22 in 2013, which translates into a 633% increase over a 9-year period. Note also that both 2012 and 2013 show the same number of ESG ETFs, since our query of ETF Global LLC database didn't generate any new ETFs for 2013. Most of the growth happened between 2005 and 2008, a period characterized by a 260% increase in their number (from 5 to 18 ETFs). While growth is now more modest by comparison, it seems to have resumed its ascending path after the 2008–2009 financial crisis, which bodes well for the future of the ESG ETF market.

10.4.3.1 ESG ETFs: Return Performance Statistics

Graph 2 shows the cumulative average returns of ESG ETFs over increasing time horizons alongside those of the SPY. The latter is a large-cap ETF issued by State Street Global Advisors on January 29, 1993, to replicate the total return of the S&P500 Index, a benchmark widely used by market participants as a barometer for the overall market. Due to its faithful mimicking of the market benchmark, the SPY will be used in this chapter as a gauge by which both performance and risk of ESG ETFs will be evaluated.

When their average cumulative returns are compared to those of the SPY as shown in Table 10.2, one doesn't fail to notice a significant difference between short- and long-term performances. Over the 1- to 2-year horizons, ESG ETFs and the SPY have more or less kept pace with each other, with a slight advantage to the SPY. The situation, however, is quite different over investment horizons longer than 2 years. The cumulative return differential in favor of the SPY then becomes quite noticeable over these longer time frames. At 45.46% (63.37–45.46%), the 4-year horizon indicates the widest spread in cumulative returns, followed by the 3-year horizon with 32.52% (46.21–13.69%) and then the 5-year horizon with a still-sobering 28.68% (106.61–77.93%).

To put things into perspective, using the largest spread, if one equally invested $10,000 on January 1, 2010, in the 21 ESG ETFs shown in Table 10.3, this initial investment would have grown to $11,791 ($10,000×1.1792) by December 31, 2013, compared to a loftier $16,337 ($10,000×1.4546) had it been allocated to the SPY instead. Although either outcome is considered highly desirable from an investment perspective, the substantial $4,545 difference ($16,337−$11,791) between the two end results is quite hard to overlook and certainly favors an investment in the SPY. For the 3-year horizon, the SPY's outperformance translates into a difference of $3252 and a slightly lower $2868 for the 5-year horizon.

A more detailed view of these outcomes can be found in Table 10.4, which shows cumulative returns associated with

---

[21]ETF Global LLC is an independent management consulting firm that offers investment advice, research support, and risk analytics services focusing on the exchange-traded funds industry (http://www.etfg.com).

TABLE 10.4 Performance Statistics: ESG ETFs versus SPY

Ticker	Product Name	Issuer	AUM of ETP (in $)	One Month (in %)	Three Month (in %)	Six Month (in %)	One Year (in %)	Two Year (in %)	Three Year (in %)	Four Year (in %)	Five Year (in %)
CGW	Guggenheim S&P Global Water Index ETF	Guggenheim	310,786,000	1.66	3.45	7.40	25.71	47.12	32.68	48.87	96.37
DSI	iShares KLD 400 Social Index Fund ETF	BlackRock	272,945,000	1.82	4.16	8.68	35.05	46.83	46.58	59.56	110.29
EAPS	ESG Shares Pax MSCI EAFE ESG Index ETF	Pax World	55,333,500	0.85	0.88	4.50	23.49	40.59	NA	NA	NA
EVX	Market Vectors Environment Index ETF Fund ETF	Van Eck	19,629,000	-0.81	1.98	4.15	29.89	40.09	25.82	51.97	92.82
FAN	First Trust ISE Global Wind Energy ETF	First Trust	80,017,523	1.80	1.34	7.70	67.36	43.06	11.85	-24.87	-8.78
FIW	First Trust ISE Water Index Fund ETF	First Trust	192,888,068	3.11	3.43	7.47	32.12	62.73	51.54	80.07	120.20
GEX	Market Vectors Global Alternative Energy ETF	Van Eck	96,995,356	1.62	1.82	5.36	71.49	70.70	-6.65	-25.80	-17.56
GIVE	AdvisorShares Global Echo ETF	AdvisorShares	6,072,000	1.74	3.52	5.32	17.91	NA	NA	NA	NA
GRID	First Trust NASDAQ Clean Edge Smart Grid Infrastructure Index Fund ETF	First Trust	12,622,122	-0.57	-0.40	4.98	23.89	44.66	13.80	11.65	0.00
HECO	Huntington EcoLogical Strategy ETF	Huntington Strategy Shares	17,317,406	0.83	1.19	5.76	29.86	NA	NA	NA	NA
ICLN	iShares S&P Global Clean Energy Index Fund ETF	BlackRock	46,980,000	-1.89	-2.26	2.57	46.20	21.55	-34.39	-53.29	-50.14
KLD	iShares KLD Select Social Index Fund ETF	BlackRock	241,693,200	1.62	3.62	7.43	30.40	38.98	39.13	54.97	103.13

Ticker	Name	Provider									
KWT	Market Vectors Solar Energy ETF	Van Eck	29,120,000	−2.61	2.01	14.66	102.67	30.98	−55.82	−68.90	−65.75
PBD	PowerShares Global Clean Energy Portfolio ETF	Invesco PowerShares	86,602,250	0.23	0.90	5.31	55.92	46.63	−13.30	−27.69	−1.65
PBW	PowerShares WilderHill Clean Energy Portfolio ETF	Invesco PowerShares	221,094,000	0.47	−0.31	1.91	58.81	26.23	−38.76	−42.60	−23.81
PHO	PowerShares Water Resource Port ETF	Invesco PowerShares	1,012,092,000	3.15	5.19	9.18	28.61	55.25	36.89	53.97	86.19
PIO	PowerShares Global Water Portfolio ETF	Invesco PowerShares	248,325,000	2.96	3.71	7.38	28.31	46.09	13.93	24.18	77.93
PUW	PowerShares WilderHill Progressive Energy Portfolio ETF	Invesco PowerShares	41,743,000	2.13	5.76	9.30	27.55	41.69	13.58	34.09	119.72
PZD	PowerShares Cleantech Portfolio ETF	Invesco PowerShares	84,190,500	2.24	5.00	8.13	38.05	46.53	19.53	28.08	78.58
QCLN	First Trust NASDAQ Clean Edge US Liquid Series Index Fund ETF	First Trust	94,080,036	3.93	5.28	10.60	93.94	87.06	8.73	11.22	65.34
TAN	Guggenheim Solar ETF	Guggenheim	357,710,240	−7.33	−2.00	6.90	135.24	44.86	−50.85	−65.13	−58.05
SPY	SPDR S&P 500 ETF	SSgA	166,937,987,635.00	1.56	4.27	8.83	31.27	46.47	46.21	63.37	106.61
Median ex SPY			86,602,250	1.62	2.01	7.38	32.12	44.86	13.69	17.92	77.93
Average ex SPY			168,011,248	0.81	2.30	6.89	47.74	46.40	6.35	8.35	42.64
High ex SPY			1,012,092,000	3.93	5.76	14.66	135.24	87.06	51.54	80.07	120.20
Low ex SPY			6,072,000	−7.33	−2.26	1.91	17.91	21.55	−55.82	−68.90	−65.75
SPY			166,937,987,635.00	1.56	4.27	8.83	31.27	46.47	46.21	63.37	106.61

Source: ETF Global LLC (as of December 31, 2013).

each one of these ESG ETFs alongside those specific to the SPY. The table shows that the average cumulative returns of ESG ETFs were stalled by the significant underperformance, to say the least, of some members of the group. If we look at the row before the last one labeled "Low ex SPY" on the table, we notice that the lowest cumulative returns over the 3-, 4-, and 5-year horizons (−55.82, −68.90, and −65.75%, respectively) were recorded by Van Eck's Market Vectors Solar Energy ETF (KWT). Unfortunately, Van Eck's ETF was not the only significant laggard among the group. Guggenheim Solar ETF (TAN) came in a close second with a disconcerting −50.85, −65.13, and −58.05%, respectively, over the same investment horizons, followed by BlackRock's iShares S&P Global Clean Energy Index Fund ETF (ICLN) with a still-alarming −34.39, −53.29, and −50.14%. Although these three showed the most disheartening results of the group, a few others contributed to the overall ESG ETFs' underperformance portrayed in Graph 2.

So far, we have evaluated these funds only in terms of their returns. In order to offer more meaningful insight on how they have really performed, it's important to also look at the risk taken in the process of managing these funds. Table 10.5 shows three measures of risk: alpha, beta, and the Sharpe ratio, each measuring a specific aspect of risk.

10.4.3.2 ESG ETFs: Risk Performance Statistics

Funds' returns consist of two components: a part that is attributable to their exposure to systematic risk, commonly referred as beta as previously discussed, and a part generated by alpha that cannot be explained by the fund's exposure to systematic risk. Starting with beta, it's important to add to our previous discussion on the topic[22] that its reliability depends on the coefficient of determination, or R^2, a tool that measures how well a fund's movements can be explained by the movements of its related index (shown in Table 10.5). R^2 ranges from 0 to 100, with 100 indicating that the movements are perfectly synchronized and 0 signifying that their variations are unrelated. Hence, the lower the value of the coefficient of variation, the less reliable the measurements of the fund's beta and alpha.[23]

It should also be noted that, from a portfolio diversification perspective, a beta that is lower than 1.0 is a desired outcome, as it reduces the overall portfolio's systematic risk. If such is the case, then it would be prudent for the fund to comprise a larger proportion of a portfolio. Table 10.5 shows a wide distribution for beta, ranging from a high of 2.62 for Van Eck's KWT to a low of 0.85 for the Guggenheim S&P Global Water Index ETF (CGW), with an average of 1.37 for the entire sample. The result for KWT indicates that the fund's returns will move up and down with the market but at a 262% greater rate. This means that, over the past three years, it has performed 262% better than the index in upmarkets and 262% worse in downmarkets.

The CGW, on the other hand, has lagged its index by 15% in upmarkets and outperformed it by 15% in downmarkets. As to 1.37, it means that on average the sample of ETFs shown in Table 10.5 has outperformed the market by 37% in upmarkets and lagged behind it by the same percentage on average in downmarkets. Based on their risk performance over the past 3 years, only three other ETFs, out of the 21 shown in Table 10.5, exhibit a beta lower than 1.0 and hence would have been able to reduce the systematic risk of an overall performance of a portfolio that includes them: DSI, EVX, and FIW with 0.95, 0.91, and 0.95, respectively. KLD and, to some extent, PHO and PIO are on the borderline in terms of their contribution to systematic risk, with beta equal to 1.0, 1.03, and 1.07, respectively. But, in view of their average (1.37) and the number of ETFs that exhibit high betas (7 out of 21), overall, ESG ETFs seem to contribute some level of systematic risk.

The distribution of alpha is even wider than that of beta, further emphasizing the volatile nature of this type of ETF. In terms of risk, alpha is a good companion to beta. If beta tells us which part of the fund's total return is attributable to its exposure to systematic risk, alpha on the other hand enables us to size up the part that is not. A positive 3-year alpha within the study's context means that the fund has earned a persistent average rate of return over a 3-year period that is independent of the fund's level of systematic risk. That represents a superior performance on the part of the fund. The fund's alpha will be negative over the same sampled period if, on the other hand, the stocks comprising the fund on average yield negative returns independently of the market movements. This means that the fund has performed poorly. If a fund's alpha is equal to zero, then the fund is not yielding either abnormal or subnormal returns relative to the market.

Table 10.5 shows both positive and negative alphas and three "NAs." The NAs result from the fact that the performance data underlying these three ETFs (EAPS, GIVE, and HECO in Table 10.5) are not sufficient for the calculation of the table's 3-year risk statistics due to their recent launch. Of the available statistics, only four exhibit positive alphas, ranging from 0.17 (DSI) to 6.44 (FIW). Negative alphas are more dominant, in terms of both the number of ETFs displaying them (14) and their size. Combined, they lowered the average sample's alpha to −9.34, which represents a significant difference between the return expected based on the exposure of these funds to systematic risk and the actual return earned over the past 3 years.

Granted that low R^2 could cast a cloud over the validity of these results, as previously mentioned, the KWT is the largest underperformer of the group with an alpha of −37.81,

---

[22] See section on "The Faith-Based ESG Market."
[23] A few sources, such as Morningstar, suggest a threshold of 70 for R^2, below which both alpha and beta become unreliable predictors, but others disagree.

TABLE 10.5 Risk Statistics: ESG ETFs versus SPY

Ticker	Product Name	Issuer	Three-Year[a]				Expense Ratio (in %)[b]
			Alpha	Beta	Sharpe Ratio	R^2 (in %)	
CGW	Guggenheim S&P Global Water Index ETF	Guggenheim	3.42	0.85	0.94	86.96	0.70
DSI	iShares KLD 400 Social Index Fund ETF	BlackRock	0.17	0.95	1.30	97.24	0.50
EAPS	ESG Shares Pax MSCI EAFE ESG Index ETF	Pax World	NA	NA	NA	NA	0.55
EVX	Market Vectors Environment Index ETF Fund ETF	Van Eck	−0.35	0.91	0.78	75.22	0.55
FAN	First Trust ISE Global Wind Energy ETF	First Trust	−6.58	1.33	0.29	59.51	0.60
FIW	First Trust ISE Water Index Fund ETF	First Trust	6.44	0.95	1.04	75.25	0.60
GEX	Market Vectors Global Alternative Energy ETF	Van Eck	−11.09	1.36	0.11	54.93	0.62
GIVE	AdvisorShares Global Echo ETF	AdvisorShares	NA	NA	NA	NA	1.50
GRID	First Trust NASDAQ Clean Edge Smart Grid Infrastructure Index Fund ETF	First Trust	−4.79	1.14	0.49	81.44	0.70
HECO	Huntington EcoLogical Strategy ETF	Huntington Strategy Shares	NA	NA	NA	NA	0.95
ICLN	iShares S&P Global Clean Energy Index Fund ETF	BlackRock	−22.58	1.62	−0.20	58.82	0.48
KLD	iShares KLD Select Social Index Fund ETF	BlackRock	−2.34	1.00	1.09	96.18	0.50
KWT	Market Vectors Solar Energy ETF	Van Eck	−37.81	2.62	−0.22	56.68	0.65
PBD	PowerShares Global Clean Energy Portfolio ETF	Invesco PowerShares	−15.86	1.55	0.00	68.17	0.75
PBW	PowerShares WilderHill Clean Energy Portfolio ETF	Invesco PowerShares	−26.71	1.72	−0.30	65.23	0.70
PHO	PowerShares Water Resource Port ETF	Invesco PowerShares	2.22	1.03	0.77	76.96	0.62
PIO	PowerShares Global Water Portfolio ETF	Invesco PowerShares	−3.41	1.07	0.47	87.80	0.75
PUW	PowerShares WilderHill Progressive Energy Portfolio ETF	Invesco PowerShares	−5.68	1.36	0.42	86.39	0.70
PZD	PowerShares Cleantech Portfolio ETF	Invesco PowerShares	−3.63	1.20	0.65	81.01	0.67
QCLN	First Trust NASDAQ Clean Edge US Liquid Series Index Fund ETF	First Trust	−8.48	1.53	0.26	61.24	0.60

(continued)

TABLE 10.5 (*Continued*)

Ticker	Product Name	Issuer	Three-Year[a]				Expense Ratio (in %)[b]
			Alpha	Beta	Sharpe Ratio	R^2 (in %)	
TAN	Guggenheim Solar ETF	Guggenheim	−30.97	2.55	−0.10	49.11	0.70
SPY	SPDR S&P 500 ETF	SSgA	−0.08	1.00	1.29	100.00	0.09
Median	ex SPY		−5.24	1.27	0.45	75.24	0.65
Average	ex SPY		−9.34	1.37	0.43	73.23	0.69
High	ex SPY		6.44	2.62	1.30	97.24	1.50
Low	ex SPY		−37.81	0.85	−0.30	49.11	0.48

[a] Yahoo Finance as of December 31, 2013.
[b] ETF Global LLC as of December 31, 2013.

followed by the TAN with −30.97 and the PowerShares WilderHill Clean Energy Portfolio ETF (PBW) with −26.71. These three ESG ETFs happen to be in the subcategory of the so-called "green ETFs."[24] With negative alphas of this magnitude, investors should justifiably wonder whether the performance of these funds' managers was sufficient enough to justify the risk they took to get the trivial 3-year cumulative returns shown in Table 10.3, −55.82, −50.85, and −38.76%, respectively, compared with a cumulative return of 46.25% for the SPY.[25]

Although in theory passively managed index funds should carry alphas of zero, a negative alpha is also conceivable due to the drag of the fund's expenses. That should explain the −0.08 alpha displayed by the SPY, caused by a very low 0.09% expense ratio; both figures are displayed in Table 10.5. The same table shows that the SPY's expense ratio is significantly lower than the average 0.69% annual cost of owning an ESG ETF. Contrary to their alpha figures, there is no reason, however, to be overly concerned, because this statistic is only marginally higher than the 0.64% average 2013 expense ratio for all ETFs recently reported by Lipper, a Thomson Reuters affiliate that provides funds data and analytical tools to the market.[26]

From a risk-adjusted returns perspective, investors in KWT, TAN, and PBW are entitled to raise the same concerns they have had with their alphas in view of the three funds' significantly low Sharpe ratios: −0.22, −0.10, and −0.30, respectively. In fact, in risk analysis, the Sharpe ratio presents a real advantage over either alpha or beta in situations where R^2 is low, as is the case here, since its relevance doesn't depend on the level of the coefficient of variation. The negative signs indicate that the three funds underperformed treasury bills on average, and for that reason, they have negative average cumulative excess returns. Hence, if investors have reasons to be alarmed based on the results associated with their alphas and betas, they should be even more concerned in view of the three funds' negative Sharpe ratios. In fact, Table 10.5 shows two other ETFs displaying negative Sharpe ratios: ICLN (−0.20) and TAN (−0.10).

Of the funds displayed in Table 10.5, most (17 out of 21) display positive Sharpe ratios ranging from 0.001 (PBD) to 1.30 (DSI). With a 3-year average Sharpe ratio of 0.43, this means that ESG ETFs have on average generated excess return per unit of risk. That is good news for these ETFs. If the results based on alpha and beta could pass for questionable based on their average level of R^2, it's certainly not the case for their Sharpe ratio, whose relevance, as mentioned earlier, doesn't depend on the latter. Under such circumstances, the use of the Sharpe ratio in our analysis offers a real advantage over that of alpha and beta and portrays ESG ETFs in a better light after a succession of risk and return misperformances.

10.5 CONCLUDING REMARKS

This chapter started with a review of the literature underlying ESG investing and a history of its evolution. It indicated that the factors fueling ESG growth haven't remained the same over time. Issues involving tobacco, human rights, and firearms, especially following several highly publicized tragic shooting incidents, certainly remain powerful in the mind of investors. They have, nonetheless, given way to environmental concerns that the investing public now deems even more pressing, following a series of unfortunate ecological disasters. The SIF seemed to agree on this point by reporting in its latest survey that the environment filter is now the most important screen that investment managers use during the process of identifying socially responsible companies.

The next section of the chapter consisted of an analysis of ESG using SIF data. It demonstrated that although this

---

[24] ETFs whose holding companies are selected based on their positive environmental characteristics.
[25] Note that this underperformance was recorded after the 2008–2009 financial crisis.
[26] http://funds.us.reuters.com/US/pdf.asp?language=UNK&docKey=1523-4489-33P0I96REDKNFQ9VASKGI0E36D

investment sector remains small in relation to the overall market, its year-to-year growth is nonetheless remarkable. This was followed by a detailed examination of faith-based funds, a subcategory that launched ESG as outlined in the section on the history of such investing. This section offered a detailed view of six clusters of faith-based funds, grouping no less than 76 funds that appeal to the specific convictions of three major religions. We observed that although this subcategory gives an impression of a busy market with its 76 funds, it's in reality highly concentrated, as it is dominated by a handful of large funds in terms of managed assets. This category of funds also provided our first opportunity to analyze this market in terms of both return performance and risk.

Swayed by the argument that the budding interest in ESG by major ETF issuers bodes well for this market in terms of its potential liquidity, a full section was dedicated to ESG ETFs. With only 21 funds in existence so far, out of the 1500 comprising the overall ETF market, investors' interest in these products seems at first even more limited than their interest in faith-based funds. A closer look, however, revealed that not only were ESG ETFs launched much more recently, but they are also growing rapidly. Not even the global crisis of 2008–2009 slowed their growth.

The risk-adjusted performance of all ESG funds appeared sturdy, for the most part, granting that their average Sharpe ratio lagged that of the Vanguard 500 in the case of the faith-based mutual funds and the SPY in the case of the ESG ETFs. A handful of the funds were, however, able to closely track the performance of those two benchmarks.

Since an analysis based on risk-adjusted performance was deemed insufficient to assess the success or failure of a fund, we also turned to alpha to determine whether the fund's return justified the risk taken. This statistic turned out to be the weakest point of ESG ETFs. The performance of these funds seemed way out of proportion to the risk taken to achieve their returns. Lastly, ESG ETFs also appeared to contribute significant systematic risk in view of their average beta, although their deviation from their benchmark in terms of beta is not nearly as drastic as that of alpha.

All in all, ESG funds are still a minute portion of the overall market. Aside from a handful of products, they still have ways to go before they can be considered viable players in terms of both their risk and their return. They will continue, however, to make important strides as long as key legal barriers such as the Retirement Income Act of 1974 abate, and as more individual and institutional investors embrace them, inspired by convictions higher than the lackluster risk-adjusted performance, these funds have on average turned so far.

APPENDIX: EXHIBITS

EXHIBIT 10.1 Southern Baptist Funds/Summary Statistics

Ticker Symbol	Fund Name	Fund Inception Date	Category	Net Assets (in $Million)	Number of Funds	YTD Return (in %)	Five-Year Average Return (in %)	Best 3-Year Total Return (in %)	Worst 1-Year Total Return (in %)	Sharpe Ratio	Alpha	Beta	Standard Deviation	R^2	Expense Ratio (in %)
GREZX	GuideStone Funds Real Estate Secs GS4	Dec 29, 2006	Real estate	244.34	1	4.65	14.38	22.40	−40.35	0.72	2.30	0.90	16.80	62.91	1.05
GDMZX	GuideStone Funds Defensv Mkt Strats GS4	Aug 31, 2011	Tactical allocation	467.09	1	15.85	N/A	N/A	8.57	N/A	N/A	N/A	N/A	N/A	1.25
GEQZX	GuideStone Funds Equity Index GS4	Aug 27, 2001	Large blend	331.62	1	24.89	15.00	14.11	−37.48	1.31	0.22	0.99	12.32	99.78	0.38
GIEYX	GuideStone Funds International Eq GS2	Aug 27, 2001	Foreign large blend	1660	1	14.26	11.73	23.94	−44.72	0.41	−0.22	1.03	17.87	98.05	0.95
GCOYX	GuideStone Funds Growth Allocation 1 GS2	Jul 1, 2003	Aggressive allocation	265	1	16.07	13.23	12.54	−32.82	0.97	−1.27	1.33	11.32	97.93	0.15
GMTZX	GuideStone Funds MyDestination 2015 GS4	Dec 29, 2006	Target date 2011–2015	426.79	1	9.11	11.90	13.45	−29.30	1.08	0.04	0.91	7.76	97.37	0.15
GNRZX	GuideStone Funds Global Ntrl Res Eq GS4	Jul 1, 2013	Natural resources	266.8	1	N/A	N/A	N/A	N/A	N/A	N/A	N/A	N/A	N/A	N/A
GEDZX	GuideStone Funds Extended-Dur Bond GS4	Aug 27, 2001	Corporate bond	276.52	1	−4.55	15.74	16.58	−8.29	0.88	0.18	2.33	8.04	67.72	0.75
GLDYX	GuideStone Funds Low- Duration Bond GS2	Aug 27, 2001	Short-term bond	833	1	−0.02	4.32	6.04	−2.76	1.5	0.85	0.30	1.16	54.41	0.36
GGIZX	GuideStone Funds Balanced Allocation GS4	Aug 27, 2001	Conservative allocation	1280	1	9.27	11.64	12.22	−24.41	1.12	0.22	0.85	7.22	98.52	0.12
GEMZX	GuideStone Funds Emerging Mkts Eq GS4	Oct 31, 2013	Diversified emerging mkts	N/A	1	N/A	N/A	N/A	N/A	N/A	N/A	N/A	N/A	N/A	N/A
GGEYX	GuideStone Funds Growth Equity GS2	Aug 27, 2001	Large growth	1340	1	27.49	16.91	17.56	−42.84	1.17	−0.59	1.06	14.00	88.30	0.87
GMIZX	GuideStone Funds MyDestination 2005 GS4	Dec 29, 2006	Target date 2000–2010	86	1	5.56	9.61	11.41	−22.88	1.15	0.44	0.62	5.31	95.75	0.20
GMWZX	GuideStone Funds MyDestination 2025 GS4	Dec 29, 2006	Target date 2021–2025	513.58	1	13.01	13.52	14.56	−35.11	1.01	−0.76	1.20	10.26	98.26	0.15
GRAZX	GuideStone Funds Real Assets GS4	Jul 1, 2013	Moderate allocation	21.83	1	N/A	N/A	N/A	N/A	N/A	N/A	N/A	N/A	N/A	N/A

GSCZX	GuideStone Funds Small Cap Equity G54	Aug 27, 2001	Small blend	542.07	1	31.12	18.59	21.36	−37.19	1.14	−1.35	1.28	16.96	88.66	1.21
GVEYX	GuideStone Funds Value Equity GS2	Aug 27, 2001	Large value	1270	1	27.71	14.80	18.38	−36.92	1.21	−1.00	1.13	14.28	97.93	0.67
GMHZX	GuideStone Funds MyDestination 2035 GS4	Dec 29, 2006	Target date 2031–2035	272.89	1	17.98	14.06	13.81	−38.86	0.9	−2.37	1.58	13.51	97.66	0.20
GMDYX	GuideStone Funds Medium-Duration Bd GS2	Aug 27, 2001	Intermediate-term bond	812.86	1	−1.07	8.90	10.74	−3.88	1.3	0.94	0.97	2.92	89.47	0.48
GGBZX	GuideStone Funds Agrsv Allocation GS4	Aug 27, 2001	World stock	848.86	1	22.07	14.20	18.71	−41.05	0.89	7.29	0.83	14.95	91.04	0.12
GFLZX	GuideStone Funds Flexible Income GS4	Jul 1, 2013	High-yield bond	106.4	1	N/A	N/A	N/A	N/A	N/A	N/A	N/A	N/A	N/A	N/A
GMDZX	GuideStone Funds Medium-Duration Bd GS4	Aug 27, 2001	Intermediate-term bond	812.86	1	−1.13	8.76	10.64	−3.97	1.25	0.79	0.97	2.91	89.71	0.63
GGBYX	GuideStone Funds Agrsv Allocation 1 GS2	Jul 1, 2003	World stock	180.57	1	22.33	14.36	12.91	−40.95	0.91	7.51	0.82	14.89	90.86	0.15
GSCYX	GuideStone Funds Small Cap Equity GS2	Aug 27, 2001	Small blend	542.07	1	31.35	18.81	21.49	−37.15	1.15	−1.05	1.28	16.91	88.52	1.01
GVEZX	GuideStone Funds Value Equity GS4	Aug 27, 2001	Large value	1270	1	27.51	14.56	18.16	−37.01	1.19	−1.31	1.14	14.36	97.84	0.92
GGBFX	GuideStone Funds Global Bond GS4	Dec 29, 2006	World bond	354.25	1	0.79	13.34	16.84	−20.12	0.85	2.75	0.88	6.28	15.73	0.83
GEMYX	GuideStone Funds Emerging Mkts Eq GS2	Oct 31, 2013	Diversified emerging mkts	N/A	1	N/A	N/A	N/A	N/A	N/A	N/A	N/A	N/A	N/A	N/A
GGIYX	GuideStone Funds Balanced Allc 1 GS2	Jul 1, 2003	Conservative allocation	384.76	1	9.34	11.80	12.39	−24.26	1.14	0.36	0.85	7.23	98.38	0.14
GGEZX	GuideStone Funds Growth Equity GS4	Aug 27, 2001	Large growth	1340	1	27.20	16.74	17.46	−42.97	1.16	−0.79	1.06	14.02	88.31	1.06
GIPZX	GuideStone Funds Inflat Prtctd Bd GS4	Jun 26, 2009	Inflation-protected bond	272.02	1	−6.15	N/A	7.92	5.82	0.51	−1.85	1.50	5.08	70.53	0.65

(*Continued*)

EXHIBIT 10.1 *(Continued)*

Ticker Symbol	Fund Name	Fund Inception Date	Category	Net Assets (in $Million)	Number of Funds	YTD Return (in %)	Five-Year Average Return (in %)	Best 3-Year Total Return (in %)	Worst 1-Year Total Return (in %)	Sharpe Ratio	Alpha	Beta	Standard Deviation	R^2	Expense Ratio (in %)
GEDYX	GuideStone Funds Extended-Dur Bond GS2	Aug 27, 2001	Corporate bond	276.52	1	−4.47	15.94	16.80	−8.12	0.91	0.43	2.32	7.99	67.84	0.52
GMFZX	GuideStone Funds MyDestination 2045 GS4	Dec 29, 2006	Target date 2041–2045	196.22	1	19.58	14.17	13.73	−40.29	0.9	−2.50	1.66	14.27	97.06	0.20
GLDZX	GuideStone Funds Low-Duration Bond GS4	Aug 27, 2001	Short-term bond	833	1	−0.18	4.12	5.83	−2.84	1.36	0.70	0.29	1.15	51.89	0.57
GIEZX	GuideStone Funds International Eq GS4	Aug 27, 2001	Foreign large blend	1660	1	14.08	11.48	23.73	−44.81	0.39	−0.49	1.03	17.88	98.06	1.19
GEQYX	GuideStone Funds Equity Index GS2	Aug 27, 2001	Large blend	331.62	1	25.03	15.17	14.31	−37.41	1.32	0.43	0.98	12.27	99.77	0.23
GDMYX	GuideStone Funds Defensv Mkt Strats GS2	Aug 31, 2011	Tactical allocation	467.09	1	16.06	N/A	N/A	8.92	N/A	N/A	N/A	N/A	N/A	0.99
GFIYX	GuideStone Funds Cnsrv Allocation I GS2	Jul 1, 2003	Conservative allocation	84.37	1	3.69	6.89	8.58	−12.97	1.23	0.61	0.40	3.54	93.26	0.15
GMGZX	GuideStone Funds MyDestination 2055 GS4	Dec 30, 2011	Target date 2051+	14	1	19.46	N/A	N/A	14.18	N/A	N/A	N/A	N/A	N/A	0.20
GCOZX	GuideStone Funds Growth Allocation GS4	Aug 27, 2001	Aggressive allocation	945.64	1	15.95	13.06	15.15	−33.06	0.95	−1.48	1.33	11.37	97.97	0.12
GFIZX	GuideStone Funds Cnsrv Allocation GS4	Aug 27, 2001	Conservative allocation	320.75	1	3.50	6.71	8.36	−13.91	1.17	0.40	0.41	3.60	93.58	0.12
Total/ Average				22,151.39	40	13.07	12.72	14.75	−24.03	1.04	0.29	1.07	10.26	86.35	0.54

Source: Yahoo Finance as of November 15, 2013.

EXHIBIT 10.2 Catholic Funds/Summary Statistics

Ticker Symbol	Fund Name	Fund Inception Date	Category	Net Assets (in $Million)	Number of Funds	YTD Return (in %)	Five-Year Average Return (in %)	Best 3-Year Total Return (in %)	Worst 1-Year Total Return (in %)	Sharpe Ratio	Alpha	Beta	Standard Deviation	R^2	Expense Ratio (in %)
LKBAX	LKCM Balanced	Dec 30, 1997	Moderate allocation	32.91	1	18.50	13.47	11.83	−19.70	1.38	2.99	1.02	9.00	91.08	0.80
LKSAX	LKCM Small Cap Equity Advisor	Jun 5, 2003	Small growth	1060	1	26.69	17.96	22.10	−39.02	1.06	−1.23	1.18	16.82	76.97	1.19
AQBLX	LKCM Aquinas Small Cap	Jan 3, 1994	Small growth	15.1	1	24.25	17.16	22.18	−37.64	0.98	−2.37	1.19	17.07	75.68	1.50
AQEGX	LKCM Aquinas Growth	Jan 3, 1994	Large growth	43.08	1	20.08	14.05	15.49	−33.07	0.89	−4.84	1.13	14.99	87.74	1.50
AQEIX	LKCM Aquinas Value	Jan 3, 1994	Large blend	56.15	1	25.95	15.22	16.12	−37.34	1.06	−2.61	1.12	14.67	91.14	1.50
LKEQX	LKCM Equity Instl	Dec 29, 1995	Large growth	305.34	1	23.27	15.57	15.61	−31.80	1.26	−0.02	1.00	12.78	96.07	0.80
LKFIX	LKCM Fixed Income	Dec 30, 1997	Corporate bond	219.24	1	0.25	6.68	8.76	−0.34	1.16	0.97	0.59	2.31	51.45	0.65
LKSCX	LKCM Small Cap Equity Instl	Jul 14, 1994	Small growth	1060	1	26.93	18.27	23.46	−38.87	1.08	−0.96	1.18	16.82	76.92	0.94
LKSMX	LKCM Small/Mid Cap Equity	Apr 29, 2011	Mid-cap growth	344.21	1	26.45	N/A	N/A	9.26	N/A	N/A	N/A	N/A	N/A	1.00
AVEWX	Ave Maria World Equity	Apr 30, 2010	World stock	36.47	1	17.80	N/A	N/A	−9.60	0.6	3.20	0.81	15.10	85.68	1.50
AVEMX	Ave Maria Catholic Values	May 1, 2001	Mid-cap blend	238.25	1	19.85	16.38	19.87	−39.83	0.95	−3.90	1.11	14.78	88.08	1.48
AVEDX	Ave Maria Rising Dividend	May 2, 2005	Large blend	632.86	1	26.76	16.23	15.62	−22.79	1.42	2.08	0.90	11.67	92.95	0.99
AVEGX	Ave Maria Growth	May 1, 2003	Mid-cap growth	265.38	1	24.76	16.47	17.14	−32.10	1.14	−0.87	1.06	14.12	86.84	1.50
AVEFX	Ave Maria Bond	May 1, 2003	Intermediate-term	143.59	1	5.07	6.75	6.69	0.30	1.65	3.73	0.14	2.51	2.64	0.70
AVESX	Ave Maria Opportunity	May 1, 2006	Mid-cap blend	48.38	1	21.53	14.75	19.34	−32.19	0.84	−4.74	1.01	13.87	83.12	1.25
Total/Average				4500.96	15	20.54	14.54	16.48	−24.32	1.11	−0.61	0.96	12.61	77.60	1.15

Source: Yahoo Finance as of November 15, 2013.

EXHIBIT 10.3 Mennonite Funds/Summary Statistics

Ticker Symbol	Fund Name	Fund Inception Date	Category	Net Assets (in $Million)	Number of Funds	YTD Return (in %)	Five-Year Average Return (in %)	Best 3-Year Total (in %)	Worst 1-Year Total (in %)	Sharpe Ratio	Alpha	Beta	Standard Deviation	R^2	Expense Ratio (in %)
MMSCX	Praxis Small Cap A	May 1, 2007	Small growth	70.55	1	26.04	17.54	21.78	−38.74	1.11	−0.24	1.14	16.29	75.48	1.72
MVIAX	Praxis Value Index A	May 11, 2001	Large value	100.06	1	25.34	12.55	16.09	−40.15	1.08	−2.62	1.05	13.33	97.16	1.21
MBAPX	Praxis Genesis Balanced A	Dec 31, 2009	Moderate allocation	48.98	1	12.45	N/A	6.77	−0.30	1.07	−0.11	0.93	7.90	98.40	0.61
MMSIX	Praxis Small Cap 1	May 1, 2007	Small growth	70.55	1	26.75	18.13	22.25	−38.53	1.14	0.34	1.14	16.34	75.53	1.05
MIIAX	Praxis Intermediate Income A	May 12, 1999	Intermediate-term bond	344.9	1	−0.92	6.80	7.87	1.82	1.1	0.22	0.84	2.45	94.98	0.96
MPLAX	Praxis International Index A	Dec 31, 10	Foreign large blend	156.88	1	12.14	N/A	N/A	−16.54	N/A	N/A	N/A	N/A	N/A	1.73
MCONX	Praxis Genesis Conservative A	Dec 31, 2009	Conservative allocation	17.61	1	5.45	N/A	6.11	2.77	1.32	1.15	0.46	4.05	90.27	0.61
MVIIX	Praxis Value Index 1	May 1, 2006	Large value	100.06	1	26.13	13.28	10.99	−39.94	1.13	−1.84	1.05	13.29	96.97	0.47
MMDEX	Praxis Growth Index 1	May 1, 2007	Large growth	138.71	1	24.64	16.23	15.87	−37.09	1.33	0.77	0.94	11.91	95.88	0.48
MGNDX	Praxis Growth Index A	May 1, 2007	Large growth	138.71	1	24.27	15.69	15.44	−37.34	1.28	0.21	0.94	11.93	95.93	1.08
MIIIX	Praxis Intermediate Income 1	May 1, 2006	Intermediate-term bond	344.9	1	−0.59	7.19	8.11	3.33	1.25	0.60	0.84	2.45	94.41	0.56
MPLIX	Praxis International Index 1	Dec 31, 2010	Foreign large blend	156.88	1	12.72	N/A	N/A	−16.05	N/A	N/A	N/A	N/A	N/A	0.84
MGAFX	Praxis Genesis Growth A	Dec 31, 2009	Aggressive allocation	41	1	17.02	N/A	7.16	−2.32	0.98	−0.99	1.24	10.64	97.12	0.61
Total/ Average				**1729.79**	**13**	**16.26**	**13.43**	**12.59**	**−19.93**	**1.16**	**−0.23**	**0.96**	**10.05**	**92.01**	**0.92**

Source: Yahoo Finance as of November 15, 2013.

EXHIBIT 10.4 Muslim Funds/Summary Statistics

Ticker Symbol	Fund Name	Fund Inception Date	Category	Net Assets (in $Million)	Number of Funds	YTD Return (in %)	Five-Year Average Return (in %)	Best 3-Year Total Return (in %)	Worst 1-Year Total Return (in %)	Sharpe Ratio	Alpha	Beta	Standard Deviation	R^2	Expense Ratio (in %)
AMAGX	Amana Growth	Feb 3, 1994	Large growth	2080	1	16.21	13.34	25.82	−29.67	0.9	−3.99	0.88	11.31	94.18	1.11
AMIGX	Amana Growth	Sep 25, 2013	Large growth	2080	1	16.25	13.34	N/A	N/A	0.9	−3.98	0.88	11.31	94.19	N/A
AMIDX	Amana Developing	Sep 25, 2013	Diversified emerging	27.79	1	4.40	N/A	N/A	N/A	0.15	−2.35	0.53	10.37	77.92	N/A
AMINX	Amana Income	Sep 25, 2013	Large blend	1570	1	25.36	13.82	N/A	N/A	1.25	0.11	0.85	11.03	92.07	N/A
AMDWX	Amana Developing	Sep 28, 2009	Diversified emerging	27.79	1	4.31	N/A	1.31	−8.01	0.14	−2.38	0.53	10.36	78.01	1.51
AMANX	Amana Income	Jun 23, 1986	Large blend	1570	1	25.33	13.82	20.30	−23.48	1.25	0.10	0.85	11.03	92.08	1.18
ADJEX	Azzad Ethical	Dec 22, 2000	Mid-cap growth	40.75	1	27.57	19.54	24.81	−43.38	0.98	−2.93	1.11	15.26	82.16	0.99
WISEX	Azzad Wise	Apr 5, 2010	Short-term bond	55.26	1	1.96	N/A	N/A	0.93	1.01	1.50	0.33	2.46	14.92	1.49
Total/ Average				**7451.59**	**8**	**15.17**	**14.77**	**18.06**	**−20.72**	**0.82**	**−1.74**	**0.75**	**10.39**	**78.19**	**1.26**

Source: Yahoo Finance as of November 15, 2013.

REFERENCES

Bauer, R., K. Koedijk and R. Otten. 2005. International evidence on ethical mutual fund performance and investment style. *Journal of Banking & Finance* 29:1751–1767.

Capelle-Blancard, G. and S. Monjon. 2012. Trends in the literature on socially responsible investment: looking for the keys under the lamppost. *Business Ethics: A European Review* 21:239–250.

Friedman, M. 1970. The Social Responsibility of a Business is to Increase its Profits. New York Times, September 13, 1970.

Galema, R., A. Plantinga and B. Scholtens. 2008. The stocks at stake: return and risk in socially responsible investment. *Journal of Banking & Finance* 32:2646–2654.

Henderson, D. 2001. *Misguided Virtue: False Notions of Corporate Social Responsibility*. New Zealand Business Roundtable, Wellington.

Hickman, K.A., W.R. Teets and J.J. Kohls. 1999. Social investing and modern portfolio theory. *American Business Review* 17:72–78.

Hoepner, A.G.F. and D.G. McMillan. 2009. Research on 'Responsible Investment': An Influential Literature Analysis Comprising a Rating, Characterization, Categorisation and Investigation. Working paper, University of Saint Andrews, School of Management, Saint Andrews.

Entine, J. 2003. The myth of social investing. *Organizations & Environment* 16:352–368.

Margolis, J.D., H.A. Elfenbein and J.P. Walsh. 2007. Does It Pay to Be Good? A Meta-Analysis and Redirection of Research on the Relationship between Corporate Social and Financial Performance. Working paper, Harvard University.

Russo, M. and P. Fouts. 1997. A resource-based perspective on corporate environmental performance and profitabilities. *Academy of Management Journal* 40(3):534–559.

Renneboog, L., J.T. Horst and C. Zhang. 2008a. The price of ethics and stakeholder governance: the performance of socially responsible mutual funds. *Journal of Corporate Finance* 14:302–322.

Renneboog, L., J.T. Horst and C. Zhang. 2008b. Socially responsible investments: institutional aspects, performance, and investor behaviour. *Journal of Banking & Finance* 32:1723–1742.

Schueth, S. 2003. Socially responsible investing in the United States. *Journal of Business Ethics* 43:189–194.

Sievanen, R., H. Rita and B. Scholtens. 2013. The drivers of responsible investment: the case of European pension funds. *Journal of Business Ethics* 117 (1):137–151.

Soros, G. 2000. *The Open Society: Reforming Global Capitalism*. Public Affairs Press, New York.

Van Dyck, T.W. and S. Paulker. 2001. Qualitative screening: looking beyond the balance sheet. *Public Retirement Journal* April:4–8.

11

THE ROLE OF PUBLIC RELATIONS AND ORGANIZATIONAL COMMUNICATION IN ENVIRONMENTAL MANAGEMENT

RICARD W. JENSEN
Department of Marketing, Montclair State University, Montclair, NJ, USA

Abstract: This chapter describes the use of public relations and organizational communication tools that governmental entities, non-profit organizations, special interest groups, and corporations are using to effectively manage issues related to the environment. Examples of the use of each of these strategies are described, and needs for additional research and scholarship are presented.

Keywords: environmental public relations, environmental organizations, marketing, stakeholders, rhetoric, Public Relations Society of America, strategic public relations, new media, corporate social responsibility, greenwashing, conservation, cause-related marketing.

11.1	Introduction	277
11.2	Public Relations	278
	11.2.1 Example: Shell oil uses Public Relations to Create Dialog and Build Relationships	279
	11.2.2 Areas for Future Research	279
11.3	Organizational Communication	279
	11.3.1 Example: How a Major Electrical Utility Used Organizational Communication to Talk with Customers during a Crisis	279
	11.3.2 Areas for Future Research	280
11.4	Best Practices in Public Relations	280
	11.4.1 Example: The United Parcel Service uses Public Relations to Communicate its Commitment to Sustainability	280
	11.4.2 Areas of Research	280
11.5	Public Relations Failures	280
	11.5.1 Example: How Public Relations was Used in Canada to Spin Climate Change	281
	11.5.2 Areas for Future Research	281
11.6	Best Practices in Organizational Communication	281
	11.6.1 Example: How a Texas Agency Used Organizational Communication to Bring Diverse Stakeholders Together	282
	11.6.2 Areas for Future Research	282
11.7	How Environmental Organizations are Using Public Relations	283
	11.7.1 Example: How Major League Soccer Supports "Green" Playing Fields	283
	11.7.2 Areas for Future Research	283
11.8	The Strategic Use of Public Relations to Achieve Environmental Goals	284
	11.8.1 Example: Australia Defeats a Drought by Teaching Residents to Conserve	284
	11.8.2 Areas for Future Research	284
11.9	Creating "Win–Win" Partnerships	284
	11.9.1 Example: Hosting the Super Bowl Benefits Local Communities, The Environment, and Locals	285
	11.9.2 Areas for Future Research	285
11.10	Summary	285
	References	286

11.1 INTRODUCTION

In this era of heightened concerns about the environment, it is more important than ever to effectively communicate the missions, goals, objectives, and activities of any organization working in the environmental sector to all affected parties, also known as stakeholders. Public relations efforts need to be strategic in nature—rather than just presenting information, the best PR efforts are persuasive and rhetorical in nature, and the goal is often to advocate causes that

---

An Integrated Approach to Environmental Management, First Edition. Edited by Dibyendu Sarkar, Rupali Datta, Avinandan Mukherjee, and Robyn Hannigan.
© 2016 John Wiley & Sons, Inc. Published 2016 by John Wiley & Sons, Inc.

your organization supports (e.g., think of cause-related marketing). Public relations professionals must manage a delicate balancing act in which they present the best face of the organization to the public and the media that makes it appealing to support the stance of your organization, but PR professionals cannot spin or misrepresent the facts to such an extent that they lose credibility. Because the public has greater power than ever to access information about the activities of virtually anyone, organizations working with the environment have to be totally transparent and honest in communicating with stakeholders; the consequences are serious: if people discover that an organization's rhetoric does not mirror how that entity conducts its daily affairs, the public will find out the discrepancy and a loss of trust may ensue that will be difficult to repair.

Similarly, it is essential that leaders of organizations communicate effectively with employees, regulatory officials, stakeholders, and the public through effective organizational communication. The practice of organizational communication focuses on the processes by which entities communicate, the dynamics associated with this communication, and the effects of communication practices on performance outcomes the organization feels are mission critical.

This chapter provides several examples of how public relations and organizational communications practice has affected the performance of environmental entities. It includes examples of successes and failures of the scenarios in which corporations, environmental organizations, and the public come together to create situations that benefit all parties.

11.2 PUBLIC RELATIONS

The practice of public relations is defined as the art and science of using strategic and rhetorical communications to build long-term mutually beneficial relationships between an organization and its various stakeholders (e.g., people who have a stake or vested interest in what the entity is doing). In practice, this means that organizations must be able to identify the stakeholders or publics that are most essential to their success and then begin to develop meaningful working relationships with them. The only way to do this is to become a responsive organization that seeks out the concerns of stakeholder groups, listens to them, and then changes the culture and behavior of the organization to meet their needs. The Public Relations Society of America (PRSA) is the lead organization that represents public relations professionals in the United States. In 2013, the PRSA defined public relations as "Strategic communication process that build mutually beneficial relationships between organizations and their publics" (PRSA, 2013). To develop these mutually beneficial long-term relationships, public relations professionals develop and implement a variety of tactics, including creating publicity, placing stories in the mass media, communicating with people within the organization, developing special events, writing press releases and fact sheets, and, increasingly, creating online content for websites, mobile phone apps, and a wide range of emerging new media and social media including (among others) Facebook, Twitter, Instagram, Vine, YouTube, etc. While defining what public relations does consist of, it is also important to clarify some misperceptions of PR practice. Public relations cannot consist merely of unethically misrepresenting the facts ("spinning the issues") to make the organization look great even when that entity has made some grievous errors. Public relations is not just about creating publicity. Public relations cannot include making false statements about how great your organization is or how bad your opponents are just to advocate your cause. When public relations is practiced properly, it must be entirely ethical.

A 2011 survey of the PRSA members in the United States provides several insights (Bortree, 2011). It found that the environmental issues most often communicated via public relations include (in order) energy efficiency, recycling, green products and services, waste management, water conservation, greenhouse gases, and environmental-friendly product packaging. Survey results suggested that the respondents (public relations practitioners) believed they were committed to communicating transparently with the public, although several critics had a decidedly different opinion. One important aspect of public relations is to develop rhetorical institutional advertising, which has traditionally been placed in print magazines and newspapers; the intent is to create a better image of the firm in the mind of the public. One way to infer the results of public relations programs is to examine image advertising in the mass media; Ahern, Bortree, and Nutter-Smith (2012) studied corporate advertisements about the environment that have appeared in the prestigious *National Geographic* magazine from 1970 to 2000; they found that the tone of image advertising in the magazine has shifted so that the more recent ads concentrated on telling consumers why it was good for the environment and society when a corporation embraced the ethos of sustainability, rather than merely describing why their product might require less energy inputs. Several recent public relations theories have been developed that focus on the need to develop stakeholder relationships, especially when focusing on environmental concerns. In a recent study that mapped the types of work done in public relations practice, ranging from the extent to which stakeholders felt they were trusted or being exploited, results show that organizations that strove to develop a sense of community with key publics were regarded most highly by stakeholders (Waters and Bortree, 2013). The extent to which the way in which public relations is practiced and how that work influences public opinion was the focus on a recent study by Roper (2008); the author found that Shell Oil only truly began to listen to its stakeholders after the company was widely criticized for environmental mismanagement in the 1980s; afterward, the corporation initiated meaningful efforts to listen to public concerns, it embraced sustainability, and its image has improved. Currently, listening to the needs of concerned publics and rapidly responding to them in a meaningful way have now become an imperative (not an option) for corporations working in the environmental sector (Frandsen and Johansen, 2013). In many cases, public relations methods can be successfully utilized to empower leaders of corporations and NGOs to assert more influence in situations involving environmental dilemmas as well as many other issues (Wakefield, 2012). But there are still important concerns about the extent to which public relations is being used optimally, especially in the environmental sector. A survey found that many corporate leaders in Australia viewed the purpose of public relations as only to gain publicity in the mass media, rather than building dialog between an organization and its stakeholders (Benn, Todd, and Pendleton, 2010).

11.2.1 Example: Shell Oil Uses Public Relations to Create Dialog and Build Relationships

The way in which Royal Dutch Shell significantly changed how it uses public relations serves as a good illustration of how organizations can best use public relations. In 1995, Shell found two ecological disasters that it handled poorly—the company disposed of Brent Spar oil drilling platform by sinking it off the coast of Scotland, and the firm was charged with not caring about the welfare of the people of Nigeria, where Shell had extracted petroleum for decades. As a result of these and other incidents, environmental NGOs such as Greenpeace and several others began boycotting and protesting the company. Shell chose to respond to these disasters by significantly rethinking how the firm viewed environmental issues and how it communicated them to the public. Shell reviewed its operations to determine if the firm was acting in an environmentally sustainable manner; Shell then took steps to improve its actions to attempt to be more ethical and environmentally responsible while still trying to make a profit. Shell began to actively seek the opinions of environmental leaders, green organizations, and other stakeholders, and the use of public relations is a key component of those efforts (Roper, 2005).

11.2.2 Areas for Future Research

- Developing and refining theoretical models about how to examine and assess how strategic public relations can be used to develop mutually beneficial relationships within the environmental sector
- Examining the extent to which different public relations strategies and tactics are being used within the environmental sector and assessing the effectiveness of these programs

11.3 ORGANIZATIONAL COMMUNICATION

In contrast to public relations, organizational communication encompasses the manner in which communications flow both inside and outside the organization and the flow of communication. For example, this discipline examines the extent to which corporate leaders issue directives without feedback or the extent to which they solicit and value comments from internal and external stakeholders. Organizational communications scholars also study the process through which effective communication occurs and the extent to which factors like "noise" and interference compromise the clarity of communications. Finally, scholars in the field research the extent to which organizations communicate in a formal or informal manner. Several recent papers have examined how organizations communicate about the environment. Linnenluecke and Griffiths (2010) studied the leadership style and motivations of leaders of corporations in Australia that affect the environment; they found that it is essential that senior corporate personnel must fully buy into corporate social responsibility and the need to communicate environmental challenges to stakeholders. Liska et al. (2012) analyzed the rhetoric emanating from coal companies and government agencies about a pollution incident using chaos theory; this pollution crisis caused the corporation to recognize that they had to totally rethink how they communicated with the public. O'Connor and Shumate (2010) assessed the ways in which different types of corporations communicate messages about corporate social responsibility; they found that businesses higher up in the value chain (those that sell directly to consumers) are often more fully vested in telling how they are socially responsible than companies lower on the value chain (those that extract natural resources). Setthasakko (2009) studied the extent to which the lack of effective organizational communication might hinder corporate environmental responsibility in Thailand. He found that corporations that do not have the leadership and buy-in from top management cannot communicate the idea of environmental sustainability to the public.

11.3.1 Example: How a Major Electrical Utility Used Organizational Communication to Talk with Customers during a Crisis

Commonwealth Edison, a major electric utility in the United States, provides an excellent example of how organizations can communicate during a crisis to meet consumer needs. In July 2011, a powerful storm swept across Chicago and more than 900,000 customers lost electric power. In order to assist as many people as possible, the utility used Facebook and Twitter in a very proactive manner—they created an "eChannels Department" focused on empowering consumers to engage with the company via the utility website, apps, social media, phone calls, and other means. Once this storm hit, the utility set up a war room that was staffed with up to 15 people around the clock to make sure that the company could immediately respond to customer needs. Their strategy was praised by the local media and many public leaders because their response to this crisis showed that the utility is concerned about its customers and other stakeholders (Diermeier and Petrella, 2009).

11.3.2 Areas for Future Research

- Examining the ways in which environmental organizations communicate with different groups of employees (e.g., executives, management, and blue-collar workers)
- Investigating the ways in which organizations communicate with regulatory agencies
- Assessing how organizations are incorporating new media and social media into the communications tool set and evaluating the effectiveness of these techniques

11.4 BEST PRACTICES IN PUBLIC RELATIONS

For companies working in any area related to the environment, the task of successfully telling stakeholders of your efforts represents both an opportunity and a threat. If your organization can convince the public that it really is socially and environmentally responsible, stakeholders will reward you and want to work with you. For many years, several scholars have advocated that senior public relations professionals be placed at level of a senior manager (equivalent to the corporate attorney or director of human resources) so that they can exercise more direct input into the daily decision making of the organization about important environmental issues; if this were to occur, public relations professionals could play a key role as the "conscience" of the organization, encouraging leaders to consistently do the right thing. Despite the need to have public relations offers in senior leadership roles, research has found that public relations professionals often do not have the support they need within their organizations to fulfill this vital role (Bowen, 2008). There are many instances in which successfully communicating an organization's work in the environment has improved the public perception and the bottom line. A comparative study of 20 corporations in the United Kingdom sought to determine the extent to which successfully practicing and communicating about CSR might satisfy the demands from several stakeholder groups while maximizing profits. Results suggest that being seen as socially responsible is essential for a firm's continued survival in this era when the public is demanding that companies are good stewards of the environment and respectful of all people (Samy, Odemilin, and Bampton, 2010). In India, research suggests that those corporations that best communicate with stakeholder groups are seen as more socially responsible than companies that do not do so (Sangle, 2010). Studies of corporate public relations campaigns in Spain and the United States confirm that corporations that communicate their environmental good deeds are most effective in building a positive brand image when they use all the tools of integrated marketing communications, not just public relations (Andreu, Mattila, and Aldás, 2011). Marketing a corporation as being socially responsible provides opportunities to focus on the best deeds of the company, thereby offsetting prevailing notions of how that firm may be negatively affecting the environment (Zyglidopoulos et al., 2012).

11.4.1 Example: The United Parcel Service Uses Public Relations to Communicate its Commitment to Sustainability

In 2007, the United Parcel Service (UPS) developed an ongoing public relations campaign to effectively communicate their green initiatives to the media and the public. The campaign was titled "Brown Goes Green" and it provided transparent information about the company's fuel use and its carbon footprint. The intent of the campaign was to highlight the reductions the UPS was achieving in gasoline use as the company developed hybrid vehicles that are more fuel efficient. In this instance, the UPS was already engaged in doing important things to incorporate best environmental practices throughout the company; it only had to find a way to tell the public its story, and public relations proved to be a great way to accomplish this. The result of the campaign is that it improved the company's environmental image (Griggs, 2008).

11.4.2 Areas of Research

- Identifying the criteria that characterize successful public relations programs
- Developing and testing theories and models that assess the extent to which public relations professionals are successful in changing the organizational culture
- Investigating the extent to which organizations working in the environment are articulating the story of their environmental good deeds in corporate sustainability reports that also address related issues (e.g., human rights, fair trade, sweatshops, etc.)

11.5 PUBLIC RELATIONS FAILURES

There are also several instances in which the use of public relations has not achieved the desired outcomes or has even worsened the image of the organizations sponsoring these programs. Many of these problems occur when organizations fail to understand or listen to the concerns of stakeholders, don't respond quickly enough, or believe that they can improve their green image by deceiving the public by "spinning" controversial issues. In the aftermath of the 2011 Fukushima nuclear disaster in Japan, many residents of the region became angry and hostile when the Tokyo Electric Power Company (TEPCO) tried to rationalize why this accident occurred and why they were not to blame, rather than focusing on the emotional impact of this tragedy on the lives of victims (Utz, Schultz, and Glock, 2013). After the Exxon Valdez oil tanker ran aground and spilled large amounts of crude oil in Alaska, corporate leaders of the oil company did not act quickly to communicate with the media and the public and did not express concern or remorse to those who were adversely affected. As a result, the reputation of Exxon and other oil companies suffered (Pauly and Hutchinson, 2005). The clear bottom line is that senior leaders need to speak up when environmental crises occur, and when they do speak, they must demonstrate that they understand the issues and the concerns of people who may be hurt by these tragedies. When the King Salmon Company in New Zealand tried to sell genetically modified salmon to the public in 1999, corporate leaders focused on one-way public relations by trying to persuade people to the merits of their argument and ignored soliciting and responding to feedback from stakeholder groups. As a result, the public opinion about GMOs has suffered in New Zealand and elsewhere (Weaver and Motion, 2002). There are also instances where failures in internal communication led in large part to environmental disasters. Many of the problems that led to the horrific Bhopal industrial pollution crisis in India in 1984 were caused in part by inadequate training of the staff working at the plant and a breakdown in how workers at the site communicated with each other (Weick, 2010). Since Bhopal, there is evidence that several corporations are trying to become more transparent and truthful about their environmental performance, even during the midst of a crisis. The South African petrochemical giant Total SA communicated immediately after one of its oil tankers sank and after an explosion at one of its refineries. The firm tried to defend its role in these accidents by trying to legitimize the actions that were taken during these events (Cho, 2009). An especially egregious negative public relations practice occurs when corporations engage in "greenwashing" to intentionally attempt to create the image that they are proenvironmental when their actions suggest otherwise (Munshi and Kurian, 2005); some prominent examples include British Petroleum's (BP) attempt to suggest its practices were "beyond petroleum" (Muralidharan, Dillistone, and Shin, 2011) and the corporate sponsorship of United Nations' environmental programs by well-known polluters (Lightfoot and Burchll, 2004). Research suggests that the widespread use of greenwashing as unethical public relations may make the public more skeptical when they see or hear honest reports about CSR. Generally speaking, much of the public strongly resents greenwashing and those organizations that engage in this type of deceptive practice (Alves, 2009).

11.5.1 Example: How Public Relations was Used in Canada to Spin Climate Change

There are many instances in which the practice of public relations appears to have been misused in order to shape public opinion about environmental issues. In Canada, a public relations firm was hired to create a front group with the intent of increasing skepticism about claims made by advocates of climate change. The front group, "Friends of Science," was funded by oil company interests in Canada, but the ties between these corporations and the organization were hidden for some time. Once it became apparent that Friends of Science was a very agenda-driven, corporate-funded entity, it faced significant opposition from the mass media and much of the public as well as scrutiny from regulatory agencies. The legacy of this ill-conceived public relations program is that it damaged relationships between the oil companies that sponsored this campaign and the stakeholders they probably wanted to build relationships with (Greenberg, Knight, and Westersund, 2011).

11.5.2 Areas for Future Research

- Assessing the extent to which environmental organizations respond rapidly during and after environmental crises and evaluating the success of these efforts
- Investigating the extent to which environmental organizations are engaging in greenwashing practices
- Carrying out long-term longitudinal studies to gauge the extent to which unethical public relations campaigns harm relationships between corporations and NGOs and valued stakeholders

11.6 BEST PRACTICES IN ORGANIZATIONAL COMMUNICATION

Organizations must have a genuine interest in listening to the needs of stakeholders, valuing their input, and then being prepared to act on the feedback they receive. Some of the strategies used to solicit feedback from the public include hosting open meetings and town hall forums (Kent, 2013). Organizational decisions about environmental decision making will be perceived as more inclusive as public meetings that foster dialog are conducted, even if this means increased conflict among participants (Jarrell, Ozymya, and McGurrin, 2012). Research shows that public relations campaigns must facilitate a thorough exchange of arguments between diverse parties if meaningful dialog between stakeholders is going to be accomplished (Mackerron and Berkhout, 2009). When corporations fully commit to actively recruiting feedback from all of their stakeholders, they often experience heightened environmental performance, compared to those companies who do not do so (Kock, Santaló, and Diestre, 2011). Organizations must be totally committed to being open and transparent in their day-to-day operations; this is more important than ever before in the Internet era, which gives the public more opportunities than ever to obtain information on their own (Waddock, 2008). Recent research shows that increased corporate transparency is producing real benefits for corporations working in the environment; mining firms found that they gained legitimacy among stakeholders and improved their brand image as they provided the public with more access to information about their activities (Haufler, 2010). Consumers are especially interested in transparency when corporate environmental activities might affect their health; it is especially important to communicate the extent to which pesticides and fertilizers that may be applied to the foods we eat (Wrigley, Ota, and Kikuchi, 2006).

An emerging trend is that many corporations are working to communicate their environmental accomplishments as part of broader efforts related to corporate social responsibility. Several companies including Patagonia, Chipotle, and Starbucks (among many others) are highlighting environmental good deeds as a keystone of their overall marketing strategy (Gopaldas, 2014). Husted and Allen (2007) suggest that CSR provides corporations with an opportunity to reconfigure the competitive landscape by branding themselves as distinctively caring about the environment and society at large. To successfully communicate the story of an organization's CSR efforts, it's essential to develop working relationships with reporters in traditional and new media. Haddock-Fraser (2012) examined the extent that various types of corporations in the United Kingdom work with newspaper reporters; they found that environmental corporations are more proactive in developing relations with newspaper reporters than other types of firms. Reilly and Weirup (2012) investigated the extent to which reporting about corporate CSR activities via new media (e.g., Twitter, Facebook, YouTube, etc.) is affecting perceptions about these firms; they suggest that most corporations recognize the imperative to create a positive presence in social media, but few of them have come to grips with how to develop ongoing positive relationships with reporters working in the new media landscape. Organizations must be knowledgeable and competent about the changing landscape of the news media as it is being transformed by social media; it's especially important for organizations to be responsive and transparent to bloggers and reporters who post on Twitter and Facebook (Waters, Tindall, and Morton, 2010).

Several researchers are working to develop theoretical models that can be applied to examine processes that occur in environmental public relations. Piercy and Lane (2009) developed a model that shows the extent to which effectively marketing the fact that corporations are engaged in CSR can positively influence the perceptions of stakeholders and can give the corporation more social credibility. Grimmelikhuijsen (2010) developed simulation models to ascertain the extent to which increased transparency might lead to higher levels of public trust; the research shows that merely providing more information will not necessarily improve a company's green brand image unless it is accompanied by good deeds.

11.6.1 Example: How a Texas Agency Used Organizational Communication to Bring Diverse Stakeholders Together

In 2000, the state of Texas significantly changed its approach to water resources planning and management to emphasize that strategies to deal with droughts, floods, and water pollution should be created by groups of stakeholders in each region of the state instead of being the work of teams of bureaucrats working for state agencies in Austin. An example of how this regional approach was best applied occurred in a large South Texas watershed near Corpus Christi in South Texas, where many people were concerned about a scarce amount of water that could be best shared by competing water users. The solution implemented by the local water management agency was to host focus groups and town hall meetings to provide all the interests in the region to voice their concerns and offer solutions, including farmers, mayors, industrial leaders, recreational water users who enjoyed rafting and fishing, and individuals interested in environmental stewardship. By creating an environment that was inclusive and by valuing public feedback, the process of developing a water management plan became a forum that facilitated goodwill and laid a groundwork for cooperation in dealing with many other environmental issues for years to come (Jensen and Uddameri, 2009).

11.6.2 Areas for Future Research

- Examining the extent to which there is a connection between environmental organizations that are open and transparent and how they are viewed by the public
- Developing strategies and tactics that have the potential for being most effective in helping environmental organizations reach public relations goals
- Creating tools that utilize new media and social media to engage with stakeholders, including such technologies as mobile phone apps, Twitter, Facebook, YouTube, and other advanced Internet-based technologies

11.7 HOW ENVIRONMENTAL ORGANIZATIONS ARE USING PUBLIC RELATIONS

Many experts believe that the practice of public relations shows its real value to a corporation or organization before, during, and after a crisis occurs. This is especially true for corporations and organizations working with environmental causes. Schultz et al. (2012) studied how BP communicate in the crisis that ensued following a massive oil spill in the Gulf of Mexico in 2010; they found that BP was effectively able to use public relations to frame itself as not being culpable for the spill while at same time being competent in managing this disaster. After a prolonged drought occurred in Australia in the early 2000s, local governments used public relations to develop an education campaign to curb water use; this public relations effort was so successful that it changed consumer behavior and significantly reduced water use over the long term (Walton and Hume, 2011). Jones, Hillier, and Comfort (2013) investigated how the leading retailing firms in the United Kingdom used public relations to combat concerns that they were not carrying about business in a sustainable manner; they found that most of these firms made the argument that they are incorporating sustainability into their business models to facilitate long-term growth. Beauchamp and O'Conner (2012) examined in CSR reports made by the chief operating officers of major US corporations; they found that the majority of CEOs made statements focusing on how corporate social responsibility was positively affecting the firm's revenues, but relatively few leaders addressed how CSR was environmentally sustainable. Rolland and Bazzoni (2009) examined environmental messages communicated by automobile manufacturers in CSR reports; they found that car makers that have a strong international presence may be more likely to stress the proenvironmental features of their vehicles in marketing and public relations materials because people throughout the world value these principles. With the rapid rise of new media technologies, it is not surprising that more and more corporations are using the Internet tools to communicate the strength of their organizations during an environmental crisis (Lovejoy, Waters, and Saxton, 2012; Waters, Tindall, and Morton, 2010).

But public relations can also be mismanaged and can have the effect of creating a worse image for corporations working in the environment. Greenberg, Knight, and Westersund (2011) examined how public relations was used to communicate the global warming crisis; they suggest that public relations campaigns often employed ethically questionable tactics to spin public opinion in favor of corporate polluters. Similarly, Miller and Horsley (2009) examined public relations tactics used by the coal mining industry during crises; they found that many coal mining firms were excellent in communicating the technical aspects of a mine disaster, but they were less able to persuade the public that they were acting in a responsibly sound manner. Durham (2005) described the dilemma that public relations professionals at Monsanto faced when trying to market genetically modified StarLink corn in 2000; messages from the public relations staff were intended to improve the image of the product and boost sales but did not include disclosing the fact that the corn would be unsafe for human consumption to the farmers who would grow the product. As a result, the StarLink brand deteriorated into a crisis, and the public relations effort was labeled as a debacle. When agricultural multinational firm ConAgra faced a health crisis associated with salmonella contamination of its peanut butter and pot pies in 2007, they responded by first denying there was a problem and later claiming they were not responsible for these outbreaks; Miller and Littlefield (2010) suggest that ConAgra's public relations efforts seemed more intent on protecting the company image when they ought to have focused on caring about consumers.

11.7.1 Example: How Major League Soccer Supports "Green" Playing Fields

In the United States, Major League Soccer (MLS) is one of the youngest and fastest-growing professional sports, and it's important for the league to create and maintain a positive brand. As a result, the league initiated its "MLS Works" public relations campaign in 2007, and a significant component of this effort is a commitment that the stadiums in the league are operated in an environmentally sustainable manner. When the league held its All-Star Game in New Jersey at The Red Bulls Arena in 2009, they made sure that every MLS team and corporate sponsor purchased renewable energy certificates to offset the amount of electrical power used during the weeklong All-Star Game festivities. The league also launched a new campaign titled "Greener Goals" that asked fans to recycle and take part in pollution cleanup efforts. The results have been very positive—now, fans can be assured that MLS cares as much about supporting green issues as it does about promoting its sport (Mallen, Chard, and Sime, 2013).

11.7.2 Areas for Future Research

- Investigating how public relations campaigns are strategically used in the midst of environmental crises and correlating why these efforts succeed or failed
- Developing theoretical and practical models to predict the elements of public relations that are most likely to change specific consumer behaviors (e.g., reducing the use of natural resources, increasing the proclivity of people to recycle more often, etc.)
- Creating scenarios that will empower environmental corporations and NGOs to communicate and share information effectively in crisis situations

11.8 THE STRATEGIC USE OF PUBLIC RELATIONS TO ACHIEVE ENVIRONMENTAL GOALS

There are several instances in which corporations, governments, and nonprofits are developing and implementing public relations campaigns to change the attitudes and behaviors of the public in order to make them more supportive of environmental stewardship. Fischer et al. (2013) described the public relations campaign created and implemented by a utility in the United States to educate consumers about energy conservation, but the program did not lower energy use. Omran, Aziz, and Robinson (2009) studied the effectiveness of a public relations campaign to increase recycling in Malaysia; they found that it is essential to increase the extent to which consumers understand the need for recycling before the behaviors of citizens can be changed, especially in developing countries. A recent study in Romania expresses similar concerns about the need to make sure that public relations campaigns are conducted that reflect the distinctive cultures, values, and languages of the nation or region in which they are being implemented; White, Vanc, and Coman (2011) examined the content and context of an effort implemented in Romania to increase public awareness about corporate social responsibility and found that challenges in communicating across cultures limited the effectiveness of this effort. A recurring theme is that the power of new media and social media is increasingly being harnessed to drive public relations and public education campaigns. Prestin and Pearce (2010) described a public relations campaign being used in public schools to motivate schoolchildren and their parents to recycle and reuse natural resources; they conclude that persuasive messages can be especially effective among "millenials" who have grown up in an era in which it's popular to be concerned about protecting the planet. Waters and Tindall (2011) investigated the reasons Americans donated funds to charities in response to the Japanese tsunami disaster of 2004; results suggest that public relations messages, as well as coverage in the mass media, influence the extent to which people are willing to donate to support environmental causes.

Scholars are also studying the extent to which public relations campaigns influence environmental activists. Lenox and Eesley (2009) examined the extent to which environmental activists develop and implement public relations campaigns; they found that several activist groups focus their message on punishing polluters rather than emphasizing motivating citizens to perform environmental good works. There are some concerns that the new media efforts of some environmental activist groups are not as effective as they could be in mobilizing supporters, in part because activist websites have been designed as passive websites that provide information rather than being instruments to engage and produce dialog (Sommerfeldt, Kent, and Taylor, 2012).

11.8.1 Example: Australia Defeats a Drought by Teaching Residents to Conserve

From 2006 through 2009, the region near Brisbane on the east coast of Australia was suffering from one of the worst droughts ever. To respond to this crisis, the Queensland Water Commission developed and implemented the "Target 140" public education and public relations campaign; the goal was to reduce household water use by 20% to 140 l per person per day. The reason this campaign worked so well is that it taught homeowners how they could conserve more water than ever by reducing the time it took to shower from 7 to 4 min; residents were given a timer to measure how long they were in the shower as well as educational materials. Over the short term, this public relations effort was so successful that it helped offset the effects of the drought; the campaign also sought to instill a change in consumer behavior that would provide long-term benefits (Walton and Hume, 2011).

11.8.2 Areas for Future Research

- Investigating the extent to which environmental public relations programs may be effective at helping nonprofit organizations raise funds, especially following natural disasters
- Examining the extent to which there may be a correlation between the success of public relations campaigns and the brand equity or value of corporations

11.9 CREATING "WIN–WIN" PARTNERSHIPS

There are several examples in which public relations programs can be the impetus that brings together corporations, not for profits, and governmental and regulatory agencies to fund and enact programs that benefit the environment. Corporations are partnering with environmental organizations to fund good works, in part because they see the benefit of being viewed as socially responsible companies. In addition, corporation are also recognizing how embracing environmental sustainability can positively affect their financial bottom line, especially if it means they can use fewer natural resources and dispose of less waste. Environmental organizations are also learning about how they can benefit by partnering with corporations to achieve environmental goals; corporations can provide the finances, marketing, and public awareness that is needed to help public education and public relations efforts succeed at a large scale (Sharma et al., 2010).

There is a richness of diversity surrounding the types of corporations that have chosen to partner with environmental organizations to achieve environmental good works and to boost their image, including (among many others) oil companies (Rondinelli and Berry, 2000); private and public utilities (Lyon and Maxwell, 2006); sports teams, events, and organizations (Trendafilova and Babiak, 2013); sports apparel makers; computer makers (Dauvergne and Lister, 2012); car makers (Aggeri, Elmquist, and Pohl, 2009); restaurants (Bitzer, Francken, and Glasbergen, 2008); retailers (Horwitz, 2009); and hotels (DeGrosbois, 2012). The reasons corporations enter into such public relations partnerships are also diverse and often include (among others) reducing the company's cost of doing business by using and disposing of fewer natural resources, promoting sustainability, helping local communities, assisting poor people in developing countries and other regions, fostering fair trade, and trying to achieve more social justice. While the idea of leveraging public relations campaigns as a way to engage corporations to fund environmental good works seems promising, it needs to be recognized that there are legitimate concerns, especially as it relates to the extent to which the private sector may be willing to embrace real environmental change, especially if it may hurt their profits (Hahn et al., 2010). Environmental organizations are also leery that corporate partners might be only interested in creating the perception of a green brand, and these NGOs do not want to be tainted by scandals associated with greenwashing (Jamali and Keshishian, 2009). In response to these concerns, scholars have developed a set of specific criteria nonprofit organizations should seek when looking for corporate partners that can enhance the social responsibility of both parties (Seitanidi and Crane, 2009).

11.9.1 Example: Hosting the Super Bowl Benefits Local Communities, the Environment, and Locals

In the United States, the National Football League's (NFL) annual championship game, the Super Bowl, is the largest-watched televised sports event. This weeklong sports spectacle draws in hundreds of thousands of tourists, a horde of reporters, and more than 80,000 fans who come to watch the game. At first glance, it might seem as though mega-events like this might be harmful to the environment. In reality, the city that earns the right to host the Super Bowl each year has to commit to a legacy project that will provide benefits to the people of the region where the game is played, including (among others) projects to benefit the environment, education, families, and public health. Babiak and Wolfe (2006) describe some of the NFL's efforts to use this special event as a way to raise the profile of the need for environmental good works, get people involved, and create a long-term shift in the minds of the public to convince them to become better stewards of the environment over the long-term. When New Orleans hosted the Super Bowl, programs were organized to clean debris that had washed ashore after Hurricane Katrina; when Detroit hosted the game, more than 2500 trees were planted to offset carbon emissions. When New York City and New Jersey hosted this event, more than $70 million was raised to support local causes, and the operations of MetLife Stadium (the site of the game) were significantly modified to improve recycling and lower natural resources use over the long term. The legacy of this partnership is that whenever a city earns the bid to host the Super Bowl, they're not just the site of one sports game. In contrast, they're really creating a significant opportunity to leverage the NFL and its corporate partners to work with local environmental groups to achieve good works.

11.9.2 Areas for Future Research

- Developing models and rubrics that NGOs can use to select the best corporate partners to facilitate their environmental programs
- Examining trends related to the extent to which corporations are partnering with environmental organizations, why they are doing so, and how these collaborations are protecting and bettering the planet
- Exploring how new media and social media are now being used in public relations campaigns to effectively communicate these programs to targeted publics across various platforms

11.10 SUMMARY

Public relations can be a very effective tool that corporations, nonprofit organizations, and governmental and regulatory agencies can use to achieve environmental goals. It is essential to regard the purpose of public relations as building long-term, mutually beneficial relations with valued publics and stakeholders. Public relations is not merely spinning an issue to make a corporation appear to be green when in fact they might be polluters; such practices are unethical and are condemned by the public relations industry. Environmental corporations and organizations also need to examine their organizational communications, ranging from the statements made by CEOs to instructions given to factory workers who need to know how to communicate during a crisis. It is obvious that many public relations practices benefit environmental corporations and organizations when communications are honest, open, and transparent; this is especially true when dealing with the public in times of crises. Finally, we are seeing an increasing number of examples where corporations are teaming up with environmental NGOs to support best practices that benefit the planet and the people who live on it. The use of solid public relations theory and technique will create opportunities for more of this sort of strategic philanthropy and will ensure that the real accomplishments of these programs can be told honestly and truthfully.

REFERENCES

Aggeri, F., M. Elmquist, and H. Pohl. 2009. Managing learning in the automotive industry—the innovation race for electric vehicles. *The International Journal of Automotive Technology and Management* 9:123–147.

Ahern, L., D. Sevick-Bortree, and A. Nutter-Smith. 2012. Key trends in environmental advertising across 30 years in National Geographic magazine. *Public Understanding of Science* 22:479–494.

Alves, I. 2009. Green spin everywhere: how greenwashing reveals the limits of the CSR paradigm. *The Journal of Global Change and Governance* 2:1–25.

Andreu, L., A. Mattila, and J. Aldás. 2011. Effects of message appeal when communicating CSR initiatives. *Advances in Advertising Research* 2:261–275.

Babiak, K., and R. Wolfe. 2006. More than just a game? Corporate social responsibility and Super Bowl XL. *Sport Marketing Quarterly* 15:214–222,

Beauchamp, L., and A. O'Connor. 2012. America's most admired companies: a descriptive analysis of CEO corporate social responsibility statements. *Public Relations Review* 38:494–497.

Benn, S., L. Todd, and J. Pendleton. 2010. Public relations leadership in corporate social responsibility. *The Journal of Business Ethics* 96:403–423.

Bitzer, V., M. Francken, and P. Glasbergen. 2008. Intersectoral partnerships for a sustainable coffee chain: really addressing sustainability or just picking (coffee) cherries? *Global Environmental Change* 18:271–284.

Bortree, D. 2011. The state of environmental communication: a survey of PRSA members. *Public Relations Journal* 5:1–17.

Bowen, S. 2008. A state of neglect: public relations as corporate conscience or ethics counsel. *Journal of Public Relations Research* 20:271–296.

Cho, C. 2009. Legitimization strategies used in response to environmental disaster: a French case study of Total SA's Erika and AZF Incidents. *The European Accounting Review* 18:33–62.

Dauvergne, P., and J. Lister. 2012. Big brand sustainability: governance prospects and environmental limits. *Global Environmental Change* 22:36–45.

DeGrosbois, D. 2012. Corporate social responsibility reporting by the global hotel industry: commitment, initiatives and performance. *The International Journal of Hospitality Management* 31:896–905.

Diermeier, D., and D. Petrella. 2009. *Commonwealth Edison: The Use of Social Media in Disaster Response.* Evanston, IL: Northwestern University.

Durham, F. 2005. Public relations as structuration: a prescriptive critique of the StarLink global food contamination case. *Journal of Public Relations Research* 17:29–47.

Fischer, R., A. Akin, B. White, D. Arant, T. Chamberlain, and S. Bolton. 2013. The smallest user campaign to decrease energy consumption in two mid-south neighborhoods: a field experiment. *Public Relations Review* 39:391–393.

Frandsen, F., and W. Johansen. 2013. Public relations and the new institutionalism: in search of a theoretical framework. *Public Relations Inquiry* 2:205–221.

Gopaldas, A. 2014. Marketplace sentiments. *The Journal of Consumer Research* 41:995–1014.

Greenberg, J., G. Knight, and E. Westersund. 2011. Spinning climate change: corporate and NGO public relations strategies in Canada and the United States. *The International Communication Gazette* 73:65–82.

Griggs, B. 2008. "EPA, UPS join to launch eco-friendly hybrid trucks," in CNN. Available at http://www.cnn.com/2008/TECH/biztech/10/28/ups.hybrid.trucks/. Accessed March 8, 2015.

Grimmelikhuijsen, S. 2010. Transparency of public decision-making: towards trust in local government? *Policy & the Internet* 2:5–35.

Haddock-Fraser, J. 2012. The role of the news media in influencing corporate environmental sustainable development: an alternative methodology to assess stakeholder engagement. *Corporate Social Responsibility and Environmental Management* 19:327–342.

Hahn, T., F. Figge, J. Pinkse, and L. Preuss. 2010. Trade-offs in corporate sustainability: you can't have your cake and eat it. *Journal of Business Strategy and the Environment* 19: 217–229.

Haufler, V. 2010. Disclosure as governance: the extractive industries transparency initiative and resource management in the developing world. *Global Environmental Politics* 10:53–73.

Horwitz, S. 2009. Wal-Mart to the rescue: private enterprise's response to Hurricane Katrina. *The Independent Review* 13:511–528.

Husted, B., and D. Allen. 2007. Strategic corporate social responsibility and value creation among large firms: lessons from the Spanish experience. *Long Range Planning* 40:594–610.

Jamali, D., and T. Keshishian. 2009. Uneasy alliances: lessons learned from partnerships between businesses and non-governmental organizations in the context of corporate social responsibility. *Journal of Business Ethics* 84:277–295.

Jarrell, M., J. Ozymya, and D. McGurrin. 2012. How to encourage conflict in the environmental decision-making process: imparting lessons from civic environmentalism to local policy-makers. *The International Journal of Justice and Sustainability* 18:184–200.

Jensen, R., and V. Uddameri 2009. Using communication research to gather stakeholder preferences to improve groundwater management models: a South Texas case study. *Journal of Science Communication* 8:1–8.

Jones, P., D. Hillier, and D. Comfort. 2013. In the public eye: sustainability and the UK's leading retailers. *Journal of Public Affairs* 13:33–40.

Kent, M. 2013. Using social media dialogically: public relations role in reviving democracy. *Public Relations Review* 39:337–345.

Kock, C., J. Santaló, and L. Diestre 2011. Corporate governance and the environment: what type of governance creates greener companies? *Journal of Brand Management* 49:492–514.

Lenox, M., and C. Eesley. 2009. Private environmental activism and the selection and response of firm targets, *Journal of Economics & Management Strategy* 18:45–73.

Lightfoot, S., and J. Burchll. 2004. Green hope or greenwash? the actions of the European Union at the World Summit on Sustainable Development. *Global Environmental Change* 14:337–334.

Linnenluecke, M., and A. Griffiths. 2010. Corporate sustainability and organizational culture. *Journal of World Business* 45:357–366.

Liska, C., E. Petrun, T. Sellnow, and M. Seeger. 2012. Chaos theory, self-organization, and industrial accidents: crisis communication in the Kingston coal ash spill. *Southern Communication Journal* 77:180–197.

Lovejoy, K., R. Waters, and G. Saxton. 2012. Engaging stakeholders through Twitter: how nonprofit organizations are getting more out of 140 characters or less. *Public Relations Review* 38:313–318.

Lyon, T., and J. Maxwell. 2006. Greenwash: corporate environmental disclosure under threat of audit. *Journal of Economics and Management* 20:3–41.

Mackerron, G., and F. Berkhout. 2009. Learning to listen: institutional change and legitimation in UK radioactive waste policy. *Journal of Risk Research* 12:989–1008.

Mallen, C., C. Chard, and I. Sime. 2013. Web communications of environmental sustainability initiatives at sport facilities hosting Major League Soccer. *Journal of Management and Sustainability* 3:115–130.

Miller, B., and S. Horsley. 2009.Digging deeper: crisis management in the coal industry. *Journal of Applied Communication Research* 37:298–316.

Miller, A., and R. Littlefield. 2010. Product recalls and organizational learning: ConAgra's responses to the peanut butter and pot pie crises. *Public Relations Review* 36:361–366.

Munshi, D., and P. Kurian. 2005. Imperializing spin cycles: a postcolonial look at public relations, greenwashing, and the separation of publics. *Public Relations Review* 31:513–520.

Muralidharan, S., K. Dillistone, and J. Shin. 2011. The gulf coast oil spill: extending the theory of image restoration discourse to the realm of social media and beyond petroleum. *Public Relations Review* 37:226–232.

O'Connor, A., and M. Shumate. 2010. An economic industry and institutional level of analysis of corporate social responsibility communication. *Management Communication Quarterly* 24:529–551.

Omran, M., A. Aziz, and G. Robinson 2009. Investigating households attitude toward recycling of solid waste in Malaysia: a case study. *The International Journal for Environmental Research* 3:275–288.

Pauly, J., and L. Hutchison. 2005. Moral fables of public relations practice: the Tylenol and Exxon Valdez cases. *Journal of Mass Media Ethics* 20:231–249.

Piercy, N., and N. Lane. 2009. Corporate social responsibility: impacts on strategic marketing and customer value. *The Marketing Review* 9:335–360.

Prestin, A., and K. Pearce. 2010. We care a lot: formative research for a social marketing campaign to promote school-based recycling. *The Journal of Resources, Conservation and Recycling* 54:1017–1026.

Reilly, A., and A. Weirup. 2012. Sustainability initiatives, social media activity, and organizational culture: an exploratory study. *The Journal of Sustainability and Green Business* 1:1–15.

Rolland, D., and J. Bazzoni. 2009. Greening corporate identity: CSR online corporate identity reporting. *Corporate Communications: An International Journal* 14:249–263.

Rondinelli, D., and M. Berry. 2000. Environmental citizenship in multinational corporations: social responsibility and sustainable development. *The European Management Journal* 18:70–84.

Roper, J. (2005). Symmetrical communication: excellent public relations or a strategy for Hegemony? *Journal of Public Relations Research* 17:69–86.

Roper, J. 2008. Symmetrical communication: excellent public relations or a strategy for hegemony? *Public Relations Review* 17:69–86.

Samy, M., G. Odemilin, and R. Bampton. 2010. Corporate social responsibility: a strategy for sustainable business success. An analysis of 20 selected British companies. *Corporate Governance* 10:213–237.

Sangle, S. 2010. Critical success factors for corporate social responsibility: a public sector perspective. *The Journal of Corporate Social Responsibility for the Environment* 17:205–214.

Schultze, F., J. Kleinnijenhuis, D. Oegema, S. Utz, and W. van u Atteveldt. 2012. Strategic framing in the BP crisis: a semantic network analysis of associative frames. *Public Relations Review* 38:97–107.

Seitanidi, M., and A. Crane. 2009. Implementing CSR through partnerships: understanding the selection, design and institutionalisation of nonprofit-business partnerships. *Journal of Business Ethics* 85:413–429.

Setthasakko, W. 2009. Barriers to implementing corporate environmental responsibility in Thailand: a qualitative approach. *International Journal of Organizational Analysis* 17:169–183.

Sharma, A., R. Gopalkrishnan, A. Mehrotra, and R. Krishnan. 2010. Sustainability and business-to-business marketing: a framework and Implications. *Industrial Marketing Management* 39:330–341.

Sommerfeldt, E., M. Kent, and M. Taylor. 2012. Activist practitioner perspectives of website public relations: why aren't activist websites fulfilling the dialogic promise? *Public Relations Review* 38:303–312.

The Public Relations Society of America. (2013). What is Public Relations? PRSA's Widely Accepted Definition. Available at http://www.prsa.org/aboutprsa/publicrelationsdefined/#.VTAd0pOnG0U. Accessed April 30, 2015.

Trendafilova, S., and K. Babiak. 2013. Understanding strategic corporate environmental responsibility in professional sport. *International Journal of Sport Management and Marketing* 13:1–26.

Utz, S., F. Schultz, and S. Glock. 2013. Crisis communication online: how medium, crisis type and emotions affected public reactions in the Fukushima Daiichi nuclear disaster. *Public Relations Review* 39:40–46.

Waddock, S. 2008. The development of corporate responsibility and corporate citizenship. *The Organization Management Journal* 5:29–39.

Wakefield, R. 2012. Personal influence and pre-industrial United States: an early relationship model that needs resurgence in U.S. public relations. *Public Relations Review* 39:131–138.

Walton, A., and M. Hume. 2011. Creating positive habits in water conservation: the case of the Queensland Water Commission and the Target 140 campaign. *The International Journal of Nonprofit and Voluntary Sector Marketing* 16:215–224.

Waters, R., and D. Bortree. 2013. Advancing relationship management theory: mapping the continuum of relationship types. *Public Relations Review* 38:123–127.

Waters, R., and N. Tindall. 2011. Exploring the impact of American news coverage on crisis fundraising: using media theory to explicate a new model of fundraising communication. *The Journal of Nonprofit & Public Sector Marketing* 23:20–40.

Waters, R., N. Tindall, and T. Morton. 2010. Media catching and the journalist–public relations practitioner relationship: how social media are changing the practice of media relations. *Journal of Public Relations Research* 22:241–264.

Weaver, K., and J. Motion. 2002. Sabotage and subterfuge: public relations, democracy and genetic engineering in New Zealand. *Media, Culture and Society* 24:325–343.

Weick, K. 2010. Reflections on enacted sensemaking in the Bhopal disaster. *Journal of Management Studies* 47:537–550.

White, C., A. Vanc, and I. Coman. 2011. Corporate social responsibility in transitional countries: public relations as a component of public diplomacy in Romania. *International Journal of Strategic Communication* 5:281–292.

Wrigley, B., S. Ota, and A. Kikuchi. 2006. Lightning strikes twice: lessons learned from two food poisoning incidents in Japan. *Public Relations Review* 32:349–357.

Zyglidopoulos, S., A. Georgiadis, C. Carroll, and D. Siegel. 2012. Does media attention drive corporate social responsibility? *Journal of Business Research* 65:1622–1627.

12

THE ECONOMICS OF ENVIRONMENTAL MANAGEMENT

DAVID TIMMONS
Department of Economics, University of Massachusetts, Boston, MA, USA

Abstract: In this chapter, we describe an economic approach to environmental management, where environmental attributes are quantified and valued in order to identify optimum environmental quality. The typical market model of marginal benefits and marginal costs is modified to account for external costs—unintentional effects of production and consumption like pollution, which are prominent in environmental economics. We introduce benefit–cost analysis as a formal method to compare benefits and costs of environmental policy options, including benefits and costs that occur at different points in time. And since many environmental goods and services do not have market prices, we describe approaches to nonmarket valuation. We also consider equity or fairness of environmental outcomes and sustainability of environmental quality over time.

Keywords: environmental economics, resource economics, ecological economics, externalities, open-access resources, public goods, benefit–cost analysis, discount rate, ecosystem services, nonmarket valuation, environmental justice, sustainability, biodiversity.

12.1 Introduction	289	
12.2 A Market Model	290	
12.2.1 Marginal Values	290	
12.2.2 Markets and Welfare	292	
12.2.3 When Markets Fail: Externalities	293	
12.2.3.1 Open-Access Resources	294	
12.2.3.2 Public Goods	295	
12.3 An Optimal Pollution Model	296	
12.4 Benefit–Cost Analysis	297	
12.4.1 Value and Time	298	
12.4.2 Ecosystem Services	299	
12.4.3 Nonmarket Valuation	300	
12.4.4 Capital Stock Values	302	
12.5 Summary	303	
References	303	

12.1 INTRODUCTION

Many people assume that economics is mostly about money, investment, and banking. While economists study these subjects, the field of economics is much broader. Economics is fundamentally about good management, and the word "economist" in fact derives from the Greek word *oikonomos*, meaning manager or steward of a household. The most basic economics, home economics, was once widely taught in US schools, instructing students in how to manage household needs like food and clothing. This chapter introduces environmental economics, the application of economic methods to managing the world environmental "household."

Economic methods can be used to study many subjects, including the environment (Fullerton and Stavins, 2012). Economists quantify problems and potential solutions, which is possible whenever costs and benefits can be measured. Of course, economics cannot answer every environmental question—it will not provide us with a good estimate of the spiritual value of nature, for example—but there are economic approaches to even such difficult questions. By the end of this chapter, you should know how economists approach environmental management, some of the methods that environmental economists use, and applications and limitations of environmental economics methods.

Many economic problems involve finding an optimum. Consider, for example, forest resources. Forests have many valuable ecological functions, including providing timber, building soil, sequestering carbon, and providing habitat for many species. Clearly, forests are valuable. Yet in much of the world, forests are also under intense pressure, especially for the development of new farmland. And farmland resources have value too. For economists, there is always an **opportunity cost**, the foregone value of a path not taken. An opportunity cost of preserving forests is not developing potential farmland, and an opportunity cost of developing more farmland is loss of forests.

A world with no farmland is not feasible with the current world population: while forests can provide some food resources from hunting and gathering, farmland provides much more food per hectare. But a world with no forests would not be good either. Preserving all potential forests and allowing no farmland is probably a bad solution, and removing all forests is also a bad solution. An economic approach would identify an optimum amount of forest preservation, which is probably neither 100% nor 0%, but somewhere in between. Exactly where depends on the relative values of forest and farmland. To compare these, an economist would attempt to assign monetary values to all the good (and bad) things that farmlands and forests provide. These values depend in part on how much farmland and forest we have already—but more on that later.

Throughout this chapter, we will explore the example of forest preservation to get a feel for how to apply environmental economics to a real-world problem. But the example is arbitrary, and you could use the same processes to consider how much coastal wetland to preserve, how much fossil fuel to extract and burn, how clean the water in a river should be, whether to build a new hydroelectric dam, or whether to impose new emission limits on diesel trucks. All of these questions involve environmental opportunity costs, and all have values that we can quantify and analyze.

In this chapter, I use the term environmental economics broadly. Some authors use the term "natural resource economics" for questions about environmental inputs to the economy and "environmental economics" for problems of pollution or outputs from the economy. While the input and output problems have some differences, similar methods are used in both subfields, so for convenience, I combine them here. The environmental economics presented here is based mostly on **microeconomics**. While some background in that subject is helpful, no background is required to understand this chapter. The "micro" of microeconomics refers to the unit of analysis, which is mostly people and firms. Microeconomics is specifically about how people, firms, and governments can make optimal decisions. **Macroeconomics** examines broader economy-wide issues like growth and development, unemployment, and inflation—issues we will not address here. The newer field of **ecological economics** ascribes a more central economic role to ecological processes (Daly and Farley, 2010). Ecological economists study whether there is some optimal size for the economy as a whole, and have also made significant contributions in valuing ecosystem services, a topic we discuss later in this chapter.

12.2 A MARKET MODEL

Ever since people have bought and sold goods in markets, goods have had prices, and what determines these prices is probably the original economic question[1]. Consider, for example, the diamond–water paradox: why does water, on which all life depends, have such a low price, while diamonds, which are mostly just ornamental, have very high prices? Adam Smith raised this question in 1776: "Nothing is more useful than water; but it will purchase scarce any thing; scarce any thing can be had in exchange for it. A diamond, on the contrary, has scarce any value in use; but a very great quantity of other goods may frequently be had in exchange for it" (Smith, 1776). Smith's explanation for this paradox was not very satisfactory, and it was more than a century after Smith before economists completely resolved this paradox.

12.2.1 Marginal Values

The answer to the diamond–water paradox relates to scarcity: diamonds are expensive because they are scarce, while water is cheap because there is so much of it (in many places). But as the abundance of something changes, its value changes also. How much would you pay for water if you were dying of thirst in a desert? Most of us would trade our diamonds (and anything else we owned) for just one cup of water in this extreme situation. Greater scarcity implies a higher price. This principle holds for most market goods—think of Picasso paintings, original editions of Superman comic books, or rare baseball cards.

More specifically, the **marginal value** of almost anything changes with the quantity available. Marginal literally means "at the edge." In an economic context, a marginal value is the value of another unit, given how much of something we already have. For me, the marginal value of cups of coffee is a good example, since I am quite addicted to coffee. The marginal value of the first cup of coffee in the morning is very high. I know that without that first cup, I will be unproductive and grumpy. If necessary, I would pay a very high price for that first cup, maybe $10. But this does not mean I would be willing to pay $10 for every cup of coffee. Having had that first cup in the morning, the marginal value of a second cup is much lower: a second cup is nice but not essential for my productivity and attitude. I would only pay a normal market price for a second cup of coffee. The marginal value of a third cup of coffee is about zero for me: I would drink a third cup to be polite if I were visiting someone and they offered me one, but I would not buy a third cup. And I would have to be paid to drink a fourth cup, since I know this would just give me a headache.

The coffee example is extreme because of the addictive nature of caffeine, but the general principle holds for most goods in the marketplace: the more of something we have already, the less valuable is having even more of it.

---

[1] Any introductory microeconomics textbook provides a longer description of this standard market model. For example, see Goodwin et al. (2014)

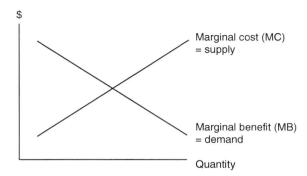

FIGURE 12.1 Marginal benefits and marginal costs.

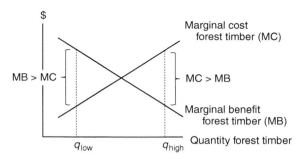

FIGURE 12.2 Forest timber marginal benefits and marginal costs.

A household's first refrigerator can be very important for quality of life, even transformational. A second refrigerator might be convenient but is not critical. A household with ten refrigerators would probably have some of them taken away (space being an opportunity cost of owning refrigerators).

With water, a small amount is necessary just to keep living, so the marginal value is very high for the initial quantity. After satisfying our thirst, we might use the next quantity of water for washing ourselves, our clothes, and our dishes. Cleanliness is good, but not as important as sustaining life. After washing, we might use water for watering a lawn or filling a swimming pool. This is not critical at all, and the marginal value of water in a pool is lower than water for drinking or washing. Marginal benefits of things decrease as we get more of them, because we fulfill our most important needs first. This is one of the most important insights in modern economics, and thinking at the margin is an important economic skill. Quantity and value are always intertwined.

The market for any good includes many people who want to buy things, and these people also have different willingness (and ability) to pay. While my marginal value for a first cup of coffee is high, a less caffeine-addicted person would have a lower willingness to pay. The market demand for coffee (or any good) includes the preferences of everybody who might buy in that market. The principle that marginal benefit decreases with quantity for individuals is also true for the market as a whole.

Graphically, as shown in Fig. 12.1, this results in a downward sloping (from left to right) **marginal benefit curve**, which we also call a market **demand curve** (Goodwin et al., 2014). As the market quantity increases (moving to the right on the horizontal axis), the marginal benefit decreases (moving down on the vertical axis). This is both because marginal benefits decline with quantity consumed for all the individuals in the market and because if prices go down, more individuals buy and quantity demanded increases. The relationship between lower prices and greater quantities demanded is one of the most reliable results in all of economics and is called the **law of demand**. This is half of the story of how prices are set in the market, the marginal benefit or demand side.

The other half is the marginal cost or supply side, which follows a similar logic. Instead of considering benefits, we consider the cost of resources it takes to provide each additional unit in the market. Marginal cost also changes with quantity, but in this case, marginal cost rises with greater quantities. Consider what it takes a city to provide drinking water. Many cities were built near springs, where clean groundwater naturally surfaces. Supplying spring water would have a low marginal cost, since the water is already at the surface and is ready to drink. If the city grew, demand for water might exceed the supply of natural springs. The city could drill wells to provide more water, but the marginal cost would be higher than for water from the springs, since drilling wells requires resources, and the city would have to pump the water to the surface. To supply still greater demand, the town could use water from a nearby river, but this would likely need to be filtered and treated before drinking, and the marginal cost would once again rise. The marginal cost to supply almost anything is rising, because we use the lowest-cost resources first. Though alternatives are usually available, they are more expensive. In Fig. 12.1, the **marginal cost curve** slopes upward (from left to right). As the quantity supplied increases (moving to the right on the horizontal axis), the marginal cost increases also (moving up on the vertical axis). The marginal cost curve is also called a market **supply curve**.

Marginal benefits and marginal costs apply in our forest example. In Fig. 12.2, the horizontal quantity axis is forest timber production, which could be measured in cubic meters or metric tons of timber. At the left side of the graph, the quantity of timber produced is very low, perhaps representing a preindustrial situation. With little timber available on the market, the marginal benefit of getting more timber is very high: timber is needed for houses, schools, and bridges. And the marginal cost of getting timber is very low, because our hypothetical preindustrial forest covers the landscape, is easily accessed, and loggers get to cut the highest-quality, easiest-to-mill trees. At q_{low}, the marginal benefit of using a low quantity of timber is greater than the marginal cost of supplying it (though we have not yet considered environmental costs—more on that shortly). On the other hand, at right side of the quantity axis, there is much forest timber on the market and correspondingly less forest cover on the landscape. At q_{high}, the marginal benefit of getting even more timber is low, since we have presumably met our most pressing timber needs already. And with a greater quantity of timber, the marginal cost of getting timber is very high:

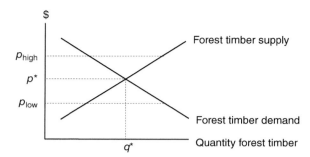

FIGURE 12.3 Forest timber optimum market quantity and price.

having used most of the forest resource, the remaining trees are in inaccessible locations and of lower quality, requiring more costly processing. At q_{high}, the marginal cost of forest harvest is greater than the marginal benefit. Marginal benefits and marginal costs of producing timber depend greatly on how much timber is already being produced. This is true for most goods and services in the economy.

12.2.2 Markets and Welfare

To return to the original economic question, it is the combination of marginal benefits (or demand) and marginal costs (or supply) in the marketplace that determines prices. Specifically, for any good, an **equilibrium price**—a price that would prevail if no market conditions changed—occurs where the marginal benefit equals the marginal cost. In Fig. 12.3, this is the price level where the marginal benefit and marginal cost curves intersect (read from the intersection across to the vertical price axis to find the equilibrium price, p^*).

Why is this so? If the price for a good is above equilibrium (too high), the quantity supplied exceeds the quantity demanded, and there will be a surplus (p_{high} in Fig. 12.3). The surplus means not all goods can be sold. To correct this, sellers reduce their prices until the quantity demanded equals the quantity supplied, and all goods sell. If the price for a good is below equilibrium (too low), the quantity demanded exceeds the quantity supplied (p_{low} in Fig. 12.3). There is a shortage, and some buyers cannot find goods to buy. Sellers look for alternative ways to bring additional goods to market (at higher cost). The quantity supplied increases and the quantity demanded decreases until they are equal, and marginal benefit equals marginal costs. The market's own mechanisms act to steer prices toward an equilibrium where supply equals demand and the marginal benefit of providing another unit equals its marginal cost.

The intersection of marginal benefit and marginal cost curves also identifies an **equilibrium quantity** (in Fig. 12.3, read down from the intersection to the horizontal quantity axis to see equilibrium quantity q^*). This is the quantity of a good provided by the market when demand equals supply and marginal benefit equals marginal cost.

Surprisingly, in a perfect market, this is also the quantity of goods that maximizes total benefit to society. To make this conclusion, assume for the moment that all marginal benefits and all marginal costs are reflected in the demand and supply curves (next we will see what happens when this is not true). If the quantity were less than q^* (like q_{low} in Fig. 12.2), the marginal benefit of another unit would exceed its marginal cost. Someone is willing to pay more for a good than it costs to provide the good, and total social welfare can be increased by providing that good. On the other hand, if the market provided more than q^* (like q_{high} in Fig. 12.2), the marginal cost is greater than the marginal benefit. Nobody is willing to pay as much as it costs to produce this quantity. Society is better off with a lower quantity. If social welfare improves by increasing quantities less than q^* and decreasing quantities greater than q^*, then q^* must be the point of greatest welfare.

This is the miracle of a perfect market: it makes society as well off as it can be with respect to a market good and does this almost by itself. A perfect market's own incentives lead to providing optimal quantities. Markets determine production and consumption of everything from candy bars to condominiums and self-correct any shortages or surpluses that may arise. Consumers and producers both use a single point of information, the price, to make decisions about how much to consume and how much to produce, and the result is that production invariably equals consumption. In 1776, Adam Smith called this the **invisible hand**, since it seems that some unseen force is causing demand to equal supply and social welfare to be maximized. This is why most economists love markets and why some people advocate little government involvement in markets. Where production and consumption decisions have been determined by a central administration, as in the former Soviet Union, results have been less than satisfactory.

But note that I carefully used the adjective "perfect" to describe welfare-maximizing markets. There are a number of technical conditions required for perfect markets, and most of these conditions do not completely hold in real markets. In environmental economics, we are particularly concerned about situations where a price, the single point of information that coordinates both consumption and production, does not provide all of the environmental information about a good. This is the case of an external cost or simply an **externality**, which we discuss in Section 12.2.3.

Even a perfect market makes no claim to **equity** or fairness. While we can prove in a mathematically rigorous way that a perfect market maximizes society's total net benefits, this proof combines all benefits regardless of who receives them. Some people can benefit much more than others. Initial resource allocations—advantages that individuals having starting out in life—have large impacts on the final distribution of resources in a market system, and free market economies often have large disparities of income and wealth. If a society values equity, it must have policies to achieve this. Environmental damage can also result from unequal income distribution (Boyce, 2002). The world's poorest people often do environmental harm through survival-level decisions they must make, for example, when subsistence farmers are forced to cultivate steep, erodible hillsides in order to eat.

As with equity in general, markets do not ensure sustainability, or equity between people living on Earth now and in the future (Costanza et al., 1997; Solow, 2012). In maximizing its own welfare, each generation can leave the world in worse environmental condition for all future generations, which is a fairness issue to unborn generations. New knowledge produced in each generation can improve the welfare of future generations: for example, the food crops developed by our ancestors have great benefit today. But there is no assurance that such benefits will always outweigh costs of a degraded environment. As with equity, if society values sustainability, it must impose conditions on the market to achieve this. The market has no self-correcting mechanisms to provide either intergenerational equity or intragenerational equity.

Thus, markets have many potential problems: they are subject to externalities and other market imperfections, they do not provide equity, they do not ensure sustainability, and many things important to society are not traded in markets. Yet environmental economists commonly use the market model described in this section, because in spite of its limitations, it is useful in many contexts. The most important elements of this model are benefits identified separately from costs, marginal benefits that decline with greater quantities, and marginal costs that increase with greater quantities.

12.2.3 When Markets Fail: Externalities

In the earlier timber market example, we assumed that all marginal benefits and all marginal costs were reflected in their respective curves. But on its own, a market recognizes only the private benefits and costs of market participants. Let's say I own a forest where I could harvest timber and you are builder who needs timber to build a school. We could reach a sales agreement based on prevailing prices. These prices only reflect benefits and costs for the two of us (and for other buyers and sellers like ourselves in the market), though nonmarket participants may also bear costs or get benefits. For example, if there is a stream running through my forest, soil erosion after I harvest could foul the stream and destroy fishing for people who live downstream. This loss of downstream fishing is a real cost to society, so the total cost of harvesting my trees is more than just my cost of cutting. We call the downstream impact an external cost since it is not a direct cost to either you or me, but to third parties (external parties) who happen to live downstream and are not involved in our timber transaction. External costs are called **negative externalities**, while external benefits are called **positive externalities** (see Harris and Roach, 2013, chapter 3).

When negative or positive externalities exist, these are not reflected in a free market price, and the free market fails to deliver a welfare-maximizing quantity at an appropriate price. In Fig. 12.4, we still have private marginal cost and benefit curves as before, but now we have a social marginal cost curve that indicates the true cost of supplying timber to the market. This includes the marginal costs of any negative externalities from harvesting timber (like loss of fishing). The social marginal cost curve lies above the private marginal cost curve, since there are social costs in addition to the private ones (true costs are higher than private costs). The optimum, welfare-maximizing quantity and price are now identified by the intersection of the *social* marginal cost curve (the true cost curve) and marginal benefit curve. As you can see from the figure, the welfare-maximizing quantity is lower, and the price is higher than market quantity and price (q_{market} and p_{market}), where only private benefits and costs are considered. This is a general result: when negative externalities are involved, a free market provides too high a quantity of goods at too low a price. Social welfare would be improved with less production and lower costs from externalities like pollution.

Environmental externalities are present in many markets. In the timber market example, other negative externalities from timber harvest include release of carbon sequestered

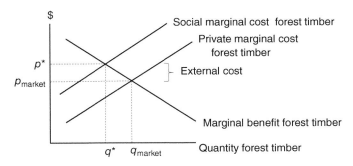

FIGURE 12.4 Forest timber optimum quantity and prince with negative externality.

in the forest, destruction of forest habitat, and negative aesthetic impacts for those who view the forest from offsite. More broadly, any cost of pollution or loss of resources borne by society as a whole rather than by those involved in a market transaction is considered a negative externality. The market fails to deliver an optimum quantity because the price does not reflect the full marginal cost of making a resource available. Adam Smith's invisible hand of the market is unable to account for effects that fall outside of the market. Some type of social intervention, usually by a government, is required to ensure that all costs are considered, since private buyers and sellers have little incentive to consider costs borne by others. Identifying and quantifying such environmental externalities is a major focus of environmental economics. In order to make appropriate environmental management decisions, we must know the true costs of producing goods and services that society needs. In some cases, these external costs are large and significant and in other cases not, but we cannot know this until such costs are quantified.

Two types of externalities are so ubiquitous in environmental management that they warrant special attention: open-access resources and public goods. Open-access resource problems are a result of negative externalities, while public goods issues stem from positive externalities.

12.2.3.1 Open-Access Resources

In 1968, Garrett Hardin wrote "The Tragedy of the Commons," now famous in the environmental literature. Hardin described a hypothetical pastoral society where grazing land was commonly owned (Hardin, 1968). Herders were free to graze as many animals as they wished on the common pasture. And for every herder, it was optimal to add more animals: someone adding a cow got the full private benefit of raising and later consuming or selling the cow. Each additional cow also degraded the pasture just slightly. But the cost of degradation was an externality for any individual herder: most of the pasture degradation cost was borne by the other herders and by the community as a whole. Of course, this was the situation for everyone, so in Hardin's parable, all the herders kept adding animals until they exceeded the carrying capacity of the pasture and the pasture grazing system collapsed (perhaps due to severe soil erosion). Instead of the invisible hand guiding the market to optimal social welfare, the unfettered market led to ecological collapse. Such are the problems of open-access resources (Goodstein, 2011, section 3.1).

We describe open-access resources as being **rival** but **nonexcludable**. Rival means that the more each person uses the resource, the greater the negative externality for others. Each user affects other users. Nonexcludable indicates that it is difficult to prevent people from using the resource, because it is freely available.

Since Hardin's essay, scholars have noted that the tragedy lay not in the common ownership of the pasture, but rather in the uncontrolled use of the resource, or in its open access. Hardin might have more accurately titled his essay "The Tragedy of Open-Access Resources." There are many examples of communally owned resources that did not in fact collapse over centuries of use, including the traditional pastures Hardin described. But societies that successfully managed common resources had mechanisms to prevent overuse, either formal quotas or informal sanctions against community members who overused the resource. Elinor Ostrom won the 2009 Nobel Prize in economics in part for her work describing ways that communities can sustainably manage commonly owned resources (Ostrom et al., 2012).

In spite of having many examples of successfully managed common resources, open-access externalities are still widespread in the world. A typical example is world fisheries, with overfishing and fishery collapse common around the world (see Box 12.1). As in Hardin's example, each fisher reaps the full benefit of every fish caught. There is a cost of leaving fewer fish in the ocean, curtailing fish reproduction, and making the remaining fish harder to catch, but this cost is an externality borne by the fishing industry as a whole. Each fisher's incentive is to catch as many fish as possible, while others suffer the consequences. These individual incentives of an open-access fishing resource lead every fisher to behave in the same way, causing a fishery to collapse, which can result in almost complete loss of fish for everyone. Today, some world fisheries are in fact sustainably managed, and a number of government policies have been effective in preventing overfishing. While these policies differ, all of them somehow limit access to potential fishers and change the ability or incentive of fishers to impose negative externalities on the rest of the fishing industry.

Forests can also be open-access resources if property rights to forest land do not exist or are not enforced. If anyone can cut trees in a forest (if there is open access), each logger has an incentive to cut as many trees as possible as quickly as possible, cutting especially the most valuable trees. Though this kind of cutting may reduce a forest's long-run yield potential (or destroy it completely), those costs are left to others. Every logger knows that if she does not cut the trees, someone else will, so no logger has an incentive to conserve or to manage the forest for optimum long-run production.

Today, the most prominent open-access externality problem is the open-access nature of the atmosphere with respect to greenhouse gas emissions. If I burn a gallon of gasoline in my car, I derive the full benefit of gasoline-powered travel, while the external cost of CO_2 emissions is spread to everybody else in the world. This is true for every individual as well as for every country.

> **BOX 12.1 Overexploitation of World Fisheries**
>
> Ocean fishing has been called the "last hunting economy" (Wasserman, 2001), with dynamics similar to traditional hunting and gathering, but using modern technology to hunt the prey. Many fisheries around the world show remarkably similar patterns of overuse, due to incentives for individual fishers to externalize costs of overfishing to others. Fishery collapse, or greatly reduced catch level, occurs because too few fish are left behind to produce succeeding generations. The United Nations' Food and Agriculture Organization (FAO) has kept fishery statistics since 1950, and an analysis of these data shows that about 24% of world fisheries have collapsed at some point, some quite suddenly (Mullon et al., 2005). While collapse is not in the interest of any of the fishers, their collective actions often lead to this outcome.
>
> For example, according to FAO data, the US and Canadian groundfish catch (cods, hakes, haddocks) in the northwest Atlantic averaged 537,000 tons per year from 1950 to 1970, peaked at 727,000 tons in 1982, then collapsed to just 82,000 tons in 1995, and stayed near that level through 2011 (http://www.fao.org/fishery/statistics/global-capture-production/en).
>
> World fisheries have been under considerable pressure since the mechanization of world fishing fleets in the twentieth century (Pauly et al., 2002). While historic sailing and rowing boats were limited in how many fish they could catch, the new motorized boats could use trawling and other methods to net numbers of fish that were previously unattainable. New sonar, satellite, and navigational technologies also made it easier to find fish in the ocean. The only way to manage fisheries sustainably is to limit the effort (equipment, labor, and technology) that goes into catching fish.
>
> Most of the collapsing fisheries are not even completely open-access resources; the great majority of fisheries are in exclusive economic zones (national waters) of fishing countries (Pauly et al., 2002). But the incentives for overfishing are so great that collapse continues to occur even with most fishing grounds now under government jurisdiction and regulation.

Each country's incentive is to continue enjoying the benefits of burning carbon fuels while letting other countries (and future generations) bear most of the cost of climate change. As in Hardin's example, unless altered by common agreement, these incentives will lead to the collapse of a climate conducive to human civilization. The "tragedy" of Hardin's title, as in an ancient Greek tragedy, alludes to the completely foreseeable consequences of a situation and the seeming inevitability of an awful conclusion. But humans have avoided the tragedy of open access in other contexts and are capable of solving the climate change crisis as well. There is a particularly urgent need for action at the international level, where coordinating policies and aligning incentives are most difficult.

12.2.3.2 Public Goods

A final important case of externalities concerns public goods (Goodstein, 2011, section 3.2). We define public goods as being nonexcludable (like open-access resources) but **nonrival**: one person's use does not prevent another's use of the same good. For example, if I own a forest, I might manage it to maximize carbon sequestration. Storing more carbon in forests can reduce the impact of carbon emissions and help to stabilize the climate. This is a positive externality: my decision to manage for carbon sequestration benefits other parties, without my necessarily intending to do so. Because the benefits of public goods are nonrival, many people can benefit from the same public good, and the total value of public goods can be very large—everyone on Earth benefits from carbon sequestration.

Note that the term public good indicates only that a good is nonexcludable and nonrival and does not indicate that the government provides the good. Public universities, for example, are often provided by the government (at least in part) but are both excludable (one cannot attend without being admitted) and rival (admitting one student means rejecting another), so public universities are not a public good. Similarly, radio broadcasts are a public good because they are nonexcludable (it is difficult to prevent someone from tuning in a radio broadcast) and nonrival (everyone can use the same radio broadcast), but most radio broadcasts are provided by the private sector rather than the government. "Public good" is an economic term for a nonexcludable, nonrival good and is not defined by who provides it.

The main problem with public goods is that there are not enough of them. There are two reasons for this. As with open-access resources, the first problem relates to economic incentives. Because a public good is nonrival, everyone benefits if anyone provides it. Instead of making an effort to sequester more carbon in my forest, I would be better off if my neighbor made the investment. I would get the carbon sequestration benefit without any of the cost. Everyone has the same incentive to wait for somebody else to provide a public good. This is known as the **free-rider problem**. Secondly, because public goods are nonexcludable (it is difficult to prevent anyone from using them), businesses cannot make a profit by selling public goods. The market depends on excludability, since people will not normally buy things they could get for free. The market does not provide public goods. Many environmental goods have a public goods character: clean air, clean water, and scenic landscapes. Public goods can have very large values but are not provided by the market—a significant market failure due to positive externalities. Box 12.2 discusses biodiversity, which may be the prime world example of an environmental public good.

BOX 12.2 Biodiversity as a World Public Good

Biodiversity of world ecosystems continues to decline at an alarming rate (Butchart et al., 2010). Species diversity has a number of tangible benefits for humans, including providing genetic material for new crop varieties, new drugs, etc. Preserving other species on Earth is also an ethical question about existence rights (see Chapter 17 in this text), and people may value biodiversity for its own sake. There is increasing evidence that biodiversity loss interferes with fundamental ecosystem functioning, reducing ecosystem services of value to humans (Cardinale et al., 2012).

To preserve biodiversity, important habitats must be conserved and not developed (an opportunity cost of biodiversity). Yet biodiversity is a classic example of a nonrival, nonexcludable good: any country that preserves habitat for biodiversity benefits people in all other countries, and countries that choose not to preserve biodiversity may still benefit from the variety of genetic material found elsewhere. While all humans depend on biodiversity, each country has an incentive to let others protect it—an international free-rider problem.

This public goods character of biodiversity is one reason to expect that biodiversity is underprovided relative to its actual value. Martin (2013) develops a dynamic economic model suggesting that losses of biodiversity in the present will make preserving biodiversity even more difficult in the future or, alternatively, that current nonsustainable behavior makes a sustainability goal even more difficult to achieve later. These results provide more evidence suggesting that biodiversity should be protected now.

12.3 AN OPTIMAL POLLUTION MODEL

The model introduced earlier of demand and supply (or marginal benefits and marginal costs) developed from observations of real working markets. But the concepts are quite general, and environmental economics extends this model into some nonmarket contexts. One example is a model of pollution reduction, which is not normally bought and sold in markets (Field, 2012, chapter 5).

How clean should the air in your region be? A typical optimum solution is neither completely clean nor completely polluted. Some pollution inevitably results from creating things that we need or want: food, houses, and concerts. Completely eliminating pollution would mean forgoing too many desirable things; in economic terms, the marginal cost would be higher than the marginal benefit. But with no pollution control at all, the marginal cost from death and disease would be higher than the marginal benefit from the last increment of production. As in the market model, the optimum amount of pollution reduction is where the marginal benefit of pollution reduction equals its marginal cost.

Figure 12.5 shows a model of pollution reduction that looks very much like our market model in Fig. 12.2. Note that we have defined the good as pollution reduction, since pollution is bad, and in order to maintain our previous marginal benefit and marginal cost curves, we need to redefine pollution as a good—pollution reduction. At the left side of the graph (q_{low}), there is little or no pollution reduction. The marginal benefit of pollution reduction would be very high in this polluted environment—people would be willing to pay a lot for a healthier and more pleasant environment. At the same time, the marginal cost of pollution reduction can be very low in this situation. Companies often have some options to reduce pollution that are very inexpensive, and may even save them money, as indicated by the below-zero marginal costs in Fig. 12.5. But as the environment gets cleaner with more pollution reduction (moving from left to right on the horizontal axis), the marginal benefits decline and the marginal costs increase. Once the environment is already fairly clean, the marginal benefit of making it even cleaner may be rather small. And after companies have used all their low-cost options to reduce pollution, the remaining options may be quite expensive. At q_{high}, the marginal cost of pollution reduction exceeds its marginal benefit. The optimum stopping point in pollution reduction is where marginal benefit and marginal cost curves intersect and

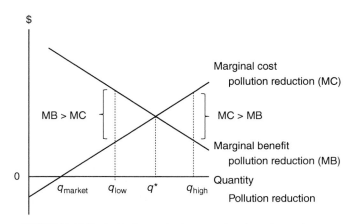

FIGURE 12.5 Pollution reduction marginal benefits and costs.

have equal values. More pollution reduction would mean cost greater than benefit for each additional unit of pollution removed, and less pollution reduction would mean that not all the potential net benefits had been gained.

Unlike the market model (Fig. 12.1), in the pollution reduction model (Fig. 12.5), the intersection of the marginal benefit and marginal costs curves does not represent a market equilibrium—an important difference. There are no market forces that would cause pollution reduction to occur at q^*, the optimum quantity for society, since all of the benefits of pollution reduction are externalities for those who create pollution. Most economists would assume that a market equilibrium would be at q_{market}—companies implement all the pollution reduction measures that save them money and no more (though this typical economic assumption likely underestimates the real-life altruistic behavior of at least some companies in most markets).

Getting to the optimal q^* level of pollution reduction requires some kind of collective action. In modern societies, this is normally accomplished by government regulation. In the United States, the Environmental Protection Agency (EPA) could simply issue rules that would limit pollution to the optimum level. A more market-based approach would be to impose a pollution fee approximately equal to the marginal cost of pollution for society. This is known as a **Pigouvian tax**, after economist Arthur Pigou who developed the idea. Paying the pollution fee is a cost of business for a company, and a company's incentive is to produce a quantity of goods where the marginal benefit of production (the sale price) equals the marginal cost of production, which now includes the cost of pollution (represented by the fee). We say the pollution fee internalizes the company's external cost of pollution, since the pollution cost now enters the decision-making process of a profit-maximizing company.

Subsidies are another policy alternative for achieving efficient environmental outcomes. Subsidies can be thought of as negative taxes, and like taxes, they can be used to represent values that would otherwise be absent from the market. Instead of taxing pollution, government could subsidize pollution reduction to encourage firms to emit the efficient quantity of pollution. Subsidies are particularly important for public goods like new environmentally friendly technology, which once developed can benefit everyone. While subsidies and taxes are market-based solutions, in that they create incentives similar to those in a free market, we must emphasize that free markets will not impose pollution fees or provide subsidies on their own. In these cases, government action is required to maximize society's welfare.

As with market benefits, environmental costs and benefits are not necessarily distributed equally or fairly throughout society—some members of society may bear much greater environmental burdens than others. For example, there is considerable evidence that areas where poor people live are more likely to be polluted than areas where rich people live. This topic is known as environmental justice (Tietenberg and Lewis, 2012, chapter 19). If society values environmental justice, government must take action to provide this. In the United States, the EPA includes an Office of Environmental Justice (www.epa.gov/environmentaljustice).

12.4 BENEFIT–COST ANALYSIS

In the market model (Fig. 12.4), maximum welfare occurs when marginal benefit equals social marginal cost, with all externalities included. While this is the ideal, much data is required to apply this standard in practice. We need to know the marginal values of all relevant goods and services, including marginal values of all externalities. And even though we recognize that marginal value changes with quantity, we rarely have estimates for full social marginal cost and marginal benefit functions at all relevant quantities.

Benefit–cost analysis is a simpler and more typical way to inform environmental management decisions, which is consistent with our welfare maximization model (Goodstein, 2011, chapters 8–10; Arrow et al., 2012). In benefit–cost analysis, we first define some policy proposal and estimate benefits and costs to society of implementing the proposal. Then we divide benefits by costs to get a benefit–cost ratio. If benefits are greater than costs, the value of the ratio is greater than one, indicating the project is worthwhile.

Though a benefit–cost ratio greater than one justifies a project, it does not reveal whether a project produces maximum benefit, that is, whether marginal benefit equals marginal cost. Thus, a further recommended step is sensitivity testing or changing the proposal slightly to see whether the benefit–cost ratio grows. For example, in a pollution control proposal, we could check benefits and costs of restricting emissions slightly more. If the benefit–cost ratio grows, this indicates stronger controls might be justified. This approximates a result where marginal cost equals marginal benefit by using a discrete number of tests at specific quantities. Benefit–cost analysis is widely used to develop environmental policies, for example, by the EPA.

With benefit–cost analysis, we must also ask who gains benefits and who incurs costs. As an extreme example, a project could have a benefit–cost ratio greater than one but have all the benefits accruing to a single individual, with costs spread out to millions of others. Most people would consider such a project unfair, even if total benefits exceeded total costs. But we do not necessarily need to weigh all benefits and costs equally.

The Stern Review (2006) was one of the first benefit–cost analyses to conclude that the benefits of controlling climate change exceed the costs of controlling climate change. The study reached this conclusion in part through its treatment of effects on people with different incomes. The benefit or utility that people derive from income was assumed to be equal to the logarithm of income. This says that utility from income increases at a decreasing rate, or that one dollar has more value to a poor person than to a rich person who already has many dollars. Thus, income losses experienced by the poor have greater weight in the analysis than losses by the rich, and potential losses from climate change have more weight than potential gains. Given these and similar assumptions, the Stern Review found unambiguous net benefits from addressing climate change now. Note that this conclusion rests in part on equity principles that were incorporated in the analysis.

12.4.1 Value and Time

As in the forest example earlier, many economic problems involve things that happen at different points in time. Costs or benefits incurred now are often not equivalent to the same costs or benefits incurred in the future. In this discussion, I am disregarding inflation, which just changes our measure of value. Like the length of a kilometer changing, with inflation, a 2010 dollar does not have the same value as a 2020 dollar, and economic studies must adjust for this. Here, I refer to something more fundamental, to differences in true value depending on when things happen.

For most people, getting something later has less value than getting the same thing now, and how much later can make a big difference in value. For example, getting $100 now is worth more than getting $100 five years from now. Why? If I had the $100 now, I could invest it to earn interest and have more than $100 five years from now. Interest rates broadly reflect what occurs with underlying physical assets: if I had a potato field now, in five years I would still have the field, plus the value of five years of potato crops. Waiting means giving up potato harvests until I get the field.

Present value is an economic tool for comparing costs or benefits that occur at different times (Conrad, 2010, section 1.3; Goodstein, 2011, sections 6.3–6.6). As the name implies, present value is what something is worth now, though the value may not be received until sometime later. It works like compound interest in reverse, essentially calculating the compounded interest given up by waiting until later to get something. The formula for present value (PV) is

$$PV = \frac{FV_t}{(1+r)^t}$$

where FV is future value, or the value that will be realized at some time in the future, t is how many years we must wait to realize the future value, and r is called the discount rate. A discount rate works just like an interest rate, except that we use the term "discount" instead of "interest" when calculating present values. At any discount rate greater than zero, present value is less than future value: we discount the future value by some amount because we have to wait to get it.

For example, at a 5% discount rate, the present value of getting $100 five years from now is

$$PV = \frac{\$100}{(1+0.05)^5} = \$78.35$$

If you try using different numbers in the present value formula, you will find that two things greatly affect the present values of future costs or benefits: how many years from now something happens and the discount rate used. Higher discount rates and more years into the future both reduce present values. For example, the present value of $100 5 years from now at a 10% discount rate is $62.09, PV of $100 in 50 years at 5% is $8.72, and PV of $100 in 50 years at 10% is just $0.85.

Returning to the forest harvest example, you can see that with conventional discount rates of perhaps 2–10%, the present value of any harvests after 50 years is very small and thus has little impact on the value of my forest asset. In fact, if the forest grows at a rate of 2% per year and I can get a 5% return on another investment (which is probably the basis for my discount rate), I would be financially ahead if I cut down all the trees and put the revenue in the other investment that grows faster than the forest.

While harvesting everything now may be financially optimal for me, it cannot be simultaneously true for every forest in the world, since society must have forest products on an ongoing basis. This is true for many kinds of environmental assets, or **natural capital**. Sustainability requires preserving natural capital, especially for renewable resources like forests and fisheries. Preserving these assets for future generations is really an ethical decision, a statement that unborn generations have rights too. While discounting future values is a reasonable way for individuals and businesses to allocate resources over their own lifetimes, effects on future generations are an externality, since unborn people have no voice in today's decisions. Again, public policy is needed to correct this externality and to ensure sustainability.

Giving equal weight to present and future generations can be accomplished with a 0% discount rate. While this is unusual in economic analysis, it can be appropriate when considering sustainability, especially for the most important natural capital stocks that can provide flows of benefits in perpetuity. With a low discount rate, I would have to consider the long-term impacts of harvesting my forest, and I would only harvest in a way that preserved the flow of forest benefits for future generations' use. While zero discount rates are not often used in practice, near-zero rates or discount rates that decline for values further in the future are increasingly recognized as necessary for sustainability.

Incorporating the concept of present value, the simple benefit–cost ratio becomes

$$\frac{B}{C} = \frac{\sum_{t=1}^{T}\left(B_t/(1+r)^t\right)}{K + \sum_{t=1}^{T}\left(C_t/(1+r)^t\right)}$$

where B is the benefit, K is the initial capital cost, C is the operating cost over time, and, as before, r is the discount rate and t is the number of years in the future. Present values for each future cost or benefit must be calculated separately for each year in which they occur. Initial capital costs K are incurred in the present, so we do not need to calculate their present value. All present values of costs and benefits are then summed, and present value of benefits is divided by present value of costs to get the benefit–cost ratio, as before. The choice of discount rate is a critical question in benefit–cost analysis, and changing discount rates often changes results. Sustainability concerns suggest using low or zero discount rates in some contexts, giving similar consideration to present and future welfare.

12.4.2 Ecosystem Services

The essence of environmental economics is applying economic theory and methods to environmental issues beyond the market where the economics discipline developed (Scorse, 2010). This usually involves assigning values to some aspects of the environment. Unless we assign values in dollars (or some other currency) to environmental goods and services, we have no way to conduct benefit–cost analysis or to optimize decisions where market transactions have environmental opportunity costs. Unless we estimate the value of forest habitat, for example, we have no way to decide whether the value of timber provided by cutting a forest exceeds the value of lost habitat or to decide under what conditions we will harvest forests.

Assigning values or **monetizing** environmental goods can be controversial for several reasons. Economists may not completely trust these environmental values, because they are not usually based on prices observed in markets (more on valuation methods in the following), and thus, these prices often have high levels of uncertainty. Environmentalists may contend that it is inappropriate or even immoral to place prices on things that are ultimately priceless. Part of this objection may stem from using dollars to measure value. It may help to think of prices as measures of importance or, in a statistical sense, as weights. The more important something is, the more weight we give it in our decisions, and the higher its price. A price is simply a device that allows us to compare disparate things and help us optimize decisions. If we did not use prices, we would still have to decide how to weigh different criteria in making environmental decisions.

For some people, including me, nature also has **intrinsic value**. This is the idea that ecosystems and species have rights to exist and have their own values that humans cannot fully comprehend and certainly cannot fully estimate. Many of the world's religious traditions support this view, though they may not express the view in economic terms. But even if we accept this idea, we must also acknowledge that all valuation is **anthropocentric** or conducted from a human point of view. We have no access to values other than the ones we create. If we ascribe intrinsic value to nature, then our environmental value estimates will always be **lower bounds**, or minimum estimates of true value, with true value always being somewhat higher but unknowable. From a practical standpoint, these lower-bound estimates are often useful in making public policy choices. Some noneconomic policy decisions are also needed to reflect values we cannot completely understand or measure: for example, wilderness preservation is a way to retain intact biomes without fully understanding their value.

The ecosystem service framework is one way to organize values of nature, which aids in assessing it value under different environmental management scenarios (Daly and Farley, 2010, chapter 6). Ecosystem services are ecological processes that provide benefits to humans—again, this is an anthropocentric approach. For example, compared to agricultural land, forest land slows rainwater runoff and increases aquifer recharge. Forests thus provide a water regulation service, making water available on a more consistent basis to society. Forest cover also filters runoff water and keeps it cleaner than some other types of land cover. These ecosystem services have real economic value, for example, in reducing water storage and filtration expenses that would otherwise be required (see Box 12.3).

Because of the importance of ecosystem services to all human societies, the United Nations Environment Programme (UNEP) conducted the Millennium Ecosystem Assessment to catalog and evaluate the state of the Earth's ecosystem services. The study took 5 years and involved more than 1360 experts in various fields. Table 12.1 shows the ecosystem service classification system used by the Millennium Ecosystem Assessment. The Millennium Ecosystem Assessment did not attempt to attach monetary values to these services, but found that many ecosystem services were significantly degraded in many parts of the world, a degradation representing real economic costs. The assessment also warned that ecosystem service degradation was accelerating in many regions, threatening the welfare of current and future generations.

As you can see from Table 12.1, some ecosystem services are easier to monetize than others. For example, it would be easier to estimate economic values for food and fiber, fresh water, and storm protection than for cultural diversity, sense of place, spiritual, and religious values. Studies of ecosystem service values have used this Millennium framework (and

BOX 12.3 Ecosystem Services Reduce New York City's Cost of Water

Metropolitan areas must provide large volumes of clean water for their residents. In the 1990s, New York City was facing an investment of $6 billion for new water filtration facilities, which would also have cost an estimated $300 million per year to operate (Postel and Thompson, 2005). Instead, the city decided to invest in protecting its main watershed in the Catskill–Delaware region.

While about 75% of the watershed was forested, with the potential to provide clean drinking water, most of it was also privately owned. New York City undertook a comprehensive watershed protection program in partnership with about 70 local governments in the watershed region. The city identified and then purchased many of the most sensitive properties for drinking water protection, allowing hunting, fishing, and similar recreation on these new public lands. In addition, New York paid for programs to help landowners manage agricultural and forestry operations in ways that preserved water quality, undertook programs to stabilize stream banks and reduce erosion, and invested in improved wastewater treatment systems in the watershed region. The combined measures cost about $1.5 billion over 10 years, or one-fourth of the cost of building new filtration facilities.

New York City water consumers got high-quality water at a low price, and residents of the watershed benefited from new recreation opportunities and the many environmental improvements in their region. The New York City program clearly demonstrated the value of forest water filtration ecosystem services—water filtration was provided much less expensively by nature than it could have been by building a filtration plant.

Source: Postel, S. L. and B. H. Thompson. 2005. Watershed protection: capturing the benefits of nature's water supply services. *Natural Resources Forum* 29:98–108.

TABLE 12.1 Ecosystem Service Typology, Millennium Ecosystem Service Assessment (2005)

Provisioning services
  Food and fiber
  Fuel
  Genetic resources
  Biochemicals, natural medicines, pharmaceuticals
  Ornamental resources
  Fresh water
Regulating services
  Air quality maintenance
  Climate regulation
  Water regulation
  Erosion control
  Water purification and waste treatment
  Regulation of human disease
  Biological control
  Pollination
  Storm protection
Cultural services
  Cultural diversity
  Spiritual and religious values
  Knowledge systems
  Educational values
  Inspiration
  Aesthetic values
  Social relations
  Sense of place
  Cultural heritage values
  Recreation and ecotourism
Supporting services (necessary for production of other ecosystem services)
  Soil formation
  Nutrient cycling
  Primary production
  Production of atmospheric oxygen
  Water cycling
  Provision of habitat

similar ones) to identify services that might be altered by human activity, which is a first step in valuing those services. We turn now to methods for estimating ecosystem service values.

12.4.3 Nonmarket Valuation

Establishing values for environmental goods and services is a primary goal of environmental economics, especially if important environmental goods are not bought or sold in markets. We can use a number of different methods to estimate these values, depending on the character of a good and the data available (Field, 2012, chapter 7). The key is finding some way to connect an environmental asset or service to people's willingness to pay for that asset or service.

Some environmental goods in fact have market prices. For example, the timber in the forest example earlier would have a clear market price. Where available, market prices are a first choice in valuation.

Often, a nonmarket environmental good is necessary for producing some market good that does have a price. If we can estimate the environment's contribution the market good, we can derive the environmental good's value. This is the **production factor method**. For example, clean water is an important in a fishery, and many fish have commercial or recreational values. Polluted water might cause fish to grow more slowly, to be less valuable, or to be absent all together. If we can estimate how fish value changes with water quality, we can use this change in the fish market value as an estimate of clean water value. Because clean water likely has other values as well, the estimate is a lower bound: we would only know the value of clean water with respect to producing more or better fish. We could say that clean water was worth at least our estimated value (and probably more).

Similarly, many environmental services or environmental damages directly impact human health, and in these cases, we can use the **cost-of-illness approach** to estimate environmental damages. For example, particulate matter from diesel engines damages human respiratory systems and may lead to hospitalizations or even premature deaths, especially in vulnerable or very exposed populations. The health-care cost for exposed individuals is one measure of the cost of particulate matter pollution. Again, it is a lower-bound estimate, since most people would value their health at far more than the cost of being treated.

Environmental damages like diesel particulate matter may also increase the probability of premature death, which has an obvious but hard-to-quantify cost. The value of human life is a difficult question, since many of us consider life to be priceless. Yet judgments about the value of life are often needed, for example, in making safety rules for consumer products. Most people would agree to spend $100 on a safety measure that would prevent one death, but most would not spend $1 trillion to prevent the same death—so we already have some (imprecise) concept of the value of a life.

There are ways to estimate how much people value their own lives. For example, if we compare wages in two similar jobs where one job is more dangerous, the wage premium for the dangerous job could be viewed as a self-valuation of life, the amount of compensation necessary for a person to accept the increased probability of dying on the job. Based on these and similar methods, the USEPA's National Center for Environmental Economics recommends $7.4 million (2006 dollars) as the **value of a statistical life** (VSL), or the value of preventing one death to a random person. While the specific value for a life may be controversial, it does provide an important valuation mechanism. Whenever an environmental service protects lives or environmental damage increases risk of death, we can use the change in probability of death along with the VSL to estimate the value of the environmental change.

Another valuation approach is based on how people's behavior regarding environmental goods is manifested in actual market transactions, a **revealed preference** method. For example, the **hedonic property value method** analyzes how property values are affected by environmental amenities or disamenities. If different houses are similar in all respects except that some are near lakes and the houses near lakes have higher values, we can assume that the lakes are the source of this added value. Summing the lake value for all affected properties gives us a total value estimate for lakes. Similarly, if some houses are near toxic waste dumps and otherwise identical houses are not, and the houses near dumps have lower values, the sum of property value losses from toxic waste proximity is one estimate of its cost. In this context, "hedonic" refers to attributes, so the hedonic property value method looks at how property attributes affect their value. All of these valuation techniques require large data sets and statistical methods to estimate values, but the hedonic property value method is particularly data intensive, since differences in property values with respect to environmental attributes may be small relative to total property value—statistically, it can be difficult to identify $300 environmental amenity differences in $300,000 homes.

One of the original revealed preference valuation methods is the **travel cost method**, where the cost of getting to a place is used to estimate the value of the place. For example, this method has been used to estimate the value of national parks: if a family spends $200 to go on vacation at a national park, the park must be worth at least $200 to the family. If the benefits received from the park were not at least $200, the family would not have made the trip. Again, this is a lower bound: the family might have been willing to pay $300 or $400 to use the park; all we know is that they were willing to pay at least $200 (since they did). In practice, this method has some difficulties. For example, it is not clear whether to count the time driving to the park as a cost and, if so, at what cost per hour—a normal wage is probably too high an estimate. Also, people may make one trip to multiple destinations, and it can be difficult to assess the independent value of each destination. But the development of the travel cost method ultimately led to other methods for nonmarket valuation, and the travel cost method is still used for some applications.

All of the methods described earlier relate primarily to **use values** of an environmental good. As the name implies, a use value is the benefit somebody gets from actually utilizing a good—like visiting a national park. But environmental economists realized that many environmental goods also have **nonuse values**. For example, people might have **existence values** for Yosemite National Park even if they do not go there, or they might have an **option value** for the possibility of going to Yosemite in the future. Somebody with a **bequest value** would benefit from the knowledge that Yosemite is being passed along to future generations. None of the valuation methods described earlier capture these nonuse values.

As discussed earlier, nonuse values like existence value for biodiversity can be very large because they are public goods: since they are nonrival, there is no limit on the number of people who can benefit from the same good. But because public goods are nonexcludable, markets rarely provide them. Estimating nonuse values is clearly important.

A group of methods known as **contingent valuation** uses surveys to estimate values for hypothetical goods that have nonuse values (Goodstein, 2011, section 8.4; Hanemann, 2012). The word "contingent" relates to the hypothetical nature of the good—a person would be willing to pay for the good, contingent on a market for it actually existing. Contingent valuation is also called a **stated preference** technique, to contrast it with the revealed preference methods discussed earlier. Stated preference values are based on what people say, while revealed preference values are based on what they do.

For example, we could use contingent valuation to estimate the value of protecting coral reefs. While a coral reef may have some market values (fish) and other use values (SCUBA diving), these do not adequately capture the value of a biologically intense and aesthetically beautiful area like a coral reef. But since protecting coral reefs is not free, we would like to have some estimate of their value to the public.

Contingent valuation researchers have found that simply asking open-ended survey questions like "How much would you pay to protect a coral reef?" does not work very well. Since coral reefs have no actual market price (nobody buys them), survey respondents do not have any frame of reference for a reasonable price. Responses might vary from $5 to $5 million, and we would have trouble making any sense of the resulting data. Also, the question as worded does not refer to any specific coral reef, and people might have different willingness to pay for coral reefs in different places. A better approach is to ask whether people would be willing to pay a specific price for a specific project. We might ask visitors to Key Largo, Florida, whether they would be willing to pay $100 to help protect the reefs at John Pennekamp Coral Reef State Park. Even better is to describe a specific approach that would solve a specific problem. We might ask whether visitors would support improving the Key Largo wastewater treatment system, if the improvement would result in a 20% increase in the number and diversity of fish on the reef but also cause hotel room rates to rise by $10 a night. Though the question is about wastewater treatment and hotel rates, we could use responses to estimate the value of underlying benefits that people derive from a coral reef. This example is purely hypothetical (like much of contingent valuation), but perhaps you can see that one key to good contingent valuation research is to make the valuation scenario as realistic as possible and as similar as possible to a real purchasing decision that people make.

Contingent valuation has a number of variations, all based on asking people direct questions in surveys. There is now a large literature on different contingent valuation methods, and if you were intending to conduct a contingent valuation study, you should review this literature carefully (e.g., Hanemann, 2012). There are a number of important methodological subtleties for contingent valuation as well as for survey research in general. How questions are asked can affect responses and can affect value estimates.

Contingent valuation is the most versatile valuation method, since it can be used to value almost anything—nonuse values, nontangible assets, as well as environmental goods or services that do not even exist yet. If we were trying to estimate the spiritual value of nature, contingent valuation would be the method to try. Yet some contingent valuation studies have also been strongly criticized: because the prices estimated are always for hypothetical goods, it can be difficult to assess the accuracy of contingent valuation studies. One important source of error in these studies is **hypothetical bias**. In actual market transactions, people trade resources they have for the goods and services they most prefer. Importantly, people are constrained in these choices by their incomes (or at least by their incomes and their credit), so market decisions reflect the difficult choices that people have made in prioritizing their needs. Most economists take this to be the best reflection of value that we can get. But contingent valuation responses are not constrained by income: I can *say* that I would spend an extra $10 per night at Key Largo to protect coral reefs, or $100 per night or $1 billion per night, since I do not actually have to pay it—the payment is hypothetical. Hypothetical bias is when the value I state is not the value I would actually pay. There are now methods in contingent valuation to control hypothetical bias, but a researcher must be attentive to these. Rigorous research methods are particularly important in contingent valuation, since the process does not include observations of real markets.

12.4.4 Capital Stock Values

A final important topic in resource valuation is the distinction between **resource stocks** and **resource flows** (Meadows, 2008, chapter 1). Returning to our earlier example, a forest has a number of stocks and flows. Perhaps the most obvious stock is the amount of timber in standing trees. The timber-related flows are annual tree growth and annual tree harvest or decay. A minimum condition for sustainable forest management is that outflow of harvest not exceed the inflow of new growth. If more timber is removed than grows, the timber stock declines. If this happens repeatedly, the timber stock is eventually depleted and the forest disappears, as surely as bathtub with more outflow than inflow must eventually empty.

A forest stock is a **capital** asset, which can yield a flow of valuable timber over time, in fact for perpetuity. The nature of any capital stock is that it generates flows of resources over time: a factory generates a flow of manufactured goods, a company stock share generates a flow of dividends, a solar panel generates a flow of electricity, and a house generates a flow of shelter services. The value of a capital asset is based primarily on the value of the flows it generates. And the owner of a capital asset can benefit from selling the flows, selling the capital stock, or some combination of both.

Again, let's say I own a forest, one that takes 50 years to grow from seedlings to maturity. One way I could receive benefits from the forest capital stock is to harvest 1/50 or 2% of the forest stock each year. Growth equals harvest, so this strategy is sustainable, and I and my successors could do this forever. The capital value of my forest never changes (in an unchanging perfect market), because the timber stock never changes and the forest's capacity to produce a flow of timber resources is undiminished. Another option is to cut the whole forest now. With this strategy, I get all the timber income now, which I may prefer, but I would then have to wait

50 years to get any more mature trees. If I sold the forest land with no trees, the new owner would also have to wait for 50 years for timber income, so the capital value of my forest asset would be much less than before I harvested. In deciding to clear cut and sell, I have simply traded lower capital value for more current income.

In some ways, this is a straightforward financial calculation. If the market for forest products is unchanging, if the capital value of my forest is based only on future timber flows, if clear cutting does not change the forest's annual growth rate, and if the land generates no other benefits, harvesting annually and harvesting all the trees at once are financially equivalent—there is no inherent reason to favor one course over the other. But of course there are complicating factors.

In addition to a flow of timber, my forest capital stock yields a flow of ecosystem services like providing habitat, increasing biodiversity, sequestering carbon, and increasing groundwater recharge. All of these are positive externalities of my forest, and as a forest owner, I normally would receive no payment for these services. Similarly, if cutting all the trees greatly reduces these ecosystem services, I bear no direct cost. Society's losses of ecosystem services are not reflected in the capital value of my land. Again, this is the nature of an externality: the capital value to society of my intact forest is greater—perhaps much greater—than it appears from the market value of forest products. The forest stock generates valuable ecosystem service flows in addition to flows of marketable timber.

Also, in many parts of the world (depending on the specific forest ecology), a decision to remove all the trees may also change possible tree growth significantly, perhaps catastrophically. Soil erosion, nutrient loss, and changes in seed germination conditions after complete tree removal may preclude a new forest from growing immediately, or ever. The capital stock may be severely degraded by a complete harvest. In this case, my complete harvest decision would cost more than just a 50-year harvest delay, and some of these costs would be borne by people in the future. Sustainability requires that such costs to future generations be considered along with current costs.

12.5 SUMMARY

An economic approach to environmental management can improve stewardship of the environment. An economic approach recognizes both that environmental goods and services have value and that protecting the environment also requires resources. The goal of good management is balancing costs and benefits to achieve the best outcomes for society.

While a free market can achieve this balancing of costs and benefits with respect to market goods, free market outcomes are not optimal when externalities are involved or when goods are not bought and sold in markets. Both of these situations are typical with environmental goods, and market failures are thus common with respect to the environment. Two of the most common types of externality-related market failure are open-access resources and public goods.

Since many environmental goods and services are not bought and sold in markets, they do not have market prices. Applying economic methods requires estimating values for environmental goods. Though there are many methods for estimating values of nonmarket goods, all of them are imprecise, and many provide only lower-bound estimates of value. But using such estimates in benefit–cost analysis has been an import tool in supporting policies for environmental protection. Many environmental economics methods are relatively new and are still improving.

Besides not accounting for external costs and benefits, markets alone cannot ensure fair outcomes in a society. Environmental justice considers how fairly environmental costs are distributed in the present, and sustainability is about how fair environmental effects are to future generations.

While environmental economics can help to identify optimum environmental solutions for society, implementing these solutions usually requires political action. Most environmental problems are the result of market failures, and markets do not normally self-correct for these failures. Yet some of the best policy solutions mimic market mechanisms and require only minimal government intervention in free markets. One example is applying Pigouvian taxes or subsidies to represent values of negative or positive externalities.

Environmental policy approaches are covered elsewhere in this text. In the United States, most significant environmental policies have an economic basis, since the EPA is required to conduct benefit–cost analysis for new environmental regulations. Economic insights can also help to inform policy approaches to environmental improvement. The most successful policies align individual incentives with social priorities, so that each person in society has an incentive to act in a way that provides the best environmental outcomes for society as a whole. This can be an effective way to enhance the environment and improve our stewardship of those natural resources on which all life ultimately depends.

REFERENCES

Arrow, K. J., M. L. Cropper, G. C. Eads, R. W. Hahn, L. B. Lave, R. G. Noll, P. R. Portney, M. Russell, R. Schmalensee, and V. K. Smith. 2012. Is there a role for benefit-cost analysis in environmental, health, and safety regulation?. In R. N. Stavins, ed., *Economics of the Environment: Selected Readings*. W.W. Norton & Company, New York, pp. 219–224.

Boyce, J. K. 2002. *The Political Economy of the Environment*. Edward Elgar Publishing, Cheltenham, UK.

Butchart, S. H., M. Walpole, B. Collen, A. van Strien, and J. P. Scharlemann. 2010. Global biodiversity: indicators of recent declines. *Science* 328(5982):1164–1168.

Cardinale, B. J., J. E. Duffy, A. Gonzalez, D. U. Hooper, C. Perrings, P. Venail, A. Narwani, G. M. Mace, D. Tilman, and D. A. Wardle. 2012. Biodiversity loss and its impact on humanity. *Nature* 486(7401):59–67.

Conrad, J. M. 2010. *Resource Economics*. Cambridge University Press, New York.

Costanza, R., J. H. Cumberland, H. Daly, R. Goodland, and R. B. Norgaard. 1997. *An Introduction to Ecological Economics*. CRC Press, Boca Raton, FL.

Daly, H. E. and J. Farley. 2010. *Ecological Economics: Principles and Applications*. Island Press, Washington, DC.

Field, B. 2012. *Environmental Economics: An Introduction*. McGraw-Hill, New York.

Fullerton, D. and R. N. Stavins. 2012. How economists see the environment. In R. N. Stavins, ed., *Economics of the Environment: Selected Readings*. W.W. Norton & Company, New York, pp. 3–8.

Goodstein, E. S. 2011. *Economics and the Environment*. John Wiley & Sons, Inc., Hoboken.

Goodwin, N., J. Harris, J. A. Nelson, B. Roach, and M. Torras. 2014. *Microeconomics in Context*. ME Sharpe, Armonk, NY.

Hanemann, M. W. 2012. Valuing the environment through contingent valuation. In R. N. Stavins, ed., *Economics of the Environment: Selected Readings*. W.W. Norton & Company, New York, pp. 148–174.

Hardin, G. 1968. The tragedy of the commons. *Science* 162:1243–1248.

Harris, J. and B. Roach. 2013. *Environmental and Natural Resource Economics: A Contemporary Approach*. M.E. Sharpe, New York.

Martin, D. 2013. A macroeconomic model of biodiversity protection. *Theoretical Economics Letters* 3:39.

Meadows, D. H. 2008. *Thinking in Systems: A Primer*. Chelsea Green Publishing, White River Junction, VT.

Millennium Ecosystem Assessment. 2005. *Ecosystems and Human Well-Being: A Framework for Assessment*. Island Press, Washington, DC.

Mullon, C., P. Fréon, and P. Cury. 2005. The dynamics of collapse in world fisheries. *Fish and Fisheries* 6(2)111–120.

Ostrom, E., C. Chang, M. Pennington, and V. Tarko. 2012. *The Future of the Commons: Beyond Market Failure and Government Regulation*. The Institute of Economic Affairs, London.

Pauly, D., V. Christensen, S. Guénette, T. J. Pitcher, U. R. Sumaila, C. J. Walters, R. Watson, and D. Zeller. 2002. Towards sustainability in world fisheries. *Nature* 418(6898):689–695.

Postel, S. L. and B. H. Thompson. 2005. Watershed protection: capturing the benefits of nature's water supply services. *Natural Resources Forum* 29:98–108.

Scorse, J. 2010. *What Environmentalists Need to Know about Economics*. Palgrave Macmillan, New York.

Smith, A. 1776. *An Inquiry into the Nature and Causes of the Wealth of Nations*, An Introduction, Notes, Marginal Summary and an Enlarged Index by Edwin Cannan. Methuen, London.

Solow, R. M. 2012. Sustainability: an economist's perspective. In R. N. Stavins, ed., *Economics of the Environment: Selected Readings*. W.W. Norton & Company, New York, pp. 543–550.

Stern, N. H. 2006. *Stern Review: The Economics of Climate Change*. HM Treasury, London.

Tietenberg, T. and L. Lewis. 2012. *Environmental and Natural Resource Economics*. Pearson, New York.

Wasserman, M. 2001. The last hunting economy. *Regional Review*(Q 2):8–17.

13

LAW AND POLICY IN ENVIRONMENTAL MANAGEMENT

DIANNE RAHM

Department of Political Science, Texas State University, San Marcos, TX, USA

Abstract: This chapter explores three large topics. First are the institutions that shape public policy. At the federal level, these include Congress, the presidency, the bureaucracy, and the courts. The chapter also explores the role of the states and the relationship between the states and the federal government. Other policy actors are described including political parties, interest groups, the media, and public opinion. Second, the chapter addresses policymaking, specifically how laws and regulations are made and implemented. The discussion addresses the issues of who is involved in policymaking, how it is made, and how it is implemented. Third, the chapter provides an in-depth overview of the substance of US laws and policies that shape environmental management. These include international treaties and federal and state laws and regulations in place regarding land, natural resource, water, air, waste, and transborder environmental issues. The chapter concludes with a discussion of the most pressing environmental management issue today—global climate change.

Keywords: public policy institutions, Congress, presidency, bureaucracy, courts, states, political parties, interest groups, the media, public opinion, policy process, rulemaking, environmental regulatory approaches, natural resource policy, energy policy, environmental policy, global climate change, sustainable energy, Clean Air Act, Clean Water Act.

13.1 Introduction	305	
13.2 What is Public Policy?	306	
13.3 Federal Public Policy Institutions: An Overview	307	
13.3.1 Congress	307	
13.3.2 The Presidency	308	
13.3.3 The Bureaucracy	309	
13.3.4 The Courts	310	
13.4 The Role of the States: An Overview	311	
13.5 Other Public Policy Actors: An Overview	312	
13.5.1 Political Parties	312	
13.5.2 Interest Groups	312	
13.5.3 The Media	312	
13.5.4 Public Opinion	313	
13.6 Institutions and Actors in Environmental Policy	313
13.7 The Policy Process	316
13.8 Rulemaking: The Process	317
13.9 Types of Environmental Regulatory Approaches	320
13.10 Evolution of Natural Resource, Energy, and Environmental Policy	321
13.10.1 Natural Resource Policy	321
13.10.2 Energy Policy	323
13.10.3 Environmental Policy	326
13.10.4 Global Climate Change	328
13.11 Conclusion	333
References	334

13.1 INTRODUCTION

What we refer to as law and policy in environmental management is really a mix of three distinct but overlapping areas of concern. Natural resource policy typically centers on the use of land, water, minerals, and other resources. The main issues in natural resource policy concern the protection or use of the land and water for development and other purposes. Energy policy is important because a great deal of the pollution that affects our world is a direct result of fossil fuel use and nuclear waste. Finally, environmental policy focuses on protection of air, water, and species as well as the safe disposal of solid and hazardous waste. Environmental policy also includes international issues, including treaties

An Integrated Approach to Environmental Management, First Edition. Edited by Dibyendu Sarkar, Rupali Datta, Avinandan Mukherjee, and Robyn Hannigan.
© 2016 John Wiley & Sons, Inc. Published 2016 by John Wiley & Sons, Inc.

for the ozone hole, global climate protection, and the maintenance of biodiversity.

This chapter explores three large topics. First are the institutions that shape public policy with special emphasis on federal, state, and other policy actors. Second, the chapter addresses policymaking, specifically how laws and regulations are made and implemented. The discussion addresses the issues of who is involved in policymaking, how it is made, and how it is implemented. Third, the chapter provides an in-depth overview of the substance of US laws and policies that shape environmental management. These include international treaties and federal and state laws and regulations in place regarding land, natural resource, water, air, waste, and transborder environmental issues. The chapter concludes with a discussion of the most pressing environmental management issue today—global climate change.

Environmental laws and policy have been phased in over time in response to various environmental concerns that have periodically arisen. Environmental laws and policies are cumulative, that is, they have added up over time to include a considerable array of laws, treaties, and regulations to address a variety of concerns. The first fears to arise in the United States were regarding poor air and water quality. Some states, like California, grew aware of poor air quality in the 1940s and 1950s and began to implement state-based laws for air emissions. The Great London Smog Event of 1952, which killed over 12,000 people, raised international concern over air quality. Air quality was poor in most urban areas of the United States but particularly so in California. Other spectacular events, such as the oil spill off the Santa Barbara coast in California and the burning of the Cuyahoga River in Cleveland, Ohio, in the 1960s, served to focus the nation's attention on water issues. Love Canal and the fear of toxic waste emerged in the 1970s as a key issue. Acid rain and the discovery of the hole in the ozone layer in the 1980s shifted the focus of the environmental movement to transborder issues. Concern today remains with transborder issues. The most prominent global environmental issue of the 2010s is climate change and its many effects.

The manner of environmental implementation has changed over time as well. Early efforts were voluntary, that is, the law merely suggested that the regulated community reduce its pollution or change its behaviors. However, these efforts proved insufficient in mitigating environmental pollution. In response to increasing concern for environmental pollutions, most environmental laws regarding air, water, solid, and hazardous waste became mandatory. Compliance was now required with fines and other punishments accruing in the event of failure. This sort of regulation, commonly called command and control, became unpopular by the 1980s. Different mechanisms of enforcement arose including market-based systems such as cap and trade. First applied to sulfur dioxide emissions under the Clean Air Act (CAA) of 1990, these market-based approaches were later accepted in the international framework and became the method used in the Kyoto Protocol, the first international treaty to deal with climate change, to reduce greenhouse gas (GHG) emissions.

Let us begin our discussion of environmental law and policy by examining the basics of the US government functions in general. We will begin by focusing on several related questions: What is public policy? What government institutions are involved in policymaking? Who has the power to influence those institutions? What is the role of the states in policymaking? What roles are played by the media, political parties, and interest groups? After we have laid this general foundation, we will then discuss the institutions and actors of policy with a particular focus on US environmental policy. We then turn to a discussion on the policy process and explore how rules and regulations are written and implemented. After the policy discussion, we move to a summary of the legal framework for natural resource policy, energy policy, and environmental policy. The chapter discusses the major federal laws for land, air, water, waste, energy, and transborder issues. We conclude with a discussion of a contemporary issue of key concern—global climate change.

13.2 WHAT IS PUBLIC POLICY?

There are many definitions of public policy. Most include some recognition that public policy in some manner involves government. For our purposes, public policy can be defined as a course of action followed by government in response to an issue of public concern. Government actions include laws passed by Congress, regulations proposed and finalized by government agencies, and court decisions. Adherence to these governmental decisions is mandatory. Public policy is backed by the coercive power of the state. Failure to comply with laws, rules, or court decrees can result in incarceration, fines, or both (Cochran, Mayer, Carr, and Cayer, 1999).

When government decides to act, the result is most often the establishment of a public program that spends government revenues to achieve a desired outcome. Government revenues are most often generated by income tax, although sometimes funds for public programs are raised through user fees or special taxes. For example, when the government decides to build a new weapons system, the funding comes from general taxation. When the government decided to provide a national repository for spent nuclear fuel, a special tax was placed on the nuclear power industry to raise money. National and state parks, which operate on sums paid by the people visiting them, are an example of programs funded by user fees.

Public policies can also be put in place by government regulation. In this case, the government determines the desired outcome and passes a law requiring companies to implement and pay for their own programs to achieve the sought-after results. For example, when the CAA was amended in 1990, it required the Environmental Protection Agency (EPA) to regulate a large number of chemicals and other pollutants that had previously been uncontrolled. After passage of the act, private industry was held responsible for putting programs in place to eliminate the offending pollutants. Thus, public policy was enacted, but the required outcome (cleaner air) was to be paid for by the private sector. The cost of implementing environmental regulations typically falls to the private sector.

Public policies can also be mandated by the courts. In 1954, in the famous case of *Brown v. Board of Education*, the US Supreme Court ruled that segregated educational facilities were a violation of the Constitution. After the decision, it fell to all other levels of government to implement and pay costs associated with the desegregation of schools. Likewise, when the courts ruled that the overcrowding of prisons violated the rights of prisoners, states were required either to release prisoners or to assume the cost of expanding their corrections facilities. The Supreme Court ruling on *Massachusetts v. Environmental Protection Agency* in 2007 affirmed that the EPA had the authority under the CAA to regulate GHG emissions. The court decision allowed the EPA to issue regulations to reduce GHG emissions.

Not all concerns raised by the public are acted upon by the government. The government may decide to do nothing about a particular issue because most people consider it to be a private, rather than public, matter. Public policy, then, also consists of what governments fail to do in response to an issue of public concern (Dye, 1995).

13.3 FEDERAL PUBLIC POLICY INSTITUTIONS: AN OVERVIEW

The legislature, executive, bureaucracy, and judiciary are the official institutions through which public policy is made on the national level. Articles I, II, and III of the Constitution are devoted to the design and role of the legislature, the executive, and the courts, respectively. The bureaucracy, which has expanded immensely since the nation's beginnings, is not given much attention in the Constitution. Regardless, the bureaucracy is a major actor in public policymaking.

The Founders of the Republic were careful to disperse power across government organizations so as to safeguard that no one unit of government would have dominance. The notion of one part of government "checking" the other, which is commonly known as checks and balances, was a fundamental part of the strategy. The Founders had an enormous fear of tyranny, and they sought to distribute power so that no one group could control the entire government. Commonly referred to as the separation of powers, the distribution of authority across the executive, the legislative, and the judiciary seeks to guarantee that no single institution will be able to dominate federal governance.

The United States is ruled by a system of laws and procedures agreed upon by the governed. The first statement of these is the Constitution. The first three words in the Preamble to the Constitution, "We the People," denote the source of authority of the laws.

The United States has three branches of government—the legislative, the executive, and the judiciary. The bureaucracy, which is part of the executive branch, consists of 14 cabinet departments as well as a host of agencies, commissions, and boards. All the organizations that comprise the bureaucracy are under the authority of the president of the United States. The sections that follow look briefly at each of these parts of the federal government. The discussion emphasizes how each institution is organized and suggests its key public policy roles.

13.3.1 Congress

At the federal level, the legislative task is given to Congress, which consists of two elected bodies: the House of Representatives and the Senate. The House of Representatives is comprised of 435 members elected for 2-year terms. The states have varying numbers of congressional representatives based upon their population. Every 10 years, the census counts the nation's population, and the new demographic figures are used to reapportion the House of Representatives. The process of reapportionment adjusts the congressional districts so that the 435 House seats best represent the population of the country. Heavily populated states, like New York and California, have a greater number of Representatives than do states with small populations, such as Wyoming. The Senate is comprised of 100 senators, two from each state, elected for terms of 6 years.

Congress has several key public policy roles. Its first major role is that it passes the laws that establish the programs that constitute much of public policy. Congress has the power of the purse, or the authority to raise money through taxation and to spend it through appropriation (Holmes, Englehardt, Elder, Zoetewey, and Ryden, 1998). In other words, Congress both establishes (or authorizes) programs to exist and appropriates the money to run them (Schick, 1980). A second and essential power of Congress is oversight. Congress has the authority to supervise the administration of programs that it has established to determine how well they are doing. Congress exercises this oversight by holding hearings, conducting investigations, and consulting with the public and constituents about government programs and issues of concern (Lowi and Ginsberg, 1998).

Congress is organized by political party. In the US two-party system, this typically means Republicans and Democrats (although from time to time it has meant Whigs, Know Nothings, Union, etc.). Voters decide whether they prefer the Republican or Democratic candidate for each seat that is up for election. The party that wins the majority of the seats in each house of Congress becomes the majority party of that house. For instance, if the Democrats hold the greatest number of seats in the House of Representatives, they are the majority party in the House of Representatives. This means that the Speaker of the House would be the ranking Democrat, and all committees would be chaired and dominated by Democrats. Likewise, the Senate is run by the leadership of the party that wins the majority of Senate seats, and all committees are chaired and dominated by members of the majority party. One party, say, the Democrats, might win the majority in the House of Representatives, while the other party, say, the Republicans, might win the majority in the Senate, or one party might win both houses of Congress.

Congress is organized into a series of committees and subcommittees, each with specific expertise. It is not possible for members of Congress to be experts on all matters. The division of Congress into committees and subcommittees allows members to focus their attention on a relatively small number of issues. This form of institutional design has several immediate consequences. First, the procedural rules followed by Congress make it difficult for legislation to be voted upon by either the House or the Senate without first being positively voted out of the committee with jurisdiction (Lindbloom, 1968). Second, Congress relies heavily on committee expertise. Most members of Congress defer to the opinions of their colleagues who serve as committee members and follow their lead in voting (Kelman, 1987). These two factors render committees very powerful.

The organization of Congress into committees and their power over policy goes back to the early Republic, when ad hoc, select, or special committees were established from time to time to deal with matters of concern. Permanent standing committees organized to deal with specific recurring matters of interest soon became a feature of Congress. Committees are typically organized to receive proposals for legislation and to produce bills ready to be voted upon. Committee jurisdiction is usually based upon the subject matter of legislation. Today, the jurisdiction of standing committees (with the exception of the House Rules Committee and the House and Senate Appropriations Committees) parallels the major agencies in the executive bureaucracy (Lowi and Ginsberg, 1998). These committees and subcommittees hold hearings and conduct investigations of programs in the corresponding executive agency. Committees are divided into subcommittees, each of which is given jurisdiction over specific matters.

Political parties play a critical role here. A member of the majority party chairs each committee or subcommittee. Members are assigned to committees by their party's leadership in the House and Senate, and they generally are given the assignments they request. Typically, members of Congress seek to be on the most prestigious committees and those that directly affect their constituencies. Status is determined largely by the extent of power of the committee. Committees of high ranking in the House are Appropriations, Budget, Commerce, Rules, and Ways and Means. Prestigious committees in the Senate include Appropriations, Armed Services, Budget, Finance, and Foreign Relations (Holmes, Englehardt, Elder, Zoetewey, and Ryden, 1998).

A large number of staff members are assigned to committees and subcommittees to assist members of Congress with their legislative work. Time pressures and the ever-increasing complexity of policy issues force members of Congress to rely heavily on staff for advice and guidance.

13.3.2 The Presidency

On the national level, the president is the leader of the executive branch. Elected for a maximum of two 4-year terms, the president has a number of roles in public policy-making. First, the president is the commander in chief and the key actor when it comes to foreign policy. Second, the president has what President Theodore Roosevelt called the "bully pulpit," the ability to set the legislative agenda by persuading Congress to pass legislation the administration desires. This power is considerable and should not be overlooked. Another way the president communicates what the administration wants to accomplish in the upcoming year is by presenting Congress with a draft presidential budget for annual federal spending. Third, the president has the power to veto legislation passed by Congress. Finally, the president is the managerial head of the executive branch of government, with extensive power over the operation of the bureaucracy.

The public policy powers of the executive rest largely on the ability of the president to set the legislative agenda (LeLoup and Shull, 1993). This occurs primarily at the beginning of a new administration, when the incoming president announces the major policy initiatives. Most presidents also avail themselves of an annual opportunity to influence the legislative agenda. Article II, Section 3, of the Constitution states that the president "shall from time to time give to the Congress information on the State of the Union and recommend to their consideration such measures" judged "necessary and expedient." The State of the Union address is typically given by modern presidents with great fanfare as they ceremoniously enter the halls of Congress and present their priorities for legislative action (Light, 1999). While the State of the Union address may not always be a succinct list of presidential desires, it does provide the administration with an opportunity to inform Congress of preferred policy directions (Kingdon, 1995).

As managerial head of the executive branch of government, the president has three key policy roles. Since 1921, agency budget requests have been reviewed and approved by the Office of Management and Budget (OMB) before being sent to Congress. This gives the president extensive power over the planned activities of the bureaucracy. A president who frowns on a particular agency or a programmatic emphasis can simply cut or refuse the budget request (Kelman, 1987). Second, presidents have the power to appoint the senior bureaucrats who head the government's agencies for the duration of the administration. In making these choices, presidents have some control over the direction bureaucracy will follow (Hooton, 1997). If a president does not particularly like the mission of an agency, such as the EPA, the president can appoint someone to head that agency who is also opposed to the mission, thus reigning in the agency. Ronald Reagan did this in his first term with both the EPA and the Department of the Interior (DOI), thus effectively crippling the agencies and their missions from being carried out. Finally, the president may issue executive orders, which are instructions to executive agencies. These instructions can be very influential in driving environmental policy. For instance, environmental justice became an important issue for all federal agencies and federal contractors as a result of President Clinton's Executive Order 12898, which instructed all federal agencies to ensure that their programs "do not unfairly inflict environmental harm on the poor and minorities."

The organization of the White House and its staff is important to understanding the public policy process. Before the 1920s, presidents relied heavily on their cabinet secretaries for policy advice. By 1939, however, managing the presidency had become far too complex, and three permanent managerial changes were instituted. The first was the creation of the Executive Office of the President (EOP) as the official staff arm of the presidency. The second was that the OMB (called the Bureau of Budget until 1970) was moved from the Treasury Department into the EOP. Finally, the White House Office (WHO) was created within the EOP to serve as a smaller, more direct staff office to provide presidents with immediate advice and information (Ragsdale, 1998).

Even with these changes, presidents struggle with the managerial task of supervising so many people. The total number of executive staffers has varied from time to time. During World War II, President Franklin D. Roosevelt (FDR) ran the war effort out of the White House, and as a consequence, the number of executive staffers grew to nearly 200,000. After the war, the number dropped rapidly. For most of Harry Truman's, Dwight Eisenhower's, and John Kennedy's administrations, the executive staff averaged about 2000 employees. Under the Great Society programs of Lyndon Johnson's administration, the staff averaged about 4000, and it grew during the Richard Nixon years to more than 5000. Gerald Ford cut the number back substantially, to less than 2000. Since the Reagan administration, presidents have had an executive staff of about 1500 (Ragsdale, 1998). The large size of the White House staff suggests a certain amount of organizational complexity. Greater size brings communication challenges and the real possibility of the emergence of competing policy goals from among the many subunits.

Presidents play a pivotal role in foreign policymaking. Both as commander in chief and as the principal diplomatic negotiator, a president can take the lead in determining the path the nation will follow internationally. At the most fundamental level, a president may engage American troops in combat. The president may seek a formal declaration of war from Congress, although modern presidents have not done this. As the figure the world looks to for the "American" position on events, presidents have enormous power. Indeed, since the Second World War, the American president has been known as the "leader of the free world." Presidents sign treaties that indicate the willingness of the nation to comply with international agreements. Presidents, of course, must submit treaties to the Senate for ratification, but presidents may heavily lobby for Senate support. A president may also announce the intention of the government to withdraw from a treaty. George W. Bush did this with the Kyoto Protocol on climate change (even though the Senate had never ratified the document President Clinton had signed).

Presidents also play a critical role in domestic policymaking. Along with the constitutional power of the veto, presidents have other ways to influence the agenda of Congress. Presidents tell Congress what legislation they would like to see passed. While presidents have no power to write legislation themselves, they are well equipped to use the media and other popular forms of communication to demand specific actions from Congress. The extent to which presidents are successful depends on their ability to persuade. If a president can rally public support for a proposal or action, Congress is likely to go along. Presidents also take advantage of the State of the Union address to directly tell Congress what they would like to see passed and to set some priorities for desired legislation.

13.3.3 The Bureaucracy

Unlike the other institutions of government that comprise our national system, the bureaucracy is not fully specified in the Constitution. While the executive, legislative, and judiciary each have a full article of the Constitution devoted to them, the bureaucracy is mentioned only in passing. For instance, under a listing of powers granted to Congress in Section 8 of Article I, the Constitution mentions certain functions of government such as providing for an army and navy, coining money, creating standards for weights and measures, establishing post offices, and protecting the rights of inventors. These functions have become institutionalized in, respectively, the Department of Defense, the Department of the Treasury, the National Institute of Standards and Technology (formerly the Bureau of Weights and Standards), the US Postal Service, and the Patent and Trademark Office. The function of performing a census of the population every 10 years (and thus the foundation for the US Census Bureau) is also specified in the Constitution (Article I, Section 2).

The bureaucracy has expanded enormously over time. As Congress passed laws to regulate areas that the Constitution specified as within its domain of authority, many government programs were established. The first major growth spurt came during the Great Depression with the many programs introduced during FDR's New Deal. These initiatives provided for the stabilization of the banking industry (the Emergency Banking Act), work and shelter for the destitute and unemployed (the Civilian Conservation Corps), regulation of agricultural prices and production (the Agricultural Adjustment Act), grants to the states so that they could help the poor (the Federal Emergency Relief Act), a major economic development program in the hard-hit southeast (the Tennessee Valley Authority), the regulation of the securities industry (the Federal Truth in Securities Act), a national system of state-run public employment offices (the Wagner–Peyser Act), financial protection for homeowners (the Home Owners' Loan Act), and a major industrial policy effort to stabilize industry (the National Industrial Recovery Act) (Robertson and Judd, 1989). Each created or expanded government bureaus to oversee and manage many new programs. The Roosevelt administration also created what is today the largest single federal program, Social Security.

The bureaucracy reached another point of great expansion in the 1960s with the establishment of President Lyndon Johnson's Great Society programs. Among these were several large health and poverty programs, including Medicare, Medicaid, and food stamps. Each necessitated the creation of large bureaus. By the late 1980s, the executive branch of the government employed about three million civilians, most working in bureaucratic positions overseeing and running government programs. While that number fell slightly in the 1990s, with the emphasis on homeland security in the aftermath of the 2001 terrorist attacks, the federal government once again began growing. The Justice Department, the intelligence agencies, and the military expanded, and the federalization of airport security alone added thousands of new employees to the federal workforce.

The bureaucracy is generally thought of as having a key role in public policymaking. Its main function is to implement the policies called for by congressional legislation, which typically involves provision of services through government programs or third-party providers overseen by bureaucrats, as well as the establishment and enforcement of rules and regulations. The rulemaking and enforcement authority granted to agencies by Congress is a powerful policy role. Moreover, Congress frequently writes vague legislation, which vastly increases the extent of administrative discretion exercised by officials. Congressional instructions to governmental organizations are frequently no more detailed than "to establish regulations suitable and necessary for carrying this law into effect; which regulations shall be binding" (Kerwin, 2003).

13.3.4 The Courts

The federal courts play a significant role in public policymaking in several ways. Their case decisions set precedents for later cases. In public policy, this role is important because interest groups bring "test" cases to court so that the court's pronouncement will set a precedent for future rulings by other judges. Second, the courts interpret the intent of congressional language in legislation. Third, while not specifically stated in the Constitution, judicial power over policymaking has been vastly enhanced by the successful attempt on the part of the courts to review acts of other branches of government and invalidate those they view as unconstitutional. This is the power of judicial review. Not only do courts invalidate legislation passed by Congress, but they also hear appeals on judgments made by regulatory agencies (concerning both individual cases and general regulations) (Smith, 1993).

The large public policy role of the courts has a long history in the United States. Courts have traditionally acted as a referee between contending actors disagreeing over controversial issues. Frequently, the argument has involved highly visible policy outcomes. Before the Civil War, for instance, the courts were involved in disputes over the appropriate role of the federal government, states' rights, and slavery. In the Progressive Era, the courts mediated the growing problems of urbanization and industrialization, often acting to protect property rights. With the passage of large volumes of Depression era legislation, the courts eventually moved from protecting property to protecting civil rights. In *Brown v. Board of Education* in 1954, they began an extensive public policy role in civil rights, which would continue and expand (Shapiro, 1995). Indeed, the agenda of the Supreme Court after *Brown* shifted away from a focus on economics to more clearly address the issues of social and moral policy.

The federal court system is composed of several types of specialty courts (such as tax courts and international trade courts) and three types of general courts. These general courts, which do most of the work of the system, are district courts, courts of appeals, and the Supreme Court. For most cases, the point of entry into the federal judicial system is the district court. Most cases go no farther, and these are the main courts of the federal system. There are 94 federal district courts, one in each judicial district. Each state has at least one judicial district, and some states have several. Cases in a district court are tried by a single judge, with or without a jury. Decisions can be appealed to the federal courts of appeals. The courts of appeals are organized based upon 12 circuits. Cases are decided by panels of three judges. The highest court is the Supreme Court. Its size, determined by Congress, has been set at nine since 1869 (Baum, 1990).

13.4 THE ROLE OF THE STATES: AN OVERVIEW

Not all areas of public policy come from the federal government. State and local governments have primary authority for large areas of public policy, including elementary and secondary education, public health and safety, disaster emergency response, and provision of such services as highway maintenance, libraries, water and sewer utilities, and public hospitals. Many municipal utilities continue to provide electricity to their localities. The States share power with the federal government in many areas of public policy, including environmental protection, social welfare, economic development, and job training.

The division of power between federal and state governments was established by the Constitution. The federal government was given the powers enumerated in the first three articles of the Constitution. The Tenth Amendment reserves all powers not clearly delegated to the federal government (or prohibited to the states) for the states. This division of power is often referred to as federalism, and the interactions between the different levels of government are referred to as intergovernmental relations (IGR). This is particularly important for environmental policy because many federal environmental laws allow delegation of federal enforcement power to the states, who then become the first line of implementation for environmental protection.

The states draw their authority from their citizens, who draft and approve the state constitutions. A state constitution establishes the executive, legislative, judicial, and bureaucratic framework for the state. Local governments within the states include municipal governments, counties, townships, school districts, and special districts (e.g., flood control districts) (De Grazia, 1957). In addition, there are a host of interstate compacts that deal with problems that cross state lines (Saffell, 1993). In areas of current policy in which both the state and federal governments have power, federal law is superior when there is a conflict between it and state law. In other words, a state law may not overturn or lessen the effect of a federal law. But states can exceed federal law (De Grazia, 1957). This is important in environmental policy largely because California typically has stronger environmental laws than does the federal government. Also, individual states and regions can advance environmental policy when the federal government fails to do so. Early climate change policy was put in place in this way due to the failure of the federal government to act (Rahm, 2010).

Federal power grew over time. After the Civil War, the Fourteenth Amendment made the federal government the guarantor of individual rights, thus expanding its role. In addition, the Commerce Clause (Article I, Section 8, Clause 3) came to be interpreted more broadly over time, and the federal government used the idea of regulating commerce between the states as an opportunity to regulate industries such as transportation, food, and drugs. To raise the money to support these new federal efforts, the Sixteenth Amendment created a national income tax.

Troubled times also increased the power of the federal government. FDR's grants-in-aid to state and local governments helped them finance Depression era programs designed at the federal level but implemented in the states. The 1960s brought another wave of increased federal support to state and local governments. To implement policies desired by the federal government that were controversial in many states, the federal government created a series of federal–local partnerships that often bypassed state governments entirely. Federal money was carefully allocated in the form of categorical grants, whereby the state or local government applying for the funding was carefully monitored by a federal agency (Cochran, Mayer, Carr, and Cayer, 1999).

This process of transferring federal revenues to states and localities in the form of grants to be spent on specific programs is a key feature of public policy. The federal government partners with other levels of government by providing full or partial funding for programs implemented and managed on the lower level. Many human service, agricultural, transportation, education, training, and environmental programs receive some, if not all, of their funding from the federal government.

If there has been a general trend in IGR over the last several decades, it has been the restructuring of intergovernmental relationships to emphasize state and local control (Jones, 1984). President Richard Nixon's revenue sharing program began the process by making federal money available to states and localities without extensive federal controls over their use. President Ronald Reagan had an even greater effect with his policy of New Federalism, which provided even greater policy flexibility and control to the states. The steady ascendancy to the presidency of former state governors (including Arkansas Governor William Clinton and Texas Governor George W. Bush) has ensured that the states receive a hearing at the highest level of the federal government.

Public policy in the last several decades has been affected by the trend of federal government downsizing. The efforts to reduce the size of the federal government have been accompanied by a movement toward devolution, the transfer of federal functions to the state or local level. As the federal presence has grown smaller, the role of the states and localities has become more important.

13.5 OTHER PUBLIC POLICY ACTORS: AN OVERVIEW

There are large numbers of other potential actors in public policymaking. Besides judges, members of Congress, the president, and the vast staffs that attend to them, there are political parties, policy advocates and interest groups, the media, and members of the public and the business community who are interested in particular issues and policy outcomes.

13.5.1 Political Parties

Political parties play an important role in policymaking. Their chief influence is on ideology and the stands that parties take regarding which policies should be promoted and which should not. Party platforms, or official statements of party opinion on particular issues, are determined by delegates at party conventions. Platforms are sometimes thought to be no more than rhetorical statements of the party's values; however, they are frequently illustrative of the true values and social preferences of influential party members. In recent decades, this has been particularly true of the Republican Party.

Parties play two roles—one in the electorate and the other in the government (Ripley, 1975). In the electorate, parties compete for dominance among voters. In the United States, third parties have been rare and short-lived, and the battle is primarily between Republicans and Democrats. The parties fight to get greater percentages of the electorate to support and vote for their candidates. In government, the battle has to do with the control of the presidency, majority control of each house of Congress, and the size of the majority party's margin in each house of Congress. The degree of control reveals how confident a party might be of moving its agenda forward.

13.5.2 Interest Groups

The First Amendment states, "Congress shall make no law … abridging the freedom of speech, or of the press; or the right of the people peaceably to assemble, and to petition the Government for a redress of grievances." Policy advocates and interest groups take this right very seriously. These groups vary widely in ideology and the means they use to achieve their desired policy outcomes. Some are firmly associated with particular industries or trade associations. Others form around shared ideology or values. Groups such as the National Rifle Association, the Sierra Club, the Mothers Against Drunk Driving, the Chemical Manufacturers Association, the League of Women Voters, the Planned Parenthood, the Cato Institute, the Heritage Foundation, and Earth First! differ greatly; yet they share the common goal of wanting to influence policy outcomes.

Most interest groups restrict their activities to peaceful participation in the democratic process. Interest groups use petition drives to influence public opinion and thus win the attention of members of Congress. They produce and distribute literature that explains their positions on policy issues. Many more professionalized interest groups hire lobbyists to lobby members of Congress and staff. By knocking on doors and talking with staff and members of Congress, they try to communicate their views and desires. Through lobbying activities, interest groups provide the information that they consider essential to understanding a policy issue. Interest groups ask the general public to write, call, or send e-mail to their representatives. Interest groups often appear at public hearings and enter formal comments about proposed policies or proposed regulations. They may raise funds and give the money to political candidates running for reelection (Jones, 1984).

Some interest groups and policy advocates engage in public protests and civil disobedience to try to influence the policy debate. The civil rights marches of the 1960s and the anti-Vietnam War protests of the 1960s and 1970s were very successful in influencing public opinion and policymakers. The protest against the World Trade Organization in Seattle in 2001 is an example of this type of civic activism. Members of environmental interest groups have chained themselves to trees to keep them from being cut down. Some militant groups like Earth First! engage in violent guerilla tactics, including driving long nails into trees so as to cause a chain saw to recoil and kill its operator, in order to save trees from loggers.

The business community can be particularly effective in influencing public policy, particularly at the state and local levels. Communities, states, and the country as a whole are very aware of the need for jobs. Government actions that might adversely affect job creation or retention are likely to cause concern. The business community also has financial resources from which to draw in its attempts to weigh in on policy approaches.

13.5.3 The Media

The First Amendment's protections of freedom of speech and the press created an environment for the development of a number of well-established written and broadcast media in the United States. The media have a long reach, and nearly all Americans have access to some of them. They affect policy debates by providing space and time to discuss policy issues, and they can give an issue high salience by providing this forum.

The media used to be dominated by a small elite group. Three major broadcast channels (ABC, NBC, and CBS) and a handful of influential national newspapers, journals, and news services (*The New York Times*, the *Wall Street Journal*, the *Washington Post*, *Time*, *Newsweek*, *US News & World Report*, and the Associated Press) took precedence in informing debate. Their supremacy has been challenged by the rise of cable broadcast news providers such as CNN and stations that deliver direct coverage of national policy events without journalistic commentary such as C-SPAN. The greatest threat to the preeminence of the elite, though, has been the emergence of news providers using the Internet as a means of delivery. Blogs and social media, in particular, have strong influence.

13.5.4 Public Opinion

Not all members of the community are involved in policy debates. Many people are simply not interested. Those who do have an interest, however, can influence policy outcomes in a variety of ways. As voters, they can alert their representatives about their views on policy issues. They can donate money to causes that interest them and to politicians who support those causes. They can participate in public forums and express their viewpoints by writing letters to newspapers and other public sources. In the era of the Internet, members of the public have access to a host of outlets where they can air their views and attempt to influence friends and followers.

Common measures of general public opinion include surveys of trust in government, ratings of presidential job handling, ratings about the most serious problems facing the nation, consumer confidence in the economic future of the country, and general public concern over specific policy issues. Public attitudes can and do affect the choices that political leaders make.

13.6 INSTITUTIONS AND ACTORS IN ENVIRONMENTAL POLICY

Who makes natural resource, energy, and environmental policy? As indicated in our overview, the formal actors at the federal level include the Congress, the president, and the courts. Congress passes laws. The Congress also has the power of oversight to review the actions of federal agencies involved in natural resource, energy, and environmental policy. The president is the executive and is charged with enforcing and implementing the laws. The president has the entire executive branch bureaucracy to accomplish this. The courts interpret laws and hear citizen suits. In this capacity, they may review laws to determine legislative intent and review executive acts (Birkland, 2005).

Several key executive agencies play a primary role in environmental, natural resource, and energy policy. The chief one for environmental policy is the EPA. The EPA was created by President Richard Nixon in 1970 as an independent federal agency. The EPA was initially constructed from the various environmental fragments that existed in other agencies. The EPA was patched together from pieces of the Department of Health, Education, and Welfare (air quality, solid waste, and drinking water), the DOI (water quality and pesticide research), the Department of Agriculture (pesticide regulation), and other agencies such as the Food and Drug Administration and the Atomic Energy Commission (AEC) (Quarles, 1976). These pieces were patched together and centered in the EPA. The creation of the EPA was vigorously opposed by several key members of the Nixon administration, including the secretaries of the DOI; Department of Health, Education, and Welfare; and Department of Agriculture. The opposition from the DOI was largely due to the fact that the Interior wanted to take over all environmental regulatory functions. The other agencies did not want to lose power to the newly created agency. The business community generally favored the creation of the EPA, believing that a single regulatory agency was preferable to multiple state-based regulatory standards (O'Leary, 1993). The EPA headquarters are in the Washington, DC area, but the EPA divides the nation into 10 geographic regions, and each region has a regional office. The mission of the EPA is to protect human health and the environment. The EPA implements and enforces legislation passed by Congress concerning air, water, waste, and transborder issues. The EPA, one of the largest regulatory agencies in the federal government, has over 17,000 employees.

Does the EPA have the ability to carry out the tasks expected of it? Both detractors and supporters of the agency have claimed that it does not. The formation of the agency from the remnants of other agencies is said to have created organizational confusion, if not dysfunction. These rocky beginnings were magnified by the Reagan administration's operation of the agency. President Reagan was openly hostile to the EPA, and while he was unable to eliminate the agency, he did cut its budget and staff and appointed managers who were equally opposed to its mission. This led to congressional micromanagement of the agency. Congress passed a series of extremely detailed and rigid laws to compel strict administrative compliance. Congress also enacted complex and ambitious new environmental laws, but the EPA's budget and staffing did not increase proportional to the new tasks.

By the mid-1990s, the EPA found itself in a highly charged political setting. Mounting dissatisfaction with its programs, from both supporters and detractors, ended in calls for reform. The 1994 elections brought to office a group of Republicans determined to lead the country down another road of regulatory reform. Congress focused on detailed scientific risk assessment, cost–benefit analysis, challenging existing rules, and finding ways to devolve authority to the states whenever possible. Threats of budget cuts also loomed. The Clinton administration's response was general agreement that the EPA needed reform.

Internal administrative reforms were undertaken under Vice President Al Gore's National Performance Review. These reforms included identification of obsolete and burdensome regulations that needed revision or elimination. There was general nonpartisan agreement about the need for a coherent mission statement to set agency priorities, an adequate budget for research and development that permits creation and defense of risk assessments, a more integrated approach to pollution management, a different way of working with the states, and an elimination of the quagmire of congressional oversight by 6 Senate committees, 7 House committees, and 31 subcommittees. The administration and Congress disagreed about which reforms would improve the EPA's performance and which were attempts to dismantle the agency and its programs.

George W. Bush's administration has made further attempts to weaken the agency. Not unlike Ronald Reagan, Bush was hostile to the EPA and its regulatory agenda. Early in his term of office, he attempted to reduce environmental protections put in place by former administrations. For instance, he called for increasing the allowable level of arsenic in water. Public concern forced the administration to abandon the plan. Bush's energy policy statements also called for drilling for oil in previously protected areas of the Alaskan wilderness. Congress failed to act to allow this. Additional controversies stemmed from the administration's attempt to roll back clean air standards so that coal-burning electric generating plants might continue to produce electricity unhindered by air quality standards. Under the Obama administration, the EPA has received at least verbal support from the president. But President Obama came to office under the pressure of the greatest economic crisis since the Great Depression and needed to keep his eyes firmly on jobs. Anything that might cost jobs was suspect, and environmental regulation was included in that concern. The economic crisis worked two ways though. The stimulus bill provided the Obama administration an opportunity to fund some environmental projects, particularly the development of renewable energy.

The EPA is not in charge of natural resource policy. That falls to a series of other agencies. The DOI is one of the largest executive agencies in the federal government, employing more than 70,000 people and having control over more than 20% of US land. DOI supplies water to 17 western states and leases oil and gas wells in the Outer Continental Shelf. The DOI has a number of bureaus that deal with natural resource and environmental issues including the Bureau of Land Management (BLM); the Bureau of Reclamation; the Bureau of Ocean Energy Management, Regulation, and Enforcement (formerly the Minerals Management Service); the National Park Service; the Office of Surface Mining Reclamation and Enforcement; the US Geological Survey (USGS); and the Fish and Wildlife Service (FWS). The US Department of Agriculture (USDA) has the Forest Service, which has a mission of preservation and conservation of the nation's forests, improving watersheds, and protecting the health of private forestlands.

Even though energy use and production are intimately tied to environmental quality, the EPA does not have the authority to regulate energy. Instead, use and production of energy are controlled by the Department of Energy (DOE).

When the DOE was formed in 1977, it took over the responsibilities of the Federal Energy Administration, the Energy Research and Development Administration, the Federal Power Commission, and parts and programs of several other agencies. The primary responsibility of the DOE is to administer the nation's nuclear weapons. It also conducts some research and development of energy technology, is involved in energy conservation and energy regulatory programs, and has a central energy data collection and analysis program. Energy use is a critical environmental issue especially when we consider the impact of the burning of fossil fuels on climate change, oil spills, mining damage, nuclear waste, and the potential of renewable energy.

A number of other agencies are also involved with the environment. The National Aeronautics and Space Administration (NASA) collects data to support climate research and to describe and measure energy and environmental phenomena that may contribute to climate variation and change. NASA's Mission to Planet Earth program plays a central role in this environmental mission. The National Oceanic and Atmospheric Administration (NOAA) gathers worldwide environmental data about the oceans, land, air, space, and Sun. The NOAA also maintains a national environmental database that combines its own data with environmental information collected by other agencies. The USGS collects and maintains data on streams, groundwater, erosion, flooding, water contamination, and sedimentation. Other agencies play tangential roles, for instance, the State Department when international environmental treaties are negotiated or the Transportation Department when fuel use standards are determined.

The Council on Environmental Quality (CEQ) created as part of the 1970 National Environmental Policy Act (NEPA) provides advice to the president on environmental matters. The CEQ has no regulatory power. Rather, it makes recommendations to the president and sometimes provides evaluations of environmental protection programs in the federal government. The CEQ's budget and influence are dependent on the current president. Under some presidents, it is powerful, while under others, it is less so (Vaughn, 2007).

The federal courts play a significant role in energy, natural resource, and environmental policy in several ways. The US Supreme Court has the power to review legislation and agency actions under the law for constitutionality. A good example of the influence of the Supreme Court on environmental policy was the court's decision in *Massachusetts v. Environmental Protection Agency* (2007, 127 S Ct 1438). In 2003, the EPA had decided that it would not regulate air emissions that contributed to climate change. Twelve states and several cities brought suit against the EPA to force it to regulate carbon dioxide and other GHG under its CAA authority to control air pollution. The court said that the EPA did indeed have the obligation to regulate GHG emissions and ordered it to do so or to explain why it failed to do so. The court thus changed US policy on the regulation of GHG emissions.

The lower federal courts play a significant role in environmental policy as a result of the use of citizen suits. A citizen suit is when an individual person sues a government agency, corporation, or another citizen to force them to enforce legislation. Many suits have been filed to compel enforcement of environmental laws. For instance, a citizen can sue a corporation for polluting water under the CAA. Most commonly, however, citizen suits are aimed at the EPA for failure to create and enforce regulations that environmental laws require it to issue. A number of environmental laws allow citizen suits including the Clean Water Act (CWA); Safe Drinking Water Act (SDWA); CAA; Resource Conservation and Recovery Act (RCRA); Comprehensive Environmental Response, Compensation, and Liability Act; Surface Mining Control and Reclamation Act; Endangered Species Act; and Emergency Planning and Community Right-to-Know Act. Because citizen suits are allowed, several large environmental advocacy organizations, such as the Sierra Club, have strategies to use litigation as a key mechanism to pursue their goals.

The states are also active players in environmental policy. Almost all states have the state equivalent of the EPA although most states have unique names for their environmental agencies. The states vary greatly in the advocacy and activism of their state agencies. For instance, California has a very activist environmental agenda, while Texas has a very minimal one. Accordingly, their state agencies reflect the overall perspective of their state's leadership. The role of the states is determined in a federal system by the Constitution. The Constitution specifies the specific role of the federal government and the states (and when there is a disagreement, the courts step in and resolve the matter). Most environmental laws allow for a process of delegation, that is, the federal government delegates to the states the authority to carry out federal legislation in their jurisdiction if the federal government lead agency approves the state's plan of operations. Most states prefer to control their own internal affairs, and so most states seek delegated authority if they have the resources to implement the policy. The federal government often lacks adequate resources itself to implement federal environmental legislation and so is eager for the states to partner with it. As long as the states are willing partners, this system works well. However, when a state fails to enforce federal law to the satisfaction of the federal government lead agency, the federal government can revoke the delegated power and assume the enforcement within that state's jurisdiction. In 2011, 2012, and 2013, the state of Texas battled the federal government's interpretation of its implementation of the CAA. The EPA briefly took back its delegated authority based on its contention that Texas' enforcement was inadequate. Regulatory federalism has inherent conflicts. As mentioned earlier, some states have a more progressive agenda and seek to exceed federal guidelines, while other states look at federal regulation as an unnecessary intrusion on their sovereignty.

A host of organizations with a mission focused on natural resource, energy, and environmental issues exist. Many of them are nonprofit organizations with an environmental advocacy mission. There is a great variety in the perspectives they have and the types of action they undertake. Some are international in focus, while others are local. Some are mainstream organizations, while others are "radical" in philosophy and tactics.

The mainstream environmental organizations include groups like the National Audubon Society, Sierra Club, and World Wildlife Fund. They have a long history of engagement. These groups are membership-based organizations drawing most of their operating funds from membership dues and contributions. They tend to engage in advocacy and lobbying activities, although more recently Sierra has changed strategy and has devoted much of its resources to litigation. Like Environmental Defense Fund, Sierra is a professionalized organization with many attorneys able to pursue the organization's goals. These groups are organized internationally or nationally and typically have regional or local chapters. Some are focused on a specific activity. Audubon, for instance, is focused on birding. Others, like Sierra Club, have more widespread interests that include opportunities for local involvement, activism, and outings. The Sierra Club also sponsors current priority campaigns such as protecting wildlands, global warming, and population stabilization. The World Wildlife Fund is devoted to wildlife ecology and management. They focus especially on threatened and endangered species. Friends of the Earth is another mainstream group that focuses on climate and energy, food and technology, oceans and forests, and the economics of environmental sustainability.

Some environmental organizations engage in what they call "direct action" tactics. Greenpeace pioneered these efforts. Greenpeace has international and national organizations. Greenpeace USA is focused on protecting forests and the oceans and stopping global warming. Greenpeace was founded in the 1970s by a group of antiwar activists that were taking nonviolent direct action against US nuclear weapons testing. Nonviolent direct action is linked to the Quaker concept of "bearing witness." As used by Greenpeace, direct action involves physically acting to stop an environmental wrong and to raise public awareness of the issue. Greenpeace famously hangs spectacular banners in public spaces and sails ships between whales and whaling vessels. While direct action tactics are not always legal, they are nonviolent.

Not all activist groups share the nonviolent commitment of Greenpeace. For instance, Earth First! uses "direct pressure" rather than direct action. They are committed to Deep Ecology action. Within the Deep Ecology perspective, all life on the planet is of sacred value. Their symbol is a clenched fist with the caption "No Compromise in the

Defense of Mother Earth!" They provide tools such as the *Wolf Hunt Sabotage Manual*, which describes how to find and destroy wolf traps and release live wolves that have been captured and the use of air-compressed horns and smoke bombs for stopping wolf hunts in the states that permit the hunts. They also provide a book called *Ecodefense: A Field Guide to Monkeywrenching*. This book explains how to engage in a variety of activities including how to spike trees to prevent logging (and likely kill potential loggers) and how to disrupt construction sites, plug waste discharge pipes, disrupt grazing, spike roads (to flatten tires), burn machinery, disable vehicles and heavy machinery, sabotage aircraft, attack an urban residence and private automobiles, trash a condo, sabotage a computer, make stink and smoke bombs, and avoid arrest. They believe that other environmental groups have sold out to corporate culture and reject them as too moderate. Their strategy crosses the line into illegal activities and promotes any and all action to protect wilderness.

A special case of environmental advocacy organization is the land trust. These organizations focus on conservation, ecology, wildlife ecology, and land management. They are typically private nonprofit organizations. Land trusts seek to acquire land and then to hold it in stewardship. Most states have a number of these that typically are organized around protecting a particular region. Some, like the Wilderness Land Trust, is national in focus and owns nearly 200 holdings in 35 national wilderness areas in the southwest and northwest.

Green Parties are another player in environmental activism. While they are more influential in Europe than in the United States, the Green Party of the United States has significant power at municipal, county, and state levels. The electoral victories of the Greens date to the mid-1980s. The Green Party has run candidates in several presidential elections; the most famous (or infamous) was Ralph Nader in 2000. Many believe that Nader took enough votes in Florida to split the state, resulting in an eventual Gore loss (after *Bush v. Gore* was settled by the Supreme Court).

A final group that we should discuss is the "permanent opposition" (Hays, 2000; Switzer, 1997). Early in the development of the environmental movement, there was a sense of bipartisanship regarding the need to clean and protect the environment. The public and its elected officials were widely supportive. That early consensus, however, did not survive long. As certain interest groups realized that there would be costs associated with a cleaner environment, they began to form opposition groups. The constituency of the opposition has changed over time, but it is generally correct to say that the main interest groups in the opposition include farmers, ranchers, business, mining, automakers and autoworkers. Those states with an economic base in extraction tend to be antienvironmental (while those states based in tourism or the service industries tend to be supportive). Sometimes the opposition came together, such as in the western movement known as the Sagebrush Rebellion. As environmental regulations increased in number, those who favored less regulation joined the opposition. By 1980, almost the entire Republican Party was antiregulatory and therefore found itself in conflict with traditional environmentalists. In the decades that followed, the Republicans grew less and less supportive of federal environmental regulations and environmental policy in general. When climate change emerged as a significant issue, most Republican candidates openly denied its existence or questioned whether it was caused by human actions (such as the burning of fossil fuels). In the 2012 campaign, many Republican candidates called for the elimination of the EPA. Democrats, however, remain generally concerned with environmental issues. The public generally mirrors the political split.

13.7 THE POLICY PROCESS

The policy process describes the making of law and policy. Generally, it is conceived of as a series of stages; however, it is important to note that each stage may not be independent of each other. There can be a certain amount of overlapping. These stages consist of agenda setting, policy formulation, policy adoption, implementation, and evaluation.

Agenda setting involves an issue getting the attention of lawmakers. Different approaches have been used to describe agenda setting. Some are sequential models of stages in that they assume that a problem causes a governmental response (Anderson, 2000). More recent research has been done on agenda setting that has revealed it to be less linear and directional or rational and comprehensive than previously thought. Using a derivation of Cohen, March, and Olsen's (1972) garbage can theory of organizational choice used to describe how complex organizations work, John Kingdon (1995) postulated an equivalent model for the political process. Kingdon describes what he calls windows of opportunities that open when a focusing event causes an issue to become salient. For example, the Love Canal event opened a window of opportunity for lawmakers to pass legislation that dealt with hazardous waste, or the burning of the Cuyahoga River provided a focusing event to pass clean water legislation. But in contrast to government solving problems when they arise in a linear model of problems to government and solutions to the people, Kingdon argued that solutions to potential problems have already been developed by individuals or groups he dubs policy entrepreneurs. The potential solutions are already formed and floating around in the ether. These policy entrepreneurs have their pet policies they wish to implement but must wait for a focusing event to open a window of opportunity. When this happens, they have a short period of time to muster support for their solution before the window closes and no policy is enacted. The length of time the window stays open is associated with what Downs (1972) calls the "issue–

attention cycle." According to Downs, an issue only remains on people's minds for a brief period of time before they move on to another issue. Kingdon refers to that time span as the window of opportunity. If an issue makes it to the agenda, it moves to the next stage.

Policy formulation occurs after an issue is on the agenda, and general discussions begin around potential solutions. In this stage, policy entrepreneurs actively advocate for support for specific policies, which may include modifying original policy proposals or competing for proposals proposed by policy entrepreneurs. The formulation of the policy represents the culmination of advocacy, negotiations, and trade-offs. Policy formulation may involve a number of actors including the president, officials from executive agencies, members of Congress, congressional staffers, the public, specific interest groups, stakeholders from the states and business community, and the nonprofit advocacy groups.

Policy adoption is the stage of acceptance of a particular policy solution as formulated. It often involves the passing of legislation and/or executive agency regulations. Political bodies may hold hearings to discuss the formulated policy. The public is often not involved at this stage, rather actors conduct closed-door conversations to finalize the proposed policy, before it is formally adopted.

Policy implementation is the stage at which the policy is put into action. The legislation is turned over to a government agency for program development and implementation. At the federal level, this would involve the EPA, DOI, or USDA. For the EPA, one of the most serious challenges for implementation includes the process of drafting and finalizing rules (described in the following text). Legislation is typically vague. While legislation may include the major goals members of Congress wish to see achieved, it rarely states specifics. Rather, legislation typically instructs the executive agency in charge of implementation to draft and finalize any and all regulations it deems necessary for the implementation of the law and its intended goals. A good example of this is the 1972 Federal Water Pollution Control Act. The main emphasis of the legislation was to give the EPA six deadlines by which it was to grant permits to water pollution sources, issue wastewater guidelines, and require polluters to install water pollution control technology to make the nation's waterways safe for fishing and swimming. But what constituted safe for fishing and swimming was left to the EPA to determine and regulate (Vaughn, 2007). Implementation also involves the daily running of programs that includes some measure of ongoing data collection, oversight, supervision, and enforcement of the regulated community.

Policy evaluation occurs as the last stage of the model. After a policy has been in place for some time, it is necessary to evaluate its success or lack thereof. Policy evaluation is the time when this occurs. There are different types of evaluations. Most evaluations look at programs that implement policy and evaluate them either for process or for outcomes. A process evaluation seeks to determine if a program is running efficiently and economically. An outcome evaluation seeks to determine if the program is meeting its desired goals. While the desired result of an evaluation is to improve program and policy performance, other possibilities exist. For instance, many states have put in place sunset laws that mandate cyclical evaluation of programs with the intent of terminating programs that are determined to be underperforming.

13.8 RULEMAKING: THE PROCESS

Large parts of environmental, energy, or natural resource policy are regulatory. That is, wide swaths of policy are implemented using rules or regulations issued by the lead agency. Rulemaking is the process that is followed by a federal government agency to issue regulations. The rulemaking process is specified in the Administrative Procedure Act of 1946. The Act specifies that an agency wishing to issue a regulation must first publish a draft regulation in the Federal Register, briefly wait for and accept comments, and then publish a final rule in the Federal Register.

Even though the law has never been changed, agencies today follow a more extensive path to new regulations. Perhaps this is due to the unpopularity of regulations and the desire of agencies to include as much feedback into their deliberation processes as possible. Today, agencies do far more than just issue a draft notice in the Federal Register. Frequently, they will begin with a scoping exercise, which usually involves the hiring of some neutral third party to contact the likely affected community and begin discussions with them regarding the forthcoming rule. They will ask how the rule could be written in such a way that it will cause the least amount of disruption. Agencies will also hold public meetings at various locations around the country to get input on the concerns of those who will be regulated. Public agencies allow communication orally, in traditional written form, and by electronic means. Agencies today may also publish in the Federal Register a discussion of their considerations regarding feedback received.

Once the final rule is published, it becomes part of the Code of Federal Regulations (CFR) and has the status of law. The CFR is organized into 50 Titles by subject area. The Titles that pertain to environmental issues include the following: Title 10, Energy; Title 18, Conservation of Power and Water Resources; Title 21, Food and Drugs; Title 30, Mineral Resources; Title 33, Navigation and Navigable Waters; Title 36, Parks, Forests, and Public Property; Title 40, Protection of Environment; Title 43, Public Lands; and Title 50, Wildlife and Fisheries.

Regulations can be quite lengthy and detailed. For instance, the procedures for implementing the NEPA consist of four long subparts. To give the readers a sense of a regulation, the policy and purpose section of subpart A and one section from subpart B that describes how to prepare an environmental impact statement are quoted below:

§ 6.100 Policy and purpose.
   a. The National Environmental Policy Act of 1969 (NEPA), 42 U.S.C. 4321 *et seq.*, as implemented by the Council on Environmental Quality (CEQ) Regulations (40 CFR Parts 1500 through 1508), requires that Federal agencies include in their decision-making processes appropriate and careful consideration of all environmental effects of proposed actions, analyze potential environmental effects of proposed actions and their alternatives for public understanding and scrutiny, avoid or minimize adverse effects of proposed actions, and restore and enhance environmental quality to the extent practicable. The U.S. Environmental Protection Agency (EPA) shall integrate these NEPA requirements as early in the Agency planning processes as possible. The environmental review process shall be the focal point to ensure NEPA considerations are taken into account.
   b. Through this part, EPA adopts the CEQ Regulations (40 CFR Parts 1500 through 1508) implementing NEPA; subparts A through C of this part supplement those regulations, for actions proposed by EPA that are subject to NEPA requirements. Subparts A through C supplement, and are to be used in conjunction with, the CEQ Regulations.

§ 6.207 Environmental impact statements.
   a. The Responsible Official will prepare an environmental impact statement (EIS) (see 40 CFR 1508.11) for major federal actions significantly affecting the quality of the human environment, including actions for which the EA analysis demonstrates that significant impacts will occur that will not be reduced or eliminated by changes to or mitigation of the proposed action.
      1. EISs are normally prepared for the following actions:
         (i) New regional wastewater treatment facilities or water supply systems for a community with a population greater than 100,000.
         (ii) Expansions of existing wastewater treatment facilities that will increase existing discharge to an impaired water by greater than 10 million gallons per day (mgd).
         (iii) Issuance of new source NPDES permit for a new major industrial discharge.
         (iv) Issuance of a new source NPDES permit for a new oil/gas development and production operation on the outer continental shelf.
         (v) Issuance of a new source NPDES permit for a deepwater port with a projected discharge in excess of 10 mgd.
      2. The Responsible Official, or other interested party, may request changes to the list of actions that normally require the preparation of an EIS (i.e., the addition, amendment, or deletion of a type of action).
      3. A proposed action normally requires an EIS if it meets any of the following criteria. (See 40 CFR 1507.3(b)(2)).
         (i) The proposed action would result in a discharge of treated effluent from a new or modified existing facility into a body of water and the discharge is likely to have a significant effect on the quality of the receiving waters.
         (ii) The proposed action is likely to directly, or through induced development, have significant adverse effect upon local ambient air quality or local ambient noise levels.
         (iii) The proposed action is likely to have significant adverse effects on surface water reservoirs or navigation projects.
         (iv) The proposed action would be inconsistent with state or local government, or federally-recognized Indian tribe approved land use plans or regulations, or federal land management plans.
         (v) The proposed action would be inconsistent with state or local government, or federally-recognized Indian tribe environmental, resource-protection, or land-use laws and regulations for protection of the environment.
         (vi) The proposed action is likely to significantly affect the environment through the release of radioactive, hazardous or toxic substances, or biota.
         (vii) The proposed action involves uncertain environmental effects or highly unique environmental risks that are likely to be significant.
         (viii) The proposed action is likely to significantly affect national natural landmarks or any property on or eligible for the National Register of Historic Places.
         (ix) The proposed action is likely to significantly affect environmentally important natural resources such as wetlands, significant agricultural lands, aquifer recharge zones, coastal zones, barrier islands, wild and scenic rivers, and significant fish or wildlife habitat.
         (x) The proposed action in conjunction with related federal, state or local government, or federally-recognized Indian tribe projects is likely to produce significant cumulative impacts.

(xi) The proposed action is likely to significantly affect the pattern and type of land use (industrial, commercial, recreational, residential) or growth and distribution of population including altering the character of existing residential areas.

4. An EIS must be prepared consistent with 40 CFR Part 1502.

b. When appropriate, the Responsible Official will prepare a legislative EIS consistent with 40 CFR 1506.8.

c. In preparing an EIS, the Responsible Official must determine if an applicant, other federal agencies or state or local governments, or federally-recognized Indian tribes are involved with the project and apply the applicable provisions of § 6.202 and Subpart C of this part.

d. An EIS must:

1. Comply with all requirements at 40 CFR parts 1500 through 1508.
2. Analyze all reasonable alternatives and the no action alternative (which may be the same as denying the action). Assess the no action alternative even when the proposed action is specifically required by legislation or a court order.
3. Describe the potentially affected environment including, as appropriate, the size and location of new and existing facilities, land requirements, operation and maintenance requirements, auxiliary structures such as pipelines or transmission lines, and construction schedules.
4. Summarize any coordination or consultation undertaken with any federal agency, state and/or local government, and/or federally-recognized Indian tribe, including copies or summaries of relevant correspondence.
5. Summarize any public meetings held during the scoping process including the date, time, place, and purpose of the meetings. The final EIS must summarize the public participation process including the date, time, place, and purpose of meetings or hearings held after publication of the draft EIS.
6. Consider substantive comments received during the public participation process. The draft EIS must consider the substantive comments received during the scoping process. The final EIS must include or summarize all substantive comments received on the draft EIS, respond to any substantive comments on the draft EIS, and explain any changes to the draft EIS and the reason for the changes.
7. Include the names and qualifications of the persons primarily responsible for preparing the EIS including an EIS prepared under a third-party contract (if applicable), significant background papers, and the EID (if applicable).

e. The Responsible Official must prepare a supplemental EIS when appropriate, consistent with 40 CFR 1502.9.

§ 6.208 Records of decision.

a. The Responsible Official may not make any decisions on the action until the time periods in 40 CFR 1506.10 have been met.

b. A record of decision (ROD) records EPA's decision on the action. Consistent with 40 CFR 1505.2, a ROD must include:

1. A brief description of the proposed action and alternatives considered in the EIS, environmental factors considered, and project impacts;
2. Any commitments to mitigation; and
3. An explanation if the environmentally preferred alternative was not selected.

c. In addition, the ROD must include:

1. Responses to any substantive comments on the final EIS;
2. The date of issuance; and
3. The signature of the Responsible Official.

d. The Responsible Official must ensure that an applicant that has committed to mitigation possesses the authority and ability to fulfill the commitment.

e. The Responsible Official must make a ROD available to the public.

f. Upon issuance of the ROD, the Responsible Official may proceed with the action subject to any mitigation measures described in the ROD. The Responsible Official must ensure adequate monitoring of mitigation measures identified in the ROD.

g. If the mitigation identified in the ROD will be included as a condition in the permit or grant, the Responsible Official must ensure that EPA has the authority to impose the conditions. The Responsible Official should ensure that compliance with assistance agreement or permit conditions will be monitored and enforced under EPA's assistance agreement and permit authorities.

h. The Responsible Official may revise a ROD at any time provided the revision is supported by an EIS. A revised ROD is subject to all provisions of paragraph (d) of this section.

§ 6.209 Filing requirements for EPA EISs.

a. The Responsible Official must file an EIS with the NEPA Official no earlier than the date the document is transmitted to commenting agencies and made available to the public. The Responsible Official must comply with any guidelines established by the NEPA Official for the filing system process and comply with 40 CFR 1506.9 and 1506.10. The review periods are computed through the filing system process and published in the FEDERAL REGISTER in the Notice of Availability.

b. The Responsible Official may request that the NEPA Official extend the review periods for an EIS. The NEPA Official will publish notice of an extension of the review period in the FEDERAL REGISTER and notify the CEQ. (GPO, n.d.)

Subpart C, Requirements for Environmental Information Documents and Third-Party Agreements for EPA Actions Subject to NEPA, and subpart D, Assessing the Environmental Effects Abroad of EPA Actions, are not shown nor are the remaining sections of subpart B. However, they are each of equal or greater length and complexity than what has been used to illustrate a regulation above. As the reader has undoubtedly determined, dealing with the CFR can be cumbersome and time-consuming! Yet much of energy, natural resource, and environmental policy is specified in regulations such as this one.

Thousands of new rules are written each year. The number of rules written by an agency varies depending on the legislation it administers. The EPA administers most of the country's environmental protection legislation and as a consequence has many rules that it has written or needs to write to implement the law. Rules can take years to draft and finalize. Agencies strive to produce rules in a timely way, but given limited institutional capacity, many agencies like the EPA struggle under the load. As a result, the production of rules to support legislation is often delayed. For the EPA, a sizeable backlog of rules at various stages of completion exists (Kerwin, 2003).

13.9 TYPES OF ENVIRONMENTAL REGULATORY APPROACHES

The first wave of laws passed in the United States during the post-World War II period typically used voluntary approaches. The fact that very little was accomplished by voluntary means lead to a mandatory approach. This approach, commonly called command and control or standards and enforcement, is used by most of the nation's environmental laws and regulations. Command and control involves five stages: goals, criteria, quality standards, emission standards, and enforcement. In stage 1, Congress determines what goals should be reached. As mentioned earlier, legislation can be quite vague, so the agencies need to provide the specifics. But the statement of goals, even if vague, sends a message to the polluters regarding what pollutants and sources will be given priority and how vigorously Congress intends to implement programs. In stage 2, the agency sets criteria. Criteria are technical data provided by scientists that indicate which substances are responsible for pollution and the various levels of concentration at which environmental damage occurs. Based on this information, the agency can move to stage 3, the setting of quality standards. Quality standards specify the maximum level of specific substances to be allowed in the air, water, soil, workplace, etc. The level of pollutants to be tolerated is associated with the risk the agency is willing to assume. Cost may be a factor in determining quality standards as well as science. Higher standards may result in millions or billions of additional dollars spent on pollution control technology and may result in job losses. Quality standards are put in place through emission standards that prescribe the acceptable pollutant discharges for key sources of pollution. Congress has used two approaches to set emission standards. One approach is to set them at the quality standard and require polluters to meet that level of cleanliness regardless of cost. The other approach is setting emission standards based on readily accessible technology, which is usually less costly even if it is less protective. The final stage is enforcement. Here, the regulatory agency typically requires the regulated community to keep track of their own emissions and self-report any accidental or intentional lapses. Clearly, this sort of system must be backed up with spot inspections to assure that the regulated community is adequately keeping records and reporting. Violators of specific regulations are fined (Rosenbaum, 2014).

Command and control approaches allow the regulatory agency the discretion to be very prescriptive regarding methods to reduce pollution. Under this system, the EPA could (and does) specify specific technology that companies must purchase and install, regardless of the cost. Companies have little freedom in choosing specific technologies.

Command and control regulatory approaches generally came under attack in the 1980s during the presidency of Ronald Reagan. Reagan came to office with the belief that he had a mandate to reduce the nation's regulatory burden, and environmental regulations were high on his list. During this era, there was a worldwide movement against tight government control. Margaret Thatcher was the prime minister of Great Britain, and she, like Ronald Reagan, championed the private sector over the government. The Soviet Union had fallen and a consensus was developing around the virtues of free enterprise and capitalism. Many suggested reforms to the regulatory system to embrace the power of the market and thereby replace command and control. The EPA experimented with several pilot programs during the 1980s. By 1990, Congress went forward with the shift when it passed the CAA Amendments of 1990.

The part of the CAA amendments that addressed acid rain was structured to use market mechanisms to reduce pollution. A cap-and-trade market was set up for sulfur emissions.

Polluters were given emission permits in the form of vouchers and trading was allowed. The idea was that those companies with modern efficient equipment would be able to do business with the vouchers allocated to them, whereas inefficient businesses would have to either update their equipment or purchase additional vouchers, thus making them less competitive. Over time, the cap would be lowered, thus reducing the overall emissions of sulfur and reducing acid rain. At the same time, companies were released from the burden of command and control that typically dictated the means of pollution reduction. With this reform, it was up to individual companies to determine how best to reduce their sulfur emissions. Cap-and-trade for sulfur was successful and nitrous oxides were later added. Polluters generally preferred the relative freedom these market-based mechanisms provided as opposed to the heavy-handedness of command and control (Vaughn, 2007). This type of regulatory approach became so popular in the 1990s that the United States insisted upon a carbon market being part of the Kyoto Protocol, the first international treaty to deal with climate change. European countries generally fought the idea, preferring traditional command and control approaches. Nevertheless, to gain US support for the treaty, a carbon market was written into the treaty. Ironically, the United States did not ratify the treaty and therefore never participated in the market (Rahm, 2010). During George Bush's years in the White House, the emphasis on voluntary approaches returned. For example, after pulling out from the Kyoto process, the United States articulated its voluntary efforts to reduce carbon emissions.

13.10 EVOLUTION OF NATURAL RESOURCE, ENERGY, AND ENVIRONMENTAL POLICY

Environmental policy is a mix of three distinct but overlapping policy areas. Natural resource policy typically centers on land use. The main issues in natural resource policy concern the protection or use of the land for development and other purposes and protection of species. Energy policy is important because a great deal of the pollution that affects our world is a direct result of fossil fuel use. Nuclear waste is also an issue. Environmental policy focuses on clean air, clean water, the disposal of solid and hazardous waste, and transborder pollution.

There have been three great waves of natural resource policy. The first dates to the beginnings of the nation when the federal government used its vast acres of western lands to lure settlers to the West. By the time of the Civil War, a movement had begun to preserve wilderness lands for future generations. This early preservationist movement was countered by a conservationist movement that encouraged the managed use of forest resources for economic development while conserving them for sustained yield. The second great wave of natural resource policy came in the 1930s when the center of attention was flood control and soil conservation. The third wave began when attention turned once again to conservation and preservation in the 1960s.

Energy policy is critical to a cleaner environment. Fossil fuel use creates air pollution, acid precipitation, particulates that are known to cause cancer, and carbon dioxide (the leading climate change gas). Finding substitutes for fossil fuels while still increasing the amount of available energy for growth and development is one of the largest challenges confronting society today.

Environmental policy burst on the national scene in the late 1960s. The first areas to receive attention in the 1970s were clean air and clean water. By 1980, hazardous waste and toxic waste were added to the list of environmental issues needing government action. The 1990s focused on the destruction of ecologically critical lands and forests, climate change, ozone depletion, loss of biodiversity, deterioration of urban quality of life, and sustainable development. Each of these areas of policy is further described later.

13.10.1 Natural Resource Policy

Natural resource policy is older than either energy or environmental policy, dating to the very beginnings of the nation. One of America's primary resources was land, and when the nation was formed, colonies with claims to western lands were required to release those lands to the federal government. As a result, the federal government became the holder of vast acres of lands. When the Jefferson administration purchased the Louisiana Territory in 1803, the size of the country doubled. As settlers moved west, more and more land was acquired and added to the national domain.

Congress used much of this land to lure development to the west. The Homestead Act of 1862, which gave 160 acres of public land to anyone willing to farm it, eventually converted 250 million acres of public lands to private farms. The federal government encouraged the development of railroads by giving land to railroad corporations in exchange for miles of track laid. The federal government also provided incentives for westward expansion and development by providing mineral and grazing rights on public lands. The Mining Law of 1872, for instance, gave miners free access to mineral deposits on public lands. In like fashion, ranchers were allowed to graze their herds on public lands (Kraft, 1996).

By the time of the Civil War, a movement had begun to preserve some wilderness lands for future generations. This movement was spearheaded by John Muir, founder of the Sierra Club, who was able to secure the protection of Yosemite Valley from economic development in 1864. Congress set aside the Yellowstone National Park in 1872 and followed with other national parks. This early preservationist movement was countered by the conservationist movement championed by Gifford Pinchot, who became head of what is today the US Forest Service in Theodore Roosevelt's administration. Pinchot encouraged the managed use of forest resources for economic development while conserving them for sustained yield. Despite the differences between conservation and preservation, both movements achieved considerable success in the early 1900s.

The National Wildlife Refuge System was created in 1903. Today, it contains 91 million acres, about 85% of which are in Alaska. There are more than 500 refuges in the system, which provide habitat for animals, as well 18 million acres of forest on which limited logging and other commercial activities such as grazing, mining, and oil drilling are permitted.

The National Forest System, which was created in 1905, includes 187 million acres of land mostly in Alaska, the Far West, and the Southeast. Fifty national forests located in eastern states are chiefly managed by the National Forest Service.

The National Park System was established in 1916. Today, it covers more than 80 million acres and contains more than 360 parks. Fifty of them are national parks. The National Park System also contains 300 national monuments, battlefields, memorials, historic sites, and recreational areas (Kraft, 1996).

The second great wave of natural resource policy came in the 1930s. During President FDR's 12 years in office, much was done to promote flood control and soil conservation in response to the drought that resulted in the Dust Bowl. The Civilian Conservation Corps and the Soil Conservation Service were established to help repair environmental damage. Controls on overgrazing were instituted through the Taylor Grazing Act of 1934. The BLM was created within the DOI in 1946 when President Harry Truman consolidated DOI's old grazing service and General Land Office (Rosenbaum, 2014).

Attention once again turned to conservation and preservation in the 1960s. The National Wilderness Preservation System was created in 1964. It contains millions of acres without any roads, permanent improvements, or sustained human habitation. In 1968, Congress created the National Wild and Scenic Rivers System, which today protects 10,500 miles of free-flowing rivers and shorelines from development. In the early 1970s, additional land was given protected status in National Marine Sanctuaries and National Estuarine Research Reserves, which contain 13 marine units of 9000 square nautical miles. These were ordered to be protected by the NOAA under the Coastal Zone Management Act of 1972.

Membership in older established conservation societies grew rapidly in the 1960s, and new environmental groups were established. Differences among these interest groups emerged. The mainstream groups, including the Sierra Club and the National Wildlife Federation, devoted their efforts to national public policy. The "Greens," including Greenpeace and Earth First!, focused mainly on local issues; emphasized public education, direct action, and social change; and worked to mobilize the grassroots. Policy analysis and research-based interest groups such as the Resources for the Future, the Union of Concerned Scientists, the Worldwatch Institute, and the World Resources Institute gained popularity. Private land conservation groups like the Nature Conservancy also emerged.

There have been conflicts over the management of federal lands since the nation's beginning. The agency at the center of this debate is the BLM, which oversees more land than any other government agency. As a result of the Alaska Land Act of 1980, BLM controls large areas of wilderness; these now constitute about half of BLM lands. The agency is also in charge of federal mineral leases. In 1976, the Federal Land Policy and Management Act formally ended the practice of transferring the national domain to private owners and gave BLM control of the remaining national domain. BLM has been moving slowly toward a policy of sustainable resource management.

As part of a review of land use in the 1970s, the Forest Service undertook an inventory of roadless land within the National Forest System for possible inclusion in the protected wilderness system. Roadless Area Review and Evaluation (RARE) I was completed in 1976 and RARE II in 1979. The Reagan administration was not sympathetic to the idea of wilderness preservation, and Interior Secretary James Watt proposed opening wilderness areas to mineral development. Congress designated lands as protected wilderness despite Reagan administration opposition. When President Reagan and Secretary Watt tried to withdraw millions of acres from possible wilderness designation, Congress objected. In 1990, Congress refused to allow logging on more than one million acres in southeast Alaska.

In its last days, the Clinton administration issued final regulations banning road construction and most logging in 60 million acres of federal land, or about one-third of the federal forests. President George W. Bush opposed the designation of land as wilderness but decided to let it stand with some modifications. The Bush administration intended to let local officials modify the ban on a case-by-case basis and allow logging, mining, and drilling. According to the Bush administration, these changes addressed federal lawsuits brought by Idaho and Boise Cascade. Environmentalists argued that the changes would damage the protections intended by the regulation. In 2002, the Bush administration proposed the Healthy Forests Initiative, which was later included in the Healthy Forests Restoration Act. This initiative excluded logging from federal review. The Bush administration also exempted Alaska's Tongass forest from the roadless rule making it more likely to be developed. While the Obama administration has been criticized by Republicans for attempting to slow mineral extraction from public lands, the president is supportive of domestic gas and oil production, even on federal land.

13.10.2 Energy Policy

We need energy to light, heat, and cool our homes and workplaces; to run our machines; and to fuel our cars, planes, and trains. Without energy, everything stops. Products cannot be manufactured; computers cannot run; and transport ceases. An advanced industrial economy such as ours is totally dependent on an uninterrupted supply of affordable energy.

Unfortunately, energy use results in pollution. Fossil fuel use is the number one cause of air pollution. The emissions from automobiles and from power plants that burn fossil fuel put millions of tons of sulfur dioxide and nitrogen dioxide into the air. These chemicals combine with water to form acid precipitation. Power plants and autos emit particulates (small solid particles) that are known to cause cancer. Fossil fuel use also results in the production of carbon dioxide, the main cause of global climate change.

To avoid the potential consequences of global climate change (e.g., rising sea levels, habitat destruction, species extinction, and increasing number and strength of disastrous storms), carbon dioxide emissions—and thus fossil fuel use—must be reduced. However, global demand for energy has tripled in the past 50 years and is likely to triple again in the next 30. In the past, the increasing demand for energy came from industrialized countries; in the future, it is likely to come from developing nations (Hinrichs and Kleinbach, 2002). Finding substitutes for fossil fuels while still providing energy for growth and development is one of the largest challenges confronting society today. Two options to replace fossil fuels are sustainable energy and nuclear power.

Sustainable energy is energy that comes from renewable sources. Often, these sources are common geological features such as sunshine, blowing wind, moving water, or heat released from the Earth. Sources of sustainable energy can be grown when agricultural production is used for energy crops. Waste can be converted into fuel when agricultural and forest wastes (corn stalks and woody plant material) or landfill gas is burned. Advanced technologies, such as wind turbines and photovoltaic (PV) panels, can be brought to bear to provide sustainable energy. Experimental technologies, such as hydrogen fuel cells, drawing on hydrogen produced from renewable sources, are under development.

Conservation and energy efficiency have a huge role to play in the movement toward a sustainable energy society. Green building designs that drastically reduce the need for heating and air conditioning, advanced designs for transport including vehicles designed to run partially or totally on ethanol or gasoline–electric hybrid designs that vastly increase mileage, as well as energy-efficient appliances and lighting are all of enormous value. The more conservation and energy efficiency are brought into play, the easier it will be to provide for remaining energy needs using renewables.

Shifting to a society that relies primarily on sustainable sources of energy makes logical sense for environmental, security, and long-term viability reasons, but making the transition is not easy. The fossil fuel economy and infrastructure are well entrenched. Diffusion of new technologies and processes takes time and often requires assistance to get over the barriers of initial adoption. This is where public policy plays a crucial role. While energy is and will remain a private sector industry, the public sector does play a role in helping sustainable energy gain a foothold in the economy.

The main thrust of federal energy policy since the Second World War has been to assure a safe and reliable supply of cheap and abundant energy. To that end, the federal government has largely supported the interests of the private sector fossil fuel industry, although providing a regulatory framework to those sectors perceived as monopolies. Another role that the federal government has played in energy policy is in the subsidy of nuclear power. In the aftermath of the invention of nuclear weaponry, the federal government played a major role in promoting the peaceful use of that technology—nuclear power.

While it holds great potential, nuclear power presents environmental challenges. Accidents at nuclear power plants (such as the meltdown at Pennsylvania's Three Mile Island, the far more catastrophic massive radioactive release at Chernobyl in Soviet Ukraine, and the destruction of the Fukushima plant in Japan) have drawn attention to the dangers of nuclear power. Another major problem is the failure of a fully developed system to deal with nuclear waste, an unfortunate but significant by-product of nuclear power. While the United States had moved toward using Yucca Mountain as a national repository, funding for that project ended under the Obama administration, and no other program has been put in place as a substitution.

The United States has abundant domestic coal and natural gas, but imports the majority of oil that is used, largely from the Arab nations of the Middle East. Keeping the oil flowing is a national security issue. In 1991, the Gulf War was in large part fought to keep friendly oil-rich nations from succumbing to regional powers unfriendly to the West. Our energy dependency also has a massive impact on our economy due to the ability of the Organization of the Petroleum Exporting Countries (OPEC) to control supply and prices. OPEC's withholding of oil, which increased prices in 1973 and in 1979, created "oil shocks" that profoundly affected our economy (Rosenbaum, 2014).

Before the energy crisis of 1973, the federal government acted more as a broker among diversified interests than as a master planner. With the exception of supporting the development of nuclear energy, long-range energy planning was largely left to the private sector, and government energy policy was little more than reliance on the private sector to provide cheap and abundant energy, primarily from fossil fuels supplemented with government-subsidized nuclear power. When the government did intervene, it was only to stabilize prices and encourage consumption. There was little concern with conservation, energy efficiency, or international dependency.

As a result of the 1973 Yom Kippur War with Israel, the Arab states implemented a 5% reduction in exports to nations that supported Israel, resulting in a rapid increase in cost of oil. The United States responded by shifting to non-Persian Gulf sources for imported oil, legislating a 55-mile-per-hour speed limit to save gas, and building the Alaskan oil pipeline. We also reacted by restructuring the government. The Energy Reorganization Act of 1974 replaced the AEC with the Nuclear Regulatory Commission (NRC) and the Energy Research and Development Administration (ERDA). The NRC was given the responsibility of licensing and regulating private nuclear facilities, and the ERDA was responsible for centralizing federal research programs and running AEC's national labs and nuclear weapon production facilities (Cochran, Mayer, Carr, and Cayer, 1999).

Further efforts to reorganize the nation's energy agencies came during the Carter administration. In 1977, President Jimmy Carter established the DOE as a cabinet-level department. The centerpiece of Carter's national energy policy, the DOE was created from the ERDA and several other agencies and was responsible for coordinating national energy policy and managing federal energy programs, including production, distribution, research, regulation, pricing, and conservation. The Carter administration also deregulated natural gas to encourage production, provided incentives to convert from oil to coal, instituted research into clean coal technologies, established tax credits to encourage use of renewables (sources of energy such as solar power and wind power), imposed a gas-guzzler tax on large cars, and introduced the Corporate Average Fuel Economy (CAFE) standards to promote automobile energy efficiency.

Passed in the wake of the oil shock of the early 1970s, the National Energy Act (NEA) of 1978 focused on decreasing dependence on foreign oil importation and increasing conservation and energy efficiency. The Public Utility Regulatory Policies Act (PURPA) was a part of the NEA. The PURPA sought to improve energy conservation and efficiency in the utility sector and had important impacts on the development of renewable energy (Energy Information Administration, 2013). Electrical power generation and delivery, since its nineteenth century origins, had been dominated by large companies who owned both power production facilities and the transmission lines. In this monopoly situation, the federal government provided regulatory controls until deregulation was introduced in 1978 with the PURPA. The PURPA initiated partial competition in the generation of electricity by requiring that monopoly utilities compare the cost of purchasing energy from other vendors rather than producing it themselves. This allowed the entrance of some independent producers into the previously monopoly controlled market. The PURPA required that utilities purchase energy from small independent producers with the price to be set at the "avoided cost"—the cost the utilities would have to pay if they expanded their facilities to generate the added energy (Hinrichs and Kleinbach, 2002). Avoided cost calculations were to be set by each state. Some states, like California and New York, set their avoided cost calculations high so that renewables would be favorably impacted. In 1995, however, the Federal Energy Regulatory Commission (FERC) took responsibility for determining avoided cost. This change resulted in lower avoided cost scales than those in states that had written their calculations to be favorable to renewables (Energy Information Administration, 2013).

The Energy Tax Act of 1978, also part of the NEA, created a series of financial incentives to promote renewables. It included a 30% investment tax credit for residential consumers for solar and wind energy equipment and a 10% investment tax credit for businesses for the installation of solar, wind, or geothermal equipment. These credits changed over time and expired in 1985 (Energy Information Administration, 2013).

In 1980, the incoming Reagan administration affirmed the use of fossil fuels to support energy use. Instead of continuing Carter's reliance on conservation and government programs to strategically manage energy policy, there was a return to full reliance on the private sector to deliver cheap energy. The Reagan administration sought to boost supply by opening federal lands to gas and oil exploration and by the further development of nuclear power. Renewable energy technologies were not given much government support. These policies have continued without major modification to today. The administration of George W. Bush relied on the private sector to generate cheap and abundant energy and denigrated conservation as a solution for the nation's energy problems. The administration favored the revitalization of the moribund nuclear energy industry as well as removal of environmental protections to allow drilling for oil and gas in areas previously designated as protected (Peters, 1998).

The 1992 Energy Policy Act (EPAct) was the most important energy bill of the 1990s, albeit the only major one, for the promotion of renewables. It expanded the PURPA to include a wider range of electricity generators, effectively creating a deregulated market for electricity across the United States. This law allowed independent power producers from any geographical region to sell electricity to industry or utilities in other regions of the country. Transmission lines were opened to allow a competitive market (Hinrichs and Kleinbach, 2002). In addition, the EPAct provided a 10-year 1.5 cent per kilowatt hour (kWh) production tax credit (PTC) for investor-owned wind turbines and biomass plants for plants brought into production between 1994 and 1999 (Energy Information Administration, 2005b). While this incentive undoubtedly provided a huge incentive to the development of renewables, several problems were associated with it. First, the PTC repeatedly expired and was subsequently reinstated by

Congress. It was most recently reinstated by the EPAct of 2005, discussed more fully in the succeeding text. The result of this instability fueled a boom and bust cycle economy that dominated (and to a large extent continues to dominate) the investor-owned renewable energy sector. Also, the PTC was initially restricted to wind and biomass. This had an adverse effect on the emergence of other renewables, like solar, until the rules were widened to allow them.

Titles III and V of the EPAct contained provisions to promote the use of alternative fuels, that is, fuels that are not derived from petroleum. The goal was to reduce dependence on foreign oil. Congress established requirements under the EPAct to build a fleet of alternative fuel vehicles (AFVs) to be used in large urban areas. The law gave the DOE the authority to manage requirements for federal, state, local, and private programs for fleet acquisition. The DOE never implemented mandated controls for local or private fleets. Federal AFV fleet requirements set by the EPAct required that beginning in FY 2000, 75% of light-duty vehicles be AFVs. Vehicles other than light-duty vehicles—those weighing 8500 pounds or more—were not covered by the law. Law enforcement, emergency, and military vehicles were also excepted. The DOE manages this AFV requirement by using a complex system of credits granted to various vehicle types and alternate fuel use. The EPAct also required that a fixed percentage of state fleet vehicles be AFV if the state operates more than 50 light-duty vehicles in its fleet. The exceptions for federal fleets also apply to state fleets; however, state fleets are also exempted for a variety of additional reasons.

During the George W. Bush administration, a great deal of controversy arose over then Vice President Dick Cheney's Energy Task Force, which recommended energy policy goals for the Bush administration. The oil and gas industry was very influential on the task force. Despite the fact that the courts ruled in favor of the White House secrecy demands, documents show that executives from both ConocoPhillips and ExxonMobil participated on the task force (Rahm, 2010). One recommendation of the Energy Task Force was that Congress exempt hydraulic fracturing from regulation under the SDWA. The National EPAct of 2005 did just that.

After 4 years in the making, the EPAct of 2005 was passed in August of 2005. While largely a law designed to support fossil fuel producers and nuclear energy, the energy bill included a number of initiatives designed to promote renewable energy. The law provided several financial incentives. It extended the PTC through 2007 for wind, biomass, geothermal, small irrigation power facilities, landfill gas, and trash combustion facilities. A new category of tax credit bonds were created by the law. These bonds, known as clean renewable energy bonds (CREBs), can be issued by governmental bodies, tribal governments, and cooperative electrical companies to finance capital expenditure for renewable energy facilities. To support energy efficiency and conservation, the law added biodiesel fuel credits of 10 cents a gallon for up to 15 million gallons of biodiesel produced before December 31, 2008. Taxpayers are permitted to claim a 30% credit for the cost of installing clean-fuel vehicle refueling properties. Clean fuels were defined by the law as those containing at least 85% volume of ethanol, natural gas, compressed natural gas, liquefied natural gas, liquefied petroleum gas and hydrogen, and any mix of diesel and biodiesel that contains at least 20% biodiesel. This credit was allowed through January 1, 2010. The law also provides a credit for business installation (by 2008) of fuel cells, stationary microturbine power plants, and solar arrays. To help create a market for some of this newly generated renewable energy, the law created federal procurement requirements. By 2013, the federal government was required to purchase 7.5% of its power needs from renewable sources, increasing from 3% in 2007 (Neff, 2005).

Building standards were affected by the law as well. Contractors receive a tax credit for the construction of energy-efficient homes purchased between 2005 and 2008. Energy-efficient commercial buildings also qualified for a tax credit. Appliance manufacturers became eligible for a tax credit for engineering more energy-efficient appliances, and the law required some new products to achieve new efficiency standards. Homeowners who installed solar energy systems were allowed to claim a tax credit (Perkins, 2005).

In terms of vehicles, the Act offered incentives for consumers to purchase energy-efficient hybrid, clean diesel, and fuel cell vehicles. While not increasing the CAFE standards, the law directed the Department of Transportation to study the impact of new fuel efficiency standards (Neff, 2005). The EPAct provided a financial incentive for AFVs equal to the percentage of any additional cost of placing such a vehicle in service. AFVs covered in the Act include compressed natural gas, liquefied natural gas, liquefied petroleum gas, hydrogen, and other liquids at least 85 percent ethanol. The law mandated the annual use of 7.5 billion gallons of ethanol by 2012 and provided refiners, blenders, and importers of gasoline a credit for any fuel that replaced or reduced the quantity of fossil fuel present in the fuel mixture (Perkins, 2005).

After its passage, the EPAct of 2005 was widely criticized for its weak provisions in regard to conservation, energy efficiency, water protection, and renewables. Despite its nearly 2000 pages of text, the law failed to address two major concerns for energy policy—the contribution of fossil fuel use to global climate change and the national security implications of continued dependence on oil importation (Crook, 2005). The Act was also criticized for the priority it gave to building new fossil fuel and nuclear plants as well as for the authority it gave the FERC to locate highly volatile liquid-natural gas ports, all potential terrorism targets, even over the opposition of state and local governments (Clarke, 2005). Many criticized the law for its glaring exemption of hydraulic fracturing from regulation under the SDWA.

While no major energy policy law was passed during the Obama administration, two major changes occurred, primarily linked to climate change (which is discussed more fully later). First, the administration was able to raise the CAFE standards substantially. A new rule issued in 2012 set the CAFE standards for cars to 54.5 mpg by 2025. Second, the historic economic stimulus bill passed early in the Obama administration contained key energy and environmental provisions to promote alternative energy and energy efficiency.

13.10.3 Environmental Policy

In 1962, Rachel Carson's influential book, *Silent Spring*, raised the issue of the potential effects of chemicals on the environment. In 1965, President Lyndon Johnson's State of the Union Address spoke of the population explosion and its potential effects on the capacity of the Earth to provide for humanity. A new consciousness brought together issues of conservation, public health, natural resources, energy use, population growth, urbanization, and consumer protection and created demand for environmental legislation.

The first areas to receive attention in the 1970s were clean air and clean water. By 1980, the problems of hazardous and toxic waste were added to the list of environmental issues needing government action. The 1990s focused on the destruction of ecologically critical lands and forests along with the loss of wilderness and wildlife. The 1990s also saw the emergence of global environmental issues, including climate change, ozone depletion, loss of biodiversity, deterioration of quality of life in urban areas, and sustainable development. Climate change remains the most pressing environmental concern today.

The first wave of environmental policy focused on the air and water. A series of laws with a primary goal of protecting human health and a secondary goal of protecting the environment were drafted. These laws generally relied on national environmental standards and national regulatory mechanisms, primarily implemented by the EPA in cooperation with the states.

The first law was the amendment of the CAA. The CAA of 1963 had modestly initiated federal air pollution control by supporting research and development and assisting the states to implement their own policies. It was generally unsuccessful, because it relied on voluntary compliance and independent state standards (Durant, 1985). The 1970 CAA amendments were a fairly radical change in policy. They established the National Ambient Air Quality Standards (NAAQSs) to be enforced by the EPA and the states. The primary standards were designed to protect human health, and the secondary standards were included to protect buildings, forests, water, and crops from damage due to air pollution. The EPA set the NAAQS for the six major "criteria" pollutants (sulfur dioxide, nitrogen dioxide, lead, ozone, carbon monoxide, and particulates) to provide an "adequate margin of safety." An "ample margin of safety" was mandated for toxic pollutants including arsenic, chromium, hydrogen chloride, zinc, pesticides, and radioactive substances. The law required setting the NAAQSs without consideration of cost of attainment. If control technologies were not available to meet the standards, Congress expected such technologies to be developed by fixed deadlines.

The 1970 CAA amendments also set national emission standards for mobile sources of air pollution (cars, buses, and trucks). They were set to promote 90% hydrocarbon emissions reduction in 1975 vehicles and 90% nitrogen oxide reductions in 1976 vehicles. The amendments also set tough emission standards for such stationary sources as refineries, chemical companies, and other industrial facilities. New sources of pollution were held to New Source Performance Standards to be set by industry, enforced by the states, and based on the best available technology, while existing sources of pollution were held to lower standards set by the states according to a State Implementation Plan (SIP). The nation was divided into 247 air quality control regions, for which the states were responsible. SIPs were to determine how much of the total pollution load within a region was the responsibility of each polluter and how much emission control that polluter must achieve (Rosenbaum, 2014).

In 1977, the CAA was amended again. This amendment eased compliance dates but strengthened the rules for nonattainment areas (areas that had been unsuccessful in reducing pollution to acceptable levels to protect human health) and added provisions for prevention of significant deterioration (PSD) in areas that were already cleaner than national standards. The nation was divided into classes, and Class I areas such as national parks were protected against further deterioration of air quality. Congress provided for protection of visibility in national parks and wilderness areas that were affected by haze and smog. The 1977 amendments also called for the use of scrubbers to remove sulfur dioxide emissions from new fossil fuel-burning power plants (Liroff, 1986).

Significant changes were ushered in with the 1990 CAA amendments. These amendments extended the act to control acid rain (sulfur dioxide and nitrogen oxides) emitted primarily by coal-burning power plants and to control CFCs (chlorofluorocarbons) that affect the ozone layer. They called for 35–60% reductions in auto emissions between 1994 and 1996 and for the development of cleaner fuels for use in nonattainment areas. The 1990 amendments established a plan to bring urban areas into compliance within 20 years, required the EPA to set emissions levels for toxic air pollutants, and listed 189 toxic chemicals that the EPA was to regulate. Title V required major stationary sources of air pollution to get EPA-issued operating permits that specify allowable emissions and necessary control measures (commonly called Title V permits) (Bryner, 1995).

One of the more innovative and controversial aspects of the 1990 amendments was the introduction of emissions trading. This policy was made part of the 1990 amendments after the success of experiments conducted by the EPA to change its traditional command and control (standards and enforcement) posture. In command and control, the federal government sets standards of acceptable pollution levels, prescribes pollution control technology, licenses and monitors polluters, and identifies and punishes violators. The experiments substituted the market incentives of netting, offsets, and bubbling for traditional command and control. In netting, firms creating new emissions sources within the same plant could reduce emissions from other plant sources so that net emissions did not significantly increase. For offsets, the EPA allowed new pollution sources to locate within a nonattainment area if they could offset new pollution emissions by reducing emissions from other sources in that area. Firms could buy pollution allowances from other local firms to accomplish this. Bubbling allowed firms operating within a "bubble" (a locally designated airshed such as the Los Angeles area) to decide on the most cost-effective methods for obtaining overall bubble compliance and allowed banking or selling of unused pollution credits (Rosenbaum, 2014).

The CAA amendments formalized these experiments in the Title IV emissions trading provisions. Title IV required the EPA to give each major coal-fired electric utility an allowance for each ton of sulfur dioxide emission permitted and to limit utility emissions to the total allowances issued. The allowances could be bought, sold, or traded by other companies or people, including by environmentalists who might want to take them out of circulation. The purpose of the program was to replace command and control with more efficient mechanisms of pollution reduction (Cook, 1988).

The regulation of water pollution followed a similar path. The first law, the Water Pollution Control Act of 1948, emphasized research on clean water issues, but the federal government had no authority over water quality and no standards were set. Some strengthening occurred in the 1965 Water Quality Act, which required the states to establish water quality standards and limit pollution discharges into interstate bodies of water. The CWA of 1972 changed things dramatically. It established national policy and deadlines for eliminating the discharge of wastes into navigable waters. It was very vague, however, stating only that all waters were to be "fishable and swimmable" by 1983, which meant that the EPA had to determine what constituted fishable and swimmable water. It made the discharge of toxic amounts of pollutants illegal. The CWA amendments of 1977 and 1987 generally strengthened water protection but postponed several deadlines for compliance (Kraft, 1996).

Like the CAA, the CWA gives states primary responsibility for implementation as long as they follow federal standards and guidelines. Discharges into navigable waters must meet federal standards, and a discharger must get a permit that specifies the type and quantity of discharge allowed. The EPA has granted authority to most states to issue the National Pollutant Discharge Elimination System (NPDES) permits. These permits apply to municipal facilities as well as to industry. Compliance is determined by self-reported discharge data and on-site state inspections. The states apply water quality criteria (WQC) that define the maximum allowable concentration of pollutants in surface waters. The EPA effluent limitations specify how much pollution a discharger may emit into the water and the specific treatment technologies to be used prior to discharge (Jasper, 1997).

The CWA gave local governments federal money to build municipal wastewater treatment facilities. The subsidies were significant: in the 1970s, the federal government assumed 75% of the capital costs, amounts in excess of $7 billion a year. These federal subsidies were reduced in the 1980s to about $2.5 billion a year (Kraft, 1996). Funding for wastewater treatment plants now relies on a system of grants and loans.

As its name suggests, the 1974 SDWA focused on tap water. The law required the EPA to set National Primary Drinking Water Standards for chemical and microbiological contamination of tap water and to regularly monitor the nation's tap water supplies. The EPA made slow progress, setting only 22 standards for 18 substances by the mid-1980s. In 1986, the law was amended, and the EPA was required to set maximum contamination level standards for 83 specific chemicals by 1989, for 25 more by 1991, and for 25 more every 3 years. The states were given primary authority for enforcing standards and monitoring levels, using the best available technology to remove pollutants. Many states saw this law as an unfunded mandate in that they receive only about half the funding they need. Funding issues are crucial for small water systems that cannot afford new water treatment technologies. In 2000, small systems were paying more than $3 billion to comply with standards and another $20 billion to repair, replace, and expand systems (Rosenbaum, 2014).

Hazardous and toxic waste policy is another critical area. As with air and water policy, legislation passed prior to the 1970s, such as the 1965 Solid Waste Disposal Act, was weak, relying on the states for standards and expecting voluntary compliance by polluters. The RCRA was a 1970 amendment to the Solid Waste Disposal Act. Its purpose was to regulate hazardous waste disposal sites and practices and to promote conservation and recovery of resources through the management of solid waste.

The RCRA required the EPA to identify and characterize hazardous wastes and develop criteria for safe waste disposal. The Department of Commerce was responsible for promoting waste recovery technologies. The EPA was to develop a cradle-to-grave system of regulation to monitor and control production, storage, transportation, and disposal of hazardous waste. The agency subsequently developed specific measures of toxicity, ignitability, corrosivity, and chemical reactivity. A substance that tests positive on any one of these indicators is governed by the RCRA. A substance may also be specifically listed by the EPA as hazardous and thus falls under the RCRA control. In an effort to eliminate midnight dumping, illegal secret dumping of wastes on public or private lands, the EPA established the national manifest system to keep track of the generation and transportation of hazardous wastes.

Implementation of the RCRA was slow. The EPA took 4 years to issue the first major regulations and 6 years to issue standards for incinerators, landfills, and surface storage tanks. Upset about the amount of time implementation was taking, Congress strictly limited the EPA's administrative discretion when it passed the 1984 Hazardous and Solid Waste Water Amendment (HSWA) Acts. One of the most detailed and restrictive environmental laws ever written, the HSWA contains 76 statutory deadlines, eight of which had provisions that would take effect if the EPA failed to meet a deadline. The HSWA sought to phase out disposal of most hazardous wastes in landfills by establishing demanding safety standards; covering more wastes, small sources previously omitted, and leaking underground storage tanks; and establishing a specific timetable. The HSWA made handling hazardous waste very costly, and the law introduced strong economic incentives for companies to adopt nonpolluting alternatives.

The Toxic Substances Control Act (TSCA) of 1976 gave the EPA authority to identify, evaluate, and regulate risks associated with commercial chemicals. The EPA was required to produce an inventory of chemicals in commerce and to regulate their manufacture, processing, use, disposition, and disposal. The agency was given the power to ban chemicals or to require special labeling. Under the TSCA, the manufacturer of a new chemical must notify the EPA and supply test data 90 days before putting the chemical on the market. The TSCA was amended by the 1986 Asbestos Hazard Emergency Response Act and the 1992 Residential Lead-Based Paint Hazard Reduction Act, which added asbestos and lead-based paint to the list of toxic substances. Implementation of the TSCA has not gone smoothly. There was business opposition to the law. Because the EPA has to prove that a chemical is unsafe before banning it, only a few chemicals have been banned. The dependence of the EPA on the regulated community for data is also considered problematic (National Research Council, 1977).

The Federal Insecticide, Fungicide, and Rodenticide Act (FIFRA) of 1947 originally dealt with labeling requirements for pesticides and was implemented through the Department of Agriculture. The FIFRA was amended in 1964, 1972, and again in 1978 as a result of increasing public awareness of the dangers of pesticides. Implementation was placed under the EPA in 1970. Today, the FIFRA allows commercial use of only those pesticides registered with the EPA. Less stringent than other environmental laws passed during the 1970s, the FIFRA clearly requires the EPA to take cost into account. The EPA must prove an existing pesticide harmful before suspending registration, but the manufacturer must demonstrate the safety of new pesticide.

The Comprehensive Environmental Response, Compensation, and Liability Act of 1980 (CERCLA, which is called Superfund) was passed as a response to the discovery of toxic waste in Love Canal, a dry canal that was not connected to upstate New York's wider canal system and that was used by the Hooker Chemical Corporation as a disposal site. Discovery of the chemicals that had leaked into the basements of houses surrounding Love Canal led to the passage of the law. While the RCRA deals with active sites (sites with current waste generation and disposal), Superfund deals with abandoned or uncontrolled waste sites. Congress initially gave the EPA $1.6 billion to identify, characterize, and clean these abandoned sites. The EPA was expected to track down the original polluters and make them pay for the final cleanup. This emphasis resulted in a great deal of litigation, and Superfund is a very controversial law not only because of liability issues but also because cleanup has been slow and the program has been expensive (Rahm, 1995).

In 1986 Superfund was amended by the Superfund Amendments and Reauthorization Act (SARA), which added $8.5 billion to the fund. Because of a 1984 accident at Bhopal, India, in which a release of toxic gas killed five thousand people, SARA initiated the Toxics Release Inventory (TRI) under the Community Planning and Right-to-Know section of the law (Rosenbaum, 2014).

13.10.4 Global Climate Change

Debate about global environmental issues began in the 1980s and grew more urgent thereafter. For more than 80 years, the United States has been a participant in international environmental agreements. The scope and pace of such agreements have increased in the last 30 years, and during that period, the United States has adopted 97 multilateral environmental treaties, bringing the total to 152. Prior to the 1980s, most dealt with access to common global resources. Since then, most have dealt with transborder issues including acid precipitation, ozone depletion, and global warming. Major meetings have been held in Stockholm, Montreal, Rio de Janeiro, Kyoto, and Copenhagen to determine a multinational response to these environmental challenges.

The theme of the 1972 Stockholm meeting was Only One Earth. The first truly international meeting on environmental issues, it was attended by 113 countries and 19 representatives from international organizations. A key outcome was the Stockholm Declaration, which stated that nations have the sovereign right to exploit their own resources in accord with their own environmental policies and that nations have the obligation to ensure that activities within their jurisdiction or control do not damage the environment of other nations or areas.

The outcome of the Montreal meeting was the 1987 Montreal Protocol on Substances That Deplete the Ozone Layer. This protocol limited domestic production and consumption of CFCs and other chemicals that were destroying the ozone layer, which blocks harmful ultraviolet radiation twenty to thirty miles above the Earth. It was signed by the United States and 46 other nations. Subsequent conferences in London in 1990 and Copenhagen in 1992 accelerated the pace of CFC reduction. The success of the protocol was due in large part to broad scientific consensus on ozone depletion as well as the development of alternatives to CFCs.

In 1992, the Conference on Environment and Development, or the Earth Summit, was held in Rio de Janeiro and was attended by 179 countries. It emphasized the growing urgency of environmental problems and focused on issues faced by developing nations. The outcome was the 28 guiding principles of the Rio Declaration on Environment and Development. Principle 16 stated that the polluter should bear the cost of cleaning up pollution. The Rio Declaration also supported the "precautionary principle," which indicated that nations should take action to abate potentially harmful pollutants even in the absence of scientific certainty about their effects (Rosenbaum, 2014).

The Kyoto Protocol grew out of an earlier series of international meetings and agreements among scientists and policy advisors. The first of these came in 1988 with the creation of the Intergovernmental Panel on Climate Change. The IPCC, organized by the World Meteorological Organization and the United Nations Environment Program at the recommendation of the United States, joined the National Academy of Sciences as an official advisor to the US government on climate change issues. The IPCC represented almost all the world's governments and their climate experts (Weart, 2003). In 1992, the United Nations Framework Convention on Climate Change was adopted at the Earth Summit, which met in Rio de Janeiro that year. The United States signed and ratified the treaty. The UNFCCC required all parties to stabilize GHG concentrations in the atmosphere at a level that would not interfere with climate, with due regard to sustainable economic development. UNFCCC also required that parties create and publish an inventory of GHG emissions. A series of other meetings, called the Conference of the Parties (COPs), followed the Rio Earth Summit. The COPs developed the legal framework for the treaty, put all UNFCCC parties on a schedule to reduce emissions, and negotiated a policy agreement that industrialized countries should be first to reduce their emissions, while developing countries would be allowed to continue to develop without having to consider GHG emissions (Vaughn, 2007).

It was this last point that created conflict in the Congress and convinced President Clinton that if he took the treaty to the Congress for ratification, it would fail. So President Clinton never presented the treaty for ratification.

Early in the George W. Bush administration, President Bush withdrew the United States from the Kyoto Protocol agreement arguing that the science was uncertain, that mandatory limits on GHG emissions would hurt the US economy, and that it was unfair to exempt developing nations from mandatory reductions while expecting binding targets from developed nations. This stand that reversed the US position on climate change begun under George W. Bush's father, George H. W. Bush, when he was the 41st president of the United States—a policy that was continued under President Bill Clinton. Both George H. W. Bush and Bill Clinton supported international engagement on the issue of climate change. Despite the stated policy position of the federal government under the George W. Bush administration, some progress was made in the United States between 2000 and 2008 to control GHG emissions. The steps forward were largely accomplished by the states acting alone or together in regional partnerships to redirect US climate change policy after the default of the federal government.

The withdrawal from the Kyoto Protocol was accompanied by increased rhetoric from the administration calling into question the scientific accuracy of global warming. The administration repeatedly questioned whether there was sufficient evidence to conclude that global warming was really occurring and, if it was, whether it was the result of human activities rather than some natural climate shift. The Bush administration used these points to frame a debate that many who opposed any policy action on climate change supported. Environmentalists and those who supported a responsible assessment of the threat of climate change were discontented with the position of the Bush administration. They not only disagreed with the US withdrawal from the Kyoto Protocol but also were concerned that the administration tried to justify its position by attacking what most considered accurate and valid scientific evidence. The Bush administration's claim that the science was unsettled fits with the claims coming from the well-organized industry-based anticlimate change lobby spearheaded by ExxonMobil and supported by many oil, coal, and fossil fuel-intensive companies.

The presence of a determined opposition to environmental advocacy is nothing new in American politics (Hays, 2000). Since the 1970s, with the beginnings of mandatory regulations to deal with air and water pollution, the environmental opposition has been a formidable obstacle that those who favor environmental policy action have had to overcome. In the debate over of climate change and what, if any, measures should be taken to address it, the Bush administration became the champion of those who favored policy inaction. Early in the tenure of the Bush administration, these opponents included many from the business community that saw climate change legislation as yet another regulatory burden. Later, this would change as the states and regions began implementing a range of plans in different geographic locations. The latter was seen as more of a burden to industry than a single mandatory national standard.

It is interesting to note that in 2001, the same year the Bush administration assumed office, IPCC scientists issued a strong statement of consensus regarding global warming and its causes. In their third draft report, they concluded, as they had similarly done in their first and second draft, that it was much more likely than not that the planet was undergoing human-induced global warming. Finally in 2004, the Bush administration reluctantly gave up its sustained opposition to the world's scientists. It sent Congress an analysis that was accompanied by supportive cover letters from the Secretaries of Energy and Commerce as well as the President's Science Advisor. The analysis at long last accepted the scientific evidence that humans were causing global warming. The report confirmed that GHG emissions were the only explanation for the atmospheric warming observed over the past several decades. The Bush administration, however, proposed no mandatory policy actions to deal with GHG emissions still arguing that the costs of reducing emissions was too high and that the United States should not be required to reduce emissions if developing countries were not also required to do so (Weart, 2003).

Rather than follow along with efforts being made internationally for the implementation of the Kyoto Protocol, the Bush administration called for a summit on climate change to be held in Washington in September of 2007. The purpose for the summit was to explore an alternative process for moving forward on the climate change issue. The conference was attended by only midlevel officials from 16 nations. Many of the European representatives feared that the Bush summit was an attempt to derail the UN-sponsored process that had culminated in the Kyoto Protocol. The parallel process the Bush administration was trying to set up relied entirely on voluntary reductions and continued noncooperation with the mandatory reductions supported within the UN–Kyoto process.

With the end of the Kyoto Protocol scheduled for 2012, the next round of UN-sponsored international talks to address climate change came with the COP in Bali, Indonesia, held in December of 2007. The goal of the Bali meeting was to establish a roadmap for negotiating a successor agreement to the Kyoto Protocol by 2009. The United States sent a representative but continued to raise issues about how strongly a successor agreement to Kyoto should demand GHG reductions on the part of developing nations, especially China. The United States insisted that no firm 2020 target numbers for GHG reductions appear in the preamble, and so a compromise was agreed to whereby the language simply recognized the need for "deep cuts" in emissions. The bitterness toward the role of the United States in the proceedings was revealed by former US Vice President Al Gore who told delegates that the United States was obstructing progress. Gore had arrived at the Bali COP after receiving the Nobel Peace Prize along with the IPCC for their efforts to alert the world of the threat of climate change. Delegates at the conference saw two sides of the United States: the official delegates representing the Bush administration and American activists, like Gore. The great divide in America over what to do about climate change was abundantly obvious to all.

In the end, the Bali Action Plan was agreed to by all parties but only after considerable drama that included booing and hissing by many delegates aimed at the US delegation. The Bali Action Plan had no binding commitments, as required by the United States, but did provide that the next 2 years of negotiation proceed on a two-track path—one for those countries not committed to mandatory limits and one that builds on the Kyoto Protocol. In the end, the United States agreed to the two tracks to avoid the total breakdown of the meeting. The roadmap thus allowed for continued negotiations for mandatory reductions to succeed Kyoto in 2012. Countries that favored the mandatory approach held out hope that with a new administration in Washington after 2008, the United States might shift its position. It seems they were correct in this expectation for the Obama administration has shifted policy.

It should be noted that the US federal government under the Bush administration did have some programs in place to address GHG emissions. Those that existed, however, were extremely limited and consisted entirely of voluntary approaches. For instance, the federal government programs Climate VISION and Energy STAR worked with industry to reduce GHG emissions voluntarily. These federal programs and clean energy R&D were coordinated by the Federal Climate Change Technology Program. The federal government also had the Federal Climate Change Initiative, which had a goal of reducing the GHG intensity of the economy by eighteen percent (from 2002 to 2012). However, none of these federal efforts called for a mandatory cap on emissions, and none directly sought to reduce the overall level of GHG emissions (Bogdonoff and Rubin, 2007). In fact, the plan to reduce GHG intensity of the economy specifically allowed for continued growth in overall GHG emissions.

The default of the federal government under the Bush administration did not result in total inaction on the part of other governmental organizations. Concern about climate change had been growing at both local and state levels for many years. In several regions of the country, the state and local response to the US withdrawal from the Kyoto process was the trigger that moved them to policy action. With inaction at the federal level, leadership fell to the states (Peterson and Rose, 2006).

What happened at the state and local levels over the climate change issue was that enterprising civil servants with experience in related areas of environmental management such as recycling or air pollution moved into action when the federal government failed to act. They drew on their experience in other related areas to build networks and develop ideas that could work to reduce GHG emissions. Operating in the absence of federal action, they crafted innovative programs that could both reduce GHG emissions and protect the economic self-interest of the state at the same time. Concerns over quality of life issues also pressured governors to adopt green policies that would continue to attract businesses and investors to their state. Early adopters of climate change policy action saw such policy as tied to long-term economic development as well as the immediate economic benefit of such things as protection of oceanfront development, tourism, water adequacy, biodiversity, or other environmentally sensitive features of state life. Those forging climate change policy at the subnational level also considered the potential for cascading impacts or cobenefits from GHG reductions including such things as overall air pollution reduction, diversification of the energy supply, less traffic congestion, and stability of the regulatory framework for regulated firms (Rabe, 2004).

A number of states have taken individual climate policy action including Oregon, California, Iowa, New Mexico, Minnesota, Massachusetts, and Maryland. Localities have also taken direct action to combat climate change. In June of 2005, the US Conference of Mayors approved the US Mayors Climate Protection Agreement. By signing the agreement, mayors pledge to reduce GHG emissions in their cities by seven percent below 1990 levels by 2012. These levels represent what the United States would have been required to do if the United States had become a signatory of the Kyoto Protocol. Over 750 US cities by the end of 2007 had signed the pledge. Localities are assisted in these efforts by a number of civil sector organizations including the Sierra Club and Environmental Defense (U.S. Conference of Mayors, 2005).

It is important to add that most states have some form of renewable portfolio standard (RPS) in place to encourage the development of renewable energy. While these RPS requirements were not generally instituted to respond to climate change issues, they serve that purpose in any event by reducing fossil fuel use, one of the leading causes of CO_2 emissions. RPS vary from state to state, but generally, they require utilities to diversify their energy portfolio to include a specific capacity from renewables. The percentage coming from renewables increases each year thus creating a stable market for clean energy technologies.

As the states developed independent policies to deal with climate change, they began to reach out to their neighbors to form regional solutions. The earliest regional agreement, the Climate Change Action Plan, was established in 2001 by several New England states and Canadian provinces. Maine, New Hampshire, Vermont, Massachusetts, Rhode Island, and Connecticut joined with the five eastern Canadian provinces of Nova Scotia, Newfoundland and Labrador, Prince Edward Island, New Brunswick, and Quebec in the agreement. This plan was a coordinated regional agreement to lower GHG emissions to 1990 levels by 2010 and 10% below 1990 levels by 2020. The parties to the agreement further pledged to eventually reduce their emissions to a level at which they would pose no threat to the climate. The plan was established based on "no regrets" measures, that is to say, using efforts that both reduce energy costs while at the same time lower GHG emissions. The agreement included provisions for establishing a standardized GHG emissions inventory, developing a plan for reductions, increasing public awareness, emphasizing conservation and energy efficiency, and exploring a market-based emissions trading scheme (Selin and Vandeveer, 2005).

To reach the commitments made in the Climate Change Action Plan, the New England governors and Eastern Canadian premiers subsequently implemented several programs. One of these programs was the Regional Greenhouse Gas Initiative (RGGI), which was initiated on December 20, 2005, by a memorandum of understanding (MOU) signed by the governors of Connecticut, Delaware, Maine, Massachusetts, New Hampshire, New Jersey, New York, and Vermont. The RGGI was the first US cap-and-trade program for power plant emissions of CO_2. The RGGI was scheduled to begin by capping emissions at 2009 levels in 2009 and then reducing them by 10% by 2019. Maryland subsequently joined the RGGI. The District of Columbia, Pennsylvania, Rhode Island, and the Eastern Canadian Provinces and New Brunswick are observers in the process. The RGGI focuses on electricity-generating power plants only. These utilities will either have to reduce operations, invest in cleaner technologies, or buy offsets to meet their emissions reduction targets (RGGI, 2007).

In 2003, Oregon, California, and Washington joined in the West Coast Governors' Global Warming Initiative (WCGGWI) and set the goal of coordinating policy across the three states. As a result of this agreement, these states agreed to collaborate on the purchase of hybrid vehicles for their fleets, increase by 1% per year retail sale of renewable energy, and expand the adoption of energy efficiency standards to those products not under federal regulatory control. In October of 2006, the governors of New York and California announced plans to link California's GHG reduction programs to the RGGI, which in turn increased the potential for greater RGGI–WCGGWI cooperation (Marris, 2007).

On February 28, 2006, Arizona and New Mexico entered into the Southwest Climate Change Initiative. This agreement fosters coordination between the states to reduce GHG emissions while promoting energy efficiency, new technologies, and clean energy sources. The expectation was that economic development would be fostered as much as climate change issues would be addressed (Pew Center on Climate Change, 2007).

The Western Climate Initiative (WCI) was signed on February 26, 2007, by the governors of Arizona, California, New Mexico, Oregon, and Washington. The WCI was a joint effort to promote mitigation strategies while at the same time reduce GHG emissions. The emissions covered by the agreement include carbon dioxide, methane, nitrous oxide, hydrofluorocarbons, perfluorocarbons, and sulfur hexafluoride. Utah later joined WCI as did British Columbia and Manitoba. The WCI obligates parties to jointly set regional emissions objectives and to establish (by August of 2008) some type of a market-based system to meet those targets. Effective August of 2007, the emissions targets were set at 15% below 2005 levels by 2020. The WCI built on efforts already taken by separate states or provinces as well as the Southwest Climate Change Initiative and the WCGGWI (Marris, 2007).

In November of 2007, the governors of Illinois, Iowa, Kansas, Michigan, Minnesota, and Wisconsin signed the Midwestern Greenhouse Gas Reduction Accord (MGGRA). By signing this agreement, the states committed themselves to set GHG reduction targets within a year, to establish a cap-and-trade system to achieve reduction targets, and to join the Climate Change Registry to track, manage, and credit entities that reduce GHG emissions. The governors also signed the Energy Security and Climate Stewardship Platform, which consigned the states to a regional goal to "[m]aximize the energy resources and economic advantages and opportunities of Midwestern states while reducing emissions of atmospheric CO_2 and other greenhouse gases." In all, the accord was a broadly structured energy sector platform for measuring goals, setting targets for energy efficiency, renewable energy, biofuels, and carbon capture. The accord included this broad range of policy tools and emphasized more than just a cap-and-trade agreement. Because it emphasized energy efficiency and renewables, it sought GHG emissions reductions independent of the cap.

A series of battles between the states and the federal government over regulation of GHGs played out in the courts in the last several years. Three of these cases involve a requested waiver from California that would allow California and subsequently other states as well to impose stricter air pollution standards than the federal government. The reason for this odd situation with California is that in the 1960s it had already implemented vehicle emission standards. This was well before the federal government passed laws to do so. When the federal government did pass such legislation, all states were preempted from adopting separate vehicle emission standards, except for California, which was required to seek a waiver of preemption from the EPA in the event it wanted to impose higher standards. The CAA was later amended to allow other states to adopt California's standards but only after California received a waiver (Adler, 2007).

In 2002, California passed a law that set limits on one source of global warming gases—automobiles. The bill empowered the California Air Resources Board to set standards for California's automobile fleet to achieve a cost-effective reduction in global warming pollution coming from automobile tailpipes. Specifically, the law required that all cars and light-duty trucks sold in California beginning in model year 2009 have 22% reduced CO_2 emissions compared to emissions in 2002 and increasing to 30% by model year 2016. Since 2002, 11 states have followed California's lead and set GHG emission standards for passenger vehicles. Three states—California, Vermont, and Rhode Island—were sued by the automobile industry to block their efforts. The fourth case was brought by several environmental groups and the state of New York in an attempt to induce the EPA to impose GHG controls on new power plants. These cases all stalled in the courts awaiting the results of yet another case that had made its way to the Supreme Court (Holtkamp, 2007).

The Supreme Court ruled on April 2, 2007, in the *Massachusetts v. Environmental Protection Agency* that the EPA had the authority under the CAA to regulate GHG emissions. The court further ruled that the EPA could not avoid its responsibility to regulate GHGs unless it could provide a scientific basis on which to do so. The case had been brought by a group of states (California, Connecticut, Illinois, Main, Massachusetts, New Jersey, New Mexico, New York, Oregon, Rhode Island, Vermont, and Washington), local governments (District of Columbia, American Samoa, New York City, and Baltimore), and private organizations (Center for Biological Diversity, Center for Food Safety, Conservation Law Foundation, Environmental Advocates, Environmental Defense, Friends of the Earth, Greenpeace, International Center for Technology Assessment, National Environmental Trust, Natural Resources Defense Council, Sierra Club, Union of Concerned Scientists, and US Public Interest Research Group).

With the Supreme Court ruling in *Massachusetts v. Environmental Protection Agency*, California urged to the EPA to take up its 2005 request for a waiver to limit GHG emissions from vehicles. California threatened to file legal action if the EPA did not respond to its request for a waiver. When the EPA did not act, California filed a law suit on November 8, 2007, to force a decision from the EPA.

The decision on California's request for a waiver was not long in coming. Only hours after the Congress passed and President Bush signed into law the Energy Independence and Security Act on December 21, 2007, EPA Administrator Stephen L. Johnson denied California's waiver request. Johnson announced that the new law, which raised the CAFE standards to 35 MPG by 2020, made the California waiver unnecessary because the primary way California would reduce GHG emissions from automobiles would be by imposing higher fuel economy standards. Indeed, if the waiver had been granted, the California standard would have required 36 MPG by 2016 a more stringent standard than the one passed by Congress. California and fifteen other states filed suit against the Bush administration for denial of the waiver on January 3, 2008, marking the next round in the 5-year battle between the states and the federal government under the Bush administration over who has the right to regulate GHG emissions.

When the Obama administration took office, however, one of the first acts of the new president was to order the EPA to reconsider its prior denial of California's waiver. The waiver was finally granted. Barack Obama, while running for the presidency in 2008, said that under his administration the United States will enter an economy-wide cap-and-trade system for controlling GHG emissions. Obama pledged to make responding to the threat of climate change a high priority of his administration. After the election of President Barack Obama, the federal government reversed the positions taken by former President Bush. President Obama called for the United States to once more assume leadership on the issue of climate change. Early in his administration, a number of steps were taken to steer the United States on the path toward climate leadership and domestic GHG emissions reduction. In his first address to a joint session of Congress, President Obama called for Congress to send him legislation imposing a market-based carbon control scheme for the United States. The Obama administration made the expansion of renewable energy a centerpiece of his initial policy goals and adopted policies that would pave the way for the creation of a new green energy sector in the United States. Clear signals were sent to the international community that the United States would once again act in partnership with them to address the climate crisis.

Within the first month of being in office, the Obama administration was able to successfully push through Congress an economic stimulus bill, the American Recovery and Reinvestment Act of 2009, aimed primarily to address the severe recession gripping the country. The historic economic stimulus bill contained key provisions to fulfill President Obama's agenda for energy and the environment—a significant effort to deal with climate change. Of the historic $787 billion bill, more than $45 billion focused on energy efficiency and alternative energy programs and tax breaks. The bill included $13 billion to make federal buildings and public housing more energy efficient and to weatherize as many as a million residential homes. More than $10 billion was provided to modernize the electricity grid by applying digital technology to create a smart grid—an interactive electrical network that is decentralized, reliable, efficient, capable of fully utilizing all sources of renewable energy, and responsive to consumer demands. Also included in the bill was $20 billion for further development of renewable energy power, $18 billion for a variety of environmental projects, and $2 billion for R&D on carbon capture and storage. The bill also provided tax credits of up to $7500 for purchasers of plug-in hybrid cars (Broder, 2009). Despite these advances, the United States had not passed a domestic climate change bill 6 years into the Obama administration.

13.11 CONCLUSION

This chapter explored three large topics. The first was policymaking. The chapter identified the key actors and institutions in the policymaking process. Second, we explored the policymaking process, specifically how laws and regulations are made and implemented in the US context. The third was the substance of natural resource, energy, and environmental policies. We concluded with a discussion of the most pressing environmental management issue—global climate change.

The complexity of environmental policymaking is rooted in the very fabric of our nation's legal system and structure. To understand natural resource, energy, and environmental policy, we must first grasp the overall nature of public policymaking and then add to it the scientific and technological issues that provide the backdrop for natural resources, energy, and the environment.

Natural resource policy is the oldest of the three policy areas explored but remains contested today. Political battles still ensue over policies for timber use, extraction of minerals (especially from public land), domestic drilling for natural gas and oil, protection of the wilderness, and development of places like ANWAR. Despite more than 100 years of policy formulation, the key issue remains the correct balance between preserving the wilderness and the responsible use of natural resources.

Energy policy, likewise, is contentious. While most actors in the policy debate agree that there is a definite downside associated with fossil fuel use, the key issue remains how we can successfully transition to a sustainable energy future. With new technologies, like fracking, opening vast new sources of domestic production of fossil fuels, this question becomes all the more important.

As this chapter showed, environmental policy started out with bipartisan agreement. The EPA was created by a Republican president and supported by Republican members of Congress. Today, however, there is a clear political divide. The anti-environment forces weigh heavily toward reducing environmental regulations, while the pro-environment forces demand protection. The debate is largely over the economic costs of regulation. While the costs are real, the value of living in a clean environment should not be forgotten. We have made great strides in cleaning up the environment since the legal push began in earnest in the 1970s.

Environmental problems never seem to completely go away. We concluded with a detailed exploration of the most critical current problem—climate change. Unfortunately, making and implementing environmental policy in a global context is more challenging than doing so only domestically. Yet this is the issue of pressing concern today, and so the challenge must be met.

REFERENCES

Adler, J.H. 2007. Can California catch a waiver? *National Review* 24, 2007.

Anderson, J.E. 2000. *Public policymaking: An introduction.* 4th edition. Houghton Mifflin, Boston.

Baum, L. 1990. *American courts: Process and policy.* 2nd edition. Houghton Mifflin, Boston.

Birkland, T.A. 2005. *An introduction to the policy process: Theories, concepts, and models of public policy making.* M.E. Sharpe, Armonk, NY/London, England.

Bogdonoff, S. and J. Rubin. 2007. The regional greenhouse gas initiative: Taking action in Maine. *Environment* 49(2):9–16.

Broder, J.M. 2009. A smaller, faster stimulus plan, but still with a lot of money. *New York Times*, February 14, A14.

Bryner, G.C. 1995. *Blue skies green politics: The clean air act of 1990 and its implementation.* 2nd edition. CQ Press, Washington, DC.

Clarke, R.A. 2005. Things left undone. *The Atlantic Monthly* 296(4):37–38.

Cochran, C.E., L.C. Mayer, T.R. Carr, and N.J. Cayer. 1999. *American public policy: An introduction.* 6th edition. St. Martins, Boston.

Cohen, M., J. March, and J. Olsen 1972. A garbage can model of organizational choice. *Administrative Science Quarterly* 17(March):1–25.

Cook, B.J. 1988. *Bureaucratic politics and regulatory reform: The EPA and emissions trading.* Greenwood Press, New York.

Crook, C. 2005. Does oil have a future? *The Atlantic Monthly* 296(3):31–32.

De Grazia, A. 1957. *The American way of government.* John Wiley & Sons, New York.

Downs, A. 1972. Up and down with ecology – The issue-attention cycle. *The Public Interest* 28(Summer):38–50.

Durant, R.F. 1985. *When government regulates itself: EPA, TVA, and pollution control in the 1970s.* University of Tennessee Press, Knoxville, TN.

Dye, T.R. 1995. *Understanding public policy.* 8th edition. Prentice-Hall, Englewood Cliffs, NJ.

Energy Information Administration. 2013. Renewable and alternative fuels. Available at http://www.eia.doe.gov/fuelrenewable.html. U.S. Department of Energy (accessed September 8, 2013).

GPO. n.d. Electronic code of federal regulations. Part 6 – Procedures for implementing the national environmental policy act and assessing the environmental effects abroad of EPA actions. Available at www.ecfr.gov (accessed September 4, 2013). U.S. Government Printing Office, Washington, DC.

Hays, S.P. 2000. *Environmental politics since 1945.* University of Pittsburg Press, Pittsburg, PA.

Hinrichs, R.A. and M. Kleinbach 2002. *Energy: Its use and the environment.* 3rd edition. Harcourt College Publishers, Ft. Worth, TX.

Holmes, J.E., M.J. Englehardt, R.E. Elder, J.M. Zoetewey, and D.M. Ryden 1998. *American government: Essentials and perspectives.* 3rd edition. McGraw-Hill, New York.

Holtkamp, J.A. 2007. Dealing with climate change in the United States: The non-federal response. *Journal of Land, Resources, & Environmental Law* (27):79–86.

Hooton, C.G. 1997. *Executive governance: Presidential administrations and policy change in the federal bureaucracy.* M.E. Sharpe, London.

Jasper, M.C. 1997. *Environmental law.* Oceana Publications, Dobbs Ferry, NY.

Jones, C.O. 1984. *An introduction to the study of public policy.* 3rd edition. Brooks/Cole, Monterey, CA.

Kelman, S. 1987. *Making public policy: A hopeful view of American government.* Basic Books, New York.

Kerwin, C.M. 2003. *Rulemaking: How government agencies write law and make policy.* 3rd edition. CQ Press, Washington, DC.

Kingdon, J.W. 1995. *Agendas, alternatives, and public policy.* 2nd edition. HarperCollins College Publishers, New York, New York.

Kraft, M.E. 1996. *Environmental policy and politics.* HarperCollins College Publishers, New York.

LeLoup, L.T. and S.A. Shull. 1993. *Congress and the president: The policy connection.* Wadsworth, Belmont, CA.

Light, P.C. 1999. *The president's agenda: Domestic policy choice from Kennedy to Clinton.* 3rd edition. Johns Hopkins University Press, Baltimore.

Lindbloom, C.E. 1968. *The policy making process.* Prentice-Hall, Englewood Cliffs, NJ.

Liroff, R.A. 1986. *Reforming air pollution regulation: The toil and trouble of EPA's bubble.* The Conservation Foundation, Washington, DC.

Lowi, T.J. and B. Ginsberg. 1998. *American government: Freedom and power.* 5th edition. W.W. Norton & Company, New York.

Marris, E. 2007. Western states launch carbon scheme. *Nature* 446:114.

National Research Council. 1977. *Decision making in the environmental protection agency.* National Academy of Sciences, Washington, DC.

Neff, S. 2005. Review of the Energy Policy Act of 2005—Summary. Center for Energy, Marine Transportation and Public Policy, Columbia University.

O'Leary, R. 1993. *Environmental change: Federal courts and the EPA.* Temple University Press, Philadelphia.

Perkins, B. 2005. Energy Policy Act of 2005. Realty Times. Available at http://realtytimes.com/printrtpages/20050809_energypolicy.htm (accessed September 8, 2013).

Peters, G.B. 1998. *American public policy: Promise and performance.* 5th edition. Seven Bridges Press, New York.

Peterson, T.D. and A.Z. Rose. 2006. Reducing conflicts between climate policy and energy policy in the US: The important role of the states. *Energy Policy* 34:619–631.

Pew Center on Climate Change. 2007. What's Being Done in the States. Available at http://www.pewclimate.org/what_s_being_done/in_the_states (accessed September 8, 2013).

Quarles, J. 1976. *Cleaning up America: An insider's view of the environmental protection agency.* Houghton Mifflin, Boston.

Rabe, B.G. 2004. *Statehouse and greenhouse: The emerging politics of American climate change policy.* Brookings Institution Press, Washington, DC.

Ragsdale, L. 1998. *Vital statistics on the presidency: Washington to Clinton.* Congressional Quarterly, Washington, DC.

Rahm, D. 1995. Controversial cleanup: Superfund and the implementation of U.S. hazardous waste policy. *Policy Studies Journal* 26(4):719–734.

Rahm, D. 2010. *Climate change policy in the United States: The science, the politics and the prospects for change*. McFarland & Company, Inc. Publishers, Jefferson, NC/London.

RGGI. 2007. Participating States. Regional Greenhouse Gas Initiative. http://www.rggi.org/ (accessed September 8, 2013).

Ripley, R.B. 1975. Policy-making: A conceptual scheme. 1–20. In Ripley, R. B. and G.A. Franklin (eds.) *Policy-making in the federal executive branch*. Collier Macmillan Publishers, London.

Robertson, D.B. and D.R. Judd. 1989. *The development of American public policy: The structure of policy restraint*. Scott, Foresman and Company, Glenview, IL.

Rosenbaum, W.A. 2014. *Environmental politics and policy*. 9th edition. CQ Press, Washington, DC.

Saffell, D.C. 1993. *State and local government: Politics and public policies*. 5th edition. McGraw-Hill, New York.

Schick, A. 1980. *Congress and money: Budgeting, spending and taxing*. The Urban Institute, Washington, DC.

Selin, H. and Vandeveer, S.D. 2005. Canadian-U.S. environmental cooperation: Climate change networks and regional action. *American Review of Canadian Studies* 35(2): 353–378.

Shapiro, M. 1995. The United States. 43–49. In Tate, C. N. and T. Vallinder (eds.) *The global expansion of judicial power*. New York University Press, New York.

Smith, C.E. 1993. *Courts and public policy*. Nelson-Hall Publishers, Chicago.

Switzer, J.V. 1997. *Green backlash: The history and politics of environmental opposition in the U.S.* Lynne Rienner Publisher, Boulder and London.

U.S. Conference of Mayors. 2005. U.S. conference of mayors climate protection agreement. http://usmayors.org/climateprotection/agreement.htm (accessed September 8, 2013).

Vaughn, J. 2007. *Environmental politics: Domestic and global dimensions*. 5th edition. Thomson Wadsworth, Belmont.

Weart, S. 2003. *The discovery of global warming*. Harvard University Press, Cambridge.

14

ENVIRONMENTAL ETHICS

Tatjana Višak

Department of Philosophy and Business Ethics, Mannheim University, Mannheim, Germany

Abstract: Environmental management decisions and the moral evaluation of such decisions depend on judgments about (i) whom we should take into account in our considerations, (ii) what is harmful or beneficial for those whom our actions concern, and (iii) what our duties are toward existing, future, and possible humans and animals, as well as toward the rest of nature. These normative ethical and value-theoretical questions are addressed in environmental ethics in a systematic, analytical, and argumentative way. Environmental ethics is a branch of applied ethics, just like business ethics and medical ethics.

This chapter provides an introduction to environmental ethics. Following upon some background explanation of what ethics is all about, the reader will be invited to be with me in actually engaging with environmental ethics. I address two major topics in environmental ethics, namely, "interspecies justice" and "intergenerational justice." I explore these topics by confronting them with questions that belong to two core areas of ethics: value theory and normative ethics. Environmental ethics as a field of application of normative ethical reasoning is very relevant to the work of environmental managers.

Keywords: value theory, normative ethics, metaethics, interspecies justice, intergenerational justice, anthropocentrism, sentientism, biocentrism, ecocentrism, welfare, harm of death, nonidentity problem, Singer, Rolston, Taylor, Kymlicka, Bradley, intrinsic value, moral status, value of existence.

14.1 Introduction	337	
14.2 Preliminaries: What is Ethics?	338	
14.2.1 Introduction	338	
14.2.2 Three Core Fields of Ethics: Value Theory, Normative Ethics, and Metaethics	338	
14.2.3 Are There Objective Moral Standards?	340	
14.3 Hands-On: Doing Environmental Ethics	340	
14.3.1 Interspecies Justice	340	
14.3.1.1 Introduction	340	
14.3.1.2 Considering Humans, Animals, and the Rest of Nature	341	
14.3.1.3 What is (Animal) Welfare and What Does the Harm of Death Consist In?	347	
14.3.1.4 How Should We Treat Nonhuman Animals and the Rest of Nature?	352	
14.3.2 Intergenerational Justice	353	
14.3.2.1 Introduction	353	
14.3.2.2 Considering Those Who Might Live	353	
14.3.2.3 Can It Be Better or Worse for an Individual to Exist than Never to Exist?	354	
14.3.2.4 Moral Duties: What Are Our Duties toward Those Who Do Not yet Exist?	357	
14.4 Conclusion	358	
References	359	

14.1 INTRODUCTION

Environmental management is the management of the interaction and impact of human societies on the environment. Within this field, the shrinking Aral Sea is known as an example of poor management of water resources diverted for irrigation. On what grounds is the shrinking of the Aral Sea bad? For whom is it bad? Does it matter morally that the Aral Sea is shrinking due to poor management? Do we have the duty to manage this water resource in a better way? These are all moral questions. Answers to these moral questions usually inform and motivate the environmental resource manager's work. Since answers to these questions are contested, the environmental resource manager does not only take moral stances

---

An Integrated Approach to Environmental Management, First Edition. Edited by Dibyendu Sarkar, Rupali Datta, Avinandan Mukherjee, and Robyn Hannigan.
© 2016 John Wiley & Sons, Inc. Published 2016 by John Wiley & Sons, Inc.

but will, at times, need to explicitly defend a normative position. This means that he or she has to engage with ethics.

Ethical questions related to the human interaction with the environment are currently discussed under the heading of "environmental ethics." This is a branch of applied ethics, just like business ethics and medical ethics. The divisions between these areas of study are, however, not fixed and are as such open to change. It is, for instance, common to talk about "population ethics," "animal ethics," and "climate ethics." These fields partly overlap with environmental ethics. One can broaden or narrow the focus, depending on one's interest and purpose. In all cases, ethical reasoning is applied to practical questions.

This chapter provides an introduction to environmental ethics. Following upon some background explanation of what ethics is all about, the reader will be invited to actually join with me in engaging with environmental ethics. I will tackle what I take to be two major topics in environmental ethics, namely, "interspecies justice" and "intergenerational justice." I will explore these topics by confronting them with questions that belong to two core areas of ethics: value theory and normative ethics. These areas, along with a third, metaethics, will be introduced in Section 14.2. My introduction to these core fields within the discipline of ethics provides a useful basis for the hands-on engagement with environmental ethics in Section 14.3. Finally, in Section 14.4, I will draw some conclusions and bring together insights related to both broad areas of justice.

Engaging with environmental ethics means taking a position on normative questions regarding human interaction with the environment. When ethicists take such a position, they do so while showing that they are well informed of the relevant arguments. Furthermore, they engage with these arguments in order to ground and justify their own position. That is what I aim to do in what follows. Thereby, I hope to strike a balance between (i) providing an overview of relevant debates, arguments, and their interrelations and (ii) offering at least a rough example of how to navigate that terrain, defending a coherent and well-grounded position of one's own. The reader will, of course, be invited to consider the various arguments, to agree or disagree, to read further, and to eventually defend his or her own position in environmental ethics.

14.2 PRELIMINARIES: WHAT IS ETHICS?

14.2.1 Introduction

This chapter is about environmental ethics in environmental management. So a natural question to start with is: What is ethics? After having provided a basic sketch of what "ethics" refers to, I will explain what characterizes environmental ethics in particular. The final question to be addressed in this preliminary section concerns the status of moral claims: Can they be true or false, and if so, how can we know that?

14.2.2 Three Core Fields of Ethics: Value Theory, Normative Ethics, and Metaethics

Ethics is an area of enquiry within the broader discipline of philosophy. Philosophy, in turn, is the study of general and fundamental problems, such as those connected with reality, existence, knowledge, values, reason, mind, and language (Grayling, 1999; Teichmann and Evans, 1999). The philosophical approach to such problems is critical and generally systematic, and it relies on rational arguments (Quinton, 1995). The word "philosophy" comes from the Ancient Greek φιλοσοφία (*philosophia*), which literally means "love of wisdom." The introduction of the terms "philosopher" and "philosophy" has been ascribed to the Greek thinker Pythagoras (see Diogenes Laertius: "Lives of Eminent Philosophers," I, 12; Cicero: "Tusculanae disputations," V, 8–9).

The field of ethics, or "moral philosophy," is vast. First of all, we need to distinguish descriptive ethics from normative ethics. Descriptive ethics is an enquiry into the moral views that people actually have. Descriptive ethicists might, for instance, explore whether people in fact think that we ought to protect the environment. Much like sociologists, empirical ethicists can produce statistics of people's opinions about various issues. Here, I will not be concerned with empirical ethics. I will, instead, focus solely on ethics as a normative enquiry. Three core areas of this broad field are value theory, normative ethics, and metaethics (Shafer-Landau, 2010, 1–2).

Value theory addresses questions about what is valuable or good. In particular, it explores what is ultimately good or intrinsically good, that is, good not only as a means, but in and of itself. Money, for instance, is usually considered an instrumental or extrinsic good. It is not good in itself, but it is good for what you can do with it. For instance, you can buy a coat with it. Is the coat good in itself? Arguably, the coat is not good in itself either. It is good for something else that it gives you: it prevents you from being cold in winter. Is being prevented from being cold in winter good in itself? Perhaps, it isn't. Perhaps, the good thing about staying warm in winter is that it prevents illnesses, such as having a cold. Having a coat thus contributes to your health. Is being healthy a good in itself? Some would argue that it is. Others would argue that it is not. Perhaps, even health is only instrumentally good. Perhaps, being healthy is only good to the extent that being healthy gives you pleasure and that being unhealthy causes suffering. Is pleasure good in itself? Pleasure might well be good in itself, just like suffering might be bad in itself. What the ultimate goods are is a matter of some controversy. There is no consensus about whether several of them exist, or just one.

When talking about ultimate goods, it is common to distinguish between what is *good for* someone and what is good *simpliciter*. A good of the first type is known as a "prudential good." We talk about prudential goodness in terms of welfare: whatever is *good for me* contributes to my welfare, and whatever is good for any individual contributes to his or her welfare. What, then, is good for me? As we have seen, pleasure looks like a plausible candidate for what is ultimately good for me. Hedonists say that pleasure is the only ultimate good and that suffering is the only ultimate bad. Hedonism is, then, an account of welfare, for it is a theory about what is ultimately good and bad for an individual.

The most well-known rival accounts of welfare are preferentialism and objective list accounts. According to preferentialism, having your preferences fulfilled is what is ultimately good for you. On that account, having money, having a coat, being warm, and being healthy are not intrinsically good. Rather, they are only good to the extent that they contribute to the satisfaction of your preferences. On this account, even pleasure is not good in itself. For pleasure is only good for you if that is what you desire.

The so-called objective list accounts of welfare, in contrast, stipulate lists of items that are considered to be intrinsically good for you. Possible items on such a list might include pleasure, health, and friendship. On such an account, some things are good for you no matter whether you actually desire them and enjoy them or not. The categorization of accounts of welfare in hedonism, preferentialism, and objective list accounts is common but also problematic. Hedonism, for instance, might also be considered an "objective list account" with one item, namely, pleasure, on the list.

Besides prudential value, other things may or may not be valuable for their own sake. Perhaps, for instance, virtue is ultimately good, even though it might not always be *good for* the individual to be virtuous. It is controversial what, if anything, is *good simpliciter*. Some argue that things can only be good if they are good for someone. Others accept that things can be good without being in any way good for anybody. In general, one conceives of ultimately good things as those whose mere existence contributes to the goodness of the world.

Another core area of moral philosophy is normative ethics. While value theory is concerned with what is good, normative ethics is concerned with what is right. The central enquiry that belongs to normative ethics is that into how we ought to live and in particular how we ought to act. Ethics can be understood as an investigation into reasons for action. Reasons, in turn, can be understood as properties of actions that count for the agent in favor of the action (Crisp, 2006a). It seems, on the face of it, that various properties of actions can provide us reasons. For instance, the action might promote my self-interest. That might provide a self-interested reason for doing the action. In addition, the action might be an act of promise keeping. That might provide a reason for the action as well, based on the fact that it is an instance of promise keeping and that keeping one's promises is somehow important. Some reasons might speak, for the agent, in favor of the action. Other might speak, for the agent, against the action. The question that will interest us most is what the agent has most reason to do, all things considered. That means that we are, strictly speaking, not particularly interested in "moral reasons"—whatever that might be—as one kind of reasons among others. After all, if we knew the agent's moral reasons, we would still not know how to weight these reasons against others in order to arrive at a judgment about what the agent has most reason to do, all things considered. That "all things considered" judgment is what we are interested in. If one wishes to call this all things considered reason the moral reason, that's fine with me. I will call it the "morally right" thing to do. Note that what the agent *has* most reason to do is not necessarily equivalent with what the agent *believes* she has most reason to do. I would say that we are primarily interested in the reasons that the agent has, not from her own perspective, but objectively speaking. Whether there are such objective reasons is a matter of debate in metaethics.

Normative ethics is concerned with morally evaluating actions. (One might also evaluate other "evaluative focal points" such as rules or attitudes, but I will focus on actions here.) Normative theories are very important in the field of normative ethics. Such theories aim primarily at telling us what is morally right and wrong and why it is so. Normative theories typically contain an account of what is good and an account of what is right. For instance, utilitarianism, one of the classical moral theories, holds that welfare is the sole ultimate good. Furthermore, it tells us that we ought to maximize welfare: the action that maximizes welfare is that which is morally right. Any other action is morally wrong. Normative theories in general aim at providing a coherent and unified account of moral rightness, which we can then apply to particular situations and actions.

Applied ethics belongs to the field of normative ethics. It is about applying normative reasoning to various fields of application, such as how we ought to deal with animals or the environment. Thus, environmental ethics is a field of applied ethics. Before engaging with environmental ethics, a few remarks are in order concerning the status of what we are actually doing when we are doing ethics.

Metaethics, the third core area of moral philosophy, concerns questions *about* the other two areas. In particular, it inquires about the status of moral claims and the possibility of moral knowledge. Can ethical theories or specific moral judgments be true? Can we know what is morally right or wrong? Can we know, as I framed it, what we have, all things considered, most reason to do? If so, how? Since the answers to these questions determine what one is actually doing when one engages with environmental ethics, I will conclude this preliminary section with some arguments for and against moral objectivity.

14.2.3 Are There Objective Moral Standards?

Ethical objectivism is the view that there are objective moral standards. There is, on this view, a correct answer to the question of what one has most reason to do, and that answer is independent of what one thinks one has most reason to do. That means that these standards apply to everyone, no matter whether people actually know them or believe in them. As Russ Shafer-Landau explains in his *The Fundamentals of Ethics*: "Moral claims are objectively true whenever they accurately tell us what these moral standards are, or tell us about what these standards require or allow us to do" (2010, 305). Those who do not believe in moral objectivity are moral skeptics. The skeptics are either moral nihilists or ethical relativists. Moral nihilists think that there are no moral truths at all. Ethical relativists accept that some moral standards are correct, but hold that this is always relative to a particular group or individual: the correct moral standards vary between groups or individuals, because they are a matter of the group's or individual's moral commitments. Note that ethical relativism is not merely the undeniable empirical claim that, as a matter of fact, different individuals and groups happen to accept different standards. Rather, it is the metaethical claim that different standards are in fact true. There are various versions and interpretations of each of these views. It is controversial which of them is right and what it implies for the moral discourse.

I will not settle this issue here, but restrict myself to discussing a couple of common arguments against moral objectivity. Since these arguments are so easy to rebut, I do not want you to buy into them and dismiss moral objectivity all too quickly. One such argument claims that if moral objectivity were true, there would be no exceptions to any moral rule: moral rules would be absolute. That, however, is mistaken. Moral objectivity is a view about the ontological *status* of moral standards, that is, whether they are ever true and whether their truth depends on what people believe or desire. Moral objectivity is not about the *stringency* of moral rules. If there are true moral rules that hold for everybody, these rules may or may not allow for exceptions. These are simply different issues. For example, there are rules in other, nonmoral areas, such as physics, which count as objective, even though they allow for exceptions. Another argument against moral objectivity is based on the idea that everyone has an equal right to an opinion. However, even if this is granted, it does certainly not follow that every opinion is equally plausible. The concern with moral objectivity expressed in such arguments relates to the fear that believing in moral objectivity may lead to dogmatism or intolerance. That, however, is not necessarily the case. The view that there are objective moral standards implies nothing about what these standards are. It implies nothing about the content of the moral standards. Moral objectivity is compatible with the claim that being open-minded or tolerant is morally right. If moral objectivism is true, it cannot be the case that all opinions concerning moral issues are equally right. Some are correct and others are wrong. That, however, does not mean that one is permitted to force one's insights upon others. Finally, moral skeptics claim that reasonable people disagree about what is morally right and that therefore there cannot be one moral truth. That, however, does not follow. Reasonable people disagree on many issues: for instance, they disagree on whether God exists, on what the Earth's average temperature will be 10 years from now, and on where the first human beings originated. In spite of that disagreement, there surely is a fact of the matter as to all of these things. The same may well be true in case of moral disagreement.

It may well be true that there are objective moral standards that hold for everybody. In any case, many often-heard arguments against moral objectivity are unconvincing. Furthermore, the skeptics' positions face serious challenges. For instance, relativists claim that standards are true simply because individuals or groups firmly believe in them. That would seem to make an individual or group infallible, even if the views are based on errors and prejudice. Moral skepticism does not allow for fundamental moral progress and not even for meaningful moral debate. Nihilistic positions come in different varieties, and there are currently heated debates about their plausibility. Objectivism faces some interesting challenges as well. For instance, how can we know of the objectivity of our moral claims? Do we have a special kind of intellect or intuition to grasp moral truths? Can we reason all the way to moral truth? Is what is morally right or wrong reducible to some facts in the world? It remains to be seen how the various positions develop and what exactly they imply concerning the standard of moral claims.

14.3 HANDS-ON: DOING ENVIRONMENTAL ETHICS

In what follows, I'd like to invite the reader to join me in actually doing environmental ethics. I will explore what I take to be two major topics in environmental ethics. These topics are interspecies justice and intergenerational justice. In order to explore these topics, I will ask relevant questions concerning value, moral status, and moral duties. I will give an overview of various answers to these questions and provide my own answer. In this way, I shall introduce these major themes in environmental ethics and invite further reflection and discussion.

14.3.1 Interspecies Justice

14.3.1.1 Introduction

An example of environmental management is the erection of fencing that separates the so-called big game from vehicles along the *Quebec Autoroute 73* in Canada. Is it a good thing that these fences are built? For whom is it good? Is it good for the drivers that are protected from large animals that otherwise may hit their cars in an effort to escape the sport hunters' guns or is it good for the animals? What will the fencing do to the animals? Will they run into the fence, rather than cross the road? Does that benefit them? Are there alternative ways of preventing accidents? Does it matter morally what the fencing does to the animals? Are these large animals rightly treated as game in the first place? Are they a resource for us to be managed? Or are they among those who should be considered when contemplating the protection of resources, such as their forest home? These are all questions with a normative aspect that environmental managers need to consider when contemplating or evaluating such measures. They concern issues of interspecies justice.

In what follows, I will discuss the moral status of humans, animals, and the rest of nature. This concerns their moral considerability, that is, whether we need to take them into account in our moral considerations (Section 14.3.1.2). Consequently, I will turn to questions concerning prudential value: What does animal or plant welfare consist in and is death harmful for an animal or plant (Section 14.3.1.3)? Finally, I will address questions concerning our moral duties toward various kinds of animals and the rest of nature (Section 14.3.1.4).

14.3.1.2 Considering Humans, Animals, and the Rest of Nature

In this section, I will introduce the concept of "moral status," along with various positions as to the moral status of humans, animals, and the rest of nature. That an individual has direct moral status means that we ought to include the individual in our moral considerations for its own sake. What we do to such an individual matters morally in a direct way. I will defend the position that all and only sentient beings have moral status.

Those to whom we direct our moral norms are called "moral agents" or "moral subjects." Moral agents are those whom we consider capable of acting morally, that is, acting for moral reasons. Typically, all mentally sane adult human beings are considered moral agents. It is these individuals that we hold morally responsible for what they are doing. The age boundary for moral agency is actually lower than 18. Teenagers and even younger children are typically considered morally responsible, at least to some extent. Clearly, newborn babies are not moral agents. I will not attempt to define here where exactly the boundary lies. As children grow older, they are socialized and expected to live increasingly in accordance with some moral norms and to do so for the right reasons. They are increasingly considered to live in accordance with what they have most reason to do.

Those whom we, moral agents, take into account in our moral considerations are called "moral objects" or "moral patients." Thus, the group of moral patients is conceptually different from the group of moral agents. It is controversial whether or not as a matter of fact both groups have the same members. In other words, do we only need to take into account other moral agents in our moral considerations, or is the scope of those whom we need to consider different, perhaps larger or narrower?

Consider human babies. They may become moral agents in the future but are not moral agents now, since they cannot act morally. Does it matter, morally speaking, what we do to them? One might argue that it matters what I do to my neighbor's baby, because if I hurt it, for example, my neighbor won't be happy. If that were the only reason for why what I did to my neighbor's baby mattered morally, the baby would not have moral status. After all, hurting the baby would not matter because of what it does to the baby. It would only matter because of what it does to someone else, in this case the neighbor. What I do to the baby would not matter directly, so to speak; it would only matter indirectly. That I ought to refrain from harming the baby would rather be because of the neighbor, not because of the baby. In the same vein, I ought to refrain from smashing my neighbor's car, not because what I do matters to the car, but because it matters to its owner. But is it true that babies do not have any moral status?

In order to decide who has moral status, we need to know the criteria on the basis of which moral status should be attributed to a particular individual. In other words, whom do we have reason to take into account when we are contemplating how to act? On what grounds can we determine whom we have reason to take into account? These are normative questions. Unsurprisingly, it is controversial how to answer them. Different moral theories have different implications as to who has moral status.

Consider, as an example, utilitarianism and the question whether nonhuman animals have moral status. Utilitarianism requires the neutral maximization of welfare. Do we have to take animals into account in our moral considerations? Since many nonhuman animals have welfare and the maximization of welfare is all that matters, their welfare counts on a par with human welfare. Thus, it follows that, according to utilitarianism, we have to take effects on the welfare of animals into account in our moral considerations. In fact, it follows that all beings that can be affected in their welfare count morally. Effects on the welfare of all beings that "have a welfare" have to be taken into account on an equal basis.

This description of utilitarianism and of who, according to this moral theory, has moral status, sounds as if the moral requirement to maximize welfare precedes the determination of the scope of moral objects. One might, indeed, understand utilitarianism in that way. One might consider an account of the good, which is defined as welfare, fundamental and consider the promotion of this good the moral aim und thus consider the requirement to count everybody's welfare equally only a by-product of this more fundamental duty to maximize welfare. This has lead to the reproach that utilitarians do not primarily care about sentient beings after all. They merely care about welfare as an abstract quantity and conceive of sentient beings as mere "receptacles" of welfare.

Another understanding of the basis of utilitarianism is possible as well. One might start with the question of whom to take into account in our actions. The answer might be those who care what happens to them. These are, arguably, sentient beings or beings with interests. One might then inquire how to treat those moral objects. The answer might be to treat them on an equal basis, to count, as Jeremy Bentham (1748–1832), one of the founding fathers of utilitarianism, put it, "each for one and no-one for more than one." The utilitarian understanding of "taking sentient beings equally into account" then might be "considering the effects of our actions on their welfare." From this might follow the utilitarian principle to neutrally maximize welfare. Maximization, on this view, arises as a by-product of the requirement to take everybody's welfare equally into account.

Which beings "have welfare" and thus deserve moral consideration according to utilitarianism depends on one's exact definition of welfare. For instance, if welfare is defined in terms of desire satisfaction and frustration, all beings that have desires need to be included in moral considerations. If, alternatively, welfare is defined in terms of pleasure and pain or, more broadly, enjoyment and suffering, then it follows that all beings that can experience these mental states count morally. Since there is a strong link between sentience and the having of desires (in the sense that all and only sentient beings seem to have desires—at least typically the desire not to feel pain), both accounts of welfare usually come to the conclusion that all and only sentient beings deserve moral consideration. This is the "sentientist" position on who has moral status. This is based on the idea that all and only sentient beings care about what happens to them. They have desires that can be fulfilled or frustrated, and they can experience enjoyment and suffering and can thus be affected in their welfare. Other accounts of welfare might imply that another group of beings has welfare, but utilitarians usually accept sentientism. Thus, they hold that the effects of our actions on all sentient beings that are affected in their welfare should be taken into account when considering what we have most reason to do.

There is disagreement between different moral theories and even within moral theories as to who has moral status. Thus, it is disputed whether nonhuman animals ought to be granted moral status. The so-called contractualist moral theories conceive of what is morally right and wrong as following from some (hypothetical) agreement between individuals. The individuals are ascribed certain characteristics, such as being free, equal, and self-interested. These individuals, at least hypothetically, agree to restrict their freedom for the mutual benefit of all. Since nonhuman animals cannot be among the hypothetical contractors, we do not need to take them directly into account, according to contractualist theories as commonly understood (Rawls, 1973). Efforts have been made, however, to incorporate sentient animals into contractualist moral theories (Rowlands, 1997) and accord them moral status on a contractualist basis.

Kantian justifications of our moral duties are based on ideas about what one can rationally will. Animals are not granted moral status in traditional interpretations of Kantian moral theories, because animals cannot live in accordance with moral rules. However, some contemporary Kant scholars argue, convincingly to my mind, that animals ought to be granted moral status even on a Kantian basis. For instance, Christine Korsgaard (2005) argues on a Kantian basis that we must regard animals as ends in themselves, that we have moral duties to animals, and that animals should have legal rights, such as the right to life. Contemporary Kantians base moral status either on the capacity of moral autonomy or on the capacity of agency. Those who want to base moral status on agency tend to focus on the capacity for moral agency. Kaldewaij (2013) argues that while being a moral agent is not necessary for moral status, being a purposive agent is sufficient. Animals are purposive agents: they have desires and act in order to get what they want. Thus, we have to take their perspective into account when determining what to do. Kaldewaij thus defends purposive agency as the most plausible Kantian basis for moral status.

It is also controversial what having moral status means *precisely*. According to utilitarianism, as I explained, that an entity has moral status means nothing more and nothing less than that effects on the entity's welfare have to be taken equally into account in the evaluation of outcomes. As such, the fact *that* an entity is accorded moral status has no implications as to *how* this entity might be treated. Utilitarianism does not accept that certain kinds of actions as such are morally right or wrong or that there are any limits as to how an entity ought to be treated. Whether an action is right or wrong and how an entity ought to be treated depends on the consequences of the action in terms of overall welfare. It depends on nothing else. Thus, every individual's welfare will be taken equally into account when considering how to act. Since, however, one ought to do whatever maximizes the overall welfare, there is no guarantee as to what every individual's welfare will be after the action. The maximization of overall welfare is compatible with unequal distributions of welfare and with some individuals being worse off than others. According to other moral theories, such as Kantianism, moral status means that the individual can rightfully make some very strong claims on others. These claims are understood as rights. The emphasis in such theories is typically on the so-called negative rights, such as the right not to be killed and certain rights of noninterference. There is typically less emphasis on the so-called positive rights, such as the right of assistance in realizing some goods, such as welfare. Those who accept rights hold, for instance, that the honoring of individuals' rights should limit the promotion of welfare.

This different understanding of "moral status" goes together with different ascriptions of moral duties. Utilitarians, who aim at the promotion of overall welfare, reject the moral relevance of the doing–allowing distinction. After all, one can affect welfare by what one does as well as by what one declines to do. For instance, one can negatively affect someone's welfare by not helping her if she needs help. Kantians, in contrast, hold that there is a morally relevant distinction between what one does and what one merely allows happening. Negative rights point out in what ways one ought not to interfere with others. They are not about how one ought to assist them. Another difference is that Kantians reject some forms of aggregation. "Aggregation" is necessary in order to compare the results of two possible actions in terms of welfare. If an action affects various individuals, one needs to calculate the action's overall effect on welfare by taking together the effects on the welfare of each concerned individual. A common way of aggregating is simply adding up the harms and benefits that an action brings about. For that reason, utilitarians typically hold that realizing slight benefits to many people can justify significantly harming one person. Kantians, in contrast, disagree. They believe, after all, that a person has certain rights not to be interfered with. Kantians reject adding up the benefits that accrue to many people and comparing it to the harm that one individual suffers. Again, that is in line with their focus on negative rights. However, since even Kantians seem to accept some aggregation, it has been argued (Norcross, 2009) that there is no plausible way to draw a line between acceptable and unacceptable aggregation, and therefore, the Kantian refusal of thoroughgoing aggregation seems to be built on weak ground.

Quite generally, according to utilitarianism, but also according to other moral theories, equal moral status is compatible with unequal treatment. For instance, if different animals have different interests, treating them differently—each according to his or her interests—is compatible with granting them equal moral consideration. For instance, some utilitarians have argued that killing a normal human teenager is worse than killing a mouse. According to that view, mice might be killed for reasons that would not justify the killing of human teenagers. If this unequal treatment is based on the idea that mice lack interest in continued life, which human teenagers have, killing the mice rather than the teenagers is compatible with the equal consideration of their interests. After all, the equal consideration of interests means that *equal* interests should get equal consideration. Unequal interests can therefore justify unequal treatment, even in case of equal moral status. (The question whether death is indeed a lesser harm to nonhuman animals will be discussed in Section 14.3.1.3.) Sentientism, the position that all and only sentient beings have moral status, does not only go together with the utilitarian moral theory. Other moral theories may also accept sentientism, albeit on different grounds, as some rights-based theories do (Cochrane, 2007, 2009). Whatever criterion one proposes for demarcating the scope of moral patients, it is hard to exclude all nonhuman animals and at the same time include all humans. This observation has been called the "argument from marginal cases." There will always be the so-called marginal humans, who score worse on certain criteria than individual animals. For instance, if intelligence is brought forward as the relevant criterion, there will always be individual nonhuman animals that are more intelligent than individual humans. For instance, certain great apes have been found to perform better in certain intelligence tests than human infants, and there will always be mentally handicapped humans or Alzheimer patients who score worse than certain intelligent animals. I do not think that intelligence matters at all for moral status. Rather, what matters is whether an individual can be affected in her welfare, that is, whether it matters to the individual what happens to her. The fact that we have moral duties toward babies, for instance, does not relate to their intelligence, but simply to the fact that it is in our power to harm and benefit them.

In order to try to avoid the "argument from marginal cases," one might simply draw the boundary between those who have moral status and those who do not on the basis of species membership. That, however, would be arbitrary and amount to unjustified discrimination. Discrimination is a matter of making distinctions. We draw lines and make distinctions all the time, for instance when we say that only people above 18 may drive a car or vote in national elections. It is also common to have separate toilets for males and females in public buildings. This is discrimination, but is it unjustified? That depends on whether a morally relevant difference can be brought forward that justifies that different treatment. Perhaps, voting and driving require certain capacities that are generally linked to age, and perhaps, while there are individual exceptions, drawing sharp lines is still the best bet in terms of policymaking. Perhaps, singling out people on the basis of individual capacities would be impractical and involve an invasion of privacy. Perhaps, it is not such a big deal to draw these rough boundaries with regard to voting and driving, since just about everyone will eventually become old enough to be allowed to do these things. On the other hand, if it is really the capacities of the individual that matter, there is something nonideal about these age boundaries and perhaps some morally better alternative could be designed. Perhaps different needs can justify differentiation in the case of male and female toilets. The question is always whether differential treatment can be morally justified. If it cannot be morally justified, it amounts to unjust discrimination. Racism and sexism are examples of unjust discrimination. The analogous word "speciesism" has been used to criticize unjust discrimination of the basis of species (Ryder, 1989). For instance, Peter Singer (1995) has forcefully pointed out that we routinely harm nonhuman animals for reasons that we would not take to justify harming humans.

While some consider sentientism too broad, others find it overly restrictive. Why should we restrict our moral consideration to those who can feel what happens to them? From the beginning of the development of environmental ethics, which arose in the counterculture of the 1960, there has been the feeling that a whole set of new values was needed to inspire a radical shift in outlook and behavior (Jamieson, 1998). Sometimes, the emerging set of values is described as "environmental ethics." This, however, is problematic, since environmental ethics, as I explained in the beginning of this chapter, should be understood as a branch of applied ethics. It is simply the application of ethical inquiry to issues of human interaction with the environment. Environmental ethics should therefore not be identified with specific substantial views about what that human–environment relationship should be. That would be just as wrong as identifying medical ethics with particular views about right conduct in the field of medicine. While environmental ethics as a field should therefore not be identified with any particular moral outlook, it is true that a particular moral outlook has been, and still is, prominently discussed within that field. By the early 1980s, three claims were seen as characteristic of the new values, which, according to some prominent figures in environmental ethics, were desperately needed (Crisp, 1998):

1. Nonsentient entities have value (nonsentientism).
2. Collective entities such as ecosystems have value (collectivism).
3. Value is mind independent: even if there were no conscious beings, aspects of nature would be valuable (mind independence).

The latter claim about the mind independence of value is a metaethical position. This metaethical position is independent of whether one is a nonsentientist or a sentientist, and it is also independent of whether one is a collectivist or an individualist. The reason why it has prominently been discussed is that its independence of the normative issues has been ill understood. Mind independence in this context is usually understood as the thesis that there can be value without valuers. It is an objectivist position about value. Mind dependence, in contrast, is understood as the thesis that value depends on and arises from valuing individuals. It is a subjectivist position about value. (Strictly speaking, even objectivists about value can hold that value is mind dependent. Those who think that pleasure is objectively valuable are objectivists, since they believe that the value of pleasure is independent of its being valued. However, obviously, pleasure cannot be mind independent, since pleasure is a state of mind.) An argument for mind dependence of value is that, intuitively, it might seem that it is valuing individuals who, by valuing, bring value into the world. For that reason, one might think that there is no value without valuers. On the other hand, if value depended on valuing and in fact if value meant nothing else than "being valued," that would have some odd implications. There would be no need for saying that anything has value. It would suffice to say that some things just happen to be valued. If, however, nothing has value, why *should* we value some things, rather than others? Why should we have reason to value some things above others?

Leaving this metaethical question about value aside, one of the two other widely discussed positions in environmental ethics was and still is nonsentientism. Richard Routley was an early defender of nonsentientism. In a 1973 paper, Routley introduced a series of thought experiments in order to show that moral status should not be restricted to sentient individuals. Routley asked us to consider a "last man" whose final act is to destroy such natural objects as mountains and salt marshes. Although these natural objects would not be appreciated by conscious beings even if they were not destroyed, Routley thought that it would still be wrong for the "last man" to destroy them. These intuitions were widely shared, and many environmental philosophers thought that they were evidence for the moral status of nonsentient objects. Biocentrism is a specific version of nonsentientism. Biocentrists ascribe moral status to all living organisms. The main justification for this extension of the scope of moral objects is the assumption that all living individuals have a "good of their own." For instance, they can be healthy or unhealthy; they can flourish or die. Human life, on that account, is not superior to other live forms. Al living organisms are considered to have "equal inherent worth." In his influential book *Respect for Nature*, Paul Taylor (1986) defends such a biocentric outlook, called "biocentric egalitarianism." It is based on the observation of interdependence of life forms. According to Taylor, all organisms (and only organisms) are teleological (i.e., goal-directed) centers of life. One might think of plants that are seeking light as an example. Such goal-directed organisms, according to Taylor, have goods of their own. Taylor calls this their "welfare interests." More recently, Robin Attfield (2014) also defends a biocentric position, which he calls "biocentric consequentialism." According to that position, the resulting harms and benefits to all living organisms determine whether an action is right or wrong. While Attfield grants moral status to all living beings, he grants a higher degree of moral status to individuals with certain capacities, such as sentience or autonomy. Mary Midgley (1983) previously defended a similar gradual account of moral status.

What to make of these biocentric arguments? Sentientists can agree that it would be wrong for the "last man" to harm animals. Would it be bad if the last man destroyed plants? Even if many people have the intuition that it would be bad, that is not enough to establish that plants have moral status.

There are, after all, different explanations for that intuition. In normal circumstances, needlessly destroying things is considered bad, for instance because it reveals aggression, which is bad because it might also be directed against sentient creatures. In addition, needlessly destroying things is bad, because sentient creatures might care about these things. So, perhaps when considering the thought experiment, our judgment is due to not abstracting from these otherwise very relevant facts. We cannot depart from what we normally think about the badness of destroying things. This psychological mechanism has been documented in thought experiments (Weijers, 2013). Furthermore, if we were asked whether it would be bad if the "last man" destroyed some great works of art, we would likely also say that it is bad. Yet, biocentrists do not wish to confer moral standing to works of art. So, simply referring to our intuition does not support the biocentrist position. Additional arguments for conferring moral standing to nonsentient living organisms are needed.

The main additional argument that biocentrists provide is the claim that living organisms have a good of their own. In a way, that is true. A tree can, for instance, be healthy and flourishing or ill and dying. Plants also have certain mechanisms that are directed toward their survival and reproduction. Flowers, for instance, turn to the sun. Plants and nonsentient animals, just like sentient beings, thrive in certain environments and not in others. They seem to have certain "needs," as in the case of a flower that needs water and sunlight. These needs can be fulfilled or frustrated. However, in opposition to sentient individuals, nonsentient organisms do not have an inner life. They are unconscious and do not have, in that sense, any perspective of their own. They do not, in that sense, care about what happens to them. But is that inner perspective really a necessary requirement for moral status? Attfield (2014) argues: "If being healthy is good for us even if we do not care for it or for anything else for which health is instrumentally valuable, then certainly health can be good for plants without them caring for it." This is an interesting argument. However, it is controversial whether health is intrinsically good for us. A common position is that health is instrumentally good for us to the extent that it contributes to our enjoyment or to our desire satisfaction. If, however, health is really an aspect of welfare, it might follow that plants, insofar as they can be healthy or unhealthy, have welfare and thus can be better or worse off (Visak and Balcombe, 2013). In that case, it seems that a utilitarian would need to take into account plant welfare as well. That would come very close to Attfield's "biocentric consequentialism."

But can things really go better or worse for a plant in the relevant sense? Being handled with care is good for my laptop, in some sense. But that does not imply that my laptop has a welfare, let alone moral status. Biocentrists reject this analogy between something being "good for" an unanimated object and something being good for a living organism. They say that in the former case, handling my laptop with care is really good for me. In case of the living organism, it is really good for the living organism, since its welfare it really its own. In what way is it its own, though? In what sense is the flower's welfare its own welfare and is my laptop's welfare *not* my laptop's own? For me, the relevant difference between flowers and laptops on the one hand and sentient beings on the other hand is that the latter but not the former have the capacity of enjoyment and suffering, desires and projects, an own perspective, and a mental life. It is possible at all to emphasize with sentient beings, but not with nonsentient organisms. There is nothing it is like to be a nonsentient object or organism. Claiming otherwise is understandable and often, I think, well intended. But it is nevertheless unwarranted projection.

If empirical science tells us that plants or any other organisms that we previously considered nonsentient do have subjective experiences after all, that will, of course, be relevant information. But the claim of the biocentrists is not that plants are sentient. It is rather that being alive rather than being sentient should be the relevant criterion for ascribing moral status. Whether they are right is one of the central debates within environmental ethics.

The second of the two abovementioned widely discussed positions within environmental ethics is holism. Holists believe that collective entities, such as ecosystems or species, have moral status. This position is opposed to individualism, which restricts moral status to individuals. Holism in a pure form is the ascription of moral status only to collectives. Holists might, for instance, be in favor of killing individual animals in order to prevent a threatened plant species from extinction. Holists might also ascribe moral status to both, individuals *and* collectives. Holists in environmental ethics are also called ecocentrists.

The debate between Peter Singer and Holmes Rolston III is illustrative (Rolston, 1999; Singer, 1999). Singer is one of the world's most famous ethicists and utilitarians. He defends a sentientist position. Holmes Rolston is one of the best-known defenders of an ecocentric position in environmental ethics, who grants moral status to collectives as well as to individual living organisms.

Singer defends sentientism as follows:

If a being suffers there can be no moral justification for refusing to take that suffering into consideration. No matter what the nature of the being, the principle of equality requires that its suffering be counted equally with the like suffering—in so far as rough comparisons can be made—of any other being. If a being is not capable of suffering, or of experiencing enjoyment or happiness, there is nothing to be taken into account. So the limit of sentience (using the term as a convenient if not strictly accurate shorthand for the capacity to suffer and/ or experience enjoyment) is the only defensible boundary of concern for the interests of others. (AL 2nd edn, 8–9, cited in Singer (1999, 328))

Rolston agrees that plants may not have wills or desires, but insists that every organism has a "good of its kind." Rolston does not take this talk about plants having their own good to be a metaphor, but "the plain fact of the matter." Singer, however, is unconvinced. He claims:

> Surely, the plain fact of the matter is that over millions of years, those plants that evolved defense mechanisms that significantly improved their chances of passing on their genes had more offspring than those plants that did not, and over time tended to supplant them. There is no talk of "standards," let alone of values or of what "ought to be." An evolutionary explanation suffices. (Singer, 1999, 329)

Does sentientism imply that there is no need to protect plants, rivers, and ecosystems? Of course, it doesn't. After all, the welfare of us, sentient beings, for now and in the future, depends crucially on intact environments and ecosystems. When I talk about our "welfare," I do not only mean our physical and mental health and survival. As should become clear from the accounts of welfare to be described in the next section, whatever is good for me in any way contributes to my welfare. This may include aesthetic, scientific, recreational, and spiritual endeavors. In that sense, protecting species and ecosystems, mountains, deserts, and individual plants and nonsentient organisms can be very important and valuable.

The flourishing of human and animal live depends a lot on supporting natural environments. There is, for instance, a growing amount of empirical evidence of the positive effects of natural environments on human welfare. Time spent in contact with the natural environment has been associated with better psychological well-being (Kaplan, 1973), superior cognitive functioning (Cimprich, 1990; Hartig et al., 1991; Tennessen and Cimprich, 1995; Kuo, 2001), fewer physical ailments (Moore, 1981; West, 1986), and speedier recovery from illness (Ulrich, 1984; Verderber, 1986; Verderber and Reuman, 1987). Due to their greater plasticity or vulnerability, exposure to nature may particularly affect *children's* functioning and welfare (Faber Taylor et al., 2001, 2002, 1998; Grahn et al., 1997; Wells, 2000). Children's preferred environments include a predominance of natural elements (Korpela, 2002). For example, in a study, where urban children aged 9–12 were asked to make a map or drawing of all their favorite places, 96% of the illustrations were of outdoor places (Moore, 1986). The welfare and survival of countless sentient animals depend on the particular ecosystems that support them. That some animal species disappear from our planet is not only bad because of the animals that are concerned but also because many humans care about those animals and feel bad about their extinction.

Biocentrists and ecocentrists argue that these kinds of sentientist (or even anthropocentric) arguments in favor of plants, species, and ecosystems are too weak. Could they provide reasons to regret the extinction of species before they ever were discovered? It seems that since nobody cares about any undiscovered species, their extinction cannot be bad for any sentient being and thus cannot be bad at all. This, however, is a mistake. First of all, even if the plant species, say, is not known by human beings, other sentient animals may in some way value or depend on this plant species. Secondly, if, counterfactually, the plant species had not disappeared, it might have been discovered and might thus have contributed to the welfare, broadly conceived, of humans, if only because the mere discovery of a new species is typically bringing some form of enjoyment. Finally, due to the complex interdependence of ecosystems, even the extinction of a single plant species usually has impacts on other plants and, either directly or indirectly, also on animals. On the other hand, if the extinction of some plant species wouldn't negatively affect animal welfare in any way, then it wouldn't be bad, after all. Biocentrists also argue that the effects of nonsentient individuals or collectives on animal welfare cannot grant them sufficient protection. After all, wouldn't it be valuable that plants flourish, even if all sentient life has long gone extinct? Wouldn't there be a reason to protect the future flourishing of nonsentient life, even if there were no sentient being around to enjoy it? Again, I think that we shouldn't place too much trust in our intuitions here. If we compare a barren and empty planet with one covered by beautiful flowers and forests, of course we are in favor of the latter. When comparing the value of these worlds, we cannot but imagine ourselves living in these worlds or at least viewing them and thus being affected by them. Furthermore, if we compare a very beautiful lifeless world, perhaps with great beaches, impressive mountains, and icebergs, to a world inhabited by some living bacteria, slimy worms or ugly bugs, we might well prefer the former to the latter. So, here again, our intuitions do not necessarily support the biocentric position.

We have seen that there are different views in ethics as to which beings have moral status. The view that all and only human beings have moral status is known as "anthropocentrism." This view has been criticized as implying unjustified discrimination on the basis of species. Sentientism is the view that all and only sentient beings have moral status. This view is also known as "zoocentrism" because it accords moral status to human and nonhuman animals. However, it should be noted that most animals are in fact insects and it is unclear whether insects are sentient. Indeed, the prevailing scientific opinion at the moment is that they are not. Mammals, birds, fish, and also some other animals are generally considered sentient. Therefore, the label "sentientism" is more adequate. Biocentrism is the view that all living organisms have moral status. Ecocentric views tend to ascribe moral status to further entities, such as species and ecosystems. While biocentrism and ecocentrism are prominently discussed in environmental ethics, I consider sentience the most plausible basis for ascribing moral status.

Let me end this discussion of whom to take into account in our actions with a note of caution. In environmental ethics, the question about who or what has moral status is frequently discussed in terms of who or what has "intrinsic value." In these discussions, the notion of "intrinsic value" is different from how I used it here. Having intrinsic value can be taken to mean "having moral standing." This line of thought usually ascribes intrinsic value to what is taken to be a due object of our moral concern, what we should directly take into account in our moral considerations. According to that line of thought, intrinsic value can be ascribed to people, animals, plants, or even collectives, depending on one's specific position. That's what I describe here as the issue of who has moral status. I do not talk about this question in terms of who or what has "intrinsic value." According to another line of thought, "intrinsic value" refers to what is good not only instrumentally but for its own sake. That might, for instance, be pleasure, knowledge, achievement, or welfare. That is how I use the notion of "intrinsic value" or "ultimate value" in this chapter. It is a very common way of using this notion. I ascribe intrinsic value, thus understood, to welfare. One can already see that the list of things that have intrinsic value according to the first line of thought is different from the list of things ascribed intrinsic value according to the second line of thought (Bradley, 2006). This reinforces the suspicion that one is talking about different things (Schroeder, 2012). One concerns the question of whom to take into account; the other concerns the question of what makes the world better. The next section will further elaborate on this latter question.

14.3.1.3 What Is (Animal) Welfare and What Does the Harm of Death Consist In?

If indeed all and only sentient animals, both human and non-human, deserve our moral consideration, we are faced with the question of what we owe to these individuals. Before addressing this question in the next section, it will be helpful to have at least a rough idea about what benefits these individuals and what harms them. As such, I will discuss the questions "What makes our lives good for us?" and "What, if anything, makes death bad for us?" in this section. I consider the self-fulfillment account of welfare and the deprivation account of the harm of death the most convincing answers to these questions. These views entail, roughly, that being happy and being successful in the projects that I consider important for who I am make my life good for me. These are both aspects of welfare, which, as I will explain in the following, should be defined as self-fulfillment. Note that any account of welfare should be applicable to all beings that have a welfare, thus to all sentient beings. Death is bad for sentient beings to the extent that it deprives them of whatever would have made the future lives good for them.

The question of what makes my life good for me and death bad for me is about prudential value: what is good or bad *for someone*. It is debated whether anything except welfare is ultimately valuable and indeed whether welfare is of ultimate value at all. Welfarism entails that welfare is the sole ultimate value, meaning that it is the only thing that is valuable for its own sake. With this in mind, we will start with a quick look into that debate and with a brief defense of welfarism.

Welfarism
Welfarism holds that the only ultimate value is welfare. On this account, if other things are valuable at all, they are valuable only to the extent that they contribute to welfare. Thus, they are valuable only as means or instrumentally. Welfare, in contrast, is the only *intrinsic* value.

A major line of criticism against welfarism claims that it is counterintuitive. Aren't other things, besides welfare, good in themselves? Isn't it good, for instance, that beauty exists, even if nobody ever contemplates it or gets pleasure from it? Isn't knowledge, truth, or virtue good for their own sake? Proponents of welfarism reply that these things are not good for their own sake, after all. While we do appreciate them, the ultimate reason for doing so is that they tend to promote our welfare. The same holds, according to welfarism, for other alleged intrinsic values, such as biodiversity, forests, rivers, or species. According to welfarism, these entities are valuable to the extent that they bring about welfare. If they do not promote the welfare of anyone, there is nothing good about them. Critics may question: "Is it always good if people are well off? Is it the more welfare the better for them? Doesn't it matter what the source of their welfare is? What if it is grounded in vicious activities?" Proponents of welfarism argue that positive welfare is always good. The bad thing about vicious activities is that they tend to have a negative effect upon welfare. That, however, doesn't make positive welfare bad as such, not even if it is based on vicious activities. Vicious activities are extrinsically bad. They are bad, because they tend to bring about negative welfare.

Another line of criticism is more structural. Isn't the striving for welfare and nothing but welfare self-defeating? Won't we end up unhappy if we spend our days preoccupied with welfare? Welfarists grant that this may well be the case. Perhaps, we will actually be happier if we do things for their own sake, rather than for their effects on our or on others' welfare. However, even if this is right, this does not rule out the thought that welfare is the sole ultimate value; it only points to the strategic advice that striving for welfare directly can be counterproductive.

Another structural criticism has it that various options, states of affairs, or lives are not comparable in terms of welfare. What does it mean to say that A's life contains more welfare than B's? Welfare according to this criticism cannot be the sole ultimate value, because of this incommensurability. Proponents of welfarism can reply, however, that we do compare things, options, and states of affairs in terms of welfare when it concerns our own welfare. For instance, we decide whether to spend our money on a new car or on a holiday. Various ways of measuring and comparing welfare across persons and across times (and across states of affairs) have been proposed, both in theory and in practice (Broome,

2004). Note that the view that only welfare affects the value of a world does not entail that everybody should only be concerned with his or her own welfare. Various moral theories accept that welfare is the sole ultimate good. Utilitarianism accepts welfarism and requires the neutral maximization of welfare. Standard egalitarian moral theories might also accept welfarism and argue that welfare is the sole ultimate good that should be distributed equally. The "prioritarian" view, according to which one should give special consideration to improving the lives of the worst off individuals, can also rest on welfarism. Also, rights views can be built on a welfarist basis.

Welfare
Most people consider the welfare of themselves and of those they care about very important. It also seems that effects of our actions on welfare provide us reasons for or against actions. That should certainly be considered in environmental management, where many possible measures affect the welfare of people or animals, for the better or for the worse. What, then, is welfare?

One of the classical accounts of welfare is hedonism. Hedonism is a mental state theory that defines welfare in terms of subjectively experienced pleasure and pain or, more broadly, enjoyment and suffering (Bentham, 1996 [1789]; Mill, 1998 [1863]; Tännsjö, 1998; Crisp, 2006b; Feldman, 2006; Bradley, 2009). A key insight of hedonism is that it acknowledges the importance of one's mental states as an aspect of welfare. But are mental states the only aspect of welfare? Against hedonism, it has been argued that if pleasant experiences were the only things that made our lives good for us, we would be as well off as we could be on a machine or on drugs that gave us endless pleasant experiences (Nozick, 1974; Sumner, 1996; Crisp, 2006b). That, however, seems implausible.

One of the major rival accounts of welfare is preferentialism (Harsanyi, 1982; Griffin, 1986; Crisp, 2006b; Sobel, 2009; Wessels, 2011). According to preferentialism, having your preferences fulfilled is what is ultimately good for you. On that account, having money, having a coat, being warm, and being healthy are not intrinsically good. Rather, they are only good to the extent that they contribute to the satisfaction of your preferences. Happiness, as well, might be prudentially good, because it is desired. A more intimate relationship between desire satisfaction and happiness is conceivable as well. The relevant notion of "desire" might be defined with recourse of the notion of happiness, such as "We desire something if and only if correctly and vividly imagining it, under certain conditions, would make us happy" (Wessels, 2011, 86). Furthermore, happiness might be defined in terms of desire fulfillment, such as "Being in a situation in which one desires to be means that one is happy" (Heathwood, 2006).

The key insight behind preferentialism is that different things seem to promote the welfare of different individuals. Preferentialism explains that this is due to their different desires. This insight lets us explain why what is good for one individual may be irrelevant or even harmful to the well-being of another—different preferences explain why religious worship matters to the well-being of many believers but not to most atheists and why fertility matters to the welfare of those who want children but is irrelevant to the welfare of those who don't.

A major criticism against the desire account of welfare is that desiring things does not make them good for us. It is rather the other way around: we desire what is good for us (Crisp, 2006b. See also Aristotle (1984 [C4BCE], Metaphysics 1072a, tr. Ross): "[…] desire is consequent on opinion rather than opinion on desire."). Further, if desire satisfaction were all that mattered for our welfare, it seems that we ought to adjust our desires constantly to the chances of fulfillment. This would certainly bring us more desire satisfaction. However, changing our desires whenever doing so would make it easier to fulfill them seems to make us lose ourselves (Barry, 1989, 280; Wessels, 2011, 132).

The third group of the classical accounts of welfare is the so-called objective list accounts. Objective list accounts of welfare stipulate lists of items that are considered to be intrinsically good for you. Typical proponents of objective list accounts hold that well-being consists in the possession of certain basic goods that are beneficial to all human beings regardless of what they desire or prefer (Moore, 1903; Nussbaum, 1988; Sen, 1993;Parfit, 2011; Hausman, 2012). The degree to which one possesses those things determines one's level of welfare. The things on that "objective list" might be health, wealth, friendship, or other items. Eudemonistic accounts of welfare, such as Nussbaum's capabilities account, conceive of welfare in terms of "nature fulfillment." At some level, they might also present a list of items whose realization is considered to be relevant for welfare. Eudemonistic accounts of welfare and objective list accounts are typically classified as objectivist accounts of welfare and contrasted with subjectivist accounts, such as hedonism and preferentialism. However, that classification is highly unclear and controversial and should not bother us here (Sumner, 1996; Crisp, 2006b, section 4.3; Haybron, 2008, 177–178; Bognar, 2010).

The key insight of objective list theorists is that it seems counterintuitive to base the notion of welfare solely on the individual's own perspective, such as their preferences. After all, people sometimes prefer what leaves them worse off. Adaptive preferences shed light on this problem of preferences that leave one worse off (Elster, 1982). Nussbaum (2001) gives examples of women in third-world countries that form adaptive preferences to accept low pay, remain with abusive partners, and prepare their daughters for marriage rather than education. These women are not getting what is best for them and their daughters; in fact, the deformation of their preferences is part of the injustice committed against them. In other cases, ignorance leads people to prefer what harms them. In still other cases, people are aware that what they prefer is bad for them, but lack the willpower to change their preferences—drug addiction is a paradigm example.

A challenge for objective list accounts of welfare is determining what counts as a primary good in an objective way. Why, for instance, is education good for me? It seems that a justification of the goodness of education cannot avoid pointing out that it ultimately leads to happiness or to desire satisfaction. Furthermore, it seems to be an open question whether anyone who fulfills the criteria on the objective list is actually faring well, from her own perspective. What if someone does not care about education and seems to be perfectly happy without it? Some alleged objective accounts of welfare are really about something else: they do not offer an account of prudential value. For instance, a definition of welfare in terms of "living according to one's excellences" conflates prudential value and perfectionist value. Perfectionist value is about how much one lives up to certain standards of perfection (Sumner, 1996, 78).

In recent years, proponents of various theories proposed revisions in order to deal with alleged counterintuitive implications. Recently, hybrid or combined accounts of welfare have been proposed in order to combine promising aspects of different theories and avoid their shortcomings (Heathwood, 2006; Wessels, 2011).

Let me finally mention a promising new account of welfare that Daniel Haybron (2008) recently proposed. Haybron's self-fulfillment account of welfare differs interestingly from the classical accounts of welfare. It is promising, because it draws on intuitions that lie at the basis of these other accounts of welfare, while it seems to be able to avoid their counterintuitive implications. Roughly, according to Haybron, welfare is self-fulfillment, which, in turn, consists of three aspects: (i) emotional flourishing; (ii) success in projects, which one considers important to who one is; and (iii) the fulfillment of aspects that do not concern the individual's personality, but that are nevertheless important for welfare, such as health and vitality. So, strictly speaking, only emotional flourishing and success in identity-related projects are aspects of self-fulfillment. However, the focus on self-fulfillment distinguishes Haybron's account of welfare from better-known varieties of eudemonism, which define welfare in terms of nature fulfillment. These other forms of eudemonism do not focus on the individual and her specific psychological makeup at all. Haybron calls them "perfectionist" and "externalist," rather than "nonperfectionist," and "internalist." The debate about what welfare ultimately consists in is ongoing.

Interventions in the environment affect both humans and nonhuman animals. All sentient animals can experience enjoyment and suffering. All sentient animals have desires that can be fulfilled or frustrated. All sentient animals care about what happens to them. Next to pleasure and pain, they can experience a wide range of emotions, such as joy, sadness, stress, and anxiety. Our actions can harm these animals or can benefit them. Effects of our actions on the welfare of animals need to be taken into account in environmental management and in fact in all our actions.

Harm of Death
Having explored what the welfare of both humans and animals might consist in, and thus what might benefit or harm them, I will now explore what if anything is harmful about death. I am interested in the harm that death is for the being that dies. Effects on others are not considered. In order to capture the harm of death as such, one should imagine that the being is killed painlessly while asleep. In that way, fear of death and other otherwise important side effects that death might have are left out of consideration. Of course, these things count in a moral evaluation of killing, but not in an account of the harm of death. Note that a view about the harm of death is not the same as a view about the wrongness of killing. When considering the wrongness of killing, considerations other than the harm of death need to be taken into account. For instance, killing can have effects on others. Furthermore, for laws about killing certain clear-cut rules might be preferable to gradual distinctions, even if the harm of death is a gradual matter. Therefore, any view concerning the harm of death does not translate directly into a view about the wrongness of killing, nor does it imply any legal position on how killing should be regulated or punished.

One influential position among the desire-fulfillment conceptions of welfare is that a being is harmed more by death if more of its desires remain unsatisfied. The thought is that the more future-oriented desires a being has when it dies, the more it is harmed by death (Singer, 1995a). This is because harm is defined as the frustration of desires. If no desires are frustrated, a being is not harmed.

Individuals who have a conception of their own existence over time usually have plans and projects for the future and desire to go on living. Those who lack a conception of their own existence over time lack a desire for continued life. They only have immediate and short-term desires, such as the desire to escape frightening or painful situations, the desire to eat, or the desire to rest. Thus, according to this view, animals can be harmed by death to different degrees. For normal adult humans, death is a great harm, insofar as they strongly wish to continue living and have all kinds of plans for the future. Great apes appear to have future-oriented projects as well and might desire their continued existence (Singer, 2011, 98–99). However, they seem to live more by the day, as do most other nonhuman animals. So, it seems that nonhuman animals are harmed significantly less by death than normal adult humans. Unborn and newborn babies and young children as well as some mentally handicapped humans are comparable to nonhuman animals in this respect. They live more by the day and seem to have only a very limited grasp of their own existence over time. Like nonhuman animals, they are harmed significantly less by death, according to that view. On this view, the harm of death consists in existing preferences being left unsatisfied.

Even those who define welfare in terms of desire satisfaction acknowledge that getting what one desires is not always beneficial and not getting what one wants is not always harmful. If one wants to marry a person and does it, this can turn out to not contribute positively to one's welfare after all. Likewise, if a suicidal teenager wants to die and is prevented from it, not getting what she wants at this point might actually be good for her, if the desire is based on mistaken beliefs. Therefore, proponents of the desire account usually stipulate that only informed and rational desires count. At this point, we encounter a dilemma, however. If the desire account defines welfare in terms of the satisfaction of actual desires, it has to be able to deal with the counterexamples we have just discussed. If, in contrast, it counts only desires about what is really good for a being, the view will not focus on what is desired but on what is valuable for a being (Bradley, 2009, 128–129).

The question is, then, a matter of how far one goes in focusing on what would be rational and informed desires rather than actual desires. Some want to go so far as arguing that a fetus would have the desire to stay alive if she were fully informed and rational (Marquis, 2009, 144). Thus, a fetus would be harmed by death, even though it may not posses any future-oriented desires. However, others have argued that this interpretation of the informed desire account stretches things too far:

> Adjusting a person's actual desires for errors is one thing; attributing a wholly new desire to a being *that is not capable of having any desires at all*, or any desires of the relevant kind is something else altogether and something for which there is no obvious motivation. Preference utilitarianism should be formulated to cover only the former, as follows: We should satisfy, to the greatest extent possible, the preferences a being has, except that we should not satisfy a preference that results from errors of reasoning or errors about matters of fact. (Singer, 2009, 156. Italics mine)

Thus, adjusting actual desires for errors is sanctioned by the informed desire account. Attributing desires to a being that is not capable of having any desires, however, goes too far. After all, the informed desire account is an account of welfare. Those who do not have feelings and desires have no welfare at all, according to that account. However, attributing the desire to stay alive to a being that *is* capable of having desires does not run into these problems. This is because it does not attribute desires to a being that has none. Even if an animal does not consciously or directly desire to stay alive because it has no concept of life and death, it might well desire other things. Staying alive might be instrumental for the fulfillment of those desires. However, the ascription of the desire to go on living goes further than the correction of existing desires for errors. For it involves taking into account a desire of which the being in question is not aware, provided that the fulfillment of that desire is instrumental for the fulfillment of the desires of which the being is aware. It seems to me that the desire-satisfaction account of welfare does not obviously rule this out.

One might oppose this with the thought that some desires do not presuppose staying alive. Examples of such desires include the desire to end hunger or the desire for the cessation of pain. If this is what an animal desires, death might serve this goal (DeGrazia, 2003, 428). It seems to me that by desiring food animals do not only desire the cessation of hunger. They also desire the positive feeling that goes together with the fulfillment of their desire for food. Fulfilling a desire for food does not only cancel out a negative feeling, it can and usually does add something positive. This holds for other seemingly "negative" experiential desires as well. Animals seem to positively enjoy what they necessarily have to do. This positive aspect of fulfillment is missed if the hunger problem is "solved" by killing the animal. Furthermore, even if some desires do not presuppose staying alive, others might. And at least for the fulfillment of those desires, staying alive would be instrumental.

Against the argument that even animals that have only short-term desires (instrumentally) desire to stay alive, it can be said that when those animals are asleep, they might not have those short-term desires. Thus, while asleep, they do not retain desires. Unlike persons, they do not desire things they might acquire after they have woken up. As such, killing such animals while they sleep would not frustrate any desires. Killing those animals painlessly while asleep would indeed not frustrate any desires if those animals do really not have any desires while asleep. Obviously, animals do not have conscious desires while they are sleeping (unless they have them in their dreams). However, desires that are not in the forefront of one's mind are also real desires. The animal might have such desires: it might desire to go on sleeping, to stay in the company of others (such as parents or offspring), or to smell or feel certain things (such as the smell of their parents or offspring or the warmth of the sun). Animals always seem to have at least some dispositional desires. Staying alive is instrumental for the fulfillment of those desires.

So far, I have described a prominent account of the harm of death, based on the desire-fulfillment account of welfare. According to that account, animals are harmed less by death than normal adult humans, because they lack the desire to go on living and have little or no other future-oriented desires. I have shown that it is questionable whether the desire account must deny that animals desire to go on living. After all, the account does not focus on actual desires, but on "ideal" desires. Thus, for instance, desires are corrected for errors of reasoning or errors about matters of fact. Furthermore, dispositional desires are taken into account along with actual desires. One might also take into account desires, the fulfillment of which is instrumental for the fulfillment of other desires that the being has. It is these that I call "implicit desires." Even animals that lack the conception of life and death but desire other things have an implicit desire to go on living that can be taken into account. As animals have at least some desires, including dispositional desires while being asleep, and—if one is willing to accept this—also an implicit desire to go on living, they are harmed by death. However, if the harm of death is defined in terms of the frustration of the desires that the being has when killed, it is true that those with fewer frustrated desires are harmed less by death. Therefore, according to that account, death is less harmful for animals (as well as for babies, infants, and some mentally handicapped humans) than for persons.

An alternative account of the harm of death does not focus on the frustration of what the being wants at the moment of its death, but on the loss of what would have been valuable for the being. That alternative account of the harm of death is compatible with the desire-satisfaction account of welfare and also with every other account of welfare. Thus, unlike the account that has been presented earlier, the alternative account to be presented in the following section is not linked to a particular account of welfare.

This alternative account of the harm of death claims that the harm of death is not the frustration of wants but the forbearance of value. This view is called the "foreclosure view" or "forbearance view" or the "deprivation view" concerning the harm of death (Kaldewaij, 2008). Instead of focusing on how much a being *wants* the future it would have had, this approach focuses on how much *value* that future would have had *for the being*. According to this view, the harm of death consists in the foreclosure of the value that the subject's future life would have brought for the subject. In other words, the harm of death for the subject can be determined by calculating the difference between the value of a being's actual life for her and the value of the being's counterfactual life for her, in case she had not died when she did (Bradley, 2009, 69–72). For instance, if a baby of 3 weeks old dies in an accident and the baby would otherwise have had a happy childhood, studied philosophy and made a good career as a philosopher, had a happy life, and died at the age of 80, then the baby is harmed a lot by death. Imagine the death of a philosophy student who had an equally happy childhood as the baby would have had and who would have had a good career as a philosopher and an equally happy life as the baby would have had. According to the foreclosure view, death is a lesser harm for the student than for the baby because the student is deprived of less value.

The foreclosure view is compatible with different accounts of welfare. For instance, a hedonist account of welfare obviously does not put any weight on future-oriented desires or the desire to stay alive. On a hedonist account, death is harmful because it takes away all the enjoyment that the being otherwise would have experienced. The more enjoyment is taken away, the greater the harm. Thus, all sentient beings that would otherwise have continued a pleasant life are harmed by death. It does not matter how future directed those beings are. On a hedonist account of the harm of death, nonhuman animals are significantly harmed by death, insofar as death deprives them of the pleasurable experiences they would otherwise have enjoyed.

If one accepts the desire-satisfaction account of welfare, the foreclosure view of the harm of death is a plausible option as well. According to the desire-satisfaction version of the forbearance view, death is harmful insofar as it precludes all future satisfaction of desires that a being would otherwise have had. The desire satisfaction that is missed consists of both the satisfaction of the desires that the being already has and the satisfaction of the desires that the being would have come to have. In a way, it is more straightforward to take into account all desire satisfaction that is lost due to death, rather than counting only the lost satisfaction of desires that existed prior to death.

According to the foreclosure view, the harm of death is not determined by how much the being *wants* the future he would have had, but by the loss of the *value* that the future would have provided for the being. Hence, according to that view, animals can be significantly harmed by death even if they do not desire to go on living and even if they have no or few future-oriented desires.

Let me finally mention a proposed modification of the deprivation view. Although according to the deprivation view the possession of future-oriented desires does not determine the harm of death, it has been suggested that death is a lesser harm for beings that have less psychological connectedness. If the harm of death for a being consists in the deprivation of that being's future welfare, it seems to matter how far the being that dies is connected to its future self. In other words, the extent to which whatever is taken away really is *her* future matters a great deal. This issue concerns the concept of personal identity, in particular personal identity over time. This is relevant in order to determine the amount of personal harm caused by death. This is what we are after when we seek to determine the harm of death: it is the harm that death is for somebody. Jeff McMahan advanced the following suggestion:

> In addition to asking how much good a person's future life would have contained, we must also ask [...] How close would the prudential unity relations have been between the individual as he was at the time of his death and himself as he would have been at those later times when the goods of his future life would have occurred? [...] The badness of the loss must be discounted for the absence of [psychological connections]. (2002, 183–184, cited in Bradley, 2009, 131)

So according to this view, which McMahan calls the *Time-Relative Interest Account*, the loss of welfare is discounted in proportion to the lack of psychological connectedness between the being's present self and its future "selves." In order to determine the degree of psychological connectedness, it matters, for instance, to what extent a person's beliefs and desires remain the same, how rich the person's mental life is, how much the person remembers things from the past, and how much a person strives to satisfy desires that occurred in the past (McMahan, 2002, 80; see also Bradley, 2009, 130).

In this account of the harm of death, animals are harmed less by death than normal adult humans. Although the degree of psychological connectedness varies per animal, animals generally live more in the moment and have fewer far-reaching memories and desires. However, even if this account of the harm of death were true and animals would indeed be harmed less by death than normal adult human beings, the harm of death for animals would not be nothing, on this account. At least mammals and birds are known to have the capacity for rich mental lives, memories, and desires for the (at least near) future. For this reason, death might still be a significant harm for those beings.

Interventions in the environment might bring about the death of animals or protect their lives. An example of environmental management that is clearly intended to benefit nonhuman animals is ecoducts that help animals cross highways or fish bypass systems that allow fish to safely pass obstacles, such as dams and hydro projects. Since death harms sentient animals that would otherwise have had an enjoyable future, this effect on animal welfare, along with other harms and benefits, needs to be taken into account in environmental management and indeed in all our actions.

14.3.1.4 How Should We Treat Nonhuman Animals and the Rest of Nature?

Just as the answer to the question of whom to take into account in our moral considerations depends on the moral theory that one accepts, so does the question of how to *treat* nonhuman animals and the rest of nature. Of course, if one thinks that only humans have moral status, then nonhuman animals and the rest of nature are seen merely as resources. Moral theories would then serve to regulate the distribution of these resources among humans. I will discuss the question of the just distribution of resources across generations in Section 14.3.2. Here, I will sketch an answer to the question of what we owe to individuals of other species. I will assume sentientism, the view that all and only sentient beings have moral status. On this view, our direct moral duties are restricted to them. I will not assume any particular moral theory here. Direct duties toward animals can be based on utilitarianism but also on a rights-based ethics. Given that we have, by now, a rough idea about what benefits animals and what harms them, I will directly jump to more practical issues concerning how to actually treat various animals.

One of the most considered and comprehensive accounts of our diverse duties to other animals can be found in Will Kymlicka and Sue Donaldson's (2010) groundbreaking book *Zoopolis: A Political Theory of Animal Rights*. Donaldson and Kymlicka accept that conscious, sentient animals have inviolable rights: "Accepting that animals are selves or persons will have many implications, the clearest of which is to recognize a range of universal negative rights—the right not to be tortured, experimented on, owned, enslaved, imprisoned, or killed. This would entail the prohibition of current practices of farming, hunting, the commercial pet industry, zoo-keeping, animal experimentation, and many others." Since respecting the basic rights of animals "need not, and indeed cannot, stop all forms of human–animal interaction," the authors insist that a more robust theoretical grounding is needed for spelling out the fair terms of interaction. Kymlicka and Donaldson go beyond the focus on negative rights that is typical of the traditional animal rights movement and also choose to emphasize our positive duties to animals:

> Different relationships generate different duties—duties of care, hospitality, accommodation, reciprocity, or remedial justice—and much of our moral life is an attempt to sort out this complex moral landscape, trying to determine which sorts of obligations flow from which types of social, political, and historical relationships. Our relations with animals are likely to have a similar sort of moral complexity, given that enormous variation in our historic relationships with different categories of animals. (2010, 6)

The authors present a particular conception of citizenship theory as the necessary extension of the animal rights framework. They draw a parallel with the human situation: while universal human rights hold for everybody, more specific rights and duties are spelled out on the basis of particular political relationships. Examples given include (i) the relationship between cocitizens, (ii) the relationship of citizens toward members of other sovereign states, and (iii) the relationship of citizens toward individuals who (temporarily) share one's territory without being cocitizens, such as tourists or illegal immigrants and guest workers. Citizenship theory is concerned with how these different kinds of relationships can be framed in a just way.

Domesticated animals, according to the authors, should be considered cocitizens. This might seem odd, as nonhuman animals cannot be politically active—a requirement that might first come to mind when thinking about citizenship. However, Donaldson and Kymlicka refer to new conceptions of citizenship, arising from disability theory, which present the following requirements as core capabilities of citizenship: (i) having and expressing a subjective good, (ii) complying with schemes of social cooperation, and (iii) participating as agents in social life. According to the authors, rather than implausibly stretching the original conception of citizenship, discussions in disability theory have contributed to a better understanding of what citizenship actually means for all of us. In the same way, they argue that citizenship theory will actually profit from the inclusion of nonhuman animals.

Kymlicka and Donaldson argue that an improved conception of citizenship can and indeed should accommodate domesticated animals as citizens. Equally, the authors argue that we can and should accord sovereignty to wild animals and denizenship to liminal animals, such as mice and pigeons with which we coexist on shared territory without interacting as closely as we do with domesticated animals. Possible doubts about the applicability of those concepts are discussed and countered by pointing toward developments within citizenship theory and toward experiences with interactions with animals.

In passing, Donaldson and Kymlicka discuss some shortcomings of animal rights theory and show how their own approach does better in those respects. For instance, the category of animals that the authors consider under the heading of "denizenship" is broadly ignored by animal rights theorists. Such theorists tend to draw a dichotomy between wild animals that ought to be left alone and domesticated animals, which should either become wild again or gradually cease to exist. The authors convincingly point out that this dichotomy is untenable, because there will always be mixed human–animal communities. Although the process of domestication has seriously infringed upon animal rights, it does not preclude the possibility of interrelating with existing animals albeit now on fair terms. The authors draw an analogy with former slaves who have rightly been offered the opportunity to become cocitizens. Furthermore, Donaldson and Kymlicka convincingly argue that our duties toward wild animals are not suitably captured by the requirement of "leaving them alone." The authors discuss in some detail the delicacies of whether and under which conditions to assist wild animals in emergency situations. Furthermore, they explore the problem of indirect effects of human activity on wild animal communities, such as pollution, habitat destruction, and climate change.

Kymlicka and Donaldson argue that approaching the animal question as, first and foremost, an issue of determining their relationship to political communities, and hence their current and potential membership status, is essential to a complete theory of animal rights. Some animals really are members of our society, others are more akin to passing visitors, and yet others should be seen as living autonomously outside of our control. The authors convincingly show that these differences generate distinctive sets of obligations, in ways that cannot be captured by an exclusive focus on animals' capacities for pain or their possession of cognitive skills. The authors persuasively show that these issues of membership status require the conceptual tools of the social sciences and of political theory, as a supplement to the longstanding concepts of both ethology and moral philosophy. The growing literature in political theory that considers nonhuman animals can be a rich source of inspiration for environmental managers who are willing to consider nonhuman animals not as resources but as moral equals.

14.3.2 Intergenerational Justice

14.3.2.1 Introduction

One of the aims of environmental management is to ensure that ecosystem services are protected and maintained for future human generations (Pahl-Wost, 2007). This assumes that we do have duties toward future people and thus that they have moral status. This section deals with the ethical foundations of this seemingly plausible assumption and draws the reader's attention to some particular ethical challenges that come up when considering our duties toward those who do not yet exist. In Section 14.3.2.2, I will present different views about the moral status of those who do exist, will exist, or might exist. I will point out that the moral status of those in the last category in particular is controversial. In Section 14.3.2.3, I will turn to value theory again, and consider whether it can be better or worse for an individual to exist than to never exist. This question is particularly pressing when we can influence who will live in the future and how many will live. Finally, in Section 14.3.2.4, I will point out a difficulty that comes up when considering our duties toward those who do not yet exist.

14.3.2.2 Considering Those Who Might Live

In Section 14.3.1.2, I explored possible candidates for criteria on the basis of which one can ascribe moral status to an individual. I criticized the appropriateness of criteria such as "species membership" and "intelligence" and defended sentientism as the most plausible demarcation of the scope of moral objects. Now, one might consider the temporal location of the sentient individual and enquire: Does it matter for the ascription of moral status *when* an individual exists? The so-called presentists claim that it matters. They accord moral status only to presently existing beings and completely leave out of consideration those who do not yet exist. This position is rarely defended. It is generally assumed, correctly I think, that temporal location does not matter. After all, if the individual will exist, and if we can affect that individual's life for the better or the worse, there seems to be no reason for disregarding this individual in our moral considerations.

The more interesting issue does not concern an individual's temporal location, but what have been called the "modal features" of an individual's life (Arrhenius, 2000, 152). That criterion is about the relation of an individual's life to reality. It is about whether a being does in fact live, will live, or has lived, as compared to whether they might live or surely will never live. Thus, whether the possible welfare of a possible being is morally relevant might depend on how likely it is that the potential being will come into existence. That kind of (modal) distinction concerns *possibilities* that might or might not come true. It is important to note that not all potential beings will ever become actual beings. A being that is a possible being at a certain time might in fact never come to live. For instance, a couple's possible "next child" might never be conceived; it might never be an actual sentient being. *Potential* beings, therefore, will not necessarily live in some other time. This is in contrast to *future* beings, which will certainly live, even though they do not live yet. Potential beings might not come to exist at all. Often, it will be uncertain whether they will come to exist. Those beings that might or might not come to exist are also called "contingent beings." They are distinguished from "necessary beings," that is, beings that live, have lived, or certainly will live. A crucial question is whether this distinction between necessary and contingent beings is morally relevant.

In what follows, I will introduce two rival views about that issue. One of those views, the Prior Existence View, only considers necessary beings and not contingent beings to be morally relevant. The Total View, on the other hand, considers both necessary and contingent beings to be relevant in this respect (Singer, 1995b, 103).[1] This distinction is particularly relevant for consequentialist moral theories. These theories, such as utilitarianism, judge an action's moral status solely on the basis of the action's consequences. For these theories, it obviously matters whether or not they consider the consequences of some action for the contingent beings it might come to affect. Whether this distinction is relevant for other moral theories depends on how they determine what makes an action right or wrong and to what extent the rightness or wrongness of an action depends on its (nonmoral) consequences, such as their consequences for someone's welfare.

Thus, the practical implications of these different views can best be explained when one assumes a utilitarian moral theory that aims at maximizing welfare. Proponents of such a moral theory need to assess the outcomes of possible actions in terms of their consequences upon welfare. Imagine that some management intervention would cause more well off individuals to exist. Would this be a good thing, all else being equal? In order to answer this question, those who aim at maximizing welfare need to know two further things. First, they need to know whether outcomes should be assessed in terms of the amount of welfare that they contain or whether they should be assessed in terms of the harms and benefits to sentient beings. Second, they need to know whether existing, as opposed to not existing, benefits a being. If the outcomes should be assessed in terms of harms and benefits, and if coming into existence can be a benefit, then bringing such a being into existence is a good thing. If, alternatively, outcomes should be assessed in terms of harms and benefits, but coming into existence cannot be a benefit, then bringing additional well-off beings into existence would not be a good thing. If, in contrast, outcomes should be assessed in terms of the overall amount of welfare that they contain, it does not matter whether coming into existence will actually benefit the individuals. Since the additional individuals would be well off, the overall quantity of welfare would be greater. For this reason, the addition of these well-off beings to the world would be a good thing.

These considerations may seem pretty abstract, but they can be of great relevance. After all, many choices that we make in terms of the management of natural resources might determine how many individuals will exist. Such measures might influence the size of the human and animal populations. I cannot go into all the theoretical complexities of that issue here, but in the next section, I will address the question whether it can be better or worse for an individual to exist than never to exist. In Section 14.3.2.4, I will discuss the major challenge to the view that outcomes should be assessed in terms of harms and benefits to individuals.

14.3.2.3 Can It Be Better or Worse for an Individual to Exist than Never to Exist?

Some forms of environmental management will affect the number of humans or animals that will exist. For instance, due to the preservation of some ecosystem, individual animals will come to be born and exist there that would not otherwise have existed. If ecosystems are destroyed or natural resources depleted, some individuals that might otherwise have existed will never be born. Can this be a good or bad thing that should be considered in our reasons for action? Can it be better or worse for an individual to exist as compared to never existing? It is controversial whether existence can be better or worse for an individual than nonexistence. This is the debate about the value of existence. It seems that existence as compared to nonexistence cannot benefit or harm an individual, since nonexistent individuals do not have any level of welfare, so existence cannot be any better or worse for them than nonexistence in terms of welfare (Broome, 1999, 168; Arrhenius, 2009). Furthermore, relaxing various conceptual requirements of the notions of harm or benefit just so that one can say that coming into existence can benefit or harm would be ad hoc and unjustified (Bykvist, 2007, 343). Some philosophers, however, disagree and the debate is ongoing (Holtug, 2001). In what follows, we will

---

[1] In addition to the Prior Existence View, there are other views that make a distinction on the basis of the modal features of a being's life. The most prominent of those is "actualism" (Reviewer comment: "a bit unclear what the difference is between actualism and the Prior Existence View"), which only counts those beings that do exist, have existed, or will exist as morally relevant and does not count "merely possible beings," that is, those who might but will in fact never be actual, as such. Furthermore, the various views can also be defended in soft versions. It can, for instance, be claimed that the welfare of possible beings or the welfare of a certain category of possible beings should count for *less*, rather than not at al. I will not consider the soft versions here and discuss only the Prior Existence View and the Total View, since those are most prominently discussed in applied ethics.

glimpse at that debate. This will also give us an idea of the seemingly arcane issues that ethicists often need to tackle in order to arrive at down to earth judgments about reasons for or against preserving an ecosystem.

How should we decide whether nonexistence has a zero value and is thus comparable in value to existence or whether it has no value at all for the individual and thus cannot be better or worse for an individual than her nonexistence? In order to decide whether something is an instance of zero value or rather of absent value, or both, it makes sense to employ some general principles of empirical inquiry. Such principles are also used in reflections about zero-value physical quantities (Balashov, 1999): the idea is, quite simply, to assemble known instances of a certain property—let us call it P—to find among them putative instances of $P=0$ and then to oppose such instances to those in which P is absent. In the case at hand, we are concerned with the property "welfare," and we want to compare instances of zero welfare with instances in which welfare is absent. Here are four general argumentative strategies along those lines.

The Argument from Composition
The argument is that a combination of two or more P-hoods (positive and negative) cannot amount to complete P-lessness. With respect to welfare, this means that we can speak of zero welfare if instances of positive and negative welfare hold the balance. If positive and negative welfare hold the balance, that is different from not having any welfare. That point can easily be granted. The dispute is about whether there are *other* instances of zero welfare, *besides* positive and negative welfare holding the balance. The following arguments are relevant to that dispute.

The Argument from Parity
This argument is that if anything else, except the aforementioned states of combinations of positive and negative P-hoods, claims to have zero P-hood, it must show relevant similarities with the known instances of zero P-hood. Thus, when one claims that the nonexistent have zero welfare, one must point to relevant similarities between the nonexistent and those in which positive and negative welfare hold the balance. In the recent debate, Bradley has brought forward an argument that is supposed to show relevant similarities between nonexistence and situations in which positive and negative welfare hold the balance.

Here is Bradley's argument: Bradley (2009, 98) claims that someone, let us call him Kris, will be indifferent, as far as his welfare is concerned, as to whether he dies immediately or lives through a period of zero welfare before he dies. This is supposed to show that putative instances of zero P-hood (nonexistence) and known instances of zero P-hood (neutral welfare of existing beings) are relevantly similar. It allegedly does not matter for Kris as far as his prudential value is concerned whether he first lives through a period of zero welfare or dies immediately.

What are we to make of Bradley's argument? Bradley argues that since Kris would value immediate death or death after a period of zero welfare equally, it follows that both future scenarios are equal as far as welfare is concerned, and this is taken to be a reason for thinking that Kris has zero welfare while dead. However, from the fact that Kris might *be indifferent* between a period of zero welfare followed by death and immediate death, it does not follow that Kris *has* zero welfare while dead. It might just mean that Kris' life is equally good for him in both cases. But that is a comparison between two lives, rather than between existence and nonexistence in terms of welfare. So in order to make that point about the value of life, it is not necessary to make any assumptions about the welfare level of dead individuals.

The Argument from Unification
This is the claim that instances of zero P-hood and instances of positive or negative P-hood must show similarities, related to the possession of P-hood. After all, they are all instances of P-hood. Furthermore, having zero P should be shown to be different from P-lessness.

One might refer to responsiveness in order to draw a relevant distinction between those who have welfare, including zero welfare, and the nonexistent to whom the concept of welfare does not even apply and hence who don't even have zero welfare (Luper, 2007, 244; Bradley, 2009, 102). Bradley challenges this view by offering a counterexample. According to Bradley, it is metaphysically possible for a dead person to be revived. Currently, reviving dead people is not among our capacities, but it might be in the future, according to Bradley. Bradley's counterargument, which adheres to the metaphysical possibility of reviving the dead, is problematic. After all, death is defined as the permanent end of all functions of life in an organism. So, if the loss appears to be reversible, the individual in question cannot have been dead. One might also grant that in a time in which "dead" people can actually be revived, those "dead" people are still responsive. This controversy shows that it needs to be clarified what responsiveness means. In what way must it be "possible" for an individual's level of welfare to change?

Bradley puts forward another argument against the claim that in order to have a welfare level, an individual must be responsive. Bradley suggests that it is possible that someone lives with a permanent welfare level of zero and that this welfare level "could not rise or fall given the person's inherent capacities." Bradley suggests the following example, assuming hedonism:

> Imagine Marsha is born without the capacity to feel pleasure or pain, and never develops that capacity; imagine Greg is born with that capacity, but due to his circumstances, he never actually feels any pleasure or pain. Given Luper's

account of responsiveness, Marsha is relevantly like a shoe—she has no well-being level at all—while Greg has a well-being level of zero. This just seems wrong. (2009, 103)

Again, assuming hedonism, it makes sense to say that a living person without any capacity for subjective feeling lacks a welfare level. Thus, Marsha might indeed be like a shoe in this respect: she lacks a welfare level. Whether Greg is also relevantly like a shoe depends on the circumstances under which he lives. If he can feel pleasure and pain, but simply happens to have only neutral feelings, he still has a welfare level. Responsive individuals are not likely to remain in a state of neutral feeling for long. If Greg's capacity to feel pleasure or pain has been knocked out, then he is relevantly like a shoe as well. The example of Marsha and Greg does not dismiss the claim that having a welfare level is linked to responsiveness.

The definition of responsiveness that Bradley offers seems to stretch things too far:

Person S is responsive at time t only if there is some world w and some time t_n such that S has a positive or negative well-being level at $<w, t_n>$. (2009, 104)

Bradley's notion of responsiveness is very inclusive. Since t_n can also be in the past, it implies that dead people are responsive. Of course, Bradley welcomes this implication, because it allows him to attribute a welfare level of zero to dead people without forcing him to do the same for shoes. However, Bradley has offered no convincing argument for this counterintuitive move. As Luper (2009) has already pointed out in reply to Bradley: "Just because something is (the sort of thing that) such that its attaining goods or evils at some time is not impossible, it does not follow that it has a welfare level at some given time, or at every time." Neither does it follow from the fact that it is metaphysically possible that someone has a welfare level at a particular time that this individual actually does have a welfare level at that time. The idea behind the concept of responsiveness seems rather to be that it must be physically possible for the individual's welfare level to go up or down when certain conditions are met, either with or without external help.

The Argument from Disparity
This is the argument that if an alleged instance of $P=0$ and an instance that has nothing to do with P-hood differ in a trait known to relate to P-hood, this strongly supports the claim that the first instance really has zero P rather than no P.

One might compare a shoe that does not experience any positive or negative welfare at time t and a person who does not experience any positive or negative values at time t. One might argue that if the person has the capacity to experience positive or negative welfare (even though those capacities might be temporarily blocked), the person still is responsive (Luper, 2009). Thus, there seems to be a relevant difference between an instance of zero P-hood and a known instance of P-lessness (the shoe). That difference (responsiveness) can be explained in terms of P-hood (i.e., welfare): one is only responsive if one's welfare level can rise or fall.

These considerations suggest that the nonexistent do not have neutral welfare. While neutral welfare is typically identified with scoring "zero" on the imaginary welfare scale, the absence of welfare might not have a place on the welfare scale. This, in turn, suggests that it is incommensurable with any welfare score and therefore that it is impossible to claim that someone's state of existence can be compared to that individual's nonexistence in terms of his or her welfare.

If, in spite of these arguments, nonexistence could be accorded a zero value on the welfare scale, a further question would come to the fore. Those who assume that nonexistence has zero value must explain for whom it has zero value. Zero value is about welfare. Welfare is a prudential value. For whom would welfare be a good in a case of nonexistence? It seems that there is nobody in the case of nonexistence who could experience this "value."

Bradley's position implies that in a case of nonexistence, the nonexistent must have zero welfare. However, it has convincingly been argued that the nonexistent cannot have any properties, not even zero welfare (Broome, 1993, 77; Bykvist, 2007, 343; Johansson, 2010). If one wants to argue that nonexistence can be neutral for a person, then one speaks about a genuine relation that holds between a state of affairs (nonexistence) and a subject (the person). For such a claim to be true, both the state of affairs and the subject must exist in the same world. But the person does not exist in the case of her nonexistence. Therefore, it cannot be true that her nonexistence has zero value for her. The nonexistent person does not stand in any relation to anything. So, nothing can be neutral for her.

Neither does it help to claim that nonexistence has zero value for the *existent*. After all, it has been argued that something (A) can only be better for somebody than something else (B), if (B) is worse for that individual than (A). Hence, if existence is said to be better for the person, this implies that nonexistence must be worse for her. However, nonexistence cannot be worse for her, as we have just seen. If a person does not exist, she does not stand in any relations. Furthermore, if something (A) is better for a person than (B), this must be true, no matter whether (A) or (B) actually obtain. So, if it is claimed that nonexistence can be better or worse for the existent, this must be true even if nonexistence obtained. However, as we have seen, nonexistence cannot be better or worse for the nonexistent, because nothing can have any value for the nonexistent.

Ascribing zero value or "neutral value" to nonexistence seems to be ruled out by these arguments. However, ascribing zero welfare to nonexistence is necessary in order to claim that it can be better or worse than existence for the individual (in terms of her welfare). So, it seems that we cannot claim that existence can be better or worse for an individual than nonexistence. As I said, that discussion might seem far removed from environmental ethics. I do, however, think that it is relevant. After all, environmental ethics is a branch of applied ethics. Thus, it is about applying all kinds of ethical insights that seem relevant for an evaluation of the human interaction with the environment. Since our interaction with the environment influence how many individuals might live at all, these considerations are relevant for an assessment of our actions.

14.3.2.4 Moral Duties: What Are Our Duties toward Those Who Do Not yet Exist?

As indicated in Section 14.3.2.3, the Person-Affecting Restriction evaluates actions in terms of the difference they make for the welfare of sentient beings. What matters, according to the Person-Affecting Restriction, is in how far sentient beings are made better or worse off. The Person-Affecting View is typically defined as follows:

The Person Affecting View
a. If outcome A is better (worse) than B, then A is better (worse) than B for at least one individual.
b. If outcome A is better (worse) than B for someone but worse (better) for no one, then A is better (worse) than B. (Arrhenius, 2003, 188)

Thus, the Person-Affecting View describes a necessary and sufficient condition for an outcome's being better than another (see also Broome, 2004; Parfit, 1984; Temkin, 1999). I will now introduce a major challenge to this particular account of the evaluation of outcomes.

Derek Parfit has famously explored what he calls the Nonidentity Problem. This problem refers to the fact that some of our choices, which affect the welfare of people, also affect their identity. In other words, the compared outcomes differ not only in how well off they leave existing people but also in who those people are. Parfit points this out in the example of the 14-year-old girl:

> The 14-year-old Girl: A 14-year-old girl has a baby. The baby has a difficult start in life and therefore also some problems later on in its life. So her child has a somewhat diminished quality of life overall, but a life that is still overall pleasant. Had the 14-year-old girl waited several years before becoming pregnant, she would have been able to give her baby a better start in life and her child would have a better life overall. (1984, 358)

If one assumes that one's personal identity depends at least in part on the genetic material of which one is made, then given the fact that sperm (and eggs) have short lives, it follows that the time of conception is relevant for who exactly will exist. In the example, the girl would have had a different child, growing from different sperm and egg, if she had delayed conception. Therefore, it seems impossible to condemn the 14-year-old girl for having the child with reference to the interests or the welfare of the child. After all, the child has a pleasant life, and if the girl had waited, this child would not exist at all. So, the Nonidentity Problem challenges our intuition that the girl ought to wait instead of having her first child now. The Nonidentity Problem consists in the fact that nobody would benefit from that seemingly morally right choice. So, either our intuitions lead us astray, or the Person-Affecting Restriction is untenable.

Here is another case that illustrates the Nonidentity Problem; let us call it the case of the Two Women:

> The case of the Two Women: Two women plan to become pregnant. Woman A is already pregnant when she receives good and bad news. The bad news is that her child will be born with a defect that will lower its quality of life, although it will still be able to have a worthwhile life. The good news is that the woman can take a pill to cure this defect. Woman B gets good and bad news when she is about to stop using contraception. The bad news is that she has a medical condition that will result in a child with a defect that will lower its quality of life, and which is just as serious as that which woman A's child would have had, if she does not take the pill. The good news is that woman B will have a healthy baby if she waits three months before becoming pregnant (Parfit, 1984, 367. See also Singer, 1995b, 123).

It seems that in order to do the best thing for her child, woman A should take the pill in order to have a healthy baby. Likewise, woman B should wait three months in order to have a healthy baby. However, the difference is that by taking the pill woman A would make her child (not yet born, but already existing in her womb) better off. Woman B, in contrast, would not benefit any baby. By delaying conception, she would have a different baby than she would otherwise have had. The Person-Affecting Restriction is focused on harms and benefits to sentient beings. It seems that in the case of woman B's delayed conception nobody would be benefitted. After all, there is no particular being made better off by having a different baby, or so it seems. Hence, it seems that the Person-Affecting Restriction would require that woman A takes the pill but not that woman B delays conception.

The Nonidentity Problem can also show up on a larger scale. Major social policy decisions that affect the welfare of future generations also affect the identity of the people who are going to exist. Consequently, the population that would come to exist after the implementation of a major social policy would be different to that which would come to exist were that policy to not be implemented. As such, the very makeup of the future population is at stake in decisions over whether or not to implement such policies. Imagine that the current population adopts a policy that causes disadvantages for the future generation, for instance in terms of depleted resources or pollution. Assuming that those future people will have lives that are pleasant overall and that they would not have existed had that policy not been adopted, it seems that adopting that policy does not harm them. (Although cases like this are commonly brought forward in order to illustrate the Nonidentity Problem, they are less clear than both of the aforementioned cases, because in such cases it is less clear that either outcome would involve the nonexistence of all people.)

Thus, the "Nonidentity Problem" refers to the observation that in choices between which of two future people to bring into existence, the outcome that is intuitively considered better (such as delaying conception in case of woman B or the 14-year-old girl) does not make any particular person worse off. In other words, it seems impossible to condemn the action that produces less welfare or to recommend the action that produces more welfare in person-affecting terms. For instance, the 14-year-old girl would not benefit any particular child by delaying conception. If she delayed conception, she would have a different child. Note that the Impersonal View, as opposed to the Person-Affecting View, would not lead to the Nonidentity Problem. The Impersonal View evaluates outcomes simply in terms of the harms of benefits that they contain, no matter whether any particular individual is harmed of benefits. Thus, for instance, the outcome in which the woman has a better off child contains more welfare than the outcome in which the woman has a different, worse-off child. Even if no particular child will be harmed, the latter outcome contains less welfare. For this reason, the Impersonal View would clearly conclude that the outcome that contains more welfare is better in terms of welfare. Even though the Impersonal View seems to yield the intuitively correct judgment here (although intuitions differ on this issue), it seems generally implausible to the extent that the Person-Affecting View seems plausible, that is, in as far as we believe that an outcome cannot be better unless it is better for somebody. The Nonidentity Problem is not only discussed with regard to consequentialist moral theory but also with regard to rights-based theory. In case of a rights-based theory, one could argue that bringing into existence individuals under less advantageous conditions violates their rights. However, to this one can reply that surely the individuals who are brought into these conditions would waive their rights, if the alternative would be not to exist at all. Thus, quite generally, the Nonidentity Problem is a challenge for the Person-Affecting View on the evaluation of outcomes (Roberts, 2009).

Proponents of the Person-Affecting View propose various strategies for dealing with this challenge. Some propose a noncomparative account of harm, which implies that those who are born into less advantageous conditions are actually harmed, even if they are not worse off than they would otherwise have been (Harman, 2009). Typically, when we say that some event harmed an individual, we think that the individual is worse off than he or she would otherwise have been. A noncomparative notion of harm rejects this comparison with a counterfactual situation. It simply asserts that some states are harmful. Bringing about these states harms the individual, according to that account. This noncomparative account of harm is itself not without problems. Alternatively, some argue that the seemingly disadvantaged individuals in the aforementioned cases are morally wronged, even if they are not harmed. That strategy loosens the connection between harm and benefit and the moral evaluation of the action. I think that the most promising strategy is to avoid the Nonidentity Problem by shifting the focus from the particular individual, such as the genetically unique child that the 14-year-old girl might have, to "her next child, whoever it will be." Different variations of this strategy have been defended, for instance, with a focus on role-based duties (Kumar, 2003) or by way of an appeal to the notion of de dicto betterness (Hare, 2007). According to these strategies, the mother's duty is toward "her next child, whoever it will be," rather than to any particular, genetically unique individual. If those strategies succeed, they can defend the Person-Affecting View against the challenge from the Nonidentity Problem.

As we have seen, difficult issues come into play when considering our moral duties toward those who do not yet exist and in particular toward those who might or might not come into existence, depending on what we are doing.

14.4 CONCLUSION

Environmental ethics as a field of application of normative ethical reasoning is very relevant to the work of environmental managers. Environmental management decisions and the evaluation of such decisions depend on judgments about (i) whom we should take into account in our moral considerations, (ii) what is harmful or beneficial for those whom our actions concern, and (iii) what our moral duties are toward existing, future, and possible humans and animals, as well as toward the rest of nature. These normative ethical and value-theoretical questions are addressed in environmental ethics in a systematic, analytical, and argumentative way. Exploring these questions sheds light on the issues of interspecies justice and intergenerational justice. The status of moral claims, whether they can be objectively true and how one can know that, is contested. I briefly discussed the metaethical debate on objectivity in ethics and rejected some common counterarguments to the existence of objective moral standards.

I discussed the moral status of humans, nonhuman animals, and the rest of nature and defended a sentientist position, arguing that we should consider all and only sentient beings in our moral deliberations. I pointed out that this holds for presently existing sentient beings, as well as for those who do not yet exist but will exist in the future. The question whether and how to take into account those who might or might not exist, depending on how we act, is more difficult to answer. I explored various positions and their challenges.

Once we know whom to take into account in our moral considerations, the question is what it actually means to harm or benefit them. I discussed the notion of (animal) welfare and the question what the harm of death consists in. I also discussed whether existence as compared to never existing might be a harm or benefit. Considering these questions leads me to favor the self-fulfillment account of welfare and the deprivation view on the harm of death. I also argued that existence can be good, bad, or neutral for an individual, but it cannot be better or worse or equally good as nonexistence. These states are not comparable in terms of what is better for the individual. These considerations should be of interest to environmental managers, because our way of dealing with scarce resources is likely to influence who and how many will live or die and will affect the welfare of sentient beings.

I also addressed the question of what our specific duties are toward other sentient beings. I introduced the idea that, given equal moral status, political theory can be helpful in determining the specific duties that we have toward others. In determining our duties of justice, one might draw on recent insights from citizenship theory and apply them to nonhuman animals as Kymlicka and Donaldson (2010) did in their groundbreaking work on that subject.

The reader may or may not be convinced by my presentation and evaluation of the relevant arguments. The aim of this chapter was to introduce the reader to some major discussions in environmental ethics and to inspire further exploration of this important field of applied ethics.

REFERENCES

Aristotle. 1984 [350BCE]. Metaphysics. Translated by W.D. Ross.

Arrhenius, G. 2000. Future generations: A challenge for moral theory. PhD Dissertation. Uppsala University, Uppsala.

Arrhenius, G. 2003. The person-affecting restriction, comparativism, and the moral status of potential people. *Ethical Perspectives*, 10 (3): 185–195.

Arrhenius, G. 2009. Can the person-affecting restriction solve the problems in population ethics? M. Roberts and D. Wasserman (eds.) *Harming future persons: ethics, genetics and the nonidentity problem*. Springer, Dordrecht, 291–316.

Arrhenius, G. Forthcoming. *Population ethics*. Oxford University Press, New York.

Attfield, R. 2014. *Environmental ethics*. Polity Press, Cambridge.

Balashov, Y. 1999. Zero-value physical quantities. *Synthese*, 119: 253–286.

Barry, B. 1989. Utilitarianism and preference change. *Utilitas*, 1 (2): 278–282.

Bentham, J. 1996 [1789]. In: J. Burns and H.L.A. Hart (eds.), *An introduction to the principles of morals and legislation*. Clarendon Press, Oxford.

Bognar, G. 2010. Authentic happiness. *Utilitas*, 22 (3): 272–284.

Bradley, B. 2006. Two concepts of intrinsic value. *Ethical Theory and Moral Practice*, 9: 111–130.

Bradley, B. 2009. *Well-being and death*. Clarendon Press, Oxford.

Broome, J. 1993. Goodness is reducible to betterness: The evil of death is the value of life. In: P. Koslowski, Y. Shionoya (eds), *The good and the economical*. Springer, Berlin, 70–86.

Broome, J. 1999. *Ethics out of economics*. Cambridge University Press, Cambridge.

Broome, J. 2004. *Weighing lives*. Oxford University Press, Oxford.

Bykvist, K. 2007. The benefits of coming into existence. *Philosophical Studies*, 135 (3): 335–362.

Cimprich, B.E. 1990. Attentional fatigue and restoration in individuals with cancer. PhD Dissertation. University of Michigan, Ann Arbor.

Cochrane, A. 2007. Animal rights and animal experiments: An interest-based approach. *Res Publica*, 13 (3): 293–318.

Cochrane, A. 2009. Ownership and justice for animals. *Utilitas*, 21 (4): 424–442.

Crisp, R. 1998. Animal liberation is not an environmental ethic. A response to Dale Jamieson. *Environmental Values*, 7 (4): 476–478.

Crisp, R. 2006a. *Reasons and the good*. Oxford University Press, Oxford.

Crisp, R. 2006b. Hedonism reconsidered. *Philosophy and Phenomenological Research*, 73 (3): 619–645.

DeGrazia, D. 2003. Identity, killing, and the boundaries of our existence. *Philosophy and Public Affairs*, 31 (4): 413–442.

Elster, J. 1982. Sour grapes—utilitarianism and the genesis of wants. In: A. Sen and B. Williams (eds), *Utilitarianism and beyond*. Cambridge University Press, Cambridge, 219–239.

Faber Taylor, A., Wiley, A., Kuo, F.E., and Sullivan, W.C. 1998. Growing up in the inner city: Green spaces as places to grow. *Environment & Behavior*, 30: 3–27.

Faber Taylor, A., Kuo, F.E., and Sullivan, W.C. 2001. Coping with ADD: The surprising connection to green play settings. *Environment & Behavior*, 33: 54–77.

Faber Taylor, A., Kuo, F.E., and Sullivan, W.C. 2002. Views of nature and self-discipline: Evidence from inner city children. *Journal of Environmental Psychology*, 22: 49–63.

Feldman, F. 2006. *Pleasure and the good life: On the nature, varieties, and plausibility of hedonism*, 2nd edition. Oxford University Press, Oxford.

Grahn, P., Mårtensson, F., Lindblad, B., Nilsson, P., and Ekman, A. 1997. *Ute på dagis. Stad and Land, 145 [Outdoor daycare. City and country]*. Norra Skåne Offset, Hässleholm, Sverige.

Grayling, A.C. 1999. Editor's introduction. In: A.C. Grayling (ed.), *Philosophy 1: A guide through the subject*, Vol. 1. Oxford University Press, Oxford.

Griffin, J. 1986. *Well-being: Its meaning, measurement, and moral importance.* Clarendon Press, Oxford.

Hare, C. 2007. Voices from another world: Must we respect the interests of people who do not, and will never exist?. *Ethics*, 117 (3): 498–523.

Harman, E. 2009. Harming as causing harm. In: M. Roberts and D. Wasserman (eds), *Harming future persons. Ethics, genethics and the nonidentity problem.* Springer, Dordrecht, 137–154.

Harsanyi, J.C. 1982. Morality and the theory of rational behavior. In, A. Sen and B. Williams (eds), *Utilitarianism and beyond.* Cambridge University Press, Cambridge, pp. 39–62.

Hartig, T., Mann, M., and Evans, G.W. 1991. Restorative effects of natural environment experiences. *Environment & Behavior*, 23: 3–26.

Hausman, D. 2012. *Preference, value, choice, and welfare.* Cambridge University Press, Cambridge.

Haybron, D.M. 2008. *The pursuit of unhappiness: The elusive psychology of well-being.* Oxford University Press, Oxford.

Heathwood, C. 2006. Desire satisfactionism and hedonism. *Philosophical Studies*, 128: 539–563.

Holtug, N. 2001. On the value of coming into existence. *The Journal of Ethics*, 5 (4): 361–384.

Jamieson, D. (1998). Animal liberation is an environmental ethic. *Environmental Values.* 7: 41–57. doi: 10.3197/096327198129341465

Jamieson, D. (ed.), 1999. *Singer and his critics.* Blackwell, Oxford.

Johansson, J. 2010. Being and betterness. *Utilitas*, 22 (3): 285–302.

Kaldewaij, F. 2008. Animals and the harm of death. In: S.J. Armstrong and R.G. Botzler (eds), *The animal ethics reader*, 2nd edition. Routledge, London.

Kaldewaij, F. 2013. The animal in morality. Justifying duties to animals in Kantian moral philosophy. PhD Dissertation. Utrecht University, Utrecht.

Kaplan, R. 1973. Some psychological benefits of gardening. *Environment & Behavior*, 5: 145–152.

Korpela, K. 2002. Children's environments. In: R.B. Bechtel and A. Churchman (eds), *Handbook of environmental psychology.* Wiley, New York, 363–373.

Korsgaard, C. 2005. Fellow creatures: Kantian ethics and our duties to animals. *The Tanner Lectures on Human Values*, 25: 77–111.

Kumar, R. 2003. Who can be wronged? *Philosophy and Public Affairs*, 31: 99–118.

Kuo, F.E. 2001. Coping with poverty: Impacts of environment and attention in the inner city. *Environment & Behavior*, 33: 5–34.

Kymlicka, W. and Donaldson, S. 2010. *Zoopolis. A political theory of animal rights.* Oxford University Press, Oxford.

Luper, S. 2007. Mortal harm. *The Philosophical Quarterly*, 57 (227): 239–251.

Luper, S. 2009. Review of Ben Bradley: Well-being and death. Notre Dame Philosophical Reviews, available at http://ndpr.nd.edu/review.cfm?id=16606. Accessed March 10, 2015.

Marquis, D. 2009. Singer on abortion and infanticide. In: J.A. Schaler (ed.), *Singer under fire. The moral iconoclast faces his critics.* Open Court, Illinois.

McMahan, J. 2002. *The ethics of killing. Problems at the margins of life.* Oxford University Press, New York.

Midgley, M. 1983. *Animals and why they matter: A journey around the species barrier.* University of Georgia Press, Georgia.

Mill, J.S. 1998 [1863]. *Utilitarianism.* In: Crisp, R. (ed). Oxford University Press, Oxford.

Moore, G.E. 1903. *Principia ethica.* Cambridge University Press, Cambridge.

Moore, E.O. 1981. A prison environment's effect on health care service demands. *Journal of Environmental Systems*, 11: 17–34.

Moore, R.C. 1986. *Childhood's domain.* Croom Helm, London.

Norcross, A. 2009. Two dogmas of deontology: Aggregation, rights and the separateness of persons. *Social Philosophy and Policy*. 26(1):76–95.

Nozick, R. 1974. *Anarchy, state and utopia.* Basic Books, New York.

Nussbaum, M. 1988. Non-relative virtues: An Aristotelian approach. *Midwest Studies in Philosophy*, 13: 32–53.

Nussbaum, M. 2001. Adaptive preferences and women's options. *Economics and Philosophy*, 17: 67–88.

Pahl-Wost, C. 2007. The implications of complexity for integrated resource management. *Environmental Modeling and Software*, 22 (5): 561–569.

Parfit, D. 1984. *Reasons and persons.* Clarendon Press, Oxford.

Parfit, D. 2011. *On what matters*, Vol. 1. Oxford University Press, Oxford.

Quinton, A. 1995. The ethics of philosophical practice. In: T. Honderich (ed.), *The Oxford companion to philosophy.* Oxford University Press, Oxford, 666.

Rawls, J. 1973. *A theory of justice.* Oxford University Press, Oxford.

Roberts, M.A. 2009. The non-identity problem. In: N. Zalta (ed), *The Stanford Encyclopedia of Philosophy*, Fall 2013 edition. Available at http://plato.stanford.edu/archives/fall2013/entries/nonidentity-problem/ (accessed on May 27, 2015).

Rolston, H. 1999. Respect for life: Counting what Singer finds of no account. In: D. Jamieson (ed.), *Singer and his critics.* Blackwell, Oxford, 247–268.

Routley (later Sylvan), R. 1973. Is there a need for a new, an environmental, ethic? Proceedings of the World Congress of Philosophy, Varna, Bulgaria, pp. 205–210.

Rowlands, M. 1997. Contractarianism and animal rights. *Journal of Applied Philosophy*, 14 (3): 235–247.

Ryder, R. 1989. *Animal revolution: Changing attitudes towards speciesism.* Cambridge University Press, Cambridge.

Schroeder, M. 2012. Value theory. In: *The Stanford encyclopedia of philosophy*, available at http://plato.stanford.edu/archives/sum2012/entries/value-theory. Accessed March 30, 2014.

Sen, A. 1993. Capability and well-being. In: M. Nussbaum and A. Sen (eds), *The quality of life.* Clarendon Press, New York, 30–53.

Shafer-Landau, R. 2010. *The fundamentals of ethics*. Oxford University Press, Oxford.

Singer, P. 1995a. *Animal liberation*, 2nd edition. Pimlico, London.

Singer, P. 1995b. *Practical ethics*, 2nd edition. Cambridge University Press, New York.

Singer, P. 1999. A response (Rolston). In: D. Jamieson (ed.), *Singer and his critics*. Blackwell, Oxford, 269–335.

Singer, P. 2009. Reply to Marquis. In: J.A. Schaler (ed.), *Singer under fire. The moral iconoclast faces his critics*. Carus Publishing Company, Illinois, 153–162.

Singer, P. 2011. *Practical ethics*, 3rd edition. Cambridge University Press, New York.

Sobel, D. 2009. Subjectivism and idealization. *Ethics*, 119: 336–352.

Sumner, L.W. 1996. *Welfare, happiness and ethics*. Clarendon Press, Oxford.

Tännsjö, T. 1998. *Hedonistic utilitarianism*. Edinburgh University Press, Edinburgh.

Taylor, P.W. 1986. *Respect for nature. A theory of environmental ethics*. Princeton University Press, Princeton.

Teichmann, J., Evans, K.C. 1999. *Philosophy: A beginner's guide*. Blackwell Publishing, Cambridge.

Temkin, L.S. 1999. Intransitivity and the person-affecting principle: A response. *Philosophy and Phenomenological Research*, 59 (3): 777–784.

Tennessen, C.M. and Cimprich, B. 1995. Views to nature: Effects on attention. *Journal of Environmental Psychology*, 15: 77–85.

Ulrich, R.S. 1984. View through a window may influence recovery from surgery. *Science*, 224: 420–421.

Verderber, S. 1986. Dimensions of person-window transactions in the hospital environment. *Environment & Behavior*, 18: 450–466.

Verderber, S. and Reuman, D. 1987. Windows, views, and health status in hospital therapeutic environments. *Journal of Architectural and Planning Research*, 4: 120–133.

Visak, T. and Balcombe, J. 2013. The applicability of the self-fulfillment account of welfare to non-human animals, babies and mentally disabled humans. *Philosophy and Public Policy*, 31 (2): 26–33.

Weijers, D. 2013. Intuitive biases in judgments about thought experiments: The experience machine revisited. *Philosophical Writings*, 41 (1): 17–31.

Wells, N.M. 2000. At home with nature. Effects of "Greenness" on Children's Cognitive Functioning. *Environment & Behavior*, 32(6):775–795.

Wessels, U. 2011. *Das gute*. Vittorio Klostermann, Frankfurt am Main.

West, M.J. 1986. Landscape views and stress responses in the prison environment. Master's Thesis. University of Washington, Seattle, 15 (3).

SECTION III

ENVIRONMENTAL MANAGEMENT: THE METHODS AND TOOLS PERSPECTIVE

15

PARTICIPATORY APPROACHES IN ENVIRONMENTAL MANAGEMENT

STENTOR B. DANIELSON
Department of Geography, Geology and the Environment, Slippery Rock University, Slippery Rock, PA, USA

Abstract: Recent decades have seen a rise in the use of participatory approaches in environmental management, which seek to give a broad set of stakeholders more of a say in decision making. Proponents of this participatory shift argue that participation produces higher-quality decisions, is more morally justified, and encourages more public acceptance than decisions made in a top-down or authoritarian way. Critics charge that participation can be undemocratic, dysfunctional, and culturally biased. The design of an effective participation process must answer questions such as who will participate, how will they interact, how will scientific and nonscientific knowledge be integrated, and how much authority will each participant have over the final decision.

Keywords: participation, public participation, community involvement, community-based natural resource management, community-based forest management, devolution, decentralization, deliberative democracy, Habermas, Juergen, Young, Iris Marion, community advisory group, deliberative polling, public hearings, citizen science, tribal sovereignty, local knowledge, cooptation.

15.1 Introduction	366	
15.2 What is Participation?	366	
15.2.1 Defining Participation	366	
15.2.2 History of Participation	366	
15.3 Arguments for Participation	367	
15.3.1 Substantive Value of Participation	367	
15.3.2 Normative Value of Participation	368	
15.3.3 Instrumental Value of Participation	368	
15.3.4 Self-Development Value of Participation	368	
15.3.5 Deliberative Democracy	368	
15.4 Arguments against Participation	369	
15.4.1 Practical Problems with Participation	369	
15.4.2 Legal Problems with Participation	369	
15.4.3 Problems with Equality and Justice in Participation	370	
15.4.4 Problems with Manipulation and Cooptation in Participation	370	
15.4.5 Discrimination Problems in Participation	370	
15.4.6 Poor Decision-Making Problems with Participation	371	
15.5 How to Conduct Participation	371	
15.5.1 How Much, and How Little, Participation is Allowed?	371	
15.5.2 How Much Participation is Needed?	372	
15.5.3 Who Will Participate?	372	
15.5.4 How Will Participants Interact?	372	
15.5.5 In What Environment Will Interaction Occur?	372	
15.5.6 Who Will Set the Agenda?	373	
15.5.7 How Will Opinions Be Aggregated?	373	
15.5.8 How Much Power Does Each Stakeholder Have over the Final Decision?	373	
15.5.9 How Will Scientific and Technical Information Be Integrated with Lay Knowledge (Including Indigenous Traditional Knowledge)?	373	
15.5.10 What are the Practical Constraints on Participation?	374	
15.5.11 Participation Mechanisms	374	
15.6 Evaluating Participation	375	
15.7 Conclusion	376	
References	376	

An Integrated Approach to Environmental Management, First Edition. Edited by Dibyendu Sarkar, Rupali Datta, Avinandan Mukherjee, and Robyn Hannigan.
© 2016 John Wiley & Sons, Inc. Published 2016 by John Wiley & Sons, Inc.

15.1 INTRODUCTION

In the early 1980s, the US Environmental Protection Agency (EPA) identified a former chemical plant in New Jersey as a potential threat to the health of the local population. The plant had contaminated the underlying soil and groundwater with volatile organic chemicals. The EPA's initial cleanup plan called for incineration of the contaminated soil and filtering of groundwater to a standard suitable for release into the environment.

A group of citizens living in the area took an interest in the cleanup and objected to the EPA's plans. These citizens organized into two community groups, whose members invested countless hours in educating themselves to understand the technical data underlying the EPA's plans. They came to public hearings held by the EPA and met privately with officials from the EPA and the chemical company that owned the plant.

On the basis of input from the citizen groups, the EPA made a series of changes to their cleanup plan. The groundwater treatment plant was moved up to the first phase of the cleanup, and the water was cleaned to a higher drinking water standard. One group acquired funding to hire its own professionals to double-check the EPA's water testing data. Data from air quality monitors around the site were published daily online. And most significantly, the incineration plan was scrapped in favor of an innovative bioremediation system in which the soil was cleaned through microbial action.

Not everyone in the surrounding community has been entirely happy with the outcome of the site cleanup, which wrapped up in the 2000s, or with the process of making decisions about how the cleanup would occur. But the scope of changes made as the result of involvement by members of the public provides a striking illustration of the movement toward participatory approaches in environmental management (Danielson et al., 2008).

Since the 1970s, the idea of "participation" has grown in importance in environmental management throughout the world. "Top-down" and "command and control" have become powerful insults used against forms of management that are viewed as outdated and undemocratic. The demand for participation has come not only from grassroots groups and ordinary people but also from law and policy decisions.

As the conventional wisdom has coalesced around the value of participation, additional complexities have emerged. What, exactly, constitutes effective participation? When is participation genuine, and when is it merely window dressing? What techniques enable good participation? And are there situations where participation can become a detriment to environmental management and even an injustice?

This chapter suggests how these questions might be answered in practical situations. It asks why participation is valuable, what its flaws are, and how it can be done better.

15.2 WHAT IS PARTICIPATION?

15.2.1 Defining Participation

In this chapter, participation refers to the broader involvement of stakeholders in decision making about the management of an environmental system. This involvement may be broader in terms of what stakeholders are involved—expanding beyond just government officials to incorporate people such as local residents, environmental and other NGOs, unions and trades groups, indigenous tribes, and industry. Participation may also broaden the power held by these groups, moving from consultation to a vote or even a veto over the decision.

A variety of terms have been used for participation. Participatory approaches are sometimes referred to as "stakeholder involvement," "community involvement," "public engagement," "collaboration," "comanagement," "civic environmentalism," "community-based natural resource management" (CBNRM) (and its specific variants such as community-based forest management (CBFM) and community-based fire management (CBFiM)), or "joint forest management" (JFM). For the purposes of this chapter, work going by all of these terms is considered together.

Most examples of participation in the literature involve a sponsoring agency setting up one or more participatory forums, into which it invites various other stakeholders. Within this forum, the stakeholders make decisions about how to manage the environmental system at issue. The sponsoring agency is typically a government agency such as Environment Canada or the US Forest Service, which holds formal authority over the management decisions to be made. Nevertheless, the ideas and concepts presented in this chapter can also apply to more radically democratic forms of decision making. These more radical forms include decentralization, in which the state hands over full control of an environmental system to a community organization, and the grassroots or bottom-up organization of management for a previously unmanaged environmental system.

Participation centers on a process of communication. The core activity of participatory decision making, whatever specific form it may take, is the exchange of information, ideas, and arguments, with the goal of persuading others to change their views and learning new information that may shift one's own views. Various aspects of the design of a participatory forum aim to enhance and regulate the sharing and processing of information while neutralizing exercises of power. How to best achieve this open communication and whether it is possible or desirable to eliminate the use of power in decision making are key debates in the literature on participation.

15.2.2 History of Participation

The early and middle twentieth century saw huge growth in bureaucracies in both the industrialized world and their poorer current and former colonies. These bureaucracies used a very top-down, authoritative approach, in which expert planners evaluated objective evidence and dictated policy (Lane, 2005; Torgerson, 1986). This top-down approach was built on advances in logistics and operations management developed during the two World Wars for military applications.

Several factors came together in the latter part of the twentieth century to create a movement away from authoritarian governance and toward participation. Among the key ones were:

1. *Failures of top-down environmental management.* Top-down approaches to decision making proved unable to deliver on their promises of rationalizing society. Planners faced serious resistance from communities who felt disempowered (Day, 1997; Roberts, 1995). Schemes that looked good on paper ended up failing in practice because planners lacked adequate information and buy-in from stakeholders (Scott, 1998). And recognition of the negative effects of technology (e.g., nuclear waste) undermined faith in experts.

2. *Grassroots demands for participation arising from "new social movements."* A common thread to the complaints of the "new social movements" of the 1960s and 1970s (Habermas, 1981; Pichardo, 1997)—such as the feminist and civil rights movements—was that certain populations were being excluded from having a say in their own fate. In addition to enhancing their ability to take part in the formal political process (e.g., through ending voter suppression practices), these groups demanded expanded forms of participation in decisions that affected them (Langton, 1978).

3. *Research demonstrating the viability of existing community-based resource management.* Researchers who examined community-based resource management systems found that in many cases they were ecologically and economically sustainable. Most notable among these researchers was political scientist Elinor Ostrom, who showed that common-pool resources could be successfully managed by a community of users without triggering a tragedy of the commons (Ostrom, 1990; Ostrom et al., 2002).

4. *Neoliberalism and the hollowing out of the state.* The economic doldrums of the 1970s brought the rise of neoliberalism, a philosophy of rolling back inefficient government management and replacing it with the private sector or civic society (Castree, 2008; Peck and Tickell, 2002). Participation fit neatly into neoliberalism, as it replaced heavy-handed state policymaking with locally based multistakeholder partnerships (Holifield, 2004; Larner and Craig, 2005). At least in the abstract, then, participatory approaches could find champions on both the right and the left (McCarthy, 2005).

As a result of these forces, participation has become an extremely popular buzzword. Governments at all levels, development agencies, nonprofits, and even corporations are quick to endorse participation. In the United States, for example, federal laws that mandate participation in environmental decision making include the National Environmental Policy Act (1970), the National Forest Management Act (1976), the Resource Conservation and Recovery Act (1976), and the Healthy Forests Restoration Act (2004). International development agencies like the World Bank, USAID, and the Asian Development Bank increasingly require participation in projects that they sponsor (Chambers, 1997). A growing industry of participation consultants turns a profit by helping to organize participatory processes (Hendriks and Carson, 2008).

15.3 ARGUMENTS FOR PARTICIPATION

A wide variety of arguments have been advanced for increasing the level of participation in environmental decision making. In some cases, the case for participation is purely self-interested—a particular stakeholder or stakeholder coalition believes that they are more likely to get the substantive outcome that they desire if the decision is made in a participatory way (Yung, Patterson, and Freimund, 2010). This section will focus instead on principled, procedural arguments in favor of participation. That is, reasons that participation is the right way to make decisions regardless of what the content of that decision might turn out to be.

The classic statement of the benefits of participation comes from Daniel Fiorino. Fiorino argues that there are three types of benefits to participation, which he labels substantive, normative, and instrumental (Fiorino, 1990; Stirling, 2006).

15.3.1 Substantive Value of Participation

The substantive value of participation is that by involving more people in the decision-making process, a larger and more diverse body of ideas and information can be accessed, leading to objectively superior decisions (Beierle, 2002). Stakeholders may have a variety of information not easily accessible to the sponsoring agency, which they can provide during a participatory process. This may include information about the environmental system at issue, information about the history of the site, or social information about local ways of life that is relevant to the feasibility and effectiveness of various possible solutions.

Beyond simply gathering a wider body of information, the process of participation can also improve the strength of reasoning by participants. Through the give-and-take of argument and refutation, participants can develop stronger and more inclusive views of the situation (Easterling, Neblo, and Lazer, 2011; Gregory, 2000). Some research suggests that deliberation can help to overcome common irrational psychological biases (Druckman, 2004).

15.3.2 Normative Value of Participation

The normative value of participation arises from the principle, common to most modern political philosophies, that legitimate government requires the consent of the governed. This has traditionally been instituted by free and fair elections that select a government, which then makes policy and appoints administrators. The movement toward participation pushes democracy farther by getting direct consent from affected parties for specific decisions on specific cases. Proponents of participation see the democratic accountability of elections as too distanced from decisions by the bureaucracy and too dominated by powerful interests. The political theories behind this shift toward participation are described in the section on deliberative democracy.

Demands for greater participation often come from activists affiliated (explicitly or implicitly) with the environmental justice movement (Gallagher, 2009; McClymont and O'Hare, 2008). A core element of the ongoing marginalization of poor, ethnic minority, indigenous, and other communities is exclusion from political decision making (Holifield, 2012). It should therefore be no surprise that principle seven from the list of principles of environmental justice developed at the First National People of Color Environmental Leadership Summit in Washington, DC, was "Environmental Justice demands the right to participate as equal partners at every level of decision-making, including needs assessment, planning, implementation, enforcement and evaluation" (First National People of Color Environmental Leadership Summit, 1991). Participation helps to equalize power relations between dominant and marginalized groups in society (Masaki, 2004; Webler, 1995).

15.3.3 Instrumental Value of Participation

The instrumental value of participation is that a decision made through a participatory process is more likely to have broad public support and thus be able to be implemented easily and without controversy. A lack of public support for a decision can stall it in a variety of ways, such as through lawsuits filed against it. In many cases, public compliance with a decision is essential, as the decision entails a requirement for action by members of the public (e.g., to clear flammable brush from around their homes) (Campbell, Koontz, and Bonnell, 2011; Pimbert and Gujja, 1997). A more participatory decision is, of course, more likely to match the preferences of the public who participated (Danielson et al., 2008; Rosener, 1982). Moreover, the process of deliberation may generate greater consensus on the decision, bringing conflicting stakeholders around to a common view they can all support (Niemeyer, Ayirtman, and HartzKarp, 2013; Stagl, 2006). Even if deliberation does not produce consensus on one option, it may at least lead to participants evaluating options on the same criteria, such that a stable compromise can be found (Dryzek and List, 2003; Farrar et al., 2010).

Empirical research has generated ambivalent findings with respect to the theory behind the instrumental value of participation. Some studies indicate that a fair and participatory process will lead to acceptance of the decision even by those who see it as substantively incorrect (Arvai, 2003; Hunt and McFarlane, 2007). Yet other studies suggest that people's acceptance of a decision is largely driven by their opinion of the outcome (Arvai and Froschauer, 2010; Earle and Siegrist, 2008). Indeed, people may decide that a process itself is unfair if they dislike the outcome it produced.

15.3.4 Self-Development Value of Participation

One final value of participation not noted by Fiorino, but which figures in many explanations of participation, is the self-development value. This idea goes back as far as Aristotle, who declared that man (and he did mean only men) is a "political animal," able to reach his full development only through participation in political life (Aristotle, 1984; Day, 1997). Through participation, one can build connections to the community, improve one's vision of the world through discussing it with others, find opportunities for ethical action, and develop one's own autonomy as an agent (Gooch, 2005; Mohan and Hickey, 2004).

Relatedly, participation can have value in self-development for the community. The process of participation can build bonds among community members, promote social cohesion, and empower the community to solve future problems (Paveglio et al., 2009; Wagner and Fernandez-Gimenez, 2008). Burns, Taylor, and Hogan (2008) went so far as to describe participation as "healing" to a community devastated by a catastrophic wildfire.

15.3.5 Deliberative Democracy

Work on participation in environmental decision making has drawn heavily on the tradition of deliberative democracy in political philosophy (Ackerman and Fishkin, 2005; Dryzek, 1995; Gutmann and Thompson, 1996). The basic idea of deliberative democracy is that traditional democracy treats citizens' policy views as existing prior to, and independent of, any public political activity—democracy simply consists in counting up how many people favor plan X versus plan Y. Deliberative democracy theory argues that legitimate preferences must be developed through discussion with fellow citizens.

The most notable and influential theorist of deliberative democracy is the German intellectual Juergen Habermas. Habermas argued that modern governments face a "legitimation crisis," meaning that citizens no longer felt that laws and policies were authoritative and legitimate, and thus people felt alienated and lacked motivation to comply with social norms. To solve this legitimation crisis, Habermas proposed that political decisions should be made through deliberation in the "ideal speech situation." In an ideal speech situation, people would be free of all influence other than the force of the better argument. This would enable dialogue to reach genuine consensus based on rational agreement and mutual understanding, rather than on coercion or manipulation. Policy decisions made in an ideal speech situation would have real force and validity, because all people affected by them would be able to consent to them. Anyone who had misgivings about the policy would be able to express those misgivings and either receive a satisfactory answer or get the policy changed (Habermas, 1984a, b; Habermas, 1996).

The ideal speech situation is a hypothetical model that could never be implemented in practice. But it offers a standard by which actual speech situations can be judged as better or worse and more or less ideal (Webler, 1995).

Habermas's ideal of communication has been expanded and challenged by feminist theorists such as Iris Marion Young (1995, 1997, 2001) and Seyla Benhabib (1986, 1992). These thinkers point out that Habermas envisions discourse as a highly rationalistic process, involving the give-and-take of abstract philosophical arguments rooted in universal reason and detached from individuals' personal experiences. But this is a very limiting notion of communication. In his zeal to exclude manipulative and harmful communication modes, Habermas has also excluded some of the normal ways that humans communicate—ways that are often associated with already marginalized groups in society such as women and people of color. In cases of long-standing structural injustice, it is necessary for oppressed groups to directly state a demand for better conditions for themselves and share their specific stories, rather than trying to speak only of the universal good.

Benhabib and Young also argue that Habermas is too quick to seek universalization. Rather than hoping that the force of undistorted communication will lead us to truth shared by all, they suggest that discourse should focus on making connections between concrete, specific people in their actual circumstances. The result would be a web of interconnected perspectives, rather than one universal truth.

15.4 ARGUMENTS AGAINST PARTICIPATION

The conventional wisdom in the field of environmental management has settled firmly in favor of participation. Most criticism directed against current participatory decision making amounts to a charge that the process is not participatory enough. Nevertheless, there are a variety of principled arguments made against participatory approaches. It is important for those proposing participatory approaches to give serious consideration to critiques such as those outlined in the succeeding text.

15.4.1 Practical Problems with Participation

Participation can be more expensive, more time-consuming, and more demanding of skills than nonparticipatory decision making, at least when the latter goes according to plan. Agencies and companies who feel short on these resources may be reluctant to expend them on a deep participatory process (Chess and Johnson, 2006; Pini and Haslam McKenzie, 2006). For seemingly straightforward decisions, the gains may simply not appear worth the cost.

Calls for participation have usually arisen from cases in which there is a high level of public engagement and controversy—but many environmental management decisions are far more mundane and excite much less public concern. A great deal of political science research has suggested that the public is for the most part apathetic and disinclined to purse greater participation in policy decision making even when the option is open to them (Eliasoph, 1998; Hibbing and Theiss-Morse, 2002).

The costs of participation may fall not just on the sponsoring agency, but also on the participants. The stakeholder groups that are formed to make participatory decisions may lack the resources that the state could bring to bear and thus be ill-equipped to carry out research, invest in monitoring, or enforce the decision. This could lead to a decision that falls short on substantive grounds (Henry, 2004).

15.4.2 Legal Problems with Participation

Participation is rejected in some cases because it violates regulatory requirements. If an agency has a mandate to carry out a certain type of regulation through a specified process, it may lack the legal and political freedom to fully open its decisions to participation. In such a situation, the agency will have trouble making a credible commitment to abide by the results of a participatory process, it may be at risk of being sued by groups unhappy with the participatory process's decision, and it may be limited in the problem framing and feasible options that it can accept (Renn, Webler, and Kastenholz, 1998). When these regulatory requirements emerged from a legitimate political system, it may even be regarded as antidemocratic to do an end run around the duly instituted laws passed by a democratically elected legislature (Coggins, 1999; McCarthy, 2006).

Related to the issue of regulatory requirements is the question of tribal sovereignty. Standard participation approaches place a high value on equality. Though it may be compromised in practice, the official position is that all stakeholders have equal input and an equal say in a decision. This principle may threaten the special status of indigenous peoples, who have a special legal and political status as a result of being the original inhabitants of the land (Holifield, 2012; Lloyd, van Nimwegen, and Boyd, 2005). Indigenous rights, such as rights secured by treaties, ought to trump the preferences of other stakeholders and should not be up for debate. For this reason, tribal organizations may refuse to join in participatory forums or may request special differentiated status or veto power (Cullen et al., 2010).

Participation can violate private rights such as the right to private property. From this perspective, certain decisions ought to rest solely with a particular person or entity, to make wisely or poorly as they see fit (Galston, 1999). Participation unfairly socializes decisions that should be part of someone's protected sphere of personal liberty.

15.4.3 Problems with Equality and Justice in Participation

Commonly, participation is implemented through face-to-face meetings—workshops, advisory committees, hearings, etc. This creates the potential for participation to favor certain stakeholders. There is a tendency for organizers to favor "local people"—people who live close to the environmental system about which the decision is being made. This tends to disempower more distant stakeholders (Sneddon and Fox, 2008). These "local people" may be dependent on exploitation of a natural resource (e.g., through logging or mining), leading to outcomes objectionable to environmentalists who had hoped that strict environmental laws would protect ecosystems (Coggins, 1999; Yung, Patterson, and Freimund, 2010).

In any participatory process, there will be a strong tendency for more input to be provided by those who are more socially privileged and who will thus have more time, resources, self-esteem, and social respect to get their voice heard. Though participatory approaches aspire to neutralize this inequality, many authors remain skeptical that power can be adequately neutralized within a participatory process (Mosse, 1994; Sultana, 2009).

Finally, participation can legitimize unjust perspectives (Benhabib, 1986; Levine and Nierras, 2007). Participatory approaches are generally built on an ideal of procedural neutrality, in which all stakeholders and perspectives are treated equally to avoid discrimination and allow the best ideas to prevail on their own merits. While equality raises the voices of the marginalized, it also secures an equal place for the oppressor. It treats the oppressor's views as legitimate and reasonable. It is unreasonable, these critics contend, to ask the oppressed to reason with their oppressors.

15.4.4 Problems with Manipulation and Cooptation in Participation

Participation has been criticized for being a form of manipulation or cooptation. This criticism is especially directed against instances in which participation is initiated by a powerful sponsoring agency such as the government, rather than arising out of the grassroots (Chilvers and Burgess, 2008; Waddington and Mohan, 2004). From this perspective, participants will come to adopt the sponsoring agency's socially dominant ideology as a result of repeated exposure to it and feelings of comradeship that develop over the course of the process (Amin, 2005; Cooke, 2004; Young, 2001). Stakeholders may refuse to participate in officially sponsored forums, in order to maintain the legitimacy of their connection to the people and interests they represent; an uncompromised critical voice from outside is thus seen as preferable to a compromise-laden seat at the table.

Manipulation may also occur when there is strong pressure for reaching consensus. Consensus is an explicit goal of many participation processes, as it seems to represent a state in which all stakeholders are in agreement and give their consent to the plan. But consensus may be reached by unfair means, such as social pressure, closing down of the agenda, or overreliance on the status quo as a default (Peterson, Peterson, and Peterson, 2005; Sanders, 1997). This need not be a nefarious process—ordinary group dynamics have a tendency to push people with unique information or divergent perspectives to keep silent, even on issues where there is a clear factual right or wrong answer (Sunstein, 2005).

Participation has been criticized for being an abdication of responsibility by the state. Functions once served by the government are increasingly turned over to the private sector or to civil society (Amin, 2005; Coggins, 1999; Larner and Craig, 2005). The government no longer stands as a guarantor of rights and welfare.

15.4.5 Discrimination Problems in Participation

The dominant form of participation has been challenged as ethnocentric. The rules of participatory systems and the way they divide up problems may disrupt cultural practices (Saito-Jensen and Jensen, 2010). In some cases, the mere fact that a model of participation is seen to originate from a particular cultural context may imbue it with a heavy political meaning. For example, Taddei (2011) found that for inhabitants of the Amazon, "participation" was seen as part and parcel of "modernity," which influenced people's desire and ability to participate fully.

Most participation processes demand that information be put into a format acceptable to a technically minded sponsoring agency, which can be a source of cultural bias. It can be difficult, distorting, or even impossible to convert lay perspectives or indigenous knowledge into a form compatible with a technical mindset (Nadasdy, 2003). This issue becomes particularly visible in the case of participatory use of Geographic Information Systems (GIS). While frequently praised as democratizing, participatory GIS can force information to be reduced to an extremely spatially precise and narratively anemic format that GIS software can handle (Brown and Knopp, 2008; Ramsey, 2008).

Other thinkers argue that the fundamental ideals of equal, open communication that lie at the root of participation are culturally specific. For example, some communities place a high value on deference to the views and decisions of elders. If the organizers of a participatory forum insist on equal participation by all members of the community, they may refuse—or may participate—and in so doing disrupt the values and authority structure of their community (Huntington et al., 2006; Sletto, 2009). Even when the antiparticipatory structure of the community (e.g., exclusion of women from decision making) is regarded by the organizer as clearly unjust, it can be a delicate situation to exploit the greater power and privilege of the organizer to impose equal participation.

Participation has been critiqued for being discriminatory against those who are incapable of participation. Models of participation assume that the participants possess a certain base level of competency to form a viewpoint or preferences, to communicate those views and preferences to others, and to respond to others' communication—sometimes loosely summarized as "the ability to speak." While advocates of participation urge every effort to include every stakeholder who has the ability to speak, it is unclear what is to be done with beings who simply cannot speak—such as infants, people with severe mental illnesses or disabilities, future generations, animals, and ecosystems (Dobson, 2010; Eckersley, 1999; Heyward, 2008). There are proposals to incorporate their interests indirectly, through some sort of appointed advocates (Stone, 1987). Other authors suggest that nature is already speaking, if we only knew how to properly listen (Plumwood, 1996; Povinelli, 1995). Finally, some suggest that to safeguard the rights of those who can't speak, the possible options to be deliberated over must be restricted, for example, by mandating use of the precautionary principle (Eckersley, 1999; Heyward, 2008).

15.4.6 Poor Decision-Making Problems with Participation

In contrast to the argument that participation will improve the quality of decisions, some argue that participation is dysfunctional. The general public possesses a wide variety of irrational beliefs, misunderstandings of science, and unreasonable expectations of government. To invite these people into the decision-making process would spoil the whole thing (Breyer, 1993; Predmore et al., 2011).

Moreover, the group interaction processes that are at the center of most participatory processes encourage a variety of biases and irrational thought patterns such as groupthink, polarization, information cascades, or the Abilene paradox (Cooke and Kothari, 2001; Mendelberg, 2002; Sunstein, 2005). Some writers prefer a hierarchical, expert-driven decision process as an alternative (Breyer, 1993), while others advocate greater reliance on the market (Somin, 2013).

Finally, some people argue that participation is an illusion. The various arguments for participation presume that some sort of sharing of ideas and perspectives, a meeting of the minds, will occur in the participatory forum. But according to some social theories, no such thing is possible. All human action is a power struggle. People may appear to be making reasoned arguments and listening to others' input—but that is simply because an appearance of reasonableness seems to be a useful strategy for advancing their interests (Chilvers and Burgess, 2008; Flyvbjerg, 1998; Shapiro, 1999). If participants believe this, they will tend to regard offers of participation as merely a way of obfuscating the power relations at issue and will then seize on the process as a way to advance their own partisan interests (Mosse, 1994; Simon, 1999). And even if some people are participating in good faith, powerful structural interests—whether within or outside the participatory process—can undercut the results of the process if they threaten the status quo (Lauria and Whelan, 1995; Tauxe, 1995).

Moreover, even this appearance of reasonableness is fragile and likely to degenerate into a shouting match between fixed ideologies. This view gains support from research suggesting that people are incredibly resistant to changing their minds when faced with evidence and opinions contrary to their preexisting beliefs. Instead, people adjust their beliefs on specific issues to be consistent with their cultural group's worldview (Earle and Siegrist, 2008; Kahan and Braman, 2003). Indeed, the process of rationalizing and explaining away conflicting information may strengthen and entrench ideological conflict.

15.5 HOW TO CONDUCT PARTICIPATION

It is not enough to simply identify a need for participation in general. Participation comes in a wide variety of forms, and executing it incorrectly or inappropriately can result in a failed decision process and wasted time and resources (Barreteau, Bots, and Daniell, 2010; Buck and Stone, 1981). It is necessary to design a specific mechanism for implementing a participatory decision-making process that fits a clear "story line" about what participation is and why it is being used in this instance (Hendriks, 2005). In designing a process, several key questions must be answered (Fung, 2003; Rowe and Frewer, 2000, 2005).

15.5.1 How Much, and How Little, Participation is Allowed?

Regulatory mandates and political realities may place upper and/or lower bounds on participation. On one hand, the law may grant final decision-making power to a particular entity.

In such a case, no matter what promises that agency may make to abide by the results of a participatory process, they can always choose—or be forced by the courts—to unilaterally reject the participatory decision (Renn, Webler, and Kastenholz, 1998). On the other hand, the law may mandate the creation of certain types of participatory forums—for example, the Healthy Forests Restoration Act in the United States makes funding for wildfire risk reduction contingent on the creation of a community wildfire protection plan through a participatory process.

15.5.2 How Much Participation is Needed?

It may seem, to an advocate of participation, that more participation is always better. But even setting practical trade-offs aside, in some cases, intensive participation is neither necessary nor even helpful (Smiley, de Loë, and Kreutzwiser, 2010; Vroom and Jago, 1978). Within a deliberative democratic system, a bureaucratic or technocratic approach still has its place, especially for routine and everyday matters in which there is no great clash of worldviews.

15.5.3 Who Will Participate?

Defining the list of stakeholders is a tricky process (Fishkin, 2006). The most common principle is that everyone who is affected by a decision should participate. But who is affected by a decision is often exactly what is in question—for example, did this pollution actually cause this person's child's cancer? Moreover, some people are affected much more significantly than others and would seem to therefore deserve a greater say (Davis, 2011, but see in contrast Nature, 2000). Is merely taking an interest in the issue, such as when an environmentalist living far away from the environmental system being managed nevertheless places great "existence value" on an endangered species, enough to get a seat at the table? And even if participation is formally open to certain stakeholders, those stakeholders may not be practically able to participate. Most participatory processes depend on some level of self-selection. This can lead to participants being a biased sample of the public that they are taken to represent (Halvorsen, 2006; Hildyard et al., 2001).

Participatory processes usually aspire to include the full breadth of people within the eligible set of stakeholders. In most cases, this is done by drawing up a list of different types of stakeholders based on the social position and relationship to the issue held by various people—a list that may include entries like local government, industry, environmental groups, nearby residents, indigenous tribes, etc. Then representatives of each of those categories are invited to participate. Social network analysis has been proposed as a more sophisticated way of identifying structurally diverse sets of stakeholders (Prell et al., 2010). Some researchers have suggested that instead organizers should seek to represent a diversity of viewpoints, regardless of the structural position of those who hold the viewpoints. This diversity of viewpoints may be based on an existing theory, or they may be empirically identified (Billgren and Holmén, 2008; Cuppen et al., 2010).

15.5.4 How Will Participants Interact?

Interaction between participants can take a wide variety of forms. Generally speaking, participation processes can range from ones that merely collect raw public opinion to ones that give extensive opportunity for opinions to be refined through debate and learning (Arnstein, 1969; Fishkin, 2006). Sometimes interaction may be greatly circumscribed, as in the case of a public hearing at which each person signs up for a 1–2 min speaking slot, allowing them to make a statement but not necessarily receive any feedback. Some participatory processes engage participants in a very structured decision-making process, guiding them through specific forms of group interaction and reflection (Arvai, Gregory, and McDaniels, 2001; Rutherford et al., 2009). Other times, there is a vision of a more freewheeling debate, with citizens able to speak their minds at length, respond to each other's comments with elaboration or refutation, and collectively workshop their ideas.

There is also much debate over the balance between reason and emotion, and between rhetoric and logic, in participatory talk. On the one hand, many theorists call for participants to stick to reasoned arguments, in order to avoid manipulation or browbeating by emotive statements, and to ensure that rational progress can be made (Chambers, 2009; Habermas, 1984a, b). Other theorists reply that the demand for cold reason shuts out the experiences of many already marginalized people and demands that they conform themselves to a white, middle-class, masculinist debate style that they lack skill at participating in (Benhabib, 1992). Researchers generally agree that interaction must be open to dialogue and counterargument, rather than just statements of fixed positions (Tuler, 2000; Webler, 1995).

Furthermore, there is a question of the number of people interacting at once. Classic ideas of participation often envision a small group (perhaps a dozen participants) who can hold an involved discussion among themselves. But the larger the number of stakeholders trying to participate, the more these traditional models of discussion break down (Cohen and Fung, 2004; Parkinson, 2003). This has led to the development of systems like deliberative polling that aim to enable hundreds of people to effectively debate an issue together (Bryson and Anderson, 2000; Weeks, 2000).

15.5.5 In What Environment Will Interaction Occur?

Face-to-face interaction is central to most proposals for participation, though there is growing interest in synchronous (e.g., videoconferencing) or asynchronous (e.g., message boards and wikis) interaction over the Internet (Rowe and Gammack, 2004). Face-to-face interaction can take place in a variety of places—hotel meeting rooms, agency offices, public buildings like schools and churches, or even outside in a field or plaza. Some spaces may be more or less welcoming to various participants. A frequent complaint is that sponsoring agencies demand that participation be done on "their turf," inhibiting other participants from attending or being comfortable expressing themselves fully (Clarke and Agyeman, 2011).

15.5.6 Who Will Set the Agenda?

A common tension in participatory processes has to do with the scope of the decision to be made. The agenda incorporates questions of problem definition, geographical and temporal scale, analytical criteria, and the menu of available policy options. In most cases, sponsoring agencies prefer to set a very circumscribed agenda—for example, this process is about deciding whether to grant a permit for this particular factory to be built. Other stakeholders want to hold a broader discussion, for example, about the need for sustainable economic development for their community or about justice for past infractions by the sponsoring agency (French and Bayley, 2011; Sze et al., 2009).

A common lesson for participation in the literature is that it is vital to involve the broadest set of stakeholders early in the process (Davis, 2011; Kasperson, 1986). When stakeholders are engaged early, they can have an influence over the agenda. When they are engaged late, they may feel that the issue has already been decided and that they are being asked to endorse a plan from a too-narrow set of options. Nevertheless, laypeople may find early participation too abstract or distant from issues that really affect them (Desai, 1989).

15.5.7 How Will Opinions Be Aggregated?

The organizer of a participatory process needs to decide how the varying views of the participants will be put together into a final decision (Hardy and Koontz, 2009). Will there be a vote (and if so, will it require a majority? A supermajority?)? Will the process continue until consensus is achieved? Will the process issue a report that describes all positions without endorsing one over the other? Or will the process simply collect data (in the form of survey responses, Q sorts, etc.) about public opinion (Lo, 2011; van den Hove, 2006)?

15.5.8 How Much Power Does Each Stakeholder Have over the Final Decision?

In many cases, the participatory process is not binding. The sponsoring agency retains full power to make whatever decision they choose, with the other participants having only persuasive power. In other cases, the participatory process's decision is binding on the sponsoring agency. In general, processes that produce definite policy rather than just advisory information tend to be more successful (Hajjar, Kozak, and Innes, 2012; Hill et al., 2012). Even when the process is not binding, it will be more successful when the sponsoring agency shows ongoing commitment to the process (Branch and Bradbury, 2006; Genskow, 2009).

Processes that give greater power to participants raise issues of representation. That is, are the individuals who are actually involved in the participatory process trustees authorized to act on behalf of a larger population that they represent or delegates implementing the will of constituents who have not had the benefit of deliberation? If the participants were self-selected or selected by the sponsoring agency, rather than being elected by their constituency, are they allowed to compromise or shift their positions due to learning and interaction with other stakeholders (Danielson et al., 2008; Parkinson, 2003)?

15.5.9 How Will Scientific and Technical Information Be Integrated with Lay Knowledge (Including Indigenous Traditional Knowledge)?

Many topics on which participation is pursued involve highly technical issues—groundwater flow models, parts per million of contamination, and so on, which may challenge the abilities of stakeholders without technical backgrounds (Robson et al., 2010). Stakeholders may be presented with educational materials introducing them to the relevant science, and they may have opportunities to ask questions and make analysis requests from experts and may even be given funding to hire their own outside experts to double-check official claims (Busenberg, 2000; Renn, Webler, and Kastenholz, 1998; Renn et al., 1993). Most advocates of participation have a high level of confidence in the public's ability to become educated about complex scientific topics if given the opportunity (Ehrenhalt, 1994).

As noted previously, nonscientific knowledge held by stakeholders may be extremely useful. Lay knowledge can provide longer-term baseline and monitoring information, social context, and local history. Official records may differ greatly from what stakeholders know to be actual practice. Communities may engage with their environment in ways that differ from the assumptions embedded in scientific risk analyses (Holifield, 2012; Wynne, 1989). In some cases, stakeholders' worldviews may be wildly at odds with the basic presuppositions of the scientific perspective—for example, indigenous worldviews that hold that game abundance is a function of divine punishment and reward rather than a function of hunting pressure (Blaeser, 2009; Ebbin, 2011).

Thus, an effective participatory process must have strong mechanisms for soliciting nonscientific knowledge and incorporating it into decision making. Techniques like participatory rural appraisal (Chambers, 1997) and participatory vulnerability scoping (Howe et al., 2013) have been developed as frameworks for letting laypeople define the priorities for policy and further research. Some participatory processes have experimented with developing integrated models of the environmental system that incorporate both scientific knowledge and laypeople's knowledge (including indigenous knowledge) (Serrat-Capdevilla et al., 2009; Walkerden, 2006). The balance between scientific and nonscientific knowledge may be especially politically fraught when indigenous people are major stakeholders, given the long history of nonnatives using science to disparage and erase indigenous perspectives (Lane and Williams, 2009).

15.5.10 What are the Practical Constraints on Participation?

Participation, especially high-quality participation, can be very costly in terms of money, time, and human capital (Rosenbaum, 1978; Steelman and DuMond, 2009). The ideal form of participation may simply be beyond the means of the sponsoring agency. In an era of widespread budget cutbacks, it is unfortunately necessary for those organizing participatory processes to weigh the gains to be made from participation against the costs incurred. Time constraints, such as when responding to an urgent crisis, may also make some forms of participation unfeasible (MacKenzie and Larson, 2010; Vroom and Jago, 1978).

15.5.11 Participation Mechanisms

A great variety of specific mechanisms for conducting participation have been proposed and tried out (Konisky and Beierle, 2001; Rowe and Frewer, 2000, 2005). It is difficult to make a comprehensive list of such mechanisms, because so many have been proposed and new names have frequently been given to substantively similar approaches. Some common mechanisms include the following:

- *Public hearings* are the most common, and most widely derided, participatory mechanism (Checkoway, 1981; Chess and Purcell, 1999). Hearings generally begin with a presentation by the sponsoring agency, followed by a series of questions or statements from the audience. Hearings provide little opportunity for any back-and-forth discussion between stakeholders, and the agenda is tightly circumscribed by the sponsoring agency. Nevertheless, strong public sentiment expressed through hearings may be able to make a difference in agency policy (Rosener, 1982).
- *Workshops or charettes* (Gute and Taylor, 2006; Wood, Good, and Goodwin, 2002) involve stakeholders being invited to brainstorm ideas for dealing with an environmental issue and then refining those ideas into a wish list of suggestions. Workshops may vary in who participates—sometimes they are invitation only in order to focus on the most important stakeholders, while other times they are open to anyone who chooses to come.
- *Community advisory groups* are ongoing panels of selected stakeholders that meet repeatedly through the course of the project to advise decision makers and sometimes to actually hold decision-making power (Branch and Bradbury, 2006; Leach, Pelkey, and Sabatier, 2002; Lynn and Busenberg, 1995). Though the CAG meetings may be open to the public, only a smaller set of official CAG members fully participate in the process. CAGs allow for focused interaction and long-term relationship building among key stakeholders who are taken as representing the interests and views of certain constituencies (Paveglio et al., 2009).
- *Negotiation or mediation* is a closed-door, generally binding, decision-making process. Key stakeholders—typically limited to those with legal standing to file a lawsuit over the resulting decision—are brought together to hash out an agreement that they will all accept and agree not to block (Fiorino, 1995; Striegnitz, 2006).
- *Polls, surveys, interviews, and public comment periods* are a fairly limited form of participation. They involve a one-way transfer of information from stakeholders to the sponsoring agency, with little opportunity for interaction among other stakeholders or real power over the final decision. Nevertheless, they are frequently used due to cost and effort (Darnall and Jolley, 2004; McComas and Scherer, 1999).
- *Deliberative polls* aim to assess what the public's views would be if they were to engage in deliberation (Fishkin, 1995; Fishkin, Luskin, and Jowell, 2000). Standard polls ask for respondents' existing attitudes, which may be shaped by misinformation, a lack of serious thought, or even just made up on the spot. A deliberative poll takes a random sample of the general population, provides them with background information on the issue to be debated, and then guides them through deliberation among the participants. Participants' attitudes are then surveyed at the end of the exercise. Deliberative polls may be done either in person or via the Internet. Generally, they are taken to have only informational or advisory power, not real decision-making power.
- *Citizens' juries* call on a panel of laypeople to question a set of experts and, on the basis of what they hear, produce a decision or plan of action (Hendriks, 2002; Keeney, Von Winterfeldt, and Eppel, 1990). Like juries in the justice system, members of citizens' juries are typically randomly selected and stand in for an undifferentiated "general public." Other stakeholders, such as industry or environmental groups, participate in the

role of experts that may be called on for evidence by the jury. The experts may be chosen by the sponsoring agency, by the jury, or both.

- *Referendums* allow public opinion to be directly binding on a decision. A referendum has a clear appearance of democracy, with everyone in the jurisdiction getting an equal vote. Nevertheless, referendums give no guarantee that a voter has engaged in any systematic thought about the issue (McDaniels and Thomas, 1999; Parkinson, 2001). Moreover, the power to frame the issue by writing the specific text of the ballot question is enormous. For example, Australia's 1999 referendum on ending the monarchy affirmed the status quo despite polls showing majority support for a republic, because the ballot question proposed a presidential appointment process that most supporters of a republic disliked (Higley and Case, 2000).

- *Citizen science projects* aim to involve lay stakeholders in the process of generating scientific information relevant to a decision (Fernandez-Gimenez, Ballard, and Sturtevant, 2008; Robertson and Hull, 2003). Rather than presenting lay stakeholders with expert-produced science to evaluate (as in a citizens' jury model), lay stakeholders are involved in the scientific process itself. They may help to frame the research question, collect data, and interpret the results. Citizen science projects may be one-off responses to a specific problem, or they may involve ongoing monitoring of environmental conditions (Whitelaw et al., 2003).

- *Decentralized resource management*, sometimes called devolution, occurs when a centralized government agency turns over control of a resource (such as a forest or fishery) to a local community (Agrawal, 2005; Logan and Moseley, 2002; Saito-Jensen and Jensen, 2010). This community is typically made up of users of the resource and neighboring residents. Usually, a formalized board or committee will hold power and may use other participatory mechanisms to gain input from a broader public. Instead of full devolution, some areas are managed through comanagement or joint management programs, in which local people and government agencies share responsibility. This is a particularly common system with respect to indigenous people (Izurieta et al., 2011; Kendrick, 2003).

Some research has suggested that the specific institutional form of a participatory process is not as important as the commitment and trustworthiness of the sponsoring agency (Chess and Purcell, 1999; Lynn and Busenberg, 1995), as well as contextual factors outside the sponsoring agency's control (Robson and Kant, 2009; Smiley, de Loë, and Kreutzwiser, 2010). Any institutional mechanism can be manipulated and undercut by a reluctant agency, and any mechanism can become a site for effective participation if the agency is enthusiastic about encouraging participation. What is perhaps most important is that the process is flexible (Bull, Petts, and Evans, 2010; Jakes et al., 2011), so that it can adapt to the particulars of the situation, the needs of stakeholders, and feedback gained through evaluation.

15.6 EVALUATING PARTICIPATION

It is not enough to design a participatory process that appears ex ante appropriate to the situation. Participation requires ongoing evaluation to ensure that it is meeting its goals (Beierle, 1999; Danielson et al., 2012; Santos and Chess, 2003).

Before evaluation begins, it is essential to determine what precisely is being evaluated. There are a wide variety of criteria and goals that an evaluator may wish to measure (Frewer, 2001; Rowe and Frewer, 2004; Webler and Tuler, 2006). Among them are the following:

- *Participant satisfaction*. The simplest (but not necessarily the most important) question for an evaluator to ask is whether the participants are happy with the process and with its outcomes (Charnley and Engelbert, 2005; Coglianese, 2003). It is vital to pay attention not just to whether people are satisfied with the process but also what reasons they have for their dissatisfaction.

- *Process fairness*. This is according to an external or theory-based standard (Laird, 1993; Tuler and Webler, 1999). For example, an evaluator may use Habermas's theory of the ideal speech situation as a yardstick for the quality of the process (Webler, 1995). In this case, the evaluator wishes to see whether the process lived up to a standard that he or she believes defines what is morally or ethically required of good participation.

- *Development of trust and social capital*. One important outcome of participation is often held to be the development of relationships among the participants. These networks can be understood as "social capital," a sort of resource that people can draw on. Interaction between stakeholders can also build trust between them, which can aid in resolving environmental conflicts (Kasperson, Golding, and Tuler, 1992).

- *Conflict resolution and consensus on a decision*. In this case, the evaluator asks whether the process led to a final decision that all participants can agree on or at least accept as legitimate (Creighton, 1991; Kellert et al., 2000). This goal may be relaxed, to call for merely accurately and fully representing all perspectives, and reducing the level of nasty (and even violent) conflict, even if there is no convergence on an agreement (Gregory, 2000; Niemeyer and Dryzek, 2007).

- *Achievement of substantive goals*. Depending on the nature of the decision being made, there may be consensus about certain substantive goals (e.g., "reduce the risk of catastrophic wildfires" or "clean up the toxic waste"), with the dispute primarily about the means to achieve them. Evaluation would ask whether the process has resulted in a plan that will (or, better, has) actually achieved this goal (Kellert et al., 2000; Muñoz-Erickson et al., 2010).

Depending on what the goals of evaluation are, there is a variety of means for conducting the evaluation. Sponsoring agencies and stakeholders generally conduct informal evaluations based on their own perceptions of the process and unstructured conversations with others (Chess, 2000; US EPA, 2001). As useful as this informal evaluation may be, it can be immensely helpful and challenging to conduct formal evaluations using accepted social science methods (Danielson et al., 2012). The evaluator may conduct a quantitative survey, for example, or hold structured focus groups to probe key questions about the effectiveness of the process.

In most cases, it is helpful for evaluation to be overseen by an outside party, such as a university researcher or a private consulting firm. Stakeholders will give more honest feedback to, and put more credence in the results obtained by, an evaluator that they perceive as unbiased. At the same time, the evaluation process itself can be made participatory, with stakeholders involved in selection of criteria, conducting analyses, and interpreting results (Izurieta et al., 2011; Plottu and Plottu, 2009; Syme and Sadler, 1994).

It is not necessary (or even possible) to implement every suggestion made during the course of the evaluation. Nevertheless, the evaluation may reveal weaknesses in the approach and ways that the participatory process is falling short of its goals. Adjusting to the results will strengthen the participatory process going forward (Measham, 2009; Torres and Preskill, 2001).

15.7 CONCLUSION

Participation is, at present, a well-entrenched part of environmental management processes across a wide variety of environmental issues and geographical locations. It is vital for contemporary environmental management professionals to understand the potential benefits and pitfalls of participatory decision-making processes, so that the merits of different approaches may be judged in specific cases. Many fundamental questions about participation remain highly disputed among researchers: questions as basic as "does participation improve the quality of information use?" and "can a participatory process create meaningful equality among different stakeholders—and should it even try?" Careful and context-aware judgment is vital to the success of any participatory process, from the design of the process to formulating the criteria by which its success can be judged.

REFERENCES

Ackerman, B. and J.S. Fishkin. 2005. *Deliberation day*. Yale University Press, New Haven.

Agrawal, A. 2005. *Environmentality: Technologies of government and the making of subjects*. Duke University Press, Durham.

Amin, A. 2005. Local community on trial. *Economy and Society* 34(4): 612–633.

Aristotle. 1984. *Politics*, trans. B. Jowett. Princeton University Press, Princeton. Available at http://classics.mit.edu/Aristotle/politics.html (Accessed March 19, 2015).

Arnstein, S.R. 1969. A ladder of citizen participation. *Journal of the American Institute of Planning* 35(4): 216–224.

Arvai, J.L. 2003. Using risk communication to disclose the outcome of a participatory decision-making process: Effects on the perceived acceptability of risk-policy decisions. *Risk Analysis* 23(2): 281–289.

Arvai, J.L. and A. Froschauer. 2010. Good decisions, bad decisions: The interaction of process and outcome in evaluations of decision quality. *Journal of Risk Research* 13(7): 845–859.

Arvai, J.L., R. Gregory, and T.L. McDaniels. 2001. Testing a structured decision approach: Value-focused thinking for deliberative risk communication. *Risk Analysis* 21(6): 1065–1076.

Barreteau, O., P.W.G. Bots, and K.A. Daniell. 2010. A framework for clarifying "participation" in participatory research to prevent its rejection for the wrong reasons. *Ecology and Society* 15(2): 1. Available at http://www.ecologyandsociety.org/vol15/iss2/art1/ (Accessed March 19, 2015).

Beierle, T.C. 1999. Using social goals to evaluate public participation in environmental decisions. *Policy Studies Review* 16(3/4): 75–103.

Beierle, T.C. 2002. The quality of stakeholder-based decisions. *Risk Analysis* 22(4): 739–749.

Benhabib, S. 1986. *Critique, norm, and utopia: A study of the foundations of critical theory*. Columbia University Press, New York.

Benhabib, S. 1992. *Situating the self: Gender, community and postmodernism in contemporary ethics*. Polity Press, Cambridge.

Billgren, C. and H. Holmén. 2008. Approaching reality: Comparing stakeholder analysis and cultural theory in the context of natural resource management. *Land Use Policy* 25: 550–562.

Blaeser, M. 2009. The threat of the yrmo: The political ontology of a sustainable hunting program. *American Anthropologist* 111(1): 10–20.

Branch, K.M. and J.A. Bradbury. 2006. Comparison of DOE and Army advisory boards: Application of a conceptual framework for evaluating public participation in environmental risk decision making. *Policy Studies Journal* 34(4): 723–753.

Breyer, S. 1993. *Breaking the vicious circle: Toward effective risk regulation*. Harvard University Press, Cambridge.

Brown, M. and L. Knopp. 2008. Queering the map: The productive tensions of colliding epistemologies. *Annals of the Association of American Geographers* 98(1): 40–58.

Bryson, J.M. and S.R. Anderson. 2000. Applying large-group interaction methods in the planning and implementation of major change efforts. *Public Administration Review* 60(2): 143–162.

Buck, J.V. and B.S. Stone. 1981. Citizen involvement in federal planning: Myth and reality. *Journal of Applied Behavioral Science* 17(4): 550–565.

Bull, R., J. Petts, and J. Evans. 2010. The importance of context for effective public engagement: Learning from the governance of waste. *Journal of Environmental Planning and Management* 53(8): 991–1009.

Burns, M.R., J.G. Taylor, and J.T. Hogan. 2008. Integrative healing: The importance of community collaboration in postfire recovery and prefire planning. p. 81–97. *In* Martin, W.E., Raish, C., Kent, B. (eds.), *Wildfire risk: Human perceptions and management implications*. Resources for the Future, Washington, DC.

Busenberg, G.J. 2000. Resources, political support, and citizen participation in environmental policy: A reexamination of conventional wisdom. *Society and Natural Resources* 13(6): 579–587.

Campbell, J.T., T.M. Koontz, and J.E. Bonnell. 2011. Does collaboration promote grass-roots behavior change? Frame adoption of best management practices in two watersheds. *Society and Natural Resources* 24(11): 1127–1141.

Castree, N. 2008. Neoliberalizing nature: The logics of deregulation and reregulation. *Environment and Planning A* 40: 131–152.

Chambers, R. 1997. *Whose reality counts? Putting the first last*. Intermediate Technology Publications, London.

Chambers, S. 2009. Rhetoric and the public sphere: Has deliberative democracy abandoned mass democracy? *Political Theory* 37(3): 323–350.

Charnley, S. and B. Engelbert. 2005. Evaluating public participation in environmental decision-making: EPA's superfund community involvement program. *Journal of Environmental Management* 77(3): 165–182.

Checkoway, B. 1981. The politics of public hearings. *Journal of Applied Behavioral Science* 17(4): 566–582.

Chess, C. 2000. Evaluating environmental public participation: Methodological questions. *Journal of Environmental Planning and Management* 43(6): 769–784.

Chess, C. and B.B. Johnson. 2006. Organizational learning about public participation: "Tiggers" and "Eeyores." *Human Ecology Review* 13(2): 182–192.

Chess, C. and K. Purcell. 1999. Public participation and the environment: Do we know what works? *Environmental Science and Technology* 33(16): 2685–2692.

Chilvers, J. and J. Burgess. 2008. Power relations: The politics of risk and procedure in nuclear waste governance. *Environment and Planning A* 40(8): 1881–1900.

Clarke, L. and J. Agyeman. 2011. Is there more to environmental participation than meets the eye? Understanding agency, empowerment and disempowerment among black and minority ethnic communities. *Area* 43(1): 88–95.

Coggins, G.C. 1999. Regulating federal natural resources: A summary case against devolved collaboration. *Ecology Law Quarterly* 25: 602.

Coglianese, C. 2003. Is satisfaction success? Evaluating public participation in regulatory policymaking. p. 69–86. *In* O'Leary, R., Bingham, L.I. (eds.), *The promise and performance of environmental conflict resolution*. Resources for the Future, Washington, DC.

Cohen, J. and A. Fung. 2004. Radical democracy. *Swiss Political Science Review* 10(4): 23–34.

Cooke, B. 2004. Rules of thumb for participatory change agents. p. 42–55. *In* Hickey, S., Mohan, G. (eds.), *Participation: From tyranny to transformation?* Zed Books, London.

Cooke, B. and U. Kothari. 2001. The case for participation as tyranny. p. 1–15. *In* Cooke, B., Kothari, U. (eds.), *Participation: The new tyranny?* Zed Books, London.

Creighton, J.L. 1991. A comparison of successful and unsuccessful public involvement: A practitioner's viewpoint. p. 135–141. *In* Zervos, C. (ed.), *Risk analysis: prospects and opportunities*. Plenum Press, New York.

Cullen, D., G.J.A. McGee, T.I. Gunton, and J.C. Day. 2010. Collaborative planning in complex stakeholder environments: An evaluation of a two-tiered collaborative planning model. *Society and Natural Resources* 23(4): 332–350.

Cuppen, E., S. Breukers, M. Hisschemöller, and E. Bergsma. 2010. Q methodology to select participants for a stakeholder dialogue on energy option from biomass in the Netherlands. *Ecological Economics* 69: 579–591.

Danielson, S., S.L. Santos, T. Webler, and S.P. Tuler. 2008. Building and breaking a bridge of trust in a Superfund site remediation. *International Journal of Global Environmental Issues* 8(1/2): 45–60.

Danielson, S., S.P. Tuler, S.L. Santos, T. Webler, and C. Chess. 2012. Three tools for evaluating participation: Focus groups, Q method, and surveys. *Environmental Practice* 14(2): 101–109.

Darnall, N. and G.J. Jolley. 2004. Involving the public: When are surveys and stakeholder interviews effective? *Review of Policy Research* 21(4): 581–593.

Davis, N.A. 2011. Broadening participation in fisheries management planning: A tale of two committees. *Society and Natural Resources* 24(2): 103–118.

Day, D. 1997. Citizen participation in the planning process: An essentially contested concept? *Journal of Planning Literature* 11(3): 421–434.

Desai, U. 1989. Public participation in environmental policy implementation: Case of the Surface Mining Control and Reclamation Act. *American Review of Public Administration* 19(1): 49–66.

Dobson, A. 2010. Democracy and nature: Speaking and listening. *Political Studies* 58: 752–768.

Druckman, J.N. 2004. Political preference formation: Competition, deliberation, and the (ir)relevance of framing effects. *American Political Science Review* 98(4): 671–686.

Dryzek, J.S. 1995. Political and ecological communication. *Environmental Politics* 4(4): 13–30.

Dryzek, J.S. and C. List. 2003. Social choice theory and deliberative democracy: a reconciliation. *British Journal of Political Science* 33: 1–28.

Earle, T.C. and M. Siegrist. 2008. On the relation between trust and fairness in environmental risk management. *Risk Analysis* 28(5):1395–414.

Easterling, K.M., M.A. Neblo, and D.M.J. Lazer. 2011. Means, motive, and opportunity in becoming informed about politics: A deliberative field experiment with members of Congress and their constituents. *Public Opinion Quarterly* 75(3): 483–503.

Ebbin, S.A. 2011. The problem with problem definition: Mapping the discursive terrain of conservation in two Pacific salmon management regimes. *Society and Natural Resources* 24(2): 148–164.

Eckersley, R. 1999. The discourse ethic and the problem of representing nature. *Environmental Politics* 8(2): 24–49.

Ehrenhalt, A. 1994. Let the people decide between spinach and broccoli. *Governing Magazine* 7(10): 6.

Eliasoph, N. 1998. *Avoiding politics: How Americans produce apathy in everyday life*. Cambridge University Press, Cambridge.

Farrar, C., J.S. Fishkin, D.P. Green, C. List, R.C. Luskin, and E.L. Paluck. 2010. Disaggregating deliberation's effects: An experiment within a deliberative poll. *British Journal of Political Science* 40: 333–347.

Fernandez-Gimenez, M.E., H.L. Ballard, and V.E. Sturtevant. 2008. Adaptive management and social learning in collaborative and community-based monitoring: A study of five community-based forestry organizations in the western USA. *Ecology and Society* 13(2): 4.

Fiorino, D.J. 1990. Citizen participation and environmental risk: A survey of institutional mechanisms. *Science, Technology, and Human Values* 15(2): 226–243.

Fiorino, D.J. 1995. Regulatory negotiation as a form of public participation. p. 223–237. In Renn, O., Webler, T., Wiedemann, P. M. (eds.), *Fairness and competence in citizen participation: Evaluating models for environmental discourse*. Kluwer, Dordrecht.

First National People of Color Environmental Leadership Summit. 1991. *Principles of environmental justice*. Available at http://www.ejnet.org/ej/principles.html (Accessed March 19, 2015).

Fishkin, J.S. 1995. *The voice of the people*. Yale University Press, New Haven.

Fishkin, J.S. 2006. Strategies of public consultation. *Integrated Assessment Journal* 6(2): 57–72.

Fishkin, J.S., R.C. Luskin, and R. Jowell. 2000. Deliberative polling and public consultation. *Parliamentary Affairs* 53(4): 657–666.

Flyvbjerg, B. 1998. Empowering civil society: Habermas, Foucault and the question of conflict. p. 185–211. In Douglass, M., Friedmann, J. (eds.), *Cities for citizens: Planning and the rise of civil society in a global age*. John Wiley & Sons, Ltd, Chichester.

French, S. and C. Bayley. 2011. Public participation: Comparing approaches. *Journal of Risk Research* 14(2): 241–257.

Frewer, L.J. 2001. Environmental risk, public trust, and perceived exclusion from risk management. *Research in Social Problems and Public Policy* 9: 221–248.

Fung, A. 2003. Recipes for public spheres: Eight institutional design choices and their consequences. *Journal of Political Philosophy* 11(3): 338–367.

Gallagher, D.R. 2009. Advocates for environmental justice: The role of the champion in public participation implementation. *Local Environment* 14(10): 905–916.

Galston, W.A. 1999. Diversity, toleration, and deliberative democracy: Religious minorities and public schooling. p. 39–48. In Macedo, S. (ed.), *Deliberative politics: Essays on Democracy and disagreement*. Oxford University Press, New York.

Genskow, K.D. 2009. Catalyzing collaboration: Wisconsin's agency-initiated basin partnerships. *Environmental Management* 43(3):411–424.

Gooch, M. 2005. Volunteering in catchment management groups: Empowering the volunteer. *Australian Geographer* 35(2): 193–208.

Gregory, R. 2000. Using stakeholder values to make smarter environmental decisions. *Environment* 42(5): 34–44.

Gute, D.M. and M.R. Taylor. 2006. Revitalizing neighbourhoods through sustainable brownfields redevelopment: Principles put into practice in Bridgeport, CT. *Local Environment* 11(5): 537–558.

Gutmann, A. and D. Thompson. 1996. *Democracy and disagreement*. Belknap Press, Cambridge.

Habermas, J. 1981. New social movements. *Telos* 49: 33–37.

Habermas, J. 1984a. *The theory of communicative action, vol. 1: Reason and the rationalization of society*. Beacon Press, Boston.

Habermas, J. 1984b. *The theory of communicative action, vol. 2: Lifeworld and system*. Beacon Press, Boston.

Habermas, J. 1996. *Between facts and norms: Contributions to a discourse theory of law and democracy*. MIT Press, Cambridge.

Hajjar, R.F., R.A. Kozak, and J.L. Innes. 2012. Is decentralization leading to "real" decision-making power for forest-dependent communities? Case studies from Mexico and Brazil. *Ecology and Society* 17(1). Available at http://www.ecologyandsociety.org/vol17/iss1/art12/ (Accessed June 12, 2013).

Halvorsen, K.E. 2006. Critical next steps in research on public meetings and environmental decision making. *Human Ecology Review* 12(2): 150–160.

Hardy, S.D. and T.M. Koontz. 2009. Rules for collaboration: Institutional analysis of group membership and levels of action in watershed partnerships. *Policy Studies Journal* 37(3): 393–414.

Hendriks, C.M. 2002. Institutions of deliberative democratic processes and interest groups: Roles, tensions and incentives. *Australian Journal of Public Administration* 61(1): 64–75.

Hendriks, C.M. 2005. Participatory storylines and their influence on deliberative forums. *Policy Sciences* 38: 1–20.

Hendriks, C.M. and L. Carson. 2008. Can the market help the forum? Negotiating the commercialization of deliberative democracy. *Policy Sciences* 41(4): 293–313.

Henry, L. 2004. Morality, citizenship, and participatory development in an indigenous development association: The case of GPSDO and the Sebat Bet Gurage of Ethiopia. p. 140–155. In Hickey, S., Mohan, G. (eds.), *Participation: From tyranny to transformation?* Zed Books, London.

Heyward, C. 2008. Can the all-affected principle include future persons? Green deliberative democracy and the non-identity problem. *Environmental Politics* 17(4): 625–643.

Hibbing, J. and E. Theiss-Morse. 2002. *Stealth democracy: Americans' beliefs about how government should work*. Cambridge University Press, Cambridge.

Higley, J. and R. Case. 2000. Australia: The politics of becoming a republic. *Journal of Democracy* 11(3): 136–150.

Hildyard, N., P. Hegde, P. Wolvekamp, and S. Reddy. 2001. Pluralism, participation, and power: Joint forest management in

India. p. 56–71. *In* Cooke, B., Kothari, U. (eds.), *Participation: The new tyranny?* Zed Books, London.

Hill, R., C. Grant, M. George, C.J. Robinson, S. Jackson, and N. Abel. 2012. A typology of indigenous engagement in Australian environmental Management: Implications for knowledge integration and social-ecological system sustainability. *Ecology and Society* 17(1). Available at http://www.ecologyandsociety.org/vol17/iss1/art23/ (Accessed June 12, 2013).

Holifield, R. 2004. Neoliberalism and environmental justice in the United States environmental protection agency: Translating policy into managerial practice in hazardous waste remediation. *Geoforum* 35: 285–297.

Holifield, R. 2012. Environmental justice as recognition and participation in risk assessment: Negotiating and translating health risk at a Superfund site in Indian country. *Annals of the Association of American Geographers* 102(3): 591–613.

Howe, P.D., B. Yarnal, A. Coletti, and N.J. Wood. 2013. The participatory vulnerability scoping diagram: Deliberative risk ranking for community water systems. *Annals of the Association of American Geographers* 103(2): 343–352.

Hunt, L.M. and B.L. McFarlane. 2007. Understanding self-evaluations of effectiveness by forestry advisory committee members: A case of Ontario's Local Citizens Committee members. *Journal of Environmental Management* 83: 105–114.

Huntington, H.P., S.F. Trainor, D.C. Natcher, O.H. Huntington, L. DeWilde, and F.S. Chapin III. 2006. The significance of context in community-based research: Understanding discussions about wildfire in Huslia, Alaska. *Ecology and Society* 11(1): 40. Available at www.ecologyandsociety.org/vol11/iss1/art40/ (verified 30 March 2014).

Izurieta, A., B. Sithole, N. Stacey, H. Hunter-Xenie, B. Campbell, P. Donohoe, J. Brown, and L. Wilson. 2011. Developing indicators for monitoring and evaluating joint management effectiveness in protected areas in the Northern Territory, Australia. *Ecology and Society* 16(3) Available at http://www.ecologyandsociety.org/vol16/iss3/art9/ (Accessed January 9, 2012).

Jakes, P.J., K.C. Nelson, S.A. Enzler, S. Burns, A.S. Cheng, V. Sturtevant, D.R. Williams, A. Bujak, R.F. Brummel, S.A. Grayzeck-Souter, and E. Staychock. 2011. Community wildfire protection planning: Is the Healthy Forests Restoration Act's vagueness genius? *International Journal of Wildland Fire* 20: 350–363.

Kahan, D.M. and D. Braman. 2003. More statistics, less persuasion: A cultural theory of gun-risk perceptions. *University of Pennsylvania Law Review* 151: 1291–1327.

Kasperson, R.E. 1986. Six propositions on public participation and their relevance to risk communication. *Risk Analysis* 6(3): 275–281.

Kasperson, R.E., D. Golding, and S.P. Tuler. 1992. Social distrust as a factor in siting hazardous facilities and communicating risks. *Journal of Social Issues* 68: 161–187.

Keeney, R.L., D. Von Winterfeldt, and T. Eppel. 1990. Eliciting public values for complex policy decisions. *Management Science* 36(9): 1011–1030.

Kellert, S.R., J.N. Mehta, S.A. Ebbin, and L.L. Lichtenfeld. 2000. Community natural resource management: Promise, rhetoric, and reality. *Society and Natural Resources* 13: 705–715.

Kendrick, A. 2003. Caribou co-management in Canada: Fostering multiple ways of knowing. p. 241–267. *In* Berkes, F., Colding, J., Folke, C. (eds.), *Navigating social-ecological systems: Building resilience for complexity and change.* Cambridge University Press, Cambridge.

Konisky, D.M. and T.C. Beierle. 2001. Innovations in public participation and environmental decision making: Examples from the Great Lakes region. *Society and Natural Resources* 14(9): 815–826.

Laird, F.N. 1993. Participatory analysis, democracy, and technological decision making. *Science, Technology, and Human Values* 18(3): 341–361.

Lane, M.B. 2005. Public participation in planning: An intellectual history. *Australian Geographer* 36(3): 283–299.

Lane, M.B. and L.J. Williams. 2009. The Natural Heritage Trust and indigenous lands: The trials and tribulations of "new technologies of governance." *Australian Geographer* 40(1): 85–107.

Langton, S. 1978. *Citizen participation in America.* D.C. Heath and Company, Lexington.

Larner, W. and D. Craig. 2005. After neoliberalism? Community activism and local partnerships in Aotearoa New Zealand. *Antipode* 37: 402–424.

Lauria, M. and R.K. Whelan. 1995. Planning theory and political economy: The need for reintegration. *Planning Theory* 14: 8–33.

Leach, W.D., N.W. Pelkey, and P.A. Sabatier. 2002. Stakeholder partnerships as collaborative policymaking: Evaluation criteria applied to watershed management in California and Washington. *Journal of Policy Analysis and Management* 21(4): 645–670.

Levine, P. and R.M. Nierras. 2007. Activists' views of deliberation. *Journal of Public Deliberation* 3(1): 4.

Lloyd, D., P. van Nimwegen, and W.E. Boyd. 2005. Letting indigenous people talk about their country: A case study of cross-cultural (mis)communication in an environmental management planning process. *Geographical Research* 43(4): 406–416.

Lo, A.Y. 2011. Analysis and democracy: The antecedents of the deliberative approach of ecosystems valuation. *Environment and Planning C: Government and Policy* 29: 958–974.

Logan, B.I. and W.G. Moseley. 2002. The political ecology of poverty alleviation in Zimbabwe's Communal Areas Management Programme for Indigenous Resources (CAMPFIRE). *Geoforum* 33: 1–14.

Lynn, F.M. and G. Busenberg. 1995. Citizen advisory committees and public policy: What we know, what's left to discover. *Risk Analysis* 15(2): 147–162.

MacKenzie, B.F. and B.M.H. Larson. 2010. Participation under time constraints: Landowner perceptions of rapid response to the emerald ash borer. *Society and Natural Resources* 23(10): 1013–1022.

Masaki, K. 2004. The "transformative" unfolding of "tyrannical" participation: The corvée tradition and ongoing local politics in western Nepal. p. 125–139. *In* Hickey, S., Mohan, G. (eds.), *Participation: From tyranny to transformation?* Zed Books, London.

McCarthy, J. 2005. Devolution in the woods: Community forestry as hybrid neoliberalism. *Environment and Planning A* 37: 995–1014.

McCarthy, J. 2006. Neoliberalism and the politics of alternatives: Community forestry in British Columbia and the United States. *Annals of the Association of American Geographers* 96(1): 84–104.

McClymont, K. and P. O'Hare. 2008. "We're not NIMBYs!" Contrasting local protest groups with idealized conceptions of sustainable communities. *Local Environment* 13(4): 321–335.

McComas, K.A. and C.W. Scherer. 1999. Providing balanced risk information in surveys used as citizen participation mechanisms. *Society and Natural Resources* 12(2): 107–119.

McDaniels, T. and K. Thomas. 1999. Eliciting preferences for land use alternatives: A structured value referendum with approval voting. *Journal of Policy Analysis and Management* 18(2): 264–280.

Measham, T.G. 2009. Social learning through evaluation: A case study of overcoming constraints for management of dryland salinity. *Environmental Management* 43(3): 1096–1107.

Mendelberg, T. 2002. The deliberative citizen: Theory and evidence. p. 151–193. *In* Delli Carpini, M., Huddy, L., Shapiro, R.Y. (eds.), *Research in micropolitics, volume 6: Political decision making, deliberation and participation*. Elsevier, New York.

Mohan, G. and S. Hickey. 2004. Relocating participation within a radical politics of development: Critical modernism and citizenship. p. 59–74. *In* Hickey, S., Mohan, G. (eds.), *Participation: From tyranny to transformation?* Zed Books, London.

Mosse, D. 1994. Authority, gender and knowledge: Theoretical reflections on the practice of participatory rural appraisal. *Development and Change* 25: 497–526.

Muñoz-Erickson, T.A., B. Aguilar-González, M.R.R. Loeser, and T.D. Sisk. 2010. A framework to evaluate ecological and social outcomes of collaborative management: Lessons from implementation with a northern Arizona collaborative group. *Environmental Management* 45: 132–144.

Nadasdy, P. 2003. Reevaluating the co-management success story. *Arctic* 56(4): 367–380.

Nature. 2000. Benefits of increased public participation. *Nature* 405(6784): 259.

Niemeyer, S. and J.S. Dryzek. 2007. The ends of deliberation: Metaconsensus and intersubjective rationality as ideal outcomes. *Swiss Political Science Review* 13(4): 497–526.

Niemeyer, S., S. Ayirtman, and J. HartzKarp. 2013. Understanding deliberative citizens: The application of Q methodology to deliberation on policy issues. *Operant Subjectivity* 36(2): 114–134.

Ostrom, E. 1990. *Governing the commons: The evolution of institutions for collective action*. Cambridge University Press, Cambridge.

Ostrom, E., T. Dietz, N. Dolšak, P.C. Stern, S. Sonich, and E.U. Weber. 2002. *The drama of the commons*. National Academy Press, Washington, DC.

Parkinson, J. 2001. Deliberative democracy and referendums. p. 131–152. *In* Dowding, K., Hughes, J., Margetts, H. (eds.), *The challenges to democracy: Ideas, involvement and institutions*. Palgrave, London.

Parkinson, J. 2003. Legitimacy problems in deliberative democracy. *Political Studies* 51: 180–196.

Paveglio, T.B., P.J. Jakes, M.S. Carroll, and D.R. Williams. 2009. Understanding social complexity within the wildland-urban interface: A new species of human habitation? *Environmental Management* 43(6): 1085–1095.

Peck, J. and A. Tickell. 2002. Neoliberalizing space. *Antipode* 34: 381–404.

Peterson, M.N., M.J. Peterson, and T.R. Peterson. 2005. Conservation and the myth of consensus. *Conservation Biology* 19(3): 762–767.

Pichardo, N.A. 1997. New social movements: A critical review. *Annual Review of Sociology* 23: 411–430.

Pimbert, M.P. and B. Gujja. 1997. Village voices challenging wetland management policies: Experiences in participatory rural appraisal from India and Pakistan. *Nature and Resources* 33(1): 34–42.

Pini, B. and F. Haslam McKenzie. 2006. Challenging local government notions of community engagement as unnecessary, unwanted, and unproductive: Case studies from rural Australia. *Journal of Environmental Policy and Planning* 8(1): 27–44.

Plottu, B. and E. Plottu. 2009. Approaches to participation in evaluation. *Evaluation* 15(3): 343–359.

Plumwood, V. 1996. Has democracy failed ecology? An ecofeminist perspective. *Environmental Politics* 4(4): 134–168.

Povinelli, E.A. 1995. Do rocks listen? The cultural politics of apprehending Australian Aboriginal labor. *American Anthropologist* 97(3): 505–518.

Predmore, S.A., M.J. Stern, M.J. Mortimer, and D.N. Seesholtz. 2011. Perceptions of legally mandated public involvement processes in the U.S. Forest Service. *Society and Natural Resources* 24(12): 1286–1303.

Prell, C., M. Reed, L. Racin, and K. Hubacek. 2010. Competing structure, competing views: The role of formal and informal social structures in shaping stakeholder perceptions. *Ecology and Society* 15(4): 34.

Ramsey, K. 2008. A call for agonism: GIS and the politics of collaboration. *Environment and Planning A* 40(10): 2346–2363.

Renn, O., T. Webler, H. Rakel, P.C. Dienel, and B.B. Johnson. 1993. Public participation in decision making: A three-step procedure. *Policy Sciences* 26: 189–214.

Renn, O., T. Webler, and H. Kastenholz. 1998. Procedural and substantive fairness in landfill siting: A Swiss case study. p. 253–270. *In* Löfstedt, R., Frewer, L. (eds.), *The Earthscan reader in risk and modern society*. Earthscan Publications, London.

Roberts, R. 1995. Public involvement: From consultation to participation. p. 221–246. *In* Vanclay, F., Bronstein, D.A. (eds.), *Environmental and social impact assessment*. John Wiley & Sons, Ltd, Chichester.

Robertson, D.P. and R.B. Hull. 2003. Public ecology: An environmental science and policy for global society. *Environmental Science and Policy* 6: 399–410.

Robson, M. and S. Kant. 2009. The influence of context on deliberation and cooperation in community-based forest management in Ontario, Canada. *Human Ecology* 37(5): 547–558.

Robson, M., J. Rosenthal, R.H. Lemelin, L.M. Hunt, N. McIntyre, and J. Moore. 2010. Information complexity as a constraint to public involvement in sustainable forest management. *Society and Natural Resources* 23(12): 1150–1169.

Rosenbaum, N.M. 1978. Citizen participation and democratic theory. p. 43–54. *In Citizen participation in America*. D.C. Heath and Company, Lexington.

Rosener, J.B. 1982. Making bureaucrats responsive: A study of the impact of citizen participation and staff recommendations on regulatory decision making. *Public Administration Review* 42(4): 339–345.

Rowe, G. and L.J. Frewer. 2000. Public participation methods: A framework for evaluation. *Science, Technology, and Human Values* 25(1): 3–29.

Rowe, G. and L. Frewer. 2004. Evaluating public-participation exercises: A research agenda. *Science, Technology, and Human Values* 29: 512–556.

Rowe, G. and L. Frewer. 2005. A typology of public engagement mechanisms. *Science, Technology, and Human Values* 30: 251–290.

Rowe, G. and J.G. Gammack. 2004. Promise and perils of electronic public engagement. *Science and Public Policy* 31(1): 39–54.

Rutherford, M.B., M.L. Gibeau, S.G. Clark, and E.C. Chamberlain. 2009. Interdisciplinary problem solving workshops for grizzly bear conservation in Banff National Park, Canada. *Policy Sciences* 42: 163–187.

Saito-Jensen, M. and C. Jensen. 2010. Rearranging social space: Boundary-making and boundary-work in a joint forest management project, Andhra Pradesh, India. *Conservation and Society* 8(3): 196.

Sanders, L.M. 1997. Against deliberation. *Political Theory* 25(3): 347–376.

Santos, S.L. and C. Chess. 2003. Evaluating citizen advisory boards: The importance of theory and participant-based criteria and practical implications. *Risk Analysis* 23(2): 269–279.

Scott, J.C. 1998. *Seeing like a state: How certain schemes to improve the human condition have failed*. Yale University Press, New Haven.

Serrat-Capdevilla, A., A. Browning-Allen, K. Lansey, T. Finan, and J.B. Valdés. 2009. Increasing social–ecological resilience by placing science at the decision table: The role of the San Pedro Basin (Arizona) decision-support system model. *Ecology and Society* 14(1): 37.

Shapiro, I. 1999. Enough of deliberation: Politics is about interests and power. p. 28–38. *In* Macedo, S. (ed.), *Deliberative politics: Essays on democracy and disagreement*. Oxford University Press, New York.

Simon, W.H. 1999. Three limitations of deliberative democracy: Identity politics, bad faith, and indeterminacy. p. 49–57. *In* Macedo, S. (ed.), *Deliberative politics: essays on democracy and disagreement*. Oxford University Press, New York.

Sletto, B.I. 2009. "We drew what we imagined": Participatory mapping, performance, and the arts of landscape making. *Current Anthropology* 50(4): 443–476.

Smiley, S., R. de Loë, and R. Kreutzwiser. 2010. Appropriate public involvement in local environmental governance: A framework and case study. *Society and Natural Resources* 23(11): 1043–1059.

Sneddon, C. and C. Fox. 2008. Struggles over dams as struggles for justice: The World Commission on Dams (WCD) and anti-dam campaigns in Thailand and Mozambique. *Society and Natural Resources* 21(7): 625–640.

Somin, I. 2013. *Democracy and political ignorance: Why smaller government is smarter*. Stanford University Press, Stanford.

Stagl, S. 2006. Multicriteria evaluation and public participation: The case of UK energy policy. *Land Use Policy* 23: 53–62.

Steelman, T.A. and M.E. DuMond. 2009. Serving the common interest in U.S. forest policy: A case study of the Healthy Forests Restoration Act. *Environmental Management* 43(3): 396–410.

Stirling, A. 2006. Analysis, participation and power: Justification and closure in participatory multi-criteria analysis. *Land Use Policy* 23: 95–107.

Stone, C.D. 1987. *Earth and other ethics: The case for moral pluralism*. Harper and Row, New York.

Striegnitz, M. 2006. Conflicts over coastal protection in a National Park: Mediation and negotiated law making. *Land Use Policy* 23: 26–33.

Sultana, F. 2009. Community and participation in water resources management: Gendering and naturing development debates from Bangladesh. *Transactions of the Institute of British Geographers* 34: 346–363.

Sunstein, C.R. 2005. Group judgments: Statistical means, deliberation, and information markets. *New York University Law Review* 80: 962.

Syme, G.J. and B.S. Sadler. 1994. Evaluation of public involvement in water resources planning: A researcher-practitioner dialogue. *Evaluation Review* 18(5): 523–542.

Sze, J., J. London, F. Shilling, G. Gambirazzio, T. Filan, and M. Cadenasso. 2009. Defining and contesting environmental justice: Socio-natures and the politics of scale in the Delta. *Antipode* 41(4): 807–843.

Taddei, R. 2011. Watered-down democratization: Modernization versus social participation in water management in Northeast Brazil. *Agriculture and Human Values* 28(1): 109–121.

Tauxe, C.S. 1995. Marginalizing public participation in local planning: An ethnographic account. *Journal of the American Planning Association* 61(4): 471–482.

Torgerson, D. 1986. Between knowledge and politics: Three faces of policy analysis. *Policy Sciences* 19: 33–59.

Torres, R.T. and H. Preskill. 2001. Evaluation and organizational learning: Past, present, and future. *American Journal of Evaluation* 22(3).

Tuler, S.P. 2000. Forms of talk in policy dialogue: distinguishing between adversarial and collaborative discourse. *Journal of Risk Research* 3(1): 1–17.

Tuler, S.P. and T. Webler. 1999. Voices from the forest: What participants expect of a public participation process. *Society and Natural Resources* 12: 437–453.

US EPA. 2001. Stakeholder involvement and public participation at the U.S. EPA: Lessons learned, barriers, and innovative approaches. United States Environmental Protection Agency. Available at http://www.epa.gov/stakeholders/pdf/sipp.pdf (Accessed June 8, 2010).

Van den Hove, S. 2006. Between consensus and compromise: Acknowledging the negotiation dimension in participatory approaches. *Land Use Policy* 23: 10–17.

Vroom, V.H. and A.G. Jago. 1978. On the validity of the Vroom-Yetton model. *Journal of Applied Psychology* 63(2): 151–162.

Waddington, M. and G. Mohan. 2004. Failing forward: Going beyond PRA and imposed forms of participation. p. 219–234. *In* Hickey, S., Mohan, G. (eds.), *Participation: From tyranny to transformation?* Zed Books, London.

Wagner, C.L. and M.E. Fernandez-Gimenez. 2008. Does community-based collaborative resource management increase social capital? *Society and Natural Resources* 21(4): 324–344.

Walkerden, G. 2006. Adaptive management planning projects as conflict resolution processes. *Ecology and Society* 11(1): 48.

Webler, T. 1995. "Right" discourse in citizen participation: An evaluative yardstick. p. 35–86. *In* Renn, O., Webler, T., Wiedemann, P.M. (eds.), *Fairness and competence in citizen participation: evaluating models for environmental discourse.* Kluwer, Dordrecht.

Webler, T. and S.P. Tuler. 2006. Four perspectives on public participation process in environmental assessment and decision making: Combined results from 10 case studies. *Policy Studies Journal* 34(4): 699–722.

Weeks, E.C. 2000. The practice of deliberative democracy: Results from four large-scale trials. *Public Administration Review* 60(4): 360–372.

Whitelaw, G., H. Vaughan, B. Craig, and D. Atkinson. 2003. Establishing the Canadian Community Monitoring Network. *Environmental Monitoring and Assessment* 88: 409–418.

Wood, N.J., J.W. Good, and R.F. Goodwin. 2002. Vulnerability assessment of a port and harbor community to earthquake and tsunami hazards: Integrating technical expert and stakeholder input. *Natural Hazards Review* 3(4): 148–157.

Wynne, B. 1989. Sheepfarming after Chernobyl: A case study in communicating scientific information. *Environment* 31(2): 10–15, 33–39.

Young, I.M. 1995. Communication and the other: Beyond deliberative democracy. p. 134–152. *In* Wilson, M., Yeatman, A. (eds.), *Justice and identity: antipodean practices.* Allen and Unwin, St. Leonards.

Young, I.M. 1997. Asymmetrical reciprocity: On moral respect, wonder, and enlarged thought. *Constellations* 3(3): 340–363.

Young, I.M. 2001. Activist challenges to deliberative democracy. *Political Theory* 29(5): 670–690.

Yung, L., M.E. Patterson, and W.A. Freimund. 2010. Rural community views on the role of local and extralocal interests in public lands governance. *Society and Natural Resources* 23(12): 1170–1186.

16

STATISTICS IN ENVIRONMENTAL MANAGEMENT

JENNIFER A. BROWN
School of Mathematics and Statistics, University of Canterbury, Christchurch, New Zealand

Abstract: One of the features underpinning good quality environmental management is that it is based on reliable, accurate, and timely information. Statistics in environmental management is a key part of a good management system. The role of statistics in environmental management is to supply information on the ecosystem and the effects of human interactions and impacts, so that appropriate management decisions can be made. Statistics is associated with how relevant data are collected, summarized, analyzed, and reported.

In this chapter, two broad areas of environmental statistics are introduced: how to collect data and how to assess trends. Data collection can be considered the fundamental part of statistics in environmental management. Data that are collected should be a good representation of the population, and we discuss some of the many data collection designs that can be used. These include simple random sampling, stratified sampling, designs with spatial balance (GRTS and BAS), and adaptive sampling. The ability to detect change and trends from data collected from environmental systems is discussed. Changes and trends can be displayed in graphics for a visual representation. We discuss a method to analyze data called linear mixed model that can be used to assess trend.

Keywords: survey sampling, simple random sampling, stratified sampling, sample mean, sample variance, sample precision, adaptive cluster sampling, spatially balanced design, generalized random tessellation stratified design, balanced acceptance sampling, detecting change and trend, detecting change, linear mixed models, repeated surveys, random effects, fixed effects, model comparison, display of the data, adequacy of the model, normal distribution.

16.1 Introduction	383	16.3 Detecting Change 390
16.2 Data Collection	384	16.3.1 Models For Detecting Change and Trends 390
16.2.1 Simple Random Sampling	384	16.4 Summary 394
16.2.2 Stratified Sampling	386	References 395
16.2.3 Other Sampling Designs	387	

16.1 INTRODUCTION

Environmental management, by its very nature, is a challenging task. Ecosystems, and of more relevance, the interactions and impacts of humans with them, are very complex. One characteristic of a reliable, accurate, and timely management system is that it is based on reliable, accurate, and timely information. This is the role of statistics in environmental management, to ensure that information on the ecosystem, the human interactions, and impacts are conveyed into the management system, and that there is an appropriate feedback system to ensure management practices and systems improve over time.

There are three key roles for statistics in environmental management (Manly, 2008). These roles are (i) to provide baseline information on the current status of the environment, (ii) to assess change from a planned or unplanned perturbation (e.g., a new method to control a pest species or an oil spill), and (iii) to provide regular updates on the change in status and trend in the environment. Although these three roles have quite different purposes and objectives, the underlying statistical theme is to provide information. This information will be used in reports and other documents produced by environmental managers and read by other agencies and the public. This level of scrutiny means that statistical methods in environmental management need to be robust and defendable.

An Integrated Approach to Environmental Management, First Edition. Edited by Dibyendu Sarkar, Rupali Datta, Avinandan Mukherjee, and Robyn Hannigan.
© 2016 John Wiley & Sons, Inc. Published 2016 by John Wiley & Sons, Inc.

In this chapter, we review some statistical ideas as applied to environmental science and discuss the role of data collection and trend analysis.

16.2 DATA COLLECTION

16.2.1 Simple Random Sampling

One of the natural tendencies when faced with the task of initiating an environmental management program is to set up a grand field design and collect a lot of data. While the intentions seem good, a well designed and planned field program of data collection will not only be more cost effective but it will provide useful information. In any monitoring design, the aims and objectives need to be clearly stated, both in a temporal and a spatial scale. These aims and objectives guide development of the survey design and data analysis (Gilbert, 1987).

A census is when all units in a population are measured. It is rare that there is sufficient time, money, or interest in conducting a census and more commonly a selection of units from a population is chosen. Only the selected units are measured. The individual elements from a population that are selected are called a sample. A sample could be the selection of people, of trees, or of the lakes and wetlands.

The survey design refers to the way the sample is selected. One of the important features of a survey design is the size of the sample that will be selected. Conducting a survey costs time and money so the size of the sample cannot be excessive. The data collected from the survey is summarized by statistics—the most common summary statistic is the sample mean. These summary statistics are used to estimate population parameters. The sample mean is an estimate of the mean of the population. It is natural that the manager would want a reliable estimate of the population mean, so choosing the best sample scheme and the most appropriate method to estimate population parameters is important.

There are many different survey designs that can be used to collect data. Whatever design is used the sample should be representative of the population and provide information relevant to the survey objective. Defining what is meant by the population in an environmental management context may not be straightforward. For biologists, the term population is usually associated with a species of interest, for example, a bird population. However, in statistical terminology, a population is the collection of units from which a sample is drawn (Thompson, 2002). It helps to think first about how data will be collected in the field and what measurements will be taken. Consider an example where information on the status of an invasive grass species is required. The biological population is clearly the grass species, but the statistical population is best thought of as the entire collection of units that could be sampled. Field surveys for this grass are usually by counting the number of grass plants, or coverage, inside 1-m^2 quadrates. The statistical population is therefore the total number of all the quadrates that could be placed within the study area. From this population, a subset, or sample, will be chosen. Within each sample unit, a measurement will be made, for example, what area within the quadrate is covered by the grass species or, if possible, the count of plants within the quadrate. Often with spreading grasses, identifying individual plants can be difficult so coverage is used as a surrogate of grass density.

As another example, 5-min bird counts of bird calls have been successfully used in New Zealand in environmental management (Hartley, 2012). The measurement is the count of bird calls of the species of interest heard within the 5-min period, the sample unit is the 5-min period and the population is all the possible 5-min periods that could have been selected in the sample (e.g., during daylight hours). With this example, there is quite a difference between the biological population (the bird species of interest) and the statistical population (the collection of sample units), and loose terminology can only lead to confusion.

It is very relevant at this stage to pause and think about the idea of a target population and a sample frame. Again, these terms can be difficult at first glance but careful attention to them avoids confusion later. The target population is the statistical population of interest, for example, the quadrates in the study area, keeping in mind that in this example, we measure the grass cover within each quadrate. The sample frame is the list or map of the quadrates that can possibly be visited and measured. Imagine that some of the study area is near the top of a steep cliff and is very inaccessible, and within the study area, there are wet, boggy areas. The sample frame is all the quadrates that could be visited and will not include the inaccessible quadrates. The quadrates in the wet, boggy area may well be selected in the sample, but when the field crew travel to the site to measure them, they chose not to (despite strict protocols). These quadrates are now not in the frame, although this was not known prior to sampling. Restrictions on access to private land can also lead to the frame being different from the target population.

In Fig. 16.1, another problem can be seen with the frame not matching the target population. Continuing with the grass example, the study area is 137,856 m^2 and has 1,378,561-m^2 quadrates overlain using a mapping software package. Figure 16.1 shows a small part of the study area. Notice the lower left hand corner, some quadrates extend outside the boundary of the study area due to the shape. If these quadrates were discarded from a sample, the sample frame would cover a smaller area than target population, yet if they were included, the frame would cover a larger area. This "edge effect" can be a serious issue for study areas that are very irregularly shaped.

The sample frame is used as the starting point to select a sample. Assume there are $N = 137,856$ quadrates in the study area. Each quadrate has a number, 1, 2, 3, ... 137,856. A sample of $n = 50$ units is required, and once the 50 quadrates

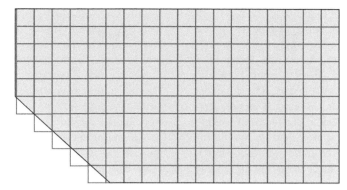

FIGURE 16.1 A portion of a large study area overlain with 1-m² quadrates. The study area is shaded, and in the lower left corner, the quadrates extend outside the study area.

are selected, each is visited and the cover of grass measured. The order the quadrates are visited in this example is not important as long as all 50 are visited and all within a relatively close time (e.g., with a day). To select the 50 quadrates, the simplest way is to draw 50 numbers, at random and without replacement, between the numbers 1 and 137,856. This can be done in a spreadsheet package or with a random number generator often found in software packages.

This sample selection method is the simplest design and is called simple random sampling. There are some important features of this design. It is a probability sampling design where the random numbers, and probability, has been used to select the sample. Nonprobability sampling would be where you subjectively chose a selection of 50 quadrates because they appeared to give a good representative sample. Probability sampling is the more preferred method.

Another feature is that the sample was selected with equal probability sampling, in other words, each possible sample of 50 quadrates was equally likely to be selected. In this example, there are too many possible samples to record, but consider that the sample of quadrates numbered 1, 2, 3, ..., 50 is as equally likely to be selected as a sample of the last 50 quadrates or indeed 50 quadrates spread roughly evenly between quadrate 1 and quadrate 137,856. Selecting a sample of quadrates 1–50 from a possible collection of 137,856 quadrates would be statistically acceptable, but as a field biologist, you may not be that comfortable about whether it were a good representation or not of the entire population. Imagine in Fig. 16.1 there was an underlying environmental trend, such as soil pH and the grasses preferred acidic soils. The soils in the upper left corner were quite acidic, and the pH increased trending toward the lower right corner. A sample that was selected from sequential quadrates in the upper left corner would not be a very good representation of the entire area because all the quadrates would have high grass cover, whereas in the field, there were many sections of the area with low cover. We discuss a few designs later that can be used to improve this.

The final feature to mention is that the sample is drawn without replacement, so that a quadrate can only appear once in the sample. The reason for this is that having 50 unique quadrates in the sample will be more informative than having the same quadrate being measured twice (or more). However, in some complex designs, sampling with replacement may be assumed to assist with the analysis. When the total frame size is far larger than the sample size, even sampling with replacement may have little real effect on having the same sample unit appear more than once.

The sample mean, $\bar{y}$, is the estimate of the population mean, μ. In this example, μ is the average grass cover per square meter for the study area:

$$\bar{y} = \frac{1}{n}\sum_{i=1}^{n} y_i = \frac{1}{n}(y_1 + y_2 + \cdots + y_n)$$

where y_i is the value (the measure of grass cover) for the ith quadrate.

The sample variance is estimated as

$$s^2 = \frac{1}{n-1}\sum_{i=1}^{n}(y_i - \bar{y})^2.$$

The estimator of the variance of the sample mean is

$$\hat{\text{var}}(\bar{y}) = \left(\frac{N-n}{N}\right)\frac{s^2}{n}.$$

This statistic, the estimated variance of the sample mean, is important to think about in designing and planning a survey. This statistic is used in calculating the confidence interval for a sample estimate (e.g., the estimate of the population mean). In this case, it is a measure of how variable, or different, all the possible sample means are. You would normally have only one sample, but thinking as a statistician, there are a finite, but very large, number of possible samples of size 50 that can be drawn from your population of 137,856 quadrates. The variance of the sample mean is a measure of how variable these are. In reality, only one sample is taken so we can only estimate this.

The estimator of the variance of the sample mean is often referred to as the sample precision, where a precise sample is a desirable aim. There are two ways to improve precision—by taking a large sample and by improving the design. The question of how large a sample is needed is probably the most common one asked of a statistician. The quickest answer is to take as large a sample as you can. There are various formulae you can use to estimate the required sample size. These use prior estimates, or approximations, of within-sample variation and for the desired precision to be stated. See Thompson (2002) and Scheaffer et al. (2011) for details.

The other component of the equation for the estimator of the variance of the sample mean is the term $((N-n)/N)$, which is often referred to as the finite population correction factor and allows for the increase in precision from sampling a large fraction of the population. The easiest way to understand this is to compare a sample of size 50 drawn from a population of 100 units. The finite population correction factor would be $((100-50)/100) = 1/2$. The effect of this is to reduce the value of the estimator. If the sample of 50 had been drawn from a population of 100,000 units, then only a small fraction of the population (0.05%) would have been sampled. The sample fraction for a sample size of 50 out of a population of size 100 is 50%, and this large fraction will improve precision.

16.2.2 Stratified Sampling

Stratified sampling is a good example of a design, where if planned well, can improve sample precision. The idea of stratified random sampling is that the population is divided into groups called strata, and these groupings are made so that the population units within a group are similar. Within each stratum, a sample is taken. The variance of the sample estimator is the weighted sum of the within-stratum variances. Because the groupings have been made so that units within a stratum are similar, strata should be less variable than the population as a whole.

Consider an example of surveying cockles on the SW coast of France. The abundance of cockles varies considerably over the estuarine area, but within sections, the cockle numbers are more similar. In the sandy flat areas, there are high numbers of cockles; in the deeper water channels, there are few cockles; and when the substrate is muddy, cockle numbers are at a medium level. Surveying cockles involves taking grab samples from the estuary from a boat where each grab is the sample unit (remember the count of cockles is the measurement from the sample unit). A simple random sample design would mean that the location of grabs was randomly placed anywhere in the estuary. In a stratified design, the area can be marked on a map as separate strata, and within each stratum, the location of the grabs is randomly placed. This has two advantages: Firstly, it means that every stratum has some survey effort so there are some statistics for each subarea (strata) of the estuary. And secondly, as long as the strata boundaries are placed in a sensible way to group areas of similar abundance, the sample estimate should be more precise than with simple random sampling. There is a third advantage, which is very practical in nature, in that the strata can be used to plan and structure the fieldwork (e.g., visit one stratum a day).

An important feature of stratified sampling to realize improvements in precision over simple random sampling is to have an understanding of how survey effort among strata should be allocated. The simplest method is to allocate effort relative to the size of each stratum.

In Fig. 16.2, the three strata are quite different in size, $N_1 = 200$, $N_2 = 150$, and $N_3 = 450$, where N_h is the total size of the stratum based on the number of sample units that would fit within it and $N = \sum_{h=1}^{L} n_h$. If the desired sample size was $n = 50$ and with

$$n_h = n \cdot \frac{N_h}{N},$$

then $n_1 = 13$, $n_2 = 9$, and $n_3 = 28$. There would be 13 quadrates randomly placed in the sandy areas, 9 in the water channels, and 28 in the mud areas. Notice in Fig. 16.2 that a stratum does not have to be a contiguous area, and the sand stratum is spread over both sides of the water channel.

There are other methods to allocate effort to improve precision by putting more effort into the more variable strata (and more into the larger strata). In the cockle example, stratum 3 had the highest sample variance. Increasing the sample size within this stratum and reducing the sample size in the strata with lower variance would improve precision.

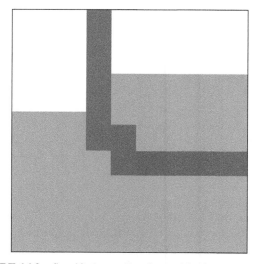

FIGURE 16.2 Stratified sampling for cockles in an estuary. The sandy areas have highest cockle densities (light areas), the muddy areas have medium densities (dark areas), and the water channels (very dark areas) have very few cockles.

Further improvements can be made by allocating more effort to the strata that have the lowest cost per unit effort. Putting all three of these terms together (stratum size, N_h, variance, σ_h^2, and sample costs, c_h) is called optimal allocation:

$$n_h = n \cdot \frac{N_h \sigma_h / \sqrt{c_h}}{\sum_{h=1}^{L} N_h \sigma_h / \sqrt{c_h}}.$$

The result of optimal allocation is that strata that are variable, large, and proportionally cheap to survey are allocated larger sample sizes compared with strata that have low variance, are small, and are expensive to survey. In our example, assuming the cost of surveying a unit was the same for all strata and that the within-stratum variances were $\sigma_1^2 = 4.5$, $\sigma_2^2 = 2.3$, and $\sigma_3^2 = 1.9$, the optimal allocation for $n = 50$ among the three strata would be then $n_1 = 22$, $n_2 = 8$, and $n_3 = 20$. Notice how in the first stratum, despite the fact it is not the largest, the most survey effort is allocated to it. This is because it was the stratum with the highest variance. In most situations, the within-stratum variance won't be known exactly, but data from previous studies or similar studies may be available.

Stratum boundaries can be defined in any way, but it makes sense to have the boundaries related in some way to different levels within the population. The idea behind stratified sampling is that units within strata should be as similar as possible so, for example, in the cockle survey, the strata were defined so that there were areas of low, medium, and high densities. The strata were defined based on information related to the underlying substrate, water depth, and currents, information that was available from maps. The assumption was that the cockle density was related in some way to these features.

16.2.3 Other Sampling Designs

One of the features of simple random sampling discussed previously is that the sample is selected with equal probability sampling; in other words, each possible sample of 50 quadrates is equally likely to be selected. What this means in practice is that in Fig. 16.1, a sample of the quadrates numbered 1, 2, 3, ..., 50 is just as likely to be chosen as a sample where the $n = 50$ quadrates are more spread over the possible 137,856 quadrates. In environmental sampling where the population of interest is related in some way to underlying environmental processes, having sample effort spread over the entire study area would be viewed more favorably than a survey where effort is concentrated in 50 adjacent quadrates. For environmental monitoring designs where there may be multiple objectives, and objectives may change, having a survey with good spatial coverage can be advantageous (McDonald, 2012). The idea of trying to ensure sample effort is spread over different parts of the environmental system is part of the motivation for stratified sampling. Stratified sampling ensures survey effort is allocated to separate strata. In the cockle example, the stratified design ensured some surveying was undertaken in all three strata: the deeper water channels, the sandy flat areas, and the mud areas.

Systematic sampling is another design that can be used to ensure the survey is spread across the underlying environmental system. For example, in a survey for roadside weeds, sample unit could be placed every 2 km to ensure a fairly even coverage of the road. The downside with this design is that information about variability among the sample units is restricted to a minimum interval of 2 km. Another way to achieve a sample that is spread across the environmental domain is to use a spatially balanced design where the study area is divided into relatively small strata and one or two units sampled from within each. A generalized random tessellation stratified (GRTS) design (Stevens and Olsen, 2004) uses an algorithm based on tessellations to spread survey effort over two-dimension space (usually a map where the two dimensions are longitude and latitude) or along a linear feature such as a river system. An invertible mapping technique converts the two-dimension space into one dimension. Then, within the one-dimension space, a systematic sample is taken, and once converted back to two-dimension space, the resultant sample is remarkably well spread over the map and without any obvious regularity in spacing. The concept of spatial balance is to have fairly even sample effort across the map while not going so far as to have a fixed regular and even distance between adjacent sample units. It is worth noting as an aside here that unequal probability sampling where sample intensity is increased in some areas and decreased in others can be accommodated in a GRTS design.

A development on the GRTS design is to extend the concept of balance beyond two dimensions and into n-dimensions. In balanced acceptance sampling (BAS) (Robertson et al., 2013), survey effort can be spread across multiple dimensions, for example, across geographic space (longitude and latitude) and along environmental gradients (e.g., soil pH), biological gradients (e.g., population genetics), and gradients related to species risk (e.g., conservation status). Any dimension can be used and the BAS design ensures survey effort is balanced. The easiest way to conceptualize this is to think in three dimensions, where two axes of a cube are the longitude and latitude of the grassland study area, and the height of the cube (the third dimension) is months in a year (Fig. 16.3). A BAS design would allow survey effort to be balanced across the geographic space and, at the same time, ensure survey effort within the year was balanced so that no one part of the map received all the survey effort in 1 month. In Fig. 16.3, each layer represents the geographic location of the survey sites in a year. Each layer has balance. Then, looking down through the layers, there is balance across time for any section of the geographic

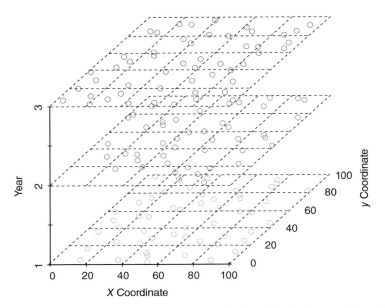

FIGURE 16.3 An illustration of a survey design that is balanced in both geographic and temporal space. The x- and y-coordinate axes are longitude and latitude (geographic space), and the third axis is time measured in years (temporal space).

space. This means that no one section of the study area is only surveyed in the early years and another section only surveyed in the later years. The figure only shows 3 years but if the design were to continue for longer, for example, 20 years, having the balance through time is important.

It is harder to conceptualize, but a fourth dimension can be added to accommodate, for example, underlying gradients in soil moisture so that survey effort was balanced in geographic space, in time, and in this environmental gradient. A fifth dimension could be added to reflect the population structure to balance effort across grass subspecies and so on. Again, a BAS design can readily accommodate unequal probability sampling, in addition to sampling across n-dimensions.

One final area to discuss here about survey sampling designs is adaptive sampling. Adaptive sampling refers to sampling designs where the protocol for data collection changes, evolves, or adapts during the course of the survey. One of the best-known adaptive designs is adaptive cluster sampling introduced by Thompson (1990). In environmental sampling, often the population of interest is clustered and rare. These two terms are widely used now to describe populations such as an invasive weed—the weed is recently invaded and is not that common (it is rare), but where it does occur, it is in a clump (it is clustered). The occurrence of an infectious disease in humans is a classic example of a clustered and rare population—few have the disease, but when someone does those in close contact with them may also be infected. A certain type of pollutant in soils in a large-scale landscape can be an example of a rare and clustered population—most of the soil is not polluted (i.e., polluted soils are rare), but when a small sample of soil is polluted, the surrounding soil is also polluted.

In these situations, a useful design is adaptive cluster sampling, or one of the many alternatives. See Brown (2011) for a recent review of adaptive designs. Adaptive cluster sampling starts with a simple random sample. Each selected sample unit is inspected. If any of the units in this initial sample meet or exceed a preset threshold, $y_i \geq C$, then neighboring units are sampled. This threshold may be something as simple as $C=1$, meaning if the sample unit has a value of 1 or more, the neighboring units are sampled. In the grassland example, a count of 1 grass plant or more (more simply if the grass is present), the neighboring units are sampled. Surveying continues and if any of these neighboring units meet this condition, their neighboring units are selected and so on. Sampling continues, in this example, as more and more of the grass-occupied sample units are found until no more occupied units are detected. This sequential process of surveying extra sample units means that once a cluster is found, the cluster size and shape can be mapped. The true boundary of a cluster may not be mapped because the sequential path traveled depends on the way the neighborhood is searched. For example, defining the neighborhood as the quadrates to the north and south will result in quite different clusters compared with when the neighborhood is defined as the eight surrounding quadrates (Brown, 2003). Another factor that will affect what is a cluster is the preset threshold value. For a very low density population where most of the quadrate counts are 0 or 1, having a threshold value of $y_i \geq 2$ will mean there will be few occasions when adaptive selection of neighboring units is initiated.

In Fig. 16.4, the population is considered rare and clustered. Most of the quadrates are unoccupied, but when one is occupied, the surrounding quadrates appear to be

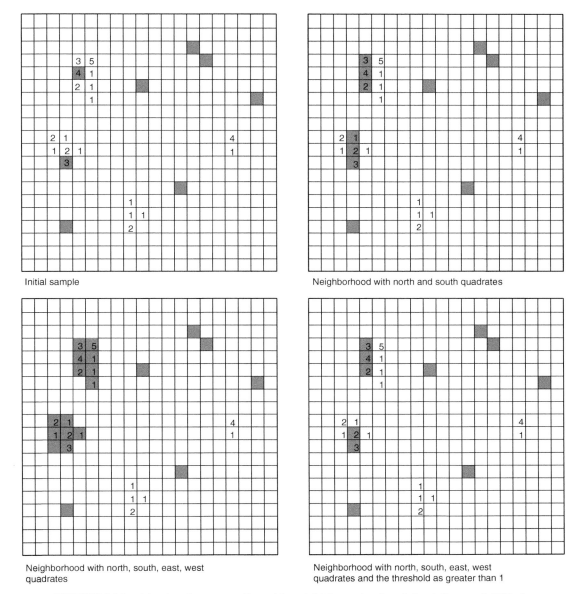

FIGURE 16.4 Adaptive cluster sampling with an initial sample of $n=8$ (top left corner). With the neighborhood defined as the north and south quadrates $n=12$ (top right corner), and $n=19$ when the neighborhood defined as the north, south, east, and west quadrates (lower left corner). With this neighborhood definition and a threshold value of greater than 1, $n=12$.

occupied. The population has four clusters, and in the initial phase of the adaptive cluster sample, two are detected (upper left corner of Fig. 16.4). The initial sample is a random sample with $n_0 = 8$. When the neighborhood is defined as the quadrates to the north and south of the one that triggered adaptive selection, the final sample size is $n=12$ (upper right corner of Fig. 16.4). With a larger neighborhood definition with north, south, east, and west quadrates searched, the final sample size is $n=19$ (lower left corner of Fig. 16.4). The shape of the network of detected quadrates within the cluster is quite different. Notice that when the threshold value is increased from $y_i \geq 1$ to $y_i \geq 2$, once again the final sample size is different, $n=12$. The shape of the network of the detected quadrates within the cluster has also changed (lower right corner of Fig. 16.4).

The idea of adaptive cluster sampling is very attractive to practicing field scientists, because once a rare plant, animal, or pollutant has been found, the natural tendency is to search in the immediate neighborhood. This design takes this intuitive behavior and puts it in the framework of probability sampling. A strict protocol of how to search for sample units that meet or exceed the threshold condition must be followed. This protocol is important to allow the probability of including a unit in the sample to be estimated.

The inclusion probability is used in estimating sample statistics such as the population total and the sample variance. The design is an adaptive survey because at the beginning the surveyor does not know what units will be included in the sample. As information becomes available, in this case a unit has a value that meets the threshold condition, it triggers searching in the neighborhood.

As another example of an adaptive design, two-phase stratified design was proposed by Francis (1984). Surveying starts with a conventional stratified sample design. After this initial phase, additional sample units are allocated one by one to an individual stratum. At each step of this sequential allocation of sample units, the stratum that is allocated the unit is chosen on the basis of where the greatest reduction in variance will be. The adaptive allocation of effort can improve sample efficiency. The design is particularly useful for field surveys where extra time for the field crew is scheduled in case of bad weather. If the weather turns out fine every day, this design is a way of allocating the extra field effort.

16.3 DETECTING CHANGE

16.3.1 Models for Detecting Change and Trends

Very often in environmental management, the interest is in how an ecosystem or environment is changing over time (e.g., the decline in abundance of a rare species) or how it has responded to some environmental perturbation (e.g., an oil spill). The discussion so far has focused on how to collect information in the field for use in assessing status or trend. We will now review a method to estimate changes in status or trends for when observations have been collected over time, for the same sample unit.

Continuing with the grassland example, in a study to assess grass coverage changes over time, 12 of the sample units (quadrates) are surveyed every year for 10 years. These data could be used to estimate trends. There are two ways to consider this data set. If we consider each sample unit separately, then we can conduct an individual analysis on each sampling unit and then summarize results across all the units. The other way is to pool the sample units and consider trends across the set of sample units. The pooled approach to analyze this would usually involve mixed models (Myers et al., 2010), and this is the focus of the remainder of this chapter.

The name mixed models comes from the idea of having both random and fixed effects. The sample units in our grassland study are a random sample from all the potential sample units that could be chosen. The repeated surveys each year are called a fixed effect. With mixed models, you can analyze the data to assess trends in the individual sample units as well as across the population of units as a whole. The models can also accommodate analysis when there are missing data such as when there are unequal numbers of observations from the sampling unit.

As with any analysis, the first part of the study should be initial data exploration, and graphs are ideal for this. Index plots (Fig. 16.5) can be useful for seeing overall trends. In the figure, the grass cover at each sample unit in each year is displayed, with the year given as a code (years 1–10). The horizontal line on the plot is at the level of the overall average cover for all 120 observations, and it appears that grass cover increases over time.

Another useful graph is shown in Fig. 16.6. The boxplots illustrate some of the spread among the 12 grass cover measurements taken each year over the 10 years, and again, the general trend that grass cover is increasing can be seen. What also can be seen is that within each year, the quadrates are quite different from each other.

One of the simplest ways to consider change is to compare the average grass cover between 2 years. In this example, it would be interesting to compare the average cover in the first year with what was observed in the last year. A simple t-test can be used for this analysis. Each quadrate was measured in year 1 and in year 10 so we use a paired t-test. With t-tests, the data from year 1 and year 10 should both be normally distributed, although in practice as long as the data set is large enough, the test is quite robust for departures from normality (Sawilowsky and Blair, 1992). The test statistic is

$$\frac{\bar{y}_{10} - \bar{y}_1}{\text{se}(\bar{y}_{10} - \bar{y}_1)} \approx t_{11}$$

where $\bar{y}_{10}$ and $\bar{y}_1$ denote the average measurements from year 10 and year 1, $\text{se}(\bar{y}_{10} - \bar{y}_1)$ is the estimated standard error of the difference in the measurements, and t_{11} is the test statistic that follows a t-distribution with 11 degrees of freedom. The test statistic is $t = 18.86$ with $p \leq 0.001$ showing that the average grass cover was significantly greater at the last year compared with the beginning of the study. The observed mean difference in cover was 168.50.

Another display of the data is shown in Fig. 16.7. We use a scatterplot graph with separate panels for each year. Grass cover appears to increase over time but now we can see that the rate of increase differs among quadrates. For example, quadrate 3 appears to have had little increase in grass cover, whereas quadrate 1 shows quite an increase in cover over the years. Also from Fig. 16.7, we can see that the quadrates differ in overall grass cover in the start of the study (compare year 1 for quadrate 1 and 5), and there is a steady rate of increase in grass cover over the years.

In a mixed model analysis, we can start by thinking about the graph in Fig. 16.7 and consider fitting a separate linear model to each quadrate. This would involve fitting 12 individual models to the data, one for each quadrate. A linear model for grass cover by year would be

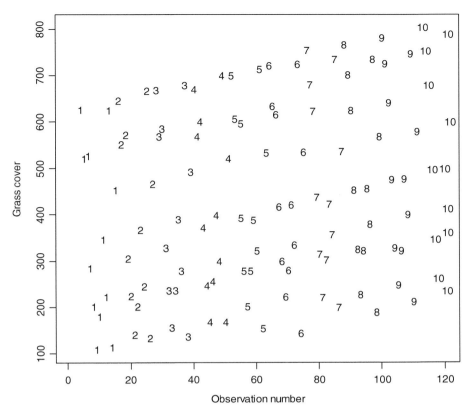

FIGURE 16.5 An index plot of grassland cover for the 12 quadrates over 10 years. There are 120 observations in total, and the points are indexed by the year of the study. Grass cover appears to have increased over the 10 years.

$$\text{cover}_{ij} = \beta_{0i} + \beta_{1i}\text{year}_j + \varepsilon_{ij}, \text{ for } i = 1, 2, \ldots, 12 \text{ and } j = 1, 2, \ldots, 10,$$

where cover_{ij} is the grass cover for the ith quadrate in the jth year. The terms β_{0i} and β_{1i} are the intercept of the y-axis and slope for the regression line for the ith quadrate. The term ε_{ij} is the model error associated with each station. In a linear regression analysis, the model errors are assumed to following a normal distribution with mean 0 and constant variance, σ^2. The y-intercept for each quadrate is the initial grass cover at the start of the study, and the slope is the change in grass cover per year (Table 16.1).

In a mixed model, we combine the information from the 12 quadrates while taking into account any random variation in the intercepts and slopes. A mixed model would be

$$\text{cover}_{ij} = \beta_0 + \delta_{0i} + \beta_1 \text{year}_j + \delta_{1i} \text{year}_j + \varepsilon_{ij}$$

where β_0 is the average y-intercept and δ_{0i} is the random variation from the overall average of the y-intercept for the ith quadrate. In the same way, β_1 is the expected change per year in grass cover and δ_{1i} is the difference between this overall average and the per year change for the ith quadrate. In linear mixed models, δ_{0i} and δ_{1i} are assumed to follow normal distributions with mean 0 and variance, $\sigma^2_{\delta_0}$ and $\sigma^2_{\delta_1}$, respectively. The random intercepts and slopes are assumed to be correlated, denoted by $\rho_{\delta_0 \delta_1}$, where a positive correlation would mean that the quadrates with higher grass cover at the start of the study would increase in cover faster than those with lower cover. The opposite would be a negative correlation where quadrates with higher grass cover at the start of the study would increase in cover slower than those with lower cover.

Usually, a step-by-step approach is used, and we start with random intercept model. The random intercept model, where only the intercepts for each sampling unit differ, is

$$\text{cover}_{ij} = \beta_0 + \delta_{0i} + \beta_1 \text{year}_j + \varepsilon_{ij}$$

This model has an AIC value of 1090.10, whereas the full model with both random intercepts and slopes AIC is 1034.34 suggesting the full model is preferable. AIC is one method for model comparison, and lower values of AIC are associated with better fitting models. The typical output from a statistics software package from the model with both random intercepts and slopes is show in Table 16.2.

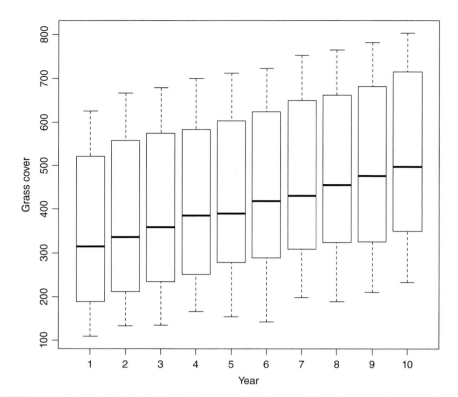

FIGURE 16.6 Boxplots for the 12 quadrates over the 10 years. There is a general trend that grass cover is increasing and, within each year, the quadrates are quite different from each other.

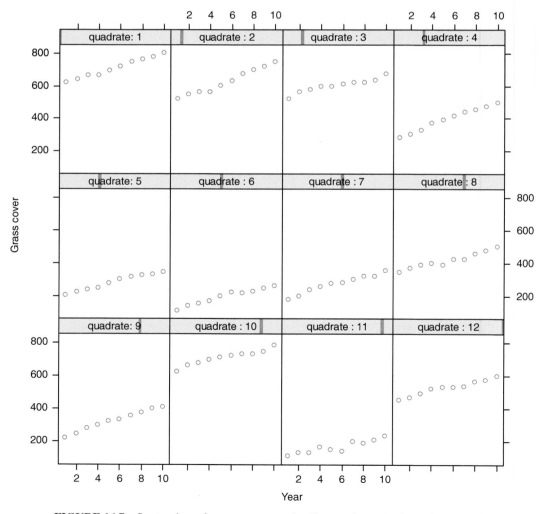

FIGURE 16.7 Scatterplots of grass cover over the 10 years for each of the 12 quadrates.

The correlation between the random intercepts and the random slopes for our data was low, 0.10 (Fig. 16.8). This suggests that there was no relationship between initial grass cover and the annual increases in cover. In some studies a high, positive correlation would suggest that the units with high initial values (intercepts) increased faster (slopes).

The other details provide information about the variance components. It assesses the relative size of variation in the grass cover that is attributable to variation across the study site (i.e., variation among quadrates) and how much of the variation is due to the within site variation (i.e., variation within quadrates). This information can be useful in planning future allocation of sampling effort (i.e., the merits of having additional quadrates).

The overall trend can be found in the results under the heading fixed effects, and there is an estimated increase of 17.75 cover per year. The initial overall grass cover is estimated as 338.71. The predicted per year change in grass cover in the mixed model ranges from 12.36 to 25.54 (Table 16.3). The coefficients for each quadrate from the mixed model differ slightly from the coefficients given in Table 16.1 where each quadrate was considered separately.

Checking for adequacy of the model is an important step. The model residuals can be used to give an indication of this. For a linear regression model, the residuals should appear to be normally distributed. Plotting the model residuals against the fitted values is one way to check for any obvious trends. The residuals in our model don't appear to have any trend up or down and are roughly an even scatter (Fig. 16.9).

In addition, because we have used a mixed model, the normality assumption applies to the random slopes and the random intercepts. A normal Q–Q plot of the predicted coefficients for the slope and the intercept can be used

TABLE 16.1 Predicted *y*-Intercept and Slope for 12 Individual Regression Lines, One for Each of Quadrates 1–12

Quadrate	Intercept	Slope
1	602.13	20.10
2	486.20	26.13
3	531.26	13.37
4	263.87	23.99
5	189.20	16.40
6	106.07	16.11
7	172.40	18.20
8	329.93	15.68
9	208.47	21.08
10	629.07	14.73
11	101.67	12.01
12	444.27	15.22

TABLE 16.2 Software Output from R (version 2.14.1) from a Mixed Effects Model

Linear mixed effects model fit by REML
Random intercepts
AIC 1090.10
Random effects: formula: ~1 | quadrate

	Intercept	Residual			
Standard deviation	195.84	16.06			

Fixed effects: cover ~ year

	Value	Standard error	DF	*t*-Value	*p*-Value
(Intercept)	338.71	56.62	107	5.98	0
Year	17.7525	0.51	107	34.378	0

Correlation
Year   −0.05

Random intercepts and slopes
AIC 1034.34
Random effects: formula: ~year | quadrate

	Standard deviation
(Intercept)	192.13
Year	4.12
Residual	10.66

Fixed effects: cover ~ year

	Value	Standard error	DF	*t*-Value	*p*-Value
(Intercept)	338.71	55.50	107	6.10	0
Year	17.75	1.24	107	14.34	0

Correlation
Year   0.10

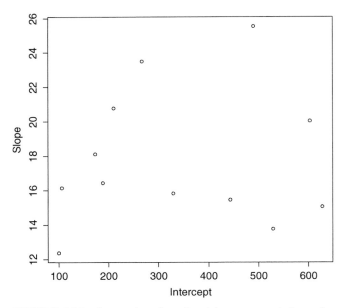

FIGURE 16.8 Scatterplot of predicted intercepts and slopes for the 12 quadrates.

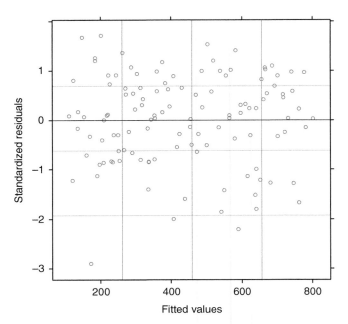

FIGURE 16.9 Standardized residuals plotted against the fitted values.

TABLE 16.3 Predicted y-Intercept and Slope for 12 Quadrates from the Mixed Model

Quadrate	Intercept	Slope
1	602.516	20.02
2	489.39	25.54
3	528.97	13.77
4	266.67	23.48
5	188.98	16.45
6	105.92	16.16
7	172.97	18.11
8	329.08	15.84
9	210.17	20.78
10	627.12	15.07
11	99.80	12.34
12	442.96	15.46

(Fig. 16.10). A straight line would suggest that the normality assumptions for the intercept and slope are satisfied. In this example, although the lines are not perfectly straight, we consider the assumptions are met.

16.4 SUMMARY

In this chapter, we have discussed two broad areas of environmental statistics: how to collect data and how to assess trends. Data collection can be considered the fundamental part of statistics in environmental management. A sample should be a good representation of the population, and there are many designs that can be used. We discussed a few deigns here, simple random sampling, stratified sampling, designs with spatial balance (GRTS and BAS), and adaptive sampling. We have discussed how to collect data, but the usefulness of the survey depends also on what data is collected. The example we used was grass cover for a grassland survey, but in other studies, it may be counts of birds, observations of an animal's behavior, pH levels of water sample, species diversity in soil samples, and so on. The timing of the survey and the repeat survey internal are all important features to consider in a survey. Sampling an animal with a marked diurnal pattern at one time of day will give quite different results from surveying at another time. Similarly, counting herbaceous plants that have a seasonal pattern will give different counts among season.

We have discussed a way to assess trend using grass cover measured over 10 years from 12 quadrates as an example. The grass cover increased over the 10 years but at a different rate in each quadrate. Different plots were used to display the changes in grass cover and help explore the data set. The graphs were helpful in seeing the overall trend and for observing the differences in starting value and rates of increase among the 12 quadrates.

We used a linear mixed model to assess trend. Initial analysis was done quadrate by quadrate and we created 12 individual liner models. In the mixed model approach, the data is considered as one pool, but we can still extract quadrate-level information. The overall trend from the mixed model analysis was estimated as an increase in 19.42 cover per year with per year changes ranging from 12.10 to 40.92. There was variation among quadrates in grass cover in the first year, and quadrates with higher

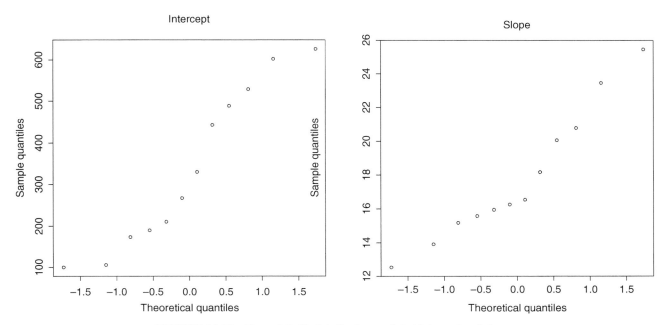

FIGURE 16.10 Normal Q–Q plots for the predicted intercept and slope.

cover tended to increase at as high of a rate than those with lower initial levels.

In this chapter, we have touched on two topics only. Within each, there are considerably more details that can be considered and, of course, many more relevant topics that could have been covered. Statistics in environmental management has a very important role. To make sound management decisions and environmental policies, we need a good understanding of the environment. Much of this understanding and knowledge comes from data and the way data is summarized, reported, analyzed, and used in models.

Collecting survey data in the field will always be an important part of environmental management. There are many other survey designs, and our understanding of how best to collect data is expanding. Data collected in the field can now be supplemented (or indeed replaced) by accessing worldwide or countrywide data sets that are becoming increasingly available as our computational power and associated technology evolve.

Another area of change in statistics for environmental management is the ever-increasing range of statistical methods and techniques. We discussed in this chapter linear mixed models for assessing trend. We had a very simple example, where there were equal numbers of repeated observations for each quadrate (other studies may have unbalanced data resulting from missing values or the way the study was designed), quadrates were visited once a year (other complex designs can have more than one visit in a season and visits in more than one season.), and the data was normally distributed (other surveys can have nonnormal data such as what you could see if the observations were counts, rates, percentages, or many other forms). Modern statistics is evolving and modeling techniques are becoming available to accommodate these complexities.

REFERENCES

Brown, J.A. 2003. Designing an efficient adaptive cluster sample. *Environmental and Ecological Statistics* 10(1):95–105.

Brown, J.A. 2011. Adaptive sampling of ecological populations. In, Y. Rong, ed., *Practical Environmental Statistics and Data Analysis*. ILM Publications, St Albans, p. 81–96.

Francis, R.I.C.C. 1984. An adaptive strategy for stratified random trawl surveys. *New Zealand Journal of Marine and Freshwater Research* 18:59–71.

Gilbert, R.O. 1987. *Statistical Methods for Environmental Pollution Monitoring*. John Wiley & Sons, Inc., New York.

Hartley, L.J. 2012. Five-minute bird counts in New Zealand. *New Zealand Journal of Ecology* 36(3):268–278.

Manly, B.F.J. 2008. *Statistics for Environmental Science and Management*, 2nd Edition. Chapman and Hall/CRC, Boca Raton, FL.

McDonald, T. 2012. Spatial sampling designs for long-term ecological monitoring. In, R.A. Gitzen, et al., ed., *Design and Analysis of Long-term Ecological Monitoring Studies*. Cambridge Press, Cambridge, p. 101–125.

Myers R.H., D.C. Montgomery, G.G. Vining, and T.J. Robinson. 2010. *Generalized Linear Models with Applications in Engineering and the Sciences*, 2nd Edition. Wiley, New York.

R Core Team. 2014. R: a language and environment for statistical computing. The R Foundation for Statistical Computing, Vienna, Austria. http://www.R-project.org/ (accessed April 30, 2015).

Robertson, B.L., J.A. Brown, T. McDonald, and P. Jaksons. 2013. BAS: balanced acceptance sampling of natural resources. *Biometrics* 63(3):776–784.

Sawilowsky, S. and R.C. Blair. 1992. A more realistic look at the robustness and type II error properties of the t test to departures from population normality. *Psychological Bulletin* 111(2): 353–360.

Scheaffer, R.L., W. Mendenhall, R.L. Ott, and K.G. Gerow. 2011. *Elementary Survey Sampling*, 7th Edition. Duxbury, Belmont.

Stevens, D.L. Jr. and A.R. Olsen. 2004. Spatially balanced sampling of natural resources. *Journal of the American Statistical Association* 99:262–278.

Thompson, S.K. 1990. Adaptive cluster sampling. *Journal of the American Statistical Association* 85:1050–1059.

Thompson, S.K. 2002. *Sampling*, 2nd Edition. John Wiley & Sons, Inc., New York.

17

REMOTE SENSING IN ENVIRONMENTAL MANAGEMENT

MARK J. CHOPPING

Department of Earth and Environmental Studies, Montclair State University, Montclair, NJ, USA

Abstract: Environmental management decisions depend on measurements: without measurement, it is not known if some environmental parameter of interest is changing importantly, or how. For many environmental issues, it is important to have information that is spatially distributed, that is, mapped. Remote sensing from space, the air, or on the ground can provide maps of many important environmental parameters—information that is not available in any other way. This chapter illustrates how remote sensing can be used to further understanding of environmental problems and aid in decision making. The science and art of remote sensing are described with special attention paid to the issues of temporal and spatial scales as these determine applications of these geospatial technologies in environmental management. The chapter provides a set of concrete examples of the use of remote sensing in a diverse set of environmental applications, before concluding with suggested readings and a sample syllabus.

Keywords: remote sensing, imaging, sensor, satellite, orbit, sampling, spatial scales, temporal scales, spectroscopy, urbanization, air pollution, aerosols, wetlands, forest, carbon, CO_2, climate, warming, ice, drought, water resources, aquifers, gravity, radiometer, radar, lidar.

17.1	Introduction	397	17.3.3	The Greenland Ice Sheet Surface Melt of July 2012	411
17.2	What is Remote Sensing?	398	17.3.4	Remote Sensing of Urbanization	414
	17.2.1 Remote Sensing in a Nutshell	398	17.3.5	Remote Sensing of Aquifers I: Gravity Remote Sensing	414
	17.2.2 What Can Remote Sensing Data Tell Us?	403	17.3.6	Remote Sensing of Aquifers II: Radar Remote Sensing	415
	17.2.3 Environmental Management at Diverse Spatial and Temporal Scales	403	17.3.7	Remote Sensing of Urban Air Pollution	416
	17.2.4 Google Earth: A Remote Sensing Panacea?	404	17.4	Concluding Remarks	419
17.3	Selected Applications of Remote Sensing	407		Acknowledgments	420
	17.3.1 Remote Sensing in Wetland Management	407		References	420
	17.3.2 Deforestation and Forest Degradation	408			

17.1 INTRODUCTION

In the business world, there is a well-known adage regarding management: "you cannot manage what you do not measure." Measurement is also paramount in environmental management: without measurement, we do not know if some environmental parameter of interest is changing importantly—but what to measure, how, and how often are key questions. Many environmental management issues—for example, the effects of industrial activities on air and water quality in adjacent areas or point source pollution events such as dilbit leaks from pipelines—cover relatively small scales in space and time and may best be measured *in situ* by going to the field and taking measurements, or collecting physical samples to bring back to the laboratory for subsequent analysis. In many cases, these methods are the only, or the best, option. However, for many of the environmental challenges that face us today, Yogi Berra's famous saying "you can learn a lot just by looking" rings true—especially if we "look" using technologically advanced sensors flying on aircraft or Earth-orbiting satellites that view over much larger areas than those we can sample on the ground. This is "remote sensing."

There are many environmental parameters that cannot be inferred using remote sensing, either because they occur

---

An Integrated Approach to Environmental Management, First Edition. Edited by Dibyendu Sarkar, Rupali Datta, Avinandan Mukherjee, and Robyn Hannigan.
© 2016 John Wiley & Sons, Inc. Published 2016 by John Wiley & Sons, Inc.

below the surface (e.g., groundwater contamination) or because they cannot be observed at the necessary spatial or temporal scales or with sufficient precision; however, there are many important issues that can *only* be addressed using remote sensing. For example, understanding phenomena that occur at regional to global scales often requires multiple, linked data sets at commensurate spatial scales and at seasonal or interannual intervals for applications (drivers) as diverse as mapping and monitoring crop yields (climate, management), carbon sequestration in forest biomass (deforestation, degradation, management), algal blooms and ocean "dead zones" (agricultural nutrient and effluent runoff), and sea level rise (ice sheet and glacier melting, thermal expansion of the oceans).

It has been said that remote sensing began with people climbing hills to get a better view of the landscape—and the location of opposing military forces. It remains true that many developments in remote sensing have been spurred by military and intelligence considerations: the field expanded rapidly during both World War II (using high-altitude aerial photography) and the Cold War (using Earth-orbiting satellites for the first time) but has seen major advances more recently with the advent of digital imaging, powerful small computers, and the Internet. These three developments have allowed many more people to use this technology or at least become aware of it, e.g., through the Google Earth™ high-resolution image browser.

This chapter will describe how remote sensing is used by environmental management practitioners and scientists to inform and aid in decision making; it is thus essential that the reader have a good understanding of remote sensing fundamentals and how these systems are constrained by physics, practicality, and cost, as well as the available sensing and computing technologies. The chapter therefore begins with a brief primer and then proceeds to examine the critical issue of scale and how this relates to our understanding of "environmental management." It includes several case studies that illustrate the application of the technology and how it is used at a wide range of scales—from local to global—often together with other geospatial technologies: Geographical Information Systems and Global Positioning Systems. The chapter concludes with suggestions for further reading so that interested readers may easily locate more in-depth information. Key terms and concepts needed by anyone learning the basics of environmental remote sensing are italicized.

17.2 WHAT IS REMOTE SENSING?

17.2.1 Remote Sensing in a Nutshell

Remote sensing is the science and art of inferring information about surfaces or features that are distant from the observer, usually using measurements of reflected or emitted electromagnetic radiation (EMR). In the *visible* wavelengths—approximately 0.4–0.7 μm—our eyes can detect this radiation and we call it "light." Remote sensing instruments are not limited to this narrow range and are capable of detecting EMR at both shorter (e.g., ultraviolet) and longer (e.g., near-infrared, thermal infrared, microwave, and radio) wavelengths. If the source of the detected EMR is natural—that is, sunlight or the longer wavelength radiation emitted from the Earth's surfaces or atmospheric elements—the remote sensing method is termed *passive*. If, on the other hand, the target surface is illuminated artificially through generation of EMR pulses as either radio waves (*r*adio *d*etection *a*nd *r*anging (radar)) or laser light (*l*ight *d*etection *a*nd *r*anging (lidar)), then the method is termed *active*. Both active remote sensing methods work at night—no sunlight required—but only radar can penetrate clouds (the radio receiver in your iPod will still receive your favorite station, even if you are indoors: radio wavelength EMR can easily penetrate the walls of buildings; this is why ground-penetrating radar techniques can be used by police to locate buried cadavers).

Instruments (*sensors*) that sense reflected sunlight in a few regions of the EM spectrum (*bands*) are termed *multispectral*, while those that sense in a hundred or more narrow bands—thus allowing *imaging spectroscopy*—are termed *hyperspectral*. Images produced using light across the visible to near-infrared range are called *panchromatic* (meaning "all colors"). Observations are usually made from above, with the remote sensing instrument located on a *platform*—balloon, helicopter, fixed-wing aircraft, or satellite—although ground-based methods with sensors located on towers or sometimes just a meter or two above the surface are also used, for example, for the collection of *ground reference data*. The phrase "ground truth" should be avoided since it is impossible to collect data on the ground with no error. Flying sensors on satellites provide the benefits in greater stability, regular repeat viewing, and global coverage.

All remote sensing systems *sample* the observed surfaces: these samples may be contiguous or overlap (in which case a 2-D *image* may be constructed from them), or they may be isolated (in which case linear profiles or grids of the point samples may be constructed and a 2-D image may only be built using *interpolation* between the grid points). Thus, while most remote sensing devices generate data that can be interpreted as an image—not all do—it must never be assumed that the sampling is complete, or regular, or uniform, or consistent over time. Some imaging sensors *scan* surface-leaving EMR using rotating mirrors to sample a *scanline* perpendicular to the platform's track direction, while others sample in continuous strips using arrays of multiple detectors (e.g., CCD arrays; Fig. 17.1). Scanning systems usually have a lower signal-to-noise ratio because the *dwell time* (the period during which light falls on the detectors) is much shorter. Both types may observe the surface far away from *nadir* (viewing straight down).

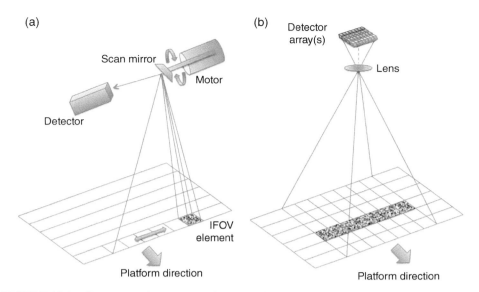

FIGURE 17.1 Conceptual diagrams showing the difference between (a) across-track scanning and (b) detector array systems that use charge-coupled devices, one linear array for each multispectral band. IFOV, instantaneous field of view: the IFOV element (or "ground-projected IFOV") is the area that is contributing EMR to the sensing device at any given moment. Note that IFOV elements are shown to be square in these diagrams but are usually elliptical (circular only at nadir, viewing straight down toward the center of the Earth).

Remote sensing systems are characterized by an instantaneous field of view (IFOV; unit, degrees or radians): they measure target-leaving EMR from an area on the surface that is termed the *ground-projected instantaneous field of view* (GIFOV; the projection of the instrument's IFOV onto the surface). If the GIFOV of an imaging system is approximately 1 ± 3 m, the system is termed *high resolution*; if it is between 5 and 50 m, it is termed *medium resolution*; and if it is larger than 50 m, it is called *moderate resolution*. Panchromatic images can have higher spatial resolution because the light incident on the detector is not divided into spectral bands. We must also note that the surface under the GIFOV that is sampled—often called a *pixel* (i.e., "picture element")—is rarely composed of one kind of material and even a 1 m pixel may contain contributions from concrete, tarmac, grass, and painted metal (e.g., for a divided highway). Thus, our samples might more accurately be called *mixels*, though unfortunately the term *pixel* has remained in common usage.

To compound matters, surfaces scatter light in different directions with respect to the source of illumination (usually the Sun). Very few natural surfaces exhibit specular reflection, where incident light is reflected as if from a mirror; most scatter light in different directions, anisotropically. *Anisotropy* is the quality of exhibiting properties with different values when measured along axes in different directions (derived from the Greek *isos* (equal) and *tropos* (way), with the prefix *an* indicating an exception). The function that describes this *reflectance anisotropy* is called the *bidirectional reflectance distribution function* (BRDF). Reflectance anisotropy arises from the 3-D structure of the surface and is partly owing to shadowing effects, though structures do not have to be large: even short grasses and bare soil surfaces demonstrate considerable anisotropy. This can be tested if the Sun is not close to *zenith* (overhead) by comparing digital photographs of a uniform lawn target taken in the solar and antisolar directions. Even with autoexposure, it will be evident that the shadows cast by grass leaves darken the picture taken looking toward the Sun; with the Sun behind the viewer, shadows are hidden by the objects casting them. Even soil elements can contribute to strong BRDF effects (Fig. 17.2).

Both the sensor design and the *platform* on which the sensing device is mounted determine the eventual spatial and temporal sampling characteristics of the remote sensing system. For example, consider the case of an imaging Earth-observing satellite in a polar orbit around the Earth (*orbital inclination* of ~98° to the equator) at an altitude of 705 km (438 miles) equipped with a sensor with a 185 km (115 miles) swath. This satellite will circle the globe every 98.9 minutes imaging the entire globe every 16 days, except for the highest polar latitudes, with approximately 14.5 orbits per day (Fig. 17.3a). These parameters describe the joint National Aeronautics and Space Administration (NASA)/US Geological Survey (USGS) Landsat 8 mission launched on February 11, 2013 (Irons et al., 2012), the latest in the Landsat series that has imaged the Earth's land surfaces since 1972. As with all remote sensing systems, Landsat 8 has some limitations: from the revisit period, it might seem feasible to construct a global 16-day image series; however,

FIGURE 17.2 Photographs of a desert grassland landscape in southern New Mexico (a) viewing in the solar direction (Sun in front of viewer) and (b) viewing in the antisolar direction (Sun behind viewer). Note the degree to which shadowing by shrubs, smaller plants, and even soil elements darkens the image in (a), even though the landscape in (b) is almost identical. Shadows are hidden behind these elements when looking 180° from the Sun direction—and they are all hidden when also viewing at zenith (i.e., as if viewing along a wire from the Sun to the target; this is called the "hot spot" geometry).

in the wavelengths in which the sensor observes, light is not capable of penetrating the turbulent collections of water droplets that we call clouds. This importantly reduces the frequency with which comprehensive, wall-to-wall image series can be constructed.

In contrast to this imaging system, consider NASA's Geoscience Laser Altimeter System (GLAS) that operated on the Ice, Cloud, and Land Elevation Satellite (ICESat) from January 2003 to October 2009. ICESat was in a near-circular, near-polar orbit (94° inclination) with an altitude

(a)

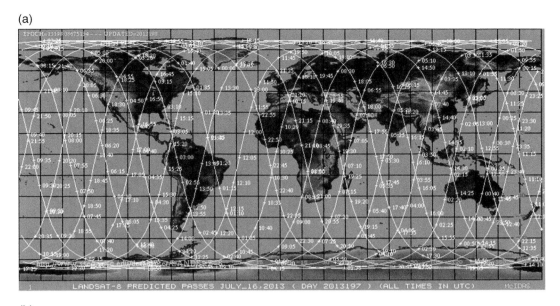

(b)

FIGURE 17.3 (a) Landsat 8 orbital tracks (courtesy of the University of Wisconsin–Madison Space Science and Engineering Center). (b) Landsat 8 Operational Land Imager true color image near the Copper River Delta, Alaska, acquired on May 28, 2013. Image width is approximately 20 km. The spring thaw swells the river with water loaded with glacial sediment from the Childs and Miles glaciers that is clearly visible. The sediment is a source of nutrients for phytoplankton and marine plants, which in turn support abundant salmon runs. Landsat imagery courtesy of NASA Goddard Space Flight Center and US Geological Survey (http://earthobservatory.nasa.gov/IOTD/view.php?id=81784).

of approximately 600 km (373 miles): not very different from the current Landsat 8 orbit. However, unlike Landsat, GLAS was neither imaging nor passive. Instead of measuring reflected sunlight, it fired green and near-infrared wavelength laser pulses toward the surface at regular intervals and sensed the reflections, resulting in a linear sampling in the satellite's along-track direction with a gap of about 170 m between each 70 m diameter shot (Fig. 17.4). The repeat period was 8 days. This profiling, "remote-sensing-with-gaps" approach may seem odd to those used

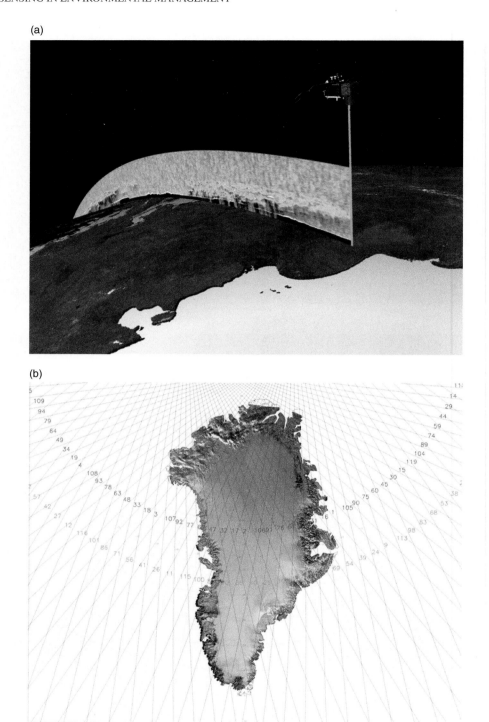

FIGURE 17.4 (a) Visualization of the ICESat satellite transmitting a green beam of laser light toward a true color surface map of the Earth and a vertical slice of the atmosphere from Geoscience Laser Altimeter System (GLAS) measurements, showing the height and thickness of clouds and aerosols. (b) Greenland 8-day repeat pattern. The returns over ice provide elevation data. Source: NASA (http://earthobservatory.nasa.gov/IOTD/view.php?id=4033; http://icesat.gsfc.nasa.gov/icesat/).

to imagery—but the ICESat/GLAS 8-day, 170m sampling was more than sufficient to address the primary objective of the mission: to measure the elevation of Earth's great ice sheets (Greenland, Antarctica) and thus track rates of change in ice sheet mass balance (Zwally et al., 2002). ICESat-2 is slated for launch in 2016; NASA is filling the 6-year gap using aerial mapping with *Operation IceBridge*, a series of aircraft campaigns.

17.2.2 What Can Remote Sensing Data Tell Us?

There are a limited number of information domains that can be exploited in environmental remote sensing, namely:

- Spatial: texture, pattern, hue, brightness, context, and range (distance)
- Spectral: variation with respect to wavelength
- Directional: variation with respect to viewing and illumination directions
- Polarization: variation with respect to horizontal or vertical polarization
- Gravity: variation with respect to gravity anomalies
- Magnetism: variation with respect to magnetic field strength and direction
- Temporal: variation in any other remotely sensed quantity over time

Space precludes an in-depth discussion of all of these sources of information but note that the most common methods involve exploiting spectral, spatial, and temporal variations, while gravity has been successfully used only recently—and magnetism is rarely used. It is often possible to exploit more than one of these sources of information simultaneously; for example, examining changes in ice sheet surface roughness (derived from analysis of surface BRDF) would exploit information in the directional and temporal domains.

Before examining other uses of remote sensing in environmental management, it is worth noting that *almost no environmental parameters are accessed directly* but are *inferred* from measurements; that these measurements are subject to error from sensor noise, orbital drift, and attenuation by the atmosphere (aerosols, water vapor, ozone, clouds, rain); and the inherent difficulty of extracting information at the appropriate spatial and temporal scales and combining it with other geospatial information—such as elevation data—in order to address the environmental problem at hand.

This is even true of high-resolution imagery such as that seen in Google Earth™: although quite small features such as automobiles, houses, and trees may be resolved, it remains difficult to derive useful information from such images. For example, how to identify, count, and measure these objects? It is too time-consuming to perform these operations manually, object by object, except for very small study areas: an automated—or at least semiautomated—method is required. Fortunately, computer hardware, software, and interconnectivity have improved rapidly over the last few decades, so that it is now possible to employ a diverse set of strategies for interpreting remote sensing imagery. Our marvelous brains can be used to obtain information on the target of interest by visually assessing hue, texture, context, shape, size, pattern, and shading but can be supported by the use of computers to implement methods that require many millions of calculations, often required when studying large areas. Clearly, it is not feasible to perform such calculations manually—if we hope to make it home for the holidays. It is now possible to transfer gigantic, multi-gigabyte files around the world with ease and store a terabyte of imagery on a hard drive the size of a smartphone. Moreover, software developments in compression mean that vast swathes of highly detailed imagery can be navigated seamlessly: the interpretation toolset has never been so advanced. However, these technological advances are not a panacea: it is still up to the remote sensing expert to use her/his knowledge to analyze (take apart) and/or synthesize (put together) appropriate data sets for each application and to use her/his expertise, discretion, and intelligence in obtaining valid and useful results.

The impact of inexpensive yet powerful computing has had another important effect: it has enabled new and better uses of data from the historical archive. Although it might seem less exciting to work with older data, observing changes through time often yields much greater understanding than a snapshot. Many activities are driven by cost and remote sensing is no exception. This is why for regional- and continental-scale land cover and vegetation mapping, moderate-resolution (1.1 km) imagery from the NOAA Advanced Very High Resolution Radiometer (AVHRR) and NASA's Moderate Resolution Imaging Spectroradiometer (MODIS) was preferred over medium-resolution (about 30 m) Landsat imagery: not only were Landsat images expensive at around $600 per 180 × 180 km scene, but they required what was considered at the time to be hefty data storage and processing capabilities. Now, the entire Landsat record has been made freely available, and new software has made it possible to process imagery consistently (e.g., the Landsat Ecosystem Disturbance Adaptive Processing System software, Masek et al., 2012). This has led to a remarkable resurgence in Landsat applications, even for global-scale studies that previously would have been too data-heavy and plagued by inconsistencies; the value of the lengthy Landsat archive is finally being realized (Kluger and Walsh, 2013, Kim et al, 2014).

17.2.3 Environmental Management at Diverse Spatial and Temporal Scales

The remote sensing sensors we have at our disposal are highly diverse: they can be passive or active, imaging or profiling, and detect reflected or emitted EMR in a varying number of broad or narrow bands in the ultraviolet to microwave wavelengths. Some systems exploit gravity (Song et al., 2013). Importantly, they are able to observe at a wide range of spatial and temporal scales. To see how useful a particular remote sensing system is in addressing a specific environmental management issue, it is useful to consider the following five characteristics:

1. Geographic coverage (site, local, regional, global)
2. Spatial resolution (grain)
3. Duration of the record (period of operation)

4. Frequency of observation (revisit time)
5. The nature of the measurement (i.e., can the required information be inferred?)

Four of these five characteristics are related to scale. Even if the measurement type is completely able to supply the required information, to be useful, it must also be available for a long enough period of time, at a suitable frequency, over a sufficiently large area, and with sufficient spatial detail for the area of interest.

The question of scale has implications for the way we regard environmental management itself. *Dictionary.com* defines "environmental management" quite broadly as "an attempt to control human impact on and interaction with the environment in order to preserve natural resources." Some practitioners consider most "environmental management" issues to be local-, city-, or at most hinterland-scale problems, perhaps because of an implicit assumption that only problems occurring at these scales can be "managed." However, we should by now be painfully aware that mankind's reach extends much further in both space and time: our collective environmental impact is global and has long-term consequences. For many of the truly momentous environmental problems of the twenty-first century—including the increase in net solar energy retained on Earth as a result of enhanced greenhouse and aerosol forcings and their proximate and indirect impacts (i.e., "global warming")—it is imperative to observe at the global scale as well as locally. For example, it is now understood that important reductions in Arctic sea ice extent at the end of the northern hemisphere summer and the resulting increased heating of the Arctic Ocean and lower atmosphere are responsible for reductions in the thermal gradient from low to high latitudes and some important changes to atmospheric circulation patterns, including sudden stratospheric warmings, a weakened polar jet stream, and unusually strong and persistent high-pressure blocking patterns, resulting in extreme weather patterns, for example, the 2011 Texas drought, the severe winter of 2012 in the United Kingdom, and the current deep California drought. In short, unlike Las Vegas, *what happens in the Arctic does not stay in the Arctic* (Francis and Vavrus, 2012). Although we have only been able to monitor sea ice extent using satellite remote sensing from 1979 (NASA, 2012a), this is long enough to allow us to appreciate the unexpected rapidity of the changes and see how far the minimum area has decreased from those of the preceding decades—and from the expectations of glaciologists and modelers (Fig. 17.5).

You may have heard the saying *all politics is local* but is the same true for environmental management? Is it reasonable to contend that *management* of global environmental issues is just not feasible? After all, many years after the gravity of the problem was recognized, we have failed spectacularly in attempts to negotiate a single international treaty to reduce the emissions and particulates that are causing rapid climate change. So, is it even possible to "manage" environmental problems at global scales? The pervading political climates of many of the major developed nations (including the United States, Canada, Australia, the United Kingdom, and Japan) do not currently seem to be conducive to such an agreement, and no binding environmental treaties have been established. However, it is not so very long ago that the Montreal Protocol on Substances that Deplete the Ozone Layer of the Vienna Convention for the Protection of the Ozone Layer entered into force (January 1, 1989); and since then, two ozone treaties have been ratified by 197 states and the European Union. International environmental management agreements have thus been shown to be possible and effective, and others are in the pipeline (see Section 17.3.2). With respect to the relationship between global environmental treaties and remote sensing, it is worth noting that high-quality, robust measurements are required, especially where there is the potential for large economic impacts of regulation (John R. Townshend, Keynote Talk at the 4th Global Vegetation Workshop at the University of Montana, Missoula, June 16–19, 2009). In the case of stratospheric ozone, measurements from NASA's Total Ozone Mapping Spectrometer (TOMS) instrument on the Nimbus 7 satellite did record the loss of stratospheric ozone over Antarctica in 1985, but the values were so low that a screening algorithm flagged them as "bad values," so they were ignored. This flagging was reasonable because atmospheric scientists did not think that such low values were possible; when it became clear that the ozone layer had in fact been severely depleted, the TOMS data were reanalyzed and found to confirm the losses seen in Sun photometer readings. Three TOMS instruments were subsequently used to monitor the extent of the "hole" in the ozone layer, until 2005 when they were replaced by the Ozone Monitoring Instrument (OMI) on the Aqua satellite and the Ozone Mapping and Profiler Suite (OMPS) on the Suomi NPP satellite.

17.2.4 Google Earth: A Remote Sensing Panacea?

As an everyday example of the importance of scale, consider *Google Earth*. While appearing to nonspecialists as a flexible and user-friendly tool for mapping the Earth's diverse environments, it quickly becomes obvious that there are numerous environmental problems that it is unable to address, for two majors reasons. First, the imagery presented is mostly *pan-sharpened true color* imagery from commercial imaging satellites such as DigitalGlobe, Inc.'s *QuickBird* series. It is called "true color" because the sensor's red, green, and blue multispectral bands are used to produce an image in which natural colors are approximated. It is "pan-sharpened" because an algorithm is employed that uses higher spatial resolution (0.6 m) panchromatic imagery to increase the spatial variation of the 2.4 m resolution multispectral imagery. To understand this, imagine a

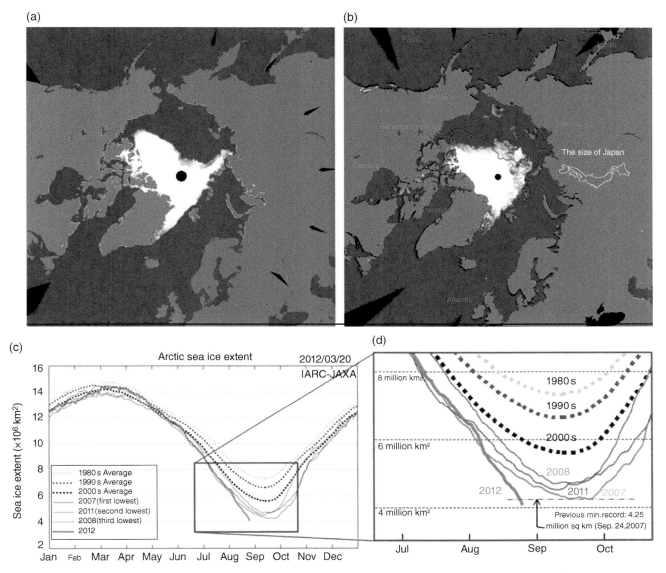

FIGURE 17.5 (a) AMSR-E Arctic sea ice extent in September 2007. (b) AMSR2 minimum in September 2012. (c) Seasonal variation of the Arctic sea ice extent (updated on August 24, 2012). (d) Detail of the area in (c). A record minimum sea ice coverage of 4.21 million km² was observed by satellite on August 24, 2012, one month earlier than previous minimum record set on September 24, 2007. Based on AMSR-E and AMSR2 microwave imagery. *Source:* Japan Aerospace Exploration Agency (JAXA). URL: http://www.eorc.jaxa.jp/en/imgdata/topics/2012/tp120825. html, last accessed September 2, 2013.

2.4 m true color image printed on a semitransparent plastic sheet that is placed on top of a print of a 0.6 m grayscale image. The detail of the finer grayscale imagery is superimposed on the color imagery. Pan sharpening achieves the same thing but using software. This provides a detailed and quite realistic pictorial representation of the landscape, so that when Google Earth was first released in 2005, there were suggestions that it would obviate the need for other remote sensing methods: anyone could now simply navigate to almost any location and determine the nature of the surface and its features. However, a moment's thought reveals the fallacy of this thinking. First, many important environmental parameters—soil moisture, atmospheric pollution, aquifer levels, ice sheet surface melt, ice sheet mass balance, sea surface temperature, sea level, and ocean salinity, among others—cannot be measured using pictorial representations, no matter how detailed. Second and perhaps more importantly, the temporal sampling provided is

FIGURE 17.6 An example of discontinuities in Google Earth imagery near the New Jersey–New York state boundary in the vicinity of Greenwood Lake. Note that these differences are owing to differential atmospheric effects, as well as the season of image acquisition, and can obscure interannual changes that we might want to isolate in an environmental management exercise.

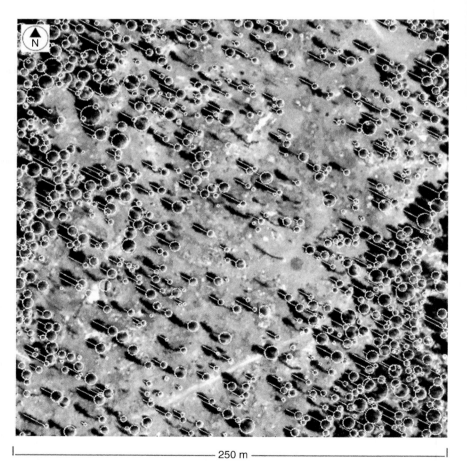

FIGURE 17.7 High-resolution (0.6 m) panchromatic QuickBird satellite image chip over forest in California's Sierra Nevada, showing tree location and shadow length measurements by the CANAPI algorithm (Chopping, 2011). The Sun is in the SE; shadows are cast toward the NW. With knowledge of the Sun elevation angle, tree heights can be estimated.

inadequate, owing to sparse data collection and the presence of clouds that obscure the surface. Google Earth imagery contains marked discontinuities at the boundaries of adjacent images (Fig. 17.6). These are owing to differing dates of acquisition, snow/no-snow conditions, differing vegetation conditions (depending on whether the imagery was acquired at the beginning, peak, or end of the growing season), and differing atmospheric conditions (haziness, variable aerosol optical depth, high thin cirrus clouds). Large discontinuities can also be caused by differences in Sun elevation that occur with changes in season and latitude: the 3-D structure of the surface—soil, grass, shrubs, and trees—results in scattering and shadowing effects that change the amount of light reflected to the sensor. Imagine the Sun is nearly overhead (zenith), as in summer time or at low latitudes: shadows will fall underneath the objects that cast them. Contrast this with the situation where the Sun is near the horizon, as in winter or at high latitudes: shadows will then fall far from the objects casting them, effectively darkening imagery. These perturbing factors not only reduce consistency in the Google Earth database—they also affect lower-resolution remote sensing—but can be addressed only partly by empirical corrections or modeling (e.g., of atmospheric scattering and absorption).

This does not mean that high spatial resolution imagery is not useful: clearly, there are many applications that can only be addressed using this kind of imagery, especially if interpretation or feature recognition can be partly automated. For example, the shadowing problem can be turned into an opportunity: although imagery is a 2-D representation of the surface, we can obtain information about 3-D structures (e.g., the locations, crown radii, and tree heights) through analysis of the patterns of sunlit and shaded features apparent in the imagery (e.g., the sunlit and shaded parts of tree or shrub crowns and the shadows cast by them). Provided the canopy is not too dense, tree number density, crown radii distributions, tree heights, and aboveground biomass estimates can be rapidly obtained using algorithms such as *CANopy Analysis with Panchromatic Imagery* (CANAPI; Chopping, 2011; Fig. 17.7). This reminds us that one person's noise can be another's signal.

17.3 SELECTED APPLICATIONS OF REMOTE SENSING

This section presents a series of applications of remote sensing that are relevant to environmental management in its broadest sense. This is far from an exhaustive collection: there is simply not enough space to include anything but a selection. The reader should note that among the areas not covered in the case studies presented, there are very important atmospheric applications of remote sensing, for example, in monitoring carbon dioxide sources and sinks with NASA's Orbiting Carbon Observatory-2 (OCO-2) that was launched on July 2, 2014 (NASA, 2014a), and tracking plumes from volcanic eruptions that are hazardous to human health and jet aircraft (NASA, 2013d). Remote sensing has also been used to evaluate water quality (eutrophication and turbidity) by estimating suspended sediment and chlorophyll concentrations (Ritchie et al., 2003).

17.3.1 Remote Sensing in Wetland Management

Hyperspectral imaging (or *imaging spectroscopy*) is an increasingly important remote sensing technology that exploits the ability to record reflected light in many narrow spectral bands (tens to hundreds), rather than just a few. This enables applications that would otherwise be difficult or impossible: the inherent dimensionality (information content) of the data is increased, and surface materials can be more finely differentiated. This is highly advantageous for applications where surface features might otherwise be confused in spectral space, such as mapping invasive species in wetlands. Invasive aquatic plants can be a major problem for managers: they compete with native plants, degrade water quality, and reduce water availability by increasing evapotranspiration. An unrestricted spread of invasive plants may lead to irreversible impacts, so there is a strong motivation for mapping efforts to aid in predicting the direction and rate of change. Researchers at the Center for Spatial Technologies and Remote Sensing (CSTARS) at the University of California, Davis, have performed this kind of wetland mapping application (e.g., Khanna et al., 2011). Spectra that are typical of the major scene components (known as *endmembers*) are either extracted from the hyperspectral imagery at locations where the surface type is known or acquired in the field (*in situ*) using a spectroradiometer (Fig. 17.8a). The endmembers are used to train specialized algorithms—such as spectral angle mapper (SAM), linear spectral unmixing (LSU), and continuum removal—or create spectral indices that are then used with a classifier or learning algorithm (e.g., maximum likelihood, decision tree, support vector machine, artificial neural network). CSTARS researchers have used imagery from the HyMap sensor—which records light in 126 bands across the 400–2400 nm region—to track the spread of highly invasive water hyacinth (*Eichhornia crassipes*) in the Sacramento–San Joaquin River Delta in California, where it co-occurs with native pennywort (*Hydrocotyle umbellata* L.) and nonnative water primrose (*Ludwigia* spp.). A decision tree classifier was trained with a set of metrics derived from the HyMap spectra to create maps for 2004, 2007, and 2008 (Fig. 17.8b). An accuracy assessment found that the three floating species were mapped with an average overall accuracy of 88% for water hyacinth, 87% for pennywort, and 71% for water primrose. This kind of application helps managers assess the rate of change and the likely impacts of future spread of species such as water

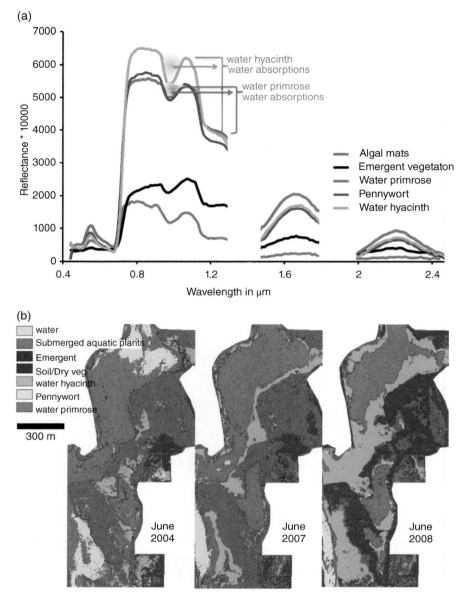

FIGURE 17.8 (a) High-resolution spectra for scene components (endmembers). (b) Mapped increase in water hyacinth cover in Stone Lake from 2004 to 2008, using HyVista HyMap imaging spectroscopy data. Credit: Shruti Khanna, Maria J. Santos, and Susan L. Ustin, Center for Spatial Technologies and Remote Sensing (CSTARS) at the University of California, Davis. By kind permission of Taylor & Francis Ltd. (www.tandfonline.com).

hyacinth without expensive field surveys: a credible cost–benefit analysis can only be accomplished if this kind information is available to managers.

17.3.2 Deforestation and Forest Degradation

The United Nations' REDD+ program is a near-future policy-driven global initiative designed to limit emissions of the greenhouse gas carbon dioxide. It will seek to do this by creating a financial value for the carbon stored in forests and offering incentives for developing countries to reduce emissions from forested lands and invest in low-carbon paths to sustainable development. It goes beyond addressing only deforestation and forest degradation and includes the role of conservation, sustainable management of forests, and active enhancement of forest carbon stocks (United Nations, 2009). In order for REDD+ to be successful, it must be seen to be equitable, so objective and verifiable forest monitoring capabilities are required. Advanced remote sensing of standing biomass from space is the only practical way of comprehensively inventorying forest stocks globally, without the use of process models that may not be trusted by all parties. Unfortunately, estimating forest biomass with satellite remote sensing is not as straightforward as it may at first appear.

Remote sensing of vegetation productivity, abundance, and/or function is often performed using what is known as a *vegetation index* (VI), a ratio of measurements of near-infrared to red light reflected from the surface (Rouse et al., 1974). These are based on the observation that red light is absorbed by green leaves, while near-infrared light is reflected from them. High (low) values of such an index reflect large (small) amounts of photosynthetically active vegetation—but also high (low) fractional cover of the nonvegetated surface, the spectral reflectance of the nonvegetated components (e.g., soil), and the depth of the foliage. There have been many studies using VIs to estimate vegetation production and carbon storage in forests (Hai-qing et al., 2007; Huete, 2012), but there are sometimes problems. For example, VI data from NASA's MODIS sensor showed an apparent slight "greening" in Amazonian rainforest in 2005—the year that parts of the Amazon experienced the deepest drought conditions in over a century. The greening seen in the satellite VI imagery was surprising to ecologists. However, subsequent investigations showed that the VI data consistently showed a pattern of negligible changes in greenness levels for 90% of the anomalies, with small random patches of temporary greening and browning (Samanta et al., 2012). The inconsistency may have been partly owing to the small signal, the presence of smoke from the many forest fires that burned during the drought, the (then) poor screening, and BRDF effects (Morton et al, 2014).

In spite of these limitations, VIs have been useful in tracking forests through time. For example, over the period 2000–2010, the MODIS enhanced vegetation index (EVI) showed declining forest cover and/or density and/or photosynthetic activity during summer in four subregions of the Eastern United States: the Upper Great Lakes, southern Appalachian, mid-Atlantic, and southeastern Coastal Plain regions. Nearly 40% of the forested area within the mid-Atlantic subregion alone showed a significant decline in forest canopy cover (Potter et al., 2012). The declining trend in greenness was consistent with increasing temperatures and increasing variability in precipitation, with strong correlations with climate, moisture index, and growing degree days.

However, since VIs are based on measurements of reflected sunlight from green leaves, woody plant parts, and soil, they retain a dependence on parameters that are not directly related to biomass (units: $Mg \cdot ha^{-1}$). Standing forest biomass is a function of tree stem density, trunk diameter, and tree height: spectral reflectance ratios do not provide this information uniquely, so we must turn to other methods: *lidar*, *radar*, and *multiangle imaging*. *Lidar* (a contraction of <u>li</u>ght <u>d</u>etection <u>a</u>nd <u>r</u>anging) employs pulses of coherent laser light (visible or near-infrared EMR): these are used to illuminate a small *footprint* on the surface and the returns detected by a sensor (Fig. 17.9). Scanning lidar systems allow construction of 3-D models of a surface or feature (e.g., Radiohead, 2007).

Lidar has emerged as the best technology for accurate forest biomass mapping, but there are no dedicated orbiting vegetation lidar systems to date. NASA's promising <u>D</u>eformation, <u>E</u>cosystem <u>S</u>tructure, and <u>Dy</u>namics of <u>I</u>ce

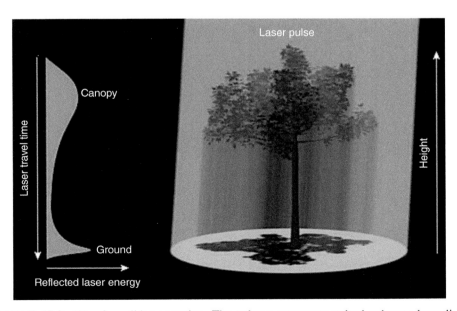

FIGURE 17.9 Waveform lidar operation. The column represents pulsed coherent laser light, resulting in a "waveform" from the returns that can be interpreted to give canopy height, crown shape, and woody biomass.

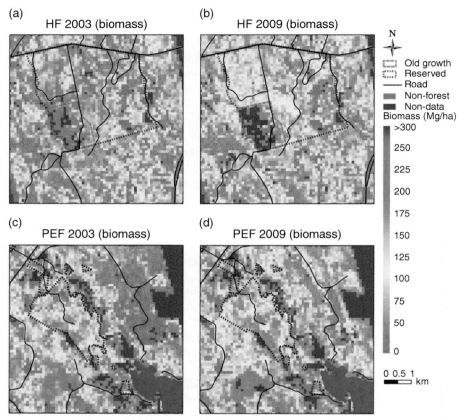

FIGURE 17.10 Aboveground biomass map for Howland Forest (HF) site (a) and (b) and Penobscot Experimental Forest (PEF) site in Maine. (c and d) In 2003 and 2009 at 1.0 ha level by the combined RH50 models. A color of orange to dark green indicates an increase of biomass. At HF site, the pink polygon is near-matured old-growth forest; and dark blue polygon is the outline of reserve area. From Huang et al. (2013).

(DESDynI) radar+lidar mission was canceled, so mapping is currently limited to aircraft campaigns (e.g., the Carnegie Airborne Observatory), although the NASA Global Ecosystem Dynamics Investigation (GEDI) advanced lidar mission will launch in 2019 (NASA 2014b). Data from ground-based (e.g., Echidna), airborne (e.g., NASA's Land, Vegetation, and Ice Sensor (LVIS)), and space-based (e.g., NASA's ICESat) lidar instruments have allowed the development of novel canopy structure and biomass measurement techniques. These instruments and associated techniques, data, and models are key in remote sensing of forest canopy structural parameters, both directly and in combination with multiangle approaches (Ni-Meister et al., 2010; Chopping et al., 2012). Huang et al. (2013) explored forest biomass prediction lidar waveform metrics using data from LVIS acquired over Howland Forest and Penobscot Experimental Forest, Maine, United States, and field-based inventory data. Two types of regression models were developed: combined models with no consideration of disturbances (fire, storms, pests, pathogens) and disturbance-specific models. The results demonstrated that the disturbance-specific models performed slightly better ($R^2 = 0.89$, RMSE = 27.9 Mg·ha^{-1}, and relative error of 22.6%) than the combined model ($R^2 = 0.86$, RMSE = 31.0 Mg·ha^{-1}, 25.1%) at the 20 m lidar footprint scale. When aggregated to 1.0 ha scale, both disturbance-specific and combined model predictions agreed well with field estimates, with $R^2 = 0.91$, RMSE = 23.1 Mg·ha^{-1}, 16.1%, and $R^2 = 0.91$, RMSE = 22.4 Mg·ha^{-1}, 15.6%. Biomass maps for 2003 and 2009 were produced (Fig. 17.10), and average annual biomass reduction rates from disturbance at Howland Forest and Penobscot Experimental Forest were calculated to be −7.0 and −6.2 Mg·ha^{-1}, while the average annual biomass accumulation rates from regrowth were +4.4 and +5.2 Mg·ha^{-1}, respectively.

Radar remote sensing employs pulses of much longer wavelength radio (microwave) waves; these interact with the large, hard structural elements of forests (stems, branches), thus providing information on canopy structure and woody biomass. While lidar and radar are active technologies, multiangle imaging exploits changes in the intensity of reflected sunlight with direction (because of shadowing and light scattering by leaves) and is thus a passive technology. In these systems, the target is observed from different angles, and information on canopy structure (canopy cover, mean canopy height) can be

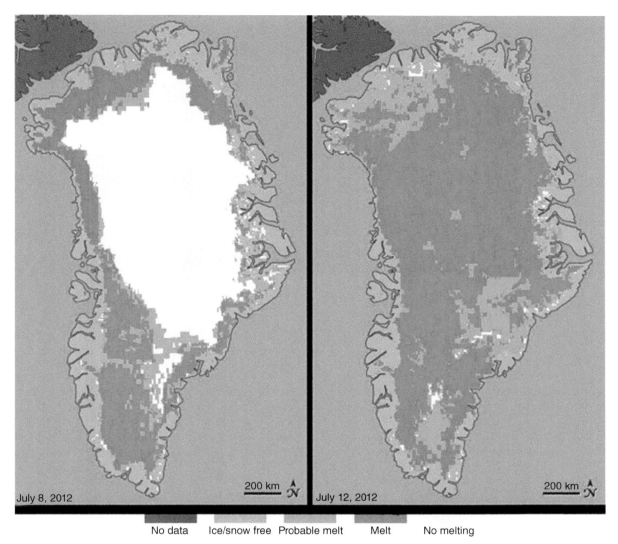

FIGURE 17.11 Extent of surface melt over Greenland's ice sheet on July 8 (40% of area) and July 12 (97% of area) 2012. Based on measurements from the Indian Space Research Organization Oceansat-2, NASA Moderate Resolution Imaging Spectroradiometer, and US Air Force Special Sensor Microwave Imager/Sounder. Credit: Nicolo E. DiGirolamo, SSAI/NASA GSFC, and Jesse Allen, NASA Earth Observatory. http://www.nasa.gov/topics/earth/features/greenland-melt.html.

inferred by using the observations to adjust the parameters of a canopy reflectance model (e.g., Wang et al., 2011).

The development and implementation of these advanced active and passive remote sensing methods are an important imperative because forest degradation is at least as important as deforestation on the most relevant timescales (decades): we must improve measurement precision and our ability to assess forest degradation if we are to inventory forest biomass and carbon pools with sufficient accuracy at regional to global scales (Houghton and Goetz, 2008). The DESDynI mission would have used both lidar and radar technologies synergistically to provide an accurate picture of the state of the world's forest stocks, including degradation (Freeman et al., 2009). It was selected as a first-tier (priority) mission by the Decadal Survey report of the US National Academy of Sciences in 2007 (National Research Council, 2007) but was shelved in 2011.

17.3.3 The Greenland Ice Sheet Surface Melt of July 2012

Understanding phenomena that occur at regional to global scales often requires *multiple, linked* data sets at commensurate spatial scales. The Greenland ice sheet surface melt event of the summer of 2012 provides a good case study. Using diverse remote sensing methods, NASA scientists determined that in July 2012, almost 97% of the surface of the Greenland ice sheet was melting, even at the higher, cooler elevations of

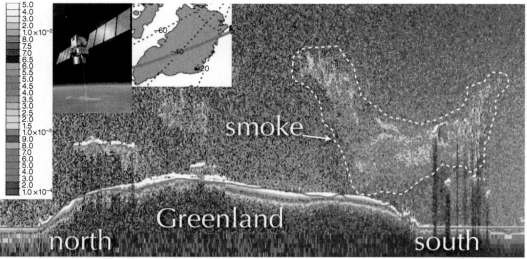

FIGURE 17.12 Latitudinal 532 nm (green laser) lidar profile through the lower atmosphere from the Cloud-Aerosol Lidar and Infrared Pathfinder Satellite Observations (CALIPSO) satellite, a joint NASA/French Space Agency mission. Map inset: satellite track over Greenland. Annotation: Jason Box (http://www.meltfactor.org/blog/?cat=74).

the center—something that had never before been observed and that NASA called "unprecedented" (Fig. 17.11; NASA, 2012b). This conclusion was based on measurements from no less than three satellite sensors: a radar instrument on the Indian Space Research Organization's Oceansat-2 satellite, thermal detectors on NASA's MODIS flown on the Terra and Aqua satellites, and the US Air Force' Special Sensor Microwave Imager/Sounder flown on Defense Meteorological Satellite Program (DMSP) satellites. Data from all three sensors showed that on July 8, about 40% of the ice sheet had undergone thawing at or near the surface, with the melting dramatically accelerating so that by July 12 an estimated 97% of the ice sheet had liquid water at the surface. The satellite data were combined in the NASA maps in the following manner: areas corresponding to those sites where at least one satellite detected surface melting were classified as "probable melt," while the areas correspond to sites where two or three satellites detected surface melting were classified as "melt."

Subsequent research suggested two proximate causes of the exceptional melting: changes in the location of the northern polar jet stream (Hanna et al., 2013) and darkening of the surface by deposition of soot particles from wildfires in the Arctic (Box et al., 2012, 2013). Such connections can be made *only* through the use of satellite remote sensing. For example, the Cloud–Aerosol Lidar and Infrared Pathfinder Satellite Observations (CALIPSO) satellite is able to show the trajectories of smoke (aerosols) entrained from wildfires in the atmospheric profiles it collects (Fig. 17.12).

When extraordinary claims are made in any area of science, extraordinary evidence is called for; scientists are, after all, trained to be skeptical. The scientists who saw melting in the satellite data did not at first trust the measurements (a good thing). However, multiple lines of independently collected evidence supported the 2012 wildfire–Greenland ice sheet melt hypothesis. Estimates of surface shortwave albedo (surface brightness) from NASA's MODIS showed a dramatic decrease in surface albedo (darkening) in the summer of 2012. This is what would be expected if important volumes of soot from distant wildfires were deposited on the ice sheet—and it is reflected in the MODIS albedo anomaly map (Fig. 17.13a) and the area-averaged values (Fig. 17.13b). This followed an accelerated darkening of the Greenland ice sheet from 2006 to 2012 compared with the previous 6 years.

In addition to the obvious decline in surface albedo, there is other supporting evidence from both remote sensing and ground observations: MODIS is able to determine the "snowline" that is a close proxy of equilibrium line altitude (ELA): the highest altitude at which winter snow survives in the summer. In 2012, this was higher than at any previous time; and the ground observed ELA was 3.7 times the standard deviation above the 21-year mean value (Box et al., 2013). Furthermore, meteorological data show that atmospheric circulation was characterized by warm air advection from the south into western Greenland, observed by the ground meteorological stations, radiosondes (instruments on weather balloons that measure temperature, humidity, and ozone), *and* satellite data. The subsequent flooding was witnessed by several imaging instruments, including the Advanced Land Imager on NASA's EO-1

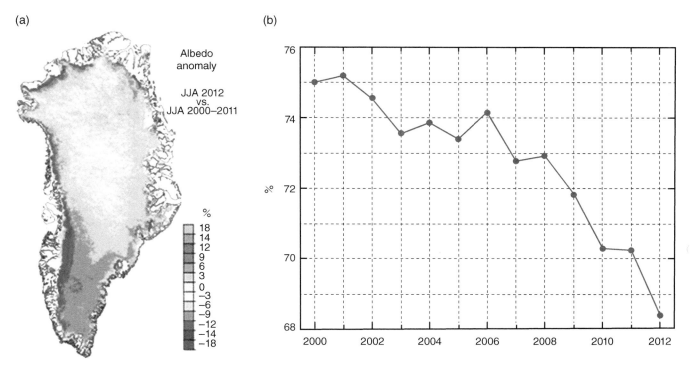

FIGURE 17.13 (a) Summer (June–July–August) albedo anomaly in 2012 relative to the 2000–2011 average. Data were derived from the MODIS MOD10A1 product. (b) Area-averaged albedo of the Greenland ice sheet during June through August each year of the period 2000–2012. Credit: Box et al. 2012, European Geosciences Union.

FIGURE 17.14 NASA Advanced Land Imager true color image of flooding in Kangerlussuaq, Greenland, on July 12, 2012. Credit: NASA Earth Observatory. http://earthobservatory.nasa.gov/NaturalHazards/view.php?id=78685 (video of this event: http://youtu.be/7SuJ1sFn_B0, http://youtu.be/RauzduvIYog last accessed 09/10/13).

satellite (Fig. 17.14). All these data—from Oceansat-2, MODIS, SSM/I, CALIPSO, radiosondes, and ground measurements—provide a coherent picture of what happened that is supported by multiple lines of evidence collected independently at different scales and using diverse technologies (active and passive microwave remote sensing, solar wavelength remote sensing, lidar remote sensing, ground and aerial sampling). It has become clear that there are important *teleconnections* between fire and ice, and we can surmise that changes in wildfire regimes will not only result in loss of lives, property, and timber—and reductions in air quality that have important health implications—but also have impacts on summer ice sheet melting.

17.3.4 Remote Sensing of Urbanization

Urbanization is a process seen in many parts of the world. It involves economic, social, demographic, resource distribution, and cultural factors. There are two types of urban expansion: the expansion of the city toward its exterior, a spatial process resulting in an increase in area, and from higher population density and greater economic activity within a city, without increasing area. The traditional study of city expansion relies on survey data, but it is difficult and expensive to conduct repeat surveys over a large region. Remote sensing can address this in a number of ways. For example, medium and high spatial resolution imaging instruments on orbiting satellites can show the extent of developed land. Another method is to examine maps of nighttime lights. With very sensitive detectors on satellites, it is possible to map artificial lighting in urban areas from street lighting, homes, factories, office buildings, shops, and vehicles. This is useful when we are interested in economic activity, since lit area is directly related to energy use. This method was used to assess increases in the extents and spatial configurations of urban areas along the Yellow River in Inner Mongolia Autonomous Region, China, over the period 1992–2003. The Chinese economy has expanded very rapidly in the last three decades, with growth rates above 10%, with much of the expansion occurring through manufacturing in the special economic zones on the eastern and southern coasts of the country. Nighttime light images from the Version 2 DMSP Operational Linescan System (OLS) were used to examine the question of economic development in more remote areas, such as this large region in Inner Mongolia (Qi and Chopping, 2007). Lit area data extracted for the county level were used with population and economic data to examine patterns of urban expansion for 1992, 1998, and 2003 (Fig. 17.15).

The results show not only a dramatic increase in urbanization and energy use but also that the region is becoming increasingly interconnected, that is, *metropolization* has occurred. This is evident from the almost continuous pattern of lights along the Hohhot–Baotou transportation corridor that links the two largest cities of the province (Fig. 17.15b). However, it has also clearly occurred all along the Yellow River and even at some distance from it: over this period, lit areas were expanding on the northern side of the city of Dongsheng toward Baotou, and vice versa. An important decline in urban population density that reflects increasing prosperity was also seen, while some of the more remote areas saw small towns disappear as people migrated to larger towns and cities (sometimes towns are covered by shifting desert sands). The use of satellite nighttime lights can thus illuminate our understanding with respect to multiple aspects of urbanization.

17.3.5 Remote Sensing of Aquifers I: Gravity Remote Sensing

Around the world, arid regions depend on groundwater from aquifers to satisfy domestic and agricultural needs. Northern India is no exception: with rates of extraction increasing as a result of population growth, economic development, and expansion of crop area, there is great concern about aquifer depletion. Until 2009, it was known that the states of Rajasthan, Punjab, and Haryana were extracting groundwater at a rate exceeding replenishment—but depletion and recharge rates were unknown (NASA, 2009a). Knowledge of these rates is critical: we would like to know if the aquifer is likely to be able to support human demands until 2020, 2050, or 2100. To address this, gravity anomalies mapped by NASA's Gravity Recovery and Climate Experiment (GRACE) mission were used (Adam, 2002; Tapley et al., 2004). GRACE consists of two spacecraft flying in the same orbit approximately 200 km apart that bounce microwave signals off each other to determine the precise distance between them. When the satellites fly over a more massive part of the Earth, they are first accelerated and then decelerated, altering this distance. In this way, maps of gravity anomalies can be obtained. Since the mass of an aquifer changes according to the volume of water stored in it, GRACE is able track water storage changes beneath the land surface. NASA researchers used 6 years of monthly GRACE data for northern India to examine depletion and recharge rates (Fig. 17.16). They found that groundwater levels have been declining by an average of $0.3\,m\cdot year^{-1}$ (1 m every 3 years; Rodell et al., 2009). According to NASA, "More than 109 km^3 of groundwater was extracted between 2002 and 2008—double the capacity of India's largest surface water reservoir, the Upper Wainganga, and triple that of Lake Mead, the largest man-made reservoir in the United States... The loss is particularly alarming because it occurred when there were no unusual trends in rainfall. In fact, rainfall was slightly above normal for the period" (NASA, 2009a). GRACE has also been used to examine groundwater storage trends and drought in the United States. In 2012, nearly 80% of the US farm, orchard, and grazing land was affected, and 28%

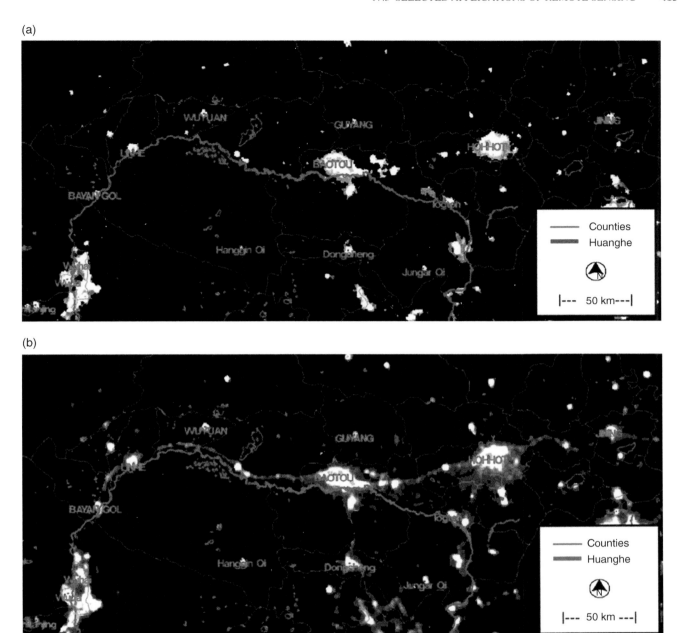

FIGURE 17.15 Extents and spatial configuration of nighttime lights in the vicinity of the Yellow River in Inner Mongolia Autonomous Region, China, in (a) 1992 and (b) 2003. These maps are from the Version 2 Defense Meteorological Satellite Program (DMSP) Operational Linescan System (OLS) nighttime lights series. Similar maps can be made with the "day–night band" of the Visible Infrared Imaging Radiometer Suite (VIIRS) on the Suomi NPP satellite launched in October 2011. Low levels of light can be detected partly because all light in the green to near-infrared wavelengths is used.

experienced extreme to exceptional drought (Fig. 17.17; NOAA, 2012; NASA, 2013a). The drought fueled wildfires, damaged crops, and caused near-record-low water levels on the Mississippi River (NASA, 2013b). California is currently in the grip of an even deeper drought and GRACE has recently shown alarming reductions in groundwater storage across the southern US and California (Famiglietti and Rodell, 2013).

17.3.6 Remote Sensing of Aquifers II: Radar Remote Sensing

The ground-penetrating capability of radar has been used for locating belowground water resources, mostly recently in Kenya, prompting the headline and subhead "Kenya water discovery brings hope for drought relief in rural north" and "Two vast underground aquifers seen by satellite

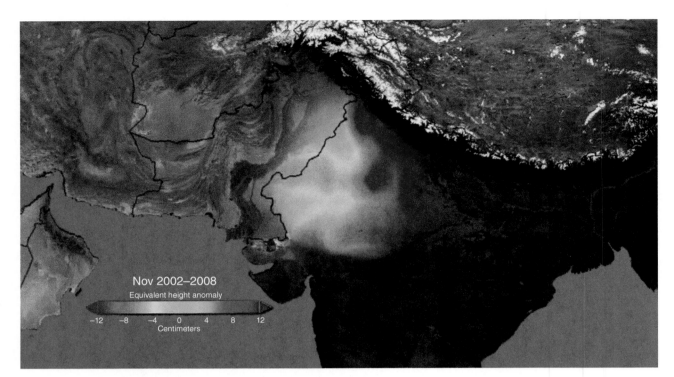

FIGURE 17.16 Groundwater storage changes in northwestern India between 2002 and 2008, relative to the mean for the period expressed as the height of an equivalent layer of water, ranging from −12 cm (deep red) to 12 cm (dark blue). These maps were derived via analysis of gravity anomalies using data from NASA's twin satellite Gravity Recovery and Climate Experiment (GRACE) system, launched in 2002. Credit: NASA/Trent Schindler and Matt Rodell.

technology in Turkana county may provide much-needed water for barren area," respectively (NASA, 2013c; Plaut, 2013). This area is the least developed and poorest area of Kenya and has suffered extended drought that impact both crop and livestock economies. The Groundwater Resources Investigation for Drought Mitigation in Africa Programme (GRIDMAP) groundwater mapping project was spearheaded by the United Nations Educational, Scientific and Cultural Organization (UNESCO) in partnership with the government of Kenya and with the financial support of the Government of Japan. By the end of the project, two vast underground aquifers storing billions of liters of water had been discovered. These aquifers contain at least 250 billion m^3 of water, or about 66 trillion gallons; they are replenished by rainfall at a rate of approximately 898 billion gallons annually. Radar Technologies International (RTI, of Tarascon en Provence, France) achieved this major success using its WATEX™ mapping system that combines passive imagery from Landsat 7 with radar, climate, and seismic data to locate potential water sources. The radar satellite remote sensing data products used have included European Space Agency European Remote Sensing 1 (ERS-1) satellite 30 m C-band synthetic aperture radar imagery, Japan Aerospace Exploration Agency (JAXA) Japanese Earth Resources Satellite 1 (JERS-1) 18 m L-band synthetic aperture radar imagery, and NASA Shuttle Radar Topography Mission (SRTM) 90 m digital topographic maps. While the Landsat imagery shows only surface features, the C-band radar sees to a depth of about 50 cm, and the L-band radar penetrates down to a maximum of 20 m. Using all three provides a cross-sectional model of the landscape and aquifer detection (Fig. 17.18). In addition to the Kenya success, RTI has applied these methods in Afghanistan, Darfur (Sudan), Chad, Angola, Iraq, and Ethiopia. According to RTI, in each case, the ability of WATEX™ to build a new large-scale map of the groundwater resource where only unreliable information had previously existed has been recognized; it has also made borehole drilling more effective, improving the rate of successful drilling by over 95% in all cases (RTI, 2013).

17.3.7 Remote Sensing of Urban Air Pollution

Air pollution has become an increasingly pressing issue in the rapidly growing cities of Asia, such as Beijing, Harbin, and Shanghai, China. This is owing to the continued

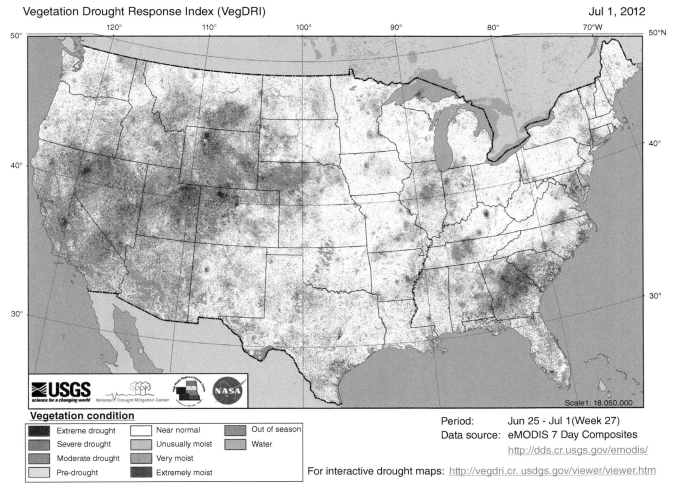

FIGURE 17.17 Vegetation Drought Response Index (VegDRI) for the conterminous United States on June 25–July 1, 2012, showing the extent of the drought (NOAA, 2012). The VegDRI integrates satellite-based observations of vegetation condition, climate data, and information on land cover/land use type, soil characteristics, and ecological setting.

burning of coal for the generation of electricity and the switch from bicycles and buses to private motorized transportation for large numbers of people, exacerbating already high airborne particulate loadings from desert dust that blows in from northwestern China. Nowhere is this problem more evident than in the capital, Beijing. The "Airpocalypse" of early 2013 was a major public health event; however, it followed other similar but less acute events. The issue of urban air quality became pressing for the authorities in 2008, when Beijing hosted the Olympic Games for the first time. Knowing that the poor air quality was likely to be noticed by athletes and spectators alike, steps were taken to reduce particulate pollution, including restrictions on personal vehicle use in the capital (reduced by about 50%) and emissions from electrical power generation and industry.

The results of these efforts can be assessed using satellite remote sensing, that allows the measurement of different constituents of the atmosphere: gases, aerosol particles, and clouds. The technologies employed include spectroscopy, lidar, sounding, and multiangle observing (looking at the target air mass from different directions with respect to the solar direction). Scientists have used images from the NASA Multi-angle Imaging Spectro-Radiometer (MISR) to show the changes in air pollution levels during the Games over East China, including Beijing. The MISR instrument has nine cameras that look through the atmosphere at different angles: four viewing ahead of the Terra spacecraft, one directly down, and four viewing to the aft. The different viewing angles allow calculation of particulate (aerosol) loadings and estimates of aerosol type. MISR saw a rapid and important reduction in air pollution during the

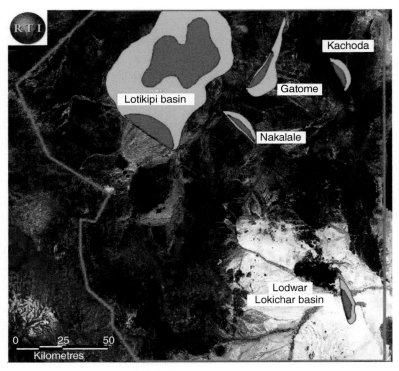

FIGURE 17.18 Regional model representation of five deep-seated structures in northern-central Turkana County, Kenya. These deep-seated aquifers (100–3000 m) were detected with the Radar Technologies International WATEX™ Deep Aquifer Model (DAM) that uses Landsat imagery, Shuttle Radar Topography Mission topographic maps, and C- and L-band synthetic aperture radar imagery.

Games (August 8–September 17, 2008), although the effects of transient weather conditions and topography are still evident (NASA, 2009b; southern and easterly winds remove pollutants, while northern or westerly winds cause accumulation of pollutants owing to the mountains). The story told by MISR follows this sequence:

July 25	Heavy particle pollution was apparent from MISR over the entire Beijing metropolitan area, even though pollution controls were in imposed on July 20 (Fig. 17.19, top left), corroborated by fine particle pollution (PM 2.5) readings on the ground that reached 140 μg m^{-3} in downtown Beijing. This was a function of both emissions and weather conditions (high temperatures and light winds from the east and southeast)
August 8–24 (Olympic Games)	MISR was unable to observe aerosol loading because of cloud cover; however, pollution restrictions remained in place for the duration of the Olympic and Paralympic Games (August 8–September 17)
September 2	According to MISR measurements, air quality improved considerably over the entire Beijing area (Fig. 17.19 upper right). This is corroborated by a daily PM 2.5 concentration measured on the ground of only 19 μg m^{-3}. The improvement may reflect the fact that air pollution control measures had been fully implemented for 3 weeks, as well as improved weather conditions (cooler temperatures, lower relative humidity, and higher wind speeds from the north)
September 4	MISR observed low aerosol concentrations over Beijing itself (Fig. 17.19 lower left), but regional pollution transport was starting began to have an influence, with aerosol concentrations over the heavily industrialized regions south of Beijing climbing rapidly
September 20	Three days after the end of the Paralympic Games, MISR observed increased aerosol concentrations over Beijing (Fig. 17.19 lower right) as anthropogenic emissions began to return to pre-Olympic levels (with daily PM2.5 concentrations at the monitoring site were 70 μg m^{-3}), even though some air pollution control measures were still in effect

This episode shows how remote sensing from orbit can inform questions of public health by providing not only measurements of city airborne particulate pollution levels but also the larger regional context; this "bigger picture" is often required in order to understand the proximate causes of threats to human health from poor air quality.

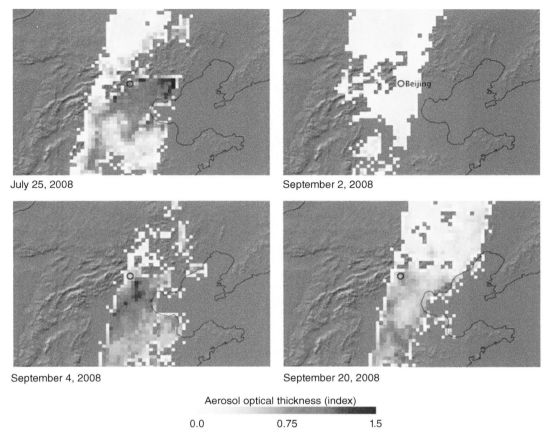

FIGURE 17.19 Aerosol loadings (including airborne particulate pollution) over Beijing and the surrounding region from the NASA/JPL Multi-angle Imaging SpectroRadiometer (MISR) flown on the Terra satellite. High aerosol concentrations (poor air quality) are brown, while low concentrations (good air quality) are cream. From NASA's Earth Observatory: http://earthobservatory.nasa.gov/IOTD/view.php?id=38290.

17.4 CONCLUDING REMARKS

Remote sensing is an extraordinarily useful tool for addressing many of the environmental problems we face today. With the advent of orbiting satellites remote sensing has become useful at a wide variety of scales, although it is still especially useful at regional to global, scales where other methods are often inadequate. It can be used to address a very wide range of applications, some quite surprising; for example, the mapping of gravity anomalies by NASA's GRACE mission allows us to infer storage in aquifers that supply drinking water to millions of people in Asia and permit extensive dryland agricultural production in (e.g.,) western Texas.

While the satellite remote sensing record length is limited to several decades at most, this is often adequate for assessing the impacts of many of the environmental problems we are facing. Moreover, new technologies are allowing applications that were not even imagined just a few years ago (e.g., mapping ocean salinity and soil moisture using microwave measurements, estimating ice sheet mass balance and aquifer depletion rates using gravity anomalies, inventorying forest biomass using radar and lidar systems, tracking black carbon aerosol transport from wildfires using lidars, spectropolarimeters, and multi-angle imagers). A recent application of a new technology is very-high-resolution spectroscopy for monitoring atmospheric CO_2. On July 2, 2014, NASA's OCO-2 was finally successfully placed in orbit, after an unfortunate launch failure in 2009 (NASA, 2014). OCO-2 is capable of measuring column CO_2 at scales on the order of 2.5 km and allows the identification of important CO_2 sources and sinks on a global scale, including identification of terrestrial biosphere sinks using the new technique of solar-induced chlorophyll fluorescence (Frankenberg et al, 2013). (i.e., forest regrowth). Other recent advances using quite different technologies include the global mapping of soil moisture and freeze-thaw cycles by NASA's Soil Moisture Active Passive (SMAP) mission launched in January 2015 (Entekhabi et al, 2015); and the deployment of a constellation of tiny cubesat imagers by Planet Labs, Inc. of San Francisco (www.planet.com). The cubesats, termed "Doves" by the firm, are about the size of a shoebox and carry an 90 mm aperture optical telescope payload, coupled with a high-resolution camera. Constellations ("Flocks of Doves") are a means to overcome the temporal sampling limitations of existing commercial high-resolution satellite imagers (Marshall, 2014). The Doves are deployed via the International Space Station and 132 had been put into low Earth orbit by the end of 2014 (Hand 2015). Planet Labs' Doves are visible light imagers but Surrey Satellite Technology Ltd has developed and deployed the international Disaster Monitoring Constellation (DMC) of medium-resolution multispectral imagers coordinated by DMCii (www.dmcii.com), in which each satellite is independently owned and controlled by a DMC Consortium member of high spatial resolution airborne and satellite imagery is cataloging changes to coastal landscapes affected by large storms, such as changes to the New Jersey shore owing to Superstorm Sandy (NASA, 2012c). Constellations of imagers will be useful in tracking the development of natural disasters in a much more timely manner than has been possible to date, potentially saving many lives.

There is almost no geoscience, geographic, environmental, or ecological discipline that has not benefited from the use of remote sensing: from wildfire tracking to oceanography, forestry to volcanology, coral reef mapping to assessing ozone depletion, crop yield prediction to coastal erosion, urban and regional planning to water quality monitoring, estimating building heat loss to invasive species mapping, and measurement of atmospheric carbon dioxide, carbon monoxide, NO_x, and aerosol loadings to algal blooms. We must make sure that we continue to develop and exploit this amazing technology and that scientists and environmental managers receive the education and training they need to make best use of this unique resource.

ACKNOWLEDGMENTS

I would like to thank the anonymous reviewer, my graduate student Rocio Duchesne, and the students in my Fall 2013 *Fundamentals of Remote Sensing of Environment* classes for providing candid criticisms of early drafts of the manuscript and notably Jeffrey Baker, Carolyn Haines, and Jeffrey Lenik. While this chapter has benefited materially from their eagle eyes and keen discernment, any errors and inaccuracies remain mine.

REFERENCES

Adam, D. 2002. Gravity measurement: Amazing grace. *Nature* 416 (6876): 10–11.

Box, J.E., X. Fettweis, J.C. Stroeve, M. Tedesco, D.K. Hall, and K. Steffen. 2012. Greenland ice sheet albedo feedback: Thermodynamics and atmospheric drivers. *The Cryosphere* 6: 821–839.

Box, J.E., J. Cappelen, C. Chen, D. Decker, X. Fettweis, T. Mote, M. Tedesco, R.S.W. van de Wal, and J. Wahr. 2013. Greenland Ice Sheet. *In* Jeffries, M.O., Richter-Menge, J.A., and Overland, J.E. (eds.) *Arctic Report Card 2012*. United States National Oceanic and Atmospheric Administration. http://www.arctic.noaa.gov/reportcard, last accessed September 2, 2013; http://www.arctic.noaa.gov/reportcard/greenland_ice_sheet.html, last accessed September 2, 2013.

Chopping, M. 2011. CANAPI: Canopy analysis with panchromatic imagery. *Remote Sensing Letters* 2(1): 21–29.

Chopping, M., M. North, J. Chen, C.B. Schaaf, J.B. Blair, J.V. Martonchik, and M. Bull. 2012. Forest cover and height from MISR in a topographically complex landscape assessed with high quality reference data. *IEEE Journal of Selected Topics in Applied Earth Observations and Remote Sensing* 5(1): 44–58.

Entekhabi, D., S. Yueh, P. E. O'Neill, E. F. Wood, E. G. Njoku, J. K. Entin, and K. H. Kellogg. 2015. The NASA Soil Moisture Active Passive (SMAP) Mission Status and Early Results, *Geophysical Research Abstracts* 17, EGU2015-5973, EGU General Assembly 2015.

Famiglietti, J.S., and M. Rodell. 2013. Water in the Balance. *Science* 340: 1300–1301.

Francis, J.A. and S.J. Vavrus. 2012. Evidence linking Arctic amplification to extreme weather in mid-latitudes. *Geophysical Research Letters* 39: L06801. doi:10.1029/2012GL051000.

Frankenberg, C., J. Berry, L. Guanter, and J. Joiner. 2013. Remote sensing of terrestrial chlorophyll fluorescence from space, SPIE Newsroom online article. DOI: 10.1117/2.1201302.004725 http://spie.org/x92267.xml last access July 7, 2015.

Freeman, A., P. Rosen, R. Jordan, W.T.K. Johnson, S. Hensley, T. Sweetser, A. Loverro, J. Smith, G. Sprague, and Y. Shen. 2009. DESDynI – A NASA Mission for Ecosystems, Solid Earth, and Cryosphere Science. *Proceedings of the 4th International Workshop on Science and Applications of SAR Polarimetry and Polarimetric Interferometry (PolInSAR 2009)*, January 26–30, 2009, Frascati, Italy. ESA Special Publication 668.

Hai-qing, H., L. Yuan-chun, and J. Yan. 2007. Estimation of the carbon storage of forest vegetation and carbon emission from forest fires in Heilongjiang Province, China. *Journal of Forestry Research* 18(1): 17–22.

Hand, E.. 2015. How tiny satellites spawned in Silicon Valley will monitor a changing Earth, *Science*, http://news.sciencemag.org/space/2015/04/feature-how-tiny-satellites-spawned-silicon-valley-will-monitor-changing-earth last access July 7, 2015.

Hanna, E., X. Fettweis, S.H. Mernild, J. Cappelen, M.H. Ribergaard, C.A. Shuman, K. Steffen, L. Wood, and T.L. Mote. 2013. Atmospheric and oceanic climate forcing of the exceptional Greenland ice sheet surface melt in summer 2012. *International Journal of Climatology* 34(4): 1022–1037.

Houghton, R.A. and S.J. Goetz. 2008. New satellites help quantify carbon sources and sinks. *Eos, Transaction, American Geophysical Union* 89(43), 417–418.

Huang, W., G. Sun, R. Dubayah, B. Cook, P. Montesano, W. Ni, and Z. Zhang. 2013. Mapping biomass change after forest disturbance: Applying LiDAR footprint-derived models at key map scales. *Remote Sensing of Environment* 134: 319–332.

Huete, A.R. 2012. Vegetation indices, remote sensing and forest monitoring. *Geography Compass* 6: 513–532.

Irons, J.R., J.L. Dwyer, and J.A. Barsi. 2012. The next Landsat satellite: The Landsat Data Continuity Mission. *Remote Sensing of Environment* 122: 11–21.

Khanna, S., M.J. Santos, and S.L. Ustin. 2011. An integrated approach to a biophysiologically based classification of floating aquatic macrophytes. *International Journal of Remote Sensing* 32(4): 1067–1094.

Kim, D-H, J. O. Sexton, P. Noojipady, C. Huang, A. Anand, S. Channan, M. Feng, and J. R. Townshend. 2014. Global, Landsat-based forest-cover change from 1990 to 2000. *Remote Sensing of Environment* 155: 178–193.

Kluger, J. and B. Walsh. 2013. Timelapse, Landsat 8/Time/Google Timelapse. http://world.time.com/timelapse/, last accessed September 2, 2013.

Marshall, W.. 2014. Tiny satellites that photograph the entire planet, every day. TED Talk at https://youtu.be/UHkEbemburs last access July 7, 2015.

Masek, J.G., E.F. Vermote, N. Saleous, R. Wolfe, F.G. Hall, F. Huemmrich, F. Gao, J. Kutler, and T.K. Lim. 2012. LEDAPS: Landsat Calibration, Reflectance, Atmospheric Correction Preprocessing Code. Model product. Oak Ridge National Laboratory Distributed Active Archive Center, Oak Ridge, TN. http://dx.doi.org/10.3334/ORNLDAAC/1080, last accessed March 11, 2015.

Morton, D. C., J. Nagol, C. C. Carabajal, J. Rosette, M. Palace, B. D., Cook, E. F. Vermote, D. J. Harding, and P. R. J. North. 2014. Amazon forests maintain consistent canopy structure and greenness during the dry season, *Nature* 506, 221–224.

NASA. 2009a. NASA Satellites Unlock Secret to Northern India's Vanishing Water. http://www.nasa.gov/topics/earth/features/india_water.html, last accessed September 30, 2013.

NASA. 2009b. Aerosols Over Beijing, NASA Earth Observatory. http://earthobservatory.nasa.gov/IOTD/view.php?id=38290, last accessed July 2, 2014.

NASA. 2012a. Arctic Sea Ice Shrinks to New Low in Satellite Era. http://www.nasa.gov/topics/earth/features/arctic-seaice-2012.html, last accessed August 23, 2013.

NASA. 2012b. Satellites See Unprecedented Greenland Ice Sheet Surface Melt. http://www.nasa.gov/topics/earth/features/greenland-melt.html, last accessed September 29, 2013.

NASA. 2012c. A Changed Coastline in Jersey, NASA Earth Observatory. http://www.earthobservatory.nasa.gov/NaturalHazards/view.php?id=79622, last accessed June 30, 2014.

NASA. 2013a. Water Storage Maps Show Improvement. http://earthobservatory.nasa.gov/IOTD/view.php?id=81408, last accessed June 29, 2014.

NASA. 2013b. Despite Rain, Drought Lingers in the United States. http://earthobservatory.nasa.gov/IOTD/view.php?id=80221, last accessed September 30, 2013.

NASA. 2013c. Vast Water Reserves Found in Drought-Prone Northern Kenya. http://landsat.gsfc.nasa.gov/?p=6404, last accessed September 30, 2013.

NASA. 2013d. NASA Satellite Data Improving Volcanic Ash Forecasts for Aviation Safety. http://www.nasa.gov/larc/volcanic-ash-aviation/#.U7RakI1vmHk, last accessed July 2, 2014.

NASA. 2014a. NASA Launches Carbon Mission to Watch Earth Breathe. http://www.jpl.nasa.gov/news/news.php?release=2014-215, last accessed July 2, 2014.

NASA. 2014b. NASA Selects Instruments to Track Climate Impact on Vegetation, NASA Press release 14-199, July 30, 2014. URL: http://www.nasa.gov/press/2014/july/nasa-selects-instruments-to-track-climate-impact-on-vegetation/ last access July 7, 2015.

National Research Council (NRC) Committee on Earth Science and Applications from Space: A Community Assessment and Strategy for the Future. 2007. *Earth Science and Applications from Space: National Imperatives for the Next Decade and Beyond*. The National Academies Press, Washington, DC.

Ni-Meister, W., S. Lee, A.H. Strahler, C.E. Woodcock, C.B. Schaaf, T. Yao, K.J. Ranson, G. Sun, and J.B. Blair. 2010. Assessing general relationships between aboveground biomass and vegetation structure parameters for improved carbon estimate from lidar remote sensing. *Journal of Geophysical Research* 115: G00E11.

NOAA. 2012. NOAA National Climatic Data Center, State of the Climate: Drought for June 2012. http://www.ncdc.noaa.gov/sotc/drought/2012/6, last accessed June 29, 2014.

Plaut, M. 2013. Kenya water discovery brings hope for drought relief in rural north. *The Guardian* (US Edition, Global Development section). http://theguardian.com, last accessed September 10, 2013.

Potter, C., S. Li, and C. Hiatt. 2012. Declining vegetation growth rates in the eastern United States from 2000 to 2010. *Natural Resources* 3: 184–190. .

Qi, X. and M. Chopping. 2007. Expansion of urban area in the Yellow River zone, Inner Mongolia Autonomous Region, China, from DMSP OLS nighttime lights data. *Proceedings of 2007 IEEE International Geoscience and Remote Sensing Symposium*, July 23–27, 2007, Barcelona, Spain, pp. 2002–2005.

Radar Technologies International (RTI). 2013. Advanced Survey of Groundwater Resources of Northern and Central Turkana County, Kenya: Final Technical Report, 78 p. http://www.rtiexploration.com/s/Turkana-Survey-Report.pdf, last access September 30, 2013.

Radiohead. 2007. House of Cards. http://youtu.be/8nTFjVm9sTQ?t=2m11s, last accessed September 2, 2013.

Ritchie, J.C., P.V. Zimba, and J.H. Everitt. 2003. Remote sensing techniques to assess water quality. *Photogrammetric Engineering & Remote Sensing* 69(6): 695–704.

Rodell, M., I. Velicogna, and J. Famiglietti. 2009. Satellite-based estimates of groundwater depletion in India. *Nature* 460: 999–1002.

Rouse, J.W., R.H. Haas, J.A. Schell, and D.W. Deering. 1974. Monitoring vegetation systems in the Great Plains with ERTS. *In* S.C. Freden, E.P. Mercanti, and M. Becker (eds) *Third Earth Resources Technology Satellite–1 Symposium*. Volume I: Technical Presentations, NASA SP-351. NASA, Washington, DC, pp. 309–317.

Samanta, A., S. Ganguly, E. Vermote, R.R. Nemani, and R.B. Myneni. 2012. Interpretation of variations in MODIS-measured greenness levels of Amazon forests during 2000 to 2009. *Environmental Research Letters* 7: 024018.

Song, C., B. Huang, and L. Ke. 2013. Modeling and analysis of lake water storage changes on the Tibetan Plateau using multi-mission satellite data. *Remote Sensing of Environment* 135: 25–35.

Tapley, B.D., S. Bettadpur, J.C. Ries, P.F. Thompson, and M. Watkins. 2004. GRACE measurements of mass variability in the earth system. *Science* 305(5683): 503–505.

United Nations. 2009. UN-REDD Program (Reducing Emissions from Deforestation and Forest Degradation) Program. http://www.un-redd.org/AboutREDD/tabid/102614/Default.aspx, last accessed September 3, 2013.

Wall, M.. 2013. Planet Labs Unveils Tiny Earth-Observation Satellite Family, Space.com article published August 31, 2013. http://www.space.com/22622-planet-labs-dove-satellite-photos.html, last access November 2, 2014.

Wang, Z., C.B. Schaaf, P. Lewis, Y. Knyazikhin, M.A. Schull, A.H. Strahler, T. Yao, R.B. Myneni, and M. Chopping. 2011. Retrieval of canopy vertical structure using MODIS data. *Remote Sensing of Environment* 115(6): 1595–1601.

Zwally, H.J., B. Schutz, W. Abdalati, J. Abshire, C. Bentley, A. Brenner, J. Bufton, J. Dezio, D. Hancock, D. Harding, T. Herring, B. Minster, K. Quinn, S. Palm, J. Spinhirne, and R. Thomas. 2002. ICESat's laser measurements of polar ice, atmosphere, ocean, and land. *Journal of Geodynamics* 34(3–4): 405–445.

18

GEOGRAPHIC INFORMATION SYSTEMS IN ENVIRONMENTAL MANAGEMENT

DANLIN L. YU[1] AND SCOTT W. BUCHANAN[2]

[1]Department of Earth and Environmental Studies, Montclair State University, Montclair, NJ, USA
[2]Department of Natural Resources Science, University of Rhode Island, Kingston, RI, USA

Abstract: The utilization of geographic information (information that is associated with specific *geographic location*) has attracted more and more attention in a wide range of fields. The application of geographic information systems (GIS) in environmental management is of special interest due to GIS's spatially explicit capability to integrate environmental data analysis and decision support. The specific characteristics of geographic information (inherently associated with set locations) make the interpretation and analysis of geographic information rather unique. Geographic information science, which develops from the applications of GIS, is experiencing rapid growth and offers great opportunities for interdisciplinary investigation of scientific endeavors such as environmental management. In this chapter, we will give some background of the development of GIScience and a case study of how GIS can be employed for a wildlife ecology practice. More importantly, however, we devote this particular chapter to the discussion of the characteristics of geographic information and introduce a few classical and new techniques in analyzing geographic information that could be of importance to environmental management. In particular, this chapter pays specific attention to the analysis of spatial heterogeneity that inherently exists in geographic information and the regressed relations among variables using geographic information. The extension of analyzing spatial heterogeneity in a temporal context is also introduced and discussed. The chapter echoes a belief that one of the major tasks of GIScience is to develop new methods to help understand the ever-increasing volume of geographic information to support environmental management. We hope the chapter will serve as a starting place for interested scholars to energetically seek better analytical techniques that, through the application of GIScience, will provide new insights and opportunities for environmental management.

Keywords: geographic information sciences, geographic information systems, environmental management, geographic information analysis, ArcGIS®, spatial reference system, geographic coordinate systems, projection, projected coordinate systems, spatial effects, spatial autocorrelation, spatial heterogeneity, geographically weighted regression, Bayesian spatially varying coefficient process, spatiotemporal process, spatiotemporal model, geographically weighted panel regression, wildlife ecology, snakes in the sand.

18.1	Geographic Information Systems in Environmental Management	424	18.5 The Next Step, Geographic Information in a Temporal Context: Spatiotemporal Analysis	431
18.2	What is GIS/GIScience and What Does it Do?	424	18.5.1 Studies on Spatiotemporal Models and Processes	431
18.3	Geographic Coordinates and Coordinate Transformation (Map Projection)	425	18.5.2 Studies on Relationships among Variables	432
	18.3.1 GCS	426	18.6 GIS in Action: Wildlife Ecology, Snakes in the Sand	433
	18.3.2 Map Projection and PCS	426	18.7 Conclusion	435
18.4	The Unique Geographic Information: Spatial Data are Special	427	References	436
	18.4.1 Types of Spatial Data	427		
	18.4.2 Spatial Effects Make Spatial Data Special	427		
	18.4.2.1 Analyzing Spatial Autocorrelation	427		
	18.4.2.2 Analyzing Spatial Heterogeneity	428		

An Integrated Approach to Environmental Management, First Edition. Edited by Dibyendu Sarkar, Rupali Datta, Avinandan Mukherjee, and Robyn Hannigan.
© 2016 John Wiley & Sons, Inc. Published 2016 by John Wiley & Sons, Inc.

18.1 GEOGRAPHIC INFORMATION SYSTEMS IN ENVIRONMENTAL MANAGEMENT

Environmental management is not really the management of the environment, as the name might suggest, but management of human activities and their impacts on the natural environment. Environmental management requires that we understand how human activities impact natural resources and ecosystems. The ultimate goal for environmental management is to seek a dynamic balance between human activities and the environment. Since humans are, in fact, an integral part of the environment, environmental management seeks to enable and support sustainable development. Locational information regarding natural resources, pollution sites, specific ecosystems, land and water resources, various human activities (residential, commercial, and industrial), and the like is crucial to the design, planning, and implementation of effective environmental management strategies. The cartographic capability of a geographic information system (GIS) provides essential support to store, retrieve, and display such information, either retrospectively or in real time.

Cartographic capability, however, has been, for quite some time, regarded as the only capability GIS offers to environmental management (Berry, 1999). Albeit central to environmental management, GIS offers much more than just visualization of location information through creation of a map. GIS allows for spatial data manipulation and analysis, such as map overlay, spatial or attribute queries, and proximity analysis. The coupled human and environmental system (CHES), which is the focus of environmental management, is essentially a complex system that is comprised of a human subsystem (broadly defined, including all human activities that both positively and negatively impact upon the environment) and an environmental subsystem (including resources, environmental carrying capacities, etc.). Maps can often be used to visualize the spatial relationships of various components of these two subsystems. The intrinsic and detailed relations are not just spatial and so cannot be fully expressed in a two-dimensional map. Statistical or empirical models needed to explore such relations. For instance, a map with well-designed color schemes and symbols will give a direct and visual record of how specific environmental management practices (such as waste water treatment and brownfield redevelopment) impact land use or land values in specific area. This map does not, however, provide detailed quantitative measures of the magnitude of the impact used in decision making.

To answer these types of question, a detailed understanding of interactions among the involved components, from both the human and environmental subsystems, is required. With such data, statistical models can be developed to support further and detailed analysis of interactions among variables across space and time. Traditional statistical modeling, however, often concentrates on relations between attributes without much consideration of the fact that those attributes are actually attached to specific geographic locations. The importance of attributes being attached to specific geographic locations is that attributes are no longer independent from one another (often referred to as spatial autocorrelation). The lack of independence reduces the effective sample size thus reducing the power and significance of a statistical model. If the issue is not taken into consideration, as is the case in most traditional statistical models, the analytical results could be rather misleading or even erroneous. GIS provides excellent tools to evaluate spatial associations among geographic objects, and newly developed analytical approaches are able to take into consideration such associations.

The remainder of this chapter will include a brief discussion of what GIS really is (other than being an excellent cartographic toolset) and what can be done with it. The chapter will elaborate the concepts of coordinates (how geographic coordinates are determined) and coordinate transformation (projection) since these two concepts constitute the fundamental building blocks of GIS. The fourth section discusses why geographic information (interchangeably called spatial data) is unique and how such uniqueness can be treated, explored, and embedded in further analyses. This is followed by an introduction to the incorporation of temporal information into geographic information analysis. Integration of temporal information is critical to contemporary environmental management practices. Such methodologies attempt to provide a holistic understanding of geographic information in a temporal context. The chapter concludes with a summary of GIS and applications to environmental management.

18.2 WHAT IS GIS/GISCIENCE AND WHAT DOES IT DO?

Without engaging in a debate about GIS is an acronym for geographic information systems or geographic information science (see Pickles, 1997; Wright & Goodchild, 1997a, 1997b), we follow common practice and use GIScience when referring to geographic information science and GIS when referring to geographic information systems. Such clarification is mainly semantic, however as modern applications of GIS as GIScience have rendered the debate moot.

GIS, as a toolset, has been defined and redefined in many ways (Burroughs, 1986; Clarke, 1997; Goodchild, 1992). A common understanding of GIS is that it is a powerful set of tools that can be used to deal with digitally stored geographic information. In this regard, GIS is often simplified to a particular software package (such as the popular ArcGIS® series). Definition of GIScience, however, proves to be rather challenging (e.g., Wright & Goodchild, 1997a,

1997b). Such challenges apparently arise from the ambiguous use of the acronym and the fact that boundaries between "doing GIS" and "researching GIS" are often blurred due to the relatively short history of converting geographic information from paper/map to digital formats. A definition that carries some consensus was given by the National Science Foundation (NSF) in 1999:

Geographic information science (GIScience) is the basic research field that seeks to *redefine* geographic concepts and their use *in the context of GIS*. GIScience also examines the impacts of GIS on individuals and society and the influences of society on GIS. GIScience reexamines some of the most fundamental themes in traditional spatially oriented fields such as geography, cartography, and geodesy while incorporating more recent developments in cognitive and information science. It also overlaps with and draws from more specialized research fields such as computer science, statistics, mathematics, and psychology, and contributes to progress in those fields. It supports research in political science and anthropology, and draws on those fields in studies of geographic information and society. (Mark, 2003, emphasis added by the author).

The most important part of the definition earlier is the implicit backbone provided by GIS, which is usually associated with some powerful computer program(s) (such as the dominant ArcGIS series packages offered by the Environmental Systems Research Institute (ESRI)). Because of the appearance and application of GIS, many geographic concepts and their traditional application are immediately changed and redefined. The most obvious change and redefinition is, of course, that those concepts are now able to be *digitally* stored and retrieved. Borrowing the power of computer science and other disciplines that take advantage of, or integrate as, standard toolsets in computer science, geographers are able to understand, utilize, and manipulate traditional geographic concepts in a never-before-seen way. All traditional geographic objects and concepts can now, through computational systems, be represented under a single umbrella: geographic information.

In addition, the NSF definition also points out that GIScience is a scientific endeavor that develops around the use of GIS (Goodchild, 1992). Applications of GIS play a significant role in the development of GIS as a scientific discipline. In this regard, GIS does draw upon quite a few new developments from a variety of scientific disciplines (e.g., computer science, statistics, mathematics, psychology, political science, and anthropology) and supports the development of these disciplines in a *new* way, by looking at geographic information as data that can be digitally stored and manipulated. More importantly, the use and in-depth development of GIS also changed geography from a primarily recording and repeating practice into a scientific discipline that not only records and repeats information but more importantly *analyzes* and presents the information with *added value*.

There are tendencies, as observed by Mark (2003), to make GIScience appear to be an all-encompassed superscientific discipline that subsumes all disciplines that use GIS. This is especially true with the recent boom of GIS applications to environmental management. While appealing in its breadth, it is dangerous to assume that if you use GIS, you are doing GIScience. Worst yet, from this application-oriented perspective, GIScience will eventually degrade to simply a pretentious name for GIS. Fortunately, the insights provided by NSF's definition suggest that GIScience, as a field of study, has its own distinct intellectual identity and is more than just the application of specific computer programs.

So, back to the ancient question—what is GIS/GIScience and what does it do? From the earlier discussion, without overgeneralizing and while admitting that GIS is a set of powerful computer tools that deals with geographic information, GIScience is a scientific discipline that studies, understands, and analyzes *digitally* stored *geographic information* in order to enrich the knowledge of and bring *new* insights to geography and other scientific disciplines including environmental management. At the center of this scientific endeavor is the digitally stored geographic information. Such information carries unique characteristics that are governed by the "First Law of Geography" (Tobler, 1970). Since the 1950s, the uniqueness of geographic information demands entirely different ways to effectively utilize such information. The development of new methods, techniques, and approaches to using these data comprise the primary focus of GIScientists. While it is both impractical to cover the breadth of GIScience in one chapter and potentially repetitive given the work of other such as Mark (2003), this chapter focuses specifically on exploring the uniqueness of geographic information and the emerging methods that are developed recently to address such uniqueness in data analysis that would enable better practices, planning, and implementation of environmental management strategies.

18.3 GEOGRAPHIC COORDINATES AND COORDINATE TRANSFORMATION (MAP PROJECTION)

Geographic information is all about location; environmental management is really but a spatial endeavor. Accurately referencing the location of environmental objects on the surface of the Earth and relative to one another requires a particular reference system. The definition of a reference system can vary depending on the text, but the essence of a reference system is defined as an arbitrary set of structure that something can be referred within. In GIS, this "something" is "location," and within the context of GIS, this reference system is sometimes called a "spatial reference system" but more commonly just "coordinate system."

For geographic information analysis, two types of coordinate systems are of particular interest: one is a three-dimensional coordinate system, the *geographic coordinate system* (acronym GCS), and the other is a two-dimensional coordinate system, which is called *projected coordinate system* (or *planar coordinate system*; either case, the acronym is PCS).

18.3.1 GCS

GCS, at first glance, is fairly simple and is more commonly referred to as the "longitude/latitude" system (Clarke, 1997). Longitude and latitude are basically angular measures for a three-dimensional location referencing system. This referencing system works well to refer to locations on the surface of a sphere-like object, like the Earth.

However, since the Earth is not a perfect sphere, we can't actually use a spherical reference system to refer to locations on the surface of Earth, at least not directly. If, however, one is willing to make compromises on accuracy, then we will be able to establish the GCS based on a spherical coordinate system developed from a perfect sphere.

In general, to make the compromise (or distortion) as small as possible, an ellipsoid (in GIS, more often referred to as a "spheroid") is used instead of a perfect sphere. An ellipsoid fits better with the shape of the Earth and hence is often used as a model of the Earth. Still with contemporary technology, especially satellite techniques, we can have an extremely close-fit spheroid for the Earth (at least at a global level); a spheroid can't model the actual fluctuation of the Earth's surface. A better (and closer) model of the surface of the Earth is called a Geoid. A Geoid is defined as the surface of the Earth if the Earth is covered entirely by water, the water remains still, and only the Earth's own gravity is influencing the water. Sometimes, this is referred to as the mean sea level (MSL). Although the Geoid is a rather good fit to the surface of the Earth, unfortunately, due to its irregular surface, a spherical coordinate system cannot be derived for the Geoid. The common practice for deriving a GCS in geographic information analysis is to combine the strength of a spheroid and the Geoid. This combination is called a Datum.

A Datum is defined as a particular spheroid that fits as close to the Geoid as possible. Apparently, for deriving a GCS for the entire globe, a Datum is really nothing but a particular spheroid, like the WGS 84. However, for deriving a local GCS, a global spheroid will not be an appropriate choice. Instead, a local spheroid will be chosen so that at the specific location where the GCS will be used, the spheroid's local surface fits as close as possible to the surface of the Geoid there. In so doing, the spherical reference system can be used to derive the geographic coordinates.

To generate actual geographic coordinates, the first task is to establish the origins of the longitude and latitude. For a spherical reference system, based on either a sphere or a spheroid, the origin of the latitude is relatively easy to determine. Often, the largest circle on the latitudinal direction, that is, the equator, is set as the origin. The origin of the longitude, however, is rather flexible, since on the longitudinal direction (or the east–west direction), there is really no difference for any longitudinal lines (because all longitudinal lines are equal length half circles or half ellipses). As a matter of fact, there were plenty of arbitrarily set origins for longitudinal lines in history; it really depended on which area was being mapped. Lately, however, it is agreed upon by the international geodetic community that the longitudinal line that crosses through Greenwich, Great Britain, be set as the internationally accepted origin of longitude. With the origins of both longitude and latitude determined, measurement will then take place and geographic coordinate are produced. If a global Datum is used, the coordinates are for measuring the entire globe; however, measurement errors could be large in places where the global Datum fits poorly to the Geoid. For these situations, a local Datum can be used, but the measurement (coordinates) can only be applied to the specific area of interest. In this case, the local measurement would be rather accurate.

18.3.2 Map Projection and PCS

Obtaining an accurate GCS is only the first step (albeit an extremely critical step) for generating an accurate map. Due to the fact that the surface of the Earth is not *developable*, meaning it cannot be flattened without introducing distortion to the four properties of an object on Earth: *shape*, *area*, *distance*, and *direction* (oftentimes, shape and direction are the same property), we'll need a *developable* surface to create a map, which is just a flat plane (a two-dimensional surface).

This is where *map projection* comes into play. Formally, a map projection is a series of mathematical operations that transfers the three-dimensional GCS to a two-dimensional PCS. The process begins via *projecting* the surface of the Datum (which is not developable) to the surface of a *developable* object. Three such objects are commonly used: the cylinder, the cone, and the plane itself (hence the names of a cylindrical projection, conic projection, and planar/azimuthal projection). We will not delve into the details on how a transformation is conducted since the transformation is standardized and computerized, and many GIS software packages can perform this action effortlessly.

However, what is important is how to choose a specific projection to generate a set of PCS that is as accurate as possible (all ensuing analyses will depend on the accuracy of the PCS). The essential principles are basically the same as when choosing an appropriate Datum to generate GCS: that is, a projection shall be chosen to reduce the distortions as much as possible. Three steps are involved in the process:

First, the GCS must be accurate. This means that the local Datum is appropriate. Second, just as an appropriate local

Datum means the surface of the local Datum fits *as closely as possible to the surface of the Geoid* of that particular area, an appropriate developable surface and the way it interacts with the Datum (either by tangent or secant) need to be chosen such that the developable surface is *as close to the surface of the Datum as possible* at the *particular area* for which the map is being generated. Third, a specific transformation to convert the GCS to PCS needs to be chosen, and all ensuing geographic analysis will be performed based on this PCS. Needless to say, the appropriateness of the aforementioned three steps will determine how appropriate the ensuing analyses will be and how effective such analyses will be to support environmental management decisions.

18.4 THE UNIQUE GEOGRAPHIC INFORMATION: SPATIAL DATA ARE SPECIAL

18.4.1 Types of Spatial Data

Before we get to the question about why spatial data is special, let's first review the types of spatial data or geographic information often used in practice (in this chapter, we use spatial data and geographic information interchangeably while admitting there are differences under various circumstances). The best place to start is to imagine a specific geography in map form, for instance, a county or a state. The most common way of representing such a geography would be using the *geographic coordinates (or projected coordinates)* as we have discussed previously to delineate its boundary. This is typically called the *vector* view of geographic information or, as in O'Sullivan and Unwin (2003), the *object* view. In this vector/object view, the geography is organized as a series of entities located in space and generally can be classified as three distinct types: the point (such as a polluting source), the line (a river), and the area (a county). Information associated with these geographies is normally stored in tabular form as attributes of the entities. In such a view, a single entity can theoretically carry an infinite amount of attribute information. Such information can then be *displayed, analyzed,* and *stored* in a GIS. In the early stages of development of GIS, this view was the typical way of representing geographic information.

The other way of representing a geography is imagining it is filled with certain properties that vary continuously across the space. The most common example is the surface of the Earth itself, where the varying property is the height above the sea level (the elevation). Of course, it is neither practical nor possible to obtain the information for the *entire* surface of the geography. The surface is always *approximated* via a raster or grid system in which the surface is divided into adequately small regular units (normally squares) called *pixels* or *cells*. For each pixel/cell, there will be one and only one value associated with the entire unit to represent the property (such as elevation, precipitation, and reflection of light). This view is called the field view in O'Sullivan and Unwin (2003). This way of representing geography is mostly in image processing, where the geographic information is obtained via remotely sensed platforms such as airplanes, satellites, or other remote sensing equipment.

18.4.2 Spatial Effects Make Spatial Data Special

No matter which way the geographic information is organized and stored, common to all representations is that the information is always associated with geography (coordinates or pixels/cells). That is, geographic information always contains positional information and hence can be used to answer the question: "where is it?" With the attribute information associated, we can answer the question: "what is where?" (Bivand, Pebesma, & Gómez-Rubio, 2008). More importantly, as aforementioned, geographic information is always dictated by the "First Law of Geography" (Tobler, 1970), which states everything is related with everything else, but near things are more related than distant things. Omitting the details of how "near" and "distant" are defined (these are extensively discussed in Anselin (1988, 2001); Bailey and Gatrell (1995); Bivand et al. (2008); Fotheringham, Brunsdon, and Charlton (2002); O'Sullivan and Unwin (2003), to name but a few), this first law immediately implies two *unique* characteristics for geographic information, or more commonly known as the spatial effects, the spatial autocorrelation and spatial heterogeneity. More importantly, per the first law, the existence of *both* autocorrelation and heterogeneity in geographic information is not accidental, but inherent. It is the existence of such unique characteristics of geographic information that makes analyzing spatial data rather different from other types of data.

18.4.2.1 Analyzing Spatial Autocorrelation

The effect of spatial autocorrelation has long been recognized, and methods have been developed to measure it in the field of regional science and econometrics (see Anselin (1988) and Baltagi (2001) for a detailed review). Anselin (2001) concisely depicts spatial autocorrelation as the "coincidence of value similarity with locational similarity." Inspired by how information is autocorrelated in time series, the statistics of measuring spatial autocorrelation are designed and widely used in current spatial data analysis practices. Among them, the two most commonly used ones are the Moran's I and Geary's C (Anselin, 1988, 2001; Cliff & Ord, 1973; Geary, 1954; Moran, 1950; Ripley, 1981). With some assumptions on spatial structure (or neighborhood definition) that defines the geographic arrangement of observations, W, whose elements w_{ij} indicate whether i and j are neighbors or is defined on some distance-decaying function using the distance between i and j (again, we omit the details of how W is defined here; interested readers are referred to Anselin (1988); Banerjee, Carlin, and Gelfand (2004) for further discussions), Moran's I takes the form

$$I = \frac{n\sum_i\sum_j w_{ij}(Y_i - \bar{Y})(Y_j - \bar{Y})}{\left(\sum_{i \neq j} w_{ij}\right)\sum_i(Y_i - \bar{Y})^2}$$

Moran (1950) shows under the null hypothesis that Y_i are i.i.d. and I is asymptotically normally distributed. However, Tiefelsdorf (2000, 2002) indicates that since the Moran's I is basically a ratio of quadratic forms in Y, the exact distribution can be obtained by applying the Imhof (1961) formula. Detailed discussion can be found at Leung, Mei, and Zhang (2003); Lieberman (1994); and Tiefelsdorf (2000).

Similarly, Geary's C takes the form

$$C = \frac{(n-1)\sum_i\sum_j w_{ij}(Y_i - \bar{Y})^2}{2\left(\sum_{i \neq j} w_{ij}\right)\sum_i(Y_i - \bar{Y})^2}$$

Apparently, C is also a ratio of quadratic forms in Y and like I can be either approximated via normal distribution or its exact distribution can be obtained via applying Imhof's formula. With their asymptotic or exact distribution available, we can use these statistics to test the null hypothesis that observations in space are randomly arranged. Though both statistics mimic somewhat the Pearson's correlation coefficient, R, Banerjee et al. (2004) recommend the use of these statistics as exploratory measures of spatial association instead of a test of spatial significance.

Furthermore, due to the multidirectional characteristics of a two-dimensional geographic space, unlike series autocorrelation, spatial autocorrelation might be quite different in different *directions*. This is especially true when the geographic information is obtained from remotely sensed imageries (for an example, see Agarwal, Gelfand, & Silander, 2002). Although such anisotropic characteristics add complexity to analyzing geographic information, more importantly, it greatly enriches our understanding of the collected geographic information. Not only are we able to understand that geographic information might be autocorrelated in space, we could choose specific directions to study such autocorrelation to see the difference and discover new patterns or trends.

18.4.2.2 Analyzing Spatial Heterogeneity

Per Tobler's Law, spatial autocorrelation only reflects the first half of the statement that "near things are more related than distant things." The second half, which means "distant things are less related," indicates the other characteristic of geographic information, the spatial heterogeneity. Apparently, spatial autocorrelation and heterogeneity are two sides of one coin. As Brunsdon, Fotheringham, and Charlton (1999) and Fotheringham et al. (2002) notice, in a regression analysis, if the actual relationships between the dependent variable and independent variables are different from place to place (one case of spatial heterogeneity) and such relationship is modeled by a global model that assumes the relationship remains the same everywhere (assumes spatial homogeneity), the residual of the global model will certainly be spatially autocorrelated.

While understanding and measuring spatial autocorrelation primarily focus on analyzing the spatial patterns existing in *one* variable (the univariate scenario), spatial heterogeneity manifests itself more in *relationships* among variables (the multivariate scenario). The most often mentioned methods of capturing such heterogeneity currently in practice include the geographically weighted regression (GWR; see Fotheringham et al., 2002) and the Bayesian theory-based spatially varying coefficient processes (SVCP) method (see Gelfand, Kim, Sirmans, & Banerjee, 2003).

The GWR technique is a recently developed GIS and spatial data analysis method designed specifically for dealing with spatial heterogeneity among regressed relationships. It has recently received increasing attention among scholars (Brunsdon, Fotheringham, & Charlton, 1996; Brunsdon et al., 1999; Fotheringham & Brunsdon, 1999; Fotheringham et al., 2002; Harris, Fotheringham, & Juggins, 2010; Huang & Leung, 2002; Leung, Mei, & Zhang, 2000a, 2000b; Páez, Uchida, & Miyamoto, 2002a, 2002b; Yu, 2006; Yu, Peterson, & Reid, 2009; Yu, Wei, & Wu, 2007; Yu & Wu, 2004). GWR develops the idea of the Casetti's (1972, 1986) expansion regression method in spatial terms. However, differing from Casetti's treatment of spatial terms, GWR allows regression coefficients to vary across space without explicitly specifying a determinant form on which the relationship drifts. In particular, GWR would assume the observation of the set of dependent and independent variables is a realization of a distance-decaying spatial process that is dictated by Tobler's first law. Empirical analyses suggest (Fotheringham et al., 2002) that Gaussian or Gaussian-like kernel functions can be used to emulate such processes sufficiently well. GWR uses the following process to calibrate the regression coefficients at particular locations.

First, for data location i, a Gaussian or Gaussian-like (such as bisquare or tricube; see Fotheringham et al. (2002) for detail) distance-decaying kernel function assigns weights to each data location in the study region. The weights act to ensure that data locations near location i have more influence than those further away. It is found through numerous experiments that the exact form of the distance-decaying function exerts very little influence on the resulting spatial pattern (Fotheringham et al., 2002; Yu, 2006; Yu et al., 2007, 2009). Second, once the weights for all the relevant data locations are generated, they will be applied to the observations at corresponding data locations, and a weighted least squares procedure takes place and produces the set of coefficients of data location i. Third, the above procedure will repeat for the rest of the data locations until the coefficients of all the data locations in the study region have been generated.

Different weighting schemes are presented and discussed in the literature (Fotheringham et al., 2002; Páez et al., 2002a, 2002b; Yu & Wu, 2004). In general, all the schemes fall within two categories, fixed or adaptive weighting (Fig. 18.1). In a fixed scheme, one optimum spatial kernel (represented by the spatial bandwidth (b)) is determined and applied uniformly across the study area (4.1a). However, such an approach, as pointed out by Fotheringham et al. (2002) and Páez et al. (2002a, 2002b), may produce a large local estimation variance in areas where data are sparse (4.1a). This in turn might exaggerate the degree of coefficient variation present in this area. In areas where data are dense, subtle coefficient variation might be masked. On the other hand, the adaptive scheme produces adaptive spatial kernels (4.1b). The kernels are able to adapt themselves in size to variations in the density of the data such that the kernels have larger bandwidths where the data are sparse and have smaller ones where the data are plentiful (4.1b). Due to this consideration, in practice, the adaptive spatial weighting scheme is often suggested. To implement the idea, a nearest neighbor approach is used to produce the adaptive spatial kernels.

In practice, the optimum number of the nearest neighbor (nnb) is obtained through an out-of-sample cross-validation (cv) procedure (Cleveland, 1979; Fotheringham et al., 2002) or through minimizing the goodness-of-fit statistics, the Akaike Information Criterion (AIC; Fotheringham et al., 2002; Hurvich, Simonoff, & Tsai, 1998). In essence, the procedure is to minimize either the cv score or the AIC statistics:

$$\text{cv} = \sum_{i=1}^{n} \left[y_i - \hat{y}_{\neq i}(\text{nnb}) \right]^2$$

$$\text{AIC} = 2n \ln(\hat{\sigma}) + n \ln(2\pi) + n \left(\frac{n + \text{tr}(\mathbf{S})}{n - 2 - \text{tr}(\mathbf{S})} \right)$$

where y_i is the observed dependent variable on location i; $\hat{y}_{\neq i}$(nnb) is the GWR fitted value of y_i using the nearest neighbor nnb, with the observation for location i omitted from the calibration process; n is the total number of observations; $\hat{\sigma}$ is the maximum likelihood estimated standard deviation of the error term; and tr($\mathbf{S}$) is the trace of the hat matrix $\mathbf{S}$ of the GWR, which maintains the following relationship (Fotheringham et al., 2002):

$$\hat{\mathbf{y}} = \mathbf{S}\mathbf{y}$$

where $\mathbf{y}$ and $\hat{\mathbf{y}}$ are the vector of the dependent variable and the GWR estimated values using a particular nearest neighbor nnb.

GWR's popularity has grown rapidly since Fotheringham and colleagues published their research in 2002 and developed an easy to use software package (Fotheringham et al., 2002). As one of the first proposed methods to deal with spatial heterogeneity in regressed relationships, GWR also receives criticism that it has to fit a model for each and every single location while at the same time repetitively use the same observations with different weights. This makes GWR somewhat "expensive" in both consuming the effective degrees of freedom and computational complexity (Fotheringham et al., 2002; Yu et al., 2007). In addition, Griffith (2008); Waller, Zhu, Gotway, Gorman, and Gruenewald (2007); Wheeler (2007); and Wheeler and Tiefelsdorf (2005) point out that the local models might suffer from enlarged multicollinearity and increased correlation among local coefficients, making local inference much less confident. Fotheringham et al. (2002) regard this as a variance–bias trade-off. As an alternative, Griffith (2008) proposed to construct interaction terms using spatial filter eigenvectors of the geographically referenced attributes. Farber and Páez (2007), on the other hand, suggest that

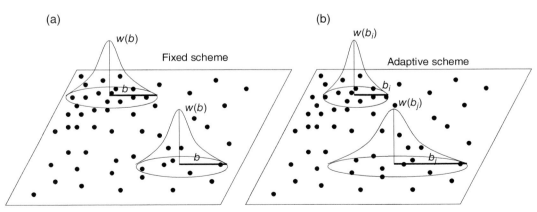

FIGURE 18.1 Fixed (a) and adaptive (b) weighting schemes in GWR.

adjusting the bandwidth in GWR's selection of local sample size (via cross-validation or minimizing AIC) might be a tenable approach to relieve the potential strong correlation among local coefficients. Except for the theoretical debate over the appropriateness of GWR, many studies inform that GWR has the potential to reveal local patterns that are globally hidden.

As Congdon (2003) points out, since GWR fits n separate regressions (with separate likelihoods) to cross-sectional data for n areas, each one regarding a particular area as the "center" for the purpose of a separate model, coefficients at specific locations are deemed somewhat "fixed," though different from one another. Under such an argument, GWR amounts to a fixed-effect method in dealing with varying coefficients. Opposite to the fixed-effect treatment, Congdon (2003, 2007), Gelfand et al. (2003), and Banerjee et al. (2004) propose to treat the variation in local coefficients as a random effect, hence the SVCP model.

When specifying an SVCP model, an analytical solution, as in the fixed-effect case (GWR), is not tenable. Instead, SVCP is normally based on a Bayesian framework in which estimation of the varying coefficients and relevant statistics (confidence interval, standard error, etc.) can be obtained via combining the prior knowledge (a prior distribution of the unknown parameters) and the data to produce a posterior distribution of the parameters (Banerjee et al., 2004). One of the attractive aspects of the Bayesian-based SVCP model is that the inference for the random spatial effects (heterogeneity) can be investigated rather formally from the posterior distribution.

In practice, the SVCP model is usually specified in a hierarchical manner. In the first stage, the distribution of the data (Y) is specified as conditional on unknown parameters (β, the coefficients, and τ^2, the residual covariance):

$$\left[Y \mid \beta, \tau^2\right] = N\left(\mathbf{X}^T \beta, \tau^2 \mathbf{I}\right)$$

The bracket notion [A|B] denotes that the distribution of A is conditional on B. $N(\bullet)$ is a Gaussian distribution. $\mathbf{X}^T$ is an n by np block diagonal matrix (n is the number of observations, and p the number of independent variables) where each row contains a row from the n by p design matrix of independent variables, with zeros in filling the rest rows (the independent variables from the design matrix are shifted p places in each subsequent row in $\mathbf{X}^T$). $\mathbf{I}$ is the n by n identity matrix. The aforementioned formula indicates that we assume the data is Gaussian conditional on the unknown parameters. Other distributions can be assumed as well for specific cases such as binary responses and categorical responses (Congdon, 2003). Based on the assumed distribution, the likelihood can then be obtained (Gelfand et al., 2003) as the input of data in the Bayesian framework.

In the second stage of the model, SVCP specifies the prior of the regression coefficients by explicitly taking into consideration the possible spatial dependence in the coefficients, β, and through a covariance matrix Σ_β:

$$\left[\beta \mid \mu_\beta, \Sigma_\beta\right] = N\left(1_{n \times 1} \otimes \mu_\beta, \Sigma_\beta\right)$$

The vector μ_β represents the means of the regression coefficients corresponding to each of the p independent variables; the symbol $\otimes$ denotes the Kronecker product operator, by which every element in $1_{n \times 1}$ is multiplied by μ_β. To take advantage of the prior knowledge that we assume the coefficients have the potential to be spatially autocorrelated, for each β_p, we can assume a priori that it follows an area unit model (such as the conditional autoregressive model or simultaneous autoregressive model, as seen in Banerjee et al. (2004), or by following Leyland, Langford, Rasbash, and Goldstein (2000), we can assume that the β_p is linked to an unstructured effect via a spatial weighting scheme (see Congdon, 2003, pp. 164–167 for technique details). The covariance matrix for the p different coefficients at each of the n locations, Σ_β, can take either a separable or nonseparable form (Banerjee et al., 2004; Gelfand et al., 2003; Wheeler & Calder, 2007). The separable form contains a spatial dependence in the regression coefficients and dependence between coefficients of the same type within the site (for technique details, see Gelfand et al. (2003) and Wheeler and Calder (2007)). The separable form is more restrictive in terms of specifying the spatial dependence since each of the p regression coefficients is assumed to have the same spatial association structure. Yet, it is more convenient computationally to adopt a separable form since it reduces the number of matrix inversions when obtaining the posterior distribution. The nonseparable form, on the other hand, would allow different spatial association structures for each regression coefficient (Banerjee, Johnson, Schneider, & Durgan, 2005); therefore, once the data volume becomes large, computation could quickly become prohibitive.

Once the aforementioned priors are specified, the Bayesian SVCP model is complete. Calibration and inference of the parameters are obtained by applying the Bayes theorem to get the posterior distribution:

$$[\theta \mid Y] \propto [Y \mid \theta] * [\theta]$$

where $[\theta \mid Y]$ is the posterior distribution of the parameters conditional on the data Y, $[Y \mid \theta]$ is the likelihood of the data, and $[\theta]$ are the priors specified for the parameters based on our a priori knowledge.

One of the major debates over using Bayesian models is that the posterior distribution is impossible to obtain via analytical means. Recent developments in simulation-based tools, such as the Gibbs sampler-based Markov chain Monte Carlo (MCMC) methods (Casella & George, 1992; Congdon, 2007; Koop, Poirier, & Tobias, 2007; Ntzoufras, 2011; Pace & LeSage, 2009), can be used to produce fairly robust and well-converged results for obtaining the posterior distribution. Algorithms that implement the Bayesian method are gradually becoming available (at the forefront is the freeware WinBUGS). This renders the Bayesian analysis to be more and more accepted in mainstream statistical and econometric data analysis.

In the past decade, the analysis of spatial heterogeneity via either GWR or Bayesian SVCP method is becoming popular. Such trends reveal the fact that *special* spatial data requires *special* treatments that are only viable with enhanced computing power. GWR and SVCP represent two viewpoints on how spatial heterogeneity is reflected in regressed relationships. The former assumes a varying but fixed effect in the coefficients, while the latter treats the spatial heterogeneity as a random effect that varies over a global mean. Researchers comparing the two methods (Congdon, 2003; Wheeler & Calder, 2007; Wheeler & Waller, 2009) often find that SVCP provides better inference capability since it essentially produces posterior *distributions* for the parameters under study. Comparing this to the relatively straightforward implementation of GWR model, however, the implementation of Bayesian SVCP model involves additional complexity and is not readily available to the general research community (Wheeler & Calder, 2007).

18.5 THE NEXT STEP, GEOGRAPHIC INFORMATION IN A TEMPORAL CONTEXT: SPATIOTEMPORAL ANALYSIS

From the earlier discussion, it is apparent that the inherent spatial effects in geographic information have recently been widely recognized in environmental management, geography, regional science, econometrics, statistics, and other fields largely due to the availability of toolsets and the recent development of geographic information sciences. On the other hand, time series analysis and the panel data analysis have seen much longer development in both environmental management practices and other fields. However, until the very recent past, combining the strength of the two has received rather limited attention. Investigations that involve both space and time are of particular importance in environmental management due to the fact that environmental management practices and implementation will yield much better strategies from a dynamic perspective.

18.5.1 Studies on Spatiotemporal Models and Processes

Spatial data analysis techniques have borrowed many ideas from time series analysis. For instance, one of the most important aspects of spatial data, spatial autocorrelation, resembles the series autocorrelation, though differs in the way **lags** are defined (Anselin, 1988; Anselin, Le Gallo, & Jayet, 2008). The fundamental similarity between spatial data and time series data is that both follow a "*neighbors are similar*" law. In spatial data, this is Tobler's (1970) "First Law of Geography." Analyzing geographic information in a temporal context has been an interesting but rather daunting task due to the potential added complexity. Such practices are usually termed spatiotemporal analysis.

There is a large body of literature in a wide range of scientific and engineering fields that addresses analyses with spatiotemporal data. Generally, studies in this genre can be divided into two categories. The first category focuses specifically on spatiotemporal clustering of observations and interpolation. For instance, Knox (1964) investigates the space–time interaction of epidemics and develops the Knox test to determine whether or not there are apparent spatiotemporal clusters (Bailey & Gatrell, 1995). Bilonick (1985); Eynon and Switzer (1983); Kyriakidis and Journel (2001); Oehlert (1993); Rouhani, Ebrahimpour, Yaqub, and Gianella (1992); and Vyas and Christakos (1997) apply the spatiotemporal models to determine space–time trends in the deposition of atmospheric pollutants. Huang and Cressie (1996) develop the spatiotemporal Kalman filter to predict the snow water equivalent. Goovaerts and Chiang (1993); Heuvelink, Musters, and Pebesma (1997); Papritz and Flühler (1994); and Snepvangers, Heuvelink, and Huisman (2003) use spatiotemporal geostatistics to model the temporal evolution of spatial patterns in soil moisture content. Armstrong, Chetboun, and Hubert (1993) and Rouhani and Wackernagel (1990) apply a spatiotemporal kriging procedure to estimate rainfall in various regions. Carroll et al. (1997); Christakos and Hristopulos (1998); Christakos and Lai (1997); and Sun, Tsutakawa, Kim, and He (2000) investigate the spatiotemporal patterns of diseases and exposure to pollutants. Hohn, Liebhold, and Gribko (1993) develop a spatiotemporal model to characterize population dynamics in ecology. Mardia and Goodall (1993), Rodríguez-Iturbe and Mejía (1974), Sampson and Guttorp (1992), and Switzer (1979) design sampling networks for monitoring spatiotemporal processes, to name but a few.

As pointed out by Kyriakidis and Journel (1999, 2001), this category of the joint analysis of space and time in a spatiotemporal framework mainly builds on the extension of established spatial analytical techniques that are widely applied in the fields of geology, forestry, and meteorology. Such extensions usually treat time as an added spatial dimension, hence enlarging the two-dimensional geographic space to a three-dimensional geographic–temporal space. However, simple extension might not be all that plausible due to the fundamental differences between geographic space and time (or ***geographic*** space and ***temporal*** space). Geographic space represents a state of coexistence, in which there can be multiple directions. While temporal space represents a state of successive existence, a nonreversible ordering in which only one direction is present (Snepvangers et al., 2003). Isotropy is well defined in **geographic** space but has no meaning in a space–time context due to the ordering and nonreversibility of time. Bilonick (1985), however, presents a fairly interesting solution to this dilemma. In Bilonick's approach, he separates the interactions among observations into three components: the geographic interaction, the temporal interaction, and the joint geographic–temporal interaction. While geographic and temporal interactions can be modeled in the regular fashion based on spatial lag (h_s) and temporal lag (h_t), the geographic–temporal interaction is modeled via the geographic–temporal lag (h_{st}). h_{st} is determined by introducing an anisotropy ratio, α, as in

$$h_{st} = \sqrt{h_s^2 + \alpha h_t^2}$$

The advantage of Bilonick's process of the spatiotemporal framework is that it contains all the possible interactions that are present in spatiotemporal datasets. However, separating those interactions might not be as straightforward as the earlier equation suggests.

In addition, the majority of the previously mentioned studies are largely confined to the field of geostatistics (Kyriakidis & Journel, 1999). The primary goals of these studies were all fairly similar (Snepvangers et al., 2003): to predict an attribute $z = \{z(s,t)\, s \in S, t \in T\}$ defined on a geographical domain $S \subset R^2$ and a time interval $T \subset R^1$, at a space–time point (s_0, t_0), where z was not measured. The prediction was to be based on n geographic measurements at t time intervals, which constitute the nt points (s_i, t_i), with $i = 1, \ldots, n$.

18.5.2 Studies on Relationships among Variables

The second category of analyses that deal with spatiotemporal data focuses on relationships expressed in regression analysis, represented mainly by the Bayesian hierarchical modeling approach (Banerjee et al., 2004, chapter 10; Congdon, 2003, chapter 6, pp. 310–317; Congdon, 2007, chapter 11) and the spatiotemporal econometric approach (Pace & LeSage, 2009, chapter 7). The Bayesian framework is basically an extension of the Bayesian SVCP model with longitudinal data (Gelfand et al., 2003). The extension to the spatiotemporal dimension is extensively discussed in Banerjee et al. (2004, chapter 10) and others (Congdon, 2003, pp. 168–169; Gelfand et al., 2003, p. 391). However, as pointed out by Wheeler and Calder (2007), this approach is relatively difficult to implement in practice and involves much more complicated model specifications and computations; thus, its empirical applications are rather limited. On the other hand, the spatiotemporal econometric approach examines specifically how spatiotemporal processes working over time can lead to equilibrium outcomes that exhibit spatial dependence (Pace & LeSage, 2009). The focus of this approach lies on extending existing spatial models to more complicated specifications that can model the spatiotemporal data generating process. Both approaches contribute significantly to the understanding of spatiotemporal processes in regional studies, epidemiological studies, and public health fields.

Other than the two approaches mentioned previously (Yu, 2010, 2013; Yu & Lv, 2009), Huang, Wu, and Barry (2010) have initiated an investigation of incorporating temporal information in the analytical framework of GWR. In Huang et al. (2010), Yu (2010), and Yu and Lv (2009), the method is a straightforward extension of the GWR analytical framework by incorporating temporal information as an added dimension other than the X and Y coordinates (hence the name geographically and temporally weighted regression (GTWR)). The treatment of the added temporal coordinate in their study follows closely the spirit proposed in Bilonick (1985). The temporal coordinate is assigned an anisotropy ratio to enable it to be measurable with the spatial coordinates. Yu (2013) and Yu and Lv (2009) assign such an anisotropy ratio arbitrarily. Huang et al. (2010), on the other hand, use an empirical approach to choose the anisotropy ratio via minimizing the cross-validation score or the AIC statistics.

Inspired by the variable coefficient panel model, Hsiao (2003) examines the concept of incorporating temporal information into the GWR analytical framework from a different perspective. Yu (2010) termed it a geographically weighted panel regression (GWPR) analysis. Panel data analysis with geographic data has only recently attracted scholarly attention (Anselin et al., 2008; Elhorst, 2001, 2003, 2010; Le Gallo & Ertur, 2003; Le Gallo, Ertur, & Baumont, 2003; Lv & Yu, 2009). The focus of this trend of *spatial panel data analysis*, as termed in both Elhorst (2001, 2003, 2010) and Anselin et al. (2008), is primarily an extension of the spatial data analysis techniques with cross-sectional data. Estimations of the parameters focus on the pooled model that incorporates either a spatial lag term in the right-hand side (RHS) of the equation or a spatial error term. The potential for heterogeneous parameters is usually overshadowed by the less accurate prediction performance than the pooled model (Baltagi, 2008) or willingness to trade bias over a reduction in variance (Toro-Vizcarrondo & Wallace, 1968).

Of course, this is not to say that panel data analysis cannot deal with heterogeneous parameters. As a matter of fact, Hsiao (2003) indicates that "when data do not support the hypothesis of coefficients being the same, yet the specification of the relationships among variables appears proper or it is not feasible to include additional conditional variables, then it would seem reasonable to allow variations in parameters across cross-sectional units and/or over time as a means to take account of the interindividual and/or interperiod heterogeneity" (p. 141). Many studies also indicate that pooling parameters over cross-sectional units might not be very tenable (Pesaran, Shin, & Smith, 1999; Pesaran & Smith, 1995; Robertson & Symons, 1992). This is especially true when the cross-sectional units are samples from geographic space, as dictated by the "First Law of Geography." The variable coefficient panel model, though can be estimated by imposing a series of restrictions (see Hsiao, 2003, p. 153), has not gained much application in empirical studies. This is largely due to the reason that the estimation of the variance–covariance matrix of the composite error term in any reasonable sized panel dataset would be computationally prohibitive, even under the current computing power.

More importantly, other than the computational consideration, it has been noted that the variable coefficient panel analysis is largely an **aspatial** approach. The unique characteristics of geographic information (autocorrelation and heterogeneity) are not quite utilized explicitly. According to Yu (2010), the central idea of GWPR analysis is an extension of GWR with panel data and panel data analytical techniques. In GWPR, it is assumed that the time series of observations at a particular geographic location is a realization of a smooth spatiotemporal process. Such spatiotemporal processes follow a distribution that closer observations (either in geography or in time) are more related than distant observations. Depending on whether the panel analysis intends to pool over geographic (cross-sectional) or temporal observations, different models can be applied. For instance, if the only concern is that the regression coefficients vary over cross-sectional units (geographic space), the spatiotemporal process is effectively reduced to a spatial process just as in a GWR analysis. Unlike GWR analysis, however, the spatial process is applicable to all the temporal observations simultaneously. The spatial process can be assumed to be temporally invariant or variant. In the latter case, temporal scalars for each period can be assigned based on exploratory analysis with the data. If we assume the spatial process is temporally invariant (as often seen in relatively short panels, with less than 10 temporal periods), the GWPR can be seen as an expanded version of the cross-sectional GWR analysis to panel data. Following similar arguments as in GWR, a bandwidth or a number of nearest neighbors can be obtained for each location to determine a set of local calibrating locations. Observations within the local calibrating locations will be weighted based on a kernel function just as in GWR (Fotheringham et al., 2002). The difference here is that the weights will be applied to each temporal period simultaneously to form a "*locally weighted panel data set.*" Within these local calibrating locations, it is assumed that the panel is poolable over geographic space. By using these "*locally weighted panel data set,*" either a fixed- or random-effect model, as detailed in Baltagi (2008), can be applied to obtain the coefficients of the explanatory variables and other statistics at that specific location. Implementation of GWPR is partially done with R scripts (R Core Team, 2013) by combining codes from the cross-sectional GWR (SPGWR, Bivand, 2013) and panel data analysis codes (PLM, Croissant & Millo, 2013). Detailed discussion and an application can be found in Yu (2010).

18.6 GIS IN ACTION: WILDLIFE ECOLOGY, SNAKES IN THE SAND

Understanding what drives spatial and temporal distributions in populations of animals is central to effective wildlife science and management. Not surprisingly, wildlife science has become increasingly dependent on the use of GIS in researching and managing species of conservation concern. GIS is used regularly by wildlife scientists to investigate aspects of spatial ecology, habitat use, population analysis, and landscape genetics, among others. Studies of spatial ecology and habitat use are particularly common as they often address fundamental ecological questions for a given species.

An investigation of spatial ecology and habitat use of eastern hognose snakes (*Heterodon platirhinos*) was carried out at Cape Cod National Seashore, Massachusetts, United States, from May 2009 to November 2010. The study site was located at the northern extent of the Cape Cod peninsula in an extensive dune ecosystem with surficial geology formed entirely of postglacial beach and dune deposits. Plant communities found in the study site included beach grass dune, scrub forest, pine forest, and cranberry wetland, among others. *H. platirhinos* is a species of increasing conservation concern, especially in the Northeast (Seburn, 2009; Therres, 1999). Sixteen snakes (10 females, 6 males) were surgically implanted with radio transmitters and relocated approximately once every 4 days during the late spring and summer months (May–August) and less frequently during the fall and winter months (September–April). A total of 413 relocations were logged over these 2 years.

By recording geographic location during each snake encounter, we were able to amass a spatial dataset for each snake upon which we calculated a number of movement parameters including average daily movements (ADMs) and home range estimates. Home range estimates are an important concept in wildlife ecology as they offer a degree of understanding of the space used by individuals for a given species over a given period of time. They are often used to test hypotheses regarding the distribution of resources in an environment (Sperry & Weatherhead, 2009) and to formulate informed decisions when attempting to determine how much land should be set aside for conservation. However, they can prove problematic to define in practice (Gregory, Macartney, & Larsen, 1987; Hemson et al., 2005). A number of different methods have been developed to obtain home range estimates, each with their own caveats. Minimum convex polygons (MCPs) and kernel density estimates

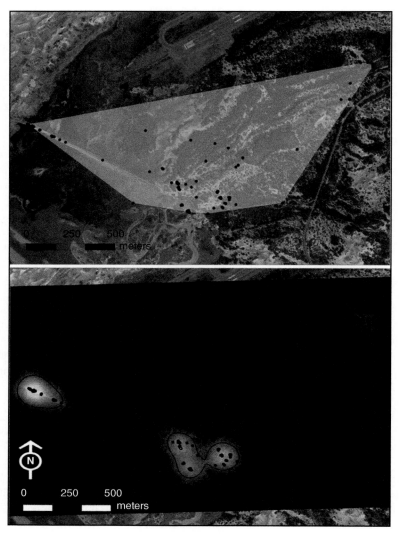

FIGURE 18.2 MCP (top) and KDEs with 95 and 50% isopleth contours (bottom) for an individual snake tracked in 2010. These home range estimates were generated using ArcGIS and the Geospatial Modeling Environment extension. The different techniques result in multiple estimates of home range (MCP = 95 ha; 50% KDE = 10 ha; 95% KDE = 61 ha) that allow for greater comparison with other studies. Each point on the maps represents a snake relocation.

(KDEs) are the two most common methods of home range estimation. We made calculations using both approaches with ArcGIS version 10.2 using the Geospatial Modeling Environment extension (Beyer, 2012). By quickly and efficiently making multiple estimates for each individual, we were able to maximize our ability to compare our results to other studies and to make ecological inferences based on our results (Fig. 18.1).

ADMs were another parameter calculated (using ArcGIS) that helped to shed light on snake ecology. ADMs are derived by dividing the calculated Euclidean distance between successive relocations by the number of days elapsed. These results can be compared over time, between sexes, and between studies to make ecological inferences. When plotted over time, ADM can reveal behavioral shifts in response to reproductive biology, as was evident during the course of our study. By periodically weighing females that we knew to be gravid, we were able to detect egg laying within a relatively narrow range of dates. By plotting ADMs over the course of an entire activity season, it became apparent that peak movements occurred immediately after or shortly after egg laying had occurred (Fig. 18.2). These long-distance movements suggest a migratory pattern to areas that are most appropriate for nesting and may help resource managers identify habitat types that are critical for conservation. In addition, simply knowing when the snakes are most likely to exhibit heightened movements could allow resource managers to make more informed decisions with respect to when to

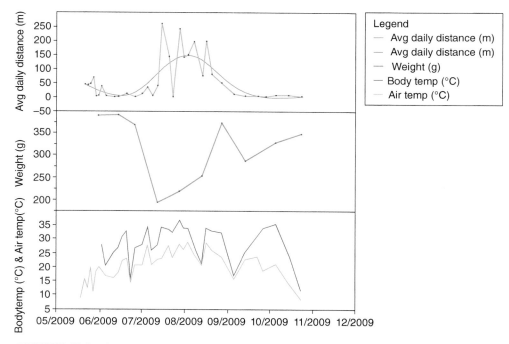

FIGURE 18.3 Average daily movement, weight and temperature data plotted over time for an individual snake tracked in 2009. Note the peak in ADM immediately following a sharp drop off in weight (indicating egg laying). This was a common phenomenon among gravid females during the course of our study.

implement protective measures (e.g., road closures) or when to most efficiently conduct surveys to detect the species (Fig. 18.3).

The tools provided by GIS analysis offer exciting opportunities to work with spatial data such as that generated by a wildlife radiotelemetry project. Merging these technologies and others is very much a contemporary theme in wildlife research (Delaney, Riley, & Fisher, 2010; Hoss, Guyer, Smith, & Shuett, 2010). The expanding field of remote sensing seems poised to elevate the potential for future synthesis with improved understanding of how animal populations function.

18.7 CONCLUSION

As Berry (1999) puts it, environmental management is essentially a spatial endeavor. Effective environmental management practices require strong spatial analytical components. Such components include not only mapping and spatial database management, often described as the primary functions of GIS, but also in-depth understanding of spatial data (geographic information) and how to best analyze such information. In this chapter, we took the liberty to deviate somewhat from traditional views of how GIS will be employed in environmental management practices (which appeared in many case studies over the past decades).

Instead, we engaged into a journey of discussing the uniqueness of geographic information and how such uniqueness demands largely unprecedented methodology to effectively analyze geographic information. We did, however, provide a case study in wildlife ecology to demonstrate how GIS toolsets can be used to utilize geographic information to understand species characteristics, which could be of particular interest to land use planning, environmental planning, and habitat preservation. The discussion in this chapter, granted, is rather limited as an attempt to address the question of how GIS facilitates environmental management.

Many of the discussed "newly developed" methodologies were not really developed in the past few years (for instance, the theoretical discussions of the Bayesian SVCP model and the variable coefficient panel analysis have both existed in statistics and econometrics for decades), but only recently emerged in empirical studies thanks to enhanced computing power. This ultimately points to the fact that the development of GIScience is closely linked to the development of computer science and technologies. Even the relatively newly emerged methods, such as the GWR/GTWR/GWPR, are only plausible with current computing power. Through the investigation and discussion of these methods, it is not surprising to find out that the popularity of a particular method is directly linked to the availability of specific computing tools (software packages) that implement the method. This is especially true in the case of GWR and Bayesian SVCP models when analyzing spatial heterogeneity in regression relationships (Wheeler & Calder, 2007).

Now back to the question again, how will GIS facilitate environmental management practices? Different practitioners will have rather varied answers. Hence, providing a comprehensive answer is not the goal of this chapter. The discussion in this chapter, however, echoes a belief that no matter how GIS is defined and employed, the essence of it is the "GI," geographic information; and one of the most important tasks of GIS in environmental management is to analyze this GI to get the most out of what it intends to offer. Our brief introduction of GIS in wildlife ecology sheds some light on the topic. This is especially true when we consider the increasingly available geographic information in the fields of environmental management, which is led by the weekly, daily, and even hourly collections of satellite imageries via remote sensing technologies. The discussion in this chapter is only a tiny portion of what is available out there. Yet we hope the discussion here would serve as a starting place for interested scholars to develop more efficient, better, and faster methodologies in analyzing geographic information for more effective environmental management practices.

REFERENCES

Agarwal, D. K., A. E. Gelfand, and J. A. Silander 2002. Investigating tropical deforestation using two-stage spatially misaligned regression models. *Journal of Agricultural, Biological, and Environmental Statistics*, 7(3):420–439.

Anselin, L. 1988. *Spatial Econometrics: Methods and Models*. Dordrecht: Kluwer Academic Publishers.

Anselin, L. 2001. Spatial econometrics. In B. H. Baltagi (Ed.), *A Companion to Theoretical Econometrics* (pp. 310–330). Oxford: Blackwell.

Anselin, L., J. Le Gallo, and H. Jayet (2008) Spatial panel econometrics. In L. Matyas and P. Sevestre (Eds.), *The Econometrics of Panel Data* (pp. 625–660). Berlin: Springer.

Armstrong, M., G. Chetboun, and P. Hubert 1993. Kriging the rainfall. In A. Soares (Ed.), *Lesotho Geostatistics Tróia'92* (pp. 661–672). Berlin: Springer.

Bailey, T. C. and A. C. Gatrell 1995. *Interactive Spatial Data Analysis*. New York: Prentice Hall.

Baltagi, B. H. 2001. *A Companion to Theoretical Econometrics*. Oxford: Blackwell.

Baltagi, B. H. 2008. *Econometric Analysis of Panel Data*. New York: Wiley.

Banerjee, S., B. P. Carlin, and A. E. Gelfand 2004. *Hierarchical Modeling and Analysis for Spatial Data*. Boca Raton: Chapman & Hall/CRC.

Banerjee, S., G. A. Johnson, N. Schneider, and B. R. Durgan 2005. Modelling replicated weed growth data using spatially-varying growth curves. *Environmental and Ecological Statistics* 12(4):357–377.

Berry, J. K. 1999. GIS technology in environmental management: a brief history, trends and probable Future. In D. L. Soden and B. S. Steel (Eds.), *Handbook of Global Environmental Policy and Administration* (pp. 49–80). New York: Marcel Dekker, Inc.

Beyer, H. L. 2012, *Geospatial Modelling Environment*, URL: http://www.spatialecology.com/gme/images/SpatialEcologyGME.pdf (accessed May 4, 2015).

Bilonick, R. A. 1985. The space-time distribution of sulfate deposition in the northeastern United States. *Atmospheric Environment* 19(11): 1829–1845.

Bivand, R. S. 2013. *SPGWR: Geographically Weighted Regression*. Retrieved from http://cran.r-project.org/web/packages/spgwr/index.html (accessed March 14, 2015).

Bivand, R. S., E. J. Pebesma, and V. Gómez-Rubio 2008. *Applied Spatial Data Analysis with R*. New York: Springer.

Brunsdon, C. F., A. S. Fotheringham, and M. E. Charlton 1996. Geographically weighted regression: a method for exploring spatial nonstationarity. *Geographical analysis* 28(4): 281–298.

Brunsdon, C. F., A. S. Fotheringham, and M. E. Charlton 1999. Some notes on parametric significance tests for geographically weighted regression. *Journal of Regional Science* 39(3):497–524.

Burroughs, P. A. 1986. *Principles of Geographic Information Systems for Land Resource Assessment* (Vol. 12). New York: Oxford Science.

Carroll, R. J., R. Chen, E. I. George, T. H. Li, H. J. Newton, H. Schmiediche, and N. Wang 1997. Ozone exposure and population density in Harris County, Texas. *Journal of the American Statistical Association* 92(438):392–404.

Casella, G. and E. I. George 1992. Explaining the Gibbs sampler. *The American Statistician* 46(3):167–174.

Casetti, E. 1972. Generating models by the expansion method: applications to geographical research. *Geographical Analysis* 4(1):81–91.

Casetti, E. 1986. The dual expansion method: an application for evaluating the effects of population growth on development. *IEEE Transactions on Systems, Man and Cybernetics* 16(1):29–39.

Christakos, G. and D. T. Hristopulos 1998. *Spatiotemporal Environmental Health Modelling: A Tractatus Stochasticus*. Boston: Kluwer Academic.

Christakos, G. and J. J. Lai 1997. A study of the breast cancer dynamics in North Carolina. *Social Science & Medicine* 45(10):1503–1517.

Clarke, K. C. 1997. *Getting Started with Geographic Information Systems* (Vol. 3). Upper Saddle River: Prentice Hall.

Cleveland, W. S. 1979. Robust locally weighted regression and smoothing scatter plots. *Journal of the American Statistical Association* 74(368):829–836.

Cliff, A. D. and J. K. Ord 1973. *Spatial Autocorrelation*. London: Pion.

Congdon, P. 2003. Modelling spatially varying impacts of socioeconomic predictors on mortality outcomes. *Journal of Geographical Systems* 5(2):161–184.

Congdon, P. 2007. *Bayesian Statistical Modelling* (Vol. 704). New York: Wiley.

Croissant, Y. and G. Millo 2013. *PLM: Linear Models for Panel Data*. Retrieved from http://cran.r-project.org/web/packages/plm/index.html (accessed March 14, 2015).

Delaney, K. S., S. P. D. Riley, and R. Fisher 2010. A rapid, strong, and convergent genetic response to urban habitat fragmentation in four divergent and widespread vertebrates. *PLoS one* 5:e12767.

Elhorst, J. P. 2001. Dynamic models in space and time. *Geographical Analysis* 33(2):119–140.

Elhorst, J. P. 2003. Specification and estimation of spatial panel data models. *International Regional Science Review* 26(3): 244–268.

Elhorst, J. P. 2010. *Spatial Panel Data Models Handbook of Applied Spatial Analysis* (pp. 377–407). Berlin: Springer.

Eynon, B. P. and P. Switzer 1983. The variability of rainfall acidity. *Canadian Journal of Statistics* 11(1):11–23.

Farber, S. and A. Páez 2007. A systematic investigation of cross-validation in GWR model estimation: empirical analysis and Monte Carlo simulations. *Journal of Geographical Systems* 9(4):371–396.

Fotheringham, A. S. and C. F. Brunsdon 1999. Local forms of spatial analysis. *Geographical Analysis* 31(4):340–358.

Fotheringham, A. S., C. F. Brunsdon, and M. E. Charlton 2002. *Geographically Weighted Regression: The Analysis of Spatially Varying Relationships*. West Sussex: Wiley.

Geary, R. C. 1954. The contiguity ratio and statistical mapping. *The Incorporated Statistician* 5(3), 115–146. doi: 10.2307/2986645

Gelfand, A. E., H. J. Kim, C. F. Sirmans, and S. Banerjee 2003. Spatial modeling with spatially varying coefficient processes. *Journal of the American Statistical Association* 98(462): 387–396.

Goodchild, M. F. 1992. Geographical information science. *International Journal of Geographical Information Systems* 6(1):31–45.

Goovaerts, P. and C. N. Chiang 1993. Temporal persistence of spatial patterns for mineralizable nitrogen and selected soil properties. *Soil Science Society of America Journal* 57(2): 372–381.

Gregory, P. T., J. M. Macartney, and K. W. Larsen 1987. Spatial patterns and movements. In R. A. Siegel, J. T. Collins, and S. S. Novak (Eds.), *Snakes: Ecology and Evolutionary Biology* (pp. 366–395). New York: Macmillan.

Griffith, D. A. 2008. Spatial-filtering-based contributions to a critique of geographically weighted regression (GWR). *Environment and Planning A* 40(11):2751.

Harris, P., A. S. Fotheringham, and S. Juggins 2010. Robust geographically weighted regression: a technique for quantifying spatial relationships between freshwater acidification critical loads and catchment attributes. *Annals of the Association of American Geographers* 100(2):286–306.

Hemson, G., P. Johnson, A. South, R. Kenward, R. Ripley, and D. McDonald 2005. Are kernels the mustard? Data from global positioning system (GPS) collars suggests problems for kernel home-range analyses with least-squares cross-validation. *Journal of Animal Ecology* 74:455–463.

Heuvelink, G. B. M., P. Musters, and E. J. Pebesma 1997. Spatio-temporal kriging of soil water content. *Geostatistics Wollongong* 96:1020–1030.

Hohn, M. E., A. M. Liebhold, and L. S. Gribko 1993. Geostatistical model for forecasting spatial dynamics of defoliation caused by the gypsy moth (Lepidoptera: Lymantriidae). *Environmental Entomology* 22(5):1066–1075.

Hoss, S. K., C. Guyer, L. L. Smith, and G. W. Shuett 2010. Multiscale influences of landscape composition and configurations on the spatial ecology of eastern diamondback rattlesnakes. *Journal of Herpetology* 44:110–123.

Hsiao, C. 2003. *Analysis of Panel Data* (Vol. 34). Cambridge: Cambridge University Press.

Huang, H. C. and N. Cressie 1996. Spatio-temporal prediction of snow water equivalent using the Kalman filter. *Computational Statistics & Data Analysis* 22(2):159–175.

Huang Y. F. and Y. Leung 2002. Analysing regional industrialisation in Jiangsu province using geographically weighted regression. *Journal of Geographical Systems* 4(2):233–249.

Huang, B., B. Wu, and M. Barry 2010. Geographically and temporally weighted regression for modeling spatio-temporal variation in house prices. *International Journal of Geographical Information Science* 24(3):383–401.

Hurvich, C. M., J. S. Simonoff, and C. L. Tsai 1998. Smoothing parameter selection in nonparametric regression using an improved Akaike information criterion. *Journal of the Royal Statistical Society: Series B* 60(2):271–293.

Imhof, J. P. 1961. Computing the distribution of quadratic forms in normal variables. *Biometrika* 48(3/4):419–426. doi: 10.2307/2332763

Knox, G. 1964. Epidemiology of childhood leukaemia in Northumberland and Durham. *British Journal of Preventive & Social Medicine* 18(1):17–24.

Koop, G., D. J. Poirier, and J. L. Tobias 2007. *Bayesian Econometric Methods* (Vol. 7). Cambridge: Cambridge University Press.

Kyriakidis, P. C. and A. G. Journel 1999. Geostatistical space–time models: a review. *Mathematical Geology* 31(6):651–684.

Kyriakidis, P. C. and A. G. Journel 2001. Stochastic modeling of atmospheric pollution: a spatial time-series framework. Part I: methodology. *Atmospheric Environment* 35(13): 2331–2337.

Le Gallo, J. and C. Ertur 2003. Exploratory spatial data analysis of the distribution of regional per capita GDP in Europe, 1980–1995. *Papers in Regional Science* 82(2):175–201.

Le Gallo, J., C. Ertur, and C. Baumont 2003. A spatial econometric analysis of convergence across European regions. In B. Fingleton (Ed.), *European Regional Growth* (pp. 99–129). Berlin: Springer-Verlag.

Leung, Y., C. L. Mei, and W. X. Zhang 2000a. Statistical tests for spatial nonstationarity based on the geographically weighted regression model. *Environment and Planning A* 32(1):9–32.

Leung, Y., C. L. Mei, and W. X. Zhang 2000b. Testing for spatial autocorrelation among the residuals of the geographically weighted regression. *Environment and Planning A* 32(5): 871–890.

Leung, Y., C.-L. Mei, and W.-X. Zhang 2003. Statistical test for local patterns of spatial association. *Environment and Planning A* 35(4):725–744.

Leyland, A. H., I. H. Langford, J. Rasbash, and H. Goldstein 2000. Multivariate spatial models for event data. *Statistics in Medicine* 19(17–18):2469–2478.

Lieberman, O. 1994. Saddle point approximation for the distribution of a ratio of quadratic forms in normal variables. *Journal of the American Statistical Association* 89(427):924–928. doi: 10.1080/01621459.1994.10476825

Lv, B. Y. and D. L. Yu 2009. Improvement of China's regional economic efficiency under the gradient development strategy: a spatial econometric perspective. *Social Science in China* 20(6):60–72.

Mardia, K. V. and C. R. Goodall 1993. Spatial-temporal analysis of multivariate environmental monitoring data. *Multivariate Environmental Statistics* 6(347–385):76.

Mark, D. M. 2003. Geographic information science: defining the field. In M. Duckham, M. F. Goodchild, and M. F. Worboys (Eds), *Foundations of Geographic Information Science* (pp. 3–18). New York: Taylor & Francis.

Moran, P. A. P. 1950. Notes on continuous stochastic phenomena. *Biometrika* 37(1/2):17–23. doi: 10.2307/2332142

Ntzoufras, I. 2011. *Bayesian Modeling Using WinBUGS* (Vol. 698). Hoboken: Wiley.

O'Sullivan, D. and D. J. Unwin 2003. *Geographic Information Analysis*. Hoboken: Wiley.

Oehlert, G. W. 1993. Regional trends in sulfate wet deposition. *Journal of the American Statistical Association* 88(422):390–399.

Pace, R. K. and J. P. LeSage 2009. *Introduction to Spatial Econometrics*. Boca Raton: Chapman & Hall/CRC.

Páez, A., T. Uchida, and K. Miyamoto 2002a. A general framework for estimation and Inference of geographically weighted regression models: 1. Location-specific kernel bandwidths and a test for local heterogeneity. *Environment and Planning A* 34:733–754.

Páez, A., T. Uchida, and K. Miyamoto 2002b. A general framework for estimation and inference of geographically weighted regression models: 2. Spatial association and model specification tests. *Environment and Planning A* 34(5):883–904.

Papritz, A. and H. Flühler 1994. Temporal change of spatially autocorrelated soil properties: optimal estimation by cokriging. *Geoderma* 62(1):29–43.

Pesaran, M. H. and R. P. Smith 1995. Estimating long-run relationships from dynamic heterogeneous panels. *Journal of Econometrics* 68(1):79–113.

Pesaran, M. H., Y. Shin, and R. P. Smith 1999. Pooled mean group estimation of dynamic heterogeneous panels. *Journal of the American Statistical Association* 94(446):621–634.

Pickles, J. 1997. Tool or science? GIS, technoscience, and the theoretical turn, Editorial, Annals of the Association of American Geographers, p. 363. Retrieved from http://search.ebscohost.com/login.aspx?direct=true&db=a9h&AN=9707110009&site=ehost-live&scope=site (accessed March 14, 2015).

R Core Team. 2013. R: a language and environment for statistical computing, Vienna, Austria. Retrieved from http://www.R-project.org/ (accessed March 14, 2015).

Ripley, B. D. 1981. *Spatial Statistics*. New York: Wiley.

Robertson, D. and J. Symons 1992. Some strange properties of panel data estimators. *Journal of Applied Econometrics* 7(2):175–189.

Rodríguez-Iturbe, I. and J. M. Mejía 1974. The design of rainfall networks in time and space. *Water Resources Research* 10(4):713–728.

Rouhani, S. and H. Wackernagel 1990. Multivariate geostatistical approach to space-time data analysis. *Water Resources Research* 26(4):585–591.

Rouhani, S., M. R. Ebrahimpour, I. Yaqub and E. Gianella 1992. Multivariate geostatistical trend detection and network evaluation of space-time acid deposition data - I. *Methodology Atmospheric Environment: Part A – General Topics* 26(14):2603–2614.

Sampson, P. D. and P. Guttorp 1992. Nonparametric estimation of nonstationary spatial covariance structure. *Journal of the American Statistical Association* 87(417):108–119.

Seburn, D. 2009. *Recovery Strategy for the Eastern Hog-Nosed Snake (Heterodon platirhinos) in Canada* (Species at risk act recovery strategy series). Ottawa: Parks Canada Agency.

Snepvangers, J. J. J. C., G. B. M. Heuvelink, and J. A. Huisman 2003. Soil water content interpolation using spatio-temporal kriging with external drift. *Geoderma* 112(3):253–271.

Sperry, J. H. and P. J. Weatherhead 2009. Does prey availabililty determine seasonal patterns of habitat selection in Texas rat snakes? *Journal of Herpetology* 43:55–64.

Sun, D., R. K. Tsutakawa, H. Kim, and Z. He 2000. Spatio-temporal interaction with disease mapping. *Statistics in Medicine* 19(15):2015–2035.

Switzer, P. 1979. Statistical considerations in network design. *Water Resources Research* 15(6):1712–1716.

Therres, G. D. 1999. Wildlife species of regional conservation concern in the northeastern United States. *Northeast Wildlife* 54:93–100.

Tiefelsdorf, M. 2000. *Modelling Spatial Processes: The Identification and Analysis of Spatial Relationships in Regression Residuals by Means of Moran's I*. New York: Springer-Verlag.

Tiefelsdorf, M. 2002. The saddle point approximation of Moran's I's and Local Moran's Ii's reference distributions and their numerical evaluation. *Geographical Analysis* 34(3):187–206. doi: 10.1111/j.1538-4632.2002.tb01084.x

Tobler, W. R. 1970. A computer movie simulating urban growth in the Detroit region. *Economic Geography* 46:234–240. doi: 10.2307/143141

Toro-Vizcarrondo, C. and T. D. Wallace 1968. A test of the mean square error criterion for restrictions in linear regression. *Journal of the American Statistical Association* 63:558–572.

Vyas, V. M. and G. Christakos 1997. Spatiotemporal analysis and mapping of sulfate deposition data over eastern USA. *Atmospheric Environment* 31(21):3623–3633.

Waller, L. A., L. Zhu, C. A. Gotway, D. M. Gorman, and P. J. Gruenewald 2007. Quantifying geographic variations in associations between alcohol distribution and violence: a comparison of geographically weighted regression and spatially varying coefficient models. *Stochastic Environmental Research and Risk Assessment* 21(5):573–588. doi: 10.1007/s00477-007-0139-9

Wheeler, D. C. 2007. Diagnostic tools and a remedial method for collinearity in geographically weighted regression. *Environment and Planning A* 39(10):2464–2481. doi: 10.1068/a38325

Wheeler, D. C. and C. A. Calder 2007. An assessment of coefficient accuracy in linear regression models with spatially varying coefficients. *Journal of Geographical Systems* 9(2):145–166. doi: 10.1007/s10109-006-0040-y

Wheeler D. C. and M. Tiefelsdorf 2005. Multicollinearity and correlation among local regression coefficients in geographically weighted regression. *Journal of Geographical Systems* 7(2):161–187.

Wheeler, D. C. and L. A. Waller 2009. Comparing spatially varying coefficient models: a case study examining violent crime rates and their relationships to alcohol outlets and illegal drug arrests. *Journal of Geographical Systems* 11(1):1–22. doi: 10.1007/s10109-008-0073-5

Wright, D. J. and M. F. Goodchild 1997a. GIS: tool or science? *Annals of the Association of American Geographers* 87(2):346.

Wright, D. J. and M. F. Goodchild 1997b. Reply: still hoping to turn that theoretical corner. *Annals of the Association of American Geographers* 87(2):373.

Yu, D. L. 2006. Spatially varying development mechanisms in the Greater Beijing Area: a geographically weighted regression investigation. *Annals of Regional Science* 40(1):173–190. doi: 10.1007/s00168-005-0038-2

Yu, D. L. 2010. Exploring spatiotemporally varying regressed relationships: the geographically weighted panel regression analysis. Paper presented at the Joint International Conference on Theory, Data Handling and Modelling in GeoSpatial Information Science, Hong Kong, China, pp. 134–139.

Yu, D. L. 2013. Understanding regional development mechanisms in Greater Beijing Area, China, 1995–2001, from a spatial-temporal perspective. *GeoJournal*. doi: 10.1007/s10708-013-9500-3

Yu, D. L. and B. Y. Lv 2009. Challenging the current measurement of China's provincial postal factor productivity: a spatial econometric perspective. *China Soft Science Magazine* 11:160–170.

Yu, D. L. and C. S. Wu 2004. Understanding population segregation from Landsat ETM+ imagery: a geographically weighted regression approach. *GIScience & Remote Sensing* 41(3):187–206.

Yu, D. L., Y. H. D. Wei, and C. S. Wu 2007. Modeling spatial dimensions of housing prices in Milwaukee, WI. *Environment and Planning B: Planning and Design* 34(6):1085–1102.

Yu, D. L., N. A. Peterson, and R. J. Reid 2009. Exploring the impact of non-normality on spatial non-stationarity in geographically weighted regression analyses: tobacco outlet density in New Jersey. *Geoscience & Remote Sensing* 46(3):329–346. doi: 10.2747/1548-1603.46.3.329

19

LIFE CYCLE ANALYSIS AS A MANAGEMENT TOOL IN ENVIRONMENTAL SYSTEMS

DIMITRIOS A. GEORGAKELLOS
Department of Business Administration, School of Economic, Business and International Studies, University of Piraeus, Piraeus, Greece

Abstract: Environmentally conscious decision making requires information about environmental consequences of alternative products, processes, or activities over their entire life cycle. Nowadays, the most recognized and well-accepted method of carrying such environmental assessments is the life cycle assessment (LCA). It is a method for evaluating the environmental impact associated with a product, process, or activity during its life cycle by identifying and describing, both quantitatively and qualitatively, its requirement for energy and materials, as well as the emissions and waste released to the environment. The LCA process consists of four components: goal definition and scoping, inventory analysis, impact assessment, and interpretation. Currently, the ISO 14040:2006 and 14044:2006 standards provide the indispensable framework for LCA, which, however, leaves the individual practitioner with a range of choices. The driving force to conduct an LCA in most cases is the benchmarking of different products, as well as the evaluation of environmental improvement options on a product-specific basis. Thus, LCA can be used in different contexts, such as for the development and optimization in (environmentally conscious) product development and design, strategic planning, policymaking, marketing, and so on. Moreover, LCA could be applied to complex government policies or business strategies relating to consumption and lifestyle options in a range of society sectors. In spite of certain weak points of the methodology, LCA is perceived as a very robust tool that offers opportunities for assisting companies and policymakers in environmental management. LCA allows a decision maker to study the entire product system, hence avoiding the suboptimization that could result if only a single process were the focus of the study. Therefore, this information can be used with other factors, such as cost and performance data, to design a product or process. It is a tool that provides indicators about the sustainability of industrial systems.

Keywords: attributional LCA, characterization, classification, consequential LCA, grouping, design for the environment (DfE), endpoints, functional unit, goal and scope definitions, impact indicators, ISO 14040, ISO 14044, LCA software, life cycle impact assessment (LCIA), life cycle interpretation, life cycle inventory analysis (LCI), life cycle stages, midpoints, normalization, Society of Environmental Toxicology and Chemistry (SETAC), system boundaries, valuation, weighting.

19.1 Introduction	442	
19.2 Brief History of LCA	443	
19.3 The Methodology of LCA	445	
19.3.1 The LCA Framework	445	
19.3.2 The Goal and Scope Definition	446	
19.3.3 The LCI	448	
19.3.4 The LCIA	451	
19.3.5 The Life Cycle Interpretation	455	
19.4 Applications of LCA	456	
19.4.1 Product Design and Development	457	
19.4.2 Strategic Planning	457	
19.4.3 Marketing	458
19.4.4 Environmental Improvement	458
19.4.5 Purchasing	459
19.4.6 Public Policymaking	459
19.4.7 Learning	459
19.5 Limitations of LCA	460
19.6 Recent Trends and Developments in LCA	461
References	462

An Integrated Approach to Environmental Management, First Edition. Edited by Dibyendu Sarkar, Rupali Datta, Avinandan Mukherjee, and Robyn Hannigan.
© 2016 John Wiley & Sons, Inc. Published 2016 by John Wiley & Sons, Inc.

19.1 INTRODUCTION

As environmental awareness increases, industries and businesses are assessing how their activities affect the environment. Society has become concerned about the issues of natural resource depletion and environmental degradation (Curran, 2006). In order to meet these challenges, environmental impacts, especially those related to goods and services, are being considered and integrated into the decision making of businesses, individuals, public administrators, and policymakers (Finnveden et al., 2009). Many businesses have responded to this awareness by providing "greener" products and using "greener" processes. The environmental performance of products and processes has become a key issue, which is why some companies are investigating ways to minimize their effects on the environment (Curran, 2006). In order to do that, knowledge must be available. When studying environmental impacts of products and services, it is vital to study them in a life cycle perspective because there is a risk of problem shifting from one part of the life cycle to another and from one geographical area to another (Finnveden et al., 2009).

Environmentally conscious decision making requires information about environmental consequences of alternative products, processes, or activities over their entire life cycle (Joshi, 2000). The most recognized and well-accepted method of carrying out environmental assessment of products and services along their life cycles is the life cycle assessment (LCA) (Sonnemann et al., 2004). LCA is a method for evaluating the environmental impact associated with a product, process, or activity during its life cycle by identifying and describing, both quantitatively and qualitatively, its requirement for energy and materials, as well as the emissions and waste released to the environment (Bersimis and Georgakellos, 2013). LCA is a structured, comprehensive, and internationally standardized method (European Commission—Joint Research Centre—Institute for Environment and Sustainability, 2010). It also considers all attributes or aspects of natural environment, human health, and resources (Finnveden et al., 2009). LCA is unique because it encompasses all processes and environmental releases beginning with the extraction of raw materials and the production of energy used to create the product through the use and final disposition of the product (Curran, 2006). Figure 19.1 illustrates the possible life cycle stages that can be considered in LCA and the typical inputs and outputs measured.

LCA was developed based on the idea of comprehensive environmental assessments of products, which was conceived in Europe and in the United States in the late 1960s and early 1970s (Ekvall et al., 2005). Since the 1990s, there has been an increased interest in the topics related to LCA. When the first scientific papers emerged, LCA was firstly considered with high expectations, but its results received some criticism. Since then, there has been a considerable amount of development and coordination that has resulted in an international standard for LCA. This along with a number of guidelines and textbooks has helped to progressively expand the maturity and methodological robustness of it (Finnveden et al., 2009).

LCA is becoming recognized as a vital and powerful decision support tool, complementing other methods, which are equally necessary to help effectively and efficiently make consumption and production more sustainable (European Commission—Joint Research Centre—Institute

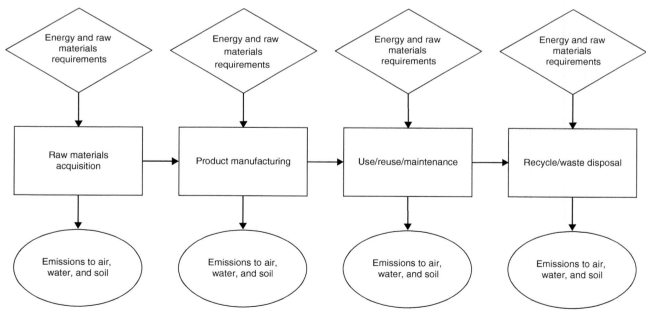

FIGURE 19.1 Life cycle stages.

for Environment and Sustainability, 2010). Today, the LCA approach is widely recognized as a useful framework, and attempts are underway to integrate life cycle thinking into business decisions (Joshi, 2000). In fact, international policy documents call for the application of LCA in the development of sustainable patterns of consumption and production (Hertwich and Peters, 2006). It should be noted that LCA is no substitute for (environmental) risk assessment (RA). Instead, the results from the LCA reflect the potential contributions to actual impacts or risks pending on the relevance and validity of the reference conditions assumed in the underlying models (Finnveden et al., 2009).

The LCA process is a systematic, phased approach and consists of four components: goal definition and scoping, inventory analysis, impact assessment, and interpretation (Curran, 2006). The goal and scope component includes the reasons for carrying out the study, the intended application, and the intended audience. It is also the place where the system boundaries of the study are described and the functional unit is defined. The functional unit is a quantitative measure of the functions that the goods (or services) provide. The result from the inventory analysis is a compilation of the inputs (resources) and the outputs (emissions) from the product over its life cycle in relation to the functional unit. The impact assessment assesses the potential human and ecological effects of energy, water, and material usage and the environmental releases identified in the inventory analysis. In the interpretation, the results from the previous phases are evaluated in relation to the goal and scope in order to reach conclusions and recommendations (Curran, 2006; Finnveden et al., 2009). Despite its apparent simplicity and attractiveness, LCA is usually difficult to carry out in real life because the final products often require a large number of diverse inputs. Besides the diversity of inputs, there are also increased interdependencies in them, making their modeling a difficult process. Thus, tracing all the direct and indirect inputs as well as the related environmental burdens all the way to basic raw material extraction is a rather daunting task (Joshi, 2000). There are, however, numerous worldwide initiatives in progress attempting to develop a consensus and provide guidance. Among the most important of them are the Society of Environmental Toxicology and Chemistry (SETAC), the Life Cycle Initiative of the United Nations Environment Program (UNEP), the European Platform for LCA of the European Commission, and the emerging International Reference Life Cycle Data System (ILCD).

During the evolution of LCA, a number of related applications emerged, of which we give some examples below (Jensen et al., 1997):

- Internal industrial use in product development and improvement
- Internal strategic planning and policy decision support in industry
- External industrial use for marketing purposes
- Governmental policymaking in the areas of ecolabeling, green procurement, and waste management opportunities.

Similar applications can be used at a strategic level, dealing with government policies and business strategies. Thus, the way an LCA project is implemented depends on the intended use of the LCA results (Guinée, 2004).

19.2 BRIEF HISTORY OF LCA

The LCA methodology had its beginnings in the 1960s but has developed and somewhat matured during the last decades (Finnveden et al., 2009). Concerns over the limitations of raw materials and energy resources sparked interest in finding ways to cumulatively account for energy use and to project future resource supplies and use (Curran, 2006). The first attempts of this kind were energy analyses of industrial systems undertaken at that time and subsequently in response to the oil crises of the early 1970s. However, interest in these studies declined in the late 1970s, and it was not until the rise of environmental awareness in the late 1980s that attention was again focused on LCA as a potentially valuable environmental management tool (McLaren, 2010).

Specifically, it is not easy to determine exactly when studies related to the LCA methodology started. One of the first studies was H. Smith's, whose calculations of energy requirements for manufacturing final and intermediate chemical products entered the public domain in 1963. Later in the 1960s, global modeling studies published in *The Limits to Growth* (1972) and *A Blueprint for Survival* (1972) resulted in predictions of the effects of the world's changing populations on the demand for finite raw materials and energy resources. These predictions, together with the oil crisis of the 1970s, encouraged more detailed studies, focused mainly on the optimum management of energy resources. During this period, about a dozen studies were performed to estimate costs and environmental implications of alternative sources of energy. Although these studies focused basically on the optimization of energy consumption, they also included estimations on emissions and releases. The foundation for the current methods of life cycle inventory analysis (LCI) in the United States is considered an internal study for The Coca-Cola Company carried out by the Midwest Research Institute (MRI) in 1963, aimed at determining the type of container with the lowest environmental effect. Other companies in both the United States and Europe performed similar comparative life cycle inventory analyses in the early 1970s. The process of quantifying the resource use and environmental releases of products became known as a Resource and Environmental Profile Analysis (REPA) in the United States and Ecobalance in Europe.

Approximately 15 REPAs were performed between 1970 and 1975. From 1975 through the early 1980s, LCI continued to be conducted, and the methodology improved through a slow stream of about two studies per year, most of which focused on energy requirements. Throughout this time, European LCA practitioners developed approaches parallel to those being used in the United States (Sonnemann et al., 2004; Curran, 2006).

In 1979, the SETAC, a multidisciplinary society of professionals with industrial, public, and scientific representatives, was founded. One of SETAC's goals was, and continues to be, the development of LCA methodology and standards. In 1984, the EMPA (Swiss Federal Laboratories for Materials Testing and Research) conducted research that added the effects on health to emission studies and took into consideration a restricted number of parameters, simplifying accordingly assessment and decision making. Products were assessed on the basis of their possible environmental impact expressed as energy consumption, air pollution, water contamination, and solid wastes (Sonnemann et al., 2004). The first international meetings for LCA researchers and practitioners were held under the auspices of the SETAC in 1990 and 1991. During the 1990s, SETAC organized working groups and published reports on various aspects of LCA methodology and application, and the SETAC annual meetings became a forum for LCA researchers and practitioners to develop a common understanding of the purpose and practice of LCA (McLaren, 2010). In 1993, the International Organization for Standardization (ISO) created the Technical Committee 207 with the goal of developing international norms and rules for environmental management. One of the six subcommittees created was assigned standardization within the field of LCA. Its aim was to prevent the presentation of partial results or data of uncertain reliability from LCA studies for marketing purposes. This ensured that each application is carried out in compliance with universally valid structure and features (Sonnemann et al., 2004). As a result, a series of four ISO LCA standards (ISO 14040 to 14043) were published between 1997 and 2000. It was also during this time that a growing number of researchers had become interested in the use of LCA as a decision support tool. In 2002, a collaborative partnership was launched between the SETAC and the UNEP: the Life Cycle Initiative (McLaren, 2010). The three programs of the initiative focused on putting life cycle thinking into practice and on improving the supporting tools through better data and indicators (Curran, 2006).

Aside from the UNEP/SETAC Life Cycle Initiative, there are also other ongoing international initiatives attempting similarly to help build consensus and provide recommendations, including the European Platform for LCA of the European Commission and the emerging ILCD (Finnveden et al., 2009). Over the last few years, the attention on carbon emissions and water consumption, through the carbon footprinting and the emerging water footprinting, has led to a broader acknowledgment of LCA's appropriateness in support of environmental management initiatives. The UK's PAS 2050 is essentially based on LCA methodology as defined in the ISO LCA standards. The ISO 14067 Carbon Footprint of Products standard seems to be also along the lines of the ISO LCA standards (McLaren, 2010). Figure 19.2 attempts to illustrate the LCA origins.

Currently, activities regarding databases, quality assurance, consistency, and coordination of methods contribute to the

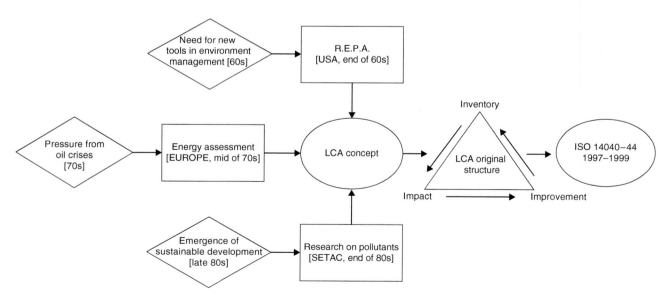

FIGURE 19.2 Life cycle assessment origins.

evolution of the LCA methodology. Simultaneously, there are several areas where the development has been strong during the last years. These include better models for impact assessment, a better understanding of the differences between discrete types of LCA (e.g., attributional and consequential LCA), methods for the combination of LCA with other environmental systems analysis tools (e.g., with cost–benefit analysis and life cycle costing), and databases for the inventory analysis (Finnveden et al., 2009).

19.3 THE METHODOLOGY OF LCA

19.3.1 The LCA Framework

As mentioned previously and shown in Fig. 19.3 in Section 19.3.1, the LCA framework is described by four phases:

1. Goal and scope definitions
2. LCI
3. Life cycle impact assessment (LCIA)
4. Life cycle interpretation.

The double arrows between the phases indicate the interactive nature of LCA (Jensen et al., 1997).

LCA demands a systematic process that involves a number of iterations. Some issues cannot be addressed initially; however, they will be examined, improved, or revised in the usually two to three iterations of almost any LCA study. The goal and the initial scope are defined at the beginning of the process and will help identify the requirements on the subsequent work. However, in the later stages—the life cycle inventory phase of data collection, the impact assessment, and the interpretation—more information becomes available, resulting typically in a revision or redefinition of the initial scope (European Commission—Joint Research Centre—Institute for Environment and Sustainability, 2010).

The development of the international standards for LCA (ISO 14040:1997, ISO 14041:1999, ISO 14042:2000, ISO 14043:2000) was a central step to consolidate procedures and methods of LCA. Their contribution to the general approval of LCA by all stakeholders and by the international community was very important (Morris, 2004; Finkbeiner et al., 2006). Currently, the ISO 14040:2006 and 14044:2006 standards provide the indispensable framework for LCA. This framework, however, leaves the individual practitioner with a range of choices, which can affect the legitimacy of the results of an LCA study (European Commission—Joint Research Centre—Institute for Environment and Sustainability, 2010). According to the ISO (2006), "the scope, including the system boundary and level of detail, of an LCA depends on the subject and the intended use of the study. The depth and the breadth of LCA can differ considerably depending on the goal of a particular LCA. The life

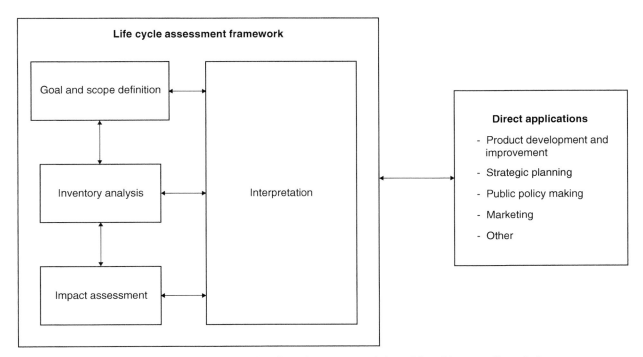

FIGURE 19.3 Framework for life cycle assessment (adapted from European Commission—Joint Research Centre—Institute for Environment and Sustainability, 2010).

cycle inventory analysis phase is the second phase of LCA. It is an inventory of input/output data with regard to the system being studied. It involves collection of the data necessary to meet the goals of the defined study. The life cycle impact assessment phase is the third phase of the LCA. The purpose of LCIA is to provide additional information to help assess a product system's LCI results so as to better understand their environmental significance. Life cycle interpretation is the final phase of the LCA procedure, in which the results of an LCI or an LCIA, or both, are summarized and discussed as a basis for conclusions, recommendations and decision-making in accordance with the goal and scope definition." It should be noted that there are cases where the goal of an LCA can be satisfied by performing only an inventory analysis and an interpretation. Such a study is usually referred to as an LCI one (International Organization for Standardization, 2006).

The following are some critical issues related to all the phases and aspects of LCA work (Finkbeiner et al., 2006; European Commission—Joint Research Centre—Institute for Environment and Sustainability, 2010).

A very crucial issue throughout the scope definition as well as in the later inventory and impact assessment phases is to ensure a high degree of consistency regarding all important methodological and data aspects of the LCA. Consistency will be examined moreover as part of the evaluation step in the interpretation phase and is to be considered in conclusions and recommendations drawing as well as in communication.

Reproducibility is another crucial requirement for the credibility of the LCA study and an important issue for review. However, in a number of published LCA studies, reproducibility has to be in balance with confidentiality. An independent and external critical review of the data guarantees the quality of LCI data sets and, especially for comparative LCA studies, the robustness—reproducibility—of their results.

The system's function and functional unit are central elements of an LCA since, without them, a consequential and valid comparison, mainly of products, is not possible. This functional unit defines what is being studied, and all subsequent analyses are then relative to it. Therefore, all inputs and outputs in the LCI and consequently the LCIA profile are related to the functional unit.

Due to the inherent complexity in LCA, transparency is an important guiding principle in executing LCAs in order to ensure a proper interpretation of the results. Moreover, LCA should consider (if possible) all attributes or aspects of natural environments, human health, and resources. By considering all attributes and aspects within one study in a cross-media perspective, potential trade-offs can be identified and assessed.

Natural science should preferably be the base for decisions within an LCA, while, if this is impracticable, other scientific approaches such as social and economic sciences or even international conventions can also be applied. If neither of these justifications or conventions is applicable, decisions could be based on value choices (Finkbeiner et al., 2006).

19.3.2 The Goal and Scope Definition

The goal and scope definition is the first phase in an LCA containing the following main issues: goal, scope, functional unit, system boundaries, data quality, and critical review process (Jensen et al., 1997). The goal and scope definition phase of an LCA includes several decisions that are important for all subsequent steps (Finkbeiner et al., 2006). During this step, the strategic aspects relating to questions to be answered and identifying the intended audience are defined. To carry out the goal and scope of an LCA study, the practitioner must follow some procedures (Sonnemann et al., 2004):

- Define the purpose of the LCA study, including the definition of the functional unit, which is the quantitative reference for the study.
- Define the scope of the study, which embraces the establishment of the spatial limits between the product system under study and its neighborhood (the generally called "environment") and the detail description of the system through drawing up its unit processes flowchart.
- Define the data required, which includes a specification of the data necessary for the inventory analysis and for the subsequent impact assessment phase.

The goal definition is the first phase of any LCA, independently, whether the LCI/LCA study is limited to the development of a single unit process data set or it is a complete LCA study of a comparative assertion to be published. The goal definition determines the level of sophistication of the study and the requirements for reporting. The goal definition also has to define the intended use of the results and users of the results. The practitioner, who has to reach the goal, needs to comprehend the detailed purpose of the study in order to make appropriate decisions all over the study. Six aspects shall be addressed and verified during the goal definition: (i) intended applications of the results; (ii) limitations due to the method, assumptions, and impact coverage; (iii) reasons for carrying out the study and decision context; (iv) target audience of the results; (v) comparative studies to be disclosed to the public; and (vi) commissioner of the study and other influential actors. The goal definition guides all the detailed aspects of the scope definition, which in turn sets the frame for the LCI and LCIA work (Jensen et al., 1997; European

Commission—Joint Research Centre—Institute for Environment and Sustainability, 2010).

The definition of the scope of the LCA sets the borders of the assessment—what is included in the system and what detailed assessment methods are to be used (Jensen et al., 1997). The main purpose of the scope definition is to define the prerequisites on methodology, quality, reporting, and review in compliance with the goal of the study such as the motives for the study, the decision context, the intended applications, and the receivers of the results. The main issues that should be clearly described or defined, when deriving the scope of an LCA study from the goal, are the following (European Commission—Joint Research Centre—Institute for Environment and Sustainability, 2010):

- The types of the deliverables of the LCA study, in line with the intend applications
- The system or process that is studied and its functions, functional unit, and reference flows
- The LCI modeling framework and handling of multifunctional processes and products
- The system boundaries, completeness requirements, and related cutoff rules
- The LCIA impact categories to be covered and selection of specific LCIA methods to be applied
- Other LCI data quality requirements regarding technological, geographical, and time-related representativeness and appropriateness
- The types, quality, and sources of required data and information and especially the required precision and maximum permitted uncertainties
- The special requirements for comparisons between systems
- Identifying critical review needs
- Planning reporting of the results

The scope should be adequately well defined to ensure that the breadth, the depth, and the detail of the study are compatible and sufficient to address the stated goal. As it has been already mentioned, LCA is an iterative technique, and consequently, the scope of the study may need to be adapted while the study is being conducted and additional information is gathered (Jensen et al., 1997).

The basis of an LCA is the definition of functional unit or performance characteristics. This is because all data collected in the inventory phase will be related to the functional unit. Additionally, the functional unit sets the scale for comparison of two or more products including improvement to one product. Thus, when comparing different products fulfilling the same function, the definition of the functional unit is of particular importance. Three aspects have to be considered when defining the functional unit: the efficiency of the product, the durability of the product, and the performance quality standard (Jensen et al., 1997). Selected examples of the functional unit are given in Table 19.1.

The system boundaries define the processes and operations (e.g., manufacturing, transport, and waste management processes) and the inputs and outputs to be taken into account in the LCA. The definition of system boundaries is a rather subjective action and includes geographical boundaries, life cycle boundaries (i.e., limitations in the life cycle), and boundaries between the technosphere and biosphere (Jensen et al., 1997). More precisely, there are three major types of system boundaries in the LCI: (i) between the technical system and the environment, (ii) between significant and insignificant processes, and (iii) between the technological system under study and other technological systems. In many instances, the system boundary linking the technical system and the environment is clear, while there are some cases (e.g., when the life cycle includes forestry, agriculture, emissions to external wastewater systems, and landfills) where the system boundary should be unambiguously defined. Furthermore, it is difficult to distinguish the system boundary between significant and insignificant processes since it is generally not known beforehand what data are insignificant and there is no reason to omit them. A common way is to include easily accessible data, examine their significance, refine when it is necessary or possible, and perform the LCI and LCIA in iterative loops until the necessary precision

TABLE 19.1 Examples of Functional Units in Life Cycle Assessment

Class of Products or Activities	System Function	Functional Unit
Goods use	Beverage container	100 l of beverage
	Laundry detergent	6 kg of washed clothes
Process	Electricity generation	1 kWh of electricity generated
	Wastewater treatment	1 kg of removed COD per m^3
Service	Road transport	1 pkm (person-km) or tkm (ton-km)
	Tourism	One guest overnight stay with breakfast

Adapted from Sonnemann et al. (2004).

has been realized. In regards to the third category, an LCA is often restricted to a specific production technology or to a level of technology, usually the best available technology. The system boundary toward other technological systems (for instance, when the LCA includes multifunctional processes) has to be defined as well. This may happen when a process is shared among several product systems, and it is not obvious to which product the environmental impacts should be allocated (Finnveden et al., 2009). Because of the subjectivity of system boundaries definition, it is of crucial importance to the transparency of the defining process and the relevant assumptions.

The quality of the final LCA is being affected by the quality of the data used in the life cycle inventory. Since this quality can be described in different ways, it is crucial that it is assessed in a systematic way that lets one to recognize, comprehend, and value the actual data quality. Additional data quality indicators that should be taken into consideration in all studies (apparently in a level of detail that depends on the goal and scope definition) are as follows: precision, completeness, representativeness, consistency, and reproducibility (Jensen et al., 1997). Based on the particular data required and the quality prerequisites, the sources for the data and information should be recognized. A wide range of prospective LCI data sources exist:

- The producers of goods and operators of processes and services, as well as their associations (primary data sources)
- Generic data, for example, national databases, consultants, and research groups or data sources that give access to primary data possibly after remodeling (secondary data sources)

Correspondingly, the choice of the sources for the other data needed (recycling rates, statistical data, etc.) is, as for the LCI data, a critical action that should be taken thoroughly (European Commission—Joint Research Centre—Institute for Environment and Sustainability, 2010).

19.3.3 The LCI

LCI is the second phase in an LCA, in which all the environmental loads or environmental effects generated by a product, process, or activity during its life cycle are identified and evaluated. In this phase, environmental loads are described as the amount of substances, noise, radiation, and so on emitted to or removed from the environment that cause actual or potential damaging effects. Thus, according to this description, it can be found in raw materials and energy consumption, as well as in environmental pollution, that is, air and water emissions, waste generation, radiation, noise, odors, and so on. Other types of effects such as aesthetic and social, although they must often be taken into account, are not considered in LCI because environmental loads must be quantifiable (Sonnemann et al., 2004). Life cycle inventory contains the following main issues: data collection, refining system boundaries, validation of data, allocation, and calculation (Jensen et al., 1997). Data must be collected based on a process flow diagram of the system under study according to the defined life cycle boundaries. A typical and rather uncomplicated example of LCA flow diagram applied to soft drink containers (bottles and cans) is shown in Fig. 19.4 (Georgakellos, 2006).

According to this figure, this specific LCA system consists of 11 subsystems that together cover the entire life cycle of the containers. These subsystems are (i) raw materials acquisition and materials manufacture (it includes all the activities required to gather or obtain a raw material or energy source from the earth and to process them into a form that can be used to fabricate a particular container), (ii) materials transportation (to the point of containers fabrication), (iii) containers fabrication (this is the process step that uses raw or manufactured materials to fabricate a container ready to be filled), (iv) containers transportation (to the point of filling), (v) filling—final product production (it includes all processes that fill the containers and prepare them for shipment), (vi) final product transportation (to retail outlets), (vii) final product use (it comprises activities such as storage of the containers for later use, preparation for use, and consumption), (viii) solid wastes collection and transportation for landfilling (it begins after the containers have served their intended purpose and enter the environment through the waste management system), (ix) solid wastes landfilling (it includes all necessary activities for the land disposal of waste), (x) used containers collection and refilling (it includes all the activities required to off-site reuse such as the return of the containers to the bottler to be refilled for their original purpose), and (xi) recycling (it encompasses all activities necessary to take the used containers out of the waste management system and deliver them to the container fabrication stage) (Georgakellos, 2006).

An LCI has increased data requirements. Setting up inventory data can take much effort and be time-consuming, being often a challenging issue due to the lack of appropriate data for the product system under examination. Many databases, such as public national or regional databases, industry databases, and consultants' databases, have been developed, in order to facilitate the LCI data collection and avoid duplication in data compilation. Such examples are the Swedish SPINE@CPM database, the German PROBAS database, the Japanese JEMAI database, the US NREL database, the Australian LCI database, the Swiss Ecoinvent database, and the European Reference Life Cycle Database (ELCD). More databases are presently under development all over the world, for example, in Germany, Canada, Brazil, China, Thailand, Malaysia, and other countries (Finnveden et al., 2009). It should be noted, however, that using data from an

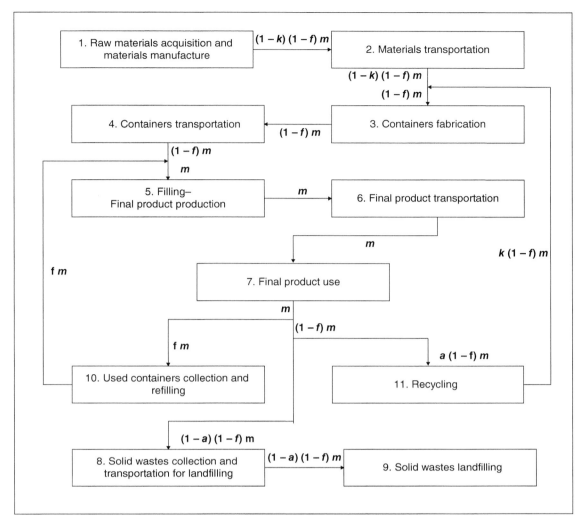

FIGURE 19.4 An LCA system illustration concerning soft drink containers (Georgakellos, 2006).

electronic database or literature, one has to make sure that they (i) concern the related processes, (ii) come from reliable sources, (iii) are updated, and (iv) are in line with the goal and the system earlier defined (Sonnemann et al., 2004). The result of the data collection can be presented in an inventory table as shown in Table 19.2 with an example from the material data concerning the same LCI example of soft drink containers (Georgakellos, 2005).

The system boundaries, defined as a part of the scope definition procedure, can be refined after the initial data collection, for example, as a result of decisions of inclusion of new unit processes shown to be noteworthy or exclusion of life stages, subsystems, or material flows. During the data collection process, the validation of data has to be performed in order to increase the overall data quality. Data validation may indicate areas where data quality must be improved or further data must be found in similar processes, unit processes, etc. On the other hand, in many LCAs, especially those of complex systems, it is not possible to manage all impacts and outputs inside the system boundaries. This problem can be solved either by expanding the system boundaries to include all the inputs and outputs or by allocating the relevant environmental impacts to the studied system. Expanding the system boundaries may have the risk of making the system too complicated, which means that the other LCA phases or steps (e.g., data collection, impact assessment, and interpretation) could take too much time and become too expensive. Thus, allocation may be a better alternative, if an appropriate method can be found (Jensen et al., 1997).

Allocation resolves the multifunctionality by separating the amounts of the particular inputs and outputs

TABLE 19.2 Life Cycle Inventory Inputs and Outputs of Soft Drink Containers from the Greek Market (Georgakellos, 2005)

Inputs and Outputs (per 1000 l of soft drinks)	Type of Container		
	Glass (0.25 l)	Aluminum (0.33 l)	PET (0.50 l)
Consumption of Energy (MJ) and Water (g)			
Total fuel plus feedstock	10,494.19	23,923.45	20,999.05
Water	42,463.33	110,241.1	97,425.44
Atmospheric Emissions (g)			
Particles	655.4028	1,694.266	377.3994
Carbon monoxide (CO)	192.1243	748.5988	1,554.898
Hydrocarbons	65.73306	1,537.727	2,166.874
Nitrogen oxides (NO_x)	2,801.414	3,088.879	2,694.381
Nitrous oxide (N_2O)	1.02563	55.80012	137.2929
Sulfur dioxide (SO_2)	5,693.381	6,724.978	4,606.816
Aldehydes	22.36611	24.65838	22.82026
Organic compounds	0.224066	3.964149	14.3779
Ammonia (NH_3)	0.054823	0.755076	0.131386
Hydrogen chloride (HCl)	0.650502	1.88769	7.454027
Fluoride[a] and hydrogen fluoride (HF)	0.252988	18.76364	0.00132
Lead (Pb)	0.162626	0	0
Volatile organic compounds (VOC)	85.2819	36.79974	40.0208
Waterborne Waste (g)			
Suspended materials	0.018285	0.226523	0.0264
Dissolved materials	32.67909	599.9079	3,487.114
BOD	0.018285	30.16529	0.0264
COD	0.054823	718.0773	0.131386
Oil	0.439812	6.531407	43.09564
Phenol	0	0	0.065386
Fluoride	0.00215	1.132614	0.250186
Ammonia (NH_3)	0.000922	0.030203	0.0924
Sulfate	0.000307	0	0.0396
Nitrate	0.000614	0	0.0396
Chloride	0.0000246	0	0.00264
Na-ions	0.000307	0	0.0264
Fe-ions	0.00000307	0	0.000396
Solid Waste (cm^3)			
Municipal waste, etc.	11,965.91	71,414.99	66,013.55

[a] HFC, CFC, HCFC, CF_4, F_2, etc.

among the cofunctions in accordance with an allocation criterion, which is a property of the cofunctions (e.g., mass, energy or element content, and market price). Allocation should be performed according to the essential causal physical, chemical, and biological connection between the different products or functions. If it is not possible to find clear common physical causal relationships among the cofunctions, then it is recommended to execute the allocation in relation to another relationship (an economic relationship or a relationship between some other properties of the cofunctions) (European Commission—Joint Research Centre—Institute for Environment and Sustainability, 2010).

After data collection and the selection of the allocation criteria, a model of environmental loads calculation is set up for the product system. No formal demands exist for calculation in LCA. However, due to the amount of data, it is recommended that a spreadsheet be developed and used for that specific purpose. Currently, a number of spreadsheet applications, together with many software programs developed especially for LCA, exist. The selection of the appropriate program depends on the kind and amount of data to be handled (Jensen et al., 1997). An example of a rather unfussy calculation model related to the LCI system presented in Fig. 19.4 is the total energy consumption of the system (E_{tot}). E_{tot} can be calculated by the following equation:

$$E_{tot} = (e_1 + e_2)(1-f)(1-k)m + (e_3 + e_4)(1-f)m$$
$$+ (e_5 + e_6 + e_7)m + (e_8 + e_9)(1-f)(1-a)m$$
$$+ e_{10}fm + e_{11}(1-f)am$$

where "e_j" is the *specific energy consumption* of the subsystem j ($j = 1, 2, \ldots, 11$), "m" is the *mass* of the functional unit of the product, "f" is the *reuse–refilling rate* of the product ($0 \leq f \leq 1$), "a" is the *recycling rate* of the product that examined ($0 \leq a \leq 1$), and "k" is the *recycled content level* of the product ($0 \leq k \leq 1$). Recycling coefficients "a" and "k" are not always the same since coefficient "k" shows the percentage of raw material used in product production that is not virgin but recycled, while coefficient "a" shows the percentage of product waste that goes for recycling. The total consumption of any material or the total release of any waste of the system can be calculated by a similar equation (Georgakellos, 2006).

It is noteworthy that, apart from the allocation or system expansion approaches, another important decision that must be made in this phase concerns the principles and method approaches that should be applied in the LCI system modeling, that is, attributional or consequential modeling. This influences many of the subsequent choices such as which inventory data are to be collected. According to the ILCD Handbook (European Commission—Joint Research Centre—Institute for Environment and Sustainability, 2010), "the attributional life cycle inventory modelling principle (also referred to as 'accounting', 'book-keeping', 'retrospective', or 'descriptive') depicts the potential environmental impacts that can be attributed to a system (e.g. a product) over its life cycle, i.e. upstream along the supply-chain and downstream following the system's use and end-of-life value chain. Attributional modelling makes use of historical, fact-based, measureable data of known (or at least knowable) uncertainty, and includes all the processes that are identified to relevantly contribute to the system being studied. In attributional modelling, the system is hence modelled as it is or was (or is forecasted to be). The consequential life cycle inventory modelling principle (also called 'change-oriented', 'effect-oriented', 'decision-based', 'market-based') aims at identifying the consequences that a decision in the foreground system has for other processes and systems of the economy, both in the analyzed system's background system and on other systems. It models the analyzed system around these consequences. The consequential life cycle model is hence not reflecting the actual (or forecasted) specific or average supply-chain, but a hypothetic generic supply-chain is modelled that is prognostizised along market-mechanisms, and potentially including political interactions and consumer behavior changes." Consequential modeling can be applied both to LCI and LCIA. Consequential assessment, and therefore also consequential modeling, is relevant in most application areas of LCA. However, there are application areas where attributional modeling is more relevant (Weidema et al., 2009).

19.3.4 The LCIA

The life cycle inventory provides environmental information consisting mostly of a quantified list of environmental loads (energy and material consumption, emissions of air and water, wastes, etc.) to be assigned to the product or process. However, the environmental damage associated with them is not yet known. Thus, the LCIA phase attempts at making the results from the inventory analysis more clear and manageable in relation to human health, natural environment, and natural resources. To achieve this, the inventory table has to be converted into a smaller number of indicators (Sonnemann et al., 2004). In other words, the purpose of the LCIA is to provide additional information to help evaluate the results from the LCI and better understand their environmental importance. Thus, the LCIA should interpret the inventory results into their potential impacts on areas of protection. These impacts are modeled applying the best available knowledge about relationships between interventions in the form of resource extractions, emissions, land and water use, and their impacts in the environment. The schematic presentation of an environmental mechanism underlying the modeling impacts and damages in LCIA is given in Fig. 19.5. In this figure, a distinction is made between midpoint and endpoint indicators, where endpoints are defined at the level of the areas of protection and midpoints specify impacts somewhere between the emission and the endpoint (Finnveden et al., 2009).

Even though midpoints and endpoints can be usually overlapped, midpoint indicators are used to determine a substance's strength of effect (mostly characterized by using a threshold) and do not consider the severity of the expected impact. This is because, normally, an indicator defined closer to the environmental intervention will result in more certain modeling, while an indicator further away from it will provide environmentally more relevant information, that is, more directly linked to society's concerns and the areas of protection (Sonnemann et al., 2004).

Impact assessment contains the following main issues:

Categories of impact indicators and models selection

Classification of environmental loads within the different environmental impact categories

Characterization of environmental loads through a reference pollutant typical of each category of environmental impact

Optional elements and information that can be used dependent on the goal and scope of the LCA study include the following:

Normalization, that is, all impact scores (contribution of a product system to one impact category) are related to a reference situation

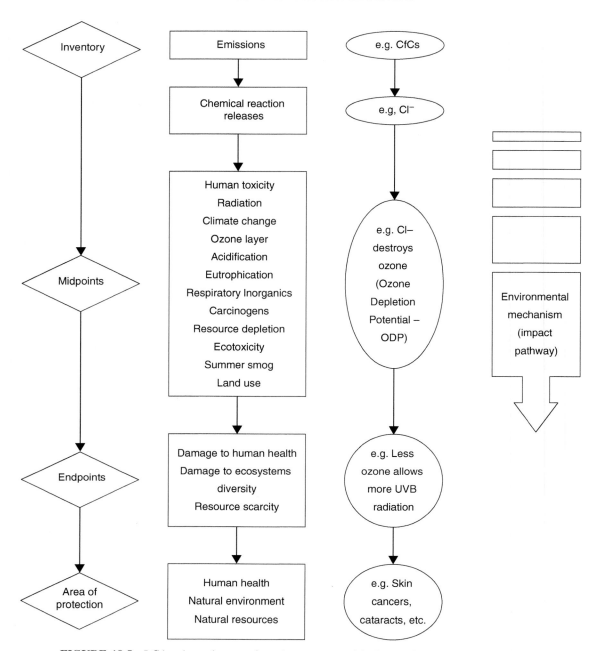

FIGURE 19.5 LCA schematic steps from inventory to midpoints and endpoints (adapted from European Commission—Joint Research Centre—Institute for Environment and Sustainability, 2010).

Grouping, that is, the indicators can be sorted and probably ranked

Weighting, that is, quantitative comparison of the seriousness of the different resource consumption or impact potentials of the product, attempting to cover and possibly aggregate indicator results across impact categories

Data quality analysis, that is, better understanding of the reliability of the LCIA results

The first step in an LCIA is to choose the impact categories that will be considered as part of the overall LCA. This is a follow-up of the decisions made in the goal and scoping phase. Based on the type of information collected in the inventory phase, the boundaries defined in the goal and scoping may be redefined. The environmental items identified in the LCI may harm human health (e.g., by causing cancer or cardiovascular diseases) and have environmental impacts (e.g., by causing global warming or ozone depletion). A number of issues have to be considered when choosing impact categories: completeness (all environmental problems of relevance should be covered), practicality (the list should not contain too many categories), independence (double counting should be prevented by choosing mutually unconnected impact categories), and relation to the characterization step (the selected impact categories should be interrelated to available characterization methods) (Jensen et al., 1997).

Some impact categories at midpoint level that should be covered by the combination of selected LCIA methods are climate change, (stratospheric) ozone depletion, human toxicity, respiratory inorganics, ionizing radiation, (ground-level) photochemical ozone formation, acidification (land and water), eutrophication (land and water), ecotoxicity, land use, and resource depletion (minerals, fossil and renewable energy resources, water). Likewise, some areas of protection are human health, natural environment, and natural resources. If possible, it is recommended to use them together with consistent impact factors on the endpoint level (European Commission—Joint Research Centre—Institute for Environment and Sustainability, 2010).

The classification assigns the emissions from the inventory to these impact categories according to the substances' ability to contribute to different environmental problems (Finnveden et al., 2009). For LCI data that contribute to only one impact category, the procedure is a clear-cut assignment. For instance, carbon dioxide emissions can be classified into the global warming category. For LCI data that contribute to two or more different impact categories, there are two ways of assigning: partition a representative portion of the LCI results to the impact categories to which they contribute (for cases when the effects are dependent on each other); assign all LCI results to all impact categories to which they contribute (when the effects are independent of each other). For example, since nitrogen dioxide could potentially affect both ground-level ozone formation and acidification simultaneously, the entire quantity of nitrogen dioxide would be assigned to both impact categories. In any case, the procedure used must be clearly documented.

Impact characterization uses science-based conversion factors (characterization factors) to convert and combine the LCI results into representative indicators of impacts to human and ecological health. Specifically, the impact of each emission is modeled quantitatively according to the environmental mechanism and expressed as an impact score in a unit common to all contributions within the impact category (e.g., kg CO_2 equivalents for greenhouse gases contributing to the impact category climate change). Characterization allows summing the contributions from all emissions and resource extractions within each impact category, translating the inventory data into a profile of environmental impact scores (Finnveden et al., 2009). In other words, characterization factors translate different inventory inputs into directly comparable impact indicators. Thus, characterization provides a way to directly compare the LCI results within each impact category. Impact indicators are typically characterized using the following equation:

Inventory data × characterization factor = impact indicators

For example, all greenhouse gases of the LCI results can be expressed in terms of CO_2 equivalents by multiplying them by a CO_2 characterization factor. Subsequently, combining the resulting impact indicators, an overall global warming potential indicator can be provided. Normalization is an LCIA tool used to express impact indicator data in a way that can be compared among impact categories. Normalization expresses the relative magnitude of the impact scores on a scale that is common to all the categories of impact (typically the background impact from society's total activities) in order to facilitate the interpretation of the results (Finnveden et al., 2009). This is found by dividing the calculated indicators by the particular reference value. Normally, the impact or damage results of the total annual territorial elementary flows in a geographic area (i.e., in a region, country, continent, or globally) or per average citizen (i.e., per capita) are used as reference values. These reference impact or damage results are called "normalization basis" and are calculated for each of the impact categories or damages, from their inventory, likewise as the impact indicators of the examined system are calculated from its LCI. Consequently, for midpoint level, the normalization basis is the overall potential impact, calculated from the annual inventory of elementary flows, while for endpoint level, the normalization basis is the overall damage to the areas of protection (European Commission—Joint Research Centre—Institute for Environment and Sustainability, 2010).

The final steps of the LCIA include grouping or weighting of the different environmental impact categories and resource consumptions reflecting the relative importance they are assigned in the study (Finnveden et al., 2009). Grouping assigns impact categories into one or more sets in order to better facilitate the interpretation of the results into specific areas of concern. In general, grouping involves sorting or ranking indicators by two possible ways: (i) by characteristics such as location (e.g., local, regional, or global) or emissions (e.g., air and water emissions) or (ii) by a ranking system (based on value choices), such as high, low, or medium priority.

Weighting (in ISO terminology) or valuation (in SETAC workgroup terminology) is the phase of LCIA that involves formalized ranking, weighting, and, possibly, aggregation of the indicator results into a final score across impact categories (Sonnemann et al., 2004). More precisely, weighting intends to rank, weight, or aggregate the results of different LCIA categories to find the relative importance of these different results. Even if the weighting process is not technical, scientific, or objective as these various LCIA results, it may be supported by applying scientifically based analytical techniques (Jensen et al., 1997). In weighting, each typically normalized indicator for the different impact categories or damages is multiplied by a specific weighting factor that reflects the relative significance of the different impact category endpoints among each other. Weighting sets are usually set by public policymakers or appropriate panels (e.g., industry panels, broader stakeholder panels, and expert panels) reflecting not only scientific knowledge but also political and other value-based concerns. However, weighting factors are fundamentally always normative/subjective and reflect value assumptions, making crucial that their identification should be well justified and documented. This should be done during the initial scope phase of the study and be in accordance with its goal, particularly the anticipated applications and target audience (European Commission—Joint Research Centre—Institute for Environment and Sustainability, 2010). It should be noted that the ISO standard for LCIA does not permit weighting to be performed in studies supporting comparative assertions disclosed to the public (Finnveden et al., 2009).

According to the ISO standard on LCA, selection of impact categories, classification, and characterization are mandatory steps in LCIA, while normalization and weighting are optional. While the ISO standard for LCIA presents the framework and some general guidelines to follow, it abstains from a standardization of more detailed methodological options. Over the last decade, a number of well-known LCIA methods have been developed filling this gap (Finnveden et al., 2009). However, the selection or development of any LCIA methods, according to the ILCD Handbook, must meet the following requirements (European Commission—Joint Research Centre—Institute for Environment and Sustainability, 2010):

> "The impact categories, category indicators and characterization models should enjoy international acceptance. LCIA methods that are endorsed by a governmental body of the relevant region where the decision is to be supported or where the reference of the accounted system is located, if available.
>
> The category indicators shall include those that are relevant for the specific LCI/LCA study performed, as far as possible. Any gaps shall be documented, and be explicitly discussed in the results interpretation.
>
> The characterization model for each category indicator shall be scientifically and technically valid, and based upon a distinct identifiable environmental mechanism or reproducible empirical observation.
>
> The entirety of characterization factors should have no relevant gaps in coverage of the impact category they relate to, as far as possible; relevant gaps shall be approximated, reported and explicitly be considered in the results interpretation.
>
> The category indicators—if endpoint level LCIA methods are included—are to represent the aggregated impacts of the related inputs and outputs of the system on the category endpoint(s).
>
> Double counting should be avoided across included characterization factors, as far as possible, and unless otherwise required by the goal of the study (e.g. as covering impacts of the same elementary flows to more than one impact categories with alternative impact pathways of the elementary flow).
>
> Value-choices and assumptions made during the selection of impact categories and LCIA methods should be minimized and shall be documented as part of the LCIA method data set documentation and preferably of a more extensive report."

Currently, several LCIA methods are normally used to quantify the different environmental indicators. Specifically, according to Pieragostini et al. (2012), some of the LCIA methods are "the CML 92 and 01 developed by the Institute of Environmental Sciences of Leiden University, the Ecoindicator 95 and 99 of PRé Consultants, the Environmental Design of Industrial Products (EDIP), EDIP 1997 and EDIP 2003 of the Danish UMIP, the IMPact Assessment of Chemical Toxics (IMPACT 2002+) proposed by the Swiss Federal Institute of Technology, the method of the International Panel on Climate Change (IPCC), the Tool for the Reduction and Assessment of Chemical and other environmental Impacts (TRACI) developed by the Environmental Protection Agency (EPA), the Critical Volume Aggregation and Polygon-based Interpretation developed by the University of Piraeus, the Custos Ambientais Associados à Geração Elétrica: Hi dreletricas x Termelétricas à Gás Natural of the Instituto Alberto Luiz Coimbra de Pó s-graduacao e pesquisa e engenheria (COPPE/UFRJ), the Environmental Priority Strategies in product design (EPS 2000) proposed by the Center for Environmental Assessment of Products and Material Systems, Chalmers University of Technology, the Externalidades na geração hidrelétrica e termelétrica, the External costs of Energy (ExternE project) developed by the European Commission, the LCA-net scheme of Mie University and the Waste Reduction Algorithm (WAR) algorithm of National Risk Management Research Laboratory."

Moreover, by applying commercial LCA software such as GaBi (Department of Life Cycle Engineering of the Chair of Building Physics at the University of Stuttgart and PE International GmbH) and SimaPro (PRé Consultants), one can easily choose among several different LCIA approaches for assessment (Chen et al., 2012). Other computational tools related to the LCA methodology include Umberto (IFU Hamburg and IFEU Heidelberg), TEAM (Ecobalance), POLCAGE (De La Salle University, Philippines, and University of Portsmouth, United Kingdom), and GEMIS (Öko-Institut). They are based on the ISO 14040 methodology, and they use general databases (apart from GEMIS and POLCAGE).

In addition, Ecoinvent database (Swiss Centre for Life Cycle Inventories) is integrated into these software tools giving the access to a range of unit processes as well as to other inventories covering various industrial areas (Pieragostini et al., 2012).

Almost all these methodologies aim to evaluate or to compare products on the aspects of climate change, ozone depletion, acidification, eutrophication, toxicity, fossil energy resource depletion, and more environmental impact categories through appropriate numerical scores. All environmental releases, fossil energy resource extractions, and land use activities that belong to a product life cycle are translated and aggregated in the right proportions to deliver an environmental profile in terms of the overall contribution of the product to a limited number of impact categories (Sleeswijk et al., 2008). This is being achieved through algorithms of calculation fed by databases (Millet et al., 2007). Normally, the relationship between inventory data and impact category indicators is linearly expressed by the characterization factors. Most of the impact assessment methods take into account environmental mechanisms to implement the assessment (Heijungs et al., 2010).

On the other hand, recently, a number of more specific impact assessment and LCI result evaluation tools have been developed and reported in the literature. Such an example is the Ecovalue08 tool, which is based on willingness-to-pay estimates for environmental quality, and market values for resource depletion (Ahlroth and Finnveden, 2011). Furthermore, it has been proposed the COMPLIMENT tool, which integrates parts of methods such as LCA, multicriteria analysis, and environmental performance indicators. The methodology is based on environmental performance indicators, expanding the scope of data collection toward a life cycle approach and including a weighting and aggregation step (Hermann et al., 2007). Likewise, Myllyviita et al. (2012) proposed a process of assessing environmental impacts of two alternative raw materials based on LCA and multicriteria decision analysis attempting to facilitate the selection of relevant impact categories with a clearer distinction between objective and subjective elements. Alternatively, Herva et al. (2012) proposed the FEcoDI ecodesign tool integrating the criteria provided by three environmental evaluation methodologies, namely, ecological footprint, LCA, and environmental RA, and constructed on the basis of fuzzy logic reasoning and features.

It is quite evident that a significant number of methodologies regarding LCIA have been developed. However, when they have been compared, little or no agreement among most of them has been reported (Millet et al., 2007; Pizzol et al., 2011), while very often they are responsible for ambiguities and uncertainties in LCA results. For that reason, the selection of the method for the environmental impact assessment involves uncertainty in LCA. As a matter of fact, each method assesses impacts using different pollutant substances and characterization factors, and a unique correct choice does not exist (Bersimis and Georgakellos, 2013).

19.3.5 The Life Cycle Interpretation

A major concern with LCAs that can have a significant impact on the conclusions and recommendations made is the assessment and interpretation of environmental impacts (Crawford, 2011). The interpretation phase aims to evaluate the results from the inventory analysis or impact assessment and compare them with the goal of the study defined in the first phase (Sonnemann et al., 2004). Thus, if the iterative steps of the LCA have resulted in the final LCI model and results, and particularly for comparative LCA studies (while, to some extent, applicable to other types of studies as well), the interpretation phase serves to obtain robust conclusions and, often, recommendations. However, this LCA phase has also one more purpose: during the iterative steps of the LCA and for all sorts of deliverables, the interpretation phase serves to guide the work in the direction of improving the LCI model to meet the requirements arisen from the study goal (European Commission—Joint Research Centre—Institute for Environment and Sustainability, 2010).

Hence, the life cycle interpretation is the LCA phase where the results of the other phases are considered, combined, and examined with regard to the achieved accuracy, completeness, and precision of the applied data and the assumptions, which have been made throughout the study. In life cycle interpretation, the results of the LCA are judged in order to meet issues posed in the goal definition. The interpretation is associated with the intended applications of the LCA study and is used to develop recommendations (European Commission—Joint Research Centre—Institute for Environment and Sustainability, 2010). The following steps can be distinguished within this phase (Sonnemann et al., 2004):

Identification of the most important LCI and LCIA results

Evaluation of the study's findings, through the application of various techniques, such as completeness and consistency check, as well as sensitivity and uncertainty analysis

Conclusions including explanation of the final outcome, comparison with the original goal of the study, recommendations, and final reporting of the results

The interpretation phase may lead to a new iteration round of the study, including a possible adaptation of the primary goal of it. The elements of the interpretation phase and their relations to other phases of the LCA are presented in Fig. 19.6.

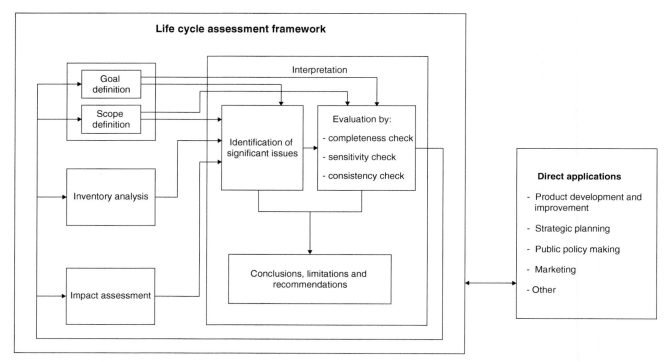

FIGURE 19.6 The elements of the life cycle interpretation phase (adapted from European Commission—Joint Research Centre—Institute for Environment and Sustainability, 2010).

The objective of the first step is to structure the information from the inventory analysis and the LCIA phase in order to determine the significant environmental issues in accordance with the goal and scope definition. Environmental issues are the results of the inventory phase, that is, inputs and outputs, and the results of the LCIA phase, that is, environmental indicators (Jensen et al., 1997). The evaluation step is performed to establish the basis for subsequently extracting the conclusions and provide recommendations throughout the interpretation of the LCA results. To facilitate the determination of the reliability and robustness of the results, the evaluation should be performed along with the identification of significant issues step. The evaluation builds on the results of the earlier LCA phases, analyzing them in an integrated perspective. In particular, it is based on the outcome of the inventory data collection and modeling and of impact assessment, and its focus is on the significant issues identified among methodological choices and data. The evaluation comprises:

Completeness check
Sensitivity check together with scenario analysis and possibly uncertainty analysis
Consistency check

The result of the evaluation step is vital in order to empower the conclusions and recommendations from the study. For that reason, it must be illustrated in a way that gives a clear understanding of the outcome (European Commission—Joint Research Centre—Institute for Environment and Sustainability, 2010).

The final step of the interpretation (conclusions and recommendations) is almost the same to any typical concluding and recommending part of a scientific and technical appraisal and so on. The purpose of this step is to arrive at conclusions and recommendations for the LCA report. This step is essential to enhance the reporting and the clearness of the study. Both are crucial for the LCA report audience. During the presentation of the conclusions and recommendations, the results of the critical review of the study should also be included (Jensen et al., 1997). Summing up or conclusions say whether the questions that were posed in the formulation of the goal definition can be answered by the LCA (European Commission – Joint Research Centre – Institute for Environment and Sustainability, 2010).

19.4 APPLICATIONS OF LCA

Product life cycle planning is one of the key factors identified to guide new design approaches. Life cycle options concern not only the postconsumption stage such as "reuse" and "recycle" but also those that influence other life stages such as design changes to increase life and repair of a product (Zwolinski and Brissaud, 2006). In order to do that, knowledge must be available. Consequently, when studying environmental impacts of products and services, it is vital to study these in a life cycle perspective. This helps to avoid problem shifting from one part of the life cycle to another, from one geographical area to another, or from one area of environmental concern to another (Finnveden et al., 2009). An LCA can help decision makers, not only select the product or process that result in the least impact to the environment, but it identifies the transfer of environmental impacts from one media to another (e.g., eliminating air emissions by creating wastewater pollutants instead) or from one life cycle stage to another (e.g., from the production of the product to the use and reuse phase). An LCA allows a decision maker to study the entire product system, hence avoiding the suboptimization that could result if only a single process were the focus of the study. Therefore, this information can be used with other factors, such as cost and performance data, to design a product or process (Curran, 2006).

The primary objectives of all LCA studies typically remain the same, besides some variations in the initial reasons and intended goals of undertaking them. These objectives are generally either:

To assist designers, manufacturers, and final consumers to recognize and evaluate the environmental performance of one or several products, processes, or services with the purpose of finding out the most environmentally friendly alternative in a life cycle perspective or

To offer a base for recognizing areas of potential improvement in the environmental performance of specific products or processes

LCA uses are broad and may include the following:

Environmental improvement targeting and pursuing
Identifying cost and environmental savings as a result of enhanced resource efficiency
Environmental knowledge managing
Fulfilling international standards and market requirements that emerge
Developing ecologically sustainable patterns of production and consumption
Public policy setting and managing
Providing information for product marketing claims and environmentally responsible design
Supporting ecolabeling programs (Crawford, 2011)

LCA can be used in different contexts, such as for the development and optimization in (environmentally conscious) product development and design, strategic planning, policy-making, marketing, and so on (Werner, 2005).

19.4.1 Product Design and Development

Product design has been recognized as a key leverage aspect for promoting a change to more sustainable business systems. Product development is mainly the evolution of information dispersed by decisions as the assurance of resources. The ultimate goal of sustainability is to redirect corporate investments toward business models that attain a better balance of environmental protection and social equity (Swarr and Hunkeler, 2008). Design for the environment (DfE) is a tool that can guide this transition. DfE is a general term in order to describe several methods for integrating environmental factors into the design process. DfE and LCA have many similarities and may not be distinguishable from each other in many cases, even if they have been developed without formal links. The main purpose of DfE is the optimization of the environmental performance of a product throughout its life cycle, integrating in manufacturing pollution prevention concepts and concerns about product energy efficiency. Another key objective of DfE is to design products with the target of minimizing afterlife impacts and costs. In other words, DfE and LCA conceptually deal with the same problem areas. Even if dedicated LCA tools were not an integral part of early DfE initiatives, the development of LCA has currently been connected to product development and, consequently, to DfE as a concept. Some important issues in DfE are presented in Table 19.3. By applying LCA in the design phase, companies and designers have the opportunity of preventing or minimizing predictable environmental impacts without negotiating the total quality of the product, constituting LCA a usual choice in product development (Jensen et al., 1997).

19.4.2 Strategic Planning

The incorporation of environmental issues in strategic business planning is becoming a common aspect in many companies. The management of environmental concerns is often formalized and supported through an environmental management scheme like the ISO 14001 standard or EMAS (environmental management and auditing scheme). However, a lot of companies still handle the issues on a one by one basis (Jensen et al., 1997). The identification of environmental impacts from a particular product or process helps decision makers to adopt improvement strategies and allocate funding in order to maximize any potential environmental benefits. These improvements usually involve replacing and modifying inefficient equipment or processes and could create value maximization. This approach is often allied with accomplishing existing

TABLE 19.3 Important Issues in Design for the Environment (Jensen et al., 1997)

Materials Selection
Minimize toxic chemical content
Incorporate recycled and recyclable materials
Use more durable materials
Reduce materials use

Production Impacts
Reduce process waste
Reduce energy consumption
Reduce use of toxic chemicals

Product Use
Energy efficiency
Reduce product emissions and waste
Minimize packaging

Design for Recycling and Reuse
Incorporate recyclable materials
Ensure easy disassembly
Reduce materials diversity
Label parts
Simplify products (e.g., number of parts)
Standardize material types

Extending the Useful Life of Products and Components
Design for remanufacture
Design for upgrades
Make parts accessible to facilitate maintenance and repairs
Incorporate reconditioned parts or subassemblies

Design for End of Life
Safe disposal

and rising compliance prerequisites and potential future market demands to ensure the viability and competitiveness of the business while focusing on environmental performance. LCA can be used to identify potential opportunities for a business that may otherwise have been neglected; define strategies for ensuring or growing market share, mainly through improving or emphasizing the environmental attributes; comply with existing or possible future environmental standards; and minimize possible future dangers, legal responsibilities, risks, or costs associated with compliance or shifting market demands (Crawford, 2011). The motivating factors behind the decision to integrate environmental issues in strategic planning could be consumer demands, compliance with legislation, community needs for environmental improvement, security of supply, product and market opportunities, and so on. The environmental performance is therefore altering from being a mandatory issue (all regulatory requirements shall be fulfilled) to being a strong positioning feature on the market. In this context, LCA is a very valuable tool (Jensen et al., 1997).

19.4.3 Marketing

The increased interest of consumers in environmental issues, accompanied by an increased demand for products that is compatible with specific environmental standards or addresses particular environmental concerns, has in its turn increased the interest of manufacturers in applying LCA to inform potential consumers of their product's environmental attributes. This practice, which can be used to preserve or even improve their competitive advantage, would be really valuable as a marketing tool. On the other hand, as "greenwash" becomes progressively prevailing, manufacturers are obliged to present consistent proof to support their declaration. Thus, many "green" marketing claims are often substantiated using the findings of studies, which are based on the LCA methodology. Additionally, LCA can be applied in order to recognize potential environmental benefits or attributes of a particular product over a rival's alternative product or process (Crawford, 2011). Consequently, at least four different uses of LCA in environmental marketing can be distinguished (Jensen et al., 1997): (i) environmental labeling, (ii) environmental claims, (iii) environmental declarations, and (iv) organizational marketing.

Particularly, the findings of an LCA are often used as the basis for environmental labeling and rating schemes by demonstrating specific product improvements in their environmental performance, for example, a reduction in carbon dioxide emissions. This information may then be presented to consumers through a product declaration, or so-called ecolabel, to provide environmental information to consumers in simple and easily comparable terms. Examples of environmental labeling include energy or water rating labels indicating the energy or water efficiency of domestic appliances or fuel consumption labels that reveal the fuel efficiency of motor vehicles (Crawford, 2011). A recent illustration of ecolabeling is the tag proposed by Bersimis and Georgakellos (2013) and presented in Fig. 19.7 for three packaging materials. It is developed by applying principal component analysis (PCA) on LCI results, providing a valuable tool for categorizing the examined products according to the impact severity as well as the impact types associated with them, in a pure mathematical and, thus, more unambiguous way (Bersimis and Georgakellos, 2013).

19.4.4 Environmental Improvement

Many considerable environmental and financial benefits for an organization can result through the improvements to the environmental performance of its existing processes or practices, which may be originated by the need to comply with external regulations or even internal company policies. LCA is frequently used to appraise the environmental

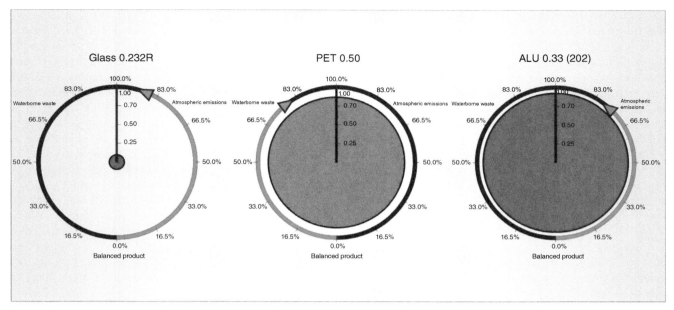

FIGURE 19.7 A proposed tag (here for three packaging materials) as an illustration of LCA-based ecolabeling (Bersimis and Georgakellos, 2013).

loadings of alternative processes within a product system as well as identify the best possible choices of action, enabling decision makers to improve existing processes (Crawford, 2011).

19.4.5 Purchasing

Purchasing (public and private) is a business area with extensive potential for environmental improvement. In private companies, incentives for green purchasing mostly involve business issues like cost saving potential, consumer demands, product design, company environmental policies, and business risk. Green public procurement is mostly motivated by policies and regulation. Green purchasing is reliant on environmental information about products and process, which is where LCA could be used as a supporting information tool (Tillman, 2010).

19.4.6 Public Policymaking

LCA can facilitate the development, implementation, and management of public policies associated with the environmental performance of the built environment, through the establishment of building regulations that deal with their operational efficiency like mandatory efficiency targets concerning, for example, cooling and heating. In such cases, LCA studies have helped to make clear where noteworthy environmental improvements are required within the building life cycle and to develop relative policies and regulations, which could assist in dealing with these issues (Crawford, 2011).

19.4.7 Learning

One of the most important applications of LCA originally identified was the identification of environmental "hot spots" and the exploration for potential improvements. The purpose of these applications was related to learning (i.e., to understand the environmental strengths and weaknesses of a product or activity from a life cycle perspective) and not so much to decision making (although they may evidently lead to that). This learning may lead to organizational or operational modifications, together with physical changes to the product or the process chain. Thus, life cycle knowledge about products and processes enables companies to face issues from customers, government, or NGOs regarding the environmental performance of their products and hence better manage the risk for exposure (Tillman, 2010).

According to international surveys, the driving force to conduct an LCA in most cases is the benchmarking of different products, as well as the evaluation of environmental improvement options on a product-specific basis (Werner, 2005). Specifically, according to such a survey, the main reasons why LCA is used are (Crawford, 2011) to support business strategy (18%) and research and development (18%), as an input to product or process design (15%), for education (13%), and for labeling or product declarations (11%). On the other hand, several surveys conducted in the region of Europe to examine the current and future use of LCA have been shown that, actually, LCA results are more probably integrated (of course together with other sources) into environmental design strategies and their tools (i.e., checklists, recommendations, guidelines, and standards). These environmental design strategies are called "Design for Environment," "Design for Recycling," "Design for Remanufacturing," or "Life Cycle Design," conditional on their focus (Werner, 2005).

Finally, LCA can implement a constructive task in both public and private environmental management relative to products, processes, and activities. This may involve environmental comparisons between existing products, comparisons with prototypes, the development of new products, and so on. Furthermore, despite direct product applications, LCA may also be used in a wider sense. For instance, LCA could be applied to complex government policies or business strategies relating to consumption and lifestyle options in a range of society sectors, rather than dealing with just specific physical goods or simple services (Guinée, 2004).

19.5 LIMITATIONS OF LCA

The core characteristic of LCA, which is its holistic nature, is its major strength and simultaneously its main limitation. This is because the broad scope of analyzing the complete life cycle of a product can only be achieved at the sacrifice of simplifying other aspects (Guinée, 2004). In other words, LCA is not a process that can be easily or rapidly implemented, as a consequence of a number of problems and limitations related to its use. A part of them result from the subjective character of the decisions made within an LCA study, such as system boundaries definition, data sources selection, and potential impacts weighting (Crawford, 2011). Some of these limitations are discussed henceforward.

LCA is very data intensive, and lack of data can restrict the conclusions that can be drawn from a specific study (Finnveden et al., 2009). More precisely, an LCI involves the collection of a wide range of data from a variety of sources. The availability of data depends on (i) the degree that data is being collected and (ii) the readiness of companies to make their data available since some data are regarded as commercially sensitive, while companies usually prefer to avoid public knowledge of their environmental impact. On the other hand, the level of confidence that can be placed on LCA outcomes is related to the availability and use of appropriate data. Many authors have recognized data availability, accuracy, and cost as major restrictions to the use of LCA (Crawford, 2011). However, as discussed previously, better databases and growing experience will help show where to focus the efforts including in terms of quality assurance (Finnveden et al., 2009).

Even if the concern for the global environment has been increased over the last decades, there is still a need to enlarge the understanding of the impacts on the environment from human activity and the actions that have been taken to minimize them (Crawford, 2011). In this context, while LCA intends to provide a comprehensive view of environmental impacts, not all types of impacts are equally well covered in a typical LCA. Moreover, human health aspects other than those connected to outdoor exposure of pollutants have generally received inadequate attention (Finnveden et al., 2009).

An LCA can include a number of methodological choices, which are uncertain and may potentially affect the results (Finnveden et al., 2009). This is because the LCA tool is even now in a condition of development worldwide. Noteworthy assumption still exists over the most appropriate assessment methodologies, data sources, and interpretation techniques that should be used (Crawford, 2011). For instance, LCIA has native limitations. Even though the LCIA process follows a systematic procedure, there are many primary assumptions, simplifications, and subjective value choices. Depending on the LCIA methodology chosen and the inventory data used, some of the main limitations may include lack of spatial resolution (e.g., a certain amount of a waterborne waste is worse in a small stream than in a large river), lack of temporal resolution (e.g., a certain amount of an air pollutant during one-hour period is worse than the same release spread through a week), inventory speciation (e.g., broad inventory listing such as "VOC," "hydrocarbons," or "metals" do not provide enough information to accurately assess environmental impacts), and threshold and nonthreshold impact (e.g., 10 tons of contamination is not necessarily 10 times worse than 1 ton of it). Other examples comprise allocation methods and time limits for the LCI, as well as the definition of the system boundaries. The most common LCA frameworks of SETAC and ISO are pushed to draw a limit around the considered processes, which potentially reduces consideration to only a small percentage of the system inputs or outputs, thereby making current assessments incomplete (Finnveden et al., 2009; Crawford, 2011).

A special type of uncertainty is related to the lack of knowledge on the actual system to be studied (Finnveden et al., 2009). This means that the life cycle of a product is a theoretical concept, which has no clearly definable temporal or spatial boundaries in a complex context. Modeling the actual very complex social, economic, and environmental systems results in temporal and spatially undifferentiated models.

Assuming, during modeling, that all other things are equal or held constant (the "ceteris paribus" assumption) to reduce the complexity of real social, economic, and environmental systems makes the validation of such models impossible. As a result, no empirical approach based on measurements is feasible to validate the models set up or used in an LCA (Werner, 2005).

LCA is a time-consuming methodology and a high-cost tool. The data collection process is usually a complex and difficult task, requiring significant time and sources. In most cases, the necessary data of an LCA study are not publicly available, affecting the effort required to conduct an LCA study. However, as data becomes more accessible through the development of appropriate databases and the progressively increasing knowledge base of the methodology from its use to a greater extent, the time and costs of conducting LCA studies will be decreased (Crawford, 2011).

LCA only accounts for impacts related to normal and abnormal operation of processes and products, but not covering, for example, impacts from accidents, spills, and similar occurrences (European Commission—Joint Research Centre—Institute for Environment and Sustainability, 2010). However, accidents might have very dramatic impacts. An example for this is the risk of a serious accident in a nuclear power plant (Frischknecht et al., 2007).

Data related to environmental or other specific impacts can differ extensively across global geographic boundaries. Variations in production and manufacturing processes and factors (e.g., types of fuels, raw materials, and transportation distances) can be quite significant not only from country to country but between areas within the same country as well. Applying nonsuitable data from other countries or areas with considerably different characteristics can result in substantial miscalculations and inaccuracies in LCA findings. Successful studies should ensure that product-, process-, technology-, and location-specific data have been used, in order to provide accurate and reliable findings (Crawford, 2011).

Nevertheless, it should be noticed that the LCA methodological development has been strong over the last decades. Thus, a number of the limitations that have been mentioned in critical reviews have also been addressed. On the other hand, several of them can never be completely resolved and are common to other tools as well, but better data and better methods are developed (Finnveden et al., 2009).

19.6 RECENT TRENDS AND DEVELOPMENTS IN LCA

LCA has currently evolved over the past several years. Thus, there are ongoing efforts to refine the way LCA performs in such a way that is better than before. Better impact assessment models are being developed and more databases are being set up (Tillman, 2010). A variety of methods, information, and data exist, and international business representatives as well as government bodies consider that there is a need for guidance on what to use, when, and how. For this purpose, several international activities have been initiated including the ILCD and the UNEP/SETAC Life Cycle Initiative. Moreover, there are some new trends in applications rising. One of them is the attention in life cycle thinking from retailers (industry has used LCA for a long time both for learning and decision making). Such an example is Wal-Mart who established targets for reduced environmental impacts of products they are trading, demanding life cycle-related information from their suppliers. A further trend is the increased promotion and use of LCA and life cycle thinking on policy planning. For instance, life cycle thinking is an important constituent of European environmental policy. Additionally, in the directive proposal on the use of energy from renewable sources, greenhouse gas emissions savings thanks to the use of renewable fuels should be calculated with a clear life cycle approach. Such legislation, on national level, is already in place in some countries. In Switzerland, for example, a new law requires a full LCA of biofuels in order to quantify the fuel tax to be paid (Finnveden et al., 2009).

A trend in the reverse direction has been the extension of the objectives of LCA by merging it with other types of tools and methodologies (Tillman, 2010). Such an illustration could be LCA and RA: a product can be analyzed using LCA, and at the same time, RA can be performed for various central processes in the chain, in which the importance is on the local environmental impacts. Both methods may well be necessary for decision making. Another interesting approach is the complementary use of LCA and substance flow analysis (SFA), if one specific substance dominates the product (e.g., cadmium in rechargeable batteries or phosphates in detergents). For a single substance, marketing mechanisms might then also become part of the analysis (Guinée, 2004). Input–output LCA (IOA–LCA), where LCA is merged with macroeconomics, is such an example. Further examples are the use of monetary weighting methods together with cost–benefit analysis and life cycle costing and the combination of LCA and strategic environmental assessment (SEA), which is a technical tool for evaluating environmental impacts of plans, policies, and programs. LCA could be used as an analytical tool within the framework of SEA (Finnveden et al., 2009; Tillman, 2010).

Another recent trend is social LCA, which widens the scope of LCA to also include social impacts. To this objective, efforts have been undertaken to develop methodologies analogous to LCA in their cradle to grave approach but accounting for social aspects as well. One of the incentives for developing such a methodology was making LCA more applicable to developing countries. However, such a methodology (social LCA) seems to be an even greater challenge. The UNEP/SETAC Life Cycle Initiative has recently published guidelines for social LCA, which set up a framework for social LCA that makes parallel with the ISO LCA framework as far as possible. However, it is apparent that many challenges remain, mostly as it concerns to impact assessment (Tillman, 2010). Currently, while LCA is identified as the best tool for assessing the life cycle impacts of products and there have been many achievements such as the international standards on LCA, there are still obstacles that slow down the broader implementation of life cycle thinking. However, the ongoing research could reduce many of them (Finnveden et al., 2009).

From all the aforementioned, it is obvious that, in spite of certain weak points of the methodology (e.g., LCA results are disputable, it is still burdened with general methodological difficulties, it has high data demand in the early stages of product development, it generally has large data requirements, it is a very heavy tool, and it is costly), LCA is perceived as a very robust tool that offers opportunities for assisting companies and policymakers in environmental management. Some of the strengths of LCA (e.g., LCA considers the whole life cycle of a product for material and process selection, it makes the connection between product features and environmental impacts understandable, it allows one to understand environmental trade-offs, it produces a learning effect on environmental matters, and it is not simply a method for calculation but, potentially, a completely new framework for business thinking) make this methodology exceptional. It is a tool that provides indicators (such as greenhouse gas emissions, climate change, and resource depletion) about the sustainability of industrial systems. In this context, companies could increase LCA application through a process that starts with pilot projects, followed by establishment of internal knowledge and leading to a more systematic and prospective way of application. Moreover, LCA could be more deeply involved in decision-making processes of companies through further integration of it into environmental management schemes. Undoubtedly, this process would generate more environmental benefits in the long run.

REFERENCES

Ahlroth, S. and G. Finnveden. 2011. Ecovalue08—A new valuation set for environmental systems analysis tools. *Journal of Cleaner Production* 19:1994–2003.

Bersimis, S. and D. Georgakellos. 2013. A probabilistic framework for the evaluation of products' environmental performance using life cycle approach and Principal Component Analysis. *Journal of Cleaner Production* 42:103–115.

Chen, Z., H.H. Ngo, and W. Guo. 2012. A critical review on sustainability assessment of recycled water schemes. *Science of the Total Environment* 426:13–31.

Crawford, R.H. 2011. *Life cycle Assessment in the built environment*. Spon Press, New York.

Curran, M.A. 2006. *Life cycle assessment: Principles and practice*. National Risk Management Research Laboratory – Office of Research and Development. U.S. Environmental Protection Agency, Cincinnati.

Ekvall, T., A.-M. Tillman, and S. Molander. 2005. Normative ethics and methodology for life cycle assessment. *Journal of Cleaner Production* 13:1225–1234.

European Commission – Joint Research Centre – Institute for Environment and Sustainability. 2010. *International Reference Life cycle data system (ILCD) handbook—General guide for Life cycle assessment—Detailed guidance*. Publications Office of the European Union, Luxembourg.

Finkbeiner, M., A. Inaba, R.B.H. Tan, K. Christiansen, and H.-J. Klüppel. 2006. The new international standards for life cycle assessment: ISO 14040 and ISO 14044. *International Journal of LCA* 11(2):80–85.

Finnveden, G., M.Z. Hauschild, T. Ekvall, J. Guinée, R. Heijungs, S. Hellweg, A. Koehler, D. Pennington, and S. Suh. 2009. Recent developments in life cycle assessment. *Journal of Environmental Management* 91:1–21.

Frischknecht R., N. Jungbluth, H.-J. Althaus, G. Doka, T. Heck, S. Hellweg, R. Hischier, T. Nemecek, G. Rebitzer, M. Spielmann, and G. Wernet. 2007. *Ecoinvent report No. 1: Overview and methodology*. Swiss Centre for Life Cycle Inventories, Dübendorf.

Georgakellos, D.A. 2005. Evaluation of life cycle inventory results using critical volume aggregation and polygon-based interpretation *Journal of Cleaner Production* 13:567–582.

Georgakellos, D.A. 2006. The use of the LCA polygon framework in waste management. *Management of Environmental Quality: An International Journal* 17:490–507.

Guinée, J.B. 2004. *Life cycle assessment operational guide to the ISO standards*. Kluwer Academic Publishers, Dordrecht.

Heijungs, R., G. Huppes, and J.B. Guinée. 2010. Life cycle assessment and sustainability analysis of products, materials and technologies. Toward a scientific framework for sustainability life cycle analysis. *Polymer Degradation and Stability* 95:422–428.

Hermann, B.G., C. Kroeze, and W. Jawjit. 2007. Assessing environmental performance by combining life cycle assessment, multi-criteria analysis and environmental performance indicators. *Journal of Cleaner Production* 15:1787–1796.

Hertwich, E.G. and G.P. Peters. 2006. Feasibility and scope of life cycle approaches to sustainable consumption. *In* D. Brissaud et al. (eds) *Innovation in life cycle engineering and sustainable development*. Springer, Dordrecht, pp. 3–16.

Herva, M., A. Franco-Urva, E.F. Carrasco, and E. Roca. 2012. Application of fuzzy logic for the integration of environmental criteria in ecodesign. *Expert Systems with Applications* 39:4427–4431.

International Organization for Standardization. 2006. *ISO 14040: Environmental management—Life cycle assessment—Principles and framework*, Second edition. ISO, Geneva.

Jensen A.A., L. Hoffman, B.T. Møller, A. Schmidt, K. Christiansen, J. Elkington, and F. van Dijk. 1997. *Life cycle assessment—A guide to approaches, experiences and information sources*. European Environment Agency, Copenhagen.

Joshi, S. 2000. Product environmental life-cycle assessment using input-output techniques. *Journal of Industrial Ecology* 3(2&3):95–120.

McLaren, S.J. 2010. Life cycle assessment (LCA) of food production and processing: An introduction. *In* U. Sonesson et al. (eds) *Environmental assessment and management in the food industry Life cycle assessment and related approaches*. Woodhead Publishing Limited, Cambridge, pp. 37–58.

Millet, D., L. Bistagnino, C. Lanzavecchia, R. Camous, and T. Poldma. 2007. Does the potential of the use of LCA match the design team needs? *Journal of Cleaner Production* 15:335–346.

Morris, A.S. 2004. *ISO 14000 environmental management standards—Engineering and financial aspects*. John Wiley & Sons, Ltd, Chichester.

Myllyviita, T., A. Holma, R. Antikainen, K. Lähtinen, and P. Leskinen. 2012. Assessing environmental impacts of biomass production chains—Application of life cycle assessment (LCA) and multi-criteria decision analysis (MCDA). *Journal of Cleaner Production* 29–30:238–245.

Pieragostini, C., M.C. Mussati, and P. Aguirre. 2012. On process optimization considering LCA methodology. *Journal of Environmental Management* 96:43–54.

Pizzol, M., P. Christensen, J. Schmidt, and M. Thomsen. 2011. Impacts of "metals" on human health: A comparison between nine different methodologies for life cycle impact assessment (LCIA). *Journal of Cleaner Production* 19:646–656.

Sleeswijk, A.W., L.F.C.M. van Oers, J.B. Guinée, J. Struijs, and M.A.J. Huijbregts. 2008. Normalisation in product life cycle assessment: An LCA of the global and European economic systems in the year 2000. *Science of the Total Environment* 390:227–240.

Sonnemann, G., F. Castells, and M. Schuhmacher. 2004. *Integrated life-cycle and risk assessment for industrial processes*. Lewis Publishers, Boca Raton.

Swarr, T. and D. Hunkeler. 2008. Life cycle costing in life cycle management. *In* D. Hunkeler et al. (eds) *Environmental life cycle costing*. SETAC and CRC Press, Boca Raton, pp. 77–90.

Tillman, A.-M. 2010. Methodology for life cycle assessment. *In* U. Sonesson et al. (eds) *Environmental assessment and management in the food industry life cycle assessment and related approaches*. Woodhead Publishing Limited, Cambridge, pp. 59–80.

Weidema, B.P., T. Ekvall, and R. Heijungs. 2009. *Guidelines for application of deepened and broadened LCA—Deliverable D18 of work package 5 of the CALCAS project*. ENEA—The Italian National Agency on new Technologies, Energy and the Environment, Rome.

Werner, F. 2005. *Ambiguities in decision-oriented life cycle inventories—The role of mental models and values*. Springer, Dordrecht.

Zwolinski, P. and D. Brissaud. 2006. Designing products that are never discarded. *In* D. Brissaud et al. (eds) *Innovation in life cycle engineering and sustainable development*. Springer, Dordrecht, pp. 225–244.

20

ENVIRONMENTAL AUDIT IN ENVIRONMENTAL MANAGEMENT

IAN T. NICOLSON
The Kadoorie Institute, The University of Hong Kong, Pokfulam, Hong Kong

Abstract: Auditing environmental management systems (EMS) requires a knowledge of environmental aspects and impacts, how EMS are structured, how they should deliver continual improvement, and how to interpret its requirements and a detailed understanding of ISO 14001 (the international standard for EMS), plus knowledge and skills on how to audit, what to audit, and how to conduct and report the audit. This chapter answers all of these needs.

Keywords: Environmental management systems, aspects and impacts, ISO14001, auditing and reporting.

20.1 Introduction to Environmental Management	466
20.1.1 What Drives Implementation of an EMS?	466
20.1.1.1 Environmental Legislation	466
20.1.1.2 Supply Chain Pressure	467
20.1.1.3 Investor Pressure	467
20.1.2 What are the Benefits of Implementing an EMS?	467
20.1.3 Aspects and Impacts	467
20.1.3.1 Key Environmental Issues	468
20.1.4 Core Elements of an EMS	469
20.1.4.1 Initial Environmental Review	469
20.1.4.2 Environmental Policy	470
20.1.4.3 Plan	471
20.1.4.4 Do	471
20.1.4.5 Check	471
20.1.4.6 Act	471
20.1.4.7 PDCA Loops	472
20.1.4.8 Conclusion	472
20.2 ISO 14001 Interpretation	473
20.2.1 Introduction	473
20.2.1.1 Section 1: Scope of ISO 14001	474
20.2.1.2 Section 2: Normative References	475
20.2.1.3 Section 3: Terms and Definitions Used in the Standard	475
20.2.1.4 Section 4: EMS Requirements	475
20.2.2 General Requirements	475
20.2.3 Environmental Policy	475
20.2.4 Planning	476
20.2.4.1 Environmental Aspects	476
20.2.4.2 Legal and Other Requirements	477
20.2.4.3 Objectives, Targets, and Program(s)	478
20.2.5 Implementation and Operation	479
20.2.5.1 Resources, Roles, Responsibility, and Authority	479
20.2.5.2 Competence, Training, and Awareness	480
20.2.5.3 Communication	481
20.2.5.4 EMS Documentation	481
20.2.5.5 Control of Documents	481
20.2.5.6 Operational Control	482
20.2.5.7 Emergency Preparedness and Response	482
20.2.6 Checking	483
20.2.6.1 Monitoring and Measurement	483
20.2.6.2 Evaluation of Compliance	484
20.2.6.3 Nonconformity, Corrective, and Preventive Actions	484
20.2.6.4 Records	485
20.2.6.5 Internal Audit	485
20.2.7 Management Review	486

An Integrated Approach to Environmental Management, First Edition. Edited by Dibyendu Sarkar, Rupali Datta, Avinandan Mukherjee, and Robyn Hannigan.
© 2016 John Wiley & Sons, Inc. Published 2016 by John Wiley & Sons, Inc.

20.3	Environmental Auditing	487
	20.3.1 Introduction	487
	20.3.1.1 What Is an Environmental Audit?	487
	20.3.1.2 Is an Audit an Inspection?	487
	20.3.2 Principles of Auditing	487
	20.3.3 Types of Environmental Audit	488
	20.3.3.1 First-Party Audit	488
	20.3.3.2 Second-Party Audit	489
	20.3.3.3 Third-Party Audit	489
	20.3.3.4 Depth of Audit	489
	20.3.4 Audit Objectives	489
	20.3.5 Audit Terminology	490
	20.3.6 Audit Preparation	490
	20.3.6.1 Why Prepare for the Audit?	491
	20.3.6.2 Initiating the Audit	491
	20.3.6.3 Initial Document Review	493
	20.3.6.4 Preparing On-Site Activities	494
	20.3.7 On-Site Activities	495
	20.3.7.1 Conducting the Opening Meeting	495
	20.3.7.2 Collect and Verify Information	496
	20.3.7.3 Prepare Audit Findings	499
	20.3.7.4 Consolidate Team Findings	500
	20.3.7.5 Closing Meeting	500
	20.3.8 Report the Audit	500
	20.3.8.1 Report Preparation	500
	20.3.8.2 Report Approval and Distribution	501
	20.3.9 Audit Completion	501
	20.3.10 Audit Follow-Up	501
	20.3.11 Auditor Competence	502
References		502
Further Reading		502

20.1 INTRODUCTION TO ENVIRONMENTAL MANAGEMENT

To get started along the pathway to environmental management, one has to ensure awareness of all the environmental issues around us. Normally, the best place to start is with a definition of environment. An easy way to visualize this is to think of primitive man and what he needed to survive (see Fig. 20.1):

- Air to breath
- Water to drink
- Some shelter
- Plants and animals for food
- Some wood or stone to make tools

FIGURE 20.1 Survival?

Not surprising then when one looks at ISO 14001 (ISO, 2004a) is that they have come up with the following definition:

Surroundings in which an organisation operates including:
Air, Land, Water, Flora and Fauna, Natural Resources and Humans and their interrelation

This is further enhanced by stating that the "Environment extends from within an organization to the global system."

The next thing one needs to understand is what is an environmental management system (EMS)? This can be visualized with the following graphic, which can be applied to ANY organization including homes, schools, or universities (Fig. 20.2).

Every organization has activities; these activities require raw materials as input and produce products and/or services as output. In order to produce these products and services, the organization may need to interact with suppliers and contractors. As a result of all these actions, environmental aspects will be created that will ultimately have an impact on the environment. The idea of an EMS is to manage all of these environmental aspects and minimize their environmental impact. So the focus of an EMS is aspects (not customer as in quality management systems).

20.1.1 What Drives Implementation of an EMS?

20.1.1.1 Environmental Legislation

Regulatory compliance is normally one of the greatest drivers, with ISO 14001 compliance a minimum requirement. Legal and regulatory requirements normally exist for air, noise, wastewater, waste, environmental impact assessment (EIA), contaminated land, and so on. Implementing an EMS will ensure a clear understanding of these requirements, appropriate environmental behavior, and

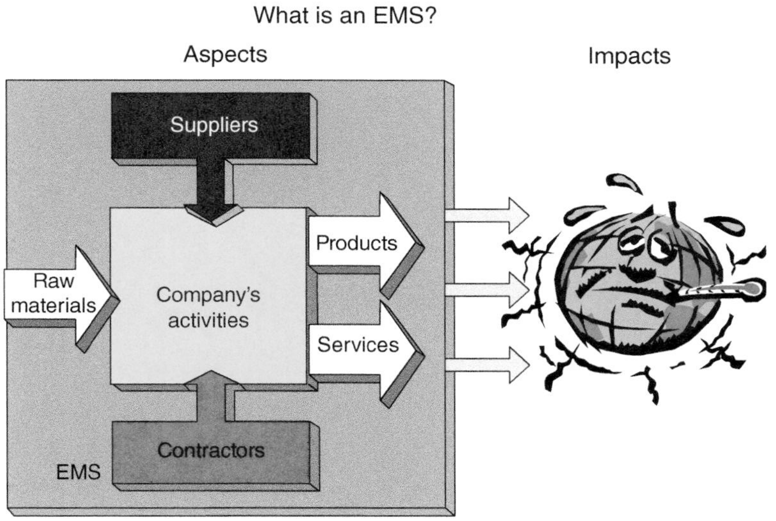

FIGURE 20.2 What is an EMS?

the need for some sort of system, training, and auditing. An EMS can provide the peace of mind that systems and controls are in place to ensure compliance with legislation.

20.1.1.2 Supply Chain Pressure

There are increasing concerns up and down the supply chain from companies, governments, and organizations regarding the environmental impacts of the products and services that they purchase. These concerns normally manifest themselves via increasing scrutiny on the environmental performance of suppliers through questionnaires or by auditing suppliers' operations.

Organizations who do not respond to these pressures will risk losing business.

20.1.1.3 Investor Pressure

Banks and institutional lenders have found themselves potentially liable for the environmental misdemeanors of their borrowers. So banks are starting to require evidence of good practice from lenders, in many cases in the form of an EMS and public reporting as part of good corporate governance. Many stock markets now rate companies based on their corporate governance score, which includes environmental performance.

20.1.2 What are the Benefits of Implementing an EMS?

Like all other management systems, an EMS ensures risk identification, control, and management. Other benefits are:

- Awareness of legal requirements and improved compliance
- Improved management and environmental performance
- Reduced incidents, prosecutions, liability, and insurance
- Financial savings through opportunities for improved efficiency and reduced operating costs
- Enhanced staff awareness, competence, ownership, and engagement
- Integration of corporate responsibility commitments
- Improved image and reputation
- Satisfies vendor/investor criteria and improves access to capital

20.1.3 Aspects and Impacts

An EMS revolves around *environmental aspects*. Using loss control principles of *cause and effect*, aspects can best be regarded as the *cause* of environmental impacts. ISO 14001 (ISO, 2004a) defines aspects as "Elements of an organisation's activities, products and services which can interact with the environment." So for a simple example of driving a car (an activity), an environmental aspect would be consumption of fossil fuel, another could be generation of exhaust gases, and another could be generation of noise. So an aspect can be thought of as the outcome from an activity, product, or service.

By comparison using loss control principles of *cause and effect*, environment impacts can best be regarded as the result (or *effect*) from environmental aspects. Similarly, ISO 14001 states, "Any change to the environment whether *adverse or beneficial* wholly or partially resulting from an organisation's activities, products or services." So using the aforementioned example of activity and aspects, the relevant impacts would be the following.

Aspects and impacts are far easier to define if one describes the associated activity with them first (Table 20.1).

Thinking of activities, products, and services leads to the following examples for clarification (Table 20.2).

TABLE 20.1

Activity	Aspect	Impact
Driving a car	Consumption of fossil fuel	Reduction of natural resources
	Generation of exhaust gases	Air pollution
	Generation of noise	Nuisance to humans

TABLE 20.2

Organization's Actions	Aspect	Impact
Activity: Handling hazardous chemicals	Potential for accidental spillage	*Soil and/or water contamination*
Product: Redesign of packaging	Reduction of packaging materials used	**Conservation of natural resources**
Service: Vehicle/plant maintenance	Lubrication oil replacement	*Consumption of natural resources*

> **Practical Example**
>
> *If you run a car and consume petrol, do you have to consider all the aspects associated with running the petrol station (e.g., delivery, storage, and spillage of fuel)? Ask the following question: if you sell the car tomorrow and have no vehicle, will the petrol station close down? An answer of "no" means that it is not your aspects; however, all the aspects associated with running your car (like use of fuel, cleaning, maintenance, etc.) are yours until you sell it!*

Remember that impacts can be **positive** as well as *negative*, although most people tend to concentrate on the latter.

One of the more difficult concepts to understand is the concept of aspects and impacts that you can control or have influence over. I prefer to use the terms *direct and indirect* aspect/impacts as was defined in the EMAS regulation.

Direct aspects/impacts are those which an organization can control. Examples include air emissions, water emissions, waste, use of natural resources and raw materials, local issues (noise, vibration, odors), land use, air emissions related to transport, and risks of environmental accidents and emergency situations.

Examples of indirect aspects/impacts are product life cycle-related issues; capital investment; insurance services; administrative and planning decisions; environmental performance of contractors, subcontractors, and suppliers; and finally choice and composition of services, for example, transport, catering, and so on.

The evaluation of indirect aspects/impacts should involve all those which the organization can control and those which it could reasonably be expected to have influence.

The simplest way to define direct and indirect aspects is to ask the question:

If the activity under review stops tomorrow, will the aspect continue? If yes, then it's not your aspect; if however the aspect stops when you stop the activity, it's your aspect.

20.1.3.1 Key Environmental Issues

Figure 20.3 (http://www.cleartheair.nsw.gov.au, 2011).

In order to ensure comprehensive coverage of aspects and impacts identification, the issues that the organizations must address are listed in the following text; further information can be found here (Coopers and Lybrand, 1994) and in the previous chapters in this book.

Atmospheric Emissions
Acid rain, local air pollution, ozone layer depletion, global warming, volatile organic compounds (VOC's), vehicle emissions, and particulates

Water Use/Discharges
Effluent discharges, sewage, water supply, groundwater pollution, aquifers, leachate, and eutrophication

Solid Waste
Packaging, paper, recycling, waste management, landfill, incineration, and producer's responsibility

Energy
Energy, energy efficiency, energy sources, coal, diesel/oil, gas, nuclear, and geothermal, wind

The Natural Environment
Deforestation, sustainable resources, damage to ecosystem, and loss of biodiversity

Human Health
Noise and nuisance

Conservation and Wildlife
Protected species, protected trees, sensitive sites, protected buildings, "scheduled" monuments, and areas of archeological importance

Resource Use
Water consumption, land use, energy use, minerals/ores, trees, and plants

Corporate/Market Pressures
Government/industry awards/grants, compulsory environmental auditing, mandatory environmental labeling, environmental taxes/economic instruments, mandatory environmental reporting, insurance risk assessments, and investors/shareholders/lenders requirements

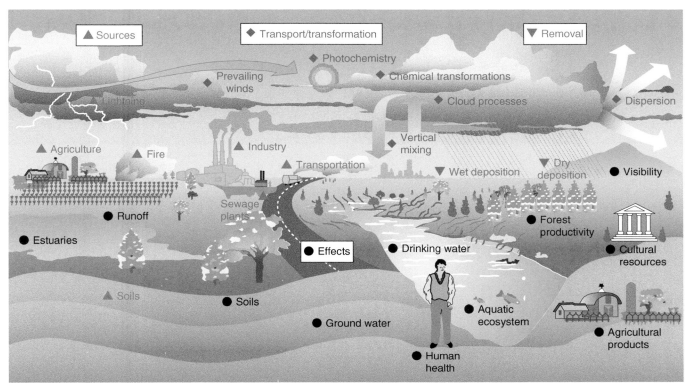

FIGURE 20.3

TABLE 20.3 Classification of Impacts

Resource Depleting Impacts	Polluting Impacts
Renewable resource depletion	Global warming
Nonrenewable resource depletion	Ozone depletion
Habitat disturbance	pH alteration
Aesthetic considerations	Eutrophication
Soil depletion	COD/BOD discharges
Landscape alteration	Photochemical smog
Hydrology alteration	Human toxicity
Habitat alteration	Environmental toxicity
(i.e., loss of biodiversity)	Hazardous wastes
	Nuisance

Environmental Impacts

Once all the aforementioned environmental issues have been addressed within the EMS, the outcome (i.e., the impacts) will normally fall under one of the following categories (Table 20.3).

It is useful to remember that there are five basic RECEIVING media that should be addressed:

- Air
- Water
- Land
- Ecological resources
- Human resources

These media should be studied in enough detail to identify and understand all their possible pathways and impacts, potential sources of loss, and potential losses occurring in the past (historic evaluation).

In addition to local issues, relevant regional and global issues regarding the receiving environment may be documented in the study (usually as part of an advanced EMS).

20.1.4 Core Elements of an EMS

20.1.4.1 Initial Environmental Review

So where do we start? Identify environmental aspects by conducting an initial environmental review (IER) to establish the organization's current position or baseline (Fig. 20.4).

The core of an EMS starts with top management commitment: no commitment = no EMS. That established, the team conducting the IER will need to look at the organization's operations and find out what current management practices/procedures are in place, gathering information and data as they go. They will need to walk around the site and identify any areas of sensitivity and then establish what legislation is relevant to their operations and which permits/licenses they should have in place. (Details of how to conduct an IER can be found in ISO 14004 (ISO, 2004b).)

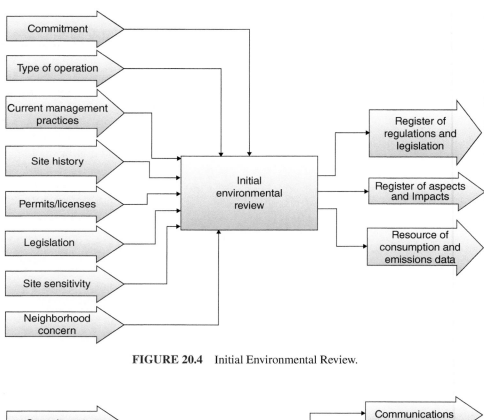

FIGURE 20.4 Initial Environmental Review.

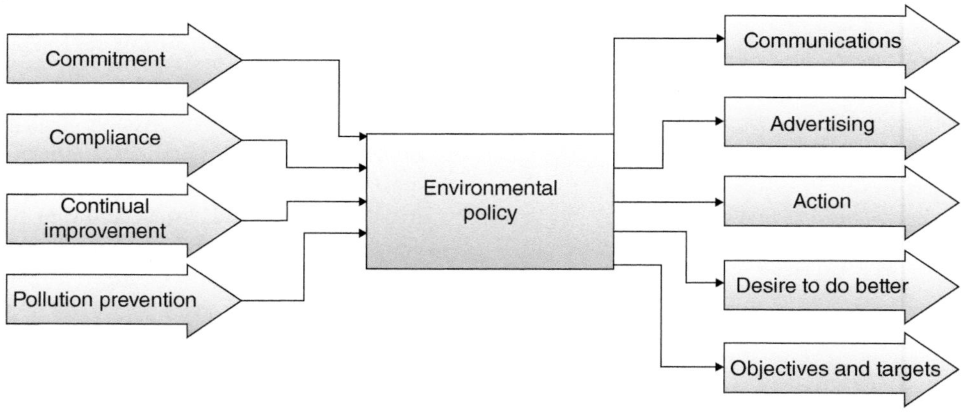

FIGURE 20.5 Environmental Policy Inputs and Outputs.

The output from this review should be a list or Register of Environmental Aspects, a Register of Legal and Other Requirements and some data on resource consumption and emissions.

20.1.4.2 Environmental Policy

Using the information from the IER, the team can prepare an initial environmental policy showing the organization's commitment to comply with regulatory requirements, prevent pollution, and continually improve their environmental performance, by setting and activating objectives and targets (O&Ts) (Fig. 20.5). This draft policy can then be used for promotion, demonstrating the organization's desire to do better.

When TC207, the technical committee within ISO, set about designing ISO 14001, they based the structure on Dr W.E. Deming's plan> do> check> act cycle of continuous improvement

Plan   *Establish performance objectives and standards*
Do   *Measure actual performance*
Check   *Compare actual performance with objectives and standards and determine the gap*
Act   *Take necessary actions to close the gap and make necessary improvements*

Dr Deming's statistical quality control (SQC) methods, popular in the United States during World War II, faded in America in the postwar boom. However, he was active in Japan since 1950, where his statistical techniques have evolved into a total management system.

While statistical methods and graphic tools remain key components of the Deming's method, it's now much more. According to Deming, the only way to become competitive is to undergo a top-to-bottom, quality-based transformation. By focusing on quality, you will improve productivity, innovation, and profitability.

What is needed, Deming maintained, was a *participative* system based on teamwork in which workers have the training and resources to make quality an ongoing daily pursuit. Deming further emphasized that quality goals are never fully achieved; process improvement is never ending.

Successful application in the quality area prompted TC207 to apply it in the environmental management area to achieve continual improvement.

20.1.4.3 Plan

This improvement cycle is readily adapted to EMS core activities by formulating a **PLAN** to control or improve significant aspects (Fig. 20.6).

Taking the outputs from the IER and applying some significance criteria, a list of significant aspects can be drawn up. Then each significant aspect is reviewed to determine whether they can be managed or have to be improved. Those that can be managed will be controlled via operational control procedures, while those needing improvement will be addressed via objectives targets and programs.

20.1.4.4 Do

The next step is to implement the PLAN through a **DO-**ing stage (Fig. 20.7).

Significant aspects need to be managed; to do this within an organization, you need to allocate responsibilities, communicate what they have to manage, and train people how to manage their aspects through procedures and work instructions or instructions. Make them aware of the legal requirements they have to comply with and drill them on how to handle emergencies. This will lead to reduced impacts, potential savings in cleanup costs, and general improvements in environmental awareness. To prove that they are managing their significant aspects, certain records will be produced.

20.1.4.5 Check

The next stage is to start CHECKING to see if significant aspects are being managed via operational control or improved via programs linked to O&T's (Fig. 20.8). If they are not, then noncompliances need to be raised and action taken to correct or prevent recurrence. The organization also needs to check regulatory compliance and keep records. Audits can be used as another method of checking conformance with in-house EMS controls and progress toward achieving continual improvement.

20.1.4.6 Act

The final stage in the plan>do>check>act approach is to conduct a management review by pulling together all the information gained from monitoring and measuring, results from internal audits, details of process changes, information on regulatory and business goal changes, results of improvement programs, and data on compliance or noncompliance (Fig. 20.9).

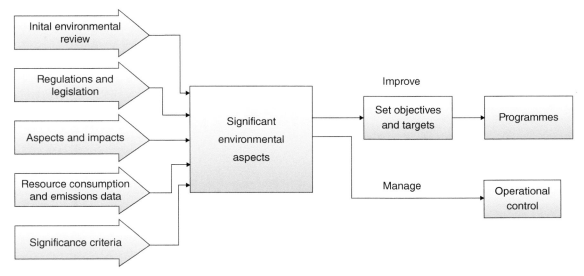

FIGURE 20.6 Plan.

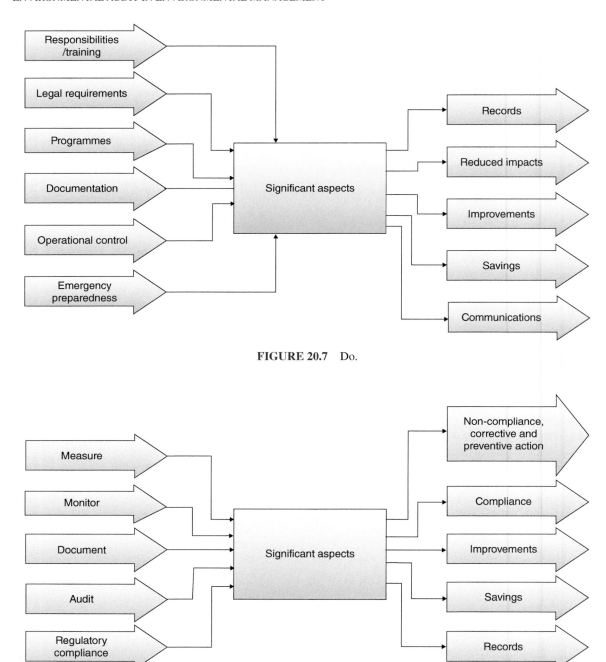

FIGURE 20.7 Do.

FIGURE 20.8 Check.

The outputs from the management review could be revision of the environmental policy, O&T's, details of the effectiveness of corrective and/or preventive action taken, and hopefully appreciation of the improvements made.

20.1.4.7 PDCA Loops

An effective EMS will have two PDCA loops, one for control and one for improvement. The input to the improvement loop will be data from the IER (Fig. 20.10). After one PDCA cycle, once O&T's have been reviewed, there will be feedback into the *policy* to update it in the light of progress or performance improvement, whereas the control loop will maintain the status quo.

20.1.4.8 Conclusion

Thus, to summarize, an EMS is based on identifying all applicable environmental aspects related to an organization's activities, products, and services, using some criteria

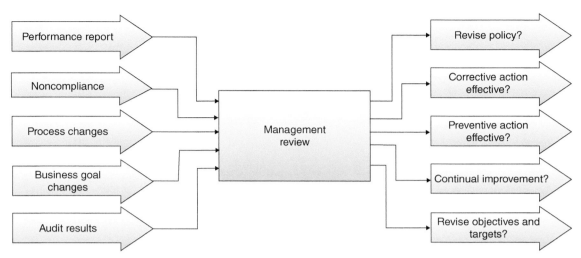

FIGURE 20.9 Act.

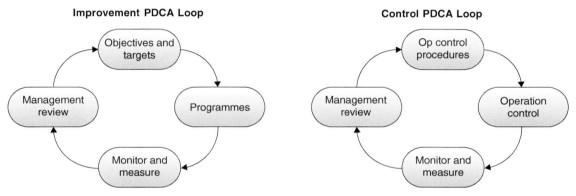

FIGURE 20.10 PDCA Loops.

to define significance, testing for significance, and then asking whether a significant aspect can be managed via operational control (including maintenance) or needs improvement via objectives, targets, and programs (Fig. 20.11). This can be referred to as the "heart of an EMS." So every significant aspect will either be managed or improved.

This on its own is not enough to prescribe an EMS as it only covers PD of PDCA. What we need is some blood circulation from the heart of an EMS via monitoring and measurement to see if the EMS complies with requirements; if not, then nonconformities need to be raised and corrective action is taken and checked for effectiveness (Fig. 20.12). This will bring us to PDCA if we subsequently have a health check via audits and take action after a management review.

20.2 ISO 14001 INTERPRETATION

20.2.1 Introduction

In order to conduct a meaningful audit against ISO 14001 (ISO, 2004), the auditor needs to know what ISO 14001

Practical Example

A bus company that operates public transport has a large fleet of buses that range from 25 years old to brand new. These buses conform to the European emission standards for buses under the classification of Euro 1, Euro 2, Euro 3, Euro 4, and Euro 5 (European Commission 1987, 1999, 2001, 2005, 2009), but they also have a large number of old buses pre-Euro 1. The bus company was getting certified to ISO 14001 and had identified their environmental aspects with respect to air emissions as follows (Table 20.4).

They then analyzed this aspect with respect to the different bus types and tried to define whether with good operational control, that is, maintenance, the emissions would comply with upcoming changes in vehicle emission regulations (VER), which would match with the Euro 1 standard, to be implemented 3 years later. The results of their analysis were as follows (Table 20.5).

So they raised an objective to improve their exhaust gases from pre-Euro 1 buses by fitting 200 catalytic converters each year over a 3-year period, such that when the regulation was enacted, they would be in compliance. For the other buses, they would continue making exhaust gas measurements and conduct regular maintenance based on the bus' running hours.

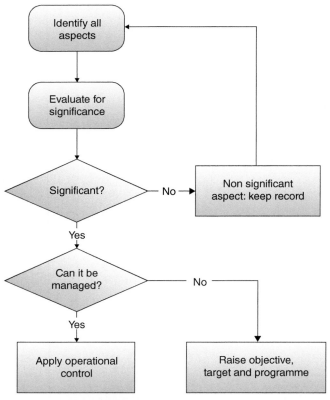

FIGURE 20.11 The heart of an EMS.

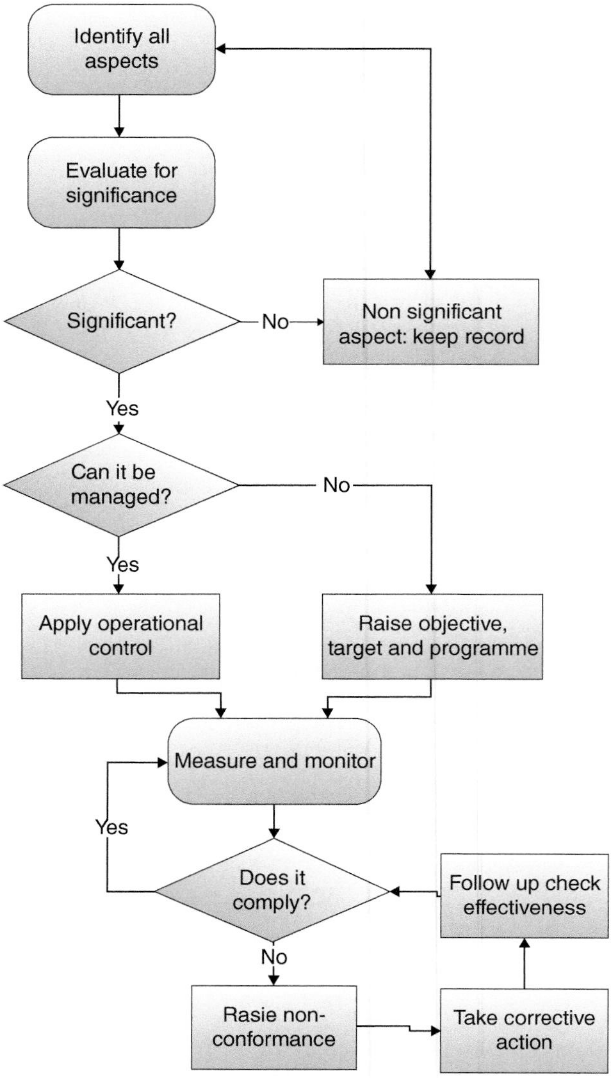

FIGURE 20.12

TABLE 20.4 Example of one aspect and impact

Activity/Product/Service	Aspect	Impact
Delivery of bus services	Generation of exhaust gases	Air pollution

TABLE 20.5 Practical Example

Service	Aspect	Impact	VER Compliance?	Management Option
Euro 4/5 bus operation	Generation of exhaust gases	Air pollution	Yes	Ensure engine management system working via maintenance
Euro 1, Euro 2, and Euro 3 bus operation	Generation of exhaust gases	Air pollution	Yes	Ensure regular maintenance
Pre-Euro 1 bus operation	Generation of exhaust gases	Air pollution	No	Raise objective and target to fit exhaust catalytic converter

requires and how to audit it. What follows in this section is a clause-by-clause breakdown for ISO 14001 as seen from an auditor's perspective. Practical examples of how the standard can be implemented will give the auditor an insight on its interpretation.

20.2.1.1 Section 1: Scope of ISO 14001

The standard literally can be applied to any organization's activities, products, and services from SME office activities to tobacco companies to nuclear power generators.

20.2.1.2 Section 2: Normative References

There are no typical standards or models related to ISO 14001; it is singular in its approach.

20.2.1.3 Section 3: Terms and Definitions Used in the Standard

Needs to be read carefully and referred to as needed, other terms relative to common management system language can be found in ISO 9000:2000 (ISO, 2000); if all else fails, check a standard dictionary!

20.2.1.4 Section 4: EMS Requirements

This section of the standard contains clauses that will be used by certification companies to check if the organization's EMS confirms to the standard. Guidance on how to interpret ISO 14001 (ISO, 2004) is given in ISO 14004 (ISO, 2004).

20.2.2 General Requirements

Environmental aspects emanate from activities, products, and services that the organization can control within an EMS but may still have an impact on the environment and certainly will have an impact if they are not controlled. This clause requires answers to the following questions:

- Is there an EMS?
- Is it implemented/supported?
- Has the scope been defined?
- Fundamentally, are significant aspects being controlled or improved?

Practical Example

A management system consultant who sets out to help companies get certified to ISO 9001, ISO 14001, and OHSAS 18001 (ISO 2008; BSI, 1999; BSI, 2007) had developed an integrated management system (IMS) to address all three management systems. When audited for ISO 14001 certification, the lead auditor found an IMS manual that addressed all the requirements of ISO 14001 (and ISO 9001/OHSAS 18001); he also found procedures addressing all the requirements of ISO 14001 (and ISO 9001/OHSAS 18001), but when asked for records of his environmental aspects plus other records to show how the system had been implemented, he could not produce any records.

The lead auditor had no choice but to issue a major nonconformance against clause ISO 14001, Clause 4.1, for lack of implementation and stopped the audit.

20.2.3 Environmental Policy

The environmental policy should be appropriate to the nature, scale, and environmental impacts of its activities, products, and services within the scope of the EMS.

The standard requires top management commitment to:

- Continual improvement
- Prevention of pollution
- Legal compliance
- Compliance to other requirements

The policy should:

- Provide a framework for setting and reviewing environmental O&T's
- Be documented, implemented, and maintained
- Be communicated to all persons working for or on behalf of the organization
- Be available to the public

Interpretation

- *Appropriate to the nature, scale, and environmental impact means that it describes who you are and what you do. So the easy test is to change the name on the policy to that of the local supermarket chain, and if the policy reads true, it is not appropriate.*
- *Commitment to continual improvement and prevention of pollution leads to an <u>environmental performance improvement</u> (NOT continual improvement of the management system). The policy must contain the words "committed to"; "will try to" and "will strive to" are not commitments.*
- *Legal compliance is interpreted as follows:*
  - *If there is noncompliance with regulations/legislation but there is evidence of a dialogue and agreement with the regulator plus an internal improvement plan, this will be seen as compliance.*
  - *No dialogue or agreement or improvement plan will be interpreted as noncompliance.*
  - *Auditors should check if the operator has signed up to, for example, ICC (International Chamber of Commerce, 2000)/Responsible Care (CIAC, 1985), etc. and should assess for compliance with the policy requirements of these principles/codes of practice (Fig. 20.13).*
- *The process of continual improvement is readily visualized with this plot of performance against time using the PDCA cycle to climb the hill toward sustainable development. The cycle is stopped from slipping back via quality or environmental controls.*

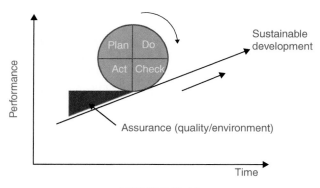

FIGURE 20.13

- *The framework for O&T's means a high-level statement, like we will reduce waste, improve energy efficiency, and so on.*
- *Documented and implemented means a hard copy on show and its concepts in evidence practically.*
- *Communicated does not mean given to all staff in their wage packet, but a session has been held where all staff, including contractors, have had the policy explained and, as a result of questions asked, are aware of its contents, not the need to earn it parrot-fashion but are aware of the major bullet points (prevention of pollution, continual improvement, legal compliance, plus some idea of O&T's).*
- *Available to the public means it's on the organization's website for all to see.*

20.2.4 Planning

The second major element or principle in ISO 14001 is the planning process and its three subelements:

1. Environmental aspects
2. Legal and other requirements
3. Objectives, targets, and programs

20.2.4.1 Environmental Aspects

Environmental aspects for activities, products, or services that the organization can control or influence have to be identified and kept up to date within the defined scope of the EMS, taking into account new developments or modification of these activities, products, or services.

Those aspects that have a significant environmental impact have to be determined and have to be considered when creating, applying, and maintaining the EMS.

Interpretation

- *There needs to be a constant review process rather than just an initial review, which is why ISO 14001 does not require an initial review. The emphasis must be on <u>updating</u>.*
- *Aspects have to be considered for all of the organization's activities, products, and services:*
  - *Within the defined scope of the EMS*
  - *Including all past, present, and future activities as well as*
  - *All operating conditions: normal, abnormal, accident, and emergency*
- *The <u>aspects</u> an organization <u>can control</u> are equivalent to direct aspects (see Section 20.1.3). Those it can <u>influence</u> are equivalent to <u>indirect</u> aspects. Care in evaluation should be taken in cases where highly polluting activities have been contracted out—they are still under the operation's <u>influence</u>!*
- *The significance method has to be documented and should be systematic and consistent. If measures of quantity like "large" have been used, there should be definitions of what "large" means. The method should be kept as simple as possible such that the auditee and the auditor can easily come up with the same answer. ISO 14004 has come up with nice examples of how to derive significance.*

Reference should be made to Fig. 20.11, "the heart of an EMS."

Practical Example

Methods to define significance could be based on environmental considerations such as:

- *Environmental concerns*
- *Regulatory requirements*
- *Scale of the impact*
- *Severity of the impact*
- *Duration of the impact*
- *Scientific evidence of risks*

And business concerns are as follows:

- *Regulatory and legal exposure*
- *Concerns of interested parties*
- *Effect on public image*
- *Compliance with industry codes of practice*

But "is our business concerned" should not be used as the organization may decide just not to do anything.

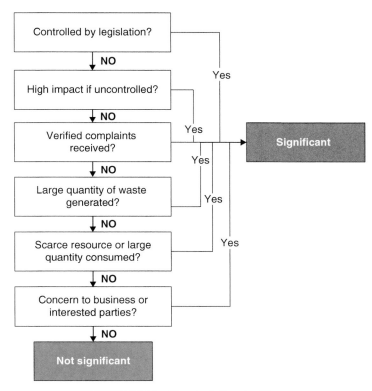

FIGURE 20.14 How to define significance.

The simplest method seen is to ask three basic questions:

1. *Is it large?* Where large is defined by *environmental criteria (such as scale, severity and duration of the impact, or type, size, and frequency)*
2. *Is it regulated?*
3. *Is it of public concern?*

Where any yes answer merits significance. This gets over the problem encountered during certification of a construction company where they asked only one question: "Is it regulated?" This managed to avoid all aspects related to raw material consumption and transportation to site. Applying the three questions above then made these aspects significant, to the auditor's satisfaction.

Sometimes, defining large is difficult, so other environmental criteria (such as scale, severity and duration of the impact, or type, size, and frequency) can be used. In the earlier example, large was defined by imagining sitting at the entrance to the site and counting all the raw material movements, in and out (Fig. 20.14).

20.2.4.2 Legal and Other Requirements

The standard requires the organization to develop and support a method to have access to applicable legal and other requirements related to environmental aspects, determine how they apply to these aspects, and take them into account when developing and supporting the EMS.

Interpretation

- *Legislation and regulations must be directly related to the operation's activities. A database on a CD-ROM of all the prevailing legislation in that country is not enough. There must be a record that shows how they relate to activities.*
- *How are they kept up to date? Once a year in the management review is too infrequent, maybe quarterly would be OK.*
- *Who keeps them up to date, are they competent to do it?*
- *Other requirements would include contracts, corporate environmental policy, insurance company requirements, industry guides, or codes of practice.*

Practical Example

Register of legal and other requirements See Table 20.6.

20.2.4.3 Objectives, Targets, and Program(s)

The standard requires the organization to develop and support documented O&T for each relevant level and function. When establishing O&T's, the following need to be considered:

- Legal and other requirements
- Significant environmental aspects
- Technological options
- Financial, operational, and other business requirements
- Views of interested parties
- Be measurable, where practicable
- Be consistent with environmental policy, including commitments to prevention of pollution and compliance with applicable legal requirements and with other requirements and continual improvement

Interpretation
Objectives should be long term and cover a period of at least 3 years. Targets should cover 1–3 years, and the program should detail activities over at least a 1-year period.

There needs to be a <u>minimum</u> of one improvement objective addressing a significant aspect per activity area in an operation.

Objectives should be SMART (Doran, 1981), that is, specific, measurable, achievable, results oriented, and time bound.

> *Objectives are broad goals arising from the policy and aspects evaluation that an organization sets itself to achieve.*
> *Targets are measurable performance requirements applicable to the organization, or parts thereof, that arise from the environmental objectives and that need to be set to achieve those objectives.*

Application of improvement O&T's if accompanied by appropriate resources given to people with the appropriate responsibility may over a defined timescale bring environmental performance improvements. The aforementioned example shows how O&T's can be used in advance of a change in regulatory requirements (the red dotted line) to improve performance (Fig. 20.15).

The organization should develop and support program(s) to achieve its O&T's. Program shall include:

- Responsibilities for achieving objectives at each relevant function and level
- Means and time frame by which they are to be achieved

TABLE 20.6   Register of legal and other requirements

Regulation	Date Enacted	Implication to the Business	Significant Aspects	Operational Control Procedure	Control Limits
Air pollution control	1999	Vehicle emissions	Truck exhausts	EMP 4.6 vehicle maintenance	<50 hartridge smoke units

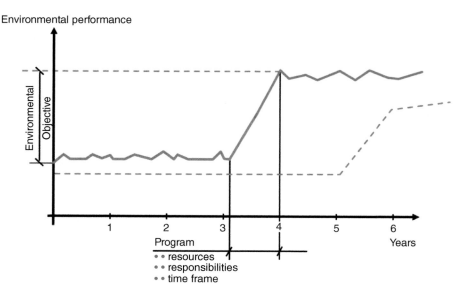

FIGURE 20.15 The use of objectives and targtets to improve performance.

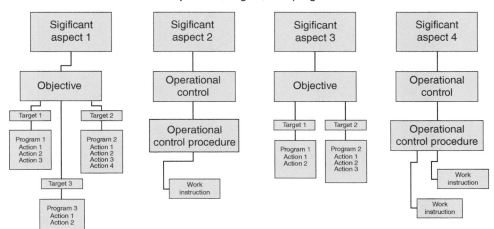

FIGURE 20.16

Interpretation

- *Responsibility cannot be shared! If the records show that a department is responsible, check to see if the department manager knows his responsibilities.*
- *Timeframe should cover at least 1 year.*

Application of objectives, targets, and programs over time will allow the organization to achieve continual improvement.

Figure 20.16 illustrates how an EMS should finally be structured, with every significant aspect either being managed or improved. Having raised the objectives, targets, and programs to improve significant aspects, it's then important to check whether the policy provides a basic framework for these objectives.

Practical Example

See Table 20.7.

TABLE 20.7 Example of an EMP

Environment Management Program for Electricity Reduction

Objective: To reduce energy consumption by 10% over the next 2 years
Target 1: Reduce energy consumption by 5% in year 1
Action 1: Replace lighting electronic ballast in area 2
Action 2: Replace 30 T8 fluorescent tubes with T5 tubes in area 2
Target 2: Reduce energy consumption by 5 % in year 2
Action 1:Replace lighting electronic ballast in areas 1 and 3
Action 2: Replace 25 T8 fluorescent tubes with T5 tubes in area 1 and 3
Responsibilities: Targets 1 and 2 electrical engineering department

20.2.5 Implementation and Operation

The third major element or principle in ISO 14001 is the implementation and operation process and its subelements. The do principle is broken down into seven subsections:

1. **Resources, roles, responsibility, and authority**
2. **Competence, training, and awareness**
3. **Communication**
4. **Documentation**
5. **Control of documents**
6. **Operational control**
7. **Emergency preparedness and response**

These clauses are normally referred to as significant aspects control elements (in red) and the supporting system elements (in blue).

20.2.5.1 Resources, Roles, Responsibility, and Authority

The organization's management shall provide resources essential to develop, support, and improve the EMS. These roles, responsibilities, and authority shall be documented and communicated to ensure effective environmental management. The relevant resources such as human resources, specialized skills, technologies, and financial resources shall be provided.

A specific management representative(s) shall be appointed who, as well as his other roles and responsibilities:

- Shall ensure that an EMS is developed and supported in accordance with the requirements of ISO 14001
- Shall reports on the performance of the EMS to top management

Interpretation

- *The need for roles and resources to be defined applies to all operations that relate to the standard.*
- *The management representative does not necessarily have to be a member of the management team.*
- *The actions required to activate the management system should not all lie on the management representative's shoulders; others throughout the organization should have defined roles and responsibilities. The auditor will be looking for a "management system" not a "man system."*

Practical Example

The organization chart previously shows how the people who can influence environmental matters (those in green) fit in to the organization (Fig. 20.17).

20.2.5.2 Competence, Training, and Awareness

Those performing tasks, for the organization or on its behalf that have potential to cause significant environmental impact, shall be competent on the basis of appropriate education, training, and/or experience.

Training needs to be associated with its environmental aspects, and its EMS shall be identified. Training and/or coaching shall be provided and records kept addressing these needs.

Staff and contractors need to know:

- Why they must comply with the environmental policy and procedures
- Significant environmental impacts of their work
- Their roles and responsibilities in complying with the environmental policy
- What will happen if they don't follow specific operating procedures

Interpretation
In the following is an example of a training needs survey that must be prepared for each employee whose work can impact on the environment. As well as a needs analysis, there needs to have actually been some training, and there should also be a record of who has done what.

Remember that training is normally one of the weakest areas of any EMS and the auditor will be looking to identify that it actually has happened and won't spend hours looking at training records, but will be checking on the effectiveness of training by talking to staff!

People who can <u>influence</u> impacts on the environment need to show that they are <u>competent</u> to do so.

All others on-site should be <u>aware</u> of the operation's environmental impact. During the first certification audit, there should be a training program evident. The company being assessed must either have trained everyone or they can prove that without training, there is no environmental risk.

Practical Example

For activities that have—or could have—a significant environmental impact, the facility shall identify the minimum E&T level and ensure that only qualified employees conduct such activities (Table 20.8).

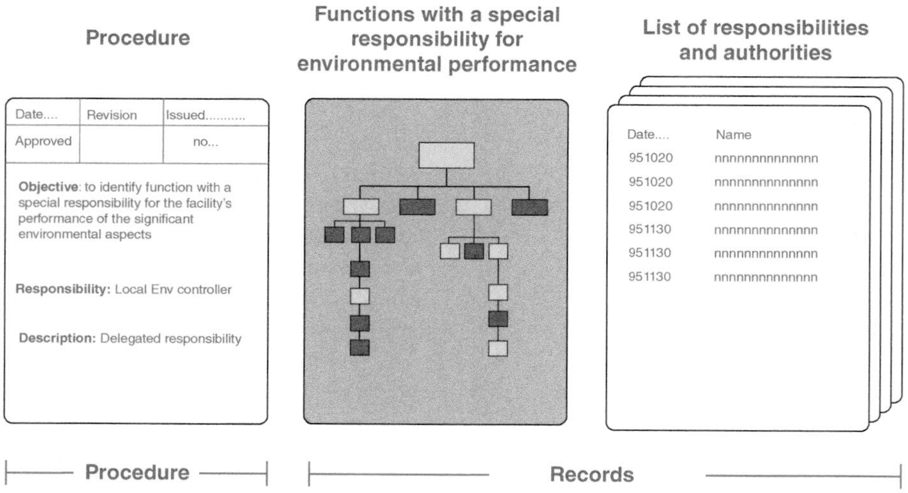

FIGURE 20.17

TABLE 20.8

Environmental Aspects	Activity	Min. Education and Training Level	Qualified Employees Name	Date for Qualification
Emission of dust	Maintenance of bag filters	Filter training course provided by the filter manufacturer	NN NN	DD MM YY DD MM YY
Noise from driven pile machine	Monitoring of noise level	Training for noise meter use from QEHS Staff	NN	DD MM YY

20.2.5.3 Communication

Organization should develop and support methods:

- For internal communication within the organization
- For responding to external communications from interested parties

The organization also needs to decide whether to PROACTIVELY communicate external on significant environmental aspects; if yes, they will need a method.

Interpretation

(a) and (b) are REACTIVE communication.

(c) is PROACTIVE communication and there needs to be a record to show how the thought process evolved.

An EMS will include the means to communicate with and provide information to interested persons inside and outside the organization.

There may also be pressure from, say, banks or financial institutions or from the neighboring community to produce information about the organization's environmental activities. An organization therefore needs to establish methods to facilitate dialogue with the public and to report environmental performance. A sense of openness encourages public acceptance and may lead to a reduction in complaints from local residents, as well as helping to increase trust between the organization and its stakeholders.

An efficient system for handling any complaints that do occur can improve local relations and provide useful information for reviewing targets and objectives.

Reporting on the improving environmental performance of the organization can:

- *Demonstrate the organization's ongoing commitment to the environment*
- *Provide a stimulus for continued commitment and improvement*
- *Act as an efficient method for dealing with concerns and queries about the environmental aspects of its activities, products, and services*

20.2.5.4 EMS Documentation

EMS documentation shall include:

- Environmental policy and O&Ts
- A description of the scope of the EMS
- A description of the core elements of EMS and their interaction
- Provide direction to related documentation
- Documents necessary to ensure the management and improvement of significant environmental aspect

Interpretation
In essence, the documentation should act as a "signpost" to the EMS. It should be helpful to both users and auditors, for example, there may be a section relating to each clause of the standard, which provides references to relevant procedures, work instructions, and so on. and cross-references to any related procedures in other parts of the management system (e.g., quality). Remember the focus of ISO 14001 is records to show that the standard has been implemented. The old ISO 9001 need for documented procedures for every clause is not there.

20.2.5.5 Control of Documents

The organization needs to develop and support a method to:

- Approve documents prior to issue
- Review, update, and reapprove documents as necessary
- Identify changes and current revision status of documents
- Ensure the right documents are in the right place
- Ensure documents remain legible and readily identifiable
- Ensure documents of external origin are identified and their distribution controlled
- Prevent the unintended use of obsolete documents and label them if they are retained for any purpose

Interpretation
Does the organization have a process for developing and maintaining EMS documentation?

*There needs to be a **master list**, or equivalent, which **identifies** the current revision status of the documents. This should be readily available.*

Unlike ISO 9001, this clause does not have the highest priority in the auditor's mind; it will only feature when there is a problem related to management of significant aspects. Auditors who start their audit on this clause should be ostracized, because the focus of the audit should be on the effective control or improvement of significant aspects.

> **Practical Example**
>
> Document Issue *The following information should be included in each document, when they are issued:*
>
> - *A document title*
> - *A reference number*
> - *A revision number, including the revision status, being replaced by the new document/copy*
> - *The recipient of the document/copy*
> - *An "acknowledgement of receipt" note/signature from the copy holder*
> - *Names/signatures of issuing and approving persons*
> - *Date of issue*

20.2.5.6 Operational Control

Organization shall identify activities that affect significant impacts and control them:

- Via documented procedures
- Within operational control limits

Organization shall identify significant aspects from suppliers and contractors and communicate these procedures with them.

Interpretation
The main theme of ISO 14001 EMS Certification is "PERFORMANCE" and "PROTECTION."

The ultimate question asked by an auditor is, "is the EMS adequately managing significant aspects, achieving regulatory compliance, and protecting the environment?" Operational control should demonstrate the organization's commitment to address this question.

The next questions the auditor will focus on are: "Have operational control limits been set? Does everyone know them? What happens when environmental performance is outside these limits?"

Having tackled management of environmental aspects via O&T's, the next most significant area that needs control is contracted services. ISO 14004 gives the following guidance.

The organization should consider the different operations and activities contributing to its significant environmental impacts when developing or modifying operational controls and procedures, for example:

- *R&D design and engineering*
- *Purchasing*
- *Contractors*
- *Raw materials handling and storage*
- *Production and maintenance processes*
- *Laboratories*
- *Storage of products*
- *Transport*
- *Marketing and advertising*
- *Customer service*
- *Acquisition or construction of property and facilities*

> **Practical Example**
>
> *Operational control will only manage significant aspects and maintain the status quo IF everyone who can control the aspects knows how to steer the process to keep the environmental performance on track. For example, if an effluent stream has a discharge consent (permit) of pH 5–9, the operator should know these limits, and when the pH approaches 5.5 or 8.5, he should take correct action to maintain compliance (Table 20.9).*

20.2.5.7 Emergency Preparedness and Response

Organization shall:

- Identify potential for accidents and emergency situations
- Prevent and mitigate environmental impacts that may be associated with them
- Review and revise emergency preparedness, especially after the occurrence of accidents or emergency situations
- Periodically test such procedures where practicable

TABLE 20.9

Significant Aspect	Process	Procedure Requirement	Operation Criteria	Monitoring Criteria
Discharge of wastewater	Effluent treatment	pH control	pH 5.5–8.5	Continuous in line
		Sludge concentration	100 g/l	1/day
		COD/BOD	COD < 200 ppm	1/day
		Total fatty matter	BOD < 40 ppm	1/week
			TFM < 150 ppm	1/day

Interpretation

Identifying and planning to avoid environmental risks should address the following:

- *What materials are handled on-site, for example, flammable liquid, storage tanks, compressed gases, and pipelines that transport hazardous materials?*
- *What is the most obvious source of an emergency situation or accident?*
- *What is the potential for an emergency situation or accident at a nearby facility (e.g., plant, road, rail line)?*
- *What are the consequences of natural disasters (e.g., flooding, earthquakes, tsunamis)?*
- *What are the consequences of a malfunction of an environmental facility (e.g., water scrubber, wastewater treatment plant)?*

Another approach is the application of the Loss Control Management "IEDIM" concepts:

- <u>*Identify*</u> *all potential emergencies.*
- <u>*Evaluate*</u> *the risk. How severe could this scenario be if it reaches its full potential? How likely is it to happen? Both risk identification and evaluation should be included in the environmental risk assessment.*
- <u>*Develop*</u> *a procedure to control the emergency and mitigate its effects. This should include defining objectives and setting success criteria against which the performance of the procedure may be judged.*
- <u>*Implement*</u> *the procedure. This involves communicating the procedure to all those who will be involved in and providing them with any necessary training to prepare them for their role. It also involves putting in place all systems, equipment, materials, procedures, practices, and activities identified as being required by the procedure.*
- <u>*Monitor*</u> *the performance of the procedure. This will involve testing the procedure at regular intervals against the performance and success criteria mentioned previously.*

For high-risk processes, application of hazard and operability (HAZOP) (British Standards Institution, 2000) studies would readily identify problems and produce solutions to mitigate any potential environmental impacts.

Nowadays, auditors will look for evidence to show that emergency drills have been conducted before certification, and organizations normally produce a set of sequenced photographs demonstrating the drills. What's often forgotten is: was the drill effective and who missed the drill?

20.2.6 Checking

The fourth major element or principle in ISO 14001 is the checking and corrective action process and its five subelements:

1. Monitoring and measurement
2. Evaluation of compliance
3. Nonconformity, corrective, and preventive action
4. Controls of records
5. Internal audit

20.2.6.1 Monitoring and Measurement

ISO 14001 requires:

- Procedures to monitor process parameters that affect environmental impacts
- Record performance against O&Ts
- Calibration records of monitoring equipment

Interpretation

Monitoring: continuous assessment of performance over time

Measuring: instantaneous measurements and discrete check of the acceptability of a parameter

Dictionary Definition

Monitor: person or device that checks, controls, warns, or keeps records of something

Measure: ascertain size, quantity of (as determined by measurements)

When defining monitoring requirements, the following need to be addressed:

- *Frequency*
- *Measuring method*
- *Equipment*
- *Calibration of equipment*
- *Qualification and responsibility*
- *Reporting*

The main environmental purpose of process monitoring is to:

a. *Check the effectiveness of the environmental control procedure*
b. *Demonstrate that the procedure is achieving compliance*
c. *Assist in improvement of the current situation*
d. *(viz., "If you can't measure, you can't manage!")*

20.2.6.2 Evaluation of Compliance

Develop and support a method for periodic evaluation of compliance with applicable legal requirements:

- Keep records.

Develop and support a method for periodic evaluation of compliance with other requirements:

- Keep records.

Interpretation
Be aware of what you have to comply with and check at a frequency that is appropriate to the magnitude of the environmental impact. Once per year for everyday activities for most organizations is not enough. Only checking for environmental prosecutions will not always guarantee regulatory compliance has been achieved.

Practical Example

A construction company with four active project sites, all at differing stages of completion, would need local staff at the project site to conduct a weekly site walk with the client to demonstrate compliance with legal and contract conditions. Head office monitoring of site compliance with legal and other requirements would probably be conducted during the monthly progress review either in head office or on-site.

20.2.6.3 Nonconformity, Corrective, and Preventive Actions

Develop and support a method to:

- Identify and correct nonconformities and take action to mitigate their environmental impacts
- Investigate nonconformities, determine their causes, and take actions in order to avoid their recurrence
- Evaluate need for action to prevent nonconformities and implement appropriate actions designed to avoid their occurrence
- Record results of corrective action and preventive action taken
- Review effectiveness of corrective action and preventive action taken
- Ensure actions taken are appropriate to magnitude of problems and environmental impacts encountered
- Ensure that any necessary changes are made to EMS documentation

Interpretation
Firstly, a definition (Table 20.10) is given as follows:
Nonconformity: Nonfulfillment of a requirement
In simple words, correction is rectifying the problem, but corrective action is to eliminate the root cause of the problem (so it does not repeat). If an NC has been detected, it should firstly be fixed, then the root cause of the problem identified, and a corrective action solution proposed and then implemented. Once activated, the effectiveness of the corrective action should be determined; if effective, then preventive action should be proposed to eliminate the original cause of the problem.

Practical Example

For each environmental aspect with prescribed maximum level, there must be a plan on how to handle nonconformance and corrective actions (Table 20.11).

The preceding text shows an example of how, with good practice, all environmental aspects limited by permits or consents can be managed if their operational control drifts outside the required limits.

Example
– Effluent discharge consent pH 6–8
Nonconformance
– Discharge for 3 h at pH 10
Nonconformance corrections
– Effluent diverted to buffer tank and pH adjusted
Corrective actions
– pH probe recalibrated, dosing system for pH adjustment recalibrated, and process control set point for corrective action adjusted

TABLE 20.10

	Nonconformity (NC) Detected	NC **Not Yet** Detected
Fix the problem	**Prevent** recurrence	*Eliminate* cause of potential NC
Correction	**Corrective action**	**Preventive action**

TABLE 20.11

Significant Aspect	Prescribed Max Level	Action If Nonconformance	Corrective and Preventive Action
Copper concentration in discharge water	0.2 mg/l	Report to supervisor	Investigate causes

Preventive actions
- Sources of effluent streams studied to see if they could be combined to reduce pH variations

20.2.6.4 Records

Records need to be maintained, as required by the system and organization to DEMONSTRATE conformance to the requirements of the EMS.

The organization needs to develop and support a method such that environmental records:

- Are legible, identifiable, and traceable to activity involved
- Are stored and maintained so they are easily retrieved and protected against spoilage
- Have retention and disposal times defined and recorded

Interpretation
Great care has to be taken with this clause because armed with this clause the auditor can expect to find records to demonstrate everything he may demand! For each "develop and support a method," expect to find a record or verbal evidence, confirmed by sampling.

When ISO 14001 was first launched in 1996, quality auditors had great difficulty trying to understand how an EMS could be audited, since there was no requirement for documented procedures everywhere, but they finally got their heads around the problem by looking for records. Records may take numerous forms and can range from handwritten notes on a notepad or in a logbook to pen recorder charts, to completed checklists, and to computer records in a database. Be wary however for computer-produced checklists as it is so easy just to change the date and produce a new record!

> **Practical Example**
>
> *A road drainage contractor was replacing foul water drains in an urban area, blocking pavements, diverting traffic, and making noise during day work hours to everyone's displeasure. Within the contract requirements, they needed to man a complaint line and keep records. When the auditor asked to see their communications log, which was a handwritten notebook, there were records for all months except October and November. When asked where the records had gone, the client stated they had fallen out of the notebook; on closer inspection, it was obvious that the pages had been ripped out. The auditor had no choice but to write an NC against lack of maintaining records properly.*

20.2.6.5 Internal Audit

The organization shall have methods in place for periodic EMS audits to determine if the EMS conforms to the plan, ISO 14001, and has been implemented and supported. Information on audit results should be provided to management.

An audit program and schedule need to be developed and implemented. The program needs to be based on the potential environmental impact of activity and results of previous audits ad using ISO 19011:2011 (ISO, 2011) as reference.

Audit procedures need to be developed to cover responsibilities and requirements for planning and conducting audits, reporting results, retaining associated records, audit criteria, scope, frequency, and methods.

Selection of auditors and conduct of audits shall ensure objectivity and impartiality of the audit process.

Interpretation
The EMS (internal) audit performs two key functions:

1. It aims to establish whether or not the organizations' EMS and activities conform to the requirements of ISO 14001.
2. It should determine whether the EMS has been properly implemented and maintained.

The audit framework should address all components of an organization's activities including its organizational structure, administration, operational procedures, documentation, and environmental performance.

The audit findings should reveal the effectiveness of the EMS and also identify and record any problems inherent in the system and make recommendations for remedial action.

The main factors to be considered when drawing up an audit program will include:

- Areas where risks of impacts are the greatest
- Where there is a history of nonconformance
- Any changes to the system or organization that could impact on the effectiveness of the system

The above should all be considered in order to provide the most coherent approach to auditing the system. It is also important, however, to consider the resources available for the audit; the organizational structure, especially the number and type of sites and departments involved; and the time required between audits for corrective action to be implemented.

ISO 14001 is not as specific on the frequency of auditing, but states that "the audit programme, including any schedule, shall be based on the environmental importance of the activity concerned and the results of previous audits."

Reference should be made to ISO 19011:2011 with respect to audit programs, procedures, and auditor competence. Auditor must not audit their own department, and they must have sufficient knowledge and experience in the audit area to be able to do their job. If there is insufficient knowledge or experience available, technical experts may be called in to support the audit.

> **Practical Example**
>
> *When auditing an infrastructure contractor, choosing project sites can be somewhat random, but normally, it would be best to choose those sites with high public concern, those sites that have shown problems before, any brand new project site, or any project where the environmentally relevant site staff have changed recently.*

20.2.7 Management Review

Top management shall review the management system at predefined intervals to check its relevance and effectiveness and to collect all relevant information.

To ensure good coverage, inputs to the management review should address:

- Results of internal audits
- Evaluations of compliance with legal and other requirements
- Communication from external interested parties, including complaints and
- environmental performance of the organization
- Extent to which O&T's have been met
- Status of corrective and preventive actions
- Follow-up actions from previous management reviews
- Changing circumstances including developments in legal and other requirements related to their environmental aspects
- Recommendations for improvement

Similarly, for a meaningful management review, the output should question whether the policy, objectives, and procedures need changing based on:

- Audit findings
- Changing operational conditions
- Commitment to continual improvement

Interpretation
Annex A.4.6 of ISO 14001 states that "The scope of the review should be comprehensive, though not all elements of an EMS need to be reviewed at once and the review process may take place over a period of time."

Review of the EMS is likely to include (Fig. 20.18):

- *Objectives, targets, and environmental performance*
- *EMS audit findings*
- *An evaluation of the effectiveness of the EMS*
- *An evaluation of the suitability of the environmental policy and the need for changes in light of:*
  - *Legislative changes*
  - *Changes in expectation or requirements of interested parties*
  - *Changes in the products, activities, or services of the organization*
  - *Advances in science and technology*
  - *Lessons learned from environmental incidents*
  - *Reporting and communication*
  - *Market considerations*

For a sound understanding of the improvement process, the definition of an EMS in ISO 14001 is important. The planning of environmental management (in this case environmental O&T's plus the environmental program) is an inseparable part of the system. Thus, the objectives are also subject to improvement (read: refinement). This refinement

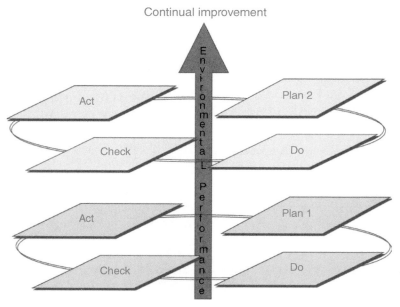

FIGURE 20.18

in principle takes place continually until the environmental policy of the company has been fully implemented.

The improvement process comes up again in the management review. In that respect, the environmental policy is from time to time reconsidered or adjusted with a view to external developments and the commitment to continual improvement. This breaks through the improvement cycle within the limits of the environmental policy established, and the continual improvement is aimed at achieving the new policy objectives at a higher level of ambition. This shows the importance of the management review in the overall improvement process.

20.3 ENVIRONMENTAL AUDITING

20.3.1 Introduction

20.3.1.1 What Is an Environmental Audit?

A number of definitions of an environmental audit have been developed. Examples include:

- **US EPA** (1986): A systematic, documented, periodic, and objective review by regulated entities of facility operations and practices related to meeting environmental requirements (US EPA, 1986)
- **International Chamber of Commerce** (1988): A management tool comprising a systematic, documented, periodic, and objective evaluation of how well environmental organization, management, and equipment are performing with the aim of helping to safeguard the environment by:
  - Facilitating management control of environmental practices
  - Assessing compliance with organization policies that would include meeting regulatory requirements (ICC, 1989)
- **ISO 14001**: EMS audit. Systematic and documented verification process to objectively obtain and evaluate evidence to determine whether an organization's EMS conforms to the EMS audit criteria set by the organization and communication of the results of the process to management (ISO, 2005)

Environmental auditing originated in the United States primarily in response to environmental laws and regulations introduced in the 1970s and 1980s. Since then, the use of environmental audits has become commonplace in Europe and the rest of the world.

The key concepts of environmental audits are that they are:

1. **Systematic**: Carried out in accordance with set protocols
2. **Documented**: Reports are prepared
3. **Periodic**: Conducted to an established schedule
4. **Objective**: Auditors have some independence
5. **Verification**: Evaluate compliance to requirements

An audit is a health check.

It seeks to verify that a system and its documentation are in compliance with the environmental management standard.

20.3.1.2 Is an Audit an Inspection?

An audit is different to an inspection. Inspection is one of the best tools available to identify problems and assess their risks before environmental accidents/incidents occur.

Informal inspections are simply the purposeful awareness of people as they go about their regular activities. Properly promoted and utilized, they can spot many potential problems as changes occur and work progresses. Informal inspections do have limitations, however, as they are not systematic and may miss things that take extra effort to find.

Planned inspections, as the name suggests, are more formalized and comprehensive; examples include:

- **General**—planned walk-through of an entire area. The inspectors look at everything to identify potential problem areas.
- **Critical parts/items**—components of machinery, equipment, materials, structures, or areas more likely than other components to result in a major problem when worn, damaged, abused, misused, or improperly applied.
- **Preuse**—many types of equipment have systems such as controls, emergency controls, and level alarms that are vital to safe operation; these systems can be damaged or become substandard between normal maintenance schedules or commissioning. For such equipment, preuse checks are essential.
- **Housekeeping**—housekeeping evaluations are a vital part of effective environmental management. This includes both *cleanliness* and *order*. Dirt and disorder are enemies of environment, safety, quality, productivity, and cost-effectiveness.

20.3.2 Principles of Auditing

ISO 19011:2011 (ISO, 2011) states that "Auditing is characterized by reliance on a number of principles. These principles should help to make the audit an effective and reliable tool in support of management policies and controls, by providing information on which an organization can act in order to improve its performance. Adherence to these principles is a prerequisite for providing audit conclusions that are relevant and sufficient and for enabling auditors, working independently from one another, to reach similar conclusions in similar circumstances."

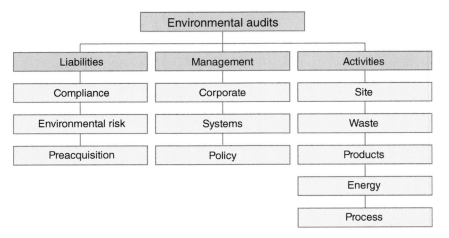

FIGURE 20.19

The six principles are:

1. **Integrity**: *the foundation of professionalism*
2. **Fair presentation**: *the obligation to report truthfully and accurately*
3. **Due professional care**: *the application of diligence and judgment in auditing*
4. **Confidentiality**: *security of information*
5. **Independence**: *the basis for the impartiality of the audit and objectivity of the audit conclusions*
6. **Evidence-based approach**: *the rational method for reaching reliable and reproducible audit conclusions in a systematic audit process*

20.3.3 Types of Environmental Audit

There are a wide variety of environmental audit types, as shown in Fig. 20.18 (ERM, 1996) (Fig. 20.19). There is often an inconsistency in terminology with many organizations referring to environmental audits as environmental reviews. Reviews tend to be the first assessment of an organization's environmental performance and compliance (see IER in Section 20.1.4.) (Thompson and Thrievel, 1986).

Liability audits check an organization's performance against any effects that are damaging to the environment and/or expose them to financial obligation. As an example, the preacquisition audit is a method of assessing the liabilities associated with a property transaction. The exercise is undertaken to ascertain what environmental liabilities and risks exist and to quantify, when appropriate, any liabilities found.

Issues typically addressed include:

- Contaminated land liability and remediation costs
- Latent environmental pollution
- Regulatory noncompliance and fines
- Costs associated with upgrading sites to achieve compliance

Management audits will address specific aspects of management.

As an example, a policy audit checks whether the policy is relevant and whether its claims are being actioned. Management system audits are the type of audit we will be focusing on in this chapter.

An **activity audit** will establish process efficiency with respect to inputs and outputs, for example, a waste audit should:

- Evaluate compliance
- Check procedure for identifying and quantifying waste
- Assess waste management practices and procedures
- Recommend improvement programs

20.3.3.1 First-Party Audit

There are three main types of management system audits: first party, second party, and logically third party (www.praxiom.com/ISO19001definitions.htm).

First party or "us on ourselves" is an INTERNAL AUDIT carried out in-house by an organization's own personnel in order to assess the implementation and effectiveness of the organization's management system.

The **primary objectives** of an internal audit are to discover whether procedures have been implemented, are understood, and are effective. The **secondary objectives** are to clear misconceptions, to train the users of the system, to identify training needs, and to improve internal communications up and down and additionally to seek improvements in procedures, forms, and methods used, thus ultimately asking the question, could the job be done better?

20.3.3.2 Second-Party Audit

Second party or "us on them or them on us" is an audit by a purchaser (or customer) on a vendor or potential vendor, that is, of one party, to a contract on the other, for the supply of goods/services. If the organization's customer is auditing the organization, the audit is EXTRINSIC (extrinsic means operating from without): "them on us."

If we are auditing an existing or potential subcontractor, the audit is EXTERNAL: "us on them."

In both cases, the basis of the audit is likely to be ISO 14001 and/or the purchasing/supply contract.

Objectives of a second-party audit are:

- To gain confidence in suppliers/contractors
- To minimize exposure to environmental risk
- To determine whether suppliers/contractors understand their environmental responsibilities
- To extend the range of environmental understanding, control, and product stewardship

20.3.3.3 Third-Party Audit

Third party or "them on us" is an audit for the purpose of EMS certification/registration. The auditor and the auditee have no contractual relationship such as purchaser or vendor of the goods and services. The basis of the audit is ISO 14001 with possible supplemental requirements from the accreditation body (e.g., RAB/UKAS or ISO/IEC 17021 (ISO, 2011)).

There may be special cases such as calibration laboratories where the basis is specified by the approval body. This is an EXTRINSIC AUDIT. Typical examples of third-party audits are first-stage assessments and certification audits as carried out by an accredited certification (or registration) body, for example, BVQI/UL.

20.3.3.4 Depth of Audit

There are generally two levels of audit:

1. Systems audit
   Carried out to establish if there is a management **system** in place and the degree to which its implementation complies with a prescribed quality or environmental management standard, for example, ISO 9001, ISO 14001, and so on.
2. Compliance audit
   Carried out against the requirements of **specific procedures** rather than system requirements to establish the degree of implementation and effectiveness of the procedures. Compliance audits are sometimes referred to as procedural audits.

Choosing the Right Level of Audit

The level of audit undertaken will, to a large extent, depend on the type of audit, for example, INTERNAL AUDITS will, on an ongoing basis, be COMPLIANCE audits, but these may be supplemented by a full SYSTEMS audit carried out on an annual basis for purposes of overall review.

EXTERNAL AUDITS may be carried out at SYSTEM level prior to award of a contract. They may become COMPLIANCE audits once a contract has been awarded and a set of operational control procedures have been agreed.

EXTRINSIC AUDITS may be initially SYSTEMS audits but may become COMPLIANCE audits on award of contract or certification.

20.3.4 Audit Objectives

An EMS audit should have defined objectives, for example:

- To determine conformance of an auditee's EMS with the EMS audit criteria
- To determine whether the auditee's EMS has been properly implemented and maintained
- To identify areas of potential improvement in auditee's EMS
- To assess the ability of the internal management review process to ensure continuing suitability and effectiveness of the EMS
- To evaluate the organization's EMS where there is a desire to establish a contractual relationship, for example, with a potential supplier or a joint venture partner

ISO 14001:2004 requires that an EMS audit performs two key functions:

1. It aims to establish whether or not the organization's EMS and activities conform to the requirements of the environmental management standard.
2. It should determine whether the EMS has been properly implemented and maintained.

The audit framework should address all components of an organization's activities including its organizational structure, administration, operational procedures, documentation, and environmental performance. The audit findings should reveal the effectiveness of the EMS and also identify and record any problems inherent in the system and make recommendations for remedial action.

As discussed earlier (Section 20.2.6.5), the audit should be used to optimize the control of significant environmental aspects and contribute to the overall aim of continual improvement (Fig 20.20).

The net result of implementing an EMS is continual improvement. The audit is part of this process as it is used to check that planned management processes are being implemented and are delivering desired results (Fig. 20.21).

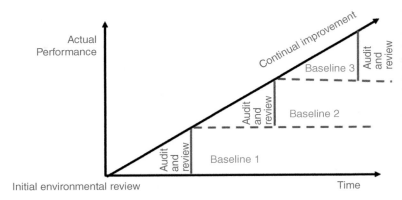

FIGURE 20.20

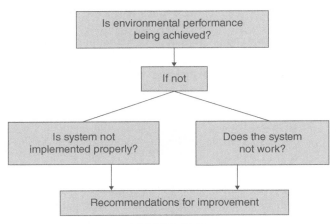

FIGURE 20.21

The audit can be seen as an information gathering exercise to ensure that the environmental management process is on track.

20.3.5 Audit Terminology

Audit objectives	Why the audit is being carried out
Audit criteria	Against what collected audit evidence is compared
Audit scope	Description of the extent and boundaries of an audit, in terms of physical locations, organization units, activities, and processes as well as the time period covered
Audit program	Arrangements for a set of one or more audits, planned for a specific time frame, and directed toward a specific purpose
Audit plan	Description of the activities and arrangements of an audit
Audit procedure	Method used by the auditor and his team to achieve the audit objectives

The audit procedure is used to guide the auditor through the on-site process. Such a guide adds consistency to the audit approach and can be used to train the audit team. The audit procedure represents the plan of what the auditor is to do to accomplish the objectives of the audit. It is an important tool since it not only serves as the auditors guide to collecting evidence but also as a record of the audit procedure completed by the teams. Audit procedures should identify the documents, records, and reports that need to be checked during the audit. The experience and qualification of auditors should be specified, in addition to means of ensuring their independence.

20.3.6 Audit Preparation

For the overview of the audit process, see Fig. 20.22.

The audit process involves seven key steps:

1. Initiating the audit
2. Initial document review
3. Preparing on-site audit activities
4. On-site activities
5. Report the audit
6. Audit completion
7. Audit follow-up

Preparation includes all the activities that take place before the actual audit. Auditing includes the activities involved in examining how well the environmental practices in the organization fit with the requirements outlined in its documented EMS and how well that system fits with the requirements of ISO 14001.

Reporting and follow-up include the steps involved in providing feedback to the auditee on the status of the system and the roles and responsibilities involved in following up and closing nonconformities.

20.3 ENVIRONMENTAL AUDITING

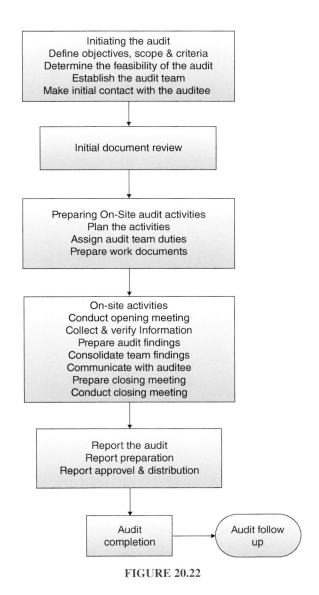

FIGURE 20.22

20.3.6.1 Why Prepare for the Audit?

Preparing helps the auditor plan the program and activities of the audit. By preparing, the auditor will ensure that any functional areas that need attention will not be missed. Without adequate preparation, a great deal of effort and time involved in the audit may be lost. Furthermore, an unprepared auditor would appear unprofessional and would lose credibility with the auditee.

20.3.6.2 Initiating the Audit

To have an effective internal auditing program, it is useful to establish and document a procedure for conducting internal environmental audits.

This internal auditing procedure should cover at least the following issues:

1. Objective, scope, criteria, and frequency of internal audits
2. What to do before, during, and after the audit
3. Who will conduct internal audits
4. Required qualifications of internal auditors

NOTE: The internal auditor must be independent of the function being audited: for example, a production manager cannot be the internal auditor for the production function.

Creating the Audit Plan

Once the overall audit program has been resolved, an audit plan for each specific audit should be drawn up; this plan should address the following issues:

   a. Audit objectives, scope, and criteria
   b. Audit locations, dates, expected times, and meetings
   c. Audit methodologies including sampling
   d. Roles and responsibilities, including guides and observers
   e. Resources for critical areas of the audit
   f. Auditee's representative
   g. Language
   h. Report topics
   i. Logistics and communication channels
   j. Measures to be taken if objectives cannot be achieved
   k. Confidentiality/security matters
      And this may also include:
   l. Any follow-up actions from a previous audit
   m. Any follow-up activities to the planned audit
   n. Coordination with other audit activities, in case of a joint audit

Remember a good audit plan should be designed to be flexible, to allow for changes in emphasis based on information gathered during the audit, and to permit effective use of resources. It should be communicated to the client organization, the audit team members, and the auditee. The client organization should then review and approve the plan.

If an organization is small and its environmental aspects/impacts few, it may be feasible to conduct the entire audit over a few days. However, for most companies, the internal audit is more likely to be part of a rolling audit program. This program should be planned with respect to the intervals between which specific parts of the organization will be audited and who will undertake the audits.

Practical Example

You have been requested to carry out a second-party audit on a potential contractor called Waste Management Co Ltd (WMC). This is one of four possible suppliers your organization is considering.

> This is a waste management organization able to remove and process food waste through recycling of a variety of waste streams and composting. You intend to select an organization to manage the waste elements of a new project your company is proposing.
>
> WMC consists of around 50 staff and undertakes a wide range of waste management contracts for a variety of commercial companies including facilities management organizations. WMC has its head office in London and supports waste management activities for commercial customers up to 50 miles from London.
>
> One of WMC's current customers, Downtown Foods Ltd (DFL), has agreed to allow you onto their site to audit WMC's activities. The DFL site is approximately 10 miles from WMC's head office.
>
> Although WMC has not sought registration to an environmental standard, they have indicated that they have an EMS that meets the requirements of ISO 14001:2004. Your senior projects manager requires you to plan for a one-day audit utilizing two auditors checking WMC (head office) and their activities at DFL on-site (Tables 20.12, 20.13, and 20.14).

Determining the Feasibility of the Audit

Before conducting the audit, the team leader should consider if there is:

- Sufficient and appropriate information for planning and conducting the audit
- Adequate cooperation from auditee
- Adequate time and resources to conduct the audit

If the audit is not feasible, an alternative should be proposed to the audit client, in agreement with the auditee.

Establishing the Audit Team

The audit team leader should establish from the scope the auditor's skills and industry experience needed to cover the environmental aspects of the organization's activities, products, and services. He should then determine if the proposed audit team members have those required competences and if not additional external resources should be drafted in to complement the team. The team leader should also determine that the team members are independent of the audit

TABLE 20.12 Practical Example: Key Personnel

Head Office	Site Staff
Managing director: CJ King	Site project manager: Y M Late
Senior projects manager (MR): Sheil B Right	Transport manager: L S Angels
HR manager: Zach M Hall	Site operative: Ruth Jones
Sales and purchasing manager: Will Buyitt	Site operative: Barry Humphreys
Office opening hours: 09:00–17:00	Site hours: 06:00–19:00

TABLE 20.13 Practical Example: Audit Plan

Auditee Organization	Waste Management Co Ltd(WMC)
Audit Reference No.: *WMC/1*	Date of Audit: *22/12/2013*

Audit objective:
To establish whether WMC is capable of working as a subcontractor to manage waste on-site and to subsequently dispose of it in an environmentally friendly way

Audit scope:
WMC's client Downtown Foods Ltd site in SE England

Audit criteria:
Local Environmental regulations and ABC Foods contract conditions

Previous audit findings/feedback/concerns/etc.:
One complaint to local press concerning littering of waste from one of WMC's disposal trucks

Specific auditor needs (equipment/facilities/passes/PPE):
Safety shoes and clean bill of health (other covering overalls, hats, and snoods will be provided on-site)

Audit team specific competency needs:
Understanding of food waste handling

Auditor(s) name:
Hyam de Boss(A), Victor Meldrew(B)

Audit schedule reference no.: *2nd-Party No3*

Prepared by: *Hyam de Boss*	Date: *12/8/2013*

TABLE 20.14 Practical Example: Audit Plan Timings

Day	Time	Auditor	Functions/Processes	Auditee	Clauses to be Audited
1	09:00	A+B	Opening meeting	All	
	09:30	A+B	Document and contract review at head office	MR	All
	10:00	B	Travel to site	YML	
	10:30	A	MR role and activities	MR	4.2, 4.3, 4.4, 4.5, 4.6
	10:30	B	Site tour and site activities	YML	4.4, 4.5
	11.30	A	Sales and purchasing	WB	4.4, 4.5
	13:30	A	HR function	ZMH	4.4, 4.5
	13:30	B	Site activities	RJ/BH	4.3, 4.4, 4.5
	14:30	A	Project reviews	MR	4.3, 4.4, 4.5, 4.6
	14:30	B	Transportation	LSA	4.3, 4.4, 4.5
	15:30	A	Senior management	CJK	4.1, 4.2, 4.3, 4.4, 4.5, 4.6
	15.30	B	Travel back to head office	YML	
	16:00	A+B	Audit team meeting and report writing		
	17.00	A+B	Closing meeting	All	

area and are capable of delivering an audit that meets the audit objectives.

Initial Contact with the Auditee
Before conducting the audit, the team leader should make contact with the client organization to:

- Establish communications with the auditee's representatives
- Confirm authority to conduct audit
- Provide information on audit objectives, scope, methods, and audit team composition, including technical experts
- Request access to relevant documents and records for planning purposes
- Determine relevant legal and contractual requirements and other requirements
- Confirm the extent of the disclosure and the treatment of confidential information
- Schedule the dates
- Determine any location-specific requirements for access, security, health and safety, or others
- Agree attendance of observers and need for guides
- Determine any areas of interest or concern to the auditee

20.3.6.3 Initial Document Review

An important part of preparing for the audit is becoming familiar with the organization's EMS documentation. The policies and procedures defined in these documents should provide a clear picture of the organization's objectives, commitments, and methods for achieving environmental goals.

Documentation should be reviewed to determine the organization's readiness for the audit. Sufficient information should be gathered to be able to prepare the audit activities and applicable work documents, for example, on processes and functions and to establish an overview of the extent of the system documentation to detect possible gaps. The review by the team leader should consider the size and complexity of the organization and may be complemented by a short site visit to confirm requirements.

The documents that should be reviewed include:

- The organization's environmental manual
- Relevant procedures, work instructions, environmental aspects and impacts, objectives, targets and programs, and any other documents relevant to the areas to be audited

EMS documents should be reviewed for compliance with the requirements set out in ISO 14001 and for clarity and comprehensiveness with regard to meeting the organization's environmental management policy, objectives, and targets that should be outlined in the environmental manual or in other procedures.

How to Review Documents
To review EMS documents, the auditor needs to refer to both the document he is reviewing and to the relevant section of ISO 14001, addressing the issues under review.

For example, if the auditor is reviewing the EMS procedure on management review, he must review it against the requirements specified in ISO 14001 under clause 4.6. The EMS procedure must, at the minimum, cover the requirements of this clause, though it may be quite a bit more detailed and should be more specific than ISO 14001.

The auditor should identify any ISO requirements that have not been addressed and point out any vague statements that need to be clarified, or that may need to be checked in other documentation, to make sure they have been adequately covered. VAGUE ISO REQUIREMENTS MUST BE DEFINED!!

Remember that the ISO standard has been written to accommodate the needs of different types of organization. Statements in the ISO standard are often open to interpretation; words such as "adequate," "as required," "appropriate," and "periodical" must be specifically defined in the organization's EMS documentation.

20.3.6.4 Preparing On-Site Activities

Planning for Site Activities   Once the organization's audit plan has been approved by management, it is ready to be implemented. The team leader has to schedule and prepare for the specific audits:

- Determine the agenda for the audit
- Confirm audit dates and times with the auditee
- Revise the program, if and when required
- Notify any other auditor on the team

The audit plan should include proposed dates and times for:

- Preaudit meeting with the auditee's management
- A tour of the facility/departments (if the auditor is not familiar with the area)
- Auditing activities in each department and/or for each clause to be audited
- Postaudit meeting with the auditee's management

The audit team leader should confirm the audit dates and times in writing at least 2–3 weeks before the proposed programmed date for the audit. If it is necessary to make changes to the audit agenda, the team leader should ask the auditee to confirm the revised program. Written confirmation of the audit agenda is preferable in order to prevent misunderstandings.

Assigning Audit Team Duties
The audit team leader should assign to each team member specific processes, activities, functions, or locations and in so doing should consider independence and competence of auditors, effective use of resources, as well as different roles and responsibilities of auditors, auditors in training, and technical experts.

These assignments should be made by the team leader in consultation with the audit team members. During the audit, the team leader may change work assignments to ensure achievement of the audit objectives.

Preparation of Work Documents
The audit team members should collect and review information relevant to their audit assignments and use this to prepare work documents; for reference and recording audit evidence, for example, checklists, audit sampling plans, and audit trails; and for recording information such as supporting evidence, audit findings, and records of meetings.

Records produced should be maintained at least until the completion of the audit. Confidential or proprietary information should be safeguarded by audit team members.

The checklist is the principal tool for use during site inspection and interview; it serves four primary purposes:

1. It guides the auditor to ensure complete coverage of audit issues.
2. It prompts the auditor as to which questions to ask.
3. It provides a means for recording observations and responses and simplifies preparation of the audit report.
4. It assures continuity from audit to audit.

The checklist cannot cover all possible situations, nor can all situations be adequately described by a "yes" or "no" response. Therefore, the auditor should supplement the checklist with written comments that record observations, response of the interviewees, or information from records or documents. The objective should be to record sufficient information during the audit, either as a checklist response or as a comment to be able to prepare the audit report. The auditor must not slavishly follow the checklist to the exclusion of making other observations or asking additional questions.

There are essentially two basic types of checklists: system-based checklists and compliance-based checklists.

Practical Example

To develop a system checklist based on ISO 14001 clauses, simply take each sentence/paragraph in the ISO standard and turn it into a question.

The following example is from ISO 14001, clause **4.4.4, EMS documentation**.

The environmental management system documentation shall include:
a. the environmental policy, objectives and targets;
b. description of the scope of the environmental management system;
c. description of the main elements of the environmental management system and their interaction and reference to related documents;
d. documents, including records, required by this International Standard; and

e. *documents, including records, determined by the organization to be necessary to ensure the effective planning, operation and control of processes that relate to its significant environmental aspects.*

Checklist Questions

1. *Does the EMS Documentation include:*
   a. *The environmental policy, objectives, and targets*
   b. *A description of the scope of the EMS*
   c. *A description of the main elements of the EMS and their interaction and any reference to related documents in other management systems or elsewhere*
   d. *Any documents, including records, required by ISO 14001*
   e. *Any documents, including records, determined by the organization to be necessary to ensure the effective planning, operation, and control of processes that relate to its significant environmental aspects*

Remember to include in the checklist questions that address requirements specified in the environmental manual and in operating procedures. Also remember to note where (what department/ function) and to whom each question on the checklist should be asked. These notes will help prepare for the audit (Table 20.15).

20.3.7 On-Site Activities

This section will go through the practical aspects of how to conduct an audit on-site.

20.3.7.1 Conducting the Opening Meeting

The opening meeting, however brief, is an important element of the audit. The meeting should be used by the auditor to explain the reasons for the audit and how the audit will be carried out and address any concerns the auditor may have.

The purpose of the opening meeting is to:

- Confirm agreement of all parties (e.g., auditee, audit team) to the audit plan
- Introduce the audit team
- Ensure that all planned audit activities can be performed

TABLE 20.15 A Sample of an EMS System Audit Checklist

Audit No: IEA001 Date: 9.3.13 Area: EMR
Auditor. WTS Audit Liaison Contact: DEMER
Audit of. ISO14001 Elements
Standards/References: ISO 14001:2004, ISO 19011:2011,

Requirement	Method of Demonstrating Requirement	(Y/N)	Evidence	Documents Examined	Contents Verified
4.2 Top management has defined the policy	Who signed the document and when?				
4.2.a The policy is appropriate to the nature, scale and environmental impacts of the organization	Does it describe all of the company's activities, products & services?				
4.2.b The company commits to continual improvement	IS there a statement committing to continual improvement?				
4.2.b The company commits to Pollution prevention	Is there a statement committing to pollution prevention?				
4.2.c The company commits to comply with relevant environmental legislation	Is there a statement committing to legislative compliance?				
4.2.d The policy provides a framework for setting objectives and targets	Are there broad statements concerning O & T?				
4.2.e The policy is documented, implemented and maintained	Is the policy visible, when was it last updated or reviewed?				
4.2.f The policy is communicated to all staff who work for on behalf of the organisation	Is there evidence of meeting where the policy has been discussed & disseminated, how are contractors briefed?				
4.2.g The Policy is available to the public	Is it their web site, or how else is it made available?				

The opening meeting is chaired by the audit team leader, covering the following:

- Introduction of the participants, including observers and guides, and an outline of their roles
- Confirmation of:
  - Audit objectives, scope, and criteria
  - Audit plan
  - Date and time for closing meeting
  - Any interim meetings
  - Any late changes
  - Presentation of methods to be used, including audit evidence, will be based on a sample of the information available
- Introduction of methods to manage risks to the organization that may result from the presence of the audit team members
- Identify communication channels between audit team and auditee
- Explain the language to be used during audit
- Acknowledge that the auditee will be kept informed of audit progress
- Ensure resources and facilities needed by the audit team are available
- Explain confidentiality and information security
- Explain the relevant health and safety and emergency and security procedures to the audit team
- Explain the method of reporting audit findings including grading
- Explain why the audit may be terminated
- Give closing meeting details
- How to deal with possible findings during audit
- How to handle feedback from the auditee on findings or conclusions of the audit, including complaints or appeals

20.3.7.2 Collect and Verify Information

Document Review The review of relevant documentation may be conducted before the interviews or site inspection or in conjunction with them. Generally, the audit team should spend their initial efforts in reviewing records and files to gain a better understanding of the key issues and any problem areas. Relevant documentation should be reviewed to determine conformity of the system, as far as is documented, with the audit criteria, and to gather information to support audit activities.

During the audit, information relevant to the audit objectives, scope, and criteria, including information relating to interfaces between functions, activities, and processes, should be collected by means of appropriate sampling and should be verified. Only information that is verifiable should be accepted as audit evidence. Audit evidence leading to audit findings should be recorded. If, during the collection of evidence, the audit team becomes aware of any new or changed circumstances or risks, these should be addressed by the team accordingly. Methods of collecting information include interviews, observations, and review of documents, including records.

Refer to Fig. 20.23 (audit method).

Interview
Many consider the interview phase of an audit to be the most important aspect of the process when auditors can obtain details not found in records, as well as get a feel for the operation. The interview is also one of the most problematic areas of the audit, involving interpersonal skills as well as technical knowledge. The findings of interviews should be verified through record checks, inspection, and verification interviews. It is important to elicit the auditee's cooperation during an audit interview, and there are certain things that can assist in obtaining this cooperation.

By offering an explanation of why certain questions are being asked, the auditee will generally be more cooperative because they understand that the auditor is not being

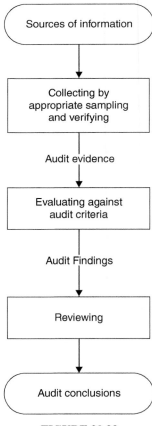

FIGURE 20.23

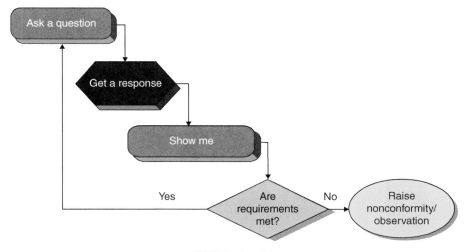

FIGURE 20.24

awkward, pedantic, and so on and because they appreciate that the auditor is considering them as a person.

It is quite common for questions to be initially misunderstood. This usually becomes clear when the expected answer is not received or if a muddled answer is received. Auditees are often unwilling to admit that they have not understood the question for fear of losing face in front of their supervisor. In this circumstance, sensitivity on the part of the auditor in recognizing this situation will help, and the question can be phrased differently or different language or jargon used. Sometimes, a question may have to be rephrased a number of times before it is understood; be patient!

Some individuals become highly stressed in an audit situation, and this can manifest itself as aggression, particularly if they feel defensive about their work/situation. If the situation appears to be getting out of hand, it is often diplomatic to break off (perhaps use a coffee break as an excuse) and return to the area at a later stage, when the auditee has had a chance to think about the questions and cooled down.

How to Ask Questions
Rudyard Kipling's poem "The Elephant Child" helps to remember one of the best techniques of asking questions:

I keep six honest serving men
(They taught me all I knew)
Their names were What and Why and When
And How and Where and Who

When used in the *open question* mode, these "six servants" elicit a response, but when used in the *closed question* style, they obtain only a partial yes/no response. The most useful questions to ask any auditee are the following:

- *What are you doing?*
- *How do you do it?*
- *Why do it that way?*
- *Where do you do it?*
- *When do you do it?*

especially if asked in this order!

Rhetorical questions like "*Are you following procedure xvz11?*" or "*Are you competent?*" should not be used.

Examples of closed questions include: "*Do you know where a copy of the documented procedure is kept?*" "*Do you know how to operate this machine?*" Both examples may elicit the answer "yes," but the auditor is left with asking further questions to verify the real facts. A more effective set of questions would be: "*Where is a copy of the documented procedure kept?*" "*How does this machine operate?*" These open questions demonstrate whether the auditee's knowledge is adequate or not.

Once the auditor has established some basic information, such as the location of the documented procedure, he/she may ask to see it or state boldly: "*Show me!*"

Alternatively, when a vague or oblique answer is received from the auditee, the auditor may ask: "*I don't understand, please explain how.*"

This concentrates the mind of the auditee to be more specific. Following a satisfactory description of the normal "process," the auditor may wish to question the auditee with regard to potential weaknesses of the system by asking: "*What if…?*" or "*Suppose that…?*"

The same approach can usefully be employed where there appears to be an apparent lack of objective evidence available. Auditing is like detective work; many skills and techniques need to be used to extract the information you require.

Communicating

In communicating the verbal message, there are five basic objectives or needs.

To ensure that the listener(s):

- HEARS what you tell them (or SEES what you show them)
- UNDERSTANDS what they have heard (or seen)
- AGREES with what they have heard or seen
- TAKES ACTION that agrees with your objectives
- PROVIDES FEEDBACK

When You Find Something That Appears to Be Wrong

- It is important not to criticize: it tends to make people defensive.
- Investigate the reasons for the noncompliance: do not make assumptions.
- Do not attempt to stop the job: this undermines supervision and is not the auditor's role.
- Do not undermine supervision: this can lead to tensions and lack of desire to cooperate and can damage future relations.
- It is vital to record specific detail of what is wrong: nonconformity notes must be accurate.
- Do record compliances as well as noncompliances: this helps in writing a positive audit report. Acknowledge the effort that has been made in developing the EMS, which helps to reduce any demotivation resulting from the identification of nonconformities.

Human Aspects of Auditing

- Be objective in your work and courteous in your manner.
- The auditor should be looking for nonconformities of the EMS: not carrying out a "witch hunt."
- Do persist if you believe that a nonconformity exists, but not to the extent of spending too much time.
- Do not get annoyed with the auditee.
- Make your own judgments.
- Do not take sides; it is better to remain independent in your judgments.
- Plan your time carefully.
- Arrive at the decision "on the spot." If you move away, you might forget the circumstances or the evidence may be "modified" in your absence.
- Be positive; do not set out to find faults in any specific area or activity.
- Ask the auditee to find the information; do not attempt to try and find the information yourself.

Site Inspection

With respect to site inspection, there are some concepts that should be addressed:

- Inspect remote areas; many sites have out of the way areas used for the storage or handling of hazardous materials.
- Make sure you walk around the outside of the site looking for fugitive emissions to air and water.
- If possible, observe operations during times of effluent discharges, during start-up/shutdown, or during maintenance periods when it is most likely that consent/permit limits are exceeded.
- Observe sampling or monitoring procedures.
- Don't let the guide misguide you around site to avoid local environmental "black holes"; you go where you want to go!

Audit Communications

The audit team should talk regularly to exchange information; assess audit progress; reassign work between audit team members, as needed; and raise any issue outside the scope that could affect the achievement of the audit objectives. The audit team leader should update the auditee with any changes to the audit plan, the objectives, scope, or the need for termination. The audit team should meet as needed to review the audit findings at appropriate stages during the audit; normal practice is to have team reviews at the end of each auditing day.

Guides and Observers

Guides and observers may accompany the audit team, but they should not influence or interfere with the conduct of audit. If this cannot be assured, the audit team leader should have the right to deny observers from taking part in certain audit activities.

Guides should assist the audit team and act on request of the audit team leader by assisting auditors in identifying individuals to participate in interviews and confirming timings, arranging access to specific locations of the auditee, and ensuring that rules concerning location safety and security procedures are known and respected by the audit team members and observers. They can also witness the audit on behalf of the auditee and provide clarification or assist in collecting information.

Collecting Audit Evidence

Sufficient audit evidence should be collected to be able to determine whether the auditee's EMS conforms to the EMS audit criteria. Audit evidence should be collected through interviews, examination of documents, and observation of activities and conditions. Indications of nonconformity to the EMS audit criteria should be recorded.

Information gathered through interviews should be verified by acquiring supporting information from independent sources, such as observations, records, and results of existing measurements. Nonverifiable statements should be identified as such.

The audit team should examine the basis of relevant sampling programs and the procedures for ensuring effective quality control of sampling and measurement processes used by the auditee as part of the organization's EMS activities.

Refer to Fig. 20.20 (how to audit).

The auditor should remember that as questions are asked and information or procedures are found missing, sufficient evidence to demonstrate that the finding is correct should be collected. A good practical tip is to try to think through how a nonconformance report would be worded as evidence is collected (see section "How to Write a Nonconformance Report").

20.3.7.3 Prepare Audit Findings

The audit team should review all of their audit evidence to determine where the EMS does not conform to the EMS audit criteria. The audit team should then ensure that audit findings are documented in a clear, concise manner and supported by audit evidence. Audit findings can indicate either conformity or nonconformity with audit criteria. When specified by the audit plan, individual audit findings should include conformity and good practices along with their supporting evidence, opportunities for improvement, and any recommendations to the auditee.

Nonconformity reports are required wherever the EMS fails in some specific way to meet any of the requirements of the standard. Corrective action is then required from the auditee, within a specified maximum time period, to rectify the detected weaknesses or omissions in the EMS. Normally, the lead auditor will discuss and agree the nonconformity with the departmental manager or supervisor and the EMS representative during the daily review of the audit.

REMEMBER: All nonconformities must be based on OBJECTIVE EVIDENCE!

How to Write a Nonconformance Report
There are three essential parts to writing a nonconformity report (NCR):

Requirement	This is the clause or subclause of the standard or the organization's own policy/codes of practice/procedure that should have been met by the EMS.
Failing	This should provide specific details of the failing, so that the auditee is completely clear about what is wrong with the system.
Evidence	These are the details of the objective evidence that is used to support the nonconformity. It should include job titles of people interviewed (but NOT individuals' names), documents, records, instruments used, and observations made.

Practical Example

This is based on Procedure for Environmental Policy, EMP4.2.
  Requirement
  • "3 Responsibility and Authority
    ○ The environmental policy shall be set by the site's senior management and shall be checked regularly, especially after environmental audits. It shall then be adapted as necessary."
  Finding
  • policy was not checked after audit IA3, IA4, IA5, but was after IA6

Based on these audit findings, the NCR could be written as follows:

Requirement: EMP4.2 requires that the environmental policy is checked regularly especially after environmental audits.
Finding: however, it had only been done once out of 4 audits
Evidence: witness record EMR 011 12/12/10 in which the update process was recorded

Another example of an audit finding:

• Effluent waters were discharged to sewer for 10 h on 12/3/13, above pH 10,
• (normal control range is specified in Operational Control Procedure EMS/OCP567 as 5.5–10, to ensure the discharge meets permit discharge level of max pH 11)

Requirement: Operational Control Procedure EMS/OCP567 specifies a normal effluent discharge pH control range of 5.5–10
Finding: Effluent waters were discharged to sewer for 10 hours on 12/3/13 above pH 10
Evidence: Witness Process Plant discharge Control chart for W/ending 15/3/13.

Thus, the NCR should answer the following questions—"who," "what," "when," "where," and "how"—substantiated with objective evidence. If the NCR is to prove useful, it must address all of the above. If insufficient information is provided, then the NCR will be of no value in terms of helping to improve the EMS. Typical information that should be included in the NCR is as follows:

• ID/title of persons involved
• Location
• Precise document references
• Process and/or equipment ID
• Clear and concise description of the nonconformance
• Auditor's name and date of audit

Classification of Nonconformities

Nonconformance (NC): nonfulfillment of a requirement as specified in a system manual, procedure, work instruction, and/or regulation

Observation/area for improvement: indication of a need for some action(s) to be taken either to prevent a situation deteriorating into a nonconformity or to promote an improvement

A major nonconformance: a systematic breakdown or a situation where key requirements of the audit criteria are missing, or a defect of significant consequence or impact, or a repeated problem in different areas, or lack of corrective action from a previous minor nonconformity (escalation). If a major is raised, it will prevent certification.

A minor nonconformance: a single or isolated lapse of control or a defect of minor consequence or impact that will not prevent certification.

Observation: a situation that could lead to a nonconformity or undesirable situation, if not addressed.

Opportunity for improvement: an improvement that could lead to enhanced effectiveness or efficiency in a process or activity.

During certification, there is no requirement from the certification body to follow up on observations, but for internal audits, they should be followed up where appropriate to bring system improvements.

20.3.7.4 Consolidate Team Findings

Before "closing the book," a quick completeness check is required to ensure the audit has been done to the required standard.

A quality check on audit trails is needed to ensure:

- All protocol questions have been asked.
- All records have been seen.
- All notes have been taken.
- All documents seen have been recorded.

In writing up the results and conclusions, care should be taken to distinguish between facts, observations, probabilities, interpretations, uncertainties, assessments, judgments, opinions, advice, and recommendations.

Good audit trail notes are essential; the quality check should ensure that someone could trace the audit from procedure → response → evidence → records → finding.

Communicate with the Auditee
Audit findings should be reviewed with the responsible auditee manager with a view to obtaining acknowledgement of the **factual basis of all findings** of nonconformity.

20.3.7.5 Closing Meeting

After completion of the audit evidence collection phase and prior to preparing an audit report, the audit team should hold a meeting with the auditee's management and those responsible for the functions audited. The main purpose of this meeting is to present audit findings to the auditee in such a manner as to obtain their clear understanding and acknowledgement of the audit findings.

Disagreements should be resolved, if possible, before the lead auditor issues the report. Final decisions on the significance and description of the audit findings ultimately rest with the lead auditor, though the auditee or client may still disagree with these findings.

Normally, the closing meeting will follow this agenda:

- Thanks
- Restate: objectives, criteria, and scope
- Present findings
- State conclusions
- Invite comments
- Agree follow-up (if any)
- Handover noncompliance forms and get them signed (retain a copy)

The findings presented in a closing meeting should not be a surprise to the auditee. To avoid arguments with a large group of senior people, it has been shown to be extremely effective to conduct a daily review with the front line management, explaining the findings and why they have arisen. This softens the blow for the closing meeting, allowing the auditee time to react, accept the findings, and prepare his defense in some cases!

20.3.8 Report the Audit

20.3.8.1 Report Preparation

The audit report is the formal output from the audit process. It should be a useful summary of what was looked at and what the findings were. The audit report is prepared under the direction of the lead auditor, who is responsible for its accuracy and completeness. The topics to be addressed in the audit report should be those determined in the audit plan. Any changes desired at the time of preparation of the report should be agreed upon by the parties concerned.

Content of Audit Report
The audit report should be dated and signed by the lead auditor. It should contain the audit findings and/or a summary thereof with reference to supporting evidence. Subject to agreement between the lead auditor and the client, the audit report may also include the following:

a. Identification of the organization audited and of the client
b. Agreed objectives, scope, and plan of the audit
c. Agreed criteria, including a list of reference documents against which the audit was conducted
d. Period covered by the audit and the date(s) the audit was conducted
e. Identification of the audit's representatives participating in the audit
f. Identification of the audit team members
g. A statement of the confidential nature of the contents
h. Distribution list for the audit report
i. A summary of the audit process including any obstacles encountered
j. Audit conclusions such as:
   - EMS conformance to the EMS audit criteria
   - Whether the system is properly implemented and maintained
   - Whether the internal management review process is able to ensure the continuing suitability and effectiveness of the EMS

The report should include accurate and complete information, positive comments, specific nonconformances, objective evidence, realistic action plans, and authorized approvals. The report should NOT include:

- Issues beyond the scope of the audit
- Sensitive or contentious issues
- Confidential information
- Issues not discussed in the audit

20.3.8.2 Report Approval and Distribution

The audit report should be sent to the client by the lead auditor. Distribution of the audit report should be determined by the client in accordance with the audit plan. The auditee should receive a copy of the audit report unless specifically excluded by the client. Additional distribution of the report outside the auditee's organization requires the auditee's permission. Audit reports are the sole property of the client; therefore, confidentiality should be respected and appropriately safeguarded by the auditors and all report recipients.

The audit report should be issued within the agreed time period in accordance with the audit plan. If this is not possible, the reasons for the delay should be formally communicated to both the client and the auditee, and a revised issue date established. If it is delayed, the reasons should be communicated to the auditee and the person managing the audit program. It should be dated, reviewed, and approved, as appropriate, in accordance with audit program procedures. It should then be distributed to the recipients as defined in the audit procedures or audit plan.

20.3.9 Audit Completion

The next stage of the audit process is to determine how the nonconformity is to be addressed. Hopefully, the auditee will undertake some form of root cause analysis to define the source of the problem. Once this has been identified, there should be some corrective action taken and ultimately some preventive measures put in place to stop the problem ever happening again. Each of these should be pinned down with appropriate implementation dates.

The audit is complete when all planned audit activities have been carried out or as otherwise agreed with the audit client (e.g., there might be an unexpected situation that prevents the audit being completed according to the plan).

Documents pertaining to the audit should be retained or destroyed by agreement between the participating parties and in accordance with audit procedures and applicable requirements.

Unless required by law, the audit team and the person managing the audit program should not disclose the contents of documents, any other information obtained during the audit, or the audit report to any other party without the explicit approval of the audit client and, where appropriate, the approval of the auditee. If disclosure of the contents of an audit document is required, the audit client and auditee should be informed as soon as possible.

Lessons learned from the audit should be entered into the continual improvement process of the management system of the audited organizations.

20.3.10 Audit Follow-Up

Either shortly after the audit or during the next cycle of audits, the auditor should try to establish the effectiveness and timeliness of any actions taken to solve problems.

How to follow-up on nonconformances:

- Examine documents
- Sample implementation/actual practice
- Check records
- Raise and report any new nonconformances (related to the original nonconformances only) and follow up again at a later date

If any corrective action has been taken that cannot be shown to be effective, then the auditor needs to follow his own procedures on what to do next. Normally, this would involve rejecting the corrective action plan if it was seen to be inadequate before any action was taken. If the action was taken and shown to be inadequate, another NC would need to be raised and the process started again.

All working documents, draft, and final reports pertaining to the audit should be retained by agreement between the client, the lead auditor, and the auditee in accordance with any applicable requirements as environmental records.

These records can subsequently be used as evidence of continual improvement.

20.3.11 Auditor Competence

Addressed in ISO 19011:2011 clause 7: Competence and evaluation of auditors. The qualification criteria addressed by the standard are applicable to both internal and external auditors, although internal auditors might not need to meet all the criteria, depending upon the nature of the organization. Auditors should be independent from the area/activity being audited and have experience and expertise including some or all of:

- Management systems and standards
- Environmental auditing
- Environmental legislation
- Technical knowledge of area/activity
- Environmental science and technology

With regard to education, the requirements are for at least secondary or equivalent with 5-year work experience or a degree with 4-year work experience. Formal training should address environmental science, technical aspects of facility operations, environmental law, EMS and standards, and audit procedures and techniques. These areas of knowledge require periodic refreshment, thus ensuring maintenance of competence.

On the job training must comprise the equivalent of 20 work days of auditing and a minimum of 4 audits within a 3-year period. Other requirements are demanded of a lead auditor, which may either be demonstrated through interviews or completion of at least three additional audits (in one acting as lead auditor).

Further details of the competence requirements for EMS and certification auditors have been recently published (ISO, 2012). This document takes the competence requirements of ISO 19011:2011 and spells out the skills and experience related to environmental aspects and other issues.

Anyone wishing to become an environmental auditor should refer to the registration schemes offered by the Institute of Environmental Management and Assessment (IEMA; IEMA.net) and the International Register of Certificated Auditors (IRCA; www.irca.org) in the United Kingdom and the National Registry of Environmental Professionals (NREP; NREP.org) in the United States. Professional qualifications for competent environmental auditors leading to chartered environmentalist (CEnv) can be pursued with the IEMA and the Chartered Institution of Water and Environmental Management (CIWEM; www.ciwem.org).

REFERENCES

British Standards Institution (2000) *BSEN: ISO9001:2000, Quality Systems*, BSI, London.

British Standards Institution (2002) *BS: IEC61882-2002, Hazard and Operability Studies—Application Guide*, BSI, Bristol.

British Standard Institution (2007) *BS: OHSAS18001-2007, Occupational Health and Safety management systems-specification*, BSI 389, London.

British Standardization Institute (1999) *Occupational Health and Safety Management Systems-Specification, OHSAS18001:2007*, BSI, London.

Chemistry Industry Association of Canada (1985) *Responsible Care*, CIAC, Ottawa.

Coopers & Lybrand (1994) *Your Business and the Environment: A DW Review for Companies*, Coopers & Lybrartd/Business in the Environment, Melbourne, Australia.

Doran, G. T. (1981) There's a S.M.A.R.T. way to write management's goals and objectives. *Manag. Rev.* 70(11): 35–36.

Environmental Resources Management (1996) Environmental audit and assessment: concepts, measures, practices and initiatives. Environmental Resources Management, SNH Review No. 46. SNH, Perth.

European Commission (1987) Council Directive 88/77/EEC. *Official J.* L36: 33–61.

European Commission (1999) Council Directive 1999/96/EC. *Official J.* L44: 1–155.

European Commission (2001) Commission Directive 2001/27/EC. *Official J.* L107: 10–23.

European Commission (2005) Commission Directive 2005/55/EC. *Official J.* L275: 1–163.

European Commission (2009) Regulation 595/2009 E.C. *Official J.* L188: 1–13.

http://www.earthlyissues.com/airpollution.htm. Accessed March 22, 2015.

International Chamber of Commerce (1989): *Environmental Auditing*, Publication 468. ICC, Paris.

International Chamber of Commerce (2000) *Business Charter for Sustainable Development*, Publication 210/356. ICC, Paris.

International Standards Organisation (2000), *ISO 9001:2000, Quality Management Systems–Requirements*. ISO, Geneva.

International Standards Organisation (2004a) *ISO 14001: Environmental Management Systems: Requirements with Guidance for Use*. ISO, Geneva.

International Standards Organization (2004b) *ISO14004:2004 Environmental Management Systems – General Guidelines on Principles, Systems and Support Techniques*. ISO, Geneva.

ISO (2011) ISO-19011 *Guidelines for Auditing Management Systems*. ISO 19011:2011(E), ISO, Geneva.

ISO (2012) *ISO/IEC TS 17021-2: 2012 Conformity Assessment - Requirements for Bodies Providing Audit and Certification of Management Systems. Part 2: Competence Requirements for Auditing and Certification of Environmental Management Systems*. ISO, Geneva.

FURTHER READING

Det Norske Veritas (1966) Loss control management. http://www.dnv.nl/binaries/managing%20risk%20through%20sms%20%20eng_tcm141-444092.pdf. Accessed March 22, 2015.

European Commission (2011) Regulation 582/2011. E.C. *Official J.* L167: 1–168.

21

RISK ASSESSMENT AS A TOOL IN ENVIRONMENTAL MANAGEMENT

KOFI ASANTE-DUAH
Environmental Protection Administration, District Department of the Environment, Washington, DC, USA

Abstract: Risk assessment generally promises a systematic way for developing technically sound strategies to aid the management of environmental contamination problems. It typically serves as a tool that can be used to properly organize, structure, and compile scientific information in order to help identify existing hazardous situations or problems, anticipate potential problems, establish priorities, and provide a basis for regulatory controls and/or corrective actions. Indeed, risk assessment represents one of the fastest evolving tools available for developing appropriate and cost-effective strategies to facilitate environmental management decisions.

This chapter provides a broad overview of the many facets of risk assessments as relevant to environmental contamination issues and related management problems. The discussion includes several important concepts, tools, and techniques/methodologies that can be used to address environment-related risk management problems in a consistent, efficient, and cost-effective manner—while also recognizing that a key underlying principle of risk assessments is that some risks are invariably tolerable.

Keywords: Acceptable risk (or *de minimis* risk), Baseline risk assessment, Comparative risk assessment, Consequence/impact assessment, Dose-response assessment, Ecological risk assessment, Effects assessment, Exposure, Exposure assessment, Exposure-response, Exposure route, Exposure scenario, Hazard, Hazard assessment, Human health risk assessment, Hazard identification, Individual risk, Risk, Risk assessment, Risk assessment process, Risk characterization, Risk-cost-benefit, Risk estimation, Risk grouping, Risk management, Societal risk, Toxicity assessment, Vulnerability analysis, Worst-case scenario.

21.1	Introduction	504
21.2	What is Risk Assessment?	504
	21.2.1 Purpose and Attributes	504
	21.2.2 Utility to Environmental Management Systems	505
21.3	Fundamental Principles and Concepts in Risk Assessment	506
	21.3.1 Key Nomenclatural Components: Hazards, Exposures, and Risks	506
	21.3.2 General Nature of Risk Measures: Qualitative versus Quantitative Descriptors	506
	21.3.3 The "Safety Paradigm" in Risk Assessment: Erring on the Side of Safety	507
	21.3.4 Risk Group Categorization: The Nature of Target Population Groups	507
	21.3.5 Risk Acceptability and Risk Tolerance Concepts in Environmental Risk Management	508
	21.3.6 Risk Assessment versus Risk Management in Environmental Management Programs	509
21.4	The Risk Assessment Framework and Paradigm	509
	21.4.1 Principal Components of a Risk Assessment	510
	21.4.2 Procedural Elements of the Risk Assessment Process	510
	21.4.2.1 Hazard Identification and Accounting	510
	21.4.2.2 Exposure–Response Evaluation	511
	21.4.2.3 Exposure Assessment and Analysis	511
	21.4.2.4 Risk Characterization and Consequence Determination	511
	21.4.3 General Types of Analytical Protocols Commonly Used in the Appraisal of Environmental Contamination Problems	511
	21.4.3.1 Technical Approach to Human Health Endangerment Assessments	511
	21.4.3.2 Technical Approach to Ecological Endangerment Assessments	513

An Integrated Approach to Environmental Management, First Edition. Edited by Dibyendu Sarkar, Rupali Datta, Avinandan Mukherjee, and Robyn Hannigan.
© 2016 John Wiley & Sons, Inc. Published 2016 by John Wiley & Sons, Inc.

21.4.4	Risk Appraisal Strategies for Effectual Risk Management	514
	21.4.4.1 Baseline Risk Assessments	515
	21.4.4.2 Comparative Risk Assessments	515
	21.4.4.3 Risk–Cost–Benefit Optimization	515
21.5	Using Risk Assessment to Shape Environmental Management Decisions	516
	21.5.1 General Scope of the Practical Application of Risk Assessments to Environmental Contamination Problems	516
	21.5.2 General Implementation Framework for the Application of Risk Assessments to Environmental Contamination Problems	517
21.6	Conclusion	518
References		519

21.1 INTRODUCTION

Environmental management systems and issues represent complex global sectors that require careful analyses, as well as the use of systematic and effective analytical tools. For instance, the matter of environmental contamination and/or impacts may pose significant risks to the general public because of the potential health and environmental effects—among several other things; in particular, risks to both human and ecological health that arise from environmental contamination and/or impacts are a matter of significant concern worldwide. Thus, the effective management of such environmental contamination and/or impact problems represents a very important environmental concern that calls for judicious and reliable methods of evaluation.

Risk assessment represents one of the best analytical and pragmatic tools available for developing appropriate and cost-effective management strategies to aid environmental protection and restoration, as well as facilitate related environmental management decisions. In fact, to ensure public health and environmental sustainability, decisions relating to environmental contamination and/or impacts management should be based on systematic and scientifically valid processes—such as can be realized through the use of some form of risk assessment.

This chapter provides a broad overview on the utility of risk assessment as an essential tool necessary to help establish effective environmental management programs—with particular emphasis on risk management and corrective action needs in relation to chemical exposure problems arising from environmental contamination and/or impacts.

21.2 WHAT IS RISK ASSESSMENT?

Several somewhat differing definitions of risk assessment have been published in the literature by various authors to describe a variety of risk assessment methods, processes, and/or protocols (see, e.g., Asante-Duah, 1996, 1998, 2002; Bates, 1994; Cohrssen and Covello, 1989; Conway, 1982; Cothern, 1993; Covello and Merkhofer, 1993; Covello and Mumpower, 1985; Covello et al., 1986; Crandall and Lave, 1981; Davies, 1996; Douben, 1998; Glickman and Gough, 1990; Gratt, 1996; Hallenbeck and Cunningham, 1988; Kates, 1978; Kolluru, 1994; Kolluru et al., 1996; LaGoy, 1994; Lave, 1982; Neely, 1994; NRC, 1982, 1983, 1994; Pollard et al., 2002; Richardson, 1990, 1992; Rowe, 1977; Turnberg, 1996; USEPA, 1984a, b; van Ryzin, 1980; Whyte and Burton, 1980). In a generic sense, risk assessment may be considered to be a systematic process for arriving at estimates of all the significant risk factors or parameters associated with an entire range of "failure modes" and/or exposure scenarios in connection with some hazard situation(s).

Risk assessment of environmental systems entails the evaluation of all pertinent scientific information to enable a description of the likelihood, nature, and extent of harm to the health of living organisms as a result of exposure to a variety of environmental stressors. The process is used to evaluate the collective demographic, geographic, physical, chemical, biological, and related factors associated with environmental stressor or impact problems; this helps to determine and characterize possible risks to public health and the environment. The overall objective of such an assessment is to determine the magnitude and probability of actual or potential harm that the environmental stressor problem poses to human health and the environment.

Almost invariably, every process for developing effectual environmental risk management programs should ideally incorporate some concepts or principles of risk assessment. For instance, decisions on restoration plans for potential environmental contamination problems will include, implicitly or explicitly, some elements or attributes of risk assessment. Ultimately, the risk assessment process seeks to estimate the likelihood of occurrence of adverse effects resulting from exposures of human and/or ecological receptors to chemical, physical, and/or biological agents or stressors present in the receptor environments.

21.2.1 Purpose and Attributes

The purpose of the risk assessment process is to provide, insofar as possible, complete information set to risk managers—so that the best possible decision can be made concerning a potentially hazardous situation. Indeed, as Whyte and Burton (1980) succinctly indicate, a major objective of risk assessment is to help develop risk management decisions that are more systematic, more comprehensive, more accountable, and more self-aware of appropriate programs than has often been the case in the past. Ultimately, the principal objective of a typical risk assessment is to provide a basis for actions aimed at minimizing the impairment of the environment and/or of public health, welfare, and safety.

Risk assessment—as used for developing appropriate strategies for environmental contamination management decisions—seeks to answer three basic questions, namely, (Asante-Duah, 1998, 2002):

- What could potentially go wrong?
- What are the chances for the "event" to happen?
- What are the anticipated consequences if the above "event" should indeed happen?

Invariably, a complete analysis of risks associated with a given situation or activity will generate answers to these questions. Subsequently, a decision can be made as to whether any existing risk is sufficiently high to represent an environmental or public health concern, and if so, to determine the nature of risk management actions. Appropriate mitigative activities can then be initiated by implementing the necessary corrective action and risk management programs.

In general, the risk assessment process provides a framework for developing the risk information necessary to assist risk management decisions; information developed in the risk assessment will typically facilitate decisions about the allocation of resources for safety improvements and hazard/risk reduction. Also, the analysis will generally provide decision makers with a more justifiable basis for determining risk acceptability, as well as aid in choosing between possible corrective measures developed for risk mitigation programs.

By and large, the information generated in a risk assessment is often used to determine the need for and the degree of mitigation required for environmental contamination and/or impact problems. Furthermore, risk assessment can be used to compare the risk reductions afforded by different remedial or risk control strategies.

The type and degree of detail of any particular risk assessment depend on its intended use. In fact, its purpose will shape the data needs, the protocol, the rigor, and related efforts. Overall, the processes involved in any risk assessment usually require a multidisciplinary approach—often covering several areas of expertise in most situations.

It is noteworthy here that, there are inherent uncertainties associated with all risk assessments. This is due in part to the fact that the analyst's knowledge of the causative events and controlling factors usually is limited, and also because, to a reasonable extent, the results obtained depend on the methodology and assumptions used. Furthermore, risk assessment can impose potential delays in the implementation of appropriate corrective measures—albeit the overall gain in program efficiency is likely to more than compensate for any delays.

In fact, the benefits of risk assessments designed to facilitate environmental risk management decisions outweigh any possible disadvantages; still, it must be recognized that this process will not be without tribulations. Its use, however, is an attempt to widen and extend the decision maker's knowledge base and thus improve the decision-making capability. Thus, the method offers a well-deserving recognition as an important environmental risk management tool.

21.2.2 Utility to Environmental Management Systems

Risk assessment has several specific applications that could affect the type of decisions to be made in relation to environmental risk management programs. The broad applications of risk assessment that are particularly relevant to the management of environmental contamination and chemical exposure problems include the following (Asante-Duah, 1998, 2002):

- Identification and ranking of all existing and anticipated potential hazards
- Explicit consideration of all current and possible future exposure scenarios
- Qualification and/or quantification of risks associated with the full range of hazard situations, system responses, and exposure scenarios
- Identification of all significant contributors to the critical pathways, exposure scenarios, and/or total risks
- Determination of cost-effective risk reduction policies via the evaluation of risk-based remedial action alternatives and/or the adoption of efficient risk management and risk prevention programs
- Identification and analysis of all significant sources of uncertainties

Appropriately applied, risk assessment techniques can indeed be used to estimate the risks posed by environmental hazards under various exposure scenarios and to further estimate the degree of risk reduction achievable by implementing various scientific remedies. For instance, the application of the risk assessment process to environmental contamination problems will generally serve to document the fact that risks to human health and the environment have been evaluated and incorporated into a set of appropriate response actions.

In its application to environmental contamination and impact problems, the risk assessment process is essentially used to organize, structure, and compile the scientific information that is necessary to support environmental risk management decisions. The approach is used to help identify existing hazardous situations or problems, anticipate potential problems, establish priorities, and provide a basis for policy decisions about regulatory controls and/or corrective actions.

Ultimately, the risk assessment process is intended to give the risk management team the best possible evaluation of all available scientific data, in order to arrive at justifiable and defensible decisions on a wide range of issues. For example, to ensure public safety in chemical exposure situations, receptor exposure point concentrations must be below some stipulated risk-based maximum exposure level—typically established through a risk assessment process. In fact, risk-based decision making will generally result in the design of better environmental risk management programs. This is because risk assessment can produce more efficient and consistent risk reduction policies. It can also be used as a screening device for setting priorities.

21.3 FUNDAMENTAL PRINCIPLES AND CONCEPTS IN RISK ASSESSMENT

In order to adequately evaluate the risks associated with a given hazard situation, several concepts are usually employed in the processes involved. In fact, risk assessment methods commonly encountered in the literature of environmental and public health risk management and/or relevant to the management of environmental contamination and impact problems require a clear understanding of several fundamental issues. Some of the key fundamental principles and concepts that will facilitate a better understanding of the risk assessment process and related application practices—and that may also affect environmental risk management decisions—are introduced in the following sections.

21.3.1 Key Nomenclatural Components: Hazards, Exposures, and Risks

Hazard is that object with the potential for creating undesirable adverse consequences, *exposure* is the situation of vulnerability to hazards, and *risk* is considered to be the probability or likelihood of an adverse effect, or an assessed threat, due to some hazardous situation.

In the typical relationship between these fundamental elements, risk represents a measure of the probability and severity of adverse consequences from an exposure of potential receptors to hazards—and may simply be represented by the measure of the frequency of an event. In fact, it is the likelihood to harm as a result of exposure to a hazard that distinguishes risk from hazard. For example, a toxic chemical that is hazardous to human health does not present a risk unless human receptors/populations are exposed to such a substance (Asante-Duah, 2002). That is, potential receptors will have to be exposed to the hazards of concern before any risk could be said to exist.

Potential risks are estimated by considering the probability or likelihood of occurrence of harm, the intrinsic harmful features or properties of specified hazards, the population at risk, the exposure scenarios, and the extent of expected harm and potential effects. The integrated assessment of hazards, exposures, and risks are indeed a very important contributor to any decision that is aimed at adequately managing a given hazardous situation. Consequently, the availability of adequate and complete information sets on all these individual components is an important prerequisite for producing sound hazard, exposure, and risk assessments.

21.3.2 General Nature of Risk Measures: Qualitative versus Quantitative Descriptors

Measuring parameters or metrics used in risk analysis take various forms, depending on the type of problem, degree of resolution appropriate for the situation on hand, and the analysts' preferences. In general, the risk parameter may be expressed in quantitative terms—in which case it could, for instance, take on values from zero (associated with certainty for "no-adverse effects") to unity (associated with certainty for "adverse effects" to occur). In several other cases, risk is only described qualitatively—such as by use of descriptors like "high," "moderate," "low," etc.; or indeed, the risk may be described in semiquantitative/semiqualitative terms.

In the risk assessment of environmental contamination or impact issues, quantitative tools are often used to better define exposures, effects, and risks in the broad context of risk analysis. Although the utility of numerical risk estimates in risk analysis has to be appreciated here, these estimates should be considered in the context of the variables and assumptions involved in their derivation—and indeed in the broader context of scientific opinions, miscellaneous "confounding" factors, and actual exposure conditions. Consequently, directly or indirectly, qualitative descriptors also become part of the quantitative risk assessment process. For instance, in evaluating the assumptions and variables relating to both toxicity and exposure conditions for a chemical exposure problem, the risk outcome may be provided in qualitative terms—albeit the risk levels are expressed in quantitative terms.

By and large, the attributable risk for any given problem situation can be expressed in qualitative, semiquantitative, and/or quantitative terms. In any event, the risk qualification or quantification process will normally rely on the use of several measures, parameters, and/or tools as reference yardsticks (Asante-Duah, 2002). Individual lifetime risk (represented by the probability that the individual will be subjected to an adverse effect from exposure to identified hazards) is about the most commonly used measure of risk in environment-related impact studies.

21.3.3 The "Safety Paradigm" in Risk Assessment: Erring on the Side of Safety

Many of the parameters and assumptions used in hazard, exposure, and risk evaluation studies tend to have high degrees of uncertainties associated with them. Thus, it is common practice for safe design and analysis to model risks such that risk levels determined for management decisions are preferably overestimated. Such conservative estimates (also, often cited as "worst-case" or "plausible upper-bound" estimates) used in risk assessment are based on the premise that pessimism in risk assessment (with resultant high estimates of risks) is more protective of public health and/or the environment. For example, in performing risk assessments, scenarios have often been developed that will reflect the worst possible exposure pattern; this notion of *"worst-case scenario"* in the risk assessment refers to the event or series of events resulting in the greatest exposure or potential exposure.

On the other hand, gross exaggeration of actual risks could lead to poor decisions being made with respect to the oftentimes very limited resources available for general risk mitigation purposes. Thus, after establishing a worst-case scenario, it is often desirable to also develop and analyze more realistic or "nominal" scenarios, so that the level of risk posed by a hazardous situation can be better bounded—by selecting "best" or "most likely" sets of assumptions for the risk assessment. But in deciding on what realistic assumptions are to be used in a risk assessment, it is imperative that the analyst chooses parameters that will, in a worst case, result in erring on the side of safety.

Notwithstanding the aforementioned discussions, it is noteworthy that there also are a variety of ways and means of making risk assessments more realistic—that is, rather than the dependence on wholesale compounded conservative assumptions (see, e.g., Anderson and Yuhas, 1996; Burmaster and von Stackelberg, 1991; Cullen, 1994; Maxim, in Paustenbach, 1988). Among other things, there usually is the need to undertake sensitivity analyses—including the use of multiple assumption sets that reflect a wider spectrum of exposure scenarios. This is important because management decisions or strategies that are based on the so-called upper-bound estimate or worst-case scenario may address risks that are almost nonexistent and impractical. Indeed, risk assessment using extremely conservative biases do not provide risk managers with the quality information needed to formulate efficient and cost-effective management strategies. Also, using plausible upper-bound risk estimates or worst-case scenarios may lead to the spending of scarce and limited resources to regulate or control insignificant risks—while more serious risks are probably being ignored. Thus, conservatism in individual assessments may not be optimal or even conservative in a broad sense if some sources of risk are not addressed simply because others receive undue attention.

Consequently, the overall recommendation is to strive for accuracy rather than conservatism. Indeed, it has been recognized that a continued reliance on conservative (viz., worse case) assumptions tends to distort risk assessment results—producing estimates that may overstate likely risks by several orders of magnitude. Thus, it becomes difficult to discern serious hazards from trivial ones—potentially distorting prioritization for resource allocation in risk management scenarios that involve decisions or choices between multiple or competing problem situations, alternative remedial actions, competing risk control or abatement measures, etc. Ultimately, the unfortunate outcome of using upper-bound estimates based on compounded conservative assumptions is that, it may lead to the regulation and control of insignificant or less significant risks whiles ignoring more serious risks.

21.3.4 Risk Group Categorization: The Nature of Target Population Groups

Broadly, three types of "risk groupings" can be identified for most risk management situations, namely,

- Risks to individuals
- Risks to the general population
- Risks to highly exposed or highly sensitive subgroups of a population

The latter two categories may indeed constitute "societal" or "group" risk—representing population risks associated with more than one person or the individual receptor.

In the application of risk assessment to environmental and public health risk management programs, it often becomes important to distinguish between "individual" and "societal" risks, in order that the most appropriate category can be used in the analysis of case-specific problems. *Individual risks* are considered to be the frequency at which a given individual could potentially sustain a given level of adverse consequence from the realization or occurrence of specified hazards. *Societal risk*, on the other hand, relates to the frequency and the number of individuals sustaining some given level of adverse consequence in a given population due to the occurrence of specified hazards. The population risk provides an estimate of the extent of harm for the population or population segment being addressed.

Individual risk estimates represent the risk borne by individual persons or receptors within a population—and are more appropriate in cases where individuals face relatively high risks. However, when individual risks are not inequitably high, then it becomes important during resource allocation to consider possible society-wide risks that might be relatively higher. Indeed, risk assessments almost always deal with more than a single individual. Frequently, individual risks are calculated for some or all of the persons or receptors in the population being studied and are then put into the context of where they fall in the distribution of risks for the entire population.

It is noteworthy that, at an individual level, the choice of whether or not to accept a risk is primarily a personal decision (in the case of human receptors). However, on a societal level (where values tend to be in conflict and decisions often produce "winners" and "losers"), the decision to accept or reject a risk is much more difficult (Cohrssen and Covello, 1989). In fact, no numerical level of risk will likely receive universal acceptance, but also eliminating all risks is an impossible task—especially for our modern society in which people have become so accustomed to numerous "hazard-generating" luxuries of life. Indeed, for many activities and technologies of today, some level of risk has to be tolerated in order to gain the benefits of the activity or technology. Consequently, levels of risk that may be considered as tolerable or relatively "safe enough" should typically be identified/defined—at least on the societal level—to facilitate rational risk management and related decision-making tasks. In this process, it must be acknowledged that individuals at the high end of a risk distribution are often of special interest to risk managers—that is, when considering various actions to mitigate the risk; these individuals often are either more susceptible to the adverse health effect than others in the population or are highly exposed individuals or both.

21.3.5 Risk Acceptability and Risk Tolerance Concepts in Environmental Risk Management

An important concept in environmental risk management is that there are levels of risk that are so great that they probably must not be allowed to occur under any set of circumstances and/or costs, and yet there are other risk levels that are so low that they are not worth bothering with even at fairly insignificant costs; these are known, respectively, as *de manifestis* and *de minimis* levels (Kocher and Hoffman, 1991; Suter, 1993; Travis et al., 1987; Whipple, 1987). Risk levels between these bounds are typically balanced against costs, technical feasibility of mitigation actions, and other socioeconomic, political, and legal considerations—in order to determine their acceptability or tolerability.

Risk is *de minimis* if the incremental risk produced by an activity is sufficiently small so that there is no incentive to modify the activity (Cohrssen and Covello, 1989; Covello et al., 1986; Fischhoff et al., 1981; Whipple, 1987). These are risk levels judged to be too insignificant to be of any social concern or to justify use of risk management resources to control them, compared with other beneficial uses for the often limited resources available in practice. In simple terms, the *de minimis* principle assumes that extremely low risks are trivial and need not be controlled. A *de minimis* risk level would therefore represent a cutoff, below which a regulatory agency or policy analyst could simply ignore alleged problems or hazards.

In fact, the concept of *de minimis* or *acceptable* risk is essentially a threshold concept, in that it postulates a threshold of concern below which there would be indifference to changes in the level of risk. In general, the selection of a *de minimis* risk level is contingent upon the nature of the risks, the stakeholders involved, and a host of other contextual variables (such as other risks being compared against).

In summary, *de minimis* is a lower bound on the range of acceptable risk for a given activity. When properly utilized, a *de minimis* risk concept can help prioritize risk management decisions in a socially responsible and beneficial way. It may also be used to define the threshold for regulatory involvement. Indeed, it is only after deciding on an acceptable risk level that an environmental or public health risk management program can be addressed in a most cost-effective manner. Ultimately, in order to make a determination of the best environmental or public health risk management strategy to adopt for a given problem situation, a pragmatic and realistic acceptable risk level ought to have been specified a priori.

It is noteworthy that, the concept of *de manifestis* risk is generally not seen as being controversial—because, after all, some hazard effects are clearly unacceptable. However, the *de minimis* risk concept tends to be controversial—in view of the implicit idea that some exposures to and effects of pollutants or hazards are acceptable (Suter, 1993). In any case, it is still desirable to use these types of criteria to eliminate obviously trivial risks from further risk management actions—considering that society cannot completely eliminate or prevent all human and environmental health effects associated with likely environmental contamination and chemical exposure problems.

Indeed, virtually all social systems have target risk levels—whether explicitly indicated or not—that represent tolerable limits to danger that the society is prepared to accept in consequence of potential benefits that could accrue from a given activity. This tolerable limit is often designated as the *de minimis* or "*acceptable*" risk level. Thus, in the general process of establishing "acceptable" risk levels, it is possible to use *de minimis* levels below which one need not be concerned (Rowe, 1983). Prevailing regulatory requirements are particularly important considerations in establishing such acceptable risk levels. In any event, with maintenance of public health and safety being a crucial goal for environmental and public health risk management, it should be acknowledged that reasons such as budgetary constraints alone could not be used as justification for establishing an acceptable risk level on the higher side of a risk spectrum.

21.3.6 Risk Assessment versus Risk Management in Environmental Management Programs

Risk assessment has been defined as "the characterization of the potential adverse health effects of [human] exposures to environmental hazards" (NRC, 1983). In a risk assessment, the extent to which a population group has been or may be exposed to certain environmental stressors is determined; the extent of exposure is then considered in relation to the kind and degree of hazard posed by the contaminant or stressor—thereby allowing an estimate to be made of the present or potential risk to the target population. Depending on the problem situation, different degrees of detail may be required for the process; however, the continuum of acute to chronic hazards and exposures should be fully investigated in a comprehensive assessment, so that the complete spectrum of risks can be defined for subsequent risk management decisions.

Risk management is a decision-making process that entails weighing policy alternatives and then selecting the most appropriate remedial action to deal with a hazard. This is accomplished by integrating the results of risk assessment with scientific data as well as with social, economic, and political concerns, in order to arrive at an appropriate decision on a potential hazard situation (Cohrssen and Covello, 1989; NRC, 1994; Seip and Heiberg, 1989; van Leeuwen and Hermens, 1995). Risk management may also include the design and implementation of policies and strategies that result from this decision process.

The risk management process—that utilizes prior-generated risk assessment information—is used in making a decision on how to protect public health and/or the environment. Overall, risk assessment is conducted to aid in risk management decisions. Whereas risk assessment focuses on evaluating the likelihood of adverse effects, risk management involves the selection of a course of action in response to an identified risk—and the latter is based on many other factors (e.g., social, legal, political, or economic) in addition to the risk assessment results. Essentially, risk assessment provides *information* on the health risk, and risk management is the *action* taken based on that information.

By and large, risk management programs are typically directed at risk reduction (i.e., taking measures to protect humans and/or the environment against previously identified risks), risk mitigation (i.e., implementing measures to remove risks), and/or risk prevention (i.e., instituting measures to completely prevent the occurrence of risks). Indeed, risk management, mitigation, and preventative programs generally can help facilitate an increase in the level of protection to public health and safety, as well as aid in the reduction of liability.

Several considerations go into the risk management decisions about environmental contamination problems. These include a consideration of issues such as the harmful effect of the environmental pollutants that need to be controlled, the risk–benefit–cost factor (which may include such items as the cost of pollution controls, the effects of alternative management practices, the relinquished benefits of using a pesticide or other toxic chemical, the danger of displacing private sector initiatives, etc.), and the measure of confidence attached to several components of the risk assessment. Ultimately, the risk management process is used to provide a context for balanced analysis and decision making.

21.4 THE RISK ASSESSMENT FRAMEWORK AND PARADIGM

Risk assessment methods may fall into several general major categories—typically under the broad umbrellas of hazard assessment, exposure assessment, consequence assessment, and risk estimation (Covello and Merkhofer, 1993; Norrman, 2001). The *hazard assessment* may consist of monitoring (e.g., source monitoring and laboratory analyses), performance testing (e.g., hazard analysis and accident simulations), statistical analyses (e.g., statistical sampling and hypotheses testing), and modeling methods (e.g., biological models and logic tree analyses). The *exposure assessment* may be comprised of monitoring (e.g., personal exposure monitoring, media contamination monitoring, biologic monitoring), testing (e.g., laboratory tests and field experimentation), dose estimation (e.g., as based on exposure time, material disposition in tissue, and bioaccumulation potentials), chemical fate and behavior modeling (e.g., food chain and multimedia modeling), exposure route modeling (e.g., inhalation, ingestion, and dermal contact), and population-at-risk modeling (e.g., general population vs. sensitive groups). The *consequence assessment* may include health surveillance, hazard screening, animal tests, human tests, epidemiologic studies, animal-to-human extrapolation modeling, dose–response modeling, pharmacokinetic modeling, ecosystem monitoring, and ecological effects modeling. The *risk estimation* will usually take such forms as relative risk modeling, risk indexing (e.g., individual risk vs. societal risk), nominal versus worst-case outcome evaluation, sensitivity analyses, and uncertainty analyses.

Most of the techniques available for performing risk assessments are structured around decision analysis procedures—in order to facilitate comprehensible solutions for even complicated problems. In fact, the risk assessment process can be used to provide a "baseline" estimate of existing risks that can be attributed to a given agent or hazard, as well as to determine the potential reduction in exposure and risk under various mitigation scenarios. Detailed listings for key elements of the principal risk assessment methods are provided elsewhere in the literature (e.g., Asante-Duah, 1998, 2002; Covello and Merkhofer, 1993; Norrman, 2001).

In general, most risk assessment methods may be classified as *retrospective*, focusing on injury after the fact (e.g., nature and level of risks at a given contaminated site; or it may be considered as *predictive*—such as evaluating possible

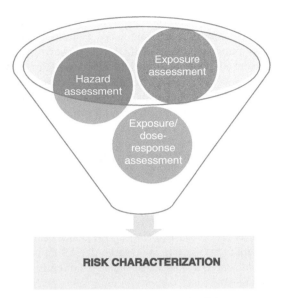

FIGURE 21.1 Principal elements of a risk assessment.

future harm to human health or the environment (e.g., risks anticipated if a newly developed food additive is approved for use in consumer food products). In the investigation of environmental contamination or impact problems, the focus of most risk assessments tends to be on a determination of potential or anticipated risks to the populations potentially at risk.

21.4.1 Principal Components of a Risk Assessment

Specific forms of risk assessment generally differ considerably in their levels of detail. Most risk assessments, however, share the same general logic—consisting of four basic elements, namely, hazard assessment, exposure/dose–response assessment, exposure assessment, and risk characterization (Fig. 21.1).

Hazard assessment describes, qualitatively, the likelihood that an environmental stressor (e.g., chemical agent) can produce adverse (health) effects under certain environmental exposure conditions. *Exposure/dose–response assessment* quantitatively estimates the relationship between the magnitude of exposure and the degree and/or probability of occurrence of a particular (health) effect. *Exposure assessment* determines the extent of receptor exposure. *Risk characterization* integrates the findings of the first three components to describe the nature and magnitude of (health) risk associated with environmental exposures. A brief discussion of these fundamental elements appears in the following—with more detailed elaboration given elsewhere in the risk analysis literature (e.g., Asante-Duah, 1998, 2002; Cohrssen and Covello, 1989; Conway, 1982; Cothern, 1993; Gheorghe and Nicolet-Monnier, 1995; Hallenbeck and Cunningham, 1988; Huckle, 1991; Kates, 1978; Kolluru et al., 1996; LaGoy, 1994; Lave, 1982; McColl, 1987; McTernan and Kaplan, 1990; Neely, 1994; NRC, 1982, 1983, 1994; Paustenbach, 1988; Richardson, 1990; Rowe, 1977; Suter, 1993; USEPA, 1984a, b, 1989a, b; Whyte and Burton, 1980).

21.4.2 Procedural Elements of the Risk Assessment Process

Procedures for analyzing hazards and risks typically will be comprised of several steps—invariably consisting of the following general elements (Asante-Duah, 1998):

- Hazard identification and accounting
  - Identify hazards (including nature/identity of hazard, location, etc.).
  - Identify initiating events (i.e., causes).
  - Identify resolutions for hazard.
  - Define exposure setting.
- Vulnerability analysis
  - Identify vulnerable zones or locales.
  - Identify concentration/impact profiles for affected zones.
  - Determine populations potentially at risk (such as human and ecological populations, as well as critical facilities).
  - Define realistic exposure scenarios.
- Consequences/impact assessment
  - Determine risk categories for all identifiable hazards and exposure scenarios.
  - Determine probability of adverse outcome (from exposures to hazards).
  - Estimate consequences (including severity, uncertainties, etc.).

Some or all of these elements may have to be analyzed in a comprehensive manner, depending on the nature and level of detail of the hazard and/or risk analysis that is being performed. The analyses typically fall into two broad categories—endangerment assessment (which may be considered as contaminant based, such as human health and ecological risk assessment (ERA) associated with chemical exposures) and safety assessment (which is system failure based, such as probabilistic risk assessment of hazardous facilities or installations).

Ultimately, the final step will be comprised of developing risk management and/or risk prevention strategies for the problem situation.

21.4.2.1 Hazard Identification and Accounting

Hazard identification and accounting involve a qualitative assessment of the presence of and the degree of hazard that an environmental stressor or contaminant could have on potential receptors. For instance, the hazard identification will typically consist of gathering and evaluating data on the types of health effects or diseases that may be produced by a chemical and the exposure conditions under which public health damage, injury, or disease will be produced. Hazard identification is not a risk assessment per se. This process involves simply determining whether it is scientifically correct to infer that toxic effects observed in one setting will occur in other settings—for example, whether substances found to be carcinogenic or teratogenic in experimental animals are likely to have the same results in humans.

21.4.2.2 Exposure–Response Evaluation

The *exposure–response evaluation* (or the *effect assessment*) is the estimation of the relationship between dose or level of exposure to a substance/stressor and the incidence and severity of an effect. It considers the types of adverse effects associated with environmental exposures, the relationship between magnitude of exposure and adverse effects, and related uncertainties.

21.4.2.3 Exposure Assessment and Analysis

An *exposure assessment* is conducted in order to estimate the magnitude of actual and/or potential receptor exposures to contaminants or chemical stressors present in the human and ecological environments. The process considers the frequency and duration of the exposures, the nature and size of the populations potentially at risk (i.e., the risk group), and the pathways and routes by which the risk group may be exposed. Indeed, several physical and chemical characteristics of the environmental stressors of concern will provide an indication of the critical exposure features.

In general, exposure assessments involve describing the nature and size of the population exposed to a substance/stressor and the magnitude and duration of their exposure. The evaluation could concern past or current exposures or exposures anticipated in the future.

21.4.2.4 Risk Characterization and Consequence Determination

Risk characterization is the process of estimating the probable incidence of adverse impacts to potential receptors under a set of exposure conditions. Typically, the risk characterization summarizes and then integrates outputs of the exposure and hazard effect assessments—in order to be able to qualitatively and/or quantitatively define risk levels. The process will usually include an elaboration of uncertainties associated with the risk estimates. Exposures resulting in the greatest risk can be identified in this process—and then mitigative measures can subsequently be selected to address the situation in order of priority and according to the levels of imminent risks.

In general, risk characterizations involve the integration of the data and analysis of the first three components of the risk assessment process (viz., hazard identification, exposure/dose–response assessment, and exposure assessment)—to determine the likelihood that receptors will experience any of the various forms of effects associated with an environmental stressor.

To the extent feasible, the risk characterization should include the distribution of risk among the target populations. Ultimately, an adequate characterization of risks from hazards associated with environmental exposure problems allows risk management and corrective action decisions to be better focused.

21.4.3 General Types of Analytical Protocols Commonly Used in the Appraisal of Environmental Contamination Problems

Variant risk assessments application schemes appear in the wide literature—utilizing a variety of risk assessment methods and/or protocols (see, e.g., Asante-Duah, 1998, 2002). Some of the key technical approaches finding common utilization in environmental contamination and/or impact problems are presented later.

21.4.3.1 Technical Approach to Human Health Endangerment Assessments

Human health risk assessment is often an integral part of most risk management and corrective action response programs that are designed to address environmental contamination problems. This section presents a general discussion on methods of approach used for completing human health risk assessments as part of the investigation and management of environmental contamination and chemical exposure problems. Specifically, it provides a procedural framework and an outline of the key elements of the health risk assessment process; the specific elements are discussed in greater detail elsewhere in the literature (see, e.g., Asante-Duah, 1998, 2002).

Human health risk assessments consist of the characterization of the potential adverse health effects associated with human exposures to environmental hazards (NRC, 1983). In a typical human health endangerment assessment process, the extent to which potential human receptors have been or could be exposed to chemical(s) associated with an environmental contamination or stressor problem is determined. The extent of exposure is considered in relation to the type and degree of hazard posed by the chemical(s) of concern, thereby permitting an estimate to be made of the present or future health risks to the populations at risk. Ultimately, the human health risk assessment technique of choice can be employed to better facilitate responsible risk management programs.

A General Framework for Human Health Risk Assessments

The basic components and steps typically involved in a comprehensive human health risk assessment designed for use in environmental contamination management programs usually will comprise the following tasks (Asante-Duah, 1998, 2002):

- Data evaluation
  - Assess the quality of available data.
  - Identify, quantify, and categorize environmental contaminants.
  - Screen and select chemicals of potential concern.
  - Carry out statistical analysis of relevant environmental or site data.

Invariably, all environmental contamination or impacts management programs start with a hazard identification/data collection and data evaluation phase. The data evaluation aspect of a human health risk assessment consists of an identification and analysis of the chemicals associated with the environmental contamination problem that should become the focus of the problem characterization and risk management program. In this process, an attempt is generally made to select all chemicals that could represent the major part of the risks associated with site-related exposures.

- Exposure assessment
  - Compile information on the physical setting of the site or problem location.
  - Identify source areas, significant migration pathways, and potentially impacted or receiving media.
  - Determine the important environmental fate and transport processes for the chemicals of potential concern, including cross-media transfers.
  - Identify populations potentially at risk.
  - Determine likely and significant receptor exposure pathways.
  - Develop representative conceptual model(s) for the site or problem situation.
  - Develop exposure scenarios (to include both the current and potential future land uses for the site or locale).
  - Estimate/model exposure point concentrations for the chemicals of potential concern found in the significant environmental media.
  - Compute potential receptor intakes and resultant doses for the chemicals of potential concern (for all potential receptors and significant pathways of concern).

The exposure assessment phase of the human health risk assessment is used to estimate the rates at which chemicals are absorbed by potential receptors. Human populations may become exposed to a variety of chemicals via several different exposure routes—represented primarily by the inhalation, ingestion, and dermal exposure routes. Since most potential receptors tend to be exposed to chemicals from a variety of sources and/or in different environmental media, an evaluation of the relative contributions of each medium and/or source to total chemical intake could be critical in a multipathway exposure analysis. In fact, the accuracy with which exposures are characterized could be a major determinant of the ultimate validity of the risk assessment. The likely types and significant categories of human exposures to a variety of chemical materials or environmental stressors that could affect public health risk management decisions are annotated further in the risk analyses literature (e.g., Al-Saleh and Coate, 1995; Asante-Duah, 2002; Corn, 1993; OECD, 1993).

- Toxicity assessment
  - Compile toxicological profiles (to include the intrinsic toxicological properties of the chemicals of potential concern, such as their acute, subchronic, chronic, carcinogenic, and reproductive effects).
  - Determine appropriate toxicity indices (such as the acceptable daily intakes or reference doses, cancer slope, or potency factors, etc.).

A toxicity assessment is conducted as part of the human health risk assessment, in order to be able to both qualitatively and quantitatively determine the potential adverse health effects that could result from a human exposure to the chemicals of potential concern. This involves an evaluation of the types of adverse health effects associated with chemical exposures, the relationship between the magnitude of exposure and adverse effects, and related uncertainties. The quantitative evaluation of toxicological effects typically consists of a compilation of toxicological profiles (including the intrinsic toxicological properties of the chemicals of concern, which may include their acute, subchronic, chronic, carcinogenic, and/or reproductive effects) and the determination of appropriate toxicity indices.

- Risk characterization
  - Estimate carcinogenic risks from carcinogens.
  - Estimate noncarcinogenic hazard quotients and indices for systemic toxicants.
  - Perform sensitivity analyses, evaluate uncertainties associated with the risk estimates, and summarize the risk information.

Finally, the risk characterization consists of estimating the probable incidence of adverse impacts to potential receptors under various exposure conditions. It involves integrating the toxicity and exposure assessments, resulting in a quantitative estimation of the actual and potential risks and/or hazards due to exposure to each key chemical constituent—and also due to the possible additive effects of exposure to mixtures of the chemicals of potential concern. Typically, the risks to potentially exposed populations resulting from exposure to the environmental contaminants are characterized through a calculation of noncarcinogenic hazard quotients and indices and/or carcinogenic risks (CAPCOA, 1990; CDHS, 1986; USEPA, 1989a). These parameters can then be compared with benchmark standards in order to arrive at risk decisions about an environmental contamination and/or chemical exposure problem.

The principal elements of the human endangerment assessment process—together with details of the requisite algorithms for estimating potential human exposures and intakes—are elaborated in greater details elsewhere in the literature (see, e.g., Asante-Duah, 1996, 1998, 2002; Hoddinott, 1992; Huckle, 1991; NRC, 1983; Patton, 1993; Paustenbach, 1988; Ricci, 1985; Ricci and Rowe, 1985; USEPA, 1984a, b, 1985, 1987, 1989a, 1991, 1992b; Van Leeuwen and Hermens, 1995).

21.4.3.2 Technical Approach to Ecological Endangerment Assessments

An ERA is comprised of a process that evaluates the likelihood that adverse ecological effects may occur or are occurring as a result of exposure to one or more environmental stressors—with such stressors usually having been imposed by human activities (Bartell *et al.*, 1992; Linthurst *et al.*, 1995; NRC, 1983; Richardson, 1995; Solomon, 1996; USEPA, 1986, 1989b, USEPA 1992a, b, c, 1994, 1995). The process evaluates the potential adverse effects of human activities on the plants and animals that make up the ecosystems. In fact, an ERA is often an integral part of most risk management and corrective action response programs that are designed to address environmental contamination problems.

This section briefly presents the major components of the ERA methodology—as may be applied to the evaluation of an environmental contamination problem (USEPA, 1988, 1992b). Specifically, it provides a procedural framework and an outline of the key elements of the ERA process. In general, ERA techniques can be employed to better develop responsible environmental risk management and corrective action response programs.

A General Framework for ERAs
The objectives of an ERA, conducted to address environmental contamination problems, consist of identifying and estimating the potential ecological impacts associated with environmental stressors found in a region or area of interest. Indeed, the procedural components of an ERA program will generally consist of several tasks—usually conducted in phases to ensure program cost-effectiveness. Where it is deemed necessary, a more detailed assessment that is comprised of biological diversity analysis and population studies may become part of the overall ERA process.

The framework for ERA is conceptually similar to the approach used for human health risk assessments (presented in preceding section) but is distinctive in its emphasis in three areas. First, ERA can consider effects beyond the individual or species level and may examine a variety of assessment end points, an entire population, community, or ecosystem. Second, the ecological values to be protected are selected from a wide range of possibilities based on both scientific and policy considerations. Finally, ERAs consider nonchemical stressors to the environment, such as loss of wildlife habitat. At any rate, a general framework for the conduct of an ERA will typically consist of the following three primary phases—*problem formulation*, *problem analysis*, and *risk characterization* (USEPA, 1992a, b, c).

Problem formulation is a formal process for generating and evaluating preliminary hypotheses about why ecological effects have occurred, or may occur, from ensuing human activities. It involves identifying and delimiting goals and assessment endpoints, preparing the conceptual model, and developing an analysis plan. Successful completion of problem formulation depends on the quality of several investigatory elements—especially relating to assessment endpoints that adequately reflect management goals and the ecosystem they represent and conceptual models that describe key relationships between a stressor and assessment endpoint or among several stressors and assessment endpoints (USEPA, 1992a, b, c). Essential to the development of these problem formulation elements are the effective integration and evaluation of available information on sources of stressors and stressor characteristics, exposure characteristics, ecosystem potentially at risk, and ecological effects.

The problem analysis phase of the ERA is composed of two principal activities—the characterization of exposure and the characterization of ecological effects—that ultimately result in the development of exposure and stressor-response profiles. It consists of hazard identification plus dose–response and exposure assessments—and that involves evaluating exposure to stressors and the relationship between stressor levels and ecological effects. This comprises the technical evaluation of data to facilitate the development of conclusions about ecological exposure and relationships between the stressor and ecological effects. During this phase, measures of exposure (e.g., source attributes, stressor levels in the environment), measures of effects (e.g., results of laboratory or field studies), and measures of ecosystem and receptor attributes (e.g., life history characteristics) are used to evaluate issues that were identified in the problem formulation phase. The analysis phase involves creation of profiles to evaluate the exposure of ecological receptors to stressors and the relationships between stressor levels and ecological effects. The end result of the problem analysis phase typically will consist of summary profiles that describe exposure and the stressor–response relationship. When combined, these profiles provide the basis for reaching conclusions about risk during the risk characterization phase.

The purpose of the *risk characterization phase* of the ERA is to evaluate the likelihood that adverse effects have occurred or will occur as a result of exposure to a stressor. Its key elements consist of estimating risk through integration of exposure and stressor–response profiles, describing risk by discussing lines of evidence, and determining ecological adversity. In general, the risk estimation is used to determine the likelihood of adverse effects to assessment endpoints by integrating exposure and effects data and then evaluating any associated uncertainties. The process uses exposure and stressor–response profiles derived a priori in the problem analysis phase of the ERA. Typically, a quantitative ecological risk characterization is accomplished by using the so-called ecological risk quotient (ErQ) method—similar to a hazard quotient employed in a human health risk characterization. In the ErQ approach, the exposure point concentration or estimated daily dose is compared to a benchmark critical toxicity parameter (e.g., a national ambient water quality criteria or a threshold reference value). The ErQ estimates the risk of an environmental contaminant to indicator species, independent of the interactions among species or between different chemicals of potential ecological concern. In general, if ErQ <1, then an acceptable risk is indicated. Conversely, an ErQ >1 calls for action or further refined investigations, due to the possibility of unacceptable levels of risk to potential ecological receptors.

Overall, the process used to evaluate environmental or ecological risk parallels that used in the evaluation of human health risks (as discussed earlier). In both cases, potential risks are determined by the integration of information on chemical exposures with toxicological data for the contaminants of potential concern. That is, the doses determined for the ecological receptors and community during the exposure assessment are integrated with the appropriate toxicity values and information derived in the ecotoxicity assessment in order to arrive at plausible ecological risk estimates. This entails both temporal and spatial components—requiring an evaluation of the probability or likelihood of an adverse effect occurring, the degree of permanence and/or reversibility of each effect, the magnitude of each effect, and the receptor populations or habitats that will be affected.

21.4.4 Risk Appraisal Strategies for Effectual Risk Management

A number of operational procedures and strategies are available for conducting and/or implementing risk assessments. The key issues requiring significant attention in the processes involved will typically consist of finding answers to the following questions (Asante-Duah, 1996, 1998, 2002):

- What environmental stressors (e.g., chemicals) pose the greatest risk?
- What are the levels of environmental stressors (e.g., concentrations of the chemicals) of concern in the exposure media?
- Which exposure pathways/routes are the most important?
- Which population groups, if any, face significant risk as a result of the possible exposures?
- What is the range of risks to the affected populations?
- What are the environmental and public health implications for any identifiable corrective action and/or risk management alternatives?

Furthermore, as a general guiding principle, risk assessments are best carried out in an iterative fashion, and then appropriately adjusted to incorporate new scientific information and regulatory changes. Typically, an iterative approach would start with relatively inexpensive screening techniques—and then for hazards suspected of exceeding the *de minimis* risk, further evaluation is conducted by moving on to more complex and resource-intensive levels of data gathering, model construction, and model application (NRC, 1994).

In general, no detailed/comprehensive evaluation is warranted if an initial screening estimate is below a preestablished reference or target level (i.e., the *de minimis* risk). On the other hand, if the screening risk estimate is above the *de minimis* risk level, then a more comprehensive/detailed evaluation (that utilizes more sophisticated and realistic data evaluation techniques than were employed in the screening) should be carried out. This next step will confirm the existence (or otherwise) of significant risks—which then forms the basis for developing any risk management action plans. The rationale for such a tiered approach is to optimize the use of resources—in that it makes efficient use of time and resources by applying more advanced and time-consuming techniques to environmental stressors of potential concern and scenarios only where necessary. Thus, a comprehensive/detailed risk assessment is performed only when truly warranted. Irrespective of the level of detail, however, a well-defined protocol should always be used to assess the potential risks. Ultimately, a decision on the level of detail (e.g., qualitative, quantitative, or combinations thereof) at which an analysis is carried out will usually be based on the complexity of the situation, as well as the uncertainties associated with the anticipated or predicted risk.

Finally, risk assessment may be conducted as a "baseline" case, in which a risk determination is made for the "as-is" problem situation, where no mitigative actions are taken, or in "comparative analysis" scenarios, in which relative risks are evaluated. A brief discussion of the distinctive nature of these types of assessments is provided in the following.

21.4.4.1 Baseline Risk Assessments

Baseline risk assessments involve an analysis of the potential adverse effects (current or future) caused by receptor exposures to hazardous substances, in the absence of any actions to control or mitigate these exposures (i.e., under an assumption of "no action"). Thus, the baseline risk assessment provides an estimate of the potential risks to the populations at risk that follows from the receptor exposure to the hazards of concern, when no mitigative actions have been considered. Because this type of assessment identifies the primary threats associated with the situation, it also provides valuable input to the development and evaluation of alternative risk management and mitigative options.

Indeed, baseline risk assessments are usually conducted to evaluate the need for and the extent of corrective action required for a hazardous situation; that is, they provide the basis and rationale as to whether or not remedial action is necessary. Overall, the results of the baseline risk assessment are typically used to (Asante-Duah, 1996, 1998, 2002):

- Document the magnitude of risk at a given locale, as well as the primary causes of the risk
- Help determine whether any response action is necessary for the problem situation
- Prioritize the need for remedial action, where several problem situations are involved
- Provide a basis for quantifying remedial action objectives
- Develop and modify remedial action goals.
- Support and justify "no further action" decisions—as appropriate, by documenting the likely insignificance of the threats posed by the hazard source(s).

In general, baseline risk assessments are designed to be case specific—and therefore may vary in both detail and the extent to which qualitative and quantitative analyses are used. The level of effort required to conduct a baseline risk assessment depends largely on the complexity and particular circumstances associated with the hazard situation under consideration.

21.4.4.2 Comparative Risk Assessments

Comparative risk assessment (CRA) has become an important aspect of risk analysis. In essence, CRA is directed at developing risk rankings and priorities that would put various kinds of hazards on an ordered scale from small to large (ACS and RFF, 1998; NRC, 1989). ACS and RFF (1998) identify two principal forms of CRA, namely:

- *Specific risk comparisons*—which involve side-by-side evaluations of distinct risks on the basis of likelihood and severity of effects. This form of CRA is comprised of a side-by-side evaluation of the risk (on an absolute or relative basis) associated with exposures to a few substances, products, or activities. Such comparisons may involve similar risk agents (e.g., the comparative cancer risks of two chemically similar pesticides) or widely different agents (e.g., the cancer risk from a particular pesticide compared with the risk of death or injury from automobile travel). Specific risk comparisons can be particularly useful when one is considering the relative importance of risks within the context of similar products, activities, or risk management actions. A more popular application has been in the area of risk communication—where such comparisons have been helpful in facilitating nontechnical audiences' understanding of the significance of varying risk levels (as, e.g., weighing the expected risks of new products or technologies against those that are already accepted or tolerated). Paired comparisons of reasonably similar risks represent the most straightforward application of comparative risk analysis; such evaluations may be conducted simply based on estimated risk levels and the extent of anticipated harm. For example, a pair of chemical pesticides might be compared with respect to their expected chronic health effects, adjusted for likelihood.
- *Programmatic CRA*—which seeks to make macrolevel (i.e., "big picture") comparisons among many and widely different types of hazards/risks. This is usually carried out in order to provide information for setting regulatory and budgetary priorities for hazard reduction. In this kind of comparison, risk rankings are based on the relative magnitude of risk (i.e., which hazards pose the greatest threat) or on relative risk reduction opportunities (i.e., the amount of risk that can be avoided with available technologies and resources). In fact, by its nature, programmatic CRA spans many dissimilar risks and provides a ready forum for value debates over what is important in gauging the seriousness of a hazard and in establishing priorities. Arguably, the major strength of programmatic CRA is the opportunity it provides for discussion and debate among various important points of view—especially those from technical experts, policy makers, and the public.

In general, risk comparisons are especially useful in situations requiring the comparison of the risks of alternative options and also to gauge the importance of different causes of the same hazard.

21.4.4.3 Risk–Cost–Benefit Optimization

Subjective and controversial as it might appear to express certain hazards in terms of cost, especially where public health and/or safety is concerned, such approach has nevertheless been used to provide an objective way of evaluating risk management action problems. This is particularly true where risk factors are considered in the overall study.

Risk–cost–benefit analysis is a generic term for techniques encompassing risk assessment and the inclusive evaluation of risks, costs, and benefits of alternative projects or policies. In performing risk–cost–benefit analysis, one attempts to measure risks, costs, and benefits, to identify uncertainties and potential trade-offs, and then to present this information coherently to decision makers. A general form of objective function for use in a risk–cost–benefit analysis that treats the stream of benefits, costs, and risks in a net present value calculation is provided in the literature elsewhere (see, e.g., Crouch and Wilson, 1982; Massmann and Freeze, 1987). In this type of approach, the risk term is defined as the expected cost associated with the probability of significant impacts or failure and is a function of the costs due to the consequences of failure at any given time. In general, trade-off decisions made in the process will be directed at improving both short- and long-term benefits of the program.

21.5 USING RISK ASSESSMENT TO SHAPE ENVIRONMENTAL MANAGEMENT DECISIONS

It is apparent that some form of risk assessment is inevitable if environmental management programs are to be conducted in a sensible and deliberate manner. In fact, risk assessment seems to be gaining greater grounds in making public policy decisions with regards to the control of risks associated with environmental management systems. This state of affairs may be attributed to the fact that the very process of performing a risk assessment does lead to a better understanding and appreciation of the nature of the risks inherent in a study and further helps develop steps that can be taken to reduce these risks.

The risk assessment process is intended to give the risk management team the best possible evaluation of all available scientific data—in order to arrive at justifiable and defensible decisions on a wide range of issues. For example, to ensure public safety in chemical exposure situations, contaminant migration beyond a compliance boundary into the public exposure domain must be below some stipulated risk-based maximum exposure level—typically established through a risk assessment process. Ultimately, based on the results of a risk assessment, decisions can be made in relation to the types of risk management actions necessary to address a given environmental contamination problem or hazardous situation. In fact, risk-based decision making will generally result in the design of better environmental management programs. This is because risk assessment can produce more efficient and consistent risk reduction policies. It can also be used as a screening device for setting priorities.

A risk assessment will typically provide the decision maker with scientifically defensible procedures for determining whether or not a potential environmental contamination problem could represent a significant adverse health and environmental risk and if it should therefore be considered a candidate for mitigative actions. In fact, the application of risk assessment can remove some of the ambiguity in the risk management decision-making process; it can also aid in the selection of prudent, technically feasible, and scientifically justifiable risk management actions that will help protect public health in a cost-effective manner. To successfully apply the risk assessment process to a potential chemical exposure problem, however, the process must be tailored to the case-specific conditions and relevant regulatory constraints. Based on the results of a risk assessment, decisions can then be made relating to the types of risk management actions needed for a given chemical exposure problem. If unacceptable risk levels are identified, the risk assessment process can further be employed in the evaluation of remedial or risk control action alternatives. This will ensure that net risks to human health are truly reduced to acceptable levels via the remedial or risk management action of choice.

In the end, the application of risk assessment to environmental contamination problems helps identify critical migration and exposure pathways, receptor exposure routes, and other extraneous factors contributing most to total risks. It also facilitates the determination of cost-effective risk reduction policies. Used in the corrective action planning process, risk assessment generally serves a useful tool for evaluating the effectiveness of remedies associated with environmental contamination problems and also for determining acceptable restoration goals. Inevitably, risk-based corrective action programs facilitate the selection of appropriate and cost-effective restoration measures.

21.5.1 General Scope of the Practical Application of Risk Assessments to Environmental Contamination Problems

Oftentimes, risk assessment is used as a management tool to facilitate effective decision making on the control of environmental pollution problems. In most applications, it is used to provide a baseline estimate of existing risks that are attributable to a specific agent or hazard; the baseline risk assessment consists of an evaluation of the potential threats to human health and the environment in the absence of any remedial or response action. Risk assessment can also be used to determine the potential reduction in exposure and risk under various corrective action scenarios, as well as to support remedy selection in risk mitigation/abatement or control programs. In particular, risk assessment can effectively be used to support remedy selection in environmental restoration programs; for instance, decisions about remediating contaminated sites are made primarily on the basis of human health and ecological risks. Overall, the risk assessment process can be used to define the level of risk, which will in turn aid in determining the level of analysis and the type of risk management actions to adopt for a given chemical exposure or environmental management problem.

In fact, risk assessment has several specific applications that could affect the type of decisions to be made in relation to environmental risk management programs. The application of the risk assessment process to environmental risk management and related problems will generally serve to document the fact that risks to human health and the environment have been evaluated and incorporated into a set of appropriate response actions. Appropriately applied, risk assessment techniques can indeed be used to estimate the risks posed by environmental hazards under various exposure scenarios and to further estimate the degree of risk reduction achievable by implementing various scientific remedies. In fact, almost invariably, every process for developing effectual environmental restoration and risk management programs should incorporate some concepts or principles of risk assessment. In particular, all corrective action decisions and restoration plans for potentially contaminated land problems will include, implicitly or explicitly, some elements of risk assessment.

A number of practical examples of the potential application of risk assessment principles, concepts, and techniques in environmental management practice—including the identification of key decision issues associated with specific problems—abound in the literature of risk analysis; some of the broad applications often encountered in chemical exposure situations are annotated in Asante-Duah (1996, 1998, 2002). For instance, among several other applications, risk assessment is generally considered an integral part of the diagnostic assessment of potentially contaminated land problems. In its application to the investigation of potentially contaminated land problems, the risk assessment process encompasses an evaluation of all the significant risk factors associated with all feasible and identifiable exposure scenarios that are the result of contaminant releases into the environment. It involves the characterization of potential adverse consequences or impacts to human and ecological receptors that are potentially at risk from exposure to site contaminants. Invariably, risk management and corrective action decisions about environmental contamination and contaminated land problems are made primarily on the basis of potential human health and ecological risks. Under such circumstances, a risk assessment process is utilized to determine whether the level of risk at a contaminated land warrants remediation and to further project the amount of risk reduction necessary to protect public health and the environment. In particular, baseline risk assessments are usually conducted to evaluate the need for and the extent of remediation required at potentially contaminated sites. That is, they provide the basis and rationale as to whether or not remedial action is necessary. Overall, the baseline risk assessment contributes to the adequate characterization of contaminated land problems. It further facilitates the development, evaluation, and selection of appropriate corrective action response alternatives.

21.5.2 General Implementation Framework for the Application of Risk Assessments to Environmental Contamination Problems

Several issues that—directly or indirectly—affect public health and environmental management programs may be addressed by using some form of risk assessment or related tools. In such applications of risk assessment, it is important to adequately characterize the exposure and physical settings for the problem situation, in order to allow for an effective application of the appropriate risk assessment methods of approach. It is noteworthy that there are several complexities involved in real-life scenarios that are unique in characterizing environmental contamination problems. Also, the populations potentially at risk from environmental contamination problems are usually heterogeneous—and this can greatly influence the anticipated impacts/consequences. Critical receptors should therefore be carefully identified with respect to numbers, locations (areal and temporal), sensitivities, etc., so that risks are neither underestimated nor conservatively overestimated.

Procedures typically used in the risk assessment process for environmental contamination problems will usually be comprised of the following tasks (Asante-Duah, 1996, 1998, 2002):

- Identification of the sources of environmental contamination and chemical exposures
- Determination of the contaminant migration pathways and chemical exposure routes
- Identification of populations potentially at risk
- Determination of the specific chemicals of potential concern
- Determination of frequency of potential receptor exposures to chemicals
- Evaluation of chemical exposure levels
- Determination of receptor response to chemical exposures
- Estimation of likely impacts or damage resulting from receptor exposures to the chemicals of potential concern

In the final analysis, potential risks are estimated by considering the probability or likelihood of occurrence of harm, the intrinsic harmful features or properties of specified hazards, the populations potentially at risk, the exposure scenarios, and the extent of expected harm and potential effects. Subsequently, the results of the risk assessment are typically used to (Asante-Duah, 1996, 1998, 2002):

- Document the magnitude of risk or threats posed by a site, including the primary causes of the risk
- Facilitate a determination as to whether further response action is necessary at a site or to support and justify a

"no further action" decision, based on both existing and anticipated exposure scenarios associated with the site
- Prioritize the need for site restoration and provide a basis for quantifying remedial action objectives

Accordingly, risk assessment is often considered an integral part of the diagnostic assessment of environmental contamination and chemical exposure problems. In its application to the investigation of environmental contamination and chemical exposure problems, the risk assessment process encompasses an evaluation of all the significant risk factors associated with all feasible and identifiable exposure scenarios. It includes a characterization of potential adverse consequences or impacts to the populations potentially at risk from the environmental contamination and chemical exposure.

Risk assessment is indeed a powerful tool for developing insights into the relative importance of the various types of exposure scenarios associated with potentially hazardous situations. But, unless care is exercised and all interacting factors are considered, risk assessments directed at single issues, followed by ill-conceived risk management strategies, can create problems worse than those the management strategies were designed to correct. To be truly instructive and constructive, therefore, risk assessments should preferably be conducted on an iterative basis—being continually updated as new knowledge and information become available. Ultimately, it is quite important to examine the total system to which a given risk assessment is being applied.

21.6 CONCLUSION

There seems to be several health, environmental, political, and socioeconomic implications associated with environmental contamination and related impact problems. It is therefore important to generally use systematic and technically sound methods of approach in the relevant risk management programs directed at addressing such issues. Risk assessment provides one of the best mechanisms for completing the tasks involved. In fact, a systematic and accurate assessment of current and future risks associated with a given environmental impact or stressor problem is crucial to the development and implementation of a cost-effective corrective action plan. Consequently, risk assessment is generally considered as an integral part of the corrective action assessment processes used in the investigation, characterization, remediation, and/or management of potential environmental impact or stressor problems.

In fact, risk assessment principles and methodologies have found extensive use in a wide variety of application scenarios. For instance, it has typically been used to evaluate many forms of new products (e.g., foods, drugs, cosmetics, pesticides, consumer products); to set environmental standards (e.g., for air and water); to predict the health threat from contaminants in air, water, and soils; to determine when a material is hazardous (i.e., to identify hazardous wastes and toxic industrial chemicals); to set occupational health and safety standards; and to evaluate soil and groundwater remediation efforts. Generally, risk assessment is considered a process used to determine the magnitude and probability of actual or potential harm that a hazardous situation poses to human health and the environment.

As a holistic approach to environmental and public health risk management, risk assessment integrates all relevant environmental and health issues and concerns surrounding a specific problem situation, in order to arrive at risk management decisions that are acceptable to all stakeholders. By and large, the encompassing process should incorporate information that helps to answer the following pertinent questions:

- Why is the project/study being undertaken?
- How will results and conclusions from the project/study be used?
- What specific processes and methodologies will be utilized?
- What are the uncertainties and limitations surrounding the study?
- What contingency plans exist for resolving newly identified issues?

Also, effective risk communication should be recognized as a very important element of the holistic approach to managing chemical exposure and related environmental hazard problems. Thus, a system for the conveying of risk information derived from a risk assessment should be considered as a very essential and integral part of the overall approach.

Ultimately, the primary focus of a risk appraisal is the assessment of whether existing or potential receptors are presently, or in the future may be, at risk of adverse effects as a result of exposure to conditions from potentially hazardous situations. This evaluation then serves as a basis for developing mitigation measures in risk management and risk prevention programs. Typically, the risk assessment will help define the level of risk as well as set performance goals for various response alternatives.

The application of risk assessment can indeed provide for prudent and technically feasible and scientifically justifiable decisions about corrective actions that will help protect public health and the environment in a most cost-effective manner. It is expected, therefore, that there will be growing applications of the risk assessment paradigm to several specific environmental contamination and impact problems, and this could affect the type of decisions made in relation to an environmental and/or public health risk management program.

Finally, it is noteworthy that to effectively utilize it as a public health and environmental management tool, risk assessment should be recognized as a multidisciplinary process that draws on data, information, principles, and expertise from many scientific disciplines.

REFERENCES

ACS and RFF, 1998. *Understanding Risk Analysis (A Short Guide for Health, Safety and Environmental Policy Making)*, A Publication of the American Chemical Society (ACS) and Resources for the Future (RFF)—written by M. Boroush, ACS, Washington, DC.

Al-Saleh, I.A. and L. Coate, 1995. Lead exposure in Saudi Arabia from the use of traditional cosmetics and medical remedies. *Environmental Geochemistry and Health*, 17: 29–31.

Anderson, P.S. and A.I. Yuhas, 1996. Improving risk management by characterizing reality: a benefit of probabilistic risk assessment. *Human and Ecological Risk Assessment*, 2: 55–58.

Asante-Duah, D.K., 1996. *Managing Contaminated Sites: Problem Diagnosis and Development of Site Restoration*. John Wiley & Sons, Ltd, Chichester, UK.

Asante-Duah, D.K., 1998. *Risk Assessment in Environmental Management: A Guide for Managing Chemical Contamination Problems*. John Wiley & Sons, Ltd, Chichester, UK.

Asante-Duah, D.K., 2002. *Public Health Risk Assessment (for human exposure to chemicals)*. Kluwer Academic Publishers, Dordrecht, The Netherlands.

Bartell, S.M, R.H. Gardner, and R.V. O'Neill, 1992. *Ecological Risk Estimation*. Lewis Publishers, Chelsea, MI.

Bates, D.V., 1994. *Environmental Health Risks and Public Policy*. University of Washington Press, Seattle, WA.

Burmaster, D.E. and K. von Stackelberg, 1991. Using Monte Carlo simulations in public health risk assessments: estimating and presenting full distributions of risk. *Journal of Exposure Analysis and Environmental Epidemiology*, 1: 491–512.

CAPCOA, 1990. *Air Toxics "Hot Spots" Program: Risk Assessment Guidelines*. California Air Pollution Control Officers Association (CAPCOA), California.

CDHS, 1986. *The California Site Mitigation Decision Tree Manual*. California Department of Health Services (CDHS), Toxic Substances Control Division, Sacramento, CA.

Cohrssen, J.J. and V.T. Covello, 1989. *Risk Analysis: A Guide to Principles and Methods for Analyzing Health and Environmental Risks*. National Technical Information Service (NTIS), US Department of Commerce, Springfield, VA.

Conway, R.A. (ed.), 1982. *Environmental Risk Analysis of Chemicals*. Van Nostrand Reinhold Co., New York.

Corn, M. (ed.), 1993. *Handbook of Hazardous Materials*. Academic Press, San Diego, CA.

Cothern, C.R. (ed.), 1993. *Comparative Environmental Risk Assessment*. Lewis Publishers/CRC Press, Boca Raton, FL.

Covello, V.T. and Merkhofer, M.W., 1993. *Risk Assessment Methods: Approaches for Assessing Health and Environmental Risks*. Plenum, New York.

Covello, V.T. and J. Mumpower, 1985. Risk analysis and risk management: an historical perspective. *Risk Analysis*, 5: 103–120.

Covello, V.T., J. Menkes, and J. Mumpower (eds.), 1986. *Risk Evaluation and Management*. Contemporary Issues in Risk Analysis, Vol. 1, Plenum Press, New York.

Crandall, R.W. and B.L. Lave (eds.), 1981. *The Scientific Basis of Risk Assessment*. Brookings Institution, Washington, DC.

Crouch, E.A.C. and R. Wilson, 1982. *Risk/Benefit Analysis*. Ballinger, Boston, MA.

Cullen, A.C., 1994. Measures of compounding conservatism in probabilistic risk assessment. *Risk Analysis*, 14(4): 389–393.

Davies, J.C. (ed.), 1996. *Comparing Environmental Risks: Tools for Setting Government Priorities*. Resources for the Future, Washington, DC.

Douben, P.E.T., 1998. *Pollution Risk Assessment and Management*. John Wiley & Sons, Ltd, Chichester, UK.

Fischhoff, B., S. Lichtenstein, P. Slovic, S. Derby, and R. Keeney, 1981. *Acceptable Risk*. Cambridge University Press, New York.

Gheorghe, A.V. and M. Nicolet-Monnier, 1995. *Integrated Regional Risk Assessment*, Volume I & II. Kluwer Academic Publishers, Dordrecht, The Netherlands.

Glickman, T.S. and M. Gough (eds.), 1990. *Readings in Risk*. Resources for the Future, Washington, DC.

Gratt, L.B., 1996. *Air Toxic Risk Assessment and Management: Public Health Risk from Normal Operations*. Van Nostrand Reinhold, New York.

Hallenbeck, W.H. and K.M. Cunningham, 1988. *Quantitative Risk Assessment for Environmental and Occupational Health*. Lewis Publishers, Inc., Chelsea, MI.

Hoddinott, K.B. (ed.), 1992. *Superfund Risk Assessment in Soil Contamination Studies*. American Society for Testing and Materials, ASTM Publication STP 1158, Philadelphia, PA.

Huckle, K.R, 1991. *Risk assessment, regulatory need or nightmare*. Shell publications, Shell Center, London, England. (8pp).

Kates, R.W., 1978. *Risk Assessment of Environmental Hazard*. SCOPE Report 8, John Wiley & Sons, New York.

Kocher, D.C. and F.O. Hoffman, 1991. Regulating environmental carcinogens: where do we draw the line? *Environmental Science & Technology*, 25: 1986–1989.

Kolluru, R. (ed.), 1994. *Environmental Strategies Handbook: A Guide to Effective Policies and Practices*. McGraw-Hill, New York.

Kolluru, R.V., S.M. Bartell, R.M. Pitblado, and R.S. Stricoff (eds.), 1996. *Risk Assessment and Management Handbook (for Environmental, Health, and Safety Professionals)*. McGraw-Hill, New York.

LaGoy, P.K., 1994. *Risk Assessment, Principles and Applications for Hazardous Waste and Related Sites*. Noyes Data Corp., Park Ridge, NJ.

Lave, L.B. (ed.), 1982. *Quantitative Risk Assessment in Regulation*. The Brooking Institute, Washington, DC.

van Leeuwen, C.J. and J.L.M. Hermens (eds.), 1995. *Risk Assessment of Chemicals: An Introduction*. Kluwer Academic Publishers, Dordrecht, The Netherlands.

Linthurst, R.A., P. Bourdeau, and R.G. Tardiff (eds.), 1995. *Methods to Assess the Effects of Chemicals on Ecosystems*. SCOPE 53/IPCS Joint Activity 23/SGOMSEC 10, John Wiley & Sons, Ltd, Chichester, UK.

Massmann, J., and R.A. Freeze, 1987. Groundwater Contamination from waste management sites: The interaction between risk-based engineering design and regulatory policy—1. Methodology & 2. Results. *Water Resources Research*, 23(2): 351–380.

McColl, R.S. (ed.), 1987. *Environmental Health Risks: Assessment and Management*. Institute for Risk Research, University of Waterloo Press, Waterloo, Ontario, Canada.

McTernan, W.F. and E. Kaplan (eds.), 1990. *Risk Assessment for Groundwater Pollution Control*. ASCE Monograph, American Society of Civil Engineers, New York.

Neely, W.B., 1994. *Introduction to Chemical Exposure and Risk Assessment*. Lewis Publishers/CRC Press, Boca Raton, FL.

Norrman, J., 2001. *Decision Analysis under Risk and Uncertainty at Contaminated Sites—A Literature Review*. SGI Varia 501, Swedish Geotechnical Institute (SGI), Linkoping, Sweden.

NRC (National Research Council), 1982. *Risk and Decision-Making: Perspective and Research. NRC Committee on Risk and Decision-Making*, National Academy Press, Washington, DC.

NRC, 1983. *Risk Assessment in the Federal Government: Managing the Process*. National Research Council, Committee on the Institutional Means for Assessment of Risks to Public Health, National Academy Press, Washington, DC.

NRC, 1989. *Improving Risk Communication*. National Research Council, Committee on Risk Perception and Communication, National Academy Press, Washington, DC.

NRC, 1994. *Science and Judgment in Risk Assessment*. Committee on Risk Assessment of Hazardous Air Pollutants, Board on Environmental Studies and Toxicology, Commission on Life Sciences, National Press Academy, Washington, DC.

OECD (Organization for Economic Cooperation and Development), 1993. *Occupational and Consumer Exposure Assessment*. OECD Environment Monograph 69, Organization for Economic Cooperation and Development (OECD), Paris, France.

Patton, D.E., 1993. The ABCs of risk assessment. *EPA Journal*, 19:10–15.

Paustenbach, D.J. (ed.), 1988. *The Risk Assessment of Environmental Hazards: A Textbook of Case Studies*. John Wiley & Sons, Inc., New York.

Pollard, S.J., R. Yearsley, et al., 2002. Current directions in the practice of environmental risk assessment in the United Kingdom. *Environmental Science & Technology*, 36(4): 530–538.

Ricci, P.F. (ed.), 1985. *Principles of Health Risk Assessment*. Prentice-Hall, Englewood Cliffs, NJ.

Ricci, P.F. and M.D. Rowe (ed.), 1985. *Health and Environmental Risk Assessment*. Pergamon Press, New York.

Richardson, M.L. (ed.), 1990. *Risk Assessment of Chemicals in the Environment*. Royal Society of Chemistry, Cambridge, UK.

Richardson, M.L. (ed.), 1992. *Risk Management of Chemicals*. Royal Society of Chemicals, Cambridge, UK.

Richardson, M. (ed.), 1995. *Environmental Toxicology Assessment*. Taylor & Francis Ltd., , London, UK.

Rowe, W.D., 1977. *An Anatomy of Risk*. John Wiley & Sons. Inc, New York.

Rowe, W.D., 1983. *Evaluation Methods for Environmental Standards*. CRC Press, Inc., Boca Raton, FL.

van Ryzin, J., 1980. Quantitative risk assessment. *Journal of Occupational Medicine*, 22: 321–326.

Seip H.M. and A.B. Heiberg (eds.), 1989. *Risk Management of Chemicals in the Environment*. NATO, Challenges of Modern Society, Vol. 12, Plenum Press, New York.

Solomon, K.R., 1996. Overview of recent developments in ecotoxicological risk assessment, *Risk Analysis,* 16(5):627–633.

Suter, G.W., 1993. *Ecological Risk Assessment*. Lewis Publishers, Boca Raton, FL.

Travis, C.C., S.A. Richter, E.A.C. Crouch, R. Wilson, and E.D. Klema, 1987. Cancer risk management. *Environmental Science & Technology*, 21:415–420.

Turnberg, W.L., 1996. *Biohazardous Waste: Risk Assessment, Policy, and Management*. John Wiley & Sons. Inc, New York.

USEPA (US Environmental Protection Agency), 1984a. *Risk Assessment and Management: Framework for Decision Making*. EPA 600/9-85-002, USEPA, Washington, DC.

USEPA, 1984b. *Approaches to risk assessment for multiple chemical exposures*. EPA-600/9-84-008, Environmental Criteria and Assessment Office, US Environmental Protection Agency (USEPA), Cincinnati, OH.

USEPA, 1985. *Principles of Risk Assessment: A Nontechnical Review*. Office of Policy Analysis, US Environmental Protection Agency (USEPA), Washington, DC.

USEPA, 1986. *Ecological Risk Assessment*. Hazard Evaluation Division Standard Evaluation Procedure, Washington, DC.

USEPA, 1987. *Handbook for Conducting Endangerment Assessments*. USEPA, Research Triangle Park, NC.

USEPA, 1988. *Superfund Exposure Assessment Manual*. EPA/540/1-88/001, OSWER Directive 9285.5-1, USEPA, Office of Remedial Response, Washington, DC.

USEPA, 1989a. *Risk Assessment Guidance for Superfund: Volume I--Human Health Evaluation Manual (Part A)*. EPA/540/1-89/002, Office of Emergency and Remedial Response, Washington, DC.

USEPA, 1989b. *Risk Assessment Guidance for Superfund. Volume II—Environmental Evaluation Manual*. EPA/540/1-89/001, Office of Emergency and Remedial Response, Washington, DC.

USEPA, 1991. *Risk Assessment Guidance for Superfund, Volume 1: Human Health Evaluation Manual (Part B, Development of Risk-Based Preliminary Remediation Goals)*. EPA/540/R-92/003, Office of Emergency and Remedial Response, Washington, DC.

USEPA, 1992a. *Framework for Ecological Risk Assessment*. EPA/630/R-92/001, February, 1992, Risk Assessment Forum, Washington, DC.

USEPA, 1992b. *Guidance for Data Useability in Risk Assessment (Parts A & B)*. Publication No. 9285.7-09A&B, Office of Emergency and Remedial Response, USEPA, Washington, DC.

USEPA, 1992c. *Guidelines for Exposure Assessment*. EPA/600/Z-92/001, Risk Assessment Forum, Office of Research and Development, Office of Health and Environmental Assessment, USEPA, Washington, DC.

USEPA, 1994. *Ecological Risk Assessment Issue Papers*. EPA/630/R-94/009, Risk Assessment Forum, Washington, DC.

USEPA, 1995. *Ecological Risk: A Primer for Risk Managers*. EPA/734/R-95/001, United States Environmental Protection Agency, Prevention, Pesticides and Toxic Substances, Washington, DC.

Whipple, C., 1987. *De Minimis Risk. Contemporary Issues in Risk Analysis*, Vol. 2, Plenum Press, New York.

Whyte, A.V., and I. Burton (eds.), 1980. *Environmental Risk Assessment*. SCOPE Report 15, John Wiley & Sons, Inc, New York.

APPENDIX A

SUPPLEMENTAL READINGS

The appended reference list to these chapters include many publications that merit further reading, including several peer-reviewed general review articles along with contributions from journalists succinctly summarizing the political and social implications of the technical issues raised. The following should be of particular interest:

CHAPTER 1

California Geological Survey. 2004. *Recommended criteria for delineating seismic hazard zones in California*, Special Publication 118, revised. State of California, the Resources Agency, Department of Conservation, Sacramento, CA.

Dean, C. 2012. Costs of shoring up coastal communities. *The New York Times*. November 5, p. D1.

Foderaro, L.W. 2013. Breach through fire island also divides opinions. *The New York Times*. April 5, p. A13.

Galbraith, K. 2013. As fracking increases, so do fears about water supply. *The New York Times*. March 8, p. A21.

Gornitz, V., S. Couch, and E.K. Hartig. 2001. Impacts of sea level rise in the New York City metropolitan area. *Global and Planetary Change*. 32(1):61–88.

Graham, B., W.K. Reilly, F. Beinecke, D.F. Boesch, T.D. Garcia, C.A. Murray, and F. Ulmer. 2011. *Deep water: The Gulf oil disaster and the future of offshore drilling. Report to the President*. National Commission on the BP Deepwater Horizon Oil Spill and Offshore Drilling, Washington, DC.

Griffith, M.B., S.B. Norton, L.C. Alexander, A.I. Pollard, and S.D. LeDuc. 2012. The effects of mountaintop mines and valley fills on the physicochemical quality of stream ecosystems in the central Appalachians: A review. *Science of the Total Environment*. 417–418:1–12.

Hubbert, M.K. 1956. *Nuclear energy and the fossil fuels*, Shell Development Company, Exploration and Production Research Division, Publication No. 95. Shell Development Company, Exploration and Production Research Division, Houston, TX.

Hunt, J.M. 1996. *Petroleum Geochemistry and Geology*, 2nd ed. Freeman, New York.

IPCC. 2007. *Climate change 2007: The Physical science basis. Contribution of Working Group I to the Fourth Assessment Report of the Intergovernmental Panel on Climate Change*. Solomon, S., D. Qin, M. Manning, Z. Chen, M. Marquis, K.B. Averyt, M. Tignor, and H.L. Miller (Eds.) Cambridge University Press, Cambridge/New York.

Johnson, D.B. and K.B. Hallberg. 2005. Acid mine drainage remediation options: a review. *Science of the Total Environment*. 338:3–14.

Joyce, C. 2012. With gas boom, Pennsylvania fears new toxic legacy. National Public Radio. Available at http://www.npr.org/2012/05/14/149631363/when-fracking-comes-to-town-it-s-water-water-everywhere. Accessed May 14, 2012.

Keefer, D.K. 2002. Investigating landslides caused by earthquakes—a historical review. *Surveys in Geophysics*. 23(6): 473–510.

Kunzig, R. 2009. Once considered too expensive, as well as too damaging to the land, exploitation of Alberta's oil sands is now a gamble worth billions. National Geographic Magazine, 1–8. March 2009. Available at http://ngm.nationalgeographic.com/2009/03/canadian-oil-sands/kunzig-text/2. Retrieved March 5, 2013.

Petroski, H. 2013. The stormy politics of building. *The International New York Times*. October 22. (http://www.nytimes.com/2013/10/23/opinion/international/the-stormy-politics-of-building.html?_r=0)

Pilkey, O.H. 2012. We need to retreat from the beach. *The New York Times*. November 14, p. A35.

Pipkin, B.W., D.D. Trent, R. Hazlett, and P. Bierman. 2008. *Geology and the Environment*, 5th ed. Thompson Brooks/Cole, Belmont, CA.

Schleisner, L. 2000. Comparison of methodologies for externality assessment. *Energy Policy*. 28(15):1127–1136.

Smith, V.H., G.D. Tilman, and J.C. Nekola. 1999. Eutrophication: Impacts of excess nutrient inputs on freshwater, marine, and terrestrial ecosystems. *Environmental Pollution*. 100:179–196.

Soeder, D.J. and W.M. Kappel. 2009. *Water resources and natural gas production from the Marcellus Shale*, U.S. Geological Survey Fact Sheet 2009-3032. U.S. Department of the Interior, U.S. Geological Survey, Reston, VA.

Zisman, E.D. 2013. The Florida sinkhole statute: its evolution, impacts and needed improvements. *Carbonates and Evaporites*. 28:95–102.

CHAPTER 2

Chapin III, FS, GP Kofinas, and C Folke (eds). 2009. *Principles of Ecosystem Stewardship: Resilience-Based Natural Resource Management in a Changing World*. Springer, New York.

Comín FA (ed.). 2010. *Ecological Restoration: A Global Challenge*. Cambridge University Press, Cambridge/New York.

Conservation International. Conservation International's "The Biodiversity Hotspots" page: http://www.conservation.org/where/priority_areas/hotspots/Pages/hotspots_main.aspx (accessed March 17, 2015).

Cornell Laboratory of Ornithology. Cornell Laboratory of Ornithology's Citizen Science program: http://www.birds.cornell.edu/page.aspx?pid=1664 (accessed March 17, 2015).

Goudie A. 2005. *The Human Impact on the Natural Environment*. The MIT Press, Cambridge, MA.

Gunderson LH and CS Holling (eds). 2001. *Panarchy, Understanding Transformations in Human and Natural Systems*. Island Press, Washington, DC.

Hobbs RJ, ES Higgs, and C Hall. 2013. *Novel Ecosystems: Intervening in the New Ecological World Order*. Wiley-Blackwell, Hoboken, NJ. 380pp.

Hughes TP, LH Gunderson, C Folke, AH Baird, D Bellwood, F Berkes, B Crona, A Helfgott, H Leslie, J Norberg, M Nystrom, P Olsson, H Osterblom, M Scheffer, H Schuttenberg, RS Steneck, M Tengo, M Troell, B Walker, J Wilson, and B Worm. 2007. Adaptive management of the Great Barrier Reef and the Grand Canyon World Heritage Areas. *Ambio* 36:586–592.

Kareiva P, H Tallis, TH Ricketts, GC Daily, and S Polasky. 2011. *Natural Capital: Theory and Practice of Mapping Ecosystem Services*. Oxford University Press, New York. 432pp.

Meridian Institute. The Collaborative Adaptive Management Network: http://www.adaptivemanagement.net/ (accessed March 17, 2015).

Millennium Ecosystem Assessment: http://www.millenniumassessment.org/en/index.html (accessed March 17, 2015).

Multi-Resolution Land Characteristics Consortium (MRLC): http://www.mrlc.gov/ (accessed March 17, 2015).

National Ecological Observatory Network: http://www.neoninc.org/ (accessed March 17, 2015).

NOAA (National Oceanographic and Atmospheric Administration). NOAA's National Climatic Data Center: http://www.ncdc.noaa.gov/ (accessed March 17, 2015).

Randolph J. 2004. *Environmental Land Use Planning and Management*. Island Press, Washington, DC. 664pp.

Stanford University. The Natural Capital Project: http://www.naturalcapitalproject.org/ (accessed March 17, 2015).

The Long Term Ecological Research Network: http://www.lternet.edu/lter-sites (accessed March 17, 2015).

Urbigkit C. 2008. *Yellowstone Wolves: A Chronicle of the Animal, the People, and the Politics*. The McDonald & Woodward Publishing Company, Blacksburg, VA. 350pp.

US Department of Interior. The Comprehensive Everglades Restoration Plan: www.evergladesrestoration.gov (accessed March 17, 2015).

Wuerthner G. 2006. *Wildfire: A Century of Failed Forest Policy*. Island Press, Washington, DC. 340pp.

CHAPTER 3

Ashman, M.R. and G. Puri. 2002. *Essential soil science: A clear and concise introduction to soil science*. Blackwell Science Ltd, Oxford/Maiden, MA.

Brady, N.C. and R.R. Weil. 2007. *The nature and properties of Soil*. 14th edition. Prentice Hall, Upper Saddle River, NJ.

Clapp, C.E., M.H.B. Hayes, N. Senesi, P.R. Bloom, and P.M. Jardine (eds). 2001. *Humic substances and chemical contaminants*. Soil Science Society of America, Madison, WI.

Dixon, J.B. and D.G. Schulze (eds). 2002. *Soil mineralogy with environmental applications*. Soil Science Society of America, Madison, WI.

Ellis, B.G., and H.D. Foth. 1996. *Soil fertility*. 2nd edition. CRC Press, Boca Raton, FL.

Essington, M.E. 2004. *Soil and water chemistry: An integrative approach*. CRC Press, Boca Raton, FL.

Gregory, P.J. 2006. Roots, rhizosphere and the soil: The route to a better understanding of soil science? *Eur. J. Soil. Sci.* 57: 2–12.

Havlin, J.L., S.L. Tisdake, W.L. Neslon, and J.D. Beaton. 2004. *Soil fertility and fertilizers: An introduction to nutrient management.* 7th edition. Prentice Hall, Upper Saddle River, NJ.

Hornsby, A.G., R.D. Wauchope, and A.E. Herner. 1995. *Pesticide properties in the environment*, Springer-Verlag, New York.

Huang, P.M., N. Senesi, and J. Buffle (eds). *Structure and surface reactions of soil particles.* John Wiley & Sons, Inc., Chichester.

IFPRI (International Food Policy Research Institute). 2013. *The 2012 global food policy report.* IFPRI, Washington, DC.

Johnston, C.T. and E. Tombcz. 2002. Surface chemistry of soil minerals, pp. 37–67. in K.J.B. Dixon and D.G. Schulze (eds). *Soil mineralogy with environmental applications.* Soil Science Society of America, Madison, WI.

Lal R. 2004. Soil carbon sequestration impacts on climate change and food security. *Science*, 304, 1623–1627.

Langmuir, D. 1997. *Aqueous environmental geochemistry.* Prentice Hall, Upper Saddle River, NJ.

Lehr J.H. 2000. *Wiley's remediation technologies handbook.* John Wiley & Sons, Inc, Hoboken, NJ.

Rathore H. and L. Nollet. 2012. *Pesticides: Evaluation of Environmental Pollution*, CRC Press, Taylor & Francis Group, Boca Raton, FL.

Rengel, Z. (ed.). 2003. *Handbook of soil acidity.* Marcel Dekker, New York.

Sharma H.D. and K. Reddy. 2004. *Geoenvironmental engineering: Site remediation, waste containment and emerging waste management technologies.* John Wiley & Sons, Inc., Hoboken, NJ.

Singh B.P., A. Cowie, and K. Chan. 2011. *Soil health and climate change.* Springer, New York.

Stumm, W. and J.J. Morgan. 1996. *Aquatic chemistry.* John Wiley & Sons, Inc., New York.

Wild, A. 1995. *Soils and the environment: An introduction.* Cambridge University Press, Cambridge/New York.

CHAPTER 4

Soli J. Arceivala and Shyam R. Asolekar (2006) *Wastewater Treatment for Pollution Control.* (3rd Edition, 8th Reprint). Tata McGraw Hill Education (India) Pvt. Ltd., New Delhi (Chapter 9, 10 and 11).

Shyam R. Asolekar, Pradip P. Kalbar, Manoj. K. M. Chaturvedi, and Krishnanand Y. Maillacheruvu (2014) Rejuvenation of Rivers and Lakes in India: Balancing Societal Priorities with Technological Possibilities. *Reference Module in Earth Systems and Environmental Sciences, from Comprehensive Water Quality and Purification*, Volume 4, Pages 181–229.

Shyam R. Asolekar and R. Gopichandran (2005) *Preventive Environmental Management—An Indian Perspective.* Foundation Books Pvt. Ltd., New Delhi (the Indian associate of Cambridge University Press, UK) (Chapters 10, 11 and 16).

Paritosh C. Deshpande, Atit K. Tilwankar, and Shyam R. Asolekar (2012) A Novel Approach To Estimating Potential Maximum Heavy Metal Exposure To Ship Recycling Yard Workers in Alang, India. *Science of the Total Environment*, Volume 438, Pages 304–311.

Paritosh C. Deshpande, Pradip P. Kalbar, Atit K. Tilwankar, and Shyam R. Asolekar (2013) A Novel Approach To Estimating Resource Consumption Rates and Emission Factors for Ship Recycling Yards in Alang, India. *Journal of Cleaner Production*, Volume 59, Pages 251–259.

Anand M. Hiremath, Atit K. Tilwankar, and Shyam R. Asolekar (2014) Significant Steps in Ship Recycling Vis-a-vis Wastes Generated in a Cluster of Yards in Alang: A Case Study. *Journal of Cleaner Production*, Volume 87, Pages 520–532.

Anand M. Hiremath, Sachin Kumar Pandey, Dinesh Kumar, and Shyam R. Asolekar (2014) Ecological Engineering, Industrial Ecology and Eco-Industrial Networking Aspects of Ship Recycling Sector in India. *APCBEE Procedia*, Volume 10, Pages 159–163.

Dinesh Kumar and Shyam R. Asolekar (2015) Experiences with Laboratory and Pilot Scale Constructed Wetlands for Treatment of Sewages and Effluents. In *SaphPani Handbook Entitled: Natural Water Treatment Systems for safe and sustainable water supply in the Indian context.* IWA Publishing, London.

Dinesh Kumar, Anand M. Hiremath, and Shyam R. Asolekar (2014) Integrated Management of Wastewater through Sewage Fed Aquaculture for Resource Recovery and Reuse of Treated Effluent: A Case Study. *APCBEE Procedia*, Volume 10, Pages 74–78.

Dinesh Kumar, Shyam R. Asolekar, and Saroj Kumar Sharma (2015) Constructed Wetlands and Other Natural Treatment Systems for Treatment of Sewages and Effluents: India. Status Report. In *SaphPani Handbook Entitled: Natural Water Treatment Systems for Safe and Sustainable Water Supply in the Indian Context.* IWA Publishing, London.

Dinesh Kumar, Shyam R. Asolekar, and Saroj Kumar Sharma (2015) Significance of Incorporating Constructed Wetlands to Enhance Reuse of Treated Sewages in India. In *SaphPani Handbook Entitled: Natural Water Treatment Systems for safe and sustainable water supply in the Indian context.* IWA Publishing, London.

CHAPTER 5

Blok, K., van Vuuren, D., van Wijk, A.J.M., and Hein, L.G. *Policies and measures to reduce CO_2 emissions by efficiency and renewable.* 1996. WWF; Zeist, the Netherlands.

Brower, M. and Union of Concerned Scientists *Cool energy: The renewable solution to global warming.* 1990. Union of Concerned Scientists; Cambridge, MA.

Dieter, S. and Walter, W. *Renewable energy, the facts.* 2010. p. 251. Earthscan; London.

Fräss-Ehrfeld, C. *Renewable energy sources: A chance to combat climate change.* 2009. Kluwer Law International; The Netherlands.

IPCC *Proceedings of the IPCC scoping meeting on renewable energy sources.* 2008. IPCC, Germany.

IPCC *Renewable energy sources and climate change mitigation: Summary for policymakers and technical summary*. 2011. Intergovernmental Panel on Climate Change; Geneva.

IPCC *Renewable energy sources and climate change mitigation: Special report of IPCC*. 2012. Cambridge University Press; Cambridge.

Martin, K., Wolfgang, S. and Andreas, W. (Eds) *Renewable energy, technology, economics & environment*. 2007. p. 564. Springer-Verlag; Berlin. 564p.

Quaschning, V. *Renewable energy and climate change*. 2010. John Wiley & Sons Ltd; Chichester, West Sussex.

Renewables 2013 Global Status Report http://www.ren21.net/Portals/0/documents/Resources/GSR/2013/GSR2013_lowres.pdf (accessed March 17, 2015).

San Cristóbal Mateo, J.R. *Multi-criteria analysis in the renewable energy industry*. 2012. Springer-Verlag; London.

Sayigh, A. *Comprehensive renewable energy, vol 1: Photovoltaic solar energy*. 2012. Elsevier; Amsterdam.

Sayigh, A. *Comprehensive renewable energy, vol 2: Wind energy*. 2012. Elsevier; Amsterdam.

Sayigh, A. *Comprehensive renewable energy, vol 3: Solar thermal systems: Components and applications*. 2012. Elsevier; Amsterdam.

Sayigh, A. *Comprehensive renewable energy, vol 4: Fuel cells and hydrogen technology*. 2012. Elsevier; Amsterdam.

Sayigh, A. *Comprehensive renewable energy, vol 5: Biomass and biofuel production*. 2012. Elsevier; Amsterdam.

Sayigh, A. *Comprehensive renewable energy, vol 6: Hydro power*. 2012. Elsevier; Amsterdam.

Sayigh, A. *Comprehensive renewable energy, vol 7: Geothermal energy*. 2012. Elsevier; Amsterdam.

Sayigh, A. *Comprehensive renewable energy, vol 8: Ocean energy*. 2012. Elsevier; Amsterdam.

Sweet, W. *Kicking the global warming and the case for renewable and nuclear energy*. 2006. Columbia University Press; New York.

Policy on renewable energy development and energy conservation: Green energy imitative. 2004. Ministry of Energy and Mineral Resource; Jakarta.

The World Wind Energy Association 2012 Annual Report http://www.wwindea.org/webimages/WorldWindEnergyReport2012_final.pdf (accessed March 17, 2015).

World Energy Outlook 2012 International Energy Agency, Global Energy Trends http://www.worldenergyoutlook.org/media/weowebsite/2012/WEO2012_Renewables.pdf (accessed March 17, 2015).

CHAPTER 6

Akimoto, H. (2003) "Global air quality and pollution," *Science*, 302(5651), 1716–1719.

Davis, M. L. and Cornwell k. (2012) Chapter 2: Materials and Energy Balances in *Introduction to Environmental Engineering* (5th edition), McGraw-Hill Science/Engineering/Math, New York.

Hertwich, E. G. and Peters, G. P. (2009) "Carbon footprint of nations: A global, trade-linked analysis," *Environmental Science & Technology*, 43, 6414–6420.

Kolpin, D. W., Furlong, E. T., Meyer, M. T., Thurman, M. E., Zaugg, S. D., Barber, L. B., and Buxton, H. T. (2002) "Pharmaceuticals, hormones, and other organic wastewater contaminants in US streams, 1999–2000: A national reconnaissance," *Environmental Science & Technology*, 36(6), 1202–1211.

Logan, B. E. and Rabaey, K. (2012) "Conversion of wastes into bioelectricity and chemicals by using microbial electrochemical technologies," *Science*, 337(6095), 686–690.

Sedlak, D. L. and von Gunten, U. (2011) "The Chlorine dilemma," *Science*, 331(6013), 42–43.

Smith, V. H., Tilman, G. D., and Nekola, J. C. (2002) "Eutrophication: Impacts of excess nutrient inputs on freshwater, marine, and terrestrial ecosystems," *Environmental Pollution*, 100(1999), 179–196.

Staley, B. and Barlaz, M. (2002) "Present and long-term composition of MSW landfill leachate: A review," *Critical Review in Environmental Science and Engineering*, 32(4), 297–336.

Staley, B. and Barlaz, M. (2009) "Composition of municipal solid waste in the united states and implications for carbon sequestration and methane yield," *Journal of Environmental Engineering*, 135(10), 901–909.

CHAPTER 7

Altomonte, S. & Schiavon, S. (2013). Occupant satisfaction in LEED and non-LEED certified buildings. *Building and Environment*, 68, 66–76.

Black, E. (2008). Green neighborhood standards from a planning perspective: the LEED for neighborhood development (LEED-ND). *Focus: Journal of the City and Regional Planning Department*, 5(1), 11.

Describes LEED for Neighborhood Development from a planning perspective.

Blumberg, D. (2012). LEED in the US Commercial Office Market: market effects and the emergence of LEED for existing buildings. *The Journal of Sustainable Real Estate*, 4(1), 23–47.

Boschmann, E. E. & Gabriel, J. N. (2013). Urban sustainability and the LEED rating system: case studies on the role of regional characteristics and adaptive reuse in green building in Denver and Boulder, Colorado. *The Geographical Journal*, 179(3), 221–233.

Bray, J. & McCurry, N. (2006). Unintended consequences: how the use of LEED can inadvertently fail to benefit the environment. *Journal of Green Building*, 1(4), 152–165.

The article discusses the challenges of LEED and examines whether the rating system creates any negative inadvertent environmental effects and, if so, what they are?

Brown, K. T. (2010). *A LEED-ND Based Methodology for the Statewide Mapping of Smart Growth Locations: A Case Study of the State of Connecticut*. University of Connecticut, Mansfield.

Discusses LEED-ND as a methodology to map smart growth in Connecticut.

Cidell, J. (2009). Building green: the emerging geography of LEED-certified buildings and professionals. *The Professional Geographer*, 61(2), 200–215.

The article examines the geography of the green building industry through a study of the spatial distribution of two different elements of that industry: green buildings and LEED-accredited professionals.

DeLisle, J., Grissom, T., & Högberg, L. (2013). Sustainable real estate: an empirical study of the behavioural response of developers and investors to the LEED rating system. *Journal of Property Investment & Finance*, 31(1), 10–40.

Dib, H., Adamo-Villani, N., & Niforooshan, R. (2012). *A Serious Game for Learning Sustainable Design and LEED Concepts*. Paper presented at the Computing in Civil Engineering: Proceedings of the 2012 ASCE International Conference on Computing in Civil Engineering, June 17–20, 2012, Clearwater Beach, FL.

Ding, G. K. (2008). Sustainable construction—the role of environmental assessment tools. *Journal of Environmental Management*, 86(3), 451–464.

An applied approach to environmental assessments

Garde, A. (2009). Sustainable by design?: insights from US LEED-ND pilot projects. *Journal of the American Planning Association*, 75(4), 424–440.

Results of a survey of LEED-ND registered pilot projects in the United States

Hess, G. (2012). Green-building standards update draws fire. *Chemical & Engineering News Archive*, 90(27), 10.

Hamedani, A. Z. & Huber, F. (2012). A comparative study of DGNB, LEED and BREEAM certificate systems in urban sustainability. *The Sustainable City VII: Urban Regeneration and Sustainability*, 1, 121–132.

Results of a survey of DGNB, LEED, and BREEAM certificate systems

Haselbach, L. (2010). *The engineering guide to LEED—New construction: Sustainable construction for engineers (GreenSource)*. McGraw-Hill, New York.

Provides an applied approach to the LEED-NC rating system, with equations and exercises.

HUD (United States Department of Housing and Urban Development). (2011). *LEED for Neighborhood Development and HUD*. Housing and Urban Development Report. Retrieved March 14, 2014, 2014, from portal.hud.gov/hudportal/documents/huddoc?id=HUDandLEED.pdf.

Keysar, E. & Pearce, A. R. (2007). Decision support tools for green building: facilitating selection among new adopters on public sector projects. *Journal of Green Building*, 2(3), 153–171.

A study of "275 design-related tools examined to address ... green building concepts and represent a range of applicability to different design tasks."

Jrade, A. & Jalaei, F. (2013). Integrating building information modelling with sustainability to design building projects at the conceptual stage. *Building Simulation*, 6(4), 429–444.

Liang, S. (2012). *International Experiences on Sustainable Scoring System: Comparisons and Applications*. Paper presented at the 2012 6th International Association for China Planning Conference (IACP), Wuhan University, Wuhan, China.

Mehdizadeh, R., Fischer, M., & Burr, J. (2013). The green housing privilege? An analysis of the connections between socioeconomic status of California communities and Leadership in Energy and Environmental Design (LEED) certification. *Journal of Sustainable Development*, 6(5), p37.

The study discusses the socioeconomic patterns of residential LEED certification in California.

Planners, C. (2012). A Methodology for Inventorying LEED-ND Location-Eligible Parcels in a Local Jurisdiction. Retrieved March 17, 2014, 2104, from http://www.usgbc.org/resources/methodology-inventorying-leed-nd-location-eligible-parcels-local-jurisdiction (accesses March 17, 2014).

A LEED-ND guidance document from Criterion Planners for the USGBC

Prum, D. A., Aalberts, R. J. & Del Percio, S. (2012). In third parties we trust—the growing antitrust impact of third-party green building certification systems for state and local governments. *Journal of Environmental Law & Litigation*, 27(1), 191–235.

Retzlaff, R. C. (2009). The use of LEED in planning and development regulation. *Journal of Planning Education and Research*, 29(1), 67.

Stoppel, C. M. & Leite, F. (2013). Evaluating building energy model performance of LEED buildings: identifying potential sources of error through aggregate analysis. *Energy and Buildings*, 65, 185–196.

Talen, E., Allen, E., Bosse, A., Ahmann, J., Koschinsky, J., Wentz, E., & Anselin, L. (2013). LEED-ND as an urban metric. *Landscape and Urban Planning*, 119, 20–34.

Zhang, B. Q. & Chen, Y. S. (2013). A Comparative Study on Different Evaluation Standards of Green Building. Paper presented at the Applied Mechanics and Materials, Switzerland.

CHAPTER 8

Business Is Beautiful: The Hard Art of Standing Apart, by Jean-Baptiste Danet, Nick Liddell, Lynne Dobney, and Dorothy MacKenzie (2013). LID, London.

Cradle to Cradle: Remaking the Way We Make Things, by William McDonough and Michael Braungart (2002). North Point Press, New York.

Ecological intelligence: How knowing the hidden impacts of what we buy can change everything, by Daniel Goleman (2009). Broadway Books, New York.

Green to Gold: How Smart Companies Use Environmental Strategy to Innovate, Create Value, and Build Competitive Advantage, 2nd edition, by Daniel Esty and Andrew Winston (2009). John Wiley & Sons, Hoboken, NJ.

Natural Capitalism: Creating the Next Industrial Revolution, by Paul Hawken, Amory Lovins, and L. Hunter Lovins (1999). Little, Brown and Co., Boston, MA.

Shopping for Good, by Dara O'Rourke (2012). MIT Press, Cambridge, MA.

Small Is Beautiful: Economics as if People Mattered, by E.F. Schumacher (1973). Harper & Row, New York.

Strategy for Sustainability: A Business Manifesto, by Adam Werbach (2009). Harvard Business Press, Boston, MA.

Sustainable Excellence: The Future of Business in a Fast-Changing World, by Aron Cramer and Zachary Karabell (2010). Rodale, Emmaus, PA.

The Upcycle: Beyond Sustainability—Designing for Abundance, by William McDonough and Michael Braungart (2013). North Point Press, a division of Farrar, Straus and Giroux, New York.

The Responsibility Revolution: How the Next Generation of Businesses Will Win, by Jeffrey Hollender and Bill Breen (2010). Jossey-Bass, San Francisco, CA.

The New Sustainability Advantage: Seven Business Case Benefits of a Triple Bottom Line, by Bob Willard (2012). New Society Publishers, Gabriola Island, BC.

What's Mine Is Yours: The Rise of Collaborative Consumption, by Rachel Botsman and Roo Rogers (2010). Harper Business, New York.

CHAPTER 9

Textbooks

Fuller, D. 1999, *Sustainable Marketing: Managerial-Ecological Issues*, Sage Publications, Thousand Oaks, CA.

Kotler, P., Roberto, N., & Lee, N. 2002, *Social Marketing: Improving the Quality of Life*, Sage Publications, Thousand Oaks, CA.

Murr, C. 2008, *Beyond Green Marketing*, VDM, Germany.

Peattie, K. 1995, *Environmental Marketing Management: Meeting the Green Challenge*, Pitman, London.

Journals and Business Articles

Banerjee, S., Gulas, C.S., & Iyer, E. (1995). Shades of green: A multidimensional analysis of environmental advertising. *Journal of Advertising*, 24(2), 21–31.

Baumgartner, R.J. & Ebner, D. (2010), Corporate sustainability strategies: Sustainability profiles and maturity levels. *Sustainable Development*, 18(2), 76–89.

Beale, J.R. & Bonsall, P.W. (2007). Marketing in the bus industry: A psychological interpretation of some attitudinal and behavioural outcomes. *Transportation Research*, Part F 10, 271–287.

Benoit-Moreau, F. & Parguel, B. (2011). Building brand equity with environmental communication: An empirical investigation in France. *EuroMed Journal of Business*, 6(1), 100–116.

Beverland, M. & Lockshin, L.S. (2004). Crafting a competitive advantage: Tempering entrepreneurial action with positioning-based values. *Qualitative Market Research*, 7(3), 172–182.

Bonn, I. & Fisher, J. (2011). Sustainability: The missing ingredient in strategy. *Journal of Business Strategy*, 32(1), 5–14.

Bush, V.D., Bush, A.J., Shannahan, K.L.J., & Dupuis, R.J. (2007). Segmenting markets based on sports orientation: An investigation of gender, race and behavioral intentions. *The Marketing Management Journal*, 17(1) 39–50.

Calonius, H. (2006). Contemporary research in marketing: A market behaviour framework. *Marketing Theory*, 6(4), 419–428.

Chamorro, A. & Bañegil, T.M. (2005). Green marketing philosophy: A study of Spanish firms with ecolabels. *Corporate Social Responsibility and Environmental Management*, 13, 11–24.

Charters, S. & Pettigrew, S. (2006). Conceptualizing product quality: The case of wine. *Marketing Theory*, 6(4), 467–483.

Cheah, I. and Phau, I. (2011). Attitudes towards environmentally friendly products: The influence of ecoliteracy, interpersonal influence and value orientation. *Marketing Intelligence & Planning*, 29(5), 452–472.

Christ, K. & Burritt, R. (2012). Environmental management accounting: The significance of contingent variables for adoption. *Journal of Cleaner Production*, 41, 163–173.

Clarke, P. & Mount, P. (2001). Nonprofit marketing: The key to marketing's "mid-life crisis"? *International Journal of Nonprofit and Voluntary Sector Marketing*, 6(1), 78–91.

Conley, C. & Friedenwald-Fishman, F. (2006). *Marketing that matters: 10 practices to profit your business and change the world*. San Francisco, CA: Berrett-Koehler Publishers.

Cordano, M., Marshall, R., & Silverman, M. (2010), How do small and medium enterprises go "green"? A study of environmental management programs in the U.S. wine industry. *Journal of Business Ethics*, 92(3), 463–478.

Craig, C.S. & Douglas, S.P. (2001). Conducting international marketing research in the twenty-first century. *International Marketing Review*, 18(1), 80–90.

Cui, Y., Trent, E.S., Sullivan, P.M., & Matiru, G.N. (2003). Cause-related marketing: How generation Y responds. *International Journal of Retail & Distribution Management*, 31(6/7), 310–320.

Darnall, N. & Kim, Y. (2012). Which types of environmental management systems are related to greater environmental improvements? *Public Administration*, 72(3), 351–365.

Daily, L., Anderson, M., Ingenito, C., Duffy, D., Krimm, P., & Thomson, S. (2006). Understanding MBA consumer needs and the development of marketing strategy. *Journal of Marketing for Higher Education*, 16(1), 143–158.

Darroch, J., Miles, M.P., Jardine, A., & Cooke, E.F. (2004). The 2004 AMA definition of marketing and its relationship to a market orientation: An extension of Cooke, Rayburn, & Abercrombie (1992). *Journal of Marketing Theory and Practice*, 12(4), 29–38.

Dawes, J.G. (1998). Doing a market assessment for an unfamiliar product. *Journal of Marketing Practice: Applied Marketing Science*, 4(8), 221–230.

De Giovanni, P. (2012). Do internal and external environmental management contribute to the triple bottom line?. *International Journal of Operations & Production Management*, 32(3), 265–290.

Delmas, M.A. & Burbano, V.C. (2011). The drivers of greenwashing, UCLA Institute of the Environment and Sustainability and the Anderson School of Management, 54 (1), 64–87.

Fay, W.B. (1992). The environment's second wave. *Marketing Research*, 4 (4), 44–45.

Frances, B. & Bettina, W. (2011). Carbon accounting: Negotiating accuracy, consistency and certainty across organisational fields. *Accounting, Auditing & Accountability Journal*, 24(8), 1022–1036.

Fuller, D.A. (1999). *Sustainable marketing: Managerial-ecological issues*. Thousand Oaks, CA: Sage Publications.

Fuller, D.A. & Ottman, J.A. (2002). Moderating unintended pollution: The role of sustainable product design. *Journal of Business Research*, 57, 1231–1238.

Garcés-Ayerbe, C., Rivera-Torres, P., & Murillo-Luna, J.L. (2012) Stakeholder pressure and environmental proactivity: Moderating effect of competitive advantage expectations. *Management Decision*, 50(2), 189–206.

Gilbert, D.C., Powell-Perry, J., & Widijoso, S. (1999). Approaches by hotels to the use of the Internet as a relationship marketing tool. *Journal of Marketing Practice: Applied Marketing Science*, 5(1), 21–38.

Goldsmith, R.E. (2004). Current and future trends in marketing and their implications for the discipline. *Journal of Marketing Theory and Practice*, 12(4), 10–17.

Gonzalez-Benito, J, Lannelongue, G, & Queiruga, D 2011, Stakeholders and environmental management systems: A synergistic influence on environmental imbalance. *Journal of Cleaner Production*, 19(14), 1622–1630.

Grönroos, C. (2006). On defining marketing: Finding a new roadmap for marketing. *Marketing Theory*, 6(4), 395–417.

Grove, S.J., Fisk, R.P., Pickett, G.M., & Kangun, N. (1996). Going green in the service sector: Social responsibility issues, implications and implementation. *European Journal of Marketing*, 30(5), 56–66.

Gummesson, E. (2005). Qualitative research in marketing: Roadmap for a wilderness of complexity and unpredictability. *European Journal of Marketing*, 39(3/4), 309–327.

Gupta, S., Czinkota, M., & Melewar, T.C. (2012). Embedding knowledge and value of a brand into sustainability for differentiation. *Journal of World Business*, 48(3), 287–296.

Hartmann, P., Ibáñez, A., & Sainz, F.J.F. (2005). Green branding effects on attitude: Functional versus emotional positioning strategies. *Marketing Intelligence & Planning*, 23(1), 9–29.

Hills, S.B. & Sarin, S. (2003). From market driven to market driving: An alternative paradigm for marketing in high technology industries. *Journal of Marketing Theory and Practice*, 11(3), 13–24.

Jantsch, J. (2006). *Duct tape marketing: The world's most practical small business marketing guide*. Nashville, TN: Thomas Nelson.

Jenkin, T.A., Webster, J., & McShane, L. (2011). An agenda for "Green" information technology and systems research. *Information and Organization*, 21(1), 17–40.

Jocumsen, G. (2002). How do small business managers make strategic marketing decisions? A model of process. *European Journal of Marketing*, 38(5/6), 659–674.

Jones, P., Clarke-Hill, C., Comfort, D., & Hillier, D. (2008). Marketing and sustainability. *Marketing Intelligence & Planning*, 26(2), 123–130.

Kärnä, J., Hansen, E., & Juslin, H. (2003). Social Responsibility in environmental marketing planning. *European Journal of Marketing*, 37(5/6), 848–871.

Kim, J & Oki, T (2011). Visioneering: An essential framework in sustainability science. *Sustainability Science*, 6(2), 247–251.

Kiron, D., Kruschwitz, N., Haanaes, K., & Von Streng Velken, I. (2012). Sustainability nears a tipping point. *MIT Sloan Management Review*, 53(2), 69–74.

Knight, K.W. & Messer, B. (2012). Environmental concern in cross-national perspective: The effects of affluence, environmental degradation, and world society. *Social Science Quarterly*, 93(2), 521–537.

Koku, P.S. (2005). Towards an integrated marketing strategy for developing markets. *Journal of Marketing Theory and Practice*, 13(4), 8.

Kotler, P. & Armstrong, G. (2008). *Principles of marketing* (12th ed.). Upper Saddle River, NJ: Prentice-Hall, Inc. (now known as Pearson Education, Inc.)

Kotler, P. & Zaltman, G. (1971). Social marketing: An approach to planned social change. *Journal of Marketing*, 35, 3–12.

Kristoffersen, L. & Singh, S. (2004). Successful application of a customer relationship management program in a non-profit organization. *Journal of Marketing Theory and Practice*, 12(2), 28–42.

Lim, S-J. & Phillips, J. (2008). Embedding CSR values: The global footwear industry's evolving governance structure. *Journal of Business Ethics*, 83, 143–156.

Loorbach, D., van Bakel, J.C., Whiteman, G., & Rotmans, J. (2010). Business strategies for transitions towards sustainable systems. *Business Strategy and the Environment*, 19(2), 133–146.

Lueneburger, C. & Goleman, D. (2010). The change leadership sustainability demands. *MIT Sloan Management Review*, 51(4), 49–55.

Maignan, I. & Ferrell, O.C. (2004). Corporate social responsibility and marketing: An integrative framework. *Journal of the Academy of Marketing Science*, 32(1), 3–19.

Maignan, I., Ferrell, O.C., & Ferrell, L. (2005). A stakeholder model for implementing responsibility in marketing. *European Journal of Marketing*, 39(9/10), 956–977.

Menon, A. & Menon, A. (1997). Enviropreneurial marketing strategy: The emergence of corporate environmentalism as market strategy. *Journal of Marketing*, 61(1), 51–67.

Menon, A., Menon, A., Chowdhury, J., & Jankovich, J. (1999). Evolving paradigm for environmental sensitivity in marketing programs: A synthesis of theory and practice. *Journal of Marketing Theory and Practice*, 7(2) 1–15.

Mercer, D. (1998). Long-range marketing. *Journal of Marketing Practice: Applied Marketing Science*, 4(6), 174–184.

Miles, M.P. & Covin, J.G. (2000). Environmental marketing: A source of reputational, competitive, and financial advantage. *Journal of Business Ethics*, 23, 299–311.

Millar, C., Hind, P., & Magala, S. (2012). Sustainability and the need for change: Organisational change and transformational vision. *Journal of Organizational Change Management*, 25(4), 489–500.

Morgan, R.M. & Hunt, S. (1999). Relationship-based competitive advantage: The role of relationship marketing in marketing strategy. *Journal of Business Research*, 46, 281–290.

Naditz, A. (2008). The green MBA: From campus to corporate. *Sustainability*, 1, 178–182.

Ottman, J.A. (1998). *Green marketing: Opportunity for innovation* (2nd ed.). New York: J. Ottman Consulting.

Oumlil, B. & Rao, C.P. (2005). Special issue on globalization and its challenges to marketing. *Journal of Marketing Theory and Practice*, 13(4), 5–7.

Paswan, A.K. & Troy, L.C. (2004). Nonprofit organization and membership motivations: An exploration in the museum industry. *Journal of Marketing Theory and Practice*, 12(2), 1–15.

Peattie, K. & Crane, A. (2005). Green marketing: Legend, myth, farce or prophesy? *Qualitative Market Research: An International Journal*, 8(4), 357–370.

Pogutz, S. & Micale, V. (2011). "Eco-efficiency, growth and the nature of corporate sustainability," submitted to the seventh international environmental management leadership symposium. Rochester, NY, USA.

Polonsky, M.J. & Mintu-Wimsatt, A.T. (eds). (1995). *Environmental marketing: Strategies, practice, theory, and research*. New York: The Hawthorne Press.

Rao, P.M. (2005). Sustaining competitive advantage in a high-technology environment: A strategic marketing perspective. *Advances in Competitiveness Research*, 13(1), 33–47.

Rothschild, M.L. (1999). Carrots, sticks, and promises: A conceptual framework for the management of public health and social issue behaviors. *Journal of Marketing*, 63, 24–37.

Rowley, J. (1998). Promotion and marketing communications in the information marketplace. *Library Review*, 47(8), 383–387.

Royne, M.B., Levy, M., & Martinez, J. (2011). The Public health implications of consumers' environmental concern and their willingness to pay for an eco-friendly product. *Journal of Consumer Affairs*, 45(2), 329–343.

Scammon, D.L. & Mayer, R.N. (1995). Agency review of environmental marketing claims: Case-by-case decomposition of the issues. *Journal of Advertising*, 24(2), 33–43.

Schalteggar, S. (2010). *Sustainability as a driver for corporate economic success consequences for the development of sustainability management control*. Luneburg: Leuphana University.

Schiederig, T., Tietze, F., & Herstatt, C. (2012). Green innovation in technology and innovation management—an exploratory literature review. *R&D Management*, 42(2), 180–192.

Shaw, E.H. & Jones, D.G.B. (2005). A history of schools of marketing thought. *Marketing Theory*, 5(3), 239–281.

Sheth, J.N. & Parvatiyar, A. (1995). The evolution of relationship marketing. *International Business Review*, 4(4), 397–418.

Shiroyana, H., Yarime, M., Matsu, M. Schroeder, H., Scholz, R., & Ulrich, A.E. (2012). Governance for sustainability: Knowledge integration and multi-actor dimensions in risk management. *Sustainability Science*, 7(Supplement 1), 45–55.

Silverman, G. (2001). *The secrets of word-of-mouth marketing: How to trigger exponential sales through runaway word of mouth*. New York: AMACOM.

Sridhar, K. (2012). Corporate conceptions of triple bottom line reporting: An empirical analysis into the signs and symbols driving this fashionable framework. *Social Responsibility Journal*, 8(3), 312–326.

Sterman, John D. (2012). "Sustaining Sustainability: Creating a Systems Science in a Fragmented Academy and Polarized World," M. Weinstein and R.E. Turner (eds.). *Sustainability science*, Chapter 2. New York: Springer. 21–58.

Stone, G.W. & Wakefield, K.L. (2000). Eco-orientation: An extension of market orientation in an environmental context. *Journal of Marketing Theory and Practice*, 8(3), 21–31.

Tate, W.L., Ellram, L.M., & Dooley, K.J. (2012). Environmental purchasing and supplier management (EPSM): Theory and practice. *Journal of Purchasing and Supply Management*, 18(3), 173–188. http://dx.doi.org/10.1016/j.pursup.2012.07.001 (accessed March 17, 2015).

Taylor, C.R. (2000). Emerging issues in marketing. *Psychology & Marketing*, 17(6), 441–447.

Thomas, J. & Gupta, R.K. (2005). Marketing theory and practice: Evolving through turbulent times. *Global Business Review*, 6(1), 95–112.

Tinsely, S.J. & Melton, K. (1997). Sustainable development and its effect on the marketing planning process. *Eco-management and Auditing*, 4, 116–126.

Tseng, M., Wang, R., Chiu, A.S.F., Geng, Y., & Lin, Y.H. (2012). Improving performance of green innovation practices under uncertainty. *Journal of Cleaner Production*, 40, 71–82.

Wagner, E.R. & Hansen, E.N. (2002). Methodology for evaluating green advertising of forest products in the United States: A content analysis. *Forest Products Journal*, 52(4), 17–23.

Weinrich, N.K. (1999). *Hands-on social marketing: A step-by-step guide*. Thousand Oaks, CA: Sage Publications.

Weng, MH & Lin, CY (2011). Determinants of green innovation adoption for small and medium-size enterprises (SMES). *African Journal of Business Management*, 5(22), 9154–9163.

Weissman, W.A & Sekutowski, J.C. (1991). Environmentally conscious manufacturing. *AT & T Technical Journal*, 70(6), 23–30.

Wilkie, W.L. & Moore, E.S. (1999). Marketing's contribution to society. *Journal of Marketing*, 63(special issue), 198–218.

Witkowski, T.H. (2005). Fair trade marketing: An alternative system for globalization and development. *Journal of Marketing Theory and Practice*, 13(4), 22–33.

Wolfe, D.B. (1998). Developmental relationship marketing (connecting messages with mind: an empathetic marketing system). *Journal of Consumer Marketing*, 15(5), 449–467.

Young, R.A., Weiss, A.M., & Stewart, D.W. (2006). *Marketing champions: Practical strategies for improving marketing's power, influence, and business impact*. Hoboken, NJ: John Wiley & Sons, Inc.

Zaltman, G. & Zaltman, L. (2008). *Marketing metaphoria: What deep metaphors reveal about the minds of consumers*. Boston, MA: Harvard Business Press.

CHAPTER 10

Abdelsalam, O., Duygun, M., Matallín-Sáez, J.C., & Tortosa-Ausina, E. (2014). Do ethics imply persistence? The case of Islamic and socially responsible funds. *Journal of Banking & Finance*, 40, 182–194.

Abul Shamsuddin, A. (2014). Are Dow Jones Islamic equity indices exposed to interest rate risk? *Economic Modelling*, 39, 273–281.

Armstrong, J.S. & Green, K.C. (2013). Effects of corporate social responsibility and irresponsibility policies. *Journal of Business Research*, 66(10), 1922–1927.

Agudo, J.M., Ayerbe, C.G., & Figueras, M.S. (2012). Social responsibility practices and evaluation of corporate social performance. *Journal of Cleaner Production*, 35(25–38), 25–38.

Aktas, N., de Bodt, E., & Cousin, J-G. (2011). Do financial markets care about SRI? Evidence from mergers and acquisitions. *Journal of Banking & Finance*, 35(7), 753–1761.

Amalric, F. (2006). Pension funds, corporate responsibility and sustainability. *Ecological Economics*, 59(4), 440–450.

Ballestero, E., Bravo, M., & Perez Gladish, B. (2012). Socially Responsible Investment: A multicriteria approach to portfolio selection combining ethical and financial objectives. *European Journal of Operational Research*, 216(2), 487–494.

Barnea, A., Heinkel, R., & Kraus, A. (2005). Green investors and corporate investment. *Structural Change and Economic Dynamics*, 16(3), 332–346.

Baron, D.P. (2008). Managerial contracting and corporate social responsibility. *Journal of Public Economics*, 92(1–2), 268–288

Bauer, R., Derwall, J., & Otten, R. (2007). The ethical mutual fund performance debate: New evidence from Canada. *Journal of Business Ethics*, 70(2), 111–124.

Bauer, R., Koedijk, K., & Otten, R. (2005). International evidence on ethical mutual fund performance and investment style. *Journal of Banking & Finance*, 29(7), 1751–1767.

Bauer, R., Otten, R., & Rad, A. T. (2006). Ethical investing in Australia: Is there a financial penalty? *Pacific-Basin Finance Journal*, 14(1), 33–48.

Becchetti, L., & Ciciretti, R. (2009). Corporate social responsibility and stock market performance. *Applied financial economics*, 19(16), 1283–1293.

Benson, K.L., & Humphrey, J.E. (2008). Socially responsible investment funds: Investor reaction to current and past returns. *Journal of Banking & Finance*, 32(9), 1850–1859.

Berry, T., Edgerton, N., & George, A. (2011). Mainstreaming Socially Responsible Investment (SRI): A role for government? Policy recommendation from the investment community. Retrieved July 16, 2014, from http://www.isf.uts.edu.au/pdfs/srireport181105.pdf.

Besley, T. & Ghata, M. (2007).Retailing public goods: The economics of corporate social responsibility. *Journal of Public Economics*, 91(9), 1645–1663.

Bian, S.X.N. & Wong, V. (2013). An examination of the diversification benefits of SRI in a portfolio context. *Journal of Energy Technologies and Policy*, 3(11), 397–407.

Bilbao-Terol, A., Arenas-Parra, M., & Cañal-Fernández, V. (2012). A fuzzy multi-objective approach for sustainable investments. *Expert Systems with Applications*, 39(12), 10904–10915.

Chami, R., Cosimano, T.F., & Fullenkamp, C. (2002). Managing ethical risk: How investing in ethics adds value. *Journal of Banking & Finance*, 26(9), 1697–1718.

Choi, D.Y. & Gray, E.R. (2008). Socially responsible entrepreneurs: What do they do to create and build their companies? *Business Horizons*, 51(4), 341–352.

Chong, J., Her, M., & Phillips, G. M. (2006). To sin or not to sin? Now that's the question. *Journal of Asset Management*, 6(6), 406–417.

Cochran, P.L. (2007). The evolution of corporate social responsibility. *Business Horizons*, 50(6), 449–454.

Consolandi, C., Innocenti, A., & Vercelli, A. (2009). CSR, rationality and the ethical preferences of investors in a laboratory experiment. *Research in Economics*, 63(4), 242–252.

Curran, M.M. & Moran, D. (2007). Impact of the FTSE4Good Index on firm price: An event study. *Journal of Environmental Management*, 82(4), 529–537.

Dam, L. & Heijdra, B.J. (2011). The environmental and macroeconomic effects of socially responsible investment. *Journal of Economic Dynamics and Control*, 35(9), 1424–1434.

Derigs, U. & Marzban, S. (2009). New strategies and a new paradigm for Shariah-compliant portfolio optimization. *Journal of Banking & Finance*, 33(6), 1166–1176.

Derwall, J., Koedijk, K., & Ter Hors, J. (2011). A tale of values-driven and profit-seeking social investors. *Journal of Banking & Finance*, 35(8), 2137–2147.

Doane, D. (2005). Beyond corporate social responsibility: Minnows, mammoths and markets. *Futures*, 37(2–3), 215–229.

El Ghoul, S., Guedhami, O., Kwok, C., & Mishra, D.R. (2011). Does corporate social responsibility affect the cost of capital? *Journal of Banking & Finance*, 35(9), 2388–2406.

El-Hawary, D., Grais, W., & Iqbal, Z. (2007). Diversity in the regulation of Islamic Financial Institutions. *The Quarterly Review of Economics and Finance*, 46(5), 778–800.

Erkens, D.H., Hung, M., & Matos, P. (2012). Corporate governance in the 2007–2008 financial crisis: Evidence from financial institutions worldwide. *Journal of Corporate Finance*, 18(2), 389–411.

Falck, O., & Heblich, S. (2007). Corporate social responsibility: Doing well by doing good. *Business Horizons*, 50(3), 247–254.

Friedman, A.L. & Miles, S. (2001). Socially responsible investment and corporate social and environmental reporting in the UK: An exploratory study. *The British Accounting Review*, 33(4), 523–548.

Galema, R., Plantiga, A., & Scholtens, B. (2008). The stocks at stake: Return and risk in socially responsible investment. *Journal of Banking & Finance*, 32(12), 2646–2654.

Gjølberg, M. (2009). Measuring the immeasurable? Constructing an index of CSR practices and CSR performance in 20 countries. *Scandinavian Journal of Management*, 25(1), 10–22.

Goss, A. & Roberts, G.S. (2011).The impact of corporate social responsibility on the cost of bank loans. *Journal of Banking & Finance*, 35(7), 1794–1810.

Gupta, P., Mehlawat, M., & Kumar, S.A. (2013). Hybrid optimization models of portfolio selection involving financial and ethical considerations. *Knowledge-Based Systems*, 37, 318–337.

Hallerbach, W., Ning, H., Soppe, A., & Spronk, J. (2004). A framework for managing a portfolio of socially responsible investments. *European Journal of Operational Research*, 153(2), 517–529.

Hamilton, S., Jo, H., & Statman, M. (1993). Doing well while doing good? The investment performance of socially responsible mutual funds. *Financial Analysts Journal*, 49(6), 62–66.

Hancock, J. (2005). *Investing in corporate social responsibility: a guide to best practice, business planning & the UK's leading companies*. Kogan Page Ltd, London/Sterling, VA.

Heinkel, R., Kraus, A., & Zechner, J. (2001). The effect of green investment on corporate behavior. *Journal of Financial and Quantitative Analysis*, 36(4), 431–450

Hickman, K.A., Teets, W.R., & Kohls, J.J. (1999). Social investing and modern portfolio theory. *American Business Review*, 17(1), 72–78.

Heslin, P.A. & Ochoa, J.D. (2008). Understanding and developing strategic corporate social responsibility. *Organizational Dynamics*, 37(2), 125–144.

Hollingworth, S. (1998). Green investing: A growing concern? *Australian CPA*, 68(4), 28–30.

Hong, H. & Kacperczyk, M. (2009). The price of sin: The effects of social norms on markets. *Journal of Financial Economics*, 93(1), 15–36.

Hoti, S., McAleer, M., & Pauwels, L.L. (2008). Multivariate volatility in environmental finance. *Mathematics and Computers in Simulation*, 78(2–3), 189–199.

Jin, H.H., Mitchell, O.S., & Piggott, J. (2006). Socially responsible investment in Japanese pensions. *Pacific-Basin Finance Journal*, 14(5), 427–438.

Kjaerheim, G. (2005). Cleaner production and sustainability. *Journal of Cleaner Production*, 13(4), 329–339.

Klemeš, J.J., Varbanov, P.S., & Huisingh, D. (2012). Recent cleaner production advances in process monitoring and optimisation. *Journal of Cleaner Production*, 34, 1–8.

Knox, S. & Maklan, S. (2004). Corporate social responsibility: Moving beyond investment towards measuring outcomes. *European Management Journal*, 22(5), 508–516.

Kurtz, L. (1997). No effect, or no net effect? Studies on socially responsible investing. *The Journal of Investing*, 6(4), 37–49.

Leite, P. & Cortez, M.C. (2014). Style and performance of international socially responsible funds in Europe. *Research in International Business and Finance*, 30, 248–267.

Levi, J. (2006). Adoption of corporate social responsibility codes by multinational companies. *Journal of Asian Economics*, 17(1), 50–55.

Luther, R. & Matatko, J. (1994). The performance of ethical unit trusts: Choosing an appropriate benchmark. *The British Accounting Review*, 26(1), 77–89.

Luther, R.G., Matatko, J., & Corner, D.C. (1992). The investment performance of UK "ethical" Unit Trusts. *Accounting, Auditing & Accountability Journal*, 5(4), 57–70.

Neal, R. & Cochran, P.L. (2008). Corporate social responsibility, corporate governance, and financial performance: Lessons from finance. *Business Horizons*, 51(6), 535–540.

Oh, C.H., Park, J.H., & Ghauri, P. N. (2013). Doing right, investing right: Socially responsible investing and shareholder activism in the financial sector. *Business Horizons*, 56(6), 703–714.

Anastasia O'Rourke, A. (2003). The message and methods of ethical investment. *Journal of Cleaner Production*, 11(6), 68–693.

Ortas, E., Moneva, J.M., & Salvador, M. (2012). Does socially Responsible Investment equity indexes in emerging markets pay off? Evidence from Brazil. *Emerging Markets Review*, 13(4), 588–597

Renneboog, L., Ter Horst, J., & Zhang, C. (2008).The price of ethics and stakeholder governance: The performance of socially responsible mutual funds. *Journal of Corporate Finance*, 14 (3), 302–322.

Sauer, D.A. (1997). The impact of social-responsibility screens on investment performance: Evidence from the Domini 400 Social Index and Domini Equity Mutual Fund. *Review of Financial Economics*, 6(2), 137–149.

Scholtens, B. (2008). A note on the interaction between corporate social responsibility and financial performance. *Ecological Economics*, 68(1–2), 46–55.

Statman, M. (2000). Socially responsible mutual funds. *Financial Analysts Journal*, 56(3), 30–39.

Székely, F. & Knirsch, M. (2005). Responsible leadership and corporate social responsibility: Metrics for sustainable performance. *European Management Journal*, 23(6), 628–647.

Tippet, J. (2001). Performance of Australia's ethical funds. *Australian Economic Review*, 34(2), 170–178.

Unerman, J. & O'Dwyer, B. (2007). The business case for regulation of corporate social responsibility and accountability. *Accounting Forum*, 31(4), 332–353.

Van den Bossche, F., Rogge, N., Devooght, K., & Van Puyenbroeck, T. (2010). Robust Corporate Social Responsibility investment screening. *Ecological Economics*, 69(5), 1159–1169.

Veith, S., Werner, J.R., & Zimmermann, J. (2009). Capital market response to emission rights returns: Evidence from the European power sector. *Energy Economics*, 31(4), 605–613.

Waddock, S.A., Graves, S.B., & Gorski, R. (2000). Performance characteristics of social and traditional investments. *The Journal of Investing*, 9(2), 27–38.

Webb, D.J, Mohr, L.A., & Harris, K.E. (2008). A re-examination of socially responsible consumption and its measurement. *Journal of Business Research*, 61(2), 91–98.

Weber, M. (2008).The business case for corporate social responsibility: A company-level measurement approach for CSR. *European Management Journal*, 26(4), 247–261.

Webley, P., Lewis, A., & Mackenzie, C. (2001). Commitment among ethical investors: An experimental approach. *Journal of Economic Psychology*, 22(1), 27–42.

Wu, M.W., & Shen, C.H. (2013). Corporate social responsibility in the banking industry: Motives and financial performance. *Journal of Banking & Finance*, 37(9), 3529–3547.

Zeng, S.X., Meng, X.H., Yin, H.T., Tam, C. M., & Sun, L. (2010). Impact of cleaner production on business performance. *Journal of Cleaner Production*, 18(10–11), 975–983.

Zhang, B., Bi, J., Yuan, Z., Ge, J., Liu, B., & Bu, M. (2008). Why do firms engage in environmental management? An empirical study in China. *Journal of Cleaner Production*, 16(10), 1036–1045.

CHAPTER 11

Benn, S. and D. Bolton. (2011). *Key Concepts in Corporate Social Responsibility*. London: Sage Press.

Ottman, Jacquelyn. (2011). *The New Rules of Green Marketing: Strategies, Tools, and Inspiration for Sustainable Branding*. San Francisco, CA: Berrett-Koehler Publishers.

CHAPTER 12

Daly, H. E. and J. Farley. 2010. *Ecological economics: principles and applications*. Washington, DC: Island Press.

Field, B. and M. Field. 2012. *Environmental Economics: an introduction*. New York: McGraw-Hill.

Goodwin, N., J. A. Nelson, F. Ackerman, and T. Weisskopf. 2009. *Microeconomics in Context*. Armonk, NY: ME Sharpe.

Harris, J. and B. Roach. 2013. *Environmental and natural resource economics: a contemporary approach*. New York: M.E. Sharpe.

Ostrom, E., C. Chang, M. Pennington, and V. Tarko. 2012. *The future of the commons: beyond market failure and government regulation*. London: The Institute of Economic Affairs.

Scorse, J. 2010. *What environmentalists need to know about economics*. New York: Palgrave Macmillan.

Tietenberg, T. and L. Lewis. 2015. *Environmental and natural resource economics*. New York: Pearson.

CHAPTER 13

Brand, Stewart. *Whole Earth Discipline: Why Dense Cities, Nuclear Power, Transgenic Crops, Restored Wildlands, and Geoengineering are Necessary*. London: Penguin Books, 2010.

Collins, Craig. *Toxic Loopholes: Failure and Future Prospects for Environmental Law*. Cambridge: Cambridge University Press, 2010.

Hays, Samuel P. *A History of Environmental Politics Since 1945*. Pittsburgh, PA: University of Pittsburgh Press, 2000.

Kraft, Michael E. *Environmental Policy and Politics*. 5th edition. Boston, MA: Longman, 2011.

Layzer, Judith A. *The Environmental Case: Translating Values into Policy*. 3rd edition. Washington, DC: CQ Press, 2012.

Reisner, Marc. *Cadillac Desert: The American West and its Disappearing Water*. New York: Penguin Books, 1993.

Smith, Laurence C. *The World in 2050: Four Forces Shaping Civilization's Northern Future*. New York: Penguin Books/Plume, 2011.

Yergin, Daniel *The Quest: Energy, Security, and the Remaking of the Modern World*. New York: Penguin Books, 2011.

CHAPTER 14

Activism

De-Shalit, Avner. *The Environment: Between Theory and Practice*. Oxford: Oxford University Press, 2000.

Francione, Gary L., and Robert Garner. *The Animal Rights Debate. Abolition or Regulation?*. New York: Columbia University Press, 2010.

List, Peter C. *Radical Environmentalism: Philosophy and Tactics*. Belmont, CA: Wadsworth Pub. Co, 1993.

Marietta, Don E., and Lester E. Embree. *Environmental Philosophy and Environmental Activism*. Lanham, MD: Rowman & Littlefield, 1995.

Munro, Lyle. *Confronting Cruelty. Moral Orthodoxy and the Challenge of the Animal Rights Movement*. Leiden: Brill, 2005.

Animals

Armstrong, Susan J., and Richard George Botzler. *The Animal Ethics Reader*, 2nd ed. London: Routledge, 2008.

Atterton, Peter, and Matthew Calarco. *Animal Philosophy*. New York: Continuum, 2004.

Cochrane, Alasdair. *Introduction to Animals and Political Theory*. New York: Palgrave MacMillan, 2010.

Garner, Robert. *Animal Ethics*. Cambridge: Polity, 2005.

Gruen, Lori. *Ethics and Animals: An Introduction*. Cambridge: Cambridge University Press, 2011.

Hargrove, Eugene C. *The Animal Rights, Environmental Ethics Debate: The Environmental Perspective*. Albany, NY: State University of New York Press, 1992.

Kymlicka, Will and Sue Donaldson. *Zoopolis. A Political Theory of Animal Rights*. Oxford: Oxford University Press.

O'Sullivan, Siobhan. *Animals, Equality and Democracy*. New York: Palgrave MacMillan, 2011.

Palmer, Clare. *Animal Ethics in Context*. New York: Columbia University Press, 2010.

Singer, Peter. *Practical Ethics*, 3rd ed. New York: Cambridge University Press.

Singer, Peter. *Animal Liberation*, 2nd ed. London: Pimlico.

Taylor, Angus, John W. Burbidge, and Angus Taylor. *Animals & Ethics: An Overview of the Philosophical Debate*. Peterborough, OT: Broadview Press, 2009.

Visak, Tatjana. *Killing Happy Animals. Explorations in Utilitarian Ethics*. New York: Palgrave MacMillan, 2013.

Climate Change

Arnold, Denis Gordon. *The Ethics of Global Climate Change*. Cambridge: Cambridge University Press, 2011.

Broome, John. *Climate Matters: Ethics in a Warming World*. New York: W.W. Norton, 2012.

Brown, Donald A. *Climate Change Ethics: Navigating the Perfect Moral Storm*. London: Routledge, 2013.

Gardiner, Stephen, Simon Caney, Dale Jamieson, and Henry Shue. *Climate Ethics: Essential Readings*. Oxford: Oxford University Press, 2010.

Garvey, James. *The Ethics of Climate Change: Right and Wrong in a Warming World*. London: Continuum, 2008.

Harris, Paul G. *World Ethics and Climate Change: From International to Global Justice*. Edinburgh: Edinburgh University Press, 2010.

Irwin, Ruth. *Climate Change and Philosophy: Transformational Possibilities*. London: Continuum Intl Pub Group, 2010.

McKibben, Bill. *The Global Warming Reader: A Century of Writing about Climate Change*. New York: Penguin Books, 2012.

Nanda, Ved P. *Climate Change and Environmental Ethics*. New Brunswick: Transaction Publishers, 2011.

Preston, Christopher J. *Engineering the Climate: The Ethics of Solar Radiation Management*. Lanham, MD: Lexington Books, 2012.

Schoenfeld, Martin. *Plan B: Global Ethics on Climate Change: The Planetary Crisis and Philosophical Alternatives*. London: Routledge, 2012.

Skrimshire, Stefan. *Future Ethics: Climate Change and Apocalyptic Imagination*. London: Continuum, 2010.

Thompson, Allen, and Jeremy Bendik-Keymer. *Ethical Adaptation to Climate Change: Human Virtues of the Future*. Cambridge, MA: MIT Press, 2012.

Emerging Technologies (Synthetic Biology, Nanotechnology, Etc.)

Allhoff, Fritz. *Nanoethics: The Ethical and Social Implications of Nanotechnology*. Hoboken, NJ: Wiley-Interscience, 2007.

Allhoff, Fritz, and Patrick Lin (eds.). *Nanotechnology & Society: Current and Emerging Ethical Issues*. Dordrecht: Springer, 2009.

Budinger, Thomas F., and Miriam D. Budinger. *Ethics of Emerging Technologies: Scientific Facts and Moral Challenges*. Hoboken, NJ: John Wiley & Sons, Inc., 2006.

David, Kenneth H., and Paul B. Thompson. *What Can Nanotechnology Learn from Biotechnology?: Social and Ethical Lessons for Nanoscience from the Debate Over Agrifood Biotechnology and GMOs*. Amsterdam: Elsevier/Academic Press, 2008.

O'Mathuna, Donal. *Nanoethics: Big Ethical Issues with Small Technology*. London: Continuum, 2009.

Ruse, Michael, and David Castle (eds.). *Genetically Modified Foods: Debating Biotechnology*. Amherst, NY: Prometheus Books, 2002.

Sherlock, Richard, and John D. Morrey (eds.). *Ethical Issues in Biotechnology*. Lanham, MD: Rowman & Littlefield, 2002.

Schmidt, Markus (ed.). *Synthetic Biology: The Technoscience and Its Societal Consequences*. Dordrecht: Springer, 2009.

Environmental Justice

Adamson, Joni, Mei Mei Evans, and Rachel Stein. *The Environmental Justice Reader: Politics, Poetics & Pedagogy*. Tucson, AZ: University of Arizona Press, 2002.

Agyeman, Julian, Peter Cole, Randolph Haluza-DeLay, and Pat O'Riley (eds.). *Speaking for Ourselves: Environmental Justice in Canada*. Vancouver, CA: UBC Press, 2009.

Cole, Luke W., and Sheila R. Foster. *From the Ground Up: Environmental Racism and the Rise of the Environmental Justice Movement*. New York: New York University Press, 2001.

Hill, Barry E. *Environmental Justice: Legal Theory and Practice*. 2nd ed. Washington, DC: ELI Press/Environmental Law Institute, 2012.

Rainey-Brown, Shirley A., and Glenn S. Johnson. *Environmental Justice Reader: Addressing the History, Issues, Policy and Change*. Deer Park, NY: Linus Publications, 2011.

Sandler, Ronald D., and Phaedra C. Pezzullo. *Environmental Justice and Environmentalism: The Social Justice Challenge to the Environmental Movement*. Cambridge, MA: MIT Press, 2007.

Shrader-Frechette, K. S. *Environmental Justice: Creating Equality, Reclaiming Democracy*. Oxford: Oxford University Press, 2002.

Stein, Rachel. *New Perspectives on Environmental Justice: Gender, Sexuality, and Activism*. New Brunswick, NJ: Rutgers University Press, 2004.

Walker, Gordon. *Environmental Justice: Concepts, Evidence and Politics*. Abingdon, Oxon: Routledge, 2012.

Food, Water, and Agriculture

Allhoff, Fritz Monroe Dave. *Food and Philosophy: Eat, Think, and Be Merry*. New York: Blackwell Publishers, 2007.

Allhoff, Fritz. *Wine & Philosophy: In Vino Veritas*. New York: Blackwell Publishers, 2008.

Brown, Peter G., and Jeremy J. Schmidt. *Water Ethics: Foundational Readings for Students and Professionals*. Washington, DC: Island Press, 2010.

Gottwald, Franz-Theo, Hans Werner Ingensiep, and Marc Meinhardt. *Food Ethics*. New York: Springer, 2010.

Kaplan, David M. *Philosophy of Food*. Berkeley: University of California Press, 2012.

Korthals, Michiel. *Before Dinner: Philosophy and Ethics of Food*. Dordrecht: Springer, 2004.

Mepham, Ben T. *Food Ethics. Professional Ethics*. New York: Routledge, 1996.

Pence, Gregory E. *The Ethics of Food: A Reader for the Twenty-First Century*. Lanham, MD: Rowman & Littlefield, 2002.

Pojman, Paul. *Food Ethics*. Boston, MA: Wadsworth Cengage Learning, 2011.

Singer, Peter, and Jim Mason. *The Ethics of What We Eat. Why Our Food Choices Matter*. Emmaus, PA: Rodale, 2006.

Telfer, Elizabeth. *Food for Thought: Philosophy and Food*. London/New York: Routledge, 1996.

Thompson, Paul B. *The Spirit of the Soil: Agriculture and Environmental Ethics*. Environmental Philosophies Series. New York: Routledge, 1994.

Thompson, Paul B. *Agricultural Ethics: Research, Teaching, and Public Policy*. New York: John Wiley & Sons, Inc., 1999.

Thompson, Paul B. *Food Biotechnology in Ethical Perspective*, 2nd ed. Dordrecht: Springer, 2007.

Future Generations

Gosseries, Axel, and Lukas H. Meyer (eds.). *Intergenerational Justice*. New York: Oxford University Press, 2009.

Partridge, Ernest (ed.). *Responsibilities to Future Generations*. Buffalo, NY: Prometheus, 1980.

Tremmel, Jorg (ed.). *Handbook of Intergenerational Justice*. Cheltenham: Edward Elgar, 2006.

Indigenous Philosophy

Berkes, Fikret. *Sacred Ecology*. 3rd ed. New York: Routledge, 2012.

Callicott, J. Baird. *Earth's Insights A Survey of Ecological Ethics from the Mediterranean Basin to the Australian Outback*. Berkeley, CA: University of California Press, 1997.

Callicott, J. Baird, and Michael P. Nelson. *American Indian Environmental Ethics: An Ojibwa Case Study*. Upper Saddle River, NJ: Prentice Hall, 2004.

Cordova, V. F., and Kathleen Dean Moore. *How It Is: The Native American Philosophy of V.F. Cordova*. Tucson, AZ: University of Arizona Press, 2007.

Dudgeon, Roy C. *Common Ground: Eco-Holism & Native American Philosophy*. Winnipeg: Pitch Black Publications, 2008.

Schweninger, Lee (ed.). *Listening to the Land: Native American Literary Responses to the Landscape*. Athens, GA: University of Georgia Press, 2008.

Public Policy

Kamieniecki, Sheldon, and Michael E. Kraft. *Business and Environmental Policy: Corporate Interests in the American Political System*. Cambridge, MA: MIT Press, 2007.

Kraft, Michael E. *Environmental Policy and Politics*. 5th ed. Boston, MA: Pearson, 2010.

Layzer, Judith A. *The Environmental Case: Translating Values into Policy*. 3rd ed. Washington, DC: CQ Press, 2012.

Vig, Norman J., and Michael E. Kraft. *Environmental Policy: New Directions for the 21st Century*. 8th ed. Los Angeles, CA: CQ Press, 2012.

Sustainability

Agyeman, Julian. *Sustainable Communities and the Challenge of Environmental Justice*. New York: New York University Press, 2005.

Becker, Christian U. *Sustainability Ethics and Sustainability Research*. New York: Springer-Verlag, 2011.

Fredericks, Sarah E. *Measuring and Evaluating Sustainability: Ethics in Sustainability Indexes*. London: Routledge, 2013.

Newton, Lisa H. *Ethics and Sustainability: Sustainable Development and the Moral Life*. Upper Saddle River, NJ: Prentice Hall, 2003.

Norton, Bryan G. *Searching for Sustainability: Interdisciplinary Essays in the Philosophy of Conservation Biology*. Cambridge: Cambridge University Press, 2003.

Norton, Bryan G. *Sustainability: A Philosophy of Adaptive Ecosystem Management*. Chicago, IL: University of Chicago Press, 2005.

Thompson, Paul B. *The Agrarian Vision: Sustainability and Environmental Ethics*. Lexington, KY: University Press of Kentucky, 2010.

Raffaelle, Ryne, Wade L. Robison, and Evan Selinger (eds.). *Sustainability Ethics: 5 Questions*. Copenhagen: Automatic Press/VIP, 2010.

Wilderness

Callicott, J. Baird, and Michael P. Nelson (eds.). *The Great New Wilderness Debate*. Athens, GA: University of Georgia Press, 1998.

Cronon, William. *Uncommon Ground: Rethinking the Human Place in Nature*. New York: W.W. Norton & Co, 1996.

Duerr, Hans Peter. *Dreamtime: Concerning the Boundary between Wilderness and Civilization*. Oxford: Basil Blackwell, 1985.

Dizard, Jan E. *Going Wild Hunting, Animal Rights, and the Contested Meaning of Nature*. rev. & expanded ed. Amherst, MA: University of Massachusetts Press, 1999.

Nash, Roderick Frazier. *Wilderness and the American Mind*. 4th ed. New Haven, CT: Yale University Press, 2001.

Nelson, Michael P., and J. Baird Callicott (eds.). *The Wilderness Debate Rages on: Continuing the Great New Wilderness Debate*. Athens, GA: University of Georgia Press, 2008.

Oelschlaeger, Max. *The Idea of Wilderness: From Prehistory to the Age of Ecology*. New Haven, CT: Yale University Press, 1991.

Anthologies

Armstrong, Adrian. *Ethics and Justice for the Environment*. Milton Park, Abingdon, Oxon: Routledge, 2012.

Armstrong, Adrian C. *Here for Our Children's Children?: Why We Should Care for the Earth*. Charlottesville, VA: Imprint Academic, 2009.

Armstrong, Susan J., and Richard George Botzler. *Environmental Ethics: Divergence and Convergence.* 3rd ed. Boston, MA: McGraw-Hill, 2003.

Attfield, Robin. *Environmental Ethics: An Overview for the Twenty-First Century.* Cambridge: Polity Press, 2003.

Benson, John. *Environmental Ethics: An Introduction with Readings.* London: Routledge, 2000.

Boylan, Michael. *Environmental Ethics (Basic Ethics in Action).* Upper Saddle River, NJ: Prentice Hall, 2001.

Brennan, Andrew, and Y. S. Lo. *Understanding Environmental Philosophy.* Durham: Acumen Publishing, Ltd., 2010.

Clowney, David, and Patricia Mosto. *Earthcare: An Anthology in Environmental Ethics.* Lanham, MD: Rowman & Littlefield Publishers, Inc, 2009.

Curry, Patrick. *Ecological Ethics: An Introduction.* 2nd ed., fully rev. and expanded. Cambridge: Polity Press, 2011.

Derr, Patrick George and Edward M. McNamara. *Case Studies in Environmental Ethics.* Lanham, MD: Rowman & Littlefield, 2003.

Des Jardins, Joseph. *Environmental Ethics.* 5th ed. Boston, MA: Wadsworth, 2012.

Donatelli, Piergiorgio. *Manuale di etica ambientale. [Handbook of environmental ethics.].* Firenze, IT: Le Lettere, 2012.

Gruen, Lori, Dale Jamieson, and Christopher Schlottmann. *Reflecting on Nature: Readings in Environmental Ethics and Philosophy.* 2nd ed. New York, NY: Oxford University Press, 2013.

Gudorf, Christine E. and James E. Huchingson. *Boundaries: A Casebook in Environmental Ethics.* 2nd ed. Washington, DC: Georgetown University Press, 2010.

Jamieson, Dale. *Ethics and the Environment: An Introduction.* Cambridge: Cambridge University Press, 2008.

Kaufman, Frederik A. *Foundations of Environmental Philosophy: A Text with Readings.* Boston, MA: McGraw-Hill, 2003.

Keller, David R. *Environmental Ethics: The Big Questions.* Chichester, West Sussex: Wiley-Blackwell, 2010.

Kernohan, Drew. *Environmental Ethics: An Interactive Introduction.* Buffalo, NY: Broadview Press, 2012.

Light, Andrew, and Holmes Rolston. *Environmental Ethics: An Anthology.* Malden, MA: Blackwell Pub, 2003.

Martin-Schramm, James B., and Robert L. Stivers. *Christian Environmental Ethics: A Case Method Approach.* Maryknoll, NY: Orbis Books, 2003.

Minteer, Ben A. *Refounding Environmental Ethics: Pragmatism, Principle, and Practice.* Philadelphia, PA: Temple University Press, 2012.

Newton, Lisa H., Catherine K. Dillingham, and Joanne Choly. *Watersheds 4: Ten Cases in Environmental Ethics.* 4th ed. Australia: Thomson Wadsworth, 2006.

O'Neill, John, Alan Holland, and Andrew Light. *Environmental Values.* London: Routledge, 2008.

Paslack, Rainer, Kees Vromans, and Gamze Yuecel Isildar (eds.). *Environmental Ethics: An Introduction and Learning Guide.* Sheffield: Greenleaf, 2012.

Pojman, Louis P. and Paul Pojman. *Environmental Ethics: Readings in Theory and Application.* 6th ed. Boston, MA: Wadsworth, 2011.

Pojman, Louis P. *Global Environmental Ethics.* Mountain View, CA: Mayfield Pub, 2000.

Rolston, Holmes III. *A New Environmental Ethics: The Next Millennium for Life on Earth.* New York: Routledge, 2011.

Sandler, Ronald L. *Character and Environment: A Virtue-Oriented Approach to Environmental Ethics.* New York: Columbia University Press, 2007.

Sarkar, Sahotra. *Environmental Philosophy: From Theory to Practice.* Malden, MA: Wiley-Blackwell, 2012.

Schmidtz, David and Elizabeth Willott. *Environmental Ethics: What Really Matters, What Really Works.* 2nd ed. New York: Oxford University Press, 2012.

Shrader-Frechette, K. S. *Environmental Justice: Creating Equality, Reclaiming Democracy.* Oxford: Oxford University Press, 2002.

Traer, Robert. *Doing Environmental Ethics,* 2nd ed. Boulder, CO: Westview Press, 2012.

VanDeVeer, Donald and Christine Pierce. *The Environmental Ethics and Policy Book: Philosophy, Ecology, Economics.* 3rd ed. Belmont, CA: Thomson/Wadsworth, 2003.

Wenz, Peter S. *Environmental Ethics Today.* New York: Oxford University Press, 2001.

Williston, Byron. *Environmental Ethics for Canadians.* Oxford: Oxford University Press, 2011.

Zimmerman, Michael E. *Environmental Philosophy: From Animal Rights to Radical Ecology.* 4th ed. Upper Saddle River, NJ: Pearson/Prentice Hall, 2005.

CHAPTER 15

Amin, A. 2005. Local community on trial. *Economy and Society* 34(4): 612–633.

Critiques the use of the concept of "community" and the reliance on community participation in decision making, arguing that this serves to coopt grassroots resistance and abdicate the state's responsibility for redistribution.

Arnstein, S.R. 1969. A ladder of citizen participation. *Journal of the American Institute of Planning* 35(4): 216–224.

Argues that forms of participation can be ranked on a "ladder" from perfunctory or manipulative approaches to ones that give real power to stakeholders.

Arvai, J.L. and A. Froschauer. 2010. Good decisions, bad decisions: the interaction of process and outcome in evaluations of decision quality. *Journal of Risk Research* 13(7): 845–859.

Reports a study showing that people see a participation process as inadequate if the outcome is not one they support.

Billgren, C. and H. Holmén. 2008. Approaching reality: comparing stakeholder analysis and cultural theory in the context of natural resource management. *Land Use Policy* 25: 550–562.

Examines the usefulness of Grid-Group Cultural Theory as a framework for ensuring a diversity of stakeholders, with mixed results.

Bora, A. 2010. Technoscientific normativity and the "iron cage" of law. *Science, Technology, and Human Values* 35(1): 3–28.

Argues that participation procedures fall short because dominant actors set the terms of the debate—focusing on "technoscientific" norms—in ways that exclude certain interests and stakeholder positions.

Byron, I. and A. Curtis. 2001. Landcare in Australia: burned out and browned off. *Local Environment* 6(3): 311–326.

Discusses the Landcare program in Australia, a nationwide participatory environmental management program, detailing how the government has displaced responsibility onto Landcare groups without giving them adequate funding.

Carter, J.L. and G.J.E. Hill. 2007. Critiquing environmental management in indigenous Australia: two case studies. *Area* 39(1): 43–54.

Compares and contrasts two projects—one successful and one unsuccessful—to gain indigenous participation in environmental management through creating sustainable trepang (sea cucumber) production systems.

Chambers, R. 1997. *Whose reality counts? putting the first last.* Intermediate Technology Publications, London.

Gives an introduction to the justifications for participation in international development projects, focusing on the methodology of Participatory Rural Appraisal.

Chess, C. and K. Purcell. 1999. Public participation and the environment: do we know what works? *Environmental Science and Technology* 33(16): 2685–2692.

Reviews research on the effectiveness of different participatory mechanisms, concluding that the precise nature of the mechanism is less important than the commitment of the sponsoring agency.

Clarke, L. and J. Agyeman. 2011. Is there more to environmental participation than meets the eye? understanding agency, empowerment and disempowerment among black and minority ethnic communities. *Area* 43(1): 88–95.

Examines how the structure of most participation processes in the United Kingdom tends to exclude black and minority ethnic people because they are based on white norms and don't engage the BME community.

Coggins, G.C. 1999. Regulating federal natural resources: a summary case against devolved collaboration. *Ecology Law Quarterly* 25: 602.

Makes a case against participation for being an abdication of government responsibility and empowering unrepresentative and antienvironmental local groups.

Cooke, B., and U. Kothari. 2001. *Participation: the new tyranny?.* Zed Books, London.

This edited volume includes a number of trenchant critiques of the dominance of participatory approaches in the context of international development.

Cullen, D., G.J.A. McGee, T.I. Gunton, and J.C. Day. 2010. Collaborative planning in complex stakeholder environments: an evaluation of a two-tiered collaborative planning model. *Society and Natural Resources* 23(4): 332–350.

Describes a participatory process in British Columbia that used a "two-table" model to give indigenous people a special say in the decision rather than treating them like just another stakeholder at the table.

Cuppen, E. 2012. Diversity and constructive conflict in stakeholder dialogue: considerations for design and methods. *Policy Sciences* 45(1): 23–46.

Explores how Q method can be used to ensure adequate diversity of perspectives in a participation process and how the process can be structured to generate constructive conflict rather than simply focusing on areas of already-existing agreement.

Daniels, S.E. and G.B. Walker. 2001. *Working through environmental conflict: the collaborative learning approach.* Praeger, Westport, CT.

Frames decision making as a social learning process, with participants gaining deeper understandings of both the facts of the issue at hand as well as the values and perspectives of other participants, leading to a better outcome and greater decision legitimacy.

Dietz, T. and P.C. Stern. 2008. Public participation in environmental assessment and decision making [Online]. Available at http://www.nap.edu/openbook.php?record_id=12434 National Academies Press, Washington, DC (accessed March 17, 2015).

Gives a positive overview of the potential for participation in decision making in the US federal government.

Dryzek, J.S. 2002. *Deliberative democracy and beyond: liberals, critics, contestations.* Oxford University Press, Oxford.

Gives a comprehensive theory of deliberative democracy.

Eckersley, R. 1999. The discourse ethic and the problem of representing nature. *Environmental Politics* 8(2): 24–49.

Argues that because nature can't participate in deliberation as envisaged by its major proponents, its rights must be safeguarded through the precautionary principle.

Fiorino, D.J. 1989. Technical and democratic values in risk analysis. *Risk Analysis* 9(3): 293–299.

Argues that democratic values should take precedence in risk management, with technical risk analysis serving an advisory function.

Fishkin, J.S. 1997. *The voice of the people: public opinion and democracy.* Yale University Press: New Haven, CT.

Makes an argument for the widespread use of deliberative polling.

Fung, A. 2003. Recipes for public spheres: eight institutional design choices and their consequences. *Journal of Political Philosophy* 11(3): 338–367.

Describes eight design choices and 10 potential outcomes to consider in designing a participatory process.

Gutmann, A. and D. Thompson. 1996. *Democracy and disagreement.* Belknap Press, Cambridge, MA.

Makes a case for deliberative democracy on political theory grounds and illustrates its application to a variety of social controversies.

Hibbing, J. and E. Theiss-Morse. 2002. *Stealth democracy: Americans' beliefs about how government should work.* Cambridge University Press, Cambridge.

Argues, on the basis of extensive survey and experimental data, that people in the United States are not interested in participating more in policymaking and would prefer if the government handled it—though they will participate if the government fails.

Hickey, S. and G. Mohan. 2004. *Participation: from tyranny to transformation? exploring new alternatives to participation in development.* Zed Books, London.

This book is a follow-up to Cooke and Kothari's *Participation: The new tyranny?* which seeks a middle ground between embrace and rejection of participation.

Innes, J.E. 2004. Consensus building: clarifications for the critics. *Planning Theory* 3(1): 5–20.

Defends consensus as a goal for participation against a variety of criticisms.

Levine, P. and R.M. Nierras. 2007. Activists' views of deliberation. *Journal of Public Deliberation* 3(1): 4.

Reports on a series of focus groups with activists and organizers of participatory processes, finding that activists see participation as biased, coopting, and conferring legitimacy on an unjust sponsoring agency.

Logan, B.I. and W.G. Moseley. 2002. The political ecology of poverty alleviation in Zimbabwe's Communal Areas Management Programme for Indigenous Resources (CAMPFIRE). *Geoforum* 33: 1–14.

Describes the failure of a community-based wildlife conservation program in Zimbabwe, attributing it to unjustified assumptions about "community," lack of real community power, and lack of resources.

Lynn, F.M. and G. Busenberg. 1995. Citizen advisory committees and public policy: what we know, what's left to discover. *Risk Analysis* 15(2): 147–162.

Reviews research on citizen advisory committees, finding that their effectiveness depends heavily on the amount of commitment and authority they get from the sponsoring agency.

Macedo, S. (Ed). 1999. *Deliberative politics: essays on "Democracy and disagreement.".* Oxford University Press, New York.

An edited collection of essays that challenge Gutmann and Thompson's case for deliberative democracy.

MacKenzie, B.F. and B.M.H. Larson. 2010. Participation under time constraints: landowner perceptions of rapid response to the emerald ash borer. *Society and Natural Resources* 23(10): 1013–1022.

Describes a case in which the need for rapid response to an invasive species made it difficult to conduct genuine participation.

McCarthy, J. 2006. Neoliberalism and the politics of alternatives: community forestry in British Columbia and the United States. *Annals of the Association of American Geographers* 96(1): 84–104.

Shows how the greater success of community forestry in Canada is due to the fact that, unlike the United States, Canada did not get locked into a structure of environmental law that focused on lawsuits as a primary form of participation in the 1970s.

McClosky, M. 1999. Local communities and the management of public forests. *Ecology Law Quarterly* 25: 624.

Argues that participation unfairly advantages those who live closer to the site of the decision and who are more socially privileged.

McComas, K.A., J.C. Besley, and C.W. Trumbo. 2006. Why citizens do and do not attend public meetings about local cancer cluster investigations. *Policy Studies Journal* 34(4): 671–698.

Reports on a survey about why local residents do or do not participate in meetings about cancer clusters, finding that participants are the curious, the fearful, and the available, while nonparticipants are the uninformed, the indifferent, the occupied, and the disaffected.

Mutz, D.C. 2008. Is deliberative democracy a falsifiable theory? *Annual Review of Political Science* 11: 521–538.

Charges that proponents of deliberative democracy ignore or evade empirical evidence that questions the effectiveness of deliberation.

Nadasdy, P. 2003. Reevaluating the co-management success story. *Arctic* 56(4): 367–380.

Sharply critiques "comanagement" partnerships between government agencies and indigenous people, using a case study from the Yukon. Argues that integration of traditional ecological knowledge with science was done in a way that was exploitative and exclusionary toward indigenous people.

Neblo, M.A., K.M. Esterling, R.P. Kennedy, D.M.J. Lazer, and A.E. Sokhey. 2010. Who wants to deliberate—and why? *American Political Science Review* 104(3): 566–583.

Challenges the idea that the public is too disengaged and apolitical for extensive participation, showing that if offered a real chance for genuine participation, lots of people—especially those disaffected from the existing political process—will take the opportunity.

Niemeyer, S. and J.S. Dryzek. 2007. The ends of deliberation: metaconsensus and intersubjective rationality as ideal outcomes. *Swiss Political Science Review* 13(4): 497–526.

Uses a case study from Queensland to show that even when deliberation can't achieve consensus, it may at least achieve metaconsensus—agreement on the problem framing and mutual respect between stakeholders.

Pini, B. and F. Haslam McKenzie. 2006. Challenging local government notions of community engagement as unnecessary, unwanted, and unproductive: case studies from rural Australia. *Journal of Environmental Policy and Planning* 8(1): 27–44.

Reports, describes, and rebuts Australian officials' views that participation is unnecessary, unwanted, and unproductive.

Prell, C., K. Hubacek, and M. Reed. 2009. Stakeholder analysis and social network analysis in natural resource management. *Society and Natural Resources* 22: 501–518.

Using a case study in the Peak District of England, explores how social network analysis can be used to identify a diverse set of stakeholders for participation.

Quaghebeur, K., J. Masschelein, and H.H. Nguyen. 2004. Paradox of participation: giving or taking part? *Journal of Community and Applied Social Psychology* 14: 154–165.

Based on a case study in Vietnam, argues that participation often involves making laypeople do extra work for the sponsoring agency's vision.

Renn, O., T. Webler and H. Kastenholz. 1998. Procedural and substantive fairness in landfill siting: a Swiss case study. p. 253–270. In Löfstedt, R., Frewer, L. (eds.), *The Earthscan reader in risk and modern society*. Earthscan Publications, London.

Gives a detailed case study of a structured decision-making process involving lay participants and expert witnesses for siting a landfill in Switzerland.

Renn, O., T. Webler, and P.M. Wiedemann. 1995. *Fairness and competence in citizen participation: evaluating models for environmental discourse*. Kluwer, Dordrecht.

Articulates standards for fair and competent participation based on Habermas's theories and then applies them to evaluating a range of common participation mechanisms.

Rowe, G., and L.J. Frewer. 2000. Public participation methods: a framework for evaluation. *Science, Technology, and Human Values* 25(1): 3–29.

Presents a framework for assessing public participation mechanisms in terms of their process and outcomes and applies it to a selection of common mechanisms, finding that community advisory groups and citizens' juries get the best marks.

Rowe, G. and L. Frewer. 2005. A typology of public engagement mechanisms. *Science, Technology, and Human Values* 30: 251–290.

Organizes participation mechanisms according to criteria of participant selection method, facilitation of elicitation of views, response mode (closed-ended or open-ended), flexibility of information input, medium of information transfer (face-to-face or mediated), and facilitation of aggregation.

Sanders, L.M. 1997. Against deliberation. *Political Theory* 25(3): 347–376.

Argues that deliberation is unable to address inequalities of power and access and therefore is an inherently conservative, rather than egalitarian, approach to decision making.

Sandman, P.M. 1990. Getting to maybe: some communications aspects of siting hazardous waste facilities. p. 233–245. In Glickman, T.S., Gough, M. (eds.), *Readings in risk*. Resources for the Future, Washington DC.

Gives nine principles for sponsoring agencies to help them remain open to public participation.

Semmens, A.A. 2005. Engendering deliberative democracy: women's environmental protection problems. *Human Ecology Review* 12(2): 96–105.

Argues that deliberative democracy is a masculine ideal that is unable to truly empower women.

Sieber, R. 2006. Public participation geographic information systems: a literature review and framework. *Annals of the Association of American Geographers* 96(3): 491–507.

Gives an overview of the history and potential pitfalls of participatory GIS.

Stokes, S.C. 1998. Pathologies of deliberation. p. 123–139. In Elster, J. (ed.), *Deliberative democracy*. Cambridge University Press, Cambridge.

Explains how deliberation can produce perverse outcomes, especially when powerful outside interests influence participants' opinions.

Sunstein, C.R. 2005. Group judgments: statistical means, deliberation, and information markets. *New York University Law Review* 80: 962.

Argues that deliberation can exacerbate biases in decision making.

Tuler, S.P. and T. Webler. 1999. Voices from the forest: what participants expect of a public participation process. *Society and Natural Resources* 12: 437–453.

Based on interviews with stakeholders in forest planning in the northeastern United States, proposes seven principles that participants think make for a good participatory process.

Vroom, V.H. and P.W. Yetton. 1973. *Leadership and decision-making*. University of Pittsburgh Press, Pittsburgh, PA.

Sets out a decision model for deciding what level of participation is necessary in a given situation.

Webler, T., and S.P. Tuler. 2006. Four perspectives on public participation process in environmental assessment and decision making: combined results from 10 case studies. *Policy Studies Journal* 34(4): 699–722.

Describes four perspectives on what makes a good participation process, found across ten case studies around the United States.

Young, I.M. 1995. Communication and the other: beyond deliberative democracy. p. 134–152. *In* Wilson, M., Yeatman, A. (eds.), *Justice and identity: antipodean practices*. Allen and Unwin, St. Leonards, NSW.

Argues that existing models of deliberation are excessively narrow, and thus exclusionary, in the types of talk that they allow.

Young, I.M. 2001. Activist challenges to deliberative democracy. *Political Theory* 29(5): 670–690.

Assesses activists' claims that deliberative democracy is just a means for perpetuating the status quo and coopting critical challenges.

CHAPTER 16

Cochran, W. G. 1977. *Sampling Techniques*. 3rd edition. John Wiley & Sons, Inc., New York.

Crawley, M. 2013. *The R Book*. 2nd edition. John Wiley & Sons, Inc., New York.

Faraway, J. J. 2005. *Extending the Linear Model with R*. Chapman Hall/CRC, London.

Faraway, J. J. 2006. *Extending the Linear Model with R*. Chapman Hall/CRC, London.

Gitzen, R. A., Millspaugh, J. J., Cooper, A. B., and D. S. Licht, (Editors) 2012. *Design and Analysis of Long-term Ecological Monitoring Studies*. Cambridge University Press, Cambridge.

Gotelli, N. J. and A. M. Ellison. 2012. *A Primer of Ecological Statistics*. 2nd edition. Sinauer, Sunderland, MA.

Lohr, S. L. 2010. *Sampling: Design and Analysis*. 2nd edition. Duxbury, Boston, MA.

Maindonald, J. and W. J. Braun. 2010. *Data Analysis and Graphics Using R: An Example-Based Approach*. 3rd edition. Cambridge University Press, Cambridge/New York.

Manly, B. F. J. 2007. *Randomization, Bootstrap, and Monte Carlo Methods in Biology*. 3rd edition. Chapman Hall/CRC, Boca Raton, FL.

Manly, B. F. J. 2008. *Statistics for Environmental Science and Management*. 2nd edition. Chapman and Hall/CRC, Boca Raton, FL.

Myers R. H., Montgomery D. C., Vining G. G., and T. J. Robinson. 2010. *Generalized Linear Models with Applications in Engineering and the Sciences*. 2nd edition. John Wiley & Sons, Inc., Hoboken, NJ.

Qian, S. S. 2010. *Environmental and Ecological Statistics with R*. Chapman Hall/CRC, Boca Raton, FL.

Scheaffer, R. L., Mendenhall, W., Ott, R. L., and K. G. Greow. 2011. *Elementary Survey Sampling*. 7th edition. Duxbury, Boston, MA.

Thompson, S. K. 2002. *Sampling*. 2nd edition. John Wiley & Sons, Inc., New York.

Thompson, W. 2004. *Sampling Rare or Elusive Species: Concepts, Designs, and Techniques for Estimating Population Parameters*. Island Press, Washington, DC.

Zuur, A. Ieno, E. N., Walker, N., Saveliev, A. A., and G. M. Smith. 2009. *Mixed Effects Models and Extensions in Ecology with R (Statistics for Biology and Health)*. Springer, New York/London.

CHAPTER 17

Societies

American Society for Photogrammetry and Remote Sensing (http://www.asprs.org)

IEEE Geoscience and Remote Sensing Society (http://www.grss-ieee.org)

The Remote Sensing and Photogrammetry Society (http://rspsoc.org.uk)

Format as a heading (like Societies", above, and 'Books' below).
Remote Sensing of Environment (Elsevier)

IEEE Transactions on Geoscience and Remote Sensing

IEEE Journal of Selected Topics in Applied Earth Observations and Remote Sensing

International Journal of Remote Sensing (*Taylor & Francis*)

Photogrammetric Engineering & Remote Sensing (*ASPRS*)

Books

Jensen, J.R. 2007. *Remote Sensing of the Environment: An Earth Resource Perspective*, 2nd edition., Prentice Hall: Upper Saddle River, NJ, 592 pages.

King, M.D., Parkinson, C.L., and K. C. Partington (eds.). 2007. *Our Changing Planet: The View from Space*. Cambridge University Press: Cambridge, UK, 400 pages.

Rees, W.G. 2006. *Remote Sensing of Snow and Ice*. CRC Press, Taylor & Francis Group: New York, 285 pages.

Sabins, F.F. 2007. *Remote Sensing: Principles and Interpretation*, 3rd edition, Waveland Press: Long Grove, IL, 512 pages.

Wulder, M.A. and S. E. Franklin (eds). 2006. *Understanding Forest Disturbance and Spatial Pattern: Remote Sensing and GIS Approaches*, CRC Press, Taylor & Francis Group: New York, 246 pages.

Xiaojun Yang (ed.). 2011. *Urban Remote Sensing: Monitoring, Synthesis and Modeling in the Urban Environment*. ISBN: 978-0-470-74958-6. John Wiley & Sons, Inc., Hoboken, NJ. 408 pages.

Web

1. NASA Earth Observatory: http://earthobservatory.nasa.gov/
2. NASA/JPL Earth: http://www.jpl.nasa.gov/earth/
3. NASA/JPL Global Climate Change: http://climate.nasa.gov/
4. NASA/JPL Eyes on the Earth: http://eyes.jpl.nasa.gov/earth/
5. NASA Land Cover Land Use Change Program: http://lcluc.umd.edu/
6. NASA Carbon Cycle and Ecosystems Program: http://cce.nasa.gov/cce/
7. Pennsylvannia State University's Geospatial Revolution: http://geospatialrevolution.psu.edu/
8. NASA/CISEIN/SEDAC Urban Remote Sensing: http://sedac.ciesin.columbia.edu/urban_rs/
9. Center for Spatial Technologies and Remote Sensing (CSTARS) at the University of California, Davis: http://cstars.metro.ucdavis.edu/
10. US Geological Survey GLOVIS: http://glovis.usgs.gov/
11. The Guardian newspaper's Satellite eye on Earth: http://www.theguardian.com/environment/series/satelliteeye

CHAPTER 18

De Smith, M. J., Goodchild, M. F., & Longley, P. A. (2007). *Geospatial analysis: a comprehensive guide to principles, techniques and software tools*. Leicester, UK: Troubador Publishing Ltd.

Longley, P. A., Goodchild, M. F., Maguire, D., & Rhind, D. W. (2005). *Geographical information systems: principles, techniques, applications and management*. Hoboken, NJ: John Wiley & Sons, Inc.

O'Sullivan, D. & Unwin, D. J. (2003). *Geographic Information Analysis*. Hoboken, NJ: John Wiley & Sons, Inc.

Text books on the principles of geographic information analysis

Clarke, Keith C. (1997). *Getting started with geographic information systems* (Vol. 3). Upper Saddle River, NJ: Prentice Hall.

Price, M. (2013). *Mastering ArcGIS* (6th ed.). New York, McGraw-Hill.

Textbooks for mastering the basic concepts and practices for Geographic Information Systems.

Goodchild, M. F., Steyaert, L. T., Parks, B. O., & Johnston, C. (1996). *GIS and environmental modeling: progress and research issues*. New York: John Wiley & Sons, Inc.

Scally, R. (2006). *GIS for Environmental Management*. Redlands, CA: ESRI Press.

Excellent texts for understanding the environmental modeling process from a GIS perspective (the Goodchild book) and a comprehensive collection of GIS for environmental management practices (Scally book).

CHAPTER 19

Arena, M., G. Azzone, and A. Conte. 2013. A streamlined LCA framework to support early decision making in vehicle development. *Journal of Cleaner Production*. 41:105–113.

Describes the streamlined Life Cycle Assessment framework as a tool in decision making.

Ayalon, O., Y. Avnimelech, and M. Shechter. 2000. Application of a comparative multidimensional Life Cycle Analysis in solid waste management policy: The case of soft drink containers. *Environmental Science and Policy*. 3(2–3):135–144.

Bersimis, S. and D. Georgakellos. 2013. A probabilistic framework for the evaluation of products' environmental performance using life cycle approach and Principal Component Analysis. *Journal of Cleaner Production*. 42:103–115.

Blengini, G.A. 2008. Using LCA to evaluate impacts and resources conservation potential of composting: A case study of the Asti District in Italy. *Resources, Conservation and Recycling*. 52(12):1373–1381.

Blengini, G.A. 2008. Applying LCA to organic waste management in Piedmont, Italy. *Management of Environmental Quality*. 19(5):533–549.

Blengini, G.A. 2009. Life cycle of buildings, demolition and recycling potential: A case study in Turin. Italy. *Building and Environment*. 44(2):319–330.

Blengini, G.A. and T. Di Carlo. 2010. The changing role of life cycle phases, subsystems and materials in the LCA of low energy buildings. *Energy and Buildings*. 42(6):869–880.

Blengini, G.A. and E. Garbarino. 2010. Resources and waste management in Turin (Italy): The role of recycled aggregates in the sustainable supply mix. *Journal of Cleaner Production*. 18(10–11):1021–1030.

Blengini, G.A. and D.J. Shields. 2010. Green labels and sustainability reporting: Overview of the building products supply chain in Italy. *Management of Environmental Quality*. 21(4):477–493.

Blengini, G.A., M. Busto, M. Fantoni, and D. Fino. 2012. Eco-efficient waste glass recycling: Integrated waste management and green product development through LCA. *Waste Management*. 32(5):1000–1008.

Blom, I., L. Itard, and A. Meijer, 2010. LCA-based environmental assessment of the use and maintenance of heating and ventilation systems in Dutch dwellings. *Building and Environment*. 45(11):2362–2372.

Cabeza, L.F., L. Rincón, V. Vilariño, G. Pérez, and A. Castell. 2014. Life cycle assessment (LCA) and life cycle energy analysis (LCEA) of buildings and the building sector: A review. *Renewable and Sustainable Energy Reviews*. 29: 394–416.

Cecchi, T., P. Passamonti, and P. Cecchi. 2009. Is it advisable to store olive oil in PET bottles? *Food Reviews International*. 25(4):271–283.

Cellura, M., F. Ardente, and S. Longo. 2012. From the LCA of food products to the environmental assessment of protected crops districts: A case-study in the south of Italy. *Journal of Environmental Management*. 93(1):194–208.

Chen, C., G. Habert, Y. Bouzidi, A. Jullien, and A. Ventura. 2010. LCA allocation procedure used as an incitative method for waste recycling: An application to mineral additions in concrete. *Resources, Conservation and Recycling*. 54(12):1231–1240.

Chiaramonti, D. and L. Recchia. 2010. Is life cycle assessment (LCA) a suitable method for quantitative CO_2 saving estimations? The impact of field input on the LCA results for a pure vegetable oil chain. *Biomass and Bioenergy*. 34(5): 787–797.

Clavreul, J., D. Guyonnet, and T.H. Christensen. 2012. Quantifying uncertainty in LCA-modelling of waste management systems. *Waste Management*. 32(12):2482–2495.

An article in the study of the uncertainty in LCA modeling.

Collado-Ruiz, D. and H. Ostad-Ahmad-Ghorabi. 2010. Comparing LCA results out of competing products: Developing reference ranges from a product family approach. *Journal of Cleaner Production*. 18(4):355–364.

Crawford, R.H. 2011. *Life Cycle Assessment in the Built Environment*. Spon Press, New York.

Curran, M.A. 2006. Life Cycle Assessment: Principles and practice. *National Risk Management Research Laboratory—Office of Research and Development*. U.S. Environmental Protection Agency, Cincinnati, OH.

Dandres, T., C. Gaudreault, P. Tirado-Seco, and R. Samson. 2011. Assessing non-marginal variations with consequential LCA: Application to European energy sector. *Renewable and Sustainable Energy Reviews*. 15(6):3121–3132.

Deng, L., C.W. Babbitt, and E.D. Williams. 2011. Economic-balance hybrid LCA extended with uncertainty analysis: Case study of a laptop computer. *Journal of Cleaner Production*. 19(11), 1198–1206.

Ekvall, T., A.-M. Tillman, and S. Molander. 2005. Normative ethics and methodology for life cycle assessment, *Journal of Cleaner Production*. 13:1225–1234.

European Commission—Joint Research Centre—Institute for Environment and Sustainability. 2010. *International Reference Life Cycle Data System (ILCD) Handbook—General Guide for Life Cycle Assessment—Detailed Guidance*. Publications Office of the European Union, Luxembourg.

Finkbeiner, M., A. Inaba, R.B.H. Tan, K. Christiansen, and H.-J. Klüppel. 2006. The new international standards for Life Cycle Assessment: ISO 14040 and ISO 14044. *International Journal of LCA*. 11(2):80–85.

Finnveden, G., M.Z. Hauschild, T. Ekvall, J. Guinée, R. Heijungs, S. Hellweg, A. Koehler, D. Pennington, and S. Suh. 2009. Recent developments in life cycle assessment. *Journal of Environmental Management*. 91:1–21.

Frischknecht R., N. Jungbluth, H.-J. Althaus, G. Doka, T. Heck, S. Hellweg, R. Hischier, T. Nemecek, G. Rebitzer, M. Spielmann, and G. Wernet. 2007. *Ecoinvent Report No. 1: Overview and Methodology*. Swiss Centre for Life Cycle Inventories, Dübendorf.

Georgakellos, D.A. 2005. Evaluation of life cycle inventory results using critical volume aggregation and polygon-based interpretation. *Journal of Cleaner Production*. 13:567–582.

Georgakellos, D.A. 2006. The use of the LCA polygon framework in waste management. *Management of Environmental Quality: An International Journal*. 17:490–507.

Georgakellos, D.A. 2012. Climate change external cost appraisal of electricity generation systems from a life cycle perspective: The case of Greece. *Journal of Cleaner Production*. 32:124–140.

Grillo Renó, M.L., E.E. Silva Lora, J.C. Escobar Palacio, O.J. Venturini, J. Buchgeister, and O. Almazan. 2011. A LCA (Life Cycle Assessment) of the methanol production from sugarcane bagasse. *Energy*. 36(6):3716–3726.

Guardigli, L., F. Monari, and M.A. Bragadin. 2011. Assessing environmental impact of green buildings through LCA methods: A comparison between reinforced concrete and wood structures in the European context. *Procedia Engineering*. 21:1199–1206.

Guinée, J.B. 2004. *Life cycle assessment operational guide to the ISO standards*. Kluwer Academic Publishers, Dordrecht.

Guo, M. and R.J. Murphy. 2012. LCA data quality: Sensitivity and uncertainty analysis. *Science of the Total Environment*. 435–436:230–243.

Heijungs, R., G. Huppes, and J.B. Guinée. 2010. Life cycle assessment and sustainability analysis of products, materials and technologies. Toward a scientific framework for sustainability life cycle analysis. *Polymer Degradation and Stability*. 95:422–428.

Hermann, B.G., C. Kroeze, and W. Jawjit. 2007. Assessing environmental performance by combining life cycle assessment, multi-criteria analysis and environmental performance indicators. *Journal of Cleaner Production*. 15:1787–1796.

Hertwich, E.G. and G.P. Peters. 2006. Feasibility and scope of life cycle approaches to sustainable consumption. *In* D. Brissaud et al.

(ed.) *Innovation in Life Cycle Engineering and Sustainable Development*. Springer, Dordrecht.

Herva, M., A. Franco-Urυa, E.F. Carrasco, and E. Roca. 2012. Application of fuzzy logic for the integration of environmental criteria in ecodesign. *Expert Systems with Applications.* 39:4427–4431.

Humpenöder, F., R. Schaldach, Y. Cikovani, and L. Schebek. 2013. Effects of land-use change on the carbon balance of 1st generation biofuels: An analysis for the European Union combining spatial modeling and LCA. *Biomass and Bioenergy.* 56:166–178.

International Organization for Standardization. 2006. *ISO 14040: Environmental Management—Life Cycle Assessment—Principles and Framework*, 2nd edition. ISO, Geneva.

Jensen A.A., L. Hoffman, B.T. Møller, A. Schmidt, K. Christiansen, J. Elkington, and F. van Dijk. 1997. *Life Cycle Assessment—A Guide to Approaches, Experiences and Information Sources*. European Environment Agency, Copenhagen.

Jeswani, H.K., A. Azapagic, P. Schepelmann, and M. Ritthoff. 2010. Options for broadening and deepening the LCA approaches. *Journal of Cleaner Production.* 18(2):120–127.

Jijakli, K., H. Arafat, S. Kennedy, P. Mande, and V.V. Theeyattuparampil. 2012. How green solar desalination really is? Environmental assessment using life-cycle analysis (LCA) approach. *Desalination.* 287:123–131.

Joshi, S. 2000. Product environmental life-cycle assessment using input-output techniques. *Journal of Industrial Ecology.* 3(2&3):95–120.

Jung, J., N. von der Assen, and A. Bardow. 2013. An uncertainty assessment framework for LCA-based environmental process design. *Computer Aided Chemical Engineering.* 32:937–942.

Kikuchi, Y., K. Mayumi, and M. Hirao. 2010. Integration of CAPE and LCA tools in environmentally-conscious process design: A case study on biomass-derived resin. *Computer Aided Chemical Engineering.* 28:1051–1056.

Li, X., Y. Zhu, and Z. Zhang, 2010. An LCA-based environmental impact assessment model for construction processes. *Building and Environment.* 45(3):766–775.

Liu, Y., V. Langer, H. Høgh-Jensen, and H. Egelyng. 2010. Life Cycle Assessment of fossil energy use and greenhouse gas emissions in Chinese pear production. *Journal of Cleaner Production.* 18:1423–1430.

Löfgren, B., A.-M Tillman, and B. Rinde, 2011. Manufacturing actor's LCA. *Journal of Cleaner Production.* 19(17–18):2025–2033.

Löfgren, B. and A.-M. Tillman. 2011. Relating manufacturing system configuration to life-cycle environmental performance: Discrete-event simulation supplemented with LCA. *Journal of Cleaner Production.* 19(17–18):2015–2024.

Martínez, E., E. Jiménez, J. Blanco, and F. Sanz. 2010. LCA sensitivity analysis of a multi-megawatt wind turbine. *Applied Energy.* 87(7): 2293–2303.

Marvuglia, A., E. Benetto, S. Rege, and C. Jury. 2013. Modelling approaches for consequential life-cycle assessment (C-LCA) of bioenergy: Critical review and proposed framework for biogas production. *Renewable and Sustainable Energy Reviews.* 25:768–781.

Menten, F., B. Chèze, L. Patouillard, and F. Bouvart. 2013. A review of LCA greenhouse gas emissions results for advanced biofuels: The use of meta-regression analysis. *Renewable and Sustainable Energy Reviews.* 26:108–134.

Mestre, A. and J. Vogtlander. 2013. Eco-efficient value creation of cork products: An LCA-based method for design intervention. *Journal of Cleaner Production.* 57:101–114.

McLaren, S.J. 2010. Life cycle assessment (LCA) of food production and processing: An introduction. *In* U. Sonesson et al. (ed.) *Environmental Assessment and Management in the Food Industry Life Cycle Assessment and Related Approaches.* Woodhead Publishing Limited, Cambridge.

Millet, D., L. Bistagnino, C. Lanzavecchia, R. Camous, and T. Poldma, T., 2007. Does the potential of the use of LCA match the design team needs? *Journal of Cleaner Production.* 15:335–346.

Morais, S.A. and C. Delerue-Matos. 2010. A perspective on LCA application in site remediation services: Critical review of challenges. *Journal of Hazardous Materials.* 175(1–3):12–22.

Morris, A.S. 2004. *ISO 14000 Environmental Management Standards—Engineering and Financial Aspects*. John Wiley & Sons, Inc., Chichester.

Nakano, K. and M. Hirao. 2011. Collaborative activity with business partners for improvement of product environmental performance using LCA. *Journal of Cleaner Production.* 19:1189–1197.

Narodoslawsky, M. 2013. From processes to life cycles to technology networks—new challenges for LCA in chemical engineering. *Current Opinion in Chemical Engineering.* 2(3):282–285.

Nessi, S., L. Rigamonti, and M. Grosso. 2012. LCA of waste prevention activities: A case study for drinking water in Italy. *Journal of Environmental Management.* 108, 73–83.

Nwe, E.S., A. Adhitya, I. Halim, and R. Srinivasan. 2010. Green supply chain design and operation by integrating LCA and dynamic simulation. *Computer Aided Chemical Engineering.* 28:109–114.

Pawelzik, P., M. Carus, J. Hotchkiss, R. Narayan, S. Selke, M. Wellisch, M. Weiss, B. Wicke, and M.K. Patel. 2013. Critical aspects in the life cycle assessment (LCA) of bio-based materials—Reviewing methodologies and deriving recommendations. *Resources, Conservation and Recycling.* 73:211–228.

Pieragostini, C., M.C. Mussati, and P. Aguirre. 2012. On process optimization considering LCA methodology. *Journal of Environmental Management.* 96:43–54.

Pizzol, M., P. Christensen, J. Schmidt, and M. Thomsen. 2011. Impacts of "metals" on human health: A comparison between nine different methodologies for life cycle impact assessment (LCIA). *Journal of Cleaner Production.* 19:646–656.

Portha, J.-F., S. Louret, M.-N. Pons, and J.-N. Jaubert. 2010. Estimation of the environmental impact of a petrochemical process using coupled LCA and exergy analysis. *Resources, Conservation and Recycling.* 54(5):291–298.

Rives, J., J. Rieradevall, and X. Gabarrell. 2010. LCA comparison of container systems in municipal solid waste management. *Waste Management*. 30(6):949–957.

Ross, S. and D. Evans. 2002. Use of Life Cycle Assessment in environmental management. *Environmental Management*. 29(1): 132–142.

Rubio Rodríguez, M.A., J. De Ruyck, P. Roque Díaz, V.K. Verma, and S. Bram. 2011. An LCA based indicator for evaluation of alternative energy routes. *Applied Energy*. 88(3):630–635.

Shen, L., E. Worrell, and M.K. Patel. 2010. Open-loop recycling: A LCA case study of PET bottle-to-fibre recycling. *Resources, Conservation and Recycling*. 55(1):34–52.

Slagstad, H. and H. Brattebø. 2012. LCA for household waste management when planning a new urban settlement. *Waste Management*. 32(7):1482–1490.

Soimakallio, S., J. Kiviluoma, and L. Saikku. 2011. The complexity and challenges of determining GHG (greenhouse gas) emissions from grid electricity consumption and conservation in LCA (life cycle assessment)—A methodological review. *Energy*. 36(12):6705–6713.

Sonnemann, G., F. Castells, and M. Schuhmacher. 2004. *Integrated Life-Cycle and Risk Assessment for Industrial Processes*. Lewis Publishers, Boca Raton, RL.

Steen, B. 2005. Environmental costs and benefits in life cycle costing. *Management of Environmental Quality: An International Journal*. 16(2):107–118.

Stewart, J.R., M.W. Collins, R. Anderson, and W.R. Murphy. 1999. Life Cycle Assessment as a tool for environmental management. *Clean Product and Processes*. 1:73–81.

Swarr, T. and D. Hunkeler. 2008. Life cycle costing in life cycle management. *In* D. Hunkeler et al. (ed.) *Environmental Life Cycle Costing*. SETAC and CRC Press, Boca Raton, RL.

Tillman, A-M. 2010. Methodology for life cycle assessment. *In* U. Sonesson et al. (ed.) *Environmental Assessment and Management in the Food Industry Life Cycle Assessment and Related Approaches*. Woodhead Publishing Limited, Cambridge.

Tingley, D.D. and B. Davison. 2012. Developing an LCA methodology to account for the environmental benefits of design for deconstruction. *Building and Environment*. 57:387–395.

Toniolo, S., A. Mazzi, M. Niero, F. Zuliani, and A. Scipioni. 2013. Comparative LCA to evaluate how much recycling is environmentally favourable for food packaging. *Resources, Conservation and Recycling*. 77:61–68.

Turconi, R., A. Boldrin, and T. Astrup: 2013. Life cycle assessment (LCA) of electricity generation technologies: Overview, comparability and limitations. *Renewable and Sustainable Energy Reviews*. 28:555–565.

Valente, C., R. Spinelli, and B.G. Hillring. 2011. LCA of environmental and socio-economic impacts related to wood energy production in alpine conditions: Valle di Fiemme (Italy). *Journal of Cleaner Production*. 19(17–18):1931–1938.

Van den Heede, P. and N. De Belie. 2012. Environmental impact and life cycle assessment (LCA) of traditional and "green" concretes: Literature review and theoretical calculations. *Cement and Concrete Composites*. 34(4):431–442.

Vinodh, S. and G. Rathod. 2010. Integration of ECQFD and LCA for sustainable product design. *Journal of Cleaner Production*. 18(8):833–842.

Virtanen, Y., S. Kurppa, M. Saarinen, J.-M. Katajajuuri, K. Usva, I. Mäenpää, J. Mäkelä, J. Grönroos, and A. Nissinen. 2011. Carbon footprint of food—Approaches from national input–output statistics and a LCA of a food portion. *Journal of Cleaner Production*. 19(16):1849–1856.

Weidema, B.P., T. Ekvall, and R. Heijungs. 2009. Guidelines for application of deepened and broadened LCA—Deliverable D18 of work package 5 of the CALCAS project. ENEA—The Italian National Agency on new Technologies, Energy and the Environment, Rome.

Werner, F. 2005. *Ambiguities in Decision-Oriented Life Cycle Inventories—The Role of Mental Models and Values*. Springer, Dordrecht.

Wilfart, A., J. Prudhomme, J.-P. Blancheton, and J. Aubin. 2013. LCA and energy accounting of aquaculture systems: Towards ecological intensification. *Journal of Environmental Management*. 121:96–109.

Yang, C.Y. and J.L. Chen. 2012. Forecasting the design of eco-products by integrating TRIZ evolution patterns with CBR and simple LCA methods. *Expert Systems with Applications*. 39(3):2884–2892.

Zafirakis, D., K.J. Chalvatzis, G. Baiocchi, and G. Daskalakis. 2013. Modeling of financial incentives for investments in energy storage systems that promote the large-scale integration of wind energy. *Applied Energy*. 105:138–154.

CHAPTER 20

Ed. Dr. John Brady, **Environmental Management in Organizations: The IEMA Handbook**, ISBN 1-85383-976-0, James & James Earthscan, available from the IEMA shop. *This is an indispensable reference and toolkit for environmental practitioners. It sets out the context and provides practical guidance to understand and tackle the issues effectively. An essential reference for anyone seeking to become a chartered environmentalist.*

Bothkin & Keller, **Environmental Science: Earth as a Living Planet**, ISBN 0471321737, John Wiley & Sons, Distribution Center, 1 Wiley Drive, Somerset, NJ 088751272, USA Tel +17324694400. *Explores vital environmental issues in a scientific approach.*

Cahill LB, 1996, **Environmental Audits, (7th ed)**, Government Institutes Tel: (301) 921 2300, ISBN 0865875251. *This manual provides guidance on how to begin and manage a successful audit program for a facility. Addresses legal issues, elements of a successful program, and more.*

Carroll B & Turpin T, 2002, **Environmental Impact Assessment Handbook; A practical guide for planners, developers and communities**, Thomas Telford Publishing, ISBN 0727727818. *This book gives straightforward guidance on the steps and studies needed to undertake an EIA and to produce a successful environmental statement. It is written for planners, developers,* *and communities who need sufficient introduction to EIA to know what needs to be prepared by whom and when.*

Crosbie & Knight, 1995, **Strategy for Sustainable Business: Environmental Opportunity and Strategic Choice**, ISBN 0077091337. McGraw-Hill Publishing Company, Shoppenhangers Road, Maidenhead, Berkshire, SL6 2QL, Tel: 01628 502500.

Glasson J, Therivel R & Chadwick A, 1999, **Introduction to Environmental Impact Assessment: principals & procedures, process, practice & prospects**, ISBN 1857289455, UCL Press Ltd, University College London, Gower Street, London, WC1E 6BT. *A comprehensive introduction to EIA. Covers existing legislation and provides case studies and examples.*

Grayson L, **Environmental Auditing: A Guide to Best Practice in the UK and Europe**, ISBN 0748718621, Earthscan (see "Howes" for contact details). *Provides an analysis of legislation and policy in the United Kingdom and Europe and an overview of the current best practice.*

Harrison RM, **POLLUTION: Causes, Effects and Control, 3rd (ed)**, ISBN 0854045341, Ref No. 1857, Royal Society of Chemistry, Burlington House, Piccadilly, London, W1J 0BA, Tel: 020 7437 8656. *Ranges from sources of pollutants and their environmental behaviour to the technologies and strategies available for control.*

Howes R, Skea J & Wheelan B, 1997, **Clean & Competitive: Motivating Environmental Performance in Industry**, ISBN 1853834904. Earthscan: 120 Pentonville Road, London, N1 9JN, Tel: 020 7278 0433. *This book explores the challenge of motivating industry to address environmental issues and recommends practical ways for addressing the complex environmental agenda.*

Humphrey N & Hadley M, 2000, **Environmental Auditing**, ISBN 190255826X, Palladian Law Publishing Ltd, Beach Road, Bembridge, Isle of Wight, PO35 5NQ. *Explains the process and techniques of auditing, in relation to manufacturing and industrial processes.*

Hyde, P & Reeve, P 2002, **Essentials of Environmental Management**, ISBN 0901357286, The Institution of Occupational Safety & Health, Tel: 01787 249 293. *Provides a comprehensive introduction to the management of environmental issues and is suitable for professionals across all sectors.*

Information for Industry, **Environmental Compliance Manual: Practical Guidance for Managers**, Gee Publishing, 100 Avenue Road, London, NW3 3PG, Fax: 207 3937 466. *All you need to know about environmental law. Regularly updated.*

Lovins LH, Lovins AB & von Weizsacker E, 1998, **Factor Four: Doubling Wealth: Halving Resource Use**, ISBN 1853834068. Earthscan (see "Howes" for contact details). *Provides practical examples, in detail, of technologies that can use the world resources more efficiently, therefore illustrating a practical model of sustainability.*

Murley, L, 2002, **2002 Pollution Handbook**, National Society for Clean Air and Environmental Protection, 136 North Street, Brighton, BN1 1RG. *This annual pollution handbook is a guide to UK and European Pollution Control Legislation.*

National Society for Clean Air and Environmental Protection (NSCA), 2001, **Pollution Handbook 2001**, ISBN 0903474492. NSCA, 44 Grand Parade, Brighton, BN2 2QA, Tel: 01273 878770. *Provides a comprehensive guide to all UK and EC pollution control legislation. New additions include the Pollution Prevention and Control Regulations and Contaminated Land Regulations 2000.*

National Society for Clean Air and Environmental Protection, 1992, **Pollution Glossary, Revised**, ISBN 09034743444. NSCA (see above for contact details). *Defines over 1000 terms in alphabetical order.*

Office for Official Publications of the European Communities, 2000, **Eurostat's Environment Statistics**, ISBN 9282890260. *Vast range of environmental statistics from data from 1980 to 1997.*

Petts J, 1999, **Handbook of Environmental Impact Assessment, Vol.** 1 & 2, ISBN 0632047720 & 0632047712, Blackwell Science Ltd, Osney Mead, Oxford, OX2 0EL, Tel: 01865 20606. *Vol 1. Addresses the role of EIA as a tool in sustainable resource management. Vol 2. EIA in practice.*

Rothery B, ISO **14000 and ISO 9000**, ISBN 056606479, Gower: Gower House, Croft Road, Aldershot, Hampshire, GU11 3HR, Tel: 01252 331551. *Practical description of how companies implement a comprehensive system to meet the requirements of ISO 14000 and ISO 9000.*

Starkey R & Welford R, 2001, **The Earthscan Reader in Business and Sustainable Development**, ISBN 1853836397. Earthscan (see "Howes" for contact details). *Comprehensive collection of recent work by authors in the field of business and sustainable development.*

Therivel R & Partidario MR, 1996, **The Practice of Strategic Environmental Assessment**, ISBN 1853833738, Earthscan (see "Howes" for contact details). *Review of international SEA guidance and regulations. A comprehensive set of case studies and constraints which effect SEA.*

The Landscape Institute & Institute of Environmental Management and Assessment, **Guidelines for Landscape and Visual Impact Assessment. 2nd Edition**, St. Nicholas House, 70 Newport, Lincoln, LN1 3DP, Tel: 01522 540069. *Provides advice on assessing the landscape and visual impacts on development projects.*

Vlaamse Instelling voor Technologisch Onderzock, **Life Cycle Assessment**, ISBN 0748721312, Earthscan (see "Howes" for contact details). *This report is an introductory guide to life cycle assessment showing the range of applications, the potential obstacles, and how to organize a project.*

Wathey D & O'Reilly M, 2000, ISO 14031: **A Practical Guide to Developing Environment Performance Indicators for your Business**, ISBN 0117024724, Order through the Stationary Office, Tel: 0870 600 5522. *Practical "how to guide." Material draws on the largest research and implementation program on ISO 14031 conducted in the United Kingdom.*

Welford R, **Corporate Environmental Management 1 & 2**, ISBN 1853835595 & 1853834122. Earthscan (see "Howes" for contact details). *1. A comprehensive analysis of the role of business in safeguarding the environment. 2. Explains the various organizational and cultural concepts, which firmly place* the corporate environmental management agenda within the human dimension.

ACBE, **Value, Growth, Success - how sustainable is your business?**, DETR Distribution, PO Box 236, Wetherby, West Yorkshire, LS23 7NB.

Croner CCH Group Ltd, **Croner's Environment Magazine**, 145 London Road, Kingston-upon-Thames, Surrey, KT2 6SR, Tel: 020 8547 3333. *All you need to know about UK environment. Regularly updated.*

Environment Data Services Ltd, **The ENDS Report**, Finsbury Business Centre, 40 Bowling Green Lane, London EC1R 0NE, Tel: 020 7814 5300. *The ENDS Report is the UK's No 1 source of intelligence across the carbon, environmental, and sustainability agenda.*

Commission of European Communities, **Green Paper: Integrated Product Policy**, http://ec.europa.eu/environment/ipp/2001developments.htm (accessed May 4, 2015).

Magazine of Forum for the Future, Green Futures, 13-17 Sturton Street, Cambridge, CB1 2SN, Tel: 01223 564334.

Institute of Environmental Management and Assessment & EIA Centre, **Environmental Assessment Yearbook**, St. Nicholas House, 70 Newport, Lincoln, LN1 3DP, Tel: 01522 540069.

Institute of Environmental Management and Assessment, **The Environmentalist**, St. Nicholas House, 70 Newport, Lincoln, LN1 3DP, Tel: 01522 540069.

Institute of Environmental Management and Assessment, Perspectives—**Guidelines on Participation in Environmental Decision-making**, St. Nicholas House, 70 Newport, Lincoln, LN1 3DP, Tel: 01522 540069.

Institute of Environmental Management and Assessment, Practitioner—**Managing Climate Change Emissions**, St. Nicholas House, 70 Newport, Lincoln, LN1 3DP, Tel: 01522 540069.

The International Journal of Life Cycle Assessment, Landsberg, Germany, Tel: +498191125318.

HKSARG Electrical & Mechanical Services Department, Promoting Energy Efficiency, http://www.emsd.gov.hk/emsd/eng/pee/index.shtml (accessed March 19, 2015).

HKSARG Environmental Protection Department, Environmental rules, regulations and guidance, http://www.epd.gov.hk (accessed March 19, 2015).

CHAPTER 21

Allen, D., 1996. *Pollution Prevention for Chemical Processes.* John Wiley & Sons, Inc., New York.

Ashford, N.A. and C.S. Miller, 1998. *Chemical Exposures: Low Levels and High Stakes.* John Wiley & Sons, Inc., New York

Barnthouse, L.W., 1994. Issues in ecological risk assessment: the Committee on Risk Assessment Methodology (CRAM) perspective. *Risk Analysis*, 14(3):251–256.

Bate, R. (ed.), 1997. *What Risk? (Science, Politics & Public Health).* Butterworth-Heinemann, Oxford.

Blancato, J.N., R.N. Brown, C.C. Dary, and M.A. Saleh (eds.), 1996. *Biomarkers for Agrochemicals and Toxic Substances: Applications and Risk Assessment.* ACS Symposium Series No. 643, American Chemical Society, Washington, DC.

British Medical Association, 1998. *Health & Environmental Impact Assessment (An Integrated Approach).* British Medical Association (BMA), England.

Bromley, D.W. and K. Segerson (eds.), 1992. *The Social Response to Environmental Risk: Policy Formulation in an Age of Uncertainty.* Kluwer Academic Publishers, Boston, MA.

Bruckner, J.V., 2000. Differences in sensitivity of children and adults in chemical toxicity. The NAS panel report. *Regulatory Toxicology and Pharmacology*, 31(3):280–285.

Burns, L.A., C.G. Ingersoll, and G.A. Pascoe, 1994. Ecological risk assessment: application of new approaches and uncertainty analysis. *Environmental Toxicology and Chemistry*, 13(12):1873–1874.

Calow, P. (ed.), 1998. *Handbook of Environmental Risk Assessment and Management.* Blackwell Science, Great Britain.

CDC (Centers for Disease Control and Prevention), 2005. *Third National Report on Human Exposure to Environmental Chemicals.* NCEH Pub. No. 05-0570, Department of Health and Human Services, Atlanta, GA.

Christen, K., 2001. The arsenic threat worsens. *Environmental Science & Technology*, 35(13):286A–291A.

Clewell, H.J. and K.S. Crump, 2005. Quantitative estimates of risk for noncancer endpoints. *Risk Analysis*, 25(2):285–289.

Cohen Hubal, E.A., P. Egeghy, K. Leovic, and G. Akland, 2006. Measuring potential dermal transfer of a pesticide to children in a child care center. *Environmental Health Perspectives*, 114(2)264–269.

Committee on Improving Risk Analysis Approaches Used by the U.S. EPA, 2009. *Science and Decisions: Advancing Risk Assessment.* National Research Council, Washington, DC.

Cooper, J.A.G. and S. McLaughlin, 1998. Contemporary multidisciplinary approaches to coastal classification and environmental risk analysis. *Journal of Coastal Research*, 14(2), 512–524.

Craun, G.F., F.S. Hauchman, and D.E. Robinson (eds.), 2001. *Microbial Pathogens and Disinfection By-products in Drinking Water—Health Effects and Management of Risks.* ILSI Press, Washington, DC.

Crawford-Brown, D.J., 1999. *Risk-Based Environmental Decisions: Methods and Culture.* Kluwer Academic Publishers, Dordrecht.

Crump, K.S., 1984. A new method for determining allowable daily intakes. *Fundamental and Applied Toxicology*, 4(5):854–871.

Cummins, S.K. and R.J. Jackson, 2001. The built environment and children's health. *Pediatric Clinics of North America*, 48(5):1241–1252.

Dakins, M.E., J.E. Toll, and M.J. Small, 1994. Risk-based environmental remediation: decision framework and role of uncertainty. *Environmental Toxicology and Chemistry*, 13(12):1907–1915.

Dakins, M.E., J.E. Toll, M.J. Small, and K.P. Brand, 1996. Risk-based environmental remediation: Bayesian Monte Carlo analysis and the expected value of sample information. *Risk Analysis*, 16(1):67–79.

Daston, G., E. Faustman, G. Ginsberg, P. Fenner-Crisp, S. Olin, B. Sonawane, J. Bruckner, W. Breslin, and T.J. McLaughlin 2004. A framework for assessing risks to children from exposure to environmental agents. *Environmental Health Perspectives*, 112(2):238–256.

DEFRA, 2000. *Guidelines for Environmental Risk Assessment and Management*. Department of Environmental Food and Rural Affairs, UK.

ECB, 2003. *Technical Guidance Document on Risk Assessment*. European Chemicals Bureau—Institute for Health and Consumer Protection, European Commission Joint Research Centre, Ispra.

ECETOC, 2003. *Derivation of Assessment Factors for Human Health Risk Assessment*. Technical Report No. 86, ECETOC, Brussels.

Erickson, R.L. and R.D. Morrison, 1995. *Environmental Reports and Remediation Plans: Forensic and Legal Review*, John Wiley & Sons, Inc., New York.

Fairman R., C.D. Mead, and W.P. Williams, 1999. *Environmental Risk Assessment—Approaches, Experiences and Information Sources*. Published by European Environment Agency—EEA Environmental Issue Report No 4, Monitoring and Assessment Research Centre, King's College, London.

Falck, F., A. Ricci, M.S. Wolff, J. Godbold, and P. Deckers, 1992. Pesticides and polychlorinated biphenyl residues in human breast lipids and their relation to breast cancer. *Archives of Environmental Health*, 47:143–146.

Fischer, F. and M. Black (eds.), 1995. *Greening Environmental Policy: The Politics of a Sustainable Future*. St. Martin's Press, New York.

Francis, B.M., 1994. *Toxic Substances in the Environment*. John Wiley & Sons, Inc., New York.

Freeze, R.A., 2000. *The Environmental Pendulum*. University of California Press, Berkeley, CA.

Gaylor, D.W. and R.L. Kodell, 2002. A procedure for developing risk-based reference doses. *Regulatory Toxicology and Pharmacology*, 35(2 Pt. 1):137–141.

Gee, G.C. and D.C. Payne-Sturges, 2004. Environmental health disparities: a framework integrating psychosocial and environmental concepts. *Environmental Health Perspectives*, 112(17): 1645–1653.

Gilbert, T., 2002. Geographic information systems based tools for marine pollution response. *Port Technology International*, 13:25–27.

Goldman, M., 1996. Cancer risk of low-level exposure. *Science*, 271(5257): 1821–1822.

Griffiths, C.W., C. Dockins, N. Owens, N.B. Simon, and D.A. Axelrad, 2002. What to do at low doses: a bounding approach for economic analysis. *Risk Analysis*, 22(4):679–688.

Hamed, M.M., 1999. Probabilistic sensitivity analysis of public health risk assessment from contaminated soil. *Journal of Soil Contamination*, 8(3):285–306.

Hamed, M.M., 2000. Impact of random variables probability distribution on public health risk assessment from contaminated soil. *Journal of Soil Contamination*, 9(2):99–117.

Hammitt, J.K., 1995. Can more information increase uncertainty? *Chance*, 8(3):15–17.

Hammitt, J.K. and A.I. Shlyakhter, 1999. The expected value of information and the probability of surprise. *Risk Analysis*, 19(1):135–152.

Hansson, S-O., 1989. Dimensions of risk. *Risk Analysis*, 9(1):107–112.

Hansson, S-O., 1996. Decision making under great uncertainty. *Philosophy of the Social Sciences*, 26(3):369–386.

Hansson, S-O., 1996. What is philosophy of risk? *Theoria*, 62:169–186.

Health Canada, 1994. *Human Health Risk Assessment for Priority Substances*, Environmental Health Directorate, Canadian Environmental Protection Act, Health Canada, Ottawa.

Hilts, S.R., 1996. A co-operative approach to risk management in an active lead/zinc smelter community, *Environmental Geochemistry and Health*, 18:17–24.

Hwang, J-S. and C-C. Chan, 2002. Effects of air pollution on daily clinic visits for lower respiratory tract illness, *American Journal of Epidemiology*, 155(1):1–10.

ILSI (International Life Sciences Institute), 2003. *Workshop to Develop a Framework for Assessing Risks to Children from Exposures to Environmental Agents*. ILSI Press, Risk Science Institute, Washington, DC.

Joffe, M. and J. Mindell, 2002. A framework for the evidence base to support health impact assessment. *Journal of Epidemiology & Community Health*, 56(2):132–138.

Kent, C., 1998. *Basics of Toxicology*. John Wiley & Sons, Inc., New York.

Kimmel, C.A. and D.W. Gaylor, 1988. Issues in qualitative and quantitative risk analysis for developmental toxicology. *Risk Analysis*, 8:15–20.

Klaassen, C.D. (ed.), 1996. *Casarett and Doull's Toxicology: The Basic Science of Poisons*, 5th edition. McGraw-Hill, New York.

Kolluru, R. (ed.), 1994. *Environmental Strategies Handbook: A Guide to Effective Policies and Practices*, McGraw-Hill, New York.

Landrigan, P., C.A. Kimmel, A. Correa, and B. Eskenazi, 2004. Children's health and the environment: public health issues and challenges for risk assessment. *Environmental Health Perspectives*, 112(2):257–265.

Lanphear, B. and R. Byrd, 1998. Community characteristics associated with elevated blood lead levels in children. *Pediatrics*, 101(2):264–271.

Linders, J.B.H.J., 2000. *Modelling of Environmental Chemical Exposure and Risk*. Kluwer Academic Publishers, Dordrecht.

Liu, L.J.S., M. Box, D. Kalman, J. Kaufman, J. Koenig, T. Larson, T. Lumley, L. Sheppard, and L. Wallace, 2003. Exposure assessment of particulate matter for susceptible populations in Seattle. *Environmental Health Perspectives*, 111(7):909–918.

MacIntosh, D.L., C.W. Kabiru, and P.B. Ryan, 2001. Longitudinal investigation of dietary exposure to selected pesticides. *Environmental Health Perspectives*, 109(2):145–150.

Mathews, J.T., 1991. *Preserving the Global Environment: The Challenge of Shared Leadership*, W.W. Norton & Co., New York.

McKone, T.E., W.E. Kastenberg, and D. Okrent, 1983. The use of landscape chemical cycles for indexing the health risks of toxic elements and radionuclides. *Risk Analysis*, 3(3):189–205.

Mitchell, P. and D. Barr, 1995. The nature and significance of public exposure to arsenic: a review of its relevance to South West England. *Environmental Geochemistry and Health*, 17:57–82.

Mitchell, P. and D. Barr, 1995. The nature and significance of public exposure to arsenic: a review of its relevance to South West England. *Environmental Geochemistry and Health*, 17:57–82.

Mitsch, W.J. and S.E. Jorgesen (eds.), 1989. *Ecological Engineering, An Introduction into Ecotechnology*, John Wiley & Sons, Inc., New York.

Moeller, D.W., 1997. *Environmental Health*, Revised edition. Harvard University Press, Cambridge, MA.

Mohamed, A.M.O. and K. Côté, 1999. Decision analysis of polluted sites—a fuzzy set approach. *Waste Management*, 19:519–533.

Morris P. and R. Therivel (eds.), 1995. *Methods of Environmental Impact Assessment*. UCL Press, University College, London.

Neumann, D.A. and C.A. Kimmel (eds.), 1999. *Human Variability in Response to Chemical Exposures (Measures, Modeling, and Risk Assessment)*. CRC Press LLC, Boca Raton, FL.

Nieuwenhuijsen, M.J. (ed.), 2003. *Exposure Assessment in Occupational and Environmental Epidemiology*. Oxford Medical Publications, Oxford.

NRC (National Research Council), 1993. *Pesticides in the Diets of Infants and Children*. National Academy Press, Washington, DC.

NRC, Committee on Human Biomonitoring for Environmental Toxicants, 2006. *Human Biomonitoring for Environmental Chemicals*. The National Academies Press, Washington, DC.

Pocock, S.J., M. Smith, and P. Baghurst, 1994. Environmental lead and children's intelligence: a systematic review of the epidemiological evidence. *British Medical Journal*, 309:1189–1197.

Preston, R.J., 2004. Children as a sensitive subpopulation for the risk assessment process. *Toxicology and Applied Pharmacology*, 199(2):132–141.

Price, P.S., C.F. Chaisson, M. Koontz, C. Wilkes, B. Ryan, D. Macintosh, and P. Georgopoulos, 2003. *Construction of a Comprehensive Chemical Exposure Framework Using Person Oriented Modeling*. Developed for The Exposure Technical Implementation Panel, American Chemistry Council, Contract #1388.

Rabl, A., J.V. Spadaro, and P.D. McGavran, 1998. Health risks of air pollution from incinerators: a perspective. *Waste Management & Research*, 16(4):365–388.

Rail, C.C., 1989. *Groundwater Contamination: Sources, Control and Preventive Measures*. Technomic Publishing Co., Inc., Lancaster, PA.

Raymond, C., 1995. New state programs encourage industrial development: balancing environmental protection and economic development. *Soil & Groundwater Cleanup*, November:10–12.

Richards, D. and W.D. Rowe, 1999. Decision-making with heterogeneous sources of information. *Risk Analysis*, 19(1):69–81.

Robson, M.G. and W.A. Toscano (eds.), 2007. *Risk Assessment for Environmental Health*. John Wiley & Sons, Inc., Jossey-Bass, San Francisco, CA.

Russell, M. and M. Gruber, 1987. Risk assessment in environmental policy-making. *Science*, 236:286–290.

Schleicher, K. (ed.), 1992. *Pollution Knows No Frontiers: A Reader*. Paragon House Publishers, New York.

Schulte, P.A., 1989. A conceptual framework for the validation and use of biologic markers. *Environmental Research*, 48(2):129–144.

Schuurmann, G. and B. Markert, 1966. *Ecotoxicology*. John Wiley & Sons, Inc., New York.

Sherman, J.D., 1994. *Chemical Exposure and Disease*, Princeton Scientific Publishing Co., New Jersey.

Silkworth, J.B. and J.F. Brown, Jr., 1996. Evaluating the impact of exposure to environmental contaminants on human health. *Clinical Chemistry*, 42:1345–1349.

Stern P.C. and H.V. Fineberg (eds.), 1996. *Understanding Risk— Informing Decisions in a Democratic Society*. Committee on Risk Characterization, Commission on Behavioural and Social Sciences and Education—National Research Council. National Academy Press, Washington, DC.

Strange, E., J. Lipton, D. Beltman, and B. Snyder, 2002. Scientific and societal considerations in selecting assessment endpoints for environmental decision making. Defining and assessing adverse environmental impact symposium 2001. *The Scientific World Journal*, 2(S1):12–20.

Suter II, G.W., B.W. Cornaby, C.T. Hadden, R.N. Hull, M. Stack, and F.A. Zafran, 1995. An approach for balancing health and ecological risks at hazardous waste sites. *Risk Analysis*, 15:221–231.

Till, J.E. and H.A. Grogan (eds.), 2008. *Radiological Risk Assessment and Environmental Analysis*. Oxford University Press, New York.

USEPA, 1996. *Radiation Exposure and Risk Assessment Manual*. US Environmental Protection Agency (USEPA), Washington, DC; EPA 402-R-96-016.

USEPA, 1997. *Ecological Risk Assessment Guidance for Superfund: Process for Designing and Conducting Ecological Risk Assessments*. US Environmental Protection Agency, Washington, DC; ERAGS, EPA 540-R-97-006, OSWER Directive # 9285.7-25, June 1997.

U.S. EPA, 1997. *Guiding Principles for Monte Carlo Analysis*. Office of Research and Development, Risk Assessment Forum, Washington, DC; EPA/630/R-97/001.

U.S. EPA, 1998. *Guidelines for Ecological Risk Assessment*. U.S. Environmental Protection Agency, Washington, DC; EPA/630/R-95/002F.

U.S. EPA, 2000. *Science Policy Handbook: Risk Characterization*. Science Policy Council, Washington, DC; EPA/100/B/00/002.

U.S. EPA, 2003. *Framework for Cumulative Risk Assessment*. Risk Assessment Forum, Washington, DC; EPA/630/P-02/001F.

U.S. EPA, 2006. *A Framework for Assessing Health Risk of Environmental Exposures to Children (Final)*. U.S. Environmental Protection Agency, Washington, DC; EPA/600/R-05/093F, 2006.

van Veen, M.P., 1996. A general model for exposure and uptake from consumer products. *Risk Analysis*, 16(3):331–338.

Vermeire, T.G., P. van der Poel, R. van de Laar, and H. Roelfzema, 1993. Estimation of consumer exposure to chemicals: application of simple models, *Science of the Total Environment*, 135:155–176.

Villa, F. and H. McLeod, 2002. Environmental vulnerability indicators for environmental planning and decision-making: guidelines and applications. *Environmental Management*, 29(3):335–348.

Washburn, S.T. and K.G. Edelmann, 1999. Development of risk-based remediation strategies. *Practice Periodical of Hazardous, Toxic, and Radioactive Waste Management*, 3(2):77–82.

White R.F., R.G. Feldman, and P.H. Travers, 1990. Neurobehavioral effects of toxicity due to metals, solvents, and insecticides. *Clinical Neuropharmacology*, 13:392–412.

Williams, F.L.R. and S.A. Ogston, 2002. Identifying populations at risk from environmental contamination from point sources. *Occupational and Environmental Medicine*, 59(1):2–8.

Woodside, G., 1993. *Hazardous Materials and Hazardous Waste Management: A Technical Guide*. John Wiley & Sons, Inc., New York.

Worster, D., 1993. *The Wealth of Nature: Environmental History and the Ecological Imagination*. Oxford University Press, New York.

Xu, L.Y. and X. Shu, 2014. Aggregate Human health risk assessment from dust of daily life in the urban environment of Beijing'. *Risk Analysis*, 34(4):670–682.

Zartarian, V.G. and J.O. Leckie, 1998. Dermal exposure: the missing link. *Environmental Science & Technology*, 32(5): 134A–137A.

APPENDIX B

MODEL SYLLABUS

CHAPTER 1: GEOLOGY IN ENVIRONMENTAL MANAGEMENT

Model Syllabus

Geology in Environmental Management
Natural hazards affecting humans
Human impacts on the environment
Volcanic hazard
    Ash eruptions
    Hazards to aviation
    Lahars—volcanic mudflows
Seismic hazards
    Earthquakes and associated tsunamis
    Seismic safety standards—California
    Liquefaction
    Earthquake-induced landslides
Coastal processes and environmental management
    Human intervention in coastal sediment transport mechanisms
    Impact of major coastal storms
Rivers, lakes, and environmental management
    Flooding
    Eutrophication of lakes
    Special considerations that apply to desert lakes

Groundwater and karst
    Groundwater overdraft
    Sinkhole hazards
Waste management
    Landfill leachate
    Nuclear waste disposal
Energy resource extraction and environmental management
    Coal and coal mining
    The petroleum system
    Petroleum and natural gas extraction

CHAPTER 2: BIOLOGY IN ENVIRONMENTAL MANAGEMENT

Model Syllabus

Instructor Information

Instructor:
Office Location:
Telephone:
E-mail:
Office Hours:

An Integrated Approach to Environmental Management, First Edition. Edited by Dibyendu Sarkar, Rupali Datta, Avinandan Mukherjee, and Robyn Hannigan.
© 2016 John Wiley & Sons, Inc. Published 2016 by John Wiley & Sons, Inc.

Course Identification

Course Number:
Course Name: Biology in Environmental Management
Course Location:
Class Times:
Prerequisites:

Course Description/Overview

Environmental management calls upon many subdisciplines of biology for solutions to complex problems: botany, landscape ecology, conservation biology, and restoration ecology. Environmental management (which also includes environmental restoration) typically involves five stages, which depend heavily on a diversity of data to understand where the ecosystem is relative to where managers want it to be. However, successful management or restoration requires that human activities and land use are managed as well, and therefore, the involvement of local communities and stakeholders is critical in all stages.

In this course, you will gain a broad overview of how biologists engage in environmental management and restoration, using their knowledge of complex dynamics, scale issues, biodiversity and community assembly, and abiotic characteristics of a region to identify feasible management or restoration targets and build a management plan to meet them. The Florida Everglades is used as a case study to demonstrate how these issues manifest themselves in the management (and then restoration) of a large, internationally famous ecosystem.

Course Learning Objectives

By the end of the course, students should:

1. Distinguish the main difference between environmental management and restoration
2. Define the five stages of environmental management and explain how biological disciplines influence their execution
3. Describe what adaptive management is and how it can be successfully implemented in environmental management and restoration
4. Build a management or restoration plan for a specific ecosystem, which includes targets, indicators, data sets, and avenues for incorporating stakeholders into the process

Course Resources

Course Website(s)
- Course website (e.g., Canvas or Blackboard)
- Personal website http://www.mtu.edu/forest/about/faculty/mayer/

Required Course Readings
A selection of readings will be assembled for each week to be read prior to class.

Grading Scheme

Grading System
Grades and explanations of them here

Grading Policy
Grades will be based on the following:

10-page research paper (100 for draft, 100 for final)	200
Peer review of draft research papers	50
Five homework assignments (20 points each)	100
Class journal	100
Class attendance/participation	200
Total points	**650**

Late Assignments
A late assignment loses 10% of the highest possible grade for each day it is late ("day" includes Saturday and Sunday). For example, the assignment would have received a 95% if it were turned in on time, and if it is turned in a day late, it will receive an 85%. This penalty will be waived if students notify me **prior** to the deadline of potentially late assignments… and the reason must be unavoidable!

Collaboration/Plagiarism Rules

Unless explicitly instructed, homework assignments are to be completed individually, with no help from or discussions with other students. We will use the homework assignments as springboards for discussion in class, and therefore, we need a variety of viewpoints and ideas. Research papers are to be completed individually, although discussion with classmates is acceptable and encouraged.

Out of consideration for your classmates, cell phones, Blackberries, iPods, PDAs, or any other electronic devices are not to be used in the classroom and must be shut off. Information exchanges on these devices during class are also prohibited and violate the Academic Integrity Code of Michigan Tech.

University Policies

Academic regulations and procedures are governed by the University policy. Academic dishonesty cases will be handled in accordance with the University's policies.

Academic Integrity: http://www.studentaffairs.mtu.edu/dean/judicial/policies/academic_integrity.html

If you have a disability that could affect your performance in this class or that requires an accommodation under the Americans with Disabilities Act, please see me as soon as possible so that we can make appropriate arrangements.

The Affirmative Action Office has asked that you be made aware of the following:

Michigan Tech complies with all federal and state laws and regulations regarding discrimination, including the Americans with Disabilities Act of 1990. If you have a disability and need a reasonable accommodation for equal access to education or services at Michigan Tech, please call the Dean of Students Office, at 487-2212. For other concerns about discrimination, you may contact your advisor, department head or the Affirmative Action Office, at 487-3310

Disability Services: http://www.mtu.edu/deanofstudents/students/disability/policy/

Michigan Tech is committed to providing a living, learning, and working environment that is free from harassment or discrimination based on race, religion, color, national origin, age, sex, sexual orientation, gender identity, height, weight, genetic information, marital status, ability, or veteran status. Michigan Tech prohibits any conduct that threatens or endangers the health or safety of any individual or group, including physical abuse, verbal abuse, threats, stalking, intimidation, harassment, sexual misconduct, coercion, and/or other communication or conduct that creates a hostile living, learning, or working environment. Any behavior that makes other students feel intimidated in this classroom will not be tolerated.

Affirmative Action: http://www.admin.mtu.edu/aao/

Equal Opportunity Statement: http://www.admin.mtu.edu/admin/boc/policy/ch3/ch3p7.htm

Course Assignments

Research Paper (200 Points)
This paper will be on a management or restoration case study of your choosing. The paper will have the following sections: (i) Biological description of the case study (what were the "historical" (however you define this) conditions of the ecosystem, what are the defining features or unique characteristics of the system, what are the natural disturbance regimes, what is the condition of the system today); (ii) Anthropogenic description of the case study (what are the types of human impacts or disturbances in the system, what ecosystem goods and services are important to local communities); (iii) Five stages of management or restoration (this will be the longest part of the paper, with different subsections for each of the five stages. In essence, you will either describe an existing management or restoration plan in these steps, or you will come up with them on your own); (iv) Conclusions, Assessment of Likely Success of the plan. The paper should be supported by an adequate number of *peer-reviewed journal articles/reports* to support your plan and descriptions.

Peer Review of Student Papers (2 @ 25 Points Each)
All students will be assigned two student draft papers to peer review (using the peer review form handed out to you). The point of this exercise is not only to help each other improve your papers, but also to learn from each other. Reviewing others' writing allows you to practice putting yourself in the shoes of a reviewer, so that you can be more critical and insightful with your own writing.

Homework Assignments (100 Points)
Five homework assignments will be distributed throughout the semester to make sure that you understand the key points in each of the five stages. These will be essays (about 2–3 pages) with prompts provided by the instructor. We will use these assignments as discussion fodder in class.

Class Journal (100 Points)
Each week, we will start a class session with a 10-min in-class writing exercise. You will need a notebook or Blue Book for this. I will give you a short question to write about or reflect upon. At random points in the semester, I will collect these notebooks and read through them, partially to get a general sense of how the class is doing as a whole, and partially to see how individual students are mastering the material. I will repeat some of these questions or topics throughout the semester, so you can see how far your understanding has come since earlier in the semester. If you are absent on the day that this question is given, please send me an email and let me know that you need the question, if a classmate doesn't remember what it was.

Attendance and Participation (200 Points)
  Attendance (100 points): Students will receive two unexcused absences without any penalty. For each absence thereafter, students will lose 10 points.
  Participation (100 points): These points will be given at my discretion; the scale goes something like this: if you speak up and participate in class discussions most of the time, you will get the full 100 points. If you speak so rarely that I don't know what your voice sounds like, you will get 30 or fewer points.

Course Schedule

Week 1: Introduction to course
Week 2: Disciplines associated with Environmental Management and Restoration
Week 3: Environmental Management and Restoration: Stage 1 (define the system)
Week 4: Stage 1 continued (system characteristics)
Assignment: Choose a case study for research paper and send e-mail to Instructor to confirm
Week 5: Environmental Management and Restoration: Stage 2 (identify target conditions)
Assignment: Homework #1 due (Stage 1)
Week 6: Stage 2 continued (identify indicators)

Week 7: Environmental Management and Restoration: Stage 3 (collecting data)

Assignment: Homework #2 due (Stage 2)

Week 8: Environmental Management and Restoration: Stage 4 (management)

Assignment: Homework #3 due (Stage 3)

Week 9: Stage 4 continued (restoration)

Week 10: Environmental Management and Restoration: Stage 5 (adaptive management)

Assignment: Homework #4 due (Stage 4)

Week 11: Case studies in Environmental Management (e.g., Greater Yellowstone Ecosystem)

Assignment: Draft research paper due to Instructor and two student peer reviewers

Week 12: Case studies in environmental restoration (e.g., Florida Everglades)

Assignment: Student peer reviews of research papers due back to authors; send copy of review to Instructor

Week 13: Integrating stakeholders

Assignment: Homework #5 due (Stage 5)

Week 14: Course wrap-up, lessons learned

Assignment: Final research paper due to Instructor

CHAPTER 3: SOIL SCIENCE IN ENVIRONMENTAL MANAGEMENT

Syllabus

About This Class

Soils have numerous interactions with the biosphere, hydrosphere, and atmosphere. In a global perspective, human life depends on soil quality and soil quality, is greatly influenced by human life. From food production and waste management to sustaining clean resources and deciphering global climate systems, soils are an intricate part of the equation and a necessary component to the solution. This class will provide an overview of some fundamental concepts of soil science and will illustrate its relationship with environmental management through several case studies, which constitute selected pressing environmental issues.

By the end of this class, the student will be able to:

- Describe fundamental physical, chemical, and biological properties and processes of soils
- Evaluate the role of soils in major environmental management problems
- Apply soil science principles to environmental management problems

Grades

Homework, 10%; term paper and class presentation, 15%; midterm exam, 35%; and final exam, 40%

Term Paper

The term papers should be done in groups of three people. They should focus on a special topic related to environmental management and soil science, review the pertinent literature, and present an overview in written and oral format. The written document should be a minimum of 15 pages of double-spaced text (1″ page margin, font Times New Roman 12 or Arial 11), including figures, tables, and references. An oral presentation with PowerPoint will be given during the last week of classes, of 10–15 min duration.

Indicative list of topics:

- Measures for CO_2 capture and control in geological systems
- Eutrophication of surface waters
- Soil erosion
- Management of brownfield sites
- Clean water supply in developing countries
- Naturally occurring arsenic contamination

You can come up with your own ideas as well. Topics are handed out on a first come, first served basis.

Class Policy

The lectures in this course build on the previous class's lecture; regular attendance is strongly recommended to understand the processes taught. The student is responsible for the material taught in a class not attended.

Handbook and Additional Reading

Brady, N.C. and R.R Weil. 2007. *The nature and properties of Soil*. 14th edition. Prentice Hall. Upper Saddle River, New Jersey.

Additional Reading for Case Studies

Gruhn, P., F. Goletti, and M. Yudelman. 2000. *Integrated nutrient management, soil fertility and sustainable agriculture: current issues and future challenges*. Food, Agriculture and the Environment Discussion Paper 32. Washington, DC: IFPRI.

U.S. Environmental Protection Agency. 2012. *Brownfields road map to understanding options for site investigation and cleanup*. 5th edition, EPA 542-R-12-001. Washington, DC: U.S. Environmental Protection Agency.

USGS (United States Geological Survey). 2007. *Pesticides in the Nation's streams and ground water, 1992–2001, circular 1291*. Reston, VA: U.S. Dept. of the Interior, U.S. Geological Survey.

Conant, R.T., M.G. Ryan, G.I.A. Gren, H.E. Birge, E.A. Davidson, P.E. Eliason, S.E. Evans, S.D. Frey, C.P. Giardina, F.M. Hopkins, R.H. Nen, M.U.F. Kirschbaum, J.M. Lavallee, J. Leifeld, W.J. Parton, J.M. Steinweg, M.D. Wallenstein, J.A°.M. Wetterstedt, and M.A. Bradford. (2011). Temperature and soil organic matter decomposition rates—synthesis of current knowledge and a way forward, *Global Change Biology*, 17, 3392–3404.

Course Outline

Week	Material
1	Introduction, soil functions in the environment
2	Soil composition—mineralogy
3	Soil composition—organic matter
4	Soil composition—soil solution and air
5	Soil properties—physical and chemical
6	Soil properties—chemical and biological
7	Soil formation and development
8	Chemical movement and sorptive processes
9	Soil transport processes
10	Case Study 1—soil fertility management
11	Case Study 2—water and soil contamination
12	Case Study 3—pesticide pollution
13	Case Study 4—global warming and soils
14	Class presentations

CHAPTER 4: GREEN CHEMISTRY AND ECOLOGICAL ENGINEERING AS A FRAMEWORK FOR SUSTAINABLE DEVELOPMENT

Curriculum

Typically, one semester teaching comprises of 40 contact hours.

Module	Short Title	Hours
Module 1	**The "environment" in the context of the twenty-first century**	 1
1.1	Which way the winds are blowing worldwide?	
1.2	Present state of world environment	
1.3	The unfinished agenda for the world	
Module 2	**Indian laws on environmental protection**	 3
2.1	Laws and regulations for safe disposal of wastewater	
2.2	Do we need to reexamine the rationale for effluent standards?	
2.3	New bacterial guidelines for treated municipal effluents	
2.4	Laws and regulations for protection of ambient air quality	
2.5	The Environmental Protection Act	
2.6	Regulations addressing hazardous materials and wastes	
2.7	Regulations for disposal of biomedical wastes	
2.8	Regulations for disposal of municipal wastes	
Module 3	**The framework and implementation of environmental policy and law**	 2
3.1	Theory and practice of environmental policy and law	
3.2	Case studies on important policies	
3.3	Important litigations	
Module 4	**System analysis: knowing the limits**	 2
4.1	What is the relationship of parts with the total picture?	
4.2	What is an environmental system?	
4.3	Role of chemical dynamics of aquatic and atmospheric systems	
4.4	Role of biological dynamics of aquatic and atmospheric systems	
4.5	Case studies	
Module 5	**Environmental audit**	 2
5.1	Significance of environmental audit as a tool	
5.2	Why different audits?	
5.3	Protocols and procedures	
5.4	Case studies	
Module 6	**Environmental impact assessment**	 3
6.1	Significance of EIA as a tool	
6.2	EIA is mandatory for all development projects	
6.3	Protocols and procedures	
6.4	A long way to go before EIAs facilitate sustainable development	
6.5	Case examples	
Module 7	**Life cycle assessment**	 3
7.1	What is LCA?	
7.2	Interpretation of environmental attributes	
7.3	Use of LCA in decision making	

Module	Short Title	Hours
Module 8	**Inventorization and environmental management system**	 3
	8.1 What are "wastes"?	
	8.2 Significance of waste inventory as a tool	
	8.3 Protocols and procedure of inventorization	
	8.4 Significance of EMS as a tool	
	8.5 Components of an environment management systems	
	8.6 ISO 14001: A model environment management system	
	8.7 Case studies	
	8.8 Issues related to ISO certification	
Module 9	**The broad debate on preventive environmental management**	 2
	9.1 What is cleaner production?	
	9.2 Cleaner production techniques	
	9.3 Evolution of industries	
	9.4 An illustrated approach for cleaner production	
	9.5 Alternate approaches to prioritizing options for pollution prevention	
Module 10	**Green chemistry and engineering applications**	 4
	10.1 Green chemistry	
	10.2 Green redox technologies	
	10.3 Green catalyst	
	10.4 Biocatalysts	
	10.5 Phase-transfer catalyst	
	10.6 Membrane separation (RO, MF, UF, NF)	
	10.7 Case studies	
Module 11	**Greener choices in process industry**	 2
	11.1 Choice of eco-friendly solvents	
	11.2 Choice of eco-friendly reactants	
	11.3 Choice of eco-friendly biofuels	
	11.4 Choice of eco-friendly fuel cells	
Module 12	**Redesigning unit operations and unit processes**	 2
	12.1 Mass exchange networks	
	12.2 Pinch technique	
	12.3 Case studies	
Module 13	**Recycle and reuse of treated wastewater**	 3
	13.1 Status of regulation and compliance	
	13.2 What are our options? Policies, regulations and controls, market mechanisms, role of technology, resource pricing	
	13.3 Framework for sustainable urban wastewater management	
	13.4 Issues in water recycling and reuse	
	13.5 Case studies	
Module 14	**Sustainable and intelligent consumption**	 1
	14.1 What is sustainable consumption?	
	14.2 Transitioning to sustainable consumption in the context of globalization	
	14.3 Why consumption pattern matters?	
	14.4 Rebound effects of consumption	
	14.5 Toward new approach	
	14.6 Case examples	
Module 15	**Ecological engineering**	 3
	15.1 What is ecological engineering?	
	15.2 Significance of ecological engineering in the context of sustainable development	
	15.3 "Design for ecology and environment" as the ultimate goal	
	15.4 Business strategies inspired by ecological engineering	
	15.5 Case studies	
Module 16	**Eco-industrial networks**	 1
	16.1 The concept of eco-industrial development	
	16.2 Risks and benefits	
	16.3 Case studies	
Module 17	**Nexus between trade and environment**	 3
	17.1 Emerging concerns about impacts of trade-related production activities	
	17.2 Trade parameters and environmental compliance dynamics	
	17.3 Typology of policy instruments in the trade–environment interface	
	17.4 Role of financial institutions	
	17.5 Environmental implications of trade governance	
	17.6 Internalizing externalities	
	17.7 Environmental assessment of trade agreements	

CHAPTER 5: GREEN ENERGY AND CLIMATE CHANGE

Recommended Curriculum

1. Introduction and structure
   1.1 Forms and functions of energy
   1.2 Energy flows and balance
      1.2.1 Solar radiation
      1.2.2 Wind flows
      1.2.3 Ocean energy
      1.2.4 Rivers and reservoirs
      1.2.5 Geothermal
   1.3 Green Energy types and characteristics
      1.3.1 Solar
      1.3.2 Wind
      1.3.3 Tidal
      1.3.4 Hydroelectric
      1.3.5 Cogeneration
      1.3.6 Biomass
      1.3.7 Geothermal
      1.3.8 Isolated and grid operations
      1.3.9 Conversions, losses, and coefficients
2. Solar energy principles and systems
   2.1 Architecture
   2.2 Solar thermal: Collectors, heating networks, power plants
   2.3 Passive solar energy
   2.4 Photovoltaic power generation: arrays
   2.5 Economic and environmental analysis
3. Biomass
   3.1 Photosynthesis
   3.2 Biogas and cogeneration
   3.3 Wood and other substrates and energy plantations
   3.4 Ethanol
   3.5 Synthetic fuels
   3.6 Economic and environmental analysis
4. Hydroelectric systems
   4.1 Hydropower plants and varieties
   4.2 Setup, intake, turbines
   4.3 Economic and environmental analysis
5. Geothermal energy
   5.1 Shallow geothermal sources and utilization
   5.2 Heat supply, hydro- and geothermal, and deep wells
   5.3 Well drilling
   5.4 Heat transfer
   5.5 Leakage monitoring
   5.6 Subsurface and aboveground systems
   5.7 Economic and environmental analysis
6. Emerging technologies
   6.1 Fuel cells
   6.2 Solar hydrogen
7. Emission offsets and protocols
   7.1 CDM methodologies
   7.2 Funding parameters and quantification of paybacks
   7.3 Carbon markets
8. Policies
   8.1 National and local imperatives and agreements
   8.2 Regional and global mechanisms
9. Green Energy initiatives
   9.1 Bilateral and multilateral processes
   9.2 Successes and challenges
   9.3 Information resources including databases
   9.4 Industry initiatives and associations

CHAPTER 6: ENGINEERING IN ENVIRONMENTAL MANAGEMENT

Course Outline

Week	Topics
1	Overview of engineering in environmental management
2	Materials balance
3	Reactor analysis and energy balance
4	Water supply subsystem: overview, traditional and emerging pollutants, and drinking water standards
5	Water supply subsystem (cont.): water treatment and distribution systems
6	Wastewater disposal subsystem: overview, pollutants, and wastewater treatment standards
7	Wastewater disposal subsystem (cont.): wastewater treatment and sewer systems
8	Midterm
9	Solid waste management: overview, generation, and characteristics of municipal solid waste
10	Solid waste management (cont.): source reduction, recycling, composting, and incineration
11	Solid waste management (cont.): landfills
12	Air resource system: overview and air pollutants
13	Air resource system (cont.): stationary source control
14	Air resource system (cont.): mobile source control
15	Final exam

CHAPTER 7: GREEN ARCHITECTURE IN ENVIRONMENTAL MANAGEMENT

Course Syllabus

What You Can Expect from the Course

Learning Outcomes and Expectations:

1. Define Green Architecture.
2. Understand why Green Architecture is an important factor in Environmental Management.
3. Quantify and measure Green Architecture.
   For example, LEED, Green Globes, Energy Star, CHPS, and STARS.
4. Discuss the integrated design process in creating successful and cost-effective Green Architecture.
5. Identify career opportunities in the green field.
   For example, LEED credentials, energy management, and carbon credit markets.
6. Beyond Green Architecture: Creating a culture of sustainability.

Textbooks

The Integrative Design Guide to Green Building by 7 Group & Bill Reed
   Our Choice by Al Gore
   LEED BD&C Reference Guide by USGBC
Readings—supplementary readings for further understanding of Green Building:

1. Professional journals
   (a) *Environmental Design and Construction*
   (b) *Environmental Building News*
   (c) *Green Builder Magazine*
2. Scholarly journals
   (a) *Journal of Green Building*
   (b) *International Journal of Sustainable Building Technology and Urban Development*
3. *Books*
   (a) **Cradle to Cradle: Remaking the Way We Make Things** by William McDonough and Michael Braungart
   (b) **Green to Gold: How Smart Companies Use Environmental Strategy to Innovate, Create Value, and Build Competitive Advantage** by Daniel Esty and Andrew Winston
   (c) **The New Natural House Book** by David Pearson
   (d) **Natural Capitalism: Creating the Next Industrial Revolution** by Paul Hawken, Amory Lovins, and L. Hunter Lovins

Course Overview

The purpose of this course is to provide you with an introduction to the concepts and techniques of Green Architecture. Buildings are responsible for approximately 39% of all carbon emissions in the United States. This is both a significant challenge and a major opportunity in creating a sustainable culture. One response to the challenge of sustainable development, the third-party rating system, has become an increasingly popular decision support tool in recent years. Many well-known third-party assessment toolsets were developed internationally, such as the International Initiative for a Sustainable Built Environmental Rating System and Sustainable Building Tool (iiSBE, SBTool 07) in Canada, the Building Research Establishment Environmental Assessment Method (BREEAM) Communities in England, and the Comprehensive Assessment System for Built Environment Efficiency (CASBEE) in Japan.

National, regional, and municipal initiatives include the US Green Building Council (USGBC) Leadership in Energy and Environmental Design (LEED™), Sustainable Sites Initiative™, which includes a set of arguments—economic, environmental, and social—for the adoption of sustainable land practices.

USGBC's most recent assessment tool, LEED™ for Neighborhood Development (LEED-ND), offers great support through enhanced access to information, increased public participation in decision making, and support for distributed collaboration between planners, stakeholders, and the public. This tool integrates the principles of smart growth, new urbanism, and green building into a system for sustainable neighborhood redevelopment and emphasizes the creation of compact, walkable, vibrant, mixed-use neighborhoods, with good connections to nearby communities.

The course has two components: learning the theories of Green Architecture (the lectures) and learning to apply certain theories via case studies and hands-on exercises.

Overall, this course will provide you with a good introduction to Green Architecture and Sustainable Development. You are expected to become familiar with the professional Green Building community including major books, journals, and websites and to become aware of the vast wealth of sustainable development data available on the Internet. Several reference books and journals will be used in this class, a list of which is given in the textbook section. Most of these references should be available in the library.

Session II

The second session of the course will cover the basic "Green Building and LEED™ Core Concepts". This course focuses on the understanding of core concepts of the LEED™ Rating Systems needed to pass the LEED™ Green Associate Study Exam (Tier I Component).

Objectives

The LEED™ Green Associate credential demonstrates comprehensive, general knowledge in green building and LEED and is for individuals who support, but may not directly apply, green building practices in their regular professional work. This study program provides foundational knowledge in green building and LEED and helps prepare for the LEED Green Associate exam. It clarifies the structure of the new LEED Professional Program and educates you in the subject matter covered by the LEED™ Green Associate exam. Participants will find this program series to be a focused study effort prior to sitting for the exam:

- Sustainable sites
- Water efficiency
- Energy and atmosphere
- Materials and resources
- Indoor environmental quality
- Innovation in design

CHAPTER 8: BUSINESS STRATEGIES FOR ENVIRONMENTAL SUSTAINABILITY

Model Syllabus

Instructor Information

Instructor:
Office Location:
Telephone:
E-mail:
Office Hours:

Course Identification

Course Number:
Course Name: Business Strategies for Environmental Sustainability
Course Location:
Class Times:
Prerequisites:

Course Description/Overview

Integrated approaches and solutions to environmental issues are necessary to overcome ever-increasing environmental problems. Environmental degradation caused by business activities is widely acknowledged all around the world. With the rising concerns over environmental destruction, individuals and businesses are becoming increasingly aware of the dangers that will occur if they do not carefully consider the needs of natural environment in every step they take. Thus, companies practice green initiatives at the firm or product level with the aim at introducing their efforts to help ease environmental impacts of their functions, such as reducing or eliminating ecologically harmful impacts of suppliers, productions, products, or end users. Developing successful business strategies in order to help sustain environmental well-being is an important way of adopting these attempts to satisfy needs of the stakeholders of business entities that can play a vital "bridging role" toward sustainability and today's necessary green lifestyles.

This course elaborates on the growing importance of environmental sustainability for businesses and identifies different ways to adopt this necessary action for the survival of companies in today's evolving business era. Furthermore, the course explores corporate social responsibility (CSR), triple bottom line, and greening the value chain, which are also important strategies for sustainable business success. In addition, the course looks at general business strategies and elaborates on the positioning businesses with sustainable practices and corporate reporting. Under every subject, the course also covers a number of relevant important industry examples and case studies.

Course Learning Objectives

By the end of the course, students should:

List and define the motivators for companies to become sustainable

Define CSR and how it impacts company success

Define the three different business strategies and explain how companies can make these strategies environmentally sustainable

Define Porter's value chain and explain how to green each stage of the value chain

Describe why companies need to participate in corporate reporting and define GRI

Develop a sustainable business strategy plan for a specific company

Course Resources

Course Website(s)
- Course website (e.g., Learning Management System)
- Personal website (instructor's personal website)

Required Course Readings
- A selection of readings will be assembled for each week to be read prior to class.

Grading Scheme

Grading System
Grades and explanations of them here

Grading Policy
Grades will be based on the following:

10-page research paper (group of two)	200
Research presentation (group of two)	100
Midterm	150
Final	150
Class attendance/participation	100
Total points	**700**

Late Assignments
Late assignment policy here

University Policies

Institutional policies here:
Academic integrity
Disability accommodations/services
Anti-discrimination
Affirmative action
Equal opportunity

Course Assignments

Research Paper (200 Points)
This paper will be on a development of a sustainable business strategy case study of your choosing. The paper will have the following sections: (i) general description of the case study (e.g., company information, current strategy, and shortcomings), (ii) development of green strategy for the company success (e.g., greening the value chain, differentiation), (iii) possible outcomes of the strategy adoption (assessment of likelihood of success of the plan), (iv) conclusion, and (v) references. The paper should be supported by an adequate number of peer-reviewed journal articles/reports to support your plan and descriptions. This is a group project with two students.

Research Presentation (100 Points)
Each group will present their papers in the classroom. The point of this exercise is not only to help each other improve your papers but also to learn from each other and develop better presentation skills.

Midterm (150 Points)
The midterm exam will be on those sections in the first half of the course. It will cover short essay and multiple-choice questions.

Final (150 Points)
Similar to the midterm exam, the final will also cover short essay and multiple-choice questions. The sections in the second half of the course will be the focus of this final exam.

Attendance and Participation (100 Points)
   Attendance (50 points): Students will receive two unexcused absences without any penalty. For each absence thereafter, students will lose 10 points.
   Participation (50 points): These points will be given at my discretion; the scale goes something like this: If you speak up and participate in class discussions most of the time, you will get the full 50 points. If you speak so rarely that I don't know what your voice sounds like, you will get 10 or fewer points.

Course Schedule

Week 1: Introduction to course, Business Strategies for Environmental Sustainability
Week 2: Business strategies (cost leadership, differentiation, and focus or niche strategy)
Week 3: Sustainability in the new business era and motivators for companies to become sustainable
Week 4: Theories of sustainable business strategies
Week 5: Sustainable business models and implementation
Week 6: The performance of sustainable corporations
Week 7: Midterm
Week 8: Greening the value chain
Week 9: Different perspectives on green markets and businesses
Week 10: Corporate reporting
Week 11: Being insincere: Greenwashing and case studies
Week 12: Student research presentations
Week 13: Student research presentations
Week 14: Final exam

CHAPTER 9: GREEN MARKETING STRATEGIES

Model Syllabus

Instructor Information

Instructor:
Office Location:
Telephone:

E-mail:
Office Hours:

Course Identification

Course Number:
Course Name: Green Marketing
Course Location:
Class Times:
Prerequisites:

Course Description/Overview

This subject examines sustainable marketing from a strategic approach. It focuses on the nature of the innovative green marketing methods and the extent that marketers should satisfy customer demand, keeping in mind the triple bottom line and the global demand for improved sustainability.

The subject deals with marketing activities and the management of those activities in marketing that is anchored on promoting sustainable environmental attributes of organizations and products. It outlines the major dimensions of the macroenvironment that impacts sustainability such as economic, social, cultural, political, legal, and financial environments while providing a set of conceptual and analytical tools to deal with sustainable issues. The content covered in the subject combines the existing strengths of conventional marketing with the sustainability perspectives of ecological and ethical marketing. It provides an overview of the long-term concepts of sustainable production and consumption and highlights strategies to create a new marketing paradigm based on sustainable, value-based relationships with consumers. This course not only equips students with sustainable skills but also engenders an appreciation and how to react to the challenges that lie ahead.

Course Learning Objectives

The unit is designed to provide students with an appreciation and understanding of managing for environmental sustainability through the green marketing initiatives and to introduce the theory and application of sustainability in business.
On completion of the unit, students should:

1. Have an overview of the main environmental issues and their relevance for the world economy and for business practice
2. Understand the nature and role of sustainability and green marketing in the modern business environment
3. Be familiar with the range of stakeholders in the field of sustainability and their links to management activities
4. Understand a variety of techniques for analyzing an organization's sustainability positioning
5. Appreciate the distinctions and interrelationships between individual, organizational, and social dimensions of sustainability
6. Demonstrate knowledge of functional activities associated with sustainability management
7. Demonstrate the ability to conduct an analysis of the sustainability management of an organization or industry

Course Resources

Prescribed Textbook(s) and Other Resources You Must Acquire or Have Access to
There will be no specific prescribed textbook for this unit. Reading materials including a compilation of textbook and journal article references will be provided to the students through the University library.

It is also expected that extensive use will be made of other learning references. It is recommended that students use the references listed below as sources of research for assignment information and to broaden understanding and learning in the unit.

The Study Guide for Managing for Environmental Sustainability and Green Marketing contains eight topics. The topics reflect the basic knowledge in environmental sustainability management. The reading materials selected for this unit reflect an up-to-date developments and thinking in environmental sustainability management and green marketing.

Course Website(s)
- Course website provides all the learning resources including the recorded seminars.

Required Course Readings
A selection of readings is assembled and will be updated progressively for each session of the course. The readings can be accessed through the university library.

LIST OF READINGS

Topic 1

1.1 The science of sustainability
Sterman, John D. (2012), "Sustaining Sustainability: Creating a Systems Science in a Fragmented Academy and Polarized World," M.P. Weinstein and Turner, R.E. *Sustainability Science*, Chapter 2, Springer, New York. 21–58.
Kim, J. and Oki, T. (2011), "Visioneering: An essential framework in sustainability science," *Sustainability Science*, 6 (2), 247–51.

1.2 The business case for sustainability

Kiron, D., Kruschwitz, N., Haanaes, K. and Von Streng Velken, I. (2012), "Sustainability nears a tipping point," *MIT Sloan Management Review*, 53 (2), 69–74.

Schalteggar, S. (2011), "Sustainability as a driver for corporate economic success consequences for the development of sustainability *management control*," *Society and Economy*, 33 (1), 15–28.

1.3 Corporate environmental management

Tate, W.L., Ellram, L.M., and Dooley, K.J. (2012), "Environmental purchasing and supplier management (EPSM): Theory and practice," *Journal of Purchasing and Supply Management*, 18 (3), 173–188.

Pogutz, S. and Micale, V. (2011), "Eco-efficiency, growth and the nature of corporate sustainability: submitted to the seventh International Environmental Management leadership Symposium." Rochester Institute of Technology.

Garcés-Ayerbe, Concepción, Rivera-Torres, Pilar, and Murillo-Luna, Josefina L. (2012), "Stakeholder pressure and environmental proactivity: Moderating effect of competitive advantage expectations," *Management Decision*, 50 (2), 189–206.

Gonzalez-Benito, J., Lannelongue, G, and Queiruga, D. (2011), "Stakeholders and environmental management systems: a synergistic influence on environmental imbalance," *Journal of Cleaner Production*, 19 (14), 1622–1630.

Topic 2

2.1 Communication versus greenwashing

Benoit-Moreau, Florence and Parguel, Béatrice (2011), "Building brand equity with environmental communication: An empirical investigation in France," *EuroMed Journal of Business*, 6 (1), 100–116.

Delmas, M.A. and Burbano, V.C. (2011), "The drivers of greenwashing," *California Management Review*, 54 (1), 64–87.

2.2 Regional and international dimensions of environmental sustainability

Knight, K.W. and Messer, B. (2012), "Environmental concern in cross-national perspective: The effects of affluence, environmental degradation, and world society," *Social Science Quarterly*, 93 (2), 521–537.

Topic 3

Shiroyana, H., Yarime, M., Matsu, M. Schroeder, H., Scholz, R., and Ulrich, A.E. (2012), "Governance for sustainability: Knowledge integration and multi-actor dimensions in risk management," *Sustainability Science*, 7 (Supplement 1), 45–55.

Baumgartner, R.J. and Ebner, D. (2010), "Corporate sustainability strategies: Sustainability profiles and maturity levels," *Sustainable Development*, 18 (2), 76–89.

Bonn, I and Fisher, J. (2011), "Sustainability: The missing ingredient in strategy," *Journal of Business Strategy*, 32 (1), 5–14

Lueneburger, C. and Goleman, D. (2010), "The change leadership sustainability demands," *MIT Sloan Management Review*, 51 (2), 49–55.

Topic 4

Gupta, S., Czinkota, M., and Melewar, T.C. (2012), "Embedding knowledge and value of a brand into sustainability for differentiation," *Journal of World Business*, 48 (3), 287–296.

Royne, M.B., Levy, M., and Martinez, J. (2011), "The public health implications of consumers' environmental concern and their willingness to pay for an eco-friendly product," *Journal of Consumer Affairs*, 45 (2), 329–343.

Cheah, I. and Phau, I. (2011), "Attitudes towards environmentally friendly products: The influence of ecoliteracy, interpersonal influence and value orientation," *Marketing Intelligence & Planning*, 29 (5), 452–72.

Topic 5

Darnall, N. and Kim, Y. (2012), "Which types of environmental management systems are related to greater environmental improvements?" *Public Administration*, 72 (3), 351–365.

Jenkin, T.A., Webster, J., and McShane, L. (2011), "An agenda for "Green" information technology and systems research," *Information and Organization*, 21 (1), 17–40.

Cordano, M., Marshall, R., and Silverman, M. (2010), "How do small and medium enterprises go "Green"? A study of environmental management programs in the U.S. wine industry," *Journal of Business Ethics*, 92 (3), 463–478.

Topic 6

Christ, K. and Burritt, R. (2012), "Environmental management accounting: The significance of contingent variables for adoption," *Journal of Cleaner Production*, 41 (1), 163–173.

De Giovanni, Pietro (2012), "Do internal and external environmental management contribute to the triple bottom line?," *International Journal of Operations & Production Management*, 32 (3), 265–290.

Kaushik S. (2012), "Corporate conceptions of triple bottom line reporting: An empirical analysis into the signs and symbols driving this fashionable framework," *Social Responsibility Journal*, 8 (3), 312–326.

Frances, B. and Bettina, W. (2011), "Carbon accounting: Negotiating accuracy, consistency and certainty across organisational fields," *Accounting, Auditing & Accountability Journal*, 24 (8), 1022–1036.

Topic 7

Millar, C., Hind, P., and Magala, S. (2012), "Sustainability and the need for change: Organisational change and transformational vision," *Journal of Organizational Change Management*, 25 (4), 489–500.

Loorbach, D., van Bakel, J.C., Whiteman, G., and Rotmans, J. 2010, "Business strategies for transitions towards sustainable systems," *Business Strategy and the Environment*, 19 (2), 133–146.

Topic 8

Tseng, M., Wang, R., Chiu, A.S.F., Geng, Y., and Lin, Y.H. (2012), "Improving performance of green innovation practices under uncertainty," *Journal of Cleaner Production*, 40 (1), 71–82.

Schiederig, T., Tietze, F., and Herstatt, C. (2012), "Green innovation in technology and innovation management—An exploratory literature review," *R&D Management*, 42 (2), 180–192.

Weng, M.H. and Lin, C.Y. (2011), "Determinants of green innovation adoption for small and medium-size enterprises (SMES)," *African Journal of Business Management*, 5 (22), 9154–9163.

Additional Texts

Moscardo, G., Lamberton, G., Wells, G., Fallon, W., Lawun, P., Rowe, A., Humphrey, J., Wiesner, R., Pettitt, B., Clifton, D., Renouf, M., and Kershaw, W. (2013), *Sustainability in Australian Business*. John Wiley & Sons, Queensland, Australia.

D'Sousa, C., Taghian, M., and Polonsky, M. (2011), *Readings and Cases in Sustainable Marketing*. Tilde University Press, Victoria, Australia

Epstein, M. (2008), *Making Sustainability Work: Best Practices in Managing and Measuring Corporate Social, Environmental and Economic Impacts*. Greenleaf Publishing, Sheffield, UK.

Friedman, T. (2008) *Hot, Flat and Crowded: Why the world needs a green revolution—And how we can renew our global future*. Camberwell, Penguin, Australia

Sroufe, R. and Sarkis, J. (2007), *Strategic Sustainability: The State of the Art in Corporate Environmental Management Systems*. Greenleaf Publishing, Sheffield, UK.

Schaltegger, S., Burritt, R., and Petersen, H. (2003), *An Introduction to Corporate Environmental Management*. Greenleaf Publishing, Sheffield, UK.

Barrow, C.J. (2006), *Environmental Management for Sustainable Development*, 2nd edition. Routledge, London, UK.

Dunphy, D., Benveniste, J., Griffiths, A., and Sutton, P. (2000), *Sustainability: The Corporate Challenge of the 21st Century*. Allen & Unwin, St Leonards, Australia

Welford, R. and Starkey, R. (1996). *The Earthscan Reader in Business and the Environment*. Earthscan Publications, London, UK.

Periodical Publications on Marketing

There is a range of periodical publications that carry reports or articles on all aspects of management studies. The following are concerned exclusively or partly with marketing and may be of interest to you:

- *Agriculture, Ecosystems and Environment*
- *Annual Review of Ecology and Systematics*
- *Australian Financial Review*
- *Australasian Journal of Market & Social Research*
- *Business Review Weekly*
- *Business Strategy and the Environment*
- *Current Opinion in Environmental Sustainability*
- *Ecological Applications*
- *Ecological Economics*
- *Environmental Quality Management*
- *Environment and Planning*
- *European Journal of Marketing*
- *Financial Review*
- *Harvard Business Review*
- *International Business Asia*
- *International Business Week*
- *International Journal of Operations and Production*
- *International Marketing Review*
- *International Journal of Accounting, Auditing and Performance Evaluation*
- *International Journal of Sustainability*
- *Journal of Cleaner Production*
- *Journal of Environmental Management*
- *Journal of Peace Research*
- *Journal of Social Issues*
- *Journal of World Business*
- *Journal of the Academy of Marketing Science*
- *Journal of Advertising Research*
- *Journal of Consumer Policy*
- *Journal of Consumer Research*
- *Journal of Macromarketing*
- *Journal of Marketing*
- *Journal of Marketing Research*
- *Journal of the Market Research Society*
- *Journal of Personal Selling and Sales Management*
- *Management Accounting Research*
- *Marketing Science Institute Occasional Papers*
- *Organization & Environment*
- *International Journal of Operations & Production Management*
- *Sustainable Development and Urban Form*
- *The Academy of Management Review*
- *The Academy of Management Journal*
- *The Economist (international version)*

Electronic Online Resources

http://www.environment.gov.au/
http://www.austrade.gov.au/studentcentre—Covers many areas of international marketing and trade
http://www.earthtimes.org/—Earth Times
http://www.eoearth.org/—Encyclopedia of Earth

http://iopscience.iop.org/—IOP science
http://www.ama.org/—American Marketing Association
http://www.abs.gov.au/—Australian Bureau of Statistics
http://www.chapmanrg.com/IMR/IMRINTER.HTM—Internet-Based International Marketing Readings
http://www.dfat.gov.au/—Department of Foreign Affairs and Trade
http://globaledge.msu.edu/ibrd/ibrd.asp—globalEDGE
http://www.imf.org/—International Monetary Fund
https://www.cia.gov/library/publications/—Central Intelligence Agency of the United States
http://www.tradeport.org—Covers international research, exporting, logistics, and financing
http://www.worldbank.org/—The World Bank
http://www.wto.org/—World Trade Organization
http://gcmd.nasa.gov/—Global Change Master Directory

Grading Scheme

Grading System

Your final mark in the unit is determined as indicated in the following assessment table:

Assessment Name	Weight (%)	Due Date[2]	Brief Description[1]
Assignment 1	40	Third week	Individual assignment Report on theoretical development of sustainability management and green marketing 3000 words
Assignment 2	60	Twelfth week	Individual assignment Practical project report—environmental sustainability and green marketing audit 4500 words

Your grade is determined from your mark according to the University scale:

Less than 50: N (fail grade)
50–59: P
60–69: C
70–79: D
80 or greater: HD.

Grading Policy
- To pass the unit, students need to complete all the assessment tasks and achieve a minimum total of 50 marks for all the assessment tasks combined.

Late Assignments

No extensions will be considered for assignment submission due date unless a written request is submitted and negotiated with the designated Unit Chair/Coordinator.

Assignments submitted late without an extension being granted will not be marked.

University Policies

Institutional policies here:
Academic integrity
Disability accommodations/services
Anti-discrimination
Affirmative action
Equal opportunity

Course Assignments

There are two assessment tasks in this unit as follows:

Assignment 1—Report on theoretical development of sustainability management and green marketing

Students are required to prepare a report on a topic in sustainability management and green marketing. This is an individual assignment.

Students will choose a topic of interest to them for the assignment. The selection of the topic should relate to one of the areas covered in the unit.

The expectation is for students to investigate and prepare a detailed yet succinct report, which demonstrates a thorough review of the extant literature. It should identify and assess the key contributions by scholars in that topic. The report requires (i) a theoretical discussion and (ii) the application of the theory, where possible, as well as (iii) the discussion of the expected future developments.

A format will be provided for students to use in preparing this assignment.

Assignment 2—Practical project report—environmental sustainability and green marketing audit

The environmental sustainability audit is a practical project to be conducted in a company.

Students need to search for and negotiate with a company to get permission for conducting the audit. The company can be either a manufacturing or a service organization. Students may choose the company of their employment for this assignment.

The audit includes a review of the organization's status on environmental sustainability. It will be a comprehensive review of the organization's strategy, objectives, actions, and achievements in planning for and implementation of environmental sustainability.

A comprehensive format will be provided to assist students to complete this assignment (audit).

Course Schedule

- **Week 1**: Introduction to course, The science of sustainability
- **Week 2**: The business case for sustainability and green marketing fundamental
- **Week 3**: An overview of corporate environmental management
- Assignment 1: Report on theoretical development of sustainability management and green marketing due to Instructor
- **Week 4**: Stakeholder issues: Spheres of influence for management actions: Greenwashing
- **Week 5**: Regional and international dimensions of environmental sustainability and green marketing
- **Week 6**: Basic sustainability strategy options and risk management
- **Week 7**: Eco-marketing, positioning, and brand management
- **Week 8**: Environmental Management systems and operations
- **Week 9**: Environmental management accounting
- **Week 10**: Sustainable business via organizational change management
- **Week 11**: Fostering "green" innovation
- **Week 12**: Course wrap-up, lessons learned
- Assignment 2: Practical project report—environmental sustainability and green marketing audit due to Instructor

CHAPTER 10: ROLE OF ENVIRONMENTAL, SOCIAL, AND GOVERNANCE (ESG) FACTORS IN FINANCIAL INVESTMENTS

Instructor Information

Instructor: A. Seddik Meziani

Office Location: Partridge Hall/Room 412, Montclair State University, Montclair, NJ 07043

Telephone: 973-655-4135

E-mail: meziania@mail.montclair.edu

Office Hours: Tuesday and Thursday, 3:00 P.M.–5:00 P.M.

Course Identification

Course Number:
Course Name: ESG Investment
Course Location:
Class Times:
Prerequisites:

Course Description/Overview

The growth of environmental, social, and governance (ESG) investment has been no less than impressive during the last two decades. Considered an alternative investment approach, ESG investment embraces companies that endorse a set of core values in the areas of environment, social commitment, human rights, labor standards, avoidance of harmful products such as tobacco, and anti-corruption. In terms of corporate governance, this course addresses the areas of investor protection. All in all, ESG investment adds concerns about social and environmental issues to standard financial concerns about risk and return as determinants of investment portfolio construction.

This course also aims to provide relevant information on the distinguishing features between conventional and ESG investment and discuss the outcomes of integrating ESG issues into investment strategies for institutional and individual investors alike. As such, its content is a multidisciplinary process that calls upon the fields of statistics, investment, and asset management. The likelihood that responsible investment can offer opportunities to generate higher returns with lower levels of investment risk is explored throughout the course.

Students will gain a broad overview of the asset management industry and on how asset managers engage in portfolio management in a capital market where all information relevant to a fair assessment of risk and return is assumed to be free flowing and known to all. In this manner, efficient capital market prices guide allocation of all resources including ESG investment.

Course Learning Objectives

By the end of the course, students should:

1. Gain a broad overview of the investment environment and the investment process
2. Learn that the risk–return trade-off is a no free lunch notion: higher expected returns come only to those who accept to bear greater investment risk
3. Devote considerable attention to the implications of diversification on portfolio risk
4. Distinguish the main differences between conventional investment and ESG investment
5. Be able to add ESG issues to standard financial concerns about risk and return as determinants of investment portfolio construction

Course Resources

Course Website(s)
- Course website (e.g., Learning ESG Investment)
- Personal website (instructor's personal website)

Required Course Readings
- A selection of readings will be assembled for each week to be read prior to class.

Grading Scheme

Grading System
Grades and explanations of them here

Grading Policy
Grades will be based on the following:

10-page research paper (100 for draft, 100 for final)	200
Peer review of draft research papers	50
Five homework assignments (20 points each)	100
Class attendance/participation	200
Total points	**550**

Late Assignments
Late assignment policy here

University Policies

Institutional policies here:
Academic integrity
Disability accommodations/services
Anti-discrimination
Affirmative action
Equal opportunity

Course Assignments

Research Paper (200 Points)
This paper will be on an ESG investment study of your choosing. The paper will have the following sections: (1) Description of the case study (what was the "historical" (however you define this) background and conditions of the case, what are the defining features/unique characteristics of the case, what prompted you to choose this case); (2) Defining influences on the subject of the case study (what are the types of human impacts in the case, what types of risk–return and diversification investors do you expect), (3) Stages of portfolio construction and management (this will be the longest part of the paper, with different subsections for each of the stages) (i) Define your chosen ESG investment and the reason for choosing it; (ii) Collect data and define the risk–return target you are willing to assume as an investor; calculate risk and return based on the historical record of your chosen ESG investment; (iii) Define and justify your capital allocation to your ESG investment; (iv) Assess the diversification impact of your ESG investment on the overall portfolio risk; (v) Define and implement your method of hedging (preservation) your overall portfolio against adverse market movements on the part of your ESG investment; (vi) Monitoring (performance attribution) and adaptive (risk adjustment) management; and (vi) Conclusion: success or failure in relation to the target risk–return you defined in 2.

The paper should be supported by an adequate number of *peer-reviewed journal articles/reports* to support your plan and descriptions.

Peer Review of Student Papers (2 @ 25 Points Each)
All students will be assigned two student draft papers to peer review (using the peer review form handed out to you). The point of this exercise is not only to help each other improve your papers but also to learn from each other. Reviewing others' writing allows you to practice putting yourself in the shoes of a reviewer, so that you can be more critical and insightful with your own writing.

Homework Assignments (100 Points)
Five homework assignments (about 2–3 pages) will be distributed by the instructor throughout the semester to make sure that you understand the key points in each of the six stages. These assignments will be used as discussion "fodder" in class.

Attendance and Participation (200 Points)
  Attendance (100 points): Students will receive two unexcused absences without any penalty. For each absence thereafter, students will lose 10 points.
  Participation (100 points): These points will be given at my discretion; the scale goes something like this: If you speak up and participate in class discussions most of the time, you will get the full 100 points. If you speak so rarely that I don't know what your voice sounds like, you will get 30 or fewer points.

Course Schedule

  Week 1: Introduction to course, ESG investment
  Week 2: Introduction to the disciplines associated with ESG investment: statistics, investment, and asset management
  Week 3: Market efficiency and ESG investment
  Week 4: Risk–return and ESG investment
  Assignment: Choose a case study for research paper; send e-mail to Instructor to confirm
  Week 5: Diversification and capital allocation to ESG investment
  Assignment: Homework #1 due
  Week 6: Diversification and capital allocation to ESG investment, continued
  Assignment: Review of Homework #1
  Week 7: Preservation: method of hedging

Assignment: Homework #2 due

Week 8: Monitoring: performance attribution

Assignment: Homework #3 due

Assignment: Review of Homework #2

Week 9: Monitoring: performance attribution, continued

Assignment: Review of Homework #3

Week 10: Tying it all together

Assignment: Homework #4 due (Stage 4)

Week 11: Case Study 1

Assignment: Draft research paper due to Instructor and two student peer reviewers

Assignment: Review of Homework #4

Week 12: Case study 2

Assignment: Student peer reviews of research papers due back to authors; send copy of review to Instructor

Homework #5 due

Week 13: Course wrap-up, lessons learned

Assignment: Review of Homework #5

Assignment: Final research paper due to Instructor

Week 14: Research papers feedback

CHAPTER 11: THE ROLE OF PUBLIC RELATIONS AND ORGANIZATIONAL COMMUNICATION IN ENVIRONMENTAL MANAGEMENT

Model Syllabus

Instructor Information

Instructor:
Office Location:
Telephone:
E-mail:
Office Hours:

Course Identification

Course Number:
Course Name: Environmental Public Relations
Course Location:
Class Times:
Prerequisites:

Course Description/Overview

It is essential for any organization dealing with environmental issues (e.g., corporations, governments, nonprofits, etc.) to understand how to successfully communicate with various publics (e.g., stakeholders, the mass media, employees, investors, etc.). This course will teach students the theories and principles associated with the use of public relations and organizational communication, specifically as it relates to issues related to the environment. The course will introduce students to contemporary real-world examples in which public relations and organizational communications have been used to address environmental issues.

Course Learning Objectives

By the end of the course, students should:

1. Understand the definition of public relations and how it can be applied to issues involving the environment
2. Understand the definition of organizational communication and how it can be applied to issues involving the environment
3. Understand how public relations can be used strategically
4. Understand how to use public relations to create "win-win" scenarios
5. Understand the extent to which decisions about the management of public relations and organizational communications led to success or failure

Course Resources

Course Website(s)
- Course website (e.g., Learning Management System)
- Personal website (instructor's personal website)

Textbooks
Benn, S. and D. Bolton. (2011) *Key Concepts in Corporate Social Responsibility*. London: Sage Press.

Ottman, Jacquelyn. (2011) *The New Rules of Green Marketing: Strategies, Tools, and Inspiration for Sustainable Branding*. San Francisco, CA: Berrett-Koehlman Publishers.

Assigned Readings
- A selection of readings will be assembled for each week to be read prior to class.

Grading Scheme

Grading System
Grades and explanations of them here

Grading Policy
Grades will be based on the following:

Two 10-page research papers	200
Ten homework assignments (20 points each)	100
Class journal	100
Class attendance/participation	100
Total points	**500**

Late Assignments
Late assignment policy here

University Policies

Institutional policies here:
Academic integrity
Disability accommodations/services
Anti-discrimination
Affirmative action
Equal opportunity

Course Assignments

Two Research Papers (100 Points Each, 200 Points Total)
Each student will write two 10-page research papers.

One research paper will focus on the use of public relations or environmental communications that focus on a specific environmental issue. It might involve (among many other topics) an environmental disaster (e.g., an oil spill or a pollution event, etc.), an instance where the environment is being threatened (e.g., habitat loss, climate change, etc.), or efforts to protect or rehabilitate ecosystems or habitats.

The second research paper will focus on the ways in which a specific organization (e.g., a company, a governmental entity, or a special interest group, etc.) are using public relations and organizational communication in their regular activities.

Each paper should be supported by an adequate number of *peer-reviewed journal articles/reports* as well as articles from the news media and trade media.

Homework Assignments (200 Points)
Students will complete 10 homework assignments during the semester. There will be one homework assignment that corresponds to each chapter in the textbook. In the homework assignments, students will be asked to examine specific issues associated with the use of public relations and organizational communications related to the environment and to apply critical thinking skills to assess the decisions made in real-world scenarios. The homework assignments will be used to stimulate vigorous discussion in the classroom.

Class Journal (100 Points)
Each week, we will start the class session with a 10-min in-class writing exercise. I will introduce a recent issue from the real world about how environmental issues are being communicated and will assign specific questions for students to respond to. At random points in the semester, I will collect these notebooks and read through them to get a sense of how the class is doing as a whole and how individual students are interpreting and mastering the material.

Attendance and Participation (200 Points)
Attendance (100 points): Students will receive two unexcused absences without any penalty. For each absence thereafter, students will lose 10 points for each additional unexcused absence.

Participation (100 points): Students will be awarded points for participation at the discretion of the instructor. Students will earn participation points when they contribute to class discussion by demonstrating they have understood the homework materials that were assigned and that they can apply them to environmental causes.

Course Schedule

Week 1: Introduction to course, Overview of Public Relations
Week 2: Overview of Organizational Communication Theory and Practice
Week 3: Examples of Public Relations Successes, Related to Environmental Issues
Week 4: Examples of Public Relations Failures, Related to Environmental Issues
Students identify the topic for Research Paper 1
Week 5: Examples of Organizational Communications Success Stories, Related to Environmental Issues
Week 6: Examples of Organizational Communication Failures, Related to Environmental Issues
Week 7: Examples of How Environmental Organizations Are Using Public Relations
Week 8: Examples of How Environmental Organizations Are Using Organizational Communication
Students identify the topic of Research Paper 2
Week 9: Learning How to Develop a Strategic Public Relations Program
Week 10: Examples of How Organizations Are Strategically Using Public Relations to Achieve Environmental Goals
Week 11: The Use of Public Relations and Organizational Communication to Create "Win-Win" Partnerships
Week 12: Ethical Issues—Greenwashing
Week 13: Ethical Issues—Corporate Social Responsibility
Week 14: Environmental Issues—Green Marketing
Week 15: Students Present Research Papers

CHAPTER 12: THE ECONOMICS OF ENVIRONMENTAL MANAGEMENT

Model Syllabus

Course Objectives: This course introduces economic approaches to solving environmental problems. The first part of the course concentrates on relevant economic theory, which is used as a framework to approach a wide range of environmental issues. The course is designed for students with minimal background in economics. It aims to enable students to appreciate of the usefulness of economic approaches to environmental and problems and to enable students to critique economic analyses that they may see in future decision-making roles. The course thus spends considerable time on the application of economic techniques to policy issues.

Assignments: Weekly case studies or problem sets, midterm and final exams, and final research project

Part I. Environmental Economics: The Framework (≈5 Weeks)

Introduction
Oates, W.E. "Forty years in an emerging field," Resources (137), Fall 1999, pp. 8–11.

Economic Theory for Environmental Economics
Field, Barry C. and K. Martha Field. *Environmental Economics*, 5th edition. McGraw-Hill Irwin, New York, 2009. chapter 3: Benefits and costs, supply and demand chapter 4: Economic efficiency and markets.
Jason, S. *What Environmentalists Need to Know about Economics*, Palgrave Macmillan, New York, 2010.

Economic Methods for Environmental Economics
Tietenberg, Tom and Lynne Lewis. *Environmental and Natural Resource Economics*, 8th edition. Pearson, Boston, MA, New York, 2009. Chapter 3: Valuing the environment: methods.
Hanemann, W.M. "Valuing the environment through contingent valuation," *Journal of Economic Perspectives* 8(4), Fall 1994, pp. 19–44.
Carson, R.T., N. Flores, and N. Meade. "Contingent valuation: controversies and evidence," *Environmental and Resource Economics* 19, 2001, pp. 173–210.
Brennan, T. "Discounting the future: economics and ethics," *Resources, Resources for the Future* 120, Summer 1995, pp. 3–6.

Market Failures and Environmental Problems
Harris, J. and B. Roach. 2013. *Environmental and Natural Resource Economics: A Contemporary Approach*, M.E. Sharpe, Armonk, New York. Chapter 4: Common Property Resources and Public Goods.
Hardin, Garrett. "The tragedy of the commons," *Science*, 162(3859), 1968, pp. 1243–1248.
Dietz, T., E. Ostram, and P.C. Stern. "The struggle to govern the commons," *Science* 302(1907), 2003, pp. 1907–1912

Environmental Sustainability
Daly, Herman E. "The economics of the steady state," *The American Economic Review*, 64(2), pp. 15–21
Solow, R. "Sustainability: An economist's perspective," Chapter 28 in *Economics of the Environment*, R.N. Stavins (ed.), 6th edition, W.W. Norton, New York, 2012.
Berrens, R.P., D.S. Brookshire, M. McKee, and C. Schmidt. "Implementing the safe minimum standard approach," *Land Economics*, 74(2), May 1998, pp. 147–161.
Howarth, R.B. "Toward an operational sustainability criterion," *Ecological Economics*, 63, 2007, pp. 656–663

Allocation of depletable and renewable resources
Field, Barry C. *Natural Resource Economics*, 2nd edition. Waveland Press, Long Grove, IL, 2008. Chapter 10: Mineral Economics.
Krautkraemer, J.A. "Nonrenewable resource scarcity," *Journal of Economic Literature*, December, 1998, pp. 2065–2107.
Tilton, J.E. "Exhaustible Resources and Sustainable Development: Two Different Paradigms," *Resources Policy*, 22, 1996, pp. 91–97.
Daly, Herman E. and Farley, Joshua. *Ecological Economics: Principles and Applications*, 2nd edition. Island Press, Washington, DC, 2010. Chapter 5: Abiotic Resources Chapter 6: Biotic Resources

Part II. Environmental Economics: Applications (≈10 Weeks)

Fishery Economics
Wasserman, Miriam. "The last hunting economy." *Regional Review* 11(2), 2001, pp. 8–17.
Hartwick, John M. and Nancy D. Olewiler. "The economics of the fishery: an introduction," Chapter 2 in *The Economics of Natural Resource Use*, 2nd edition. Prentice Hall, Upper Saddle River, NJ, 1998.
Brodziak, John, Steve X. Cadrin, Christopher M. Legault, and Steve Murawski. "Goals and strategies for rebuilding New England groundfish stocks." *Fisheries Research* 94, 2008, pp. 355–366.
Arnason, Ragnar. "The Icelandic individual transferable quota system: a descriptive account." *Marine Resource Economics* 8, 1993, pp. 201–218.
Grafton, R., D. Squires, and J.F. Kirkley, "Private property rights and crises in world fisheries: turning the tide?" *Contemporary Economic Policy* XIV, October 1996, pp. 90–99.
Repetto, Robert. "A natural experiment in fisheries management." *Marine Policy* 25, 2001, pp. 251–264.
Feeny, D., S. Hanna, and A. McEvoy. "Questioning assumptions of the "Tragedy of the Commons" model of fisheries." *Land Economics* 72(2), May 1996, pp. 187–205.

Oil Spills
Roach, B.J.M. Harris, and A. Williamson. *The gulf oil spill: economics and policy issues*. Tufts University, Global Development and Environment Institute, Medford, MA, 2010.
Duffield, John. "Nonmarket valuation and the courts: the case of the Exxon Valdez." *Contemporary Economic Policy* 15, October 1997, pp. 98–110.
Kim, Inho. "Ten years after the enactment of the oil pollution act of 1990: a success or a failure." *Marine Policy* 26(197), 2002, pp. 197–207.
McCrea-Strub, A. "Potential impact of the deepwater horizon oil spill on commercial fisheries in the Gulf of Mexico." *Fisheries* 36(7), July 2011, pp. 332–336.
Nash, S. "Oil and water, economics and ecology in the Gulf of Mexico." *Bioscience* 61(4), April 2011, pp. 259–263.
Costanza, R., D. Batker, J.W. Day, R.A. Feagin, M.L. Martinez, and J. Roman "The perfect spill: solutions for averting the next Deepwater Horizon." *Solutions* 1(5), 2010, pp. 17–20.

Renewable Energy Economics

Heal, Geoffrey. "Reflections—the economics of renewable energy in the United States." *Review of Environmental Economics and Policy* 4(1), 2010, pp. 139–154.

Jacobson, M.Z. and M.A. Delucchi. "Providing all global energy with wind, water, and solar power, part I: technologies, energy resources, quantities and areas of infrastructure, and materials." *Energy Policy* 39, 2011, 1154–1169.

Darmstadter, Joel. "The economic and policy setting of renewable energy". Discussion paper, Resources for the Future, Washington, D.C., 2003.

Bergmann, Ariel, Nick Hanley, and Robert Wright. "Valuing the attributes of renewable energy investments." *Energy Policy* 34, 2006, pp. 1004–1014.

Snyder, Brian and Mark J. Kaiser. "Ecological and economic cost-benefit analysis of offshore wind energy." *Renewable Energy* 34, 2009. pp. 1567–1578.

Musial, W. and S. Butterfield. "Future of offshore wind energy in the United States." National Renewable Energy Laboratory, conference paper for EnergyOcean, Palm Beach, FL, 2004

Shale Gas Extraction

Kargbo, D.M., R.G. Wilhelm, and D.J. Campbell. "Natural gas plays in the Marcellus Shale: challenges and potential opportunities." *Environmental Science and Technology* 44(15), 2010, pp. 5679–5684.

Zoback, M., S. Kitasei, and B. Copithorne. *Addressing the environmental risks from shale gas development*. Worldwatch Institute, Washington, DC, Briefing paper 1, July 2010.

Schmidt, C.W. "Blind rush? Shale gas proceeds amid human health questions." *Environmental Health Perspectives* 119(8), August 2011, pp. A348–A353.

Osborn, S.G., A. Vengosh, N.R. Warner, and R.B. Jackson. "Methane contamination of drinking water accompanying gas-well drilling and hydraulic fracturing." *Proceedings of the National Academy of Sciences of the United States of America, PNAS* 108(20), May 2011, pp. 8172–8176.

Jiang, M., W.M. Griffin, C. Hendrickson, P. Jaramillo, J. VanBriesen, and A. Venkatesh. "Life cycle greenhouse gas emissions of Marcellus shale gas." *Environmental Research Letters* 6, 2011.

Howarth, R.W., R. Santoro, and A. Ingraffea. "Methane and the greenhouse-gas footprint of natural gas from shale formations." *Climatic Change* 106(4), 2011, pp. 679–690.

Climate Change Economics

Harris, J.M. and B. Roach. *The economics of global climate change*. Tufts University, Global Development and Environment Institute, Medford, MA, 2009.

Stern, Nicholas. *The economics of climate change: the stern review*. Cambridge University Press, Cambridge/New York, 2007.

Weitzman, M.L. "A review of the Stern Review on the economics of climate change." *Journal of Economic Literature* 65, September 2007, pp. 703–724.

Newell, R. and W. Pizer. "Discounting the benefits of climate change policies using uncertain rates." *Resources*, 146, winter 2002, pp. 15–20.

Ostrom, E. "A polycentric approach for coping with climate change." The World Bank, Washington, DC background paper to the 2010 World Development Report, October 2009.

Gupta, Joyeeta. "A history of international climate change policy." *Wiley Interdisciplinary Reviews: Climate Change* 1(5), September/October 2010, pp. 636–653.

CHAPTER 13: LAW AND POLICY IN ENVIRONMENTAL MANAGEMENT

Model Syllabus

Course Objectives: This course provides an interdisciplinary introduction to the US policy for energy, the environment, and global sustainability. Emphasis will be placed on understanding the laws, regulations, and treaties that oversee air and water pollution, solid waste, hazardous waste, energy use, natural resources, climate change, and global governance for energy, environment, and sustainability. Topics include:

- The environmental legacy
- The evolution of American environmentalism
- The process of policy formulation, implementation, and enforcement
- The institutions and politics of policymaking
- Environmental regulation for land, air, water, solid, and toxic waste
- Nuclear energy, fossil fuels, and renewable energy
- Climate change and the challenge of global policymaking
- Global sustainability
- International environmental institutions and regimes

Course Schedule

Week 1: Class introduction
Week 2: Overview of environmentalism in America
Week 3: Public policy institutions
Week 4: The policy process—the policy cycle
Week 5: The policy process—interest group politics
Week 6: Public lands and natural resources policy
Week 7: Air and water regulation
Week 8: Toxic and hazardous substances
Week 9: Energy policy
Week 10: Risk assessment and environmental justice
Week 11: Regulatory economics
Week 12: Global sustainability

CHAPTER 14: ENVIRONMENTAL ETHICS

Syllabus

About This Class

The theory and practice of environmental management is not just concerned with what is but also very much with what ought to be. The guiding ideal of sustainability is a normative one, and fundamental questions such as "Who or what counts as a resource?", "Whom should we take into account in our moral considerations?", and "What is a fair distribution of resources within and across generations?" are normative questions. In this class, students will be introduced to environmental ethics, the field of application of ethics that systematically addresses these questions.

By the end of this class, a successful student will be able to:

- Explain what ethics is and what value theory, normative ethics, and metaethics are about
- Identify core questions and concepts of the field of value theory and apply them to environmental ethics
- Identify core questions and concepts of the field of normative ethics and apply them to environmental ethics
- Defend own position with regard to central issues in environmental ethics

Assignments

Your grade will be based on three assignments:

1. *Class participation (20%)*

   Each week, you will have several readings, listed in the syllabus. You are expected to read these articles carefully and come to the week's seminar prepared to discuss them. The questions that come with the weekly readings are meant to facilitate this preparation. They will not be marked.

2. *Paper (60%)*

   One of the main controversies in environmental ethics concerns our moral duties to nonhuman animals and the rest of nature. This issue will be discussed in the first half of the class. The student is expected to engage with the various positions on this issue and defend his or her own position in a paper. In addition to submitting a written paper, each student will give a 5-min oral presentation of his or her research during the last week's class.

3. *Exam (20%)*

   The student's understanding of core notions and arguments that have been discussed in this class will be tested in a brief exam at the end of the term. The exam consists of five open questions, each of which requires a brief answer of no more than five sentences.

Schedule of Topics and Readings

Week 1: Introduction to ethics
What are ethics and philosophy all about? What are the three core fields of ethics? What is environmental ethics? What is the relevance of ethics for environmental management?
Shafer-Landau, R. (2010). *The Fundamentals of Ethics*, Oxford University Press, Oxford. (Introduction)

Singer, P. (2011). *Practical Ethics*, 3rd edition. Cambridge University Press, Cambridge. (Chapter 1 "About Ethics" and Chapter 10 "The Environment")

Week 2: The status of moral claims
What is metaethics? Are there any objective moral standards? If so, how can we know them?
Shafer-Landau, R. (2010). *The Fundamentals of Ethics*, Oxford University Press, Oxford. (Chapters 19–21 on "Ethical Relativism", "Moral Nihilism" and "Ten Arguments Against Moral Objectivity")
Singer, P. (2011). *Practical Ethics*, 3rd edition. Cambridge University Press, Cambridge. (Chapter 12, "Why act morally?")

Week 3: Moral status: humans and animals
Whom should we take into account in our moral considerations and on what grounds? Is it defendable to restrict our moral considerations to members of our own species?
Gruen, L. (2003). "The Moral Status of Animals", in *Stanford Encyclopedia of Philosophy*, Stanford University, Stanford. (http://plato.stanford.edu/entries/moral-animal/).

Week 4: Moral status: nonsentient nature
Do we have any direct moral duties toward plants, rivers, and ecosystems? If so, on what grounds? If not, why not and does it imply that these things are unimportant?
Rolston, H. (1999). "Respect for Life: Counting what Singer Finds of no Account," in D. Jamieson (ed.) *Singer and his Critics*, Blackwell, Oxford.
Singer, P. (1999). "A Response (Rolston)," in D. Jamieson (ed.) *Singer and his Critics*, Blackwell, Oxford.

Week 5: Intrinsic value
What, if anything, is valuable in and of itself? Can something be good without being good for someone?
Bradley, B. (2006). "Two Concepts of Intrinsic Value," in: *Ethical Theory and Moral Practice*, 9 pp. 111–130.
Schroeder, M. (2012). "Value Theory," in E.N. Zalta (ed.) *Stanford Encyclopedia of Philosophy*, Standford University, Standford. (http://plato.stanford.edu/entries/value-theory/) (accessed March 20, 2015).

Week 6: Animal welfare
What is good and bad for animals? What benefits or harms them?
Sandoe, P. and Christiansen, S.B. (2008). *Ethics of Animal Use*. Blackwell, Oxford. (Chapter 3 "What is a Good Animal Life?")

Week 7: The harm of death
What, if anything, makes it the case that death is bad for us? Is death a greater harm for humans than for nonhuman animals?
Bradley, B. (forthcoming). "Is Death Bad for a Cow?" In Višak, T. and Garner, R. (eds.), *The Ethics of Killing Animals*, Oxford University Press, Oxford.

Week 8: Duties and conflicts concerning interspecies justice
How to solve conflicts between individuals and species or ecosystems?
Jamieson, D. (1997). "Animal Liberation is an Environmental Ethics," in: *Environmental Values*, 7, pp. 41–57.

Crisp, R. (1998). "Animal Liberation is not an Environmental Ethic. A Response to Dale Jamieson," in: *Environmental Values*, 7, 4, pp. 476–478.

Week 9: Considering those who might live

Should we consider those who do not yet exist? Should we consider those whose existence depends on how we act?

Roberts, M.A. (2009b). "The Non-Identity Problem," in E.N.Zalta (ed.), *Stanford Encyclopedia of Philosophy*. (http://plato.stanford.edu/archives/fall2015/entries/nonidentity-problem/).

Week 10: The value of existence

Can it benefit (or harm) someone to exist as compared to never existing at all?

Matheney, G. and Chan, K.M.A. (2005). "Human Diet and Animal Welfare: The Illogic of the Larder," in: *Journal of Agricultural and Environmental Ethics*, 18, pp. 579–597.

Week 11: Duties and conflicts concerning intergenerational justice

How to solve conflicts between current needs and possible future needs?

Jamieson, D. (2002). "Ethics, Public Policy and Global Warming," in Light, A. and Rolston, H. (eds.), *Environmental Ethics. An Anthology*, Blackwell, UK.

Week 12: Student presentations

CHAPTER 15: PARTICIPATORY APPROACHES IN ENVIRONMENTAL MANAGEMENT

Syllabus

About This Class

"Participation" has become a major buzzword in environmental management. Nearly everyone endorses participation today—but how many of them really carry it out effectively? In this class, students will be introduced to the participatory trend in environmental decision making, exploring its complexities and its critics.

By the end of this class, a successful student will be able to:

1. Describe participation and identify its common manifestations in the industrialized and developing worlds
2. Make a strong argument for or against the use of participatory approaches
3. Apply the major design questions to determining the appropriate form of participation for a given issue
4. Evaluate common participation mechanisms such as public hearings, community advisory groups, citizen science projects, and comanagement

Assignments

Your grade will be based on three assignments:

1. Class participation (20%)

    Each week, you will have several readings, listed in the syllabus. You are expected to read these articles carefully and come to the week's seminar prepared to discuss them. A good model for careful reading is the "yes, no, hmmm" method:

    - *Yes*: One important thing you think the article got right
    - *No*: One important thing you disagree with in the article
    - *Hmmm*: One thing in the article that sparked you to think beyond what was in the article, generating new ideas or connections to other materials

2. Deliberative democracy paper (20%)

    In weeks 4 and 5, we will be studying theories of deliberative democracy. The various theorists we will read each propose a model of incorporating deliberation into democratic governance. By week 7, you will write a paper of 1500–2000 words presenting your own model of democracy. You should indicate in your paper how you draw on or rebut the views of the authors we have read. No additional research is required for this paper, though of course if you do draw on outside sources you should cite them.

3. Participation case study (60%)

    During week 4, all students must meet with the professor to select a case study for their major paper. The case study should be of a specific environmental management issue in a specific place. You will then write a longer paper (3000–4000 words) that answers the following questions:

    - What is the nature of the environmental management issue?
    - What, if any, types of participatory approaches have been tried in your case? How have they worked?
    - What form of participation would you recommend in this case? What factors influence your decision?

In addition to submitting a written paper, each student will give a 5-min oral presentation of their research during the last week's class.

Schedule of Topics and Readings

Week 1: What is participation?

What is participation? Why is participation so popular today in environmental management contexts?

Lynn, F.M. 1987. Citizen involvement in hazardous waste sites: two North Carolina success stories. *Environmental Impact Assessment Review* 7: 347–361.

Logan, B.I. and W.G. Moseley. 2002. The political ecology of poverty alleviation in Zimbabwe's Communal Areas Management Programme for Indigenous Resources (CAMPFIRE). *Geoforum* 33: 1–14.

Dietz, T. and P.C. Stern. 2008. Public participation in environmental assessment and decision making [Online]. Available at http://www.nap.edu/openbook.php?record_id=12434 National Academies Press, Washington DC. Chapter 1 (accessed March 20, 2015).

Week 2: Why participation?
What is the case for participation? What reasons justify participatory approaches?
Fiorino, D.J. 1990. Citizen participation and environmental risk: a survey of institutional mechanisms. *Science, Technology, and Human Values* 15(2): 226–243.
Hunold, C. and I.M. Young. 1998. Justice, democracy, and hazardous siting. *Political Studies* 46: 82–95.
Dietz, T. and P.C. Stern. 2008. Public participation in environmental assessment and decision making [Online]. Available at http://www.nap.edu/openbook.php?record_id=12434 National Academies Press, Washington DC. Chapter 2 (accessed March 20, 2015).

Week 3: Legal mandates and barriers for participation
How does current law enable participation? Can current laws get in the way of effective participation?
[NB: This week's lesson plan is focused on the US federal government. Classes held in other countries should adjust their materials to reflect their own domestic legal landscape. A class may also wish to study the law in their specific state or province.]
Poisner, J. 1996. A civic republican perspective on the National Environmental Policy Act's process for citizen participation. *Environmental Law* 26: 53.
Branch, K.M. and J.A. Bradbury. 2006. Comparison of DOE and Army advisory boards: application of a conceptual framework for evaluating public participation in environmental risk decision making. *Policy Studies Journal* 34(4): 723–753.
McCarthy, J. 2006. Neoliberalism and the politics of alternatives: community forestry in British Columbia and the United States. *Annals of the Association of American Geographers* 96(1): 84–104.
Jakes, P.J., K.C. Nelson, S.A. Enzler, S. Burns, A.S. Cheng, V. Sturtevant, D.R. Williams, A. Bujak, R.F. Brummel, S.A. Grayzeck-Souter, and E. Staychock. 2011. Community wildfire protection planning: is the Healthy Forests Restoration Act's vagueness genius? *International Journal of Wildland Fire* 20: 350–363.

Week 4: Deliberative democracy I: Pragmatism and discourse ethics
What is the theory of deliberative democracy? How do key theorists like John Dewey and Juergen Habermas envision democracy?
Addams, J. 2002. *Democracy and social ethics.* University of Illinois Press, Urbana, IL. Chapter 1.
Dewey, J. 1954. *The public and its problems.* Swallow Press, Chicago, IL. Chapter 1.
Habermas, J. 2002. Deliberative politics. p. 107–125. *In Democracy.* Blackwell, Malden, MA.
Webler, T. 1995. "Right" discourse in citizen participation: an evaluative yardstick. pp. 35–86. *In* Renn, O., Webler, T., and Wiedemann, P.M. (eds.), *Fairness and competence in citizen participation: evaluating models for environmental discourse.* Kluwer, Dordrecht.

Week 5: Deliberative democracy II: Feminist revisions
Can emotion and narrative play a role in deliberative democracy? Are deliberative ideals biased toward rich white Western men?
Benhabib, S. 1992. *Situating the self: gender, community and postmodernism in contemporary ethics.* Polity Press, Cambridge. Chapter 5.
Mouffe, C. 1999. Deliberative democracy or agonistic pluralism? *Social Research* 66(3): 745–758.
Young, I.M. 1989. Polity and group difference: a critique of the ideal of universal citizenship. *Ethics* 99(2): 250–274.
Young, I.M. 1995. Communication and the other: beyond deliberative democracy. p. 134–152. *In* Wilson, M. and Yeatman, A. (eds.), *Justice and identity: antipodean practices.* Allen and Unwin, St. Leonards, NSW.

Week 6: Deliberative democracy III: Bringing nature in
Can nature be taken into account in deliberative democracy? How do we handle future generations and nonhuman creatures who can't speak for themselves?
Plumwood, V. 1996. Has democracy failed ecology? an ecofeminist perspective. *Environmental Politics* 4(4): 134–168.
Johnson, G.F. 2007. Discursive democracy in the transgenerational context and a precautionary turn in public reasoning. *Contemporary Political Theory* 6: 67–85.
O'Neill, J. 2001. Representing people, representing nature, representing the world. *Environment and Planning C* 19: 483–500.
Heyward, C. 2008. Can the all-affected principle include future persons? Green deliberative democracy and the non-identity problem. *Environmental Politics* 17(4): 625–643.

Week 7: Designing a participatory process
What design considerations should go into a participatory process? What are the issues that must be considered in deciding between models such as public hearings, citizens' juries, or community advisory groups?
Fung, A. 2003. Recipes for public spheres: eight institutional design choices and their consequences. *Journal of Political Philosophy* 11(3): 338–367.
Rowe, G. and L. Frewer. 2005. A typology of public engagement mechanisms. *Science, Technology, and Human Values* 30: 251–290.
Vroom, V.H. and A.G. Jago. 1978. On the validity of the Vroom-Yetton model. *Journal of Applied Psychology* 63(2): 151–162.
Pritikin, T.T. 1998. A citizen's view: the nuts and bolts of co-partnerships. *Human Ecology Review* 5(1): 51–53.

Week 8: Who allows participation and who participates?
What are the characteristics of sponsoring agencies who embrace participation? What kind of laypeople will be willing to participate?
McComas, K.A., J.C. Besley, and C.W. Trumbo. 2006. Why citizens do and do not attend public meetings about local cancer cluster investigations. *Policy Studies Journal* 34(4): 671–698.

Neblo, M.A., K.M. Esterling, R.P. Kennedy, D.M.J. Lazer, and A.E. Sokhey. 2010. Who wants to deliberate—and why? *American Political Science Review* 104(3): 566–583.

Chess, C. and B.B. Johnson. 2006. Organizational learning about public participation: "Tiggers" and "Eeyores." *Human Ecology Review* 13(2): 182–192.

Sultana, F. 2009. Community and participation in water resources management: gendering and naturing development debates from Bangladesh. *Transactions of the Institute of British Geographers* 34: 346–363.

Week 9: Knowledge I: Making science accessible

Can laypeople learn to understand complex technical issues? Can science be made accessible to nonexperts?

Serrat-Capdevilla, A., A. Browning-Allen, K. Lansey, T. Finan, and J.B. Valdés. 2009. Increasing social–ecological resilience by placing science at the decision table: the role of the San Pedro Basin (Arizona) decision-support system model. *Ecology and Society* 14(1): 37.

Busenberg, G.J. 2000. Resources, political support, and citizen participation in environmental policy: a reexamination of conventional wisdom. *Society and Natural Resources* 13(6): 579–587.

Renn, O., T. Webler, and H. Kastenholz. 1998. Procedural and substantive fairness in landfill siting: a Swiss case study. p. 253–270. *In* Löfstedt, R. and Frewer, L. (eds.), *The Earthscan reader in risk and modern society*. Earthscan Publications, London.

Petersen, D., M. Minkler, V.B. Vásquez, and A.C. Baden. 2006. Community-based participatory research as a tool for policy change: a case study of the Southern California Environmental Justice Collaborative. *Review of Policy Research* 23(2): 339–353.

Week 10: Knowledge II: Traditional, indigenous, and lay knowledge

What kind of knowledge do nonscientific stakeholders bring to the table? Can this knowledge be incorporated into decision making without disrespecting or invalidating it?

Nursey-Bray, M., H. Marsh, and H. Ross. 2010. Exploring discourses in environmental decision making: an indigenous hunting case study. *Society and Natural Resources* 23(4): 366–382.

Berkes, F. 1998. Indigenous knowledge and resource management systems in the Canadian subarctic. p. 98–128. *In* Berkes, F., Colding, J., and Folke, C. (eds.), *Linking social and ecological systems: management practice and social mechanisms for building resilience*. Cambridge University Press, Cambridge.

Nadasdy, P. 2003. Reevaluating the co-management success story. *Arctic* 56(4): 367–380.

Huntington, H.P., S.F. Trainor, D.C. Natcher, O.H. Huntington, L. DeWilde, and F.S. Chapin III. 2006. The significance of context in community-based research: understanding discussions about wildfire in Huslia, Alaska. *Ecology and Society* 11(1): 40.

Week 11: Social learning

What is social learning? How can participation foster social learning? What benefits will social learning bring to a participatory process?

Reed, M.S., A.C. Evely, G. Cundill, I. Fazey, J. Glass, A. Laing, J. Newig, B. Parrish, C. Prell, C. Raymond, and L.C. Stringer. 2010. What is social learning? *Ecology and Society* 15(4): r1.

Pahl-Wostl, C., M. Craps, A. Dewulf, E. Mostert, D. Tàbara, and T. Taillieu. 2007. Social learning and water resources management. *Ecology and Society* 12(2): 5.

Daniels, S.E. and G.B. Walker. 1996. Collaborative learning: improving public deliberation in ecosystem-based management. *Environmental Impact Assessment Review* 16: 71–102.

Muro, M. and P. Jeffrey. 2008. A critical review of the theory and application of social learning in participatory natural resource management processes. *Journal of Environmental Planning and Management* 51(3): 325–344.

Week 12: Critics of participation I: Practical considerations

What practical difficulties—with respect to time, money, and knowledge—might make participation a bad idea? Is participation naïve about the realities of politics?

Laurian, L. 2004. Public participation in environmental decision making: findings from communities facing toxic waste cleanup. *Journal of the American Planning Association* 70(1): 53–66.

Pini, B. and F. Haslam McKenzie. 2006. Challenging local government notions of community engagement as unnecessary, unwanted, and unproductive: case studies from rural Australia. *Journal of Environmental Policy and Planning* 8(1): 27–44.

Breyer, S. 1993. *Breaking the vicious circle: toward effective risk regulation*. Harvard University Press, Cambridge, MA. Chapter 1.

Cooke, B. 2001. The social psychological limits of participation? p. 102–121. *In* Cooke, B. and Kothari, U. (eds.), *Participation: the new tyranny?* Zed Books, London.

Week 13: Critics of participation II: Radical voices

Is participation simply wrong in principle? Does participation give power to the wrong people?

Amin, A. 2005. Local community on trial. *Economy and Society* 34(4): 612–633.

McClosky, M. 1999. Local communities and the management of public forests. *Ecology Law Quarterly* 25: 624.

Sanders, L.M. 1997. Against deliberation. *Political Theory* 25(3): 347–376.

Shapiro, I. 1999. Enough of deliberation: politics is about interests and power. p. 28–38. *In* Macedo, S. (ed.), *Deliberative politics: essays on Democracy and disagreement*. Oxford University Press, New York.

Week 14: Participatory GIS

What is participatory GIS (PGIS)? Can PGIS solve the problems of facilitating participation and integrating science, or will it simply coopt or exclude some voices?

Morehouse, B.J., S. O'Brien, G. Christopherson, and P. Johnson. 2010. Integrating values and risk perceptions into a decision support system. *International Journal of Wildland Fire* 19: 123–136.

Ramsey, K. 2008. A call for agonism: GIS and the politics of collaboration. *Environment and Planning A* 40(10): 2346–2363.

Sieber, R. 2006. Public participation geographic information systems: a literature review and framework. *Annals of the Association of American Geographers* 96(3): 491–507.

Rundstrom, R.A. 1995. GIS, indigenous peoples, and epistemological diversity. *Cartography and Geographic Information Systems* 22(1): 45–57.

Week 15: Evaluating participation

Why should participatory processes be evaluated? How should evaluation be carried out?

Beierle, T.C. 1999. Using social goals to evaluate public participation in environmental decisions. *Policy Studies Review* 16(3/4): 75–103.

Santos, S.L. and C. Chess. 2003. Evaluating citizen advisory boards: the importance of theory and participant-based criteria and practical implications. *Risk Analysis* 23(2): 269–279.

Danielson, S., S.P. Tuler, S.L. Santos, T. Webler, and C. Chess. 2012. Three tools for evaluating participation: Focus groups, Q method, and surveys. *Environmental Practice* 14(2): 101–109.

Measham, T.G. 2009. Social learning through evaluation: a case study of overcoming constraints for management of dryland salinity. *Environmental Management* 43(6): 1096–1107.

Week 16: Student presentations

CHAPTER 16: STATISTICS IN ENVIRONMENTAL MANAGEMENT

1. Environmental sampling designs
   (a) Simple random sampling
   (b) Stratified sampling
   (c) Two-phased sampling
   (d) Systematic sampling
   (e) Unequal probability sampling
   (f) Cluster sampling
   (g) Adaptive sampling
   (h) Nonsampling errors
   (i) Missing data
   (j) Designs for questionnaires
   (k) Designs for long-term sampling including rotating panel designs
2. Models for data
   (a) Graphical display of data
   (b) Models for data
   (c) Analysis of variance (ANOVA)
   (d) Simple linear regression
   (e) Multiple linear regression
   (f) Generalized linear models (GLM)
   (g) Linear mixed models
   (h) Generalized estimating equations
   (i) Generalized linear mixed models (GLMM)
   (j) Multivariate statistics
   (k) Generalized additive models
   (l) Time series analysis
   (m) Randomization, Bootstrap, and Monte Carlo Methods
   (n) Spatial statistics
3. Risk assessment

CHAPTER 17: REMOTE SENSING IN ENVIRONMENTAL MANAGEMENT

Course Outline

The following is an idealized semester course outline (weekly sessions) for a midlevel introductory course in applied environmental remote sensing that indicates practical, hands-on activities in parentheses; a selection of these is described briefly on the following page. Each 2.5 h session comprises lecture, discussion, and the practical exercise. While students have found this to be quite demanding, they leave the course with a broad, comprehensive view of the technology and its applications, as well as a wide set of skills in interpreting and manipulating imagery.

Session 1:	Overview, applications, and images ("Google Earth Me!")
Session 2:	EMR: image visualization and interpretation ("what in the world?")
Session 3:	Sensors and systems, data types and formats (topography from lidar))
Session 4:	High-resolution imagery for forestry and urban/suburban landscapes (CANAPI)
Session 5:	Geometric correction of remote sensing imagery (warped)
Session 5:	Vegetation at continental to global scales I (NDVI I)
Session 6:	Vegetation at continental to global scales II (NDVI II)
Session 7:	Global Albedo from BRDF ("brighter later?")
Session 8:	Applications of thermal infrared remote sensing (warming lakes)
Session 9:	New Lidar technologies (multiple return lidar over Queens, NY)
Session 10:	Remote sensing of natural hazards ("don't get burned!")
Session 11:	Land cover from remote sensing I (high-resolution/color IR)

Session 12: Land cover from remote sensing II (Landsat TM/multispectral)
Session 13: Land cover from remote sensing III (accuracy assessment)
Session 14: Remote sensing in hydrology and GRACE ("water, water everywhere?")
Session 15: Hyperspectral imaging (invasion of the water hyacinths)
Session 16: Solid Earth, ocean, and land applications of radar remote sensing (deforestation in tropical rainforest)
Session 17: Remote sensing indicators of recent climate change ("fire and ice")

Outlines of Selected Practical Exercises

Google Earth Me! In this exercise, students use panchromatic and multispectral (blue, green, red, near-infrared band) imagery from the QuickBird satellite to explore image visualizations: grayscale imagery, true color composites, false color composites, data scaling, and contrast stretches. They also use these images to perform a Brovey transform (pan-sharpening operation) and are asked to comment on the result (How successful was it? Are the colors natural? Why do you think this might be? Why do the panchromatic images have higher spatial resolution than the multispectral images—and what are the implications of this for hyperspectral remote sensing with hundreds of bands?).

Topography from Lidar Students use single/discrete-return lidar data points to explore the morphology of a mesquite shrub dune landscape in southern New Mexico. They must first import the data from a plain text format (text file with X, Y coordinates and Z elevation values) and then perform an interpolation operation to obtain a wall-to-wall raster digital terrain model. Pseudo 3-D fly-through visualizations can also be made.

Multiple Return Lidar Students use multiple return lidar to explore the various features that can be extracted over this Queens, NY, urban/suburban landscape near JFK International Airport.

CANAPI Students are presented with a 500 × 500 cell 0.6 m QuickBird panchromatic image acquired over sparse forest in the Sierra Nevada of California. They must first attempt to count the number of trees by eye and—having failed within the allotted 5 min—are asked to obtain the canopy cover, mean crown radius, and mean tree shadow length—in 10 min. Having failed again, they will discover the power of image processing and automation by using the CANAPI algorithm in an ImageJ macro to calculate these parameters in just a few seconds.

CHAPTER 18: GEOGRAPHIC INFORMATION SYSTEMS IN ENVIRONMENTAL MANAGEMENT

Model Syllabus

Brief Description and Objectives of the Course

The development of Geographic Information Science (GIS) is greatly enriched by the recent achievements and advances in computer science, statistics, and regional sciences. One of the key milestones in the development of GIS is the analysis of geographic information assisted with knowledge developed in statistics, econometrics and other quantitative analysis disciplines, and application of GIS in a wide range of studies, including environmental management. Apparently, the fast development of computing technologies that are embedded in either standard commercial software packages or stand-alone public packages lends indispensable hands to the ever-growing demand for GIS scientists. In the meantime, new developments in environmental management brought new topics and issues that feed the growth of GIS. In this GIS course, you will be introduced to the field of intense Geographic Information Analysis. You will be required to explore a variety of options for data collection, compilation/organization, analysis, and presentation using either standard commercial GIS software packages (such as ArcGIS) or freely available packages (such as the SPDEP package in R (http://cran.r-project.org/web/packages/spdep/index.html) or OpenGeoDa (http://geodacenter.asu.edu/ogeoda)). Licensing issues for ArcGIS might prevent you from taking full advantage of this software. It is hence recommended for you to familiarize yourselves with SPDEP or OpenGeoDa as soon as possible.

Required Text

O'Sullivan, David and Unwin, David J., 2010, *Geographic Information Analysis* 2nd Edition, John Wiley & Sons, Inc., Hoboken, NJ.

Manuals for SPDEP and OpenGeoDa can be found on their corresponding web pages. SPDEP is much more versatile but has a relatively steep learning curve. OpenGeoDa is good enough for everyday uses and is facilitated by Luc Anselin's GeoDa workbook.

General Policies

1. Everyone must use their MSU e-mail account. All announcements and communications will be sent via e-mail (and in blackboard as well). Hence, checking your e-mail in a regular fashion is highly recommended. You are responsible for all information given in class, including changes in class schedule and any additional reading assignments. You are responsible

for acquiring your own notes to make up for lecture material missed due to an absence.
2. Readings are expected to be completed **PRIOR** to the next class meeting.
3. Participation in discussions is mandatory.
4. Academic misconduct in any form, especially plagiarism, will not be tolerated—it will result in an immediate fail grade of the course.

Grading Policies

The goal of the study is to produce a publishable final report based on a topic that you chose. The grades will be based upon the quality of your final report. If the final report is sent out to a journal for publication consideration (after I've reviewed it), you'll automatically get an A. For this matter, it is a good practice to find appropriate journals that might publish your study as soon as possible. Otherwise, the following grading scales apply.

Grade Scale:

Final grade	Credits earned (%)	Final grade	Credits earned (%)
A	93–100	C+	70–72.99
A–	88–92.99	C	67–69.99
B+	83–87.99	C–	64–66.99
B	78–82.99	D	60–63.99
B–	73–77.99	F	<60%

Reading Schedules

Week 1, Chapter 1: Geographic information and geographic information analysis

Tasks: Choose either SPDEP or OpenGeoDa (or both if you are up to the task) and get into the manuals

Week 2, Chapter 2: Pitfalls and potential of geographic information

Tasks: Familiarize yourselves with SPDEP or OpenGeoDa for data analysis

Assignment 1: Presenting your idea (start to form an abstract, outline of the final project)

Week 3, Chapter 3: Presenting your ideas: mapping it out

Tasks: Compare mapping options in SPDEP and OpenGeoDa, how are they different from ArcGIS?

Week 4, Chapter 4: Maps again: what maps are good maps

Tasks: Play around with SPDEP's plot functions or map directly in OpenGeoDa with your own data

Assignment 2: Data collecting started. Once you are clear about what you intend to do, start the data collection process immediately. Assignment 1 is due—everyone shall submit to me an abstract that is no more than 250 words and no less than 150 words summarizing the general idea of what you intend to do. In addition, an outline with rough information about each section (a title would suffice at the moment) is required as well

Week 5, Chapter 5: Analyzing point patterns

Tasks: Try to analyze a set of point data that are provided in either SPDEP package or OpenGeoDa and explore how the ideas presented in your textbook can be done in these packages

Week 6, Chapter 6: Point pattern further explored

Tasks: Continue what you've done last week

Week 7, Chapter 7: Spatial autocorrelation

Tasks: Explore the concept of spatial autocorrelation in either SPDEP or OpenGeoDa

Week 8, Chapter 8: Spatial autocorrelation: local

Tasks: Continue the exploration of spatial autocorrelation

Week 9, Chapter 9: Fields: spatial interpolation

Tasks: Explore the ideas presented in this chapter in either SPDEP or OpenGeoDa

Assignment 3: Form your methodology: what will you use to analyze your data? Assignment 2 due—present me your data sets, I need to verify them for the validity of your analysis. The report shall be no less than 2 pages, describing the data (you might want to use maps), the metadata, and potential ways to analyze them (for Assignment 3)

Week 10, Chapter 10: Statistics of the fields

Tasks: Continue what you've been doing last week

Week 11, Chapter 11: Map overlay

Tasks: Attempt to explore the ideas presented in this chapter in SPDEP, OpenGeoDa, or ArcGIS

Assignment 3 due: in no less than 5 pages, double space, present your choice of methods that you plan to analyze your data

Week 12, Chapter 12: Student presentation

Final report combining the previous three assignments is due at the end of the course

CHAPTER 19: LIFE CYCLE ANALYSIS AS A MANAGEMENT TOOL IN ENVIRONMENTAL SYSTEMS

Syllabus

About This Class

Nowadays, "Life Cycle Assessment" or "LCA" has become one of the most significant tools in environmental management. LCA is the systematic analysis of global, regional, and local material and energy flows and uses that are associated with products, processes, industrial sectors, and economies. The course covers primary and key features of LCA, including the history and aim of LCA, a concise presentation of LCA methodology, the main applications of LCA, etc., providing students with analytical tools and methods for implementing principles of life cycle analysis.

The purpose of this course is to give students the opportunity to start thinking over how LCA can be used in various fields of industry and society. Thus, by the end of this class, a successful student will be able to:

- Describe the complete function and theory of LCA
- Explain the subject matter and give details about the purpose of the LCA steps
- Consider potential applications and limitations of LCA
- Achieve a complete LCA of a product or service system

Assignments

Class participation	15%
Midterm exam	25%
Project report and presentation	25%
Final exam	35%

Course Outline

Weeks 1 and 2: The life cycle thinking approach and the history of LCA

Basic and key features of life cycle thinking and a concise overview of the LCA history

Finkbeiner, M., A. Inaba, R.B.H. Tan, K. Christiansen, and H.-J. Klüppel. 2006. The new international standards for life cycle assessment: ISO 14040 and ISO 14044. *International Journal of LCA*. 11(2):80–85.

International Organization for Standardization. 2006. *ISO 14040: Environmental management—Life cycle assessment—Principles and framework*, Second edition. ISO, Geneva.

Joshi, S. 2000. Product environmental life-cycle assessment using input-output techniques. *Journal of Industrial Ecology*. 3(2&3):95–120.

Weeks 3 and 4: The general LCA framework—The goal and scope definition

The components of the LCA methodology. Setting up and designing an LCA study

Curran, M.A. 2006. *Life cycle assessment: Principles and practice*. National Risk Management Research Laboratory—Office of Research and Development. U.S. Environmental Protection Agency, Cincinnati, OH.

European Commission—Joint Research Centre—Institute for Environment and Sustainability. 2010. *International Reference Life cycle data system (ILCD) handbook—General guide for Life cycle assessment—Detailed guidance*. Publications Office of the European Union, Luxembourg.

Finnveden, G., M.Z. Hauschild, T. Ekvall, J. Guinée, R. Heijungs, S. Hellweg, A. Koehler, D. Pennington, and S. Suh. 2009. Recent developments in life cycle assessment. *Journal of Environmental Management*. 91:1–21.

Week 5: The life cycle inventory (LCI)

Constructing a flow model and collecting data of the technical system, as defined in the goal and scope

Curran, M.A. 2006. *Life cycle assessment: Principles and practice*. National Risk Management Research Laboratory—Office of Research and Development. U.S. Environmental Protection Agency, Cincinnati, OH.

European Commission—Joint Research Centre—Institute for Environment and Sustainability. 2010. *International Reference Life cycle data system (ILCD) handbook—General guide for Life cycle assessment—Detailed guidance*. Publications Office of the European Union, Luxembourg.

Finnveden, G., M.Z. Hauschild, T. Ekvall, J. Guinée, R. Heijungs, S. Hellweg, A. Koehler, D. Pennington, and S. Suh. 2009. Recent developments in life cycle assessment. *Journal of Environmental Management*. 91:1–21.

Weeks 6 and 7: The life cycle impact assessment (LCIA) and interpretation

The principles and methods available for describing the environmental consequences of the environmental loads of a technical system. Reporting and critical review

Georgakellos, D.A. 2005. Evaluation of life cycle inventory results using critical volume aggregation and polygon-based interpretation. *Journal of Cleaner Production*. 13:567–582.

Pieragostini, C., M.C. Mussati, and P. Aguirre. 2012. On process optimization considering LCA methodology. *Journal of Environmental Management*. 96:43–54.

Pizzol, M., P. Christensen, J. Schmidt, and M. Thomsen. 2011. Impacts of "metals" on human health: a comparison between nine different methodologies for life cycle impact assessment (LCIA). *Journal of Cleaner Production*. 19:646–656.

Weeks 8 and 9: Applications of LCA

Principles and applications of LCA in product design and development, marketing and ecolabelling, strategic planning and public policy making, green purchasing, etc.

Bersimis, S. and D. Georgakellos. 2013. A probabilistic framework for the evaluation of products' environmental performance using life cycle approach and Principal Component Analysis. *Journal of Cleaner Production*. 42:103–115.

Herva, M., A. Franco-Urva, E.F. Carrasco, and E. Roca. 2012. Application of fuzzy logic for the integration of environmental criteria in ecodesign. *Expert Systems with Applications*. 39:4427–4431.

Millet, D., L. Bistagnino, C. Lanzavecchia, R. Camous, and T. Poldma, 2007. Does the potential of the use of LCA match the design team needs? *Journal of Cleaner Production*. 15:335–346.

Weeks 10 and 11: Limitations of LCA—Recent trends and developments in LCA

Presenting of the limitations mentioned in critical reviews, together with the ongoing efforts to refine the LCA methodology

European Commission—Joint Research Centre—Institute for Environment and Sustainability. 2010. *International Reference Life cycle data system (ILCD) handbook—General guide for Life cycle assessment—Detailed guidance.* Publications Office of the European Union, Luxembourg.

Finnveden, G., M.Z. Hauschild, T. Ekvall, J. Guinée, R. Heijungs, S. Hellweg, A. Koehler, D. Pennington, and S. Suh. 2009. Recent developments in life cycle assessment. *Journal of Environmental Management.* 91:1–21.

Pieragostini, C., M.C. Mussati, and P. Aguirre. 2012. On process optimization considering LCA methodology. *Journal of Environmental Management.* 96:43–54.

Week 12: Student project presentations

CHAPTER 20: ENVIRONMENTAL AUDIT IN ENVIRONMENTAL MANAGEMENT

Model Syllabus

Lecture 1: Introduction to Environmental Management

- Introduction to Environmental Management
- Introduction to aspects and impacts
- Environmental Issues
- Core elements of an EMS

Lecture 2: Aspects, Migration Paths, and Impacts

- Migration of pollution—sources, pathways, and receptors
- How to identify environmental aspects and impacts
- *Case Study 1: Aspects Identification*

Lecture 3: ISO14001 Part 1

- General requirements
- Environmental policy
- Plan: planning

Lecture 4: ISO14001 Part 2

- Do: implementation and operation
- Check: checking
- Act: management review
- *Case Study 2: ISO14001 Familiarization*

Lecture 5: Intro to Environmental Auditing

- Introduction to auditing, audit structure, and skills
- Preaudit activities—preparing for the audit
- *Case Study 3: Audit Preparation*

Lecture 6: The Audit Process

- On-site Activities—Conducting the Audit
- Following an audit trail
- Postaudit activities—reporting
- *Case Study 4: Audit Reporting*

Note: These are 3 h lectures with a 15 min break in the middle; case studies are introduced in class but done at home. This is based on my own program at HKU.

CHAPTER 21: RISK ASSESSMENT IN ENVIRONMENTAL CONTAMINATION PROBLEM MANAGEMENT

Model Syllabus

Outline for Proposed Course

This course will introduce risk assessment principles, concepts, methods, and tools that may find useful applications in the management of environmental contamination and chemical exposure problems. The main emphasis of the course will be on the use of human and ecological health risk assessment principles and methods to support risk management and corrective action decisions associated with contaminated site restoration programs. The course will also provide a broad understanding of the scientific basis of risk assessment and its applicability within the environmental industry. The proposed course content/outline is as follows:

Part I

- A general overview
  - Types and sources of environmental pollutants
  - Nature of environmental contamination problems
  - Health, environmental, and socioeconomic implications of environmental contamination problems
  - Legislative–regulatory controls affecting environmental contamination problems
- Contaminant fate and transport in the environment
  - Factors affecting contaminant migration
  - Important fate and transport properties, processes, and parameters
  - Cross-media transfer of contaminants between environmental compartments
  - Contaminant fate and transport assessment
- Conceptualization of contaminated sites
  - Elements of a conceptual site model
  - Design of conceptual site models
  - Development of exposure scenarios

Part II

- Fundamentals of hazard, exposure, and risk assessment
  - Nature of hazard, exposure, and risk
  - Purpose and attributes of risk assessment
  - Risk acceptability and risk tolerance considerations
  - Uncertainty issues in risk assessments
  - Risk-based decision making
- Elements of a risk assessment
- The risk assessment process
  - Risk assessment techniques and methods of approach
  - Data collection and evaluation for risk assessments
  - Exposure assessments
  - Toxicity assessment of environmental contaminants
  - Risk characterization
- Determination of acceptable risk-based contaminant levels
  - Development of site restoration goals and risk-based cleanup criteria for remedial action programs
  - Determination of health-protective chemical concentrations
- Risk assessment applications to environmental contamination problems
  - Conducting contaminated site risk assessment
  - Risk assessment applications to the management of contaminated land problems

APPENDIX C

MODEL ENVIRONMENTAL MANAGEMENT CURRICULA (BS, MS, PHD)

ENVIRONMENTAL MANAGEMENT, BACHELOR OF SCIENCE

2014–2015 Degree Plans

University Core Requirements (42 h)	UHCL Course Title	TCCNS	CODE
Communications (6 h)	WRIT 1301—Composition I (*grade must be "C-" or better*)	ENGL 1301	010
	WRIT 1302—Composition II (*grade must be "C-" or better*)	ENGL 1302	010
Mathematics (3 h)	MATH 1314—College Algebra	MATH 1314	020
Life and Physical Sciences (6 h)	CHEM 1311—Chemistry I	CHEM 1311	030
	CHEM 1312—Chemistry II	CHEM 1312	030
Language, Philosophy, and Culture (3 h)	Choose ONE course from the core approved list		040
Creative Arts (3 h)	Choose ONE course from the core approved list		050
US History (6 h)	HIST 1301—US History I	HIST 1301	060
	HIST 1302—US History II	HIST 1302	060
Government/Political Science (6 h)	POLS 2305—Federal Government	GOVT 2305	070
	POLS 2306—Texas Government	GOVT 2306	070
Social/Behavioral Science (3 h)	Choose ONE course from the core approved list	ECON 2301	080
Component Area Option (6 h)	COMM 1315—Public Speaking	SPCH 1315	090
	PSYC 1100—Learning Frameworks	PSYC 1100	090
	CHEM 1111—Chemistry I lab	CHEM 1111	030
	CHEM 1112—Chemistry II lab TWO 1-h Life/Physical Science Labs	CHEM 1112	030

An Integrated Approach to Environmental Management, First Edition. Edited by Dibyendu Sarkar, Rupali Datta, Avinandan Mukherjee, and Robyn Hannigan.
© 2016 John Wiley & Sons, Inc. Published 2016 by John Wiley & Sons, Inc.

Environmental Management Core Requirements (18 h)	UHCL Course Title	TCCNS
	ACCT 2301—Principles of Accounting I	ACCT 2301
	ACCT 2302—Principles of Accounting II	ACCT 2302
	(*ACCT 2301 and/or 2302 can be substituted with additional natural science*)	
	ECON 2301—Principles of Macroeconomics	ECON 2301
	ECON 2302—Principles of Microeconomics	ECON 2302
	LEGL 2301—Legal Environment of Business	BUSI 2301 or 2302
	MATH 1324—Finite Math	MATH 1324
	(*MATH 1324 can be substituted with MATH 1325, 2412, or 2413*)	

Environmental Management Major Requirements (45 h)	UHCL Course Title
Core Requirements (30 h)	CHEM 3333—Environmental Chemistry
	DSCI 3321—Statistics
	ENVR 3311—Foundations of Environmental Management
	ENVR 4313—Techniques of Environmental Assessment
	ENVR 4315—Introduction to Environmental Law
	ENVR 4332—Process of Environmental Permitting
	ENVR 4333—Introduction to Pollution Control Technology
	ENVR 4336—Administrative Practice and Ethical Considerations
	MGMT 3301—Management Theory and Practice
	WRIT 3315—Technical Writing
Select two of the following courses (6 h)	ENVR 4311—Principles of Air Quality
	ENVR 4312—Water Management Principles
	ENVR 4317—Solid Waste Management
Approved special topics courses—must be approved by faculty coordinator	
Select two of the following courses (6 h)	MGMT 3313—Organizational Communication
	MGMT 3331—Human Resource Management
	MGMT 4326—Effective Negotiations
	MGMT 4327—Leadership
	MGMT 4353—International Business Management
	MGMT 4354—Organizational Behavior
	MGMT 4357—Governmental Budget Planning and Assessment
	GEOG 4301—Urban Geography
	GEOG 4312—Human Geography
	LEGL 3351—Legal Research
	LEGL 4353—Dispute Resolution
Select one of the following courses (3 h)	BIOL 3311—Marine Biology
	BIOL 3333—Environmental Biology
	CHEM 4355—Environmental Sampling and Monitoring
	GEOG 4321—Fundamentals of Geographic Information Systems
	GEOL 3333—Environmental Geology
	GEOL 4323—Soils in the Environment
	INDH 3333—Environmental Safety and Health

Environmental Management Upper-Level Elective Requirements (15 h)	UHCL Course Title
ENVR Elective (3 h)	33XX or 43XX ENVR course not already taken
General Electives (6 h)	33XX or 43XX courses offered by Schools of Business, Human Sciences and Humanities, Science and Computer Engineering, or Education
Nonbusiness Electives (6 h)	Same as General Elective ***EXCEPT***: ACCT, DSCI, FINC, ISAM, MGMT, and MKTG courses ***cannot*** fulfill nonbusiness elective

GENERAL DEGREE REQUIREMENTS

- Students must complete at least 120 semester credit hours. A minimum of 60h of the 120 semester hours must be advanced (3000–4000 level) coursework according to the requirements of the respective major.
- Students must complete the University Core Curriculum requirements (refer to Core Curriculum Requirements section of this catalog).
- Students must fulfill the statutory requirements of the Texas State Education Code, including the following:
  - Six hours of US History (three hours may be Texas History)
  - Six hours of Constitutions of the United States and Texas
- Students must demonstrate writing proficiency by completing 9h of lower-level (1000–2000 level) and upper-level (3000–4000 level) English composition course credit with a minimum grade of "C-" or better. Some majors may require higher grades in English composition.
- Students must complete at least 25% of the credit hours required for the degree (i.e., 30 semester credit hours for a 120 credit hour program) through instruction offered by UHCL to fulfill the Southern Association of Colleges and Schools (SACS) residency requirements.
- Students must complete the final 30 semester hours of 3000 and 4000 level coursework in residence at UHCL.
- Students must complete a minimum of 12 semester credit hours of upper-level (3000–4000 level) coursework in the major in residence at UHCL.
- Students must have a cumulative GPA of 2.000 on coursework completed at UHCL with grades of "C" or better on at least 30h of resident upper-level work. Grades of "C-" or below cannot be applied toward the 30h of resident upper-level work.

MSEM: MASTER OF SCIENCE IN ENVIRONMENTAL MANAGEMENT UNIVERSITY OF SAN FRANCISCO

Launched in 1977, the Masters of Science in Environmental Management (MSEM) program at the University of San Francisco (USF) is one of the first graduate environmental management programs in the United States and the longest running in the West. Strongly rooted in the natural sciences, the MSEM interdisciplinary curriculum connects ecology and chemistry-based courses with environmental policy, law, and management. We now offer the richest mix of curriculum in the program's history. The program now has more than 85 students enrolled each year and is supported by 11 full-time faculty and nearly 25 adjunct faculty and practicing professionals. MSEM alumni number more than 600, working in government, consulting, industry, and nonprofit organizations in California and around the world.

Features of the MSEM Degree

- 30 units total for the MSEM degree.
- Use of Special Topics courses, to continue to bring current material and professional practitioners to the program, and promote flexibility in curriculum content and course scheduling.
- A diversity of courses to meet the professional and educational objectives of our diverse student population.
- Option to select an area of concentration within the degree. Concentrations are in:
  - Ecology
  - Environmental Health and Hazards
  - Water Management
  - Energy and Climate Change
- **Three required courses** for all students (3 courses × 2 units each) with **waiver option** for each:
  - ENVM 611 Ecology
  - ENVM 601 Environmental Chemistry
  - ENVM 603 Quantitative Methods
- **Required research capstone course**: ENVM 698 Master's Project. Typically 4 units (option of 1–4 units).[1]
- If a concentration is selected, it can be accomplished with completion of five 2-unit courses. Lists of courses for each concentration and sample schedules are included below.
- ENVM 690 Research Methods is recommended, but not required.
- Remaining units are open electives for the degree.

COURSE LISTS FOR AREAS OF CONCENTRATION

Ecology

- ENVM620 Applied Ecology
- ENVM621 Restoration Ecology
- ENVM622 Restoration Ecology Lab
- ENVM626 Wetland Ecology
- ENVM627 Wetland Ecology Lab
- ENVM628 Riparian Ecology
- ENVM636 Resource Management
- ENVM671 Climate Change: Global Processes and Ecological Impacts
- ENVM680 Wetland Delineation I
- ENVM680 Wetland Delineation II
- ENVM680 California Ecosystems
- ENVM680 Field Botany
- ENVM680 Field Survey Management

---

[1] MSEM also includes a Thesis Option as an alternative to the Master's Project. This option is not common and is only possible if student interest aligns with faculty research, and faculty member is willing to serve as a mentor for the thesis research project.

- EVM680 Conservation Biology
- ENVM68X Special Topics in Ecology (future offerings)

Environmental Health and Hazards

- ENVM633 Air Quality Assessment and Management
- ENVM641 Environmental Health and Safety
- ENVM643 Environmental Health
- ENVM644 Environmental Toxicology
- ENVM647 Environmental Risk Management
- ENVM648 Environmental Risk Assessment
- ENVM654 Environmental Engineering I: Contaminant Transport in Surface Water and Air
- ENVM655 Environmental Engineering II: Contaminant Transport in Ground Water
- ENVM656 Hazardous Waste Engineering
- ENVM680 Aquatic Pollution
- ENVM680 Water Treatment
- ENVM680 Environmental Emergency Management
- ENVM680 Solid Waste Management
- ENVM68X Special Topics in Environmental Health and Hazards (future offerings)

Water Management

- ENVM626 Wetland Ecology
- ENVM627 Wetland Ecology Lab
- ENVM628 Riparian Ecology
- ENVM654 Environmental Engineering I: Contaminant Transport in Surface Water and Air
- ENVM655 Environmental Engineering II: Contaminant Transport in Ground Water
- ENVM680 Marine Resources
- ENVM680 Aquatic Pollution
- ENVM680 Water Treatment
- ENVM680 Watershed Management
- ENVM680 Water Policy
- ENVM680 Hydrology and Watersheds
- ENVM680 Watershed Monitoring
- ENVM68X Special Topics in Water Management (future offerings)

Energy and Climate Change

- ENVM633 Air Quality Assessment and Management
- ENVM651 Energy Resources and Environment
- ENVM671 Climate Change: Global Processes and Ecological Impacts
- ENVM672 Climate Change Mitigation
- ENVM680 Renewable Energy
- ENVM680 Energy Auditing
- ENVM680 Green Building
- ENVM680 Urban Adaptation to Climate Change
- ENVM680 Sustainable Design (4 units, cross-listed with ARCD)
- ENVM680 Building Environmental Control Systems (4 units, cross-listed with ARCD)
- ENVM68X Special Topics in Energy and Climate Change (future offerings)

ADDITIONAL COURSES NOT SPECIFIED FOR AN AREA OF CONCENTRATION

ENVM605	Environmental Ethics
ENVM607	Env. Policy: Design and Implementation
ENVM613	Environmental Law I
ENVM680	Environmental Law II
ENVM680	Intro Geospatial Analysis
ENVM637	Accelerated Intro to GIS for ENVS
ENVM680	Advanced/Applied GIS
ENVM661	Environmental Accounting
ENVM614	Environmental Economics I
ENVM680	Environmental Economics II—Markets
ENVM680	Resource Economics
ENVM680	Environmental Finance
ENVM680	Sustainable Business
ENVM680	Process of Urban Plan and Design
ENVM680	Instrumental Analysis

SAMPLE CURRICULUM MAPS

The examples below show how each concentration could be accomplished with existing courses, including Special Topics courses.

General MSEM course plan with the concentration option [total = 30 units]:

Fall 1 [8 units]	Ecology [2], Environmental Chemistry [2], Quantitative Methods [2], one concentration course [2]
Spring 1 [8 units]	Two concentration courses [2×2], two electives [2×2]
Fall 2 [8 units]	Research Methods [2], two concentration courses [2×2], one elective [2]
Spring 2 [6 units]	One elective [2], Master's Project [4]

MSEM curriculum concentrations—sample course plan for each concentration

Total units for the degree	30
Required units	10
Concentration units	10
Elective units	10

Note: in the tables below, "req/c" denotes required courses or the number of concentration courses (5 courses × 2 units/course = 10 units).

Environmental Health and Hazards

Fall 1

req/c	#	Course Name	Units
req	611	Ecology	2
req	603	Quant. Methods	2
req	611	Env. Chemistry	2
1	643	Env. Health	2

Fall 2

req/c	#	Course Name	Units
	654	Env. Engineering I	2
4	680	Env. Emergency Management	2
	655	Env. Engineering II	2
	690	Research Methods	2

Spring 1

req/c	#	Course Name	Units
2	644	Env. Toxicology	2
3	680	Aquatic Pollution	2
	680	Solid Waste Management	2
	680	Instrumental Analysis	2

Spring 2

req/c	#	Course Name	Units
5	648	Env. Risk Assessment	2
req	698	Master's Project	4

Ecology

Fall 1

req/c	#	Course Name	Units
req	611	Ecology	2
req	603	Quant. Methods	2
req	611	Env. Chemistry	2
1	680	Wetland Ecology	2

Fall 2

req/c	#	Course Name	Units
4	680	Wetland Delineation	2
5	671	Climate Change: Global and Ecological	2
	680	California Ecosystems	2
	690	Research Methods	2

Spring 1

req/c	#	Course Name	Units
	613	Env. Law I	2
2	628	Riparian Ecology	2
	680	Env. Law II	2
3	680	Conservation Biology	2

Spring 2

req/c	#	Course Name	Units
	621	Restoration Ecology	2
req	698	Master's Project	4

Water Management

Fall 1

req/c	#	Course Name	Units
req	611	Ecology	2
req	603	Quant. Methods	2
req	611	Env. Chemistry	2
1	680	Watershed Management	2

Fall 2

req/c	#	Course Name	Units
4	680	Marine Resources	2
5	680	Water Policy	2
	607	Env. Policy	2
	690	Research Methods	2

Spring 1

req/c	#	Course Name	Units
2	680	Aquatic Pollution	2
	648	Env. Risk Assessment	2
3	680	Hydrology and Watersheds	2
	647	Env. Risk Management	2

Spring 2

req/c	#	Course Name	Units
	637	Accel Intro to GIS	2
req	698	Master's Project	4

Energy and Climate Change

	Fall 1				Fall 2		
req/c	#	Course Name	Units	req/c	#	Course Name	Units
req	611	Ecology	2		654	Env. Engineering I	2
req	603	Quant. Methods	2	5	671	Climate Change: Global and Ecological	2
req	611	Env. Chemistry	2		655	Env. Engineering II	2
	607	Env. Policy	2		690	Research Methods	2

	Spring 1				Spring 2		
req/c	#	Course Name	Units	req/c	#	Course Name	Units
1	680	Urban Adaptation to Climate Change	2		680	Env. Finance	2
2	651	Energy and Environment	2	req	698	Master's Project	4
3	672	Climate Change Mitigation	2				
4	680	Renewable Energy	2				

MONTCLAIR STATE UNIVERSITY ENVIRONMENTAL MANAGEMENT PHD CURRICULUM

Complete the following eight requirements:

CORE COURSES

Complete four courses for **12 semester hours**:

BIOL	570	Ecology	3
EAES	760	Organizational Environmental Management	3
EAES	700	Earth Systems Science	3
EAES	561	Environmental Law and Policy	3

REQUIRED RESEARCH COURSES

Complete two courses for **6 semester hours**:

ENVR	895	Research Project in Environmental Management I	3
ENVR	896	Research Project in Environmental Management II	3

PERSPECTIVE COURSES

Complete the following four requirements for **12 semester hours**:

Methods Perspective

STAT	595	Topics in Statistics—Environmental Statistics	3
OR			
STAT	541	Applied Statistics	3

Natural Science Perspective

EAES	505	Environmental Geoscience	3
OR			
EAES	533	Water Resource Management	3

Social Science Perspective

EAES	792	Special Topics—Environmental Economics	3
OR			
ANTH	522	Environment and Community	3

Business Perspective

MGMT	563	Sustainability and Corporate Responsibility	1.5
AND			
MGMT	565	Project Management	1.5

RESEARCH REQUIREMENTS

Complete the following two requirements for a minimum of 36 semester hours:

Colloquium

Complete for a minimum of **6 semester hours**:

EAES	790	Colloquium in Environmental Management	1

Dissertation

Complete for a minimum of **30 semester hours**:

EAES	900	Dissertation Advisement	3–12
EAES	901	Dissertation Extension 1 (After completion of 30 h of ENVR 900)	

ELECTIVES

Complete **6 semester hours** from the following list:

ANTH	529	Building Sustainable Communities	3
BIOL	520	Plant Physiology	3
BIOL	543	Advances in Immunology	3
BIOL	547	Molecular Biology I	3
BIOL	550	Topics in Microbiology	3
BIOL	553	Microbial Ecology	4
BIOL	554	Microbial Physiology	3
BIOL	571	Physiological Plant Ecology	4
BIOL	572	Wetland Ecology	4
BIOL	573	Shoreline Ecology	4
BIOL	595	Conservation Biology: The Preservation of Biological Diversity	3
CHEM	510	Hazardous Materials Management	3
CHEM	525	Bioinorganic Chemistry	3
CHEM	534	Chromatographic Methods: Theory and Practice	3
CNFS	505	Society and Natural Environment	2
CNFS	510	Environmental Impact of Recreation on Natural Areas	2
CNFS	525	Field/Lab Experiment—Society and Natural Environment	1
EAES	566	Environmental Problem Solving	3
EAES	565	Environmental Change and Communication	3
EAES	509	Current Issues in Sustainability Science	3
EAES	563	Natural Resource Management	3
EAES	531	Hydroclimatology	3
EAES	660	Seminar in Environmental Management	3
EAES	611	Advanced Environmental Remote Sensing and Image Processing	3
EAES	791	Research Methods	3
EAES	701	Modeling in Environmental Science	3
EAES	792	Special Topics	1–4
EAES	610	Spatial Analysis	3
EAES	710	Advanced Geographic Information Systems	3
EAES	569	Air Resource Management	3
EAES	562	Waste Management	3
EAES	525	X-ray Microanalysis	3
EAES	532	Applied Groundwater Modeling	4
EAES	550	Advanced Marine Geology	3
EAES	535	Geophysics	3
EAES	526	Geochemistry	3
EAES	527	Organic Geochemistry	3
EAES	528	Environmental Forensics	3
EAES	529	Instrumental Environmental Analysis	3
HLTH	502	Determinants of Environmental Health	3
HLTH	565	Foundations of Epidemiology	3
MKTG	561	Applied Marketing Management	1.5
MKTG	577	Special Topics in Marketing: Green Marketing Strategies for a Sustainable Future	1.5
PHMS	565	Tidal Marsh Ecology	4
SOCI	581	Sociological Perspectives on Health and Medicine	3
STAT	547	Design and Analysis of Experiments	3
STAT	548	Applied Regression Analysis	3
STAT	601	Statistical Methods for Research Workers II	3

QUALIFYING EXAMINATION/ASSESSMENT

Successfully complete the qualifying examination or assessment requirement.

ADMISSION TO CANDIDACY

Admission to candidacy follows completion of predissertation research courses and qualifying exam.

DISSERTATION REQUIREMENT

Complete a dissertation in accordance with Doctoral Program requirements.

The major **milestones** for the degree program include

Milestone	Recommended Within Semester*	
	Full-Time	Part-Time
Identification of a Chair of the Dissertation Committee	1	2
Identification and approval of a complete Dissertation Committee	2	3
Completion of the research course sequence	2	4
Completion of core courses	2	4
Qualifying written and oral examination administered by Dissertation Committee	3	5
Defense of Dissertation proposal	4	6
Enrollment in Dissertation courses	5	7
Completion of perspective and elective courses, and Admission to Candidacy	6	8
Enrollment in Sustainability Seminar Series	Every semester in residence**	
Submission and acceptance of at least one manuscript for publication	7–8	10–12
Dissertation defense and Dissertation approval	8	12

*"Semester" indicates Spring or Fall, not Summer, that is, for students enrolling in Fall 2009, Semester 2 is Spring 2010, and Semester 3 is Fall 2010.
** Waived only for approved "leave of absence."

INDEX

Note: Page numbers in *italics* refer to Figures; those in **bold** to Tables.

activated sludge process, 114, 117, *162*, 162–3
adaptive cluster sampling, 388–90, *389*
adaptive management
    Florida Everglades ecosystem, 68
    obstacles, 67
    preservation and restoration projects, 67
    successes and failures patterns, 67
Administrative Procedure Act, 1946, 317
ADMs *see* average daily movements (ADMs)
advance metering infrastructure (AMI), 139
AEC *see* anion exchange capacity (AEC)
AFVs *see* alternative fuel vehicles (AFVs)
agriculture and food processing
    baking technique, 103
    cooking of, food products, 106
    non-point source pollution, 103
    organic farming, 103
    pesticides and toxic materials, 103
air resource system
    mobile source control, 170–171
    pollutants, 168–9
    pollution management, 168
    stationary source control, 169–70
Alang ship recycling yard
    ecological engineering, 118
    geographical and coastal conditions, 120
    number of ships recycled, 119, *119*
    plate cutting activity, 120

quantity of steel recycled, 119, *120*
    recycling, 118, 119
alternative fuel vehicles (AFVs), 325
Amana Income Investor (AMANX), 259
American Recovery and Reinvestment Act, 2009, 333
AMI *see* advance metering infrastructure (AMI)
anion exchange capacity (AEC), 81, 83, 313, 324
anthropocentrism, 346
applied ethics, 338–9, 344, 357
Aquarius Platinum Limited, mining, 218
aquatic systems
    environmental management problems, 118
    JalMahal Tourism Project, Jaipur, 117, 118
    Mansagar Lake, 117
    natural resources, overexploitation, 117
audit completion, 501
audit follow-up, 501–2
auditing, environmental
    audit preparation, 490–495
    audit terminology, 490
    compliance audit, 489
    concepts, 487
    definitions, 487

EMS audit, 489
    first-party audit, 488
    informal inspections, 487
    International Chamber of Commerce, 487
    ISO 14001, 487
    liability audits, 488
    management audits, 488
    objectives, 489, *490*
    planned inspections, 487
    principles, 487–8
    second party audit, 489
    systems audit, 489
    third party audit, 489
    types, 488, *488*
    US EPA, 487
auditor competence, 502
audit preparation
    audit plan, 491, *492*
    checklist, 494, **495**
    documentation, 493
    document review, 493, 494
    internal auditing procedure, 491
    key personnel, 491, **492**
    on-site activities preparation, 494–5
    preparation, 491
    process, 490, *491*
    team establishment, 492
    timings, 492, **493**

audit report
    content, 500–501
    preparation, 500
    report approval and distribution, 501
audit terminology, 490
Ave Maria Catholic Values (AVEMX), 259
Ave Maria Rising Dividend (AVEDX), 259
average daily movements (ADMs), 433–5

backwashing, 158
baking technique, 103
balanced acceptance sampling (BAS), 387–8, *388*, 394
Bali Action Plan, 330
Bank of America Bank of America Corporation, 217
BAS *see* balanced acceptance sampling (BAS)
baseline risk assessments, 515–17
batch reactor (BR), 154, *154*
benefit–cost analysis
    benefit–cost ratio, 297
    capital stock values
        capital asset, 302–3
            resource stocks and resource flows, 302
    ecosystem services, 299–300
    nonmarket valuation, 300–302
    pollution control proposal, 297
    Stern Review, 297–8
    value and time
        natural capital, 298
        present value (PV), 298–9
bidirectional reflectance distribution function (BRDF), 399, 403
Bilonick's approach, 431–2
biocentric consequentialism, 344–5
biocentric egalitarianism, 344
biocentrism, 344, 346
biochemical oxygen demand (BOD), 30, 153, 160, 162
biodiversity and ecosystem processes
    ecosystem services, concept, 53
    endemic species, 52
    hot spots, ecosystems, 52–3
    organizational levels, 52
    primary productivity, 53
    Shannon index, 52
biology
    abiotic conditions, 51–2, *52*
    adaptive management, 48–9
    anthropogenic disturbances, 53
    biodiversity and ecosystem processes, 52–3
    collaborative and participative process, 49
    diversification, 47–8
    ecologically compatible boundaries, 48, *49*
    ecological processes, sustainability of, 48
    GIS software, 55–6
    harvests detection, nonindustrial private forests, 56, *57*
    landscapes and scale, 50–51, *51*
    press disturbances, 53
    pulse disturbances, 53
    rapid assessment approach, 50
    remote sensing data, 56
    spatial and temporal scales, 48
    stages of, 50, *50*
    "the study of life," 47
    systems thinking, 48
    target identification, 53–5
    trend map, volunteer observations, 55, *55*
    types of data, 55
A Blueprint for Survival (1972), 443
BOD *see* biochemical oxygen demand (BOD)
Bombardier Transportation, railroad, 218–19
BR *see* batch reactor (BR)
BRDF *see* bidirectional reflectance distribution function (BRDF)
British Petroleum's (BP), 281
built environment
    climate change, 174
    green buildings
        description, 174
        Holmes–Rulli residence, 174, *175*, *176*
        Morgan residence, 175, *176*
        solar canopy at TD bank, 174, *174*
        tree and we, 175, *177*
    role, 174
    warmer air, 174
"bully pulpit," 308
bureaucracy
    congress, 310
    constitution, 309
    public policymaking, 310
    Roosevelt administration, 310
Bureau of Land Management (BLM), 322
business sustainability, environmental
    added value, 200
    building consumer trust and loyalty, 200
    competitive advantage, 199
    concepts, 197
    corporate behavior
        emergence of, 204
        GHG emissions, 204
        high and low sustainability corporations, 204
    corporate engagement, 203
    corporate reputation and credibility, 200
    CSP and CFP, 205–7
    CSR *see* corporate social responsibility (CSR)
    development, 207–8
    employee motivation, 199
    employment and goods and services, 203
    environmental protection, 199
    financial environmental and social capital, 205
    high sustainability, 205
    incentive systems, 203
    increased consumer awareness, 200
    innovation and long-term growth, 199
    leapfrog innovations, 204
    management and strategic concepts, 197
    management systems, 204
    market opportunity, 199
    meeting regulations, 200
    Michael Porter's generic strategies
        cost leadership, 197–8
        description, 197
        differentiation, 198
        focus on, 198
    in new business era
        aspects, 199
        components, 198–9
        "thriving in perpetuity," 198
    not-for-profit organization, Club of Rome, 197
    performance of, 205
    products, 204
    reduced costs, 199
    responsiveness, 200
    SD, 197
    source of differentiation, 199
    theft and political subversion, 203
    time horizon of, 205
    value-addition processes, 204
    waste reduction, 199

CAA *see* Clean Air Act (CAA)
CAFE *see* Corporate Average Fuel Economy (CAFE)
California Air Resources Board, 332
California Geological Survey (CGS)
    liquefaction hazards, 10
    Seismic Hazards Mapping Act, 10
California seismic safety standards
    CGS, 10
    seismic hazard maps, 10
    tsunami evacuation route, San Francisco, 10, *10*
    urban clusters, 10
CALIPSO *see* Cloud–Aerosol Lidar and Infrared Pathfinder Satellite Observations (CALIPSO)
Campbell Soup Company
    environmental sustainability policy, 223–4
    New Jersey, United States, 223

carbon cycle (C4 models), 93
The Carlsberg Group, 217
Catholic funds/summary statistics, 273
cation exchange capacity (CEC), 80, 81
cause-related marketing (CRM), 243
CEC *see* cation exchange capacity (CEC)
CEQ *see* Council on Environmental Quality (CEQ)
CFR *see* Code of Federal Regulations (CFR)
CGS *see* California Geological Survey (CGS)
chemical oxygen demand (COD), 160
citizen science projects, 375
Clariant, chemical industry, 215
Clean Air Act (CAA), 307
Clean Water Act (CWA), 315
climate change *see* renewable energy
Cloud–Aerosol Lidar and Infrared Pathfinder Satellite Observations (CALIPSO), 412, *412*
CMFR *see* completely mixed flow reactor (CMFR)
coagulation process, 157–8
coal mining
  bituminous resources, 32–3
  hazardous trace elements, West Virginia, 34, *35*
  high-sulfur content, acid mine drainage, 33
  layered sedimentary rocks, 32
  mountaintop removal–valley fill, 34, *35*
  open-pit/strip mines, 33, *33*
  physical, chemical and biological treatments, 34
  surface subsidence, 34–5, *36*
  Will Scarlet open-pit coal mine, Williamson County, 34, *34*
coastal hazards
  Atlantic shores of Long Island and New Jersey, 13, *16*
  barrier islands, 16, *17*, 18
  beach nourishment and dune enhancement, 19
  budget, 15
  complex environmental management issues, 13
  crushed recycled glass, sand substitute, 18
  extratropical cyclone, 18
  Gateway National Seashore, 16
  inlets, 16
  large-scale hard defenses, 19
  lighthouse at Montauk Point, 16, *17*
  oblique low-altitude airphoto, Mantoloking, *18*, 19
  posttropical cyclone phase, 18
  rolling easements, 19

  seaside vacation, residents, 13
  spits and barrier islands, 13
COD *see* chemical oxygen demand (COD)
Code of Federal Regulations (CFR), 166, 317, 318
combined sewer overflows (CSOs), 22
commissioning agent (CxA)
  applications, 183
  description, 182
  green building rating systems, 182
  system integration testing, 182
community advisory groups, 374
comparative risk assessment (CRA), 515
completely mixed flow reactor (CMFR), 154, *154*
congress
  committees and subcommittees, 308
  Democrats, 308
  House of Representatives, 307
  political parties, 308
  Senate, 307
  staff members, 308
constructed wetland, Ropar, 115–16
continent-scale field data collection networks
  long-term monitoring stations, 57
  LTER, 58
  NEON, 59
  NOAA weather monitoring stations, 57–8
  USGS and USEPA water stations, 58
contingent valuation, 301–2
continuous flow stirred tank reactors (CSTRs), 154
Corporate Average Fuel Economy (CAFE), 170
corporate financial performance (CFP) *see* corporate social performance (CSP)
corporate social performance (CSP)
  businesses and policymakers, 205
  environmental responsibility, 206
  financial performance, 206
  inconsistencies, 205
  industry classification, 206–7
  influencing factor, 207
  meta-analyses, 207
  objectives, 205
  philanthropic activities, 206
  reputation, 206
  ROAE and ROAA measures, 207
  social science research, 206
corporate social responsibility (CSR)
  business executives, 200
  convergence strategy, 201
  corporate philanthropy, 200
  firms' assessment, 200
  increased, 199–200
  social norms, 200
  social responsibility, 201

corporate sustainability reporting
  GRI, 214
  industry cases
    agriculture, 214
    automotive, Tata Motors, 214–15
    aviation, Qantas, 215
    chemicals, Clariant, 215
    commercial services, 215–16
    computers, Lenovo, 216
    construction materials, Kawasaki Heavy Industries, 216
    consumer durables, LG Electronics, 216–17
    energy, US-based Hess Corporation, 217
    financial services, 217
    food and beverage products, 217
    household and personal products, 217–18
    logistics, Panalpina, 218
    mining, Aquarius Platinum Limited, 218
    railroad, Bombardier Transportation, 218–19
    retail, H&M, 219
    textiles and apparel, Viyellatex Group, 219
    tourism/leisure, Royal Caribbean Cruises Limited, 219–20
    US-based Waste Management (WM), 220
cost-of-illness approach, 301
Council on Environmental Quality (CEQ), 314, 318, 320
courts
  congressional language, 310
  federal court system, 310
  public policy, 310
CRA *see* comparative risk assessment (CRA)
CSP *see* corporate social performance (CSP)
CSR *see* corporate social responsibility (CSR)
customer relationship management (CRM), 243
CWA *see* Clean Water Act (CWA)
CxA *see* commissioning agent (CxA)

decentralized resource management, 375
deforestation and forest degradation
  active and passive remote sensing methods, 411
  howland forest (HF) site, 410, *410*
  radar remote sensing, 410
  REDD+ program, 408
  vegetation productivity, 409
  waveform lidar operation, 409

"Deming cycle," 245–6
de minimis, 508
descriptive ethics, 338
desertification
  description, 23
  meromixis recurrence, 24
  Mono Lake, California
    fluctuations in, 23, *23*
    satellite view, *22, 23*
    stratification, 23
Duckweed pond, Ludhiana, 116

earthquake-related hazards
  California seismic safety standards, 10, *10*
  East Japan earthquake and tsunami, 7–9, *9*
  explosive eruptions at volcanoes, 7
  liquefaction hazards, San Francisco, 10, *11, 12*
  slope failures, San Francisco Bay area, 10–15
East Japan earthquake and Tsunami
  disaster management, 8–9
  plate boundary at Pacific, 7
  radioactive materials and damage costs, 8
  regional planning, 8
  tsunami following 2011 Tohoku earthquake, 8, *9*
ecocentrists, 345
Ecodefense: A Field Guide to Monkeywrenching, 316
ecolabel, 458
ecological economics, 290
ecological engineering
  definition, 113
  self-designed system, 113
ecological marketing *see* green marketing strategies
ecological risk assessment, 55
economics
  benefit–cost analysis
    capital stock values, 302–3
    ecosystem services, 299–300
    nonmarket valuation, 300–302
    value and time, 298–9
  costs and benefits, 289
  ecological economics, 290
  economic problems, 289–90
  macroeconomics, 290
  market model
    externalities, 294–5
    marginal values, 290–292
    markets and welfare, 292–3
  microeconomics, 290
  opportunity cost, 290
ecosystem services
  anthropocentric, 299

assigning values or monetizing environmental goods, 299
ecosystem service typology, 299–300
importance of, 299
intrinsic value, 299
lower bounds, 299
ELA *see* equilibrium line altitude (ELA)
electrical double layer or diffuse double layer, 81
electrostatic precipitators (ESPs), 169
emission reduction
  cogeneration, 130
  description, 129
  fossil fuels, 130
  hydro systems, 130
  oil reservation, 130
  publication agencies, 129–30
  renewable energy resources, 130
emitted electromagnetic radiation (EMR), 398
EMR *see* emitted electromagnetic radiation (EMR)
energy balances
  closed system, 154, 155
  definition, 154
  open system, 154
  work, 155
energy policy, 305
  alternative fuel vehicles (AFVs), 325
  conservation and energy efficiency, 323
  federal energy policy, 323
  fossil fuel, 323
  global climate change, 323
  nuclear power, 323
  Obama administration, 326
  Organization of the Petroleum Exporting Countries (OPEC), 323
  production tax credit (PTC), 324–5
  sustainable energy, 323
Energy Policy Act (EPAct), 324–5
Energy Tax Act, 1978, 324
engineering *see also* wastewater disposal subsystem; water resource management system
  branches, 152
  description, 151
  energy balance, 154–5
  environmental management settings, 152
  fundamental principles and canons, 152, **152**
  materials balance, 152–4
  PEs, 151–2
engineers, 151
environmental management system (EMS), 245
  act approach, 471, *473*
  aspects and impacts, 467, **468**
  atmospheric emissions, 468

cause and effect, 467
conservation and wildlife, 468
continual improvement process, 246
corporate/market pressures, 468
"Deming cycle," 245–6, *246*, 471, *472*
energy, 468
environmental impacts, 469
environmental legislation, 466–7
global system, 466, *466*
"heart of an EMS," 473, *474*
human health, 468
IER, 469–70, *470*
implementation, benefits, 467
improvement cycle, 471
natural environment, 468
organization, 466, *467*
PDCA loops, 472, *473*
policy, 470–471
RECEIVING media, 469
resource use, 468
solid waste, 468
supply chain, 467
water use/discharges, 468
environmental marketing *see* green marketing strategies
environmental policy
  air and water, 326
  Clean Air Act (CAA), 326
  Federal Insecticide, Fungicide, and Rodenticide Act (FIFRA), 328
  hazardous and toxic waste policy, 327
  Resource Conservation and Recovery Act (RCRA), 327–8
  Toxic Substances Control Act (TSCA), 328
  water pollution, 327
Environmental Protection Agency (EPA), 307
environmental regulatory approaches
  energy policy, 321
  fossil fuel, 321
  natural resource policy, 321–2
environmental, social, and governance (ESG) investments
  ETFs and ESG market, 261–2
  exchange-traded funds (ETFs) providers, 255–6
  faith-based ESG market, 256, **259**, 260
  global warming and ozone depletion, 259
  history, 256–7
  literature, review of, 257
  return performance statistics, 261
  risk performance statistics, 261
  Social Investment Forum Foundation (SIF), 258
  stock of market, 259–60
  total US-domiciled ESG funds, 258
  total US-domiciled funds, 258

Environmental Systems Research Institute (ESRI), 425
EOP *see* Executive Office of the President (EOP)
EPA *see* Environmental Protection Agency (EPA)
equilibrium line altitude (ELA), 412
equilibrium price, 292
Ernst & Young USA, commercial services, 215–16
ESPs *see* electrostatic precipitators (ESPs)
ESRI *see* Environmental Systems Research Institute (ESRI)
ETFs and ESG market
　authorized participants, or APs, 263
　process of creation and redemption, 263
　return performance statistics, 263, **262–5**, 266
　risk performance statistics, **267–8**, 266, 268
　US investors, 263
ethical relativism, 340
ethics, environmental *see also* interspecies justice
　applied ethics, 339
　business and medical ethics, 338
　descriptive ethics, 338
　ethical objectivism, 340
　ethical relativism, 340
　metaethics, 339
　normative ethics, 339
　objectivism, 340
　philosophy, 338
　preferentialism, 339
　prudential good, 339
　water resources, 337
eutrophication
　beneficial use, 23
　fossil fuels combustion, 21–2
　municipal wastewater effluents, 22
　nitrogen and phosphorus, 21
　noxious algal blooms, 19
　point/nonpoint sources, 22
　water quality issue, 22
Executive Office of the President (EOP), 309
exposure assessment and analysis, 511
exposure–response evaluation, 511
externalities
　Adam Smith's invisible hand, 294
　biodiversity, 296
　environmental, 293–4
　negative, 293, *293*
　open-access resources, 294–5
　positive, 293
　public goods, 295
　World Fisheries, overexploitation, 295

face-to-face interaction, 372–3
faith-based ESG market
　ESG funds in the United States, **258**
　funds/summary statistics, 259, **260**
　return performance statistics
　　domestic equity funds, 261
　　Sharpe ratio, 261
　　Vanguard 500 Index (VFINX), 261
　risk performance statistics, 261
　summary statistics, 259, **260**
　taking stock
　　Amana funds, 259
　　Ave Maria funds, 259
　　GuideStone Funds, 259
　　Islamic funds, 259
　　Luther King Capital Management (LKCM) funds, 259
　　Praxis' 13 funds, 259
　　shunning companies, 259
FEcoDI ecodesign tool, 455
Federal Insecticide, Fungicide, and Rodenticide Act (FIFRA), 328
filter plants
　coagulation process, 157–8
　disinfection, 158–9
　filtration, 158
　flocculation process, 158
　screen, 157
　sedimentation, 158
　surface water treatment, 157
fixed effect, 390
flocculation process, 158
flooding hazards
　engineered flood control measures, 21
　flooded industrial zone, Naugatuck River, 19, *20*
　flood plain, Ansonia, 19–21, *20*
foreclosure view, 351
free-rider problem, 295
fuel cells
　solar energy applications, 140–141
　varieties, 141

GCS *see* geographic coordinate system (GCS)
generalized random tessellation stratified (GRTS) design, 387, 394
generic strategies, Porter's
　cost leadership, 197–8
　description, 197
　differentiation, 198
　focus on, 198
geographically and temporally weighted regression (GTWR), 432, 435
geographically weighted regression (GWR), 428–33, 436
geographic coordinate system (GCS), 426–7

geographic information science (GIScience), 424–5, 431, 574
geographic information systems (GIS) *see also* spatial data
　coordinates and coordinate transformation, 425
　environmental management, 424
　Environmental Systems Research Institute (ESRI), 425
　geographic coordinate system (GCS), 426
　GIScience, 425
　map projection and PCS, 426–7
　projected coordinate system (PCS), 426
　spatial reference system, 425
Getgo Car Share or Flexicar, 238
GHG *see* greenhouse gas (GHG)
GLC *see* Global Land Cover (GLC) facility
global climate change
　American Recovery and Reinvestment Act, 2009, 333
　Bali Action Plan, 330
　greenhouse gas (GHG), 330–331, 332–3
　Kyoto Protocol, 329
　Montreal Protocol, 329
　multilateral environmental treaties, 328
　Regional Greenhouse Gas Initiative (RGGI), 331
　renewable portfolio standard (RPS), 331
　Rio Declaration, 329
　Stockholm meeting, 328
　United Nations' "Framework Convention on Climate Change" (UNFCCC), 329
　Western Climate Initiative (WCI), 332
Global Land Cover (GLC) facility, 56, **58**
Global Positioning System (GPS), 56, 241
Global Reporting Initiative (GRI)
　CSR and financial performance, 214
　frameworks, 214
　global nonprofit organization, 214
　guidelines, 214
global warming
　causes, 232
　and soil management
　　carbon depletion, 94
　　climate change models, 93
　　global carbon cycle, 93
　　land management practices, 94
　　organic carbon transformations, 93, *93*
　　plant residues, organic carbon, 94
　　respiration, 93
Google Earth™, 403
　discontinuities, *406*, 407
　QuickBird satellite image, *406*, 407
　remote sensing methods, 405
　"true color," 404

gravity remote sensing
  gravity recovery and climate experiment (GRACE), 414
  groundwater storage, 414, *416*
  vegetation drought response index (VegDRI), 415, *417*
green buildings
  description, 174
  energy
    description, 183
    increase efficiency of systems, 186, *187*
    light shelfs at School of the Future, Philadelphia, 184, *185*
    net zero energy, 183, *184*
    offset the rest, 186–8, *187, 188*
    orientation, 184, *184*
    R30 walls, R50 roofs and R5 windows, 184
    Thomas Edison State College, Trenton, 184, *186*
    USEPA green building working group, 184, *184*
  Holmes–Rulli residence, 174, *175, 176*
  indoor environmental quality, 189, 191
  materials, *188–90*, 189
  Morgan residence, 175, *176*
  solar canopy at TD bank, 174, *174*
  tree and we, 175, *177*
  water, 188–9, *189*
green chemistry
  biomass, 113
  chemical sector, 98
  definition, 98
  engineering applications, 99, **100–101**
  gasifiers, types of, 113, *113*
  growth and development, 98
  low chemical potential substances
    ecology integration technology, $CaSO_4$ decomposition, 111, *112*
    "reuse, recycle and reduce," 111
  principles of, 98–9
  raw materials and resources, 99
  research and innovation, 102
  risks and hazard, 99
  SAICM and REACH, 98
  solvent substitution, 102
  "sustainable development" concept, 97
green consumers and purchase decisions
  definition, 236
  demographic factors, 236
  green consumption behavior, 236
  Greendex survey, 2012, 235, **235**
  measurement problem, 236
  segmentation activities, 236
green energy *see also* public policy
  emission reduction, 129–30
  features of, 128

fossil carbon, 128
primary energy carriers, 128
renewable, 128
solar energy systems, 128
green engineering
  advance wastewater treatment technologies, 103, **104–6**
  agriculture and food processing, 103, 106
  GC&E applications, 106, **107–8**
  principles of, 102–3
  wastewater treatment and recycling, 103
"Greener Goals," 283
green ETFs, 268
green financing, 208–9
greenhouse gas (GHG), 307
Greenland ice sheet surface melt
  Cloud–Aerosol Lidar and Infrared Pathfinder Satellite Observations (CALIPSO), 412, *412*
  ELA, 412
  MODIS albedo anomaly map, 412, *413*
  NASA's EO-1 satellite, 412–14
  teleconnections, 414
  unprecedented, 411–12
green logistics and distribution
  ecological supply chain practices, 240, *240*
  efficiency of transportation, 241
  greenhouse gas emission, 241
  green purchasing, 240
  green suppliers, economic impact, 240
  green supply chain management, definition, 239
  high-quality planning, 240
  process of planning, 241
  reverse logistics, 241
green marketing strategies
  achieving organizational goals, 234
  business benefits, 232
  company and executive perspectives, 213
  consumer perspectives, 212–13
  consumers and purchase decisions, 235–6
  continuous improvement, 245–6
  corporate social responsibility (CSR) reports, 246
  definition, 233
  environmental accounting, 247
  environmental audit, 246
  environmental change, 232
  future
    carbon taxes, 248
    environmental change, 248
    home solar systems, 248
    public policy, 248
    system-wide changes, 248–9
  governance, 245
  green chain, suppliers, 240
  implementation, 234–5, 244–5

ISO series, advantages, 246
level of consumption, 232–3
mix components, 236–7
natural resource accounting, 246
physical distribution and logistics, 241
positioning, 235
prices, 238–9
products, 237–8
promotion *see* green promotion
role of marketing function, 234
social accounting, 246–7
supply issues, 239–40
TBL, 234, 247–8
technological developments, 233
theory and practice, 233–4
green marketing theory
  environmental change and human consumption, 233
  green and conventional products, 233
  informing and educating consumers, 234
  product selection, 234
Greenpeace USA, 315
green playing fields, 283
green prices
  clean development mechanism (CDM), 239
  consumer expectations, 238
  cost saving, 238
  economies of scale firms, 239
  green product innovation, investments, 238
  indirect environmental benefits, 238–9
  market-based mechanisms (carbon taxes), 238
  for new green product, 239
  price-value relationship, 238
  product life cycle, stages, 238
green products
  definition, 237
  existing conventional product (continuous innovation), 237
  features and attributes, 237
  life cycle, 237
  new product concept (discontinuous innovation), 237–8
  removing CFCs, 237
  use of renewable resources, 237
green promotion
  advertising, 241
  communication tools, 241
  environmental labels and declarations, 241
  environmental logos/third-party endorsements, 242–3
  greenwash, 242
  integrated marketing communication, 242
  Part 260-guides, **242**
  personal selling, 243–4

publicity/public relations, 244
  sales promotion (cause-related
      marketing), 243
  social media, 244
green value chain
  consumption, 212
  financing, 208–9
  innovations with sustainable technology,
      209–10
  manufacturing, 211
  marketing and sales, 211–12
  products and services development, 210
greenwashing, 281, 458
  company's reputation, 221
  consumer demand, 220
  definition, 220
  FTC and EPA, 221
  law and of societal expectations, 221
  public, NGOs and governmental
      agencies, 221
  US economy, competition, 220–221
GRI *see* Global Reporting Initiative (GRI)
GRIHA *see* US Green Building Council and
      the Green Buildings Rating System
      India (GRIHA)
groundwater overdraft
  center pivot irrigation
    arid southwestern United States, 25, *26*
    cone of depression, Jameco aquifer, 27
    hydrologic imbalance, 26
    in operation, 25, *25*
    problems on Long Island, 26, *26*
    southwestern Kansas farms, 24, 25
  freshwater, sources of, 24
  vadose/unsaturated zone, 24
  vast grain fields, productivity, 25
  water table, 24
GRTS *see* generalized random tessellation
      stratified (GRTS) design
GuideStone Equity Index Fund (GEQZX),
      259
GWR *see* geographically weighted
      regression (GWR)

hazard identification
  and accounting, 510–511
  dose–response and exposure
      assessments, 513
hazardous and toxic waste policy, 326–7
Healthy Forests Restoration Act, 322,
      367, 372
hedonic property value method, 301
hedonism, 339, 348, 355–6
Hennes & Mauritz Hennes & Mauritz
      (H&M), 219
heterotrophs, 81
HEVs *see* hybrid electric vehicles (HEVs)
HF site *see* howland forest (HF) site

Holmes–Rulli residence
  outdoor courtyard, 174, *175*
  rainwater harvesting, 175, *175*
  recycled glass backsplash, 189, *190*
  residence net zero electric solar, 187, *187*
  residence rainwater catchment for
      irrigation, 188, *188*
  VOC free paint, daylighting and natural
      ventilation, 175, *176*
Howland Forest (HF) site, 410, *410*
Hurricane Katrina, 196, 285
hybrid electric vehicles (HEVs), 171
hypothetical bias, 302

IBM *see* International Business
      Machines (IBM)
IER *see* initial environmental review (IER)
IHA *see* International Hydropower
      Association (IHA)
implicit desires, 351
Incineration (combustion), 111, 166,
      169–70, 222, 366, 468
individual risks, 507–9
infiltration capacity, 85
initial environmental review (IER),
      469–70, *470*
INM *see* integrated nutrient management
      (INM)
integrated design process
  architect, 180
  certification systems, 178
  circular nature, open communication,
      177, *178*
  civil engineer and landscape
      architect, 180
  commitment maintenance, 179
  CxA, 182–3
  electrical engineer, 181–2
  facility management staff, end users,
      owner and contractor, 183
  goals, importance of, 178
  green champion, 179–80
  LEED™ method, 178
  mechanical engineer, 181
  MEP engineering, 181
  open communication and conduct
      charrettes, 179
  opportunity, 183, *183*
  plumbing engineer, 182
  project team, 177
  right team assemble, 178
  specialized individuals and skill sets, 176
  stakeholders, 176
  structural engineer, 180–181
  sustainability goals, 178
  synergies identification, 179
  typical, 176, *177*
  value engineering suggestions, 183

integrated nutrient management (INM)
  BMPs, 88
  fertilization, 88
  fertilizer sources, 88
  nutrient conservation, 88
  physical barriers against erosion, 88
  wind and water erosion, prevention, 88
integrated plant nutrition systems (IPNS)
      approach *see* integrated nutrient
      management (INM)
integrated solid waste management (ISWM)
  composting, 165–6
  description, 164
  landfilling and combustion disposal, 166
  recycling, 164–5, *165*
  source reduction, 164
intergenerational justice
  argument, composition, 355
  ecosystems, 354
  modal features, 353
  neutral welfare, 356
  nonidentity problem, 357–8
  person-affecting view, 357
  responsiveness, 356
  zero *P*-hood, 355
Intermediate Income A (MIIAX), 259
International Business Machines (IBM)
  climate protection, 222
  conservation, 222
  corporate policy, 222
  New York, United States, 222
  pollution prevention, 222
  product stewardship, 222
International Chamber of Commerce,
      475, 487
International Hydropower Association
      (IHA), 141
International Solar Energy Society
      (ISES), 141
interspecies justice
  anthropocentrism, 346
  attributing desires, 350
  biocentric consequentialism, 344
  biocentrism, 344, 346
  biocentrists and ecocentrists, 346
  discrimination, 343
  ecosystems, 346
  foreclosure view, 351
  harm of death, 349–50
  hedonism, 348
  holism, 345
  human–environment relationship, 344
  ideal desires, 350
  implicit desires, 351
  Kantian justifications, 342
  marginal humans, 343
  metaethical position, 344
  moral duties, 342

interspecies justice (cont'd)
    moral objects, 341
    moral status, 341, 343
    objective list accounts, 348–9
    positive and negative rights, 342–3
    sentient animals, 349
    sentientism, 343
    utilitarianism, 341
    welfare, 342, 348
    welfarism, 347–8
    zoocentrism, 346
inventory management software, 241
invisible hand, Smith, 292
ion exchange
    aluminosilicate structures, 83
    cation and anion capacity, 81
    characteristics, 83–4
    definition, 83
ion sorption
    definition, 84
    Freundlich and Langmuir isotherms, 84, *84*
    vs. ion exchange, 84
    precipitation, 84
    retention of, 84
ISES *see* International Solar Energy Society (ISES)
ISO 14001 interpretation, 487
    advantages, 246
    checking and corrective action process
        evaluation of compliance, 484
        internal audit, 485–6
        monitoring and measurement, 483
        nonconformity and preventive actions, 484–5, **484**
        records, 485
    EMS requirements, 475
    environmental policy, 475–6, *476*
    implementation and operation
        communication, 481
        competence, training and awareness, 480–481, **481**
        control of documents, 481–2
        emergency preparedness and response, 482–3
        EMS documentation, 481
        operational control, 482, **482**
        principle, 479
        roles and responsibilities, 479–80, *480*
    normative references, 475
    planning process
        environmental aspects, 475–7, *477*
        legal and requirements, 475–8
        O&T's, 478
    requirements, 475
    scope, 474
    terms and definitions, 475

issue–attention cycle, 316–17
ISWM *see* integrated solid waste management (ISWM)

Karst hazards
    bedrock, types of, 24
    and sinkhole hazards, 27–9
Kawasaki Heavy Industries, 216
KDE *see* kernel density estimates (KDE)
kernel density estimates (KDE), 433–4, *434*
Kimberly–Clark Corporation, 217–18
Kraft Foods, Inc., 222–3
Kyoto Protocol, 306, 309, 321, 329–31

lahar hazards
    description, 4
    disaster preparedness plans, 7
    liquid lava and solid pyroclastics, eruptions of, 7
    precursor earthquake activity, 7
    Ruapehu lahar 2007, New Zealand, 7, *9*
    south of Seattle, Washington, 7, *8*
land classification systems
    environmental management projects, 56
    GLC facility, 56, **58**
    LCCS, 56–7
    National Land Cover Data system, 56, **58**
Land Cover Classification System (LCCS), 56–7
landfill(s)
    environmental consideration, 166
    landfill leachate, *167*, 167–8
    leachate, 30, 31, 152, 166–8
    LFG, 166–7
    mining
        definition, 111
        disposal of wastes, 110
        integrated waste to energy, framework, 111, *112*
        stock availability, 111
        technology innovation, 111
        valorization technologies, 111
    sanitary, 166
    site selection, operation and structure, 166
landscapes and scale
    ecological processes, 51, *51*
    pattern/process of interest, 50
law of demand, 291
laws and policy, environmental
    air quality, 306
    business community, 312
    Council on Environmental Quality (CEQ), 314
    definitions, 306
    Department of the Interior (DOI), 313
    energy policy, 305
    environmental management, 305

Environmental Protection Agency (EPA), 313
    federal courts, 314–15
    green parties, 316
    Greenpeace USA, 315
    internal administrative reforms, 313–14
    international treaty, 306
    media, 312–13
    National Oceanic and Atmospheric Administration (NOAA), 314
    organizations, 315
    policy process, 316–17
    political parties, 312
    public opinion, 313
    public policy, 306–7
    public policy institutions, 307–11
    regulatory approaches, 332–21
    states, 311, 315
LCCS *see* Land Cover Classification System (LCCS)
Leadership in Energy and Environment Design (LEED), 131, 178–82, *185*, 189, 191
LEED *see* Leadership in Energy and Environment Design (LEED)
Lenovo computers, 216
LG Electronics, consumer durables, 216–17
lidar scanners, 409
life cycle assessments (LCAs) *see also* life cycle inventory (LCI) analysis
    applications
        environmental improvement, 458–9
        learning, 459–60
        marketing, 458, *459*
        product design and development, 457, **458**
        public policymaking, 459
        purchasing, 459
    characteristic, 460
    cycle inventory analysis phase, 446
    data sources, 448
    developments, 461–2
    environmental awareness, 442
    evolution, 443
    functional units, 447, **447**
    GHG emissions, 143–4
    goal and scope definition, 446–7
    life cycle impact assessment (LCIA), 446, 451–5
    life cycle interpretation, 445–6
    limitations, 460–461
    methodology, 443
    natural science, 446
    origins, 444, *444*
    phases, 445
    process, 443
    reproducibility, 446

resource and environmental profile analysis (REPA), 443–4
Society of Environmental Toxicology and Chemistry (SETAC), 443, 444
stages, 442, *442*
system's function and functional unit, 446
life cycle impact assessment (LCIA)
categories, 454
characterization, 453
FEcoDI ecodesign tool, 455
greenhouse gases, 453
impact assessment, 451
impact categories, 451–2
ISO standard, 454
midpoints and endpoints, 451
phase, 446, 451
software tools, 454–5
weighting, 454
life cycle interpretation
elements, 445–6, *446*
evaluation, 446
life cycle inventory (LCI) analysis
allocation criteria, 450
data requirements, 448
energy consumption, 450–451
environmental effects, 448
inputs and outputs of soft drink containers, 449, **450**
life cycle impact assessment (LCIA), 451
soft drink containers, 448, *449*
The Limits to Growth (1972), 443
liquefaction hazards, San Francisco
aid workers, 10, *12*
water-saturated soils, 10
zones, 10, *11*
LKCM Small Cap Equity Advisor (LKSAX), 259
Long-Term Ecological Research (LTER), 58–9
LTER *see* Long-Term Ecological Research (LTER)
Luther King Capital Management (LKCM) funds, 259

macroeconomics, 290, 461
Major League Soccer (MLS), 283
management review, 486, *486*
map projection and PCS, 426–7
marginal humans, 343
marginal values
caffeine, nature of, 290–291
demand curve, 291, *291*
law of demand, 291
marginal benefit curve, 291, *291*
marginal benefits and costs, 291–2
marginal cost curve, 291, *291*
marginal cost or supply side, 291
supply curve, 291, *291*

market model
externalities, 294–5
marginal values, 290–292
markets and welfare, 292–3
markets and welfare
equilibrium price, 292
equilibrium quantity, 292, *292*
equity or fairness, 292–3
externality, 292
invisible hand, 292
potential problems, 293
Marks and Spencer (M&S), 224–5
mass balance, 152, 153, *153*
material efficiency, 103
description, 109
flow chart, 109, *110*
life extension, 109
material scarcity, 109
policy measures, 109, 110
unit processes, 106, 109
value addition, 109
winning policies and technologies, 109
materials balance
calculus-based equation, 153
conservative *vs.* nonconservative substances, 153
control volume, 152–3
equation, 153
mass balance, 152, 153, *153*
particulate matter (PM), 152
properties, 152
reactors, 154, *154*
steady *vs.* unsteady state, 153
mechanical, electrical and plumbing (MEP) engineering, 180, 181
media, 312–13
megasites, 85
Mennonite funds/summary statistics, 274
MEP *see* mechanical, electrical and plumbing (MEP) engineering
metaethics, 338–40, 344, 358
MetLife Stadium, 285
microeconomics, 290
mineral (inorganic) constituents, 77
MISR instrument *see* multi-angle imaging spectroradiometer (MISR) instrument
mixels, 399
mobile source control
emission control, 170–171
fuels and new-generation automobiles, 171
institutional issues and regulation strategies, 170
vehicle-induced air pollution, 170
modal features, 353–4
Montreal Protocol, 147, 329, 404
moral objects, 341–2, 344, 353

Mt. Vesuvius valconic eruption, Pompeii
climate and good harbors, 4
destructive pyroclastic flows, 5
false alarm mobilization, 6
phases of, eruption, 4–5
proximity of, 5, *5*
public imagination during, 4, *5*
multi-angle imaging spectroradiometer (MISR) instrument, 407, 417–19
municipal solid wastes (MSWs)
composition of MSU generated within, 163, *164*
description, 163
generation rates, United States, 163, *164*
ISWM, 164–6
landfilling, 166–8
nonhazardous wastes, 163
RCRA, 163
municipal (centralized) wastewater treatment systems
activated sludge process, *162*, 162–3
classifications, 161
degrees of, 161, *162*
primary treatment, 162
secondary treatment, 162
tertiary (advanced) treatment, 163
Muslim funds/summary statistics, 275

National Ecological Observatory Network (NEON), 59
National Energy Act (NEA), 324
National Forest System, 322
National Land Cover Data system, 56, **58**
National Oceanic and Atmospheric Administration (NOAA), 57–8, 314, 322, 415
National Park System, 322
National Wildlife Refuge System, 322
natural attenuation (NA), 90, 168
natural fire disturbance regimes
characteristics, 67
landscape-scale heterogeneity, 66
pyrogenic ecosystems, 67
natural gas extraction
cleanburning, 39
deepwater horizon offshore oil platform, 37, *37*, 38
Marcellus Shale tight gas exploitation, 38, *39*
nonrenewable resource, 38
oil sand deposit, Alberta, 39, *40*
onshore exploration and production, 37
overlying freshwater aquifers, 39
secondary and tertiary recovery, 38
world petroleum production history and predictions, 38, *38*

natural resource policy
  Bureau of Land Management (BLM), 322
  Clinton administration, 322
  Congress, 321
  federal government, 321
  membership, 322
  National Forest System, 322
  National Park System, 322
  National Wildlife Refuge System, 322
  Roadless Area Review and Evaluation (RARE), 322
natural treatment systems (NTSs) *see also* aquatic systems
  constructed wetland, Ropar, 115–16
  Duckweed pond, Ludhiana, 116
  Karnal technology (KT), Ujjain, 117
  low O&M costs, 114
  minimum energy and degree of treatment, 114–15
  opportunities and constraints, 117
  polishing pond, Agra, 116
  reuse in India, 115, **115**
  sewage fed aquaculture, India, 116
  waste stabilization pond, Mathura, 116
NBPOL *see* New Britain Palm Oil Limited New Britain Palm Oil Limited (NBPOL)
negative rights, 342
NEON *see* National Ecological Observatory Network (NEON)
New Britain Palm Oil Limited New Britain Palm Oil Limited (NBPOL), 214
nitrification/denitrification biological process, 163
NOAA *see* National Oceanic and Atmospheric Administration (NOAA)
nonmarket valuation
  bequest value, 301
  contingent valuation, 301–2
  cost-of-illness approach, 301
  existence values, 301
  hedonic property value method, 301
  hypothetical bias, 302
  option value, 301
  production factor method, 300
  revealed preference method, 301
  stated preference technique, 301
  travel cost method, 301
  use and nonuse values, 301
  value of a statistical life (VSL), 301
non-point source pollution, 103, 397
The North American Breeding Bird Survey, 55, *55*
North American Islamic Trust (NAIT), 259
NTSs *see* natural treatment systems (NTSs)
nuclear waste disposal
  constituent isotopes, 31–2

radioactive wastes, 32
Yucca Mountain, high-level nuclear waste storage site, 32, *32*

objective list accounts, 339, 348–9
objectivism, 340
on-site activities
  audit communications, 498
  audit evidence, 498–9
  audit findings preparation, 499–500
  closing meeting, 500
  guides and observers, 498
  human aspects of auditing, 498
  interview phase, 496
  nonconformities classification, 500
  opening meeting, 495–6
  open question mode, 497
  site inspection, 498
  team findings, 500
on-site (decentralized) disposal systems, 161, *161*
opportunity cost, 290–291, 299
optimal allocation, 387
organic farming, 103
organizational communication
  areas for future research, 280, 282
  corporate transparency, 282
  electrical utility, 279–80
  environmental decision making, 281
  leadership style and motivations, 279
  organization's CSR efforts, 282
  Texas Agency, 282

Panalpina logistics, 218
participatory approaches
  citizen science projects, 375
  citizens' juries, 374–5
  community advisory groups, 374
  conventional wisdom, 366
  decentralized resource management, 375
  decision-making problems, 371
  deliberative democracy, 368–9
  deliberative polls, 374
  discrimination, 370–371
  equality and justice, 370
  evaluation
    participant satisfaction, 375
    process fairness, 375
    sponsoring agencies and stakeholders, 376
    trust and social capital development, 375
  face-to-face interaction, 372–3
  group interaction processes, 371
  Healthy Forests Restoration Act, 372
  history, 366–7
  instrumental value, 368
  legal problems, 369–70

  manipulation and cooptation, 370
  negotiation/mediation, 374
  normative value, 368
  organizer, 373
  participants interaction, 372
  polls, surveys, interviews, and public comment periods, 374
  practical constraints, 374
  practical problems, 369
  public hearings, 374
  referendum, 375
  scientific and technical information, 373–4
  self-development value, 368
  social network analysis, 372
  sponsoring agency, 373
  stakeholders, 366, 372
  substantive value, 367–8
  workshops/charettes, 374
PCS *see* projected coordinate system (PCS)
PEs *see* professorial engineers (PEs)
Pesticide National Synthesis Project, 91
pesticide pollution
  BMPs, 93
  classification, 91
  degradation, 92
  ecological impacts, 92
  environmental quality issue, 91
  health and ecological impacts, 91
  Pesticide National Synthesis Project, 91
  Stockholm Convention, 91
  transport behavior, classes, 92
  US streams and groundwater, 92
petroleum system *see also* natural gas extraction
  clay-rich sediments, 36
  components/phases, 36
  conventional and nonconventional oil and gas resources, 36
  diagenesis process, 36
  hydrocarbons, 37
  and natural gas formation, 36
PEVs *see* plug-in electric vehicles (PEVs)
PFRs *see* plug flow reactors (PFRs)
photovoltaic (PV) systems
  advantages, 133
  applications, 133, *134*
  basic electrification, 135
  cost trends, 133, *134*
  cumulative market growth, 135, *135*
  evolution of, 135, *136*
  grid parity, 133
  policy drivers, 133
  RECs mechanisms, 135
  RPOs, 133, 135
  solar radiation, 131
Pigouvian tax, 297, 303
pixels, 56–7, 399, 427

plan–do–check–act (PDCA) cycle, 245–6, 246, 472–3
plug flow reactors (PFRs), 154, *154*
plug-in electric vehicles (PEVs), 171
policy adoption, 316–17
policy process
  adoption, 317
  evaluation, 317
  formulation, 317
  implementation, 317
political parties, 306, 308, 312
pollutants
  asbestos, 169
  carbon monoxide (CO), 169
  lead (Pb), 169
  PM, 168
  primary and secondary, 168
pollution
  control proposal, 297
  management, 168
  non-point source, 103, 397
  optimal model
    environmental costs and benefits, 297
    marginal costs, *296*, 296–7
    market model, 297
    Pigouvian tax, 297
    subsidies, 297
  pesticide *see* pesticide pollution
  prevention, 222
  urban air
    aerosol loadings, 418, *419*
    multi-angle imaging spectroradiometer (MISR) instrument, 417
  vehicle-induced air, 170
  water, 327
positive rights, 342
preferentialism, 339, 348
preservation
  best management practices, 59
  conservation, 59
  description, 59
  gap analysis, 60
  habitat requirements, 62–3
  hysteresis, between different biomes, 59, *60*
  IUCN category, 60, **61–2**
  protected area networks, 63
  resilience of system, 59
  thresholds, 59
presidency
  bully pulpit, 308
  environmental justice, 309
  Executive Office of the President (EOP), 309
  president, 309
  public policy powers, 308
production factor method, 300
production tax credit (PTC), 324–5

professorial engineers (PEs)
  definition, 151
  skills improvement, 151–2
projected coordinate system (PCS), 426–7
Project FeederWatch program, 55, *55*
PTC *see* production tax credit (PTC)
public opinion, 278, 281, 283, 312, 313, 372, 375
public policy
  Clean Air Act (CAA), 307
  definitions, 306
  Environmental Protection Agency (EPA), 307
  government institutions, 306
  government regulation, 306–7
  government revenues, 306
  and green energy
    carbon-based systems, 128
    coupling energy efficiency enhancement and emission reduction, 128
    initiatives, 129
    local-level socioeconomic development, 128
    policy considerations, 128, **129**
  greenhouse gas (GHG), 307
  institutions
    bureaucracy, 309–10
    congress, 307–8
    constitution, 307
    courts, 310–11
    presidency, 308–9
    United States, 307
public relations (PR)
  definition, 278
  environmental goals, strategic use, 284
    Australia, residents to conserve, 284
    future research, 284
    public education campaigns, 284
  environmental organizations
    ConAgra's public relations, 283
    education campaign, 283
    future research, 283
    global warming crisis, 283
    Major League Soccer (MLS), 283
  failures
    Canada to spin climate change, 281
    future research, 281
    greenwashing, 281
    outcomes, 280–281
  future research, 279
  organization supports, 277–8
  practices
    CSR, 280
    research, 280
    United Parcel Service, 280
  Public Relations Society of America (PRSA), 278

  shell oil, 279
  stakeholder relationships, 278–9
  survey results, 2011, 278
Public Relations Society of America (PRSA), 278
Public Utility Regulatory Policies Act (PURPA), 324
PV *see* photovoltaic (PV) systems

Qantas, aviation industry, 215

radar remote sensing, 410
  cross-sectional model, 416, *418*
  ground-penetrating capability, 415
  WATEX™ mapping system, 416
radio-frequency identification (RFID) deployment, 241
RARE *see* Roadless Area Review and Evaluation (RARE)
RCRA *see* The Resource Conservation and Recovery Act (RCRA)
REACH *see* Registration Evaluation & Authorization of Chemicals (REACH)
REC *see* renewable energy certificate (REC) mechanisms
reclamation
  aquatic organisms, 65
  bioremediation and/or phytoremediation, 65
  mining activities, sites, 65
  SMCRA, 65
  underground mining methods, 65
Regional Greenhouse Gas Initiative (RGGI), 331
Registration Evaluation & Authorization of Chemicals (REACH), 98
rehabilitation
  brownfields, 66
  ecological processes and productivity, 65
remote sensing
  applications
    deforestation and forest degradation, 408–11
    gravity remote sensing, 414–17
    Greenland ice sheet surface melt, 411–14
    radar remote sensing, 415–16, *418*
    urban air pollution, 416–19
    urbanization, 414, *415*
    wetland management, 407–8
  bidirectional reflectance distribution function (BRDF), 399
  desert grassland landscape, 399, *400*
  detector array systems, 398–9, *399*
  domains, 403
  dwell time, 398
  electromagnetic radiation (EMR), 398

remote sensing (cont'd)
  environmental management, 397
  Geoscience Laser Altimeter System (GLAS), 401
  global scales, 398
  Google Earth, 404–7
  ground reference data, 398
  ICESat satellite, 401–2, *402*
  instantaneous field of view (IFOV), 399
  interpolation, 398
  landsat 8 orbital tracks, 399–400, *401*
  reflectance anisotropy, 399
  spatial and temporal scales, environmental management, 403–4, *405*
  surfaces scatter light, 399
The REN Alliance, 141
renewable energy
  Ad Hoc Working Group, 144
  commercial airlines, 145
  country attractiveness indices, 145
  energy efficiency enhancement, 143
  GHG emissions, 143
  GHG-intensive conventional sources, 145
  industry forums, 141, **142**
  IPCC 2011, 143
  national emissions trading systems, 144
  non-OECD countries, 144
  sources, 143–4
  transitions in India, 145–6
  UNEP 2013, 143
renewable energy certificate (REC) mechanisms, 135
renewable portfolio standard (RPS), 331
REPA *see* Resource and Environmental Profile Analysis (REPA)
Resource and Environmental Profile Analysis (REPA), 443–4
The Resource Conservation and Recovery Act (RCRA), 163, 166, 327–8
Respect for Nature, 344
restoration
  characteristics, SER, 66
  description, 63
  Florida Everglades Ecosystem, 63, **64**
  plans, 63–4
  reclamation, 65
  reference target, 63, 65
  rehabilitation, 65–6
revealed preference method, 301
RGGI *see* Regional Greenhouse Gas Initiative (RGGI)
Rio Declaration, 329
risk assessment
  acceptability and tolerance concepts, 508
  applications, 505
  assessment *vs.* management, 509
  characterization, 511
  conservatism, 507
  cost-effective management, 504
  decision making, 516
  de minimis, 508
  environmental contamination, 506
  environmental management systems, 504
  estimation, 509
  exposure assessment and analysis, 511, 512
  exposure–response evaluation, 511
  fundamental elements, 506
  hazard assessment, 509
  hazard identification and accounting, 510–511
  health effect, 508
  human health endangerment assessment, 511
  individual risks, 507
  parameters/metrics, 506
  principal elements, 510, *510*
  principles and methodologies, 518
  procedural elements, 510
  purpose and attributes, 504–5
  qualification/quantification process, 506
  retrospective and predictive, 508–9
  societal risk, 507
  techniques, 505–6
  toxicity assessment, 512
risk–cost–benefit optimization, 515–16
risk management process, 509
rivers and lakes
  artificial canals and reservoirs, 19
  episodic flooding, 19
  eutrophication, 19, 21–3
  flooding hazards, 19–21, *20, 21*
  hydropower source, 19
  lakes in arid regions, *23,* 23–4
Roadless Area Review and Evaluation (RARE), 322
Royal Caribbean Cruises Limited, tourism/leisure, 219–20
RPS *see* renewable portfolio standard (RPS)
rulemaking process
  Administrative Procedure Act, 1946, 317
  Code of Federal Regulations (CFR), 317
  environmental impact statement (EIS), 318
  National Environmental Policy Act (NEPA), 319–20
  policy and purpose, 318
  records of decision (ROD), 319
  regulations, 317–18
  Responsible Official, 318–19

SAICM *see* Strategic Approach to International Chemicals Management (SAICM)
San Andreas Fault Zone (SAFZ), 11–13
sanitary sewers, 22, 160–161, *161*
SETAC *see* Society of Environmental Toxicology and Chemistry (SETAC)
ship recycling
  Alang ship recycling yard, 118–20, 119, *119, 120*
  complete recycling, 121–2
  dismantling, 121
  "end-of-life" ship, auction, 121
  ferrous/non-ferrous metals, 118
  hazardous and nonhazardous wastes, 121
  interventions, 121
  reusable materials, 121
  workers safety, 121
simple random sampling
  collecting survey data, 395
  edge effect, 384
  finite population correction, 386
  mapping software package, 384–5
  probability sampling, 385
  sample selection method, 385
  sample variance, 385–6
  summary statistics, 384
  survey design, 384
sinkhole hazards
  boreholes, 28
  calcite, 27
  evaporite minerals, 28
  formation mechanisms, 27
  geophysical methods and surveys, 28, 29, *30*
  ruined building in Clermont, 28, *28*
  Summer Bay Resort collapse site, 28, *29*
  synthetic aperture radar, 28–9
  Winter Haven, satellite view, 27, *27*
slope failures, San Francisco Bay area
  enlargements of Daly City, 13, *14*
  factor of safety, 10
  landslide-prone coastal district of Daly City, 11, *12*
  missing houses, 13, *15*
  SAFZ, 11, 13
  zones, 10, *11*
smart grids
  AMI, 139
  drivers, classification of, 139
  modernized electricity system, 139
  RF wireless, GPRS and PLC, 139
  smartening of, 139–40
  technology areas, 139, *140*
SMCRA *see* Surface Mining Control and Reclamation Act (SMCRA)
social accounting
  benefits, 247
  corporate social reporting (CSR), 247
  definition, 246
  key criticism, 247
social network analysis, 372
social sustainability

approaches, 113–14
case studies, 114
economic and environmental aspects, 113
factors, 113
societal risk, 507
Society of Environmental Toxicology and Chemistry (SETAC), 443
softening plants
carbonate and noncarbonate hardness, 159
hardness removal, raw water, 159
soil buffering capacity, 80
soil bulk density, 79
soil color
agents of, 78
Munsell color charts, 78
varieties, 78
soil contamination
acidification and mobilization, 91
biodegradation process, 90
chemical and/or waste handling, storage and disposal, 89
common soil remediation technologies, 90, *90*
definition, 88–9
degradation, 89
interaction processes, 89, *90*
management of, 89
reducing compounds, 89
soil pollutants in Europe, 89
soil exchange capacity, 81
soil fertility management
clayey textural class, 87
definition, 87
global issues, 87
INM, 88
natural resources and genetic diversity, 87
nutrient resources, 87–8
soil organic matter (SOM)
humic and nonhumic substances, 77
humus, chemical structure, 77
living organisms and nonliving organic components, 77
management issues, 77
soil particle density, 79
soil science
atmosphere modification, 76
biological properties
flora (plants) and fauna (animals), 81
metabolic activity, 81
organism biomass, 81
organism diversity, 81, **81**
characteristics, 81–2
chemical movement and sorptive processes
adequate plant growth, 83
ion exchange, 83–4
ion sorption, 84, *84*
soil charge, 83

chemical properties
acidity, 80
AEC, 81
alkalinity, 80
buffering capacity, 80
CEC, 81
ion exchange capacity, 81
soil pH, 80
civilizations, 76
compositions
air, 78
behavior in environment, 76–7
description, 76
mineral (inorganic) constituents, 77
solution, 78
SOM, 77–8
US Department of Agriculture (USDA), 76
development processes, 83
engineering medium, human-built systems, 76
of fertilizers, 76
in situ and *ex situ*, 81
living organisms, habitat, 76
management and use, 76
physical properties
bulk density, 79
color, 78
particle density, 79
strength and consolidation process, 80
structure, 79
texture and particle size distribution, 78–9
total porosity, 79
water holding capacity, 79–80
plant debris accumulation, 81
plant growth, 76
profile formation, *82*, 82–3
quality and fate of water, 76
recycling system, 76
rocks and minerals weathering, 82
terrestrial life interfaces with, 76
transport processes, 85–7
solar energy systems
active, 131
biomass, 137
central receiver technology, 131, *132*
deployment, 136
fuel cells, 140–141
geothermal systems, 137–8
linear Fresnel lens, 131, *132*
market penetration trend, 135, *136*
nuclear energy, 138
opportunities and challenges, 139, **139**
parabolic dish technology, 131, *133*
passive, 130–131
PV systems *see* photovoltaic (PV) systems
research studies, 135–6

smart grids, 139–40
solar parabolic trough, 131, *132*
symbolizing, 130, *131*
turbine's frictional resistance, 136–7
wind availability and velocity, 136
wind power global capacity, 136, *137*
working fluids, 131
solid waste disposal
abandoned Malanka Landfill, Secaucus, 29, *31*
aerobic degradation, 30
groundwater contamination, prevention, 29, *31*
Karst terrain, 30
microbial methane accumulations, 30
SOM *see* soil organic matter (SOM)
Southern Baptist funds/summary statistics, 270–272
spatial data
autocorrelation, 427–8
directions, 428
geographically weighted regression (GWR), 428
geographic information, 427
heterogeneity, 428
spatial effects, 427
weighting schemes, 429, *429*
spatially varying coefficient processes (SVCP) model, 428, 430–432, 435–6
spatial reference system, 425, *426*
spatiotemporal analysis
aspatial approach, 433
average daily movements (ADMs), 434–5, *435*
Bilonick's approach, 432
econometric approach, 432
geographically and temporally weighted regression (GTWR), 432
geographic information, 431
geographic space, 431–2
kernel density estimates (KDE), 433–4, *434*
long-distance movements, 434
spatial data analysis techniques, 431
spatial panel data analysis, 432
SQC methods *see* statistical quality control (SQC) methods
SSCM *see* sustainable supply chain management (SSCM)
stakeholder theory
Freeman's argument, 201
globalization, 202
logical gap, 201
nondeceptive and legal, 201
in real-life situations, 202
social responsibility, 201, 202
sustainability theory of businesses, 203
TBL approach, 203

stated preference technique, 301
stationary source control
　absorption, 169
　adsorption, 169
　combustion, 169–70
　cyclones, 169
　ESPs, 169
statistical quality control (SQC) methods, 471
statistics
　change and trends, models for
　　grassland cover, 390–391
　　linear regression model, 393
　　mixed effects model, **393**
　　mixed model, 393–4, *394*
　　predicted *y*-intercept and slope, 391, 393
　　quadrates, 390, *392*
　　random and fixed effects, 390–391
　　scatterplot of intercepts and slopes, 393, *394*
　　scatterplots of grass cover, 391, *392*
　　variation among quadrates, 393
　data collection
　　adaptive sampling, 388–90, *389*
　　balanced acceptance sampling (BAS), 387–8, *388*
　　generalized random tessellation stratified (GRTS) design, 387
　　simple random sampling, 384–6
　　stratified sampling, 386–7
　　systematic sampling, 387
　　two-phase stratified design, 390
　roles of, 383
Stockholm Convention, 91
Strategic Approach to International Chemicals Management (SAICM), 98
stratified sampling, 386–8
Sudan Accountability and Divestment Act, 256
Super Bowl, 285
supply chain, 467
Surface Mining Control and Reclamation Act (SMCRA), 65
sustainable development
　anthropogenic activities, 122
　bilateral and multilateral institutions, 123
　Global Chemicals Outlook, 122
　transitioning, 122, **123**
　US President Green Chemistry Awards, 122
sustainable energy, 122, 137, 144, 146, 323
sustainable marketing *see* green marketing strategies
sustainable supply chain management (SSCM)
　environmental management goals, 210

firms, 211
organization's reputation, 211
proactive, 211
reverse logistics, 211
source materials, 210
SVCP model *see* spatially varying coefficient processes (SVCP) model
systematic sampling, 387

Target identification
　alternative futures scenarios, 54
　DDT exposure, 55
　ecological risk assessment, 55
　economic and noneconomic values, 54
　indicators, 54–5
　land cover change, Biome, 53, **54**
　land degradation mapping, 54
　land use, 53–4
　management decision making, 54
　pristine ecosystems concept, 53
Tata Motors, 214–15
textural classes, soil
　determination, 79
　textural triangle for, classification, 79, *79*
tight gas exploitation, Shale, 38, *39*
Time-Relative Interest Account, 351
Tokyo Electric Power Company (TEPCO), 281
total porosity, soil, 79
total suspended solids (TSSs), 160
Toxic Substances Control Act (TSCA), 328
trade and environment connectivity
　environmental regulations, 122
　liberalization, 122
　resources, 122
transport processes, soil science
　advection–dispersion equation, 86, *86*
　Darcy's law, 86
　groundwater flow, 86
　infiltration and evapotranspiration, 85
　overview of, 85, *85*
　quantitative models, 85
　total, pressure and elevation, 85, *85*
　vapor extraction and air sparging, 86–7
travel cost method, 301
triple bottom line (TBL)
　criticism, 247–8
　stakeholder theory, 247
TSSs *see* total suspended solids (TSSs)
tsunami *see* East Japan earthquake and tsunami
two-phase stratified design, 390

United Nations' "Framework Convention on Climate Change" (UNFCCC), 128, 143–4, 329
United Parcel Service (UPS), 280
urbanization, 13, 310, 326, 414, *415*

US-based Hess Corporation, 217
US-based Waste Management (WM), 220
US Environmental Protection Agency (USEPA), 58, 86, 89–91, 156, 159, 161, 163–5, 189, 297, 303, 513
USEPA *see* US Environmental Protection Agency (USEPA)
US Green Building Council and the Green Buildings Rating System India (GRIHA), 131
utilitarianism, 201, 339, 341–3, 348, 352, 354

value chain
　business actions, 208, *208*
　greening *see* green value chain
　Michael Porter's, 208, *208*
　SSCM, 210–211
value of a statistical life (VSL), 301
vegetation drought response index (VegDRI), 415, *417*
vegetation index (VI), 409
Vietnam War, 256, 312
Viyellatex Group, textiles and apparel, 219
Volcanic Ash Advisory Centers (VAACs), 7
volcanic hazards
　eruptions and aviation
　　European air travel, Eyjafjallajökull on Iceland, 6, 6–7
　　health issues, 7
　　northern European capitals, 6
　　VAACs, 7
　lahar hazards, Northwestern Washington State, 7, *8, 9*
　Mt. Vesuvius volcanic eruption in Pompeii, 4–6, *5, 6*
　pyroclastic materials, 4

waste management
　nuclear waste disposal, 31–2, *32*
　solid waste disposal, 29–31, *30, 31*
wastewater disposal subsystem
　flowchart of, 159–60, *160*
　industrial/domestic sewage, 159
　pollutants
　　BOD and COD, 160
　　color and pathogens, 160
　　nitrogen (N) and phosphorus (P), 160
　　TSS, 160
　sanitary sewers, 160–161, *161*
　treatment standards, 163, **163**
　treatment systems
　　municipal (centralized), 161–3
　　on-site (decentralized) disposal systems, 161, *161*
wastewater treatment systems
　conventional mechanized, 114
　minimum quality standards, 114

NTSs, 114–15
    regulatory agencies, 114
    sector-based regulations, 114
    The Water Act, 114
The Water Act, 114
water distribution systems, 116, 158–9
water resource management system
    description, 155, *155*
    distribution systems, 159
    drinking water standards, 159
    pollutants
        color, 156
        emerging contaminants, 156
        hardness, 156
        particles, 155–6
        pathogens, 156
        salts, 156
    sources, 156
    supply subsystem, 155, *155*
    treatment systems *see* water treatment systems
water treatment residuals (WTR), 158
water treatment systems
    available resources, 156
    economic consideration, 156
    environmental factors, 156
    filter plants, 157–9
    flowcharts, 157, *157*
    softening plants, 159
water quality criteria (WQC), 327
WATEX™ mapping system, 416
waveform lidar operation, *409*
WBA *see* World Bioenergy Association (WBA)
WCI *see* Western Climate Initiative (WCI)
weather monitoring stations, NOAA, 57–8
welfarism, 347–8
western climate initiative (WCI), 332

wetland management
    cost–benefit analysis, 408
    deforestation and forest degradation, 408–11
    hyperspectral imaging, 407
    scene components, 407–8, *408*
"WIN–WIN" partnerships
    areas for future research, 285
    environmental organizations, 284
    National Football League's (NFL), 285
    public relations campaigns, 285
    richness of diversity, 284–5
WM *see* US-based Waste Management (WM)
Wolf Hunt Sabotage Manual, 316
World Bioenergy Association (WBA), 141

zoocentrism, 346
Zoopolis: A Political Theory of Animal Rights, 352